Urban Forests, Climate Change and Environmental Pollution

Hukum Singh
Editor

Urban Forests, Climate Change and Environmental Pollution

Physio-Biochemical and Molecular Perspectives to Enhance Urban Resilience

Springer

Editor
Hukum Singh
Plant Physiology Discipline
Forest Research Institute
Dehradun, Uttarakhand, India

ISBN 978-3-031-67836-3 ISBN 978-3-031-67837-0 (eBook)
https://doi.org/10.1007/978-3-031-67837-0

This Springer imprint is published by the registered company Springer Nature Switzerland AG
The registered company address is: Gewerbestrasse 11, 6330 Cham, Switzerland

Preface

Urban forests play a crucial role in enhancing the resilience of cities. However, they encounter significant challenges due to climate change and environmental pollution, threatening their survival and functionality amidst rapid economic growth and urbanization. Cities serve as the economic powerhouses of nations that necessitate sustainable, resilient ecosystems to safeguard the well-being of their residents. Urban forests provide a range of invaluable services and benefits to city dwellers. They improve air quality, mitigate the urban heat island effect, foster biodiversity, and ultimately contribute to the health and well-being of urban populations. Recognized as "Green Guardians," urban forests act as nature-based solutions (NbS) to combat global air pollution and climate change impacts.

This book explores the complex morpho-physio-biochemical and molecular changes that enable urban forests to flourish amidst climate change and environmental pollution. It examines how urban trees modulate their morphological structures, including root systems and physiological and biochemical functioning, to withstand changing urban climatic and environmental conditions. Advancements in molecular biology provide insights into the genetic traits associated with stress tolerance, guiding the selection and breeding of resilient tree species. This book explains the molecular mechanisms and practical strategies for enhancing urban forest resilience through urban planning, adopting stress-tolerant species, and sustainable management practices.

Moreover, the book emphasizes the importance of public awareness and engagement in fostering environmental stewardship and ensuring the long-term success of urban forest initiatives. It aims to bridge the gap between scientific research and practical applications, offering a comprehensive foundation for understanding the mechanisms underpinning urban tree responses to environmental changes. This book is valuable for researchers, academicians, urban planners, policymakers, and practitioners. It provides valuable insights and strategies for creating and sustaining resilient urban forests that mitigate climate change impacts, improve air quality, and enhance the well-being of urban communities in a rapidly changing world.

Dehradun, India

Hukum Singh

Contents

About the Editor

Dr. Hukum Singh is a senior scientist specializing in Plant Physiology and Climate Change at the Plant Physiology Discipline of ICFRE-Forest Research Institute, Dehradun, Uttarakhand, India. He holds master's degree in three major subjects: Botany, Environmental Science, and Economics. Dr. Singh obtained his doctoral degree from the Department of Plant Physiology, G. B. Pant University of Agriculture and Technology, Pantnagar, India. His current research areas are forests and climate change, forest fire, urban forests, urban climate and environmental pollution, plant functional traits, ecophysiology and urban plant physiology, plant phenology, climate-smart forestry, greenhouse gas emissions, carbon-water-energy flux, climate change modeling, and plant response to abiotic and biotic stress. Dr. Singh has authored over 100 publications, including research publications, review articles, books and book chapters, and conference proceedings with journals and publishers of international repute.

Climate Change and Environmental Pollution: Biological Consequences on Urban Forests

Hukum Singh

Abstract Climate change and environmental pollution are formidable threats to the integrity and functionality of urban forests. This chapter provides a detailed exploration of the biological ramifications of these phenomena on urban forests, with a specific focus on phenological, physiological, biochemical, molecular, and gene expression aspects, elucidating their profound effects on plant growth and development. The influence of climate change on urban forests is evident in disrupted phenological patterns, induced physiological stress, biochemical imbalances, and molecular-level alterations in gene expression, all of which hinder plant growth and development. Similarly, environmental pollution instigates morphological, physiological, biochemical, and molecular changes in urban forest ecosystems, jeopardizing their overall health and resilience. This chapter delves into adaptation and mitigation strategies to boost urban resilience against these challenges. These strategies encompass selecting climate-resilient tree species, enhancing urban forest management practices, improving soil health and fertility, and implementing green infrastructure projects. Furthermore, it addresses the policy implications of these challenges, providing recommendations for policymakers to devise and implement effective strategies for alleviating the adverse impacts of climate change and environmental pollution on urban forests. The insights derived from this chapter have the potential to assist urban planners and policymakers in safeguarding the indispensable ecological functions and services provided by urban forests. Ensuring the sustained contribution of urban forests to the well-being of urban ecosystems and communities paves the way for a more resilient and sustainable future.

Keywords Biological consequences · Climate change · Environmental pollution · Urban forests

H. Singh (✉)
Plant Physiology Discipline, ICFRE-Forest Research Institute, P. O. New Forest Dehradun, Dehradun, Uttarakhand 248006, India
e-mail: hukumsingh97@yahoo.com

H. Singh (ed.), *Urban Forests, Climate Change and Environmental Pollution*,
https://doi.org/10.1007/978-3-031-67837-0_1

1 Introduction

Climate change presents an urgent global emergency unfolding at a pace exceeding predictions. The gradual elevation in global temperatures arises from the amplification of the greenhouse effect caused by the excessive release of greenhouse gases (GHGs). These gases absorb infrared radiation, producing atmospheric heat (Kweku et al. 2018; Kumar et al. 2022; Singh 2021). Urban forests are critical components of urban ecosystems, providing numerous ecological, social, and economic benefits. However, they face significant threats from climate change and environmental pollution, which have profound biological consequences (Lu et al. 2018; Kumar et al. 2024; Singh et al. 2024a). Climate change exacerbates existing stressors on urban forests, leading to altered precipitation patterns, increased temperatures, and more frequent extreme weather events. These changes disrupt the phenology of urban trees, affecting flowering, leaf emergence, and fruiting, consequently impacting wildlife populations dependent on these resources (Kumar et al. 2021a, b, c, d, e, f; Singh et al. 2024b). Furthermore, rising temperatures and prolonged drought periods contribute to increased tree mortality rates and reduced canopy cover, diminishing urban forests' overall health and resilience.

Environmental pollution, including air pollution and soil contamination, poses additional challenges to urban forest ecosystems. Elevated levels of ozone, nitrogen oxides, and particulate matter impair photosynthesis and nutrient uptake and induce morphological, physiological, and biochemical changes, impacting tree growth and development. Soil pollution from industrial activities and improper waste disposal lead to nutrient imbalances, soil compaction, and toxicity, inhibiting root growth and microbial activity essential for tree health (Singh et al. 2023; Dhyani et al. 2021). The biological consequences of climate change and environmental pollution on urban forests extend beyond individual trees to entire ecosystems (Singh and Verma 2013a, b; Prakash et al. 2022). Declines in tree health and biodiversity have cascading effects on associated fauna, including birds, insects, and mammals, disrupting ecological interactions and ecosystem functioning. Moreover, compromised urban forests provide reduced ecosystem services, such as air purification, carbon sequestration, and stormwater management, affecting urban residents' well-being and quality of life (Rahim et al. 2024).

Additionally, climate change may increase the risk of wildfires in regions where urban forests interface with wild lands. Warmer temperatures and drier conditions create favorable conditions for fire ignition and spread, posing significant threats to human communities and urban forest ecosystems (Radeloff et al. 2018; Singh 2021; Sharma and Singh 2021). Importantly, these changes in urban forest dynamics have implications for ecosystem services provided by urban forests, such as improved air quality, carbon sequestration, and urban cooling (Gupta et al. 2019b). Climate change alters urban forests' capacity to deliver these services, potentially impacting human health and well-being (Escobedo et al. 2011; Singh et al. 2021). Therefore, urban planners and policymakers must implement various adaptation strategies to mitigate climate change and environmental pollution impacts. These strategies include selecting climate-resilient tree species, improving soil management practices, enhancing green infrastructure, and promoting community engagement in urban forest stewardship (McDonald et al. 2020; Singh et al. 2020; Kumar et al. 2021a).

2 Climate Change Effects on Urban Forests

Climate change influences tree health, biodiversity, and ecosystem services of urban forests. Rising temperatures induce heat stress, reducing growth rates, heightening vulnerability to pests and diseases, and elevated mortality rates (Allen et al. 2010; Kumar et al. 2021b). Besides, it disrupts pest and disease dynamics, influencing the distribution and behavior of urban forest pests and pathogens. Rising temperatures may facilitate the expansion of pest ranges, while changes in precipitation patterns create conditions favorable for disease outbreaks (Pureswaran et al. 2015; Kumar et al. 2021c, d). Furthermore, shifts in climate may lead to changes in urban forest composition, favoring species more tolerant to warmer conditions or altered precipitation regimes. This could result in biodiversity loss and reduced ecosystem resilience as native species struggle to adapt or migrate (Kumar et al. 2022).

2.1 Phenological Modulations

Urban forests, crucial components of urban ecosystems, provide many ecological, social, and economic benefits to urban communities. However, these forests are increasingly susceptible to the impacts of climate change, particularly in terms of phenological shifts—the timing of critical biological events in plant and animal life cycles (Pureswaran et al. 2015; Yadav et al. 2019a, b; Gupta et al. 2019a). Understanding the phenological responses of urban forests to climate change is essential for effective management and conservation efforts (Singh and Kumar 2022a, b).

Climate change influences various aspects of phenology, i.e., leaf emergence, flowering, fruiting, and leaf senescence. Besides, the temperatures and precipitation patterns change photoperiods, triggering phenological events (Kumar et al. 2019b). The urban heat island effect causes earlier leaf emergence and keeps their leaves for extended periods in response to rising temperatures. Similarly, variations in precipitation patterns influence flowering and fruiting, potentially altering pollination and seed dissemination processes in urban forests (Gupta et al. 2019a; Kumar et al. 2019c; Tiwari et al. 2021). For example, the imbalance between the timing of plant flowering and the emergence of pollinators may lead to reduced reproductive success for both plants and pollinators. Furthermore, changes in leaf emergence and senescence timing alter nutrient cycling and carbon sequestration rates in urban forests, affecting ecosystem productivity and resilience (Polgar and Primack 2011; Kumar et al. 2019c).

2.2 Physiological Modulations

Temperatures induce heat stress in urban trees, disrupting physiological processes like photosynthesis and transpiration. This stress is compounded by water scarcity in urban areas, exacerbating tree challenges, reducing growth rates and leaf scorch, and heightened susceptibility to pests and diseases (Kumari and Singh 2018; Sharma et al. 2018; Singh et al. 2018). Furthermore, precipitation patterns, such as increased droughts and heavy rainfall, disrupt the water balance of urban trees. Drought conditions lead to water scarcity, causing wilting, leaf loss, and even mortality in vulnerable tree species. Conversely, intense rainfall triggers soil erosion, root damage, and nutrient leaching, compromising tree health (Wikle et al. 2021; Gupta et al. 2018a).

Moreover, rising atmospheric CO_2 levels in urban premises, while potentially stimulating photosynthesis and growth in some species, alter leaf chemistry and nutrient content, influencing tree-insect interactions and ecosystem dynamics (Kumar et al. 2021a; Singh 2024). In addition, climate change impacts the distribution and abundance of pests and pathogens affecting urban trees. Urban heat and milder winters facilitate the spread of pests like bark beetles and pathogens such as fungal diseases, leading to outbreaks and decline of susceptible tree species (Raffa et al. 2015; Gupta et al. 2018b). Mismatches between leaf emergence and peak insect activity impact herbivory levels and tree fitness (Meineke et al. 2021). Hence, climate-induced shifts in the phenology of urban forests and insect pest incidences ultimately affect the overall functioning of the urban ecosystem.

2.3 Biochemical Modulations

Climate change poses significant challenges to the biochemical processes of plants, with potential implications for ecosystem functioning, productivity, and global food security. These changes in biochemical processes have direct and indirect consequences on human health, including disease distribution shifts and productivity disruptions, leading to nutritional deficiencies (Haines et al. 2006).

Elevated atmospheric carbon dioxide (CO_2) levels stimulate photosynthesis in plants, as they take up more CO_2 for photosynthetic processes. However, prolonged exposure to high CO_2 levels over threshold value leads to a down-regulation of photosynthetic enzymes, reducing the efficiency of photosynthesis over time (Dusenge et al. 2019).

Water stress in urban areas disrupts biochemical processes like photosynthesis and respiration, as well as nutrient uptake and transport. Water stress produces reactive oxygen species (ROS), causing oxidative damage to plant cells (Zeppel et al. 2014). In addition, high temperatures denature enzymes involved in photosynthesis, respiration, and other metabolic pathways, reducing their efficiency (Manoj et al. 2021). Conversely, some plants may experience cold stress if temperatures fluctuate unpredictably, leading to altered enzyme activity and metabolic rates.

Furthermore, it is very interesting to note that climate change influences the production of secondary metabolites in plants, including phenolics, flavonoids, and terpenoids. These compounds are essential in plant defense against biotic and abiotic stresses (Qaderi et al. 2023). Changes in temperature, water availability, and CO_2 levels alter the synthesis and accumulation of these compounds, impacting plant resilience to environmental challenges (Kumar et al. 2021b; Joshi et al. 2020).

Changing climatic conditions cause nutrient imbalance, disrupting nutrient cycling and leading to nutrient availability imbalances in the soil. This affects plants' essential nutrient uptake, impacting biochemical processes such as nitrogen fixation, protein synthesis, and enzyme activity. Nutrient imbalances in urban areas compromise plant health and productivity, making plants more susceptible to pests, diseases, and other stressors (Sistla and Schimel 2012; Singh and Verma 2013a, b; Kumar et al. 2017).

Changing climatic variability in urban premises alters the symbiotic relationship of organisms with plants. The modulations in symbiotic relationships directly impact plant biochemistry through mechanisms such as herbivory, pathogen infection, and mycorrhizal symbiosis (Van der Putten et al. 2010). Further, changes in biotic interactions influence plant metabolism, defense mechanisms, and adaptive resilience of plants to urban climate.

2.4 Molecular and Gene Expression

Climatic variability impacts urban forests, modulating molecular and gene expression dynamics. Elevated temperatures prompt the upregulation of genes associated with heat shock proteins, which is crucial for thermal tolerance. Concurrently, alterations in precipitation patterns induce drought conditions, activating genes linked to drought tolerance mechanisms, including the synthesis of osmoprotectants and antioxidants, crucial for water retention and stress mitigation (Lamelas Penas 2023). Moreover, elevated atmospheric CO_2 levels directly influence gene expression, potentially altering photosynthetic rates and carbon assimilation processes (Kumar et al. 2022). Changes in phenology, driven by climate-induced shifts in environmental cues, disrupt the timing of biological events, influencing gene expression patterns related to developmental stages (Heard et al. 2012). Climate-driven changes in pest and pathogen distributions affect gene expression in plant defense mechanisms, including the modulation of genes encoding pathogenesis-related proteins (Pureswaran et al. 2015). Furthermore, climate-induced stressors trigger epigenetic modifications, altering gene expression without changing DNA sequences, thus contributing to phenotypic plasticity in urban forest trees. Therefore, a strong understanding of these molecular responses is pivotal for formulating effective strategies to support the resilience of urban forests in the wake of climate change.

2.5 *Plant Growth and Development*

Urban forests, though vital for mitigating the impacts of urbanization. The urban climate affects various processes, impacting the growth and development of urban forests. Temperature-induced plant heat stress affects photosynthesis, water uptake, and nutrient absorption (Nowak and Greenfield 2012; Gupta and Singh 2017). Altered precipitation patterns may stress urban trees by affecting soil moisture levels. Drought stress leads to leaf scorch, premature leaf drop, and decreased growth, particularly in species with shallow root systems.

Conversely, heavy rainfall and flooding inundate urban soils, leading to water logging and root suffocation (Rosenzweig et al. 2011). Flood stress in urban areas during the rainy season causes tree decline, root rot, and even mortality in some cases. Increased atmospheric CO_2 concentrations stimulate photosynthesis and plant growth to some extent, known as the carbon fertilization effect (Sharma et al. 2018). However, the importance of this benefit is limited by other factors such as nutrient availability, water stress, and temperature extremes.

Urban temperatures facilitate the range expansion of pests and pathogens that affect urban trees, increasing the prevalence of diseases (Raum et al. 2023; Singh et al. 2017a). Further, urban forests face new challenges from invasive species or novel disease outbreaks. Climate stressors weaken the immune systems of plants, making them more susceptible to pest and disease attacks (Tubby and Webber 2010). Stressed trees exhibit symptoms such as dieback, canopy thinning, and increased vulnerability to secondary pests.

Urban soils often suffer from compaction, contamination, and nutrient depletion, impairing plant root growth and nutrient uptake. Climate stressors exacerbate soil degradation, compromising urban trees' health and resilience (Kumar et al. 2020a, b; Singh et al. 2017b). Intense rainfall events associated with climate change increase soil erosion in urban forests, leading to topsoil loss, tree roots destabilization, and overall soil structure degradation, resulting in reduced growth and development of urban forests (Peterman and Bachelet 2012; Meena et al. 2011; Kumar et al. 2019c).

3 Environmental Pollution Impacts on Urban Forests

Environmental pollution and climate change often interact synergistically to impact urban forest ecosystems. For example, air pollution exacerbates climate change's effects on plant physiology, making trees more susceptible to heat stress, drought, and other environmental challenges (Gupta et al. 2018a, b; Kumar et al. 2020a, b; Singh and Kumar 2022a, b). Conversely, climate change influences pollutants' dispersion and deposition, altering their impacts on urban forest biochemistry.

3.1 Morphological Changes

Urban environmental pollution is a significant issue that globally impacts the morphology of forests, leading to changes in their structure, growth patterns, and overall health. Air pollutants such as ozone, sulfur dioxide, nitrogen oxides, and particulate matter cause visible damage to leaves, including necrosis (dead tissue), chlorosis (yellowing), and stippling (tiny spots). These pollutants interfere with photosynthesis and chlorophyll production, reducing leaf size, altered shape, and abnormal vein patterns (Thomas and Hendricks 1961; Singh 2024). High levels of air pollution inhibit plant growth by disrupting metabolic processes and nutrient uptake. Pollutants like ozone and sulfur dioxide penetrate leaf stomata, interfering with gas exchange and reducing carbon dioxide uptake, essential for photosynthesis (Kumar et al. 2022; Yadav et al. 2019a, b). As a result, plants may exhibit reduced shoot elongation, smaller overall size, and delayed or suppressed flowering and fruiting. Soil pollution, often associated with heavy metals, industrial contaminants, and excess salts, harms root systems and inhibits water and nutrient uptake (Sandeep et al. 2019; Jangra et al. 2024).

Compacted urban soils, a common issue in urban areas, restrict root growth and oxygen availability, further exacerbating plant stress. Damaged or restricted root systems can lead to poor anchorage, reduced stability, and increased susceptibility to drought and disease (Bauhus et al. 2017). Pollution-induced stress causes canopy thinning and dieback in urban trees, reducing foliage density and canopy cover (Kumar et al. 2022; Kanta et al. 2024). This may display cascade effects on urban microclimates, increasing solar radiation and surface temperatures, further stressing plants, and exacerbating environmental conditions. It has been reported that environmental pollution disrupts plant phenology, affecting the timing of leaf emergence, flowering, and fruiting. For example, exposure to air pollutants may delay budburst or flowering in certain species, leading to mismatches with pollinators or changes in seed dispersal patterns (Neil and Wu 2006). Pollution-tolerant species may dominate urban forests, replacing sensitive or less-adapted species. This leads to species composition shifts and reduced biodiversity, potentially impacting ecosystem stability and resilience. Chronic exposure to pollution demonstrates cumulative and long-lasting effects on plant morphology and fitness. Over time, repeated exposure may lead to genetic changes within plant populations, favoring traits that confer tolerance or resistance to pollutants.

3.2 Physiological Changes

The urban environmental pollution impacts various aspects of plants. Few changes in urban plants are pollutant-specific. Elevated ozone levels in urban areas lead to plant oxidative stress, damaging cell membranes and disrupting photosynthesis

(Tiwari et al. 2021). This leads to lower chlorophyll concentration, lower photosynthetic rates, and slower growth. Besides, particulate matter from vehicle emissions and industrial activities settles on leaf surfaces, blocking stomata and reducing gas exchange, hindering photosynthesis, limiting carbon dioxide uptake, and increasing plant water stress (Sandeep et al. 2019).

Nitrogen deposition from vehicle exhaust and fertilizer runoff on leaf and soil surfaces, respectively causes nitrogen saturation. Nitrogen deposition disturbs nutrient balance, alters root morphology, and makes plants susceptible to diseases and pests (Singh and Verma 2013a, b; Malik et al. 2024). Urban soils contain elevated levels of heavy metals from industrial activities, vehicle emissions, and improper waste disposal (Kumar et al. 2022). Accumulation of heavy metals such as lead, cadmium, and mercury inhibits root growth, impairs nutrient uptake, and disrupts physiological processes within plants (Meena et al. 2011; Singh et al. 2017b).

Soil compaction is a crucial problem in urban areas. Urbanization often leads to soil compaction due to construction activities, foot traffic, and heavy machinery. Compacted soils have reduced porosity, limiting oxygen availability to roots and impeding water infiltration (Peterman and Bachelet, 2012). All these changes result in root asphyxiation, reduced nutrient uptake, and stunted growth.

In addition, water sources in urban areas are contaminated with pollutants such as heavy metals, pesticides, and industrial chemicals (Singh et al. 2022a, b). Irrigation with polluted water introduces toxins to plant roots, leading to physiological stress, reduced growth, and leaf damage (Kumar et al. 2021f; Kumar and Singh 2024). Furthermore, nutrient runoff from urban landscapes leads to eutrophication of water bodies within or adjacent to urban forests. Excessive nutrient levels promote algal blooms, deplete oxygen levels, and disrupt aquatic ecosystems, indirectly impacting plant health through altered water quality and availability (Janke et al. 2017). This problem is more prominent in blue-green urban areas. Urban areas are warmer than surrounding rural areas due to heat-absorbing pollutants and surfaces (Haines et al. 2006). Elevated temperatures can accelerate plant metabolism, increase water loss through transpiration, and exacerbate heat stress, causing physiological drought, leaf wilting, and ultimately declining growth rates of urban vegetation (Kumar et al. 2022).

3.3 Biochemical Changes

Environmental pollution leads to biochemical changes in urban forests, affecting the health and functioning of trees and other vegetation. Urban air pollutants directly damage plant tissues and disrupt biochemical processes (Chmielewski et al. 1998). For example, ozone impairs photosynthesis by damaging chloroplasts and interfering with the functioning of photosynthetic enzymes.

Heavy metals accumulate in plant tissues and soil, disrupting enzyme activity in various metabolisms and impairing plant growth and development. Soil pollution alters soil microbial communities and nutrient cycling processes, affecting the availability of essential nutrients to plants. Plants absorb contaminants and translocate to aboveground tissues, impacting biochemical pathways and plant health. Further, water pollution leads to nutrient imbalances, salinity stress, and the accumulation of toxic substances in plant tissues (Meuser 2010). Contaminated water in urban areas modulates physiological processes such as osmosis, nutrient uptake, and transpiration, affecting overall plant health and vitality.

Environmental pollution influences the production of secondary metabolites in urban vegetation. Pollution exposure may induce the synthesis of defensive compounds such as phenolics, flavonoids, and antioxidants (Lahiri and Krishna 2024). However, chronic exposure to high levels of pollutants may also suppress secondary metabolite production, compromising plant resilience to stressors such as pests, pathogens, and environmental extremes.

3.4 *Molecular and Gene Expression*

Pollutants such as heavy metals, ozone, nitrogen oxides, and volatile organic compounds (VOCs) incite oxidative stress in plants by generating reactive oxygen species (ROS) within their cells. Plants activate diverse defense mechanisms to counteract oxidative stress, including regulating genes involved in antioxidant production, detoxification pathways, and repair mechanisms (Garg et al. 2019). Pollution disrupts plants' normal regulation of gene expression, resulting in modified gene transcription and translation patterns. Genes associated with stress responses, such as those coding for heat shock proteins, pathogenesis-related proteins, and stress-responsive transcription factors, may undergo upregulation in reaction to pollution-induced stress (Rehem et al. 2012). The disruption in photosynthesis triggers alterations in the expression of genes responsible for carbon fixation, photosystem protection, and stomatal regulation (Heber et al. 1995). Pollutants potentially impede plants' uptake and assimilation of essential nutrients, leading to nutrient imbalances and deficiencies. Genes responsible for nutrient transport, uptake, and metabolism may exhibit differential expression in response to pollution stress as plants endeavor to maintain nutrient homeostasis. Plants react to pollution stress by triggering defense and detoxification pathways to alleviate the adverse effects of pollutants. Genes encoding enzymes involved in detoxification processes, such as cytochrome P450 monooxygenases, glutathione S-transferases, and ABC transporters, may be induced to detoxify and eliminate pollutants from plant tissues (Aioub et al. 2021). Environmental pollution induces epigenetic alterations in plants, such as changes in DNA methylation, histone modifications, and small RNA expression. These epigenetic modifications influence gene expression patterns and phenotypic adaptability, enabling plants to adapt to pollution stress across successive generations.

3.5 Plant Growth and Development

Elevated particulate matter (PM) levels and gaseous pollutants directly damage plant tissues, particularly leaf surfaces of urban vegetation. This damage can impair photosynthesis, reduce the plant's ability to uptake nutrients and weaken its overall health. Elevated ground-level ozone induces oxidative stress in plants, causing chlorosis (yellowing) and necrosis (death) of leaf tissues. Ozone damage reduces photosynthetic efficiency, inhibits growth, and makes plants more susceptible to diseases and pests, hindering plant growth and development.

Contamination of urban soils with heavy metals adversely impacts plant growth and development. Heavy metals like lead, cadmium, and mercury accumulate in plant tissues, inhibiting root growth, nutrient uptake, and photosynthesis growth and productivity (Escobedo et al. 2011). Moreover, acid deposition in urban areas guides soil acidification, affecting nutrient availability and soil microbial activity. Acidic soils hinder plant root growth and nutrient uptake, leading to stunted growth and nutrient deficiencies. Urban forests are exposed to water pollution from runoff containing heavy metals, pesticides, fertilizers, and other chemicals (Chatterjee et al. 2017). The contaminated water harms plants by affecting root health, inhibiting water uptake, and causing toxicity symptoms such as leaf discoloration, wilting, and necrosis.

The synthesis of global research presented in the abovementioned sections suggested that pollution exacerbates climate change's impacts on urban forests, making plants more susceptible to heat waves, drought, and extreme weather events. However, combined effects of pollution and climate change alter plant phenology, distribution patterns, and species composition in urban forests, affecting ecosystem dynamics and biodiversity of urban areas (Allen et al. 2010).

4 Resilience Strategies and Adaptation Measures

4.1 Selection of Climate-Resilient Plantation Species

Selecting climate-resilient plantation species is crucial in adapting and mitigating urban climate change and environmental pollution through forestry interventions. This approach may guide urban planners and policymakers in maintaining the health and functionality of forest urban ecosystems.

Understanding the climate variable-plant interactions is crucial before selecting climate-resilient urban plantation species. Climate models and in-situ experiments (Open top chamber approach, free air CO_2 enrichment experimentation, temperature gradient tunnel, etc.) provide valuable insights into future impacts of changing climatic and environmental scenarios on plant species, helping to identify suitable species that thrive under changing climatic and environmental circumstances.

Identifying resilient plantation species based on plant functional traits is suggested. Traits to consider include drought tolerance, heat tolerance, resistance to pests and diseases, ability to withstand flooding or waterlogging, and flexibility in phenological timing. The selected plantation species, based on functional traits, possess adaptive characteristics that enable them to tolerate or adapt to environmental stresses associated with climate change (Kumar et al. 2022).

Assessing species diversity is essential for ecosystem resilience, increasing the likelihood of some species adapting to changing conditions. Selecting a diverse mix of tree species with various adaptive traits may enhance resilience and reduce the risk of widespread losses due to climate change impacts, pollution, pest outbreaks, etc. (Thompson et al. 2009).

Promoting native species well-adapted to local climatic conditions and soil types makes them inherently more resilient to regional and urban environmental changes. Promoting the use of native species helps preserve biodiversity, support ecosystem functions, and maintain the genetic diversity necessary for adaptation to changing conditions (Keenan 2015).

While native species are preferred for their ecological benefits, incorporating non-native species in some cases exhibits desirable resilience traits that make them suitable for planting in urban and disturbed environments (Keenan 2015). Urban planners, policymakers, and land managers are advised to consider the risks of introducing non-native species, such as invasiveness or disruption of native plant communities (Kumar et al. 2019a).

Utilizing climate-resilience tools and resources are available to assist in selecting climate-resilient tree species, including climate suitability maps, species distribution models, and databases of species traits and tolerances (Singh et al. 2023). These tools help land managers and urban planners decide which species best suit specific environmental conditions and climate scenarios.

Monitoring the plantation performance of selected tree species over time is essential for assessing their resilience and identifying any unforeseen challenges or opportunities (Keenan 2015). Adaptive management practices allow for adjustments to planting strategies based on empirical observations and feedback, ensuring urban forests' continued success and resilience in ongoing environmental changes.

Hence, prioritizing the selection of climate-resilient tree species and incorporating them into forest management and urban planning efforts offers stakeholders a way to enhance the resilience of forest ecosystems, support biodiversity conservation, and mitigate the impacts of climate change and environments in urban areas.

4.1.1 Adaptation Strategies

Planting climate-resilient tree species adapted to local conditions may enhance the long-term viability of urban forests. Moreover, implementing soil conservation practices such as mulching, aeration, and amendment improves soil health and supports plant growth. Installing irrigation systems or rainwater harvesting techniques helps mitigate the impacts of drought stress and waterlogging on urban trees (Esperon-Rodriguez et al. 2022). Adopting integrated pest management strategies reduces reliance on chemical pesticides and minimizes the spread of pests and diseases in urban forests. Integrating green infrastructure and urban forestry principles into urban planning promotes resilient urban landscapes that support plant growth and biodiversity while mitigating climate change and environmental impacts (Mell 2009). Implementing these adaptation strategies and promoting sustainable management practices in urban forests better withstand the challenges posed by climate change, ensuring their contribution to urban ecosystems, human well-being, and urban environmental quality.

4.2 Improving Urban Forest Management Practices

Urban forests are critical in enhancing cities' livability, sustainability, and resilience. Effective management practices are essential to ensure the health and vitality of urban forests in the era of climate change, pollution, and urbanization. Developing and implementing comprehensive urban forest management plans that integrate ecological, social, and economic objectives is a prerequisite for adaptive mechanisms. Establish policies and regulations to protect existing plantations, promote tree planting initiatives, and guide sustainable land use practices in urban areas (Ottitsch and Krott 2005). Similarly, collaboration with relevant stakeholders, including local governments, community organizations, and urban planners, is required to implement urban forest management strategies effectively. Prioritize selecting climate-resilient tree species that are well-adapted to local environmental conditions, including temperature, soil type, and moisture availability. Different species and age classes enhance species richness, genetic diversity, and ecosystem resilience within urban forests (Singh 2013). Implement proper planting techniques, including appropriate spacing, soil preparation, and mulching, to promote healthy root development and long-term tree growth. Establish regular maintenance programs for pruning, watering, fertilizing, and pest management to ensure the health and vitality of urban trees.

Apart from this, integrating urban forestry principles into green infrastructure planning and design significantly maximizes the benefits of trees and vegetation in urban environments. Green infrastructure strategically uses natural systems and vegetation to provide environmental, social, and economic benefits in urban and rural areas. Implementing green infrastructure projects is crucial for enhancing resilience to climate change and promoting sustainable development (Abunnasr 2013). Incorporating trees, green spaces, and vegetative buffers into urban development projects helps mitigate the impacts of heat islands, stormwater runoff, air pollution, and noise pollution. There is a need to design multifunctional green spaces that provide recreational opportunities, habitat for wildlife, and ecosystem services while enhancing urban aesthetics and quality of life (Abunnasr 2013).

Soil amendment is a critical activity that significantly contributes toward sustainable forest management. Implement soil conservation practices, such as mulching, composting, and soil aeration, to improve soil health, structure, and fertility in urban forests (Meuser 2010). Restore degraded urban soils through remediation techniques, including phytoremediation, biochar application, and green infrastructure development (Garg et al. 2019). Further, minimizing soil compaction, erosion, and pollution through sustainable land management practices, including limiting impervious surfaces and reducing vehicle traffic in urban areas, is another management practice enhancing ecosystem resilience. Recycling organic waste materials through composting may enrich soil with essential nutrients, improve soil structure, and enhance water retention capacity (Chatterjee et al. 2017). Applying organic mulches such as wood chips, straw, or leaf litter to the soil surface helps suppress weeds, conserve soil moisture, regulate soil temperature, and promote the decomposition of organic matter. Applying soil amendments such as gypsum, lime, or biochar adjusts soil pH, improves nutrient availability, and enhances soil structure. Implementing soil restoration techniques rebuild degraded soils, enhance biodiversity, and restore ecosystem services (Fajeriana et al. 2024).

Refining and implementing nutrient management strategies is crucial for maintaining the health and sustainability of urban forests. Precision management techniques, including soil testing and optimized fertilizer application, play pivotal roles in this endeavor. Furthermore, nutrient management in urban forests should prioritize using organic and slow-release fertilizers, which offer sustained nutrient availability while reducing the risk of leaching and runoff. Compost and other organic amendments in soil management plans enhance soil structure and fertility, fostering healthier root systems and tree vigor (Chaudhari et al. 2020).

5 Policy Implications and Recommendations

Urban forests play a crucial role in mitigating the impacts of climate change and environmental pollution while providing numerous benefits to urban dwellers. However, they face significant challenges due to the adverse effects of climate change and pollution, which disrupt ecosystem functioning and compromise the health and resilience of urban forests. Addressing these challenges requires comprehensive policies and strategic interventions.

5.1 Policy Implications

There is a need to implement comprehensive policies for urban forest management that prioritize preserving and expanding green spaces within urban areas. Establish sustainable land use regulations to prevent deforestation and ensure the integration of green infrastructure in urban development projects (Nowak and Greenfield 2010). Strengthen air quality standards to decrease pollutants like nitrogen oxides, sulfur dioxide, and particulate matter, negatively impacting urban forest ecosystems

(McPherson et al. 2005). Besides, policies should be implemented to promote the adoption of cleaner energy sources and technologies to mitigate air pollution. This approach may reduce greenhouse gas emissions to alleviate the effects of climate change on urban forests. Reforestation and aforestation initiatives as nature-based solutions should also be promoted to sequester carbon dioxide and enrich urban biodiversity (Escobedo et al. 2011; Naumann et al. 2011a, b). It is very worthwhile to develop strategies for biodiversity conservation to safeguard urban forest ecosystems and shield vulnerable species from the detrimental effects of climate change and pollution (Givnish 2015).

5.2 Recommendations

- Reduction of GHG emissions by implementing policies to reduce greenhouse gas emissions and air pollutants. These policies include transitioning to renewable energy sources, promoting energy efficiency measures, and adopting stricter industrial and vehicle emissions regulations in urban areas (Escobedo et al. 2011).
- Incorporate green infrastructure and urban forestry into city planning and development processes. Designating green spaces, planting trees, and preserving existing urban forests can help mitigate the urban heat island effect, improve air quality, and enhance biodiversity (Benedict and McMahon 2002). Municipalities should prioritize and integrate green infrastructure projects into long-term urban planning strategies.
- Implement sustainable forest management practices in urban areas to enhance the resilience of urban forests to climate change and pollution. This includes tree planting programs, regular maintenance of urban green spaces, and removing invasive species (Keenan 2015).
- Promoting diverse tree species composition and age structure may improve urban forests' adaptive capacity.
- Launch public awareness campaigns and educational programs to engage communities in environmental conservation efforts (Ottitsch and Krott 2005). Increasing public understanding of the importance of urban forests for mitigating climate change and pollution fosters support for policy initiatives and encourages individual actions such as tree planting and sustainable lifestyle choices.
- Investing in scientific research and monitoring programs to better understand the biological consequences of climate change and pollution on urban forests is strongly recommended. This includes studying the impacts of temperature fluctuations, air pollution exposure, and extreme weather events on tree health and ecosystem dynamics (Tiwari et al. 2021). The research findings strongly inform evidence-based policy decisions and adaptive management strategies.
- Encourage collaboration among government agencies, non-profit organizations, academic institutions, and community groups to address urban forests' complex

challenges (Tiwari et al. 2021). Collective expertise and resources leverage stakeholders to develop holistic approaches to climate change adaptation and pollution mitigation in urban environments.

- Integrate urban forestry goals into broader climate action and environmental policies at the local, regional, and national levels. Policymakers and urban planners should recognize urban forestry as an essential component of climate adaptation and mitigation strategies, with dedicated funding and institutional support to implement effective policies and programs.
- Promote the development of green economy initiatives that support sustainable land management practices and create green jobs in urban forestry and ecological restoration. Investing in green infrastructure projects and nature-based solutions generates economic benefits while enhancing urban ecosystems' resilience (Jim 2013).

6 Conclusion

The synthesis underscores the profound and multifaceted biological impacts of climate change and environmental pollution on urban forests. From alterations in phenology and physiology to shifts in biochemistry and molecular processes, these stressors pose significant challenges to the health and vitality of urban green spaces. However, within these challenges lie opportunities for resilience and adaptation. Urban planners and policymakers may adopt strategies such as carefully selecting climate-resilient tree species, enhancing urban forest management practices, and implementing green infrastructure projects to fortify urban forests against the adverse effects of climate change and environmental pollution. Moreover, robust policy interventions such as emission reduction, integrating urban forestry into urban planning, public awareness, education initiatives, research and monitoring investments, collaborative partnerships, and support for green economy endeavors are imperative to mitigate the biological consequences of climate change and pollution on urban forests. Furthermore, implementing resilience strategies, adaptation measures, and policy recommendations may guide urban planners and land managers toward fostering more resilient and sustainable urban forest ecosystems. Ultimately, these interventions are required in urban ecosystems to continue to provide invaluable ecological, social, and economic benefits to urban communities, ensuring a greener and healthier environment for future generations.

References

Abunnasr Y (2013) Climate change adaptation: a green infrastructure planning framework for resilient urban regions. University of Massachusetts Amherst

Aioub AA, Zuo Y, Li Y, Qie X, Zhang X, Essmat N, Wu W, Hu Z (2021) Transcriptome analysis of Plantago major as a phytoremediator to identify some genes related to cypermethrin detoxification. Environm Sci Pollut Res 28:5101–5115

Allen CD, Macalady AK, Chenchouni H, Bachelet D, McDowell N, Vennetier M, Kitzberger T, Rigling A, Breshears DD, Hogg ET, Gonzalez P (2010) A global overview of drought and heat-induced tree mortality reveals emerging climate change risks for forests. Forest Ecol Manag 259(4):660–684

Bauhus J, Forrester DI, Gardiner B, Jactel H, Vallejo R, Pretzsch H (2017) Ecological stability of mixed-species forests. Mixed Species for Ecol Manag 32:337–382

Benedict MA, McMahon ET (2002) Green infrastructure: smart conservation for the 21st century. Renew Resour J 20(3):12–17

Chatterjee R, Gajjela S, Thirumdasu RK (2017) Recycling of organic wastes for sustainable soil health and crop growth. Int J Waste Resour 7(03):296–302

Chaudhari SK, Biswas PP, Kapil H (2020) Soil health and fertility. The Soils of India, pp 215–231

Dhyani K, Kuniyal HB, Singh H, Sobha S (2021) Growth and physiological potential of *Terminalia arjuna* under elevated CO_2 levels in open top chamber condition. J Appl Nat Sci 13(3):1121–1126

Dusenge ME, Duarte AG, Way DA (2019) Plant carbon metabolism and climate change: elevated CO_2 and temperature impacts on photosynthesis, photorespiration and respiration. New Phytol 221(1):32–49

Escobedo FJ, Kroeger T, Wagner JE (2011) Urban forests and pollution mitigation: analyzing ecosystem services and disservices. Environ Pollut 159(8–9):2078–2087

Esperon-Rodriguez M, Rymer PD, Power SA, Barton DN, Cariñanos P, Dobbs C, Eleuterio AA, Escobedo FJ, Hauer R, Hermy M, Jahani A (2022) Assessing climate risk to support urban forests in a changing climate. Plants People Planet 4(3):201–213

Fajeriana N, Ali A, Rini RP (2024) Soil tillage and planting along the contour on sloping land to minimize the potential for erosion and surface runoff. Sarhad J Agricult 40(1):82–93

Garg N, Saroy K, Cheema A, Bisht A (2019) Microbial diversity in soil: biological tools for abiotic stress management in plants. Plant Biotic Interact State Art 12:283–321

Givnish TJ (2015) Adaptation to sun and shade: a whole-plant perspective. Austr J Plant Physiol 22(6):825–847

Gupta SK, Singh H (2017) Observations on sporadic flowering in *Ficus benjamina* L. and *Peltophorum africanum* L. in new forest. Indian for 143:290–291

Gupta SK, Ram J, Singh H (2018a) Comparative study of transpiration in cooling effect of tree species in the atmosphere. J Geosci Environ Protect 6:151–166. https://doi.org/10.4236/gep.2018.68011

Gupta SK, Ram J, Singh H, Kumar P (2018b) Comparative study of cooling under shades of some forestry tree species with respect to ambient temperatures. J Geosci Environ Protect 6:51–66. https://doi.org/10.4236/gep.2018.611004

Gupta SK, Ram J, Singh H (2019a) Relationship between leaf area index and atmospheric cooling of tree species. Indian J for 42(3):203–208

Gupta SK, Singh H, Ram J (2019b) Need for a novel paradigm in forestry research: perspectives of comparative atmospheric cooling by different forestry effect tree species. Indian for 145(6):509–515

Heard MJ, Riskin SH, Flight PA (2012) Identifying potential evolutionary consequences of climate-driven phenological shifts. Evolut Ecol 26:465–473

Heber U, Kaiser W, Luwe M, Kindermann G, Veljovic-Javonovic S, Yin Z, Pfanz H, Slovik S (1995) Air pollution, photosynthesis and forest decline: interactions and consequences. Ecophysiology of photosynthesis. Springer, Berlin, pp 279–296

Jangra A, Kumar K, Maikhuri S, Bhandari MS, Pandey S, Singh H, Barthwal S (2024) Unveiling stress-adapted endophytic bacteria: characterizing plant growth-promoting traits and assessing cross-inoculation effects on *Populus deltoides* under abiotic stress. Plant Physiol Biochem 210:108610. https://doi.org/10.1016/j.plaphy.2024.108610

Janke BD, Finlay JC, Hobbie SE (2017) Trees and streets as drivers of urban stormwater nutrient pollution. Environ Sci Technol 51(17):9569–9579

Jim CY (2013) Sustainable urban greening strategies for compact cities in developing and developed economies. Urban Ecosyst 16:741–761

Joshi R, Singh H, Chhetri R, Yadav K (2020) Assessment of carbon sequestration potential in degraded and non-degraded community forests in Terai region of Nepal. J Forest Environ Sci 36(2):13–121. https://doi.org/10.7747/JFES.2020.36.2.113

Kanta C, Kumar A, Chauhan A, Singh H, Sharma IP (2024). The interplay between plant functional traits and climate change. In: Kumar N, Singh H (eds) Plant functional traits for improving productivity. Springer, Singapore. https://doi.org/10.1007/978-981-97-1510-7_3

Keenan RJ (2015) Climate change impacts and adaptation in forest management: a review. Ann for Sci 72:145–167

Kumar M, Singh MP, Singh H, Dhakate PM, Ravindranath NH (2019a) Forest working plan for the sustainable management of forest and biodiversity in India. J Sustain for 39:1–22. https://doi.org/10.1080/10549811.2019.1632212

Kumar N, Jeena N, Singh H (2019b) Elevated temperature modulates rice pollen structure: a study from the foothill of the Himalayan agro-ecosystem in India. 3 Biotech 9(5):175. https://doi.org/10.1007/s13205-019-1700-1

Kumar N, Jeena N, Kumar N, Gangola S, Singh H (2019c) Phytoremediation facilitating enzymes: an enzymatic approach for enhancing remediation process. In: Bhatt P (ed) Smart bioremediation technologies. Academic Press, Cambridge, pp 289–306

Kumar N, Singh H, Sharma SK (2020a) Antioxidants: responses and importance in plant defense system. In: Roychowdhury R, Choudhury S, Hasanuzzaman M, Srivastava S (eds) Sustainable agriculture in the era of climate change. Springer, Cham, pp 251–264. https://doi.org/10.1007/978-3-030-45669-6_11

Kumar P, Singh R, Singh H, Chand T, Bala N (2020b) Assessment of soil carbon dioxide efflux and its controlling factors in moist temperate forest of West Himalayas. Curr Sci 19(4):661–669

Kumar A, Kumar P, Singh H, Kumar N (2021a) Modulation of plant functional traits under essential plant nutrients during seasonal regime in natural forests of Garhwal Himalayas. Plant Soil 465:197–212. https://doi.org/10.1007/s11104-021-05003-x

Kumar A, Kumar P, Singh H, Kumar N (2021b) Adaptation and mitigation potential of roadside trees with bio-extraction of heavy metals under vehicular emissions and their impact on physiological traits during seasonal regimes. Urban for Urban Green 58:126900. https://doi.org/10.1016/j.ufug.2020.126900

Kumar A, Kumar P, Singh H, Kumar N (2021c) Impact of plant functional traits on infiltration rate under different forests of Kempty watershed in Garhwal Himalaya. Indian J Soil Conserv 49(1):38–44

Kumar A, Kumar P, Singh H, Bisht S, Kumar N (2021d) Relationship of physiological plant functional traits with soil carbon stock in the temperate forest of Garhwal Himalaya. Curr Sci 120(8):1368–1373

Kumar M, Phukon SN, Singh H (2021e) The role of communities in sustainable land and forest management. In: Forest resources resilience and conflicts. Academic Press, New York. https://doi.org/10.1016/B978-0-12-822931-6.00024-1

Kumar M, Phukon SN, Paygude AC, Tyagi K, Singh H (2021f) Mapping phenological functional types (PhFT) in the Indian Eastern Himalayas using machine learning algorithm in Google Earth Engine. Comput Geosci 158:104982. https://doi.org/10.1016/j.cageo.2021.104982

Kumar A, Singh H, Kumari G, Bisht S, Malik A, Kumar A, Singh M, Raturi BS, Thakur A, Kaushal R (2022) Adaptive resilience of roadside trees to vehicular emissions via leaf enzymatic, physiological, and anatomical trait modulations. Environ Pollut 313:120191. https://doi.org/10.1016/j.envpol.2022.120191

Kumar N, Singh H, Giri K, Kumar A, Joshi A, Yadav A, Singh R, Bisht S, Kumari R, Jeena N, Khairakpam R, Mishra G (2024) Physiological and molecular insights into the allelopathic

effects on agroecosystems under changing environmental conditions. Physiol Mol Biol Plants 30:417–433. https://doi.org/10.1007/s12298-024-01440-x
Kumar N, Singh H (2024) Plant functional traits for improving productivity. Springer, Singapore. https://doi.org/10.1007/978-981-97-1510-7
Kumar N, Suyal DC, Sharma IP, Verma A, Singh H (2017) Elucidating stress proteins in rice (*Oryza sativa* L.) genotype under elevated temperature: a proteomic approach to understand heat stress response. 3 Biotech 7:205. https://doi.org/10.1007/s13205-017-0856-9
Kumari A, Singh H (2018) Impact of heat and drought stress on physiological response of *Terminalia arjuna* grown under elevated temperature condition: an adaptive mechanism to climate change. J Pharm Phytochem 7(1):1374–1378
Kweku DW, Bismark O, Maxwell A, Desmond KA, Danso KB, Oti-Mensah EA, Quachie AT, Adormaa BB (2018) Greenhouse effect: greenhouse gases and their impact on global warming. J Sci Res Rep 17(6):1–9
Lahiri M, Krishna K (2024) Effect of air pollution on plant secondary metabolites in selected trees of Delhi. Environ Qual Manag 33(3):399–409
Lamelas Penas L (2023) Systems biology-derived identification of novel heat stress biomarkers pine species (Bachelor's thesis)
Lu Y, Yuan J, Lu X, Su C, Zhang Y, Wang C, Cao X, Li Q, Su J, Ittekkot V, Garbutt RA (2018) Major threats of pollution and climate change to global coastal ecosystems and enhanced management for sustainability. Environ Pollut 239:670–680
Malik A, Kumar K, Kumari G, Kumar N, Singh H (2024) Application of functional traits in modelling productivity and resilience under climate change. In: Kumar N, Singh H (eds) Plant functional traits for improving productivity. Springer, Singapore. https://doi.org/10.1007/978-981-97-1510-7_5
McDonald RI, Mansur AV, Ascensão F, Colbert ML, Crossman K, Elmqvist T, Gonzalez A, Güneralp B, Haase D, Hamann M, Hillel O (2020) Research gaps in knowledge of the impact of urban growth on biodiversity. Nat Sustain 3(1):16–24
McPherson G, Simpson JR, Peper PJ, Maco SE, Xiao Q (2005) Municipal forest benefits and costs in five US cities. J for 103(8):411–416
Meena D, Singh H, Chaudhari SK (2011) Elucidating strontium response on growth dynamics and biochemical change in *Phaseolus mungo* L. Int J Agricult Environ Biotechnol 4:107–113
Meineke EK, Davis CC, Davies TJ (2021) Phenological sensitivity to temperature mediates herbivory. Glob Change Biol 27(11):2315–2327
Mell IC (2009) Can green infrastructure promote urban sustainability? In: Proceedings of the institution of civil engineers-engineering sustainability, vol 162. Thomas Telford Ltd., pp 23–34
Meuser H (2010) Contaminated urban soils, vol 18. Springer, New York
Naumann S, Anzaldua G, Berry P, Burch S, Davis M, Frelih-Larsen A, Gerdes H, Sanders M (2011a) Assessment of the potential of ecosystem-based approaches to climate change adaptation and mitigation in Europe. Final report to the European Commission, DG Environment
Naumann S, Davis M, Kaphengst T, Pieterse M, Rayment M (2011b) Design, implementation and cost elements of Green infrastructure projects. Final Report, European Commission, Brussels, p 138
Neil K, Wu J (2006) Effects of urbanization on plant flowering phenology: a review. Urban Ecosyst 9:243–257
Nowak DJ, Greenfield EJ (2010) Evaluating the national land cover database tree canopy and impervious cover estimates across the conterminous United States: a comparison with photo-interpreted estimates. Environ Manag 46:378–390
Nowak DJ, Greenfield EJ (2012) Tree and impervious cover change in US cities. Urban for Urban Green 11(1):21–30
Ottitsch A, Krott M (2005) Urban forest policy and planning. Urban forests and trees: a reference book. Springer, Berlin, pp 117–148
Peterman W, Bachelet D (2012) Climate change and forest dynamics: a soils perspective. Soils Food Sec Environ Sci Technol 35:158–182

Polgar CA, Primack RB (2011) Leaf-out phenology of temperate woody plants: from trees to ecosystems. New Phytol 191(4):926–941

Prakash V, Hunney K, Singh H (2022) Effect of elevated carbon dioxide on growth and development of *Santalum album* L. seedlings inoculated with plant growth promoting microorganisms in Open Top Chambers. Int J Environ Agricult Biotechnol 7(5):149–159

Pureswaran DS, De Grandpre L, Pare D, Taylor A, Barrette M, Morin H, Régniere J, Kneeshaw DD (2015) Climate-induced changes in host tree–insect phenology may drive ecological state-shift in boreal forests. Ecology 96(6):1480–1491

Qaderi MM, Martel AB, Strugnell CA (2023) Environmental factors regulate plant secondary metabolites. Plants 12(3):447

Raffa KF, Aukema BH, Bentz BJ, Carroll AL, Hicke JA, Kolb TE (2015) Responses of tree-killing bark beetles to a changing climate. Clim Change Insect Pests 7:173–201

Rahim NH, Razak SA, Chang X, Saun FC, Khan MN, Hamzah SN, Rohman F, Ali B, Kaplan A, Iqbal M, Bozhuyuk MR (2024) Ecosystem services by urban forest (UF) towards climate change adaptation: a review. Polish J Environ Stud 33:3503–3513

Raum S, Collins CM, Urquhart J, Potter C, Pauleit S, Egerer M (2023) Tree insect pests and pathogens: a global systematic review of their impacts in urban areas. Urban Ecosyst 26(2):587–604

Rehem BC, Bertolde FZ, de Almeida AAF (2012) Regulation of gene expression in response to abiotic stress in plants. Cell Metabol Cell Homeo Stress Res 12:13–38

Rosenzweig C, Solecki WD, Hammer SA, Mehrotra S (eds) (2011) Climate change and cities: first assessment report of the urban climate change research network. Cambridge University Press, Cambridge

Sandeep G, Vijayalatha KR, Anitha T (2019) Heavy metals and its impact in vegetable crops. Int J Chem Stud 7(1):1612–1621

Sharma R, Singh H (2021) Alteration in biochemical constituents and nutrients partitioning of *Asparagus racemosus* in response to elevated atmospheric CO_2 concentration. Environ Sci Pollut Res 29:6812–6821. https://doi.org/10.1007/s11356-021-16050-3

Sharma R, Singh H, Kaushik M, Nautiyal R, Singh O (2018) Adaptive physiological response, carbon partitioning, and biomass production of *Withania somnifera* (L.) Dunal grown under elevated CO_2 regimes. 3 Biotech 8:267. https://doi.org/10.1007/s13205-018-1292-1

Singh HCP (2013) Adaptation and mitigation strategies for climate-resilient horticulture. Climate-resilient horticulture: adaptation and mitigation strategies. Springer, New York, pp 1–12

Singh H (2021) An integrated approach considering physiological and biophysical based indicators for assessing tolerance of roadside plantations of *Alstonia scholaris* towards urban roadside air pollution: an assessment of adaptation of plantations for mitigating urban roadside air pollution. Trees. https://doi.org/10.1007/s00468-021-02179-8

Singh H (2024) Exploring adaptive modulation in plant functional traits and its impact on the productivity of *Acacia auriculiformis* under CO_2-enriched environment. Ind Crops Products 210:118186. https://doi.org/10.1016/j.indcrop.2024.118186

Singh H, Verma A (2013a) Physiological responses of rice cultivars to various nitrogen levels. Int J Agricult Environ Biotechnol 6(3):383–388

Singh H, Verma A (2013b) Characterization and screening of high nitrogen efficient rice genotype to curtail environmental pollution. Int J Agricult Environ Biotechnol 6(1):23–30

Singh H, Kumar M (2022a) Climate change and its impact on Indian Himalayan forests: current status and research needs. Climate change: impacts, responses and sustainability in the Indian Himalaya. Springer, Cham, pp 223–242

Singh H, Kumar M (2022b) Climate change and its impact on Indian Himalayan forests: current status and research needs. In: Rani S, Kumar R (eds) Climate change. Springer, Cham, pp 223–242. https://doi.org/10.1007/978-3-030-92782-0_11

Singh H, Savita Sharma R, Sinha S, Kumar M, Kumar P, Verma A, Sharma SK (2017a) Physiological functioning of *Lagerstroemia speciosa* L. under heavy roadside traffic: an approach to screen

potential species for abatement of urban air pollution. 3 Biotech 7(61):1–10. https://doi.org/10.1007/s13205-017-0690-0
Singh H, Verma A, Kumar M, Sharma R, Gupta R, Kaur M, Negi M, Sharma SK (2017b) Phytoremediation: a green technology to clean up the sites with low and moderate level of heavy metals. Austin Biochem 2:1012
Singh H, Sharma R, Savita R, Singh MP, Kumar M, Verma A, Ansari MW, Sharma SK (2018) Adaptive physiological response of *Parthenium hysterophorus* to elevated atmospheric CO_2 concentration. Indian for 144:1–14
Singh H, Yadav M, Kumar N, Kumar A, Kumar M (2020) Assessing adaptation and mitigation potential of roadside trees under the influence of vehicular emissions: a case study of *Grevillea robusta* and *Mangifera indica* planted in an urban city of India. PLoS ONE 15(1):e0227380. https://doi.org/10.1371/journal.pone.0227380
Singh S, Singh H, Sharma SK, Nautiyal R (2021) Seasonal variation in biochemical responses of bamboo clones in the sub-tropical climate of Indian Himalayan foothills. Heliyon 7(4):e06859. https://doi.org/10.1016/j.heliyon.2021.e06859
Singh H, Bandyopadhyaya S, Kumar A, Singh M, Malik A, Kumari G, Raturi A, Yadav SK, Kumar M, Barthwal S, Thakur A, Rawat PS (2022a) Understanding the physiological and biophysical response of urban roadside plantations for assessing adaptation and mitigation mechanisms toward vehicular emissions. Urban Clim 44:101183. https://doi.org/10.1016/j.uclim.2022.101183
Singh N, Poonia T, Siwal SS, Srivastav AL, Sharma HK, Mittal SK (2022b) Challenges of water contamination in urban areas. In: Current directions in water scarcity research, vol 6. Elsevier, pp 173–202
Singh H, Kumar N, Singh R, Kumar M (2023) Assessing the climate change impact on the habitat suitability of the range restricted bird species (*Catreus wallichii*) in the Indian Himalayan ecosystem. Environ Sci Pollut Res 12:789. https://doi.org/10.1007/s11356-023-30789-x
Singh H, Kumar N, Kumar A (2024a) Enhancing resource use efficiency in crops through plant functional traits. In: Kumar N, Singh H (eds) Plant functional traits for improving productivity. Springer, Singapore. https://doi.org/10.1007/978-981-97-1510-7_6
Singh M, Singh H, Kumar A, Kumar M, Barthwal S, Thakur A (2024b) Soil nitrogen availability determines the CO_2 fertilization effect on tree species (*Neolamarckia cadamba*): growth and physiological evidence. Environ Sustain 7:53–60. https://doi.org/10.1007/s42398-023-00300-w
Sistla SA, Schimel JP (2012) Stoichiometric flexibility as a regulator of carbon and nutrient cycling in terrestrial ecosystems under change. New Phytol 196(1):68–78
Thompson I, Mackey B, McNulty S, Mosseler A (2009) Forest resilience, biodiversity, and climate change. In: A synthesis of the biodiversity/resilience/stability relationship in forest ecosystems. Secretariat of the convention on biological diversity, Montreal. Technical Series, pp 1–67
Tiwari P, Verma P, Raghubanshi AS (2021) Forest phenology as an indicator of climate change: impact and mitigation strategies in India. India: climate change impacts, mitigation and adaptation in developing countries. Springer, Cham, pp 185–205
Tubby KV, Webber JF (2010) Pests and diseases threatening urban trees under a changing climate. Forestry 83(4):451–459
Van der Putten WH, Macel M, Visser ME (2010) Predicting species distribution and abundance responses to climate change: why it is essential to include biotic interactions across trophic levels. Philos Trans Royal Soc B Biol Sci 365(1549):2025–2034
Wikle J, Higgins H, Clark P, Cook D, Dutton D, Garton J, Kosiba A, Schadler E (2021) Climate change in forests. The Vermont Climate Assessment 2021:82
Yadav SK, Singh H, Ginwal HS, Barthwal S (2019a) Elevated CO_2 enhanced growth and physiological process of *Populus deltoides* Bartr. ex Marsh. Indian for 145(1):23–27

Yadav SK, Singh H, Nautiyal R, Ginwal HS, Ansari SA, Barthwal S (2019b) Modulation of morpho-physiological responses in *Populus deltoides* by elevated carbon dioxide and temperature. For Sci 66:105–118. https://doi.org/10.1093/forsci/fxz048

Zeppel MJB, Wilks JV, Lewis JD (2014) Impacts of extreme precipitation and seasonal changes in precipitation on plants. Biogeosciences 11(11):3083–3093

Urban Forests: Importance, Challenges and Opportunities

Alok Kumar Singh, Saurbh Shukla, M. K. Singh, Abhinav Singh, Rupali Singh, and Bipin Kumar Singh

Abstract Urban forests, comprising the collective green spaces and trees in cities, are recognized for their multifaceted benefits. They contribute significantly to environmental health by mitigating air pollution, regulating temperatures, and enhancing biodiversity. Furthermore, urban forests play an important role in promoting mental well-being and providing recreational spaces for residents. However, the burgeoning urbanization presents a series of challenges to the sustainability of urban forests. The limited available space in urban areas, coupled with issues like soil compaction and competition for resources, poses threats to the health and longevity of urban trees. Additionally, the unequal distribution of urban forests can exacerbate environmental injustices, impacting communities disparately. Despite the challenges, the chapter highlights opportunities for enhancing urban forest sustainability. Integrating green infrastructure into urban planning, implementing effective tree care practices, and fostering community engagement are identified as key strategies. The chapter explores innovative approaches, such as vertical forests and urban agroforestry, that offer potential solutions to maximize the benefits of urban forests in constrained spaces. It provides a nuanced understanding of the socio-ecological dynamics within urban environments and offers insights for policymakers, urban planners, and researchers. The comprehensive exploration of urban forest dynamics,

A. K. Singh · A. Singh
Rajiv Gandhi South Campus, Banaras Hindu University, Mirzapur, Uttar Pradesh 231001, India

S. Shukla
Dr. Yashwant Singh Parmar University of Horticulture and Forestry, Solan, Himachal Pradesh 173230, India

M. K. Singh
Department of Agronomy, Institute of Agricultural Sciences, Banaras Hindu University, Varanasi 221005, India

R. Singh (✉)
Forest Research Institute Deemed to be University, Dehradun, Kaulagarh 248003, India
e-mail: srupali333@gmail.com

B. K. Singh
Department of Silviculture and Agroforestry, Acharya Narendra Deva University of Agriculture and Technology, Kumarganj, Ayodhya, Uttar Pradesh 224 229, India

H. Singh (ed.), *Urban Forests, Climate Change and Environmental Pollution*,
https://doi.org/10.1007/978-3-031-67837-0_2

from the ecological to the social dimensions, ensures a holistic perspective for addressing the challenges and harnessing the opportunities presented by urban forests. This chapter is a valuable resource for those interested in understanding and advancing the role of urban forests. It navigates the complex interplay of ecological, social, and economic factors, providing a foundation for informed urban planning and environmental management decision-making. The chapter ultimately advocates integrating urban forests as integral components of sustainable and resilient cities.

Keywords Urban forestry · Sustainability challenges · Green infrastructure · Socio-ecological dynamics · Resilient cities

1 Introduction

Urbanization of the globe represents a massively accelerating trend concerning population growth and expansion of cities. The United Nations has pointed out that half of the world's population currently lives in urban areas, estimated to reach 64% in developing countries and 86% in developed countries by the end of 2050. As per the report of the World Bank (2022), by the end of 2050, 7 out of 10 people will live in cities. Urbanization has led to tangible advantages for human societies, including dense populations and labor forces that support large-scale production, extensive transportation networks that encourage trade and economic growth, advanced communication technologies that enhance education and scientific pursuits, improved healthcare and public services, expanded employment opportunities, and increased cultural diversity, ultimately elevating overall living standards but simultaneously it results in some pronounced negative impact on the environment and society as a whole. The ever-growing population and urbanization lead to land cover changes, natural resource depletion, and environmental degradation, which enormously impact the city's vegetation and surrounding areas (Sun et al. 2011; Sirakayaet al. 2018; Kumar et al. 2022, 2024; Singh and Kumar 2022). Despite occupying a mere fraction of less than 2%—of the Earth's land area, urban areas are responsible for a disproportionate share of global environmental impact. They contribute 78% of carbon emissions, utilize 60% of residential water resources, and consume 76% of the wood allocated for industrial purposes (Brown 2001). It leads to elevated temperature, increased air pollution, depletion of soil quality, reduction of native biodiversity, encroachment upon farmlands, and alteration to hydrological systems in the cities (Güneralpet al. 2020; Singh et al. 2024a; Singh 2024). There is an urgent need to overcome these complexities in cities, which is closely associated with achieving the United Nations Sustainable Development Goal 11: "making cities and human settlement more resilient, safe and sustainable. Amidst these complexities, urban forests are crucial components of sustainable urban development (Dhyani et al. 2021; Kumar et al. 2021a, b, c, d, e; Prakash et al. 2022). An urban forest is a harmonious blend of living organisms and their non-living surroundings, encompassing a substantial area exceeding 0.5 ha, predominantly populated by trees (Liu et al. 2003). They help provide a range of multiple benefits and ecosystem services to the urban environment, including temperature regulation (Oliveira et al. 2011), moisture conservation (Shaffer 2018), air pollutant removal, water quality improvement, improvement in biodiversity of native species, reduce the impact of wind and solar flux (O'Dell et al. 1977), and mitigation of climate change through carbon sequestration (Davies et al. 2011). Urban forest benefits are not just ecological or economic. Still, they are also associated with improving the urban society's physical and mental or psychological well-being (Takayama et al. 2014; Singh 2021; Sharma and Singh 2021; Singh et al. 2021). Urban trees possess a multitude of benefits that are often underestimated or overlooked. The power of the tree is depicted in Fig. 1. Acknowledging the myriad benefits urban trees offer, it's crucial to address the challenges they face, including insufficient maintenance, urban development pressures, pollution, and the impacts of climate change. These challenges hinder the ability of urban trees to thrive and fulfill their potential as contributors to sustainable, resilient, and liveable cities for both present and future generations.

One of the primary obstacles to the health and longevity of urban trees is the limited availability of space in urban areas. This constraint is compounded by issues such as soil compaction and resource competition, which further impede the growth and vitality of urban trees. Recognizing and mitigating these challenges can effectively harness the power of urban trees to enhance urban environments, improve public health, mitigate climate change, and foster community well-being (Singh et al. 2020;

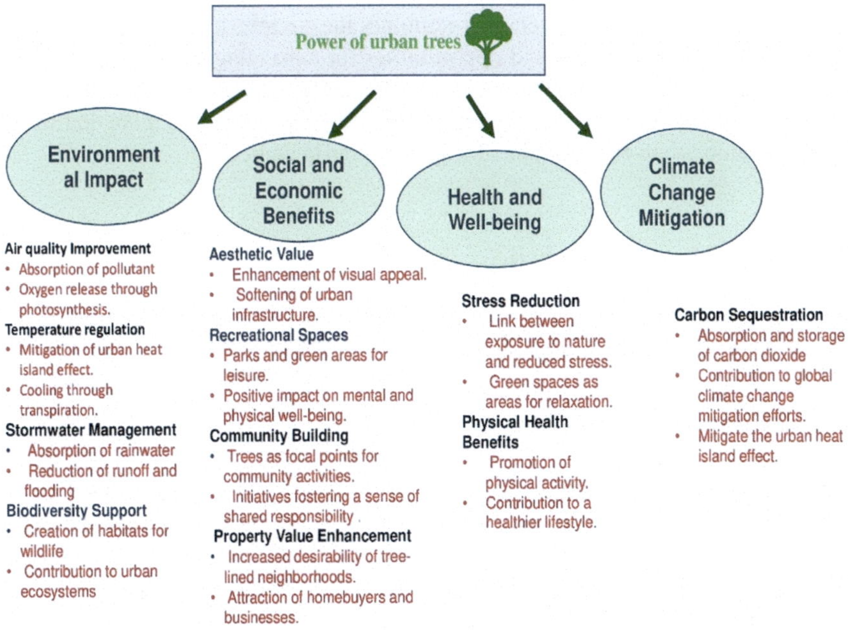

Fig. 1 Power of urban trees

Kumar et al. 2021a, b). It requires comprehensive urban planning and management strategies that prioritize preserving, maintaining, and expanding urban green spaces to ensure urban tree populations' sustained health and resilience (Kumar et al. 2021c, d). Through collaborative efforts among policymakers, urban planners, community organizations, and residents, we can create environments where urban trees thrive, providing invaluable benefits to cities worldwide. Addressing the challenges posed by the development of urban trees stands as a paramount concern for experts in urban forestry (Wirtz et al. 2021), who tackle the issue through various tactics such as initiatives for planting trees, regulations safeguarding trees, and objectives aimed at enhancing green infrastructure (Salisbury et al. 2022).

Moreover, the unequal distribution of urban forests can exacerbate existing environmental injustices, disproportionately affecting marginalized communities. Many studies indicate that numerous cities are evolving with disparities in urban residents' access to green spaces, potentially resulting in divergent living environments and social well-being, consequently affecting the physical and mental health of urban dwellers (Kim et al. 2021; Yang et al. 2020; and Chen et al. 2020). Despite these challenges, the chapter also identifies opportunities for enhancing the sustainability of urban forests (Yadav et al. 2019a, b; Gupta et al. 2019a). Through a comprehensive examination of existing research and case studies, we explore innovative strategies and practices (vertical forests and urban agroforestry) to maximize the benefits of urban forests within constrained urban spaces. These include integrating green infrastructure into urban planning, implementing effective tree care techniques, and fostering community engagement in tree stewardship initiatives (Kumar et al. 2019; Kumar et al. 2019). This chapter examines the complexities of urban forestry, exploring both the challenges and opportunities for enhancing the sustainability of urban forests in the face of rapid urbanization. Innovative strategies and collaborative efforts harness the full potential of urban forests to create healthier, more equitable, and sustainable urban environments for generations to come.

2 Importance of Urban Forests

2.1 Environmental Contributions

2.1.1 Mitigation of Air Pollution

As per Landrigan et al. (2019) and WHO (2019), approximately 91% of the world's population live in regions where the levels of pollutants are higher than the limit established by the WHO for a healthy environment, and it has led to mortality of nearly 4.2 million people every year mainly in developing countries. The higher traffic rate in urban areas is one of the primary sources of air pollution in urban areas, including industries and factories (Jung et al. 2015; Sharma et al. 2018; Singh et al. 2018). Air pollution is the precursor of many other problems, such as damage to vegetation, reduced visibility, climate change, acidic deposition, human and animal diseases, etc. (Fenger et al. 1999). The primary air pollutants in urban areas include sulphur dioxide (SO_2), nitrogen oxides (NO_x) resulting from the combustion of fossil fuels and automobiles, ozone (O_3) produced by the interaction of volatile organic compounds with NO_x, polycyclic aromatic hydrocarbons (PAH's) and particulate matter produced by vehicular emissions. Urban forests have great potential to improve the quality of urban air (Gupta et al. 2018a, b). Trees act as a barrier and deposition surface for particulate matter, change wind velocity and direction, and modify air pollutants' dispersion (Janhall 2015). Trees can absorb gaseous pollutants through stomatal pores (Leaves and bark) and CO_2 during photosynthesis (Beckett et al. 1998; Singh and Verma 2013; Kumar et al. 2017). Stomata can also have hydrophilic regions that could serve as sites for depositing hydrophilic pollutants (Burkhardt 2010; Gupta and Singh 2017; Singh et al. 2017a, b). Trees possess the unique feature of transpiration, which has the potential to reduce the temperature, limit the rate of chemical reactions that produce secondary pollutants, and aid in the settlement of particles due to increased humidity (Akbari et al. 2001; Grey and Deneke 1986). However, these effects depend upon the type of species, their growth rate, the density of plant species, the type and characteristics of pollutants, the distance between vegetation and pollutant sources, and meteorological conditions (Zhou 1993; Fowler 2002). Trees whose leaves have parallel venation, grooves, or scaly leaves can absorb large amounts of air pollutants. Trees with large coarse leaves having complex (trichomes, pubescence, cuticle thickness) structures, varying shapes, and a large number of pores on the surface are the most desirable characteristics for trees suitable for urban areas (Sgrignaet al. 2020). Trichomes increase the leaf area severalfold, providing extra surfaces for PM retention, while cuticle thickness and esterification regulate the PAH retention and translocation in subcuticular tissues (Prigionieroet al. 2023). Conifers and broadleaf are experimentally proven more effective in absorbing PM of larger aerodynamic diameter over broadleaf (Liu et al. 2018). More giant trees are more efficient in controlling pollution than smaller trees, and trees with higher leaf areas have greater potential for air phytoremediation. It is estimated that Adama Urban Forest removed 188 thousand tons of air pollutants per year, resulting from O_3, CO, NO_2, $PM_{2.5}$, and SO_2 (Korichoet al. 2022). Another study by Tan et al. (2021) showed annual air pollution of 178.26 kg (121.79 kg for O_3, 30.82 kg for NO_2, 10.90 kg for $PM_{2.5}$, and 14.75 kg for SO_2) from 1230 sample street trees in Kyoto city of Japan. A study by Nowak (1994) in Chicago's urban forests was found to remove pollutants at rates of 0.7 kg ha^{-1} $year^{-1}$ for carbon monoxide (CO), 2.1 kg ha^{-1} $year^{-1}$ for sulphur dioxide (SO_2), 2.4 kg ha^{-1} $year^{-1}$ for nitrogen dioxide (NO_2), 5.5 kg ha^{-1} $year^{-1}$ for particulate matter less than 10 μm in diameter (PM_{10}), and 6.0 kg ha^{-1} $year^{-1}$ for ozone (O_3). This removal primarily occurred during the leafed season, with a total removal capacity of up to 87.5 kg ha^{-1} $year^{-1}$. A comparative study in Frankfurt, Germany, revealed that streets lined with trees exhibited lower particulate levels, with 3000 dust particles per liter of air, compared to neighboring treeless streets where levels ranged from 10,000 to 30,000 particulates per liter of air (Mink and Witter 1982).

2.1.2 Temperature Regulation

Building materials such as concrete, asphalt, glass, and steel have high thermal capacities, which can trap large insolation during the day while stored heat is dissipated during the night hours, which results in elevated temperatures of urban landscapes. Greening urban areas through rooftop gardening, establishing urban forests and parks, and roadside/street vegetation is one of the favoured approaches for reducing the temperature variations in urban areas (Chen and Ng. 2012; Meena et al. 2011; Kumar et al. 2019, 2020). The increased vegetation density in urban landscapes performs the function of temperature reduction through an increased rate of surface evaporation and vegetation transpiration and through the effect of shading of plant canopies, which reduces the amount of insolation that reaches the earth surface (Spronken-Smith et al. 2000; Singh and Kumar 2022; Singh et al. 2024b), and acts as a windbreak to modify ambient conditions around the building. Effectively oriented trees in cities can prevent solar radiation from directly striking buildings, leading to reduced heat storage and cooling the urban environment. Shading prevents the intense solar radiation from heating the shaded ground below the vegetation. Thus, it prevents air heating below the canopy and limits the heating of the top layer of the canopy through sensible heat flow (Shashua-Bar et al. 2009). The biophysical mechanism of transpiration from vegetation and evaporation from the surface can reduce the summer air temperature as it leads to lowering the Bowen ratio between sensible and latent heat fluxes. The increased humidity following higher evapotranspiration in summer also reduces the air temperature (Chow et al. 2016; Ren et al. 2018). Among evapotranspiration and shading, shading is the principal way of cooling air in the urban environment, contributing to 80% of the cooling effect in tropical and subtropical climatic regions (Shashua-Bar and Hoffman 2000). On average, urban centers are 0.5–1.5 warmer compared to adjacent rural areas (Chen 2008), and this temperature rise can be up to 4 in large metropolitan cities (Rosenzweig et al. 2006). The cooling effect of vegetation cover in urban areas is not limited only to the area where vegetation cover is present but also expands to neighboring areas. The cooling effect of a large park can extend up to 200 m on the downwind side in adjoining residential areas (Sugawara et al. 2016). The impact of evaporation on urban air cooling is driven by the choice of species and their density. A developed model claims that evapotranspiration can reduce the urban temperature by 1 °C. However, the number of trees for 1 reduction varies with species, as 63 mature *Eucalyptus camaldulensis* and 24 *Liquidambar styraciflua* are required (Ballinas and Barradas (2016).

2.1.3 Biodiversity Enhancement

Traditionally, urban areas are sought to be of low diversity and dominated by non-native species. Introducing non-native species due to urbanization can lead to biological invasion into rural regions, causing loss of local biodiversity. Amid expansive cities and towering skyscrapers, urban forests emerge as green sanctuaries, offering a shelter for diverse life forms. These pockets of nature within urban landscapes play a crucial role in supporting biodiversity, enriching ecosystems, and enhancing the quality of life for city inhabitants (Gyllin and Grahn 2005; Yadav et al. 2019a, b; Jangra et al. 2024; Kanta et al. 2024). Despite space and human activity constraints, these green spaces host a rich array of flora and fauna, ranging from towering trees to tiny insects. The diverse vegetation in urban forests provides food, shelter, and breeding grounds for countless species, contributing to the overall biodiversity of the area (Conway and Bourne 2013; Malik et al. 2024; Kumar and Singh 2024). High species diversity can ensure high security against catastrophic losses to pests and diseases and environmental shifts such as climate change (Kendal et al. 2014). A technique has been developed for improving and maintaining the high species diversity in urban forests and is known by the name 10/12/30 rule of thumb. As per this rule, an urban forest should consist of not more than 10% of any particular species, 20% of any one genus, and 30% of any single family (Santamour 1990; Kendal et al. 2014). Urban forests help to maintain the diversity of native species, as Zhang et al. (2016) found 90% of native species and only 9% of exotic species in Changchun City, China. Urban forests serve as habitats for endangered or threatened species. As evidenced by Howenstine (1993), Cook County, the most densely populated county within the Chicago Metropolitan Area, supports 20 fauna species facing endangerment or threat, alongside a listing of 130 plant species. Urban areas can contain higher levels of biodiversity than unpopulated areas (Kühn et al. 2004; Araújo 2003). However, many studies have also shown reduced species richness levels due to loss and habitat fragmentation of natural vegetation (McKinney 2006).

2.2 *Social Benefits*

2.2.1 Promotion of Physical and Mental Well-Being

Earlier sections covered the role of urban forests in purifying the air, absorbing harmful pollutants, and emitting oxygen, thus serving as the lungs of cities. They mitigate the urban heat, regulate microclimate, and preserve biodiversity. Access to urban forests encourages physical activity and promotes public health. Green spaces offer opportunities for recreational pursuits such as walking, jogging, cycling, birdwatching, and outdoor sports, thus facilitating exercise and reducing sedentary lifestyles (Fig. 2). Regular engagement in these activities lowers the incidence of obesity, cardiovascular diseases, and other lifestyle-related ailments, thereby improving overall physical well-being. Exposure to natural environments has been linked to reduced stress levels, improved mood, and enhanced cognitive function. Time spent in urban forests fosters relaxation, mindfulness, and connection to nature, thereby alleviating symptoms of anxiety, depression, and mental fatigue (Bowler et al. 2010; Takayama et al. 2014). Several theories suggest that the improvement of the physical health and mental well-being of societies is influenced by exposure to the natural environment. The Restoration theory proposed by Kaplan and Kaplan (1989) posits that nature provides the specific environmental stimuli necessary for restoring attentional resources depleted during tasks requiring sustained, directed attention. By facilitating relaxation, stress reduction, positive affective experiences, and attentional scope expansion, exposure to natural environments promotes cognitive restoration and enhances overall well-being. Another psycho-evolutionary theory proposed by Ulrich (1983) highlights how nature elicits innate, adaptive responses that facilitate psychophysiological stress recovery. The green infrastructure (GI) enhances mental health, indicating enhancements in overall mental well-being and reduced likelihood of experiencing stress, anxiety, and mood disorders among adults (Gascon et al. 2015).

Additionally, there is suggestive evidence of diminished behavioral and emotional issues in children. Moreover, recent research has linked long-term proximity to green spaces with improved cognitive development, including enhancing working memory and attention, mainly observed in children (Dadvandet al. 2015). Studies have proved that creating green belts in urban areas reduces the risk of cardiovascular attack, diabetes, asthma, and obesity in adults (Maas et al. 2009; Lachowycz and Jones 2011a, b; Gascon et al. 2016). Well-planned and managed urban forests can be a central point for tourist attraction. This green infrastructure provides recreational spaces for the city residents and helps city dwellers' economic development. Hansmann et al. (2016) conducted a study highlighting that urban forest partnerships, which are gaining traction across Europe, often prioritize tourism as one of their objectives. Urban parks offer recreational opportunities for residents and attract visitors from outside the area. Particularly renowned parks like Central Park in New York City are significant tourist destinations (Konijnendijk et al. 2013). The research conducted in Denmark and Finland has highlighted the favorable effect of having forests nearby on property values (Konijnendijk and Randrup 2004). Nevertheless,

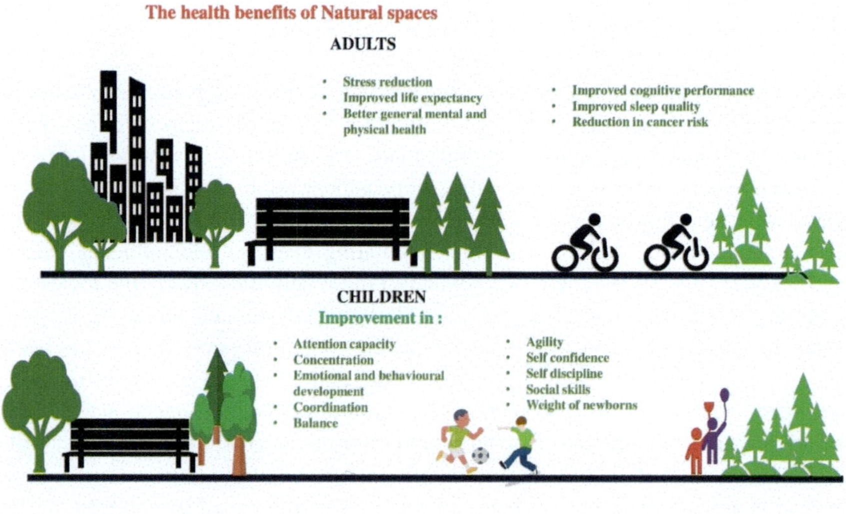

Fig. 2 Health benefits of natural spaces

the study in Europe and North America indicates the necessity of incorporating monetary and non-monetary considerations into environmental decision-making processes (Ward Thompson 2004). Such investigations would probably reveal substantial advantages linked to tourism expenditures if the impact of Urban Green Infrastructure (UGI) on tourist experiences and destinations.

3 Recreational Spaces for Residents

3.1 Challenges of Urbanization

3.1.1 Limited Space Issue

Urbanization, characterized by the rapid expansion of cities and increasing population density, brings about a fundamental challenge for urban forestry: limited space for trees and green spaces (Yokohariet al. 2000). As cities grow vertically and horizontally, land available for planting trees diminishes. This scarcity of space affects the quantity of trees and their quality and functionality within the urban landscape. One significant impact of limited space on urban trees is the reduction in canopy cover. With less space for trees to grow and spread their branches, canopy coverage may be compromised, resulting in diminished shade provision, decreased air quality improvement, and reduced habitat for urban wildlife. Moreover, limited space constrains tree growth, often resulting in stunted or poorly developed trees, which are more susceptible to environmental stressors such as pests, diseases, and extreme weather events. Urban forest management strategies must address the challenge of limited space by embracing innovative approaches to maximize tree planting opportunities. Vertical greening initiatives, such as green walls and facades, offer a solution to space constraints by utilizing vertical surfaces for vegetation.

Similarly, rooftop gardens and green roofs transform underutilized spaces into green infrastructure, providing multiple benefits such as stormwater management, energy conservation, and biodiversity enhancement. Furthermore, urban planners and landscape architects are crucial in integrating trees and green spaces into urban design. By prioritizing the preservation of existing green spaces, incorporating trees into streetscapes and urban developments, and creating multifunctional green corridors, cities can optimize limited space for urban forestry. Additionally, community engagement and participatory approaches can empower residents to advocate for protecting and expanding green areas, fostering a sense of ownership and stewardship towards urban trees.

3.1.2 Impact on Urban Trees

The ecological benefit of urban trees largely depends upon the tree growth (Cheng et al. 2022). The polluted urban air, soil contamination with heavy metals, and elevated temperature of cities present various challenges to tree growth, although each factor has detrimental and beneficial effects. Limited space restricts the ability of trees to establish and grow, leading to smaller tree sizes, reduced canopy cover, and decreased biodiversity in urban areas.

3.1.3 Soil Compaction Challenges

The growth of plants, whether trees or crops, depends mainly upon the ecosystem services provided by the soil (structure, composition, biodiversity of organisms inside soil). The physical properties of soil regulate the infiltration, percolation, and drainage and act as a storehouse of nutrients for plant growth and development. The construction of buildings, streets, and roads causes urban disturbances. Site leveling, either by removing topsoil or adding soil, leads to disturbances in the soil's ecosystem. The extensive use of heavy construction equipment, vehicle movement, transport, and storage of materials leads to soil compaction, negatively impacting the soil's root penetration, porosity, and biological properties (Pagliai et al. 2003). The ecosystem may also suffer from soil compaction since reduced infiltration in compacted soils can result in enhanced runoff and erosion, improving the amount of nutrients and other contaminants delivered to surrounding water bodies (Shober and Denny 2010). Soil compaction can be caused intentionally to provide a stable and solid foundation for rods and buildings (Schueler 2000) or unintentionally for residential movement such as walking or playing in parks (Kissling et al. 2009). However, soil compaction in urban environments is not homogenous and can vary temporally or spatially depending on work intensity and temporal changes (Edmondson et al. 2011). A bulk density over 1.4 g cm^{-3} is considered detrimental to plant growth, and Shober and Denny (2010) found that bulk density value in urban landscapes may range from 1.4 to 2.2 g cm^{-3}. These all-detrimental effects of urbanization on urban soil will reduce the growth and development of young trees, especially when they do not have robust root systems. Trees require water, oxygen, nutrients, a soil matrix, and temperature within a suitable range, but these conditions are unlikely to be met in compacted urban soil (Ruark et al. 2012; Layman et al. 2016). Survived individuals high mortality rate and slow growth were noticed under these conditions (Nowak et al. 1990). Soil compaction is reduced by tillage practices (deep or shallow), plug aeration, air tilling, use of organic amendments, and soil replacement (Shober and Denny 2010). According to Somerville et al. (2018), soil compactions were significantly reduced by applying tillage practices of adding organic amendments and supporting plant growth.

3.1.4 Resource Competition

Resource competition among trees, infrastructure, vegetation, and human activities is a pervasive challenge in urban environments, where space and resources are limited. Trees in urban areas must compete for essential resources such as water, sunlight, nutrients, and space, often facing constraints that compromise their growth and vitality. One of the primary resources that trees compete for in urban settings is water. Urbanization alters the hydrological cycle, increasing surface runoff, soil moisture variability, and water stress for trees. Competing demands for water from municipal utilities, landscaping irrigation, and impervious surfaces further exacerbate water scarcity for urban trees, particularly during drought.

Similarly, trees must contend with limited access to sunlight, a critical resource for photosynthesis and growth. Tall buildings, dense urban infrastructure, and shading from adjacent structures can cast shadows and reduce the amount of sunlight reaching trees, inhibiting their ability to thrive and achieve their full potential. Furthermore, nutrient availability and soil quality are vital in determining tree health and resilience in urban environments. Soil degradation, pollution, and nutrient depletion resulting from human activities can limit the availability of essential nutrients such as nitrogen, phosphorus, and potassium, impairing tree growth and physiological functioning. Implementing management strategies that optimize resource allocation and promote ecosystem resilience is necessary to address resource competition in urban forestry. Sustainable urban design practices, such as green infrastructure planning, tree canopy prioritization, and integrated water management, can enhance resource efficiency and support healthy tree growth. Additionally, community engagement and public education initiatives can raise awareness about the importance of urban trees and encourage stewardship practices that reduce resource competition and promote coexistence among trees and other urban elements.

3.1.5 Unequal Distribution and Environmental Injustices

Urbanization often exacerbates socio-economic disparities and inequalities in access to green spaces and environmental amenities, resulting in environmental injustices that disproportionately affect marginalized communities and vulnerable populations. One of the critical challenges associated with unequal distribution in urban forestry is the spatial disparity in access to green spaces and urban forests. Wealthier and affluent neighborhoods tend to have better-maintained parks, larger tree canopy cover, and more extensive green infrastructure than low-income communities and inner-city neighborhoods. This disparity in green space distribution contributes to environmental inequities, as residents in underserved areas are deprived of urban trees' numerous health, social, and economic benefits (Chen et al. 2024). Moreover, environmental injustices related to urban forestry extend beyond access to green spaces to encompass issues such as air quality, heat stress, and environmental hazards. Historically marginalized communities and minority populations are often disproportionately exposed to environmental pollutants, industrial emissions, and urban heat islands, which can have adverse effects on respiratory health, quality of life, and overall well-being. Community-driven initiatives, participatory design processes, and equitable resource allocation can empower marginalized communities to advocate for their environmental rights and influence policy decisions that prioritize environmental justice and social equity.

4 Opportunities for Sustainability

The long-term goals of urban forest managers are creating and managing urban forests for the public benefit and achieving sustainability. The urban forest presents many opportunities for fostering sustainability within densely populated areas. Leveraging these opportunities requires a multifaceted approach integrating green infrastructure, promoting effective tree care practices, and engaging communities meaningfully.

4.1 Green Infrastructure Integration

Integrating the urban forest into green infrastructure initiatives is paramount for sustainable urban development. Green infrastructure refers to strategically planned networks of natural and semi-natural elements that provide ecological, social, and economic benefits. By incorporating trees, parks, green spaces, and vegetated rooftops into urban planning and design, cities can mitigate the adverse effects of urbanization, such as air and water pollution, heat islands, and biodiversity loss. Trees play a pivotal role in green infrastructure by providing essential ecosystem services such as carbon sequestration, air purification, stormwater management, and temperature regulation. Strategic placement of trees in urban areas can significantly reduce energy consumption for heating and cooling buildings, lowering carbon emissions and mitigating the urban heat island effect.

Furthermore, integrating green infrastructure with existing urban infrastructure can enhance resilience to climate change impacts such as extreme weather events and flooding. Trees and vegetation act as natural buffers, absorbing excess rainwater and reducing the risk of urban flooding while improving soil health and promoting biodiversity. Municipalities can promote green infrastructure integration through policies and regulations that incentivize sustainable development practices, such as green building certifications, tree planting mandates, and tax incentives for green roofs and permeable pavements. Collaborative efforts between government agencies, urban planners, landscape architects, and community stakeholders are essential for realizing the full potential of green infrastructure in creating sustainable and resilient cities.

4.2 Effective Tree Care Practices

Effective tree care practices are essential for maintaining urban trees' health, vitality, and longevity. Urban environments present unique challenges to tree growth and survival, including soil compaction, pollution, limited space, and resource competition. Implementing sustainable tree care practices can mitigate these challenges and ensure the continued benefits of the urban forest. Proper tree selection and placement are critical considerations in urban forestry initiatives. Native and climate-adapted tree species are better suited to thrive in urban environments and require less maintenance than exotic species. Careful attention should be paid to selecting trees with diverse characteristics such as size, shape, and tolerance to environmental stressors to maximize ecosystem services and minimize risks associated with monoculture planting. Regular maintenance activities such as pruning, watering, and mulching are essential for promoting tree health and reducing the risk of pest infestations and diseases. Adopting mulching practices can improve soil moisture retention, suppress weed growth, and enhance soil fertility, reducing the need for irrigation and chemical fertilizers. Furthermore, implementing integrated pest management strategies can minimize reliance on pesticides and herbicides, reduce environmental contamination, and protect beneficial insects and wildlife. Community-based tree stewardship programs can empower residents to actively care for urban trees through tree planting events, educational workshops, and volunteer maintenance activities.

4.3 Community Engagement Strategies

Community engagement is vital to fostering a sense of ownership and stewardship of the urban forest among residents. Engaging communities in decision-making processes and encouraging participation in tree planting and maintenance activities can enhance social cohesion, improve public perception of urban forestry initiatives, and strengthen the resilience of urban ecosystems. Effective community engagement strategies involve fostering partnerships between local government agencies, nonprofit organizations, businesses, educational institutions, and community groups. Collaborative approaches that prioritize inclusivity, diversity, and equity ensure that the benefits of urban forestry are equitably distributed among all residents, particularly in underserved and marginalized communities. Educational outreach programs are vital in raising awareness about the importance of trees and green spaces in urban environments and empowering residents with the knowledge and skills to contribute to urban forestry efforts actively. Schools, community centers, and online platforms can serve as hubs for environmental education and capacity-building initiatives, providing opportunities for hands-on learning and skill development.

Moreover, leveraging social media and digital technology platforms can enhance outreach efforts and facilitate community engagement by providing access to information, resources, and interactive tools for tree mapping, monitoring, and reporting.

Citizen science initiatives enable residents to collect valuable data on tree health, biodiversity, and environmental quality, supporting evidence-based decision-making and adaptive management practices. Harnessing the urban forest's opportunities for sustainability requires a holistic approach that integrates green infrastructure, promotes effective tree care practices, and engages communities in meaningful ways.

5 Innovative Approaches

5.1 Vertical Forest

Site availability is one of the major problems in urban green infrastructure development, which ultimately depends upon urban morphology. Urban morphology is the city's structural development, including buildings, streets, parking lots, and sidewalks. With these structures already present in the cities, it is very challenging for the municipalities to create urban parks or forests. The denser the city, the less space is available for tree growth. Vertical forests existed when it was impossible to expand the urban forest area horizontally. The Vertical Forest concept represents an innovative approach to sustainable urban development, where residential buildings are hubs for metropolitan reforestation. This model facilitates environmental regeneration and enhances urban biodiversity without necessitating urban sprawl. It achieves vertical densification of natural elements within the cityscape, aligning with policies aimed at reforestation and integrating nature into expansive urban and metropolitan zones (Kucherova and Narvaez 2018). The vertical forest concept has ancient roots, as seen in the Babylonian Hanging plant relic, but recently came into more prominence as a measure to mitigate climate change and its potential environmental, social, and economic value. Stefano Boeri, an Italian architect, uses this concept in the Bosco Verticale project. The Bosco Verticale project aimed to rejuvenate the urban landscape of the heavily polluted Porta Nuova area by creating a comfortable living environment. The façade of Bosco Verticale is adorned with a vegetation structure consisting of over 13,000 trees representing more than 90 species. These include tall tree species reaching up to six meters strategically positioned on balconies. The project's implementation involves collaboration among arboriculturists, horticulturists, landscapers, and botanists. Bosco Verticale introduces the concept of a green skyscraper that functions as a building and a new ecosystem within the city. It aims to enhance environmental quality and microclimate, mitigate air pollution, increase humidity levels, and improve urban thermal comfort. The Bosco Verticale in Milan is the world's inaugural vertical forest (Pribadiet al. 2021). The plant system within the Vertical Forest plays a vital role in shaping a microclimate, generating moisture, capturing CO_2 and particulate matter, and releasing oxygen into the atmosphere. Only two vertical forest towers in Milan can absorb 19 tons of CO_2 annually and produce comparable oxygen (Kucherova and Narvaez 2018).

5.2 Urban Agroforestry

Urban agroforestry or urban food forest refers to integrating trees and other perennial vegetation into urban landscapes to support agricultural production, ecosystem services, and human well-being. It involves intentionally designing and managing trees, shrubs, and other vegetation in urban areas to provide food, fibre, and other resources while enhancing environmental quality, biodiversity, and aesthetics (Maes et al. 2015). Urban agroforestry systems can include community gardens, orchards, food forests, green roofs, and urban farms, all integrated into the city's fabric to create multifunctional green spaces that benefit both people and the environment (Fig. 3).

Species commonly grown and considered appropriate for many areas of the United States include cultivated varieties of pear (*Pyrus* spp.), apple (*Malus* spp.), cherry (*Prunus* spp.), plum (*Prunus* spp.), and blueberry (*Vaccinium corymbosum* L.). However, this list can be expanded to include lesser-known native species such as American hazelnut (*Corylus americana* Marshall), black chokeberry (*Aronia melanocarpa* (Michx.) Elliott), black walnut (*Juglans nigra* L.), persimmon (*Diospyros virginiana* L.), pawpaw [*Asimina triloba* (L.) Dunal], mulberry (*Morus* spp.), wild plum (*P. americana* Marshall), wild cherry [*P. avium* (L.) L.], pecan [*Carya illinoensis* (Wangenh.) K. Koch], and American black elderberry [*Sambucus canadensis* L]. Some berry-producing shrubs in Rubus, Ribes, and Vaccinium genera could also be included in urban gardens and orchards. These food forests produce safe food that is less contaminated by heavy metals and Poly-acrylic-hydrocarbons unless

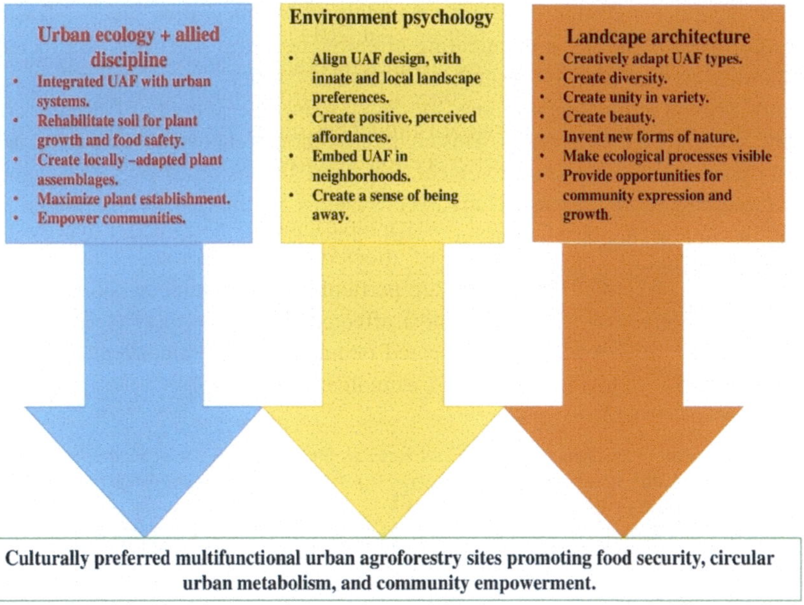

Fig. 3 Multidisciplinary design principles for multifunctional urban agroforestry

the location is very close to the pollutant source in the city (Romanova and Lovell 2021). Recently, many countries have been promoting food production within urban areas. For example, in Amsterdam, more than 350 ha of land have been allocated to urban gardens to cultivate fresh food and other commodities for the city's inhabitants (van Leeuwen 2010). Similarly, in Dar es Salaam, Tanzania, approximately 90% of leafy vegetables and 60% of the city's milk supply originate from production within or near the urban area (Lee-Smith and Prain 2006). Tropical home gardens, a well-known agroforestry system, currently face many challenges regarding sustainable production due to the load of urbanization (Thaman 1992). However, these practices can still contribute significantly to local food security and income generation from the sale of garden production (Montagnini 2006). A highly referenced case is found in Havana, Cuba, where, according to Altieri et al. (1999), home gardens in 1996 supplied a substantial quantity of fresh food to the urban populace, comprising 8500 tons of agricultural produce, 7.5 million eggs, and 3650 tons of meat.

6 Conclusion

Urban forests are vital components of sustainable and resilient cities, offering many benefits ranging from environmental to social and economic realms. Despite the challenges posed by urbanization, such as limited space and resource competition, there are ample opportunities to enhance the sustainability of urban forests through strategic planning, innovative approaches, and community involvement. Cities may nurture healthier and more resilient urban forests by integrating green infrastructure into urban planning, implementing effective tree care practices, and fostering community engagement. Innovative solutions like vertical forests and urban agroforestry present promising avenues for maximizing the benefits of urban forests within constrained urban spaces. This chapter has comprehensively examined urban forest dynamics, delving into ecological and social dimensions. Doing so has underscored the interconnectedness of urban ecosystems and human well-being, emphasizing the importance of adopting a holistic approach to urban forest management and policy-making.

References

Akbari H, Pomerantz M, Taha H (2001) Cool surfaces and shade trees to reduce energy use and improve air quality in urban areas. Sol Energy 70(3):295–310. https://doi.org/10.1016/S0038-092X(00)00089-X

Altieri MA, Companioni N, Cañizares K, Murphy C, Rosset P, Bourque M, Nicholls CI (1999) The greening of the "barrios": urban agriculture for food security in Cuba. Agric Hum Values 16(2):131–140. https://doi.org/10.1023/A:1007545304561

Araújo MB (2003) The coincidence of people and biodiversity in Europe. Glob Ecol Biogeogr 12(1):5–12. https://doi.org/10.1046/j.1466-822X.2003.00314.x

Ballinas M, Barradas V (2016) The urban tree as a tool to mitigate the urban heat island in Mexico City: a simple phenomological model. J Environ Qual 45:157–166. https://doi.org/10.2134/jeq2015.01.0056

Beckett KP, Freer-Smith PH, Taylor G (1998) Urban woodlands: their role in reducing the effects of particulate pollution. Environ Pollut 99(3):347–360. https://doi.org/10.1016/S0269-7491(98)00016-5

Bowler DE, Buyung-Ali LM, Knight TM, Pullin AS (2010) A systematic review of evidence for the added benefits to health of exposure to natural environments. BMC Public Health 10(1):1–10. https://doi.org/10.1186/1471-2458-10-456

Brown LR (2001) Eco-economy: building an economy for the earth, 1st edn. Norton, New York, p 352. https://doi.org/10.4324/9781315071893

Burkhardt J (2010) Hygroscopic particles on leaves: nutrients or desiccants. Ecol Monogr 80:369–399. https://doi.org/10.1890/09-1988.1

Chen L, Ng E (2012) Outdoor thermal comfort and outdoor activities: a review of research in the past decade. Cities 29(2):118–125. https://doi.org/10.1016/j.cities.2011.08.006

Chen Y, Yue W, La Rosa D (2020) Which communities have better accessibility to green space: an investigation into environmental inequality using big data. Landsc Urban Plan 204:103919. https://doi.org/10.1016/j.landurbplan.2020.103919

Chen J, Kinoshita T, Li H, Luo S, Su D (2024) Which green is more equitable: a study of urban green space equity based on morphological spatial patterns. Urban for Urban Green 91:128178. https://doi.org/10.1016/j.ufug.2023.128178

Cheng XL, Nizamani MM, Jim CY, Qureshi S, Liu S, Zhu ZX, Wang HF (2022) Response of urban tree DBH to fast urbanization: case of coastal Zhanjiang in south China. Urban Ecosyst 12:1–12. https://doi.org/10.1007/s11252-021-01167-7

Chow WT, Akbar SNABA, Heng SL, Roth M (2016) Assessment of measured and perceived microclimates within a tropical urban forest. Urban for Urban Green 16:62–75. https://doi.org/10.1016/j.ufug.2016.01.010

Conway TM, Bourne KS (2013) A comparison of neighborhood characteristics related to canopy cover, stem density and species richness in an urban forest. Landsc Urban Plan 113:10–18. https://doi.org/10.1016/j.landurbplan.2013.01.005

Dadvand P, Nieuwenhuijsen MJ, Esnaola M, Forns J, Basagaña X, Alvarez-Pedrerol M, Sunyer J (2015) Green spaces and cognitive development in primary schoolchildren. Proc Natl Acad Sci 112(26):7937–7942. https://doi.org/10.1073/pnas.1503402112

Davies ZG, Edmondson JL, Heinemeyer A, Leake JR, Gaston KJ (2011) Mapping an urban ecosystem service: quantifying above-ground carbon storage at a city-wide scale. J Appl Ecol 48(5):1125–1134. https://doi.org/10.1111/j.1365-2664.2011.02021.x

Dhyani K, Kuniyal HB, Singh H, Sobha S (2021) Growth and physiological potential of *Terminalia arjuna* under elevated CO_2 levels in Open top chamber condition. J Appl Nat Sci 13(3):1121–1126

Edmondson JL, Davies ZG, McCormack SA, Gaston KJ, Leake JR (2011) Are soils in urban ecosystems compacted: a citywide analysis. Biol Lett 7(5):771–774. https://doi.org/10.1098/rsbl.2011.0260

Fenger J, Hertel O, Palmgren F (1999) Urban air pollution-European aspects. Springer, New York, pp 1–321

Fowler AM (2002) Assessment of the validity of using mean potential evaporation in computations of the longterm soil water balance. J Hydrol 256:248–263

Gascon M, Triguero-Mas M, Martínez D, Dadvand P, Forns J, Plasència A, Nieuwenhuijsen MJ (2015) Mental health benefits of long-term exposure to residential green and blue spaces: a systematic review. Int J Environ Res Public Health 12(4):4354–4379. https://doi.org/10.3390/ijerph120404354

Gascon M, Triguero-Mas M, Martínez D, Dadvand P, Rojas-Rueda D, Plasència A, Nieuwenhuijsen MJ (2016) Residential green spaces and mortality: a systematic review. Environ Int 86:60–67. https://doi.org/10.1016/j.envint.2015.10.013

Grey GW, Deneke FJ (1986) Urban forestry, 2nd edn. Wiley, New York

Güneralp B, Reba M, Hales BU, Wentz EA, Seto KC (2020) Trends in urban land expansion, density and land transitions from 1970 to 2010: a global synthesis. Environ Res Lett 15(4):044015

Gupta SK, Singh H (2017) Observations on sporadic flowering in *Ficus benjamina* L. and *Peltophorum africanum* L. in new forest. Indian for 143:290–291

Gupta SK, Ram J, Singh H (2018a) Comparative study of transpiration in cooling effect of tree species in the atmosphere. J Geosci Environ Protect 6:151–166. https://doi.org/10.4236/gep.2018.68011

Gupta SK, Ram J, Singh H, Kumar P (2018b) Comparative study of cooling under shades of some forestry tree species with respect to ambient temperatures. J Geosci Environ Protect 6:51–66. https://doi.org/10.4236/gep.2018.611004

Gupta SK, Ram J, Singh H (2019a) Relationship between leaf area index and atmospheric cooling of tree species. Indian J for 42(3):203–208

Gyllin M, Grahn P (2005) A semantic model for assessing the experience of urban biodiversity. Urban for Urban Green 3:149–161. https://doi.org/10.1016/j.ufug.2005.04.004

Hansmann R, Whitehead I, Krajter Ostoić S, Zivojinović I, Stojanovska M, Jones N, Bernasconi A, Benamar S, Lelieveld C, Barstad J (2016) Partnerships for urban forestry and green infra structure delivering services to people and the environment: a review on what they are and aim to achieve. South East Eur for 7(1):9–19

Howenstine WL (1993) Urban forests as part of the whole ecosystem. In: Proceedings of the 6th national urban forestry conference. American Forests, Washington, DC, pp 118–120

Jangra A, Kumar K, Maikhuri S, Bhandari MS, Pandey S, Singh H, Barthwal S (2024) Unveiling stress-adapted endophytic bacteria: characterizing plant growth-promoting traits and assessing cross-inoculation effects on *Populus deltoides* under abiotic stress. Plant Physiol Biochem 210:108610. https://doi.org/10.1016/j.plaphy.2024.108610

Janhall S (2015) Review on urban vegetation and particle air pollution-deposition and dispersion. Atmos Environ 105:130–137. https://doi.org/10.1016/j.atmosenv.2015.01.052

Jung DY, Leem JH, Kim HC, Kim JH, Hwang SS, Lee JY, Kim BJ, Hong YC, Hong SJ, Kwon HJ (2015) Effect of traffic-related air pollution on allergic disease: results of the children's health and environmental research. Allergy Asthma Immunol Res 7(4):359–366. https://doi.org/10.4168/aair.2015.7.4.359

Kanta C, Kumar A, Chauhan A, Singh H, Sharma IP (2024) The interplay between plant functional traits and climate change. In: Kumar N, Singh H (eds) Plant functional traits for improving productivity. Springer, Singapore. https://doi.org/10.1007/978-981-97-1510-7_3

Kaplan R, Kaplan S (1989) The experience of nature: a psychological perspective. Cambridge University Press, Cambridge, pp 12–17

Kendal D, Dobbs C, Lohr VI (2014) Global patterns of diversity in the urban forest: is there evidence to support the 10/20/30 rule. Urban for Urban Green 13:411–417. https://doi.org/10.1016/j.ufug.2014.04.004

Kim J, Lee KJ, Thapa B (2021) Visualizing fairness: distributional equity of urban green spaces for marginalized groups. J Environ Plan Manag 65(5):833–851

Kissling M, Hegetschweiler KT, Rusterholz HP, Baur B (2009) Short-term and long-term effects of human trampling on above-ground vegetation, soil density, soil organic matter and soil microbial processes in suburban beech forests. Appl Soil Ecol 42(3):303–314. https://doi.org/10.1016/j.apsoil.2009.05.008

Konijnendijk CC, Randrup TB (2004) Urban forestry. In: Burley J, Evans J, Youngquist JA (eds) Encyclopedia of forest sciences. Elsevier, Amsterdam, pp 471–478

Konijnendijk CC, Annerstedt M, Busse Nielsen A, Maruthaveeran S (2013) Benefits of urban parks a systematic review. International Federation of Parks and Recreation Administration, Copenhagen, pp 1–70

Koricho HH, Seboka AD, Fufa F, Gebreyesus T, Song S (2022) Study on the ecosystem services of urban forests: implications for climate change mitigation in the case of Adama City of Oromiya Regional Sate, Ethiopia. Urban Ecosyst 25:1–10. https://doi.org/10.1007/s11252-021-01152-0

Kucherova A, Narvaez H (2018) Urban forest revolution. E3S Web Confer 33:01013

Kühn I, Brandl R, Klotz S (2004) The flora of German cities is naturally species rich. Evol Ecol Res 6(5):749–764

Kumar N, Jeena N, Kumar N, Gangola S, Singh H (2019) Phytoremediation facilitating enzymes: an enzymatic approach for enhancing remediation process. In: Bhatt P (ed) Smart bioremediation technologies. Academic Press, Cambridge, pp 289–306

Kumar A, Kumar P, Singh H, Kumar N (2021a) Modulation of plant functional traits under essential plant nutrients during seasonal regime in natural forests of Garhwal Himalayas. Plant Soil 465:197–212. https://doi.org/10.1007/s11104-021-05003-x

Kumar A, Kumar P, Singh H, Kumar N (2021b) Adaptation and mitigation potential of roadside trees with bio-extraction of heavy metals under vehicular emissions and their impact on physiological traits during seasonal regimes. Urban for Urban Green 58:126900. https://doi.org/10.1016/j.ufug.2020.126900

Kumar A, Kumar P, Singh H, Kumar N (2021c) Impact of plant functional traits on infiltration rate under different forests of Kempty watershed in Garhwal Himalaya. Indian J Soil Conser 49(1):38–44

Kumar A, Kumar P, Singh H, Bisht S, Kumar N (2021d) Relationship of physiological plant functional traits with soil carbon stock in the temperate forest of Garhwal Himalaya. Curr Sci 120(8):1368–1373. https://doi.org/10.18520/cs/v120/i8/1368-1373

Kumar M, Phukon SN, Paygude AC, Tyagi K, Singh H (2021e) Mapping phenological functional types (PhFT) in the Indian eastern Himalayas using machine learning algorithm in Google Earth Engine. Comput Geosci 158:4982. https://doi.org/10.1016/j.cageo.2021.104982

Kumar A, Singh H, Kumari G, Bisht S, Malik A, Kumar A, Singh M, Raturi S, Barthwal S, Thakur A, Kaushal R (2022) Adaptive resilience of roadside trees to vehicular emissions via leaf enzymatic, physiological, and anatomical trait modulations. Environ Pollut 313:120191. https://doi.org/10.1016/j.envpol.2022.120191

Kumar N, Singh H, Giri K, Kumar A, Joshi A, Yadav A, Singh R, Bisht S, Kumari R, Jeena N, Khairakpam R, Mishra G (2024) Physiological and molecular insights into the allelopathic effects on agroecosystems under changing environmental conditions. Physiol Mol Biol Plants 30:417–433. https://doi.org/10.1007/s12298-024-01440-x

Kumar N, Singh H (2024) Plant functional traits for improving productivity. Springer, Singapore. https://doi.org/10.1007/978-981-97-1510-7

Kumar N, Suyal DC, Sharma IP, Verma A, Singh H (2017) Elucidating stress proteins in rice (*Oryza sativa* L.) genotype under elevated temperature: a proteomic approach to understand heat stress response. 3 Biotech 7:205. https://doi.org/10.1007/s13205-017-0856-9

Kumar N, Singh H, Sharma SK (2020) Antioxidants: responses and importance in plant defense system. In: Roychowdhury R, Choudhury S, Hasanuzzaman M, Srivastava S (eds) Sustainable agriculture in the era of climate change. Springer, Cham. https://doi.org/10.1007/978-3-030-45669-6_11

Lachowycz K, Jones AP (2011a) Greenspace and obesity: a systematic review of the evidence. Obes Rev 12(5):183–189. https://doi.org/10.1111/j.1467-789X.2010.00827.x

Lachowycz K, Jones AP (2011b) Green space and obesity: a systematic review of the evidence. Obes Rev 12:183–189. https://doi.org/10.1111/j.1467-789X.2010.00827.x

Landrigan PL, Fuller R, Fisher S, Suk WA, Sly P, Chiles TC, O'Reilly S (2019) Pollution and children's health. Sci Total Environ 650:2389–2394. https://doi.org/10.1016/j.scitotenv.2018.09.375

Layman RM, Day SD, Mitchell DK, Chen YJ, Harris JR, Daniels WL (2016) Below ground matters: urban soil rehabilitation increases tree canopy and speeds establishment. Urban for Urban Green 16:25–35. https://doi.org/10.1016/j.ufug.2016.01.004

Lee-Smith D, Prain G (2006) Understanding the links between agriculture and health. Focus 13. International Food Policy Research Institute, Washington, DC

Liu CF, Li MM, He XY, Chen W, Xu WY, Zhao GL, Ning ZH (2003) Concept discussion and analysis of urban forest. Chin J Ecol 22(5):146–149

Liu J, Cao Z, Zou S, Liu H, Hai X, Wang S, Jia Z (2018) An investigation of the leaf retention capacity, efficiency and mechanism for atmo spheric particulate matter of five greening tree species in Beijing, China. Sci Total Environ 616:417–426. https://doi.org/10.1016/j.scitotenv.2017.10.314

Maas J, Verheij RA, de Vries S, Spreeuwenberg P, Schellevis FG, Groenewegen PP (2009) Morbidity is related to a green living environment. J Epidemiol Commun Health 63(12):967–973. https://doi.org/10.1136/jech.2008.079038

Maes J, Barbosa A, Baranzelli C, Zulian G, Batista SF, Vandecasteele I et al (2015) More green infrastructure is required to maintain ecosystem services under current trends in land-use change in Europe. Lands Ecol 30:517–534. https://doi.org/10.1007/s10980-014-0083-2

Malik A, Kumar K, Kumari G, Kumar N, Singh H (2024). Application of functional traits in modelling productivity and resilience under climate change. In: Kumar N, Singh H (eds) Plant functional traits for improving productivity. Springer, Singapore. https://doi.org/10.1007/978-981-97-1510-7_5

McKinney ML (2006) Urbanization as a major cause of biotic homogenization. Biol Conserv 127(3):247–260. https://doi.org/10.1016/j.biocon.2005.09.005

Meena D, Singh H, Chaudhari SK (2011) Elucidating strontium response on growth dynamics and biochemical change in *Phaseolus mungo* L. Int J Agricult Environ Biotechnol 4:107–113

Mink G, Witter G (1982) Haeuser mitGruenem Pelz, Ein HandbuchzurHausbegruenung. Verlag Dieter Fricke GmbH, Frankfurt, p 498

Montagnini F (2006) Homegardens of Mesoamerica: biodiversity, food security, and nutrient man agement. In: Kumar BM, Nair PKR (eds) Tropical homegardens: a time-tested example of sustainable agroforestry. Springer, Dordrecht, pp 61–84

Nowak DJ, McBride JR, Beatty RA (1990) Newly planted street tree growth and mortality. J Arboric 16:124–129

Nowak DJ (1994) Air pollution removal by Chicago's urban forest. Chicago's urban forest ecosystem: results of the Chicago urban forest climate project. General Technical Report NE-186, US Department of Agriculture, Forest Service, Radnor, pp 63–81

O'Dell R, Taheri M, Kabel R (1977) A model for uptake of pollutants by vegetation. J Air Pollut Control Assoc 27:1104–1109

Oliveira S, Andrade H, Vaz T (2011) The cooling effect of green spaces as a contribution to the mitigation of urban heat: a case study in Lisbon. Build Environ 46:2186–2194. https://doi.org/10.1016/j.buildenv.2011.04.034

Pagliai M, Marsili A, Servadio P, Vignozzi N, Pellegrini S (2003) Changes in some physical properties of a clay soil in Central Italy following the passage of rubber tracked and wheeled tractors of medium power. Soil Tillage Res 73(1):119–129. https://doi.org/10.1016/S0167-1987(03)00105-3

Prakash V, Hunney K, Singh H (2022) Effect of elevated carbon dioxide on growth and development of *Santalum album* L. seedlings inoculated with plant growth promoting microorganisms in Open Top Chambers. Int J Environ Agricult Biotechnol 7(5):149–159

Pribadi MA, Septina AD, Lugina M, Indartik I (2021) Vertical forest: green open space alternative in urban area development. IOP Confer Ser Earth Environ Sci 909(1):012012

Prigioniero A, Postiglione A, Zuzolo D, Niinemets Ü, Tartaglia M, Scarano P, Guarino C (2023) Leaf surface functional traits influence particulate matter and polycyclic aromatic hydrocarbons air pollution mitigation: insights from Mediterranean urban forests. J Clean Prod 418:138158. https://doi.org/10.1016/j.jclepro.2023.138158

Ren Z, He X, Pu R, Zheng H (2018) The impact of urban forest structure and its spatial location on urban cool island intensity. Urban Ecosyst 21:863–874. https://doi.org/10.1007/s11252-018-0776-4

Romanova O, Lovell S (2021) Food safety considerations of urban agroforestry systems grown in contaminated environments. Urban Agricult Reg Food Syst 6(1):e20008. https://doi.org/10.1002/uar2.20008

Rosenzweig C, Solecki WD, Slosberg R (2006) Mitigating New York City's heat island with urban forestry, living roofs, and light surfaces. In: A report to the New York state energy research and development authority, pp 1–5

Ruark GA, Mader DL, Tattar TA (2012) The influence of soil compaction and aeration on the root growth and vigour of trees-a literature review. Arboric J 6:251–265

Salisbury AB, Koeser AK, Hauer RJ, Hilbert DR, Abd-Elrahman AH, Andreu MG, Thorn H (2022) The legacy of hurricanes, historic land cover, and municipal ordinances on urban tree canopy in Florida (United States). Front for Glob Change 5(1):742157. https://doi.org/10.3389/ffgc.2022.742157

Santamour FS (1990) Trees for urban planting: diversity, uniformity, and common sense. In: Proceedings of the seventh conference of the metropolitan tree improvement alliance (METRIA), pp 57–65

Schueler T (2000) The compaction of urban soil: the practice of watershed protection. Center for Watershed Protection, Ellicott City, pp 210–214

Sgrigna G, Baldacchini C, Dreveck S, Cheng Z, Calfapietra C (2020) Relationships between air particulate matter capture efficiency and leaf traits in twelve tree species from an Italian urban-industrial environment. Sci Total Environ 718:137310. https://doi.org/10.1016/j.scitotenv.2020.137310

Shaffer HB (2018) Urban biodiversity arks. Nat Sustain 1(12):725–727. https://doi.org/10.1038/s41893-018-0193-y

Sharma R, Singh H, Kaushik M, Nautiyal R, Singh O (2018) Adaptive physiological response, carbon partitioning, and biomass production of *Withania somnifera* (L.) Dunal grown under elevated CO_2 regimes. 3 Biotech 8:267. https://doi.org/10.1007/s13205-018-1292-1

Sharma R, Singh H (2021) Alteration in biochemical constituents and nutrients partitioning of *Asparagus racemosus* in response to elevated atmospheric CO_2 concentration. Environ Sci Pollut Res 29:6812–6821. https://doi.org/10.1007/s11356-021-16050-3

Shashua-Bar L, Hoffman ME (2000) Vegetation as climatic component in the design of an urban street-an empirical model for predicting the cooling effect of urban green areas with trees. Energ Build 31:221–235. https://doi.org/10.1016/S0378-7788(99)00018-3

Shashua-Bar L, Pearlmutter D, Erell E (2009) The cooling efficiency of urban landscape strategies in a hot dry climate. Landsc Urban Plan 92:179–186. https://doi.org/10.1016/j.landurbplan.2009.04.005

Shober AL, Denny GC (2010) Soil compaction in the urban landscape EDIS SL 317/SS529, 3/2010. EDIS, 2010(2)

Singh H (2021) An integrated approach considering physiological and biophysical based indicators for assessing tolerance of roadside plantations of *Alstonia scholaris* towards urban roadside air pollution: an assessment of adaptation of plantations for mitigating urban roadside air pollution. Trees. https://doi.org/10.1007/s00468-021-02179-8

Singh H (2024) Exploring adaptive modulation in plant functional traits and its impact on the productivity of *Acacia auriculiformis* under CO_2-enriched environment. Ind Crops Prod 210:118186

Singh H, Verma A (2013) Characterization and screening of high nitrogen efficient rice genotype to curtail environmental pollution. Int J Agricult Environ Biotechnol 6(1):23–30

Singh H, Savita Sharma R, Sinha S, Kumar M, Kumar P, Verma A, Sharma SK (2017a) Physiological functioning of *Lagerstroemia speciosa* L. under heavy roadside traffic: an approach to screen potential species for abatement of urban air pollution. 3 Biotech 7(61):1–10. https://doi.org/10.1007/s13205-017-0690-0

Singh H, Verma A, Kumar M, Sharma R, Gupta R, Kaur M, Negi M, Sharma SK (2017b) Phytoremediation: a green technology to clean up the sites with low and moderate level of heavy metals. Aust Biochem 2:1012

Singh H, Sharma R, Savita SMP, Kumar M, Verma A, Ansari MW, Sharma SK (2018) Adaptive physiological response of *Parthenium hysterophorus* to elevated atmospheric CO_2 concentration. Indian for 144:1–14

Singh H, Yadav M, Kumar N, Kumar A, Kumar M (2020) Assessing adaptation and mitigation potential of roadside trees under the influence of vehicular emissions: A case study of *Grevillea robusta* and *Mangifera indica* planted in an urban city of India. PLoS ONE 15(1):e0227380

Singh S, Singh H, Sharma SK, Nautiyal R (2021) Seasonal variation in biochemical responses of bamboo clones in the sub-tropical climate of Indian Himalayan foothills. Heliyon 7(4):e06859. https://doi.org/10.1016/j.heliyon.2021.e06859

Singh H, Kumar N, Kumar A (2024a) Enhancing resource use efficiency in crops through plant functional traits. In: Kumar N, Singh H (eds) Plant functional traits for improving productivity. Springer, Singapore. https://doi.org/10.1007/978-981-97-1510-7_6

Singh M, Singh H, Kumar A, Kumar M, Barthwal S, Thakur A (2024b) Soil nitrogen availability determines the CO_2 fertilization effect on tree species (*Neolamarckia cadamba*): growth and physiological evidence. Environ Sustain 7:53–60. https://doi.org/10.1007/s42398-023-00300-w

Singh H, Kumar M (2022) Climate change and its impact on Indian Himalayan forests: current status and research needs. In: Rani S, Kumar R (eds) Climate change. Springer, Cham, pp 223–242. https://doi.org/10.1007/978-3-030-92782-0_11

Sirakaya A, Cliquet A, Harris J (2018) Ecosystem services in cities: Towards the international legal protection of ecosystem services in urban environments. EcosystServ 29:205–212. https://doi.org/10.1016/j.ecoser.2017.01.001

Somerville PD, May PB, Livesley SJ (2018) Effects of deep tillage and municipal green waste compost amendments on soil properties and tree growth in compacted urban soils. J Environ Manage 227:365–374. https://doi.org/10.1016/j.jenvman.2018.09.004

Spronken-Smith RA, Oke TR, Lowry WP (2000) Advection and the surface energy balance across an irrigated urban park. Int J Climatol 20(9):1033–1047

Sugawara H, Shimizu S, Takahashi H, Hagiwara S, Narita K, Mikami T, Hirano T (2016) Thermal influence of a large green space on a hot urban environment. J Environ Qual 45(1):125–133. https://doi.org/10.2134/jeq2015.01.0049

Sun J, Wang X, Chen A, Ma Y, Cui M, Piao S (2011) NDVI indicated characteristics of vegetation cover change in China's metropolises over the last three decades. Environ Monit Assess 179:1–14. https://doi.org/10.1007/s10661-010-1715-x

Takayama N, Korpela K, Lee J, Morikawa T, Tsunetsugu Y, Park BJ, Li Q, Tyrvainen L, Miyazaki Y, Kagawa T (2014) Emotional, restorative and vitalizing effects of forest and urban environments at four sites in Japan. Int J Environ Res Public Health 11(7):7207–7230. https://doi.org/10.3390/ijerph110707207

Tan X, Hirabayashi S, Shibata S (2021) Estimation of ecosystem services provided by street trees in Kyoto, Japan. Forests 12(3):311. https://doi.org/10.3390/f12030311

Thaman RR (1992) Agro-deforestation as a major threat to sustainable development. Box 19:194–195

Ulrich RS (1983) Aesthetic and affective response to natural environment. Behavior and the natural environment New York. Plenum Press Altman I, Wohlwill JF, pp 85–125

Van Leeuwen B (2010) Dealing with urban diversity: promises and challenges of city life for intercultural citizenship. Polit Theor 38(5):631–657. https://doi.org/10.1177/0090591710372869

Ward Thompson C (2004) Forest amenity planning approaches. Encyclopedia of forest sciences. Elsevier, Amsterdam, pp 478–486

WHO (2019) World health organization. Air pollution and child health: prescribing clean air. http://www.who.int/ceh

Wirtz Z, Hagerman S, Hauer RJ, Konijnendijk CC (2021) What makes urban forest governance successful: a study among Canadian experts. Urban for Urban Green 58:126901. https://doi.org/10.1016/j.ufug.2020.126901

World Bank (2022) Urban development

Yadav SK, Singh H, Ginwal HS, Barthwal S (2019a) Elevated CO_2 enhanced growth and physiological process of *Populus deltoides* Bartr. ex Marsh. Indian for 145(1):23–27

Yadav SK, Singh H, Nautiyal R, Ginwal HS, Ansari SA, Barthwal S (2019b) Modulation of morpho-physiological responses in *Populus deltoides* by elevated carbon dioxide and temperature. For Sci 66:105–118

Yang G, Zhao Y, Xing H, Fu Y, Liu G, Kang X, Mai X (2020) Understanding the changes in spatial fairness of urban greenery using time-series remote sensing images: a case study of Guangdong-Hong Kong-Macao Greater Bay. Sci Total Environ 715:136763. https://doi.org/10.1016/j.scitotenv.2020.136763

Yokohari M, Takeuchi K, Watanabe T, Yokota S (2000) Beyond greenbelts and zoning: a new planning concept for the environment of Asian mega-cities. Landsc Urban Plann 47(3–4):159–171. https://doi.org/10.1016/S0169-2046(99)00084-5

Zhang D, Zheng H, He X, Ren Z, Zhai C, Yu X, Wang P (2016) Effects of forest type and urbanization on species composition and diversity of urban forest in Changchun, Northeast China. Urban Ecosyst 19:455–473

Zhou HX (1993) Brownian dynamics study of the influences of electrostatic interaction and diffusion on proteinprotein association kinetics. Biophys J 64:1711–1726

Urbanization and Climate Change: Impacts, Adaption and Mitigation Strategies

Rishita Pakhira, Subhankar Biswas, and Hukum Singh

Abstract This chapter explores the multifaceted relationship between urbanization and climate change, highlighting their combined impacts on urban environments and proposing effective strategies for adaptation and mitigation. Urbanization accelerates environmental changes, exacerbating the effects of climate change through phenomena such as the Urban Heat Island effect and alterations in air quality. These changes pose significant challenges to urban ecosystems, including urban forests, soil quality, and biodiversity. However, strategic mitigation measures and adaptive strategies help alleviate these challenges, emphasizing the importance of green infrastructure, urban planning, and ecosystem-based management approaches. Integrating ecological principles, understanding local contexts, and implementing sustainable practices lead us to enhance resilience and encourage sustainable urban development in the face of urbanization and climate change. Further, this chapter provides insights and recommendations for policymakers, urban planners, and stakeholders to navigate the complexities of urbanization and climate change while promoting the preservation and prosperity of urban environments.

Keywords Air quality · Green Infrastructure · Mitigation · Urban ecosystem · Urban Heat Islands

1 Introduction

As humanity continues its relentless march towards urbanization, the intricate interplay between this phenomenon and climate change emerges as a paramount concern of the twenty-first century. The nexus between urbanization and climate change is a

R. Pakhira (✉) · H. Singh
ICFRE-Forest Research Institute, PO New Forest Dehradun, Dehradun, Uttarakhand 248006, India
e-mail: rishitapakhira@gmail.com

S. Biswas
Bidhan Chandra Krishi Viswavidyalaya, P.O. Krishi Viswavidyalaya, Mohanpur, Nadia, West Bengal 741252, India

H. Singh (ed.), *Urban Forests, Climate Change and Environmental Pollution*,
https://doi.org/10.1007/978-3-031-67837-0_3

dynamic and multifaceted relationship that shapes the fabric of our urban environments, influencing everything from air quality to biodiversity (Kumar et al. 2022; Singh et al. 2022; Kumar et al. 2024). In this introductory exploration, we delve into the profound implications of urbanization on climate change and vice versa, shedding light on the synergistic effects that amplify the challenges faced by urban ecosystems worldwide.

Urbanization, the process by which populations concentrate in urban areas, has become a defining characteristic of the modern era. The rapid expansion of cities and towns is accompanied by many environmental changes, profoundly altering local climates and ecosystems (Singh et al. 2024a; Singh 2024). Among the most conspicuous changes is the Urban Heat Island (UHI) effect, wherein urban areas experience significantly higher temperatures than their rural surroundings. This phenomenon, fuelled by the proliferation of impervious surfaces and the scarcity of green spaces, exacerbates the impacts of climate change, magnifying heat-related health risks and straining energy demands for cooling (Dhyani et al 2021; Kumar et al. 2021a, b, c, d, e; Prakash et al. 2022).

Moreover, the transformation wrought by urbanization extends beyond mere temperature fluctuations, permeating the air we breathe. Alterations in land use and transportation patterns contribute to the degradation of air quality, with implications for human health and environmental sustainability. The intricate web of interactions between urbanization and climate change also manifests in shifts in precipitation patterns, exacerbating the frequency and intensity of extreme weather events such as floods and droughts, further imperilling urban resilience (Singh 2021; Sharma and Singh 2021; Singh et al. 2021).

Amidst these challenges, however, lie opportunities for innovation and adaptation. Strategic mitigation measures and adaptive strategies offer pathways to mitigate the adverse impacts of urbanization on climate change and enhance the resilience of urban ecosystems (Singh et al. 2020; Kumar et al. 2021a, b). Embracing green infrastructure, integrating ecological principles into urban planning, and fostering ecosystem-based management approaches emerge as critical pillars in sustainable urban development.

This chapter unravels the complexities of the urbanization-climate change nexus, offering insights and recommendations for policymakers, urban planners, and stakeholders. By understanding the intricate dynamics and embracing holistic, sustainable practices, we can pave the way toward resilient, thriving urban environments in the face of unprecedented environmental challenges.

2 Global Climate Change Scenario

The Intergovernmental Panel on Climate Change (IPCC 2021) projects a significant global surface air temperature (GSAT) increase between 2021 and 2040, with an expected rise of 1.5 °C. There is a high probability that individual years may exceed this 1.5 °C threshold by 40–60% by 2030, relative to the 1850–1900 baselines.

Looking further ahead to 2081–2100, GSAT is anticipated to be 0.2–1.0 °C higher compared to the recent past (1995–2014).

Throughout the twenty-first century, an increase in annual global land precipitation is expected alongside rising GSAT. Projections indicate a range of −0.2 to +4.7% changes in globally averaged annual land precipitation between 2081 and 2100, relative to 1995–2014. These projections reveal significant regional variations and seasonal differences in precipitation changes linked to the increasing GSAT. However, uncertainties persist in near-term precipitation changes due to natural internal variability, model uncertainty, and ambiguity in natural and anthropogenic aerosol forcing, with a medium confidence level in these projections.

The Arctic Ocean is forecasted to become practically ice-free in September (annual minimum sea ice area less than 1 million km_2), averaging over 2081–2100. Moreover, global mean sea level (GMSL) will continue rising throughout the twenty-first century. Additionally, there is a very likely scenario that the cumulative carbon uptake by both the ocean and land will increase until the conclusion of the twenty-first century.

3 Global Urban Expansion

Urban population expansion is a worldwide phenomenon. Most of the urban growth and land-cover changes will occur in certain regions. Globally, more than 5.87 million km^2 of land has a positive likelihood (>0%) of becoming urban by 2030, with 20% (1.2 million km^2) having a high probability (>75%) of urban expansion. If all locations with a high chance (>75%) of urban land conversion occur, the global urban extent will rise by an astounding 185% compared to levels around 2000 (Seto et al. 2012).

4 Urban Forests and Their Importance

Urban forests are ecosystems comprised of trees, vegetation, and green spaces within urban areas. These areas provide various environmental, social, and economic benefits to the communities they serve. Urban forests encompass parks, street trees, green belts, and other planted areas within cities and towns (Kumar et al. 2021c, d). Urban forests offer many goods and services, and their value to the human community is contingent on their physical attributes and functions. As per the Millennium Ecosystem Assessment, environmental values are categorized into four main groups: (a) Provisioning services (such as food, water, and fibre); (b) Regulating services (including climate and water regulation, pollination); (c) Cultural services (such as recreation and education); (d) Supporting services (including nutrient cycling). The significance of urban forests lies in their essential qualities, and their positive contributions depend on how effectively these services fulfill their roles within the urban environment (Jansson 2014).

4.1 Ecological Services and Human Benefits of Urban Forests

4.1.1 Emotional and Psychological Services

Urban planners should prioritize the emotional and psychological connection between humans and nature, known as biophilia, to reconnect the growing population with the natural environment. Studies show exposure to nature can reduce stress, enhance mental well-being, and alleviate conditions like anxiety and depression. Incorporating biophilia into urban design, such as creating parks and planting vegetation, is essential for positive emotional and psychological health. Additionally, urban forests offer opportunities for community engagement, fostering a sense of shared identity and connection with nature (Wolf et al. 2020).

Moreover, urban forests play a crucial role in urban development, reducing environmental degradation and improving human well-being (Gwedla and Shackleton 2019). More excellent tree coverage near homes is associated with better health and lifestyle choices, including lower rates of obesity and improved neighbourhood social cohesion. Trees in urban areas have the potential to mitigate harm from air pollutants, excess heat, and crime while enhancing bodily capacities such as the immune system and cardiovascular function (Wolf et al. 2020). Effective design of urban forests should consider residents' preferences for ecosystem services, making them accessible to all.

4.1.2 Monetary Services

Property values across urban gradients vary based on available amenities. The Hedonic Pricing model, used to estimate the market value of urban trees and their ecosystem services, indicates that residents value tree coverage within certain distances of their property (Pandit et al. 2012). An increase in tree cover is linked to higher property sale prices. Despite the benefits, some property owners weigh negative aspects of urban forestry, such as blocked views and maintenance issues, when assessing preferences. Urban planners should consider the economic value of trees in shaping urban landscapes.

4.1.3 Hydrologic Services

Urbanization has decreased forest cover and increased impervious surfaces, disrupting hydrologic processes. This alteration results in higher stormflow volumes, peak flows, and stream flooding in urban watersheds than rural ones. Urban streams experience increased erosion, bank erosion, and channel enlargement, threatening hydrological stability. Land managers should monitor and assess stream conditions to prevent further degradation and protect hydrologic systems (Berland et al. 2017; Phillips et al. 2019).

4.1.4 Thermal Services

Urban heat island (UHI) effects, causing elevated air and surface temperatures, result from factors like surface types, building structures, vegetation coverage, and human activities. Urban planning decisions should focus on city design, improving transportation methods, and incorporating cooling measures like green spaces. Urban forests offer an efficient solution to mitigate rising temperatures, reducing surface and air temperatures through shading effects (Mohajerani et al. 2017; Brown et al. 2018).

4.1.5 Air Quality Services

Urban forests mitigate air pollution by absorbing pollutants through leaf stomata, plant surfaces, and interception. Their efficiency in pollutant removal is influenced by tree coverage and pollutant concentrations. Proper urban planning should involve incorporating native tree species, strategically placing street trees to avoid pollutant trapping, and spacing trees to allow for atmospheric mixing. This approach is essential for reducing anthropogenic air pollution and promoting efficient carbon fixation (Yadav et al. 2019; Gupta et al. 2019a, b).

4.1.6 Synurbization Services

As global development transforms natural landscapes, wildlife species, including birds and mammals, exhibit synurbization—adapting to urban environments by altering behaviour and utilizing new food resources. Parker and Nilon (2012) provide a model for assessing factors influencing synurbization, aiding in urban planning for coexistence.

4.1.7 Adaptation Mechanisms

Wildlife, such as white-crowned sparrows and blackbirds, adapt to urban noise by selecting high-frequency song elements and adjusting their minimum frequency, respectively, for optimal signal transmission. Understanding their adaptive strategies is crucial to managing species in evolving urban environments. Recommendations include using forest stands and green spaces as sound barriers to reduce urban noise exposure (Margaritis and Kang 2017; Kumar et al. 2019a; 2019b; c).

4.1.8 Population and Community Services

As urbanization encroaches on wildlife habitat, managing wildlife populations becomes vital. Anthropogenic stressors prompt wildlife species' movement, diet, reproduction, and survival changes. Urban forests play a crucial role in providing habitat and landscape connectivity for wildlife, emphasizing the importance of preserving scattered urban trees (Le Roux et al. 2018). Effective wildlife habitat design in urban settings involves recognizing and managing various habitats patches, including Remnant, Successional, and Managed Habitat Patches (Adams and Lindsey 2010).

5 Interaction of Urban Forests, Climate Change and Urbanization

Climate change severely threatens natural ecosystems, intricate systems influenced by environmental changes affecting abiotic and biotic factors. The repercussions include shifts in heat waves, precipitation patterns, CO_2 concentration, and temperature, leading to the emergence of new pests, weeds, and pathogens. These changes impose stresses on plants, both biotic and abiotic (Sharma et al. 2018; Singh et al. 2018; Gupta et al. 2018a, b; Choudhary et al. 2022) (Fig. 1).

The Intergovernmental Science-Policy Platform on Biodiversity as well as Ecosystem Services (IPBES) report warns that the unsustainable growth of metropolitan areas may put one million species at risk of extinction due to climate change (IPBES 2019; McDonald et al. 2020). Urbanization, through land cover conversion, disrupts biological migration by significantly altering habitat suitability and the provision of ecological services (Wang et al. 2019).

Increased demand for dwelling space and agricultural land, driven by an increasing urban population, changes land use and cover. This transformation necessitates more crop and food production, resulting in biodiversity loss and climate regulation disturbances (Mononen et al. 2016; Liu et al. 2019). Urban development disrupts entire ecosystems and species, with avian species due to a drastic reduction in vegetation cover. Species richness is positively correlated with vegetation and open areas but

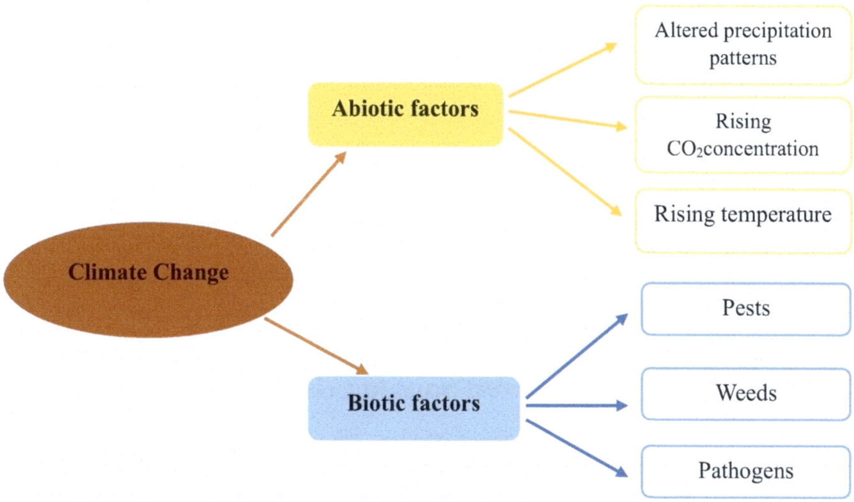

Fig. 1 Climate change impacts on abiotic and biotic factors

negatively correlated with built-up areas and road density, accounting for 40–81% variation in avian species.

Urban forests impact avian species. In highly and moderately urbanized areas, there is a median decline in species richness of 52.2 and 27.4%, respectively. The decline is attributed to species inadequately adapting to urban environments and avoiding associated risks (Sol et al. 2014). Abnormal climatic conditions have significantly decreased the reproductive capability of various species, and predictions based on biotic models indicate that most common species will be under significant stress, leading to a considerable shift in assemblage composition (Nelson et al. 2009).

5.1 Direct and Indirect Effects of Urbanization on Urban Forest Soil

Soils in urban areas undergo modifications resulting from direct and indirect effects, often coinciding. Direct effects are typically related to soil alterations during the intense disturbance phase of the Urban Soil Continuum. In most urban settings, these impacts occur concurrently with land development processes, involving levelling and relocating soil using heavy equipment, excavation, and backfilling, or shaping and constructing urban landscape features. Post-development soil management practices, including fertilization and irrigation, are also categorized as direct effects. Human-initiated management addresses limitations frequently caused by direct effects and often leads to the developing of highly productive soils (Singh and Verma 2013; Kumar et al. 2017; Gupta and Singh 2017).

In contrast, changes indirectly impact the abiotic and biotic environment caused by human activities, like the emission of pollutants. While these effects are related to some degree throughout the Urban Soil Continuum, they have a proportionately more significant impact on undisturbed soils associated with remaining parcels. Urban environmental factors contributing to indirect effects include the urban heat island (Hall et al. 2015), the introduction of exotic plant and animal species (Szlavecz et al. 2011; Singh et al. 2017a, b), and atmospheric deposition of pollutants such as nitrogen (N) and sulfur (S) (Huang et al. 2015; Decina et al. 2017), heavy metals (Werkenthin et al. 2014), and potentially toxic organic chemicals (Jensen et al. 2007).

5.2 *Response of Urban Forest Soils to Climate Change*

Human-induced changes to the global climate are already significantly evident in urban areas. Urban environments experience up to 30% higher air temperatures, and atmospheric carbon dioxide (CO_2) and methane (CH_4) concentrations are approximately 40% higher than rural or global averages. These environmental alterations reflect the anticipated overall increases in atmospheric CO_2 projected globally over the next 30 years or more (Meinshausen et al. 2011).

The impacts of expected increases in air temperature, especially at night, rainfall intensity, and the atmosphere CO_2 concentrations from global climate changes may be best understood by examining their influences on the direct and indirect effects of urban land use discussed in earlier sections. Indirect effects are linked to environmental, chemical, and species changes, while direct effects encompass alterations associated with landscape, physical, and resource changes occurring during and after urban development. Notably, soils least affected along the Urban Soil Continuum (USC), such as remnant forest soils, are particularly susceptible to indirect effects and, consequently, are asked to be most affected by long-term changes in global climate. In contrast, soils that is managed or more impacted by direct effects are likely to experience less influence from climate change.

6 The Urban Heat Island Effect

The Urban Heat Island (UHI) is an urban area experiencing significantly higher temperatures than its rural surroundings due to artificial infrastructure and human activities (Isola et al. 2023). UHIs adversely affect public health (Heaviside et al. 2017) and contribute to increased energy consumption for cooling purposes (Santamouris et al. 2015). One of the primary contributors to elevated temperatures in urban areas and the form of UHIs is the existence of impervious surfaces (Irmak et al. 2017), leading to diminished thermal comfort in cities (Morris et al. 2017). Implementing measures that specifically address the surface cover of buildings can significantly enhance the thermal behaviour of urban structures (Icaza et al. 2016).

The UHI effect primarily stems from the high density of buildings and impervious materials, significantly impacting the energy balance within a city (Masson et al. 2020). Common building materials like concrete and asphalt typically have a lower albedo than vegetation, reflecting less incoming radiation. Urban materials also exhibit higher heat capacity than vegetation, allowing them to store more absorbed solar energy. Additionally, the limited presence of vegetation in current urban landscapes reduces heat dissipation through evapotranspiration. These factors collectively result in the absorption and retention of significantly more thermal energy within the urban environment, causing temperatures to rise more rapidly than in natural areas (Kolokotroni and Giridharan 2008).

6.1 Impacts on Urban Forests

Excessive heat, a common feature in urban areas, coupled with other urban conditions, can impact the physiological functioning of trees, leading to a range of outcomes from enhanced growth to premature senescence, branch die-back, and even mortality (Hilbert et al. 2019). Elevated air temperatures resulting from Urban Heat Islands (UHIs) can influence tree growth by altering various physiological processes directly or indirectly (Dusenge et al. 2019). At the cellular level, reaction times increase with temperature up to a maximum, after which enzymatic activity drops, resulting in a species-dependent optimum curve (Arcus et al. 2016). In leaves, this optimum response is reflected in the net assimilation rate of carbohydrates, a balance of photosynthesis and respiration, with losses exceeding gains more rapidly at higher temperatures. These responses vary between species and intraspecifically due to local acclimation, a shift of optimum temperature responses after prolonged exposure, and threshold temperatures before tissue damage occurs (Geange et al. 2021).

High temperatures in temperate areas often coincide with low relative air humidity, resulting in a significant vapor pressure deficit, which can decrease stomatal conductance, governing the majority of gas exchange in leaves (Grossiord et al. 2020). Prolonged stomatal closure or decreased conductance with high temperatures may lead to reduced tree growth rate in subsequent years or even starvation, as carbohydrate reserves are depleted without sufficient replenishment. Moreover, air and soil temperatures influence cambial activity initiation, speed, and cessation, affecting radial growth throughout a growing season (Rathgeber et al. 2016). Radial growth is limited by wood formation dynamics and their relation to environmental drivers rather than solely by photosynthetic activity (Körner 2015). The availability of soil water is vital for cell expansion, limiting radial growth before photosynthesis, and is linked to local climate, as higher temperatures drive evaporation and may contribute to the depletion of soil water storage, impeding growth (Fatichi et al. 2014).

Several urban tree species have demonstrated tendencies for enhanced growth rates and/or productivity compared to rural counterparts, often attributed to increased temperatures (Jia et al. 2018). However, a wide range of effect sizes and signs (i.e., reduced growth) is specific to species and location (Berland 2020). Zhao et al. (2016) found increased productivity rates within urban clusters as urbanization intensifies, as remotely sensed vegetation indices indicate. Positive associations between air temperature and radial growth were identified for two species (*Tilia cordata* Mill., *Rubiniapseudoacacia*) commonly chosen by urban planners in Munich (Moser-Reischl et al. 2019). In Berlin, Dahlhausen et al. (2018) observed enhanced growth in highly urbanized environments for *T. cordata*, attributed to the UHI effect, while intermediate development intensity was detrimental to tree growth. Conversely, Gillner et al. (2014) noted decreased growth for Acer species, *Platanus* × *hispanica,* and *Quercus rubra* with higher summer temperatures in another major German city, particularly when compounded with drought. These growth variations may stem from species-specific responses to increased temperatures and are influenced by other factors such as water availability, pollution, road-salt loading, and structural impedance by infrastructure, or management practices.

6.2 Mitigation Strategies of UHI Effects

Green Infrastructure (GI) is rapidly being recognized as an important method for decreasing the impacts of urban heat islands (UHI). Vegetation in GI regulates microclimates by mechanisms like evapotranspiration from plant leaves and shade surfaces, improving human thermal comfort. Most surveyed studies support a comprehensive GI approach to lower temperatures, with only a handful focused on a single GI technique. The research reveals that the combined deployment of different GI solutions gives the most successful outcomes in boosting microclimates and thermal comfort, whereas scenarios with no greening at all are deemed the least advantageous option (Herath et al. 2018).

Various varieties of GI cover trees, grass, bushes, greening on buildings (such as green roofs and walls), and parks. Balany et al. (2020) indicated that trees are the most often adopted GI method, followed by grass and green roofs. Various GI are discussed in Table 1.

6.3 Climate Change and Urbanization-Induced Stress

Climate change and urbanization create a complex and dynamic environmental system that significantly impacts biotic and abiotic elements (Kumar and Verma 2018; Chaudhry and Sidhu 2022). Abiotic stress has a substantial impact on plant growth and viability markers. Plants are constantly exposed to changing environments, which need adaptive responses to numerous biotic and abiotic stressors (Meena et al. 2011; Kumar et al. 2019c; 2020; Balfagón et al. 2020). Abiotic stress emerges as the principal restriction for plants in urban and peri-urban environments

Table 1 Various green infrastructure strategies and their effects

GI type	Parameters measured	Main findings	References
Tree	MRT, UTCI	Shrub or vertical greening can prevent airflow from decreasing, affecting thermal comfort conditions	Perini et al. (2017)
Tree, grass	Ta	Significant cooling may be produced with the proper positioning. Clustered green space improves local cooling, but a scattered layout results in superior regional cooling	Zhang et al. (2017)
Tree, grass	Ta, RH, MRT	Green and blue regions show the possibility of heat reduction. Grey areas are more likely to provide an unpleasant temperature environment	Zhao and Fong (2017)
Tree	Ta	The daytime air temperature lowers as the aspect ratio (AR) increases. The tree canopy impacts pedestrian PET	Jamei et al. (2017)
Tree, green roofs	Ta, MRT	Green roofs provide cooling both during the day and at night. Removing trees significantly raises the air temperature throughout the day	Roth and Lim (2017)
Tree, grass, shrubs	Ta, MRT, MOCI	Significant reduction in MOCI with a combo scenario (vegetation, cool roof, cool pavement). The worst-case situation is utilising asphalt without vegetation	Salata et al. (2017)
Tree, grass	Ta	With more trees, Ta may be reduced by up to 0.6 °C and asphalt Ts by up to 15 °C	Tsoka et al. (2018)
Tree, green roofs	Ta, RH	Combining trees on curbsides, 50% green roofs, and green walls resulted in a maximum temperature drop of 1.9 °C	Herath et al. (2018)
Tree	Ta, RH, MRT, WS, PET	Dense trees improve thermal comfort. PET is more susceptible to MRT than wind speed. Low-rise structures cause increased heat stress	Yahia et al. (2018)
Tree	Ta, MRT, WS, PET	The impact of vegetation on thermal comfort depends on tree arrangement, LAI, crown width, and height. Trees with an aspect ratio <2 improve PET significantly	Zhang et al. (2018)
Tree, grass	Ta, RH, WS	A larger proportion of deciduous trees and grass will raise temperatures and improve thermal comfort in the winter	Afshar et al. (2018)
Tree	Ta, MRT, PMV	Vegetation improves thermal comfort in hot summer circumstances	Karakounoset al. (2018)
Tree	Ta, Ts	Zones with considerable vegetation do not exhibit a big average Urban Heat Island (UHI) effect; nonetheless, the impact is significant in densely urbanized zones	Manandhar et al. (2019)
Tree	Ta	Tree parameters like leaf albedo, LAD, and crown shape all substantially impact temperature	Shinzato et al. (2019)

(continued)

Table 1 (continued)

GI type	Parameters measured	Main findings	References
Tree	Ta	Trees with a height of 5 or 15 m can raise air temperature during the warmest hours of the day by impeding wind flow	Battista et al. (2020)
Green roofs	Ta, RH	Intensive green roofs are the most efficient form of lowering temperature, especially in the morning rather than in the afternoon	Kim et al. (2020)

Here, MRT—mean residence time, UTCI—universal thermal climate index, ambient air temperature (Ta), MOCI—multivariate ocean climate indicator, ws—wind speed, Potential evapotranspiration, or PET—predicted mean vote (PMV)

(Leotta et al. 2023), resulting in various physiological and biochemical alterations that impede osmotic adaptation (Ozturk et al. 2021).

6.4 *Drought*

Drought is unique among extreme weather events because of its substantial influence on soil fauna and plants (Zia et al. 2021; Bozca and Leblebici 2022). Drought reduces yields and alters urban woods' character. Predictions predict a global rise in tree death owing to drought, posing health concerns to urban trees and reducing ecosystem services, resulting in rising financial expenses (Petruzzellis et al. 2022). Climate change forecasts for 2070 indicate that 73% of investigated species in urban areas would lose climatically appropriate places, with 18% losing more than half of their habitat (Burley et al. 2019). Drought affects plants' phenological phases, resulting in shorter phases and, eventually, lower carbon dioxide assimilation (Singh and Kumar 2022; Singh et al. 2024b).

Drought stress has a physiological effect on leaf stomata, resulting in diminished photosynthetic capability and inefficient water usage (Driesen et al. 2020). During stress, reactive oxygen species (ROS) levels rise, causing oxidative damage and disrupting plants' antioxidant defence systems (Parvin et al. 2019).

Prolonged droughts caused by climate change present obstacles for afforestation in urban areas, making it more difficult to find acceptable seed sources depending on current climate conditions (Kamakura et al. 2021). The choice of abiotic stress resistance becomes critical for urban organisms. Monitoring plant performance with hyperspectral imaging can help pick more robust and climate-resistant plants.

According to Asrar et al. (2012), inoculating Antirrhinum majus seeds with mycorrhizal fungus improved drought stress tolerance, highlighting the positive impact of mycorrhizal fungi in increasing drought resistance in plants. Similarly, Battacharyya et al. (2015) discovered a similar response in plants treated with extracts of *Ascophyllum nodosum*. The use of these extracts improves drought stress tolerance in plant species. These findings demonstrate the possible use of mycorrhizal fungus and *A. nodosum* extracts to enhance drought tolerance in various plant species. Understanding and fostering these adaptation methods is critical in establishing more resilient and drought-tolerant urban forests, lessening climate change's negative consequences.

6.5 High Concentration of Salt

Salt stress is a severe worry intimately linked to the adverse effects of climate change, resulting in stunted development and various physiological diseases in plants. Elevated Na and Cl ion concentrations disturb osmotic activities crucial for plant health (Majeed and Muhammad 2019).

Regarding physiological factors, the spread of salinized soils highlights the rising need to produce cultivars that are tolerant or resistant to salt stress. Genome editing methods provide intriguing prospects to improve tree salt tolerance (Guo et al. 2022). Plants respond to salt and heavy metal stress by activating transporters in plasma and vascular membranes, which import toxic materials into vacuoles and transfer them to root tips and shoots. In contrast, during drought, cold, and heat stress, these transporters increase water and sugar levels in all plant organs (Gill et al. 2021).

6.6 Heavy Metal Concentration

"Heavy metal" refers to certain elements based on their toxicological, physical, and biological properties (Pourret 2018). Heavy metals pose a substantial environmental concern since they are poisonous and resistant to biodegradation (El-Naggar et al. 2018). Addressing this issue requires plants that absorb and remove heavy metals from urban surroundings.

Regarding physiological characteristics, heavy metal stress has a deleterious influence on all phases of plant growth, from seed germination to seed production. Several mechanisms have been discovered to improve plant tolerance to heavy metal pollution, including chelation, enzymatic defence systems (phenols, flavonoids, essential oils, and alkaloids), and protein repair (Antoniadis et al. 2017). Gladkov et al. (2021) demonstrated the generation of copper-tolerant cultivars by cell selection, highlighting its importance in resistance breeding. Additionally, genomic approaches help to advance more accurate breeding procedures. Transcriptomics, homologous recombination, allele research, and association mapping can be used to identify functional markers (Salgotra and Stewart 2020).

7 Air Quality and Trees Health

Air pollution has become a significant global concern in developed and developing nations. The rise in urban populations has led to a high prevalence of private vehicles in urban transportation systems, contributing to severe air pollution that affects the surrounding environment (Gulia et al. 2015). The re-suspension of road dust due to heightened traffic and the wear and tear of tires and brakes further contributes to elevated particulate matter (PM) concentrations in urban areas (Amato et al. 2014). The growing urbanization of cities, characterized by increased population and physical size, has raised travel demand and driven changes in travel behaviour (Chaudhuri and Chowdhury 2018). Cities account for more than 70% of worldwide carbon dioxide emissions. Studies show that increased carbon dioxide levels have a favourable impact on climate change, encouraging biomass, leaf area, and dry mass growth (Gurney et al. 2015). Urban vegetation has varied levels of tolerance for pollution gases. Small-leaf species, trichomes, and furrows may suggest greater urban tolerance (Barwise and Kumar 2020). However, Przybysz et al. (2021) discovered no significant effect of trichomes or leaf size on the formation of bound dust in herbaceous plants. As a result, selecting appropriate plant species is critical for reducing pollen emissions and moderating the effects of air pollution on ecosystems. Tropospheric ozone and nitrogen oxide are among the most damaging pollutants in urban environments. Urban trees and forests operate as biological filters; significantly reducing airborne dust particles (PM), particularly in highly populated urban areas (Zhang et al. 2019; Dadkhah-Aghdash et al. 2022).

7.1 Air Pollution Mitigation Strategies

7.1.1 Green Walls

Airborne particles cling to the leaves of green walls, reducing ambient concentrations. However, the pollutant removal capability of green walls in terms of varied PM size fractions and ideal species composition has not been thoroughly investigated (Weerakkody et al. 2017). Green walls are more effective at removing particulate matter (PM) than gaseous pollutants. PM10 clearance rates of 1.37 mg/year were found within 1216 ha of metropolitan Toronto and 314 kg/year in Melbourne, Australia (Jayasooriya et al. 2017).

Green walls operate as solid barriers in open road circumstances, boosting on-road particle concentration through recirculation while reducing the concentration beyond the barrier (Tong et al. 2016). The reduction in pollutant concentration behind green walls is due to dispersion and deposition.

7.1.2 Vertical Green Screens

Green screens represent an innovative concept, distinct from traditional approaches like trees or hedges, consisting of a steel or plastic mesh that supports the growth of climbing vegetation. They are designed to offer diverse advantages, including capturing air pollutants, supporting biodiversity, reducing noise, and attenuating rainfall. Green screens also serve additional purposes, such as providing privacy and tranquillity, concealing graffiti, acting as fencing, or greening noise barriers (Dover and Phillips 2015).

These screens can serve as an alternative to trees and hedges in urban areas, addressing the challenge of limited planting space. Moreover, they find applications in domestic gardens, indoor environments, or open roadsides, particularly around schools or playgrounds. Green screens are relatively cost-effective, easy to install, and require less maintenance than green walls. Climbing plants for green roofs are either planted and pre-grown in metal mesh panels supported by poles or other structures in greenhouse nurseries for approximately two years before being transplanted on-site (Dover and Phillips 2015), or they are directly planted on-site and grown in place until reaching maturity in about two and a half years (Tremper and Green 2018).

7.1.3 Green Roofs

Green roofs, also known as eco-roofs or living roofs, involve partial or complete coverage of building roofs with vegetation. These roofs offer environmental and economic benefits, enhancing building sustainability and compensating for vegetation loss during construction (Berardi et al. 2014). Critical components of green roofs include vegetation, growing substrate, filter layer, drainage layer, waterproofing membrane, and root barrier (Vijayaraghavan 2016).

Green roofs offer various benefits, including air pollution mitigation, stormwater attenuation, reduced energy consumption, and increased water quality, mitigation of the urban heat island effect, ecological preservation, sound insulation, and noise reduction (Berardi et al. 2014; Francis and Jensen 2017). Despite all these advantages, some studies have thoroughly explored the potential of green roofs to improve air quality (Francis and Jensen 2017).

8 Changes in Species Composition and Communities

Urbanization, marked by land cover changes, is a crucial driver of biodiversity loss, particularly impacting plant communities. Numerous studies (Anderson et al. 2020; Fekete et al. 2020; Heneidy et al. 2021; Salgado et al. 2021) have consistently reported decreased species richness in response to urbanization. Land cover changes, converting natural landscapes to impervious surfaces, creating a more uniform physical environment, reducing species richness, and altering community composition (Dolan et al. 2017; Freitas et al. 2020).

Functional-trait studies reveal specific features of plants favored in urban environments, contributing to habitat homogenization and lower functional-trait diversity in communities (Anderson et al. 2020). Responses to habitat homogenization include increased presence of generalist, disturbance-tolerant, ruderal, and pioneer species and a decline in phylogenetic richness. Urban habitats exhibit a convergence of traits such as shifts to self- and wind-pollination, fast growth rates, long flowering periods, and specific resource-use strategies. These changes in functional characteristics impact community stability and functioning (Dolan et al. 2017; Anderson et al. 2020).

While some native plant communities demonstrate resilience to fragmentation, maintaining their diversity, others succumb to the effects of urbanization (Fekete et al. 2020). The persistence of native species in urban areas suggests alternatives to mitigate the consequences of urbanization. Green areas with stress-adapted native species, supported by protected forest areas near urban environments, contribute to biodiversity conservation (Fekete et al. 2020; Heneidy et al. 2021).

Urbanization affects plant-pollinator interactions, influencing pollinator services and reproductive success. Fragmented urban environments may lead to a decreased level of pollinator diversity and effectiveness, impacting plant reproductive success. However, some studies suggest that certain pollinators can move between urban fragments, maintaining reproductive success for some plant species (Irwin et al. 2018; Oliveira et al. 2019).

Genetic studies provide insights into how plant populations respond to habitat fragmentation, with factors such as limited gene flow affecting genetic structure. Some species show resilience to fragmentation, maintaining genetic diversity (Emel et al. 2021). Understanding the reproductive systems of species and their genetic responses aids in conservation efforts and choosing suitable species for urban greening.

Climate change, accompanied by urban heat island effects, elevated carbon dioxide, ozone levels, and increased drought stress, further impacts urban vegetation. Phenological changes, such as advanced and extended flowering periods, are observed, influencing plant-animal interactions and resource availability (Fisogni et al. 2020; Kondratyeva et al. 2020).

Human activities, including intentional migration and ornamental plant introductions, contribute to the spread of non-native invasive species. Human-assisted dispersal through roads and railways facilitates the implementation of exotic species (Oliveira et al. 2020; Rivkin et al. 2021). Environmental education and promoting the use of native species for ornamentation are essential to prevent the entry of invasive species and restore local ecosystems' productivity.

9 Urban Forest Ecosystem-Based Management

Urban forests, as integral components of urban ecosystems, require a comprehensive approach to management that considers ecological and social dimensions. Grumbine's (1994) list of ten dominant themes in Ecosystem-Based Management (EBM) is a foundational framework that addresses scientific and social aspects (Table 2).

Table 2 Ten dominant themes of ecosystem-based management

Original EBM theme	Revised UF-EBM theme in UF context
Ecological integrity	Urban social-ecological systems with diverse structures have a high capacity to recover from disturbance or maintain function while stressed. However, trade-offs between ecological and social functions exist
Hierarchical context	Management accounts for the hierarchical organization of urban forests and the need to account for multiple spatial levels and cross-level dynamics
Ecological boundaries	Disparate social and ecological boundaries in urban forests are accounted for in the delineation and classification of management units
Data collection	The sources, formats, collection, management, analysis/visualization, and data sharing account for technological advances and emergent trends in citizen science, mobile technology, and open data
Monitoring	Frameworks of indicators that integrate social values have appropriate resources allocated and implementation priority assigned to understand ecosystem change and management efficacy
Adaptive management	Adaptive management admits huge uncertainties around urban forest management's environmental, social, and economic consequences and calls for designing management interventions as explicit learning experiments
Interagency cooperation	There is coordination between municipal departments to sustain urban forests while accommodating other needs like development and infrastructure renewal. At the same time, higher levels of government contribute to regional and national governance issues
Partnerships and civic engagement	Municipal governments partner with NGOs, community groups, and industry for program delivery while fostering civic engagement through urban forest outreach, education, and stewardship initiatives
Organizational change	Urban forestry departments shift away from individual tree management through institutional support, funding, and coordination with other departments
Social change	City property owners are aware of urban forest benefits and guidelines for effective management for planting and maintaining trees
Humans embedded in nature	Equity in the circulation of trees and their beneficial ecosystem services across the urban landscape is integrated into management
Values	Values held by the public about the urban forest are included in the planning processes

10 Conclusion

Urban forests, integral to mitigating environmental impacts in urban areas, face significant challenges to soil quality due to urbanization. The Urban Heat Island (UHI) effect exacerbates these issues, directly impacting urban forest health. However, strategic mitigation measures can enhance the adverse effects of UHI on urban trees, underscoring the importance of urban planning and green infrastructure. Climate change and urbanization induce stress, including drought, high salt levels, and heavy metal exposure, emphasizing the necessity for adaptive strategies to safeguard urban forests. These strategies must address the specific vulnerabilities of urban trees, ensuring resilience to changing climatic conditions. Air quality is a critical factor influencing urban tree health, exacerbated by urbanization and further compounded by climate change. Changes in species composition and biodiversity loss threaten urban forests, influenced by urbanization and climate change. Adopting an ecosystem-based management approach for urban forests is vital to navigate these complexities. This approach integrates ecological principles, local context understanding, and sustainable practices to enhance urban forest resilience. Integrating adaptive strategies, mitigation measures, and ecosystem-based management principles may promote the coexistence of urbanization and climate change with the preservation and prosperity of urban forests. These practices may lead to sustainable urban development while nurturing urban forests' essential ecological services.

References

Adams CE, Lindsey KJ (2010) Wildlife Management Education Needs to Go Urban. In: UENR biennial conference, session innovations in pedagogy, Engel, 27th March, 2010

Afshar NK, Karimian Z, Doostan R, Nokhandan MH (2018) Influence of planting designs on winter thermal comfort in an urban park. J Environ Eng Landsc Manag 26(3):232–240

Amato F, Cassee FR, van der Gon HAD, Gehrig R, Gustafsson M, Hafner W, Harrison RM, Jozwicka M, Kelly FJ, Moreno T, Prevot ASH, Schaap M, Sunyer J, Querol X (2014) Urban air quality: the challenge of traffic non-exhaust emissions. J Hazard Mater 275:31–36. https://doi.org/10.1016/j.jhazmat.2014.04.053

Anderson P, Charles-Dominique T, Ernstson H, Andersson E, Goodness J, Elmqvist T (2020) Post-apartheid ecologies in the City of Cape Town: An examination of plant functional traits in relation to urban gradients. Landsc Urban Plan 193:103662

Antoniadis V, Levizou E, Shaheen SM, Ok YS, Sebastian A, Baum C et al (2017) Trace elements in the soil-plant interface: phytoavailability, translocation, and phytoremediation—a review. Earth Sci Rev 171:621–645

Arcus VL, Prentice EJ, Hobbs JK, Mulholland AJ, Van der Kamp MW, Pudney CR et al (2016) On the temperature dependence of enzyme-catalyzed rates. Biochemistry 55(12):1681–1688

Asrar AA, Abdel-Fattah GM, Elhindi KM (2012) Improving growth, flower yield, and water relations of snapdragon (Antirhinum majus L.) plants grown under well-watered and water-stress conditions using arbuscular mycorrhizal fungi. Photosynthetica 50:305–316. https://doi.org/10.1007/s11099-012-0024-8

Balany F, Ng AW, Muttil N, Muthukumaran S, Wong MS (2020) Green infrastructure as an urban heat island mitigation strategy—a review. Water 12(12):3577

Balfagón D, Zandalinas SI, Mittler R, Gómez-Cadenas A (2020) High temperatures modify plant responses to abiotic stress conditions. Physiol Plant 170(3):335–344

Barwise Y, Kumar P (2020) Designing vegetation barriers for urban air pollution abatement: a practical review for appropriate plant species selection. NPJ Clim Atmosp Sci 3(1):12

Battacharyya D, Babgohari MZ, Rathor P, Prithiviraj B (2015) Seaweed extracts as biostimulants in horticulture. Sci Horticult 196:39–48

Battista G, Evangelisti L, Guattari C, De Lieto Vollaro E, De Lieto Vollaro R, Asdrubali F (2020) Urban heat Island mitigation strategies: experimental and numerical analysis of a university campus in Rome (Italy). Sustainability 12(19):7971

Berardi U, GhaffarianHoseini A, GhaffarianHoseini A (2014) State-of-the-art analysis of the environmental benefits of green roofs. Appl Energy 115:411–428

Berland A (2020) Urban tree growth models for two nearby cities show notable differences. Urban Ecosyst 23(6):1253–1261

Berland A, Shiflett SA, Shuster WD, Garmestani AS, Goddard HC, Herrmann DL, Hopton ME (2017) The role of trees in urban stormwater management. Landsc Urban Plan 162:167–177

Bozca FD, Leblebici S (2022) Interactive effect of boric acid and temperature stress on phenological characteristics and antioxidant system in *Helianthus annuus* L. South Afr J Bot 147:391–399

Brown H, Proust K, Newell B, Spickett J, Capon T, Bartholomew L (2018) Cool communities—urban density, trees, and health. Int J Environ Res Public Health 15(7):1547

Burley H, Beaumont LJ, Ossola A, Baumgartner JB, Gallagher R, Laffan S et al (2019) Substantial declines in urban tree habitat predicted under climate change. Sci Total Environ 685:451–462

Chaudhry S, Sidhu GPS (2022) Climate change regulated abiotic stress mechanisms in plants: a comprehensive review. Plant Cell Rep 41(1):1–31

Chaudhuri S, Chowdhury AR (2018) Air quality index assessment prelude to mitigate environmental hazards. Nat Hazards 91:1–17

Choudhary JR, Get S, Tripathi A, Kaldate R, Rana M, Mehta S et al (2022) Breeding efforts for crop productivity in abiotic stress environment. Augmenting crop productivity in stress environment. Springer, Singapore, pp 63–103

Dadkhah-Aghdash H, Rasouli M, Rasouli K, Salimi A (2022) Detection of urban trees sensitivity to air pollution using physiological and biochemical leaf traits in Tehran, Iran. Sci Rep 12(1):15398

Dahlhausen J, Rötzer T, Biber P, Uhl E, Pretzsch H (2018) Urban climate modifies tree growth in Berlin. Int J Biometeorol 62:795–808

Decina SM, Templer PH, Hutyra LR, Gately CK, Rao P (2017) Variability, drivers, and effects of atmospheric nitrogen inputs across an urban area: emerging patterns among human activities, the atmosphere, and soils. Sci Total Environ 609:1524–1534

Dhyani K, Kuniyal HB, Singh H, Sobha S (2021) Growth and physiological potential of *Terminalia arjuna* under elevated CO_2 levels in Open top chamber condition. J Appl Nat Sci 13(3):1121–1126

Dolan RW, Aronson MF, Hipp AL (2017) Floristic response to urbanization: filtering of the bioregional flora in Indianapolis, Indiana, USA. Am J Bot 104(8):1179–1187

Dover J, Phillips S (2015) Particulate pollution capture by green screens along the A38 Bristol street in Birmingham. Green Wall Centre Staffordshire Univ Final Rep 3:16–111

Driesen E, Van den Ende W, De Proft M, Saeys W (2020) Influence of environmental factors light, CO_2, temperature, and relative humidity on stomatal opening and development: a review. Agronomy 10(12):1975

Dusenge ME, Duarte AG, Way DA (2019) Plant carbon metabolism and climate change: elevated CO_2 and temperature impacts on photosynthesis, photorespiration and respiration. New Phytol 221(1):32–49

El-Naggar A, Shaheen SM, Ok YS, Rinklebe J (2018) Biochar affects the dissolved and colloidal concentrations of Cd, Cu, Ni, and Zn and their phytoavailability and potential mobility in a mining soil under dynamic redox-conditions. Sci Total Environ 624:1059–1071

Emel SL, Wang S, Metz RP, Spigler RB (2021) Type and intensity of surrounding human land use, not local environment, shape genetic structure of a native grassland plant. Mol Ecol 30(3):639–655
Fatichi S, Leuzinger S, Körner C (2014) Moving beyond photosynthesis: from carbon source to sink-driven vegetation modeling. New Phytol 201(4):1086–1095
Fekete R, Bódis J, Fülöp B, Süveges K, Urgyán R, Malkócs T et al (2020) Roadsides provide refuge for orchids: characteristic of the surrounding landscape. Ecol Evolut 10(23):13236–13247
Fisogni A, Hautekèete N, Piquot Y, Brun M, Vanappelghem C, Michez D, Massol F (2020) Urbanization drives an early spring for plants but not for pollinators. Oikos 129(11):1681–1691
Francis LFM, Jensen MB (2017) Benefits of green roofs: a systematic review of the evidence for three ecosystem services. Urban For Urban Green 28:167–176
Freitas ÉVD, das Dores Magalhães Veloso M, de Araújo WS (2020) Urbanization alters the composition, but not the diversity and structure, of Neotropical savanna woody plant communities. Folia Geobot 55:95–108
Geange SR, Arnold PA, Catling AA, Coast O, Cook AM, Gowland KM et al (2021) The thermal tolerance of photosynthetic tissues: a global systematic review and agenda for future research. New Phytol 229(5):2497–2513
Gill RA, Ahmar S, Ali B, Saleem MH, Khan MU, Zhou W, Liu S (2021) The role of membrane transporters in plant growth and development, and abiotic stress tolerance. Int J Mol Sci 22(23):12792
Gillner S, Bräuning A, Roloff A (2014) Dendrochronological analysis of urban trees: climatic response and impact of drought on frequently used tree species. Trees 28:1079–1093.
Gladkov EA, Tashlieva II, Gladkova OV (2021) Ornamental plants adapted to urban ecosystem pollution: lawn grasses and painted daisy tolerating copper. Environ Sci Pollut Res 28(11):14115–14120
Grossiord C, Buckley TN, Cernusak LA, Novick KA, Poulter B, Siegwolf RT et al (2020) Plant responses to rising vapor pressure deficit. New Phytol 226(6):1550–1566
Grumbine RE (1994) What is ecosystem management?. Conserv Biol 8(1):27–38. https://doi.org/10.1046/j.1523-1739.1994.08010027.x
Gulia S, Nagendra SS, Khare M, Khanna I (2015) Urban air quality management: a review. Atmosp Pollut Res 6(2):286–304
Guo J, Shan C, Zhang Y, Wang X, Tian H, Han G et al (2022) Mechanisms of salt tolerance and molecular breeding of salt-tolerant ornamental plants. Front Plant Sci 13:854116
Gupta SK, Singh H (2017) Observations on sporadic flowering in *Ficus benjamina* L. and *Peltophorum africanum* L. in new forest. Indian For 143:290–291
Gupta SK, Ram J, Singh H (2018a) Comparative study of transpiration in cooling effect of tree species in the atmosphere. J Geosci Environ Protect 6:151–166. https://doi.org/10.4236/gep.2018.68011
Gupta SK, Ram J, Singh H, Kumar P (2018b) Comparative study of cooling under shades of some forestry tree species with respect to ambient temperatures. J Geosci Environ Protect 6:51–66. https://doi.org/10.4236/gep.2018.611004
Gupta SK, Ram J, Singh H (2019a) Relationship between leaf area index and atmospheric cooling of tree species. Indian J For 42(3):203–208
Gupta SK, Singh H, Ram J (2019b) Need for a novel paradigm in forestry research: perspectives of comparative atmospheric cooling by different forestry effect tree species. Indian For 145(6):509–515
Gurney KR, Romero-Lankao P, Seto KC, Hutyra LR, Duren R, Kennedy C et al (2015) Climate change: track urban emissions on a human scale. Nature 525(7568):179–181
Gwedla N, Shackleton CM (2019) Perceptions and preferences for urban trees across multiple socio-economic contexts in the Eastern Cape, South Africa. Landsc Urban Plan 189:225–234
Hall SJ, Learned J, Ruddell B, Larson KL, Cavender-Bares J, Bettez N, Groffman PM, Grove JM, Heffernan JB, Hobbie SE, Larson K, Morse JL, Neill C, Nelson KC, O'Neil-Dunne J, Ogden LA, Pataki DE, Pearse WD, Polsky C, Roy Chowdhury R, Steele MK, Trammell TL (2015)

Convergence of microclimate in residential landscapes across diverse cities in the United States. Landsc Ecol 31:101–117. https://doi.org/10.1007/s10980-015-0297-y

Heaviside C, Macintyre H, Vardoulakis S (2017) The urban heat island: implications for health in a changing environment. Curr Environ Health Rep 4:296–305

Heneidy SZ, Halmy MW, Toto SM, Hamouda SK, Fakhry AM, Bidak LM et al (2021) Pattern of urban flora in intra-city railway habitats (Alexandria, Egypt): a conservation perspective. Biology 10(8):698

Herath HMPIK, Halwatura RU, Jayasinghe GY (2018) Evaluation of green infrastructure effects on tropical Sri Lankan urban context as an urban heat island adaptation strategy. Urban For Urban Green 29:212–222

Hilbert DR, Roman LA, Koeser AK, Vogt J, van Doorn NS (2019) Urban tree mortality: a literature review. Arboricult Urban For 45(5):167–200

Huang J, Zhang W, Mo J, Wang S, Liu J, Chen H (2015) Urbanization in China drives soil acidification of *Pinus massoniana* forests. Sci Rep 5(1):13512

Icaza LE, Van der Hoeven F, Van den Dobbelsteen A (2016) Surface thermal analysis of North Brabant cities and neighbourhoods during heat waves. TeMA J Land Use Mobil Environ 9(1):63–87

IPBES (2019) Summary for policymakers of the global assessment report on biodiversity and ecosystem services of the intergovernmental science-policy platform on biodiversity and ecosystem services. IPBES secretariat, Bonn, p 56

IPCC (2021) Summary for policymakers. In: Climate change 2021: the physical science basis. Contribution of working group I to the sixth assessment report of the intergovernmental panel on climate change. Cambridge University Press, Cambridge

Irmak MA, Yilmaz S, Dursun D (2017) Effect of different pavements on human thermal comfort conditions. Atmósfera 30(4):355–366

Irwin RE, Warren PS, Adler LS (2018) Phenotypic selection on floral traits in an urban landscape. Proceed Royal Soc B 285(1884):20181239

Isola F, Leone F, Pittau R (2023) Evaluating the urban heat island phenomenon from a spatial planning viewpoint: a systematic review. TeMA J Land Use Mobil Environ 32:75–93

Jamei E, Ossen DR, Rajagopalan P (2017) Investigating the effect of urban configurations on the variation of air temperature. Int J Sustain Built Environ 6(2):389–399

Jansson M (2014) Green space in compact cities: the benefits and values of urban ecosystem services in planning. Nord J Archit Res 2(2014):139–160

Jayasooriya VM, Ng AWM, Muthukumaran S, Perera BJC (2017) Green infrastructure practices for improvement of urban air quality. Urban For Urban Green 21:34–47. https://doi.org/10.1016/j.ufug.2016.11.007

Jensen H, Reimann C, Finne TE, Ottesen RT, Arnoldussen A (2007) PAH-concentrations and compositions in the top 2 cm of forest soils along a 120 km long transect through agricultural areas, forests and the city of Oslo, Norway. Environ Pollut 145(3):829–838. https://doi.org/10.1016/j.envpol.2006.05.008

Jia W, Zhao S, Liu S (2018) Vegetation growth enhancement in urban environments of the Conterminous United States. Global Change Biol 24(9):4084–4094

Kamakura RP, DeWald LE, Sniezko RA, Elliott M, Chastagner GA (2021) Using differences in abiotic factors between seed origin and common garden sites to predict performance of Pacific madrone (*Arbutus menziesii* Pursh). For Ecol Manag 497:119487

Karakounos I, Dimoudi A, Zoras S (2018) The influence of bioclimatic urban redevelopment on outdoor thermal comfort. Energy Build 158:1266–1274

Kim J, Lee SY, Kang J (2020) Temperature reduction effects of rooftop garden arrangements: a case study of Seoul national university. Sustainability 12(15):6032

Kolokotroni M, Giridharan R (2008) Urban heat island intensity in London: An investigation of the impact of physical characteristics on changes in outdoor air temperature during summer. Sol Energy 82(11):986–998. https://doi.org/10.1016/j.solener.2008.05.004

Kondratyeva A, Knapp S, Durka W, Kühn I, Vallet J, Machon N et al (2020) Urbanization effects on biodiversity revealed by a two-scale analysis of species functional uniqueness versus redundancy. Front Ecol Evolut 8:73

Körner C (2015) Paradigm shift in plant growth control. Curr Opin Plant Biol 25:107–114

Kumar A, Verma JP (2018) Does plant—microbe interaction confer stress tolerance in plants: a review? Microbiol Res 207:41–52

Kumar M, Singh MP, Singh H, Dhakate PM, Ravindranath NH (2019a) Forest working plan for the sustainable management of forest and biodiversity in India. J Sustain For 39:1–22. https://doi.org/10.1080/10549811.2019.1632212

Kumar N, Jeena N, Singh H (2019b) Elevated temperature modulates rice pollen structure: a study from the foothill of the Himalayan agro-ecosystem in India. 3 Biotech 9(5):175. https://doi.org/10.1007/s13205-019-1700-1

Kumar N, Jeena N, Kumar N, Gangola S, Singh H (2019c) Phytoremediation facilitating enzymes: an enzymatic approach for enhancing remediation process. In: Bhatt P (ed) Smart bioremediation technologies. Academic Press, Cambridge, pp 289–306

Kumar A, Kumar P, Singh H, Kumar N (2021a) Modulation of plant functional traits under essential plant nutrients during seasonal regime in natural forests of Garhwal Himalayas. Plant Soil 465:197–212. https://doi.org/10.1007/s11104-021-05003-x

Kumar A, Kumar P, Singh H, Bisht S, Kumar N (2021b) Relationship of physiological plant functional traits with soil carbon stock in the temperate forest of Garhwal Himalaya. Curr Sci 120(8):1368–1373. https://doi.org/10.18520/cs/v120/i8/1368-1373

Kumar A, Kumar P, Singh H, Kumar N (2021c) Adaptation and mitigation potential of roadside trees with bio-extraction of heavy metals under vehicular emissions and their impact on physiological traits during seasonal regimes. Urban For Urban Green 58:126900. https://doi.org/10.1016/j.ufug.2020.126900

Kumar A, Kumar P, Singh H, Kumar N (2021d) Impact of plant functional traits on infiltration rate under different forests of Kempty watershed in Garhwal Himalaya. Indian J Soil Conserv 49(1):38–44

Kumar M, Phukon SN, Paygude AC, Tyagi K, Singh H (2021e) Mapping phenological functional types (PhFT) in the Indian Eastern Himalayas using machine learning algorithm in Google Earth engine. Comput Geosci 158:104982. https://doi.org/10.1016/j.cageo.2021.104982

Kumar A, Singh H, Kumari G, Bisht S, Malik A, Kumar A, Singh M, Raturi BS, Thakur A, Kaushal R (2022) Adaptive resilience of roadside trees to vehicular emissions via leaf enzymatic, physiological, and anatomical trait modulations. Environ Pollut 313:120191. https://doi.org/10.1016/j.envpol.2022.120191

Kumar N, Singh H, Giri K, Kumar A, Joshi A, Yadav A, Singh R, Bisht S, Kumari R, Jeena N, Khairakpam R, Mishra G (2024) Physiological and molecular insights into the allelopathic effects on agroecosystems under changing environmental conditions. Physiol Mol Biol Plants 30:417–433. https://doi.org/10.1007/s12298-024-01440-x

Kumar N, Suyal DC, Sharma IP, Verma A, Singh H (2017). Elucidating stress proteins in rice (*Oryza sativa* L.) genotype under elevated temperature: a proteomic approach to understand heat stress response. 3 Biotech 7:205. https://doi.org/10.1007/s13205-017-0856-9

Kumar N, Singh H, Sharma SK (2020) Antioxidants: responses and importance in plant defense system. In: Roychowdhury R, Choudhury S, Hasanuzzaman M, Srivastava S (eds) Sustainable agriculture in the era of climate change. Springer, Cham. https://doi.org/10.1007/978-3-030-45669-6_11

Le Roux DS, Ikin K, Lindenmayer DB, Manning AD, Gibbons P (2018) The value of scattered trees for wildlife: contrasting effects of landscape context and tree size. Divers Distrib 24(1):69–81. https://doi.org/10.1111/ddi.12658

Leotta L, Toscano S, Ferrante A, Romano D, Francini A (2023) New Strategies to Increase the Abiotic Stress Tolerance in Woody Ornamental Plants in Mediterranean Climate. Plants 12(10):2022. https://doi.org/10.3390/plants12102022

Liu W, Zhan J, Zhao F, Yan H, Zhang F, Wei X (2019) Impacts of urbanization-induced land-use changes on ecosystem services: A case study of the Pearl River Delta Metropolitan Region, China. Ecol Indic 98:228–238. https://doi.org/10.1016/j.ecolind.2018.10.054

Majeed A, Muhammad Z (2019) Salinity: a major agricultural problem—causes, impacts on crop productivity and management strategies. In: Hasanuzzaman M, Hakeem KR, Nahar K, Alharby HF (eds) Plant Abiotic Stress Tolerance: Agronomic, Molecular and Biotechnological Approaches, 83–99. https://doi.org/10.1007/978-3-030-06118-0_3

Margaritis E, Kang J (2017) Relationship between green space-related morphology and noise pollution. Ecol Indic 72:921–933. https://doi.org/10.1016/j.ecolind.2016.09.032

Manandhar P, Bande L, Tsoupos A, Marpu PR, Armstrong P (2019) A study of local climate zones in Abu Dhabi with urban weather stations and numerical simulations. Sustainability 12(1):156. https://doi.org/10.3390/su12010156

Masson V, Lemonsu A, Hidalgo J, Voogt J (2020) Urban climates and climate change. Annu Rev Environ Resour 45:411–444. https://doi.org/10.1146/annurev-environ-012320-083623

McDonald RI, Mansur AV, Ascensão F, Colbert ML, Crossman K, Elmqvist T, Gonzalez A, Güneralp B, Haase D, Hamann M, Hillel O, Huang K, Kahnt B, Maddox D, Pacheco A, Pereira HM, Seto KC, Simkin R, Walsh B, Werner AS, Ziter C (2020) Research gaps in knowledge of the impact of urban growth on biodiversity. Nat Sustain 3(1):16–24. https://doi.org/10.1038/s41893-019-0436-6

Meena D, Singh H, Chaudhari SK (2011) Elucidating strontium response on growth dynamics and biochemical change in *Phaseolus mungo* L. Int J Agricult Environ Biotechnol 4:107–113

Meinshausen M, Smith SJ, Calvin K, Daniel JS, Kainuma ML, Lamarque JF, Matsumoto K, Montzka SA, Raper SCB, Riahi K, Thomson A, Velders GJM, van Vuuren DPP (2011) The RCP greenhouse gas concentrations and their extensions from 1765 to 2300. Clim Change 109:213–241. https://doi.org/10.1007/s10584-011-0156-z

Mohajerani A, Bakaric J, Jeffrey-Bailey T (2017) The urban heat island effect, its causes, and mitigation, with reference to the thermal properties of asphalt concrete. J Environ Manag 197:522–538. https://doi.org/10.1016/j.jenvman.2017.03.095

Mononen L, Auvinen AP, Ahokumpu AL, Rönkä M, Aarras N, Tolvanen H, Kamppinen M, Viirret E, Kumpula T, Vihervaara P (2016) National ecosystem service indicators: Measures of social–ecological sustainability. Ecol Indic 61:27–37. https://doi.org/10.1016/j.ecolind.2015.03.041

Morris KI, Chan A, Morris KJK, Ooi MC, Oozeer MY, Abakr YA, Nadzir MSM, Mohammed IY, Al-Qrimli HF (2017) Impact of urbanization level on the interactions of urban area, the urban climate, and human thermal comfort. Appl Geogr 79:50–72. https://doi.org/10.1016/j.apgeog.2016.12.007

Moser-Reischl A, Rahman MA, Pauleit S, Pretzsch H, Rötzer T (2019) Growth patterns and effects of urban micro-climate on two physiologically contrasting urban tree species. Landsc Urban Plan 183:88–99. https://doi.org/10.1016/j.landurbplan.2018.11.004

Nelson KC, Palmer MA, Pizzuto JE, Moglen GE, Angermeier PL, Hilderbrand RH, Dettinger M, Hayhoe K (2009) Forecasting the combined effects of urbanization and climate change on stream ecosystems: from impacts to management options. J Appl Ecol 46(1):154–163. https://doi.org/10.1111/j.1365-2664.2008.01599.x

Oliveira W, e Silva JLS, de Oliveira MTP, Cruz-Neto O, da Silva LAP, Borges LA, Sobrinho MS, Lopes AV (2019) Reduced reproductive success of the endangered tree brazilwood (Paubrasilia echinata, Leguminosae) in urban ecosystem compared to Atlantic forest remnant: lessons for tropical urban ecology. Urban For Urban Green 41:303–312. https://doi.org/10.1016/j.ufug.2019.04.020

Oliveira MT, Silva JL, Cruz-Neto O, Borges LA, Girao LC, Tabarelli M, Lopes AV (2020) Urban green areas retain just a small fraction of tree reproductive diversity of the Atlantic forest. Urban For Urban Green 54:126779. https://doi.org/10.1016/j.ufug.2020.126779

Ozturk M, Unal BT, García-Caparrós P, Khursheed A, Gul A, Hasanuzzaman M (2021) Osmoregulation and its actions during the drought stress in plants. Physiol Plant 172(2):1321–1335. https://doi.org/10.1111/ppl.13297

Pandit R, Polyakov M, Sadler R (2012) The importance of tree cover and neighbourhood parks in determining urban property values. In: 56th AARES annual conference, Fremantle, Western Australia, February 7–10, 2012. http://dx.doi.org/10.22004/ag.econ.124357

Parker TS, Nilon CH (2012) Urban landscape characteristics correlated with the synurbization of wildlife. Landsc Urban Plan 106(4):316–325. https://doi.org/10.1016/j.landurbplan.2012.04.003

Perini K, Chokhachian A, Dong S, Auer T (2017) Modeling and simulating urban outdoor comfort: Coupling ENVI-Met and TRNSYS by grasshopper. Energy Build 152:373–384. https://doi.org/10.1016/j.enbuild.2017.07.061

Parvin K, Nahar K, Hasanuzzaman M, Bhuyan MB, Fujita M (2019) Calcium-mediated growth regulation and abiotic stress tolerance in plants. In: Hasanuzzaman M, Hakeem KR, Nahar K, Alharby HF (eds) Plant Abiotic Stress Tolerance: Agronomic, Molecular and Biotechnological Approaches, 291–331. https://doi.org/10.1007/978-3-030-06118-0_13

Petruzzellis F, Tordoni E, Di Bonaventura A, Tomasella M, Natale S, Panepinto F, Bacaro G, Nardini A (2022) Turgor loss point and vulnerability to xylem embolism predict species-specific risk of drought-induced decline of urban trees. Plant Biol 24(7):1198–1207. https://doi.org/10.1111/plb.13355

Phillips TH, Baker ME, Lautar K, Yesilonis I, Pavao-Zuckerman MA (2019) The capacity of urban forest patches to infiltrate stormwater is influenced by soil physical properties and soil moisture. J Environ Manag 246:11–18. https://doi.org/10.1016/j.jenvman.2019.05.127

Pourret O (2018) On the necessity of banning the term "heavy metal" from the scientific literature. Sustainability 10(8):2879. https://doi.org/10.3390/su10082879

Prakash V, Hunney K, Singh H (2022) Effect of elevated carbon dioxide on growth and development of *Santalum album* L. seedlings inoculated with plant growth promoting microorganisms in Open Top Chambers. Int J Environ Agricult Biotechnol 7(5):149–159

Przybysz A, Popek R, Stankiewicz-Kosyl M, Zhu CY, Małecka-Przybysz M, Maulidyawati T, Mikowska K, Deluga D, Griżuk K, Sokalski-Wieczorek J, Wolszczak K, Wińska-Krysiak M (2021) Where trees cannot grow–Particulate matter accumulation by urban meadows. Sci Total Environ 785:147310. https://doi.org/10.1016/j.scitotenv.2021.147310

Rathgeber CB, Cuny HE, Fonti P (2016) Biological basis of tree-ring formation: a crash course. Front Plant Sci 7:734. https://doi.org/10.3389/fpls.2016.00734

Rivkin LR, Johnson RA, Chaves JA, Johnson MT (2021) Urbanization alters interactions between Darwin's finches and Tribulus cistoides on the Galápagos Islands. Ecol Evol 11(22):15754–15765. https://doi.org/10.1002/ece3.8236

Roth M, Lim VH (2017) Evaluation of canopy-layer air and mean radiant temperature simulations by a microclimate model over a tropical residential neighbourhood. Build Environ 112:177–189. https://doi.org/10.1016/j.buildenv.2016.11.026

Salata F, Golasi I, Petitti D, de Lieto Vollaro E, Coppi M, de Lieto Vollaro A (2017) Relating microclimate, human thermal comfort and health during heat waves: An analysis of heat island mitigation strategies through a case study in an urban outdoor environment. Sustain Cit Soc 30:79–96. https://doi.org/10.1016/j.scs.2017.01.006

Salgotra RK, Stewart Jr CN (2020) Functional markers for precision plant breeding. Int J Mol Sci 21(13):4792. https://doi.org/10.3390/ijms21134792

Salgado K, Martínez ML, Álvarez-Molina LL, Hesp P, Equihua M, Mariño-Tapia I (2022) Impact of urbanization and landscape changes on the vegetation of coastal dunes along the Gulf of Mexico. Écoscience 29(2):103–116. https://doi.org/10.1080/11956860.2021.1934299

Santamouris M, Cartalis C, Synnefa A, Kolokotsa D (2015) On the impact of urban heat island and global warming on the power demand and electricity consumption of buildings—A review. Energy Build 98:119–124. https://doi.org/10.1016/j.enbuild.2014.09.052

Seto KC, Güneralp B, Hutyra LR (2012) Global forecasts of urban expansion to 2030 and direct impacts on biodiversity and carbon pools. Proceed Natl Acad Sci 109(40):16083–16088. https://doi.org/10.1073/pnas.1211658109

Sharma R, Singh H (2021) Alteration in biochemical constituents and nutrients partitioning of *Asparagus racemosus* in response to elevated atmospheric CO_2 concentration. Environ Sci Pollut Res 29:6812–6821. https://doi.org/10.1007/s11356-021-16050-3

Sharma R, Singh H, Kaushik M, Nautiyal R, Singh O (2018) Adaptive physiological response, carbon partitioning, and biomass production of *Withania somnifera* (L.) Dunal grown under elevated CO_2 regimes. 3 Biotech 8:267. https://doi.org/10.1007/s13205-018-1292-1

Shinzato P, Simon H, Duarte DHS, Bruse M (2019) Calibration process and parametrization of tropical plants using ENVI-met V4–Sao Paulo case study. Archit Sci Rev 62(2):112–125. https://doi.org/10.1080/00038628.2018.1563522

Singh H (2021) An integrated approach considering physiological and biophysical based indicators for assessing tolerance of roadside plantations of *Alstonia scholaris* towards urban roadside air pollution: an assessment of adaptation of plantations for mitigating urban roadside air pollution. Trees. https://doi.org/10.1007/s00468-021-02179-8

Singh H (2024) Exploring adaptive modulation in plant functional traits and its impact on the productivity of *Acacia auriculiformis* under CO_2-enriched environment. Ind Crops Prod 210:118186. https://doi.org/10.1016/j.indcrop.2024.118186

Singh H, Verma A (2013) Characterization and screening of high nitrogen efficient rice genotype to curtail environmental pollution. Int J Agricult Environ Biotechnol 6(1):23–30

Singh H, Savita Sharma R, Sinha S, Kumar M, Kumar P, Verma A, Sharma SK (2017a) Physiological functioning of *Lagerstroemia speciosa* L. under heavy roadside traffic: an approach to screen potential species for abatement of urban air pollution. 3 Biotech 7(61):1–10. https://doi.org/10.1007/s13205-017-0690-0

Singh H, Verma A, Kumar M, Sharma R, Gupta R, Kaur M, Negi M, Sharma SK (2017b) Phytoremediation: a green technology to clean up the sites with low and moderate level of heavy metals. Aust Biochem 2:1012

Singh H, Sharma R, Savita, Singh MP, Kumar M, Verma A, Ansari MW, Sharma SK (2018) Adaptive physiological response of *Parthenium hysterophorus* to elevated atmospheric CO_2 concentration. Indian For 144:1–14

Singh H, Yadav M, Kumar N, Kumar A, Kumar M (2020) Assessing adaptation and mitigation potential of roadside trees under the influence of vehicular emissions: a case study of *Grevillea robusta* and *Mangifera indica* planted in an urban city of India. PLoS ONE 15(1):e0227380. https://doi.org/10.1371/journal.pone.0227380

Singh S, Singh H, Sharma SK, Nautiyal R (2021) Seasonal variation in biochemical responses of bamboo clones in the sub-tropical climate of Indian Himalayan foothills. Heliyon 7(4):e06859. https://doi.org/10.1016/j.heliyon.2021.e06859

Singh H, Bandyopadhyaya S, Kumar A, Singh M, Malik A, Kumari G, Raturi A, Yadav SK, Kumar M, Barthwal S, Thakur A, Rawat PS (2022) Understanding the physiological and biophysical response of urban roadside plantations for assessing adaptation and mitigation mechanisms toward vehicular emissions. Urban Climate 44:101183. https://doi.org/10.1016/j.uclim.2022.101183

Singh H, Kumar N, Kumar A (2024a) Enhancing resource use efficiency in crops through plant functional traits. In: Kumar N, Singh H (eds) Plant functional traits for improving productivity. Springer, Singapore. https://doi.org/10.1007/978-981-97-1510-7_6

Singh M, Singh H, Kumar A, Kumar M, Barthwal S, Thakur A (2024b) Soil nitrogen availability determines the CO_2 fertilization effect on tree species (*Neolamarckia cadamba*): growth and physiological evidence. Environ Sustain 7:53–60. https://doi.org/10.1007/s42398-023-00300-w

Singh H, Kumar M (2022) Climate change and its impact on Indian Himalayan forests: current status and research needs. In: Rani S, Kumar R (eds) Climate change. Springer, Cham, pp 223–242. https://doi.org/10.1007/978-3-030-92782-0_11

Sol D, González-Lagos C, Moreira D, Maspons J, Lapiedra O (2014) Urbanisation tolerance and the loss of avian diversity. Ecol Lett 17(8):942–950. https://doi.org/10.1111/ele.12297

Szlavecz K, McCormick M, Xia L, Saunders J, Morcol T, Whigham D, Filley T, Csuzdi C (2011) Ecosystem effects of non-native earthworms in Mid-Atlantic deciduous forests. Biol Invasions, 13:1165–1182. https://doi.org/10.1007/s10530-011-9959-0

Tong Z, Baldauf RW, Isakov V, Deshmukh P, Zhang KM (2016) Roadside vegetation barrier designs to mitigate near-road air pollution impacts. Sci Total Environ 541:920–927. https://doi.org/10.1016/j.scitotenv.2015.09.067

Tremper AH, Green DC (2018) The Impact of a Green Screen on Concentrations of Nitrogen Dioxide at Bowes Primary School, Enfield. Environmental Research Group, King's College, London

Tsoka S, Tsikaloudaki A, Theodosiou T (2018) Analyzing the ENVI-met microclimate model's performance and assessing cool materials and urban vegetation applications–A review. Sustain Cit Soc 43:55–76. https://doi.org/10.1016/j.scs.2018.08.009

Vijayaraghavan K (2016) Green roofs: A critical review on the role of components, benefits, limitations and trends. Renew Sustain Energy Rev 57:740–752. https://doi.org/10.1016/j.rser.2015.12.119

Wang J, Zhou W, Pickett ST, Yu W, Li W (2019) A multiscale analysis of urbanization effects on ecosystem services supply in an urban megaregion. Sci Total Environ 662:824–833. https://doi.org/10.1016/j.scitotenv.2019.01.260

Weerakkody U, Dover JW, Mitchell P, Reiling K (2017) Particulate matter pollution capture by leaves of seventeen living wall species with special reference to rail-traffic at a metropolitan station. Urban For Urban Green 27:173–186. https://doi.org/10.1016/j.ufug.2017.07.005

Werkenthin M, Kluge B, Wessolek G (2014) Metals in European roadside soils and soil solution–A review. Environ Pollut 189:98–110. https://doi.org/10.1016/j.envpol.2014.02.025

Wolf KL, Lam ST, McKeen JK, Richardson GR, Van Den Bosch M, Bardekjian AC (2020) Urban trees and human health: A scoping review. Int J Environ Res Pub Health 17(12):4371. https://doi.org/10.3390/ijerph17124371

Yadav SK, Singh H, Nautiyal R, Ginwal HS, Ansari SA, Barthwal S (2019) Modulation of morpho-physiological responses in *Populus deltoides* by elevated carbon dioxide and temperature. For Sci 66:105–118. https://doi.org/10.1093/forsci/fxz048

Yahia MW, Johansson E, Thorsson S, Lindberg F, Rasmussen MI (2018) Effect of urban design on microclimate and thermal comfort outdoors in warm-humid Dar es Salaam, Tanzania. Int J Biometeorol 62:373–385. https://doi.org/10.1007/s00484-017-1380-7

Zhao TF, Fong KF (2017) Characterization of different heat mitigation strategies in landscape to fight against heat island and improve thermal comfort in hot–humid climate (Part I): Measurement and modelling. Sustain Cit Soc 32:523–531. https://doi.org/10.1016/j.scs.2017.03.025

Zhang Y, Murray AT, Turner Ii BL (2017) Optimizing green space locations to reduce daytime and nighttime urban heat island effects in Phoenix, Arizona. Landsc Urban Plan 165:162–171. https://doi.org/10.1016/j.landurbplan.2017.04.009

Zhang L, Zhan Q, Lan Y (2018) Effects of the tree distribution and species on outdoor environment conditions in a hot summer and cold winter zone: A case study in Wuhan residential quarters. Build Environ 130:27–39. https://doi.org/10.1016/j.buildenv.2017.12.014

Zhang L, Zhang Z, Chen L, McNulty S (2019) An investigation on the leaf accumulation-removal efficiency of atmospheric particulate matter for five urban plant species under different rainfall regimes. Atmos Environ 208:123–132. https://doi.org/10.1016/j.atmosenv.2019.04.010

Zhao S, Liu S, Zhou D (2016) Prevalent vegetation growth enhancement in urban environment. Proceed Natl Acad Sci 113(22):6313–6318. https://doi.org/10.1073/pnas.1602312113

Zia R, Nawaz MS, Siddique MJ, Hakim S, Imran A (2021) Plant survival under drought stress: Implications, adaptive responses, and integrated rhizosphere management strategy for stress mitigation. Microbiol Res 242:126626. https://doi.org/10.1016/j.micres.2020.126626

Biological Insights into the Adaptive Responses of Medicinal Plants to Changing Urban Environments

Anjali Chauhan and Hukum Singh

Abstract The transformation of natural ecosystems by urban development presents significant challenges for plant species, particularly those with medicinal properties. These plants, rich in bioactive compounds with therapeutic effects, have been integral to traditional healing practices and remain central to modern pharmacology. However, adapting medicinal plants to urban environments characterized by pollution, climate fluctuations, and habitat fragmentation is crucial for maintaining their medicinal efficacy. In response to urban stressors, medicinal plants undergo diverse biological adaptations spanning morphology, physiology, biochemistry, gene expression, and epigenetics. These adaptations enable them to thrive amidst urban challenges and ensure the preservation of their medicinal qualities. Understanding these adaptive mechanisms is vital for cultivating high-quality medicinal plants in urban settings. This chapter delves into the biological responses of medicinal plants to urbanization, shedding light on how these responses influence their therapeutic potential. Integrating ecological and biological perspectives contributes to developing sustainable urban plant practices and conserving medicinal plant biodiversity.

Keywords Biological adaptations · Medicinal plants · Traditional · Therapeutic properties · Urban environment

1 Introduction

As urban areas rapidly expand, the abundance of organisms within cities significantly impacts plant growth and function. Despite this profound influence, our comprehension of the relationship between urban plants and their surroundings lags behind

A. Chauhan (✉)
Department of Botany, School of Applied and Life Science, Uttaranchal University, Dehradun, Uttarakhand 248001, India
e-mail: canjali714@gmail.com

H. Singh
Plant Physiology Discipline, Forest Research Institute, PO New Forest Dehradun, Dehradun, Uttarakhand 248006, India

H. Singh (ed.), *Urban Forests, Climate Change and Environmental Pollution*,
https://doi.org/10.1007/978-3-031-67837-0_4

that of other ecosystems (Wang and Pataki 2010; Kumar et al. 2022; Singh et al. 2022; Kumar et al. 2024). Urban environments present distinct abiotic conditions, including elevated carbon dioxide levels, increased temperatures, and heightened deposition of inorganic nitrogen pollutants. The urban heat island effect, characterized by elevated nighttime temperatures due to the release of stored heat from materials like concrete, exacerbates these conditions (Arnfield 2003; Singh et al. 2024a; Singh 2024). While nitrogen, carbon dioxide, and temperature boosts benefit photosynthesis, rising temperatures elevate respiration rates and soil water evaporation, challenging the net carbon gain (Dhyani et al. 2021; Kumar et al. 2021; Prakash et al. 2022).

Furthermore, the interactions among these abiotic factors yield complex outcomes, and studying urban plant ecology is crucial for comprehending the intricate relationships between plants and climate change (Carreiro and Tripler 2005). Understanding the intricate biological mechanisms by which medicinal plants adapt to evolving urban environments is paramount for sustainable urban development, biodiversity conservation, and achieving various UN Sustainable Development Goals (SDGs). In today's era of rapid urbanization and environmental flux, the resilience and adaptability of plant species, especially medicinal ones, have garnered significant scientific interest. Medicinal plants have provided empirical therapeutic benefits for millennia, and their history is rooted in traditional medicine across diverse cultures. Traditional herbal medicine is a rich source of pharmaceutical compounds underpinning modern pharmacotherapy (Patwardhan et al. 2008).

In urban settings, medicinal plants face myriad stressors, including air pollution, temperature fluctuations, altered soil composition, and habitat fragmentation (Singh 2021; Sharma and Singh 2021). Understanding how these plants navigate and flourish amidst such challenges is pivotal for biodiversity conservation and the sustainable utilization of these valuable botanical resources (Singh et al. 2020, 2021; Kumar et al. 2021a). Investigating the adaptive mechanisms of medicinal plants in urban environments is crucial for informing conservation strategies, urban planning, and cultivation practices aimed at bolstering their resilience.

Plant resilience, the capacity to rebound from adversity and thrive in challenging conditions, is especially critical given the inherent limitations of sessile organisms like plants in adapting to extreme temperatures. However, human activities such as resource overexploitation, land use changes, and pollution often compromise the resilience of flora and fauna in ecosystems (Iqbal et al. 2000). Climate change and environmental extremes further exacerbate these challenges, affecting plant growth and productivity (Gray and Brady 2016).

This chapter examines the intricate interplay between medicinal plants and their evolving environment by exploring their morphological, physiological, and biochemical adaptations to urban settings. Analyzing factors influencing medicinal plants and molecular aspects such as gene expression and epigenetic modifications in response to urban environments advances scientific understanding and contributes to sustainable urban ecosystems' development.

2 Medicinal Plants and Urban Environment

Medicinal plants, crucial for maintaining health and combating various diseases, are integral to traditional medicine practices globally. Even those who exclusively rely on allopathic medicine indirectly benefit from medicinal plants, as approximately 20–25% of prescribed medications originate from plant-based sources (Rates 2001). The World Health Organization (WHO) estimates that around 80% of the world's population primarily depends on traditional treatments for their healthcare needs. It is assumed that as many as two billion individuals heavily rely on medicinal plants for their health and well-being. These plants play a vital role in enhancing the livelihood and welfare of populations in both developed and developing countries. The healthcare sector increasingly recognizes the importance of medicinal plants, as evidenced by discussions on the contribution of traditional medicine to achieving health-related Millennium Development Goals (MDGs), three of which are directly linked to health.

Urban environments harbor various plant species, ranging from hundreds to thousands, providing ample opportunities to study how these plants respond to stressful conditions. Exploring plant responses to urban conditions, known as 'urban plant physiology,' offers valuable insights into immediate physiological responses, plant tolerances, and short- and long-term adaptations (Kumar et al. 2021b, c, d). Woody species are particularly significant due to their long lifespan, offering insights into long-term adaptation mechanisms. Moreover, based on tree ring analysis, dendrochronology provides a unique opportunity to retrospectively explore how historical urban conditions influenced plant growth, considering factors like water availability and temperature (Gillner et al. 2014).

Urbanization poses numerous challenges, including stress, pollution, and sedentary lifestyles, leading to increased health problems (Yadav et al. 2019a, b; Gupta et al. 2019a). Medicinal plants, renowned for their therapeutic properties, offer promising ways to mitigate urban health challenges. Urban green spaces serve as vital reserves for the biodiversity of medicinal plants, providing opportunities for their cultivation and conservation. These spaces create urban refuges that support biodiversity while enhancing the overall health of urban ecosystems. Medicinal plants are strategically cultivated in parks, gardens, and even rooftops, establishing accessible green spaces that promote physical and mental well-being. The presence of medicinal plants in urban environments introduces the concept of 'green medicine,' offering relief from the fast-paced urban lifestyle.

3 Factors Affecting Medicinal Plants in Urban Areas

3.1 Climate Change

The habitat of an organism comprises four primary components: the Atmosphere, Lithosphere, Hydrosphere, and Biosphere. Various external factors continually impact the life of an organism, with these environmental factors undergoing regular changes due to factors such as population growth, pollution, industrialization, and the unsustainable utilization of natural resources (Gupta et al. 2019a; Kumar et al. 2019a, b). The United Nations projects a population of 9.8 billion by 2050, a significant increase from 7.6 billion in 2017, driving further urbanization, immigration, and globalization. Industrial activities contribute to the continuous release of effluents containing diverse contaminants into the air, water, and soil, exacerbating environmental degradation and pollution. Emissions from industrial processes, forest fires, agricultural residues, waste incineration, and fossil fuel combustion introduce harmful gases into the atmosphere, altering its natural composition. Greenhouse gases like CO_2, methane, NOx, and CFCs exacerbate environmental degradation by trapping infrared radiation in the Earth's atmosphere.

Urban development and infrastructure evolve in tandem, with both mutually dependent and co-evolutionary relationships (Neuman and Smith 2010). Urban residents and ecosystems face threats from various factors, among which extreme heat stands out as particularly severe. Even slight changes in climatic conditions, influenced by impervious surface area and vegetation factors, can have significant effects (Larsen 2015; Sharma et al. 2018; Singh et al. 2018). The significance of medicinal plants underscores the urgency of understanding the mounting pressures from climate change, which increasingly impact every facet of species survival and conservation (Gupta et al. 2018a, b). Like other organisms, medicinal plants are vulnerable to shifting climatic conditions, with noticeable effects on their life cycles and global distribution. Medicinal plants endemic to specific regions or ecosystems are especially susceptible to climate change induced by human activities. These factors significantly impede the growth and development of plant species, as they often rely on specific temperature regimes (Nazir et al. 2018). In recent years, there has been a marked increase in the frequency and intensity of heat waves, significantly impacting the yield of medicinal plants when temperatures rise abruptly by as much as 5 degrees Celsius above ambient levels. The synthesis of secondary metabolites crucial for the medicinal properties of these plants is highly contingent on diverse environmental conditions, with temperature fluctuations emerging as a pivotal factor affecting secondary metabolite production.

Contrary to cold stress conditions, water-stressed environments stimulate the synthesis of specific chemicals, as evidenced by research on *Rehmanniaglutinosa* (Garten) by Chung et al. (2006). This underscores the profound impact of environmental factors on producing beneficial compounds in medicinal plants. However, certain species, such as *Withaniasomnifera*, exhibit an increased accumulation of essential bioactive compounds, such as steroid lactones (particularly withanolides), in response to cold stress (Mir et al. 2015). Changing climatic and environmental conditions modulate phenology—the timing of plant life cycle events (IPCC 2007). In response to shifting weather patterns and rising temperatures, early flowering and alterations in the geographical ranges of some medicinal and aromatic plants have been observed. While seemingly inconsequential at first, these changes pose significant challenges to the overall survival of these species. Researchers have documented a noticeable shift towards earlier spring onset since the 1960s, accompanied by observable changes in various phenological events, including plant bud break, leaf and flower emergence, and fruit and seed dispersal (Walther et al. 2002). Early flowering due to temperature fluctuations threatens cold-sensitive regions in late spring, as premature buds or fruits may succumb to freezing conditions, endangering or even causing the loss of economically valuable crops (Zobayed et al. 2005). Varieties of medicinal plants, including horse chestnut (*Aesculus hippocastanum*) and hawthorn (*Crataegus monogyna*), have exhibited earlier blooming in the United Kingdom under climate change (Cavaliere 2009). Irregular monsoon patterns disrupt the annual life cycle of medicinal plants, with increasing rainfall frequency and intensity hindering harvesting and production of critical medicinal compounds. These environmental challenges underscore the imperative of adopting sustainable practices to safeguard the medicinal potential of these plants (Goswami et al. 2006).

3.2 Air Pollution

Urban areas are major contributors to climate change, accounting for over 70% of the world's total carbon dioxide emissions. Elevated carbon dioxide levels can stimulate biomass growth, leading to increased leaf area and dry mass, potentially beneficial for the environment (Przybysz et al. 2021; Singh and Verma 2013; Kumar et al. 2017). However, medicinal plants exhibit diverse responses to environmental stresses, with variations observed among species. The impact of air pollutants, influenced by climatic conditions, depends on their physicochemical properties and residence time in a given area (Gupta and Singh 2017; Singh et al. 2017a, b). Industrialization and anthropogenic activities have introduced significant pollutants into the atmosphere, land, and water ecosystems, including sulphur, nitrogen oxides, carbon oxides, organic and inorganic compounds, radioactive isotopes, and toxic heavy metals (Agbaire and Akporhonor 2014).

Airborne pollutants enter plant ecosystems through dry and wet deposition mechanisms and are assimilated into metabolic pathways, directly affecting leaves and indirectly through soil acidification. Some plant species exposed to various pollutants can suffer extreme physical damage, leading to visible leaf injury (Liu and Ding 2008; Meena et al. 2011; Kumar et al. 2019c, 2020). Accumulation of nitrogen oxides, sulphur, carbon, and trace elements on plant leaves influences the physiological behaviour of plant species. Susceptible plant species may experience reductions in photosynthetic activity, growth, yield, leaf injury, stomatal damage, senescence, and membrane permeability (Tiwari et al. 2018).

For instance, medicinal plants like *Morinda lucida* and *Chromolaena odorata* collected from dumpsites in Ibadan, Nigeria, exhibited significant reductions in chlorophyll content (up to 60%) and ascorbic acid (up to 30%) due to air pollutant deposition from municipal solid waste (Falusi et al. 2016). These pollutants adversely affect plant life's physiological and biochemical aspects, influencing periodicity, phenology, fruiting, blossoming, leaf wax, biomass production, seed germination, and seed growth. Increased pollution levels rapidly generate reactive oxygen species (ROS), inducing stress responses within the plant's ascorbate–glutathione cycle (Singh and Kumar 2022; Singh et al. 2024c).

The phytotoxicity of pollutants depends on their atmospheric concentrations. Low levels of certain pollutants, like sulphur dioxide, can benefit plants by contributing to amino acid, protein, vitamin, and chlorophyll synthesis and aiding in nitrogen fixation and nodule development. However, excessive exposure can lead to adverse effects such as necrosis, leaf chlorosis, growth suppression, and even plant death (Migliore et al. 2003). Figure 1 illustrates the impact of atmospheric pollutants on synthesizing secondary metabolites (SMs) in medicinal plants.

Airborne pollutants, including sulphur dioxide, nitrogen oxides, ozone, and particulate matter, disrupt metabolic processes in medicinal plants, reducing photosynthesis, altering nutrient uptake, and affecting overall plant health. This interference can lead to a decline in the synthesis of bioactive compounds crucial for medicinal properties, directly damaging plant tissues and limiting their growth and reproductive capabilities. Consequently, urban air pollution poses a significant threat to the quality and availability of medicinal plants, potentially restricting their use in traditional medicine and pharmaceutical applications.

3.3 Habitat Fragmentation

Habitat fragmentation, a consequence of urbanization, profoundly impacts ecosystems and biodiversity, including medicinal plants. Urban expansion often leads to the fragmentation and isolation of natural habitats, creating small, disconnected green patches. This fragmentation can have various detrimental effects on medicinal plants within urban settings. Rapid human-induced environmental changes disrupt ecological processes and evolutionary pathways, potentially driving population and species extinction and contributing to global biodiversity loss (Tilman and Lehman 2001).

The size and fragmentation of habitats significantly influence the likelihood of stochastic extinction, as environmental, demographic, and genetic factors interplay (Young et al. 1996). Medicinal plants, integral to complex ecosystems, rely on diverse interactions with other organisms for vital processes like reproduction, pollination, and dispersal. Loss of these interactions can reduce genetic diversity, rendering medicinal plants more susceptible to diseases and environmental shifts. The expected decrease in genetic variation within populations, coupled with increased inbreeding and distinctiveness between populations, underscores the genetic consequences of habitat fragmentation (Leimu et al. 2006).

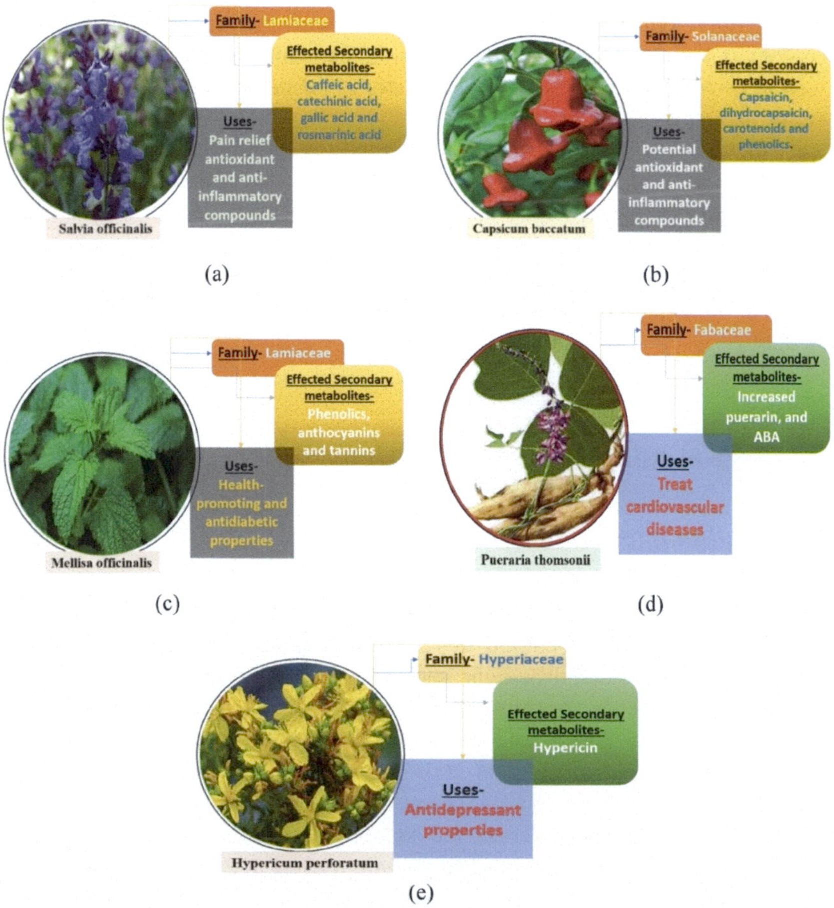

Fig. 1 Effects of some atmospheric pollutants on synthesizing secondary metabolites (SMs) in medicinal plants. (**a** Bortolin et al. 2016; **b** Pellegrini et al. 2015; **c** Tonelli et al. 2015; **d** Wong et al. 2011; **e** Stein et al. 2012)

Additionally, habitat fragmentation affects demography and ecology. Small populations face heightened extinction risks due to their limited numbers, exacerbated by isolation in fragmented landscapes, hindering demographic rescue through immigration. Ecologically, fragmentation disrupts crucial biological interactions, such as plant-pollinator mutualisms (Olesen and Jain 1994). Pollination limitations in small and fragmented populations can lead to low reproductive success, with changes in pollinators' grazing behaviour and activities further exacerbating these challenges (Kolb 2005).

Moreover, fragmentation threatens uninterrupted interactions between plants and seed-dispersing agents, potentially altering population dynamics and fitness. As plant communities fragment, the expansion of edges triggers microclimate changes, impacting light, temperature, and brightness variations, subsequently altering plant phenology (Camargo et al. 2011). Habitat fragmentation can reduce plant species' flowering periods and increase local pollinator extinctions. While evidence on the effect of habitat fragmentation on flowering phenology is limited, observations on species like Echinacea angustifolia demonstrate notable differences in flowering periods, with edge plants exhibiting extended periods compared to central habitat plants (Ison and Wegenius 2014).

These impacts extend to pollinators, either directly through population structure and distribution changes or indirectly via plant resource composition and distribution alterations, potentially reshaping pollinator assemblages and behaviours (Cagnolo et al. 2009).

4 Biological Adaptations of Medicinal Plants

4.1 Morphological Changes

Changes in morphological characteristics stem from alterations in the physical traits of plants. Medicinal plants exhibit morphological changes due to environmental conditions, nutrient availability, light intensity, temperature variations, and biotic stresses. Optimizing leaf area is crucial for enhancing photosynthesis and maximizing dry matter yield. Water stress notably impacts plant growth across various species, including Populus (Wullschlager et al. 2005), soybean (Zhang et al. 2008), and others. In urban environments, where space for plant growth may be limited, medicinal plants might adapt by reducing overall size and adopting a more compact growth pattern. Drought stress significantly reduces leaf growth, leading to diminished leaf area.

Wullschlager et al. (2005) observed significant intraspecific leaf number, area, and biomass variations between coexisting Populus species. While water stress prompts increased root growth in sunflower and Catharanthus roseus (Jaleel et al. 2008), Wullschlager et al. (2005) noted a decrease in root dry weight under similar conditions. Drought triggers adaptive physiological responses in plants, including increased thickness of the upper epidermal waxy cuticle, thickened cuticular edges, enhanced lignification, thickening of scalariform in xylem tracheids, and larger sub-stomal cavities (Gomes and Prado 2007).

Urban air pollution and dust can alter leaf morphology, leading to the evolution of thicker cuticles or denser epidermal layers to mitigate pollutant impact. Medicinal plants in urban settings may produce smaller leaves, potentially in response to space limitations, shading, or heightened air pollution levels (Yadav et al. 2019a, b; Jangra et al. 2024; Kanta et al. 2024). Some plants undergo morphological changes to mimic other species or environmental attributes as a survival strategy. Exhibiting phenotypic plasticity, plants adjust their morphology in response to environmental cues to optimize growth and reproduction.

Therefore, understanding these morphological adaptations is crucial for urban planners, ecologists, and those conserving and cultivating medicinal plants in urban areas. Such insights into plant resilience and adaptability under changing environmental conditions are invaluable for sustainable urban ecosystems.

4.2 *Physiological Changes*

Medicinal plants within urban environments undergo distinctive physiological adaptations due to the unique challenges urban living conditions pose. Pollution, altered microclimates, and limited space are among the factors influencing the growth, development, and chemical composition of medicinal plants in urban settings (Malik et al. 2024; Kumar and Singh 2024). These environmental stressors often impede plant performance, subsequently impacting productivity (Singh et al. 2017; Liu et al. 2018). Climate change exacerbates water scarcity, leading to drought stress in plants, which disrupts crucial physiological functions (Liu et al. 2018), particularly during the critical stages of seed germination, where water deficit sensitivity poses a significant challenge.

Severe water scarcity hampers cell expansion in higher plants by disrupting water flow from the xylem to expanding cells (Nonami 1998). Conversely, water logging initially reduces stomatal water flow, reducing plant water consumption and causing internal water loss. Insufficient oxygen levels under waterlogged conditions inhibit root penetration ability, leading to reduced water-borne conductivity and decreased rates of photosynthesis and transpiration (Ashraf et al. 2011). This reduction in physiological processes is linked to stomatal closure, decreased chlorophyll content, leaf senescence, and reduced leaf volume, further diminishing transpiration rates.

Air pollutants can alter soil chemistry, disrupting nutrient availability and leading to imbalances that affect plant health and growth. These changes may impact essential nutrients like nitrogen and phosphorus, which are vital for water uptake and balance, exacerbating dehydration issues or waterlogging depending on the plant's response (Driesen et al. 2020). Additionally, ozone, sulphur dioxide, and nitrogen oxides can hinder photosynthesis, reducing carbon dioxide uptake and interfering with energy production and compound synthesis. Under stress, elevated levels of reactive oxygen species contribute to oxidative damage, disrupting antioxidant defense systems.

Heat stress further compounds these challenges, impacting plant growth rates and metabolic regulation, leading to chlorophyll reduction and disruption of grain-filling mechanisms (Masouleh et al. 2020). Preserving grain mass under heat stress indicates potential heat tolerance (Ul-Hassan et al. 2021). In response, plants increase carotenoid content to protect chlorophyll from damage and produce phenolic compounds, damaging cellular structures. Activation of genes and metabolites like ascorbic acid, citric acid, glutathione, and melatonin plays a crucial role in protecting plants from heat stress. Assessments reveal that combined drought and heat stresses cause the most damage to plant species, leading to protein denaturation, enzyme inactivation, lipid membrane fluidity increase, and ROS generation (Aishwarya and

Kumar 2021). Metabolites form protective chelates with metals to mitigate these adverse effects (Dadkhah-Aghdash et al. 2022).

Hence, medicinal plants undergo significant physiological changes in response to urban environmental factors like light intensity, air pollution, temperature, water availability, and soil structure, which shape their bioactive compounds and medicinal properties.

4.3 Biochemical Changes

In urban settings, medicinal plants undergo nuanced molecular alterations influenced by the distinct conditions of urban life. These transformations typically arise in response to environmental stressors and anthropogenic activities. Urban stressors such as pollution and temperature fluctuations serve as stimuli for enhancing the production of secondary metabolites in medicinal plants. These bioactive compounds, encompassing antioxidants and phytochemicals, play pivotal roles in the plant's defense mechanisms and potentially contribute to their medicinal properties.

Many studies underscore the significance of plant gene expression in orchestrating transcriptional and post-transcriptional processes in the face of salt stress. Despite its importance, comprehending the molecular mechanisms underlying such responses remains challenging, as elucidated by prior research (Munns and Tester 2008). Notably, the upregulation of genes like SOS1 emerges as critical under salt-stress conditions, facilitating stress mitigation. While certain plants, such as *Thellungiella,* exhibit minimal changes in photosynthesis-associated genes in response to salt stress, others, like *Oryza sativa,* rely on these genes for stress recovery. Extensive research has delved into characterizing gene and protein expression under salt stress across various plant species, revealing insights into adaptive mechanisms like the role of high-affinity potassium ion transporter 1 (HKT1) in reducing sodium ion accumulation in shoot tissue.

Various proteins hold promise in enhancing stress resilience. For instance, genes encoding enzymes involved in osmolyte biosynthesis enable the synthesis of osmotic compounds counteracting stress. Examples include proline accumulation via the transformation of L-glutamic acid to proline, regulated by enzymes like P5C synthase and P5C reductase. In response to drought stress, repression of proline dehydrogenase occurs, leading to proline accumulation. Conversely, flood stress triggers significant gene expression regulation, offering avenues for identifying waterlogging stress resilience-associated genes.

To mitigate temperature stress, plants employ diverse genetic strategies, including regulating specific gene expression and the synthesis of protective proteins like heat shock proteins/factors (HSPs/HSFs), NAC genes, LEA proteins, and antioxidant enzymes. Molecular alterations in medicinal plants under urban factors underscore the intricate interplay between plant physiology and urban environments. Exposure to pollutants, altered light, and soil contamination prompt tailored molecular responses in these plants, often involving stress-related gene regulation, antioxidant pathways, and secondary metabolite production. While some medicinal plants exhibit adaptive resilience to urban stressors, others may experience diminished therapeutic potential. Understanding the molecular dynamics of medicinal plants in urban environments is crucial for sustainable urban planning and conservation efforts aimed at preserving invaluable plant resources.

5 Molecular Insights

5.1 Gene Expression

At the core of medicinal plant biology lies gene expression, the intricate process through which DNA-encoded genetic information is transcribed into RNA and translated into functional proteins. This regulatory machinery, highly responsive to environmental cues and physiological needs, governs the biosynthesis of secondary metabolites crucial for medicinal properties. Understanding the impact of environmental conditions on the quality and quantity of these metabolites is paramount. Insights from gene expression analysis under abiotic stress shed light on optimizing secondary metabolite production.

Ramakrishna and Ravishankar (2011) elucidated abiotic stresses' role as signaling factors influencing secondary metabolism. Pateraki and Kanellis (2010) documented terpenoid biosynthesis in *C. creticus*, showcasing its cytotoxic effects against cancer cells. Gene expression patterns related to terpenoid biosynthesis vary with tissues and developmental stages. Sanchita et al. (2015) analyzed *W. somnifera* gene expression under drought stress, covering diverse gene categories. Zhou et al. (2015) investigated RGHBNG gene responses in *R. glutinosa* tuberous roots under abiotic stresses, implicating its role in growth and stress response.

Artemisia annua displayed increased transcription of ADS and CYP71AV1 genes under cold, heat, or UV light. Chaudhary and Sharma (2015) utilized deep SAGE technology to analyze cold-related genes in *Hippophaerhamnoides*. Binder et al. (2009) studied UV-B effects on *C. roseus* hairy root cultures. Elevated temperatures reduced photosynthesis in Panax quinquefolius but increased ginsenoside concentrations.

Singh et al. (2002) identified the SMO1 gene in *A. annua*, conferring artemisinin production and drought tolerance. Fujimoto et al. (2000) identified AP2/ERF transcription factors' involvement in stress responses. Mishra et al. (2015) demonstrated AP2/ERF TFs' response to various stimuli in *P. somniferum*, highlighting their versatility. Mishra et al. (2013) explored WRKY TFs' role in secondary metabolite accumulation in *P. somniferum*. Basu and Roychoudhury (2014) investigated stress-related gene expression in rice, offering insights into salt tolerance.

Kohli et al. (2013) studied phytohormone signalling during abiotic stress, which is crucial for plant resilience. ABA, a key phytohormone, regulates water balance by closing stomata. Wani et al. (2016) highlighted phytohormones' roles in stress signalling. Studying gene expression in medicinal plants in urban environments offers insights into their therapeutic mechanisms. The intricate interplay of genes orchestrates bioactive compound synthesis, enhancing medicinal efficacy. Understanding these dynamics paves the way for optimizing plant resilience and medicinal potency in challenging environments.

5.2 Epigenetic Modifications

With its origins in ancient Greek, the prefix "Epi" denotes "above" or "beyond." Coined by Conrad Waddington in the 1940s, "epigenetics" originally transcended genetic perspectives (Waddington 2012), emphasizing the interplay between genes and environment to shape phenotype (Waddington 1977). Epigenetics encompasses inherited changes in gene function independent of alterations in the DNA sequence, influenced by environmental factors that modulate various physiological processes in plants. Understanding these modifications in medicinal plants is crucial for optimizing cultivation, enhancing therapeutic properties, and fostering sustainable agricultural practices.

DNA methylation yields dual consequences: methyl-cytosines within the gene body regulate gene expression, while those in repetitive sequences, like transposable elements, safeguard genome integrity. Enhanced genomic DNA methylation represses gene expression, conserving energy under biotic or abiotic stress. In contrast, reduced methylation of resistance-associated genes promotes chromatin activation and novel gene expression, conferring long-term stress resistance.

Contemporary understanding characterizes epigenetics as a mechanism influencing gene expression independently of DNA sequence changes (Martin and Zhang 2007), intricately regulated by genetic information and environmental factors. Epigenetic modifications significantly contribute to plant adaptation to various stresses, reliably inherited across generations, potentially allowing plants to anticipate future conditions based on past experiences (Latzel et al. 2016).

Epigenetic phenomena involve regulatory processes, including histone modifications, DNA methylation, and microRNAs (miRNAs)-mediated DNA methylation, which is crucial in gene expression regulation. DNA methylation, facilitated by RNA-directed DNA methylation (RDDM), inhibits transcriptional activity, while histone modification is essential for post-translational gene regulation. Chromatin structure stability depends on enzyme activity that introduces covalent changes in DNA and histone proteins or ATP-dependent energy that disrupts histone-DNA interactions.

Epigenetic modifications are vital in regulating gene expression in plants, including medicinal plants. The interplay between DNA methylation, histone modifications, and non-coding RNAs shapes gene expression diversity, impacting the bioactive compounds present in medicinal plants. Manipulating these mechanisms holds promise for enhancing therapeutic potential and optimizing bioactive compound production, ultimately advancing pharmacological interventions in plant-based medicine.

6 Conservation and Sustainable Practices

Conservation efforts for medicinal plants and their sustainable utilization have garnered considerable attention (Uprety et al. 2012). Various guidelines have been established to facilitate the conservation of medicinal plants, encompassing the development of systems for species inventory and status monitoring. Moreover, there is a pressing need for coordinated conservation endeavours that integrate both in-situ and ex-situ strategies (Hamilton 2004). As wild resources face increasing scarcity, sustainable utilization emerges as a viable conservation approach for medicinal plants. Over the past few decades, there has been a notable surge in the harvesting of medicinal plant resources, particularly from wild populations, with annual demand for forest resources in Europe, North America, and Asia escalating by 8–15%.

The viability of a species is imperiled when its fertility dips below a critical threshold (Semwal et al. 2007). Conservation recommendations have evolved to encompass strategies that emphasize both in-situ and ex-situ conservation approaches. Conservation efforts often entail preserving the medicinal properties of plants in their natural habitats by establishing nature reserves and wild nurseries. Meanwhile, ex-situ conservation practices, including botanical gardens and seed banks, along with potential replanting initiatives, play a pivotal role (Coley et al. 2003) (Fig. 2). To guide conservation endeavours effectively, a comprehensive understanding of the geographic distribution and biological characteristics of medicinal plants is imperative. This knowledge aids in determining whether a species should be conserved in its native habitat or within a controlled nursery environment.

Overharvesting medicinal plants with restricted abundance and slow growth often results in resource depletion and, in severe cases, species extinction (Baker et al.

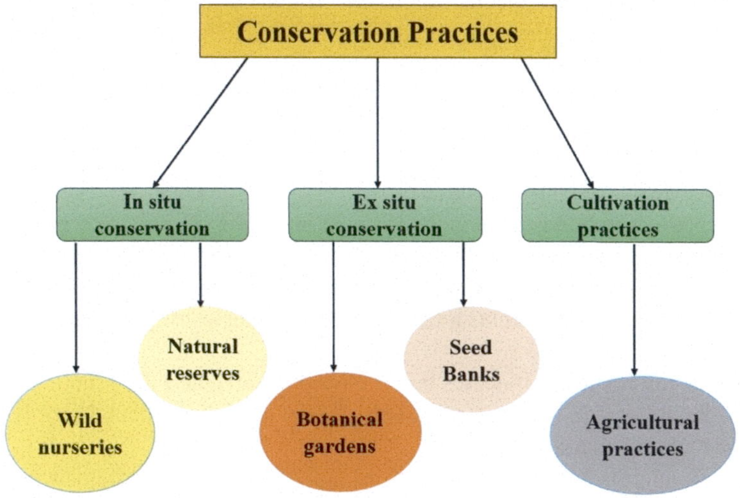

Fig. 2 Various cultivation practices for the conservation of medicinal plants

2007). Hence, prioritizing the sustainable utilization of medicinal plants underscores the necessity for well-defined harvesting practices. Root and whole plant harvesting pose more significant risks to medicinal flora, such as herbs, shrubs, and trees, than the collection of leaves, flowers, or buds. When herbal medicines are derived from whole plants or roots, utilizing their leaves as a treatment presents a gentler alternative. For example, Wang et al. (2009) discovered that extracts from both ginseng leaf-stem and roots exhibit comparable medicinal effects, yet ginseng leaf-stem serves as a more environmentally sustainable source.

7 Challenges and Future Research

The adaptive responses of medicinal plants to changing urban environments present many challenges in biological research. The interaction of environmental factors, including pollution and altered climatic conditions, complicates unraveling the intricate mechanisms controlling plant adaptability. Further complicating the scenario is examining how urbanization impacts plants' medicinal properties. There is a need to comprehend medicinal plants' molecular, physiological, and ecological adaptations in response to urban stresses. Additionally, assessing the implications of these adaptive responses on bioactive compounds produced by medicinal plants adds another layer of complexity to the research landscape.

Key research aspects that need to be addressed include:

- How do specific urban stressors, such as air pollution and soil contamination, influence gene expression and biochemical pathways associated with medicinal properties in plants?
- What are the primary molecular mechanisms underlying the adaptive responses of medicinal plants to the changing urban climate, and how do these mechanisms vary across plant species?
- How does urbanization impact the overall ecological interactions of medicinal plants, including their symbiotic relationships with microorganisms and interactions with other plant species?
- To what extent do urban-induced adaptive responses influence the yield and quality of bioactive compounds in medicinal plants, and how can cultivation practices be adjusted to enhance these properties?
- Can innovative technologies, such as advanced imaging techniques and omics approaches, offer new insights into the adaptive strategies employed by medicinal plants in urban settings?
- How can conservation strategies be developed to safeguard the genetic diversity of medicinal plants, ensuring the sustainability of their medicinal properties in the face of urbanization?
- How can interdisciplinary collaborations among biologists, ecologists, pharmacologists, and urban planners be encouraged to tackle the multifaceted challenges associated with the adaptive responses of medicinal plants to urban environments?

8 Conclusion

It is concluded that exploring biological mechanisms underlying the adaptive responses of medicinal plants to urbanization yields a comprehensive understanding of the intricate interplay between plant ecology and urban development. This synthesis underscores the remarkable ability of medicinal plants to adapt to and thrive amidst the challenges posed by urban environments, elucidating complex mechanisms at molecular, physiological, and ecological levels. Exploration of critical biological factors such as genetic diversity, epigenetic modifications, and physiological adaptations unveils the resilience of medicinal plants in urban settings. Furthermore, the nuanced interactions between plants and their urban habitats underscore the necessity of considering immediate impacts and long-term consequences on medicinal plant populations. This knowledge holds significant importance in ensuring the sustainable availability and conservation of medicinal resources within urban landscapes. In rapid urbanization, the insights garnered contribute to the enhancing capability necessary to address the challenges of sustaining medicinal plant resources. Ultimately, nurturing a balance between urban developments and preserving medicinal plant diversity is indispensable for human well-being and safeguarding these invaluable biological assets in the evolving world.

References

Agbaire PO, Akporhonor EE (2014) The effects of air pollution on plants around the vicinity of the Delta Steel Company, Ovwian-Aladja, Delta State, Nigeria. IOSR J Environ Sci Toxicol Food Technol 8:61–65

Aishwarya R, Kumar M (2021) Urban tree carbon density and CO_2 equivalent of national zoological park, Delhi. Environ Monit Assess 193(12):854

Arnfield AJ (2003) Two decades of urban climate research: a review of turbulence, exchanges of energy and water, and the urban heat island. Int J Climatol J Royal Meteorol Soc 23(1):1–26

Ashraf MA, Ahmad MSA, Ashraf M, Al-Qurainy F, Ashraf MY (2011) Alleviation of waterlogging stress in upland cotton (*Gossypium hirsutum* L.) by exogenous application of potassium in soil and as a foliar spray. Crop Pasture Sci 62(1):25–38

Baker DD, Chu M, Oza U, Rajgarhia V (2007) The value of natural products to future pharmaceutical discovery. Nat Prod Rep 24(6):1225–1244

Basu S, Roychoudhury A (2014) Expression profiling of abiotic stress-inducible genes in response to multiple stresses in rice (*Oryza sativa* L.) varieties with contrasting level of stress tolerance. BioMed Res Int 2014:5142

Binder BY, Peebles CA, Shanks JV, San KY (2009) The effects of UV-B stress on the production of terpenoid indole alkaloids in *Catharanthus roseus* hairy roots. Biotechnol Prog 25(3):861–865

Bortolin RC, Caregnato FF, Junior AMD, Zanotto-Filho A, Moresco KS, de Oliveira Rios A et al (2016) Chronic ozone exposure alters the secondary metabolite profile, antioxidant potential, anti-inflammatory property, and quality of red pepper fruit from *Capsicum baccatum*. Ecotoxicol Environ Saf 129:16–24

Cagnolo L, Valladares G, Salvo A, Cabido M, Zak M (2009) Habitat fragmentation and species loss across three interacting trophic levels: effects of life-history and food-web traits. Conser Biol 23(5):1167–1175

Camargo MGG, Souza RM, Reys P, Morellato LP (2011) Effects of environmental conditions associated to the cardinal orientation on the reproductive phenology of the cerrado savanna tree *Xylopiaaromatica* (Annonaceae). An Acad Bras Ciênc 83:1007–1020

Carreiro MM, Tripler CE (2005) Forest remnants along urban-rural gradients: examining their potential for global change research. Ecosystems 8:568–582

Cavaliere C (2009) The effects of climate change on medicinal and aromatic plants. Herbal Gram 81:44–57

Chaudhary S, Sharma PC (2015) DeepSAGE based differential gene expression analysis under cold and freeze stress in seabuckthorn (*Hippophaerhamnoides* L.). PLoS ONE 10(3):e0121982

Chung IM, Kim JJ, Lim JD, Yu CY, Kim SH, Hahn SJ (2006) Comparison of resveratrol, SOD activity, phenolic compounds and free amino acids in *Rehmanniaglutinosa* under temperature and water stress. Environ Exper Bot 56(1):44–53

Coley PD, Heller MV, Aizprua R, Araúz B, Flores N, Correa M et al (2003) Using ecological criteria to design plant collection strategies for drug discovery. Front Ecol Environ 1(8):421–428

Dadkhah-Aghdash H, Rasouli M, Rasouli K, Salimi A (2022) Detection of urban trees sensitivity to air pollution using physiological and biochemical leaf traits in Tehran, Iran. Sci Rep 12(1):15398

Dhyani K, Kuniyal HB, Singh H, Sobha S (2021) Growth and physiological potential of *Terminalia arjuna* under elevated CO_2 levels in Open top chamber condition. J Appl Nat Sci 13(3):1121–1126

Driesen E, Van den Ende W, De Proft M, Saeys W (2020) Influence of environmental factors light, CO_2, temperature, and relative humidity on stomatal opening and development: a review. Agronomy 10(12):1975

Falusi BA, Odedokun OA, Abubakar A, Agoh A (2016) Effects of dumpsites air pollution on the ascorbic acid and chlorophyll contents of medicinal plants. Cogent Environ Sci 2:1170585. https://doi.org/10.1080/23311843.2016.1170585

Fujimoto SY, Ohta M, Usui A, Shinshi H, Ohme-Takagi M (2000) Arabidopsis ethylene-responsive element binding factors act as transcriptional activators or repressors of GCC box–mediated gene expression. Plant Cell 12(3):393–404

Gillner S, Bräuning A, Roloff A (2014) Dendrochronological analysis of urban trees: climatic response and impact of drought on frequently used tree species. Trees 28:1079–1093

Gomes FP, Prado CH (2007) Ecophysiology of coconut palm under water stress. Braz J Plant Physiol 19:377–391.

Goswami BN, Venugopal V, Sengupta D, Madhusoodanan MS, Xavier PK (2006) Increasing trend of extreme rain events over India in a warming environment. Science 314(5804):1442–1445

Gray SB, Brady SM (2016) Plant developmental responses to climate change. Develop Biol 419(1):64–77

Gupta SK, Singh H (2017) Observations on sporadic flowering in *Ficus benjamina* L. and *Peltophorum africanum* L. in new forest. Indian For 143:290–291

Gupta SK, Ram J, Singh H (2018a) Comparative study of transpiration in cooling effect of tree species in the atmosphere. J Geosci Environ Protect 6:151–166. https://doi.org/10.4236/gep.2018.68011

Gupta SK, Ram J, Singh H, Kumar P (2018b) Comparative study of cooling under shades of some forestry tree species with respect to ambient temperatures. J Geosci Environ Protect 6:51–66. https://doi.org/10.4236/gep.2018.611004

Gupta SK, Ram J, Singh H (2019a) Relationship between leaf area index and atmospheric cooling of tree species. Indian J For 42(3):203–208

Hamilton AC (2004) Medicinal plants, conservation and livelihoods. Biodiver Conser 13:1477–1517

IPCC (2007) Climate change 2007: synthesis report. In: Core Writing Team, Pachauri RK, Reisinger A (eds) Contribution of working groups I, II and III to the fourth assessment report of the intergovernmental panel on climate change. IPCC, Geneva

Iqbal M, Srivastava PS, Siddiqi TO (2000) Environmental hazards plants and people. CBS Publishers

Ison JL, Wagenius S (2014) Both flowering time and distance to conspecific plants affect reproduction in *Echinacea angustifolia*, a common prairie perennial. J Ecol 102(4):920–929

Jangra A, Kumar K, Maikhuri S, Bhandari MS, Pandey S, Singh H, Barthwal S (2024) Unveiling stress-adapted endophytic bacteria: characterizing plant growth-promoting traits and assessing cross-inoculation effects on *Populus deltoides* under abiotic stress. Plant Physiol Biochem 210:108610. https://doi.org/10.1016/j.plaphy.2024.108610

Jaleel CA, Gopi R, Sankar B, Gomathinayagam M, Panneerselvam R (2008) Differential responses in water use efficiency in two varieties of *Catharanthus roseus* under drought stress. C R Biol 331(1):42–47

Kanta C, Kumar A, Chauhan A, Singh H, Sharma IP (2024) The interplay between plant functional traits and climate change. In: Kumar N, Singh H (eds) Plant functional traits for improving productivity. Springer, Singapore. https://doi.org/10.1007/978-981-97-1510-7_3

Kohli A, Sreenivasulu N, Lakshmanan P, Kumar PP (2013) The phytohormone crosstalk paradigm takes center stage in understanding how plants respond to abiotic stresses. Plant Cell Rep 32:945–957

Kolb A (2005) Reduced reproductive success and offspring survival in fragmented populations of the forest herb *Phyteuma spicatum*. J Ecol 93(6):1226–1237

Kumar M, Singh MP, Singh H, Dhakate PM, Ravindranath NH (2019a) Forest working plan for the sustainable management of forest and biodiversity in India. J Sustain For 39:1–22. https://doi.org/10.1080/10549811.2019.1632212

Kumar N, Jeena N, Singh H (2019b) Elevated temperature modulates rice pollen structure: a study from the foothill of the Himalayan agro-ecosystem in India. 3 Biotech 9(5):175. https://doi.org/10.1007/s13205-019-1700-1

Kumar N, Jeena N, Kumar N, Gangola S, Singh H (2019c) Phytoremediation facilitating enzymes: an enzymatic approach for enhancing remediation process. In: Bhatt P (ed) Smart bioremediation technologies. Academic Press, Cambridge, pp 289–306

Kumar A, Kumar P, Singh H, Kumar N (2021a) Modulation of plant functional traits under essential plant nutrients during seasonal regime in natural forests of Garhwal Himalayas. Plant Soil 465:197–212. https://doi.org/10.1007/s11104-021-05003-x

Kumar A, Kumar P, Singh H, Kumar N (2021b) Adaptation and mitigation potential of roadside trees with bio-extraction of heavy metals under vehicular emissions and their impact on physiological traits during seasonal regimes. Urban for Urban Green 58:126900. https://doi.org/10.1016/j.ufug.2020.126900

Kumar A, Kumar P, Singh H, Kumar N (2021c) Impact of plant functional traits on infiltration rate under different forests of Kempty watershed in Garhwal Himalaya. Indian J Soil Conserv 49(1):38–44

Kumar A, Kumar P, Singh H, Bisht S, Kumar N (2021d) Relationship of physiological plant functional traits with soil carbon stock in the temperate forest of Garhwal Himalaya. Curr Sci 120(8):1368–1373. https://doi.org/10.18520/cs/v120/i8/1368-1373

Kumar A, Singh H, Kumari G, Bisht S, Malik A, Kumar A, Singh M, Raturi BS, Thakur A, Kaushal R (2022) Adaptive resilience of roadside trees to vehicular emissions via leaf enzymatic, physiological, and anatomical trait modulations. Environ Pollut 313:120191. https://doi.org/10.1016/j.envpol.2022.120191

Kumar N, Singh H, Giri K, Kumar A, Joshi A, Yadav A, Singh R, Bisht S, Kumari R, Jeena N, Khairakpam R, Mishra G (2024) Physiological and molecular insights into the allelopathic effects on agroecosystems under changing environmental conditions. Physiol Mol Biol Plants 30:417–433. https://doi.org/10.1007/s12298-024-01440-x

Kumar N, Singh H (2024) Plant functional traits for improving productivity. Springer Nature, Singapore. https://doi.org/10.1007/978-981-97-1510-7

Kumar N, Suyal DC, Sharma IP, Verma A, Singh H (2017) Elucidating stress proteins in rice (*Oryza sativa* L.) genotype under elevated temperature: a proteomic approach to understand heat stress response. 3 Biotech 7:205. https://doi.org/10.1007/s13205-017-0856-9

Kumar N, Singh H, Sharma SK (2020) Antioxidants: responses and importance in plant defense system. In: Roychowdhury R, Choudhury S, Hasanuzzaman M, Srivastava S (eds) Sustainable

agriculture in the era of climate change. Springer, Cham. https://doi.org/10.1007/978-3-030-45669-6_11
Larsen L (2015) Urban climate and adaptation strategies. Front Ecol Environ 13(9):486–492
Latzel V, Rendina González AP, Rosenthal J (2016) Epigenetic memory as a basis for intelligent behavior in clonal plants. Front Plant Sci 7:1354
Leimu R, Mutikainen PIA, Koricheva J, Fischer M (2006) How general are positive relationships between plant population size, fitness and genetic variation? J Ecol 94(5):942–952
Liu YJ, Ding HUI (2008) Variation in air pollution tolerance index of plants near a steel factory: Implication for landscape-plant species selection for industrial areas. WSEAS Trans Environ Develop 4(1):24–32
Liu S, Yang R, Tripathi DK, Li X, He W, Wu M et al (2018) RETRACTED: the interplay between reactive oxygen and nitrogen species contributes in the regulatory mechanism of the nitro-oxidative stress induced by cadmium in Arabidopsis
Malik A, Kumar K, Kumari G, Kumar N, Singh H (2024) Application of functional traits in modelling productivity and resilience under climate change. In: Kumar N, Singh H (eds) Plant functional traits for improving productivity. Springer, Singapore. https://doi.org/10.1007/978-981-97-1510-7_5
Martin C, Zhang Y (2007) Mechanisms of epigenetic inheritance. Curr Opin Cell Biol 19(3):266–272
Masouleh SSS, Aldine NJ, Sassine YN (2020) The role of organic solutes in the osmotic adjustment of chilling-stressed plants (vegetable, ornamental and crop plants). Ornam Horticult 25:434–442
Meena D, Singh H, Chaudhari SK (2011) Elucidating strontium response on growth dynamics and biochemical change in *Phaseolus mungo* L. Int J Agricult Environ Biotechnol 4:107–113
Migliore L, Cozzolino S, Fiori M (2003) Phytotoxicity to and uptake of enrofloxacin in crop plants. Chemosphere 52(7):1233–1244
Mir BA, Mir SA, Khazir J, Tonfack LB, Cowan DA, Vyas D, Koul S (2015) Cold stress affects antioxidative response and accumulation of medicinally important withanolides in *Withaniasomnifera* (L.) Dunal. Ind Crops Prod 74:1008–1016
Mishra S, Triptahi V, Singh S, Phukan UJ, Gupta MM, Shanker K, Shukla RK (2013) Wound induced tanscriptional regulation of benzylisoquinoline pathway and characterization of wound inducible PsWRKY transcription factor from Papaver somniferum. PLoS ONE 8(1):e52784
Mishra S, Phukan UJ, Tripathi V, Singh DK, Luqman S, Shukla RK (2015) PsAP2 an AP2/ERF family transcription factor from *Papaver somniferum* enhances abiotic and biotic stress tolerance in transgenic tobacco. Plant Mol Biol 89:173–186
Munns R, Tester M (2008) Mechanisms of salinity tolerance. Annu Rev Plant Biol 59:651–681
Nazir N, Bilal S, Bhat KA, Shah TA, Badri ZA, Bhat FA et al (2018) Effect of climate change on plant diseases. Int J Curr Microbiol App Sci 7(6):250–256
Neuman M, Smith S (2010) City planning and infrastructure: once and future partners. J Plan Hist 9(1):21–42
Nonami H (1998) Plant water relations and control of cell elongation at low water potentials. J Plant Res 111:373–382
Olesen JM, Jain SK (1994) D: fragmented plant populations and their lost interactions. Conserv Genet 15:417–426
Pateraki I, Kanellis AK (2010) Stress and developmental responses of terpenoid biosynthetic genes in *Cistus creticu* s subsp. creticus. Plant Cell Rep 29:629–641
Patwardhan B, Vaidya AD, Chorghade M, Joshi SP (2008) Reverse pharmacology and systems approaches for drug discovery and development. Curr Bioact Comp 4(4):201–212
Pellegrini E, Francini A, Lorenzini G, Nali C (2015) Ecophysiological and antioxidant traits of *Salvia officinalis* under ozone stress. Environ Sci Pollut Res 22:13083–13093
Prakash V, Hunney K, Singh H (2022) Effect of elevated carbon dioxide on growth and development of *Santalum album* L. seedlings inoculated with plant growth promoting microorganisms in Open Top Chambers. Int J Environ Agricult Biotechnol 7(5):149–159

Przybysz A, Popek R, Stankiewicz-Kosyl M, Zhu CY, Małecka-Przybysz M, Maulidyawati T et al (2021) Where trees cannot grow–Particulate matter accumulation by urban meadows. Sci Total Environ 785:147310

Ramakrishna A, Ravishankar GA (2011) Influence of abiotic stress signals on secondary metabolites in plants. Plant Sign Behav 6(11):1720

Rates SMK (2001) Plants as source of drugs. Toxicon 39(5):603–613

Sanchita S, Singh R, Mishra A, Dhawan SS, Shirke PA, Gupta MM, Sharma A (2015) Physiological performance, secondary metabolite and expression profiling of genes associated with drought tolerance in *Withaniasomnifera*. Protoplasma 252:1439–1450

Semwal DP, Saradhi PP, Nautiyal BP, Bhatt AB (2007) Current status, distribution and conservation of rare and endangered medicinal plants of Kedarnath Wildlife Sanctuary, Central Himalayas, India. Curr Sci 12:1733–1738

Sharma R, Singh H, Kaushik M, Nautiyal R, Singh O (2018) Adaptive physiological response, carbon partitioning, and biomass production of *Withania somnifera* (L.) Dunal grown under elevated CO_2 regimes. 3 Biotech 8:267. https://doi.org/10.1007/s13205-018-1292-1

Sharma R, Singh H (2021) Alteration in biochemical constituents and nutrients partitioning of *Asparagus racemosus* in response to elevated atmospheric CO_2 concentration. Environ Sci Pollut Res 29:6812–6821. https://doi.org/10.1007/s11356-021-16050-3

Singh H (2021) An integrated approach considering physiological and biophysical based indicators for assessing tolerance of roadside plantations of *Alstonia scholaris* towards urban roadside air pollution: an assessment of adaptation of plantations for mitigating urban roadside air pollution. Trees. https://doi.org/10.1007/s00468-021-02179-8

Singh H (2024) Exploring adaptive modulation in plant functional traits and its impact on the productivity of *Acacia auriculiformis* under CO_2-enriched environment. Ind Crops Prod 210:118186. https://doi.org/10.1016/j.indcrop.2024.118186

Singh H, Verma A (2013) Characterization and screening of high nitrogen efficient rice genotype to curtail environmental pollution. Int J Agricult Environ Biotechnol 6(1):23–30

Singh KB, Foley RC, Oñate-Sánchez L (2002) Transcription factors in plant defense and stress responses. Curr Opin Plant Biol 5(5):430–436

Singh H, Savita Sharma R, Sinha S, Kumar M, Kumar P, Verma A, Sharma SK (2017a) Physiological functioning of *Lagerstroemia speciosa* L. under heavy roadside traffic: an approach to screen potential species for abatement of urban air pollution. 3 Biotech 7(61):1–10. https://doi.org/10.1007/s13205-017-0690-0

Singh H, Verma A, Kumar M, Sharma R, Gupta R, Kaur M, Negi M, Sharma SK (2017b) Phytoremediation: a green technology to clean up the sites with low and moderate level of heavy metals. Aust Biochem 2:1012

Singh H, Sharma R, Savita SMP, Kumar M, Verma A, Ansari MW, Sharma SK (2018) Adaptive physiological response of *Parthenium hysterophorus* to elevated atmospheric CO_2 concentration. Indian For 144:1–14

Singh H, Yadav M, Kumar N, Kumar A, Kumar M (2020) Assessing adaptation and mitigation potential of roadside trees under the influence of vehicular emissions: a case study of *Grevillea robusta* and *Mangifera indica* planted in an urban city of India. PLoS ONE 15(1):e0227380. https://doi.org/10.1371/journal.pone.0227380

Singh S, Singh H, Sharma SK, Nautiyal R (2021) Seasonal variation in biochemical responses of bamboo clones in the sub-tropical climate of Indian Himalayan foothills. Heliyon 7(4):e06859. https://doi.org/10.1016/j.heliyon.2021.e06859

Singh H, Bandyopadhyaya S, Kumar A, Singh M, Malik A, Kumari G, Raturi A, Yadav SK, Kumar M, Barthwal S, Thakur A, Rawat PS (2022) Understanding the physiological and biophysical response of urban roadside plantations for assessing adaptation and mitigation mechanisms toward vehicular emissions. Urban Clim 44:101183. https://doi.org/10.1016/j.uclim.2022.101183

Singh H, Kumar N, Kumar A (2024a) Enhancing resource use efficiency in crops through plant functional traits. In: Kumar N, Singh H (eds) Plant functional traits for improving productivity. Springer, Singapore. https://doi.org/10.1007/978-981-97-1510-7_6

Singh H, Kumar M (2022) Climate change and its impact on Indian Himalayan forests: current status and research needs. In: Rani S, Kumar R (eds) Climate change. Springer, Cham, pp 223–242. https://doi.org/10.1007/978-3-030-92782-0_11

Stein AC, Viana AF, Müller LG, Nunes JM, Stolz ED, Do Rego JC et al (2012) Uliginosin B, a phloroglucinol derivative from *Hypericum polyanthemum*: a promising new molecular pattern for the development of antidepressant drugs. Behav Brain Res 228(1):66–73

Tilman D, Lehman C (2001) Human-caused environmental change: impacts on plant diversity and evolution. Proceed Natl Acad Sci 98(10):5433–5440

Tiwari S, Agrawal M, Tiwari S, Agrawal M (2018) Effect of ozone on physiological and biochemical processes of plants. In: Tropospheric ozone and its impacts on crop plants: a threat to future global food security, pp 65–113

Ul-Hassan M, Rasool T, Iqbal C, Arshad A, Abrar M, Abrar MM et al (2021) Linking plants functioning to adaptive responses under heat stress conditions: a mechanistic review. J Plant Growth Regul 32:1–18

Uprety Y, Asselin H, Dhakal A, Julien N (2012) Traditional use of medicinal plants in the boreal forest of Canada: review and perspectives. J Ethnobiol Ethnomed 8:1–14

Waddington CH (1977) The evolution of an evolutionist. J Hist Biol 10(2):5874

Waddington CH (2012) The epigenotype. Int J Epidemiol 41(1):10–13

Walther GR, Post E, Convey P, Menzel A, Parmesan C, Beebee TJ et al (2002) Ecological responses to recent climate change. Nature 416(6879):389–395

Wang W, Pataki DE (2010) Spatial patterns of plant isotope tracers in the Los Angeles urban region. Landsc Ecol 25:35–52

Wang H, Peng D, Xie J (2009) Ginseng leaf-stem: bioactive constituents and pharmacological functions. Chin Med 4(1):1–8

Wani SH, Kumar V, Shriram V, Sah SK (2016) Phytohormones and their metabolic engineering for abiotic stress tolerance in crop plants. Crop J 4(3):162–176

Wong KH, Li GQ, Li KM, Razmovski-Naumovski V, Chan K (2011) Kudzu root: traditional uses and potential medicinal benefits in diabetes and cardiovascular diseases. J Ethnopharmacol 134(3):584–607

Wullschleger SD, Yin TM, DiFazio SP, Tschaplinski TJ, Gunter LE, Davis MF, Tuskan GA (2005) Phenotypic variation in growth and biomass distribution for two advanced-generation pedigrees of hybrid poplar. Can J For Res 35(8):1779–1789

Yadav SK, Singh H, Ginwal HS and Barthwal S (2019a) Elevated CO_2 enhanced growth and physiological process of *Populus deltoides* Bartr. ex Marsh. Indian For 145(1):23–27

Yadav SK, Singh H, Nautiyal R, Ginwal HS, Ansari SA, Barthwal S (2019b) Modulation of morpho-physiological responses in *Populus deltoides* by elevated carbon dioxide and temperature. For Sci 66:105–118. https://doi.org/10.1093/forsci/fxz048

Young A, Boyle T, Brown T (1996) The population genetic consequences of habitat fragmentation for plants. Trends Ecol Evolut 11(10):413–418

Zhang B, Pan X, Stellwag EJ (2008) Identification of soybean microRNAs and their targets. Planta 229:161–182

Zhou Y, Zhang Y, Wei J, Zhang Y, Li J, Wang W et al (2015) Cloning and analysis of expression patterns and transcriptional regulation of RghBNG in response to plant growth regulators and abiotic stresses in *Rehmanniaglutinosa*. Springerplus 4:1–8

Zobayed SMA, Afreen F, Kozai T (2005) Temperature stress can alter the photosynthetic efficiency and secondary metabolite concentrations in St John's Wort. Plant Physiol Biochem 43(10–11):977–984

Unraveling Urban Plant Strategies for Heavy Metal Uptake and Detoxification

Urvashi Gupta, Smriti Shukla, Laxmi Kant Bhardwaj, Prangya Rath, Divya Upadhyay, Bhawana Sharma, Rupam, Harshita Asthana, Tanu Jindal, and Hukum Singh

Abstract Heavy metal pollution in urban environments is increasingly prevalent, threatening environmental quality and human health. Urban plants facilitate mitigating this pollution by absorbing and detoxifying heavy metals from their surroundings. This chapter explores how urban plants absorb and detoxify heavy metals to maintain cellular equilibrium. Heavy metal uptake in urban plants primarily occurs through root absorption from the soil and aerial deposition on plant surfaces. Heavy metals are transported throughout the plant once absorbed and accumulated in various organs. Urban plants employ various detoxifying mechanisms to mitigate heavy metal toxicity. These include chelation and complexation with organic molecules, activation of antioxidant defense systems to counteract oxidative stress, and efflux and sequestration processes to regulate intracellular metal concentrations. Understanding urban plants' absorption and detoxification processes is crucial for developing strategies to reduce heavy metal pollution in urban settings. The phytoremediation potential of urban plants holds promise for improving environmental quality, safeguarding human health, and promoting sustainable urban ecosystems. Urban planners and policymakers may use this knowledge to identify appropriate plant species for urban greening initiatives, thereby enhancing environmental quality and minimizing human

U. Gupta (✉) · S. Shukla · L. K. Bhardwaj · T. Jindal
Amity Institute of Environmental Toxicology, Safety and Management (AIETSM), Amity University, Noida 201303, India
e-mail: gpt.urvashi@gmail.com

P. Rath · D. Upadhyay
Amity Institute of Environmental Sciences (AIES), Amity University, Noida 201303, India

B. Sharma · Rupam
Department of Bioscience and Biotechnology, Banasthali Vidyapith, Rajasthan 304022, India

H. Asthana
Department of Environment Science, St. Josephs University, Bengaluru 560027, India

H. Singh
ICFRE-Forest Research Institute, PO New Forest, Dehradun, Uttarakhand 248006, India

H. Singh (ed.), *Urban Forests, Climate Change and Environmental Pollution*,
https://doi.org/10.1007/978-3-031-67837-0_5

health risks in urban settlements. Incorporating phytoremediation into urban planning and landscaping efforts may effectively address heavy metal pollution while creating healthier and more resilient urban environments.

Keywords Detoxification mechanisms · Heavy metal pollution · Metal uptake · Urban forests · Plant strategy

1 Introduction

Heavy metal uptake and detoxification mechanisms in urban plants are vital in their ability to thrive in environments contaminated with high concentrations of toxic heavy metals, which are common in urban areas owing to industrial activities, traffic emissions, and other anthropogenic or man-made sources (Kumar et al. 2022, 2024; Singh and Kumar 2022; Singh et al. 2024a; Singh 2024). Polluting soil, aquatic waste streams, and groundwater with toxic metals pose significant health and environmental risks (Memon et al. 2001). Despite the widespread ecological damage caused by heavy metals from natural and artificial sources, effective disposal methods remain relatively inexpensive.

Heavy metals are typically defined as metals with a density greater than 5 g/cm^3, which can harm the environment and living organisms (Järup 2003). These metals, including both essential ones like iron (Fe), zinc (Zn), and copper (Cu), and nonessential ones, pose toxicity determined by factors such as concentration, reactivity, and oxidation capability (Järup 2003; Singh et al. 2020; Dhyani et al. 2021; Kumar et al. 2021; Prakash et al. 2022).

Phytoremediation stands out as a perpetual and widely recognized method due to its low cost, ease of usage, environmental safety, and effectiveness in environmental rehabilitation (Bhardwaj et al. 2023a, b). This environmentally beneficial method involves regenerating heavy metal-polluted soil cost-effectively (Yan et al. 2020), thereby promoting sustainable use of contaminated land by cultivating energy plants to hinder heavy metal entry into the food chain.

Heavy metal stress triggers the production of reactive oxygen species (ROS) in plant cells, leading to oxidative damage. Urban plants employ antioxidant defense mechanisms, such as the production of enzymes like superoxide dismutase and catalase, to scavenge ROS and counteract oxidative stress (Mansoor et al. 2023). Some urban plants possess natural phytoremediation capabilities known as hyperaccumulators, which extract or accumulate high concentrations of metals without experiencing significant toxicity. Techniques involving hyperaccumulative plants can thus be utilized to remediate contaminated urban environments and enhance soil and air quality (Singh 2021; Sharma and Singh 2021; Singh et al. 2021; Skuza et al. 2022).

Plants employ various mechanisms to absorb, translocate, and detoxify heavy metals, distributing them throughout their tissues. These mechanisms include mycorrhizal association, binding with plant cell walls, chelation by phytochelatins, efflux from the plasma membrane, and vacuolar compartmentalization (Kushwaha et al.

2015). Understanding these mechanisms is crucial for developing strategies to mitigate heavy metal pollution in urban areas and harnessing the phytoremediation potential of plants for environmental restoration (Singh et al. 2020; Kumar et al. 2021a, b).

This chapter highlights an eco-friendly and cost-effective approach to address heavy metal stress, offering a novel perspective on existing phytoremediation applications. It provides insights into heavy metal contamination, uptake and detoxification mechanisms in plants, plant responses to heavy metal stress, and techniques for mitigating heavy metal pollution in urban environments (Kumar et al. 2021c, d).

2 Sources and Types of Heavy Metals in Urban Environments

Urban environments are shaped by a complex interplay of anthropogenic and natural factors, leading to the accumulation of heavy metals in various matrices such as soil, water, air, and urban plants. Understanding the sources and types of heavy metals in urban areas is essential for effective pollution management and environmental protection. Natural processes such as seepage, rock weathering, volcanic eruptions, forest fires, and dust contribute significantly to heavy metal presence in the environment (Cimboláková et al. 2019; Yadav et al. 2019a, b; Gupta et al. 2019a, b). Conversely, human activities such as mining ores, industrial processes, tanneries, battery production, and electroplating are primary sources of heavy metal contamination (Arora et al. 2023). Other contributors include industrial and municipal wastewater, landfill leachates, urban runoff, and agricultural chemicals like fertilizers and pesticides. Heavy metals enter the water supply through industrial and consumer waste and acid rain, exacerbating contamination in urban environments. Common heavy metals associated with industrial activities include lead (Pb), cadmium (Cd), mercury (Hg), chromium (Cr), arsenic (As), nickel (Ni), and zinc (Zn) (Gavhane et al. 2021).

Heavy metals are pervasive environmental pollutants in urban areas, originating from anthropogenic and natural sources. Understanding their sources and types is crucial for effective pollution control and risk mitigation. Untreated releases from industrial facilities and agricultural runoff into nearby water bodies render the water unsafe for human consumption, endangering aquatic organisms and the surrounding ecosystem (Kumar et al. 2019b, c).

Anthropogenic contributors to heavy metals in urban soils and road dust include vehicular emissions, industrial emissions, domestic pollutants, weathering of buildings and pavement surfaces, and atmospheric deposition. Conversely, anthropogenic sources of heavy metals in agricultural soils stem from mineral extraction, waste disposal, municipal discharge, sewage sludge, and pesticide and fertilizer applications (Wei and Yang 2010; Sharma et al. 2018; Singh et al. 2018; Gupta et al. 2018a, b). Heavy metals are transported over long distances through atmospheric deposition, settling onto urban surfaces, soil, water bodies, and vegetation. Discharged into

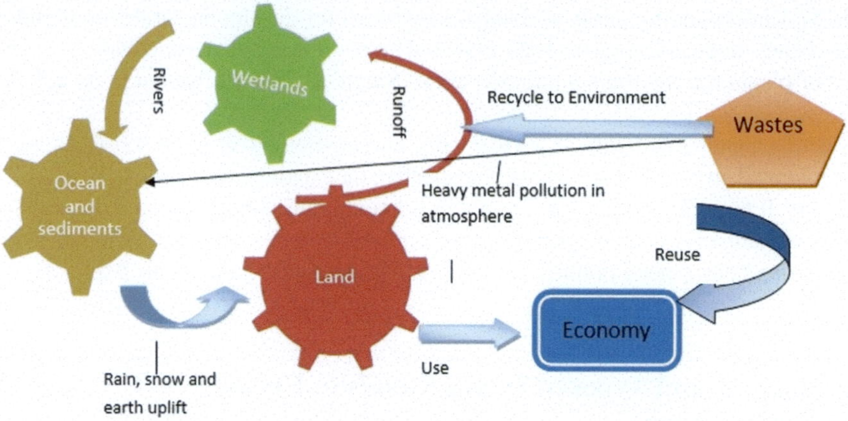

Fig. 1 Heavy metals in the environment. *Source* Odum (2016)

the atmosphere, heavy metals eventually contaminate other environmental elements, such as the hydrosphere and lithosphere, impacting the plants and wildlife that inhabit these areas. Some heavy metals pose health risks at low exposure levels, with metals like nickel, cadmium, and chromium potentially causing cancer in humans. Meteorological and topographical conditions heavily influence heavy metal dispersion and deposition in a given area (Singh and Verma 2013; Kumar et al. 2017; Gupta and Singh 2017; Gupta 2020).

Highly polluting industries, including thermal power plants, coalfields, concrete, scrap metal, and hydrocarbons, emit dust, smoke, fumes, and hazardous gases, causing significant harm to the environment and ecosystems. Often clustered in industry-specific zones, these activities surpass the environment's carrying capacity, leading to severe environmental degradation (Singh et al. 2017a, b).

Figure 1 illustrates the cycle of earth, water, and heavy metals, depicting how the economy utilizes water and mined metals, resulting in byproduct waste. Dilute contaminants are released into the environment through air and water, which wetlands can filter, while a significant percentage of concentrated wastes is reused (Odum 2016).

3 Heavy Metal Uptake Mechanisms in Plants

Plants have evolved highly specific and very effective mechanisms to obtain essential micronutrients from the environment. Heavy metal uptake mechanisms in plants involve several processes that allow plants to absorb these potentially toxic elements from the soil (Bhardwaj et al. 2023a, b). Heavy metals such as cadmium (Cd), lead (Pb), mercury (Hg), arsenic (As), and chromium (Cr) are often present in soil due to natural processes like the weathering of rocks, as well as anthropogenic activities like industrial pollution and the use of agrochemicals. Several factors, such as plant species, properties of the medium, root zone, vegetative uptake, and addition of chelating agents, affect the uptake mechanism of heavy metals.

3.1 Uptake Mechanism

Mobilization, root uptake, xylem loading, accumulation, root-to-shoot transport, and cellular compartmentation are involved in this procedure (Bhardwaj et al. 2023a, b). Most heavy metals are present in the soil in insoluble form and are not readily available. Plants tend to increase the bioavailability of heavy metals by releasing root exudates (Dalvi and Bhalerao 2013). Heavy metals are absorbed by plant roots from the soil. Metal ions in the soil diffuse or are actively transported across the root cell membranes. Complexing agents and metal ion transporters mediate this process. H^+-coupled carrier proteins or transporters are present in the root cell membrane and play a significant role in the uptake of metal ions.

Once inside the root, metals may bind to metal transporters or chelators, which facilitate their movement across the plasma membrane and into the vascular system of the plant, primarily the xylem and phloem. Two pathways absorb heavy metals into the roots (Singh et al. 2020). The apoplastic pathway is known as passive transport and involves the movement of heavy metals into roots through simple diffusion driven by concentration gradients. This process is vital for metals present at high concentrations in the soil. The symplastic pathway is active transport and depends on energy (Peer et al. 2006). This pathway uses energy in the form of ATP to move metals against their concentration gradient, allowing plants to accumulate metals even when present at low concentrations in the soil (Meena et al. 2011; Kumar et al. 2019c, 2020; Singh and Kumar 2022; Singh et al. 2024b).

Heavy metals can form complexes with chelators and are precipitated and restrained in the intracellular or extracellular spaces (Ali et al. 2013). These complexes, including sulfate, phosphate, and carbonate, act as metal ligands to facilitate the chelation of metallic ions. Citrate is known as the chief chelator for metals. From roots, these heavy metals are transferred to the other parts of the plants (Meagher 2000). Thakur et al. (2016) studied the mechanism of the uptake of heavy metals in the plant and stated that metal ions are transferred from the vacuoles of the root to the xylem and then transferred from the xylem to the shoot via xylem vessels. Heavy metals are transported and distributed in the leaves through symplast or apoplast (Tong et al. 2004; DalCorso et al. 2019).

4 Heavy Metal Detoxification Mechanisms in Plants

Heavy metal contamination poses significant threats to environmental and human health due to its persistence and toxicity (Alam et al. 2023; Bhardwaj et al. 2023a, b). As primary producers in ecosystems, plants play crucial roles in mitigating heavy metal pollution through various detoxification mechanisms (Cobbett 2002). This section explores the diverse strategies employed by plants to detoxify heavy metals, highlighting their importance in environmental remediation and sustainable agriculture.

4.1 Heavy Metal Uptake and Accumulation in Plants

Plants absorb or uptake heavy metals from the soil, water, and air through their roots and leaves. Once inside the plant, these metals can accumulate in different tissues, posing risks to plant health and ecosystem integrity. However, plants have evolved sophisticated mechanisms to tolerate and detoxify heavy metals, minimizing their adverse effects (Yaashikaa et al. 2022).

4.2 Chelation and Complexation

Chelation involves the formation of complexes between heavy metal ions and organic ligands within plant tissues. Phytochelatins (PCs) and metallothioneins (MTs) are two major classes of metal-binding peptides synthesized by plants in response to heavy metal stress. These peptides bind to metal ions, reducing their toxicity and facilitating their sequestration into vacuoles for storage or detoxification (Carrillo and Borthakur 2021).

4.3 Phytoextraction

Phytoextraction is a phytoremediation technique wherein plants can absorb and accumulate heavy metals from contaminated soils into their tissues. Hyperaccumulator plants, metastasized by their ability to accumulate exceptionally high concentrations of heavy metals, play vital roles in phytoextraction. These plants often possess specialized metal transporters, chelators, and detoxification enzymes that enhance their metal uptake and tolerance (Awa and Hadibarata 2020).

4.4 Rhizofiltration

Rhizofiltration involves using plant roots to filter heavy metals from highly polluted water sources. Aquatic plants, such as water hyacinths and ferns, are commonly employed in rhizofiltration systems due to their efficient metal uptake capabilities (Ali et al. 2020). The roots of these plants absorb dissolved heavy metals, which are sequestered in their tissues or excreted into the surrounding environment (Bakshe and Jugade 2023).

4.5 Phytostabilization

Phytostabilization aims to immobilize heavy metals in the soil, reducing their bioavailability and potential for leaching into groundwater or plant uptake. Certain plant species, known as metal-tolerant or metallophyte plants, can establish stable ecosystems in heavy metal-contaminated soils. These plants produce extensive root systems that bind soil particles and secrete compounds, such as organic acids and polysaccharides, that promote metal immobilization (Bakshe and Jugade 2023).

4.6 Phytofiltration

Phytofiltration involves using plants to remove heavy metals from contaminated air or gases. Plants capture airborne heavy metal particles through transpiration and foliar uptake and absorb them into their tissues. Indoor plants, such as spider plants and peace lilies, are often utilized in phytofiltration systems to improve indoor air quality and mitigate heavy metal exposure (Nwogwu et al. 2022).

4.7 Metabolic Transformation and Sequestration

Plants possess enzymatic machinery capable of metabolizing and detoxifying heavy metals through various biochemical pathways. Metallothioneins, glutathione S-transferases, and phytochelatin synthases are enzymes involved in heavy metal detoxification. These enzymes catalyze the conjugation, oxidation, reduction, and sequestration of heavy metal ions, rendering them less toxic or facilitating their compartmentalization within plant cells.

4.8 Microbial Interactions

Plant-associated microorganisms, including bacteria, fungi, and algae, contribute to heavy metal detoxification through symbiotic or associative relationships. Mycorrhizal fungi enhance plant metal tolerance and uptake by facilitating nutrient acquisition and providing metal-binding sites in their hyphal networks. Additionally, rhizospheric bacteria can promote plant growth and metal immobilization by producing chelating agents, siderophores, and exopolysaccharides.

4.9 Genetic Engineering

Advances in genetic engineering have enabled the manipulation of plant genomes to enhance their heavy metal tolerance and detoxification capabilities. Transgenic plants expressing metal-binding proteins, transporters, and detoxification enzymes show improved resistance to heavy metal stress and enhanced phytoremediation potential. However, carefully considering environmental safety and regulatory concerns is necessary when deploying genetically modified organisms for remediation (Khan et al. 2021a, b).

4.10 Environmental and Agricultural Implications

Plant detoxification mechanisms have profound implications for environmental remediation, sustainable agriculture, and human health. Phytoremediation technologies offer cost-effective and environmentally friendly solutions for cleaning contaminated sites and restoring ecosystem health (Ghuge et al. 2023). Furthermore, the cultivation of metal-tolerant crops and the development/construction of phytoremediation-based agroecosystems can help mitigate the impacts of toxicity of heavy metal pollution on food security and agricultural productivity.

5 Heavy Metals Transport and Sequestration

Heavy metal transport and sequestration in urban plants are complex processes governed by physiological, biochemical, and molecular mechanisms. Understanding heavy metals transport and sequestration in urban plants has significant implications for environmental remediation strategies, particularly in urban environments. Selecting suitable plant species with high metal tolerance and accumulation capacities, optimizing growth conditions, and enhancing metal sequestration mechanisms are crucial to developing green space for cleaning the urban environment and improving urban health (Yadav et al. 2019a, b; Jangra et al. 2024; Kanta et al. 2024; Malik et al. 2024).

Plants uptake heavy metals from contaminated environments using various mechanisms, i.e., passive and active mechanisms. Passive diffusion allows metals to move along a concentration gradient across the root cell membrane. On the other hand, active transport involves specific metal transporters and ion channels that actively pump metals into root cells against the concentration gradient (Kumar and Singh 2024). It is reported that mycorrhizal associations enhance metal uptake by plants, as mycorrhizal fungi extend the root surface area and increase nutrient and water uptake, thereby facilitating metal uptake. However, the extent of metal uptake varies depending on factors such as metal speciation, soil pH, and the presence of competing ions.

Once the plants absorb heavy metals, they are translocated within the plant and distributed to various tissues and organs. Translocation occurs primarily through the plant's vascular system, including the xylem and phloem. Factors like metal speciation, plant species, physiological response, and environmental conditions influence translocation. While metals accumulate in tissues such as leaves and stems, plants have developed mechanisms to minimize the translocation of heavy metals to sensitive organs. This includes sequestration of metals in root tissues, binding to metal chelators such as phytochelatins, and compartmentalization in vacuoles.

Plants sequester and detoxify heavy metals to reduce toxicity and protect cellular structures and functions. Metal chelation involves the binding of metals to peptides such as phytochelatins, forming non-toxic complexes that are stored safely in vacuoles. Compartmentalization in vacuoles and cell walls further helps isolate and detoxify heavy metals in plant tissues. Additionally, plants employ antioxidant defense systems to mitigate oxidative damage induced by metal stress, such as producing enzymes like superoxide dismutase and catalase to scavenge reactive oxygen species (ROS).

6 Plant Responses to Heavy Metal Stress

6.1 Morphological Responses

Morphological responses of plants to heavy metal stress encompass a spectrum of visible changes that reflect the plant's adaptation and defense mechanisms against metal toxicity. Understanding these morphological responses provides valuable insights into plants' physiological and adaptive coping strategies for heavy metal stress (Singh et al. 2020a; Singh et al. 2022).

Chlorosis, the yellowing of leaves due to reduced chlorophyll content, is a common morphological response to heavy metal stress. Heavy metals disrupt chlorophyll synthesis, leading to chlorotic leaves. Additionally, leaf necrotic lesions may develop due to metal-induced cellular damage, further impacting photosynthetic efficiency and overall plant health (DalCorso et al. 2019).

Heavy metal stress often inhibits plant growth, resulting in stunted growth and reduced plant size. This reduction in growth may be evident in both shoot and root systems. Further, metal toxicity impairs cell elongation and division, leading to shorter shoots and decreased root length.

The root system undergoes significant morphological changes in response to heavy metal stress. Plants may exhibit reduced root biomass and root elongation in contaminated soils. However, in some cases, plants may display increased branching in the root system to enhance nutrient uptake and explore a larger soil volume for resources.

Heavy metal exposure induces variations in leaf morphology, including size, shape, and surface characteristics. These alterations may be adaptive responses aimed at minimizing metal uptake or enhancing tolerance to metal stress. For example, smaller leaf sizes and thicker leaf cuticles may reduce metal uptake through leaf surfaces.

The epidermal layer of leaves and roots may undergo structural modifications in response to heavy metal stress. Plants may develop thicker cell walls or produce specialized cells, such as trichomes or mucilage-secreting cells, to sequester or excrete heavy metals and protect underlying tissues from damage.

Heavy metal stress often results in lower shoot height and decreased biomass production in affected plants. This reduction in shoot growth is attributed to impaired photosynthetic activity, nutrient imbalance, and overall physiological disturbances induced by heavy metal toxicity.

Besides vegetative morphological changes, heavy metal stress affects plant reproductive structures and processes. Reduced flower production, abnormal floral morphology, and impaired seed set or germination rates may occur under prolonged exposure to high levels of heavy metals.

6.2 Physiological Responses

Heavy metal stress induces many physiological and metabolic alterations in plants, from impaired water and nutrient uptake to disruption of photosynthetic and respiratory processes. These adverse effects collectively hinder plant growth and development, posing significant challenges to plant survival and productivity in metal-contaminated environments. Understanding the intricate mechanisms underlying plant responses to heavy metal stress is crucial for developing strategies to mitigate the detrimental impacts and enhance plant tolerance to metal toxicity.

Plants exposed to harmful concentrations of heavy metals undergo a cascade of physiological and metabolic changes to cope with the stress imposed by metal toxicity. One notable consequence is the decreased water uptake, which impedes the plant's ability to absorb heavy metals from the soil system, exacerbating their accumulation within plant tissues. This disturbance in water uptake also disrupts nutrient absorption, leading to an imbalance in essential nutrients critical for plant growth and development (Cseh 2002; Ghnaya et al. 2007).

Furthermore, heavy metals interfere with the translocation of nutrients within the plant, inhibiting their movement to vital organs and disrupting metabolic processes. This translocation inhibition exacerbates nutrient deficiencies and hampers the plant's ability to sustain growth and productivity. Additionally, heavy metal stress affects stomatal conductance, leading to alterations in water uptake and transpiration rates. Plants may modulate the stomatal aperture to regulate water loss and maintain cellular hydration under metal stress conditions.

Heavy metals also impair photosynthetic activity by disrupting chlorophyll synthesis, inhibiting electron transport, and damaging photosystem proteins. This disruption manifests in various ways, including generating reactive oxygen species (ROS) and inhibiting photosystem II, leading to diminished photosynthetic capacity and reduced carbon assimilation. To mitigate the adverse effects on photosynthesis, plants respond by adjusting their photosynthetic apparatus, including synthesizing photoprotective pigments like anthocyanins and carotenoids.

Additionally, heavy metals damage the plant's respiratory system and mitochondrial function, compromising energy metabolism and cellular respiration. This impairment further contributes to the overall decline in plant growth and development, as energy allocation towards essential physiological processes is hampered.

Understanding these complex responses may guide researchers to develop targeted approaches for alleviating the negative impacts of heavy metal stress on plants, ultimately enhancing their resilience in contaminated environments.

6.3 Biochemical Responses

Urban plants face significant challenges due to exposure to heavy metals from various sources. These heavy metals accumulate in urban soils and threaten plant health. In response to heavy metal stress, urban plants exhibit various biochemical responses aimed at detoxification, tolerance, and adaptation. Urban plants alleviate the contamination of heavy metals and enhance their tolerance to metal stress by employing these biochemical responses. However, the effectiveness of these responses can fluctuate depending on factors such as plant species, metal concentration, and duration of exposure, highlighting the importance of understanding the specific biochemical mechanisms underlying plant responses to heavy metal stress in urban environments.

Plants produce metal-binding compounds such as phytochelatins and metallothioneins to chelate and sequester heavy metals, reducing their toxicity. These chelators bind to heavy metals, forming stable complexes sequestered in vacuoles to prevent their interaction with essential cellular components (Hossain et al. 2012).

Heavy metals induce oxidative stress in plant cells by generating reactive oxygen species (ROS) such as superoxide radicals, hydrogen peroxide, and hydroxyl radicals. In response, plants activate antioxidant defense mechanisms, including enzymes such as superoxide dismutase, catalase, peroxidase, and glutathione reductase, as well as non-enzymatic antioxidants like glutathione, ascorbate, and tocopherol. These antioxidants scavenge ROS and mitigate oxidative damage to cellular structures.

Heavy metal stress can modulate the levels and activities of phytohormones such as abscisic acid (ABA), ethylene, and salicylic acid, which play crucial roles in plant growth, development, and stress responses. Phytohormones help regulate various biochemical pathways in metal detoxification, ion transport, and stress signaling.

Plants synthesize metallothioneins, small cysteine-rich proteins, in response to heavy metal stress. Metallothioneins bind to heavy metals and participate in metal detoxification and homeostasis by sequestering metals in cells and regulating their intracellular distribution.

Glutathione, a tripeptide (γ-glutamyl-cysteinyl-glycine), plays a crucial role in metal detoxification by forming metal complexes and facilitating their transport and excretion from cells (Cobbett 2002). Plants regulate glutathione metabolism to maintain cellular redox balance and alleviate heavy metal-induced oxidative stress.

Phytochelatins, small cysteine-rich peptides, are synthesized in response to heavy metal stress and play a vital role in metal detoxification. Phytochelatins bind to heavy metals, forming complexes sequestered in vacuoles, thereby reducing metal toxicity in the cytoplasm. Heavy metals inhibit the activity of various enzymes involved in essential metabolic pathways, including photosynthesis, respiration, and nitrogen metabolism. This enzyme inhibition disrupts cellular metabolism and adversely affects plant growth and productivity (Cobbett 2002).

6.4 Molecular Responses

Understanding the molecular responses of urban plants to heavy metal stress provides insights into their mechanisms of tolerance and adaptation in contaminated environments. Elucidating e molecular pathways may assist researchers in developing strategies to enhance plant resilience to heavy metal pollution and improve the sustainability of urban ecosystems.

Plants respond to heavy metal stress by altering the expression of numerous genes involved in metal detoxification, stress signaling, and defense responses. Transcription factors such as WRKY, NAC, and bZIP family members play pivotal roles in regulating the expression of genes encoding metal transporters, chelators, and detoxification enzymes.

Heavy metal stress triggers the activation of plant detoxification pathways, including synthesizing enzymes such as metallothioneins, glutathione S-transferases, and phytochelatin synthases. These enzymes facilitate the conjugation, chelation, and sequestration of heavy metals, aiding in their detoxification and compartmentalization.

Plants modulate the expression and activity of metal transporters and channels to regulate heavy metals' uptake, translocation, and sequestration. Transporters such as ATP-binding cassette (ABC) transporters, cation diffusion facilitators (CDFs), and heavy metal ATPases (HMAs) play crucial roles in mediating the movement of metals across cellular membranes.

Heavy metal stress activates signaling pathways such as mitogen-activated protein kinase (MAPK) cascades, calcium-dependent signaling, and reactive oxygen species (ROS)-mediated signaling (Jonak et al. 2004). These pathways coordinate the plant's responses to heavy metal stress, including activating defense genes and regulating physiological processes.

Heavy metal stress may induce epigenetic modifications in plants, including changes in DNA methylation patterns, histone modifications, and microRNA expression profiles. These epigenetic changes influence gene expression patterns and phenotypic responses to heavy metal stress, contributing to plant adaptation and tolerance.

It is essential to comprehend these various responses at various levels to create plans to increase plant tolerance to heavy metals or choose plants appropriate for phytoremediation. Furthermore, it advances our general knowledge of plant stress physiology. The tolerance of plant response to heavy metals is discussed below (Table 1).

7 Adaptation Mechanisms in Plants to Heavy Metal Stress

7.1 Proline Synthesis as a Defense Mechanism:

Research highlights the crucial role of proline, an organic compound, in mitigating heavy metal toxicity in plants. Proline's involvement in stabilizing proteins, scavenging free radicals, and maintaining cellular osmotic balance contributes to plant resilience against heavy metal stress. It reduces metal toxicity when applied exogenously or through genetic manipulation to enhance its biosynthetic pathways (Singh et al. 2015).

7.2 Morphological Adaptations

Studies on *Phaseolus mungo* and *Lens culinaris* have revealed morphological adaptations like the development of trichomes and increased stomata on leaf surfaces as responses to lead (Pb) pollution. These adaptations help maintain efficient photosynthesis and serve as a protective barrier against Pb, indicating a vital survival strategy (Emamverdian et al. 2015). *Pinus brutia, Robinia pseudoacaccia, Quercus trojana, Fraxinus ornus,* and *P. brutia* have the greatest copper, zinc, and cobalt concentrations. *Pinus brutia*'s needles last two to five years, like those of other coniferous trees. As a result, they are exposed to contaminated air for more extended periods than deciduous tree leaves. *Pinus brutia*'s depressed stomata in longitudinal bands

Table 1 Tolerance of plant growth reaction to HMs/metalloids in soil

Plant	Heavy metal	Plant response and related parameter	References
Euphorbia pulcherrima	Arsenic, lead, chromium, copper, zinc, mercury	The plant showed significant accumulations of Zn, Pb, and Cu in roots and removal and transportation of Hg in leaves; it was also found to tolerate As and Cr	Xiao et al. (2021)
Mirabilis jalapa	Cadmium	The plant successfully removed As, Pb, and Cd from contaminated soil without experiencing any harmful effects from phytotoxins	Li et al. (2022)
Eucalyptus nitens	Arsenic	For the plant treatment with 100 and 200 μg/mL. Roots may concentrate to 69.8 and 133 μg/g levels and leaves between 9.48 μg/g (200 As) and 15.9 μg/g (100 As) without morphological damage or toxic symptoms. Two different interferences were found in the Asimpact on the accumulation and translocation of metals such as Ca, Fe, K, and Zn	Ramalho et al. (2023)
Arabidopsis thaliana	Aluminium	Lipid peroxidation, growth inhibition, and an increase in ROS	Kochian et al. (2015)
Zinnia elegans	Chromium	The plant reached a 50 mg/kg Cr stress as it grew	Panda et al. (2020)
Hydrangea	Lead	The plant showed no signs of heavy metal toxicity under Pb treatments (0–1600 mg/ kg)	Ma et al. (2022)
Lavandula angustifolia	Nickel	The plant developed healthily up to 40 mg/kg. Ni of the ground	Barouchas et al. (2019)

and rough surface are among the explanations for the high heavy metal content in its leaves. *Quercus trojana* leaves had the greatest manganese and nickel concentrations. Other tree species in urban environments, such as *Cupressus arizonica*, *Albizia julibrissin*, *Platanus orientalis*, *Celtis australis*, and *Ligustrum japonicum*, have the potential to capture heavy metals.

7.3 *Anatomical Adaptations*

Plants exhibit anatomical changes, such as modifications in the structure of roots, xylem, and leaves, to combat environmental stress. These include reduced cell elongation and alterations in cell differentiation, which are crucial for adapting to challenging conditions like heavy metal contamination, drought, and extreme temperatures (Babst-Kostecka et al. 2016; Hoque et al. 2021). For example, heavy metal detoxification is rapidly reported in various plants, including *Oryza sativa*, *Zea mays*, *Arabidopsis*, and *Populus species*.

7.4 Physiological Adaptations

Certain plant species in ecosystems affected by industrial pollution demonstrate remarkable tolerance, suggesting their potential for phytoremediation. Research within the Marble Waste Polluted Ecosystem (MWPE) in northern Pakistan identified species like *Adiantum capillusveneris* with significant phytoremediation abilities, showcasing their importance in environmental sustainability (Wani et al. 2023). *Solanum nigrum*, *Brassica juncea*, *Alnus firma*, *Brassica napus*, etc., are a few examples under heavy metals stress for Cd, Pb, Zn.

7.5 Biochemical Adaptations

Heavy metal contamination leads to oxidative stress in plants, characterized by producing reactive oxygen species (ROS). Plants adapt through biochemical strategies, including the increased activity of ROS-detoxifying enzymes, to mitigate heavy metal toxicity and maintain cellular homeostasis (Grigore and Vicente 2023). *Ailanthus altissima*, *Phragmites karka*, *Solanum lycopersicum* L. are some examples of antioxidant defense in Plants Under Abiotic Stress.

7.6 Molecular Adaptations

At the molecular level, plants respond to heavy metal stress by modulating the expression of genes involved in metal transport, sequestration, and detoxification. Advancements in genomics have unveiled numerous metal-responsive genes, highlighting the intricate molecular mechanisms plants employ to cope with heavy metal toxicity. Plants such as *Raphanus sativus*, *Sesbania rostrata*, *Spartina alterniflora*, *Vigna radiata*, *Pinus sylvestris*, *Medicago sativa*, *Z. mays*, *Arabidopsis thaliana*, *Triticum aestivum*, and *Lupinus luteus* are some of the instances under heavy metal stress.

8 Factors Influencing Heavy Metal Uptake and Detoxification

Various biological and environmental factors influence heavy metal uptake and plant detoxification. These factors include the bioavailability of heavy metals in soil, plant characteristics, soil properties like pH, organic matter, clay content, redox conditions, and the biomass of plants (Kushwaha et al. 2015). Additionally, biological and environmental factors affect foliar uptake of heavy metals. Strategies to reduce heavy metal accumulation in crops involve selecting low-metal cultivars, physiological blocking water management, soil amendment, and phytoremediation techniques. These factors play a crucial role in plants' heavy metal detoxification and tolerance mechanisms.

8.1 Biological Factors

A plant's ability to accumulate and detoxify heavy metals depends on its pant characteristics, such as species, genotypes, and growth stage. Some plants have evolved specific mechanisms to tolerate and detoxify heavy metals, such as the production of phytochelatins (PCs) and metallothioneins (MTs) (Yan et al. 2020). The root characteristics play a crucial role in heavy metal uptake and transport. The root represents the first barrier to heavy metal accumulation and is responsible for the selective absorption of ions in soil solution (Yaashikaa et al. 2022). Furthermore, plant metabolic processes, such as ion uptake and transport, influence heavy metal uptake and detoxification (Ghori et al. 2019). Besides, soil microorganisms, including bacteria and fungi, can form symbiotic relationships with plant roots and influence heavy metal uptake and detoxification. Mycorrhizal associations, for example, may enhance nutrient uptake while potentially reducing the uptake of certain heavy metals (Bhantana et al. 2021). The genetic makeup of plants influences their ability to tolerate and detoxify heavy metals. Some plant species and varieties possess specific genetic traits that enhance their resistance to metal stress.

8.2 Environmental and Climatic Factors

Soil factors like pH, organic matter, clay content, and redox conditions can significantly influence plants' heavy metal availability and uptake. Soil processes, such as mineral weathering and microbial activity, also affect the solubility and form of available metal species in soil (Abdu et al. 2017). Climatic factors, such as temperature and precipitation, can impact heavy metal mobility in soil. Changes in climate patterns may affect the bioavailability and transport of heavy metals. Environmental factors, such as temperature, humidity, and atmospheric pollutants, influence foliar uptake of heavy metals. Water availability in the soil can affect the solubility of heavy metals and influence plants' uptake. It has been reported that implementing pollution control measures and sustainable land management practices significantly contributed to reducing the introduction of heavy metals into the environment.

Researchers may develop holistic approaches to manage and mitigate the impact of heavy metal contamination in plants and soils by considering the abovementioned factors. Implementing sustainable practices and developing resilient plant varieties is crucial for ensuring food security and environmental health. Various strategies, such as selecting low-metal cultivars, physiological blocking, water management, soil amendment, and phytoremediation, are employed to mitigate the adverse effects of heavy metal accumulation in plants. These strategies aim to reduce heavy metal uptake and enhance plant detoxification mechanisms.

9 Techniques for Studying Heavy Metal Uptake and Detoxification

Understanding the mechanisms of heavy metal uptake and detoxification in urban plants is crucial for assessing their role in mitigating metal pollution in urban environments. Various techniques have been developed to study these processes, ranging from traditional methods such as plant tissue analysis to advanced molecular techniques like transcriptomics and metabolomics. Different techniques and procedures commonly used to study heavy metal uptake and detoxification in urban plants are briefly discussed below.

9.1 Plant Sampling and Tissue Analysis

Plant sampling and tissue analysis are fundamental for studying urban plants' heavy metal uptake and accumulation. Soil and plant tissue samples are collected from contaminated environments, and metal concentrations are measured using analytical methods such as atomic absorption spectroscopy (AAS), inductively coupled plasma mass spectrometry (ICP-MS), or X-ray fluorescence (XRF) spectroscopy. These methods provide valuable information on metal uptake, translocation, and accumulation in different plant organs.

9.2 Microscopy and Imaging Techniques

Microscopy and imaging techniques offer insights into the spatial distribution of heavy metals within plant tissues. Scanning electron microscopy (SEM) and transmission electron microscopy (TEM) allow visualization of metal nanoparticles and their high-resolution interaction with plant cells. Confocal laser scanning microscopy (CLSM) and fluorescence microscopy enable the visualization of metal-binding ligands and cellular compartments involved in metal detoxification.

9.3 Physiological and Biochemical Assays

Physiological and biochemical assays are used to assess the impact of heavy metal stress on plant metabolism and defense mechanisms. These assays include chlorophyll content measurements, photosynthetic activity, antioxidant enzyme activity, and stress-responsive gene expression. Techniques such as enzyme-linked immunosorbent assay (ELISA) and Western blotting quantify metal-binding proteins and signaling molecules involved in metal detoxification pathways.

9.4 Isotope Tracing and Metabolomics

Isotope tracing techniques, such as stable isotope labeling and radiotracer experiments, track heavy metals' uptake, translocation, and fate within plants. These techniques provide quantitative information on metal fluxes and turnover rates in different plant compartments. Metabolomics approaches enable the identification and quantifying of metabolites in metal detoxification pathways, including gas chromatography-mass spectrometry (GC–MS) and liquid chromatography-mass spectrometry (LC–MS).

9.5 Transcriptomics and Proteomics

Transcriptomics and proteomics techniques provide insights into the molecular mechanisms underlying heavy metal uptake and detoxification in urban plants. RNA sequencing (RNA-seq) and microarray analyse profile changes in gene expression in response to metal stress, identifying key genes and pathways involved in metal uptake, transport, and detoxification. Proteomics approaches, such as two-dimensional gel electrophoresis (2D-PAGE) and mass spectrometry-based proteomics, allow the identification and quantification of metal-binding proteins and regulatory factors.

9.6 *Advanced Imaging and Spectroscopy*

Advanced imaging and spectroscopy techniques offer non-destructive and high-resolution characterization of heavy metal uptake and distribution in urban plants. X-ray microtomography (XMT) and synchrotron X-ray fluorescence (SXRF) spectroscopy enable three-dimensional imaging and elemental mapping of metal distribution in plant tissues with sub-micron resolution. Fourier-transform infrared (FTIR) spectroscopy and Raman spectroscopy provide information on metal speciation and chemical interactions within plant cells.

Despite the advances in techniques for studying heavy metal uptake and detoxification in urban plants, several challenges remain. These include the need for integration across different disciplines, standardization of experimental protocols, and validation of results under field conditions. Hence, future efforts should develop multi-omics approaches that integrate transcriptomics, proteomics, metabolomics, and imaging techniques to understand plant-metal interactions in urban environments comprehensively.

10 Implications for Urban Ecosystem Management

Urban ecosystems are shaped by a dynamic interplay of physical and socio-economic factors, presenting unique challenges and opportunities for management. Integrating ecological considerations into urban planning and design processes is paramount to fostering sustainable urban ecosystems (Kumar et al. 2019a). This entails incorporating green spaces, biodiversity corridors, and natural habitats into urban landscapes to support ecological functions and enhance urban biodiversity (Kumar et al. 2021e; Semeraro et al. 2021).

However, rapid urbanization often transforms natural ecosystems into human-dominated systems, impacting soil properties, air quality, and water quality. The proliferation of heavy metals in urban environments due to industrialization and urbanization poses significant environmental concerns globally.

In response to these challenges, utilizing urban plants for pollution mitigation has emerged as a promising strategy for environmental management. Phytoremediation involves using plants to extract, immobilize, or degrade contaminants from soil and water and offers a practical and affordable approach to managing heavy metal contamination in urban settings (DalCorso et al. 2019).

Incorporating phytoremediation into urban green infrastructure projects allows for synergistic benefits, where plants beautify the environment and actively contribute to pollution mitigation and ecosystem restoration. Urban green spaces, such as parks, green roofs, and roadside plantings, provide suitable sites for phytoremediation activities.

Moreover, the redevelopment of degraded and abandoned land presents an opportunity for urban planning and revitalization. Strategically planting urban plants capable of hyperaccumulating heavy metals in brownfield sites and industrial areas can remediate soil pollution, making these spaces suitable for new uses such as parks, green corridors, or residential developments (Nissim and Labrecque 2021).

Through their ability to absorb ions from the soil and create rhizosphere ecosystems, plants play a crucial role in recovering contaminated soil and stabilizing soil fertility. The potential of urban plants and integrating phytoremediation into urban planning and management practices may address environmental challenges, promote sustainability, and create healthier and more resilient urban ecosystems.

11 Future Directions and Challenges

In recent years, research efforts have focused on understanding how urban plants interact with heavy metals regarding uptake and detoxification mechanisms. This aims to provide insights into how plants are utilized for phytoremediation and mitigate heavy metal pollution's impacts in urban settings. The exploration of heavy metal uptake and detoxification mechanisms in urban plants represents a critical area of research with implications for environmental sustainability and human health. Looking ahead, several future directions and challenges are poised to shape this field as.

11.1 Interdisciplinary Collaboration

Integrating expertise from diverse fields such as plant biology, microbiology, soil science, and environmental chemistry will be essential for addressing the multifaceted nature of heavy metal pollution in urban environments. Collaborative efforts may provide holistic insights into the complex interactions between plants, soil, microbes, and pollutants.

11.2 Field Validation and Implementation

While laboratory studies offer valuable insights, field-based research is crucial for validating findings and assessing the effectiveness of phytoremediation strategies in real-world urban settings. Future research should prioritize long-term field trials to evaluate various remediation approaches' scalability, efficiency, and ecological impacts.

11.3 Climate Change Considerations

Climate change is anticipated to influence heavy metal dynamics in urban ecosystems through precipitation patterns, temperature regimes, and atmospheric deposition alterations. Understanding how climate change impacts the bioavailability and toxicity of heavy metals and their interactions with urban plants is essential for developing adaptive strategies to mitigate pollution risks.

11.4 Genetic Engineering and Biotechnological Solutions

Advancements in genetic engineering techniques offer opportunities to develop urban plants with enhanced metal tolerance and remediation capabilities. Future research should focus on identifying key genes and pathways involved in metal stress responses and harnessing this knowledge to engineer more effective plants detoxifying heavy metals from contaminated soils.

11.5 Microbiome-Mediated Remediation

Investigating the role of plant-associated microbes in facilitating heavy metal detoxification presents promising avenues for biotechnological interventions. Future studies should explore microbial community dynamics in urban plant-soil systems and identify microorganisms capable of enhancing metal immobilization, chelation, or volatilization to improve phytoremediation efficiency.

11.6 Technological Advancements

Leveraging omics-based approaches, remote sensing, and advanced imaging techniques provide unprecedented insights into heavy metal-plant interactions at molecular, cellular, and ecosystem scales. Future research should embrace these technological advancements to unravel the complexities of urban metal pollution and inform remediation strategies.

11.7 Policy and Governance

Effective management of heavy metal pollution in urban environments requires robust policy frameworks and governance mechanisms. Future research should engage policymakers, urban planners, and stakeholders to develop evidence-based policies that promote sustainable urban development and prioritize protecting human health and the environment.

12 Conclusion

Exploring urban plant mechanisms for heavy metal absorption and detoxification presents a promising avenue for mitigating pollution and nurturing sustainable urban ecosystems. This knowledge offers valuable insights for researchers and policymakers to devise effective strategies for leveraging urban plants in heavy metal pollution mitigation. Delving into plant processes and their interactions with heavy metals guides scientists to unlock the potential of phytoremediation, thus alleviating the impacts of heavy metal pollution in urban settings. This chapter highlights the potential of plants in phytoremediation efforts and their role in alleviating heavy metal pollution in urban environments. Moreover, insights gleaned from heavy metal-tolerant plants can inform strategies to minimize heavy metal absorption and accumulation in food plants, thereby reducing their adverse effects on human health. However, further research is essential to fully understand the intricate mechanisms involved and optimize urban plants' utilization for heavy metal remediation in urban environments.

References

Abdu N, Abdullahi AA, Abdulkadir A (2017) Heavy metals and soil microbes. Environ Chem Lett 15(1):65–84

Alam S, Bhardwaj LK, Mallick R, Rai S (2023) Estimation of heavy metals and fluoride ion in vegetables grown nearby the stretch of River Yamuna, Delhi (NCR), India. Indian J Environ Protect 43(1):64–73

Ali H, Khan E, Sajad MA (2013) Phytoremediation of heavy metals—concepts and applications. Chemosphere 91(7):869–881

Ali S, Abbas Z, Rizwan M, Zaheer IE, Yavaş İ, Ünay A et al (2020) Application of floating aquatic plants in phytoremediation of heavy metals polluted water: a review. Sustainability 12(5):1927

Arora V, Bithel N, Singh R (2023) A study on heavy metal sources and pollution: challenge to biological and ecosystem. Bull Pure Appl Sci Bot 1:11541

Awa SH, Hadibarata T (2020) Removal of heavy metals in contaminated soil by phytoremediation mechanism: a review. Water Air Soil Pollut 231(2):47

Babst-Kostecka AA, Waldmann P, Frérot H, Vollenweider P (2016) Plant adaptation to metal polluted environments—physiological, morphological, and evolutionary insights from *Biscutella laevigata*. Environ Exper Bot 127:1–13

Bakshe P, Jugade R (2023) Phytostabilization and rhizofiltration of toxic heavy metals by heavy metal accumulator plants for sustainable management of contaminated industrial sites: a comprehensive review. J Hazard Mater Adv 32:100293

Barouchas PE, Akoumianaki-Ioannidou A, Liopa-Tsakalidi A, Moustakas NK (2019) Effects of vanadium and nickel on morphological characteristics and on vanadium and nickel uptake by shoots of mojito (Mentha× villosa) and lavender (*Lavandula anqustifolia*). Notulae Botanicae Horti Agrobotanici Cluj-Napoca 47(2):487–492

Bhantana P, Rana MS, Sun XC, Moussa MG, Saleem MH, Syaifudin M et al (2021) Arbuscular mycorrhizal fungi and its major role in plant growth, zinc nutrition, phosphorous regulation and phytoremediation. Symbiosis 84:19–37

Bhardwaj LK, Kumar D, Kumar A (2023a) Phytoremediation potential of ocimum sanctum: a sustainable approach for remediation of heavy metals. In: Phytoremediation potential of medicinal and aromatic plants. CRC Press, pp 46–57

Bhardwaj LK, Sharma S, Jindal T (2023b) Estimation of physico-chemical and heavy metals in the lakes of Grovnes and Broknes Peninsula, Larsemann Hill, East Antarctica. Chem Afr 6(5):2677–2694

Carrillo JT, Borthakur D (2021) Methods for metal chelation in plant homeostasis. Plant Physiol Biochem 163:95–107

Cimboláková I, Uher I, Laktičová KV, Vargová M, Kimáková T, Papajová I (2019) Heavy metals and the environment. Environ Fact Affect Hum Health 1:29

Cobbett C (2002) Heavy metal detoxification mechanisms in plants. In: Proceedings of the 9th new phytologist symposium. Heavy metals and plants. Philadelphia

Cseh E (2002) Metal permeability, transport and efflux in plants. Physiology and biochemistry of metal toxicity and tolerance in plants. Springer, Dordrecht, pp 1–36

DalCorso G, Fasani E, Manara A, Visioli G, Furini A (2019) Heavy metal pollutions: state of the art and innovation in phytoremediation. Int J Mol Sci 20:3412. https://doi.org/10.3390/ijms20143412

Dalvi AA, Bhalerao SA (2013) Response of plants towards heavy metal toxicity: an overview of avoidance, tolerance and uptake mechanism. Ann Plant Sci 2(9):362–368

Dhyani K, Kuniyal HB, Singh H, Sobha S (2021) Growth and physiological potential of *Terminalia arjuna* under elevated CO_2 levels in Open top chamber condition. J Appl Nat Sci 13(3):1121–1126

Emamverdian A, Ding Y, Mokhberdoran F, Xie Y (2015) Heavy metal stress and some mechanisms of plant defense response. Sci World J

Gavhane SK, Sapkale JB, Susware NK, Sapkale SJ (2021) Impact of heavy metals in riverine and estuarine environment: a review. Res J Chem Environ 25(5):226–233

Ghnaya T, Slama I, Messedi D, Grignon C, Ghorbel MH, Abdelly C (2007) Effects of Cd^{2+} on K^+, Ca^{2+} and N uptake in two halophytes *Sesuvium portulacastrum* and *Mesembryanthemum crystallinum*: consequences on growth. Chemosphere 67(1):72–79

Ghori NH, Ghori T, Hayat MQ, Imadi SR, Gul A, Altay V, Ozturk M (2019) Heavy metal stress and responses in plants. Int J Environ Sci Technol 16:1807–1828

Ghuge SA, Nikalje GC, Kadam US, Suprasanna P, Hong JC (2023) Comprehensive mechanisms of heavy metal toxicity in plants, detoxification, and remediation. J Hazard Mater 450:131039

Grigore MN, Vicente O (2023) Wild halophytes: tools for understanding salt tolerance mechanisms of plants and for adapting agriculture to climate change. Plants 12(2):221

Gupta V (2020) Vehicle-generated heavy metal pollution in an urban environment and its distribution into various environmental components. Environ Conc Sustain Develop 1:113–127

Gupta SK, Singh H (2017) Observations on sporadic flowering in *Ficus benjamina* L. and *Peltophorum africanum* L. in new forest. Indian for 143:290–291

Gupta SK, Ram J, Singh H (2018a) Comparative study of transpiration in cooling effect of tree species in the atmosphere. J Geosci Environ Protect 6:151–166. https://doi.org/10.4236/gep.2018.68011

Gupta SK, Ram J, Singh H, Kumar P (2018b) Comparative study of cooling under shades of some forestry tree species with respect to ambient temperatures. J Geosci Environ Protect 6:51–66. https://doi.org/10.4236/gep.2018.611004

Gupta SK, Ram J, Singh H (2019a) Relationship between leaf area index and atmospheric cooling of tree species. Indian J for 42(3):203–208

Gupta SK, Singh H, Ram J (2019b) Need for a novel paradigm in forestry research: perspectives of comparative atmospheric cooling by different forestry effect tree species. Indian for 145(6):509–515

Hoque MN, Tahjib-Ul-Arif M, Hannan A, Sultana N, Akhter S, Hasanuzzaman M et al (2021) Melatonin modulates plant tolerance to heavy metal stress: morphological responses to molecular mechanisms. Int J Mol Sci 22(21):11445

Hossain MA, Piyatida P, da Silva JAT, Fujita M (2012) Molecular mechanism of heavy metal toxicity and tolerance in plants: central role of glutathione in detoxification of reactive oxygen species and methylglyoxal and in heavy metal chelation. J Bot

Järup L (2003) Hazards of heavy metal contamination. Br Med Bull 68(1): 167–182

Jangra A, Kumar K, Maikhuri S, Bhandari MS, Pandey S, Singh H, Barthwal S (2024) Unveiling stress-adapted endophytic bacteria: characterizing plant growth-promoting traits and assessing cross-inoculation effects on *Populus deltoides* under abiotic stress. Plant Physiol Biochem 210:108610. https://doi.org/10.1016/j.plaphy.2024.108610

Jonak C, Nakagami H, Hirt H (2004) Heavy metal stress: activation of distinct mitogen-activated protein kinase pathways by copper and cadmium. Plant Physiol 136(2):3276–3283

Kanta C, Kumar A, Chauhan A, Singh H, Sharma IP (2024) The interplay between plant functional traits and climate change. In: Kumar N, Singh H (eds) Plant functional traits for improving productivity. Springer, Singapore. https://doi.org/10.1007/978-981-97-1510-7_3

Khan AHA, Kiyani A, Mirza CR, Butt TA, Barros R, Ali B et al (2021a) Ornamental plants for the phytoremediation of heavy metals: present knowledge and future perspectives. Environ Res 195:110780

Khan MIR, Chopra P, Chhillar H, Ahanger MA, Hussain SJ, Maheshwari C (2021b) Regulatory hubs and strategies for improving heavy metal tolerance in plants: chemical messengers, omics and genetic engineering. Plant Physiol Biochem 164:260–278

Kochian LV, Piñeros MA, Liu J, Magalhaes JV (2015) Plant adaptation to acid soils: the molecular basis for crop aluminum resistance. Ann Rev Plant Biol 66:571–598

Kumar N, Suyal DC, Sharma IP, Verma A, Singh H (2017) Elucidating stress proteins in rice (*O. sativa* L.) genotype under elevated temperature: a proteomic approach to understand heat stress response. 3 Biotech 7:205. https://doi.org/10.1007/s13205-017-0856-9

Kumar M, Singh MP, Singh H, Dhakate PM, Ravindranath NH (2019a) Forest working plan for the sustainable management of forest and biodiversity in India. J Sustain for 39:1–22. https://doi.org/10.1080/10549811.2019.1632212

Kumar N, Jeena N, Singh H (2019b) Elevated temperature modulates rice pollen structure: a study from the foothill of the Himalayan agro-ecosystem in India. 3 Biotech 9(5):175. https://doi.org/10.1007/s13205-019-1700-1

Kumar N, Jeena N, Kumar N, Gangola S, Singh H (2019c) Phytoremediation facilitating enzymes: an enzymatic approach for enhancing remediation process. In: Bhatt P (ed) Smart bioremediation technologies. Academic Press, Cambridge, pp 289–306

Kumar A, Kumar P, Singh H, Kumar N (2021a) Modulation of plant functional traits under essential plant nutrients during seasonal regime in natural forests of Garhwal Himalayas. Plant Soil 465:197–212. https://doi.org/10.1007/s11104-021-05003-x

Kumar A, Kumar P, Singh H, Bisht S, Kumar N (2021b) Relationship of physiological plant functional traits with soil carbon stock in the temperate forest of Garhwal Himalaya. Curr Sci 120(8):1368–1373

Kumar A, Kumar P, Singh H, Kumar N (2021c) Adaptation and mitigation potential of roadside trees with bio-extraction of heavy metals under vehicular emissions and their impact on physiological traits during seasonal regimes. Urban for Urban Green 58:126900. https://doi.org/10.1016/j.ufug.2020.126900

Kumar A, Kumar P, Singh H, Kumar N (2021d) Impact of plant functional traits on infiltration rate under different forests of Kempty watershed in Garhwal Himalaya. Indian J Soil Conser 49(1):38–44

Kumar M, Phukon SN, Paygude AC, Tyagi K, Singh H (2021e) Mapping phenological functional types (PhFT) in the Indian Eastern Himalayas using machine learning algorithm in Google Earth engine. Comput Geosci 158:104982. https://doi.org/10.1016/j.cageo.2021.104982

Kumar A, Singh H, Kumari G, Bisht S, Malik A, Kumar A, Singh M, Raturi S, Barthwal S, Thakur A, Kaushal R (2022) Adaptive resilience of roadside trees to vehicular emissions via leaf enzymatic, physiological, and anatomical trait modulations. Environ Pollut 313:120191. https://doi.org/10.1016/j.envpol.2022.120191

Kumar N, Singh H, Giri K, Kumar A, Joshi A, Yadav A, Singh R, Bisht S, Kumari R, Jeena N, Khairakpam R, Mishra G (2024) Physiological and molecular insights into the allelopathic effects on agroecosystems under changing environmental conditions. Physiol Mol Biol Plants 30:417–433. https://doi.org/10.1007/s12298-024-01440-x

Kumar N, Singh H (2024) Plant functional traits for improving productivity. Springer, Singapore. https://doi.org/10.1007/978-981-97-1510-7

Kumar N, Singh H, Sharma SK (2020) Antioxidants: responses and importance in plant defense system. In: Roychowdhury R, Choudhury S, Hasanuzzaman M, Srivastava S (eds) Sustainable agriculture in the era of climate change. Springer, Cham, pp 251–264. https://doi.org/10.1007/978-3-030-45669-6_11

Kushwaha A, Rani R, Kumar S, Gautam A (2015) Heavy metal detoxification and tolerance mechanisms in plants: implications for phytoremediation. Environ Rev 24(1):39–51

Li X, Chang Z, Lian X, Meng G, Ma J, Guo R, Wang Y (2022) Phytoremediation of cadmium contaminated alkaline soil using the ornamental hyperaccumulator *Mirabilis jalapa* L. enhanced by double harvesting: a field study. Environ Sci Pollut Res 29(22):33506–33513

Ma W, Zhao B, Lv X, Feng X (2022) Lead tolerance and accumulation characteristics of three Hydrangea cultivars representing potential lead-contaminated phytoremediation plants. Horticult Environ Biotechnol 63(1):23–38

Malik A, Kumar K, Kumari G, Kumar N, Singh H (2024) Application of functional traits in modelling productivity and resilience under climate change. In: Kumar N, Singh H (eds) Plant functional traits for improving productivity. Springer, Singapore. https://doi.org/10.1007/978-981-97-1510-7_5

Mansoor S, Ali A, Kour N, Bornhorst J, AlHarbi K, Rinklebe J et al (2023) Heavy metal-induced oxidative stress mitigation and ROS scavenging in plants. Plants 12(16):3003

Meagher RB (2000) Phytoremediation of toxic elemental and organic pollutants. Curr Opin Plant Biol 3(2):153–162

Meena D, Singh H, Chaudhari SK (2011) Elucidating strontium response on growth dynamics and biochemical change in *P. mungo* L. Int J Agricult Environ Biotechnol 4:107–113

Memon AR, Aktoprakligil D, Özdemir A, Vertii A (2001) Heavy metal accumulation and detoxification mechanisms in plants. Turk J Bot 25(3):111–121

Nissim WG, Labrecque M (2021) Reclamation of urban brownfields through phytoremediation: implications for building sustainable and resilient towns. Urban for Urban Green 65:127364

Nwogwu NA, Ajala OA, Ajibade FO, Ajibade TF, Adelodun B, Lasisi KH et al (2022) Phytoremediation mechanisms of heavy metal removal: a step towards a green and sustainable environment. In: Innovative bio-based technologies for environmental remediation. CRC Press, pp 207–236

Odum HT (2016) Heavy metals in the environment: using wetlands for their removal. CRC Press

Panda A, Patra DK, Acharya S, Pradhan C, Patra HK (2020) Assessment of the phytoremediation potential of *Zinnia elegans* L. plant species for hexavalent chromium through pot experiment. Environ Technol Innov 20:101042

Peer WA, Baxter IR, Richards EL, Freeman JL, Murphy AS (2006) Phytoremediation and hyperaccumulator plants. In: Molecular biology of metal homeostasis and detoxification: from microbes to man, pp 299–340

Prakash V, Hunney K, Singh H (2022) Effect of elevated carbon dioxide on growth and development of *Santalum album* L. seedlings inoculated with plant growth promoting microorganisms in Open Top Chambers. Int J Environ Agricult Biotechnol 7(5):149–159

Ramalho JC, Pelica J, Lidon FC, Silva MM, Simões MM, Guerra M, Reboredo FH (2023) Soil arsenic toxicity impact on the growth and C-assimilation of *Eucalyptus nitens*. Sustainability 15(8):6665

Semeraro T, Scarano A, Buccolieri R, Santino A, Aarrevaara E (2021) Planning of urban green spaces: an ecological perspective on human benefits. Land 10(2):105

Sharma R, Singh H (2021) Alteration in biochemical constituents and nutrients partitioning of *Asparagus racemosus* in response to elevated atmospheric CO_2 concentration. Environ Sci Pollut Res 29:6812–6821. https://doi.org/10.1007/s11356-021-16050-3

Sharma R, Singh H, Kaushik M, Nautiyal R, Singh O (2018) Adaptive physiological response, carbon partitioning, and biomass production of *Withania somnifera* (L.) Dunal grown under elevated CO_2 regimes. 3 Biotech 8:267. https://doi.org/10.1007/s13205-018-1292-1

Singh H (2021) An integrated approach considering physiological and biophysical based indicators for assessing tolerance of roadside plantations of *Alstonia scholaris* towards urban roadside air pollution: an assessment of adaptation of plantations for mitigating urban roadside air pollution. Trees 21:2179. https://doi.org/10.1007/s00468-021-02179-8

Singh H (2024) Exploring adaptive modulation in plant functional traits and its impact on the productivity of *Acacia auriculiformis* under CO_2-enriched environment. Ind Crops Prod 210:118186. https://doi.org/10.1016/j.indcrop.2024.118186

Singh H, Verma A (2013) Characterization and screening of high nitrogen efficient rice genotype to curtail environmental pollution. Int J Agricult Environ Biotechnol 6(1):23–30

Singh M, Kumar J, Singh S, Singh VP, Prasad SM, Singh MPVVB (2015) Adaptation strategies of plants against heavy metal toxicity: a short review. Biochem Pharmacol 4(161):2167–2501

Singh H, Savita Sharma R, Sinha S, Kumar M, Kumar P, Verma A, Sharma SK (2017a) Physiological functioning of *Lagerstroemia speciosa* L. under heavy roadside traffic: an approach to screen potential species for abatement of urban air pollution. 3 Biotech 7(61):1–10. https://doi.org/10.1007/s13205-017-0690-0

Singh H, Verma A, Kumar M, Sharma R, Gupta R, Kaur M, Negi M, Sharma SK (2017b) Phytoremediation: a green technology to clean up the sites with low and moderate level of heavy metals. Aust Biochem 2:1012

Singh H, Sharma R, Savita SMP, Kumar M, Verma A, Ansari MW, Sharma SK (2018) Adaptive physiological response of *Parthenium hysterophorus* to elevated atmospheric CO_2 concentration. Indian for 144:1–14

Singh H, Yadav M, Kumar N, Kumar A, Kumar M (2020a) Assessing adaptation and mitigation potential of roadside trees under the influence of vehicular emissions: a case study of *Grevillea robusta* and *Mangifera indica* planted in an urban city of India. PLoS ONE 15(1):e0227380. https://doi.org/10.1371/journal.pone.0227380

Singh S, Yadav V, Arif N, Singh VP, Dubey NK, Ramawat N et al (2020b) Heavy metal stress and plant life: uptake mechanisms, toxicity, and alleviation. In: Plant life under changing environment. Academic Press, pp 271–287

Singh S, Singh H, Sharma SK, Nautiyal R (2021) Seasonal variation in biochemical responses of bamboo clones in the sub-tropical climate of Indian Himalayan foothills. Heliyon 7(4):e06859. https://doi.org/10.1016/j.heliyon.2021.e06859

Singh H, Bandyopadhyaya S, Kumar A, Singh M, Malik A, Kumari G, Raturi A, Yadav SK, Kumar M, Barthwal S, Thakur A, Rawat PS (2022) Understanding the physiological and biophysical response of urban roadside plantations for assessing adaptation and mitigation mechanisms toward vehicular emissions. Urban Clim 44:101183. https://doi.org/10.1016/j.uclim.2022.101183

Singh M, Singh H, Kumar A, Kumar M, Barthwal S, Thakur A (2024a) Soil nitrogen availability determines the CO_2 fertilization effect on tree species (*Neolamarckia cadamba*): growth and physiological evidence. Environ Sustain 7:53–60. https://doi.org/10.1007/s42398-023-00300-w

Singh H, Kumar N, Kumar A (2024b) Enhancing resource use efficiency in crops through plant functional traits. In: Kumar N, Singh H (eds) Plant functional traits for improving productivity. Springer, Singapore. https://doi.org/10.1007/978-981-97-1510-7_6

Singh H, Kumar M (2022) Climate change and its impact on Indian Himalayan Forests: current status and research needs. In: Rani S, Kumar R (eds) Climate change
Skuza L, Szućko-Kociuba I, Filip E, Bożek I (2022) Natural molecular mechanisms of plant hyperaccumulation and hypertolerance towards heavy metals. Int J Mol Sci 23(16):9335
Thakur S, Singh L, Wahid ZA, Siddiqui MF, Atnaw SM, Din MFM (2016) Plant-driven removal of heavy metals from soil: uptake, translocation, tolerance mechanism, challenges, and future perspectives. Environ Monit Assess 188:1–11
Tong YP, Kneer R, Zhu YG (2004) Vacuolar compartmentalization: a second-generation approach to engineering plants for phytoremediation. Trends Plant Science 9(1):7–9
Wei B, Yang L (2010) A review of heavy metal contaminations in urban soils, urban road dusts and agricultural soils from China. Microchem J 94(2):99–107
Wani ZA, Ahmad Z, Asgher M, Bhat JA, Sharma M, Kumar A, Sharma V, Kumar A, Pant S, Lukatkin AS, Anjum NA (2023) Phytoremediation of potentially toxic elements: role, status and concerns. Plants 12(3): 429
Xiao F, Gu Z, Sarkissian A, Ji Y, Ruonan Y, Yang L, Zeng Q, Huang P, Chen H (2021) Phytoremediation of potentially toxic elements in a polluted industrial soil using Poinsettia. Physiol Mol Biol Plants 27:675–686
Yaashikaa PR, Kumar PS, Jeevanantham S, Saravanan R (2022) A review on bioremediation approach for heavy metal detoxification and accumulation in plants. Environ Pollut 301:119035
Yadav SK, Singh H, Ginwal HS, Barthwal S (2019a) Elevated CO2 enhanced growth and physiological process of *Populus deltoides* Bartr. ex Marsh. Indian for 145(1):23–27
Yadav SK, Singh H, Nautiyal R, Ginwal HS, Ansari SA, Barthwal S (2019b) Modulation of morpho-physiological responses in *Populus deltoides* by elevated carbon dioxide and temperature. For Sci 66:105–118. https://doi.org/10.1093/forsci/fxz048
Yan A, Wang Y, Tan SN, Ghosh S, Chen Z (2020) Phytoremediation: a promising approach for revegetation of heavy metal-polluted land. Front Plant Sci 11:513099

Tree Species Suitability for Adapting to Urban Air Pollution and Climate Change

Km Nargis and Hukum Singh

Abstract Assessing species' suitability for adapting to urban air pollution and climate change involves comprehensively examining their resilience and adaptive capacities in the face of environmental stressors. Key factors influencing species adaptability include physiological tolerance to air pollutants, ability to cope with temperature fluctuations, and capacity for behavioral adjustments. Species with inherent tolerance mechanisms, such as efficient detoxification processes and adaptive respiratory systems, are more likely to withstand urban air pollution. Additionally, the ability to acclimate to changing climate conditions, such as altered precipitation patterns and temperature extremes, is crucial for long-term survival in urban environments. These efforts often combine field observations, laboratory experiments, and modeling techniques to evaluate the adaptability of different plantation species to urban climate and environmental pollution. Identifying indicator species with inherent adaptive traits helps guide conservation strategies and urban planning initiatives, fostering the development of resilient ecosystems capable of thriving amidst urban air pollution and climate change. Overall, this knowledge contributes to the sustainable coexistence of diverse species with increasingly urbanized landscapes.

Keywords Air pollution · Climate change · Resilience · Tree species · Urban environment

1 Introduction

An era of increasing urbanization has begun for human society. More than half of the world's population now lives in cities. By 2050, this will rise to roughly 68% or 6.7 billion people (Zhang et al. 2022; Kumar et al. 2022; Singh et al. 2022;

K. Nargis (✉)
Department of Botany, School of Applied and Life Sciences (SALS), Uttaranchal University, Premnagar, Dehradun, Uttarakhand 248007, India
e-mail: n1995mehtab@gmail.com

H. Singh
ICFRE-Forest Research Institute, PO New Forest, Dehradun, Uttarakhand 248006, India

H. Singh (ed.), *Urban Forests, Climate Change and Environmental Pollution*,
https://doi.org/10.1007/978-3-031-67837-0_6

Kumar et al. 2024). The process of urbanization and human activity drastically alter the ecology of landscapes. Unplanned and inappropriate land use, especially for urban uses, reduces land use for agriculture, forests, and wetlands in urban areas, degrading ecosystem services flow and environmental sustainability (Arfanuzzaman and Dahiya 2019; Singh et al. 2024a; Singh 2024). Rapid urbanization has resulted in a large-scale conversion of the vegetated regions to impervious surfaces, which has drastically changed the atmospheric and climatic conditions of cities, causing environmental problems in urban areas, including air pollution and urban emissions of greenhouse gases, which comprise more than 70% of global emissions, elevated CO_2 concentration, and the urban heat island (UHI), degradation of soil, including erosion, compaction, sealing and biodiversity loss. Trees play a crucial role in urban areas, providing various environmental, social, and economic benefits. For centuries, urban design and landscape architecture have incorporated urban trees into parks, yards, streets, and abandoned parcels and continue to be essential elements of civic spaces valued by the general public (Pataki et al. 2021; Dhyani et al. 2021). To support the ecology of metropolitan cities, plants play an indispensable role that includes temperature regulation (trees lower temperature and provide shade, which helps to reduce the urban heat island effect) and air quality improvement (trunks absorb pollutants like carbon dioxide, Sulphur dioxide, and nitrogen dioxide). Trees provide various benefits to people, including stormwater management, livability, urban planning, economic benefits, biodiversity support, aesthetic and recreational value, noise reduction, human health and well-being, and climate change mitigation. However, many metropolitan areas lack awareness about street trees' value to wildlife, which creates a knowledge gap regarding urban ecosystem conservation (Wood and Esaian 2020; Kumar et al. 2021a, b, c, d, e; Prakash et al. 2022). Plant adaptation is essentially required in urban areas due to the unique and challenging conditions urban environments present. Urbanization brings about several factors that can stress and impact plant life and adaptation becomes essential for their survival and overall well-being. It is required to choose tree species that are suitable for urban climates (Singh 2021a, b; Sharma and Singh 2021). The considerable characteristics: plants that are resilient to temperature fluctuations and heat stress, adapted to local water conditions and considered drought tolerant species, good air purifying capabilities, thrive in urban soils, Prioritize native plants that are adapted to the local climate and support local wildlife, adjust to seasonal changes, plant varieties that are known to thrive in the specific climate conditions of urban areas (Singh et al. 2021).

2 Global Issue: Air Pollution and Climate Change in Urban Areas

Air pollution and climate change are the biggest threats to rapidly expanding cities in the present day. Developing countries such as India, transitioning from predominantly rural to urban areas, face crucial challenges regarding sustainable development and climate action (Kaur and Pandey 2021; Singh et al. 2020a, b; Kumar et al. 2021a). The air pollution issues have already raised grave concerns for human and plant health. The main factors contributing to air pollution in urban areas include industrial emissions, vehicular pollution, construction emissions, particulate matter, gaseous pollutants, and so on. Primary or secondary air pollutants can be the leading cause of indoor and outdoor pollutants in urban areas. Primary air pollutants emitted directly contain particulate matter {$PM_{2.5}$, PM_{10}, suspended particulate matter (SPM), respirable particulate matter (RPM)}, SO_X, NO_X, CO Ammonia, and dust particles (Kumar et al. 2021b, c).

In contrast, secondary air pollutants include smog, ozone peroxyacyl nitrates (PANS), etc. (Kaur and Pandey 2021). Urbanization leads to decreased cropland, fallow land, and vegetated areas, contributing to climate change. The emission of greenhouse gases (GHG) is one of the most significant factors in altering climate by changing the atmospheric concentration of various gases. Additionally, scientists carefully examine how water vapors affect the climate (Kaur and Pandey 2021; Kumar et al. 2021d). Not only this, scientists are also investigating the role of black carbon in climate change due to its ability to absorb incoming solar radiations strongly. It is well known that climate change affects the planet's temperature, precipitation patterns, and solar insolation. The correlation between air pollution, climate change, and urbanization has been proposed as air pollutants, like aerosols, have a direct impact on earth's climate through absorption and scattering, as well as an indirect one through changes to cloud properties and radiation transfer processes (Kaur and Pandey 2021;Yadav et al. 2019a, b; Gupta et al. 2019). Insufficient infrastructure, including improper transportation management, unplanned industrial distribution, and paved and unpaved roads, makes it difficult for urban growth. Thus, a rise in pollution levels results from all of these (Rumana et al. 2014; Gupta et al. 2019; Kumar et al. 2019).

3 Urban Air Pollution and Its Impact on Urban Trees

In a growing urbanized world, air pollution mitigation is regarded as one of the most crucial issues in city planning. Urban trees provide large surface areas, influence the microclimate and air turbulence, and facilitate the widespread deposition of various gases and particles, all of which help to improve air quality (Grote et al. 2016a, b; Kumar et al. 2019). The worsened air quality of urban areas affects biophysical, phenological (leaf initiation, leaf development, flowering, fruiting, and other phenophases), morphological (leaf area and number, leaf thickness, leaf frequency, stomata density, and growth dynamics, etc.), biochemical (pigmentations, proline, sugars, ascorbic acid, proteins, enzymes, heavy metals etc.), and physiological (stomata conductance, carbon assimilation, transpiration, water use efficiency etc.) characteristics of urban roadside plantation species (Singh et al. 2017a, b, c; Singh et al. 2020a, b; Singh 2021a, b). The traits above are known as functional traits that govern a plant's overall functioning, which affects its growth and survival. Any changes in a plant's functional traits were considered a crucial sign for assessing the adaptation and mitigation response of urban plantations towards roadside air pollution. Vital biological processes, such as photosynthesis and other metabolic processes like protein synthesis, are impacted when the foliage of urban roadside plantations is exposed to vehicle emissions (Sharma et al. 2018; Singh et al. 2018, 2020a, b), ultimately affecting plant growth and development (Locosselli et al. 2019). Pollutants are persistently absorbed by the leaf surface, which seriously alters the stomatal behavior and gas exchange processes vital to plant life (Locosselli et al. 2019). The leaves cuticle surface is sticky, encouraging the deposition of dust and other particulate matter. Eventually, this can block stomatal opening, impacting the plant's overall physiology. The accumulation of dust particlesand heavy metals (Pb, Cd, Cr, Cu, and Fe) on the foliage results from vehicular emissions (Sevik et al. 2019; Singh et al. 2020a, b). This remarkably subtracts metabolic and physiological processes of plant growth and development (Singh et al. 2017a, b, c, 2020a, b). Pollutants may build up over time in a concentration that may adversely affect biological processes. While some plant species may develop a particular defense mechanism to keep pollutants out, others may accumulate toxic elements in different plant tissues. However, the availability or lack of these elements also affects physiological and metabolic functions such as respiration, photosynthesis, and the degeneration of cell organelles, ultimately leading to plant deaths (Gupta et al. 2018a, b; Hosseini et al. 2020).

3.1 Sources of Air Pollution in Urban Areas

3.1.1 Industrial Emissions

Emissions of greenhouse gases (GHGs) from industry are rapidly increasing and higher than those from other end-use sectors. About 47% of energy-related carbon dioxide emissions come from the global industrial sector, and large amounts of additional greenhouse gases (GHGs) are released as industrial process gases (Moomaw 1996). About one-third of the world's carbon dioxide (CO_2) emissions are caused by energy use in the industrial sector. The two main ways industrial sources contribute to greenhouse gas (GHGs) emissions are energy-related emissions from fossil fuel combustion and process-related emissions. (1) CO_2 from the production of lime and cement (calcination process), steel (cock and pig iron production), aluminum (oxidation and electrodes), hydrogen (refineries and the chemical industry) and ammonia (fertilizers and chemicals), (2) halocarbons (CFCs), hydro chlorofluorocarbons (HCFCs), and hydrofluorocarbons (HFCs) utilized as solvents, coolants, aerosol propellants, refrigerants and foam expanders, (3) CH_4 from miscellaneous industrial processes (iron and steel, oil refining, ammonia and hydrogen). (4) NO_2 from nitric acid and adipic acid (nylon) production (5) carbon tetrafluoride or CFCs-14 (CF 4) and hexaflouroethylene or CFC-16 (C2F6) from aluminum production (electrolysis).

3.1.2 Vehicular Pollution

Approximately two-thirds of urban air pollution is caused by vehicular emissions, especially automobiles. The major pollutants released by automobiles, such as lead (Pb), CO, NO_x, Sulphur oxides (SO), HC, and suspended particulate matter (SPM), have damaging effects on the environment and human health (Bhandarkar 2013; Singh and Verma 2013; Kumar et al. 2017). In gasoline engines, the gases comprise a combination of carbon monoxide (CO), nitrogen oxides (NO_X), and unburnt hydrocarbons (HC). Diesel vehicle emissions are characterized by relatively low CO concentration and unburnt hydrocarbons in the exhaust, which are compensated by high concentrations of NO_X and CO_2. Diesel cars emit 7.5 times more toxic particulate matter (PM) than comparable petrol cars (Gupta and Singh 2017; Singh et al. 2017a).

3.1.3 Construction Emission

By contributing its material inputs, the construction industry interacts with urbanization (Lin and Liu 2015). According to reports, the construction sector is the primary source of CO_2 emissions (Lin and Liu 2015). Furthermore, increasing urbanization will increase demand for infrastructure and buildings, resulting in enhanced CO_2 emissions from the construction industry (Singh et al. 2017b; Zhang et al. 2021). Construction-related activities, including developing infrastructure, buildings, and roads in urban areas (Zhang et al. 2021), often entail using equipment, materials, and energy sources that produce greenhouse gases (GHGs) and pollutants. The production of construction materials like steel, bricks, and cement is energy-intensive and can release greenhouse gases (GHGs). Construction produces a lot of waste, including discarded materials and construction debris. When this waste is disposed of, especially if burned openly, pollutants may be released into the atmosphere.

3.1.4 Particulate Matter and Gaseous Pollutants

Amidst heavy-duty vehicle emissions, particulate matter (PM) is expected to range from 1.9 to 10 μm in size (10 μm). Diesel soot particles are released into the atmosphere as both liquid and solid particles, and the surfaces of these particles allow them to absorb many chemical substances (Fang et al. 2003). Particulate matter is primarily composed of hydrophilic hydrocarbons. Another component of ambient particulate matter is black carbon (Novakov 1984), which comprises SO_4^{2-}, NO_3^-, and NH_4^+ (Cui et al. 2006). Additionally, aerosols from chemical reactions involving gases like SO_2, NO_X, and VOC can form PM2.5 in the atmosphere (Bhandarkar 2013; Meena et al. 2011; Kumar et al. 2019).

3.2 *Effects of Air Pollution on Tree Health*

Growing urban pollution levels are predicted to have significant effects on plant's biological clocks, which will then influence their phenological (flowering, fruiting, etc.), physiological (leaf thickness, leaf development, senescence, etc.), and biochemical (osmolyte accumulation, etc.) characteristics. These phenomena ultimately harm vegetation's capacity to absorb carbon dioxide, mitigate its effect, and adapt to it (Liu and Ding 2008; Singh et al. 2017a, b, c).

3.2.1 Hindrance in Physiological Functioning

Airborne pollutants cause damage to chloroplasts, which impedes photosynthesis (Singh et al. 2017a, b, c; Kumar et al. 2020; Singh and Kumar 2022a, b). As a result, the plant's capacity to generate energy and synthesize vital substances is diminished. The closure of stomata, microscopic pores on plant leaves essential for gas exchange, is caused by ozone and particulate matter (Singh et al. 2024b). This closure decreases the plant's ability to absorb carbon dioxide for photosynthesis. Air pollutants like nitrogen oxides (NO_X) and Sulphur dioxide (SO_2) can be conferred to the formation of acid rain, which is responsible for changing the pH of the soil, removing vital nutrients from it, and interfering with plant root's ability to absorb nutrients (Yadav et al. 2019a, b; Jangra et al. 2024). This can result in nutrient imbalances in plants. Industrial activities release volatile organic compounds (VOCs), which can disrupt plant metabolism.

3.2.2 Foliage Damage

Urban dust reduces stomatal diffusive resistance that gives rise to increased leaf temperature. Higher air and leaf temperatures harm leaves' internal and external structure, causing transpiration and photosynthesis rates to drop and ultimately lowering plant productivity in a polluted environment. Decreased photosynthetic pigment concentration is a marker for air pollution-induced leaf damage (Wedler et al. 1995; Cui et al. 2006). Exposure to pollution may cause compensatory growth, as indicated by SLA, which also verified foliar damage (Cui et al. 2006; Kanta et al. 2024; Malik et al. 2024).

3.2.3 Impact on Growth and Development

Toxins in the air, such as heavy metal and organic compounds, can be absorbed by plants and accumulate in tissues, affecting growth and development. This may decrease yields, stunted growth, and aberrant plant morphology (Kumar and Singh 2024). Pollen viability can be impacted by air pollution, which lowers plant's capacity for reproduction. This may impact the genetic diversity and reproductive success of plant populations. Prolonged exposure to air pollutants can weaken a plant's immune response, making it more vulnerable to pests and diseases.

4 Urban Climate Change and Its Impacts on Tree Species

Tree biology and distribution are impacted by rapid climate change. Cities can plant tree species in their urban forests that will thrive in climates predicted to occur in the future as a proactive adaptation measure. Yang (2009) indicates significant alterations in wind and precipitation patterns. Due to their long lifespans and slow rate of genetic adaptation to environmental change, trees will be impacted by this rapid climate change. The effects of climate change include changes in species range shift, growth patterns, phenological changes, host-parasite interactions, insect pest incidence, habitat adaptability, biogeochemical interactions, interactions between animals and resources, hydrological behavior, and more (Singh and Kumar 2022a, b). Trees constitute the most vulnerable and visible component of the garden that faces the impact of climate change (Bisgrove and Hadley 2002). This presents a problem for cities hoping to build and manage long-term urban forests. Furthermore, because trees live for a long time, planting a species that does not adapt well to the climate predicted for the future can have unfavorable effects on survival decades. Thus, urban foresters should consider and use species appropriate for future climates in their current urban greening initiatives.

4.1 Urban Heat Island Effect

The urban heat island (UHI) effect is the phenomenon of heat accumulation within urban areas due to human activity and urban construction. It is acknowledged as the most prominent feature of an urban environment. The urban heat island (UHI) effect will undoubtedly raise land surface temperatures, changing the structure and functions of urban ecological systems and affecting material and energy flow. These changes will have various ecological and environmental effects on urban climates, urban hydrologic situations, soil properties, atmospheric environments, biological habits, material cycles, energy metabolism, and the population's health (Yang et al. 2016). The increased temperature induced by UHI can affect plant phenology or the timing of developmental processes like leaf emergence and senescence in and around urban areas (Zipper et al. 2016).

4.1.1 Rising Temperatures in the Urban Environment

Because urban areas are warmer than the surrounding rural areas, an urban heat island, or "UHI" puts urban residents at a higher risk of heat stress than rural residents. It is now well known that in urban areas, less green space and more impervious surfaces divide more energy into sensible heat by lowering evapotranspiration and latent heat flux (Grimmond and Oke 1991). Urban heat islands (UHI) are also caused by anthropogenic heat release or heat release from human activities like traffic and building climate control (Allen et al. 2011). Another significant component is weather; low wind and cloud cover are associated with the highest UHIs.

4.2 *Effects on Tree Physiology*

Urban conditions near the roadside affect various biological processes, including gas exchange, stomatal physiology, respiration-related activities, and enzymatic activities at the metabolic level (Singh and Kumar 2022a, b).

4.2.1 Impact on Water Availability for Growth and Development of the Tree

Water stress is a significant factor limiting trees' growth and productivity. At the tree level, it can impact the rate and timing of other physiological processes, such as the production of flowers and fruits, lower the tree's capacity to withstand external stresses, and decrease the diameter and height growth of the tree. It can cause a reduction in leaf area, an increase in mortality, or the extinction of one species by another. Such stressful conditions can arise at any time of the year (for example, summer drought, winter desiccation) during the day (e.g., high evaporation demand) and in both excess and deficiency of soil moisture (Teskey and Hinckley 1986).

4.2.2 Impact of Drought Stress on Growth and Tree Mortality

Tree mortality in urban areas may result from drought stress. The impact varies depending on the severity of the stress (water availability and demand) and the individual's vulnerability. However, at the ecosystem level, the characteristics of individual trees that are significant for drought sensitivity or resistance differ depending on the availability of resources. This results from varying environmental circumstances for every tree, which in turn causes various individual growth histories to produce distinct anatomical and physiological traits. Thus, stand scale sensitivity and individual drought stress sensitivity may differ significantly. The impact of drought stress depends upon the tree's social position, characterized by the degree of suppression. Still, according to empirical data, it is correlated with resource availability, stress sensitivity, and stress exposure (Grote et al. 2016a, b).

5 Criteria for Assessing Tree Species Suitability

According to forestry experts, the recommendation for managing green spaces is to ensure the sustainability of urban trees. Proper management and planning of urban trees can offer many significant advantages to city inhabitants. These advantages are physical, energetic, biological, social, and economic (Christopoulou et al. 2007; Notaro and De Salvo 2010). Because native species are more adapted to the environment in which they live, native species tend to be healthier than non-native ones. Particular urban plant characteristics include limited space for roots and compacted soil (Watson and Kelsey 2006), warmer temperatures, or drier conditions. This may promote the development of non-native species better suited to these adverse environments. When selecting species, two fundamental presumptions are that they can thrive in high-stress environments and increase the specific diversity and coverage of trees in the area (Garcia-Garcia et al. 2016).

5.1 Characteristics of Resilient Tree Species

5.1.1 Tolerance to Air Pollutants

Pollutants from motor vehicles include carbon monoxide, nitrogen monoxide, Sulphur dioxide, and particulate matter. By gas exchange mechanisms, vegetation can absorb these pollutants. Several factors influence tree tolerance to air pollution, including leaf surface, wax coating, stomatal density, metabolic adaptation, overall plant health, and species-specific tolerance. Roadside vegetation could lower NO_2 concentrations to 3.5 and 2.3 parts per billion, respectively, based on level road structures 10 and 150 m from the road (Nasrullah et al. 1994). Through gaseous exchange processes, vegetation absorbs NO_2, which is subsequently assimilated into the nitrogenous compound and used as the primary compound in the plant to promote growth (Rowland 1986). Plants acclimating to air pollutants may exhibit morphological changes, such as longer trichomes and thicker epidermal cells. The stomata control mechanism demonstrated a physiological response by shutting down to noninjurious SO_2 levels and opening in response to noninjurious O_3 levels (Olszyk and Tibbitts 1981). The air pollution tolerance index (APTI) is derived from four physiological properties: relative water content (RWC), pH of leaf extract, total ascorbate (AsA), and chlorophyll (Chl) content. The degree of plant susceptibility to air pollutants was assessed using the APTI value. Ascorbate was recognized as an air pollution-detoxifying antioxidant molecule.

Additionally, it can regulate cell division and growth (Conklin et al. 2000). In green plants, photosynthesis is a vital process which depends on chlorophyll. Leaf chlorophyll variations can be relative indicators of environmental quality. Another rationale for incorporating pH into the APTI components was its significance in modulating physiological reactions to stress (Hartung et al. 1988).

5.1.2 Drought Resistance

Trees are in unprecedented danger of water scarcity, which can result in drought-induced tree mortality (Greenwood et al. 2017). A species' susceptibility to drought and its eventual recovery from it are explained by several physiological processes, including both long-term (adaptation) and short-term (acclimation) responses (Ruehr et al. 2019). Various plant characteristics, such as leaf water potential, leaf gaseous exchange, chlorophyll fluorescence, and leaf surface temperature, can be applied to determine the appropriate response in drought conditions (Gillner et al. 2017). Theoretically, phenotypic plasticity and interspecific genetic diversity (genetic diversity) are correlated with a species niche breadth; that is, a genotype with high plasticity permits a species to adapt its traits to various local environmental conditions and consequently achieve a trait environment match (Fajardo and Siefert 2019). One physiological response to drought is the accumulation of soluble carbohydrates, which prevents tissue desiccation by encouraging osmotic adjustment (Arndt et al. 2001). Drought-induced tree resilience, or a tree's ability to quickly re-establish its functioning and resume growth after a drought event, has been linked to C reserves because they provide soluble sugars with osmotic potential, fuel growth recovery or compensation demands, and heal any hydraulic damage (Körner 2019). Stomatal closure is a typical response of trees to drought, which forces trees to store carbon to meet osmotic and metabolic demands. Trees appear to have developed a non-water conservative drought coping mechanism based on starch conversion into soluble sugars in response to drought. This strategy was common to species with varying niche breadths (Fajardo and Piper 2021).

5.1.3 Heat Tolerance

From the molecular level to biology, processes are influenced by heat stress. Some trees have xerophytic characteristics, which make them better suited for hot climates. These characteristics include reduced leaf surface area, thick cuticles, and water-storing tissues. Trees with deep root systems can access water from deeper soil layers, which is especially helpful during hot and dry periods (Ilyas et al. 2021). Heat-tolerance trees have leaves modified to lose less water through transpiration. Water is conserved by leaf characteristics like pubescence or fine hair on the leaf surface (Ilyas et al. 2021). Deciduous trees conserve water and reduce heat stress by shedding their leaves in the hot and dry season. Heat-resistant bark shields the cambium layer and tree trunk from heat stress. To combat heat stress, trees may use biochemical techniques like synthesizing antioxidant and heat shock proteins, which shield cells from the damaging effects of high temperatures. From the cellular level up to the level of the entire tree, abnormal and prolonged heat stress negatively impacts tree biology (Wahid et al. 2007). These include thylakoid and cell membrane fluidity, electron transport, photochemistry or photosystem II, and ribulose bisphosphate carboxylase-oxygenase (RuBisCo) impairment activity (Teskey et al. 2015). Heat stress is responsible for rising respiration rates and synthesizing reactive oxygen species (ROS). Reduced growth, reproduction, stomatal conductance, photosynthesis, and other processes result in early leaf abscission, yellowing, and necrotic leaves (Percival 2023).

5.1.4 Disease Tolerance

Every plant has an inbuilt defense mechanism that shields it from most attackers. Additionally, plants may develop resistance after observing certain environmental cues. This acquired resistance (AR) is usually long-lasting and can even be passed down to future generations via alteration in DNA methylation and related chromatin density (Luna et al. 2012). Genetic mutations and the rate at which new defense genes are involved can be influenced by epigenetic processes (Wilkinson et al. 2019). Plants can coordinate multitrophic ecological interactions from these defense mechanisms to ensure survival. These external strategies entail recruiting microbes and beneficial insects both above and below ground. Beneficial microbial recruitment can develop disease-suppressive microbiomes that protect individual plants and their offspring (Hu et al. 2018). Planting nonnative species is a popular way to avoid contact with their particular pathogen (Garcia-Garcia et al. 2016).

6 Urban Planning and Policy Implications

Urban planning plays a vital role in shaping the development and functioning of cities. Efficient urban planning can contribute to sustainable, livable, and resilient urban environments. Some critical implications and considerations for urban planning and policies are sustainability (land use planning, green spaces, energy efficiency), transportation, housing, resilient and disaster preparedness, technology and innovation, community engagement, health and well-being, zoning and land use regulations, etc.

6.1 Integrating Suitable Tree Species into Urban Planning

Creating sustainable and livable cities requires incorporating suitable species of trees into urban planning. Trees have numerous environmental, social, and economic advantages, such as improved aesthetics, temperature regulation, carbon sequestration, wildlife habitat, and water purification. When selecting appropriate tree species, it's essential to consider several factors, such as the local climate's temperature, humidity, and precipitation patterns. Consider pH levels, soil conditions, and drainage to ensure the chosen trees can flourish in an urban setting. Give preference to native species because they are typically more adapted to the environment and contribute to the local biodiversity, and select adaptable species tolerant of urban stresses like pollution, compacted soil, and space constraints. Avoid planting monoculture trees to lower the danger of disease and pest outbreaks (Garcia-Garcia et al. 2016).

6.1.1 Zoning Regulations

Zoning is a government planning tool that controls how private property is used. It is mainly used in the United States and Western Europe (Smith 2021). Its goal is to advance people's general welfare, safety, and health of their community's all-encompassing municipal plan. It outlines the purpose of both current and future community development by separating communities into districts or zones and controlling land use and development intensity. Zoning for various purposes is common in urban and rural areas. These uses include historic preservation, subdivision development, agricultural development, forestry, recreation, floodplain management, and airport space (Wilson 1999).

6.1.2 Green Infrastructure Initiative

Urban Green Infrastructure (UGI) provides a doorway to sustainable urban development. More land is being developed than ever before. The primary issue facing conservation is the heavy use and fragmentation of open spaces. A framework for future development may foster various ecological, social, and economic benefits. Green infrastructure systems contribute to preserving and restoring naturally functioning ecosystems. These include increased recreational opportunities, enhanced habitat and biodiversity, conservation of natural landscape processes, cleaner air and water, better health, and intensive sense of place and connection to the natural world. In amplifying rising property values, green space can lower the cost of providing public utilities and infrastructure, such as stormwater management, water treatment, and flood control. Green infrastructure should be;

- It is designed to link elements into a system that functions as a whole rather than separate, unrelated parts.
- Our green space systems must be planned to include ecological, social, and economic benefits, functions, and values.
- Green space systems need to be laid out strategically to cross multiple jurisdictions and incorporate green space elements at each level of government.

- It should be planned and implemented with input from the public, including community organizations and private landowners.
- It must be funded as a primary public investment rather than with money left over after providing all other services (Benedict and McMahon 2002).

7 Community Engagement

Mechanisms are needed for community engagement, involvement, and development to guarantee that the needs and goals of local users in the community are met (Vargas-Hernández et al. 2017). It is critical to comprehend public support for and perception of the effectiveness of local government initiatives related to urban greening, particularly regarding improving residential roadside verges. The government often lacks the funding necessary for successful urban greening; therefore, it may rely on the assistance of neighborhood associations (Stenhouse 2004). Involving the community in planning urban green spaces is crucial because they understand how the areas will be utilized (such as parks or thoroughfares) and can offer suggestions on creating the area better for everyone.

7.1 Creating Awareness Among the Public on the Benefits of Adapted Tree Species

Creating awareness among the public on the benefits of tree species can be achieved through various methods. One approach is through creating documentaries. Kacprzyk et al. (2023) have concluded through their research work that documentaries are proven to generate increased awareness and audience engagement for plants. Analyses demonstrate that, with screen time driving audience reaction, Green Planet raised audience awareness and encouraged audience interaction for further information about the plants featured in the show. Another effective method is community awareness and awareness campaigns, which are more fruitful in developing countries for forest conservation efforts.

Additionally, online and social media platforms effectively create awareness programs. Millard et al. (2020) have shown in their studies that incorporating Wikipedia in biodiversity creates a stronger connection to biodiversity. The Species Awareness Index (SAI) is an effective metric for quantifying public biodiversity awareness at a global level.

7.2 *Involvement of the Community in Urban Greening Efforts*

Urban green spaces are community and local government programs with social, economic, and environmental goals. Local government agencies are responsible for designing institutional frameworks for urban green areas to satisfy users' needs. It should be done in such a way that it promotes stronger ties with the environment (Vargas-Hernández et al. 2017). Community support plays a vital role in ensuring the success of local urban greening initiatives. While making strategies for greening initiatives, involving local citizens as their network in co-creating strategies for sustainable cities proves to be beneficial. Houtstra et al. (2022) have discussed three effective pathways to enhance social capital for citizen initiatives and make the impact of citizen activities more explicit. Case studies demonstrated the effectiveness of these pathways. The community must be incorporated into urban green infrastructure planning and design to achieve sustainable and inclusive results. Identifying social and ecological goals through organized community participation processes and integrating community needs and values into the planning and design phases.

8 Future Direction and Research Needs

Future research directions and needs for studies on resistant plants to air pollution should be prioritized to further our understanding of how plants respond to various air pollutants and aid in developing mitigation methods for their detrimental effects on plant health. Barnes et al. (1999) mentioned the possibilities of using contemporary molecular techniques to detect, isolate, and modify air pollution resistance in fresh stock and the prospects for breeding plants with increased resistance to air pollutants. The challenges of integrating sophisticated technologies into traditional plant breeding for tolerance to air pollution are also discussed in the research. Ultimately, their study highlights the necessity for more investigation and the creation of mitigation plans for the effects of air pollution on plant life. A few potential directions and regions that require investigation are as follows: examine and pinpoint the precise processes that give plants resistance against air pollution. Part of this involves comprehending the physiological, biochemical, and genetic components that allow plant species or genotypes to characterize the genes that confer air pollution resistance on plants. This information makes plant varieties more resistant to specific pollutants that need to be bred or produced. Extensive research on the mutualistic or antagonistic relationship between various contaminants and plant species can aid in our comprehension of how plants react to their surroundings. The network of ecosystem communities is susceptible to air pollution; hence, co-occurrence network structure analysis is a valuable tool for evaluating how disturbance affects ecosystems since it offers an innovative way to gauge how well an ecosystem is operating as well as an integrated perspective on the reaction of the ecosystem (Karimi et al. 2016). We should investigate the molecular and biochemical reactions of the plant to air pollutants to better understand the metabolic pathways involved in pollution tolerance. Identify and develop bioindicators that might serve as early warning systems for how air pollution affects plant health. Following specific physiological or biochemical markers in plants can help assess the level and impact of pollution. Urban forest management practices include the selection of appropriate tree species, the application of sustainable urban design techniques, and the optimization of green infrastructure. Cities contribute significantly to the services provided by urban ecosystems. Cities' environmental efforts as their primary means of enhancing urban sustainability usually lessen the adverse effects of growing urbanization on the environment (Smith et al. 2019). Establish long-term monitoring systems to track plant populations' responses to air pollution over an extended period.

8.1 Emerging Technologies for Monitoring Tree Health

Some cutting-edge technologies are being developed or have been used to track the health of trees. The following technologies were becoming more and more popular. Large-scale forest monitoring can benefit from using high-resolution satellite imaging and airborne drones fitted with various sensors, which can offer valuable data. These technologies are used to measure changes in tree canopy density, spot stress indicators, and evaluate the general health of the forest. Forests may be precisely mapped in three dimensions using remote sensing technologies like LiDAR (light detection and ranging). This aids in measuring the structure of the forest canopy and detecting individual tree health. LiDAR can also be installed on drones or airplanes for more specialized evaluations. With machine learning and remote sensing technologies (such as LiDAR and hyperspectral images), tree species and structural characteristics have been mapped and evaluated (Alonzo et al. 2014). Extremely spatial, a thorough examination of the health of plants and trees is made possible by imaging with hyperspectral cameras, which record a broad spectrum of electromagnetic wavelengths. Researchers can identify diseases, nutritional deficits, and stress markers by examining the reflected light. Machine learning algorithms can analyze large datasets, such as sensor and satellite footage, to find patterns linked to stressed and healthy trees. Thanks to wireless sensor networks, multiple sensors can be installed throughout a forest, forming a network that continuously gathers and sends data on environmental conditions. This aids in tracking variations in humidity, temperature, and other elements that affect the health of trees.

8.2 Continuing Research on Climate-Resilient Tree Species

Certain characteristics allow trees that are climate-resilient species to adjust to shifting environmental conditions and flourish. These characteristics help them to resist different stresses brought on by climate change. Deswal et al. (2019) the anticipated pollution index (API) and air pollution tolerance index (APTI) of 40 plants in Rohtak City, Haryana, India. The plants showing higher air pollution resistance are *Calotropis gigantean*, *Syzygium cumini*, *Plumeria rubra*, *Albizia lebbeck*, and *Delonix regia*.

8.3 Policy Recommendation for Sustainable Urban Forestry

Urban gardening has received institutional support in Europe, mainly through national and local laws and initiatives to encourage urban vegetation growth (Calvet-Mir and March 2019). Laws known as 'tree ordinances' are meant to reflect the ideals of the community and the significance that its trees, or urban forest, hold. Tree laws can be put into place to promote tree planting and upkeep for the public's safety and well-being (Moran et al. 2020). Large trees are protected from removal by ordinance in some towns, and strict tree preservation laws, when put in place, show an increase in the amount of urban tree canopy cover (Landry and Pu 2010). These urban forestry policies can be implemented using an integrative approach considering social, economic, and environmental aspects. Here are a few suggestions for sustainable urban forestry policies.

- Create a thorough plan to regulate urban forests, including objectives, approaches, and steps for sustainable urban forestry. The urban environment's particular requirements and features should be considered in this strategy.
- In urban forests, biodiversity is increased by planting various native tree species. To sustain various species, encourage the installation of habitat features like insect-friendly plants and bird cages.
- Encourage cooperation between stakeholders, such as corporations, nonprofit organizations, community organizations, and government agencies. Partnerships guarantee a comprehensive approach and can increase the efficiency of urban forestry projects. Incentivize private property owners to grow and maintain trees on their land. Tax incentives, subsidies, and other monetary and non-monetary are examples of this.
- Provide training courses to help create a competent workforce for managing and maintaining urban forests. This may support economic growth and the creation of jobs.
- Establish a data collection and monitoring system to evaluate the state and health of urban trees. This data can be used to monitor developments, spot new problems, and make well-informed decisions.

9 Conclusion

It is concluded that prioritizing the selection of plant species with heightened tolerance is imperative to ensure the sustainability of urban trees amidst the challenges posed by urban settings, climate change, and escalating pollution levels. This chapter endeavors to scrutinize the adaptability of various tree species to urban air pollution and climate fluctuations. It underscores the necessity of a thorough suitability assessment, encompassing factors such as climate projections and local environmental conditions underpinned by sustained long-term monitoring efforts. Such a comprehensive approach is vital for informed decision-making in urban forestry management, facilitating the preservation and flourishing of green spaces amid evolving urban landscapes.

References

Allen L, Lindberg F, Grimmond CSB (2011) Global to city scale urban anthropogenic heat flux: model and variability. Int J Climatol 31(13):1990–2005

Alonzo M, Bookhagen B, Roberts DA (2014) Urban tree species mapping using hyperspectral and lidar data fusion. Remote Sens Environ 148:70–83

Arfanuzzaman M, Dahiya B (2019) Sustainable urbanization in Southeast Asia and beyond: challenges of population growth, land use change, and environmental health. Growth Change 50(2):725–744

Arndt SK, Clifford SC, Wanek W, Jones HG, Popp M (2001) Physiological and morphological adaptations of the fruit tree *Ziziphus rotundifolia* in response to progressive drought stress. Tree Physiol 21(11):705–715

Barnes J, Bender J, Lyons T, Borland A (1999) Natural and man-made selection for air pollution resistance. J Exper Bot 50(338):1423–1435

Benedict MA, McMahon ET (2002) Green infrastructure: smart conservation for the 21st century. Renew Resour J 20(3):12–17

Bhandarkar S (2013) Vehicular pollution, their effect on human health and mitigation measures. VE 1(2):3340

Bisgrove R, Hadley P (2002) Gardening in the global greenhouse: the impacts of climate change on gardens in the UK. UKCIP, Oxford

Calvet-Mir L, March H (2019) Crisis and post-crisis urban gardening initiatives from a Southern European perspective: the case of Barcelona. Eur Urban Reg Stud 26(1):97–112

Christopoulou O, Polyzos S, Minetos D (2007) Peri-urban and urban forests in Greece: obstacle or advantage to urban development? Manag Environ Qual Int J 18(4):382–395

Conklin PL, Saracco SA, Norris SR, Last RL (2000) Identification of ascorbic acid-deficient *Arabidopsis thaliana* mutants. Genetics 154(2):847–856

Cui HX, Jiang GM, Niu SL, Jiang CD, Liu MZ, Yu SL, Gao LM (2006) Ecophysiological response of plants to combined pollution from heavy-duty vehicles and industrial emissions in higher humidity. J Integr Plant Biol 48(12):1391–1400

Deswal M, Deswal P, Duhan SS, Laura JS (2019) An assessment of anticipated pollution index of some common plants and trees of Rohtak city of Haryana (India).

Dhyani K, Kuniyal HB, Singh H, Sobha S (2021) Growth and physiological potential of *Terminalia arjuna* under elevated CO_2 levels in Open top chamber condition. J Appl Nat Sci 13(3):1121–1126

Fajardo A, Piper FI (2021) How to cope with drought and not die trying: drought acclimation across tree species with contrasting niche breadth. Funct Ecol 35(9):1903–1913

Fajardo A, Siefert A (2019) The interplay among intraspecific leaf trait variation, niche breadth and species abundance along light and soil nutrient gradients. Oikos 128(6):881–891

Fang GC, Chang CN, Chu CC, Wu YS, Fu PPC, Yang IL, Chen MH (2003) Characterization of particulate, metallic elements of TSP, PM2.5 and PM2.5–10 aerosols at a farm sampling site in Taiwan, Taichung. Sci Total Environ 308(1–3):157–166

Garcia-Garcia MJ, Sánchez-Medina A, Alfonso-Corzo E, Garcia CG (2016) An index to identify suitable species in urban green areas. Urban For Urban Green 16:43–49

Gillner S, Korn S, Hofmann M, Roloff A (2017) Contrasting strategies for tree species to cope with heat and dry conditions at urban sites. Urban Ecosyst 20:853–865

Greenwood S, Ruiz-Benito P, Martínez-Vilalta J, Lloret F, Kitzberger T, Allen CD et al (2017) Tree mortality across biomes is promoted by drought intensity, lower wood density and higher specific leaf area. Ecol Lett 20(4):539–553

Grimmond CSB, Oke TR (1991) An evapotranspiration-interception model for urban areas. Water Resour Res 27(7):1739–1755

Grote R, Gessler A, Hommel R, Poschenrieder W, Priesack E (2016a) Importance of tree height and social position for drought-related stress on tree growth and mortality. Trees 30:1467–1482

Grote R, Samson R, Alonso R, Amorim JH, Cariñanos P, Churkina G et al (2016b) Functional traits of urban trees: air pollution mitigation potential. Front Ecol Environ 14(10):543–550

Gupta SK, Singh H (2017) Observations on sporadic flowering in *Ficus benjamina* L. and *Peltophorum africanum* L. in New Forest. Indian For 143:290–291

Gupta SK, Ram J, Singh H (2018a) Comparative study of transpiration in cooling effect of tree species in the atmosphere. J Geosci Environ Protect 6:151–166. https://doi.org/10.4236/gep.2018.68011

Gupta SK, Ram J, Singh H, Kumar P (2018b) Comparative study of cooling under shades of some forestry tree species with respect to ambient temperatures. J Geosci Environ Protect 6:51–66. https://doi.org/10.4236/gep.2018.611004

Gupta SK, Ram J, Singh H (2019) Relationship between leaf area index and atmospheric cooling of tree species. Indian J For 42(3):203–208

Hartung W, Radin JW, Hendrix DL (1988) Abscisic acid movement into the apoplastic solution of water-stressed cotton leaves: role of apoplastic pH. Plant Physiol 86(3):908–913

Hosseini NS, Sobhanardakani S, Cheraghi M, Lorestani B, Merrikhpour H (2020) Heavy metal concentrations in roadside plants (*Achillea wilhelmsii* and *Cardaria draba*) and soils along some highways in Hamedan, west of Iran. Environ Sci Pollut Res 27:13301–13314

Houtstra M, Stuiver M, Spijker J, McDonnell C (2022) Chapter 11: civic engagement as the corner stone of symbiotic cities. In: The symbiotic city: voices of nature in urban transformations. Wageningen Academic Publishers, p 196

Hu L, Robert CA, Cadot S, Zhang XI, Ye M, Li B et al (2018) Root exudate metabolites drive plant-soil feedbacks on growth and defense by shaping the rhizosphere microbiota. Nat Commun 9(1):2738

Ilyas M, Nisar M, Khan N, Hazrat A, Khan AH, Hayat K et al (2021) Drought tolerance strategies in plants: a mechanistic approach. J Plant Growth Regul 40:926–944

Jangra A, Kumar K, Maikhuri S, Bhandari MS, Pandey S, Singh H, Barthwal S (2024) Unveiling stress-adapted endophytic bacteria: characterizing plant growth-promoting traits and assessing cross-inoculation effects on *Populus deltoides* under abiotic stress. Plant Physiol Biochem 210:108610. https://doi.org/10.1016/j.plaphy.2024.108610

Kacprzyk J, Clune S, Clark C, Kane A (2023) Making a greener planet: nature documentaries promote plant awareness. Ann Bot 131(2):255–260

Kanta C, Kumar A, Chauhan A, Singh H, Sharma IP (2024) The interplay between plant functional traits and climate change. In: Kumar N, Singh H (eds) Plant functional traits for improving productivity. Springer, Singapore. https://doi.org/10.1007/978-981-97-1510-7_3

Karimi B, Meyer C, Gilbert D, Bernard N (2016) Air pollution below WHO levels decreases by 40% the links of terrestrial microbial networks. Environ Chem Lett 14:467–475
Kaur R, Pandey P (2021) Air pollution, climate change, and human health in Indian cities: a brief review. Front Sustain Cit 3:705131
Körner C (2019) No need for pipes when the well is dry—a comment on hydraulic failure in trees. Tree Physiol 39(5):695–700
Kumar N, Suyal DC, Sharma IP, Verma A, Singh H (2017) Elucidating stress proteins in rice (*Oryza sativa* L.) genotype under elevated temperature: a proteomic approach to understand heat stress response. 3 Biotech 7:205. https://doi.org/10.1007/s13205-017-0856-9
Kumar N, Jeena N, Kumar N, Gangola S, Singh H (2019) Phytoremediation facilitating enzymes: an enzymatic approach for enhancing remediation process. In: Bhatt P (ed) Smart bioremediation technologies. Academic Press, Cambridge, pp 289–306
Kumar A, Kumar P, Singh H, Kumar N (2021a) Modulation of plant functional traits under essential plant nutrients during seasonal regime in natural forests of Garhwal Himalayas. Plant Soil 465:197–212. https://doi.org/10.1007/s11104-021-05003-x
Kumar A, Kumar P, Singh H, Kumar N (2021b) Adaptation and mitigation potential of roadside trees with bio-extraction of heavy metals under vehicular emissions and their impact on physiological traits during seasonal regimes. Urban for Urban Green 58:126900. https://doi.org/10.1016/j.ufug.2020.126900
Kumar A, Kumar P, Singh H, Kumar N (2021c) Impact of plant functional traits on infiltration rate under different forests of Kempty watershed in Garhwal Himalaya. Indian J Soil Conser 49(1):38–44
Kumar A, Kumar P, Singh H, Bisht S, Kumar N (2021d) Relationship of physiological plant functional traits with soil carbon stock in the temperate forest of Garhwal Himalaya. Curr Sci 120(8):1368–1373
Kumar M, Phukon SN, Paygude AC, Tyagi K, Singh H (2021e) Mapping phenological functional types (PhFT) in the Indian Eastern Himalayas using machine learning algorithm in Google Earth engine. Comput Geosci 158:104982. https://doi.org/10.1016/j.cageo.2021.104982
Kumar A, Singh H, Kumari G, Bisht S, Malik A, Kumar A, Singh M, Raturi BS, Thakur A, Kaushal R (2022) Adaptive resilience of roadside trees to vehicular emissions via leaf enzymatic, physiological, and anatomical trait modulations. Environ Pollut 313:120191. https://doi.org/10.1016/j.envpol.2022.120191
Kumar N, Singh H, Giri K, Kumar A, Joshi A, Yadav A, Singh R, Bisht S, Kumari R, Jeena N, Khairakpam R, Mishra G (2024) Physiological and molecular insights into the allelopathic effects on agroecosystems under changing environmental conditions. Physiol Mol Biol Plants 30:417–433. https://doi.org/10.1007/s12298-024-01440-x
Kumar N, Singh H (2024) Plant functional traits for improving productivity. Springer, Singapore. https://doi.org/10.1007/978-981-97-1510-7
Kumar N, Singh H, Sharma SK (2020) Antioxidants: responses and importance in plant defense system. In: Roychowdhury R, Choudhury S, Hasanuzzaman M, Srivastava S (eds) Sustainable agriculture in the era of climate change. Springer, Cham, pp 251–264. https://doi.org/10.1007/978-3-030-45669-6_11
Landry S, Pu R (2010) The impact of land development regulation on residential tree cover: an empirical evaluation using high-resolution IKONOS imagery. Landsc Urban Plann 94(2):94–104
Lin B, Liu H (2015) CO_2 mitigation potential in China's building construction industry: a comparison of energy performance. Build Environ 94:239–251
Liu YJ, Ding HUI (2008) Variation in air pollution tolerance index of plants near a steel factory: Implication for landscape-plant species selection for industrial areas. WSEAS Trans Environ Develop 4(1):24–32
Locosselli GM, de Camargo EP, Moreira TCL, Todesco E, de Fátima Andrade M, de André CDS et al (2019) The role of air pollution and climate on the growth of urban trees. Sci Total Environ 666:652–661

Luna E, Bruce TJ, Roberts MR, Flors V, Ton J (2012) Next-generation systemic acquired resistance. Plant Physiol 158(2):844–853

Malik A, Kumar K, Kumari G, Kumar N, Singh H (2024) Application of functional traits in modelling productivity and resilience under climate change. In: Kumar N, Singh H (eds) Plant functional traits for improving productivity. Springer, Singapore. https://doi.org/10.1007/978-981-97-1510-7_5

Meena D, Singh H, Chaudhari SK (2011) Elucidating strontium response on growth dynamics and biochemical change in *Phaseolus mungo* L. Int J Agricult Environ Biotechnol 4:107–113

Millard JW, Gregory RD, Jones K, Freeman R (2020) The species awareness index (SAI): a Wikipedia-derived conservation culturomics metric for public biodiversity awareness. BioRxiv, 2020–2008

Moomaw WR (1996) Industrial emissions of greenhouse gases. Energy Policy 24(10–11):951–968

Moran R, Allison T, Bell B, Brown E, Dews M, Foster S et al (2020) Urban forest and the tree canopy: a pathway to climate resilience. Int Undergrad J Serv Learn Lead Soc Change 10(1):10–25

Nasrullah N, Tatsumoto H, Misawa AK (1994) Effect of roadside planting and road structure on NO_2 concentration near road. Eisei Kagaku 40(4):328–337

Notaro S, De Salvo M (2010) Estimating the economic benefits of the landscape function of ornamental trees in a sub-Mediterranean area. Urban for Urban Green 9(2):71–81

Novakov T (1984) The role of soot and primary oxidants in atmospheric chemistry. Sci Total Environ 36:1–10

Olszyk DM, Tibbitts TW (1981) Stomatal response and leaf injury of *Pisum sativum* L. with SO_2 and O_3 exposures: I. Influence of pollutant level and leaf maturity. Plant Physiol 67(3):539–544

Pataki DE, Alberti M, Cadenasso ML, Felson AJ, McDonnell MJ, Pincetl S et al (2021) The benefits and limits of urban tree planting for environmental and human health. Front Ecol Evolut 9:603757

Percival GC (2023) Heat tolerance of urban tree species: a review. Urban for Urban Green 36:128021

Prakash V, Hunney K, Singh H (2022) Effect of elevated carbon dioxide on growth and development of *Santalum album* L. seedlings inoculated with plant growth promoting microorganisms in Open Top Chambers. Int J Environ Agricult Biotechnol 7(5):149–159

Rowland AJ (1986) Nitrogen uptake, assimilation and transport in barley in the presence of atmospheric nitrogen dioxide. In: Fundamental, ecological and agricultural aspects of nitrogen metabolism in higher plants: proceedings of a symposium organized by the department of plant physiology, University of Groningen and the Institute for Soil Fertility, Haren, 9–12 April 1985. Springer, pp 211–214

Ruehr NK, Grote R, Mayr S, Arneth A (2019) Beyond the extreme: recovery of carbon and water relations in woody plants following heat and drought stress. Tree Physiol 39(8):1285–1299

Rumana HS, Sharma RC, Beniwal V, Sharma AK (2014) A retrospective approach to assess human health risks associated with growing air pollution in urbanized area of Thar Desert, western Rajasthan, India. J Environ Health Sci Eng 12:1–9

Sevik H, Ozel HB, Cetin M, Özel HU, Erdem T (2019) Determination of changes in heavy metal accumulation depending on plant species, plant organism, and traffic density in some landscape plants. Air Qual Atmos Health 12:189–195

Sharma R, Singh H (2021) Alteration in biochemical constituents and nutrients partitioning of *Asparagus racemosus* in response to elevated atmospheric CO_2 concentration. Environ Sci Pollut Res 29:6812–6821. https://doi.org/10.1007/s11356-021-16050-3

Sharma R, Singh H, Kaushik M, Nautiyal R, Singh O (2018) Adaptive physiological response, carbon partitioning, and biomass production of *Withania somnifera* (L.) Dunal grown under elevated CO_2 regimes. 3 Biotech 8:267. https://doi.org/10.1007/s13205-018-1292-1

Singh H (2021a) An integrated approach considering physiological and biophysical based indicators for assessing tolerance of roadside plantations of *Alstonia scholaris* towards urban roadside air pollution: an assessment of adaptation of plantations for mitigating urban roadside air pollution. Trees 36:2179. https://doi.org/10.1007/s00468-021-02179-8

Singh H (2021b) An integrated approach considering physiological-and biophysical-based indicators for assessing tolerance of roadside plantations of *Alstonia scholaris* towards urban roadside air pollution: an assessment of adaptation of plantations for mitigating roadside air pollution. Trees 36:1–15

Singh H (2024) Exploring adaptive modulation in plant functional traits and its impact on the productivity of *Acacia auriculiformis* under CO_2-enriched environment. Ind Crops Prod 210:118186

Singh H, Kumar M (2022a) Climate change and its impact on Indian Himalayan forests: current status and research needs. In: Rani S, Kumar R (eds) Climate change. Springer, Cham, pp 223–242. https://doi.org/10.1007/978-3-030-92782-0_11

Singh H, Kumar M (2022b) Climate change and its impact on Indian Himalayan forests: current status and research needs. Climate change: impacts, responses and sustainability in the Indian Himalaya. Springer, Cham, pp 223–242

Singh H, Verma A (2013) Characterization and screening of high nitrogen efficient rice genotype to curtail environmental pollution. Int J Agricult Environ Biotechnol 6(1):23–30

Singh H, Savita Sharma R, Sinha S, Kumar M, Kumar P, Verma A, Sharma SK (2017a) Physiological functioning of *Lagerstroemia speciosa* L. under heavy roadside traffic: an approach to screen potential species for abatement of urban air pollution. 3 Biotech 7(61):1–10. https://doi.org/10.1007/s13205-017-0690-0

Singh H, Verma A, Kumar M, Sharma R, Gupta R, Kaur M, Negi M, Sharma SK (2017b) Phytoremediation: a green technology to clean up the sites with low and moderate level of heavy metals. Aust Biochem 2:1012

Singh H, Savita S, Sharma R, Sinha S, Kumar M, Kumar P et al (2017c) Physiological functioning of *Lagerstroemia speciosa* L. under heavy roadside traffic: an approach to screen potential species for abatement of urban air pollution. 3 Biotech 7:1–10

Singh H, Sharma R, Savita SMP, Kumar M, Verma A, Ansari MW, Sharma SK (2018) Adaptive physiological response of *Parthenium hysterophorus* to elevated atmospheric CO_2 concentration. Indian For 144:1–14

Singh H, Yadav M, Kumar N, Kumar A, Kumar M (2020a) Assessing adaptation and mitigation potential of roadside trees under the influence of vehicular emissions: a case study of *Grevillea robusta* and *Mangifera indica* planted in an urban city of India. PLoS ONE 15(1):e0227380. https://doi.org/10.1371/journal.pone.0227380

Singh H, Yadav M, Kumar N, Kumar A, Kumar M (2020b) Assessing adaptation and mitigation potential of roadside trees under the influence of vehicular emissions: a case study of *Grevillea robusta* and *Mangifera indica* planted in an urban city of India. PLoS ONE 15(1):e0227380

Singh S, Singh H, Sharma SK, Nautiyal R (2021) Seasonal variation in biochemical responses of bamboo clones in the sub-tropical climate of Indian Himalayan foothills. Heliyon 7(4):e06859. https://doi.org/10.1016/j.heliyon.2021.e06859

Singh H, Bandyopadhyaya S, Kumar A, Singh M, Malik A, Kumari G, Raturi A, Yadav SK, Kumar M, Barthwal S, Thakur A, Rawat PS (2022) Understanding the physiological and biophysical response of urban roadside plantations for assessing adaptation and mitigation mechanisms toward vehicular emissions. Urban Clim 44:101183. https://doi.org/10.1016/j.uclim.2022.101183

Singh H, Kumar N, Kumar A (2024a) Enhancing resource use efficiency in crops through plant functional traits. In: Kumar N, Singh H (eds) Plant functional traits for improving productivity. Springer, Singapore. https://doi.org/10.1007/978-981-97-1510-7_6

Singh M, Singh H, Kumar A, Kumar M, Barthwal S, Thakur A (2024b) Soil nitrogen availability determines the CO_2 fertilization effect on tree species (*Neolamarckia cadamba*): growth and physiological evidence. Environ Sustain 7:53–60. https://doi.org/10.1007/s42398-023-00300-w

Smith H (2021) Citizen's guide to zoning. Routledge

Smith IA, Dearborn VK, Hutyra LR (2019) Live fast, die young: accelerated growth, mortality, and turnover in street trees. PLoS ONE 14(5):e0215846

Stenhouse RN (2004) Local government conservation and management of native vegetation in urban Australia. Environ Manag 34:209–222

Teskey R, Wertin T, Bauweraerts I, Ameye M, McGuire MA, Steppe K (2015) Responses of tree species to heat waves and extreme heat events. Plant Cell Environ 38(9):1699–1712

Teskey RO, Hinckley TM (1986) Moisture: effects of water stress on trees. In: Stress physiology and forest productivity: proceedings of the physiology working group technical session. Society of American Foresters National Convention, Fort Collins, Colorado, USA, July 28–31, 1985. Springer, Dordrecht, pp 9–33

Vargas-Hernández JG, Pallagst K, Zdunek-Wielgołaska J (2017) Urban green spaces as a component of an ecosystem. Funct Serv Users Commun Involv Initiat Act Rev Urban 37:1–26

Wahid A, Gelani S, Ashraf M, Foolad MR (2007) Heat tolerance in plants: an overview. Environ Exper Bot 61(3):199–223

Watson GW, Kelsey P (2006) The impact of soil compaction on soil aeration and fine root density of *Quercus palustris*. Urban For Urban Green 4(2):69–74

Wedler M, Weikert RM, Lippert M (1995) Photosynthetic performance, chloroplast pigments and mineral content of Norway spruce (*Picea abies* (L.) Karst.) exposed to SO_2 and O_3 in an open-air fumigation experiment. Plant Cell Environ 18(3):263–276

Wilkinson SW, Magerøy MH, López Sánchez A, Smith LM, Furci L, Cotton TA et al (2019) Surviving in a hostile world: plant strategies to resist pests and diseases. Annu Rev Phytopathol 57:505–529

Wilson D (1999) Zoning regulations. In: Encyclopedia of environmental science. Kluwer Academic Publishers, pp 711–713

Wood EM, Esaian S (2020) The importance of street trees to urban avifauna. Ecol Appl 30(7):e02149

Yadav SK, Singh H, Ginwal HS, Barthwal S (2019a) Elevated CO_2 enhanced growth and physiological process of *Populus deltoides* Bartr. ex Marsh. Indian For 145(1):23–27

Yadav SK, Singh H, Nautiyal R, Ginwal HS, Ansari SA, Barthwal S (2019b) Modulation of morpho-physiological responses in *Populus deltoides* by elevated carbon dioxide and temperature. For Sci 66:105–118. https://doi.org/10.1093/forsci/fxz048

Yang J (2009) Assessing the impact of climate change on urban tree species selection: a case study in Philadelphia. J For 107(7):364–372

Yang L, Qian F, Song DX, Zheng KJ (2016) Research on urban heat-island effect. Proced Eng 169:11–18

Zhang S, Li Z, Ning X, Li L (2021) Gauging the impacts of urbanization on CO_2 emissions from the construction industry: evidence from China. J Environ Manag 288:112440

Zhang L, Yang L, Zohner CM, Crowther TW, Li M, Shen F et al (2022) Direct and indirect impacts of urbanization on vegetation growth across the world's cities. Sci Adv 8(27):eabo0095

Zipper SC, Schatz J, Singh A, Kucharik CJ, Townsend PA, Loheide SP (2016) Urban heat island impacts on plant phenology: intra-urban variability and response to land cover. Environ Res Lett 11(5):054023

Phenological Shifts in Forest Ecosystems: A Strategic Response to Climate Change and Environmental Stress

R. K. Dwivedi and Pawanika Chandola

Abstract As integral components of terrestrial ecosystems, forests play a crucial role in global climate regulation and biodiversity conservation, providing essential ecosystem services. Climate change significantly affects forest ecosystems worldwide, influencing their phenology—the timing of seasonal events such as bud burst, leaf fall, and flowering. This chapter explores the dynamic relationship between forest phenology and climate change, examining the mechanisms behind timing shifts and their ecological consequences. Additionally, it investigates adaptive strategies employed by forests to mitigate climate change effects, including natural adaptations among species and human-assisted adaptations through forest management and afforestation. The chapter covers current methods to monitor and predict forest phenological changes, highlighting ongoing research and advancements. Challenges and limitations in understanding and managing forest phenology are discussed, with case studies from various regions illustrating real-world implications of phenological adaptation. Besides, this chapter addresses policy implications and conservation efforts related to forest ecosystems and phenology in the context of climate change. Conclusively, this chapter underscores the critical role of forest phenology in climate change adaptation and emphasizes the need for comprehensive research and adaptive strategies to sustain forest ecosystems. It serves as a call to action for continued exploration of emerging trends and innovations in forest phenology research and management, stressing the urgency of these efforts in the face of a changing climate.

Keywords Forest physiology · Phenology · Climate change · Adaptive strategy

R. K. Dwivedi · P. Chandola (✉)
Department of Botany, B. D. Govt. P. G. College, Jaihairkhal, Pauri Garhwal, Uttarakhand 246 193, India
e-mail: pawanika.chandola25@gmail.com

H. Singh (ed.), *Urban Forests, Climate Change and Environmental Pollution*,
https://doi.org/10.1007/978-3-031-67837-0_7

1 Introduction

As an essential part of the terrestrial ecosystem, forests sustain much of human progress, productivity, and biodiversity. The implications of our over-reliance and overexploitation on this ecosystem include deforestation, fragmentation, edge effects, and small-scale changes in humidity, temperature, and light (Singh et al. 2022; Singh et al. 2023; Raturi et al. 2024). All of these factors accelerate the process of climate change, and it has an impact on the essential functions of forest ecosystems and communities (Kumar et al. 2022, 2024; Singh and Kumar 2022; Singh et al. 2024a; Singh 2024). Plant phenology patterns can vary in response to variations in temperature and precipitation (Cleland et al. 2007; Meng et al. 2019). Phenological pattern changes may harm wildlife that depends on seasonal food supply because they are believed to be planted' first responses to climate change (Corlett and LaFrankie 1998; Dhyani et al 2021; Kumar et al. 2021a, b, c, d, e). The timing of flowering and fruiting patterns and developmental phases are altered due to climate change, and these changes negatively impact forest species' survival. There is mounting evidence that changes in the global climate have disrupted the ecological equilibrium, causing variations in the distribution and temporal patterns of vegetation cover. Monitoring phenology can help comprehend how vegetation and climate interact (Kumar et al. 2019a). However, certain species exhibit some resilience to climate change, and the degree of this resilience depends significantly on the species and the climate (Singh 2021; Sharma and Singh 2021; Singh et al. 2021; Prakash et al. 2022).

Phenology has proven to be an accurate and thorough indicator of climate variability as well as the responses of vegetation growth to it (Richardson et al. 2009). Phenology knowledge sheds essential light on the relationships between vegetation and the climate and how these relationships affect various temporal and spatial scales (Singh et al. 2020; Kumar et al. 2021a). Since it has been shown that there has been a significant shift in plant species as a result of the relatively low levels of climate warming and variability in the local environment, phenology has recently gained a lot of attention (Wang et al. 2017; Kumar et al. 2021b, c). Comprehending the phenology is crucial to adjusting to the changing climate.

2 Understanding Forest Phenology

With increased temperatures, vegetation phenology has significantly advanced, changing the seasonality of ecosystem processes and the feedback between the biosphere and atmosphere (Meng et al. 2021; Kumar et al. 2021d). Plant phenology patterns are characterized by the emergence and fall of leaves, blooming, and fruiting in duration, frequency, and production quantity (Newstrom et al. 1994). Understanding phenology is also helpful when choosing which species of trees to plant and when to harvest seeds for a seed bank. Environmental elements like temperature and precipitation impact plant phenological patterns (Cleland et al. 2007; Tang et al.

2023). Conversely, changes in forest phenology significantly impact the distribution and fitness of forests, plant-plant interactions among species, and the structure and functioning of ecosystems (Yadav et al. 2019a; b; Gupta et al. 2019a, b). The diversity of pollinators and plants is directly affected by blooming phenology, and leaf emergence cycles are important for many ecosystem functions.

The type of forest also affects phenology. Most plants on Earth depend on animal pollination to reproduce successfully, particularly in the tropics, where animals pollinate an estimated 94% of the plant species. Some plants found in tropical forests have extremely specialized phenologies. When certain things happen, like the first significant rainfall after a drought, they might only flower for 1 or 2 days (Kumar et al. 2019b, c). This reaction can occasionally be highly coordinated, with hundreds of the trees in a forest blossoming simultaneously and lasting for just one or two days. These patterns may present particular evolutionary obstacles for pollinators, herbivores, nectarivores, and frugivores (Lieth 1974). Rainfall-related occurrences are predominant in tropical forests. In tropical woods, flowering and fruiting are frequently controlled by the amount of rainfall or drought. Tropical woods have very little temperature seasonality, whereas temperate weather patterns exhibit a wide range of phenological fluctuations. There is variation in rainfall for certain plants, and trees in tropical forests react seasonally (Lieth 1974).

Trees often flower during periods of yearly drought and then bear fruit after the rains have returned. In woods with noticeable dry spells, this pattern is far more prevalent. Forest animals mostly rely on the fruits produced in these sequences that correspond with dry periods for sustenance. Fruit abundance peaks annually due to numerous trees' simultaneous flowering and fruiting (Sharma et al. 2018; Singh et al. 2018). This indicates that the food supply for rainforest animals in these environments peaks and declines over an extended period. Importantly, fruiting and flowering may fail in years with exceptionally high precipitation levels during the dry season. As a result, fewer animal food supplies are available, which could lead to hunger (Brunner et al. 2015).

3 Climate Change and Forest Phenology

Plant phenology and growth are impacted by climate change. However, the kind of plant and the environment in which it grows determine the impact's degree and intensity. One of the most trustworthy biological markers for assessing the effects of changing climate conditions is phenology, which studies the timing of repeated phases in an organism influenced by climatic influences (Gupta et al. 2018a, b). The fate of all forest dwellers is determined by trees, which are an essential component of forest ecosystems (Cleland et al. 2007). Therefore, the gross commuting in the forest during extreme weather conditions is determined by forest phenology. The local climate influences the plant growth season, culminating in the region's aberrant abiotic elements (Singh and Verma 2013; Kumar et al. 2017). Droughts often occur at the height of the growing season due to the effects of climate change. A "false autumn" phenomenon occurs when plants experience extreme water stress due to drought, which causes them to shed their leaves or change colour prematurely.

The past 50 years of satellite data are a valuable resource for observing variations in the length of the growing season for vegetation. Scientists are using this data to understand better how plants react to our warming environment and quantify the extent and effects of these changes (Gupta and Singh 2017; Singh et al. 2017a). Early springs in central and northern Europe encouraged increased vegetation growth, which caused the soil moisture to evaporate quickly and contributed to severe summer dry conditions that increased the risk of forest fires. Early spring snowmelt and higher spring and summer temperatures were directly associated with this shift. The circumstance was reported to impact plant phenology and growth (Gomory et al. 2020).

Longer growing seasons, or extended periods between the start of spring and the conclusion of fall, have been observed in most parts of the world in recent decades. The altered equilibrium of the atmosphere and climate is the primary cause of most of these phenological disturbances (Singh et al. 2017b). In turn, variations in the length of the growing season influence the cycling of carbon and water in temperate forests. A longer growing season may have contributed to higher plant growth in the northern hemisphere, according to observations from phenological investigations (Meena et al. 2011; Kumar et al. 2019c, 2020). This could have a negative feedback loop on rising atmospheric carbon dioxide levels and climate change. Thus, we must advance scientific knowledge of the connection between phenological variation and forest production to comprehend forest trees and their impact on the global climate.

Extreme weather events impact the timing of forest leaf colouration and fall. However, there is still too little information about the precise effects of each major environmental stressor on the phenology of autumn in forests and the variety of phenological sensitivities found in various forest communities and species (Singh and Kumar 2022; Singh et al. 2024b). Although contradictory, reactions to the interplay between weather and climatic conditions can be significant. Furthermore, distinct forest tree species and communities can exhibit differing responses to an identical environmental stressor, and the same species and communities might exhibit varying phenological sensitivity at various places.

Fall forest phenology significantly impacts processes such as nitrogen cycling and drift in the forest ecosystem. There is still much to learn about the drivers of fall phenology (Yadav et al. 2019a; b; Jangra et al. 2024). Beyond the impact of decreasing photoperiod and temperature during the summer and fall, it is well known that extreme weather events, such as heat stress and drought, can affect leaf senescence or the colour and drop of leaves. These events are becoming more frequent and intense due to global climate change (Kanta et al. 2024; Malik et al. 2024). While there is diversity in responses and sensitivities to the seasonal changes and extreme events from different tree species and ecosystems, the relative roles of seasonal climate variations and extreme weather events in causing leaf senescence remain unknown (Winarni et al. 2016).

4 Factors Influencing Forest Phenology

Temperature is the primary element impacting the phenology of all plant types due to changing climate conditions; since 1981, temperatures have grown globally by 0.18 °C per 10 years. Accordingly, a later arrival of autumn and an earlier start to spring are most likely caused by the warming temperature. In most of Asia, rising temperatures cause the vegetation growing season to lengthen, but rising precipitation has the reverse effect. Temperature increases cause the spring phenological stages of many tree species to accelerate and the autumn phenological stages to lag. Many temperate trees and herbs in India have different flowering phenologies, which are caused by variations in temperature. Due to warming temperatures, Rhododendron and Myrica species are experiencing early flowering and fruit sets (Gaira et al. 2014). Warming temperatures worldwide have confused plants into blooming earlier in the spring, which also throws off the temporal relationships of pollinator species.

Except in the northernmost reaches of the tropics, research conducted in Asia has revealed few reactions to temperature variations. Similarly, in regions without a noticeable dry season, there is no significant phenological signal at the individual or community levels (Tiwari et al. 2021). These regions have certain multi-year cycles, but these have not been conclusively connected to the temperature or other climatic signals. On the other hand, distinct flowering and fruiting phenology is observed in Asian woods whenever there is a noticeable dry season, which has consequent ecological ramifications for animals. However, depending on the kind of plant, temperature may have a different effect on how long the growth season lasts. Warmer temperatures can boost photosynthesis and vegetation productivity in settings where forests predominate. Conversely, a warmer temperature causes excess water to evaporate from the Earth's surface, hindering the growth of plants with shallow roots that grow on the surface, such as grasses, sedges, and some herbaceous plants.

Precipitation and global carbon are significant factors that influence the phenological shift and growth events in addition to temperature. One aspect of climate change that goes hand in hand with rising temperatures is changed precipitation patterns. The seasonality of precipitation determines the impact of changed precipitation on biodiversity and ecosystem functioning (Potts et al. 2019). Precipitation has been found to influence a plant's senescence and rejuvenation significantly. A pre-season drought significantly shortens the growing season and delays the greening-up process. Extremely low spring and summer precipitation is associated with late rejuvenation of woody species and early senescence of tiny plants (Currier and Sala 2022). Temperature and precipitation harm the phenology of flowering in herbs. However, these phenological events benefit from delayed senescence, which may allow for a longer time window for vital physiological functions. On the other hand, the nature of the influence of precipitation changed; it notably harmed the fruiting stage. The exact time and degree of CO_2 rise determine the impact of CO_2 on blooming.

5 Long-Term Climate Change Effects on Forest Phenology

The long-term effects of climate change on long-lived tree species are recorded as slow growth due to additional stresses. These consist of storms, invading species, wildfires, and pest outbreaks. Certain disturbances happen quickly, such as a wildfire. Others take decades or even centuries, such as shifts in the populations of plants or animals. In other cases, a forest may suffer lasting impacts. Canopy phenology responds to annual and decadal changes in climate, and changes in canopy phenology contribute to altered species ranges and potential increases in water and nutrient limitations. Plant species range contraction and expansion into new territories by clement climatic conditions are evidenced by the shift of some boreal species poleward as reproduction fails in their range in the southern part (Fei et al. 2017; Shay et al. 2021).

Additionally, disturbances can compound to raise the threats to forests. Drought, for instance, can weaken trees and increase a forest's susceptibility to insect or wildfire outbreaks. Likewise, wildfires can increase a forest's susceptibility to pests. Introducing new species in climate change-affected forests causes competition to native plant species, alteration, and changes in forest community structure (Osland et al. 2013).

Changes in climatic conditions and biotic interactions inside the forests also change and cause long-term effects on plant species distribution. Plant species are impacted by interactions that change the limits of their warm and cold climates (Kumar and Singh 2024). These interactions include mutualism, facilitation, pollination, herbivory, and parasitism (Benning et al. 2019; Benning and Moeller 2021). In California, the plant *Clarkia xanthiana* (Onagraceae) shifted beyond its cold limit due to climate change, resulting in the lack of pollinators and increased herbivore populations (Benning and Moeller 2021). Thus, positive and negative interactions affect plant distributions under stressful conditions and can attenuate climate stress on plants (Bulleri and Chapman 2010). Pollinator species distributions are generally strongly linked to their visiting plant geographic ranges (Jiménez-Valverde et al. 2021). Plant communities' structure, function, and makeup changes can also affect the interaction between plants and microbes (Rudgers et al. 2020). Consequently, the distribution of microbial communities' spatial variation impacts plant local adaption patterns and determines the extent of a species range. (Rolshausen et al. 2020).

The interplay between climate impacts and other stressors, such as land development, might reduce a forest's capacity for adaptation. Temperature elevation generally decreases forest growth in areas where water is limited. Increased temperature and CO_2 can also impact nutrient cycling and forest productivity in the long term (Fei et al. 2017). Thus, the impacts of climate change are variable for forest communities. Climate change affects the physiology and defenses of the host, the relationships between pests and other species, including natural enemies, competitors, mutualists, and the environment (Tang et al. 2016).

6 Observing and Monitoring Phenology

There is mounting evidence that shifts in the vegetation cover and the global climate have affected the ecological equilibrium. Phenology is crucial for energy exchange, vegetation dynamics, and ecosystem functions. As a result, phenology monitoring can be a valuable tool for comprehending how vegetation and climate interact, and the process of transitioning from plant-specific to regional-scale phenology studies is made possible by repeated observations from satellite sensors. Vegetation phenology has traditionally been tracked by direct plant observations, with phenological records going back to the ninth century (e.g., Japanese cherry trees). Typically, these indices are derived from satellite imagery (Richardson et al. 2013). Scientists usually conduct visual assessments (Menzel et al. 2006), but volunteers interested in the natural world can also do so (Berra et al. 2017). This entails monitoring the plant and documenting its phenological phase—the day the first bud or bloom emerges—per a predetermined procedure or standard (Streher et al. 2017).

Plant phenology modeling is the simplest and most effective method for evaluating phenological alterations. Water availability, temperature, and solar radiation are thought to be the main variables influencing plant phenology. Mechanistic, statistical, and theoretical approaches have been widely used for parameterizing plant phenology models. The statistical methods establish a relationship between environmental conditions and the timing of phenological phenomena. The methods contain accurate phenological mechanism assumptions and straightforward computation processes, but their applicability and prediction power are restricted. The "cause-effect relationships" that are known or presumed to exist between biological processes and many other essential driving factors are used by mechanical techniques to explain plant phenology. Mechanistic techniques offer enhanced parameter processing, practical assumptions, wide-ranging applications, and accurate forecasts (Zhang et al. 2020).

Currently, measurements from remote sensing and conventional ground observations are the primary methods used to monitor vegetation phenology. Conventional on-site observations typically document particular phenological occurrences for individuals within a group, but large-scale remote sensing is generally used to investigate vegetation phenology patterns. Research on global change is increasingly focusing on how vegetation phenology responds to climate change since vegetation phenology is highly responsive to climatic conditions. Comprehending the spatiotemporal variance in vegetation phenology is essential to estimate ecosystem carbon exchange and reveal ecological responses to local climate change. On the other hand, satellite observations at regional and global scales have yielded databases of land surface phenology (LSP) with a wide range of applications. These LSP datasets' physical meanings and accuracy vary significantly because they are sourced from various satellite datasets and methodologies. Satellite-based and aircraft-based sensors can provide remote sensing data, such as satellite photos, to track and identify changes in the vegetation cover. Mapping and monitoring accuracy is increased by fusing ground data with remotely sensed imagery, such as aircraft and satellites. These methods use long-term and real-time LSP data recordings, monitoring, and LSP metrics and detection algorithms to identify phenological changes in vegetation. Additionally, changes in vegetation cover at the species level can be detected by merging data from satellite imagery with additional information on aspects, slopes, altitudes, water bodies, and soil qualities. The varied effects of climatic variables like precipitation, temperature, solar radiation, etc., under changed environmental conditions on vegetation phenology in various geographical regions at both global and regional scales can be tracked using ground-based or remote sensing approaches.

7 Impact on Ecosystem Services

Ecosystem services comprise the natural goods and processes now affected by climate change. The change in species composition, development, reproduction, mortality, and ecosystem structures, including the capacity of forest ecosystems to operate as carbon sinks, significantly impact forest ecosystems and ecosystem services. Due to longer growing seasons, more significant atmospheric CO_2, and forest regeneration, terrestrial primary production increased globally (Campbell et al. 2009). Local elements, however, might have suppressive effects. Enhancing biogeochemical processes such as carbon and nitrogen cycling can impact terrestrial output; warming and rising atmospheric CO_2 may also affect the soil ecosystem (Melillo et al. 2017).

Depending on the kind and height of the forest, changes in primary output may result from climate change. Primary output is expected to decline in forests where soil water is scarce during the growing season (Latta et al. 2010). Mixed results have been shown in how climate change has affected primary output. However, changes in primary production are anticipated to be magnified at higher trophic levels. Changes in climate, such as fire, drought, species invasions, pest and disease outbreaks, and storms like hurricanes, are upsetting the structure, composition, and function of forest ecosystems and reducing the number of resources available for civilization (Shen et al. 2015).

Interacting species react to variations in time within an ecosystem caused by environmental influences and the timing of recurrent life-cycle stages. Plants typically use physical alterations as indicators for the next phase of their seasonal cycle. One of the most apparent effects of global climate change is that many plants respond to an earlier arrival of spring by altering their phenological events to withstand the climatic upheaval (Lieth 1974). Species that rely on distinct habitats to support phenological transitions to the subsequent life-cycle stage are eventually impacted by phenological reactions to changed environmental conditions. Inadequate phenological transitions may result in extensive disturbances to the food chain and environmental ramifications (Ehrlén and Valdés 2020).

8 Adaptive Strategies in Forest Ecosystems

Most scientists and professionals agree that site characteristics like temperature, water availability, and fertilizer supply will change due to climate change, requiring adaptive management of forests (Gebeyehu and Hirpo 2019). The process of altering ecological, social, and economic systems in reaction to the consequences of climate change is known as adaptation (Smit et al. 2000). A good adaptation plan must be adaptable enough to change course in various social, political, and environmental contexts. Plans for sustainable forest management might include climate change adaptation techniques as a risk management element. The precautionary principle encourages action by putting policies that are helpful now and would lower the possibility of unacceptably significant losses. It is necessary to evaluate forest policy to make sure that it promotes adaptability (Parker et al. 2000). Passive and active adaptation are the two key adaptive strategies in managing forest ecosystems.

The concept of passive adaptation, which involves not managing forest management, suggests utilizing forest ecosystems' natural resilience and succession mechanisms to reduce or eliminate the amount of silvicultural input (Kijowska-Oberc et al. 2020). Passive adaptation, associated with non-management, aims to apply natural forest dynamics for forest adaptation (Bolte et al. 2009). In reaction to climate change, natural ecosystem dynamics are expected to enforce dynamic ecosystem self-regulation. It appears very feasible that forests could respond to climate change through passive adaptation, which uses potential succession and natural forest dynamics. Resistance acts protect the ecosystem from outside influences, keeping circumstances intact.

The ecosystems less vulnerable to climate change's effects, such as old-growth forests, would benefit the most from this. One feature shared by all old-growth woods is heterogeneity. It is believed that old-growth forests' unique complexity—caused by the heterogeneity and diversity of stand structures—makes them capable of adapting to unexpected disturbances. Hence, unmanaged forests can develop into multiple states, while managed forests have fewer development options (Puettmann et al. 2009). While this method may work in the short term, it is more likely to contribute to the survival of the current environment. This viewpoint is increasingly prevalent among private landowners of small forested areas occasionally managed by non-professionals. (Mostegl et al. 2019).

By using silvicultural techniques (such as tending, thinning, stand conversion, and tree species enrichment) and the resilience of the forests, active adaptation refers to altering stand structures and the composition of tree species in a way that improves the forest's climate tolerance over its lifetime (Bolte et al. 2009; Millar et al. 2007). The definition of resilience is the ability to adapt to some degree of change while promoting a return to a preferred reference state or earlier state following a disturbance. Resilience measures improve the system's ability to recover from disturbances and adjust to changing environmental conditions.

Warmth, little precipitation, and forest fires can all promote resilience and facilitate the adjustment to impending disruptions in forest ecosystems that have adapted to them. Systems get more robust until they can no longer bear the quantity of change; at that point, they transform into a new state. Like the resistive approach, this one can be more risky in the long run but more successful in the short term. Active adaptation is changing stand structures and the species composition of trees in a forest to improve the forest's tolerance to climate change over time. This is done by utilizing the resilience of the forests and silvicultural practices (such as thinning, tending, stand conversion, and tree species enrichment). It depends on both the fire's intensity and temperature. The western Himalayas of India are seeing a shift in forest regime due to climate change, with temperature, precipitation, and forest fires being highlighted as important drivers (Tyagi and Kumar 2022). Systems become more resilient until they can no longer withstand the amount of change; at that moment, they transition into a new state. This strategy, like the resistive one, may be more dangerous in the long run and more effective in the short term.

Long-term efficacy is usually taken into consideration while designing transitional solutions. Their gradual introduction is common in more extensive management plans concentrating on short-term resilience initiatives. Regarding climate change adaptation strategies, these options offer the most incredible variety (Swanston and Janowiak 2012). Adaptive actions get increasingly specialized along this continuum. Ecosystems can better adapt to new conditions when transition activities proactively anticipate and accommodate change. Conversely, behaviors characterized by resistance and resilience help the existing ecosystem to survive. Under certain conditions, such as when the danger of implementing resistance and resilience tactics is deemed too high, these measures might be justified in ecosystems considered particularly vulnerable to a range of possible future climates.

9 Ecological and Environmental Implications

Fundamentally, a phenological change could give an organism flexibility in the occurrence of life-cycle events, enabling it to maintain its past climatic niche. The life-cycle event will, in theory, take place in the same or comparable climatic conditions, even though it will happen at separate periods. Such circumstances likewise impact plant-pollinator interactions. The mismatches in plant-pollinator interactions can be attributed to plants' greater sensitivity to high temperatures than insects. Higher parasite loads, pollinator demography, and morphology were the outcomes of this decline in the overall fitness and survival rate of pollinators' gender distributions and population viability (Miller-Rushing et al. 2010). Conversely, asymmetric pollinator-plant interactions affect the process of fertilization and pollination and, eventually, the plant's ability to survive (Schenk et al. 2018).

In order to survive, synchronize sexual reproduction within a population, and prevent unfavorable seasonal oscillations, plants respond phenologically to their changing surroundings. The interplay between seasonality in some tropical forests and forest fragmentation may have significant ecological ramifications. The forest margins adjacent to deforested areas could become drier than those with a more evenly distributed forest cover. This might emphasize the dry season by influencing phenology. At the same time, areas with high deforestation reduce the number of forest animals that may search for food. Moving over the landscape reduces species' ability to eat seeds or fruits to make up for their lack of food. The combination of more dramatic feast and famine brought on by more significant phenological reactions and a lower capacity to smooth feast and famine by relocating may result in severe stress for birds and mammals living in forests (Iler et al. 2021).

10 Challenges and Limitations

Climate change has hastened the reduction in habitat variety and productivity of forest ecosystems worldwide that has been caused by forestry activities in recent decades (Fischer and Lindenmayer 2007; Lindenmayer and Fischer 2007; Schütz 1997). There is still uncertainty about trees' ability to respond (Julius and West 2007). Because forests are intricate ecosystems, they are presumably more resilient to slight alterations in their surroundings. Tropical forests—predominantly intact forests—can tolerate certain degrees of climate stress (Malhi et al. 2008). However, many experts are concerned that forests won't be able to adjust to the unprecedented rates of climate change. However, many experts are worried that forests won't be able to adapt to the unprecedented rates of climate change. (Gitay et al. 2002).

Species can adjust to climate change through migration to suitable locations, adaptive evolution, or phenotypic flexibility, i.e., acclimatization (Markham 1996; Bawa and Dayanandan 1998). Without these decisions, species will decline and ultimately go extinct (Noss 2001). According to evidence from combined climate and vegetation models, global warming may necessitate migration rates far quicker than during postglacial times. This could limit biodiversity by favoring highly mobile and opportunistic species (Malcolm et al. 2002).

Defining technical adaptation metrics for forests is difficult since they depend on many contextual elements (e.g., forest types, management goals, climatic threats, and non-climatic pressures). Furthermore, although modeling has been used to examine how vulnerable forests are to climate change, forest managers and policymakers may find it challenging to employ ecosystem models and climatic scenarios due to their inherent uncertainties (Millar et al. 2007). The goal of forest ecosystem-based management, or FEM, is to preserve ecosystems' biodiversity and ecological integrity by bridging the gap between natural and managed forests. In light of climate change, FEM is a viable approach to attaining sustainable forest management (Grenon et al. 2010). Cooperation among scientists is crucial at this crucial juncture to modify forest management techniques for the future.

11 Future Research Directions

The economies and ecology of human societies will be significantly impacted by climate change because these changes will affect disturbances, change the distribution of forest species, modify tree growth, and promote the entrance of invasive species. The economies and ecology of human societies will be significantly impacted by climate change because these changes will affect disturbances, change the distribution of forest species, modify tree growth, and promote the entrance of invasive species. To ensure the application of sustainable forest management in the context of the most significant challenge facing humanity today—climate change—it is imperative that (1) long-term monitoring networks be established to document forest functioning across various forest landscapes and environmental conditions, and (2) new decision-support tools capable of predicting future scenarios in forest ecosystems be developed quickly. (3) To develop reliable indicators of climate change to decide shifts in forest ecosystems, (4) to build governance policies suggesting a legislative framework to ensure the application of sustainable adaptive forest management, (5) Finally, international collaboration among the countries must be encouraged for learning experiences and management for adaptive strategies.

12 Conclusion

Climate change has accelerated forest deterioration, impacting ecosystems and communities. Forest phenology, influenced by temperature and precipitation variations, reflects plants' responses to climate change, potentially affecting wildlife reliant on seasonal food. Understanding phenology illuminates vegetation-climate relationships. Longer growing seasons globally stem from atmospheric and climate shifts, affecting carbon and water cycles. Extreme weather alters leaf coloration and fall timing, yet effects on autumn phenology remain unclear. Rising temperatures lengthen growing seasons, while increased precipitation may counteract this. Unique flowering and fruiting phenology affect ecosystems in regions lacking a distinct dry season. Climate change's long-term impacts on trees include slowed growth and susceptibility to stressors like insects, wildfires, and storms. Climate interactions with stressors like land development may hinder forests' adaptive capacity. Monitoring phenology through remote sensing and ground observations is crucial for understanding vegetation-climate dynamics. Adaptive strategies, including active and passive adaptation, are vital for forest resilience. Forest ecosystem-based management promotes sustainable practices, but uncertainties challenge implementation. Future research should focus on monitoring networks, decision-support tools, governance policies, and international collaboration. Cooperation is critical to adapting forest management for climate change mitigation and ecosystem preservation.

References

Bawa KS, Dayanandan SD (1998) Global climate change and tropical forest genetic resources. Clim Change 39(2):473–485

Benning JW, Moeller DA (2021) Microbes, mutualism, and range margins: testing the fitness consequences of soil microbial communities across and beyond a native plant's range. New Phytol 229:2886–2900

Benning JW, Eckhart VM, Geber MA, Moeller DA (2019) Biotic interactions contribute to the geographic range limit of an annual plant: herbivory and phenology mediate fitness beyond a range margin. Am Nat 193:786–797

Berra EF, Gaulton R, Barr S (2017) Commercial off-the-shelf digital cameras on 1016 unmanned aerial vehicles for multitemporal monitoring of vegetation reflectance and 1017 NDVI. IEEE Trans Geosci Remote Sens 55:4878–4886

Bolte A, Ammer C, Löf M, Madsen P, Nabuurs GJ, Schall P et al (2009) Adaptive forest management in central Europe: climate change impacts, strategies and integrative concept. Scand J For Res 24(6):473–482

Brunner I, Herzog C, Dawes AMA, Sperisen C (2015) How tree roots respond to drought. Front Plant Sci 6:547

Bulleri F, Chapman MG (2010) The introduction of coastal infrastructure as a driver of change in marine environments. J Appl Ecol 47(1):26–35

Campbell JL, Rustad LE, Boyer EW, Christopher SF, Driscoll CT, Fernandez IJ, Groffman PM, Houle D, Kiekbusch J, Magill AH, Mitchell MJ, Ollinger SV (2009) Consequences of climate change for biogeochemical cycling in forests of northeastern North America. Can J for Res 39:264–284

Cleland EE, Chuine I, Menzel A, Mooney HA, Schwartz MD (2007) Shifting plant phenology in response to global change. Trends Ecol Evolut 22(7):357–365

Corlett RT, LaFrankie JV (1998) Potential impacts of climate change on tropical Asian forests through an influence on phenology. Clim Change 39(2–3):439–453

Currier CM, Sala OE (2022) Precipitation versus temperature as phenology controls in drylands. Ecology 103:37–93

Dhyani K, Kuniyal HB, Singh H, Sobha S (2021) Growth and physiological potential of *Terminalia arjuna* under elevated CO_2 levels in Open top chamber condition. J Appl Nat Sci 13(3):1121–1126

Ehrlén J, Valdés A (2020) Climate drives among-year variation in natural selection on flowering time. Ecol Lett 23(4):653–662

Fei S, Desprez KM, Potter I, Knott JA, Oswalt CM (2017) Divergence of species responses to climate change. Sci Adv 3(5):e1603055

Fischer J, Lindenmayer DB (2007) Landscape modification and habitat fragmentation: a synthesis. Global Ecol Biogeogr 16:265–280

Gaira KS, Rawal RS, Rawat B, Bhatt ID (2014) Impact of climate change on the flowering of *Rhododendron arboreum* in central Himalaya. India. Curr Sci 106(12):1735–1738

Gebeyehu NM, Hirpo FH (2019) Review on effect of climate change on forest ecosystem natural resource management specialized in forest and nature management. Int J Environ Sci Nat Resour 17(4):126–129

Gitay H, Suarez A, Watson RT (2002) Climate change and biodiversity: IPCC Technical Paper V. In: Intergovernmental panel on climate change. Geneva, p 77

Gömöry D, Krajmerová D, Hrivnák M, Longauer R (2020) Assisted migration versus close-to-nature forestry: what are the prospects for tree populations under climate change? Cent Eur for J 66(2):63–70

Grenon F, Jetté J-P, Leblanc M (2010) Manuel de référence pour l'aménagement écosystémique des forêts au Québec Ministère des Ressources Naturelles et de la Faune, Direction de l'environnement et de la protection des forêts, Gouvernement du Québec, pp 1–60. https://mffp.gouv.qc.ca/documents/forets/amenagement/manuel-reference-ecosystemique.pdf

Gupta SK, Singh H (2017) Observations on sporadic flowering in *Ficus benjamina* L. and *Peltophorum africanum* L. in new forest. Indian for 143:290–291

Gupta SK, Ram J, Singh H (2018a) Comparative study of transpiration in cooling effect of tree species in the atmosphere. J Geosci Environ Protect 6:151–166. https://doi.org/10.4236/gep.2018.68011

Gupta SK, Ram J, Singh H, Kumar P (2018b) Comparative study of cooling under shades of some forestry tree species with respect to ambient temperatures. J Geosci Environ Protect 6:51–66. https://doi.org/10.4236/gep.2018.611004

Gupta SK, Ram J, Singh H (2019a) Relationship between leaf area index and atmospheric cooling of tree species. Indian J for 42(3):203–208

Gupta SK, Singh H, Ram J (2019b) Need for a novel paradigm in forestry research: perspectives of comparative atmospheric cooling by different forestry effect tree species. Indian for 145(6):509–515

Iler AM, CaraDonna P, Forrest JRK, Post E (2021) Demographic consequences of phenological shifts in response to climate change. Annu Rev Ecol Evolut Syst 52:221–245

Jangra A, Kumar K, Maikhuri S, Bhandari MS, Pandey S, Singh H, Barthwal S (2024) Unveiling stress-adapted endophytic bacteria: characterizing plant growth-promoting traits and assessing cross-inoculation effects on *Populus deltoides* under abiotic stress. Plant Physiol Biochem 210:108610. https://doi.org/10.1016/j.plaphy.2024.108610

Jiménez-Valverde A, Aragón P, Lobo JM (2021) Deconstructing the abundance–suitability relationship in species distribution modelling. Global Ecol Biogeogr 30:327–338

Julius SH, West JM (2007) Preliminary review of adaptation options for climate sensitive ecosystems and resources, synthesis and assessment product 4.4. U.S. Climate Change Science Program, draft for public comment. https://www.govinfo.gov/content/pkg/GOVPUB-PREX23-PURL-LPS96474/pdf/GOVPUB-PREX23-PURL-LPS96474.pdf

Kanta C, Kumar A, Chauhan A, Singh H, Sharma IP (2024) The interplay between plant functional traits and climate change. In: Kumar N, Singh H (eds) Plant functional traits for improving productivity. Springer, Singapore. https://doi.org/10.1007/978-981-97-1510-7_3

Kijowska-Oberc J, Staszak AM, Kaminski J, Ratajczak E (2020) Review adaptation of forest trees to rapidly changing climate. Forests 11(2):123

Kumar N, Suyal DC, Sharma IP, Verma A, Singh H (2017) Elucidating stress proteins in rice (*Oryza sativa* L.) genotype under elevated temperature: a proteomic approach to understand heat stress response. 3 Biotech 7:205. https://doi.org/10.1007/s13205-017-0856-9

Kumar M, Singh MP, Singh H, Dhakate PM, Ravindranath NH (2019a) Forest working plan for the sustainable management of forest and biodiversity in India. J Sustain for 39:1–22

Kumar N, Jeena N, Singh H (2019b) Elevated temperature modulates rice pollen structure: a study from the foothill of the Himalayan agro-ecosystem in India. Biotech 9(5):175

Kumar N, Jeena N, Kumar N, Gangola S, Singh H (2019c) Phytoremediation facilitating enzymes: an enzymatic approach for enhancing remediation process. In: Bhatt P (ed) Smart bioremediation technologies. Academic Press, Cambridge, pp 289–306

Kumar A, Kumar P, Singh H, Kumar N (2021a) Modulation of plant functional traits under essential plant nutrients during seasonal regime in natural forests of Garhwal Himalayas. Plant Soil 465:197–212. https://doi.org/10.1007/s11104-021-05003-x

Kumar A, Kumar P, Singh H, Bisht S, Kumar N (2021b) Relationship of physiological plant functional traits with soil carbon stock in the temperate forest of Garhwal Himalaya. Curr Sci 120(8):1368–1373

Kumar A, Kumar P, Singh H, Kumar N (2021c) Adaptation and mitigation potential of roadside trees with bio-extraction of heavy metals under vehicular emissions and their impact on physiological traits during seasonal regimes. Urban for Urban Green 58:126900. https://doi.org/10.1016/j.ufug.2020.126900

Kumar A, Kumar P, Singh H, Kumar N (2021d) Impact of plant functional traits on infiltration rate under different forests of Kempty watershed in Garhwal Himalaya. Indian J Soil Conser 49(1):38–44

Kumar M, Phukon SN, Paygude AC, Tyagi K, Singh H (2021e) Mapping phenological functional types (PhFT) in the Indian Eastern Himalayas using machine learning algorithm in Google Earth engine. Comput Geosci 158:4982. https://doi.org/10.1016/j.cageo.2021.104982

Kumar A, Singh H, Kumari G, Bisht S, Malik A, Kumar A, Singh M, Raturi S, Barthwal S, Thakur A, Kaushal R (2022) Adaptive resilience of roadside trees to vehicular emissions via leaf enzymatic, physiological, and anatomical trait modulations. Environ Pollut 313:120191. https://doi.org/10.1016/j.envpol.2022.120191

Kumar N, Singh H, Giri K, Kumar A, Joshi A, Yadav A, Singh R, Bisht S, Kumari R, Jeena N, Khairakpam R, Mishra G (2024) Physiological and molecular insights into the allelopathic effects on agroecosystems under changing environmental conditions. Physiol Mol Biol Plants 30:417–433. https://doi.org/10.1007/s12298-024-01440-x

Kumar N, Singh H (2024) Plant functional traits for improving productivity. Springer, Singapore. https://doi.org/10.1007/978-981-97-1510-7

Kumar N, Singh H, Sharma SK (2020) Antioxidants: responses and importance in plant defense system. In: Roychowdhury R, Choudhury S, Hasanuzzaman M, Srivastava S (eds) Sustainable agriculture in the era of climate change. Springer, Cham, pp 251–264. https://doi.org/10.1007/978-3-030-45669-6_11

Latta G, Temesgen H, Adams DT (2010) Barrett Analysis of potential impacts of climate change on forests of the United States Pacific Northwest. For Ecol Manag 259:720–729

Lieth H (1974) Purposes of a phenology book. In: Lieth H (ed) Phenology and seasonality modelling. Ecological studies (analysis and synthesis). Springer, Berlin. https://doi.org/10.1007/978-3-642-51863-8_1

Lindenmayer DB, Fischer J (2007) Tackling the habitat fragmentation panchreston. Trends Ecol Evol 22(3):127–132

Malcolm JR, Markham A, Neilson RP, Garaci M (2002) Estimated migration rates under scenarios of global climate change. J Biogeogr 29:835–849

Malhi Y, Roberts JT, Betts RA, Killeen TJ, Li W, Nobre CA (2008) Climate change, deforestation, and the fate of the Amazon. Science 319:169–172

Malik A, Kumar K, Kumari G, Kumar N, Singh H (2024) Application of functional traits in modelling productivity and resilience under climate change. In: Kumar N, Singh H (eds) Plant functional traits for improving productivity. Springer, Singapore. https://doi.org/10.1007/978-981-97-1510-7_5

Markham A (1996) Potential impacts of climate change on ecosystems: a review of implications for policy makers and conservation biologists. Clim Res 6:179–191

Meena D, Singh H, Chaudhari SK (2011) Elucidating strontium response on growth dynamics and biochemical change in *Phaseolus mungo* L. Int J Agricult Environ Biotechnol 4:107–113

Melillo JM, Frey SD, DeAngelis KM, Werner WJ, Bernard MJ, Bowles FP, Pold G, Knorr MA, Grandy AS (2017) Long-term pattern and magnitude of soil carbon feed-back to the climate system in a warming world. Science 358(6359):101–105

Meng L, Zhou Y, Lia X, Asrar GR, Mao J, Wanamaker AD et al (2019) Divergent responses of spring phenology to daytime and night time warming. Agr for Meteorol 281:107832

Meng L, Zhou Y, Gu L, Richardson AD, Penuelas J, Fu Y, Wang Y, Asrar GR, De Boeck HJ, Zhang Y, Wang Z (2021) Photoperiod decelerates the advance of spring phenology of six deciduous tree species under climate warming. Glob Chang Biol 27:2914–2927

Menzel A, Sparks TH, Estrella N, Koch E, Aasa A, Ahas R et al (2006) European phenological response to climate change matches the warming pattern. Global Change Biol 12(10):1969–1976

Millar CI, Stephenson NL, Stephens SL (2007) Climate change and forests of the future: managing in the face of uncertainty. Ecol Appl 8:2145–2151

Miller-Rushing AJ, Høye TT, Inouye DW, Post E (2010) The effects of phenological mismatches on demography. Phil Trans R Soc 365(1555):3177–3186

Mostegl NM, Pröbstl-Haider U, Jandl R, Haider W (2019) Targeting climate change adaptation strategies to small-scale private forest owners. Forest Policy Econ 99:83–99

Newstrom LE, Frankie GW, Baker HG (1994) A new classification for plant phenology based on flowering patterns in lowland tropical rain forest trees at La Selva. Costa Rica. Biotropica 26(2):141–159

Noss R (2001) Beyond Kyoto: forest management in a time of rapid climate change. Conser Biol 15(3):578–590

Osland MJ, Enwright N, Day RH, Doyle TW (2013) Winter climate change and coastal wetland foundation species: salt marshes versus mangrove forests in the Southeastern United States. Global Change Biol 19(5):1482–1494

Parker WC, Colombo SJ, Cherry ML, Greifenhagen S, Papadopol C, Flannigan MD et al (2000) Third millennium forestry: what climate change might mean to forests and forest management in Ontario. Forest Chronicle 76(3):445–463

Potts DL, Barron-Gafford GA, Butterfield BJ, Fay PA, Hultine KR (2019) Bloom and bust: ecological consequences of precipitation variability in aridlands. Plant Ecol 220:135–139

Prakash V, Hunney K, Singh H (2022) Effect of elevated carbon dioxide on growth and development of *Santalum album* L. seedlings inoculated with plant growth promoting microorganisms in Open Top Chambers. Int J Environ Agricult Biotechnol 7(5):149–159

Puettmann KJ, D'Amato AW, Kohnle U, Bauhus J (2009) Individual-tree growth dynamics of mature Abies alba during repeated irregular group shelterwood (Femelschlag) cuttings. Can J For Res 39(12):2437–2449

Raturi A, Singh H, Kumar P, Chanda A, Raturi A (2024) Spatiotemporal patterns of greenhouse gas fluxes in the subtropical wetland ecosystem of Indian Himalayan foothill. Environ Monit Assess 196(10):882

Richardson AD, Hollinger DY, Dail DB, Lee JT, Munger JW, Okeefe J (2009) Influence of spring phenology on seasonal and annual carbon balance in two contrasting New England forests. Tree Physiol 29(3):321–331

Richardson AD, Keenan TF, Migliavacca M, Ryu Y, Sonnentag O, Toomey M (2013) Climate change, phenology, and phenological control of vegetation feedbacks to the climate system. Agricult for Meteorol 169:156–173

Rolshausen R, Hallman U, Dal Grande F, Otte J, Knudsen K, Schmitt I (2020) Expanding the mutualistic niche: parallel symbiont turnover along climatic gradients. Proc R Soc B Biol Sci 287:20192311

Rudgers JA, Afkhami ME, Bell-Dereske L, Chung YA, Crawford KM, Kivlin SN et al (2020) Climate disruption of plant-microbe interactions. Ann Rev Ecol Evol Syst 51(1):561–586

Schenk M, Krauss J, Holzschuh A (2018) Desynchronizations in bee-plant interactions cause severe fitness losses in solitary bees. J Anim Ecol 87(1):139–149

Schütz JP (1997) Sylviculture 2: La gestion des forêts irrégulières et mélangéesPresses Polytechniques et Universitaires Romandes

Sharma R, Singh H (2021) Alteration in biochemical constituents and nutrients partitioning of *Asparagus racemosus* in response to elevated atmospheric CO_2 concentration. Environ Sci Pollut Res 29:6812–6821. https://doi.org/10.1007/s11356-021-16050-3

Sharma R, Singh H, Kaushik M, Nautiyal R, Singh O (2018) Adaptive physiological response, carbon partitioning, and biomass production of *Withania somnifera* (L.) Dunal grown under elevated CO2 regimes. 3 Biotech 8:267. https://doi.org/10.1007/s13205-018-1292-1

Shay JE, Pennington LK, Mandussi S, Montiel-Molina JA, Toews DJ, Hendrickson BT, Sexton JP (2021) Rules of plant species ranges: applications for conservation strategies. Front Ecol Evol 9:700962

Shen M, Piao S, Dorji T, Liu Q, Cong N, Chen X, Shuai A, Wang S, Wang T, Zhang G (2015) Plant phenological responses to climate change on the Tibetan Plateau: research status and challenges. Geosciences 2(4):454–467

Singh H (2021) An integrated approach considering physiological and biophysical based indicators for assessing tolerance of roadside plantations of *Alstonia scholaris* towards urban roadside air pollution: an assessment of adaptation of plantations for mitigating urban roadside air pollution. Trees 36:179. https://doi.org/10.1007/s00468-021-02179-8

Singh H (2024) Exploring adaptive modulation in plant functional traits and its impact on the productivity of *Acacia auriculiformis* under CO_2-enriched environment. Ind Crops Prod 210:118186

Singh H, Verma A (2013) Characterization and screening of high nitrogen efficient rice genotype to curtail environmental pollution. Int J Agricult Environ Biotechnol 6(1):23–30

Singh H, Savita Sharma R, Sinha S, Kumar M, Kumar P, Verma A, Sharma SK (2017a) Physiological functioning of *Lagerstroemia speciosa* L. under heavy roadside traffic: an approach to screen potential species for abatement of urban air pollution. 3 Biotech 7(61):1–10. https://doi.org/10.1007/s13205-017-0690-0

Singh H, Verma A, Kumar M, Sharma R, Gupta R, Kaur M, Negi M, Sharma SK (2017b) Phytoremediation: a green technology to clean up the sites with low and moderate level of heavy metals. Aust Biochem 2:1012

Singh H, Sharma R, Savita SMP, Kumar M, Verma A, Ansari MW, Sharma SK (2018) Adaptive physiological response of *Parthenium hysterophorus* to elevated atmospheric CO_2 concentration. Indian for 144:1–14

Singh H, Yadav M, Kumar N, Kumar A, Kumar M (2020) Assessing adaptation and mitigation potential of roadside trees under the influence of vehicular emissions: a case study of *Grevillea robusta* and *Mangifera indica* planted in an urban city of India. PLoS ONE 15(1):e0227380

Singh S, Singh H, Sharma SK, Nautiyal R (2021) Seasonal variation in biochemical responses of bamboo clones in the sub-tropical climate of Indian Himalayan foothills. Heliyon 7(4):e06859. https://doi.org/10.1016/j.heliyon.2021.e06859

Singh H, Bandyopadhyaya S, Kumar A, Singh M, Malik A, Kumari G, Raturi A, Yadav SK, Kumar M, Barthwal S, Thakur A, Rawat PS (2022) Understanding the physiological and biophysical response of urban roadside plantations for assessing adaptation and mitigation mechanisms toward vehicular emissions. Urban Clim 44:101183. https://doi.org/10.1016/j.uclim.2022.101183

Singh H, Kumar M (2022) Climate change and its impact on Indian Himalayan forests: current status and research needs. In: Rani S, Kumar R (eds) Climate change. Springer, Cham, pp 223–242. https://doi.org/10.1007/978-3-030-92782-0_11

Singh H, Kumar N, Singh R, Kumar M (2023) Assessing the climate change impact on the habitat suitability of the range-restricted bird species (*Catreus wallichii*) in the Indian Himalayan ecosystem. Environ Sci Pollut Res 30(57):121224–121235

Singh H, Kumar N, Kumar A (2024a) Enhancing resource use efficiency in crops through plant functional traits. In: Kumar N, Singh H (eds) Plant functional traits for improving productivity. Springer, Singapore. https://doi.org/10.1007/978-981-97-1510-7_6

Singh M, Singh H, Kumar A, Kumar M, Barthwal S, Thakur A (2024b) Soil nitrogen availability determines the CO_2 fertilization effect on tree species (*Neolamarckia cadamba*): growth and physiological evidence. Environ Sustain 7:53–60. https://doi.org/10.1007/s42398-023-00300-w

Smit B, Burton I, Klein RJ, Wandel J (2000) An anatomy of adaptation to climate change and variability. Springer Netherlands, pp 223–251

Streher AS, Sobreiro J, Ferreira F, Patrıcia L, Morellato C, Sanna T, Silva F (2017) Surface phenology in the tropics: the role of climate and topography in a snow-free mountain. Ecosystems 20:1436–1453

Swanston CW, Janowiak MK (2012) Forest adaptation resources: climate change tools and approaches for land managers. General technical report NRS-87. U.S. Department of Agriculture, Forest Service, Northern Research Station, Newtown Square, PA

Tang J, Körner C, Muraoka H, Piao S, Shen M, Thackeray SJ, Yang X (2016) Emerging opportunities and challenges in phenology: a review. Ecosphere 7(8):e01436

Tang Y, Zhou W, Du Y (2023) Effects of temperature, precipitation, and CO_2 on plant phenology in China: a circular regression approach. Forests 14(9):1844

Tiwari P, Verma P, Raghubanshi AS (2021) Forest phenology as an indicator of climate change: impact and mitigation strategies in India. In: Islam MN, van Amstel A (eds) India: climate change impacts, mitigation and adaptation in developing countries. Springer, Cham. https://doi.org/10.1007/978-3-030-67865-4_8.

Tyagi K, Kumar M (2022) The resilience of Indian western Himalayan forests to regime shift: are they reaching towards no return point? Eco Inform 69:101644

Wang X, Gao Q, Wang C, Yu M (2017) Spatiotemporal patterns of vegetation phenology change and relationships with climate in the two transects of East China. Glob Ecol Conser 10:206–219

Winarni NL, Kurniasari DR, Hartiningtias D, Nusalawo M (2016) Phenology, climate, and adaptation: how does Dipterocarps respond to climate? Indonesian J for Res 3(2):129–141

Yadav SK, Singh H, Ginwal HS, Barthwal S (2019a) Elevated CO_2 enhanced growth and physiological process of *Populus deltoides* Bartr. ex Marsh. Indian For 145(1):23–27

Yadav SK, Singh H, Nautiyal R, Ginwal HS, Ansari SA, Barthwal S (2019b) Modulation of morpho-physiological responses in *Populus deltoides* by elevated carbon dioxide and temperature. For Sci 66:105–118. https://doi.org/10.1093/forsci/fxz048

Zhang X, Wang J, Henebry GM, Gao F (2020) Development and evaluation of a new algorithm for detecting 30 m land surface phenology from VIIRS and HLS time series. ISPRS J Photogramm Remote Sens 161:37–51

Understanding the Impact of Climate Change and Environments on Urban Forests: Insights from Morphological, Physiological, and Molecular Perspectives

Apurva Malik, Hukum Singh, Kishan Kumar, Garima Kumari, and Ajay Sojitra

Abstract As global climate patterns undergo unprecedented shifts, urban ecosystems, particularly urban forests, face increasing vulnerability, demanding a comprehensive exploration of their responses across various levels. This chapter synthesizes research findings on urban forests' morphological, physiological, biochemical, and molecular adaptations to changing climatic and environmental conditions. Understanding plant responses to urban pollutants is vital for mitigating adverse effects and fostering sustainable urban ecosystems. Morphologically, urban trees exhibit altered growth patterns, canopy structures, and phenological shifts in response to temperature variations, altered precipitation regimes, and urban heat island effects. Physiologically, stress-induced responses such as transpiration rates, stomatal conductance, and photosynthetic efficiencies serve as crucial indicators of adaptive strategies urban trees employ to cope with environmental challenges. Biochemical studies highlight the role of secondary metabolites, antioxidants, and stress-responsive proteins in mitigating the adverse effects of climate change. Advances in molecular biology techniques unravel intricate mechanisms of molecular pathways, including gene expression, epigenetic modifications, and signal transduction, elucidating underlying adaptive responses in urban forest species. This chapter underscores the urgency of comprehending the multifaceted impacts of climate change and urbanization on urban forests, emphasizing the need for informed conservation and management strategies.

Keywords Urban forestry · Sustainability · Physiological and molecular adaptations · Climate change

A. Malik (✉) · H. Singh · K. Kumar · G. Kumari · A. Sojitra
ICFRE-Forest Research Institute, PO New Forest, Dehradun, Uttarakhand 248006, India
e-mail: apurva.malik14@gmail.com

H. Singh (ed.), *Urban Forests, Climate Change and Environmental Pollution*,
https://doi.org/10.1007/978-3-031-67837-0_8

1 Introduction

Air is a valuable natural resource that is the foundation for human existence on Earth. The environment, particularly the atmosphere, plays a crucial role in the well-being of urban residents. Climatic conditions directly impact both life and property. Air pollution stands out as a considerable global problem, being the most significant environmental threat to human health and responsible for approximately one in every nine deaths annually. Its prevalence in urban areas is particularly worrisome, where high pollutant concentrations and potential health issues converge (Joshi et al. 2009; Kumar et al. 2022, 2024; Singh and Kumar 2022). This is exacerbated by anticipated global population growth, increased urbanization, and the effects of climate change on atmospheric conditions and weather patterns. Rapid urbanization has increased air pollution, excessive noise, and traffic congestion, especially in major cities in developing countries (Sanesi and Chiarello 2006; Singh et al. 2024a; Singh 2024). The associated environmental problems extend beyond air pollution to include water and noise pollution, as well as challenges in waste management.

The accelerated pace of urbanization has also led to a reduction in vegetation cover, with tall buildings and narrow roads creating street canyon-like conditions that impede the dispersion of air pollutants within cities. Road traffic is recognized as a significant contributor to air and noise pollution, adversely affecting human and plant health (Dhyani et al. 2021; Kumar et al. 2021a, b, c, d, e; Prakash et al. 2022). While concentrations of traditional pollutants such as sulphur dioxide (SO_2) and black smoke have decreased significantly, emissions from road traffic have emerged as the primary cause of poor air quality. Changes in the thermal properties of surface materials and the lack of proper evapotranspiration in urban areas may contribute to the urban heat island (UHI) effect (Wong et al. 2010).

Diesel and petrol-powered vehicles contribute to the emission of various pollutants, and the types and proportions of these pollutants depend on the vehicle's technology and operating conditions. Nitric oxide (NO) and nitrogen dioxide (NO_2) are particularly significant phytotoxic pollutants associated with road transport due to their high concentrations of exhaust emissions and adverse effects on plants. The sensitivity and response of different plant species to air pollution vary (Stevovi 2010).

Understanding the adaptive responses of urban forest species, where the delicate balance between natural ecosystems and anthropogenic effects provides particular challenges, is where this information is most essential (Singh 2021; Sharma and Singh 2021). The growth of cities and other human settlements drastically changed the environment. Once natural places for biodiversity, urban forests are now under constant stress from various factors, such as pollution, habitat fragmentation, and climate change (Singh et al. 2021). Plants' reactions to environmental changes are closely tied to their structural and functional characteristics (Gostin 2009; Wang et al. 2011). Environmental alterations lead to anatomical and morphological changes in plant organs. Numerous studies have explored the structure and anatomy of vegetative organs under conditions of environmental pollution (Chaturvedi et al. 2013; Kardel et al. 2010).

Certain plant species resist air pollution, displaying adaptations that allow them to grow. The resistance varies among tree species, with some capable of withstanding pollutants and maintaining average, healthy growth (Magtoto et al. 2013). Among plant organs, leaves are susceptible to environmental stress because most physiological processes, including photosynthesis, occur through leaves.

Eliminating harmful pollutants from the urban environment has become a top priority. In this context, establishing urban green belts is a highly cost-effective biological approach to address and alleviate air pollution in rapidly expanding urban areas by fostering a collaborative relationship between nature and the city (Singh et al. 2020a, b; Kumar et al. 2021a, b). Additionally, urban afforestation holds the potential to tackle urban climate issues through adaptation and mitigation, contributing to the attainment of sustainable development goals. This chapter explores the reactions of urban vegetation to human-induced environmental pollution, examining various aspects of plant responses such as morphological, physiological, biochemical, and molecular dimensions.

Morphological responses include changes in plant growth and development, leaf morphology, and anatomy, whereas physiological investigations uncover the impact on photosynthesis, water relations, and nutrient uptake (Kumar et al. 2021c, d). Biochemical responses reveal metabolic alterations, secondary metabolite production, detoxification processes, and plant defense mechanisms. At the same time, molecular studies shed light on gene expression changes, epigenetic modifications, and signaling pathways involved in stress responses.

2 Morphological Responses of Plants to Urban Environmental Pollution

As stationary organisms, plants display a spectrum of morphological and anatomical responses to grapple with the challenges imposed by urban pollution (Kumar et al. 2019, c). This section delves into the intricate changes in plant growth, leaf morphology and anatomy, and root architecture as a response to the challenges posed by urban environmental pollution.

2.1 *Changes in Plant Growth and Development*

Urban trees, morphologically, exhibit changes in growth patterns, canopy structures, and phenological shifts in response to dynamic factors like temperature variations, altered precipitation patterns, and the urban heat island effect (Yadav et al. 2019a, b; Gupta et al. 2019a, b). The morphology of plant species in urban settings is influenced by the heightened toxicity of the urban environment (Bell et al. 2011). Reduction in leaf size serves as an indicator of a polluted urban environment. Parameters such as the Length Breadth ratio (L/B) and vein-to-vein difference become crucial in assessing air pollution. Analysis of urban plant species reveals physical and potentially functional changes, indicating their adaptation to tolerate air pollution. The changes in morphological parameters disclose the impact of air pollution on plants' photosynthetic activity, leading to a decrease in leaf growth and size. Leaf characteristics, such as the petiole length, are also impacted by air pollution, with studies suggesting a reduction in petioles due to diminished plant biomass (Xu et al. 2009; Tiwari et al. 2006).

Leaf length reflects the plant's ability to withstand stress (Assadi et al. 2011). A decrease in leaf's surface area and petiole length results in less contact with natural toxins, especially air pollution, enhancing the resistance of plants to pollution (Sharma et al. 2018; Singh et al. 2018). Physiological disturbances may contribute to reducing plant species' morphological and anatomical characteristics.

Leaf thickness, a sensitive biophysical property, responds to urban pollution, other abiotic stimuli, and changing climatic conditions (Gupta et al. 2018a, b). Singh et al. (2020a, b) noted an increase in leaf thickness in roadside trees due to the waxy cuticle and the sticky nature of the leaves, allowing dust particles to adhere. This alteration affects the transpiration rate, impacting the leaf's stomatal behaviour.

2.2 *Alterations in Leaf Morphology and Anatomy*

The foliar epidermis is considered a bio-indicator of environmental quality, significantly influencing overall plant development (Singh and Verma 2013; Kumar et al. 2017). The cuticle and epidermal features play a crucial role in determining the resistance/sensitivity of each species to environmental pollutants. Stomatal and epidermal cells' size, lower recurrence, thickening of cell walls, epicuticular wax deposition alterations, and chlorosis are among the auxiliary alterations in leaves subjected to air pollution (Gupta and Singh 2017; Singh et al. 2017a).

Studies report that the size of stomata is smaller and thicker, and the number of stomata is higher in plant species at contaminated sites compared to unpolluted urban areas (El-Khatib 2012). The response of stomata to stress and mechanisms for reducing and controlling pollutant entry into leaves are vital features of plants under stressful conditions (Singh et al. 2017b). Stomatal density, a significant eco-physiological parameter, influences gas exchange between the atmosphere and the plant (Kumar et al. 2020; Singh and Kumar 2022). Plants tend to reduce stomatal size to minimize pollutant entry. Plants resistant to air pollution can compensate for reduced gas exchange and photosynthetic deficiency by increasing pore density. Under stressful conditions, plants adapt to decrease gaseous exchange with the environment, minimizing contact with polluted surroundings (Arriaga et al. 2014; Balasooriya et al. 2009).

Plants also acclimate to air pollution by altering their anatomical structures, such as thickening epidermal cells and increasing the density of stomatal, trichomes, and epidermal cells. A significant increase in the thickness of the palisade and elastic

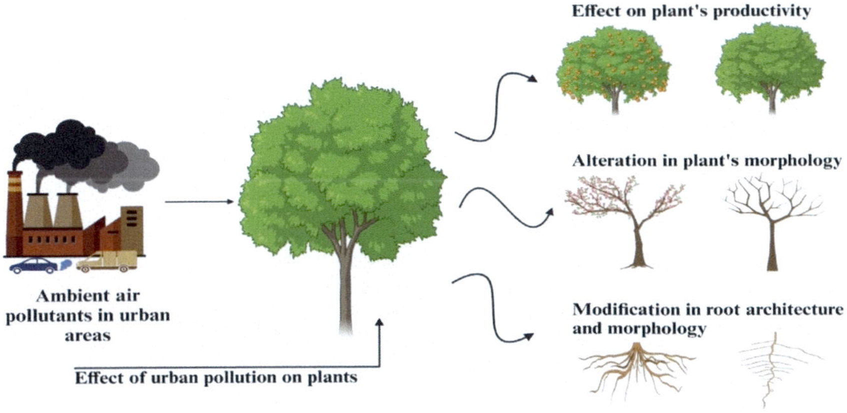

Fig. 1 Impact of air pollutants on development and productivity of urban plant species

mesophyll, lamina, and upper epidermis tissue are other adaptive changes in urban species. Longer trichomes also result from plants' mitigation strategies for urban pollution.

These leaf morphology and anatomy alterations can compromise the plant's ability to capture sunlight and involvement in essential physiological processes (Sadhna et al. 2023). Production of reactive oxygen species (ROS) causes oxidative stress, which results in cellular damage and influences leaf morphology. Studies have demonstrated a direct correlation between air pollution levels, leaf injuries, and deformities in urban plants (Fig. 1).

2.3 *Modifications in Root Architecture and Morphology*

Roots, the critical and only source for nutrient absorption and water uptake, undergo significant modifications in response to urban pollution. Altered soil conditions and pollutants prompt adaptive changes in root architecture. Root systems may become shallower, with a more considerable proliferation of lateral roots, as plants optimize resource acquisition in the face of environmental stress (Yang and Zhang, 2015; Singh et al. 2024b). In addition to changes in architecture, pollution-induced modifications in root morphology can include alterations in root hair density and length. These adaptations aim to enhance nutrient and water absorption efficiency, enabling plants to withstand the challenges posed by urban pollution (Nitoslawskiet al. 2017).

3 Physiological Responses of Plants to Urban Environmental Pollution

The physiological responses of urban trees to climate change encompass intricate adjustments in processes such as photosynthesis, transpiration, and water-use efficiency. These responses provide insights into the mechanisms trees employ to cope with changing environmental conditions (Meena et al. 2011; Kumar et al. 2019c). Examining physiological adaptations is essential for unraveling the complex interplay between urban trees and their environment.

3.1 Photosynthetic Adaptations and Limitations

The physiological well-being of urban plants is adversely affected by the toxic elements in environmental pollution. Various pollutants, such as those emitted from vehicles, industries, and the burning of solid waste, significantly impact the chlorophyll content of plants (Yadav et al. 2019a, b; Jangra et al. 2024; Kanta et al. 2024). Chlorophyll, a fundamental feature in plants, regulates various physiological activities, and its reduction due to pollution-induced toxicity results in environmental stress (Malik et al. 2024; Kumar and Singh 2024). This stress negatively influences the chloroplast machinery, subsequently hindering the growth and development of the plants.

Sulphur dioxide (SO_2), known for its phytotoxic nature, plays a role in leaching chlorophyll pigment, leading to the formation of pheophytin. This process, documented by Jyothi and Jaya (2010), contributes to the deterioration of photosynthetic rate, stomatal conductance, transpiration, and other physiological processes under adverse environmental conditions.

Particulate matter, including dust from traffic and industrial pollutants in the air, can accumulate on leaf surfaces. This deposition obstructs stomatal apertures, limiting oxygen uptake and disrupting light absorption. Combining these mechanisms inhibits energy dissipation in the antenna protein's light-harvesting system, thus impeding the energy needed for photosynthetic reactions (Horton and Ruban 2005). The clogged stomatal pores reduce transpiration and elevate CO_2 concentration within the leaf (Singh et al. 2022).

Stomatal conductance, a conceptually important factor, relies on various environmental conditions and the position and age of tree leaves. Any disruption in stomatal behaviour alters gaseous exchange, reducing the rate of CO_2 assimilation and water exchange. This interruption in physiological processes underscores the vulnerability of urban plants to the detrimental effects of environmental pollution.

3.2 *Impact on Plant Water Relations and Water Use Efficiency*

The impact of urban environmental pollution on plant-water relations and stomatal behaviour is significant. The presence of pollutants on leaf surfaces disrupts stomatal function, resulting in changes to the dynamics of water exchange within the plant. In polluted urban environments, membranous structures are disrupted, leading to alterations in stomatal morphology, and the impairment of the electron transport chain has been observed (Kulshrestha and Saxena 2016). This reduction in stomatal conductance under polluted conditions contributes to a decline in water uptake by the roots, posing a challenge to overall plant hydration, particularly in urban areas where water availability is limited.

Under stressful conditions, plants adopt strategies to conserve water in storage organs, compensating for the compromised water uptake through the roots. The relative water content, a crucial parameter in plant physiology, plays a pivotal role in maintaining a balance between water uptake and release. Higher relative water content in plants becomes instrumental in preserving physiological equilibrium during conditions of contamination and toxicity, providing resilience to the adverse effects of pollutants. Additionally, the reduced transpiration rate resulting from suppressed stomatal conductance, induced by the deposition of particulate matter on leaves, enhances physiological water use efficiency. This mechanism contributes to the plant's coping with water stress (Singh et al. 2020a, b).

Studies have indicated that elevated levels of air pollutants, such as ozone and sulfur dioxide, can lead to stomatal closure, further intensifying plant water stress (Kollist et al. 2014). These findings underscore the intricate relationship between urban pollution and plant-water relations, highlighting the multifaceted impacts of pollutants on stomatal behaviour and overall water dynamics within the plant.

3.3 *Effects on Nutrient Uptake and Assimilation*

The impact of urban pollutants extends to nutrient uptake and assimilation in plants, disrupting essential physiological processes. Contaminants in soil and air can modify nutrient availability, affecting the plant's capacity to absorb crucial elements. Common urban pollutants, including heavy metals, tend to accumulate in plant tissues, impeding nutrient uptake and causing imbalances in nutrient concentrations.

Plants also contribute significantly to monitoring and maintaining ecological balance by actively participating in nutrient and gas cycling, such as carbon dioxide and oxygen. Air pollutants increase cell permeability, resulting in water and dissolved nutrient loss and premature leaf senescence (Bakiyaraj and Ayyappan, 2014).

Furthermore, pollutants directly influence nutrient assimilation pathways within plant cells. For example, nitrogen dioxide (NO_2) and sulfur dioxide (SO_2) have demonstrated interference with nitrogen and sulfur metabolism in plants, impacting protein synthesis and overall plant growth (Luo et al. 2018). Recognizing and comprehending these disruptions are pivotal for developing strategies to mitigate the adverse effects of urban pollution on plant nutrient dynamics.

4 Biochemical Responses of Plants to Urban Environmental Pollution

Urban environmental pollution poses a significant threat to plant life, triggering biochemical responses indicative of stress and adaptation mechanisms. Exposure to pollutants like heavy metals and particulate matter induces oxidative stress, producing reactive oxygen species (ROS). Plants activate antioxidant enzymes such as superoxide dismutase, catalase, and peroxidase to counteract this. Changes in metabolic pathways and synthesizing specialized compounds, including phytochelatins and metallothioneins, further contribute to detoxification. Additionally, modifications in cell wall composition, such as increased lignification, enhance structural integrity against pollutants. This comprehensive understanding emphasizes the multifaceted impact of urban pollutants on plant biochemistry, ultimately compromising essential physiological functions and impeding overall plant growth and productivity.

4.1 Metabolic Alterations and Secondary Metabolite Production

Metabolic adjustments and the production of secondary metabolites are pivotal biochemical responses of plants confronted with urban environmental pollution. Faced with prevalent pollutants in urban environments, plants undergo notable alterations in their metabolic pathways, influencing secondary metabolites like phenolic compounds, flavonoids, and alkaloids. Phenol, known for its role as a free radical scavenger, shields plants from oxidative stress-induced damage. Chandawat et al. (2014) reported the significance of phenol in plants' resistance mechanisms against air pollution, highlighting higher total phenol levels in leaves from polluted sites compared to controls. Furthermore, air pollution has been linked to increased accumulation of phenolic content in leaves. Qayoom Mir et al. (2009) noted that alterations in concentrations of total flavonoids and phenolics in *Catharanthus roseus* L. and *Ocimum sanctum* L. could potentially serve as biomarkers for urban automobile pollution, showing a positive correlation with vehicular pollution load across different sites.

4.2 Detoxification Processes and Defense Mechanisms

Plants employ sophisticated detoxification processes and defense mechanisms in urban environmental pollution to cope with harmful pollutants. Detoxification involves the conversion or sequestration of toxic substances, and plants have evolved various mechanisms to achieve this. One prominent strategy is the activation of detoxifying enzymes, such as cytochrome P450s and glutathione S-transferases, which facilitate the breakdown and conjugation of pollutants. Additionally, plants enhance their defense mechanisms by producing specialized proteins and peptides that can bind to, sequester, or neutralize pollutants. This includes metallothioneins that bind heavy metals and phytochelatins that chelate metal ions, preventing their toxic effects. Reinforcing cell walls and forming reactive oxygen species (ROS) are also integral parts of the plant's defense arsenal.

These secondary metabolites assume diverse roles, functioning as defense mechanisms against environmental stress and signalling molecules in interactions between plants and pollutants. The heightened production of secondary metabolites often correlates with the plant's adaptive response to counter pollutants-induced oxidative stress. Enhanced tolerance in plants exposed to SO_2, NO_2, and O_3, whether individually or in combination, has been linked to increased activities of superoxide dismutase (SOD) and peroxidase (POD), indicating the interlinked primary protective role of the SOD and CAT/POD enzyme system in mitigating cellular injury. Studies have reported a decrease in cellular damage mediated by oxy-radicals in plants with elevated SOD levels. Additionally, it has been documented that catalase (CAT) activity increases in plants exposed to polluted areas (Vollenweider et al. 2003).

4.3 Changes in Enzyme Activities and Protein Expression

Urban environmental pollution induces notable changes in plant enzyme activities and protein expression as part of the intricate physiological responses to stress. Elevated levels of pollutants prompt alterations in enzyme activities, particularly those involved in antioxidative defense. Enzymes such as superoxide dismutase (SOD), catalase (CAT), and peroxidase become upregulated to counteract oxidative stress caused by pollutants, highlighting the plant's adaptive strategy. Stress-responsive proteins, including heat shock and pathogenesis-related proteins, are often overexpressed in response to urban pollution. These proteins play crucial roles in cellular protection, repair, and signalling.

Pukacka and Pukacki (2000) demonstrated that activities of SOD, POX, and APX were notably elevated in Scots pine (*Pinus sylvestris* L.) needles obtained from polluted areas, specifically around a phosphate fertilizer factory and a copper foundry, compared to a control area throughout all seasons. The needle around the fertilizer factory indicated prolonged exposure to high sulfur dioxide (SO_2) concentrations. Conversely, leaves from the copper foundry area are influenced by elevated concentrations of SO_2 and Nitrogen oxides (NOx) in the air. The investigation by Dziri and Hosni (2012) explored the impact of cement dust on the antioxidative metabolism of Aleppo pine (*Pinus halepensis*) needles. The study revealed a substantial increase in the activities of all antioxidant enzymes, consistent with the earlier findings. The authors observed positive correlations between enzyme activities, indicating a cooperative action of these antioxidant enzymes in responding to induced oxidative stress.

Accordingly, several studies have documented decreased soluble protein content when different organisms are exposed to stress (Doğanlar and Atmaca, 2011). In addition, Sairam et al. (2003) found the highest protein content in the youngest leaf (1st), whereas decreased in the older leaves (10th) of *Helianthus annuus* exposed to metal contamination. This might be due to continuous and linear increase in oxidative stress from the youngest to the oldest leaf. On the other hand, as noted by Radotic et al. (2000), there was an increase in total protein concentration in long-term cadmium-treated spruce needles compared to the short-term treated ones. The remarkable growth in total protein contents in Oztetik's (2015) study supports the idea that the high protein content reflects a better protection capacity of plants and a defensive mechanism recruited to overcome severe environmental conditions (Fig. 2).

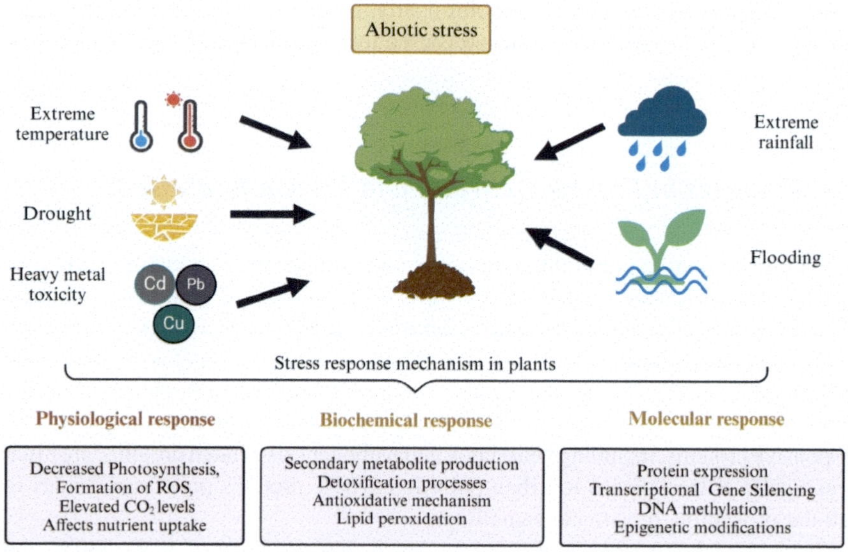

Fig. 2 Plant's responses to abiotic stressful urban environmental conditions

5 Molecular Responses of Plants to Urban Environmental Pollution

Pollutants such as sulphur dioxide (SO_2), nitrogen oxides (NO), particulate matter (PM), and heavy metals are profound in higher amounts in urban settings. These contaminants can cause various chemical reactions in plants, impacting several biological functions such as photosynthesis, transpiration, conductance, etc. According to a study conducted by Smith et al., urban-adapted plant species exposed to high PM levels reveal substantial changes in gene expression associated with photosynthesis, indicating a biological response to environmental stress.

The chloroplast's unique component of plants is the organelle responsible for photosynthesis, intracellular signalling, and synthesizing several molecules essential to metabolism, including lipids, vitamins, hormones, amino acids, and nucleotides. In addition, chloroplasts function as environmental sensors through plastid-to-nucleus during stress, maximizing nuclear gene expression following physiological needs (Chan et al. 2016). Reactive oxygen species (Pogson et al. 2008), intermediates in tetrapyrrole biogenesis (Larkin, 2016), the redox state of plastids (Piippo et al. 2006), secondary metabolites in chloroplasts (Estavillo et al. 2011; Xiao et al. 2012) and chloroplast gene expression (Jung and Chory, 2010; Leister et al. 2011) are among the several potential retrograde signalling pathways.

The genes encoding chloroplast must be expressed correctly for chloroplast formation and photosynthesis. Many studies have been conducted in the last decade to investigate the molecular processes controlling the expression of the chloroplast gene by genetic techniques. It has been determined that a large number of nucleus-encoded proteins control the expression of chloroplast genes. These protein mutants are nonetheless susceptible to various environmental stressors (Chi et al. 2013; Tadini et al. 2016; Nawaz and Kang, 2017; Robles and Quesada, 2019). These results imply that chloroplast gene expression and responses to environmental stress are related, but this connection has not received as much attention.

Urban environmental pollution poses complex challenges for plants, necessitating intricate molecular adaptations to thrive in adverse conditions. This section explores the molecular responses of plants to urban pollution, focusing on gene expression changes and regulatory mechanisms, epigenetic modifications, DNA methylation, and the signalling pathways involved in stress responses.

5.1 Gene Expression Changes and Regulatory Mechanisms

Plants respond to urban pollution at the molecular level through dynamic changes in gene expression patterns and regulatory mechanisms. The upregulation or down-regulation of specific genes is a common molecular strategy plants employ to adapt to environmental stressors. When faced with stress, plants initiate various regulatory systems. An initial response involves an increase in active oxygen forms, which, when elevated, can lead to cell death by negatively affecting proteins, lipids, and nucleic acids. To counteract this, the plant activates a complex antioxidant system to balance the pro-oxidant system. Studies have identified critical genes associated with pollutant detoxification, antioxidant defense, and regulating metabolic pathways in response to urban pollution.

Regulatory elements such as transcription factors (TFs) are pivotal in orchestrating gene expression changes for protective proteins. The activation or repression of these transcription factors modulates the plant's response to environmental stress. TFs can be categorized into four major groups based on the homology of their DNA-binding domain structure. For instance, MYB, WRKY, and NAC transcription factors have regulated genes involved in stress responses, including those related to detoxification processes and antioxidant defense mechanisms. Notably, the APETALA2/ethylene-responsive factor (AP2/ERF), WRKY, myeloblastosis (MYB), NAM/ATAF/CUC (NAC), and basic leucine zipper (bZIP) families are among the first to be activated during abiotic stress responses (Wu et al. 2022).

Also, the impact of heavy metals activates numerous gene and protein-binding signal pathways that contribute to developing resistance against heavy metals. This activation involves the expression of many protective genes facilitated by transcription factors (TF). Studies revealed that the chlorophyll content of transgenic poplar leaves was higher than that of the wild type, reflecting a protective mechanism of the PyWRKY71 gene. Additionally, the activities of other antioxidants, including POD, SOD, CAT, and MDA, were elevated in transgenic poplars, bolstering their tolerance to Cd stress (Chen et al. 2023).

5.2 *Epigenetic Modifications and DNA Methylation*

Epigenetic modifications, such as DNA methylation, play a significant role in shaping the molecular responses of plants to urban pollution. Interestingly, exposure to heat stress in urban areas can lead to increased genetic instability and higher rates of somatic homologous recombination. Chromatin configuration at specific locations in the DNA influences somatic homologous recombination, and heat stress may affect genetic stability by modifying chromatin structure and accessibility of DNA for repair and recombination. DNA methylation involves adding a methyl group to the DNA molecule, influencing gene expression without altering the underlying DNA sequence. It is indicated that urban pollution can induce changes in DNA methylation patterns, leading to altered gene expression profiles in plants (Cortijo et al. 2017).

Recent findings highlight a specific variant of histone H2A as a thermosensor, regulating temperature-dependent gene expression. The effects of heat-induced acclimation may pass to future generations through epigenetic mechanisms. However, somatic recombination rates do not increase in subsequent heat-exposed generations.

Pecinka et al. (2010) identified that extended exposure to high temperatures induces a temporary increase in the transcriptional activity of transgenic and specific endogenous repeats regulated by transcriptional gene silencing (TGS). These changes occur independently of senescence, DNA repair, and heat stress signalling. Surprisingly, transcription activation under heat stress does not necessitate DNA demethylation.

The epigenetic regulation of stress-responsive genes allows plants to fine-tune their responses to environmental challenges. DNA methylation patterns can serve as epigenetic markers, reflecting the plant's exposure to urban pollutants and providing insights into its adaptive strategies. Investigating the interplay between epigenetic modifications and gene expression dynamics enhances our understanding of the molecular basis of plant responses to urban environmental pollution.

5.3 Signalling Pathways Involved in Stress Responses

Plants employ intricate signalling pathways to perceive and respond to environmental stressors associated with urban pollution. Key signalling molecules such as salicylic acid (SA), jasmonic acid (JA), and ethylene play crucial roles in orchestrating stress responses (Pieterse et al. 2009). These signalling pathways activate a cascade of molecular events that lead to the induction of stress-responsive genes and the synthesis of defense-related proteins.

Mitogen-activated protein kinase (MAPK) cascades are crucial in transferring information from sensors to cellular responses, representing a conserved signalling mechanism in higher trees. These cascades serve as convergence points in plant defense signalling under biotic and abiotic conditions. The MPAK cascades initiate the activation of MAPKKK in response to stimuli. MAPKKK then phosphorylates into MAPKK, which, in turn, phosphorylates MAPK. Notably, plants exhibit many MAPK components, with over 20 MAPKs, 10 MAPKKs, and 60 MAPKKKs. Plants respond to elevated copper (Cu) and cadmium (Cd) stress levels by activating distinct MAPK pathways. In *Medicago sativa* L. plants, exposure to excess Cu or Cd ions triggers the activation of four specific MAPKs: SIMK (salt stress-induced MAPK), MMK2 (Medicago MAPKK), MMK3 (Medicago MAPKKK), and SAMK (stress-activated MAPK). The kinetics of MAPK activation show that SIMK, MMK2, MMK3, and SAMK respond rapidly to Cu ions, while Cd ions induce delayed MAPK activation (Jonak et al. 2004) (Fig. 3).

Mitogen-activated protein kinase (MAPK) cascades also act as central components in transmitting stress signals from the cell surface to the nucleus, triggering the appropriate transcriptional responses (Nakagami et al. 2005). Unraveling the intricacies of these signalling pathways provides a comprehensive view of how plants perceive and integrate signals in response to urban pollution. Stress-responsive genes, activated through these signalling pathways, often encode proteins involved in antioxidant defense, detoxification, and the repair of cellular damage. The modulation of these signalling networks allows plants to mount effective defense responses against the oxidative stress and toxicity induced by urban pollutants.

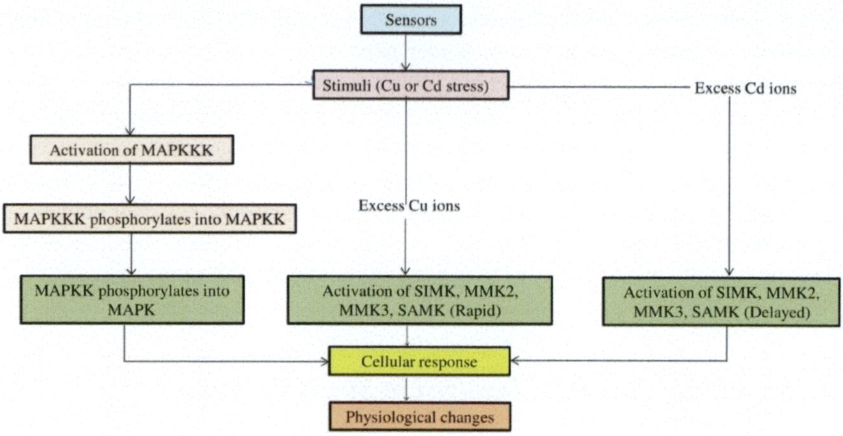

Fig. 3 Mitogen-activated protein kinase (MAPK) cascade in plants under Cu and Cd stress

6 Implications for Urban Environmental Management and Sustainable Green Spaces

As urbanization reshapes landscapes, compelling environmental management and creating sustainable green spaces become imperative. This chapter explores critical implications for urban environmental management, focusing on plant selection for urban greening projects, the potential use of plants as bio-indicators of pollution, and the vital role of plants in mitigating and improving urban air and soil quality.

6.1 Importance of Plant Selection for Urban Greening Projects

The success of urban greening projects hinges on thoughtful plant selection, considering various factors such as adaptability to urban conditions, resistance to pollutants, and contribution to biodiversity. Carefully chosen plant species enhance aesthetic appeal and play a pivotal role in improving overall environmental quality. Native and resilient plant species should be prioritized for urban green spaces, as they are better adapted to local climates, require less maintenance, and contribute to local biodiversity conservation.

Furthermore, selecting a diverse range of plant species with varying heights, textures, and growth habits can create a more resilient and visually appealing urban landscape. This diversity promotes ecological balance, supports local wildlife, and enhances the overall functionality of green spaces. Sustainable urban design should integrate the principles of biophilic cities, emphasizing the symbiotic relationship between urban environments and nature to improve residents' well-being and foster environmental sustainability (Beatley 2011).

6.2 *Potential for Using Plants as Bio-indicators of Pollution*

Plants can serve as bio-indicators, offering valuable insights into pollution in urban environments. Plant species exhibit visible changes in morphology, physiology, and biochemistry in response to environmental stressors, making them adequate air and soil quality indicators. Monitoring these changes can help identify pollution sources, assess the effectiveness of pollution control measures, and guide urban environmental management strategies.

Leaf damage, changes in chlorophyll content, and alterations in plant growth patterns are commonly observed responses to pollution. Combining traditional monitoring methods with modern molecular techniques enables a comprehensive understanding of how plants respond to different pollutants. Integrating plant-based bio-indicators into urban environmental monitoring programs can provide real-time data, aiding in timely interventions to mitigate pollution and improve overall environmental quality.

6.3 *Role of Plants in Mitigating and Improving Urban Air and Soil Quality*

Plants are crucial in mitigating and improving urban air and soil quality through various ecological processes. Trees and vegetation act as natural air filters by capturing and trapping particulate matter, absorbing pollutants, and releasing oxygen. The urban heat island effect can be alleviated by strategically planting trees to provide shade and reduce surface temperatures.

In addition to improving air quality, plants contribute to soil health and structure. Their root systems stabilize soil, preventing erosion, and promote microbial activity that enhances nutrient cycling. Phytoremediation, using plants to remove pollutants from soil, is an innovative approach to address contaminated urban soils. Certain plant species accumulate and detoxify heavy metals, hydrocarbons, and other contaminants, offering a sustainable and cost-effective solution to soil remediation.

7 Future Directions and Research Gaps

This section explores future directions and research gaps in these fields, highlighting emerging areas of research and technologies and identifying gaps and unresolved questions that require further investigation.

7.1 Emerging Areas of Research and Technologies

7.1.1 Smart Green Infrastructure

Integrating technology into urban green spaces, often called smart green infrastructure, represents a promising avenue for future research. Utilizing sensors, IoT devices, and data analytics can enhance the efficiency of green space management. Smart technologies can monitor environmental parameters, such as air and soil quality, plant health, and water usage, providing real-time data to inform decision-making and optimize resource allocation. Research in this area can explore the potential of smart green infrastructure to create more adaptive and resilient urban ecosystems.

7.1.2 Biodiversity Conservation in Urban Settings

Understanding and promoting biodiversity within urban green spaces is an emerging focus for research. Investigating the impact of different design elements, vegetation types, and management practices on urban biodiversity can inform strategies for creating ecologically vibrant and sustainable green spaces. Incorporating native plant species and designing interconnected green corridors may contribute to maintaining biodiversity and supporting essential ecosystem services in urban environments.

7.1.3 Green Roofs and Vertical Gardens

Exploring the potential of green roofs and vertical gardens as innovative solutions for urban environmental management is an evolving area of research. These features can mitigate the urban heat island effect, improve air quality, and contribute to building energy efficiency. Research could delve into the optimal plant species, substrates, and maintenance practices for maximizing the ecological and aesthetic benefits of green roofs and vertical gardens in different urban contexts.

7.2 *Identification of Gaps and Unresolved Questions*

7.2.1 Long-Term Impact of Urban Greening

While many studies have demonstrated the immediate benefits of urban greening, there is a gap in understanding the long-term impact of these interventions. Research should focus on the sustained ecological, social, and economic benefits of green spaces over extended periods. Examining the longevity of plantings, the evolution of biodiversity, and the influence of green spaces on community well-being over time will provide valuable insights for urban planning and management.

7.2.2 Socioeconomic Equity in Green Space Access

Research gaps exist in understanding the equitable distribution of green spaces across different socioeconomic groups within urban areas. Investigating the accessibility, quality, and usability of green spaces for diverse communities can shed light on potential disparities. This research can inform policies and interventions to ensure that all residents enjoy the benefits of green spaces, regardless of their socioeconomic status.

7.2.3 Integration of Cultural Ecosystem Services

While the environmental benefits of green spaces are well-documented, there is a research gap in understanding the cultural ecosystem services they provide. Exploring how green spaces contribute to cultural identity, community cohesion, and mental well-being will enhance our understanding of the multifaceted value of urban green spaces. This knowledge may guide the integration of cultural ecosystem services into urban planning frameworks.

8 Conclusion

In conclusion, urban forests, vital elements of metropolitan landscapes, are increasingly vulnerable to the complex impacts of climate change and environmental stressors. Morphological changes in urban tree species, influenced by shifting temperature regimes, precipitation patterns, and urban heat island effects, signify their adaptation to urban environments. Biochemical responses, including alterations in secondary metabolite production, antioxidant systems, and stress-responsive proteins, reveal the adaptive strategies employed by trees in urban settings. Understanding detoxification processes and defense mechanisms is crucial for developing strategies to enhance phytoremediation and foster sustainable urban ecosystems. Molecular responses of plants to urban environmental pollution involve intricate signaling pathways, emphasizing the need for effective urban environmental management and sustainable green spaces to create livable and resilient cities. Integrating plant selection for urban greening projects, utilizing plants as bio-indicators of pollution, and recognizing their multifaceted role in mitigating pollution are critical for achieving these goals. Incorporating these principles into urban planning and management practices may cultivate healthier environments, enhance biodiversity, and promote the well-being of inhabitants, contributing to a sustainable and vibrant urban future. Identifying gaps and unresolved questions in future research directions is essential for expanding scientific knowledge and creating healthier, more resilient urban environments. The insights from exploring plant responses to urban environmental pollution offer valuable knowledge for mitigating pollution impacts and promoting healthier urban ecosystems.

References

Arriaga MO, Stampacchio ML, Pepi MGF, Perelman PE, Faggi AM (2014) Use of epidermal characters as bioindicators of environmental pollution. Multequina 23:41–53

Assadi A, Pirbalouti AG, Malekpoor F, Teimori N, Assadi L (2011) Impact of air pollution on physiological and morphological characteristics of *Eucalyptus camaldulensis*. J Food Agricult Environ 9(2):676–679

Bakiyaraj R, Ayyappan D (2014) Air pollution tolerance index of some terrestrial plants around an industrial area. Int J Mod Res Rev 2(1):1–7

Balasooriya BLWK, Samson R, Mbikwa F, Boeckx P, Van Meirvenne M (2009) Biomonitoring of urban habitat quality by anatomical and chemical leaf characteristics. Environ Exper Bot 65(2–3):386–394

Beatley T (2011) Biophilic cities: integrating nature into urban design and planning. Island Press

Bell JNB, Honour SL, Power SA (2011) Effects of vehicle exhaust emissions on urban wild plant species. Environ Pollut 159(8–9):1984–1990

Chan KX, Phua SY, Crisp P, McQuinn R, Pogson BJ (2016) Learning the languages of the chloroplast: retrograde signaling and beyond. Annu Rev Plant Biol 67:25–53

Chandawat DK, Verma PU, Solanki HA, Patel YM (2014) Role of total phenol in the resistance mechanism of plants against air pollution. Biolife 2(2):586–592

Chaturvedi RK, Prasad S, Rana S, Obaidullah SM, Pandey V, Singh H (2013) Effect of dust load on the leaf attributes of the tree species growing along the roadside. Environ Monit Assess 185:383–391

Chen X, Wu X, Han C, Jia Y, Wan X, Liu Q, Zhang F (2023) A WRKY transcription factor, PyWRKY71, increased the activities of antioxidant enzymes and promoted the accumulation of cadmium in poplar. Plant Physiol Biochem 205:108163

Chi W, Sun X, Zhang L (2013) Intracellular signaling from plastid to nucleus. Annu Rev Plant Biol 64:559–582

Cortijo S, Charoensawan V, Brestovitsky A, Buning R, Ravarani C, Rhodes D et al (2017) Transcriptional regulation of the ambient temperature response by H2A.Z nucleosomes and HSF1 transcription factors in Arabidopsis. Mol Plant 10(10):1258–1273

Dhyani K, Kuniyal HB, Singh H, Sobha S (2021) Growth and physiological potential of *Terminalia arjuna* under elevated CO_2 levels in Open top chamber condition. J Appl Nat Sci 13(3):1121–1126

Doğanlar ZB, Atmaca M (2011) Influence of airborne pollution on Cd, Zn, Pb, Cu, and Al accumulation and physiological parameters of plant leaves in Antakya (Turkey). Water Air Soil Pollut 214:509–523

Dziri S, Hosni K (2012) Effects of cement dust on volatile oil constituents and antioxidative metabolism of Aleppo pine (*Pinus halepensis*) needles. Actaphysiologiaeplantarum 34(5):1669–1678

El-Khatib A (2012) Morpho-anatomical characteristics of olive (*Olea europaea* L.) trees leaf as bio-indicator of cement dust air pollution in Libya. J Environ Stud 9(1):65–72

Estavillo GM, Crisp PA, Pornsiriwong W, Wirtz M, Collinge D, Carrie C, Pogson BJ (2011) Evidence for a SAL1-PAP chloroplast retrograde pathway that functions in drought and high light signaling in Arabidopsis. Plant Cell 23(11):3992–4012

Gostin IN (2009) Structural modification induced by air pollutants in *Plantago lanceolata* leaves. AnaleleUniversitătii Din Oradea FasciculaBiologie 16(1):61–65

Gupta SK, Singh H (2017) Observations on sporadic flowering in *Ficus benjamina* L. and *Peltophorum africanum* L. in new forest. Indian for 143:290–291

Gupta SK, Ram J, Singh H (2018a) Comparative study of transpiration in cooling effect of tree species in the atmosphere. J Geosci Environ Protect 6:151–166. https://doi.org/10.4236/gep.2018.68011

Gupta SK, Ram J, Singh H, Kumar P (2018b) Comparative study of cooling under shades of some forestry tree species with respect to ambient temperatures. J Geosci Environ Protect 6:51–66. https://doi.org/10.4236/gep.2018.611004

Gupta SK, Ram J, Singh H (2019a) Relationship between leaf area index and atmospheric cooling of tree species. Indian J for 42(3):203–208

Gupta SK, Singh H, Ram J (2019b) Need for a novel paradigm in forestry research: perspectives of comparative atmospheric cooling by different forestry effect tree species. Indian for 145(6):509–515

Horton P, Ruban A (2005) Molecular design of the photosystem II light-harvesting antenna: photosynthesis and photoprotection. J Exper Bot 56(411):365–373

Jangra A, Kumar K, Maikhuri S, Bhandari MS, Pandey S, Singh H, Barthwal S (2024) Unveiling stress-adapted endophytic bacteria: characterizing plant growth-promoting traits and assessing cross-inoculation effects on *Populus deltoides* under abiotic stress. Plant Physiol Biochem 210:108610. https://doi.org/10.1016/j.plaphy.2024.108610

Jonak C, Nakagami H, Hirt H (2004) Heavy metal stress: activation of distinct mitogen-activated protein kinase pathways by copper and cadmium. Plant Physiol 136(2):3276–3283

Joshi N, Chauhan A, Joshi PC (2009) Impact of industrial air pollutants on some biochemical parameters and yield in wheat and mustard plants. Environmentalist 29:398–404

Jung HS, Chory J (2010) Signaling between chloroplasts and the nucleus: can a systems biology approach bring clarity to a complex and highly regulated pathway? Plant Physiol 152(2):453–459

Jyothi SJ, Jaya DS (2010) Evaluation of air pollution tolerance index of selected plant species along roadsides in Thiruvananthapuram, Kerala. J Environ Biol 31(3):379–386

Kanta C, Kumar A, Chauhan A, Singh H, Sharma IP (2024) The interplay between plant functional traits and climate change. In: Kumar N, Singh H (eds) Plant functional traits for improving productivity. Springer, Singapore. https://doi.org/10.1007/978-981-97-1510-7_3

Kardel F, Wuyts K, Babanezhad M, Wuytack T, Potters G, Samson R (2010) Assessing urban habitat quality based on specific leaf area and stomatal characteristics of *Plantago lanceolata* L. Environ Pollut 158(3):788–794

Kollist H, Nuhkat M, Roelfsema MRG (2014) Closing gaps: linking elements that control stomatal movement. New Phytol 203(1):44–62

Kulshrestha U, Saxena P (eds) (2016) Plant responses to air pollution. Springer, Singapore, pp 1–195

Kumar N, Singh H (2024) Plant functional traits for improving productivity. Springer, Singapore

Kumar N, Suyal DC, Sharma IP, Verma A, Singh H (2017) Elucidating stress proteins in rice (*Oryza sativa* L.) genotype under elevated temperature: a proteomic approach to understand heat stress response. 3 Biotech 7:205. https://doi.org/10.1007/s13205-017-0856-9

Kumar N, Jeena N, Singh H (2019) Elevated temperature modulates rice pollen structure: a study from the foothill of the Himalayan agro-ecosystem in India. 3 Biotech 9(5):175. https://doi.org/10.1007/s13205-019-1700-1

Kumar N, Jeena N, Kumar N, Gangola S, Singh H (2019c) Phytoremediation facilitating enzymes: an enzymatic approach for enhancing remediation process. In: Bhatt P (ed) Smart bioremediation technologies. Academic Press, Cambridge, pp 289–306

Kumar A, Kumar P, Singh H, Kumar N (2021a) Modulation of plant functional traits under essential plant nutrients during seasonal regime in natural forests of Garhwal Himalayas. Plant Soil 465:197–212. https://doi.org/10.1007/s11104-021-05003-x

Kumar A, Kumar P, Singh H, Bisht S, Kumar N (2021b) Relationship of physiological plant functional traits with soil carbon stock in the temperate forest of Garhwal Himalaya. Curr Sci 120(8):1368–1373

Kumar A, Kumar P, Singh H, Kumar N (2021c) Adaptation and mitigation potential of roadside trees with bio-extraction of heavy metals under vehicular emissions and their impact on physiological traits during seasonal regimes. Urban for Urban Green 58:126900. https://doi.org/10.1016/j.ufug.2020.126900

Kumar A, Kumar P, Singh H, Kumar N (2021d) Impact of plant functional traits on infiltration rate under different forests of Kempty watershed in Garhwal Himalaya. Indian J Soil Conser 49(1):38–44

Kumar M, Phukon SN, Paygude AC, Tyagi K, Singh H (2021e) Mapping phenological functional types (PhFT) in the Indian Eastern Himalayas using machine learning algorithm in Google Earth Engine. Comput Geosci 158:104982. https://doi.org/10.1016/j.cageo.2021.104982

Kumar A, Singh H, Kumari G, Bisht S, Malik A, Kumar A, Singh M, Raturi S, Barthwal S, Thakur A, Kaushal R (2022) Adaptive resilience of roadside trees to vehicular emissions via leaf enzymatic, physiological, and anatomical trait modulations. Environ Pollut 313:120191. https://doi.org/10.1016/j.envpol.2022.120191

Kumar N, Singh H, Giri K, Kumar A, Joshi A, Yadav A, Singh R, Bisht S, Kumari R, Jeena N, Khairakpam R, Mishra G (2024) Physiological and molecular insights into the allelopathic effects on agroecosystems under changing environmental conditions. Physiol Mol Biol Plants 30:417–433. https://doi.org/10.1007/s12298-024-01440-x

Kumar N, Singh H, Sharma SK (2020) Antioxidants: responses and importance in plant defense system. In: Roychowdhury R, Choudhury S, Hasanuzzaman M, Srivastava S (eds) Sustainable agriculture in the era of climate change. Springer, Cham

Larkin RM (2016) Tetrapyrrolesignaling in plants. Front. Plant Sci 7:1586

Leister D, Wang X, Haberer G, Mayer KF, Kleine T (2011) Intracompartmental and intercompartmental transcriptional networks coordinate the expression of genes for organellar functions. Plant Physiol 157(1):386–404

Luo M, Hou X, Gu Y, Lau NC, Yim SHL (2018) Trans-boundary air pollution in a city under various atmospheric conditions. Sci Total Environ 618:132–141

Magtoto LM, Mones DG, Lomahan FP (2013) Morpho-anatomical characterization of *Tithonia diversifolia* (Hemsl.) gray growing on sites exposed to vehicular emissions. Int J Plant Anim Environ Sci 3(3):168–172

Malik A, Kumar K, Kumari G, Kumar N, Singh H (2024) Application of functional traits in modelling productivity and resilience under climate change. In: Kumar N, Singh H (eds) Plant functional traits for improving productivity. Springer, Singapore

Meena D, Singh H, Chaudhari SK (2011) Elucidating strontium response on growth dynamics and biochemical change in *Phaseolus mungo* L. Int J Agricult Environ Biotechnol 4:107–113

Nakagami H, Pitzschke A, Hirt H (2005) Emerging MAP kinase pathways in plant stress signalling. Trends Plant Sci 10(7):339–346

Nawaz G, Kang H (2017) Chloroplast-or mitochondria-targeted DEAD-box RNA helicases play essential roles in organellar RNA metabolism and abiotic stress responses. Front Plant Sci 8:871

Nitoslawski SA, Steenberg JW, Duinker PN, Bush PG (2017) Assessing the influence of location attributes on urban forest species composition in suburban neighbourhoods. Urban for Urban Green 27:187–195

Oztetik E (2015) Biomarkers of ecotoxicological oxidative stress in an urban environment: using evergreen plant in industrial areas. Ecotoxicology 24:903–914

Pecinka A, Dinh HQ, Baubec T, Rosa M, Lettner N, Scheid OM (2010) Epigenetic regulation of repetitive elements is attenuated by prolonged heat stress in Arabidopsis. Plant Cell 22(9):3118–3129

Pieterse CM, Leon-Reyes A, Van der Ent S, Van Wees SC (2009) Networking by small-molecule hormones in plant immunity. Nat Chem Biol 5(5):308–316

Piippo M, Allahverdiyeva Y, Paakkarinen V, Suoranta UM, Battchikova N, Aro EM (2006) Chloroplast-mediated regulation of nuclear genes in Arabidopsis thaliana in the absence of light stress. Physiol Genom 25(1):142–152

Pogson BJ, Woo NS, Förster B, Small ID (2008) Plastid signalling to the nucleus and beyond. Trends Plant Sci 13(11):602–609

Prakash V, Hunney K, Singh H (2022) Effect of elevated carbon dioxide on growth and development of *Santalum album* L. seedlings inoculated with plant growth promoting microorganisms in Open Top Chambers. Int J Environ Agricult Biotechnol 7(5):149–159

Pukacka S, Pukacki PM (2000) Seasonal changes in antioxidant level of Scots pine (*Pinus sylvestris* L.) needles exposed to industrial pollution. I. Ascorbate and thiol content. Acta PhysiologiaePlantarum 22:451–456

Qayoom-Mir A, Yazdani T, Ahmad S, Yunus M (2009) Total flavonoids and phenolics in *Catharanthus roseus* L. and *Ocimum sanctum* L. as biomarkers of urban auto pollution. Casp J Environ Sci 7(1):9–16

Radotić K, Dučić T, Mutavdžić D (2000) Changes in peroxidase activity and isoenzymes in spruce needles after exposure to different concentrations of cadmium. Environ Exper Bot 44(2):105–113

Robles P, Quesada V (2019) Transcriptional and post-transcriptional regulation of organellar gene expression (OGE) and its roles in plant salt tolerance. Int J Mol Sci 20(5):1056

Sadhna PBD, Tripathi S, Devi RS, Singh R, Bhadouria R (2023) 7 Ecophysiological responses of tropical plants to rising air pollution. Ecophysiol Trop Plants Recent Trends Fut Perspect

Sairam RK, Singh DV, Srivastava GC (2003) Changes in activities of antioxidant enzymes in sunflower leaves of different ages. BiologiaPlantarum 47:61–66

Sanesi G, Chiarello F (2006) Residents and urban green spaces: the case of Bari. Urban for Urban Green 4(3–4):125–134

Sharma R, Singh H, Kaushik M, Nautiyal R, Singh O (2018) Adaptive physiological response, carbon partitioning, and biomass production of *Withania somnifera* (L.) dunal grown under elevated CO_2 regimes. 3 Biotech 8:267. https://doi.org/10.1007/s13205-018-1292-1

Sharma R, Singh H (2021) Alteration in biochemical constituents and nutrients partitioning of *Asparagus racemosus* in response to elevated atmospheric CO_2 concentration. Environ Sci Pollut Res 29:6812–6821. https://doi.org/10.1007/s11356-021-16050-3

Singh H (2021) An integrated approach considering physiological and biophysical based indicators for assessing tolerance of roadside plantations of *Alstonia scholaris* towards urban roadside air pollution: an assessment of adaptation of plantations for mitigating urban roadside air pollution. Trees 36:2179. https://doi.org/10.1007/s00468-021-02179-8

Singh H (2024) Exploring adaptive modulation in plant functional traits and its impact on the productivity of *Acacia auriculiformis* under CO_2-enriched environment. Ind Crops Prod 210:118186. https://doi.org/10.1016/j.indcrop.2024.118186

Singh H, Kumar M (2022) Climate change and its impact on Indian Himalayan forests: current status and research needs. In: Rani S, Kumar R (eds) Climate change. Springer, Cham, pp 223–242

Singh H, Verma A (2013) Characterization and screening of high nitrogen efficient rice genotype to curtail environmental pollution. Int J Agricult Environ Biotechnol 6(1):23–30

Singh H, Savita Sharma R, Sinha S, Kumar M, Kumar P, Verma A, Sharma SK (2017a) Physiological functioning of *Lagerstroemia speciosa* L. under heavy roadside traffic: an approach to screen potential species for abatement of urban air pollution. 3 Biotech 7(61):1–10. https://doi.org/10.1007/s13205-017-0690-0

Singh H, Verma A, Kumar M, Sharma R, Gupta R, Kaur M, Negi M, Sharma SK (2017b) Phytoremediation: a green technology to clean up the sites with low and moderate level of heavy metals. Aust Biochem 2:1012

Singh H, Sharma R, Savita SMP, Kumar M, Verma A, Ansari MW, Sharma SK (2018) Adaptive physiological response of *Parthenium hysterophorus* to elevated atmospheric CO_2 concentration. Indian for 144:1–14

Singh H, Yadav M, Kumar N, Kumar A, Kumar M (2020a) Assessing adaptation and mitigation potential of roadside trees under the influence of vehicular emissions: a case study of *Grevillea robusta* and *Mangifera indica* planted in an urban city of India. PLoS ONE 15(1):0227380. https://doi.org/10.1371/journal.pone.0227380

Singh H, Yadav M, Kumar N, Kumar A, Kumar M (2020b) Assessing adaptation and mitigation potential of roadside trees under the influence of vehicular emissions: a case study of *Grevillea robusta* and *Mangifera indica* planted in an urban city of India. PLoS ONE 15(1):e0227380

Singh S, Singh H, Sharma SK, Nautiyal R (2021) Seasonal variation in biochemical responses of bamboo clones in the sub-tropical climate of Indian Himalayan foothills. Heliyon 7(4):e06859. https://doi.org/10.1016/j.heliyon.2021.e06859

Singh H, Bandyopadhyaya S, Kumar A, Singh M, Malik A, Kumari G, Raturi A, Yadav SK, Kumar M, Barthwal S, Thakur A, Rawat PS (2022) Understanding the physiological and biophysical response of urban roadside plantations for assessing adaptation and mitigation mechanisms toward vehicular emissions. Urban Clim 44:101183. https://doi.org/10.1016/j.uclim.2022.101183

Singh H, Kumar N, Kumar A (2024a) Enhancing resource use efficiency in crops through plant functional traits. In: Kumar N, Singh H (eds) Plant functional traits for improving productivity. Springer, Singapore

Singh M, Singh H, Kumar A, Kumar M, Barthwal S, Thakur A (2024b) Soil nitrogen availability determines the CO_2 fertilization effect on tree species (*Neolamarckia cadamba*): growth and physiological evidence. Environ Sustain 7:53–60. https://doi.org/10.1007/s42398-023-00300-w

Stevovi S (2010) Environmental impact on morphological and anatomical structure of Tansy. Afr J Biotechnol 9(16):2413–2421

Tadini L, Pesaresi P, Kleine T, Rossi F, Guljamow A, Sommer F, Leister D (2016) GUN1 controls accumulation of the plastid ribosomal protein S1 at the protein level and interacts with proteins involved in plastid protein homeostasis. Plant Physiol 170(3):1817–1830

Tiwari S, Agrawal M, Marshall FM (2006) (2006) Evaluation of ambient air pollution impact on carrot plants at a sub urban site using open top chambers. Environ Monit Assess 119:15–30

Vollenweider P, Ottiger M, Günthardt-Goerg MS (2003) Validation of leaf ozone symptoms in natural vegetation using microscopical methods. Environ Pollut 124(1):101–118

Wang H, Shi H, Li Y (2011) Leaf dust capturing capacity of urban greening plant species in relation to leaf micromorphology. In: Proceedings of the 2011 international symposium on water resource and environmental protection, vol 3. IEEE, pp 2198–2201

Wong NH, Tan AYK, Chen Y, Sekar K, Tan PY, Chan D et al (2010) Thermal evaluation of vertical greenery systems for building walls. Build Environ 45(3):663–672

Wu X, Chen Q, Chen L, Tian F, Chen X, Han C, Zhang F (2022) A WRKY transcription factor, PyWRKY75, enhanced cadmium accumulation and tolerance in poplar. Ecotoxicol Environ Saf 239:113630

Xiao Y, Savchenko T, Baidoo EE, Chehab WE, Hayden DM, Tolstikov V, Dehesh K (2012) Retrograde signaling by the plastidial metabolite MEcPP regulates expression of nuclear stress-response genes. Cell 149(7):1525–1535

Xu F, Guo W, Xu W, Wei Y, Wang R (2009) Leaf morphology correlates with water and light availability: what consequences for simple and compound leaves. Prog Nat Sci 19(12):1789–1798

Yadav SK, Singh H, Ginwal HS, Barthwal S (2019a) Elevated CO_2 enhanced growth and physiological process of *Populus deltoides* Bartr. ex Marsh. Indian For 145(1):23–27

Yadav SK, Singh H, Nautiyal R, Ginwal HS, Ansari SA, Barthwal S (2019b) Modulation of morpho-physiological responses in *Populus deltoides* by elevated carbon dioxide and temperature. For Sci 66:105–118. https://doi.org/10.1093/forsci/fxz048

Yang JL, Zhang GL (2015) Formation, characteristics and eco-environmental implications of urban soils: a review. Soil Sci Plant Nutr 61(sup1):30–46

Unveiling the Arboreal Responses to Edaphic Factors in Urban Forest Ecosystems

Ayushman Malakar, Anshuman Das, and Subhajit Chanda

Abstract The rise of urbanization in recent years has brought about significant challenges for global urban landscapes, particularly in terms of environmental and climate-related issues. Habitat fragmentation caused by urbanization has resulted in a decline in biodiversity. One effective way to address these challenges is by strategically enhancing the green tree cover in urban areas. Urban trees are vital as nature-based solutions (NbS) to combat climate change and various types of pollution in urban areas and provide multiple ecosystem services. Additionally, they help mitigate the urban heat island effect, minimize the risk of flooding and water pollution, replenish groundwater resources, and contribute to carbon sequestration. This chapter systematically explores how urban trees adapt to the altered structure and composition of urban soil and environment. It was found that the true potential of urban trees is severely hindered by soil compaction, pollution, heavy metal accumulation, and lack of essential nutrients. In order to sustain the growth and health of urban forest ecosystems, the study elaborates on the significance of comprehending urban nutrient cycles and plant-tree interactions. The unique morpho-physiological responses and the symbiotic associations with soil micro-biome exhibited by the roots are discussed in this chapter to highlight the exquisite adaptive capability of urban trees. This chapter contemplates bridging the research gap by focusing on the responses of urban trees to the prevailing edaphic factors to revitalize and maintain healthy urban tree-soil relationships. The chapter's findings will undoubtedly contribute to the achievement of Sustainable Development Goal (SDG) 11, which calls for creating inclusive, resilient, safe, and sustainable cities and human settlements, along with significantly committing to attaining many other SDGs, such as 6, 7, 12, 13, and 15. Subsequently, the chapter will also help forestry professionals and environmental planners implement the appropriate and holistic management practices to create more resilient and sustainable urban forests that are better adapted to the constraints of the cityscapes.

A. Malakar (✉) · A. Das
ICFRE—Institute of Forest Productivity, Ranchi, Jharkhand, India
e-mail: ayushmanmalakar@gmail.com

S. Chanda
The Energy and Resources Institute, New Delhi, India

H. Singh (ed.), *Urban Forests, Climate Change and Environmental Pollution*,
https://doi.org/10.1007/978-3-031-67837-0_9

Keywords Urban trees · Urban soils · Tree health · Nutrient cycle · Plant-soil interaction · Nature-based solutions (NbS)

1 Introduction

Urbanization has grown dramatically worldwide due to the Industrial Revolution (Locosselli and Buckeridge 2023). Most of the human population is inclined to relocate to large cities or metropolitan regions for a higher standard of living due to easy access to employment opportunities and basic amenities (Kumar et al. 2022; Singh et al. 2022). According to Baklanov et al. (2016), more than half of the world's population has already moved into urban areas in 2007, and by 2050, it is expected that this percentage will have risen to about 70% (United Nations 2019). There is little doubt that the current urban ecology is paying a significant price for this rapid urbanization. Metropolitan areas are expected to experience an 80% rise in land usage by 2030, with most of this expansion occurring in developing nations (Mahendra and Seto 2019). Given that just two percent of the earth's surface is inhabited by over fifty percent of the people on earth, the fundamental challenges in urban landscapes have become exacerbated more than ever (Pincetl 2017).

Urban landscapes face a myriad of environmental and climate-related challenges. Air pollution is one of the most significant concerns in the cities. Primary sources like vehicular and industrial emissions emit harmful pollutants, leading to terrible air quality indices (AQI) in every large city around the globe (Kumar et al. 2024; Singh et al. 2024a; Singh 2024). Being '*concrete-jungles*', the impervious surfaces of the buildings absorb and retain heat, creating the urban heat-island effect. Stormwater management in urban areas is another critical concern, as impervious surfaces lie under most cities (Singh et al. 2020; Kumar et al. 2021a). As a result, the cities become prone to flash flooding and severe water pollution due to heightened runoff. In addition, urban landscapes are one of the leading causes of rapid climate change, as they release massive amounts of greenhouse gases into the atmosphere (Kumar et al. 2021b, c, d). Almost two-thirds of the world's greenhouse gas emissions can be sourced from the combustion of fossil fuels in major cities worldwide as the consumption of natural resources and energy (Crippa et al. 2021). The urbanization process causes habitat fragmentation, eventually leading to a decline in biodiversity (Singh 2021; Sharma and Singh 2021). Moreover, the absence of greenery in many metropolitan places can severely impact the inhabitants' physical and mental health.

These diverse challenges can effectively be tackled by strategically improving the green tree cover in urban environments (Fig. 1) (Singh et al. 2021). Numerous literature and research consistently affirm that urban tree cover is essential for sustaining ecological harmony and enriching the general well-being of urban populations (Oldfield et al. 2014). Forests are crucial in mitigating global climate change by acting as one of the most oversized carbon sinks. Although urban trees cannot match the carbon storage capacity of natural forests as they only cover a fraction of the land surface, they serve as the cornerstone of nature-based solutions (NbS) for addressing climate change and pollution in metropolitan areas (Pataki et al. 2021).

In addition to their aesthetic value, the urban tree ecosystem plays a crucial role in delivering various ecosystem services. Trees in urban spaces primarily absorb toxic gasses like carbon monoxide, sulfur dioxide, nitrogen oxides, particulate matter, and carcinogenic hydrocarbons from the atmosphere and purify the surroundings by producing oxygen (Tallis et al. 2011; Yadav et al. 2019a, b; Gupta et al. 2019). Besides, they reduce surface and air temperatures through crown-shading and evapotranspiration, effectively mitigating the urban heat island phenomenon (Kumar et al. 2019a). Urban forests are essential for replenishing groundwater supplies and regulating storm water as they slow down the flow of rainwater. They also reduce the risk of flooding and water contamination significantly. These trees play key roles in

Fig. 1 Urban trees can act as a remedial approach, but effectively using them requires balancing their services and disservices to address various urban landscape problems. Adapted from Locosselli and Buckeridge (2023), created in BioRender

carbon sequestration, noise-pollution reduction, and improvement of water quality in the cities (Wang et al. 2024; Kumar et al. 2019b). As a home for various plant and animal species, urban trees can promote biodiversity in the area. They offer urban dwellers access to much-needed greenery, improving their physical, mental, and financial well-being (Qin et al. 2013). Throughout the COVID-19 epidemic, these urban trees have gained widespread recognition for their numerous recreational benefits, offering city inhabitants physical and mental respite (Bamwesigye et al. 2023). Urban trees are also economically important in cities and towns since studies have shown that having urban greenery nearby may raise property prices (Escobedo et al. 2015; Sharma et al. 2018; Singh et al. 2018).

Nonetheless, the true potential of urban trees for providing ecosystem services is not often fully realized as they encounter multiple challenges due to the cityscape that make their survival and growth far more complex than trees in natural forests (Clemente 2023; Gupta et al. 2018a, b). Urban trees generally endure warmer and more humid city environments due to the perceived heat island effect. Additionally, high particulate materials in the air and air pollution from toxic fumes of vehicles and industries significantly reduce the photosynthetic capacity of urban trees (Baek and Woo 2010). Urban trees in numerous cities may also experience drought stress due to changed hydrological circumstances, which reduces their productivity (Bialecki et al. 2018). Moreover, urban trees are also susceptible to anthropogenic complications due to some disservices associated with them, such as limitation of space, conflicts with city infrastructure, release of allergy-causing pollen during certain seasons, need for regular maintenance, and the occasional occurrence of trees falling accidentally (Lyytimäki and Sipilä 2009; Singh and Verma 2013; Kumar et al. 2017). The soil is a major concern when it comes to urban trees. In urban areas, the soil is often compacted and lacks basic nutrients, making it difficult for trees to thrive (Locosselli et al. 2019; Gupta and Singh 2017; Singh et al. 2017a). The limited space for their roots to spread and grow properly hampers their overall development. Heavy metals and other soil contaminants have also been discovered to accumulate in urban soils in several parts of the world.

Therefore, an in-depth understanding of nutrient cycles and plant-soil interaction is necessary to ensure the growth and health of urban forest ecosystems. Nutrient cycles involve the continuous movement of essential elements such as nitrogen, phosphorus, and potassium, among many others, from the plants to the soil organisms and back to the environment (Singh et al. 2017b). Unfortunately, urban soils often lack organic matter and beneficial microorganisms due to disturbances and disorganization in the urban soils. This deficiency has adverse outcomes for trees as it affects their structural integrity, density, and overall health. Moreover, anthropogenic activities also disrupt the nutrient cycles (Meena et al. 2011; Kumar et al. 2019c). As a result, the trees suffer from imbalances and deficiencies that impede their growth and vigor. Plant-soil interactions also play a critical role in maintaining the health and vitality of urban trees. The interconnectedness of roots, microorganisms, and organic matter forms a complex network where nutrients are exchanged (Kumar et al. 2020; Singh and Kumar 2022; Singh et al. 2024b). Frequent disruptions of these relations in urban areas make it difficult for trees to access the required nutrients. Suppose appropriate management practices such as mulching, organic amendments, and soil inoculation are integrated. In that case, it is possible to rejuvenate and maintain a healthy nutrient exchange and plant-soil relationships for urban trees. Incorporating these aspects, urban foresters and landscape practitioners can develop more resilient and sustainable urban forests that are better suited to endure the challenges of the city's environment.

2 Effect of Urban Soil Properties on Urban Tree Health

The survival of urban trees is the most challenging factor in any urban plantation program. They face various stressors where urban soil properties and nutrient status are crucial in their growth and development. Urban trees may develop more quickly than rural trees but exhibit a lower survival rate (Yadav et al. 2019a, b; Jangra et al. 2024). According to Smith et al. (2019), urban trees face almost double the mortality rate than trees in natural forests or rural areas. The lifespan of urban trees is also reported to be much shorter, necessitating more frequent gap-filling and leading to the expense of tree management. Urban soils possess a different composition and quality than rural or forest soils, as they are highly affected by anthropogenic factors. The parent material of urban soils is a mixture of residual natural soils and other synthetic materials that influence the formation of soils and the development of soil characteristics (Pouyat et al. 2010). Urban disturbances such as the construction of high-rises, surface paving, and excessive traffic pressure prohibit soil formation from compaction and significantly change soil structure.

Additionally, introducing chemicals, fertilizers, and other toxic substances in urban soils alters the levels of nutrients and biological activities. Although the urban soils may have the necessary chemical and physical properties to foster tree growth, high concentrations of pollutants like heavy metals are often found in these soils, which can severely impact the urban tree's health. Even partially unaffected urban soil profiles can be indirectly influenced in many different ways through the urban heat island, invasive species, and other human activities (Decina et al. 2017).

Although indices for assessment of soil properties in forests and agricultural fields are well-established, the concept of creating indicators for evaluating urban soil properties is still at its juvenile stage. The Urban Site Index (USI) and Rapid Urban Site Index (RUSI) are the most widely used methodologies for quick field-based soil evaluations in urban soils. Between the two, RUSI has been proposed to be the most robust index since it evaluates 15 parameters grouped into five criteria, i.e., climatic, urban city, physical, chemical, and biological (Scharenbroch et al. 2023). The model identified "estimated rooting area" and "soil structure" as the most critical of these fifteen soil factors. Further improvements in the model revealed that it is one of the most stable approaches for urban soil assessment and capable of successfully predicting urban tree health and growth.

2.1 Soil Compaction in Urban Landscapes

Paved surfaces are the predominant feature of urban landscapes, restricting trees' planting space and soil volume. Compaction of urban soils, particularly those along roadsides, is frequently caused by building operations, paving or sealing of the soil's surface, and high traffic volume. Jim and Ng (2018) state that soil compaction weakens the soil's structure by reducing porosity and raising bulk density. These modifications restrict the flow of nutrients, air, and water through the soil, which prevents the development of roots and tree growth. Compaction in the top 30 inches of soil depth, where roots are most dense, is especially detrimental to urban trees. Because of the linked effects on physical root impedance, aeration, water drainage, etc., compaction is recognized to limit plant development. Still, the precise processes underlying this restriction remain unclear to date. A bulk density greater than 1.4–1.6 g/cm^3 is often considered the crucial threshold for significant root development (McGrath and Henry 2016). Studies in several urban areas have found bulk densities frequently exceeding this threshold (Kim and Yoo 2021). Compared to non-compacted soils, plants in compacted soils have smaller and shallower roots and stunted development (Fair et al. 2012). Poor aeration around roots results from compaction, which breaks down soil aggregates and reduces porosity and consistency. Rainwater infiltration and discharge through soil profiles diminish with increasing bulk density. Reduced infiltration can cause flooding during storms and increase root damage during droughts. Jim (1993) states an infiltration rate of 5 cm/h is vital for tree development. Also, urban regions often experience water stress, which is expected to worsen the growth and development of urban trees (Dai 2013).

Several methods have been investigated to reduce soil compaction and enhance the rooting environment for urban trees. Additional root space can be provided while retaining load-bearing capability by using "structural soils" and "structural cells" under urban pavements (Deeb et al. 2020). Compared to conventional pavements, trees nurtured in these systems show greater root and shoot proliferation (Bühler et al. 2017). Surface solutions for tree pits, such as porous pavements, may improve aeration and water penetration. Mechanical approaches such as air injection, soil loosening, and backfilling have had varied results in their usefulness in alleviating compaction. However, several recent studies have consistently demonstrated that adding organic matter (OM), such as compost, significantly improves soil compaction and encourages improved tree health (Sax et al. 2017). The healthy development of urban trees depends greatly on avoiding and minimizing soil compaction.

2.2 Contaminants in Urban Soils

Urban soils are vastly different in chemical composition, primarily affected by land use and land cover (LULC) policies. Roadside soils have more alkaline pH, less nutrient content, and more salts and heavy metals than natural soils. Urban soil alkalinity is widespread due to using cement and other calcareous materials in construction (Kim and Yoo 2021). Alkalized soils can create a shortage of some micronutrients in trees, as these nutrients are mainly insoluble at higher pH and, thus, unavailable for plants (Brady and Weil 2008). Maintaining nutrient contents seems to be a critical component for the proper growth and development of urban trees. Soil salinity and airborne salt deposition also damage the growth of urban trees across metropolitan areas. The effect of this high salinity can be seen in urban trees through discoloration of the leaves, chlorosis, necrosis, early leaf loss, dieback post-flushing, and retarded development (Cekstere et al. 2009). Furthermore, leaf necrosis is positively correlated to salt and chloride accumulation in the leaves of urban trees (Cekstere et al. 2016).

Additionally, several heavy metals accumulate in the trees of urban areas from various sources of environmental contamination. Studies conducted in Poland on linden trees (*Tilia* spp.) found in parks and streets have shown a strong correlation between contamination of heavy metals and tree viability (Olchowik et al. 2021). Heavy metals, such as lead and cadmium, have been shown to have detrimental effects on tree development in pot tests and urban forest park investigations. Nonetheless, certain tree species can evolve defensive mechanisms, such as limiting heavy metal deposition in their roots.

3 Role of Nutrient Cycling in Urban Tree Growth

Under urban ecosystems, "nutrient cycling" describes the complex mechanisms that recycle and redistribute essential nutrients like phosphorus, nitrogen, and carbon (Reisinger et al. 2016). Urban landscapes provide distinct problems cycling these nutrients due to their dense population, infrastructure, and restricted green space (Kanta et al. 2024; Malik et al. 2024). Therefore, knowledge of the urban nutrient cycle is imperative for prosperous urban tree plantations. Most nutrients are released into urban environments by decaying organic materials from urban vegetation, such as food waste and foliage. When this organic material is mineralized, its nutrients are converted into inorganic forms that microorganisms and plants take through immobilization processes.

Nonetheless, nutrient loss from urban soils can also occur when heavy rains or irrigation cause leaching or when pollutants like ammonia and nitrogen oxides are produced (van de Vlasakker et al. 2022). Human interventions considerably impact nutrient dynamics in urban environments by significantly speeding up the cycle. The natural equilibrium of the nutrient cycle gets altered by the overuse of fertilizer, although organic waste management techniques like composting and recycling increase the availability of nutrients (Kumar and Singh 2024). The rapid growth of urban infrastructure significantly modifies the flow of nutrients. Urban nutrient cycling may be improved with smart urban development. Natural nutrient cycle processes are encouraged by incorporating green places such as parks, urban woods, and green roofs (Calheiros et al. 2022). Besides, urban agriculture and soil amendment are practices that may increase the fertility of the soil and nutrient retention within the cities.

Atmospheric depositions are essential in providing necessary nutrients to urban ecosystems worldwide. In cities with low nutrient inputs, air deposition can be an important source of nutrients for the ecosystem. Nitrogen and sulfur compounds are the principal nutrients deposited on land and in aquatic bodies via atmospheric deposition. Nitrogen deposition in urban ecosystems results from gaseous ammonia (NH_3) emissions and nitrogen oxides (NO_x) from burning fossil fuels, industry, and agriculture. The same is true for sulfur deposition, caused by sulfur dioxide (SO_2) emissions from burning fossil fuels and industrial processes. Although in lower amounts, trace elements, including potassium, calcium, magnesium, and phosphorus, can also be delivered via air deposition. Atmospheric deposition may have both positive and negative impacts on urban ecosystems.

On the one hand, inputting nutrients such as nitrogen and sulfur can improve urban soil fertility, encourage urban tree growth, and enhance primary production and function. Studies have indicated that leaf fall from trees may considerably boost the amount of atmospheric nutrients incorporated into urban soil. These inputs can be increased two to eightfold in late spring, and the groundwater quality in metropolitan areas can be enhanced by the direct reach of this nutrient-rich throughfall into drainage systems (Decina et al. 2020).

On the other hand, significant SO_2 and NO_x deposition can cause acid rain, which lowers soil pH and interferes with nutrient cycling (Bhardwaj and Saxena 2023). This may exert a negative impact on soil microbial populations and the availability of nutrients for urban vegetation. Sustainable management strategies and monitoring programs are crucial to balance the effects of these depositions on urban tree ecosystems and preserve nutrient balance in urban soils.

Another critical procedure in urban environments is fertilization, which supplies nutrients to sustain urban tree growth and preserve soil fertility (Venkata Mohan et al. 2020). Applying fertilizer is vital to maintaining healthy urban trees in areas where human activity heavily influences soil. Along with inorganic fertilizers, organic fertilizers like compost and manure also enhance soil structure by increasing aggregation, water retention, and aeration, and they serve as nutrient stores that are progressively released over time. Urban soils also receive essential nutrients from the natural deposition of organic matter. Organic matters are transferred from plants, animals, and other sources to the earth's surface via litterfall, decomposition, and other methods. Organic matter deposition has a variety of consequences on the ecology and soil of metropolitan areas. Soil organisms decompose organic materials, which promote microbial activity, beneficial interactions, and soil biological diversity.

Furthermore, the accumulation of organic matter lowers erosion and increases the water-holding capacity of urban soils (Canedoli et al. 2020). Organic matter contains vital elements such as carbon, nitrogen, phosphorus, potassium, sulfur, calcium, magnesium, and other micronutrients. These are made available to the urban vegetation through decomposition, thereby encouraging the growth of urban trees.

Nutrient depletion in urban soils, like in agricultural or forest soils, happens primarily through leaching and biomass loss (Pouyat et al. 2020). Due to heavy rainfall or irrigation, leaching occurs when water-soluble nutrients are transported beyond plants' root zone. Several variables, including soil texture and the amount of rainfall, significantly impact the amount of leaching. Due to their high compaction, urban soils are more prone to leaching, which might cause nutrient deficits in urban plants. Furthermore, removing organic matter or biomass can exacerbate nutrient depletion, especially in urban ecosystems. Urban biomass comprises fallen leaves, branches, and other organic residue that decompose to provide nutrients in urban soils. Regularly removing this organic matter from urban areas by grass cutting, branch trimming, and leaf raking might interfere with the soil's natural ability to replace organic matter. As a result, over time, the soil's ability to hold onto moisture, preserve structural integrity, and support microbial activity—all crucial for the cycling of nutrients—diminishes, endangering the health of urban trees (Chaurasia et al. 2023).

4 Tree-Soil Interactions in Urban Environments

4.1 Morphological and Physiological Responses of Trees to Urban Soil Conditions

Roots are specialized multicellular organs that help in water and solute uptake and movement, mainly characterized by endogenous branching, root hairs, protective root caps, and gravitropic response (Kenrick and Strullu-Derrien 2014). Macronutrient availability in the urban soil changes the adaptive responses of the root systems of urban trees. Root systems fulfill functions such as adequate tree anchorage and stability (Dumroese et al. 2019) and adapt to their surroundings by changing morphologically, physiologically, and molecularly (Montagnoli 2022).

The root elongation process can be affected by numerous soils' physical stresses acting in combination or independently. Bengough et al. (2006) explained that the growth of root tips depends on multiple factors such as local soil conditions, communication with the rest of the plant through a supply of nutrients (e.g., carbon compounds), and responses (e.g., chemical signals). Generally, soil classifications are based on their composition, which is the relative proportions of sand, silt, and clay particles. These particle sizes and their relative proportion determine the pore space between the particles. Where clayey soils reduce the water movement, sandy soils provide more than enough space to flow the water, reducing the water holding capacity (WHC). In their study, Rellán-Álvarez et al. (2016) stated the compensatory uptake mechanism, in which substantial soil conductance heterogeneity arises in the immediate root environment, modifying and improving the water flow path. Thus, the water uptake sites shift from the dry to the wet root system portion. Since the distribution of water and nutrients in the soil is not uniform, root architecture is essential at the macro level in addition to soil parameters. Nutrients such as Nitrogen (N) follow similar principles to water as Nitrogen (N) is loosely bound to the particles, whereas phosphorus (P) is present but mostly in unavailable chemical forms. However, lateral root growth or elongation rates target nitrogen (N) and phosphorus (P) nutrient-rich regions.

Craul (1985) reported that urban soil characteristics include lack of aeration, enhanced nutrient content variability, higher pH, hard pan formation on the surface of bare soils, and high pollution concentration. Compaction of soils is also an inherent problem in urban soils due to disturbance-related mechanical actions involved in the breakdown of smaller aggregates. Urban soils with salinity also negatively affect tree and root growth. When compared to drought-sensitive varieties and drought-resistant rice varieties, it is noticed that they have a deeper and more highly branched root system (Grime et al. 1986).

4.2 Symbiotic Associations in Urban Soil Environment

The cycling of nitrogen in urban soil is significantly influenced by soil microbes (Fig. 2). As a crucial building block of proteins and nucleic acids, nitrogen (N) is a nutrient that is necessary for all living things. Furthermore, in many natural ecosystems, the supply of N to organisms controls primary productivity. In nature, N exists in various oxidation states and chemical forms, and microbes usually catalyze the transitions between these forms (Hayatsu et al. 2008). A significant amount of soil biodiversity comprises microorganisms, which are also essential to soil processes that eventually impact terrestrial ecosystem function (Philippot et al. 2013). Microbial community dynamics may be linked to temporal and regional change in N processes and rates, given the biogeochemical contributions of bacteria. The progressive aerobic oxidation of ammonia (NH_3) to nitrite is known as nitrification. Heterotrophic and autotrophic nitrification are the two types of nitrifications that occur in soils. The former is mainly performed by nitrite-oxidizing bacteria (NOB, e.g. *Nitrobacter*, *Nitrospina*, *Nitrococcus*, and *Nitrospira*), chemoautotrophic ammonium-oxidizing bacteria (AOB, e.g. *Nitrosomonas*, *Nitrosospira*, and *Nitrosococcus*), and archaea (AOA, e.g. *Nitrososphaera* and *Nitrosotalea*).

Certain heterotrophic bacteria and fungi that oxidize organic and inorganic N molecules carry out heterotrophic nitrification (Hayatsu et al. 2008). Despite the consensus that autotrophic nitrification is the most common type of nitrification in soils, high rates of nitrification in tropical and subtropical acidic soils indicate that heterotrophic nitrification, which is less affected by low pH than autotrophic nitrification, may be significant in these soils (Pett-Ridge et al. 2013). Through a greater absorbing surface area, the mycorrhizal hyphae network boosts the foraging capabilities and N acquisition for roots, especially when N is scarce (Simon et al. 2017). Mycorrhizae are, therefore, essential to the acquisition of N by trees and, as a result,

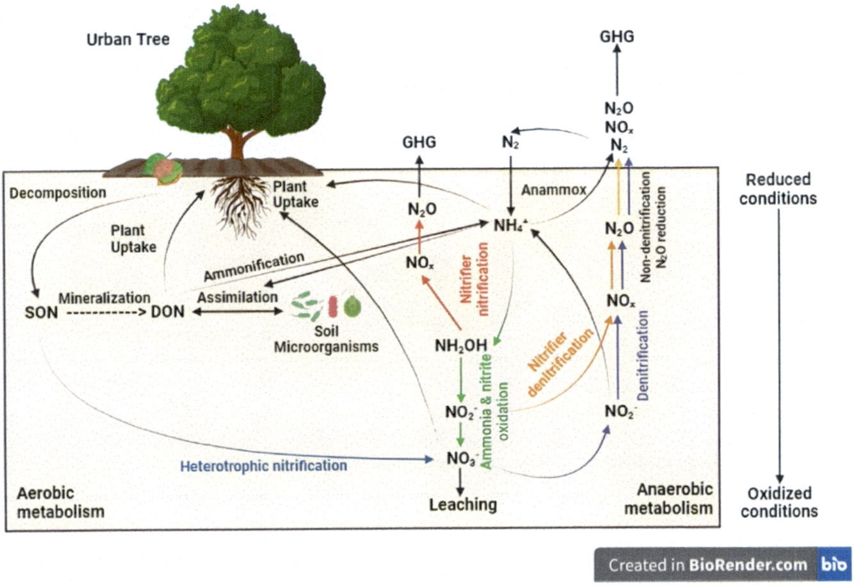

Fig. 2 Nitrogen cycle in the urban soil. Adapted from Pajares and Bohannan (2016), created in BioRender

their capacity to compete. Tree mycorrhizal relationships come in several forms, including ectomycorrhiza (EM) and arbuscular mycorrhiza (AM). Both mycorrhizal types take up both organic and inorganic N sources, and their acquisition is controlled by environmental factors like water availability and the presence or absence of specific N sources (Bukovska et al. 2018).

Research carried out in controlled environments has examined the impact of mycorrhizal type on nutrient uptake, particularly nitrogen and phosphorus (Phillips et al. 2013). Through the supply of nutrients and controlling the plant-soil microbial interactions, mycorrhizal species directly or indirectly affect the composition of plant communities (Tedersoo et al. 2020). However, understanding of species-specific tree-soil interactions remains restricted due to complex sets of drivers in urban ecosystems (Fig. 3). Competition for N is projected to be strongest at the seedling stage in urban trees due to urban soils' restricted N storage capacity (Simon et al. 2017).

The physical modifications to the soils in urban environments significantly influence the organisms present in the soil. Compared to rural soil environments, urban soil environments have a low number, diversity, and biomass of organisms (Guilland et al. 2018). Reeves et al. (1979) proved that soil disturbance influences mycorrhizal fungi. Cairney and Meharg (1999) provided evidence of soil pollution due to anthropogenic activities, and Egerton-Warburton and Allen (2000) demonstrated nitrogen (N) deposition in urban environments has considerable influence on the mycorrhizal fungi in urban environments. Bainard et al. (2011) compared mycorrhizal status

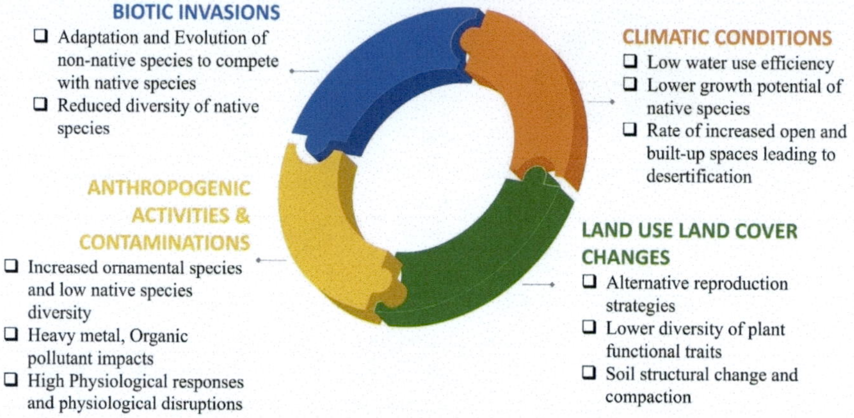

Fig. 3 Impact of urban drivers on plant-soil interaction. Authors' creation

and colonization of 26 tree species growing in urban and rural environments. The study found that the trees growing in urban environments had reduced arbuscular mycorrhizal colonization.

5 Adaptations of Urban Trees to Major Urban Stressors

5.1 Response to Drought Conditions

The global climate change scenario may cause drought to become a frequent event shortly. Huang et al. (2023) extensively studied five urban agglomerations in China's Yangtze River and stated that urbanization has become a prominent anthropogenic activity globally. Such urbanization activities modify temperature and rainfall in and around metropolitan areas. Thus, the urbanization process can be attributed to drought or decreased water levels compared to normal conditions. The conditions may be either from less rainfall or less water storage in the soil. Allen et al. (2010) linked urban tree mortality with drought events in major cities worldwide. Low soil water potential leads to varied physiological responses of urban trees, and it is well characterized. The physiological responses of trees vary from lower leaf water potential or anisohydric response to maintain leaf water potential over threshold with varying strategies (Ryan 2011).

Brunner et al. (2015) demonstrated that tree roots obtain various strategies to respond to drought stress. Such responses can be asserted to root biomass adjustments, physiological acclimations, and anatomical alterations. A study conducted by Kozlowski and Pallardy (2002) provides evidence regarding the impact of drought on both coarse and fine roots. In general, where there is less water penetration in urban environments, plants and trees suffer more or less drought-like conditions. Also, excess water from groundwater tables, where the groundwater recharge rate is low, is extracted due to urban construction plants suffering a similar drought stress event. In response, urban trees avoid drought-like conditions by avoiding low water potentials to tolerate dehydration (Verslues et al. 2006). Such strategies include maintaining required water potential by maintaining water loss rates through limiting stomatal closure and restricting shoot growth. That is why the increased root-shoot ratio of urban trees can often be observed under drought-like conditions in metropolitans. Increased fine roots and elongated deep taproot systems are also responsive mechanisms for drought tolerance of urban trees (Brunner et al. 2015). Based on these studies, multiple insights can be drawn on how drought conditions influence urban tree root morphology (Table 1).

5.2 Response to Soil Contaminations

Urban environments exert stressful conditions on plants, which affect them. Antonovics et al. (1971) primarily demonstrated the recognition of plant accumulation in metalliferous soils. The contamination of the sites can be due to metal and non-metal components. Baker (1981) grouped plant responses to heavy metals in the soil into three significant pathways. The capacity of accumulators to transfer metals from low to high soil levels makes them helpful in concentrating metals in their aboveground plant sections. The indicators reflect the internal concentration at external levels because the uptake and transport of metals to the shoot are regulated. Another is excluders, where metal concentrations in the shoot are kept low and stable over a broad range of soil concentrations up to a specific soil value at which the system malfunctions and unfettered transport occurs. The urban environment significantly impacts the physical environment, ecology of urban tree populations, and ecosystem processes. Besides heavy metal pollution, several contamination factors affect the tree's adaptation to urban soil environments. Few urban trees adapt in the contaminated sites by accumulating the contaminants in their body parts, which is referred to as phytoextraction (Dadea et al. 2017), or by immobilizing and stabilizing the contaminants *in-situ*, reducing their mobility and bioavailability, which is referred to as phytostabilization (Brunner et al. 2009). In phytoextraction, urban soil contaminants (generally heavy metals and other toxic substances) are absorbed by the roots, transported through the vascular system, and finally sequestered in the aboveground biomass of the urban trees, effectively removing the contaminants from the soil.

On the other hand, trees that are tolerant to specific contaminants execute phytostabilization processes in response to the accumulation. Microorganisms residing in the rhizosphere of urban tree roots play a crucial role in immobilizing the pollutants (Fig. 4). Applying biochar and biosolids is often recommended to support tree growth and improve soil quality in urban landscapes (Scharenbroch et al. 2013).

Table 1 Effect of urban drought conditions on root morphology of urban trees (Based on Brunner et al. 2015)

Trait categories	Root traits	Effect of urban drought conditions
Growth	Biomass	Negative
	Lifespan	Predominant negative trend
	Mortality	Predominant positive trend
	Production	Predominant positive trend
	Turnover rate	Predominant negative trend
Anatomical	Conduits	Predominant positive trend
morphological	Diameter	Predominant negative trend
	Root Area Index (RAI)	No effect
	Root Tissue Density (RTD)	No effect
	Specific Root Length (SRL)	No effect
Architectural	Branching	No effect
	Length	Predominant negative trend
	Rooting depth	Positive
	Tip frequency	Predominant negative trend
Biotic	Mycorrhizas	Predominant positive trend
chemical	Carbon (C)	Predominant negative trend
	Nitrogen (N)	Predominant positive trend
	Phosphorous (P)	Predominant negative trend
	C/N ratio	Predominant negative trend
	N/P ratio	Predominant positive trend
Biochemical	Cellulose	No effect
	Lignin/phenolics	Predominant positive trend
	Suberin/aliphatics	Predominant positive trend
Physiological	Abscisic acid (ABA)	Positive
	Exudates	Predominant negative trend
	Proline	Positive
	Respiration	Negative
	Starch	Predominant negative trend
	Sugars (non-structural carbohydrates)	No effect

6 Case Studies in the Science of Urban Tree Plantations

The first definition of "urban forestry" was put forth by Jorgensen (1986), who stated it as "a specialized branch of forestry and has as its objective the cultivation and management of trees for their present and potential contributions to the physiological, sociological, and economic well-being of urban society." He also emphasized that, in addition to their recreational and primary amenity benefits, urban trees directly

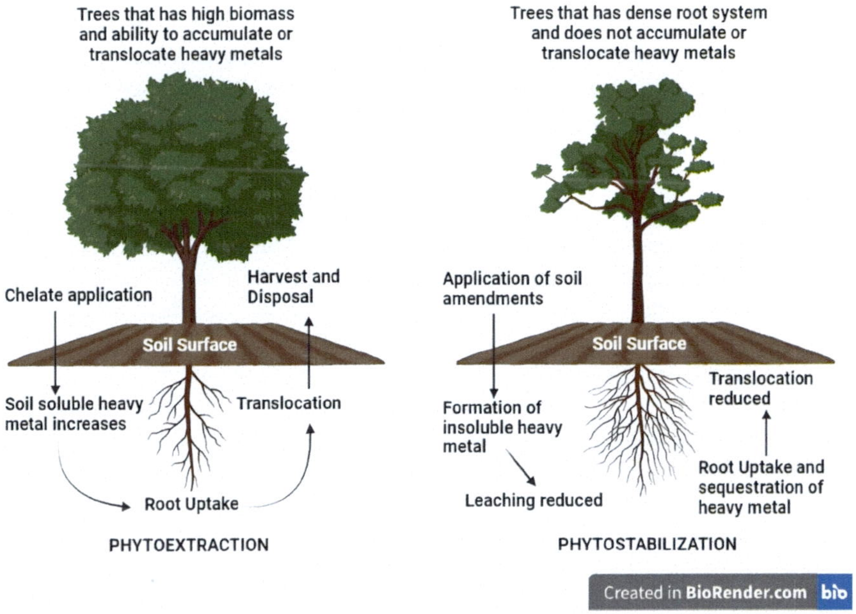

Fig. 4 Flow chart of the general processes for phytoextraction and phytostabilization. Adapted from Cunningham and Berti (2020), created in BioRender

contribute to the environment of cities. DeGraaf (1975) asserted that urban trees are a human environment component and offer habitat to a greater variety of urban wildlife.

Considering the enormous benefits of urban trees, all major cities worldwide are investing in substantial urban greening programs. Numerous established economic, ecological, and social advantages of urban forests—such as energy savings, the reduction of urban heat islands, and restorative advantages for city dwellers—are the driving forces behind these planting initiatives. In several metropolitans, these initiatives have progressed from street plantations towards afforestation and reforestation operations to establish urban forest stands consisting of native tree species (Oldfield et al. 2013).

World Resources Institute (WRI) provided great insight into three cities, Tokyo in Japan, Belfast in Northern Ireland, and Washington, DC, which are taking advancements in urban forestry to the next level (Weber 2016). During World War II, Tokyo lost many trees, prompting efforts to restore vegetation. A 1946 proposal was aimed to earmark 10% of urban lands for green spaces. By 1980, urban trees had surpassed 235,000, and by 1990, Tokyo's green space had expanded to more than 21,630 ha, supporting groundwater conservation and urban resilience. Belfast has seen nearly 200,000 trees planted in urban areas since the end of 1998. The initiative, supported by European funds, involved local and central governments, environmental organizations, and citizen volunteers called "Tree Wardens." Later, in a span of only three years (2013 to 2015), 90,000 trees were planted by such volunteers, demonstrating dedication to sustainable urban forestry. The District of Columbia, Washington D.C., also has a rich history of planning and maintaining its urban forest since 1872, with Governor Alexander Shepherd's directive to plant 60,000 street trees. The city now holds a tree canopy of 2 million trees, removing 526,000 tons of CO_2 and saving almost 1 million dollars of annual emissions.

Over nine years, the MillionTreesNYC initiative, launched in 2007 in New York City, has planted one million trees around the city. Although several Asian and African countries lag behind urban forestry initiatives, recent major movements in this sector have been observed. An ambitious initiative was launched in Singapore in 1967 to transform the concrete jungle into a city teeming with greenery. With a unique approach to the "Park Adopt-a-Plot" scheme, which encourages communities to participate in the care and maintenance of green spaces, the city successfully incorporated greenery into all aspects of urban design, from parks and gardens to streetscapes and buildings.

India, the fifth largest economy in the world, has also embarked on several urban forestry programmes that significantly impact greening the country's cities. The National Mission on Green India (NMGI) is a government-led initiative to increase India's forest cover by 5 million hectares, which includes a significant focus on planting and promoting urban trees. Some successful urban forestry programmes in the nation are worth mentioning. The Warje Urban Forest in Pune stands out as a role model. This public–private partnership transformed a barren 16-hectare area into a thriving forest with over 10,000 native plant species. It's a haven for birds, butterflies, and even small mammals. The country's capital, Delhi, also boasts two successful urban greening projects—Aravalli Biodiversity Park and Yamuna Biodiversity Park. These projects revived a degraded, mined landscape. They recreated the natural ecosystems of the Aravalli hills and Yamuna River region within the city limits, providing vital habitat for native flora and fauna of more than 200 species. Many cities like Ahmedabad and Bengaluru implement the Miyawaki method to plant species of Acacia, which creates dense urban forests in smaller spaces much faster. Additionally, several non-governmental organizations (NGOs) like 'Thuvakkam' (https://thuvakkam.org) in South India and 'Give Me Trees' (https://www.givemetrees.org/) in North India have showcased their exemplary impact in greening the urban landscapes.

7 Conclusion

Population in urban areas is indeed driving the expansion of cities and accelerating the process of urbanization. This rapid urban growth transforms natural landscapes into built-up areas, accompanied by industrialization and infrastructure development. However, this transformation is not without consequences for the environment. As urbanization alters land use patterns, it significantly impacts soil properties and plant root systems. The proliferation of high-rises and other constructions replaces open surfaces with impervious concrete and asphalt, often accumulating waste materials on soil surfaces. Consequently, this results in soil compaction, reduced infiltration rates, and pollution. Compacted soil limits root penetration and growth, hindering the development of extensive root systems essential for urban trees. Moreover, soil contaminants like heavy metals, pesticides, and hydrocarbons pose risks to the growth of urban trees, contributing to decreased soil biodiversity, reduced fertility, and increased erosion and runoff. Despite these challenges, there is a lack of comprehensive studies on the long-term effectiveness of urban forestry initiatives. The variability in urban soils and its impact on large-scale tree plantings within urban landscapes remain largely unexplored. Furthermore, funding and long-term maintenance present significant barriers to widespread urban forestry implementation. Surprisingly, little attention has been given to trees' physiological responses and overall performance in urban soils. Therefore, there is a need to systematically investigate how urban trees respond to the altered structure and composition of urban soils. Effective policies and guidelines for urban planning and green infrastructure initiatives are crucial to mitigate these impacts and maintain healthy soil for plant growth in urban environments. Urban trees may take time to provide their ecosystem services fully, but enriching urban soil from the seedling stage is essential for their survivability and proper development. Future research directions in urban forestry should focus on large-scale implications through rigorous component modeling. This approach can provide an in-depth understanding of how urbanization affects urban tree root systems and soil composition. Additionally, field experiments and long-term monitoring are necessary to assess the details and impacts of impervious surfaces, pollutant concentrations, and urban planning strategies on soil health and plant growth. Interdisciplinary research integrating ecology, soil science, urban planning, and engineering fields holds the potential for exploring innovative nature-based solutions (NbS) to mitigate the negative impacts of urbanization on soil and urban vegetation. Areas of focus could include studying soil microbiomes in urban environments and their relation to plant health parameters, soil-less cultivation systems, green infrastructure, nature-based remediation techniques, urban agroecology, and permaculture.

References

Allen CD, Macalady AK, Chenchouni H, Bachelet D, McDowell N, Vennetier M, Kitzberger T, Rigling A, Breshears DD, Hogg ET, Gonzalez P, Fensham R, Zhang Z, Castro J, Demidova N, Lim J, Allard G, Running SW, Semerci A, Cobb N (2010) A global overview of drought and heat-induced tree mortality reveals emerging climate change risks for forests. For Ecol Manage 259(4):660–684.https://doi.org/10.1016/j.foreco.2009.09.001

Antonovics J, Bradshaw AD, Turner RG (1971) Heavy metal tolerance in plants. In: Advances in ecological research. Academic Press, p 7. https://doi.org/10.1016/S0065-2504(08)60202-0

Baek SG, Woo SY (2010) Physiological and biochemical responses of two tree species in urban areas to different air pollution levels. Photosynthetica 48(1):23–29. https://doi.org/10.1007/s11099-010-0005-8

Bainard LD, Klironomos JN, Gordon AM (2011) The mycorrhizal status and colonization of 26 tree species growing in urban and rural environments. Mycorrhiza 21(2):91–96. https://doi.org/10.1007/s00572-010-0314-6

Baker AJM (1981) Accumulators and excluders-strategies in the response of plants to heavy metals. J Plant Nutrit 3(1–4):643–654. https://doi.org/10.1080/01904168109362867

Baklanov A, Molina LT, Gauss M (2016) Megacities, air quality and climate. Atmos Environ 126:235–249. https://doi.org/10.1016/j.atmosenv.2015.11.059

Bamwesigye D, Fialova J, Kupec P, Yeboah E, Łukaszkiewicz J, Fortuna-Antoszkiewicz B, Botwina J (2023) Urban forest recreation and its possible role throughout the COVID-19 pandemic. Forests 14(6):1254. https://doi.org/10.3390/f14061254

Bengough AG, Bransby MF, Hans J, McKenna SJ, Roberts TJ, Valentine TA (2006) Root responses to soil physical conditions; growth dynamics from field to cell. J Exp Bot 57(2):437–447. https://doi.org/10.1093/jxb/erj003

Bhardwaj A, Saxena A (2023) A review: acid rain; its emission, deposition and effect on ecosystem. In: AIP conference proceedings. AIP Publishing, vol 2735, no 1. https://doi.org/10.1063/5.0141305

Bialecki MB, Fahey RT, Scharenbroch B (2018) Variation in urban forest productivity and response to extreme drought across a large metropolitan region. Urban Ecosyst 21(1):157–169. https://doi.org/10.1007/s11252-017-0692-z

Brady NC, Weil RR (2008) The nature and properties of soils, 14th edn. Prentice Hall

Brunner I, Pannatier EG, Frey B, Rigling A, Landolt W, Zimmermann S, Dobbertin M (2009) Morphological and physiological responses of Scots pine fine roots to water supply in a dry climatic region in Switzerland. Tree Physiol 29(4):541–550. https://doi.org/10.1093/treephys/tpn046

Brunner I, Herzog C, Dawes MA, Arend M, Sperisen C (2015) How tree roots respond to drought. Front Plant Sci 6:547. https://doi.org/10.3389/fpls.2015.00547

Bühler O, Ingerslev M, Skov S, Schou E, Thomsen IM, Nielsen CN, Kristoffersen P (2017) Tree development in structural soil—An empirical below-ground in-situ study of urban trees in Copenhagen, Denmark. Plant Soil 413(1–2):29–44. https://doi.org/10.1007/s11104-016-2814-4

Bukovská P, Bonkowski M, Konvalinková T, Beskid O, Hujslová M, Püschel D, Řezáčová V, Gutiérrez-Núñez MS, Gryndler M, Jansa J (2018) Utilization of organic nitrogen by arbuscular mycorrhizal fungi—Is there a specific role for protists and ammonia oxidizers? Mycorrhiza 28(3):269–283. https://doi.org/10.1007/s00572-018-0825-0

Cairney JWG, Meharg AA (1999) Influences of anthropogenic pollution on mycorrhizal fungal communities. Environ Pollut 106(2):169–182. https://doi.org/10.1016/S0269-7491(99)00081-0

Calheiros CSC, Castiglione B, Palha P (2022) Nature-based solutions for socially and environmentally responsible new cities: the contribution of green roofs. In: Circular economy and sustainability. Elsevier, pp 235–255. https://doi.org/10.1016/B978-0-12-821664-4.00015-7

Canedoli C, Ferrè C, El Khair DA, Padoa-Schioppa E, Comolli R (2020) Soil organic carbon stock in different urban land uses: high stock evidence in urban parks. Urban Ecosyst 23(1):159–171. https://doi.org/10.1007/s11252-019-00901-6

Cekstere G, Osvalde A, Nikodemus O (2009) Influence of de-icing salt on K supply and street trees ecological status in Riga, Latvia. In: Rauch S, Morrison GM, Monzón A (eds), Highway and urban environment, vol 17. Springer Netherlands. https://doi.org/10.1007/978-90-481-3043-6_36

Cekstere G, Osvalde A, Vollenweider P (2016) De-icing salt impact on leaves of street trees (Tilia x vulgaris H.) in Riga, Latvia. Acta Biol Univ Daugavpiliensis 16(1):31–38

Chaurasia M, Patel K, Rao KS (2023) Impact of anthropogenic land uses on soil microbiological activity in a peri-urban landscape. Environ Monit Assess 195(10):1233. https://doi.org/10.1007/s10661-023-11822-5

Clemente M (2023) Rethinking "Streetline forestscapes" in a broader context of urban forestry: in-between ecological services and landscape design, with some evidence from Rome, Italy. Sustainability 15(4):3435. https://doi.org/10.3390/su15043435

Craul PJ (1985) Urban soils. In: Proceedings of the fifth conference of the metropolitan tree improvement alliance (METRIA), p 5

Crippa M, Guizzardi D, Pisoni E, Solazzo E, Guion A, Muntean M, Florczyk A, Schiavina M, Melchiorri M, Hutfilter AF (2021) Global anthropogenic emissions in urban areas: patterns, trends, and challenges. Environ Res Lett 16(7):074033. https://doi.org/10.1088/1748-9326/ac00e2

Cunningham SD, Berti WR (2020) Phytoextraction and phytostabilization: Technical, economic, and regulatory considerations of the soil-lead issue. In: Phytoremediation of contaminated soil and water. CRC Press, pp 359–376

Dadea C, Russo A, Tagliavini M, Mimmo T, Zerbe S (2017) Tree species as tools for biomonitoring and phytoremediation in urban environments: a review with special regard to heavy metals. Arboricult Urban Forest 3(4):155–167. https://doi.org/10.48044/jauf.2017.014

Dai A (2013) Increasing drought under global warming in observations and models. Nat Clim Change 3(1):52–58. https://doi.org/10.1038/nclimate1633

Decina SM, Templer PH, Hutyra LR, Gately CK, Rao P (2017) Variability, drivers, and effects of atmospheric nitrogen inputs across an urban area: emerging patterns among human activities, the atmosphere, and soils. Sci Total Environ 609:1524–1534. https://doi.org/10.1016/j.scitotenv.2017.07.166

Decina SM, Ponette-González AG, Rindy JE (2020) Urban tree canopy effects on water quality via inputs to the urban ground surface. In: Levia DF, Carlyle-Moses DE, Iida S, Michalzik B, Nanko K, Tischer A (eds) Forest-water interactions. Ecological studies, vol 240. Springer, pp 433–457. https://doi.org/10.1007/978-3-030-26086-6_18

Deeb M, Groffman PM, Blouin M, Egendorf SP, Vergnes A, Vasenev V, Cao DL, Walsh D, Morin T, Séré G (2020) Using constructed soils for green infrastructure—Challenges and limitations. Soil 6(2):413–434. https://doi.org/10.5194/soil-6-413-2020

DeGraaf RM (1975) Wildlife considerations in metropolitan environments. In: Forestry issues in Urban America. Proceedings of the national convention of the society of American foresters, pp 97–102

Dumroese RK, Terzaghi M, Chiatante D, Scippa GS, Lasserre B, Montagnoli A (2019) Functional traits of Pinus ponderosa coarse roots in response to slope conditions. Front Plant Sci 10:947. https://doi.org/10.3389/fpls.2019.00947

Egerton-Warburton LM, Allen EB (2000) Shifts in arbuscular mycorrhizal communities along an anthropogenic nitrogen deposition gradient. Ecol Appl 10(2):484–496. https://doi.org/10.1890/1051-0761(2000)010[0484:SIAMCA]2.0.CO;2

Escobedo FJ, Adams DC, Timilsina N (2015) Urban forest structure effects on property value. Ecosyst Serv 12:209–217. https://doi.org/10.1016/j.ecoser.2014.05.002

Fair BA, Metzger JD, Vent J (2012) Response of eight maple cultivars (Acer spp.) to soil compaction and effects of two rates of pre-plant nitrogen on tree establishment and aboveground growth. Arboricult Urban For 38(2):64–74. https://doi.org/10.48044/jauf.2012.011

Grime JP, Crick JC, Rincon JE (1986) The ecological significance of plasticity. In: Symposia of the society for experimental biology, vol 40. PubMed: 3544310

Guilland C, Maron PA, Damas O, Ranjard L (2018) Biodiversity of urban soils for sustainable cities. Environ Chem Lett 16(4):1267–1282. https://doi.org/10.1007/s10311-018-0751-6

Gupta SK, Singh H (2017) Observations on sporadic flowering in Ficus benjamina L. and Peltophorum africanum L. in new forest. Indian Forester 143:290–291

Gupta SK, Ram J, Singh H (2018a) Comparative study of transpiration in cooling effect of tree species in the atmosphere. J Geosci Environ Prot 6:151–166. https://doi.org/10.4236/gep.2018.68011

Gupta SK, Ram J, Singh H, Kumar P (2018b) Comparative study of cooling under shades of some forestry tree species with respect to ambient temperatures. J Geosci Environ Prot 6:51–66. https://doi.org/10.4236/gep.2018.611004

Gupta SK, Ram J, Singh H (2019) Relationship between leaf area index and atmospheric cooling of tree species. Indian J for 42(3):203–208

Hayatsu M, Tago K, Saito M (2008) Various players in the nitrogen cycle: Diversity and functions of the microorganisms involved in nitrification and denitrification. Soil Sci Plant Nutrit 54(1):33–45. https://doi.org/10.1111/j.1747-0765.2007.00195.x

Huang S, Gan Y, Zhang X, Chen N, Wang C, Gu X, Ma J, Niyogi D (2023) Urbanization amplified asymmetrical changes of rainfall and exacerbated drought: analysis over five urban agglomerations in the Yangtze River Basin, China. Earth's Fut 11(2):EF003117, e2022. https://doi.org/10.1029/2022EF003117

Jangra A, Kumar K, Maikhuri S, Bhandari MS, Pandey S, Singh H, Barthwal S (2024) Unveiling stress-adapted endophytic bacteria: characterizing plant growth-promoting traits and assessing cross-inoculation effects on Populus deltoides under abiotic stress. Plant Physiol Biochem 210:108610. https://doi.org/10.1016/j.plaphy.2024.108610

Jim CY, Ng YY (2018) Porosity of roadside soil as indicator of edaphic quality for tree planting. Ecol Eng 120:364–374. https://doi.org/10.1016/j.ecoleng.2018.06.016

Jim CY (1993) Soil compaction as a constraint to tree growth in tropical and subtropical urban habitats. Environ Conserv 20(1):35–49. https://doi.org/10.1017/S0376892900037206

Jorgensen E (1986) Urban forestry in the rearview mirror. Arboricult J 10(3):177–190. https://doi.org/10.1080/03071375.1986.9746750

Kanta C, Kumar A, Chauhan A, Singh H, Sharma IP (2024) The interplay between plant functional traits and climate change. In: Kumar N, Singh H (eds) Plant functional traits for improving productivity. Springer, Singapore. https://doi.org/10.1007/978-981-97-1510-7_3

Kenrick P, Strullu-Derrien C (2014) The origin and early evolution of roots. Plant Physiol 166(2):570–580. https://doi.org/10.1104/pp.114.244517

Kim YJ, Yoo G (2021) Suggested key variables for assessment of soil quality in urban roadside tree systems. J Soils Sed 21(5):2130–2140. https://doi.org/10.1007/s11368-020-02827-5

Kozlowski TT, Pallardy SG (2002) Acclimation and adaptive responses of woody plants to environmental stresses. Bot Rev 68(2): 270–334. https://doi.org/10.1663/0006-8101(2002)068[0270:AAAROW]2.0.CO;2

Kumar N, Singh H (2024) Plant functional traits for improving productivity. Springer Nature, Singapore. https://doi.org/10.1007/978-981-97-1510-7

Kumar N, Suyal DC, Sharma IP, Verma A, Singh H (2017) Elucidating stress proteins in rice (Oryza sativa L.) genotype under elevated temperature: a proteomic approach to understand heat stress response. 3 Biotech 7:205. https://doi.org/10.1007/s13205-017-0856-9

Kumar M, Singh MP, Singh H, Dhakate PM, Ravindranath NH (2019a) Forest working plan for the sustainable management of forest and biodiversity in India. J Sustain for 39:1–22. https://doi.org/10.1080/10549811.2019.1632212

Kumar N, Jeena N, Singh H (2019b) Elevated temperature modulates rice pollen structure: a study from the foothill of the Himalayan agro-ecosystem in India. 3 Biotech 9(5):175. https://doi.org/10.1007/s13205-019-1700-1

Kumar N, Jeena N, Kumar N, Gangola S, Singh H (2019c) Phytoremediation facilitating enzymes: an enzymatic approach for enhancing remediation process. In: Bhatt P (ed) Smart bioremediation technologies. Academic Press, Elsevier. 50 Hampshire Street, 5th Floor, Cambridge, MA 02139, United States, pp 289–306

Kumar N, Singh H, Sharma SK (2020) Antioxidants: responses and importance in plant defense system. In: Roychowdhury R, Choudhury S, Hasanuzzaman M, Srivastava S (eds) Sustainable agriculture in the era of climate change. Springer, Cham, pp251–264. https://doi.org/10.1007/978-3-030-45669-6_11

Kumar A, Kumar P, Singh H, Kumar N (2021a) Modulation of plant functional traits under essential plant nutrients during seasonal regime in natural forests of Garhwal Himalayas. Plant Soil 465:197–212. https://doi.org/10.1007/s11104-021-05003-x

Kumar A, Kumar P, Singh H, Bisht S, Kumar N (2021b) Relationship of physiological plant functional traits with soil carbon stock in the temperate forest of Garhwal Himalaya. Curr Sci 120(8):1368–1373. https://doi.org/10.18520/cs/v120/i8/1368-1373

Kumar A, Kumar P, Singh H, Kumar N (2021c) Adaptation and mitigation potential of roadside trees with bio-extraction of heavy metals under vehicular emissions and their impact on physiological traits during seasonal regimes. Urban for Urban Green 58:126900. https://doi.org/10.1016/j.ufug.2020.126900

Kumar A, Kumar P, Singh H, Kumar N (2021d) Impact of plant functional traits on infiltration rate under different forests of Kempty watershed in Garhwal Himalaya. Indian J Soil Conserv 49(1):38–44

Kumar A, Singh H, Kumari G, Bisht S, Malik A, Kumar A, Singh M, Raturi BS, Thakur A, Kaushal R (2022) Adaptive resilience of roadside trees to vehicular emissions via leaf enzymatic, physiological, and anatomical trait modulations. Environ Pollut 313:120191. https://doi.org/10.1016/j.envpol.2022.120191

Kumar N, Singh H, Giri K, Kumar A, Joshi A, Yadav A, Singh R, Bisht S, Kumari R, Jeena N, Khairakpam R, Mishra G (2024) Physiological and molecular insights into the allelopathic effects on agroecosystems under changing environmental conditions. Physiol Mol Biol Plants 30:417–433. https://doi.org/10.1007/s12298-024-01440-x

Locosselli GM, de Camargo EP, Moreira TCL, Todesco E, de Fátima Andrade M, de André CDS, de Andre PA, Singer JM, Ferreira LS, Saldiva PHN, Buckeridge MS (2019) The role of air pollution and climate on the growth of urban trees. Sci Total Environ 666:652–661. https://doi.org/10.1016/j.scitotenv.2019.02.291

Locosselli GM, Buckeridge MS (2023) The science of urban trees to promote well-being. Trees 37(1):1–7. https://doi.org/10.1007/s00468-023-02389-2

Lyytimäki J, Sipilä M (2009) Hopping on one leg–The challenge of ecosystem disservices for urban green management. Urban For Urban Green 8(4):309–315. https://doi.org/10.1016/j.ufug.2009.09.003

Mahendra A, Seto KC (2019) Upward and outward growth: managing urban expansion for more equitable cities in the global south. World Resources Report. World Resources Institute

Malik A, Kumar K, Kumari G, Kumar N, Singh H (2024) Application of functional traits in modelling productivity and resilience under climate change. In: Kumar N, Singh H (eds) Plant functional traits for improving productivity. Springer, Singapore. https://doi.org/10.1007/978-981-97-1510-7_5

McGrath D, Henry J (2016) Organic amendments decrease bulk density and improve tree establishment and growth in roadside plantings. Urban For Urban Green 20:120–127. https://doi.org/10.1016/j.ufug.2016.08.015

Meena D, Singh H, Chaudhari SK (2011) Elucidating strontium response on growth dynamics and biochemical change in Phaseolus mungo L. Int J Agric Environ Biotechnol 4:107–113

Montagnoli A (2022) Adaptation of the root system to the environment. Forests 13(4):595. https://doi.org/10.3390/f13040595

Olchowik J, Suchocka M, Jankowski P, Malewski T, Hilszczańska D (2021) The ectomycorrhizal community of urban linden trees in Gdańsk, Poland. PLOS ONE 16(4):e0237551. https://doi.org/10.1371/journal.pone.0237551

Oldfield EE, Warren RJ, Felson AJ, Bradford MA (2013) FORUM: challenges and future directions in urban afforestation. J Appl Ecol 50(5):1169–1177. https://doi.org/10.1111/1365-2664.12124

Oldfield EE, Felson AJ, Wood SA, Hallett RA, Strickland MS, Bradford MA (2014) Positive effects of afforestation efforts on the health of urban soils. For Ecol Manage 313:266–273. https://doi.org/10.1016/j.foreco.2013.11.027

Pajares S, Bohannan BJ (2016) Ecology of nitrogen fixing, nitrifying, and denitrifying microorganisms in tropical forest soils. Front Microbiol 7:1045. https://doi.org/10.3389/fmicb.2016.01045

Pataki DE, Alberti M, Cadenasso ML, Felson AJ, McDonnell MJ, Pincetl S, Pouyat RV, Setälä H, Whitlow TH (2021) The benefits and limits of urban tree planting for environmental and human health. Front Ecol Evol 9:603757. https://doi.org/10.3389/fevo.2021.603757

Pett-Ridge J, Petersen DG, Nuccio E, Firestone MK (2013) Influence of oxic/anoxic fluctuations on ammonia oxidizers and nitrification potential in a wet tropical soil. FEMS Microbiol Ecol 85(1):179–194. https://doi.org/10.1111/1574-6941.12111

Philippot L, Spor A, Hénault C, Bru D, Bizouard F, Jones CM, Sarr A, Maron PA (2013) Loss in microbial diversity affects nitrogen cycling in soil. ISME J 7(8):1609–1619. https://doi.org/10.1038/ismej.2013.34

Phillips RP, Brzostek E, Midgley MG (2013) The mycorrhizal-associated nutrient economy: a new framework for predicting carbon–nutrient couplings in temperate forests. New Phytol 199(1):41–51. https://doi.org/10.1111/nph.12221

Pincetl S (2017) Cities in the age of the Anthropocene: climate change agents and the potential for mitigation. Anthropocene 20:74–82. https://doi.org/10.1016/j.ancene.2017.08.001

Pouyat RV, Szlavecz K, Yesilonis ID, Groffman PM, Schwarz K (2010) Chemical, physical, and biological characteristics of urban soils. Urban Ecosyst Ecol 55:119–152. https://doi.org/10.2134/agronmonogr55.c7

Pouyat RV, Day SD, Brown S, Schwarz K, Shaw RE, Szlavecz K, Trammell TL, Yesilonis ID (2020) Urban soils. In: Forest and rangeland soils of the United States under changing conditions: a comprehensive science synthesis, pp 127–144

Qin J, Zhou X, Sun C, Leng H, Lian Z (2013) Influence of green spaces on environmental satisfaction and physiological status of urban residents. Urban For Urban Green 12(4):490–497. https://doi.org/10.1016/j.ufug.2013.05.005

Reeves FB, Wagner D, Moorman T, Kiel J (1979) The role of endomycorrhizae in revegetation practices in the semi-arid west. I. A comparison of incidence of mycorrhizae in severely disturbed vs. natural environments. Am J Bot 66(1):6–13. https://doi.org/10.1002/j.1537-2197.1979.tb06186.x

Reisinger AJ, Groffman PM, Rosi-Marshall EJ (2016) Nitrogen-cycling process rates across urban ecosystems. FEMS Microbiol Ecol 92(12):fiw198. https://doi.org/10.1093/femsec/fiw198

Rellán-Álvarez R, Lobet G, Dinneny JR (2016) Environmental control of root system biology. Annu Rev Plant Biol 67:619–642. https://doi.org/10.1146/annurev-arplant-043015-111848

Ryan MG (2011) Tree responses to drought. Tree Physiol 31(3):237–239. https://doi.org/10.1093/treephys/tpr022

Sax MS, Bassuk N, van Es H, Rakow D (2017) Long-term remediation of compacted urban soils by physical fracturing and incorporation of compost. Urban For Urban Green 24:149–156. https://doi.org/10.1016/j.ufug.2017.03.023

Scharenbroch BC, Meza EN, Catania M, Fite K (2013) Biochar and biosolids increase tree growth and improve soil quality for urban landscapes. J Environ Qual 42(5):1372–1385. https://doi.org/10.2134/jeq2013.04.0124

Scharenbroch BC, Scheberl L, Gebhard JC, Prater JR, Werner LP (2023) Towards an improved rapid urban site index. Arboricult Urban For 49(2):90–104. https://doi.org/10.48044/jauf.2023.008

Sharma R, Singh H, Kaushik M, Nautiyal R, Singh O (2018) Adaptive physiological response, carbon partitioning, and biomass production of Withania somnifera (L.) Dunal grown under elevated CO_2 regimes. 3 Biotech 8:267. https://doi.org/10.1007/s13205-018-1292-1

Sharma R, Singh H (2021) Alteration in biochemical constituents and nutrients partitioning of asparagus Racemosus in response to elevated atmospheric CO_2 concentration. Environ Sci Pollut Res 29:6812–6821. https://doi.org/10.1007/s11356-021-16050-3

Simon J, Dannenmann M, Pena R, Gessler A, Rennenberg H (2017) Nitrogen nutrition of beech forests in a changing climate: Importance of plant–soil-microbe water, carbon, and nitrogen interactions. Plant Soil 418(1–2):89–114. https://doi.org/10.1007/s11104-017-3293-y

Singh H (2021) An integrated approach considering physiological and biophysical based indicators for assessing tolerance of roadside plantations of Alstonia scholaris towards urban roadside air pollution: an assessment of adaptation of plantations for mitigating urban roadside air pollution. Trees. https://doi.org/10.1007/s00468-021-02179-8

Singh H (2024) Exploring adaptive modulation in plant functional traits and its impact on the productivity of Acacia auriculiformis under CO_2-enriched environment. Industr Crops Prod 210:118186. https://doi.org/10.1016/j.indcrop.2024.118186

Singh H, Kumar M (2022) Climate change and its impact on Indian Himalayan forests: current status and research needs. In: Rani S, Kumar R (eds) Climate change. Springer Climate. Springer, Cham, pp 223–242. https://doi.org/10.1007/978-3-030-92782-0_11

Singh H, Verma A (2013) Characterization and screening of high nitrogen efficient rice genotype to curtail environmental pollution. Int J Agric Environ Biotechnol 6(1):23–30

Singh H, Savita Sharma R, Sinha S, Kumar M, Kumar P, Verma A, Sharma SK (2017a) Physiological functioning of Lagerstroemia speciosa L. under heavy roadside traffic: an approach to screen potential species for abatement of urban air pollution. 3 Biotech 7(61):1–10. https://doi.org/10.1007/s13205-017-0690-0

Singh H, Verma A, Kumar M, Sharma R, Gupta R, Kaur M, Negi M, Sharma SK (2017b) Phytoremediation: a green technology to clean up the sites with low and moderate level of heavy metals. Austin Biochem 2:1012

Singh H, Sharma R, Savita SMP, Kumar M, Verma A, Ansari MW, Sharma SK (2018) Adaptive physiological response of Parthenium hysterophorus to elevated atmospheric CO_2 concentration. Indian Forester 144:1–14

Singh H, Yadav M, Kumar N, Kumar A, Kumar M (2020) Assessing adaptation and mitigation potential of roadside trees under the influence of vehicular emissions: A case study of Grevillea robusta and Mangifera indica planted in an urban city of India. PLoS ONE 15(1):e0227380. https://doi.org/10.1371/journal.pone.0227380

Singh S, Singh H, Sharma SK, Nautiyal R (2021) Seasonal variation in biochemical responses of bamboo clones in the sub-tropical climate of Indian Himalayan foothills. Heliyon 7(4):e06859. https://doi.org/10.1016/j.heliyon.2021.e06859

Singh H, Bandyopadhyaya S, Kumar A, Singh M, Malik A, Kumari G, Raturi A, Yadav SK, Kumar M, Barthwal S, Thakur A, Rawat PS (2022) Understanding the physiological and biophysical response of urban roadside plantations for assessing adaptation and mitigation mechanisms toward vehicular emissions. Urban Clim 44:101183. https://doi.org/10.1016/j.uclim.2022.101183

Singh M, Singh H, Kumar A, Kumar M, Barthwal S, Thakur A (2024a) Soil nitrogen availability determines the CO_2 fertilization effect on tree species (Neolamarckia cadamba): growth and physiological evidence. Environ Sustain 7:53–60. https://doi.org/10.1007/s42398-023-00300-w

Singh H, Kumar N, Kumar A (2024b) Enhancing resource use efficiency in crops through plant functional traits. In: Kumar N, Singh H (eds) Plant functional traits for improving productivity. Springer, Singapore. https://doi.org/10.1007/978-981-97-1510-7_6

Smith IA, Dearborn VK, Hutyra LR (2019) Live fast, die young: Accelerated growth, mortality, and turnover in street trees. PLOS ONE 14(5):e0215846. https://doi.org/10.1371/journal.pone.0215846

Tallis M, Taylor G, Sinnett D, Freer-Smith P (2011) Estimating the removal of atmospheric particulate pollution by the urban tree canopy of London, under current and future environments. Landsc Urban Plan 103(2):129–138.https://doi.org/10.1016/j.landurbplan.2011.07.003

Tedersoo L, Bahram M, Zobel M (2020) How mycorrhizal associations drive plant population and community biology. Science 367(6480):eaba1223. https://doi.org/10.1111/nph.12221

United Nations (2019) World urbanization prospects 2018: highlights (ST/ESA/SER. A/421). Retrieved 16 Feb 2024

Venkata Mohan S, Amulya K, Annie Modestra J (2020) Urban biocycles–Closing metabolic loops for resilient and regenerative ecosystem: a perspective. Bioresour Technol 306:123098. https://doi.org/10.1016/j.biortech.2020.123098

Verslues PE, Agarwal M, Katiyar-Agarwal S, Zhu J, Zhu JK (2006) Methods and concepts in quantifying resistance to drought, salt and freezing, abiotic stresses that affect plant water status. Plant J Cell Mol Biol 45(4):523–539. https://doi.org/10.1111/j.1365-313X.2005.02593.x

van de Vlasakker PCH, Tonderski K, Metson GS (2022) A review of nutrient losses to waters from soil- and ground-based urban agriculture—More nutrient balances than measurements. Front Sustain Food Syst 6:842930. https://doi.org/10.3389/fsufs.2022.842930

Wang R, Liu L, Tao Z, Wan B, Wang Y, Tang X, Li Y, Li X (2024) Effect of urbanization and urban forests on water quality improvement in the Yangtze River Delta: a case study in Hangzhou, China. J Environ Manage 351:119980. https://doi.org/10.1016/j.jenvman.2023.119980

Weber S (2016) 3 cities taking urban forestry to the next level. World Resources Institute. https://www.wri.org/insights/3-cities-taking-urban-forestry-next-level

Yadav SK, Singh H, Ginwal HS and Barthwal S (2019a) Elevated CO_2 enhanced growth and physiological process of Populus deltoides Bartr. ex Marsh. Indian Forester 145(1):23–27

Yadav SK, Singh H, Nautiyal R, Ginwal HS, Ansari SA, Barthwal S (2019b) Modulation of morpho-physiological responses in Populus deltoides by elevated carbon dioxide and temperature. For Sci 66:105–118. https://doi.org/10.1093/forsci/fxz048

Alteration in Biochemical and Molecular Expression of Trees in Response to Changing Climatic and Environmental Pollution

Ravita Tadiya, Rimpee Garg, Kishan Kumar, Hukum Singh, and Santan Barthwal

Abstract The chapter provides a comprehensive overview of the intricate relationship between climate change, pollution, and tree biology, emphasizing the critical need to understand how climate and environmental factors impact tree functionality. Trees respond to these stressors through diverse biochemical and molecular adaptations. They adjust photosynthetic efficiency, antioxidant defense systems, and secondary metabolite production to counteract oxidative stress and maintain cellular balance. Additionally, trees modify gene expression, activate stress-responsive pathways, and undergo epigenetic changes to enhance resilience in changing climatic conditions. Environmental pollution disrupts biochemical pathways and compromises tree health, yet some species exhibit resilience by deploying detoxification mechanisms and activating defense genes in polluted environments. This knowledge synthesis will guide conservation efforts and aid in developing sustainable practices to protect tree health amidst climate change and environmental pollution.

Keywords Biochemical response · Climate change · Environment pollution · Molecular and gene expression · Tree functionality

1 Introduction

Environmental pollution and climate change considerably impact natural ecosystems, human health, and agricultural output (Noyes et al. 2009; Noya et al. 2018; Richardson et al. 2018). Climate change and pollution are closely related variables that threaten Earth's ecosystems and significantly challenge global sustainability (Noyes et al. 2009; Ummenhofer and Meehl 2017; Kumar et al. 2022; Singh et al. 2022; Kumar et al. 2024). While these two phenomena are distinct, they commonly interchange and promote each other. Greenhouse gases (GHGs) such as carbon

R. Tadiya (✉) · R. Garg · K. Kumar · H. Singh · S. Barthwal
Genetics and Tree Improvement Division, ICFRE-Forest Research Institute, P.O. New Forest, Dehradun, Uttarakhand 248006, India
e-mail: ravita.tadiya@gmail.com

H. Singh (ed.), *Urban Forests, Climate Change and Environmental Pollution*,
https://doi.org/10.1007/978-3-031-67837-0_10

dioxide and methane are emitted during the combustion of fossil fuels and industrial activities, contributing to climate change (Yoro and Daramola 2020; Singh et al. 2024a; Singh 2024). Furthermore, rising temperatures also trigger extreme weather events, such as altered precipitation patterns, resulting in heavier rainfall and eventual flooding. It also accelerates evaporation, resulting in more water loss from the soil, contributing to drought (Trenberth 2005; Ummenhofer and Meehl 2017). Furthermore, in dry and semi-arid areas, high temperatures and increased evaporation contribute to soil salinity by accumulating salts (Brindha and Schneider 2019; Corwin 2021).

These environmental fluctuations have a substantial impact on plant development and survival via a variety of processes. According to the Food and Agriculture Organization (FAO), the decline in the production of crops leads to a significant impact on climate change, especially in the context of the world's rapid population growth (Rogelj et al. 2016; Dhyani et al. 2021; Kumar et al. 2021a, b, c, d, e; Prakash et al. 2022). Forest trees are crucial in mitigating climate change and environmental pollution while providing a broad range of products essential for human needs, such as wood, paper, biomass, fuel, etc. (Harfouche et al. 2014). Trees consistently face environmental stresses such as temperature fluctuations, altered rainfall patterns, and increased frequency of extreme weather events like elevated CO_2, all indicative symptoms of climate change (Barlow et al. 2015; Hatfield and Prueger 2015; Salehi-Lisar and Bakhshayeshan-Agdam 2016). Temperature, precipitation, and atmospheric CO_2 concentration substantially influence plant physiology and distribution, impacting the developmental processes of trees like seed germination and flowering. These processes are seasonal, and the temporal fluctuation in plant activity provides strong evidence that changes in the global environment impact species and ecosystems (Cleland et al. 2007).

Trees are distinguished from other plants due to their long lifespan, slower growth, prolonged vegetative periods, and complex life cycles (Fladung and Hoenicka 2012). Trees respond in various ways to ensure their survival and productivity in response to climate-induced stress. They continuously adapt to altering environmental conditions and varying growing seasons. Molecular networks regulate the adaptability of trees. Advances in transcriptomics, proteomics, and metabolomics have considerably contributed to identifying several genes and proteins associated with environmental stress (Lee et al. 2023; Raza et al. 2019; Thornton et al. 2014; Zandalinas et al. 2018). Understanding how forest trees adapt to harsh environmental conditions is critical for maintaining productivity and meeting the demand for future commercial products.

This chapter aims to comprehensively assess the biochemical and molecular changes occurring in trees in response to major abiotic stresses caused by climate change, such as elevated CO_2 and temperature. Exploring molecular signaling cascades, gene expression patterns, and metabolic pathways may help to reveal the processes that plants utilize to deal with the challenges posed by a rapidly changing climate. Assessing the intricate mechanisms involved in these processes enhances understanding of plant biology and holds applications in sustainable agriculture, ecosystem preservation, and climate change mitigation (Singh 2021; Sharma and Singh 2021; Singh et al. 2021).

2 Elevated Atmospheric CO_2 Induced Biochemical and Molecular Adaptations

Rising atmospheric CO_2 concentrations are a primary contributor to climate change. Carbon dioxide levels are expected to exceed 500 ppm by the end of the century, leading to a global temperature increase exceeding two degrees Celsius (Lee et al. 2023). Multiple studies have highlighted the significant influence of elevated CO_2 levels on biochemical and molecular processes, influencing the expression of genes involved in metabolic processes and stress signaling pathways (Roy and Mathur 2021).

The CO_2 fertilization effect alters photosynthesis, resulting in changes in the activity of essential enzymes like ribulose-1,5-bisphosphate carboxylase/oxygenase (Rubisco) (Priyadharsini et al. 2022). Protein structure and content are also affected by alterations in nitrogen metabolism (Conroy 1992; Taub and Wang 2008). Genes linked to nitrogen metabolism and photorespiratory pathways are downregulated, while genes related to photosynthesis, growth, development, and stress responses are upregulated (Adavi and Sathee 2021; Huang and Xu 2015). In *Betula pendula*, increased CO_2 levels enhanced the activity of Rubisco and overall photosynthesis despite a decrease in photochemical parameters and Rubisco concentration (Eichelmann et al. 2004; Riikonen et al. 2005). An investigation into aspen trees (*Populus tremuloides*) exposed to elevated CO_2 levels showed heightened photosynthesis despite the downregulation of many transcripts linked to chloroplast biosynthesis and function. This includes genes related to photosynthetic proteins like Rubisco and Rubisco activase, photosystem I and II proteins, the light harvest complex, and chlorophyll biosynthesis (Cseke et al. 2009; Taylor et al. 2005). The accumulation of sugars in the leaves may trigger negative feedback processes, potentially linked to observed transcriptional patterns at the gene expression level (Cseke et al. 2009).

Plants respond to increasing CO_2 levels by regulating stomatal conductance and optimizing water usage efficiency via systems that involve signaling molecules like abscisic acid (Negin and Moshelion 2016; Li et al. 2020). Stomatal closure in response to increased CO_2 conditions is promoted by activating ABA biosynthesis genes and ABA receptors (PYR/RCAR family) (Li et al. 2020). Under elevated CO_2, the expression of guard cell aquaporin genes is suppressed, resulting in reduced stomatal conductance and increased water-use efficiency. The efficiency is improved by the upregulation of aquaporin genes, such as CsPIP2-7, in stems and leaves (Huang and Xu 2015; Li et al. 2020). The genes ALDH, PK, PDC, GDH, ACAS, AdSS, AS, and NiR play roles in various metabolic processes in Triploid White Poplar (Liu et al. 2014). Additionally, CO_2 sensing in guard cells is mediated by carbonic anhydrases, which induce stomatal closure either through the direct activation or by influencing protein kinases (MPK, HT1, CBC, OST1, etc.) and S-type (SLAC1) anion channels (Ma and Bai 2021).

Stomatal density, stomatal conductance, transpiration of leaves, and evapotranspiration by the canopy/ecosystem are often decreased in response to elevated CO_2 levels (Singh et al. 2020; Kumar et al. 2021a, b). Importantly, though, some research challenges this conclusion, suggesting that the outcome could be the opposite when increased CO2 interacts with other climate factors (Taylor et al. 2005; Xu et al. 2016). According to a study done in poplar trees by (Yadav et al. 2020), variations in the expression of the genes SPCH, SDD1, and DAG appeared to have an impact on the number and size of stomata, which may raise transpiration and photosynthetic rates. The HIC (high CO_2) gene, involved in the signal transduction pathway leading to stomatal production, demonstrated increased expression (Xu et al. 2016; Kumar et al. 2021c, d).

The antioxidant defence system becomes crucial as CO_2 levels rise, stimulating the production of reactive oxygen species (ROS), which causes the overexpression of enzymes such as superoxide dismutase (SOD) and catalase (CAT) (Xu et al. 2015). Secondary metabolite synthesis, particularly phenolics and terpenoids, is disrupted, affecting plant defence systems (Jamloki et al. 2021). These biochemical adaptations are essential for anticipating and regulating the effects of rising CO_2 levels on plant ecosystems. Kim et al. (2022) observed that higher CO_2 levels promoted the level of catalase, MDAR, superoxide dismutase [Fe] 2 (FSD2), L-ascorbate peroxidase (APXS, APX5), and tocopherol O-methyltransferase (VTE4) in *Pinus densiflora*. Moreover, microRNAs are acknowledged for their role in plant responses to elevated CO_2 levels. The miR156/157 and miR172 microRNAs play a crucial role in modulating pathways concerning auxin signaling, stress responses, and cell wall biosynthesis (May et al. 2013).

3 Heat Stress Induced Biochemical and Molecular Responses

Heat stress, often linked to heat shock, significantly impacts various biochemical processes in plants, including enzyme activity, metabolic pathways, and biochemical reactions. As temperatures increase, protein denaturation becomes more prevalent, impacting enzymes and structural proteins within plant cells (Li et al. 2021). High temperatures activate stress-responsive pathways that harm cell components and impede plant development and seed formation. Heat stress causes stomatal closure, which reduces water loss through transpiration. Prolonged heat stress, on the other hand, may cause water scarcity, affecting total plant hydration and growth (Ohama et al. 2017; Zhao et al. 2020).

The buildup of soluble osmolytes in plant cells can lower osmotic potential, lessening the harmful effects of stress. In response to HS, plants accumulate a range of soluble osmolytes, including sugar, sugar alcohol (polyol), proline, and soluble proteins (Sairam and Tyagi 2004). Two genes, GOLS1 and HsfA2, were found to be up-regulated in grapevine berries, improving heat tolerance in heat stress. Galactinol synthesis is caused by the GOLS1 gene, which is activated by HsfA2 (Pillet et al. 2012). Similarly, even a short heat treatment stimulates the synthesis of substantial levels of proline in the leaves of *Prunus domestica L* to assist it in coping with heat stress conditions (Viljevac et al. 2022).

Elevated temperature also increases the formation of reactive oxygen species (ROS) and prompts plants to use antioxidant defense systems, including both enzymatic and non-enzymatic antioxidants, to combat possible oxidative damage (Xu et al. 2015; Zandalinas et al. 2018). In addition to ROS-scavenging enzymes like SOD, CAT, APX, monodehydroascorbate reductase (MDHAR), DHAR, and glutathione reductase (GR), the non-enzymatic mechanisms like ascorbate—ASA, glutathione—GSH, α-tocopherol, and flavonols, are involved in detoxifying the excess ROS produced by cells under stress. Song et al. (2014) conducted a study in which they exposed numerous plants to SOD, CAT, POD, and APX at high temperatures and determined that these oxidants convert the hazardous O_2^- and H_2O_2 to water and molecular oxygen (O_2), thereby safeguarding the cell from oxidative stress.

The gene responsive to heat stress can be categorized into three distinct groups. The first group encompasses proteins that probably contributed to heat stress tolerance, such as RNA helicases, heat shock proteins (HSPs), ion transporters, and aquaporins (AQPs) (Feng et al. 2019; Kim et al. 2019; Wang et al. 2016). Heat shock proteins (HSPs) and heat shock factors (HSFs) form complexes with key heat-responsive proteins to prevent their denaturation. Moreover, heat shock proteins (HSPs) serve as molecular chaperones, controlling the ubiquitin-mediated autophagy and/or 26S proteasome processes that break down misfolded or damaged proteins (Liu and Charng 2012; Ohama et al. 2017). Heat stress promotes the expression of Hsps in poplar, including Hsp90, and several are co-expressed with Heat-Shock Factors A and B. Several Hsps, including Hsp60, 70, and 100, are activated in *P. trichocarpa* during the temporary expression of HSFA2, HSFA6a, and HSFB2a (Zhang et al. 2013; Song et al. 2014).

The second group of proteins consists of regulatory proteins such as calcium-dependent protein kinases (CDPKs), mitogen-activated protein kinases (MAPK), and receptor-like protein kinases (RLKs) (Ohama et al. 2017; Zhao et al. 2020). In response to heat stress, Ca^{2+} signals are subsequently transduced by Ca^{2+} sensors, commencing signal transduction pathways (Hu et al. 2023). Under heat stress conditions in plants, hormones at low concentrations can function as signaling molecules, regulating a range of physiological and biochemical processes. The transcription factor Phytochrome Interacting Factor 4 (PIF4) functions as a central regulatory hub that integrates thermo-responsive pathways at increased temperatures by interacting with light, hormones, and circadian signaling (Choi and Oh 2016; Quint et al. 2016).Rising temperatures promote the biosynthesis pathways of hormones, including auxins, salicylic acid (SA), ABA, brassinosteroids (BRs), cytokinin (CK), jasmonate (JA), and ethylene (ET). This results in an enhanced accumulation of these hormones in plant cells, serving to cope with heat stress (Li et al. 2021). TF families also regulate heat-responsive genes, including NAC, WRKY, bZIP, and MYB (Zhao et al. 2020).

Furthermore, non-coding RNAs (ncRNAs) comprise the third group, encompassing long non-coding RNAs (lncRNAs) and microRNAs (miRNAs), both of which play a role in controlling the heat stress response. MiRNAs and lncRNAs interact with other molecules, such as proteins, RNA, and DNA, to build complex regulatory lncRNA-miRNA-mRNA networks, enhancing heat tolerance in plants (Ohama et al. 2017; Zhao et al. 2020; He et al. 2020; Song et al. 2020).

4 Drought Stress Induced Biochemical and Molecular Responses

Drought stress inhibits plant growth by interfering with a variety of physiological and biochemical processes such as respiration, translocation, ion uptake, water potential, stomatal closure, photosynthesis, sugar and nutrient metabolism, antioxidant systems, and phytohormones (Trenberth 2005; Salehi-Lisar and Bakhshayeshan-Agdam 2016). Drought stress activates multiple genes with varied functions of plants, resulting in physiological and biochemical reactions (Raza et al. 2019; Yadav et al. 2019; Gupta et al. 2019a, b; Wang et al. 2021). Under drought conditions, plants produce reactive oxygen species (ROS), which disrupt essential physiological and metabolic processes in plants (Singh and Kumar 2022; Singh et al. 2024b). Antioxidants and osmotic regulation are the key defense systems that allow plants to endure water-deprived stress (Krasensky and Jonak 2012).

Abscisic acid (ABA), a key component in stress signalling, is produced in response to increased ROS levels during stress. It plays a crucial role in regulating gene expression to facilitate biochemical reactions such as antioxidant enzymes CAT, peroxidase (POD), SOD, glutathione reductase (GR), and ascorbate peroxidase (APX) Additionally, ABA contributes to the modulation of non-enzymatic antioxidants such as phenolic compounds (Ahmad et al. 2008; Hasanuzzaman et al. 2020). In reaction to drought stress, the plant's defense mechanism includes the accumulation of soluble molecules in the cytoplasm, such as proline, glycine-betaine, and sugars. These metabolites function as Osmo protectants, maintaining water flow and molecular stability and reducing the accumulation of stress-induced free radicals (Ghosh et al. 2021). In a study conducted by Yang et al. (2009a, b), *Populus kangdingensis* from high altitude showed tolerance to drought by increasing soluble protein, sugar, free proline, and the activity of antioxidant enzymes.

The expression of genes responsive to drought is facilitated by signal transduction pathways within the complex network of drought response, which also influence osmotic regulation and stress-related chemical synthesis (Todaka et al. 2015; Meena et al. 2011; Kumar et al. 2019c, 2020). Drought-inducible genes are governed by transcriptional cascades, with resulting proteins such as protein kinases, protein phosphatases, late embryogenesis abundant (LEA) proteins, antioxidants, osmotic, aquaporin, and sugar transporters linked to water and ion uptake (Kaur and Asthir 2017). LEA proteins have been suggested as a unique form of molecular chaperones preventing protein aggregation induced by water stress. Veeranagamallaiah et al. (2011) and Kosová et al. 2014 reported the induction of several dehydrin-type proteins (class of LEA proteins), including dehydrins (Dhn1, Dhn3, Dhn5, Dhn7, and Dhn9), under terminal drought. In poplar and Norway spruce, induction of dehydrins, LEA family proteins, was seen in response to osmotic pressure (Blödner et al. 2007; Caruso et al. 2002). Two types of water-transport proteins, plasma layer characteristic proteins (PIPs) and tonoplast film natural proteins, are required to maintain the water status of plants. A decrease in PIPs resulted in the suppression of development in *Eucalyptus* spp. (Tsuchihira et al. 2010). Berta et al. (2010a; b) identified five aquaporin membrane proteins in white poplar that were elevated under drought.

The stress hormone abscisic acid (ABA) undergoes rapid synthesis in response to different stresses, such as drought. This causes stomatal closure, which limits plant development and improves resistance to drought in plants (Zeevaart 1980; Zhu 2016). When the PYR/PYL/RCAR receptors detect ABA, it triggers downstream cascade kinase processes that trigger the transcription factors (TFs) known as ABA-responsive element (ABRE)-binding proteins (AREBs) and ABRE-binding factors (ABFs). Through the induction of stress-responsive genes, activated ABA signaling improves resistance to drought (Berta et al. 2010b; Saadaoui et al. 2017).

The interplay between JA and ABA signalling in response to drought stress is demonstrated by the interaction between the ABA-responsive ABF3 and the JA-regulated MYC2 (Liu et al. 2018). This interaction highlights the crosstalk between signalling pathways. Furthermore, BES1 from BR signalling interacts with WRKY, AP2/ERF, and NAC TFs to control the response to drought (Nolan et al. 2020). The dynamic interplay between ABA-dependent and ABA-independent pathways regulates the response to drought (Yoshida et al. 2015). Furthermore, RLKs have been linked to plant responses to various abiotic stresses, including drought (Marshall et al. 2012). According to Berta et al. (2010a; b), temporary water deficiency in poplar tissues caused the production of a particular receptor-like kinases (RLKs) gene.

5 Flooding Stress Induced Biochemical and Molecular Responses

Flooding occurs when soil becomes highly saturated with water, usually due to variable or unpredictable rainfall patterns caused by meteorological circumstances (Trenberth 2005). Morales-Olmedo et al. (2015) state that floods reduce air in soil pores, preventing gas exchange between the soil and the atmosphere. Under saturated conditions, plants turn to other energy generation strategies, such as fermentation (Davies 2004; Vartapetian 1991; Singh et al. 2017a, 2017b). Plants adopt a variety of morphological modifications to ease respiratory depression in roots and offset the damage resulting from disrupted energy metabolism as a defense against water-logging issues (Meena et al. 2011; Kumar et al. 2019c, 2020). The development of air films in the upper cuticle, the prompt elongation of apical meristematic tissue, the synthesis of adventitious roots (ARs) or other specialized aeration tissues, and the creation of barriers to radial oxygen loss (ROL) are examples of morphological adaptations in response to waterlogging challenges (Ejiri et al. 2021; Loreti et al. 2016; Visser and Voesenek 2005).

In response to floods, many flood-tolerant trees establish a broad adventitious root system and show stem hypertrophy, frequently linked with lenticels. According to Sauter (2013), adventitious roots and stem lenticels work together to form a gas transport system that allows roots to respire in flood situations. Eucalyptus trees can generate a wide range of characteristics that enable them to withstand flooding (Argus et al. 2015). A study conducted by Lopez and Kursar (1999) on four tropical tree species observed that large lenticels on Primara plants in flood conditions contributed to significant root growth during flooded stress. According to Calvo-Polanco et al. (2012), the Tamarack *(Larix laricina)* tree exhibited tolerance seedlings generated adventitious roots in flood circumstances that increased tolerance potential. Many genes have been identified as adventitious root (AR) formation promoters. (Geldner et al. 2003; Steinmann et al. 1999). Modulating auxin polar transport via the activation of auxin-associated genes such as GNOM1, PIN1, and PIN2 can alter the establishment of ARs in plants (Liu et al. 2009). Ethylene may trigger the development of AR in cortical cells, which subsequently requires reactive oxygen species (ROS) as a signalling mechanism to activate the process of programmed cell death (Steffens et al. 2012; Steffens and Sauter 2009).

Plants develop radial oxygen loss (ROL) barriers in their roots to limit oxygen leakage, and numerous genes have been identified as key mediator in their formation (Githiri et al. 2006). Suberin is assumed to be required for the development of the ROL barrier and is produced by suberin biosynthesis genes such as KCS, CYP, GPAT, ABC, LTP, HD, and POD (De Simone et al. 2003; Shiono et al. 2011). Numerous genes have been linked to aerenchyma development. In a hypoxic environment, the development of lysigenous aerenchyma depends on genes such as LSD1, EDS1, and PAD4, which regulate the upstream ethylene signal and ROS generation (Yamauchi et al. 2011). Plant hormones have an important role in the response to waterlogging stress. Immature, delicate stems and internodes can rapidly elongate to reach the air and initiate respiration (Kuroha et al. 2018). Waterlogging enhances gibberellin (GA) synthesis and ethylene (ET) buildup, enhancing internode elongation. Auxin promotes cell division and the formation of adventitious roots (ARs) (Kuroha et al. 2018). Increased ABA levels activate the antioxidant defense system and stomatal closure, reducing oxidative damage and significantly improving the plant's tolerance to waterlogging (Shimamura et al. 2014). SA levels enhance AR production, improve gas exchange, and increase plant tolerance to flooding (Bai et al. 2010). Exogenous SA application to plants lowers the oxidative stress damage induced by hypoxia and improves hypoxia tolerance in *Malus robusta* (Bai et al. 2009). Under waterlogging stress, the interaction between Jasmonic Acid (JA) and ethylene (ET) plays a critical role in the formation and development of the root system and aerenchyma (Pan et al. 2021).

Plants activate internal protective mechanisms by antioxidants to counteract the damaging effects of ROS, which indicate oxidative damage (Ahmad et al. 2008). High activity of antioxidant enzymes such as superoxide dismutase (SOD), ascorbate peroxidase (APX), glutathione reductase (GR), and catalase (CAT) in trees is associated with a higher tolerance to oxygen deficiency by flooding (Arbona and Gómez-Cadenas 2008). The increase of SOD, APX, and CAT enzyme activity in Citrumelo trees permits them to sustain a temporary tolerance under hypoxia (Hossain et al. 2009). Under hypoxia, similar outcomes were noted in the species of *Prunus* and *Malus* (Bai et al. 2010; Radmann et al. 2018).

6 Salinity Induced Biochemical and Molecular Responses

A diverse range of biochemical and molecular reactions contribute to how plants adapt to and endure the impacts of salt stress. According to Hasanuzzaman and Fujita (2022), salt stress alters gene expression of various genes involved in molecular, cellular, biochemical, physiological, and morphological pathways. These changes are brought about by genes and transcriptional variables that regulate signalling, ion transport, antioxidant system, compatible solute production, senescence-associated response, hormone regulation, and SOS signalling pathways (Hasegawa et al. 2000).

When plants are subjected to high concentrations of salt, the cells in their bodies can transport sodium into a structure known as the vacuole and store it there. This mechanism aids in preventing the accumulation of excessive sodium in the central part of the cell, thereby mitigating the potential damage to the cell's metabolism (Song and Wang 2015; Kumar et al. 2019a, b). Additionally, this mechanism enables the plant cells to absorb more water due to lower osmotic pressure. Therefore, the balance of potassium and sodium in the central part of the cell is crucial for the plant to withstand high salt stress (Yang et al. 2017; Sharma et al. 2018; Singh et al. 2018; Gupta et al. 2018a, b). The gene responsible for the transport of sodium and hydrogen (Na^+/H^+) ions in and out of the vacuole is AtNHX1. The plasma membrane (SOS1) plays a crucial role in regulating the correct levels of ions inside the cells (Li et al. 2019; Yang et al. 2009a, b). In a study conducted by Shi et al. (2002) on poplar, it was observed that the overexpression of the Na^+/H^+ antiporter AtNHX1 gene occurred under saline conditions, leading to the accumulation of certain types of sugar alcohols. Wu et al. (2007) cloned a gene encoding the Na^+/H^+ antiporter PeSOS1 from *P. euphratica*. They demonstrated that PeSOS1 may be responsible for the efflux of Na^+ at the plasma membrane. Accordingly, it is believed that PeSOS1 plays a vital role in the salt stress tolerance of *P. euphratica*.

To reduce osmotic stress under saline conditions, plants produce and accumulate unique molecules called compatible solutes, which play an essential role in their response to stress (Singh and Verma 2013; Kumar et al. 2017; Gupta and Singh 2017; Negrão et al. 2017). These solutes include glycine-betaine, proline, trehalose, sucrose, fructose, glucose, mannitol, and sorbitol (types of sugar alcohol). These compatible solutes are synthesized and stored in the cells of plants under salt stress, significantly contributing to the plant's ability to tolerate the stress (Méndez-Alonzo et al. 2016; Negrão et al. 2017). The genetically modified hybrids of *Populus nigra* and *P. tomentosa* produce mannitol by introducing a gene from *Escherichia coli* called mannitol-1-phosphate dehydrogenase (mtlD). This enzyme converts fructose into mannitol, increases mannitol content, and improves the plant's ability to tolerate salt (Lin et al. 2016). Guan et al. (2018) identified four genes that have the potential to serve as indicators of salt stress in grapevine plants. One of these genes, a proline-rich cell wall protein-like (PRPs) gene, was the most promising candidate.

The activity of antioxidant enzymes like glutathione reductase (GR), ascorbate peroxidase (APX), glutathione peroxidize (GPX), catalase (CAT), and superoxide dismutase (SOD) as well as the presence of non-enzymatic antioxidant compounds, exhibit a positive correlation with salinity tolerance (Bai et al. 2013; Yang et al. 2010). The *Tamarix androssowii*TaMnSOD gene in transgenic poplar trees increased their tolerance to salt stress. The TaMnSOD gene in the poplar trees increased their tolerance to salt stress (Wang et al. 2010). The Late Embryogenesis Abundant Protein, also known as LEA, helps protect plants from dehydration (Goyal et al. 2005).

The DREB, which stands for "dehydration-responsive element", plays a role in plants' resilience to stress (Zhou et al. 2012). Lu et al. (2014), reported that DREB plays an essential role in salt tolerance in a study of DREB-transgenic *Populus.* Furthermore, Yang et al. (2017) found that the ThDREB gene from *Tamarix hispida* increased the salt and drought tolerance of transgenic tobacco and *T. hispida.* Phytohormones are crucial in signalling and regulating plant responses to salt stress. Fricke et al. (2004) detected a significant elevation in ABA levels in both roots and shoots in response to stress conditions. According to Li et al. (2009), the expression of the jasmonic ethylene responsive factor gene in transgenic poplar trees leads to increased salt tolerance.

7 Conclusion

The chapter concludes that environmental pollution and climate change alterations modulate the molecular and biochemical responses, leading to changes in tree functionality to adapt to shifting growing conditions. This exploration illuminates the interactions among various environmental stressors—ranging from elevated CO_2 levels to heatwaves, droughts, floods, and salinity—and their profound impacts on trees' biochemical and molecular dynamics. Emphasizing the critical importance of understanding these dynamics, it underscores the necessity for pioneering methodologies to preserve the irreplaceable biodiversity and ecological harmony of forest ecosystems within the overarching climate change mitigation and adaptation framework.

References

Adavi SB, Sathee L (2021) Elevated CO_2 alters tissue balance of nitrogen metabolism and down-regulates nitrogen assimilation and signalling gene expression in wheat seedlings receiving high nitrate supply. Protoplasma 258(1):219–233

Ahmad P, Sarwat M, Sharma S (2008) Reactive oxygen species, antioxidants and signaling in plants. J Plant Biol 51(3):167–173. https://doi.org/10.1007/BF03030694

Arbona V, Gómez-Cadenas A (2008) Hormonal modulation of citrus responses to flooding. J Plant Growth Regul 27(3):241–250

Argus RE, Colmer TD, Grierson PF (2015) Early physiological flood tolerance is followed by slow post-flooding root recovery in the dryland riparian tree E ucalyptus camaldulensis subsp. refulgens. Plant Cell Environ 38(6):1189–1199

Bai T, Li C, Ma F, Shu H, Han M (2009) Exogenous salicylic acid alleviates growth inhibition and oxidative stress induced by hypoxia stress in Malus robusta Rehd. J Plant Growth Regul 28(4):358–366

Bai T, Li C, Ma F, Feng F, Shu H (2010) Responses of growth and antioxidant system to root-zone hypoxia stress in two Malus species. Plant Soil 327(1–2):95–105

Bai JH, Liu JH, Zhang N, Yang JH, Sa RL, Lan WU (2013) Effect of alkali stress on soluble sugar, antioxidant enzymes and yield of oat. J Integr Agric 12(8):1441–1449. https://doi.org/10.1016/S2095-3119(13)60556-0

Barlow KM, Christy BP, O'Leary GJ, Riffkin PA, Nuttall JG (2015) Simulating the impact of extreme heat and frost events on wheat crop production: a review. Field Crops Res 171:109–119

Berta M, Giovannelli A, Sebastiani F, Camussi A, Racchi ML (2010a) Transcriptome changes in the cambial region of poplar (Populus alba L.) in response to water deficit. Plant Biol 12(2):341–354

Berta M, Giovannelli A, Sebastiani F, Camussi A, Racchi ML (2010b) Transcriptome changes in the cambial region of poplar (Populus alba L.) in response to water deficit. Plant Biol 12(2):341–354

Blödner C, Majcherczyk A, Kües U, Polle A (2007) Early drought-induced changes to the needle proteome of Norway spruce. Tree Physiol 27(10):1423–1431

Brindha K, Schneider M (2019) Impact of urbanization on groundwater quality. In: Venkatramanan S, Prasanna MV, Chung, SY (eds). Elsevier, pp 179–196. https://doi.org/10.1016/B978-0-12-815413-7.00013-4

Calvo-Polanco M, Señorans J, Zwiazek JJ (2012) Role of adventitious roots in water relations of tamarack (Larix laricina) seedlings exposed to flooding. BMC Plant Biol 12(1):99. https://doi.org/10.1186/1471-2229-12-99

Caruso A, Morabito D, Delmotte F, Kahlem G, Carpin S (2002) Dehydrin induction during drought and osmotic stress in Populus. Plant Physiol Biochem 40(12):1033–1042

Choi H, Oh E (2016) PIF4 integrates multiple environmental and hormonal signals for plant growth regulation in Arabidopsis. Mol Cells 39(8):587

Cleland EE, Chuine I, Menzel A, Mooney HA, Schwartz MD (2007) Shifting plant phenology in response to global change. Trends Ecol Evol 22(7):357–365

Conroy JP (1992) Influence of elevated atmospheric CO_2 concentrations on plant nutrition. Aust J Bot 40(5):445–456

Corwin DL (2021) Climate change impacts on soil salinity in agricultural areas. Eur J Soil Sci 72(2):842–862

Cseke LJ, Tsai C, Rogers A, Nelsen MP, White HL, Karnosky DF, Podila GK (2009) Transcriptomic comparison in the leaves of two aspen genotypes having similar carbon assimilation rates but different partitioning patterns under elevated [CO_2]. New Phytol 182(4):891–911

Davies PJ (2004). Plant hormones: biosynthesis, signal transduction, action! Springer Science & Business Media.

De Simone O, Haase K, Muller E, Junk WJ, Hartmann K, Schreiber L, Schmidt W (2003) Apoplasmic barriers and oxygen transport properties of hypodermal cell walls in roots from four Amazonian tree species. Plant Physiol 132(1):206–217

Dhyani K, Kuniyal HB, Singh H and Sobha (2021). Growth and physiological potential of Terminalia arjuna under elevated CO_2 levels in Open top chamber condition. J Appl Natl Sci 13(3):1121–1126. https://doi.org/10.31018/jans.v13i3.2490. ISSN 2231-5209.

Eichelmann H, Oja V, Rasulov B, Padu E, Bichele I, Pettai H, Möls T, Kasparova I, Vapaavuori E, Laisk A (2004) Photosynthetic parameters of birch (Betula pendula Roth) leaves growing in normal and in CO_2-and O_3-enriched atmospheres. Plant, Cell Environ 27(4):479–495

Ejiri M, Fukao T, Miyashita T, Shiono K (2021) A barrier to radial oxygen loss helps the root system cope with waterlogging-induced hypoxia. Breed Sci 71(1):40–50. https://doi.org/10.1270/jsbbs.20110

Feng Z-J, Liu N, Zhang G-W, Niu F-G, Xu S-C, Gong Y-M (2019) Investigation of the AQP family in soybean and the promoter activity of TIP2; 6 in heat stress and hormone responses. Int J Mol Sci 20(2):262

Fladung M, Hoenicka H (2012) Fifteen years of forest tree biosafety research in Germany. Iforest-Biogeosci for 5(3):126

Fricke W, Akhiyarova G, Veselov D, Kudoyarova G (2004) Rapid and tissue-specific changes in ABA and in growth rate in response to salinity in barley leaves. J Exp Bot 55(399):1115–1123

Geldner N, Anders N, Wolters H, Keicher J, Kornberger W, Muller P, Delbarre A, Ueda T, Nakano A, Jürgens G (2003) The Arabidopsis GNOM ARF-GEF mediates endosomal recycling, auxin transport, and auxin-dependent plant growth. Cell 112(2):219–230

Ghosh UK, Islam MN, Siddiqui MN, Khan MAR (2021) Understanding the roles of osmolytes for acclimatizing plants to changing environment: a review of potential mechanism. Plant Signal Behav 16(8):1913306

Githiri SM, Watanabe S, Harada K, Takahashi R (2006) QTL analysis of flooding tolerance in soybean at an early vegetative growth stage. Plant Breeding 125(6):613–618

Goyal K, Walton LJ, Tunnacliffe A (2005) LEA proteins prevent protein aggregation due to water stress. Biochem J 388(1):151–157

Guan L, Haider MS, Khan N, Nasim M, Jiu S, Fiaz M, Zhu X, Zhang K, Fang J (2018) Transcriptome sequence analysis elaborates a complex defensive mechanism of grapevine (Vitis vinifera L.) in response to salt stress. Int J Mol Sci 19(12):4019

Gupta SK, Singh H (2017) Observations on Sporadic Flowering in Ficus benjamina L. and Peltophorum africanum L. in New Forest. Indian Forester 143:290–291

Gupta SK, Ram J, Singh H (2018a) Comparative study of transpiration in cooling effect of tree species in the atmosphere. J Geosci Environ Prot 6:151–166. https://doi.org/10.4236/gep.2018.68011

Gupta SK, Ram J, Singh H, Kumar P (2018b) Comparative study of cooling under shades of some forestry tree species with respect to ambient temperatures. J Geosci Environ Prot 6:51–66. https://doi.org/10.4236/gep.2018.611004

Gupta SK, Ram J, Singh H (2019a) Relationship between leaf area index and atmospheric cooling of tree species. Indian J for 42(3):203–208

Gupta SK, Singh H, Ram J (2019b) Need for a novel paradigm in forestry research: perspectives of comparative atmospheric cooling by different forestry effect tree species. Indian Forester 145(6):509–515

Harfouche A, Meilan R, Altman A (2014) Molecular and physiological responses to abiotic stress in forest trees and their relevance to tree improvement. Tree Physiol 34(11):1181–1198. https://doi.org/10.1093/treephys/tpu012

Hasanuzzaman M, Fujita M (2022) Plant responses and tolerance to salt stress: physiological and molecular interventions. Int J Mol Sci 23(9):1–6. https://doi.org/10.3390/ijms23094810

Hasanuzzaman M, Bhuyan MHMB, Zulfiqar F, Raza A, Mohsin SM, Mahmud JA, Fujita M, Fotopoulos V (2020) Reactive oxygen species and antioxidant defense in plants under abiotic stress: revisiting the crucial role of a universal defense regulator. Antioxidants 9(8):681

Hasegawa PM, Bressan RA, Zhu J-K, Bohnert HJ (2000) Plant cellular and molecular responses to high salinity. Annu Rev Plant Biol 51(1):463–499

Hatfield JL, Prueger JH (2015) Temperature extremes: effect on plant growth and development. Weather Clim Extrem 10:4–10

He X, Guo S, Wang Y, Wang L, Shu S, Sun J (2020) Systematic identification and analysis of heat-stress-responsive lncRNAs, circRNAs and miRNAs with associated co-expression and ceRNA networks in cucumber (Cucumis sativus L.). Physiol Plantarum 168(3):736–754

Hossain Z, López-Climent MF, Arbona V, Pérez-Clemente RM, Gómez-Cadenas A (2009) Modulation of the antioxidant system in citrus under waterlogging and subsequent drainage. J Plant Physiol 166(13):1391–1404

Hu W, Liu J, Liu T, Zhu C, Wu F, Jiang C, Wu Q, Chen L, Lu H, Shen G (2023) Exogenous calcium regulates the growth and development of Pinus massoniana detecting by physiological, proteomic, and calcium-related genes expression analysis. Plant Physiol Biochem 196:1122–1136

Huang B, Xu Y (2015) Cellular and molecular mechanisms for elevated CO_2–regulation of plant growth and stress adaptation. Crop Sci 55(4):1405–1424

Jamloki A, Bhattacharyya M, Nautiyal MC, Patni B (2021) Elucidating the relevance of high temperature and elevated CO_2 in plant secondary metabolites (PSMs) production. Heliyon 7(8)

Kaur G, Asthir B (2017) Molecular responses to drought stress in plants. Biol Plant 61:201–209

Kim JH, Lim SD, Jang CS (2019) Oryza sativa heat-induced RING finger protein 1 (OsHIRP1) positively regulates plant response to heat stress. Plant Mol Biol 99:545–559

Kim T-L, Lim H, Chung H, Veerappan K, Oh C (2022) Elevated CO_2 alters the physiological and transcriptome responses of Pinus densiflora to long-term CO_2 exposure. Plants 11(24):3530

Kosová K, Vítámvás P, Prášil IT (2014) Wheat and barley dehydrins under cold, drought, and salinity–what can LEA-II proteins tell us about plant stress response? Front Plant Sci 5:343

Krasensky J, Jonak C (2012) Drought, salt, and temperature stress-induced metabolic rearrangements and regulatory networks. J Exp Bot 63(4):1593–1608

Kumar N, Suyal DC, Sharma IP, Verma A, Singh H (2017) Elucidating stress proteins in rice (Oryza sativa L.) genotype under elevated temperature: a proteomic approach to understand heat stress response. 3 Biotech 7:205. https://doi.org/10.1007/s13205-017-0856-9

Kumar M, Singh MP, Singh H, Dhakate PM, Ravindranath NH (2019a) Forest working plan for the sustainable management of forest and biodiversity in India. J Sustain for 39:1–22. https://doi.org/10.1080/10549811.2019.1632212

Kumar N, Jeena N, Singh H (2019b) Elevated temperature modulates rice pollen structure: a study from the foothill of the Himalayan agro-ecosystem in India. 3 Biotech 9(5):175. https://doi.org/10.1007/s13205-019-1700-1

Kumar N, Jeena N, Kumar N, Gangola S, Singh H (2019c) Phytoremediation facilitating enzymes: an enzymatic approach for enhancing remediation process. In: Bhatt P (ed) Smart bioremediation technologies. Academic Press, Elsevier. 50 Hampshire Street, 5th Floor, Cambridge, MA 02139, United States, pp 289–306

Kumar A, Kumar P, Singh H, Kumar N (2021a) Modulation of plant functional traits under essential plant nutrients during seasonal regime in natural forests of Garhwal Himalayas. Plant Soil 465:197–212. https://doi.org/10.1007/s11104-021-05003-x

Kumar A, Kumar P, Singh H, Bisht S, Kumar N (2021b) Relationship of physiological plant functional traits with soil carbon stock in the temperate forest of Garhwal Himalaya. Curr Sci 120(8):1368–1373. https://doi.org/10.18520/cs/v120/i8/1368-1373

Kumar A, Kumar P, Singh H, Kumar N (2021c) Adaptation and mitigation potential of roadside trees with bio-extraction of heavy metals under vehicular emissions and their impact on physiological traits during seasonal regimes. Urb Forest Urb Green 58:126900. https://doi.org/10.1016/j.ufug.2020.126900

Kumar A, Kumar P, Singh H, Kumar N (2021d) Impact of plant functional traits on infiltration rate under different forests of Kempty watershed in Garhwal Himalaya. Indian J Soil Conserv 49(1):38–44

Kumar M, Phukon SN, Paygude AC, Tyagi K, Singh H (2021e) Mapping Phenological Functional Types (PhFT) in the Indian Eastern Himalayas using machine learning algorithm in Google Earth Engine. Comput Geosci 158:104982. https://doi.org/10.1016/j.cageo.2021.104982

Kumar A, Singh H, Kumari G, Bisht S, Malik A, Kumar A, Singh M, Raturi BS, Thakur A, Kaushal R (2022) Adaptive resilience of roadside trees to vehicular emissions via leaf enzymatic, physiological, and anatomical trait modulations. Environ Pollut 313:120191. https://doi.org/10.1016/j.envpol.2022.120191

Kumar N, Singh H, Giri K, Kumar A, Joshi A, Yadav A, Singh R, Bisht S, Kumari R, Jeena N, Khairakpam R, Mishra G (2024) Physiological and molecular insights into the allelopathic effects on agroecosystems under changing environmental conditions. Physiol Mol Biol Plants 30:417–433. https://doi.org/10.1007/s12298-024-01440-x

Kumar N, Singh H, Sharma SK (2020) Antioxidants: responses and importance in plant defense system. In: Roychowdhury R, Choudhury S, Hasanuzzaman M, Srivastava S (eds) Sustainable agriculture in the era of climate change. Springer, Cham, pp 251–264. https://doi.org/10.1007/978-3-030-45669-6_11

Kuroha T, Nagai K, Gamuyao R, Wang DR, Furuta T, Nakamori M, Kitaoka T, Adachi K, Minami A, Mori Y (2018) Ethylene-gibberellin signaling underlies adaptation of rice to periodic flooding. Science 361(6398):181–186

Lee H, Calvin K, Dasgupta D, Krinner G, Mukherji A, Thorne P, Trisos C, Romero J, Aldunce P, Barret K (2023) IPCC, 2023: climate change 2023: synthesis report, summary for policymakers. In: Contribution of working groups I, II and III to the sixth assessment report of the intergovernmental panel on climate change [Core Writing Team, Lee H, Romero J (eds)]. IPCC, Geneva, Switzerland

Li Y, Su X, Zhang B, Huang Q, Zhang X, Huang R (2009) Expression of jasmonic ethylene responsive factor gene in transgenic poplar tree leads to increased salt tolerance. Tree Physiol 29(2):273–279

Li H, Xu G, Yang C, Yang L, Liang Z (2019) Genome-wide identification and expression analysis of HKT transcription factor under salt stress in nine plant species. Ecotoxicol Environ Saf 171:435–442

Li S, Li X, Wei Z, Liu F (2020) ABA-mediated modulation of elevated CO_2 on stomatal response to drought. Curr Opin Plant Biol 56:174–180

Li N, Euring D, Cha JY, Lin Z, Lu M, Huang L-J, Kim WY (2021) Plant hormone-mediated regulation of heat tolerance in response to global climate change. Front Plant Sci 11:627969

Lin K-H, Lin M-Y, Wang T-C, Wu Y-S (2016) Efficient agrobacterium-mediated transformation of Populus nigra x Populus tomentosa expressing the MtlD gene improves salt tolerance. Taiwan J Forest Sci 31:19–35

Liu H, Charng Y (2012) Acquired thermotolerance independent of heat shock factor A1 (HsfA1), the master regulator of the heat stress response. Plant Signal Behav 7(5):547–550

Liu S, Wang J, Wang L, Wang X, Xue Y, Wu P, Shou H (2009) Adventitious root formation in rice requires OsGNOM1 and is mediated by the OsPINs family. Cell Res 19(9):1110–1119

Liu S, Lv Z, Liu Y, Li L, Zhang L (2018) Network analysis of ABA-dependent and ABA-independent drought responsive genes in Arabidopsis thaliana. Genet Mol Biol 41:624–637

Liu J, Zhang J, He C, Duan A (2014) Genes responsive to elevated CO_2 concentrations in triploid white poplar and integrated gene network analysis. PLoS ONE 9(5). https://doi.org/10.1371/journal.pone.0098300

Lopez OR, Kursar TA (1999) Flood tolerance of four tropical tree species. Tree Physiol 19(14):925–932. https://doi.org/10.1093/treephys/19.14.925

Loreti E, van Veen H, Perata P (2016) Plant responses to flooding stress. Curr Opin Plant Biol 33:64–71

Lu N, Wei B, Sun Y, Liu X, Chen S, Zhang W, Zhang Y, Li Y (2014) Field supervisory test of DREB-transgenic Populus: salt tolerance, long-term gene stability and horizontal gene transfer. Forests 5(5):1106–1121

Ma X, Bai L (2021) Elevated CO2 and reactive oxygen species in stomatal closure. Plants 10(2):410

Marshall A, Aalen RB, Audenaert D, Beeckman T, Broadley MR, Butenko MA, Caño-Delgado AI, de Vries S, Dresselhaus T, Felix G (2012) Tackling drought stress: receptor-like kinases present new approaches. Plant Cell 24(6):2262–2278

May P, Liao W, Wu Y, Shuai B, Richard McCombie W, Zhang MQ, Liu QA (2013) The effects of carbon dioxide and temperature on microRNA expression in Arabidopsis development. Nat Commun 4(1):2145

Meena D, Singh H, Chaudhari SK (2011) Elucidating strontium response on growth dynamics and biochemical change in Phaseolus mungo L. Int J Agric Environ Biotechnol 4:107–113

Méndez-Alonzo R, López-Portillo J, Moctezuma C, Bartlett MK, Sack L (2016) Osmotic and hydraulic adjustment of mangrove saplings to extreme salinity. Tree Physiol 36(12):1562–1572

Morales-Olmedo M, Ortiz M, Sellés G (2015) Effects of transient soil waterlogging and its importance for rootstock selection. Chil J Agric Res 75:45–56

Negin B, Moshelion M (2016) The evolution of the role of ABA in the regulation of water-use efficiency: from biochemical mechanisms to stomatal conductance. Plant Sci 251:82–89

Negrão S, Schmöckel SM, Tester M (2017) Evaluating physiological responses of plants to salinity stress. Ann Bot 119(1):1–11. https://doi.org/10.1093/aob/mcw191

Nolan TM, Vukašinović N, Liu D, Russinova E, Yin Y (2020) Brassinosteroids: multidimensional regulators of plant growth, development, and stress responses. Plant Cell 32(2):295–318

Noya I, González-García S, Bacenetti J, Fiala M, Moreira MT (2018) Environmental impacts of the cultivation-phase associated with agricultural crops for feed production. J Clean Prod 172:3721–3733

Noyes PD, McElwee MK, Miller HD, Clark BW, Van Tiem LA, Walcott KC, Erwin KN, Levin ED (2009) The toxicology of climate change: Environmental contaminants in a warming world. Environ Int 35(6):971–986. https://doi.org/10.1016/j.envint.2009.02.006

Ohama N, Sato H, Shinozaki K, Yamaguchi-Shinozaki K (2017) Transcriptional regulatory network of plant heat stress response. Trends Plant Sci 22(1):53–65

Pan J, Sharif R, Xu X, Chen X (2021) Mechanisms of waterlogging tolerance in plants: research progress and prospects. Front Plant Sci 11:627331

Pillet J, Egert A, Pieri P, Lecourieux F, Kappel C, Charon J, Gomès E, Keller F, Delrot S, Lecourieux D (2012) VvGOLS1 and VvHsfA2 are involved in the heat stress responses in grapevine berries. Plant Cell Physiol 53(10):1776–1792

Prakash V, Hunney K, Singh H (2022) Effect of elevated carbon dioxide on growth and development of Santalum album L. seedlings inoculated with plant growth promoting microorganisms in open top chambers. Int J Environ Agric Biotechnol 7(5):149–159

Priyadharsini P, Nirmala N, Dawn SS, Baskaran A, SundarRajan P, Gopinath KP, Arun J (2022) Genetic improvement of microalgae for enhanced carbon dioxide sequestration and enriched biomass productivity: review on CO_2 bio-fixation pathways modifications. Algal Res 66:102810

Quint M, Delker C, Franklin KA, Wigge PA, Halliday KJ, Van Zanten M (2016) Molecular and genetic control of plant thermomorphogenesis. Nature Plants 2(1):1–9

Radmann EB, Klumb EK, Deuner S, Bianchi VJ (2018) Antioxidant capacity in leaf and root tissues of Prunus spp under flooding. J Exp Agric Inte 26(3):1–10

Raza A, Razzaq A, Mehmood SS, Zou X, Zhang X, Lv Y, Xu J (2019) Impact of climate change on crops adaptation and strategies to tackle its outcome: a review. Plants 8(2):34

Richardson KJ, Lewis KH, Krishnamurthy PK, Kent C, Wiltshire AJ, Hanlon HM (2018) Food security outcomes under a changing climate: impacts of mitigation and adaptation on vulnerability to food insecurity. Clim Change 147:327–341

Riikonen J, Holopainen T, Oksanen E, Vapaavuori E (2005) Leaf photosynthetic characteristics of silver birch during three years of exposure to elevated concentrations of CO_2 and O_3 in the field. Tree Physiol 25(5):621–632

Rogelj J, Den Elzen M, Höhne N, Fransen T, Fekete H, Winkler H, Schaeffer R, Sha F, Riahi K, Meinshausen M (2016) Paris Agreement climate proposals need a boost to keep warming well below 2 C. Nature 534(7609):631–639

Roy S, Mathur P (2021) Delineating the mechanisms of elevated CO_2 mediated growth, stress tolerance and phytohormonal regulation in plants. Plant Cell Rep 40(8):1345–1365. https://doi.org/10.1007/s00299-021-02738-w

Saadaoui E, Ben Yahia K, Dhahri S, Ben Jamaa ML, Khouja ML (2017) An overview of adaptative responses to drought stress in spp. Forest Stud 67(1):86–96. https://doi.org/10.1515/fsmu-2017-0014

Sairam RK, Tyagi A (2004) Physiology and molecular biology of salinity stress tolerance in plants. Curr Sci 407–421

Salehi-Lisar SY, Bakhshayeshan-Agdam H (2016) Drought stress in plants: causes, consequences, and tolerance. In: Drought stress tolerance in plants, Vol 1: Physiology and biochemistry, pp 1–16

Sauter M (2013) Root responses to flooding. Curr Opin Plant Biol 16(3):282–286. https://doi.org/10.1016/j.pbi.2013.03.013

Sharma R, Singh H, Kaushik M, Nautiyal R, Singh O (2018) Adaptive physiological response, carbon partitioning, and biomass production of Withania somnifera (L.) Dunal grown under elevated CO_2 regimes. 3 Biotech 8:267. https://doi.org/10.1007/s13205-018-1292-1

Sharma R, Singh H (2021) Alteration in biochemical constituents and nutrients partitioning of asparagus racemosus in response to elevated atmospheric CO_2 concentration. Environ Sci Pollut Res 29:6812–6821. https://doi.org/10.1007/s11356-021-16050-3

Shi H, Quintero FJ, Pardo JM, Zhu JK (2002) The putative plasma membrane Na(+)/H(+) antiporter SOS1 controls long-distance Na(+) transport in plants. Plant Cell 14(2):465–477. https://doi.org/10.1105/tpc.010371

Shimamura S, Yoshioka T, Yamamoto R, Hiraga S, Nakamura T, Shimada S, Komatsu S (2014) Role of abscisic acid in flood-induced secondary aerenchyma formation in soybean (Glycine max) hypocotyls. Plant Prod Sci 17(2):131–137

Shiono K, Ogawa S, Yamazaki S, Isoda H, Fujimura T, Nakazono M, Colmer TD (2011) Contrasting dynamics of radial O_2-loss barrier induction and aerenchyma formation in rice roots of two lengths. Ann Bot 107(1):89–99

Singh H (2021) An integrated approach considering physiological and biophysical based indicators for assessing tolerance of roadside plantations of Alstonia scholaris towards urban roadside air pollution: an assessment of adaptation of plantations for mitigating urban roadside air pollution. Trees. https://doi.org/10.1007/s00468-021-02179-8

Singh H, Verma A (2013) Characterization and screening of high nitrogen efficient rice genotype to curtail environmental pollution. Int J Agric Environ Biotechnol 6(1):23–30

Singh H, Verma A, Kumar M, Sharma R, Gupta R, Kaur M, Negi M, Sharma SK (2017b) Phytoremediation: a green technology to clean up the sites with low and moderate level of heavy metals. Austin Biochem 2:1012

Singh H, Sharma R, Savita SMP, Kumar M, Verma A, Ansari MW, Sharma SK (2018) Adaptive physiological response of Parthenium hysterophorus to elevated atmospheric CO_2 concentration. Indian Forester 144:1–14

Singh S, Singh H, Sharma SK, Nautiyal R (2021) Seasonal variation in biochemical responses of bamboo clones in the sub-tropical climate of Indian Himalayan foothills. Heliyon 7(4):e06859. https://doi.org/10.1016/j.heliyon.2021.e06859

Singh H, Bandyopadhyaya S, Kumar A, Singh M, Malik A, Kumari G, Raturi A, Yadav SK, Kumar M, Barthwal S, Thakur A, Rawat PS (2022) Understanding the physiological and biophysical response of urban roadside plantations for assessing adaptation and mitigation mechanisms toward vehicular emissions. Urb Clim 44:101183. https://doi.org/10.1016/j.uclim.2022.101183

Singh M, Singh H, Kumar A, Kumar M, Barthwal S, Thakur A (2024a) Soil nitrogen availability determines the CO_2 fertilization effect on tree species (Neolamarckia cadamba): growth and physiological evidence. Environ Sustain 7:53–60. https://doi.org/10.1007/s42398-023-00300-w

Singh H, Kumar M (2022) Climate change and its impact on Indian Himalayan forests: current status and research needs. In: Rani S, Kumar R (eds) Climate change. Springer Climate. Springer, Cham, pp 223–242. https://doi.org/10.1007/978-3-030-92782-0_11

Singh H, Savita Sharma R, Sinha S, Kumar M, Kumar P, Verma A, Sharma SK (2017a) Physiological functioning of Lagerstroemia speciosa L. under heavy roadside traffic: an approach to screen potential species for abatement of urban air pollution. 3 Biotech 7(61):1–10. https://doi.org/10.1007/s13205-017-0690-0

Singh H, Yadav M, Kumar N, Kumar A, Kumar M (2020). Assessing adaptation and mitigation potential of roadside trees under the influence of vehicular emissions: a case study of Grevillea robusta and Mangifera indica planted in an urban city of India. PLoS ONE 15(1):e0227380. https://doi.org/10.1371/journal.pone.0227380

Singh H, Kumar N, Kumar A (2024b) Enhancing resource use efficiency in crops through plant functional traits. In: Kumar N, Singh H (eds) Plant functional traits for improving productivity. Springer, Singapore. https://doi.org/10.1007/978-981-97-1510-7_6

Singh H (2024b) Exploring adaptive modulation in plant functional traits and its impact on the productivity of Acacia auriculiformis under CO_2-enriched environment. Industr Crops Prod 210:118186. https://doi.org/10.1016/j.indcrop.2024.118186

Song J, Wang B (2015) Using euhalophytes to understand salt tolerance and to develop saline agriculture: Suaeda salsa as a promising model. Ann Bot 115(3):541–553

Song Y, Chen Q, Ci D, Shao X, Zhang D (2014) Effects of high temperature on photosynthesis and related gene expression in poplar. BMC Plant Biol 14:1–20

Song Y, Chen P, Liu P, Bu C, Zhang D (2020) High-temperature-responsive poplar lncRNAs modulate target gene expression via RNA interference and act as RNA scaffolds to enhance heat tolerance. Int J Mol Sci 21(18):6808

Steffens B, Sauter M (2009) Epidermal cell death in rice is confined to cells with a distinct molecular identity and is mediated by ethylene and H_2O_2 through an autoamplified signal pathway. Plant Cell 21(1):184–196

Steffens B, Kovalev A, Gorb SN, Sauter M (2012) Emerging roots alter epidermal cell fate through mechanical and reactive oxygen species signaling. Plant Cell 24(8):3296–3306

Steinmann T, Geldner N, Grebe M, Mangold S, Jackson CL, Paris S, Galweiler L, Palme K, Jurgens G (1999) Coordinated polar localization of auxin efflux carrier PIN1 by GNOM ARF GEF. Science 286(5438):316–318

Taub DR, Wang X (2008) Why are nitrogen concentrations in plant tissues lower under elevated CO_2? A critical examination of the hypotheses. J Integr Plant Biol 50(11):1365–1374

Taylor G, Street NR, Tricker PJ, Sjödin A, Graham L, Skogström O, Calfapietra C, Scarascia-Mugnozza G, Jansson S (2005) The transcriptome of Populus in elevated CO_2. New Phytol 167(1):143–154

Thornton PK, Ericksen PJ, Herrero M, Challinor AJ (2014) Climate variability and vulnerability to climate change: a review. Glob Change Biol 20(11):3313–3328

Todaka D, Shinozaki K, Yamaguchi-Shinozaki K (2015) Recent advances in the dissection of drought-stress regulatory networks and strategies for development of drought-tolerant transgenic rice plants. Front Plant Sci 6:84

Trenberth KE (2005) The impact of climate change and variability on heavy precipitation, floods, and droughts. Encyclopedia Hydrol Sci 17:1–11

Tsuchihira A, Hanba YT, Kato N, Doi T, Kawazu T, Maeshima M (2010) Effect of overexpression of radish plasma membrane aquaporins on water-use efficiency, photosynthesis and growth of Eucalyptus trees. Tree Physiol 30(3):417–430

Ummenhofer CC, Meehl GA (2017) Extreme weather and climate events with ecological relevance: a review. Philos Trans Roy Soc B Biol Sci 372(1723):20160135

Vartapetian BB (1991) Flood-sensitive plants under primary and secondary anoxia: ultrastructural and metabolic responses. In: Plant life under oxygen deprivation. SPB Academic Publishing, The Hague, The Netherlands, pp 201–216

Veeranagamallaiah G, Prasanthi J, Reddy KE, Pandurangaiah M, Babu OS, Sudhakar C (2011) Group 1 and 2 LEA protein expression correlates with a decrease in water stress induced protein aggregation in horsegram during germination and seedling growth. J Plant Physiol 168(7):671–677

Viljevac Vuletić M, Mihaljević I, Tomaš V, Horvat D, Zdunić Z, Vuković D (2022) Physiological response to short-term heat stress in the leaves of traditional and modern plum (Prunus domestica L.) cultivars. Horticulturae 8(1):72

Visser EJW, Voesenek LACJ (2005) Acclimation to soil flooding–sensing and signal-transduction. Plant Soil 274(1):197–214

Wang YC, Qu GZ, Li HY, Wu YJ, Wang C, Liu GF, Yang CP (2010) Enhanced salt tolerance of transgenic poplar plants expressing a manganese superoxide dismutase from Tamarix androssowii. Mol Biol Rep 37(2):1119–1124

Wang D, Qin B, Li X, Tang D, Zhang Y, Cheng Z, Xue Y (2016) Nucleolar DEAD-box RNA helicase TOGR1 regulates thermotolerant growth as a pre-rRNA chaperone in rice. PLoS Genet 12(2):e1005844

Wang X, Niu Y, Zheng Y (2021) Multiple functions of MYB transcription factors in abiotic stress responses. Int J Mol Sci 22(11):6125

Wu Y, Ding N, Zhao X, Zhao M, Chang Z, Liu J, Zhang L (2007) Molecular characterization of PeSOS1: the putative Na^+/H^+ antiporter of Populus euphratica. Plant Mol Biol 65(1):1–11

Xu Z, Jiang Y, Zhou G (2015) Response and adaptation of photosynthesis, respiration, and antioxidant systems to elevated CO_2 with environmental stress in plants. Front Plant Sci 6(September):1–17. https://doi.org/10.3389/fpls.2015.00701

Xu Z, Jiang Y, Jia B, Zhou G (2016) Elevated-CO_2 response of stomata and its dependence on environmental factors. Front Plant Sci 7:657

Yadav SK, Singh H, Nautiyal R, Ginwal HS, Ansari SA, Barthwal S (2019) Modulation of morphophysiological responses in *Populus deltoides* by elevated carbon dioxide and temperature. For Sci 66:105–118. https://doi.org/10.1093/forsci/fxz048

Yadav SK, Singh H, Nautiyal R, Ginwal HS, Ansari SA, Barthwal S (2020) Modulation of morphophysiological responses in *Populus deltoides* by elevated carbon dioxide and temperature. For Sci 66(1):105–118

Yamauchi T, Rajhi I, Nakazono M (2011) Lysigenous aerenchyma formation in maize root is confined to cortical cells by regulation of genes related to generation and scavenging of reactive oxygen species. Plant Sign Behav 6(5):759–761

Yang F, Xu X, Xiao X, Li C (2009a) Responses to drought stress in two poplar species originating from different altitudes. Biol Plant 53(3):511–516

Yang Q, Chen ZZ, Zhou XF, Yin HB, Li X, Xin XF, Hong XH, Zhu JK, Gong Z (2009b) Overexpression of SOS (salt overly sensitive) genes increases salt tolerance in transgenic Arabidopsis. Mol Plant 2(1):22–31. https://doi.org/10.1093/mp/ssn058

Yang G, Yu L, Zhang K, Zhao Y, Guo Y, Gao C (2017) A ThDREB gene from *Tamarix hispida* improved the salt and drought tolerance of transgenic tobacco and T. hispida. Plant Physiol Biochem 113:187–197

Yang Y, Shi R, Wei X, Fan Q, An L (2010) Effect of salinity on antioxidant enzymes in calli of the halophyte *Nitraria tangutorum* Bobr. Plant Cell Tissue Organ Culture (PCTOC) 102(3):387–395

Yoro KO, Daramola MO (2020) CO_2 emission sources, greenhouse gases, and the global warming effect. In: Advances in carbon capture. Elsevier, pp 3–28

Yoshida T, Fujita Y, Maruyama K, Mogami J, Todaka D, Shinozaki K, Yamaguchi-Shinozaki K (2015) Four A rabidopsis AREB/ABF transcription factors function predominantly in gene expression downstream of SnRK2 kinases in abscisic acid signalling in response to osmotic stress. Plant, Cell Environ 38(1):35–49

Zandalinas SI, Mittler R, Balfagón D, Arbona V, Gómez-Cadenas A (2018) Plant adaptations to the combination of drought and high temperatures. Physiol Plant 162(1):2–12

Zeevaart JAD (1980) Changes in the levels of abscisic acid and its metabolites in excised leaf blades of *Xanthium strumarium* during and after water stress. Plant Physiol 66(4):672–678

Zhang H, Yang J, Chen Y, Mao X, Wang Z, Li C (2013) Identification and expression analysis of the heat shock transcription factor (HSF) gene family in Populus trichocarpa. Plant Omics 6(6)

Zhao J, Lu Z, Wang L, Jin B (2020) Plant responses to heat stress: physiology, transcription, noncoding RNAs, and epigenetics. Int J Mol Sci 22(1). https://doi.org/10.3390/ijms22010117

Zhou M-L, Ma J-T, Zhao Y-M, Wei Y-H, Tang Y-X, Wu Y-M (2012) Improvement of drought and salt tolerance in Arabidopsis and *Lotus corniculatus* by overexpression of a novel DREB transcription factor from *Populus euphratica*. Gene 506(1):10–17

Zhu J-K (2016) Abiotic stress signaling and responses in plants. Cell 167(2):313–324

Plant Growth Regulators/promoters Mediated Adaptation and Tolerances in Urban Plants for Enhancing Resilience Towards Multiple Stresses

Muhammad Ahsan Altaf, Sandeep Adavi, Priyanka Lal, Lopamudra Nayak, Laxmipriya Behera, Ujala Rashmi Sahu, Awadhesh Kumar, Ravinder Kumar, Rahul Kumar Tiwari, and Milan Kumar Lal

Abstract Urban agriculture plays a vital role in ensuring several cities' food and nutrition security and is gaining significance, particularly in the face of societal impacts like climate change. Environmental conditions that fluctuate globally substantially threaten the survival of natural ecosystems. Urban plants, especially trees, are well recognized for their ability to provide many ecological services across diverse facets of life. Various approaches are used to enhance plant responses to environmental stress, including phytohormones, biostimulants, and plant growth regulators (PGRs). Urban plants play a significant role in ensuring the population's mental health. It is known that PGRs significantly impact the hormone levels of plants in modern agriculture and horticulture. PGRs effectively employ many developmental processes in urban plants, such as induction of flowering and fruiting, modifying root architecture and system, and promoting fruit setting. In this chapter, PGRs

M. A. Altaf
Key Laboratory for Quality Regulation of Tropical Horticultural Crops of Hainan Province, School of Breeding and Multiplication (Sanya Institute of Breeding and Multiplication), Hainan University, Sanya 572025, China

S. Adavi
School of Crop Resistance System Research, ICAR-National Institute of Biotic Stress Management, Baronda, Raipur, Chhattisgarh 493225, India

P. Lal
Lovely Professional University, Jalandhar GT Road (NH1), Phagwara, Punjab, India

L. Nayak · L. Behera · U. R. Sahu · A. Kumar · M. K. Lal (✉)
Division of Crop Physiology and Biochemistry, ICAR—National Rice Research Institute (ICAR-NRRI), Cuttack, Odisha 753006, India
e-mail: milan2925@gmail.com

R. Kumar
ICAR-Indian Agricultural Research Institute, New Delhi, India

R. K. Tiwari · M. K. Lal
ICAR-Central Potato Research Institute (ICAR-CPRI), Shimla, Himachal Pradesh 171001, India

H. Singh (ed.), *Urban Forests, Climate Change and Environmental Pollution*,
https://doi.org/10.1007/978-3-031-67837-0_11

have gained significant attention as a viable and environmentally friendly option to mitigate the adverse impacts of environmental stresses on urban plants.

Keywords Environmental conditions · Plant growth regulators · Root architecture · Phytohormone · Biostimulant

1 Introduction

Urban areas sustain a wide range of plant species, ranging from hundreds to thousands. This diverse collection of species offers sufficient opportunities to investigate how plants respond to stressful circumstances (Calfapietra et al. 2015; Kumar et al. 2022a, b; Singh et al. 2022). Studying plant adaptations to urban environments provides a valuable chance to understand the immediate physiological responses, plant tolerance, and the scope and processes of short-term and long-term plant adaptations (Chen et al. 2011; Kumar et al. 2024; Singh et al. 2024). In this regard, woody species are crucial due to their durability and the potential to investigate the mechanisms underlying long-term adaptation. Woody species play a vital role in these circumstances due to their extended longevity and the opportunity to examine long-term adaptation processes (Grote et al. 2016; Singh 2024; Dhyani et al 2021).

Over the last several decades, there has been a noticeable rise in the frequency and intensity of severe events associated with climate change on a global scale. Climate change has enormously affected several living things, including plants (Kumar et al. 2021a, b, c, d, e; Prakash et al. 2022; Campos et al. 2023). Climate change exacerbates the abiotic stressors that affect plants, resulting in changes in precipitation patterns that cause more frequent droughts and floods, extreme temperatures, and high wind speeds (Espeland and Kettenring 2018; Shahid et al. 2021). Furthermore, many abiotic stressors may simultaneously manifest in the environment, and their effects on urban plants are predicated upon their relationship (Ju et al. 2022). Environmental stresses considerably damaged urban plant growth and development (Gill and Tuteja 2010; Kumar et al. 2021b, c).In this chapter, we described the application of PGRs in cultivating urban plants (Rademacher 2015; Singh 2021; Singh et al. 2020; Kumar et al. 2021a).

Despite the recognized impacts of exogenously applied PGRs to enhance plant resilience against environmental stresses (Sajjad et al. 2017; Sharma and Singh 2021). The primary adverse effects of environmental stresses are disruption of growth and yield, which are directly regulated by PGRs (Akram and Ashraf 2013). Diverse PGR classes, including ethylene (ET), abscisic acid (ABA), cytokinins (CKs), gibberellins (GAs), and auxin (IAA) are associated with improving plants' ability to respond to a variety of environmental stresses (Sabagh et al. 2021a, b). These PGRs are critical in maintaining stress adaption by modulating physiological, biochemical, and molecular processes. They activate the plant defence system, upregulated transcript levels, and stimulate transcription factors, metabolic genes, and stress proteins at the cellular level (Asgher et al. 2015; Singh et al. 2021). The current chapter provides

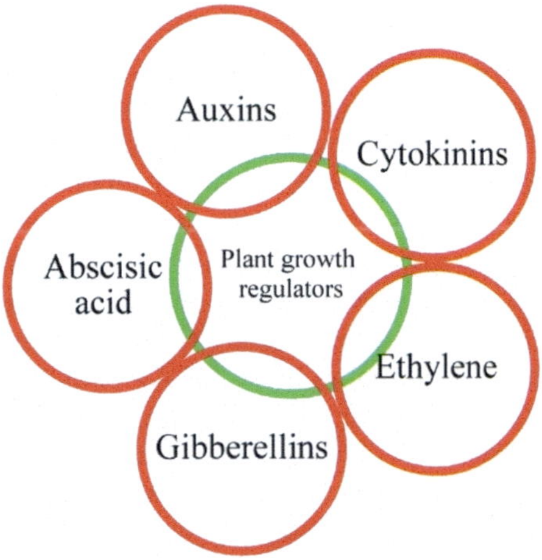

Fig. 1 Plant growth regulators described in this chapter

a comprehensive overview of the latest advancements in various PGRs to enhance plant resilience to abiotic stressors.

2 Plant Growth Regulators Enhanced Stress Tolerances in Urban Plants

Exogenous PGRs are crucial in modulating plant responses to environmental stresses by enhancing growth and developmental processes. Multiple approaches are used to improve urban plant development in unfavorable environmental circumstances (Sabagh et al. 2021a, b; Singh et al. 2020; Kumar et al. 2021a). This chapter provides a comprehensive overview of the latest advancements in various PGRs that enhance urban plant resilience to environmental stresses. Furthermore, it emphasizes the areas of study that need to be addressed regarding the production of PGRs under stressful situations and their possible functions in enhancing tolerance to unfavorable growth conditions (Upreti and Sharma 2016; Kumar et al. 2021b; Kumar et al. 2024) (Fig. 1).

3 Auxin

Auxin is a traditional plant growth regulator essential for regulating plant development and growth under various environmental circumstances (Gomes and Scortecci 2021; Kumar et al. 2021c, d). Auxin regulated seed germination, protected the photosynthetic system, balanced oxidative stress biomarkers, maintained mineral nutrient accumulation, upregulated antioxidant enzymes system, and increased plant secondary metabolites production (Yadav et al. 2019; Gupta et al. 2019; Mathur et al.

2022). Foliar application of auxin (indole acetic acid; IAA) significantly increased fruit weight and promoted fruit-set percentage in snake fruit (Rai et al. 2021). Auxin significantly enhanced growth and development in response to environmental stresses such as salinity (Abdel Latef et al. 2021), drought (Shi et al. 2014), heat (Jing et al. 2023), cold (Rahman 2013), heavy metals (Mathur et al. 2022). Foliar application auxin (NAA) considerably decreased salinity's adverse effects and increased rosemary's growth status (Abou-El-Ghit 2015). Auxin application effectively regulated the growth of Garden sage (*Salvia officinalis*). In addition, lower auxin concentrations positively regulated sage plant vegetative propagation (Koleva Gudeva et al. 2017; Gupta et al. 2019; Kumar et al. 2019a). Ault and Siqueira (2008) reported that auxin applications significantly stimulated lily root, shoot, and callus production. Indole-3-butyric acid significantly increased root traits and regulated cuttings of Dahila (Singh et al. 2023). Auxin (NAA) application enhanced root size, flower size, leaf size, branches per plant, and root per plant in marigolds (Ullah et al. 2013). Auxin promotes root length and increases root regeneration in *Chrysanthemum morifolium* (Chae 2016; Kumar et al. 2019b). Auxin application effectively improved fruit growth, early fruit development, and delayed fruit ripening in strawberries (Liu et al. 2011; Sharma et al. 2018). Low-temperature stress considerably decreased the growth and development of the tulip plant. In contrast, Auxin application markedly increased the number of leaves, plant height, leaf area, and flower diameter in tulips (Rietveld et al. 2000; Ahmad et al. 2013). Furthermore, auxin-regulated cell wall expansion and cell wall division development in rose (Jing et al. 2023).

4 Gibberellic Acid

Gibberellic acid (GA_3) is a plant hormone that promotes urban plant growth and development (Gupta and Chakrabarty 2013). GA_3 promoted seed germination, cell division, cell elongation, root development, later root development, antioxidant enzyme activity, pigments content, photosynthetic assimilation rate, proline accumulation, ion homeostasis, redox homeostasis, flowering, and seedling growth and development (Sharma et al. 2018; Singh et al. 2018; Shah et al. 2023). Foliar application of gibberellic acid (GA_3) increased the growth, flowering, and other morphological traits of freesia plants (AL-Chalabi et al. 2016). In another study, GA_3 treatment remarkably increased the physiological and morphological traits of the freesia plant (Ahmed and Saleh 2023).GA_3 application significantly improved growth characteristics, increased flowering growth-related traits, and promoted seed germination in realted marigolds (Gupta et al. 2018a, b; Kumar et al. 2020a, b). In another study, the foliar application of GA_3 improved the yield-related characteristics in marigolds (Sathappan 2018).GA_3 plays a crucial function in alleviating environmental stresses such as salinity (Shahzad et al. 2021), drought (Akter et al. 2014), heat (Nagar et al. 2021), cold (Zhu et al. 2016), and heavy metals toxicity (Ghani et al. 2021). Furthermore, GA_3 application increased the number of flowers and the size of orchid flowers under water regime environment (Cardoso et al. 2010). Matsumoto (2006) reported that GA3 application effectively increased vegetative growth and promoted early flowering of hybrid orchids. GA_3 supplementation significantly improved germination time, percentage, and seedling health index in Cyclamen species under short- and long-term conditions (Cornea-Cipcigan et al. 2020). Foliar application of GA_3 considerably improved different physiological parameters such as flower length and diameter, bud length and diameters, flower stalk length, and the number of cut flower stalks, intermodal length incarnation plants under naturally ventilated polyhouse (Benny et al. 2017). Further, GA_3 application effectively regulated the growth and flowering of Philodendron's Black Cardinal (Chen et al. 2003). Dhaduk et al. (2007) reported that GA_3 application enhanced anthurium growth attributes and flowering-related characteristics.

5 Cytokinin

Cytokinins (CKs) are crucial plant hormones that promote plant growth and development and have significant functions in stress response and nutrient metabolism in urban plants (Singh and Verma 2013; Kumar et al. 2017; Prasad 2022). CKs regulated seed germination, cell formation, bud elongation, photosynthetic system, flowering, root cell division and elongation, root formation, redox balance, mineral nutrient flux, and root development in urban plants (Werner and Schmülling 2009; Sosnowski et al. 2023). Furthermore, CKs enhanced growth and development in response to environmental stresses such as salinity (Kaya et al. 2009), drought (Pospíšilová et al. 2016), heat (Yang et al. 2016), cold (Xia et al. 2009), heavy metals toxicity (Khalil et al. 2017). CK-mediated growth stimulation in *Epipremnumaureum* L. cuttings is associated with alterations in leaf size and leaf production rate (Benedetto et al. 2013). CKs application significantly enhanced growth, net photosynthetic rate, and leaf area development in *Epipremnumaureum* (Benedetto et al. 2015). In another study, Wu et al. (2021) reported that CKs potentially enhanced the leaf area development in many plants. Foliar application of CKs increased photosynthetic assimilation rate, growth rate, leaf expansion, and whole plant growth rate in benjamina (Benedetto et al. 2020). Evidence suggests CKs protect photosynthetic apparatus under intense light stress (Cortleven and Schmülling 2015). The findings on the aspect of hormonal application demonstrate that CKs are crucial as primary hormonal regulators necessary for cambium formation (Nieminen et al. 2008).

6 Ethylene

Ethylene is a gaseous hormone that regulates various plant developmental processes by stimulating and inhibiting growth (Dugardeyn and Van Der Straeten 2008). Ethylene regulates multiple plant developmental processes such as seed germination, photosynthetic efficiency, root initiation, redox balance, mineral accumulation, vegetative growth and development, floral induction, protected photosynthetic system, maintained ROS accumulation, reaction wood formation in urban plants (Wheeler et al. 2004; Dubois et al. 2018). Ethylene significantly inhibits auxiliary buds (Reid 1995), meristems root and shoot (Iqbal et al. 2017), cell division (Jackson 2018), and DNA synthesis (Van de Poel et al. 2015). Ethylene also actions a ripening hormone (Wang et al. 2013). Ethylene positively responds to environmental stresses and protects urban plants during adverse environmental conditions (Kazan 2015; Riyazuddin et al. 2020). Ethylene actively responds during flowering opening in peony trees (Zhou et al. 2010). Ethylene is essential in coordinating and regulating senescence processes in carnation flowers (Van Altvorst and Bovy 1995). Employing functional genomics approaches will provide valuable information on the regulatory network between the ethylene signaling system and the ovule developmental program in orchids (Tsai et al. 2008). Ethylene regulates the developmental adaptability and growth stability between the shoot and root systems in vitro-cultivated epiphytic orchids (Rodrigues et al. 2014; Gupta and Singh 2017; Singh et al. 2017a). Several distinct ethylene regulatory routes are implicated in figs' maturation and ripening process, with each pathway reliant upon the individual inducer or stimulus (Owino et al. 2006). Furthermore, ethylene enhanced environmental stress tolerance such as salinity (Tao et al. 2015), drought (Sharp 2002), heat (Poór et al. 2022), cold (Eremina et al. 2016), and heavy metals (Asgher et al. 2014) toxicity. Ethylene is a phytohormone crucial in several physiological processes and influences plant growth and development throughout its life cycle.

7 Abscisic Acid

Abscisic acid (ABA) is a crucial plant growth regulator that modulates plant growth, development, and how urban plants respond to environmental stresses (Emenecker and Strader 2020). ABA plays a significant role in many physiological processes, such as seedling growth, seed development, seed germination, flowering, cell expansion rate, photosynthetic process, vegetative growth, cell division, and dormancy (Xiong and Zhu 2003; Singh and Kumar 2022). ABA is recognized for potentially enhancing stress tolerance in urban plants (Kishor et al. 2022). ABA concentrations in plants have been beneficial in two areas: seedless fruit production and extending the shelf life of green food and decorative plants (Duan et al. 2006). Drought stress considerably decreased the growth and development of bedding plants such as pansies and marigolds.

In contrast, ABA application to bedding plants significantly reduced the leaf chlorosis damage and maintained the shelf life of these plants (Waterland et al. 2010; Singh et al. 2017b). Kumar et al. (2022a, b) reported that ABA is applied exogenously to increase plant tolerance to heavy metal toxicity by enhancing secondary metabolite production, osmolyte accumulation, and antioxidant enzyme profile. The application of foliar ABA as a signal for root development considerably impacts the regulation

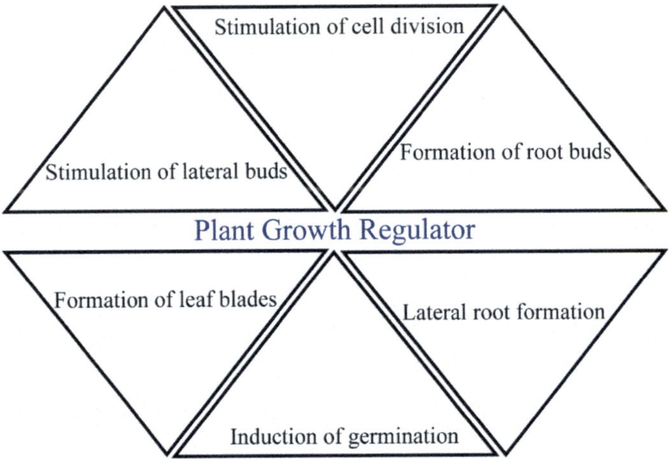

Fig. 2 Functions of plant growth regulator in plants

of root-to-shoot growth under normal circumstances. This finding shows that the hydration of leaves, rather than roots, is the primary signal for controlling plant responses to moisture (Meena et al. 2011; Kumar et al. 2019c, 2020a, b; McAdam et al. 2016). ABA enhanced the growth status of many plants in response to environmental stresses such as salinity, drought, heat, cold, and heavy metals toxicity (Sah et al. 2016; Dar et al. 2017). ABA is a crucial regulator for plant growth and development (Fig. 2).

8 Conclusion and Future Prospects

The role of urban agriculture in enhancing food security and nutritional resilience within cities is increasingly critical, particularly in light of the challenges posed by climate change. Urban plants, notably trees, are vital in providing ecological services by mitigating environmental stressors. This research underscores the potential for enhancing urban plant responses to such stressors by applying plant growth regulators (PGRs), thereby benefiting both urban ecosystems and the mental well-being of residents. Authorities may effectively manage environmental challenges and promote sustainable urban agriculture practices by leveraging PGRs to modulate plant development processes—such as flowering, fruiting, and root system architecture. Integrating innovative approaches, like PGR utilization, is imperative to bolster the resilience of urban ecosystems in the face of environmental change.

References

Abdel Latef AAH, Akter A, Tahjib-Ul-Arif M (2021) Foliar application of auxin or cytokinin can confer salinity stress tolerance in Vicia faba L. Agronomy 11(4):790

Abou-El-Ghit HM (2015) Effect of naphthalene acetic acid (NAA) on growth and yield of rosemary (*Rosemarinus officinalis* L.) under salinity stress. Egypt J Bot 56(2):303–317

Ahmed EE, Saleh YM (2023) Effect of spraying with Gibberellic and ascorbic acids on growth, flowering and corm production of two freesia cultivars Freesia Hybrid. In: AIP conference proceedings, vol. 2839, no. 1. AIP Publishing

Ahmed Z, Sheikh MQ, Siddique MAA, Jeelani MI, Singh A, Nazir G, Laishram N, Rehman SI (2013) Enhancing early blooming and flower quality of tulip (*Tulipa gesneriana* Linn.) through application of plant growth regulators. Afr J Agric 8(38):4780–4786

Akram NA, Ashraf M (2013) Regulation in plant stress tolerance by a potential plant growth regulator, 5-aminolevulinic acid. J Plant Growth Regul 32:663–679

Akter N, Rafiqul Islam M, Abdul Karim M, Hossain T (2014) Alleviation of drought stress in maize by exogenous application of gibberellic acid and cytokinin. J Crop Sci Biotechnol 17:41–48

AL-Chalabi SK, Al-Khafaji SMS (2016) Influence of foliar application of GA3 and CPPU on growth and flowering of Freesia plants" goldenmelody. Euphrates J Agric Sci 8(2):30–23

Asgher M, Khan NA, Khan MIR, Fatma M, Masood A (2014) Ethylene production is associated with alleviation of cadmium-induced oxidative stress by sulfur in mustard types differing in ethylene sensitivity. Ecotoxicol Environ Saf 106:54–61

Asgher M, Khan MIR, Anjum NA, Khan NA (2015) Minimising toxicity of cadmium in plants—role of plant growth regulators. Protoplasma 252:399–413

Ault JR, Siqueira SS (2008) Morphogenetic response of Lilium michiganense to four auxin-type plant growth regulators in vitro. HortScience 43(6):1922–1924

Benedetto AD, Galmarini C, Tognetti J (2013) Changes in leaf size and in the rate of leaf production contribute to cytokinin-mediated growth promotion in *Epipremnum aureum* L. cuttings. J Hortic Sci Biotechnol 88(2):179–186

Benedetto AD, Galmarini C, Tognetti J (2015) Effects of combined or single exogenous auxin and/or cytokinin applications on growth and leaf area development in Epipremnumaureum. J Hortic Sci Biotechnol 90(6):643–654

Benedetto AD, Galmarini C, Tognetti J (2020) Differential growth response of green and variegated Ficus benjamina to exogenous cytokinin and shade. Ornament Hortic 26:259–276

Benny JC, Devi S, Fatmi U, Jose DA (2017) Effect of plant growth regulators, Gibberellic acid (GA3) and Naphthalene Acetic Acid (NAA) on growth and yield of carnation (Dianthus caryophyllus L.) under naturally ventilated polyhouse. Plant Arch 17(2):803–812

Calfapietra C, Peñuelas J, Niinemets Ü (2015) Urban plant physiology: adaptation-mitigation strategies under permanent stress. Trends Plant Sci 20(2):72–75

Campos EV, do ES Pereira A, Aleksieienko I, Do Carmo GC, Gohari G, Santaella C, Fraceto LF, Oliveira HC (2023) Encapsulated plant growth regulators and associative microorganisms: Nature-based solutions to mitigate the effects of climate change on plants. Plant Sci, p 111688

Cardoso JC, Ono EO, Rodrigues JD (2010) Gibberellic acid and water regime in the flowering induction of Brassocattleya and Cattleya hybrid orchids. Hortic Bras 28:395–398

Chae SC (2016) Influence of auxin concentration on in vitro rooting of Chrysanthemum morifolium Ramat. Biosci, Biotechnol Res Asia 13(2):833–837

Chen J, Henny RJ, McConnell DB, Caldwell RD (2003) Gibberellic acid affects growth and flowering of Philodendron 'Black Cardinal.' Plant Growth Regul 41:1–6

Chen Z, He X, Cui M, Davi N, Zhang X, Chen W, Sun Y (2011) The effect of anthropogenic activities on the reduction of urban tree sensitivity to climatic change: dendrochronological evidence from Chinese pine in Shenyang city. Trees 25:393–405

Cornea-Cipcigan M, Pamfil D, Sisea CR, Mărgăoan R (2020) Gibberellic acid can improve seed germination and ornamental quality of selected Cyclamen species grown under short and long days. Agronomy 10(4):516

Cortleven A, Schmülling T (2015) Regulation of chloroplast development and function by cytokinin. J Exp Bot 66(16):4999–5013

Dar NA, Amin I, Wani W, Wani SA, Shikari AB, Wani SH, Masoodi KZ (2017) Abscisic acid: a key regulator of abiotic stress tolerance in plants. Plant Gene 11:106–111

Dhaduk BK, Kumari S, Singh A, Desai JR (2007) Response of gibberellic acid on growth and flowering attributes in anthurium (*Anthurium andreanum* Lind.). J Ornament Hortic 10(3):187–189

Dhyani K, Kuniyal HB, Singh H, Sobha (2021) Growth and physiological potential of *Terminalia arjuna* under elevated CO_2 levels in Open top chamber condition. J Appl Nat Sci 13(3):1121–1126. https://doi.org/10.31018/jans.v13i3.2490. ISSN 2231-5209

Duan H, Li Y, Pei Y, Deng W, Luo M, Xiao Y, Luo K, Lu L, Smith W, Mcavoy RJ, Zhao D (2006) Auxin, cytokinin and ab sei sic acid: biosynthetic and catabolic genes and their potential applications in ornamental crops. J Crop Improv 18(1–2):347–364

Dubois M, Van den Broeck L, Inzé D (2018) The pivotal role of ethylene in plant growth. Trends Plant Sci 23(4):311–323

Dugardeyn J, Van Der Straeten D (2008) Ethylene: fine-tuning plant growth and development by stimulation and inhibition of elongation. Plant Sci 175(1–2):59–70

Emenecker RJ, Strader LC (2020) Auxin-abscisic acid interactions in plant growth and development. Biomolecules 10(2):281

Eremina M, Rozhon W, Poppenberger B (2016) Hormonal control of cold stress responses in plants. Cell Mol Life Sci 73:797–810

Espeland EK, Kettenring KM (2018) Strategic plant choices can alleviate climate change impacts: a review. J Environ Manage 222:316–324

Ghani MA, Abbas MM, Ali B, Aziz R, Qadri RWK, Noor A, Azam M, Bahzad S, Saleem MH, Abualreesh MH, Alatawi A (2021) Alleviating role of gibberellic acid in enhancing plant growth and stimulating phenolic compounds in carrot (*Daucus carota* L.) under lead stress. Sustainability 13(21):12329

Gill SS, Tuteja N (2010) Reactive oxygen species and antioxidant machinery in abiotic stress tolerance in crop plants. Plant Physiol Biochem 48(12):909–930

Gomes GLB, Scortecci KC (2021) Auxin and its role in plant development: structure, signalling, regulation and response mechanisms. Plant Biol 23(6):894–904

Grote R, Samson R, Alonso R, Amorim JH, Cariñanos P, Churkina G, Fares S, Thiec DL, Niinemets Ü, Mikkelsen TN, Paoletti E (2016) Functional traits of urban trees: air pollution mitigation potential. Front Ecol Environ 14(10):543–550

Gupta R, Chakrabarty SK (2013) Gibberellic acid in plant: still a mystery unresolved. Plant Signal Behav 8(9):e25504

Gupta SK, Singh H (2017) Observations on Sporadic Flowering in *Ficus benjamina* L. and *Peltophorum africanum* L. in New Forest. Indian Forester 143:290–291

Gupta SK, Ram J, Singh H (2018a) Comparative study of transpiration in cooling effect of tree species in the atmosphere. J Geosci Environ Prot 6:151–166. https://doi.org/10.4236/gep.2018.68011

Gupta SK, Ram J, Singh H, Kumar P (2018b) Comparative study of cooling under shades of some forestry tree species with respect to ambient temperatures. J Geosci Environ Prot 6:51–66. https://doi.org/10.4236/gep.2018.611004

Gupta SK, Ram J, Singh H (2019) Relationship between leaf area index and atmospheric cooling of tree species. Indian J for 42(3):203–208

Iqbal N, Khan NA, Ferrante A, Trivellini A, Francini A, Khan MIR (2017) Ethylene role in plant growth, development and senescence: interaction with other phytohormones. Front Plant Sci 8:475

Jackson MB (2018) Ethylene in root growth and development. In: The plant hormone ethylene. CRC Press, pp 159–181

Jing H, Wilkinson EG, Sageman-Furnas K, Strader LC (2023) Auxin and abiotic stress responses. J Exp Bot 74(22):7000–7014

Ju JH, Yoon YH, Shin SH, Ju SY, Yeum KJ (2022) Recent trends in urban agriculture to improve bioactive content of plant foods. Horticulturae 8:767. https://doi.org/10.3390/horticulturae8090767

Kaya C, Tuna AL, Yokaş I (2009) The role of plant hormones in plants under salinity stress. In: Salinity and water stress: improving crop efficiency, pp 45–50

Kazan K (2015) Diverse roles of jasmonates and ethylene in abiotic stress tolerance. Trends Plant Sci 20(4):219–229

Khalil RR, Moustafa AN, Bassuony FM, Haroun SA (2017) Kinetin and/or calcium affect growth of *Phaseolus vulgaris* L. plant grown under heavy metals stress. J Environ Sci 46(2):103–120

Kishor PBK, Tiozon RN, Fernie AR, Sreenivasulu N (2022) Abscisic acid and its role in the modulation of plant growth, development, and yield stability. Trends Plant Sci 27:1283–1295

Koleva Gudeva L, Trajkova F, Mihajlov L, Troiciki J (2017) Influence of different auxins on rooting of rosemary, sage and elderberry. Annu Res Rev Biol 12(5):1–8

Kumar N, Suyal DC, Sharma IP, Verma A, Singh H (2017) Elucidating stress proteins in rice (*Oryza sativa* L.) genotype under elevated temperature: a proteomic approach to understand heat stress response. 3 Biotech 7:205. https://doi.org/10.1007/s13205-017-0856-9

Kumar M, Singh MP, Singh H, Dhakate PM, Ravindranath NH (2019a) Forest working plan for the sustainable management of forest and biodiversity in India. J Sustain for 39:1–22. https://doi.org/10.1080/10549811.2019.1632212

Kumar N, Jeena N, Singh H (2019b) Elevated temperature modulates rice pollen structure: a study from the foothill of the Himalayan agro-ecosystem in India. 3 Biotech 9(5):175. https://doi.org/10.1007/s13205-019-1700-1

Kumar N, Jeena N, Kumar N, Gangola S, Singh H (2019c) Phytoremediation facilitating enzymes: an enzymatic approach for enhancing remediation process. In: Bhatt P (ed) Smart bioremediation technologies. Academic Press, Elsevier, pp 289–306. 50 Hampshire Street, 5th Floor, Cambridge, MA 02139, United States

Kumar N, Singh H, Sharma SK (2020a) Antioxidants: responses and importance in plant defense system. In: Roychowdhury R, Choudhury S, Hasanuzzaman M, Srivastava S (eds) Sustainable agriculture in the era of climate change. Springer, Cham, pp 251–264. https://doi.org/10.1007/978-3-030-45669-6_11

Kumar P, Singh A, Laishram N, Pandey RK, Dogra S, Jeelani MI, Sinha BK (2020b) Effects of plant growth regulators on quality flower and seed production of marigold (Tagetes erecta L.). Bangladesh J Bot 49(3):567–577

Kumar A, Kumar P, Singh H, Kumar N (2021a) Modulation of plant functional traits under essential plant nutrients during seasonal regime in natural forests of Garhwal Himalayas. Plant Soil 465:197–212. https://doi.org/10.1007/s11104-021-05003-x

Kumar A, Kumar P, Singh H, Bisht S, Kumar N (2021b) Relationship of physiological plant functional traits with soil carbon stock in the temperate forest of Garhwal Himalaya. Curr Sci 120(8):1368–1373. https://doi.org/10.18520/cs/v120/i8/1368-1373

Kumar A, Kumar P, Singh H, Kumar N (2021c) Adaptation and mitigation potential of roadside trees with bio-extraction of heavy metals under vehicular emissions and their impact on physiological traits during seasonal regimes. Urban for Urban Green 58:126900. https://doi.org/10.1016/j.ufug.2020.126900

Kumar A, Kumar P, Singh H, Kumar N (2021d) Impact of plant functional traits on infiltration rate under different forests of Kempty watershed in Garhwal Himalaya. Indian J Soil Conserv 49(1):38–44

Kumar M, Phukon SN, Paygude AC, Tyagi K, Singh H (2021e) Mapping Phenological Functional Types (PhFT) in the Indian Eastern Himalayas using machine learning algorithm in Google Earth Engine. Comput Geosci 158:104982. https://doi.org/10.1016/j.cageo.2021.104982

Kumar A, Singh H, Kumari G, Bisht S, Malik A, Kumar A, Singh M, Raturi BS, Thakur A, Kaushal R (2022a) Adaptive resilience of roadside trees to vehicular emissions via leaf enzymatic, physiological, and anatomical trait modulations. Environ Pollut 313:120191. https://doi.org/10.1016/j.envpol.2022.120191

Kumar S, Shah SH, Vimala Y, Jatav HS, Ahmad P, Chen Y, Siddique KH (2022b) Abscisic acid: metabolism, transport, crosstalk with other plant growth regulators, and its role in heavy metal stress mitigation. Front Plant Sci 13:972856

Kumar N, Singh H, Giri K, Kumar A, Joshi A, Yadav A, Singh R, Bisht S, Kumari R, Jeena N, Khairakpam R, Mishra G (2024) Physiological and molecular insights into the allelopathic effects on agroecosystems under changing environmental conditions. Physiol Mol Biol Plants 30:417–433. https://doi.org/10.1007/s12298-024-01440-x

Liu DJ, Chen JY, Lu WJ (2011) Expression and regulation of the early auxin-responsive Aux/IAA genes during strawberry fruit development. Mol Biol Rep 38:1187–1193

Mathur P, Tripathi DK, Baluška F, Mukherjee S (2022) Auxin-mediated molecular mechanisms of heavy metal and metalloid stress regulation in plants. Environ Exp Bot 196:104796

Matsumoto TK (2006) Gibberellic acid and benzyladenine promote early flowering and vegetative growth of Miltoniopsis orchid hybrids. HortScience 41(1):131–135

McAdam SA, Brodribb TJ, Ross JJ (2016) Shoot-derived abscisic acid promotes root growth. Plant, Cell Environ 39(3):652–659

Meena D, Singh H, Chaudhari SK (2011) Elucidating strontium response on growth dynamics and biochemical change in *Phaseolus mungo* L. Int J Agric Environ Biotechnol 4:107–113

Nagar S, Singh VP, Arora A, Dhakar R, Singh N, Singh GP, Meena S, Kumar S, Shiv Ramakrishnan R (2021) Understanding the role of gibberellic acid and paclobutrazol in terminal heat stress tolerance in wheat. Front Plant Sci 12:692252

Nieminen K, Immanen J, Laxell M, Kauppinen L, Tarkowski P, Dolezal K, Tähtiharju S, Elo A, Decourteix M, Ljung K, Bhalerao R (2008) Cytokinin signaling regulates cambial development in poplar. Proc Natl Acad Sci 105(50):20032–20037

Owino WO, Manabe Y, Mathooko FM, Kubo Y, Inaba A (2006) Regulatory mechanisms of ethylene biosynthesis in response to various stimuli during maturation and ripening in fig fruit (*Ficus carica* L.). Plant Physiol Biochem 44(5–6):335–342

Poór P, Nawaz K, Gupta R, Ashfaque F, Khan MIR (2022) Ethylene involvement in the regulation of heat stress tolerance in plants. Plant Cell Rep 41(3):675–698

Pospíšilová H, Jiskrova E, Vojta P, Mrizova K, Kokáš F, Čudejková MM, Bergougnoux V, Plíhal O, Klimešová J, Novák O, Dzurova L (2016) Transgenic barley overexpressing a cytokinin dehydrogenase gene shows greater tolerance to drought stress. New Biotechnol 33(5):692–705

Prakash V, Hunney K and Singh H (2022). Effect of elevated carbon dioxide on growth and development of *Santalum album* L. seedlings inoculated with plant growth promoting microorganisms in Open Top Chambers. Int J Environ Agric Biotechnol 7(5):149–159

Prasad R (2022) Cytokinin and its key role to enrich the plant nutrients and growth under adverse conditions-an update. Front Genet 13:883924

Rademacher W (2015) Plant growth regulators: backgrounds and uses in plant production. J Plant Growth Regul 34:845–872

Rahman A (2013) Auxin: a regulator of cold stress response. Physiol Plant 147(1):28–35

Rai IN, Wiraatmaja IW, Astiari NK, Dwiyani R (2021) Increasing off-season snake fruit (*Salacca zalacca*) yield by preventing fruit-set failure with different exogenous auxin spray. J Appl Horticult 23(3)

Reid MS (1995) Ethylene in plant growth, development, and senescence. Plant hormones: physiology, biochemistry and molecular biology. Springer, Netherlands, Dordrecht, pp 486–508

Rietveld PL, Wilkinson C, Franssen HM, Balk PA, van der Plas LH, Weisbeek PJ, Douwe de Boer A (2000) Low temperature sensing in tulip (*Tulipa gesneriana* L.) is mediated through an increased response to auxin. J Exp Botany 51(344):587–594

Riyazuddin R, Verma R, Singh K, Nisha N, Keisham M, Bhati KK, Kim ST, Gupta R (2020) Ethylene: a master regulator of salinity stress tolerance in plants. Biomolecules 10(6):959

Rodrigues MA, Freschi L, Purgatto E, Lima VFGAP, Kerbauy GB (2014) Ethylene modulates the developmental plasticity and the growth balance between shoot and root systems in the in vitro grown epiphytic orchid Catasetum fimbriatum. J Plant Growth Regul 33:513–525

Sabagh AE, Hossain A, Islam MS, Iqbal MA, Amanet K, Mubeen M, Nasim W, Wasaya A, Llanes A, Ratnasekera D, Singhal RK (2021a) Prospective role of plant growth regulators for tolerance to abiotic stresses. In: Plant growth regulators: signalling under stress conditions, pp 1–38

Sabagh AE, Mbarki S, Hossain A, Iqbal MA, Islam MS, Raza A, Llanes A, Reginato M, Rahman MA, Mahboob W, Singhal RK (2021b) Potential role of plant growth regulators in administering crucial processes against abiotic stresses. Front Agron 3:648694

Sah SK, Reddy KR, Li J (2016) Abscisic acid and abiotic stress tolerance in crop plants. Front Plant Sci 7:571

Sajjad Y, Jaskani MJ, Asif M, Qasim M (2017) Application of plant growth regulators in ornamental plants: a review. Pakistan J Agric Sci 54(2)

Sathappan CT (2018) Effect of plant growth regulators and pinching on growth and flower yield of African marigold (*Tagetes erecta* L.). J Horticult Sci 13(1):42–47

Shah SH, Islam S, Mohammad F, Siddiqui MH (2023) Gibberellic acid: a versatile regulator of plant growth, development and stress responses. J Plant Growth Regul, pp 1–22

Shahid R, Shijie L, Shahid S, Altaf MA, Shahid H (2021) Determinants of reactive adaptations to climate change in semi-arid region of Pakistan. J Arid Environ 193:104580

Shahzad K, Hussain S, Arfan M, Hussain S, Waraich EA, Zamir S, Saddique M, Rauf A, Kamal KY, Hano C, El-Esawi MA (2021) Exogenously applied gibberellic acid enhances growth and salinity stress tolerance of maize through modulating the morpho-physiological, biochemical and molecular attributes. Biomolecules 11(7):1005

Sharma R, Singh H, Kaushik M, Nautiyal R and Singh O (2018) Adaptive physiological response, carbon partitioning, and biomass production of *Withania somnifera* (L.) Dunal grown under elevated CO_2 regimes. 3 Biotech 8:267. https://doi.org/10.1007/s13205-018-1292-1

Sharma R, Singh H (2021) Alteration in biochemical constituents and nutrients partitioning of *Asparagus racemosus* in response to elevated atmospheric CO_2 concentration. Environ Sci Pollut Res 29:6812–6821. https://doi.org/10.1007/s11356-021-16050-3

Sharp RE (2002) Interaction with ethylene: changing views on the role of abscisic acid in root and shoot growth responses to water stress. Plant, Cell Environ 25(2):211–222

Shi H, Chen L, Ye T, Liu X, Ding K, Chan Z (2014) Modulation of auxin content in Arabidopsis confers improved drought stress resistance. Plant Physiol Biochem 82:209–217

Singh H (2021) An integrated approach considering physiological and biophysical based indicators for assessing tolerance of roadside plantations of *Alstonia scholaris* towards urban roadside air pollution: an assessment of adaptation of plantations for mitigating urban roadside air pollution. Trees. https://doi.org/10.1007/s00468-021-02179-8

Singh H (2024) Exploring adaptive modulation in plant functional traits and its impact on the productivity of *Acacia auriculiformis* under CO_2-enriched environment. Industr Crops Prod 210:118186. https://doi.org/10.1016/j.indcrop.2024.118186

Singh H, Kumar M (2022) Climate change and its impact on Indian Himalayan forests: current status and research needs. In: Rani S, Kumar R (eds) Climate change. Springer Climate. Springer, Cham, pp 223–242. https://doi.org/10.1007/978-3-030-92782-0_11

Singh H, Verma A (2013) Characterization and screening of high nitrogen efficient rice genotype to curtail environmental pollution. Int J Agric Environ Biotechnol 6(1):23–30

Singh H, Savita Sharma R, Sinha S, Kumar M, Kumar P, Verma A, Sharma SK (2017a) Physiological functioning of *Lagerstroemia speciosa* L. under heavy roadside traffic: an approach to screen potential species for abatement of urban air pollution. 3 Biotech 7 (61):1–10. https://doi.org/10.1007/s13205-017-0690-0

Singh H, Verma A, Kumar M, Sharma R, Gupta R, Kaur M, Negi M, Sharma SK (2017b) Phytoremediation: a green technology to clean up the sites with low and moderate level of heavy metals. Austin Biochem 2:1012

Singh H, Sharma R, Savita SMP, Kumar M, Verma A, Ansari MW, Sharma SK (2018) Adaptive physiological response of *Parthenium hysterophorus* to elevated atmospheric CO_2 concentration. Indian Forester 144:1–14

Singh H, Yadav M, Kumar N, Kumar A, Kumar M (2020) Assessing adaptation and mitigation potential of roadside trees under the influence of vehicular emissions: A case study of Grevillea robusta and Mangifera indica planted in an urban city of India. PLoS ONE 15(1): e0227380. https://doi.org/10.1371/journal.pone.0227380

Singh S, Singh H, Sharma SK, Nautiyal R (2021) Seasonal variation in biochemical responses of bamboo clones in the sub-tropical climate of Indian Himalayan foothills. Heliyon 7(4):e06859. https://doi.org/10.1016/j.heliyon.2021.e06859

Singh H, Bandyopadhyaya S, Kumar A, Singh M, Malik A, Kumari G, Raturi A, Yadav SK, Kumar M, Barthwal S, Thakur A, Rawat PS (2022) Understanding the physiological and biophysical response of urban roadside plantations for assessing adaptation and mitigation mechanisms toward vehicular emissions. Urban Clim 44:101183. https://doi.org/10.1016/j.uclim.2022.101183

Singh S, Singh I, Miller CT, Dhatt KK, Dubey RK (2023) Increasing basal dose of indole-3-butyric acid improve rooting and growth of different cutting types in Dahlia. Rhizosphere 27:100729

Singh M, Singh H, Kumar A, Kumar M, Barthwal S, Thakur A (2024) Soil nitrogen availability determines the CO_2 fertilization effect on tree species (*Neolamarckia cadamba*): growth and physiological evidence. Environ Sustain 7:53–60. https://doi.org/10.1007/s42398-023-00300-w

Sosnowski J, Truba M, Vasileva V (2023) The impact of auxin and cytokinin on the growth and development of selected crops. Agriculture 13(3):724

Tao JJ, Chen HW, Ma B, Zhang WK, Chen SY, Zhang JS (2015) The role of ethylene in plants under salinity stress. Front Plant Sci 6:1059

Tsai WC, Hsiao YY, Pan ZJ, Kuoh CS, Chen WH, Chen HH (2008) The role of ethylene in orchid ovule development. Plant Sci 175(1–2):98–105

Ullah Z, Abbas SJ, Naeem N, Lutfullah G, Malik T, Khan MAU, Khan I (2013) Effect of indole butyric acid (IBA) and naphthalene acetic acid (NAA) plant growth regulators on Marigold (*Tagetes erecta* L.). Afr J Agric Res 8(29):4015–4019

Upreti KK, Sharma M (2016) Role of plant growth regulators in abiotic stress tolerance. In: Abiotic stress physiology of horticultural crops, pp 19–46

Van Altvorst AC, Bovy AG (1995) The role of ethylene in the senescence of carnation flowers, a review. Plant Growth Regul 16:43–53

Van de Poel B, Smet D, Van Der Straeten D (2015) Ethylene and hormonal cross talk in vegetative growth and development. Plant Physiol 169(1):61–72

Wang F, Cui X, Sun Y, Dong CH (2013) Ethylene signaling and regulation in plant growth and stress responses. Plant Cell Rep 32:1099–1109

Waterland NL, Campbell CA, Finer JJ, Jones ML (2010) Abscisic acid application enhances drought stress tolerance in bedding plants. HortScience 45(3):409–413

Werner T, Schmülling T (2009) Cytokinin action in plant development. Curr Opin Plant Biol 12(5):527–538

Wheeler RM, Peterson BV, Stutte GW (2004) Ethylene production throughout growth and development of plants. HortScience 39(7):1541–1545

Wu W, Du K, Kang X, Wei H (2021) The diverse roles of cytokinins in regulating leaf development. Hortic Res 8:118

Xia J, Zhao H, Liu W, Li L, He Y (2009) Role of cytokinin and salicylic acid in plant growth at low temperatures. Plant Growth Regul 57:211–221

Xiong L, Zhu JK (2003) Regulation of abscisic acid biosynthesis. Plant Physiol 133(1):29–36

Yadav SK, Singh H, Nautiyal R, Ginwal HS, Ansari SA, Barthwal S (2019) Modulation of morpho-physiological responses in Populus deltoides by elevated carbon dioxide and temperature. For Sci 66:105–118. https://doi.org/10.1093/forsci/fxz048

Yang D, Li Y, Shi Y, Cui Z, Luo Y, Zheng M, Chen J, Li Y, Yin Y, Wang Z (2016) Exogenous cytokinins increase grain yield of winter wheat cultivars by improving stay-green characteristics under heat stress. PLoS ONE 11(5):e0155437

Zhou L, Dong L, Jia PY, Wang WR, Wang LY (2010) Expression of ethylene receptor and transcription factor genes, and ethylene response during flower opening in tree peony (*Paeonia suffruticosa*). Plant Growth Regul 62:171–179

Zhu Z, Ding Y, Zhao J, Nie Y, Zhang Y, Sheng J, Tang X (2016) Effects of postharvest gibberellic acid treatment on chilling tolerance in cold-stored tomato (*Solanum lycopersicum* L.) fruit. Food Bioprocess Technol 9:1202–1209

Enhancing Adaptation and Stress Tolerance in Urban Plants Through Plant Growth-Promoting Rhizobacteria (PGPR)

Sagar Vishwakarma, Vikas Verma Patel, and Soban Prakash

Abstract Urban plants are subjected to many biotic and abiotic stresses. These stresses compromise crop health and lead to substantial urban agricultural losses globally. To mitigate these challenges, researchers worldwide actively pursue innovative management strategies that integrate traditional and advanced plant breeding approaches. Urban environments, characterized by diverse stress factors, offer a fertile ground for investigating the interplay between microorganisms, plant nutrition, and plant growth and development. Among these, plant growth-promoting rhizobacteria (PGPR) emerge as crucial players in biotic and abiotic stress mitigation. PGPR enhances tolerance to pesticide and heavy metal contamination, temperature extremes, salinity, and drought by colonizing the rhizosphere and endo-rhizosphere of plants. The strategic introduction of stress-tolerant PGPR into the plant rhizosphere holds promise for enhancing plant development and boosting productivity in stress-prone urban environments. This chapter explores a range of stress-tolerant PGPRs derived from urban plant varieties, elucidating their impact on various crop types and their potential for promoting plant growth under adverse conditions.

Keywords Rhizobacteria · Biotic and abiotic stress · Stress tolerance

S. Vishwakarma (✉)
Department of Botany and Microbiology, Gurukula Kangri (Deemed to be University), Haridwar, Uttarakhand 249404, India
e-mail: sagarvishwakarmarksh@gmail.com

V. V. Patel
Department of Botany, VRAL Government Girls Degree College, Bareilly, India

S. Prakash
Department of High Altitude Plant Physiology Research Centre, H.N.B. Garhwal University, Srinagar, Garhwal, Uttarakhand 246174, India

H. Singh (ed.), *Urban Forests, Climate Change and Environmental Pollution*,
https://doi.org/10.1007/978-3-031-67837-0_12

1 Introduction

Various bacteria, fungi, algae, actinomycetes, and fungi form a complex ecosystem in the soil, contributing to the biogeochemical cycling of nutrients and the decomposition of organic matter, vital processes that sustain soil productivity and health. Soil characteristics support plant growth and development as plants rely on soil for essential nutrients, water, anchoring, and other necessities throughout their lifecycle. However, plants constantly face a barrage of abiotic and biotic stresses, including variations in water availability, temperature extremes, mineral deficiencies, radiation, and microbial pathogens, which can disrupt their biosynthesis and development, leading to increased oxidative stress (Gill and Tuteja 2010; Yadav et al. 2019a, b; Jangra et al. 2024; Kanta et al. 2024; Malik et al. 2024; Kumar and Singh 2024).

The rhizosphere, predominantly inhabited by fungi and bacteria, is a hotspot for microbial activity. Among these microbes, plant growth-promoting rhizobacteria (PGPR) play a pivotal role in enhancing plant growth by colonizing roots and exerting beneficial effects such as synthesizing plant growth regulators, facilitating nutrient cycling, and suppressing pathogens (Ahmad et al. 2008). PGPR possess intrinsic traits that enable them to competitively colonize root systems, adapt to root-associated microhabitats, and stimulate plant growth.

Microbes themselves are not immune to the stresses present in their environment. Some microbes, equipped with enzymes like cellulase, can penetrate plant roots to access resources within the apoplast and vascular bundles, surviving and thriving amidst intense competition in the rhizosphere. However, the widespread adoption of synthetic insecticides and fertilizers since the mid-1900s, often called the "green revolution," has significantly increased crop yields and led to soil fertility issues and environmental contamination.

Addressing biotic and abiotic stresses in plants has traditionally relied on methods like breeding, genetic engineering, and agricultural management practices, which can be costly and time-consuming. Harnessing the multifaceted capabilities of microorganisms presents a promising, cost-effective, and sustainable approach. Research efforts increasingly focus on developing microbial strategies to manage abiotic stresses globally, with a growing recognition of bacteria's pivotal role in addressing biotic and abiotic challenges (Meena et al. 2011; Kumar et al. 2019c; 2020; Singh and Kumar 2022; Singh et al. 2024).

Arbuscular mycorrhizal fungi (AMF) exemplify beneficial soil microbes, forming symbiotic associations with various agricultural plant species, enhancing plant nutrition, adaptability, and stress resistance (Singh et al. 2012). Specific PGPR strains have shown efficacy in promoting plant growth and alleviating salt stress in salt-rich soils. Harnessing the potential of PGPR systems to mitigate both biotic and abiotic stressors holds promise for enhancing crop production and resilience. This chapter delves into the impacts of PGPR under diverse stress conditions, shedding light on their potential applications in crop management and stress mitigation.

2 Drought Stress

Urbanization, industrialization, and the excessive use of chemical fertilizers are exacerbating climate change, posing significant threats to global food security. With the world's population soaring, sustaining agricultural productivity is urgently needed. Agriculture, being highly susceptible to climate variations, faces the challenge of drought-induced yield losses in many regions worldwide (Naveed et al. 2014). Drought is a primary abiotic stressor impeding agricultural productivity globally. It's estimated that drought has caused a 9–10% reduction in cereal production in some countries. Predictions suggest that by 2050, half of the arable land will be affected by drought, posing severe threats to crop growth. As global warming exacerbates drought frequency and severity, there's a looming risk of declining food production. Therefore, urgent strategies are needed to mitigate the effects of water scarcity on food security. Enhancing plant drought tolerance is crucial for thriving in water-stressed environments, and modern agro-biotechnological tools, like genetic engineering, offer avenues to develop drought-resistant crops.

3 Drought Tolerance

Drought severely impacts plant productivity by disrupting vital processes like photosynthesis and metabolic activities, leading to decreased crop yields globally. Morphological and physiological changes in drought alter plant water potential and turgor, hampering normal functions (Kamara et al. 2003; Singh et al. 2020; Kumar et al. 2021a, b, c, d). Drought stress reduces both the quantity and quality of water uptake by roots, affecting nutrient availability in the soil. This carbon fixation and utilization imbalance disturbs sugar concentrations and osmotic potentials, triggering oxidative stress by producing reactive oxygen species (ROS) (Fang et al. 2015). Chlorophyll degradation and leaf senescence further impair photosynthetic efficiency. Ethylene overproduction during drought inhibits root growth, exacerbating plant stress.

Developing drought-tolerant crop varieties through genetic modification and breeding is a promising strategy to enhance agricultural productivity and water efficiency. Plants under stress exhibit increased reliance on Plant Growth-Promoting Rhizobacteria (PGPR), which enhance metabolic activities. Introducing specific PGPR strains like Enterobacter sp. FD17 and *Burkholderia phytofirmans* strain PsJN into crops during drought conditions positively influences photosynthetic activity, water status, and overall plant development (Kasim et al. 2013). PGPR-mediated breakdown of the ethylene precursor, ACC, alleviates plant stress and promotes average growth (Mayak et al. 2004). Such approaches hold immense potential for bolstering agricultural resilience against drought and water scarcity. The role of PGPR in drought stress in presented in Table 1.

Table 1 Role of PGPR in drought stress

S. no.	PGPR	Crop	Mechanism	References
1	A. brasilense	Tomato	Nitric oxide was a signalling molecule in the IAA-induced pathway that promoted root hair growth and lateral roots	Creus et al. (2005)
2	Rhizobium sp. YAS34	Sunflower	Raises the volume of soil macropores and the amount of root-adhering soil (RAS) per root dry mass	Alami et al. (2000)
3	Azospirillum spp.	Wheat	Plant growth	Arzanesh et al. (2011)
4	Pseudomonas fluorescens or P. putida	Rapeseed	Pseudomonas fluorescens or *P. putida* inoculation increased yield, 1000-grain weight, grains/pod, pods, and branches/ plant in rapeseeds	Arvin et al. (2012)
5	Azospirillum lipoferum	Maize	Gibberellins reduced drought stress and raised ABA levels	Cohen et al. (2009)
6	Burkholderia phytofirmans strain PsJN and Enterobacter spp. FD17	Maize	Water condition, growth, and photosynthetic activity	Naveed et al. (2014)
7	Bacillus licheniformis K11	Pepper	Increased stress gene expression CaPR-10, sHSP, VA, and Cadhn	Lim and Kim (2013)
8	Rhizobium etli overexpressing trehalose-6-phosphate synthase gene	Phaseolus vulgaris	Trehalose functions as a signalling molecule that upregulates the expression of genes related to carbon and nitrogen metabolism, stress tolerance, and	Suárez et al. (2008)
9	Azoapirillum brasilense Az39	Rice	The synthesis of cadaverine by PGPR enhanced root development and reduced osmotic stress	Cassán et al. (2009)

4 Heavy Metals Stress

Heavy metals are significant inorganic contaminants affecting urban plants and soil health. Heavy metals can positively and negatively affect soil microorganisms and climatic factors depending on their nature and bioavailability. While certain heavy metals like zinc and nickel can be harmful in high concentrations, their phytotoxic levels may accumulate in soil due to the indiscriminate use of untreated municipal waste and agrochemicals. As plants absorb these metals, they pose risks to human and animal health upon entering the food chain, potentially impacting all ecosystem components (Singh and Verma 2013; Kumar et al. 2017; Gupta and Singh 2017; Singh et al. 2017a, b).

Soil contamination by heavy metals is a pressing concern due to its direct effects on the biotic components of the ecosystem. Various human activities, such as mining, industrial processes, improper waste disposal, and agricultural practices, contribute to heavy metal pollution in soil, altering its pH, organic matter content, and other properties. These alterations can significantly affect soil microbial development and plant growth (Hall 2002; Sharma et al. 2018; Singh et al. 2018; Gupta et al. 2018a, b). Heavy metals exist in soil as elemental forms or compounds, and through biological amplification, they accumulate in living organisms, posing risks to higher trophic levels.

Plants have evolved molecular and physical defense mechanisms to cope with heavy metal stress, including physiological barriers such as trichome production, thick cuticles, and cell wall modifications. These adaptations help plants either sequester heavy metals or counteract their adverse effects. However, when heavy metals penetrate plant tissues, plants activate various defense mechanisms to mitigate their harmful effects.

5 Effects of PGPR on Crops Under Heavy Metal Stress

With increasing urbanization and industrialization, heavy metal contamination in the environment has become a significant issue, impacting plant metabolism, ecosystem dynamics, and human health. Plant growth-promoting rhizobacteria (PGPR) are crucial in mitigating heavy metal toxicity and promoting plant growth. PGPR acts through various mechanisms, including reducing ethylene synthesis, enhancing root proliferation, and converting inaccessible forms of nutrients into accessible forms for plants in the rhizosphere (Nadgórska-Socha et al. 2013).

PGPR facilitates heavy metal extraction through biotransformation, chelation, extrusion, and volatilization (Gupta and Diwan 2017). They also induce the expression of antioxidant enzymes and non-enzymatic antioxidants in plants under heavy metal stress, thereby enhancing their tolerance. Certain PGPR strains have been identified for tolerating and mobilizing heavy metals like cadmium, lead, zinc, and copper (Ma et al. 2015). These findings underscore the importance of PGPR in mitigating heavy metal stress and preserving plant health in contaminated environments (Table 2).

6 Thermal Stress and Adaptation

Temperature-induced stress alters hormone synthesis, particularly ethylene, often impeding plant development (Islam et al. 2014). High temperatures lead to protein denaturation and aggregation, potentially resulting in cell necrosis if untreated. Discrepancies between ABA and cytokinins, stemming from prolonged extreme temperatures during reproduction, can cause grain abortion. Heat responses include suppression of standard translation and transcription and heightened activation of genes coding for proteins implicated in thermotolerance and heat shock. Freezing stress can induce cell desiccation, starvation, and mortality, altering nutrient and water absorption rates. Conversely, low temperatures impair metabolic processes, modify membrane characteristics, alter protein structures, and inhibit enzyme reactions (Heino and Palva 2003).

Understanding how cereals adapt to such stressors remains limited, with few established plant growth-promoting rhizobacteria (PGPR) relationships at suboptimal temperatures. Bacteria's geographical origin is speculated to dictate their optimal development range for advantageous interactions with plants. For instance, bacteria isolated from cold climates were found to colonize wheat plant rhizospheres, enhancing their resistance to low temperatures (Yadav et al. 2019a, b; Gupta et al. 2019a, b; Kumar et al. 2019a, b). In contrast, those isolated from warm environments bolstered their high-temperature tolerance. Similarly, inoculating sorghum and wheat seedlings with Pseudomonas sp. strain AKM-P7 increased their heat tolerance by inducing higher-molecular-weight protein synthesis and elevating cellular metabolite levels (Zulfikar Ali et al. 2011). Another study identified a thermotolerant *Pseudomonas* spp. strain (AKM-P6) from the rhizosphere of Indian desert and semi-arid pigeon pea plants, which, when inoculated, boosted sorghum seedlings' heat tolerance through increased high-molecular-weight protein biosynthesis and metabolite content (Ali et al. 2009).

Similarly, inoculating wheat with various cold-resistant Pseudomonas species enhanced root development, nutrient uptake, and overall cold adaptation. In grapevines (Vitis vinifera), inoculation with *Burkholderia phytofirmans* PsJN strains mitigated heat shock-induced cell membrane damage by reducing biomass loss and osmolyte leakage (Bensalim et al. 1998). However, further research is needed to understand the mechanisms governing PGPR-plant interactions under these conditions.

Table 2 Role of PGPR in heavy metal stress

S. No.	PGPR	Heavy Metal	Effect	Plant	References
1	Enterobacter sp.	Cd	Metal-induced growth disruption was recovered by inoculation by reducing metal absorption in mung beans	Pea	Naveed et al. (2020)
2	Kocuria sp. CRB15	Cu	Brassica nigra, root and shoot development, is improved, and the harmful effects of copper on plants are reduced	Brassica nigra	Swapnil et al. (2016)
3	*Kluyvera ascorbate* SUD165, *Kluyvera ascorbate* SUD165/26	Pb	An increase in Indian mustard's dry weight	*Brassica napus* (canola); *Brassica juncea* (Indian mustard)	Burd et al. (2018)
4	Bacillus, Klebsiella, Leifsonia and Enterobacter	Cd	Significant rise in dry biomass under Cd stress, shoot and root development, and the relative water content and membrane permeability of maize leaves following inoculation	Zea mays	Ahmad et al. (2016)
5	P. aeruginosa ZN3	Zn	Increased synthesis of plant biomass, total soluble protein, and chlorophyll in the leaves of wheat plants, along with improved absorption of P and N	Wheat	Islam et al. (2014)
6	Glomusmosseae, Rhizobium trifolii and Brevibacillusbrevis	Ni	A higher biomass of shoots and roots, as well as a higher N and P content	Clover	Vivas et al. (2006)

(continued)

Table 2 (continued)

S. No.	PGPR	Heavy Metal	Effect	Plant	References
7	*Micrococcus* sp. *MU1, Klebsiella* sp. *BAM1*	Cd	Increased Cd mobilization in the impacted soil and the removal of Cd ions from an aqueous solution	*Helianthus annuus* (sunflower)	Prapagdee et al. (2013)
8	Achromobacter sp. E4L5, Bacillus sp. E4S1, Bacillus sp. E1S2, Bacillus pumilus E2S2 and Stenotrophomonas sp. E1L	Zn, Cd, Pb	Effective inoculation increased the plants' dry biomass and the development of roots and shoots. An increase in the amount of water-extractable zinc and copper in soil	Sedum plumbizincicola	Adediran et al. (2015)

7 Salinity Stress

Several factors, including climate, stress severity, and genotype tolerance, influence crop responses to salt. Soil salinity, characterized by excess soluble salts in the root zone, reduces water uptake and other essential plant functions, elevating osmotic pressure and impeding plant growth. Salinization, the accumulation of accessible salts like carbonate, calcium, potassium, bicarbonate, chloride, sodium, sulfate, and magnesium, is exacerbated by mineral weathering, altering soil chemistry and moisture levels (Farooq et al. 2015).

Salt toxicity manifests in plants through osmotic imbalance, ion toxicity, oxidative stress, and reduced photosynthesis, negatively impacting crop yield. High salt concentrations affect seed germination, root length, plant height, and fruiting, ultimately diminishing total crop output (Liang et al. 2018). Despite these challenges, plants can develop mechanisms to endure salt stress and complete their life cycle (Parida and Das 2005). PGPR possesses sophisticated physiological, chemical, and biological mechanisms that may synthesize compounds and initiate processes to fortify plants against soluble salts (Parida and Das 2005).

8 PGPR Effects on Salt-Tolerant Crops

The presence of PGPR significantly enhances rice plants' root length, shoot length, and dry weight under both normal and saline conditions. Studies have shown that PGPR induces various plant hormones, such as auxins, hydrogen cyanide (HCN), phosphate solubilization compounds, and siderophores (Sarkar et al. 2018). Auxins, like indole-3-butyric acid (IBA) and indole-3-acetic acid (IAA), promote plant growth by increasing shoot-to-root ratio, hypocotyl length, lateral root formation, and inhibiting root elongation (Zhao et al. 2001; Loper 1986).

Salinity stress is a common issue in soil, but PGPR strains isolated from the rhizosphere demonstrate salt tolerance and beneficial interactions with plants under various stress conditions. Exogenous application of PGPR in both saline and nonsaline environments enhances plant tolerance to stressors and promotes growth (Egamberdieva et al. 2017). For instance, PGPR strains reduce tomato foot and root rot in saline conditions and enhance stress tolerance by balancing reactive oxygen species (ROS) levels and increasing tomato proline content (Egamberdieva et al. 2017).

Exogenous application of PGPR also improves rice growth, nutrient uptake, and antioxidant production under salt-stress conditions. Furthermore, PGPR-treated plants exhibit reduced expression of the K^+ ion transporter (HKT1), leading to decreased sodium (Na^+) accumulation in shoots, thus mitigating salt stress (Zhang et al. 2008).

While soil bacteria exhibit lower salt tolerance compared to rhizospheric bacteria, certain PGPR strains, such as *Pseudomonas chlororaphis* TSAU13, *Pseudomonas extremorientalis* TSAU6, *Pseudomonas fluorescens* WCS356, *Pseudomonas putida* TSAU1, and *Pseudomonas extremorientalis* TSAU20, can tolerate up to 3% NaCl (Egamberdieva et al. 2017). Inoculation with *Bacillus lentus* and Pseudomonas sp. decreases salinity stress on basil plants by enhancing antioxidant enzyme activity, nutrient availability, and photosynthesis (Nabti et al. 2010). The halotolerant strain *Azospirillum brasilense* NH also improves wheat development and germination in salinized soil (Nabti et al. 2010).

PGPR induces hormone production in plants and releases exopolysaccharides, phytohormones, volatile chemicals, osmolytes, antioxidants, and ACC deaminase under stress conditions, contributing to plant growth promotion and stress tolerance (Vurukonda et al. 2016). PGPR synthesizing these growth-promoting substances is crucial for sustainable saline agriculture. PGPR strains also produce other phytohormones like gibberellins, cytokinins, and IAA, further alleviating stress (Cassán et al. 2014). These findings suggest the potential of exogenous application of PGPR and their metabolites to enhance plant resistance to abiotic stress (Table 3).

Table 3 Role of PGPR in Salt-tolerant

S. No.	PGPR	Plant	Effect	References
1	Bacillus megaterium A12	Lycopersicon esculentum	Expression of PIP aquaporin is up-regulated	Akram et al. (2019)
2	*Pseudomonas* sp.	Sunflower	Production of IAA, HCN, and siderophores; solubilization of phosphate; fixation of nitrogen; control of chitinase and β-1–3 glucanase	Tewari and Arora (2016)
3	Glomusmosseae and Pseudomonas mendocina	Lactuca sativa	Shoot biomass	
4	P. fluorescens, P. aeruginosa, and P. stutzeri	Solanum lycopersicum	Leaf size, number of leaves, branch counts, and lateral root counts	Tank and Saraf (2010)
5	Pseudomonas sp. PMDZnCd 2003	Oryza sativa	Production of indole-3-acetic acids (IAA), fixation of nitrogen, and solubilization of phosphate	Nakbanpote et al. (2014)
6	*Pseudomonas fluorescens* SA8	Black gram	Enhanced photosynthetic content, gas exchange, and water quality	Nakbanpote et al. (2014)
7	Bacillus spp.	Radish	Plant growth	Yildrim et al. (2008)
8	Glomusfasciculatum	Acacia nilotica	In addition to the intake of phosphorus (P), nitrogen (N), and potassium (K), total, shoot, and root biomass	Chandrasekaran and Boughattas (2014)

9 Pesticide Stress

Pesticides, whether applied aerially or by ground methods, undergo degradation influenced by various physicochemical properties and environmental factors such as organic matter content, pH, and moisture (Beigel and Di Pietro 1999). Temperature fluctuations, like abrupt increases of 5–7 °C, induce plant heat stress, affecting photosynthesis, water balance, flowering, fruiting, and susceptibility to pests and diseases. Pesticides exacerbate cellular stress by elevating reactive oxygen species (ROS), impacting plant metabolism and physiology. Soil-applied pesticides disrupt vital soil microorganisms, affecting nitrogen-fixing and phosphate-solubilizing bacteria, thereby diminishing crop productivity and soil fertility.

10 Pesticide Tolerance

Plant growth-promoting rhizobacteria (PGPR) are pivotal in enhancing plant resistance to biotic stressors and pesticide tolerance. PGPR-mediated systemic resistance (ISR) aids plants in combating biotic stress. Certain bacteria, like Pseudomonas and Flavobacterium spp., resist various pesticides (Nazarian and Mousawi 2005). To mitigate pesticide toxicity, PGPR employs diverse strategies, including biodegradation and enzymatic hydrolysis (Yang and Lee 2008). These bacteria also suppress pest populations directly, offering biocontrol benefits.

11 Plant Growth Promotion

PGPR enhances plant growth through nutrient fixation, disease suppression via volatile organic compounds (VOCs) and enzymes, and mitigation of biotic and abiotic stresses. They produce phytohormones like indole acetic acid and cytokinins and enzymatically degrade the ethylene precursor ACC, alleviating stress (Patten and Glick 2011). PGPR also synthesizes antioxidants and siderophores, aiding plant development by modulating iron availability and mitigating pathogens (García de Salamone et al. 2001).

12 Cytokinins

Cytokinins, a subclass of phytohormones, are pivotal in stimulating cell division, cell enlargement, and tissue growth in specific plant areas. Roots and shoots exhibit meristem activity, vascular development, nodule formation, root elongation, lateral rooting, and apical dominance in response to environmental stimuli, all influenced by cytokinins (Chen et al. 2022). These phytohormones also regulate axillary bud growth, leaf senescence, apical dominance, and cell division in both roots and shoots, contributing to cell proliferation and differentiation.

Given that many plant growth-promoting rhizobacteria (PGPRs) produce cytokinins, selecting microbial inoculants capable of generating high levels of cytokinins holds promise for mitigating salt stress. Studies have shown that up to ninety percent of rhizobacteria cultured in vitro release cytokinin-like growth stimulants (Patel and Saraf 2017). For example, rhizobacteria associated with *Coleus forskohlii*, such as *Pseudomonas stutzeri* MTP40, *Stenotrophomonas maltophilia* MTP42, and *Pseudomonas putida* MTP50, have been found to produce cytokinins that promote plant growth.

Furthermore, auxins and cytokinins collaborate to regulate plant growth, promoting xylem formation through various chemical pathways, while cytokinins stimulate phloem development. Several PGPRs can produce cytokinins, encouraging plants to release root exudates and thereby enhancing interactions between the plant and PGPRs. For instance, the symbiotic nitrogen-fixing relationship of Rhizobium is regulated by auxin and cytokinin. Additionally, cytokinins regulate root meristem differentiation and promote root hair growth, although they may also inhibit lateral root development and the spread of the primary root (Akhtar et al. 2019).

13 Conclusion

In the face of ever-changing climatic conditions and global warming concerns, urban agricultural production methods must adapt to mitigate their severe impacts on crops. Employing microbial bioresources, particularly plant growth-promoting rhizobacteria (PGPR), offers a promising solution. PGPR enhances plant growth and development, confers resistance to biotic and abiotic stresses, and reduces the reliance on chemical fertilizers. Over the past four decades, PGPR has played a crucial role in supporting sustainable agriculture, with ongoing research focusing on understanding their mechanisms and developing stress-tolerant strains. Future efforts will involve creating diverse bio-formulations, such as granular or liquid inoculants, and exploring combinations of stress-tolerant PGPRs for biocontrol against plant diseases. By reclaiming wastelands and improving soil health, PGPR-based approaches increase food production while safeguarding the environment. It is suggested that leveraging microbial bioresources like PGPR offers a promising avenue for enhancing urban agricultural productivity and ensuring food security in the face of mounting environmental challenges.

References

Adediran GA, Ngwenya BT, Mosselmans JFW, Heal KV, Harvie BA (2015) Mechanisms behind bacteria induced plant growth promotion and Zn accumulation in Brassica Juncea. J Hazard Mater 283:490–499

Ahmad F, Ahmad I, Khan MS (2008) Screening of free-living rhizospheric bacteria for their multiple plant growth promoting activities. Microbiol Res 163(2):173–181. https://doi.org/10.1016/j.micres.2006.04.001

Ahmad I, Javed M, Hafiz A, Asghar N (2016) Differential effects of plant growth-promoting rhizobacteria on maize growth and cadmium uptake. J Plant Growth Regul 35(2):303–315. https://doi.org/10.1007/s00344-015-9534-5

Akhtar S, Mekureyaw MF, Pandey C, Roitsch T (2019) Role of cytokinins for interactions of plants with microbial pathogens and pest insects. https://doi.org/10.3389/fpls.2019.01777

Akram W, Aslam H, Ahmad SR, Anjum T, Yasin A, Khan WU, Ahmad A, Guo J, Wu T (2019) Bacillus megaterium strain A12 ameliorates salinity stress in tomato plants through multiple mechanisms. https://doi.org/10.1080/17429145.2019.1662497

Alami Y, Achouak W, Marol C, Heulin T (2000) Rhizosphere soil aggregation and plant growth promotion of sunflowers by an exopolysaccharide-producing Rhizobium sp. strain isolated from sunflower roots. Appl Environ Microbiol 66(8):3393–3398. https://doi.org/10.1128/AEM.66.8.3393-3398.2000

Ali SZ, Sandhya V, Grover M, Kishore N, Rao LV, Venkateswarlu B (2009) Pseudomonas sp. strain AKM-P6 enhances tolerance of sorghum seedlings to elevated temperatures. Biol Fert Soils 46(1):45–55. https://doi.org/10.1007/s00374-009-0404-9

Arvin P, Vafabakhsh J, Mazaheri D (2012) Plant growth promoting rhizobacteria (PGPR) on yield, yield components and seed oil content of different cultivars and species of brassica oilseed rape. Scholarsresearchlibrary.Com 3(9):4444–4451. http://scholarsresearchlibrary.com/ABR-vol3-iss9/ABR-2012-3-9-4444-4451.pdf

Arzanesh MH, Alikhani HA, Khavazi K, Rahimian HA, Miransari M (2011) Wheat (*Triticum aestivum* L.) growth enhancement by Azospirillum sp. under drought stress. World J Microbiol Biotechnol 27(2):197–205. https://doi.org/10.1007/s11274-010-0444-1

Beigel C, Di Pietro L (1999) Transport of triticonazole in homogeneous soil columns influence of nonequilibrium sorption. Soil Sci Soc Am J 63(5):1077–1086

Bensalim S, Nowak J, Asiedu SK (1998) A plant growth promoting rhizobacterium and temperature effects on performance of 18 clones of potato. Am J Potato Res 75(3):145–152. https://doi.org/10.1007/bf02895849

Burd GI, Dixon DG, Glick BR (2018) Plant growth-promoting bacteria that decrease heavy metal toxicity in plants. https://doi.org/10.1139/w99-143

Cassán F, Maiale S, Masciarelli O, Vidal A, Luna V, Ruiz O (2009) Cadaverine production by *Azospirillum brasilense* and its possible role in plant growth promotion and osmotic stress mitigation. Eur J Soil Biol 45(1):12–19. https://doi.org/10.1016/j.ejsobi.2008.08.003

Cassán F, Vanderleyden J, Spaepen S (2014) Physiological and agronomical aspects of phytohormone production by model plant-growth-promoting Rhizobacteria (PGPR) belonging to the genus Azospirillum. J Plant Growth Regul 33(2):440–459. https://doi.org/10.1007/s00344-013-9362-4

Chandrasekaran M, Boughattas S (2014) A meta-analysis of arbuscular mycorrhizal effects on plants grown under salt stress. Mycorrhiza 3:611–625. https://doi.org/10.1007/s00572-014-0582-7

Chen L, Jameson GB, Guo Y, Song J, Jameson PE (2022) The LONELY GUY gene family: from mosses to wheat, the key to the formation of active cytokinins in plants. Plant Biotechnol J 20(4):625–645. https://doi.org/10.1111/pbi.13783

Cohen AC, Travaglia CN, Bottini R, Piccoli PN (2009) Participation of abscisic acid and gibberellins produced by endophytic Azospirillum in the alleviation of drought effects in maize. Botany 87(5):455–462. https://doi.org/10.1139/B09-023

Creus CM, Graziano M, Casanovas EM, Pereyra MA, Simontacchi M, Puntarulo S, Barassi CA, Lamattina L (2005) Nitric oxide is involved in the *Azospirillum brasilense*-induced lateral root formation in tomato. Planta 221(2):297–303. https://doi.org/10.1007/s00425-005-1523-7

Egamberdieva D, Davranov K, Wirth S, Hashem A, Abd Allah EF (2017) Impact of soil salinity on the plant-growth—promoting and biological control abilities of root associated bacteria. Saudi J Biol Sci 24(7):1601–1608. https://doi.org/10.1016/j.sjbs.2017.07.004

Fang Y, Liao K, Du H, Xu Y, Song H, Li X, Xiong L (2015) A stress-responsive NAC transcription factor SNAC3 confers heat and drought tolerance through modulation of reactive oxygen species in rice. J Exp Bot 66(21):6803–6817. https://doi.org/10.1093/jxb/erv386

Farooq M, Hussain M, Wakeel A, Siddique KHM (2015) Salt stress in maize: effects, resistance mechanisms, and management. A review. Agron Sustain Dev 35(2):461–481. https://doi.org/10.1007/s13593-015-0287-0

García de Salamone IE, Hynes RK, Nelson LM (2001) Cytokinin production by plant growth promoting rhizobacteria and selected mutants. Can J Microbiol 47(5):404–411. https://doi.org/10.1139/cjm-47-5-404

Gill SS, Tuteja N (2010) Reactive oxygen species and antioxidant machinery in abiotic stress tolerance in crop plants. Plant Physiol Biochem 48(12):909–930. https://doi.org/10.1016/j.plaphy.2010.08.016

Gupta P, Diwan B (2017) Bacterial Exopolysaccharide mediated heavy metal removal: a review on biosynthesis, mechanism and remediation strategies. Biotechnol Rep 13:58–71. https://doi.org/10.1016/j.btre.2016.12.006

Gupta SK, Singh H (2017) Observations on sporadic flowering in *Ficus benjamina* L. and *Peltophorum africanum* L. in New Forest. Indian Forester 143:290–291

Gupta SK, Ram J, Singh H (2018a) Comparative study of transpiration in cooling effect of tree species in the atmosphere. J Geosci Environ Prot 6:151–166. https://doi.org/10.4236/gep.2018.68011

Gupta SK, Ram J, Singh H, Kumar P (2018b) Comparative study of cooling under shades of some forestry tree species with respect to ambient temperatures. J Geosci Environ Prot 6:51–66. https://doi.org/10.4236/gep.2018.611004

Gupta SK, Ram J, Singh H (2019a) Relationship between leaf area index and atmospheric cooling of tree species. Indian J for 42(3):203–208

Gupta SK, Singh H, Ram J (2019b) Need for a novel paradigm in forestry research: perspectives of comparative atmospheric cooling by different forestry effect tree species. Indian Forester 145(6):509–515

Hall JL (2002) Cellular mechanisms for heavy metal detoxification and tolerance. J Exp Bot 53(366):1–11. https://doi.org/10.1093/jxb/53.366.1

Heino P, Palva ET (2003) Signal transduction in plant cold acclimation, vol 4, pp 151–186. https://doi.org/10.1007/978-3-540-39402-0_7

https://ageconsearch.umn.edu/record/288998%0A, http://www.fao.org/economic/esa

Islam F, Yasmeen T, Ali Q, Ali S, Saleem M, Hussain S, Rizvi H (2014) Influence of *Pseudomonas aeruginosa* as PGPR on oxidative stress tolerance in wheat under Zn stress. Ecotoxicol Environ 104:285–293

Jangra A, Kumar K, Maikhuri S, Bhandari MS, Pandey S, Singh H, Barthwal S (2024) Unveiling stress-adapted endophytic bacteria: Characterizing plant growth-promoting traits and assessing cross-inoculation effects on *Populus deltoides* under abiotic stress. Plant Physiol Biochem 210:108610. https://doi.org/10.1016/j.plaphy.2024.108610

Kamara AY, Menkir A, Badu-Apraku B, Ibikunle O (2003) The influence of drought stress on growth, yield and yield components of selected maize genotypes. J Agric Sci 141(1):43–50. https://doi.org/10.1017/S0021859603003423

Kanta C, Kumar A, Chauhan A, Singh H, Sharma IP (2024) The interplay between plant functional traits and climate change. In: Kumar N, Singh H (eds) Plant functional traits for improving productivity. Springer, Singapore. https://doi.org/10.1007/978-981-97-1510-7_3

Kasim WA, Osman ME, Omar MN, Abd El-Daim IA, Bejai S, Meijer J (2013) Control of drought stress in wheat using plant-growth-promoting bacteria. J Plant Growth Regul 32(1):122–130. https://doi.org/10.1007/s00344-012-9283-7

Kumar N, Singh H (2024) Plant functional traits for improving productivity. Springer Nature, Singapore. https://doi.org/10.1007/978-981-97-1510-7

Kumar N, Suyal DC, Sharma IP, Verma A, Singh H (2017) Elucidating stress proteins in rice (*Oryza sativa* L.) genotype under elevated temperature: a proteomic approach to understand heat stress response. 3 Biotech 7:205. https://doi.org/10.1007/s13205-017-0856-9

Kumar M, Singh MP, Singh H, Dhakate PM, Ravindranath NH (2019a) Forest working plan for the sustainable management of forest and biodiversity in India. J Sustain for 39:1–22. https://doi.org/10.1080/10549811.2019.1632212

Kumar N, Jeena N, Singh H (2019b) Elevated temperature modulates rice pollen structure: a study from the foothill of the Himalayan agro-ecosystem in India. 3 Biotech 9(5):175. https://doi.org/10.1007/s13205-019-1700-1

Kumar N, Jeena N, Kumar N, Gangola S, Singh H (2019c) Phytoremediation facilitating enzymes: an enzymatic approach for enhancing remediation process. In: Bhatt P (ed) Smart bioremediation technologies. Academic Press, Elsevier, pp 289–306. 50 Hampshire Street, 5th Floor, Cambridge, MA 02139, United States

Kumar N, Singh H, Sharma SK (2020) Antioxidants: responses and importance in plant defense system. In: Roychowdhury R, Choudhury S, Hasanuzzaman M, Srivastava S (eds) Sustainable agriculture in the era of climate change. Springer, Cham, pp 251–264. https://doi.org/10.1007/978-3-030-45669-6_11

Kumar A, Kumar P, Singh H, Kumar N (2021a) Modulation of plant functional traits under essential plant nutrients during seasonal regime in natural forests of Garhwal Himalayas. Plant Soil 465:197–212. https://doi.org/10.1007/s11104-021-05003-x

Kumar A, Kumar P, Singh H, Bisht S, Kumar N (2021b) Relationship of physiological plant functional traits with soil carbon stock in the temperate forest of Garhwal Himalaya. Curr Sci 120(8):1368–1373. https://doi.org/10.18520/cs/v120/i8/1368-1373

Kumar A, Kumar P, Singh H, Kumar N (2021c) Adaptation and mitigation potential of roadside trees with bio-extraction of heavy metals under vehicular emissions and their impact on physiological traits during seasonal regimes. Urban for Urban Green 58:126900. https://doi.org/10.1016/j.ufug.2020.126900

Kumar A, Kumar P, Singh H, Kumar N (2021d) Impact of plant functional traits on infiltration rate under different forests of Kempty watershed in Garhwal Himalaya. Indian J Soil Cons 49(1):38–44

Liang W, Ma X, Wan P, Liu L (2018) Plant salt-tolerance mechanism: a review. Biochem Biophys Res Commun 495(1):286–291. https://doi.org/10.1016/j.bbrc.2017.11.043

Lim JH, Kim SD (2013) Induction of drought stress resistance by multi-functional PGPR *Bacillus licheniformis* K11 in pepper. Plant Pathol J 29(2):201–208. https://doi.org/10.5423/PPJ.SI.02.2013.0021

Loper (1986) Phyto76n04_386.PDF

Ma Y, Oliveira RS, Wu L, Luo Y, Rajkumar M, Rocha I, Freitas H (2015) Inoculation with metal-mobilizing plant-growth-promoting rhizobacterium bacillus sp. SC2b and its role in rhizoremediation. J Toxicol Environ Health Part A Curr Issues 78(13–14):931–944. https://doi.org/10.1080/15287394.2015.1051205

Malik A, Kumar K, Kumari G, Kumar N, Singh H (2024) Application of functional traits in modelling productivity and resilience under climate change. In: Kumar N, Singh H (eds) Plant functional traits for improving productivity. Springer, Singapore. https://doi.org/10.1007/978-981-97-1510-7_5

Mayak S, Tirosh, T, Glick BR (2004) Plant growth-promoting bacteria confer resistance in tomato plants to salt stress. Plant Physiol Biochem 42(6):565–572

Meena D, Singh H, Chaudhari SK (2011) Elucidating strontium response on growth dynamics and biochemical change in *Phaseolus mungo* L. Int J Agric Environ Biotechnol 4:107–113

Nabti E, Sahnoune M, Ghoul M, Fischer D, Hofmann A, Rothballer M, Schmid M, Hartmann A (2010) Restoration of growth of durum wheat (*Triticum durum* var. waha) under saline conditions due to inoculation with the rhizosphere bacterium *Azospirillum brasilense* NH and extracts of the marine alga Ulva lactuca. J Plant Growth Regul 29(1):6–22. https://doi.org/10.1007/s00344-009-9107-6

Nadgórska-Socha A, Kafel A, Kandziora-Ciupa M, Gospodarek J, Zawisza-Raszka A (2013) Accumulation of heavy metals and antioxidant responses in *Vicia faba* plants grown on monometallic contaminated soil. Environ Sci Pollut Res 20(2):1124–1134. https://doi.org/10.1007/s11356-012-1191-7

Nakbanpote W, Panitlurtumpai N, Sangdee A, Sakulpone N, Sirisom P, Pimthong A (2014) Salt-tolerant and plant growth-promoting bacteria isolated from Zn/Cd contaminated soil: identification and effect on rice under saline conditions. J Plant Interact 9(1):379–387. https://doi.org/10.1080/17429145.2013.842000

Naveed M, Mitter B, Reichenauer TG, Wieczorek K, Sessitsch A (2014) Increased drought stress resilience of maize through endophytic colonization by Burkholderia phytofirmans PsJN and Enterobacter sp. FD17. Environ Exp Bot 97:30–39. https://doi.org/10.1016/j.envexpbot.2013.09.014

Naveed M, Mustafa A, Majeed S, Naseem Z, Saeed Q (2020) Enhancing cadmium tolerance and pea plant health through Enterobacter sp. MN17 inoculation together with biochar and gravel sand

Nazarian A, Mousawi M (2005) Study of bacterial resistance to organophosphorous, vol 2, no 3, pp 207–211

Parida AK, Das AB (2005) Salt tolerance and salinity effects on plants: a review. Ecotoxicol Environ Saf 60(3):324–349. https://doi.org/10.1016/j.ecoenv.2004.06.010

Patel T, Saraf M (2017) Biosynthesis of phytohormones from novel rhizobacterial isolates and their in vitro plant growth-promoting efficacy. J Plant Interact 12(1):480–487. https://doi.org/10.1080/17429145.2017.1392625

Patten CL, Glick BR (2011) Inspection Non-destructive inspection with shearography. Proc SPIE Int Soc Opt Eng 2248(8):1–34. https://doi.org/10.1128/AEM.68.8.3795

Prapagdee B, Chanprasert M, Mongkolsuk S (2013) Chemosphere Bioaugmentation with cadmium-resistant plant growth-promoting rhizobacteria to assist cadmium phytoextraction by Helianthus annuus. Chemosphere 92(6):659–666. https://doi.org/10.1016/j.chemosphere.2013.01.082

Sarkar A, Ghosh PK, Pramanik K, Mitra S, Soren T, Pandey S, Mondal MH, Maiti TK (2018) A halotolerant Enterobacter sp. displaying ACC deaminase activity promotes rice seedling growth under salt stress. Res Microbiol 169(1):20–32. https://doi.org/10.1016/j.resmic.2017.08.005

Sharma R, Singh H, Kaushik M, Nautiyal R, Singh O (2018) Adaptive physiological response, carbon partitioning, and biomass production of *Withania somnifera* (L.) Dunal grown under elevated CO_2 regimes. 3 Biotech 8:267. https://doi.org/10.1007/s13205-018-1292-1

Singh H, Kumar M (2022) Climate change and its impact on Indian Himalayan forests: current status and research needs. In: Rani S, Kumar R (eds) Climate change. Springer Climate. Springer, Cham, pp 223–242. https://doi.org/10.1007/978-3-030-92782-0_11

Singh H, Verma A (2013) Characterization and screening of high nitrogen efficient rice genotype to curtail environmental pollution. Int J Agric Environ Biotechnol 6(1):23–30

Singh AK, Hamel C, DePauw RM, Knox RE (2012) Genetic variability in arbuscular mycorrhizal fungi compatibility supports the selection of durum wheat genotypes for enhancing soil ecological services and cropping systems in Canada. Can J Microbiol 58(3):293–302. https://doi.org/10.1139/W11-140

Singh H, Savita Sharma R, Sinha S, Kumar M, Kumar P, Verma A, Sharma SK (2017a) Physiological functioning of *Lagerstroemia speciosa* L. under heavy roadside traffic: an approach to screen potential species for abatement of urban air pollution. 3 Biotech 7(61):1–10. https://doi.org/10.1007/s13205-017-0690-0

Singh H, Verma A, Kumar M, Sharma R, Gupta R, Kaur M, Negi M, Sharma SK (2017b) Phytoremediation: a green technology to clean up the sites with low and moderate level of heavy metals. Austin Biochem 2:1012

Singh H, Sharma R, Savita SMP, Kumar M, Verma A, Ansari MW, Sharma SK (2018) Adaptive physiological response of *Parthenium hysterophorus* to elevated atmospheric CO_2 concentration. Indian Forester 144:1–14

Singh H, Yadav M, Kumar N, Kumar A and Kumar M (2020). Assessing adaptation and mitigation potential of roadside trees under the influence of vehicular emissions: A case study of Grevillea robusta and Mangifera indica planted in an urban city of India. PLoS ONE 15(1):e0227380. https://doi.org/10.1371/journal.pone.0227380

Singh H, Kumar N, Kumar A (2024) Enhancing resource use efficiency in crops through plant functional traits. In: Kumar N, Singh H (eds) Plant functional traits for improving productivity. Springer, Singapore. https://doi.org/10.1007/978-981-97-1510-7_6

Suárez R, Wong A, Ramírez M, Barraza A, Orozco MDC, Cevallos MA, Lara M, Hernández G, Iturriaga G (2008) Improvement of drought tolerance and grain yield in common bean by overexpressing trehalose-6-phosphate synthase in rhizobia. Mol Plant Microbe Interact 21(7):958–966. https://doi.org/10.1094/MPMI-21-7-0958

Swapnil IG, Anubha S, Sharad S (2016) Alleviation of mercury toxicity in wheat by the interaction of mercury-tolerant plant growth-promoting Rhizobacteria. J Plant Growth Regul 35(4):1000–1012. https://doi.org/10.1007/s00344-016-9598-x

Tank N, Saraf M (2010) Salinity-resistant plant growth promoting rhizobacteria ameliorates sodium chloride stress on tomato plants. J Plant Interact 9145. https://doi.org/10.1080/17429140903125848

Tewari S, Arora NK (2016) Fluorescent pseudomonas sp . PF17 as an efficient plant growth regulator and biocontrol agent for sunflower crop under saline conditions. Symbiosis 68:99–108. https://doi.org/10.1007/s13199-016-0389-8

Vivas A, Biró B, Németh T, Barea JM, Azcón R (2006) Nickel-tolerant Brevibacillus brevis and arbuscular mycorrhizal fungus can reduce metal acquisition and nickel toxicity effects in plant growing in nickel supplemented soil. Soil Biol Biochem 38(9):2694–2704

Vurukonda SSKP, Vardharajula S, Shrivastava M, SkZ A (2016) Enhancement of drought stress tolerance in crops by plant growth promoting rhizobacteria. In: Microbiological research, vol 184. Elsevier GmbH. https://doi.org/10.1016/j.micres.2015.12.003

Yadav SK, Singh H, Ginwal HS, Barthwal S (2019a) Elevated CO_2 enhanced growth and physiological process of *Populus deltoides* Bartr. ex Marsh. Indian Forester 145(1):23–27

Yadav SK, Singh H, Nautiyal R, Ginwal HS, Ansari SA, Barthwal S (2019b) Modulation of morpho-physiological responses in *Populus deltoides* by elevated carbon dioxide and temperature. For Sci 66:105–118. https://doi.org/10.1093/forsci/fxz048

Yang CF, Lee CM (2008) Enrichment, isolation, and characterization of 4-chlorophenol-degrading bacterium Rhizobium sp. 4-CP-20. Biodegradation 19(3):329–336. https://doi.org/10.1007/s10532-007-9139-1

Yildrim E, Donmez MF, Turan M (2008) Use of bioinoculants in ameliorative effects on radish plants under salinity stress use of bioinoculants in ameliorative effects. J Plant Nutr 4167. https://doi.org/10.1080/01904160802446150

Zhang H, Kim MS, Sun Y, Dowd SE, Shi H, Paré PW (2008) Soil bacteria confer plant salt tolerance by tissue-specific regulation of the sodium transporter HKT1. Mol Plant Microbe Interact 21(6):737–744. https://doi.org/10.1094/MPMI-21-6-0737

Zhao Y, Christensen SK, Fankhauser C, Cashman JR, Cohen JD, Weigel D, Chory J (2001) A role for flavin monooxygenase-like enzymes in auxin biosynthesis. Science 291(5502):306–309. https://doi.org/10.1126/science.291.5502.306

Zulfikar Ali S, Sandhya V, Grover M, Linga VR, Bandi V (2011) Effect of inoculation with a thermotolerant plant growth promoting *Pseudomonas putida* strain AKMP7 on growth of wheat (Triticum spp.) under heat stress. J Plant Interact 6(4):239–246. https://doi.org/10.1080/17429145.2010.545147

Enhancing Urban Sustainability Through Heavy Metal Extraction: Mechanisms, Challenges, and Environmental Impact

Manpreet Kaur, Nemit Verma, and Hukum Singh

Abstract Urban sustainability is paramount for developing cities to demonstrate environmental stewardship, social equity, economic robustness, and resilience in emerging challenges. Rapid urbanization, industrial expansion, and intensified agricultural practices driven by growing populations have significantly escalated the risk of heavy metal contamination, posing grave environmental and public health threats. The unchecked discharge of toxic heavy metals like lead, mercury, cadmium, and arsenic imperils urban sustainability by their pervasive presence. These metals can induce toxicity in humans and wildlife, even at minimal concentrations. Phytoremediation emerges as a promising solution to counteract this menace. This eco-friendly approach harnesses the natural capabilities of plants to extract and accumulate heavy metals from their surroundings. In contrast to conventional methods, which often exacerbate environmental degradation, phytoremediation offers a sustainable alternative by leveraging plants' inherent mechanisms for metal extraction. This chapter delves into the intricate processes through which plants facilitate the remediation of heavy metal contamination in urban environments.

Keywords Urban sustainability · Urban forests · Heavy metal extraction · Phytoremediation · Environmental impact

1 Introduction

Urban sustainability refers to a city or region's ability to maintain and improve residents' quality of life while reducing environmental impact, conserving resources for future generations, and fostering environmental, economic, and social sustainability.

M. Kaur (✉) · H. Singh
ICFRE-Forest Research Institute, P. O. New Forest, Dehradun, Uttarakhand 248006, India
e-mail: manpreetchinna@gmail.com

N. Verma
Advisor under project 'Restore, Conserve, and Protect Forest and Tree Cover for NDC Implementation in India (RECAP4NDC)' Project, Deutsche Gesellschaft Für Internationale Zusammenarbeit (GIZ) GmbH, New Delhi, India

H. Singh (ed.), *Urban Forests, Climate Change and Environmental Pollution*,
https://doi.org/10.1007/978-3-031-67837-0_13

Heavy metal pollution presents a significant challenge to urban sustainability, as these contaminants are well-known for their toxic properties, long-lasting presence in the environment, and ability to accumulate in organisms through bioaccumulation (Mitra et al. 2022). While certain heavy metals like copper, zinc, and nickel are essential for human health (Engwa et al. 2019), most contribute to environmental and atmospheric pollution, posing risks to human health. Interactions with environmental elements such as water, soil, and air can exacerbate their toxicity, leading to exposure through the food chain (Mitra et al. 2022).

Heavy metals' persistence and non-biodegradable nature prolong their environmental presence, increasing health risks (Ali et al. 2020; Alengebawy et al. 2021). Prolonged exposure to heavy metals through air, water, soil, and food can result in various diseases, including neurological disorders, cancer, hypertension, myocardial infarction, skin ailments, and complications in urinary, reproductive, and respiratory systems (Li et al. 2019).

Removing heavy metals from the environment poses significant challenges due to their resistance to degradation, akin to other biological and chemical pollutants. Various technologies are employed for on-site (in situ) and off-site (ex-situ) remediation of heavy metal-contaminated soil, including chemical reduction, electrophoresis, excavation, pneumatic fracturing, soil washing, flushing, solidification, and nitrification (Dada et al. 2015). However, these conventional methods, often called "pump and treat" and "dig and dump," have limitations, particularly concerning small areas. Challenges associated with traditional approaches include incomplete removal, high energy requirements, significant production of toxic sludge, limited applicability to small areas, and high costs (Li et al. 2019; Zamora-Ledezma et al. 2021).

In recent years, there has been growing interest in an emerging, economically viable, and environmentally friendly alternative to traditional remediation practices known as "phytoremediation." This method leverages the natural capabilities of plants to cleanse the environment by extracting, accumulating, and detoxifying contaminants from soil, air, and water through physical, chemical, or biological mechanisms (Sharma et al. 2023). Several factors, including soil and plant characteristics, influence the effectiveness of phytoremediation, such as soil chemistry and physics, exudates released by plants and microbes, metal availability, and plant capacity for uptake, accumulation, translocation, sequestration, and detoxification (Wang et al. 2020). Phytoremediation encompasses various processes, including phytostabilization (reducing metal bioavailability), phytoextraction (extracting and accumulating), phytovolatilization (extracting and releasing as volatile compounds), and phytofiltration (using aquatic plants).

2 Urban Sustainability

The global ecological equation often neglects contemporary industrialized societies' over-exploitation of natural resources (Kumar et al. 2024). The rapid pace of urbanization posed numerous risks, such as population overload, resource and space constraints, and inequality, highlighting the potential dangers of urban living and posing challenges to achieving sustainable and desirable lifestyles (Wang et al. 2023). This gives rise to a need to plan a sustainable city to protect the environment and the social, economic, and physical well-being (Roy et al. 2023).

Consequently, urban sustainability emerged as a central focus in sustainable development. The United Nations introduced the Sustainable Development Goals (SDGs) 2015 to confront global challenges. These goals comprise 17 overarching goals and 169 sub-goals intending to steer the 2030 Global Framework for Sustainable Development. Urban populations constitute a significant portion of the global populace, so cities are pivotal in advancing the SDGs agenda. Hence, SDG11 (sustainable cities and communities) stands out as directly tied to fostering the sustainable growth of urban areas.

Moreover, numerous other SDGs bear relevance to urban contexts, such as SDG6 (clean water and sanitation), SDG7 (affordable and clean energy), SDG12 (sustainable consumption and production), and SDG13 (climate action), among others. These objectives are interconnected with the previously outlined urban risks, emphasizing the importance of considering all 17 SDGs in examining sustainable urban challenges. The environment is one of the critical factors for sustainable urban development (Wang et al. 2023).

Urban sustainability encompasses the principles and strategies within urban planning aimed at constructing and enhancing cities without causing permanent depletion of resources. The concept revolves around three core pillars: social sustainability, which promotes social well-being and health; economic sustainability, emphasizing efficient resource usage and economic viability; and environmental sustainability, ensuring resource utilization while minimizing environmental impacts (Roy et al. 2023; Sharma et al. 2018; Singh et al. 2018).

Environmental sustainability involves engaging with the environment responsibly to mitigate the degradation of natural resources (Kumar et al. 2019a). It preserves the environment, intersects with public health, and maintains long-term environmental quality (Thiel et al. 2015; Malik et al. 2024). Practicing environmental sustainability entails meeting current societal needs without compromising the rights of future generations. It also involves assessing the potential future impacts of human activities and implementing strategies for their sustainable management (Elleuch et al. 2018; Kumar et al. 2024; Sherman et al. 2016; World Commission on Environment and Development 1987).

3 Sources of Heavy Metal Pollution in the Environment

Heavy metal pollution includes both natural and human-induced sources, with the latter contributing significantly to the escalation of metallic substances in the environment. Various anthropogenic activities such as metal mining, smelting, foundries, and metal-based industries, as well as the leaching of metals from multiple sources like landfills, waste dumps, excretions, and agricultural practices such as the use of pesticides and fertilizers, constitute primary contributors. Additionally, natural phenomena like volcanic activity, metal corrosion, and soil erosion also contribute to heavy metal pollution in the environment (He et al. 2005; Tchounwou et al. 2012; Singh 2024; Gautam et al. 2016; Masindi and Muedi 2018; Bhat et al. 2022). The dissemination of pollutants within ecosystems is influenced by factors such as temperature, surface water movements, air mass circulation, wind speed, partition coefficient, polarity, vapour pressure, and molecular stability (Walker et al. 2012; Gupta et al. 2018a, b).

Heavy metals, being non-degradable, persist in the soil for extended periods as they cannot undergo microbial or chemical degradation. This leads to the integration of heavy metals into the food chain, posing risks to the ecosystem. Heavy metals also hinder the biodegradation of organic pollutants, exacerbating environmental pollution (Gupta et al. 2019a, b). These metals in the soil pose risks to the biosphere by being ingested directly or absorbed by plants, which can be hazardous to both the plant and the organisms in the food chain that consume it. Additionally, heavy metals alter soil properties such as pH, color, porosity, and natural chemistry, impacting soil quality and contaminating water sources (Muchuweti et al. 2006; Gupta et al. 2012; Singh et al. 2024a; Masindi and Muedi 2018).

4 Heavy Metal Pollution, a Threat to Environmental Sustainability

Cities, however, comprising a mere ~5% of the global land surface, significantly influence the entire biosphere (Angeoletto et al. 2015). Anthropogenic activities within urban areas present challenges that inadvertently impact biodiversity, ecosystem functioning, and environmental quality, subsequently leading to adverse effects on human health and well-being.

Urban areas consist of natural and built environments characterized by a higher density of human population and intricate interplay between socioeconomic elements and biological phenomena (MEA 2005; Schaefer 2011; Kumar et al. 2019c). The advancement of urban development poses risks to human health and specific aspects of biodiversity, primarily driven by factors such as climate change, notably the urban heat island effect, environmental pollution, and habitat fragmentation (Kumar et al. 2022; Mollashahi and Szymura 2022). Soil, air, water, and stream sediments are vital components of the environment that undergo alteration due to the uncontrolled release of potentially hazardous metals (Chibuike and Obiora 2014; Xie et al. 2016; Singh et al. 2022; Kumar et al. 2020). These metals enter the soil and remain contained without being exposed or washed away, as the soil can retain them (Zhang et al. 2010). Subsequently, these contained metals can leach into groundwater from the soil, posing risks to public health by cultivating vegetables that easily bioaccumulate these metals and become a significant environmental threat along the food chain (Ji et al. 2018).

5 Mechanism of Heavy Metal Extraction from the Environment

Phytoremediation, a novel technique employing plants to clean up contaminated environments, garners attention among government and industrial professionals, including environmental experts, scientists, and remediation engineers (Bhargava et al. 2012). It is considered cost-effective and environmentally friendly, as in-situ remediation using plants incurs lower installation and maintenance costs than other current technologies, with the added benefit of detoxifying hazardous waste and restoring sites (Van Aken 2009; Bhargava et al. 2012). Plants aid in conserving soil functions without disturbing topsoil, and organic matter input may also enhance soil fertility (Mench et al. 2009; Yadav et al. 2019a, b).

Different types of phytoremediation technologies include phytoextraction, phytostabilization, phytovolatilization, phytofiltration, and phytodegradation (Marques et al. 2009a, b; Fulekar et al. 2009).

5.1 Phytoextraction

Plants play a vital role in remediation efforts by absorbing contaminants from soil, water, or sediments through their roots, a process known as phytoextraction (Sarwar et al. 2017; Yanitch et al. 2020). Specifically, they translocate and accumulate these contaminants in their aboveground, harvestable parts. This process is pivotal for reducing soil pollutants, especially in polluted sites where plants capable of accumulating heavy metals are cultivated. Notably, transferring metals to aboveground biomass, particularly shoots, is a critical physiological process, given the ease of collection compared to roots.

Phytoextraction is economically viable and is the most effective approach for removing heavy metals and metalloids from disturbed soils. However, its efficacy depends on heavy metal bioavailability, soil characteristics, metal speciation, and the plant's capacity to absorb and accumulate metals in aboveground components (Yan et al. 2020; Singh et al. 2020).

The suitability of a plant for phytoextraction hinges on its ability to produce substantial biomass, actively accumulate heavy metals in easily harvestable parts, and tolerate toxic metals (Vangronsveld et al. 2009; Suman et al. 2018). Critical determinants of phytoextraction potential include shoot metal concentration and biomass (Li et al. 2010). Hyperaccumulators, which can accumulate significant metal amounts, are particularly valuable, with approximately 450–500 plant species identified as such (Chaudhary et al. 2018).

Certain plant species, like Trifolium spp., capable of multiple harvesting, exhibit considerable potential for metallic substance phytoremediation. Grasses are favored over trees and shrubs due to their rapid growth, high biomass production, and stress tolerance, and they are particularly desirable (Malik et al. 2010).

The core strategy of phytoextraction involves cultivating suitable plant species in situ, harvesting metal-containing biomass, and processing it to reduce mass and size. Techniques such as composting, compressing, dehydrating, and thermal decomposition are employed. The resulting metal-enriched biomass, containing high contaminants, can be utilized for trace element re-extraction or carefully disposed of as hazardous waste if economically feasible (Sheoran et al. 2009; Suman et al. 2018). Consequently, phytoextraction emerges as the most promising commercial application for remediating soils or water contaminated with heavy metals and metalloids (Milic et al. 2012; Sun et al. 2011).

5.2 *Phytofilteration*

Phytofiltration refers to the method through which plants eliminate pollutants from contaminated surfaces or wastewater via absorption or adsorption. This process, which includes rhizofiltration, caulofiltration, or blastofiltration, employs plant roots, shoots, or seedlings to purify contaminated water sources or surfaces (Mesjasz-Przybyłowicz et al. 2004).

In rhizofiltration, heavy metals are either adsorbed onto the root surface or absorbed by the roots themselves. The pH of the rhizosphere can be altered by root exudates, leading to the precipitation of heavy metals on plant roots (Javed et al. 2019). This mechanism minimizes the migration of heavy metals into groundwater. Trees with extensive root systems, particularly those with long roots, are pivotal in absorbing significant amounts of water. Acting as natural pumps, these trees draw up large volumes of water from the underground water table (Muthusaravanan et al. 2018), consequently bringing contaminants along with the water. Root exudates like citric acid and malic acid aid in scavenging or enhancing pollutants' absorption, adsorption, or sedimentation (Banerjee and Roychoudhury 2022; Singh et al. 2017b).

For rhizofiltration, plants are grown hydroponically in clean water to develop robust root systems. Subsequently, the clean water is replaced with polluted water to acclimate the plants. Once acclimated, the plants are moved to the contaminated site for heavy metal removal. Once the roots become saturated, they are harvested and disposed of (Wuana and Okieimen 2011).

Ideal plants selected for rhizofiltration should possess dense root systems, demonstrate high biomass production, and exhibit tolerance to heavy metals. Both terrestrial and aquatic plants can be utilized for rhizofiltration. In remediating wetland water, aquatic species like hyacinth, azolla, duckweed, cattail, and poplar are commonly employed due to their high accumulation of heavy metals, tolerance, rapid growth, and high biomass production (Hooda 2007). Terrestrial plants such as sunflower (Helianthus annuus) possess long, hairy root systems and significant heavy metal accumulation capacity during rhizofiltration (Tomé et al. 2008; Dhanwal et al. 2017).

5.3 *Phytovolatilization*

Plants can remarkably uptake contaminants from soil, transforming them into less harmful volatile compounds, which are then released into the atmosphere through phytovolatilization (Yan et al. 2020). This natural mechanism proves particularly effective in detoxifying organic pollutants and certain heavy metals like selenium (Se), mercury (Hg), and arsenic (As) (Mahar et al. 2016).

One of the key advantages of phytovolatilization lies in its ability to remove heavy metal and metalloid contaminants from the site, dispersing them as gaseous compounds without the necessity for plant harvesting and disposal. However, it's crucial to recognize that while phytovolatilization offers a remedial strategy, it does not entirely eradicate pollutants. Instead, it transfers them from the soil to the atmosphere, where volatile compounds may contaminate the surrounding air. Furthermore, these pollutants may return to the soil through precipitation (Vangronsveld et al. 2009; Jangra et al. 2024).

Given these factors, a thorough risk assessment becomes imperative before considering the implementation of phytovolatilization in the field. This evaluation should encompass the immediate impact on air quality and the potential for secondary contamination of soil and water sources.

5.4 *Phytostabilization*

Certain plant species tolerant to metals can sequester heavy metals underground, reducing their bioavailability and preventing their migration into the ecosystem and potential entry into the food chain, a phenomenon known as phytostabilization (Marques et al. 2009a, b). Phytostabilization operates through several mechanisms, including the precipitation of heavy metals or the reduction of metal valence in the rhizosphere, absorption, and sequestration within root tissues, or adsorption onto root cell walls (Ginn et al. 2008; Gerhardt et al. 2017).

Vegetation growth is crucial in maintaining soil health in heavy metal-polluted areas. Established vegetation cover stabilizes heavy metals underground, reduces leaching into groundwater, and prevents the dispersion of heavy metal-containing soil particles by wind (Vangronsveld et al. 2009; Mench et al. 2010). A notable advantage of phytostabilization is its elimination of the need to dispose of hazardous biomass, as required in phytoextraction (Wuana and Okieimen 2011).

For phytostabilization to be effective, plants must tolerate heavy metal conditions and possess dense rooting systems capable of immobilizing heavy metals, stabilizing soil structure, and preventing erosion. Additionally, they should demonstrate rapid biomass production and facilitate the establishment of vegetation cover. Furthermore, the plant cover should be easily maintained under field conditions (Marques et al. 2009a, b).

Interestingly, microorganisms in the rhizosphere, such as bacteria and mycorrhiza, can contribute to phytostabilization. These microorganisms enhance heavy metal immobilization by adsorbing metals onto their cell walls, producing chelators, and promoting precipitation processes (Göhre and Paszkowski 2006; Mastretta et al. 2009; Ma et al. 2011). Moreover, they increase plant roots' surface area and depth, facilitating phytostabilization, and act as a filtration barrier against the translocation of heavy metal ions from roots to shoots (Göhre and Paszkowski 2006; Singh et al. 2024b).

6 Challenges Associated with Heavy Metal Extraction from Urban Environment

Extracting heavy metals from urban environments poses several challenges owing to the intricate nature of these settings and the unique properties of heavy metals. Here, we outline several critical challenges linked to the extraction of heavy metals from urban environments:

6.1 Complexity of Urban Environments

Urban environments are highly complex, with diverse sources of heavy metal contamination, such as industrial activities (Li et al. 2022; Singh et al. 2021), traffic emissions (Suryawanshi et al. 2016; Kumar et al. 2022; Prakash et al. 2022), waste disposal sites, and historical pollution. Identifying the categories of pollution sources is challenging (Tan et al. 2021; Singh et al. 2022).

6.1.1 Variability and Heterogeneity

Heavy metal contamination in urban environments often exhibits spatial and temporal variability (Huang et al. 2023; Kumar et al. 2021b, c), making it difficult to assess and remediate accurately. The distribution of heavy metals can vary widely within a relatively small area due to factors such as land use, soil characteristics, and proximity to pollution sources (Gupta and Singh 2017; Kumar et al. 2017).

6.1.2 Technological Limitations

Refining and augmenting traditional extraction techniques becomes imperative when addressing the extraction of heavy metals from urban soils. Conventional methods often fall short due to hurdles such as incomplete metal removal, excessive energy consumption, generation of hazardous sludge, restricted applicability to localized areas, and substantial expenses (Li et al. 2019; Zamora-Ledezma et al. 2021; Kumar et al. 2021a). Thus, a pressing concern emerges the demand for discerning or innovative extraction approaches that are both potent and economical and capable of precisely targeting particular heavy metals without exacerbating environmental degradation (Singh and Verma 2013; Meena et al. 2011).

6.1.3 Risk of Mobilization and Redistribution

Extracting heavy metals from urban soils risks mobilizing and redistributing contaminants, potentially exacerbating environmental pollution and posing risks to human health. Careful management is required to avoid the leaching and mobilization of heavy metals during remediation to prevent further contamination (Kapahi and Sachdeva 2019).

6.1.4 Cost and Resource Intensity

Heavy metal extraction from urban environments using traditional techniques can be costly and not feasible, as it is resource-intensive and requires advanced equipment, specialized expertise, and significant time and effort (Masindi and Muedi 2018). Economic constraints may limit the scale and scope of remediation efforts, particularly in low-resource settings.

6.1.5 Regulatory and Legal Considerations

Compliance with regulatory requirements and navigating legal frameworks can present challenges for heavy metal extraction projects in urban environments. Remediation activities must adhere to environmental regulations and may require permits or approvals from relevant authorities.

6.1.6 Community Engagement and Stakeholder Management

Urban heavy metal extraction projects often involve multiple stakeholders, including local communities, government agencies, industry, and environmental organizations. Effective communication and engagement with stakeholders are essential to address concerns, build trust, and ensure the success of remediation efforts.

6.1.7 Long-Term Sustainability

Achieving long-term sustainability in heavy metal extraction projects requires considering the broader environmental and social implications, such as potential impacts on ecosystems, land use, and socioeconomic dynamics. Implementing measures to monitor and mitigate future contamination risks is critical for sustainable urban development (Dhyani et al. 2021; Kanta et al. 2024).

6.2 *Environmental Impact*

The extraction of heavy metals from soil can significantly impact urban sustainability. Heavy metals often accumulate in soil due to various human activities such as industrial processes, mining, and improper waste disposal. When these metals are extracted from the soil through remediation efforts, several environmental concerns arise;

6.2.1 Soil Degradation

The process of extracting heavy metals can further degrade soil quality, making it less fertile and unable to support plant growth. This degradation can have cascading effects on urban ecosystems, including biodiversity loss and reduced soil stability (Kumar et al. 2021d).

6.2.2 Water Contamination

Heavy metals extracted from soil can leach into groundwater or surface water, leading to contamination of water sources. This contamination can threaten aquatic ecosystems and human health if the contaminated water is used for drinking or irrigation (Kumar et al. 2021e).

6.2.3 Air Pollution

During extraction, dust and particulate matter containing heavy metals can be released into the air, contributing to air pollution. Inhalation of these particles can lead to respiratory and other health problems for nearby residents (Singh et al. 2017a, 2021; Sharma and Singh 2021).

6.2.4 Habitat Destruction

Remediation activities, such as excavation or soil removal, can disrupt natural habitats and ecosystems in urban areas. This destruction can negatively impact wildlife populations and reduce overall ecological resilience.

6.2.5 Energy Consumption

Heavy metal extraction processes often require significant energy inputs, contributing to greenhouse gas emissions and exacerbating climate change. High energy consumption can also undermine the overall sustainability of remediation efforts.

7 Conclusion

The chapter concludes that extracting heavy metals from soil is a pivotal endeavor within urban sustainability initiatives, aiming to counteract the environmental and health hazards pervasive in contaminated urban landscapes. While imperative for enhancing soil quality and safeguarding human well-being, it is paramount to acknowledge the potential ecological repercussions of such extraction processes. Efforts towards heavy metal extraction necessitate meticulous planning and execution to mitigate adverse impacts on urban ecosystems and overall sustainability. Embracing sustainable remediation practices, including adopting eco-friendly techniques and reducing energy consumption, is a pivotal strategy in curbing the environmental footprint of extraction activities. Furthermore, effective communication and community involvement emerge as linchpins for ensuring the triumph of remediation endeavors. Engaging local communities in decision-making processes and furnishing them with comprehensive information regarding the risks and advantages of remediation engenders support and collaboration, thus fostering more sustainable outcomes. In essence, while heavy metal extraction poses formidable challenges to urban sustainability, it concurrently offers avenues for enhancing environmental integrity, safeguarding public health, and cultivating resilient urban habitats when approached judiciously and responsibly.

References

Alengebawy A, Abdelkhalek ST, Qureshi SR, Wang M-Q (2021) Heavy metals and pesticides toxicity in agricultural soil and plants: ecological risks and human health implications. Toxics 9:42. https://doi.org/10.3390/toxics9030042

Ali S, Awan Z, Mumtaz S, Shakir HA, Ahmad F, Ulhaq M, Tahir HM, Awan MS, Sharif S, Irfan M, Khan MA (2020) Cardiac toxicity of heavy metals (cadmium and mercury) and pharmacological intervention by vitamin C in rabbits. Environ Sci Pollut Res 27(23):29266–29279. https://doi.org/10.1007/s11356-020-09011-9

Angeoletto F, Essy C, Ruiz Sanz JP, Silva FFD, Albertin RM, Santos JWMC (2015) Ecología Urbana: La CienciaInterdisciplinaria del Planeta Ciudad. DQuestão 13(32):6. https://doi.org/10.21527/2237-6453.2015.32.6-20

Banerjee A, Roychoudhury A (2022) Assessing the rhizofiltration potential of three aquatic plants exposed to fluoride and multiple heavy metal polluted water. Vegetos 35:1158–1164. https://doi.org/10.1007/s42535-022-00405-3

Bhargava A, Carmona FF, Bhargava M, Srivastava S (2012) Approaches for enhanced phytoextraction of heavy metals. J Environ Manag 105:103–120

Bhat SA, Bashir O, Ul Haq SA, Amin T, Rafiq A, Ali M, Américo-Pinheiro JHP, Sher F (2022) Phytoremediation of heavy metals in soil and water: an eco-friendly, sustainable and multidisciplinary approach. Chemosphere 303:134788

Chaudhary K, Agarwal S, Khan S (2018) Role of phytochelatins (PCs), metallothioneins (MTs), and heavy metal ATPase (HMA) genes in heavy metal tolerance. In: Mycoremediation and environmental sustainability. Springer, Cham, pp 39–60. https://doi.org/10.1007/978-3-319-77386-5_2

Chibuike GU, Obiora SC (2014) Heavy metal polluted soils: effect on plants and bioremediation methods. Review article. Appl Environ Soil Sci 12(Article ID 752708)

Dada EO, Njoku KI, Osuntoki AA, Akinola MO (2015) A review of current techniques of physico-chemical and biological remediation of heavy metals polluted soil. Ethiopian J Environ Stud Manage 8:606–615. https://doi.org/10.4314/ejesm.v8i5.13

Dhanwal P, Kumar A, Dudeja S, Chhokar V, Beniwal V (2017) Recent advances in phytoremediation technology. In: Kumar R, Sharma AK, Ahluwalia SS (eds) Advances in environmental biotechnology. Springer, Singapore, pp 227–241. https://doi.org/10.1007/978-981-10-4041-2_14

Dhyani K, Kuniyal HB, Singh H, Sobha (2021) Growth and physiological potential of *Terminalia arjuna* under elevated CO_2 levels in Open top chamber condition. J Appl Nat Sci 13(3):1121–1126. https://doi.org/10.31018/jans.v13i3.2490. ISSN 2231-5209

Elleuch B, Bouhamed F, Elloussaief M, Jaghbir M (2018) Environmental sustainability and pollution prevention. Environ Sci Pollut Res 25:18223–18225

Engwa GA, Ferdinand PU, Nwalo FN, Unachukwu MN (2019) Metal toxicity in human. In Mechanism and health effects of heavy metal toxicity in humans. In: Karcioglu O, Arslan B (eds) Poisoning modern world—new tricks an old dog? https://doi.org/10.5772/intechopen.82511

Fulekar MH, Singh A, Bhaduri AM (2009) Genetic engineering strategies for enhancing phytoremediation of heavy metals. Afric J Biotech 8:529–535

Gautam PK, Gautam RK, Chattopadhyaya MC, Banerjee S, Chattopadhyaya MC, Pandey JD (2016) Heavy metals in the environment: fate, transport, toxicity and remediation technologies. In: Thermodynamic profiling of pollutants view project materials for solid oxide fuel cells view project heavy metals in the environment: fate, transport, toxicity and rem.

Gerhardt KE, Gerwing PD, Greenberg BM (2017) Opinion: taking phytoremediation from proven technology to accepted practice. Plant Sci 256:170–185. https://doi.org/10.1016/j.plantsci.2016.11.016

Ginn BR, Szymanowski JS, Fein JB (2008) Metal and proton binding onto the roots of *Fescue rubra*. Chem Geol 253:130–135. https://doi.org/10.1016/j.chemgeo.2008.05.001

Gohre V, Paszkowski U (2006) Contribution of the arbuscular mycorrhizal symbiosis to heavy metal phytoremediation. Planta 223:1115–1122. https://doi.org/10.1007/s00425-006-0225-0

Gupta SK, Singh H (2017) Observations on Sporadic Flowering in *Ficus benjamina* L. and *Peltophorum africanum* L. in New Forest. Indian Forester 143:290–291

Gupta N, Khan DK, Santra SC (2012) Heavy metal accumulation in vegetables grown in a long-term wastewater-irrigated agricultural land of tropical India. Environ Monit Assess 184:6673–6682

Gupta SK, Ram J, Singh H (2018a) Comparative study of transpiration in cooling effect of tree species in the atmosphere. J Geosci Environ Prot 6:151–166. https://doi.org/10.4236/gep.2018.68011

Gupta SK, Ram J, Singh H, Kumar P (2018b) Comparative study of cooling under shades of some forestry tree species with respect to ambient temperatures. J Geosci Environ Prot 6:51–66. https://doi.org/10.4236/gep.2018.611004

Gupta SK, Ram J, Singh H (2019a) Relationship between leaf area index and atmospheric cooling of tree species. Indian J for 42(3):203–208

Gupta SK, Singh H, Ram J (2019b) Need for a novel paradigm in forestry research: perspectives of comparative atmospheric cooling by different forestry effect tree species. Indian Forester 145(6):509–515

He ZL, Yang XE, Stoffella PJ (2005) Trace elements in agroecosystems and impacts on the environment. J Trace Elem Med Biol 19:125–140

Hooda V (2007) Phytoremediation of toxic metals from soil and waste water. J Environ Biol 28(2 Suppl.):367–376

Huang C, Gou Z, Huang S, Ma X, Shi Z, Liao G, Deng O, Luo L, Zhou W, Cheng Z, Zhang Y, Yang Y, Fang C, Chen C (2023) Analysis of the spatial and temporal characteristics and sources of heavy metal pollution in arable soils in rapidly urbanizing cities in western China. https://doi.org/10.22541/au.168424806.63401240/v1 (Preprint)

Jangra A, Kumar K, Maikhuri S, Bhandari MS, Pandey S, Singh H, Barthwal S (2024) Unveiling stress-adapted endophytic bacteria: characterizing plant growth-promoting traits and assessing cross-inoculation effects on *Populus deltoides* under abiotic stress. Plant Physiol Biochem 210:108610. https://doi.org/10.1016/j.plaphy.2024.108610

Javed MT, Tanwir K, Akram MS, Shahid M, Niazi NK, Lindberg S (2019) Phytoremediation of cadmium-polluted water/sediment by aquatic macrophytes: role of plant-induced pH changes. In: Hasanuzzaman M, Prasad MNV, Fujita M (eds) Cadmium Toxicity and Tolerance in Plants. Academic Press, London, pp 495–529. https://doi.org/10.1016/B978-0-12-814864-8.00020-6

Ji Y, Wu P, Zhang J, Zhang J, Zhou Y, Peng Y, Zhang S, Chai G, Gao G (2018) Heavy metal accumulation, risk assessment and integrated biomarker responses of local vegetables: a case study along the Le'an river. Chemosphere 199:361–367

Kanta C, Kumar A, Chauhan A, Singh H, Sharma, IP (2024) The interplay between plant functional traits and climate change. In: Kumar N, Singh H (eds) Plant functional traits for improving productivity. Springer, Singapore. https://doi.org/10.1007/978-981-97-1510-7_3.

Kapahi M, Sachdeva S (2019) Bioremediation options for heavy metal pollution. J Health Pollut 9(24):191203. https://doi.org/10.5696/2156-9614-9.24.191203

Kumar N, Suyal DC, Sharma IP, Verma A, Singh H (2017) Elucidating stress proteins in rice (*Oryza sativa* L.) genotype under elevated temperature: a proteomic approach to understand heat stress response. 3 Biotech 7:205. https://doi.org/10.1007/s13205-017-0856-9

Kumar M, Singh MP, Singh H, Dhakate PM, Ravindranath NH (2019a) Forest working plan for the sustainable management of forest and biodiversity in India. J Sustain for 39:1–22. https://doi.org/10.1080/10549811.2019.1632212

Kumar N, Jeena N, Kumar N, Gangola S, Singh H (2019c) Phytoremediation facilitating enzymes: an enzymatic approach for enhancing remediation process. In: Bhatt P (ed) Smart bioremediation technologies. Academic Press, Elsevier, pp 289–306. 50 Hampshire Street, 5th Floor, Cambridge, MA 02139, United States

Kumar N, Singh H, Sharma SK (2020) Antioxidants: responses and importance in plant defense system. In: Roy Chowdhury R, Choudhury S, Hasanuzzaman M, Srivastava S (eds) Sustainable agriculture in the era of climate change. Springer, Cham, pp 251–264. https://doi.org/10.1007/978-3-030-45669-6_11

Kumar M, Phukon SN, Paygude AC, Tyagi K, Singh H (2021a) Mapping Phenological Functional Types (PhFT) in the Indian Eastern Himalayas using machine learning algorithm in Google Earth Engine. Comput Geosci 158:104982. https://doi.org/10.1016/j.cageo.2021.104982

Kumar A, Kumar P, Singh H, Kumar N (2021b) Modulation of plant functional traits under essential plant nutrients during seasonal regime in natural forests of Garhwal Himalayas. Plant Soil 465:197–212. https://doi.org/10.1007/s11104-021-05003-x

Kumar A, Kumar P, Singh H, Bisht S, Kumar N (2021c) Relationship of physiological plant functional traits with soil carbon stock in the temperate forest of Garhwal Himalaya. Curr Sci 120(8):1368–1373. https://doi.org/10.18520/cs/v120/i8/1368-1373

Kumar A, Kumar P, Singh H, Kumar N (2021d) Adaptation and mitigation potential of roadside trees with bio-extraction of heavy metals under vehicular emissions and their impact on physiological traits during seasonal regimes. Urban for Urban Green 58:126900. https://doi.org/10.1016/j.ufug.2020.126900

Kumar A, Kumar P, Singh H, Kumar N (2021e) Impact of plant functional traits on infiltration rate under different forests of Kempty watershed in Garhwal Himalaya. Indian J Soil Cons 49(1):38–44

Kumar A, Singh H, Kumari G, Bisht S, Malik A, Kumar A, Singh M, Raturi A, Barthwal S, Thakur A, Kaushal R (2022) Adaptive resilience of roadside trees to vehicular emissions via leaf enzymatic, physiological, and anatomical trait modulations. Environ Pollut 313:120191. https://doi.org/10.1016/j.envpol.2022.120191

Kumar N, Singh H, Giri K, Kumar A, Joshi A, Yadav A, Singh R, Bisht S, Kumari R, Jeena N, Khairakpam R, Mishra G (2024) Physiological and molecular insights into the allelopathic effects on agroecosystems under changing environmental conditions. Physiol Mol Biol Plants 30:417–433. https://doi.org/10.1007/s12298-024-01440-x

Li JT, Liao B, Lan CY, Ye ZH, Baker AJM, Shu WS (2010) Cadmium tolerance and accumulation in cultivars of a high-biomass tropical tree (*Averrhoa carambola*) and its potential for phytoextraction. J Environ Qual 39:1262–1268

Li C, Zhou K, Qin W, Tian C, Qi M, Yan X, Han W (2019) A review on heavy metals contamination in soil: effects, sources, and remediation techniques. Soil Sed Contamination Int J 28:380–394. https://doi.org/10.1080/15320383.2019.1592108

Li F, Yang H, Ayyamperumal R, Liu Y (2022) Pollution, sources, and human health risk assessment of heavy metals in urban areas around industrialization and urbanization-Northwest China. Chemosphere 308(2):136396. ISSN 0045-6535, https://doi.org/10.1016/j.chemosphere.2022.136396

Ma Y, Prasad M, Rajkumar M, Freitas H (2011) Plant growth promoting rhizobacteria and endophytes accelerate phytoremediation of metalliferous soils. Biotechnol Adv 29:248–258. https://doi.org/10.1016/j.biotechadv.2010.12.001

Mahar A, Wang P, Ali A, Awasthi MK, Lahori AH, Wang Q, Li R, Zhang Z (2016) Challenges and opportunities in the phytoremediation of heavy metals contaminated soils: a review. Ecotox Environ Safe 126:111–121. https://doi.org/10.1016/j.ecoenv.2015.12.023

Malik RN, Husain SZ, Nazir I (2010) Heavy metal contamination and accumulation in soil and wild plant species from industrial area of Islamabad, Pakistan. Pak J Bot 42(1):291–301

Malik A, Kumar K, Kumari G, Kumar N, Singh H (2024). Application of functional traits in modelling productivity and resilience under climate change. In: Kumar N, Singh H (eds) Plant functional traits for improving productivity. Springer, Singapore. https://doi.org/10.1007/978-981-97-1510-7_5

Marques AP, Rangel AO, Castro PM (2009a) Remediation of heavy metal contaminated soils: phytoremediation as a potentially promising clean-up technology. Crit Rev Env Sci Technol 39:622–654. https://doi.org/10.1080/10643380701798272

Marques APGC, Rangel AOSS, Castro PML (2009b) Remediation of heavy metal contaminated soils: phytoremediation as a potentially promising clean-up technology. Crit Rev Environ Sci Tech 39:622–654

Masindi V, Muedi KL (2018) Environmental contamination by heavy metals. In: Saleh HEDM, Aglan RF (eds) Heavy metals. InTech

Mastretta C, Taghavi S, Van Der Lelie D, Mengoni A, Galardi F, Gonnelli C, Barac T, Boulet J, Weyens N, Vangronsveld J (2009) Endophytic bacteria from seeds of Nicotiana tabacum can reduce cadmium phytotoxicity. Int J Phytoremediat 11:251–267. https://doi.org/10.1080/15226510802432678

MEA (2005) Urban systems. Ecosystems and human well-being: current state and trends. Island Press, Washington, DC, pp 795–825

Meena D, Singh H, Chaudhari SK (2011) Elucidating strontium response on growth dynamics and biochemical change in *Phaseolus mungo* L. Int J Agric Environ Biotechnol 4:107–113

Mench M, Lepp N, Bert V, Schwitzguebel JP, Gawronski SW, Schroder P, Jaco V (2010) Successes and limitations of phytotechnologies at field scale: outcomes, assessment and outlook from COST Action 859. J Soil Sediment 10:1039–1070. https://doi.org/10.1007/s11368-010-0190-x

Mench M, Schwitzguebel JP, Schroeder P, Bert V, Gawronski S, Gupta S (2009) Assessment of successful experiments and limitations of phytotechnologies: contaminant uptake, detoxification and sequestration, and consequences for food safety. Environ Sci Pollut Res 16:876–900

Mesjasz-Przybyłowicz J, Nakonieczny M, Migula P, Augustyniak M, Tarnawska M, Reimold U, Koeberl C, Przybyłowicz W, Głowacka EB (2004) Uptake of cadmium, lead nickel and zinc from soil and water solutions by the nickel hyperaccumulator *Berkheya coddii*. Acta Biol Cracoviensia Ser Bot 46:75–85

Milic D, Lukovic J, Ninkov J, Zeremski-Skoric T, Zoric L, Vasin J, Milic S (2012) Heavy metal content in halophytic plants from inland and maritime saline areas. Cent Eur J Biol 7:307–317

Mitra S, Chakraborty AJ, Tareq AM, Emran TB, Nainu F, Khusro A, Idris AM, Khandaker MU, Osman H, Alhumaydhi FA, Simal-Gandara J (2022) Impact of heavy metals on the environment and human health: novel therapeutic insights to counter the toxicity. J King Saud Univ Sci 34(3):101865. ISSN 1018-3647. https://doi.org/10.1016/j.jksus.2022.101865

Mollashahi H, Szymura M (2022) Urban ecosystem: an interaction of biological and physical components. In: Biodiversity of ecosystems. IntechOpen. Available at https://doi.org/10.5772/intechopen.97742

Muchuweti M, Birkett JW, Chinyanga E, Zvauya R, Scrimshaw MD, Lester JN (2006) Heavy metal content of vegetables irrigated with mixtures of wastewater and sewage sludge in Zimbabwe: implications for human health. Agric Ecosyst Environ 112:41–48

Muthusaravanan S, Sivarajasekar N, Vivek JS, Paramasivan T, Naushad M, Prakashmaran J, Gayatri V, Al-duaij OK (2018) Phytoremediation of heavy metals: mechanisms, methods and enhancements. Environ Chem Lett 16:1339–1359. https://doi.org/10.1007/s10311-018-0762-3

Prakash V, Hunney K, Singh H (2022) Effect of elevated carbon dioxide on growth and development of *Santalum album* L. seedlings inoculated with plant growth promoting microorganisms in Open Top Chambers. Int J Environ Agric Biotechnol 7(5):149–159

Roy A, Garai N, Biswas JK (2023) Exploration of urban sustainability in India through the lens of sustainable development goals. Discov Sustain 4:41. https://doi.org/10.1007/s43621-023-00158-2

Sarwar N, Imran M, Shaheen MR, Ishaque W, Kamran MA, Matloob A, Rahim A, Hussain S (2017) Phytoremediation strategies for soils contaminated with heavy metals: modifications and future perspectives. Chemosphere 171:710–721. https://doi.org/10.1016/j.chemosphere.2016.12.116

Schaefer VH (2011) Remembering our roots: a possible connection between loss of ecological memory, alien invasions and ecological restoration. Urban Ecosyst 14:35–44

Sharma R, Singh H (2021) Alteration in biochemical constituents and nutrients partitioning of *Asparagus racemosus* in response to elevated atmospheric CO_2 concentration. Environ Sci Pollut Res 29:6812–6821. https://doi.org/10.1007/s11356-021-16050-3

Sharma JK, Kumar N, Singh NP, Santal AR (2023) Phytoremediation technologies and their mechanism for removal of heavy metal from contaminated soil: an approach for a sustainable environment. Front Plant Sci 14:1076876. https://doi.org/10.3389/fpls.2023.1076876

Sharma R, Singh H, Kaushik M, Nautiyal R, Singh O (2018). Adaptive physiological response, carbon partitioning, and biomass production of *Withania somnifera* (L.) Dunal grown under elevated CO_2 regimes. 3 Biotech 8:267. https://doi.org/10.1007/s13205-018-1292-1

Sheoran V, Sheoran AS, Poonia P (2009) Phytomining: a review. Minerals Eng 22:1007–1019. https://doi.org/10.1016/j.mineng.2009.04.001

Sherman J, McGain F, Fanzca F (2016) Environmental sustainability in anesthesia. Poll Prev Patient Saf 34:47–61

Singh H (2024) Exploring adaptive modulation in plant functional traits and its impact on the productivity of *Acacia auriculiformis* under CO_2-enriched environment. Industr Crops Prod 210:118186. https://doi.org/10.1016/j.indcrop.2024.118186

Singh H, Verma A (2013) Characterization and screening of high nitrogen efficient rice genotype to curtail environmental pollution. Int J Agric Environ Biotechnol 6(1):23–30

Singh H, Savita Sharma R, Sinha S, Kumar M, Kumar P, Verma A, Sharma SK (2017a) Physiological functioning of *Lagerstroemia speciosa* L. under heavy roadside traffic: an approach to screen potential species for abatement of urban air pollution. 3 Biotech 7(61):1–10. https://doi.org/10.1007/s13205-017-0690-0

Singh H, Verma A, Kumar M, Sharma R, Gupta R, Kaur M, Negi M, Sharma SK (2017b) Phytoremediation: a green technology to clean up the sites with low and moderate level of heavy metals. Austin Biochem 2:1012

Singh H, Sharma R, Savita SMP, Kumar M, Verma A, Ansari MW, Sharma SK (2018) Adaptive physiological response of *Parthenium hysterophorus* to elevated atmospheric CO_2 concentration. Indian Forester 144:1–14

Singh H, Yadav M, Kumar N, Kumar A, Kumar M (2020) Assessing adaptation and mitigation potential of roadside trees under the influence of vehicular emissions: A case study of *Grevillea robusta* and *Mangifera indica* planted in an urban city of India. PLoS ONE 15(1):e0227380. https://doi.org/10.1371/journal.pone.0227380

Singh S, Singh H, Sharma SK, Nautiyal R (2021) Seasonal variation in biochemical responses of bamboo clones in the sub-tropical climate of Indian Himalayan foothills. Heliyon 7(4):e06859. https://doi.org/10.1016/j.heliyon.2021.e06859

Singh H, Bandyopadhyaya S, Kumar A, Singh M, Malik A, Kumari G, Raturi A, Yadav SK, Kumar M, Barthwal S, Thakur A, Rawat PS (2022) Understanding the physiological and biophysical response of urban roadside plantations for assessing adaptation and mitigation mechanisms toward vehicular emissions. Urban Clim 44:101183. https://doi.org/10.1016/j.uclim.2022.101183

Singh M, Singh H, Kumar A, Kumar M, Barthwal S, Thakur A (2024a) Soil nitrogen availability determines the CO_2 fertilization effect on tree species (*Neolamarckia cadamba*): growth and physiological evidence. Environ Sustain 7:53–60. https://doi.org/10.1007/s42398-023-00300-w

Singh H, Kumar N, Kumar A (2024b) Enhancing resource use efficiency in crops through plant functional traits. In: Kumar N, Singh H (eds) Plant functional traits for improving productivity. Springer, Singapore. https://doi.org/10.1007/978-981-97-1510-7_6

Suman J, Uhlik O, Viktorova J, Macek T (2018) Phytoextraction of heavy metals: a promising tool for clean-up of polluted environment? Front Plant Sci 9:1476. https://doi.org/10.3389/fpls.2018.01476

Sun Y, Zhou Q, Xu Y, Wang L, Liang X (2011) The role of EDTA on cadmium phytoextraction in a cadmium-hyperaccumulator *Rorippa globosa*. J Environ Chem Ecotoxicol 3:45–51

Suryawanshi PV, Rajaram BS, Bhanarkar AD, Rao CVC (2016) Determining heavy metal contamination of road dust in Delhi, India. Atmosphera 29(3):221–234

Tan B, Wang H, Wang X, Ma C, Zhou J, Dai X (2021) Health risks and source analysis of heavy metal pollution from dust in Tianshui, China. Minerals 11(5):502. https://doi.org/10.3390/min11050502

Tchounwou PB, Yedjou CG, Patlolla AK, Sutton DJ (2012) Heavy metal toxicity and the environment. EXS 101(2012):133–164

Thiel CL, Eckelman M, Guido R, Huddleston M, Landis AE, Sherman J, Shrake SO, Copley-Woods N, Bilec MM (2015) Environmental impacts of surgical procedures: life cycle assessment of hysterectomy in the United States. Environ Sci Technol 49(3):1779–1786

Tome FV, Rodriguez PB, Lozano J (2008) Elimination of natural uranium and 226Ra from contaminated waters by rhizofiltration using *Helianthus annuus* L. Sci Total Environ 393:351–357. https://doi.org/10.1016/j.scitotenv.2008.01.013

Van Aken B (2009) Transgenic plants for enhanced phytoremediation of toxic explosives. Curr OpinBiotechnol 20:231–236

Vangronsveld J, Herzig R, Weyens N, Boulet J, Adriaensen K, Ruttens A, Thewys T, Vassilev A, Meers E, Nehnevajova E, Lelie D, Mench M (2009) Phytoremediation of contaminated soils and groundwater: lessons from the field. Environ Sci Pollut Res 16:765–794. https://doi.org/10.1007/s11356-009-0213-6

Walker CH, Sibly RM, Hopkin SP (2012) D.B.P. In: Group T. And F. (ed) Principles of ecotoxicology, 4th edn. CRC Press

Wang L, Hou D, Shen Z, Zhu J, Jia X, Ok YS, Tack FMG, Rinklebe J (2020) Field trials of phytomining and phytoremediation: a critical review of influencing factors and effects of additives. Crit Rev Environ Sci Technol 50:2724–2774. https://doi.org/10.1080/10643389.2019.1705724

Wang D, Dong L, Di S (2023) Data-driven comparison of urban sustainability towards sustainable urban development under sustainable development goals (SDGs): a review based on bibliometric analysis. Front Energy Res 11:1168126. https://doi.org/10.3389/fenrg.2023.1168126

World Commission on Environment and Development (1987) Our common future. Oxford University Press, Oxford, p 27

Wuana RA, Okieimen FE (2011) Heavy metals in contaminated soils: a review of sources, chemistry, risks and best available strategies for remediation. Isrn Ecol 2011:402647. https://doi.org/10.5402/2011/402647

Xie Y, Fan J, Zhu W, Amombo E, Lou Y, Chen L, Fu J (2016) Effect of heavy metals pollution on soil microbial diversity and bermudagrass genetic variation. Front Plant Sci J 7:755. https://doi.org/10.3389/fpls.2016.00755

Yadav SK, Singh H, Nautiyal R, Ginwal HS, Ansari SA, Barthwal S (2019a) Modulation of morpho-physiological responses in *Populus deltoides* by elevated carbon dioxide and temperature. For Sci 66:105–118. https://doi.org/10.1093/forsci/fxz048

Yadav SK, Singh H, Ginwal HS, Barthwal S (2019) Elevated CO_2 enhanced growth and physiological process of *Populus deltoides* Bartr. ex Marsh. Indian Forester 145(1):23–27

Yan A, Wang Y, Tan SN, Mohd Yusof ML, Ghosh S, Chen Z (2020) Phytoremediation: a promising approach for revegetation of heavy metal-polluted land. Front Plant Sci 11:359. https://doi.org/10.3389/fpls.2020.00359

Yanitch A, Kadri H, Frenette-Dussault C, Joly S, Pitre FE, Labrecque M (2020) A four-year phytoremediation trial to decontaminate soil polluted by wood preservatives: phytoextraction of arsenic, chromium, copper, dioxins and furans. Int J Phytoremediation 22:1505–1514. https://doi.org/10.1080/15226514.2020.1785387

Zamora-Ledezma C, Negrete-Bolagay D, Figueroa F, Zamora-Ledezma E, Ni M, Alexis F, Guerrero VH (2021) Heavy metal water pollution: a fresh look about hazards, novel and conventional remediation methods. Environ Technol Innov 22:101504. https://doi.org/10.1016/j.eti.2021.101504

Zhang MK, Liu ZY, Wang H (2010) Use of single extraction methods to predict bioavailability of heavy metals in polluted soils to rice. Commun Soil Sci Plant Anal 41(7):820–831

Urban Tree Responses to Climate Change and Environmental Pollution

Rahul Gaur and Hukum Singh

Abstract The escalating impacts of climate change and urbanization on ecosystems necessitate a comprehensive understanding of how urban trees respond to the complex interplay of environmental stresses, particularly climate change and environmental pollution. Urban trees are integral components of urban ecosystems that play a crucial role in mitigating environmental stressors and enhancing the resilience of urban landscapes. Morphological adaptations, such as alterations in leaf structure, root architecture, and overall growth patterns, indicate a tree's response to changing environmental conditions. Concurrently, physiological adaptations, including modifications in water use efficiency, gas exchange, and nutrient uptake, help the trees to thrive in the face of urban challenges. Biochemical responses of urban trees involve metabolic adjustments to cope with elevated levels of pollutants and changing climate conditions. Enzymatic activities, antioxidants, and secondary metabolite production are pivotal biochemical component stress employ to deal with various environmental stresses. Moreover, molecular insights into gene expression, signaling pathways, and epigenetic modifications provide a deeper understanding of the mechanisms that govern the adaptive responses of urban trees. This study synthesizes recent research findings to underscore the interconnectedness of morpho-physio-biochemical and molecular responses in urban trees. Understanding the adaptive mechanisms of urban plants may help researchers and urban planners develop targeted strategies for enhancing the resilience of urban tree populations in the context of a rapidly changing climate. A holistic understanding of urban tree responses to climate and pollution is essential for increasing urban green spaces, sustainable urban planning, and ecosystem conservation.

Keywords Climate change · Environmental stresses · Pollution · Sustainable urban planning · Urban resilience

R. Gaur (✉)
Department of Botany, School of Applied and Life Sciences (SALS), Uttaranchal University, Dehradun, Uttarakhand 248007, India
e-mail: rahulgaursince1998@gmail.com

H. Singh
Plant Physiology Discipline, ICFRE-Forest Research Institute, PO New Forest, Dehradun, Uttarakhand 248006, India

H. Singh (ed.), *Urban Forests, Climate Change and Environmental Pollution*,
https://doi.org/10.1007/978-3-031-67837-0_14

1 Introduction

Urban trees are vital for cities' well-being, offering many environmental, social, and economic benefits. They enhance air quality by absorbing pollutants and releasing oxygen, regulate temperatures by providing shade and reducing the urban heat island (UHI) effect, and manage storm water runoff, mitigating the risk of flooding (Wang et al. 2021a, b). Beyond their ecological impact, urban trees contribute to biodiversity, provide aesthetic appeal, help in recreational spaces, and establish a community well. They act as natural noise barriers (Barber et al. 2021), play a role in carbon sequestration, which aids in the fight against climate change (Ariluoma et al. 2021; Wang et al. 2021a, b), and offer economic advantages through increased property values. Furthermore, the presence of trees in urban areas is associated with improved mental health and overall quality of life (Wolf et al. 2020). Climate change and pollution have significant adverse effects on urban trees, and elevated temperature is a threat by inducing heat stress, which alters the traditional growing seasons and impacts the reproductive patterns of trees (Woodward et al. 2023; Singh et al. 2017b).

Additionally, air pollution, stemming from vehicular emissions and industrial activities, introduces harmful pollutants like ozone (O_3) and particulate matter (PM_1, $PM_{2.5}$, and PM_{10}), which can damage the photosynthesis and nutrient absorption process (Han et al. 2020). These pollutants weaken the tree's immune system, making them more susceptible to diseases and pests. The combined impact of climate change and pollution often leads to premature aging, reduced canopy density, and increased mortality rates among urban trees.

Urban environmental conditions pose various challenges to the trees, altering the morpho-physio-biochemical characteristics and hindering molecular functioning through factors like limited space, soil compaction, air pollution, elevated temperatures, altered water availability, etc. In adapting to these stresses, urban trees often exhibit distinct morphological characteristics. The morphological responses of urban trees are the structural and physical changes of the trees in response to the unique environmental conditions of urban settings. For instance, they may develop a more compact and vertical growth habit to optimize space utilization or display leaf size, shape, or density changes to enhance resilience against air pollution. Additionally, root systems may undergo adjustments to navigate compacted soils (Locosselli et al. 2019; Sabrin et al. 2021). Urban trees undergo a series of physiological changes as they adapt to the unique challenges of urban environments. In response to urban heat islands, trees exhibit thermoregulatory mechanisms, such as increased transpiration and leaf cooling, to mitigate temperature stress (Kabisch et al. 2021). However, prolonged exposure to air pollutants like ozone (O_3) and particulate matter (PM_1, $PM_{2.5}$, andPM_{10}) can compromise photosynthesis and overall metabolic processes, affecting the tree's health.

Furthermore, the limited soil volume in urban settings leads to competition for nutrients and water. Some urban plants reduce the growth rate to cope with this challenge (Simon et al. 2019). They also follow biochemical adaptations like exposure to elevated levels of air pollutants, such as nitrogen oxides (NO_x) and particulate matter (PM_1, $PM_{2.5}$, and PM10), that activate biochemical defense mechanisms. One notable adaptation involves upregulating antioxidant enzymes, such as superoxide dismutase and catalase, to mitigate oxidative stress caused by pollutants (Garcia-Caparros et al. 2021). Additionally, urban trees may exhibit changes in secondary metabolite production, including phenolic compounds and flavonoids, which play a role in scavenging free radicals and enhancing overall stress tolerance (Isah 2019). Water stress is another significant concern in urban areas, leading to the induction of osmo protectants like proline to maintain cell turgor and mitigate drought-related damage (Miller et al. 2020). Examining the empirical insights of urban trees provides a deeper understanding of their adaptability and resilience in urban climate and pollution. At the molecular level, these trees undergo complex physiological processes that enable them to thrive in pollution, limited space, and fluctuating climatic conditions. Studies have revealed that certain genes are activated in response to urban stresses, triggering biochemical pathways that enhance the tree's ability to detoxify pollutants and cope with environmental challenges. Additionally, molecular investigations shed light on the signaling mechanisms involved in stress responses, including synthesizing protective compounds and antioxidants (Slezina et al. 2021; Kumar et al. 2024).

Studies have shown that many commonly planted trees in India, such as Neem (*Azadirachta indica* A. Juss.) and Peepal (*Ficusreligiosa* L.), are exhibiting signs of heat stress, including reduced leaf size, earlier leaf senescence, and increased susceptibility to pests and diseases (Gajjar et al. 2022; Sathiameena et al. 2020). Increased frequency and intensity of droughts are impacting tree growth and survival, particularly in drier regions. Species with deep root systems and drought-tolerant adaptations, such as Pongamia (*Pongamia pinnata* L.) and Prosopis (*Prosopis juliflora* Sw. Dc.), are faring better (Rajarajan et al. 2022; Sharma et al. 2022). Changes in rainfall patterns disrupt the natural water balance in urban ecosystems, leading to waterlogging in some areas and water scarcity in others. Trees that can adapt to both wet and dry conditions, such as Mango (*Mangifera indica* L.) and Jamun (*Syzygium-cumini* L.),are more resilient (Zafar et al. 2021). High air pollutants, such as ozone and particulate matter, can damage tree leaves, reduce photosynthesis, and impair growth. Species with waxy leaves or hairy leaf surfaces, such as Banyan (*Ficus benghalensis* L.) and Mahua (*Madhucalongifolia* Koen. Macbride),are better at tolerating air pollution (Kaziet al. 2022). Contamination of soil with heavy metals and other pollutants can limit nutrient uptake and hinder root growth. Trees with deep root systems and the ability to chelate metals, such as Shisham (*Dalbergia sissoo* Roxb.) and Golden Shower (*Cassia fistula* L.), are more tolerant of soil pollution (Fig. 1) (Nawaz et al. 2022; Bhutianiet al. 2022).

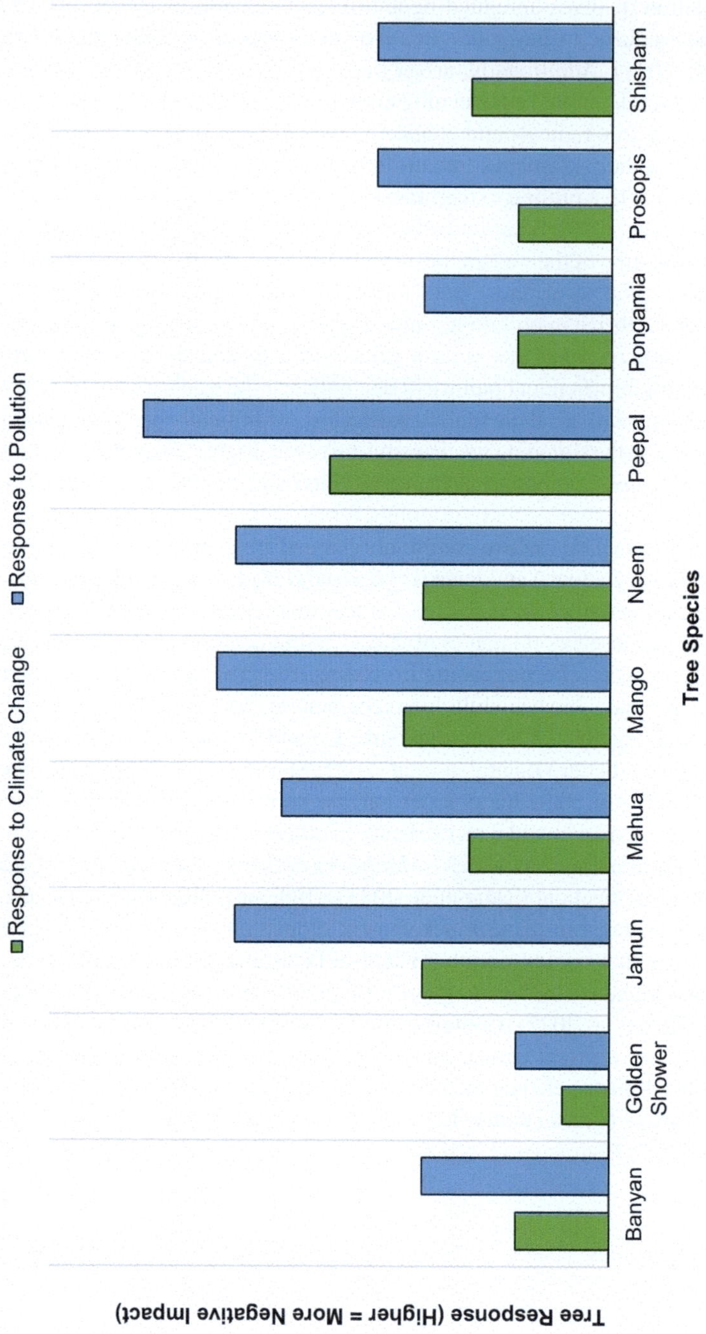

Fig. 1 Commonly planted urban trees response to climate change and pollution

2 Urban Climate and Pollution

Urbanization has been a hallmark of human civilization, transforming small settlements into metropolises. While this process has brought economic development and improved living standards for many, it has also given rise to significant environmental challenges, particularly in the urban climate and pollution. The interplay between these two phenomena has far-reaching consequences for the inhabitants of cities and the planet.

2.1 Urban Climate

The urban climate is a microcosm of global climate changes, with cities acting as localized hubs that magnify and reflect broader environmental trends. The UHI effect is one of the most noticeable aspects of urban climate (Table 1). This occurs when the concentration of buildings, roads, and other impervious surfaces in urban areas absorbs and re-emits solar radiation, leading to higher temperatures than surrounding rural areas (Cheela et al. 2021; Kumar et al. 2019b, 2020; Singh et al. 2022). The UHI effect exacerbates the warming global climate and poses significant challenges for urban residents. They contribute to high energy consumption as residents use air conditioning to combat the heat. This, in turn, intensifies greenhouse gas emissions, creating a feedback loop that further exacerbates climate change.

Additionally, higher temperatures can impact human health, leading to heat-related illnesses and straining healthcare systems. Vulnerable populations, such as the elderly and low-income communities, often bear the worst impact of these climate-related health risks. Urban climates are also marked by altered precipitation patterns, with increased frequency and intensity of extreme weather events such as heavy rainfall and storms. This can result in urban flooding, disrupting daily life, damaging infrastructure, and compromising water quality. Precipitation patterns can affect water availability, leading to water stress in some metropolitan areas (Masson et al. 2020).

2.2 Pollution

The rapid urbanization witnessed in recent decades become a significant reason for pollution. Urban environments are hotspots for various types of pollution, including air, water, and soil pollution (Table 1). These pollutants degrade the quality of life and pose significant risks to ecosystems and global environmental stability.

Table 1 Factors influencing urban climate and pollution with their mitigation strategies

Factors	Impact on urban climate	Impact on pollution	Potential mitigation strategies
Urban Heat Island (UHI) Effect	Temperature increases (5–10 °C higher)	Air pollution increases (stagnation of pollutants)	Through green spaces, cool roofs, permeable pavements
Buildings and Infrastructure	Wind speed and ventilation decreases	Air pollution increases (emissions from energy production)	Through renewable energy sources, energy-efficient buildings
Transportation	Traffic congestion increases	Air and noise pollution increases (vehicle emissions)	Through public transportation, cycling infrastructure, electric vehicles
Water management	The chances of urban flooding increases	Water quality decreases (contaminated runoff)	Through green spaces, rainwater harvesting, wastewater treatment
Waste management	Land fills increase	Air and water pollution increases (emissions from waste disposal)	Through recycling, composting, waste reduction initiatives
Light pollution	Sky visibility decreases, and disruption of circadian rhythms increases	Increases	Through shielding lights, using energy-efficient bulbs
Urban sprawl	Loss of natural habitats and fragmentation of ecosystems increases	Increases	Through smart growth planning, compact and sustainable cities

Sources Wang et al. (2011), Takahashi et al. (2020), Ngarambe et al. (2021), Gavrilescu (2022)

2.2.1 Air Pollution

It is the most visible and widely studied form of environmental pollution in urban areas. Emissions from vehicles, industrial activities, and other sources release pollutants such as particulate matter (PM_1, $PM_{2.5}$, and PM_{10}), nitrogen oxides (NO_x), sulfur dioxide (SO_2), and volatile organic compounds into the atmosphere. These pollutants have the worst effects on respiratory health, cardiovascular function, and overall well-being. Long-term exposure to air pollution is linked to a range of chronic diseases, namely asthma, lung cancer, and cardiovascular diseases. Beyond these immediate health impacts, urban air pollution contributes to global climate change. Greenhouse gas emissions from urban areas, particularly carbon dioxide (CO_2), methane (CH_4), and nitrous oxide (N_2O), contribute to the enhanced greenhouse effect, driving global warming (Takahashi et al. 2020; Gupta et al. 2018; Sharma et al. 2018; Singh et al. 2018). The challenge is not only in controlling the emissions but also in adapting urban infrastructure to mitigate the impacts of climate change, such as rising sea levels and more frequent and severe heat waves (Singh et al. 2017a).

2.2.2 Water Pollution

Urbanization often leads to the contamination of water bodies, with untreated wastewater, industrial discharges, and runoff from roads and impervious surfaces carrying pollutants into rivers, lakes, and ultimately to oceans. This not only threatens aquatic ecosystems but also poses risks to human health through consuming contaminated water. The degradation of water quality in urban areas necessitates comprehensive water management strategies, including improved sewage treatment, storm water management, and the protection of natural water sources (Wang et al. 2011).

2.2.3 Soil Pollution

Urban industrial activities, improper waste disposal, and the use of chemical substances contribute to soil contamination. Heavy metals, pesticides, and other pollutants that can accumulate in the soil threaten human health and the environment. Contaminated soils can impact the safety of urban agriculture and even lead to the transfer of pollutants through the food chain (Gavrilescu 2022; Singh and Verma 2013).

2.3 The Intersection of Urban Climate and Pollution

The complicated interactions between urban climate and pollution are evident in various ways. For instance, air pollution not only degrades air quality but also plays a role in climate change by contributing to the greenhouse effect. Simultaneously, climate change influences air quality through altered weather patterns and increased temperatures, which can enhance the formation of ground-level ozone (O_3) and other pollutants (Cavender and Donnelly 2019). Similarly, the UHI effect, a sign of urban climate, contributes to increased energy consumption for cooling purposes, leading to higher greenhouse gas emissions and exacerbating climate change. Furthermore, the altered precipitation patterns associated with climate change can influence the transport and fate of pollutants, affecting water and soil quality in urban areas (Pataki et al. 2021). Addressing this combination of interplay requires a holistic and integrated approach to urban planning, policy-making, and infrastructure development. Sustainable urban design prioritizes green spaces, promotes energy efficiency, and incorporates climate-resilient infrastructure that can mitigate the UHI effect and certain forms of environmental pollution. Additionally, transitioning to cleaner energy sources, implementing strict emission controls, and investing in public transportation can significantly reduce urban air pollution and its contribution to climate change (Dai et al. 2022).

3 Morphological Responses of Urban Trees

Urban trees often exhibit phenotypic plasticity, which means the ability to alter their morphology in response to environmental conditions. This plasticity allows trees to adjust their growth patterns and physiological processes to suit the urban environment better. As cities expand and human populations increase, urbanization stresses trees. These stresses include altered soil composition, limited space for root expansion, increased pollution, and the UHI effect. These factors collectively shape the morphological responses of urban trees, influencing their growth patterns, canopy structure, and overall health (Gupta et al. 2018; Kumar et al. 2019c).

3.1 Root System

One of the primary morphological responses of urban trees is observed in their root system. Trees often develop shallow and wide-ranging root structures in compacted urban soils to maximize water and nutrient absorption. However, this adaptation may compromise the tree's stability, increasing susceptibility to wind throw and other environmental hazards (Karlova et al. 2021). For example, the Banyan tree, known for its extensive aerial roots, is well-adapted to tropical climates, providing stability in swampy areas. On the other hand, Golden Shower trees develop a robust taproot system, enhancing their ability to withstand dry conditions.

Additionally, Jamun and mango trees exhibit an adaptable taproot root system, aiding in efficient nutrient absorption when changing climactic conditions. Neem trees are renowned for their pollution-tolerant nature, with their robust root systems contributing to soil detoxification (Omar et al. 2023). Peepal and Shisham trees, with their deep and expansive root networks, play a crucial role in stabilizing soil and mitigating the impact of pollution on the surrounding environment. Pongamia and Prosopis trees are known for their nitrogen-fixing abilities, enhancing soil fertility and combating pollution effects. Mahua trees, with their well-adapted root systems, contribute to the restoration of polluted areas. Understanding the morphological changes in root systems is crucial for selecting appropriate tree species for urban planting and designing effective urban green spaces (Chaudhry and Sidhu 2022).

3.2 Canopy Structure and Growth Patterns

The canopy structure of urban trees undergoes distinct morphological changes in response to urbanization. Limited space for lateral growth often results in narrower canopies and vertical growth (Singh et al. 2024b). This adaptation allows trees to thrive in confined urban spaces, providing shade and aesthetic appeal. However, the vertical growth may lead to imbalanced crown structures, impacting the overall stability and health of the tree. Proper pruning and canopy management are essential to mitigate these effects and maintain the structural integrity of urban trees (Sbrissia et al. 2020; Kimm et al. 2021).

3.3 *Tolerance to Pollution*

Urban environments are characterized by increased levels of air and soil pollution, posing a challenge to the health of trees. Some tree species exhibit morphological adaptations that enhance their tolerance to pollutants. For example, certain trees develop specialized leaf structures that minimize the absorption of contaminants or can shed contaminated leaves, promoting new growth. Understanding these adaptive mechanisms is crucial for selecting resilient tree species for urban planting and improving city air quality (Manjunath and Reddy 2019; Shrestha et al. 2021).

3.4 *Influence of Urban Heat Island (UHI) Effect*

The UHI effect, characterized by higher temperatures in urban areas than surrounding rural areas, significantly affects the morphological responses of trees. Trees in urban environments may develop adaptations to cope with increased temperatures, such as modified leaf structures to minimize water loss through transpiration (Kim and Brown 2021). Various studies have shown that the Banyan tree, known for its expansive canopy, exhibits increased leaf surface area to enhance transpiration and cooling. Golden Shower trees develop deep and extensive root systems to access water resources efficiently. Jamun trees have drought-resistant features, such as reduced leaf size to minimize water loss through transpiration. Mahua trees undergo alterations in leaf orientation or shape to mitigate heat stress, while Mango trees exhibit mechanisms for efficient water usage, such as stomatal regulation. Neem trees adapted by producing compounds with sun-protective properties, safeguarding against excessive radiation. Peepal trees utilize their large, heart-shaped leaves to provide shade and reduce direct exposure to sunlight. Shisham trees adjust their leaf morphology to balance water conservation and photosynthetic efficiency. Urban planners need to consider these adaptations when selecting tree species and planning green infrastructure to mitigate the impact of the heat island effect (Han et al. 2023).

4 Physiological Responses of Urban Trees

4.1 Photosynthesis and Carbon Sequestration

Photosynthesis is a tree's most important physiological process, by which trees convert sunlight into energy, producing oxygen and sequestering carbon dioxide. In urban settings, trees face challenges such as reduced green space, air pollution, and increased temperatures. Despite these challenges, urban trees exhibit remarkable adaptability, adjusting their photosynthetic rates to optimize resource utilization (Gayathri et al. 2021; Singh et al. 2021; Kumar et al. 2021a, b, c, d, e).

Elevated levels of air pollutants such as nitrogen dioxide (NO_2), sulfur dioxide (SO_2), and ozone (O_3), common in urban areas, interfere with photosynthetic processes. These pollutants damage the chloroplasts and reduce the efficiency of photosynthesis in urban trees (Singh et al. 2022; Yadav et al. 2019a, b; Gupta et al. 2019a). Moreover, the urban heat island (UHI) effect exacerbates tree stress (Kumar et al. 2022). Increased temperatures accelerate water loss through transpiration and affect the enzymes involved in photosynthesis. As a result, urban trees may experience a decline in photosynthetic rates, limiting their ability to produce energy and sustain healthy growth (Wang et al. 2019; Kumar et al. 2024).

4.2 Transpiration and Water Uptake

Transpiration is the process by which trees absorb water through their roots and release it into the atmosphere through pores called stomata on their leaves. In urban areas, trees often contend with limited soil space, soil compaction, and competition for water. Despite these challenges, urban trees demonstrate efficient water uptake and transpiration, contributing to the overall cooling of the urban environment. Their ability to regulate temperature through transpiration helps mitigate the urban heat island effect, improving city inhabitants' overall comfort and well-being (Jamalluddin et al. 2019).

4.3 Soil Health and Nutrient Cycling

The health of urban trees is intricately linked to soil quality. Urban soils are often compacted, contaminated, and subjected to various disturbances. Trees, however, have evolved mechanisms to cope with these adversities. Their roots release exudates that promote beneficial microbial activity, enhancing soil structure and nutrient availability. Urban trees contribute to nutrient cycling by absorbing, storing, and releasing essential elements, playing a vital role in maintaining the overall health of urban ecosystems (Yu et al. 2022; Singh et al. 2024a).

4.4 Stress Tolerance and Adaptation

Urban trees face various stress, including pollution, soil compaction, and limited space for root expansion. In response, trees employ multiple strategies to enhance stress tolerance and adaptation. Some species exhibit drought tolerance, while others have developed mechanisms to withstand pollution, such as accumulating heavy metals in their tissues without adverse effects (Raza et al. 2020; Gupta and Singh 2017; Kumar et al. 2017; Jangra et al. 2024).

4.5 Phenology and Seasonal Changes

The phenology of urban trees refers to their seasonal cycles, including leaf emergence, flowering, fruiting, and leaf fall. Urbanization, with its altered microclimates and heat island effects, can influence the phenological patterns of trees. Studies have shown that urban trees may exhibit earlier leaf emergence and extended growing seasons compared to their rural environment. These changes can have cascading effects on ecosystem dynamics, impacting interactions with pollinators, pests, and other wildlife (Piao et al. 2019; Singh et al. 2021; Kanta et al. 2024; Malik et al. 2024; Kumar et al. 2024).

4.6 Air Quality Improvement

One of the most tangible benefits of urban trees is their role in improving air quality. Trees act as natural air purifiers, filtering pollutants and particulate matter from the atmosphere. Through phytoremediation, certain tree species can absorb and store contaminants in their tissues, helping to detoxify the air. The presence of urban trees is particularly crucial in mitigating the adverse effects of vehicular emissions and industrial activities on air quality, contributing to a healthier living environment for urban residents (Tomson et al. 2021).

5 Biochemical Responses of Urban Trees

5.1 Antioxidant Defense Mechanisms

Trees deploy antioxidant defense mechanisms to combat the oxidative stress induced by urban pollutants. Urban trees activate enzymes such as superoxide dismutase, catalase, and peroxidase to neutralize reactive oxygen species (ROS) generated during stress conditions. These enzymes play a vital role in protecting cellular structures and maintaining the integrity of cell membranes. However, prolonged exposure to high levels of pollutants can overcome the antioxidant defense system, leading to oxidative damage in tree tissues. This is the main reason for leaf necrosis, reduced growth, and an overall decline in tree health. Understanding the balance between antioxidant defense mechanisms and oxidative stress is crucial for assessing the resilience of urban trees (Ejaz et al. 2020; Meena et al. 2011; Singh et al. 2024c).

5.2 *Soil Quality and Nutrient Uptake*

The urban environment significantly alters soil composition, posing challenges to nutrient availability for trees. Soil compaction from construction activities and concrete surfaces limit root growth and reduce the tree's ability to access essential nutrients. Pollutants from various sources, including vehicle emissions and industrial activities, contaminate urban soils. Urban trees respond to these challenges by adjusting their nutrient uptake strategies. Some species exhibit enhanced nutrient-use efficiency, optimizing the allocation of nutrients to essential physiological processes. However, prolonged exposure to poor soil conditions leads to nutrient deficiencies, negatively impacting tree health and growth (Dubeyet al. 2021).

5.3 *Hormonal Regulation in Urban Trees*

Hormones are pivotal in regulating various physiological processes in trees, including growth, development, and stress responses. Trees undergo complex hormonal changes in urban environments to cope with their challenges. Abscisic acid (ABA), a stress hormone, is often upregulated in response to environmental stresses such as drought, pollution, and high temperatures. However, auxins and cytokinins, hormones responsible for growth and development, may experience altered levels in urban trees. The balance between stress-related and growth-promoting hormones determines how well urban trees adapt to their surroundings (Łopucki et al. 2019).

5.4 *Epigenetic Modifications*

Epigenetic modifications play a crucial role in the biochemical responses of trees to urban stress. Epigenetic changes, such as DNA methylation and histone modifications, can influence gene expression without altering the underlying DNA sequence. Urban trees may undergo epigenetic modifications in response to environmental stresses, allowing for rapid adaptation to changing environmental conditions. Understanding the epigenetic mechanisms involved in the biochemical responses of urban trees provides insights into their long-term adaptability and evolution in urban environments. Research in this field is essential for developing strategies to enhance the resilience of urban tree populations (Chang et al. 2020).

6 Molecular Mechanisms and Gene Expression

The molecular mechanisms and gene expression involved in the responses of urban trees to various environmental stresses are complex and multifaceted. One of the primary stresses that urban trees face is climate change, characterized by rising temperatures, altered precipitation patterns, and increased frequency of extreme weather events. In response to these changes, trees undergo a series of molecular mechanisms to maintain cellular homeostasis and ensure survival (Lahalle et al. 2021; Ouyang et al. 2021). One key aspect of this response is the activation of stress-responsive genes. Several classes of genes are involved in the stress response of urban trees. Heat shock proteins (HSPs) are a well-studied group that plays a crucial role in protecting cells from damage caused by high temperatures. When temperatures rise, trees upregulate the expression of HSP genes, producing heat shock proteins that help prevent protein denaturation and maintain cellular integrity (Percival 2023). In addition to temperature-related stress, urban trees face challenges related to altered precipitation patterns, including drought stress. Genes associated with drought tolerance are upregulated in response to water scarcity, triggering physiological changes that enable trees to conserve water and withstand periods of drought. Aquaporins, for example, are water channel proteins whose expression increases during drought stress, facilitating water movement across cell membranes (Table 2) (Manna et al. 2021; Nazet al. 2022).

Climate change often exacerbates pollution in urban environments, further impacting tree health. Air pollutants, such as ozone (O_3) and particulate matter (PM_1, $PM_{2.5}$, and PM_{10}), can harm trees, leading to oxidative stress. Molecular mechanisms in antioxidant defense systems become activated to counteract the damaging effects of reactive oxygen species generated by pollution. Essential genes involved in antioxidant defense include superoxide dismutase (SOD), catalase (CAT), and glutathione peroxidase (GPX) (Dumanović et al. 2021). These enzymes work together to neutralize reactive oxygen species and protect cellular components from oxidative damage. The upregulation of these antioxidant genes is a crucial part of the adaptive response of urban trees to pollution stress.

Beyond individual stress response genes, a network of signaling pathways coordinates the overall stress response in trees. For example, the mitogen-activated protein kinase (MAPK) cascade is a signaling pathway that plays a central role in transducing stress signals and activating downstream stress response genes. The activation of MAPKs in response to environmental stresses is a conserved mechanism across urban trees. Epigenetic modifications also contribute to the adaptive responses of

Table 2 Molecular mechanisms and gene expressions involved in response to climate and pollution of urban trees

Climate/ pollution factor	Molecular mechanisms	Gene expression changes	Physiological/ phenotypic effects
High temperature	Heat shock proteins (HSPs) protect and refold denatured proteins, Activate heat shock element (HSE) binding proteins, and increase the production of antioxidants	Upregulation of HSP genes (HSP70, HSP90), Downregulation of photosynthesis genes, Increased expression of antioxidant enzymes (superoxide dismutase, catalase)	Reduced photosynthesis, Increased respiration, Wilting and leaf senescence, Decreased growth and productivity
Drought stress	Accumulation of abscisic acid (ABA), Activation of ABA-responsive genes, Closure of stomata to conserve water	Upregulation of stress-related genes (LEA proteins, aquaporins), Downregulation of growth and metabolism genes	Reduced stomatal conductance, Decreased transpiration, Stunted growth and delayed development, Leaf shedding
Air pollution (Ozone)	Production of reactive oxygen species (ROS), Activation of antioxidant defense system, Signaling through MAPK pathways	Upregulation of antioxidant genes (superoxide dismutase, catalase, peroxidase), Downregulation of photosynthesis genes, Induction of defense-related genes	Chlorophyll degradation, Reduced photosynthesis, Cell membrane damage, Leaf necrosis, and premature senescence
Air pollution (particulate matter)	Physical blockage of stomata, Induction of inflammatory response, Disruption of hormonal signaling	Downregulation of photosynthesis genes, Upregulation of stress-related genes, Dysregulation of gene expression patterns	Reduced light interception, Impaired gas exchange, Increased susceptibility to pathogens, Chlorosis, and stunted growth
Acid rain	Disruption of membrane transport, Leaching of essential nutrients, Activation of stress response pathways	Downregulation of nutrient transporter genes, Upregulation of stress-related genes, Dysregulation of gene expression patterns	Reduced nutrient uptake, Imbalanced ion homeostasis, Chlorosis, and leaf die-back, Decreased growth and productivity

Sources Manna et al. (2021), Naz et al. (2022), Percival (2023)

urban trees to climate and pollution stress. DNA methylation and histone modifications can regulate the expression of stress-responsive genes by modulating chromatin structure (Liu et al. 2022). These epigenetic changes allow trees to adapt to changing climatic conditions over time. Also, the microbiome associated with urban trees has influenced their responses to environmental stress. The rhizosphere, the soil region influenced by tree roots, contains a diverse community of microorganisms that can enhance the stress tolerance of trees. Certain microbial symbionts can promote nutrient uptake, produce plant growth-promoting hormones, and induce systemic resistance against pathogens, contributing to the overall resilience of urban trees (Trivedi et al. 2022).

7 Interactions and Trade-Offs

7.1 Interactions Between Climate and Urban Trees

Urban areas experience the effect, with higher temperatures than surrounding rural areas. This poses challenges for urban trees, affecting their physiological processes. Increased temperatures lead to higher evapotranspiration rates, impacting water availability for trees. Simultaneously, elevated temperatures also alter metabolic processes and growth patterns. Climate change results in altered precipitation patterns, leading to drought or excessive rainfall in certain regions. Urban trees, already stressed by the UHI effect, may face increased water stress during droughts. So, the excess rain can lead to saturated soils, affecting root health and nutrient uptake. Striking a balance in water management becomes crucial for the survival and well-being of urban trees. Increased atmospheric carbon dioxide (CO_2) levels, a key component of climate change, can influence photosynthetic rates in urban trees (Sharma and Singh 2021). While elevated CO_2 levels may initially enhance photosynthesis, sustained exposure may lead to acclimation, potentially affecting long-term carbon sequestration. Understanding the dynamics of this interaction is essential for predicting the future carbon sequestration potential of urban trees (Prakash et al. 2022; Dhyani et al. 2021; Yadav et al. 2019a, b).

7.2 Trade-Offs in Urban Tree Responses

7.2.1 Ozone Exposure and Respiratory Stress

Urban areas often face high levels of air pollution, including ground-level ozone. While trees play a crucial role in air purification by absorbing pollutants, prolonged exposure to ozone can result in respiratory stress for trees. This trade-off between pollution absorption and respiratory stress highlights the complexity of the urban tree response to air quality.

7.2.2 Adaptation to Urban Soils

Urban soils are often compacted and contaminated, posing challenges for tree root development and nutrient uptake. Trees that adapt to these conditions may exhibit altered growth patterns, but such adaptations can come at the cost of reduced overall tree health. Urban planners must carefully balance soil management practices to promote immediate tree survival and long-term ecological resilience.

7.2.3 Biotic Interactions and Pests

Urban environments introduce novel challenges for trees through altered biotic interactions. Increased fragmentation and isolation may reduce genetic diversity, making urban tree populations susceptible to pests and diseases. While certain tree species may thrive in urban settings, the trade-off may involve increased vulnerability to specific pests adapted to urban conditions.

7.2.4 Human-Tree Interaction

The presence of trees in urban areas has numerous benefits for human well-being, including improved mental health and air quality. However, the proximity of trees to human activities can also result in conflicts, such as root damage to infrastructure or the need for regular maintenance. Balancing maximizing the benefits of urban trees and minimizing potential conflicts requires careful urban planning and community engagement.

8 Urban Planning and Management Strategies

As urban areas grow, the challenges related to climate change and pollution become more urgent. Urban areas contribute significantly to climate change and environmental degradation, posing risks to human health and the ecosystem. Urban areas are characterized by a high concentration of human activities, industries, and infrastructure, leading to increased emissions of greenhouse gases (GHGs) and pollutants. The urban heat island (UHI) effect exacerbates climate-related issues, and pollution from vehicular emissions, industrial activities, and improper waste disposal further degrades air and water quality, impacting the health of urban residents and ecosystems (Singh et al. 2020). Effective urban planning and management strategies are crucial to addressing these challenges and creating sustainable, resilient cities (Sturiale and Scuderi 2019; Hurlimann et al. 2021).

8.1 Sustainable Urban Planning

Effective urban planning is foundational to building climate-resilient and low-carbon cities. It involves designing cities that prioritize environmental conservation, energy efficiency, and the well-being of their inhabitants (Jain 2023).

8.1.1 Compact City Design

Encouraging compact and mixed-use developments minimizes urban sprawl, reduces the demand for transportation, and lowers energy consumption.

8.1.2 Green Infrastructure

Incorporating green spaces, parks, and urban forests helps combat the heat island effect, enhances air quality, and provides recreational areas for residents.

8.1.3 Sustainable Transportation

Promoting public transportation, cycling, and walking reduce the use of private vehicles, thus lowering emissions and traffic-related pollution.

8.1.4 Renewable Energy Integration

Incorporating renewable energy sources, such as solar and wind, into urban infrastructure helps reduce the carbon footprint of cities.

8.2 Climate-Responsive Urban Management

Urban management involves the ongoing operations and maintenance of urban infrastructure. Climate-responsive management strategies aim to enhance a city's resilience to climate change impacts.

8.2.1 Climate-Adaptive Infrastructure

Designing and retrofitting infrastructure to withstand extreme weather events, such as floods and heat waves, is crucial for ensuring the longevity and functionality of urban systems (Vahedifard et al. 2021; Gupta et al. 2019b).

8.2.2 Water Management

Implementing sustainable water management practices, such as rainwater harvesting, storm water management, and efficient water use, helps address water scarcity and reduce the risk of flooding (Nguyen et al. 2019).

8.2.3 Waste Management

Developing comprehensive waste management systems, including recycling and waste-to-energy initiatives, minimizes the environmental impact of urban waste (Ijgosse 2019).

8.2.4 Emergency Preparedness

Establishing robust emergency response plans and early warning systems helps urban areas cope with and recover from climate-related disasters (Raikes et al. 2019).

8.3 Pollution Control and Abatement

Effective control and reduction measures are essential to combating pollution in urban areas. Urban planning and management strategies should focus on:

8.3.1 Air Quality Monitoring

Implementing air quality monitoring systems helps identify pollution sources and formulate targeted interventions to reduce emissions (Kim et al. 2020).

8.3.2 Sustainable Mobility

Promoting electric vehicles, investing in public transportation, and implementing emission standards contribute to reducing air pollution from vehicular sources (Porru et al. 2020).

8.3.3 Industrial Regulations

Enforcing strict regulations on industrial emissions and promoting cleaner production technologies are crucial for minimizing pollution from industrial activities (Divan and Rosencranz 2022).

8.3.4 Waste Reduction and Recycling

Encouraging waste reduction, recycling, and proper disposal practices help mitigate the impact of urban waste on air, water, and soil quality (Rajmohan et al. 2019).

8.4 Community Engagement and Education

Effective urban planning and management strategies require the active involvement and awareness of the community. Engaging residents through education and participation programs can foster a sense of responsibility and contribute to sustainable urban development (Rall et al. 2019; Kumar et al. 2019a).

8.4.1 Environmental Education

Incorporating environmental education into school curriculums and community programs raises awareness about the importance of sustainable living.

8.4.2 Citizen Science

Involving citizens in data collection and monitoring initiatives empowers communities to contribute to actively identifying and addressing local environmental issues.

8.4.3 Green Initiatives

Encouraging community-led green projects, such as tree planting, community gardens, and clean-up campaigns, fosters a sense of ownership and pride for city inhabitants.

9 Future Challenges and Research Directions

One of the critical challenges is revealing the complex web of morpho-physio-biochemical and molecular responses of urban trees to the urban climate and pollution. This multifaceted research demands an integrative approach that combines traditional ecological methods with advanced molecular techniques. Urban trees must adapt morphologically to withstand these changes as cities experience rising temperatures, altered precipitation patterns, and increased pollution levels. So, understanding how trees modify their physical structures, such as leaf size, shape, and root architecture, is pivotal for predicting their resilience in urban landscapes. Physiological processes like photosynthesis, transpiration, and nutrient uptake are central to a tree's ability to thrive in urban environments. Future research should focus on elucidating these processes under varying urban conditions. How do elevated temperatures affect photosynthetic rates? How does pollution alter nutrient availability in urban soils? These questions are needed for comprehensive physiological studies that can be used for urban planning and tree management strategies. Urban pollution introduces many pollutants, including heavy metals and air-borne toxins, which can trigger biochemical changes in trees. Research should aim to reveal the biochemical pathways activated in response to urban stressors and their implications for tree health. Identifying specific biomarkers indicative of stress can aid in developing early detection mechanisms for assessing tree well-being in urban settings.

Understanding how biochemical responses vary across different tree species is crucial for selecting resilient species for urban afforestation projects. As molecular biology techniques advance, researchers should focus on the genetic and molecular mechanisms governing tree responses to urban stresses. Genomic studies can identify essential genes associated with stress tolerance, providing valuable information for breeding programs to develop urban-adapted tree varieties. Epigenetic studies can uncover how environmental factors influence gene expression in urban trees, shedding light on the mechanisms of phenotypic plasticity and bridging the gap between morpho-physio-biochemical observations and molecular-level insights crucial for developing a holistic understanding of urban tree responses at different scales. Collaborative efforts between ecologists, physiologists, biochemists, and molecular biologists are essential to tackle this challenge effectively. Another critical aspect for future research involves exploring the long-term impacts of climate and pollution on urban tree populations. While short-term studies provide valuable information on tree responses, it is crucial to understand how these reactions translate into sustained survival and growth over the long term. This requires establishing long-term monitoring programs in urban areas to track the performance of tree populations over multiple years. Such studies can provide valuable information into the adaptive capacity of urban trees and inform urban planning strategies for creating resilient and sustainable green spaces.

10 Conclusion

This chapter has provided valuable insights into the interactions between trees and their urban environments. The chapter has illuminated urban trees' adaptive mechanisms to cope with changing climatic conditions and increasing pollution levels through comprehensive morpho-physio-biochemical and molecular analyses. These findings underscore the importance of understanding the resilience and vulnerability of urban trees, given their critical role in mitigating environmental stresses and enhancing the well-being of urban ecosystems. The molecular insights gained from this study have deepened our understanding of the genetic and biochemical processes that govern these responses, aiding the development of targeted urban forestry and environmental management strategies. As cities expand and face escalating environmental pressures, the knowledge derived from this research is essential for creating sustainable urban forestry practices. Implementing informed policies and practices can help preserve and enhance urban green spaces, improve air quality, mitigate the urban heat island (UHI) effect, and maintain overall ecological balance in urban areas. Future research will be crucial for refining scientific understanding of the dynamic relationships between urban trees and their environments. This knowledge will be vital for shaping urban landscapes that can thrive in the face of climate change and pollution, fostering a healthier and more sustainable coexistence between humans and nature. Ultimately, this promotes the health and well-being of people living in urban areas, ensuring that urban environments remain livable and resilient in the face of ongoing environmental challenges.

References

Ariluoma M, Ottelin J, Hautamäki R, Tuhkanen EM, Mänttäri M (2021) Carbon sequestration and storage potential of urban green in residential yards: a case study from Helsinki. Urban for Urban Green 1(57):126939. https://doi.org/10.1016/j.ufug.2020.126939

Barber A, Haase D, Wolff M (2021) Permeability of the city–Physical barriers of and in urban green spaces in the city of Halle, Germany. Ecol Ind 1(125):107555. https://doi.org/10.1016/j.ecolind.2021.107555

Bhutiani R, Tiwari RC, Chauhan P, Ahamad F, Sharma VB, Tyagi I, Singh P (2022) Potential of *Cassia fistula* pod-based absorbent in remediating water pollutants: an analytical study. In: Sustainable materials for sensing and remediation of noxious pollutants, pp 261–272. https://doi.org/10.1016/B978-0-323-99425-5.00001-3

Cavender N, Donnelly G (2019) Intersecting urban forestry and botanical gardens to address big challenges for healthier trees, people, and cities. Plants People Planet 1(4):315–322. https://doi.org/10.1002/ppp3.38

Chang YN, Zhu C, Jiang J, Zhang H, Zhu JK, Duan CG (2020) Epigenetic regulation in plant abiotic stress responses. J Integr Plant Biol 62(5):563–580. https://doi.org/10.1111/jipb.12901

Chaudhry S, Sidhu GP (2022) Climate change regulated abiotic stress mechanisms in plants: a comprehensive review. Plant Cell Rep 41(1):1–31. https://doi.org/10.1007/s00299-021-02759-5

Cheela VS, John M, Biswas W, Sarker P (2021) Combating urban heat island effect—A review of reflective pavements and tree shading strategies. Buildings 11(3):93. https://doi.org/10.3390/buildings11030093

Dai T, Zheng X, Yang J (2022) A systematic review of studies at the intersection of urban climate and historical urban landscape. Environ Impact Assess Rev 1(97):106894. https://doi.org/10.1016/j.eiar.2022.106894

Dhyani K, Kuniyal HB, Singh H, Sobha (2021) Growth and physiological potential of *Terminalia arjuna* under elevated CO_2 levels in Open top chamber condition. J Appl Nat Sci 13(3):1121–1126. https://doi.org/10.31018/jans.v13i3.2490. ISSN 2231-5209

Divan S, Rosencranz A (2022) Environmental law and policy in India: cases and materials. Oxford University Press

Dubey RS, Srivastava RK, Pessarakli M (2021) Physiological mechanisms of nitrogen absorption and assimilation in plants under stressful conditions. In: Handbook of plant and crop physiology. CRC Press, pp 579–616

Dumanović J, Nepovimova E, Natić M, Kuča K, Jaćević V (2021) The significance of reactive oxygen species and antioxidant defense system in plants: a concise overview. Front Plant Sci 6(11):552969. https://doi.org/10.3389/fpls.2020.552969

Ejaz S, Fahad S, Anjum MA, Nawaz A, Naz S, Hussain S, Ahmad S (2020) Role of osmolytes in the mechanisms of antioxidant defense of plants. Sustain Agric Rev 39:95–117. https://doi.org/10.1007/978-3-030-38881-2_4

Gajjar HH, Jeyaraman JD, Kaul DS (2022) Role of *Azadirachta indica* (Neem) and *Polyalthia longifolia* (Asopalav) trees for improving outdoor thermal environment in unorganized urban settings. Int J Biometeorol 66(10):2055–2067. https://doi.org/10.1007/s00484-022-02340-z

Garcia-Caparros P, De Filippis L, Gul A, Hasanuzzaman M, Ozturk M, Altay V, Lao MT (2021) Oxidative stress and antioxidant metabolism under adverse environmental conditions: a review. Bot Rev 87:421–466. https://doi.org/10.1007/s12229-020-09231-1

Gavrilescu M (2022) Enhancing phytoremediation of soils polluted with heavy metals. Curr Opin Biotechnol 1(74):21–31. https://doi.org/10.1016/j.copbio.2021.10.024

Gayathri R, Mahboob S, Govindarajan M, Al-Ghanim KA, Ahmed Z, Al-Mulhm N, Vodovnik M, Vijayalakshmi S (2021) A review on biological carbon sequestration: a sustainable solution for a cleaner air environment, less pollution and lower health risks. J King Saud Univ Sci 33(2):101282. https://doi.org/10.1016/j.jksus.2020.101282

Gupta SK, Singh H (2017) Observations on sporadic flowering in *Ficus benjamina* L. and *Peltophorum africanum* L. in New Forest. Indian for 143:290–291

Gupta SK, Ram J, Singh H (2018a) Comparative study of transpiration in cooling effect of tree species in the atmosphere. J Geosci Environ Prot 6:151–166. https://doi.org/10.4236/gep.2018.68011

Gupta SK, Ram J, Singh H (2019a) Relationship between leaf area index and atmospheric cooling of tree species. Indian J for 42(3):203–208

Gupta SK, Singh H, Ram J (2019b) Need for a novel paradigm in forestry research: perspectives of comparative atmospheric cooling by different forestry effect tree species. Indian for 145(6):509–515

Han D, Shen H, Duan W, Chen L (2020) A review on particulate matter removal capacity by urban forests at different scales. Urban for Urban Green 1(48):126565. https://doi.org/10.1016/j.ufug.2019.126565

Han D, Zhang T, Qin Y, Tan Y, Liu J (2023) A comparative review on the mitigation strategies of urban heat island (UHI): a pathway for sustainable urban development. Climate Dev 15(5):379–403. https://doi.org/10.1080/17565529.2022.2092051

Hurlimann A, Moosavi S, Browne GR (2021) Urban planning policy must do more to integrate climate change adaptation and mitigation actions. Land Use Policy 1(101):105188. https://doi.org/10.1016/j.landusepol.2020.105188

Ijgosse J (2019) Waste incineration and informal livelihoods: a technical guide on waste-to-energy initiatives. In: WIEGO technical brief, p 11

Isah T (2019) Stress and defense responses in plant secondary metabolites production. Biol Res 52. https://doi.org/10.1186/s40659-019-0246-3

Jain AK (2023) Climate resilient, green and low carbon built environment. Springer Nature

Jamalluddin N, Massawe FJ, Symonds RC (2019) Transpiration efficiency of Amaranth (*Amaranthus* sp.) in response to drought stress. J Hortic Sci Biotechnol 94(4):448–459. https://doi.org/10.1080/14620316.2018.1537725

Jangra A, Kumar K, Maikhuri S, Bhandari MS, Pandey S, Singh H, Barthwal S (2024) Unveiling stress-adapted endophytic bacteria: characterizing plant growth-promoting traits and assessing cross-inoculation effects on *Populus deltoides* under abiotic stress. Plant Physiol Biochem 210:108610. https://doi.org/10.1016/j.plaphy.2024.108610

Kabisch N, Pueffel C, Masztalerz O, Hemmerling J, Kraemer R (2021) Physiological and psychological effects of visits to different urban green and street environments in older people: a field experiment in a dense inner-city area. Landsc Urban Plan 1(207):103998. https://doi.org/10.1016/j.landurbplan.2020.103998

Kanta C, Kumar A, Chauhan A, Singh H, Sharma IP (2024) The interplay between plant functional traits and climate change. In: Kumar N, Singh H (eds) Plant functional traits for improving productivity. Springer, Singapore. https://doi.org/10.1007/978-981-97-1510-7_3

Karlova R, Boer D, Hayes S, Testerink C (2021) Root plasticity under abiotic stress. Plant Physiol 187(3):1057–1070. https://doi.org/10.1093/plphys/kiab392

Kazi ER, Kulkarni S, Shaikh YI, Shaikh VS, Prasad N (2022) A case study on estimation of air quality in Hadapsar Suburban of Pune India. Oriental J Chem 38(1):118. https://doi.org/10.13005/ojc/380115

Kim SW, Brown RD (2021) Urban heat island (UHI) intensity and magnitude estimations: a systematic literature review. Sci Total Environ 20(779):146389. https://doi.org/10.1016/j.scitotenv.2021.146389

Kim J, Jeong U, Ahn MH, Kim JH, Park RJ, Lee H, Song CH, Choi YS, Lee KH, Yoo JM, Jeong MJ (2020) New era of air quality monitoring from space: Geostationary Environment Monitoring Spectrometer (GEMS). Bull Am Meteor Soc 101(1):E1-22. https://doi.org/10.1175/BAMS-D-18-0013.1

Kimm H, Guan K, Burroughs CH, Peng B, Ainsworth EA, Bernacchi CJ, Moore CE, Kumagai E, Yang X, Berry JA, Wu G (2021) Quantifying high-temperature stress on soybean canopy photosynthesis: The unique role of sun-induced chlorophyll fluorescence. Glob Change Biol 27(11):2403–2415. https://doi.org/10.1111/gcb.15603

Kumar N, Suyal DC, Sharma IP, Verma A, Singh H (2017) Elucidating stress proteins in rice (*Oryzasativa* L.) genotype under elevated temperature: a proteomic approach to understand heat stress response. 3 Biotech 7:205. https://doi.org/10.1007/s13205-017-0856-9

Kumar M, Singh MP, Singh H, Dhakate PM, Ravindranath NH (2019a) Forest working plan for the sustainable management of forest and biodiversity in India. J Sustain for 39:1–22. https://doi.org/10.1080/10549811.2019.1632212

Kumar N, Jeena N, Singh H (2019b) Elevated temperature modulates rice pollen structure: a study from the foothill of the Himalayan agro-ecosystem in India. 3 Biotech 9(5):175. https://doi.org/10.1007/s13205-019-1700-1

Kumar N, Jeena N, Kumar N, Gangola S, Singh H (2019c) Phytoremediation facilitating enzymes: an enzymatic approach for enhancing remediation process. In: Bhatt P (ed) Smart bioremediation technologies. Academic Press, Elsevier, pp 289–306. 50 Hampshire Street, 5th Floor, Cambridge, MA 02139, United States

Kumar N, Singh H, Sharma SK (2020) Antioxidants: responses and importance in plant defense system. In: Roy Chowdhury R, Choudhury S, Hasanuzzaman M, Srivastava S (eds) Sustainable agriculture in the era of climate change. Springer, Cham, pp 251–264. https://doi.org/10.1007/978-3-030-45669-6_11

Kumar A, Kumar P, Singh H, Kumar N (2021a) Modulation of plant functional traits under essential plant nutrients during seasonal regime in natural forests of Garhwal Himalayas. Plant Soil 465:197–212. https://doi.org/10.1007/s11104-021-05003-x

Kumar A, Kumar P, Singh H, Bisht S, Kumar N (2021b) Relationship of physiological plant functional traits with soil carbon stock in the temperate forest of Garhwal Himalaya. Curr Sci 120(8):1368–1373. https://doi.org/10.18520/cs/v120/i8/1368-1373

Kumar A, Kumar P, Singh H, Kumar N (2021c) Adaptation and mitigation potential of roadside trees with bio-extraction of heavy metals under vehicular emissions and their impact on physiological traits during seasonal regimes. Urban for Urban Green 58:126900. https://doi.org/10.1016/j.ufug.2020.126900

Kumar A, Kumar P, Singh H, Kumar N (2021d) Impact of plant functional traits on infiltration rate under different forests of Kempty watershed in Garhwal Himalaya. Indian J Soil Cons 49(1):38–44

Kumar M, Phukon SN, Paygude AC, Tyagi K, Singh H (2021e) Mapping phenological functional types (PhFT) in the Indian Eastern Himalayas using machine learning algorithm in Google Earth Engine. Comput Geosci 158:104982. https://doi.org/10.1016/j.cageo.2021.104982

Kumar A, Singh H, Kumari G, Bisht S, Malik A, Kumar A, Singh M, Raturi A, Barthwal S, Thakur A, Kaushal R (2022) Adaptive resilience of roadside trees to vehicular emissions via leaf enzymatic, physiological, and anatomical trait modulations. Environ Pollut 313:120191. https://doi.org/10.1016/j.envpol.2022.120191

Kumar N, Singh H, Giri K, Kumar A, Joshi A, Yadav A, Singh R, Bisht S, Kumari R, Jeena N, Khairakpam R, Mishra G (2024) Physiological and molecular insights into the allelopathic effects on agroecosystems under changing environmental conditions. Physiol Mol Biol Plants 30:417–433. https://doi.org/10.1007/s12298-024-01440-x

Lahalle A, Lacroix M, De Blasio C, Cissé MY, Linares LK, Le Cam L (2021) The p53 pathway and metabolism: the tree that hides the forest. Cancers 13(1):133. https://doi.org/10.3390/cancers13010133

Liu Y, Wang J, Liu B, Xu ZY (2022) Dynamic regulation of DNA methylation and histone modifications in response to abiotic stresses in plants. J Integr Plant Biol 64(12):2252–2274. https://doi.org/10.1111/jipb.13368

Locosselli GM, de Camargo EP, Moreira TC, Todesco E, de Fátima AM, de André CD, de André PA, Singer JM, Ferreira LS, Saldiva PH, Buckeridge MS (2019) The role of air pollution and climate on the growth of urban trees. Sci Total Environ 20(666):652–661. https://doi.org/10.1016/j.scitotenv.2019.02.291

Łopucki R, Klich D, Ścibior A, Gołębiowska D (2019) Hormonal adjustments to urban conditions: stress hormone levels in urban and rural populations of *Apodemus agrarius*. Urban Ecosyst 14(22):435–442. https://doi.org/10.1007/s11252-019-0832-8

Malik A, Kumar K, Kumari G, Kumar N, Singh H (2024) Application of functional traits in modelling productivity and resilience under climate change. In: Kumar N, Singh H (eds) Plant functional traits for improving productivity. Springer, Singapore. https://doi.org/10.1007/978-981-97-1510-7_5

Manjunath BT, Reddy J (2019) Comparative evaluation of air pollution tolerance of plants from polluted and non-polluted regions of Bengaluru. J Appl Biol Biotechnol 7(3):63–68. https://doi.org/10.7324/JABB.2019.70312

Manna M, Thakur T, Chirom O, Mandlik R, Deshmukh R, Salvi P (2021) Transcription factors as key molecular target to strengthen the drought stress tolerance in plants. Physiol Plant 172(2):847–868. https://doi.org/10.1111/ppl.13268

Masson V, Lemonsu A, Hidalgo J, Voogt J (2020) Urban climates and climate change. Annu Rev Environ Resour 17(45):411–444

Meena D, Singh H, Chaudhari SK (2011) Elucidating strontium response on growth dynamics and biochemical change in *Phaseolus mungo* L. Int J Agric Environ Biotechnol 4:107–113

Miller DL, Alonzo M, Roberts DA, Tague CL, McFadden JP (2020) Drought response of urban trees and turfgrass using airborne imaging spectroscopy. Remote Sens Environ 1(240):111646. https://doi.org/10.1016/j.rse.2020.111646

Nawaz R, Aslam M, Nasim I, Irshad MA, Ahmad S, Latif M, Hussain F (2022) Air pollution tolerance index and heavy metals accumulation of tree species for sustainable environmental management in megacity of Lahore. Air 1(1):55–68. https://doi.org/10.3390/air1010004

Naz H, Akram NA, Ashraf M, Hefft DI, Jan BL (2022) Leaf extract of neem (*Azadirachta indica*) alleviates adverse effects of drought in quinoa (*Chenopodium quinoa* Willd.) plants through alterations in biochemical attributes and antioxidants. Saudi J Biol Sci 29(3):1367–74. https://doi.org/10.1016/j.sjbs.2022.01.038

Ngarambe J, Joen SJ, Han CH, Yun GY (2021) Exploring the relationship between particulate matter, CO, SO_2, NO_2, O_3, and urban heat island in Seoul, Korea. J Hazard Mater 5(403):123615. https://doi.org/10.1016/j.jhazmat.2020.123615

Nguyen TT, Ngo HH, Guo W, Wang XC, Ren N, Li G, Ding J, Liang H (2019) Implementation of a specific urban water management—Sponge City. Sci Total Environ 20(652):147–162. https://doi.org/10.1016/j.scitotenv.2018.10.168

Omar SA, Ashokhan S, Yaacob JS (2023) Salinity-induced modulation of growth and accumulation of phytochemicals composition in in vitro root cultures of *Azadirachta indica*. Biocatal Agric Biotechnol 1:102748. https://doi.org/10.1016/j.bcab.2023.102748

Ouyang SN, Gessler A, Saurer M, Hagedorn F, Gao DC, Wang XY, Schaub M, Li MH, Shen WJ, Schönbeck L (2021) Root carbon and nutrient homeostasis determines downy oak sapling survival and recovery from drought. Tree Physiol 41(8):1400–1412. https://doi.org/10.1093/treephys/tpab019

Pataki DE, Alberti M, Cadenasso ML, Felson AJ, McDonnell MJ, Pincetl S, Pouyat RV, Setälä H, Whitlow TH (2021) The benefits and limits of urban tree planting for environmental and human health. Front Ecol Evol 8(9):603757. https://doi.org/10.3389/fevo.2021.603757

Percival GC (2023) Heat tolerance of urban tree species-a review. Urban for Urban Green 8:128021. https://doi.org/10.1016/j.ufug.2023.128021

Piao S, Liu Q, Chen A, Janssens IA, Fu Y, Dai J, Liu L, Lian XU, Shen M, Zhu X (2019) Plant phenology and global climate change: current progresses and challenges. Glob Change Biol 25(6):1922–1940. https://doi.org/10.1111/gcb.14619

Porru S, Misso FE, Pani FE, Repetto C (2020) Smart mobility and public transport: opportunities and challenges in rural and urban areas. J Traff Transp Eng (english Ed) 7(1):88–97. https://doi.org/10.1016/j.jtte.2019.10.002

Prakash V, Hunney K, Singh H (2022) Effect of elevated carbon dioxide on growth and development of *Santalum album* L. seedlings inoculated with plant growth promoting microorganisms in Open Top Chambers. Int J Environ Agric Biotechnol 7(5):149–159

Raikes J, Smith TF, Jacobson C, Baldwin C (2019) Pre-disaster planning and preparedness for floods and droughts: a systematic review. Int J Disaster Risk Reduction 38:101207

Rajarajan K, Sakshi S, Taria S, Prathima PT, Radhakrishna A, Anuragi H, Ashajyothi M, Bharati A, Handa AK, Arunachalam A (2022) Whole plant response of *Pongamia pinnata* to drought stress tolerance revealed by morpho-physiological, biochemical and transcriptome analysis. Mol Biol Rep 49(10):9453–9463. https://doi.org/10.1007/s11033-022-07808-0

Rajmohan KV, Ramya C, Viswanathan MR, Varjani S (2019) Plastic pollutants: effective waste management for pollution control and abatement. Curr Opin Environ Sci Health 1(12):72–84. https://doi.org/10.1016/j.coesh.2019.08.006

Rall E, Hansen R, Pauleit S (2019) The added value of public participation GIS (PPGIS) for urban green infrastructure planning. Urban for Urban Green 1(40):264–274. https://doi.org/10.1016/j.ufug.2018.06.016

Raza A, Ashraf F, Zou X, Zhang X, Tosif H (2020) Plant adaptation and tolerance to environmental stresses: mechanisms and perspectives. In: Plant ecophysiology and adaptation under climate change: mechanisms and perspectives I: general consequences and plant responses, pp 117–145. https://doi.org/10.1007/978-981-15-2156-0_5

Sabrin S, Karimi M, Nazari R, Pratt J, Bryk J (2021) Effects of different urban-vegetation morphology on the canopy-level thermal comfort and the cooling benefits of shade trees: case-study in Philadelphia. Sustain Cities Soc 1(66):102684. https://doi.org/10.1016/j.scs.2020.102684

Sathiameena K, Narasimhan D, Pari R (2020) Role of trees in mitigating urban temperature. Plant Arch 20:4622–4932

Sbrissia AF, Schmitt D, Duchini PG, Da Silva SC (2020) Unravelling the relationship between a seasonal environment and the dynamics of forage growth in grazed swards. J Agron Crop Sci 206(5):630–639. https://doi.org/10.1111/jac.12402

Sharma R, Singh H (2021) Alteration in biochemical constituents and nutrients partitioning of *Asparagus racemosus* in response to elevated atmospheric CO_2 concentration. Environ Sci Pollut Res 29:6812–6821. https://doi.org/10.1007/s11356-021-16050-3

Sharma R, Singh H, Kaushik M, Nautiyal R, Singh O (2018) Adaptive physiological response, carbon partitioning, and biomass production of *Withania somnifera* (L.) Dunal grown under elevated CO_2 regimes. 3 Biotech 8:267. https://doi.org/10.1007/s13205-018-1292-1

Sharma M, Dinesh R, Sen S (2022) Sustainable livelihoods study and salt tolerance effects on two important arid region tree species *Prosopis cineraria* (L.) Druze and *Prosopis juliflora* (Sw.) DC. Vegetos, pp 1–0. https://doi.org/10.1007/s42535-022-00528-7

Shrestha S, Baral B, Dhital NB, Yang HH (2021) Assessing air pollution tolerance of plant species in vegetation traffic barriers in Kathmandu Valley, Nepal. Sustain Environ Res 31:1–9. https://doi.org/10.1186/s42834-020-00076-2

Simon H, Fallmann J, Kropp T, Tost H, Bruse M (2019) Urban trees and their impact on local ozone concentration—a microclimate modeling study. Atmosphere 10(3):154. https://doi.org/10.3390/atmos10030154

Singh H, Verma A (2013) Characterization and screening of high nitrogen efficient rice genotype to curtail environmental pollution. Int J Agric Environ Biotechnol 6(1):23–30

Singh H, Savita Sharma R, Sinha S, Kumar M, Kumar P, Verma A and Sharma SK (2017a) Physiological functioning of *Lagerstroemia speciosa* L. under heavy roadside traffic: an approach to screen potential species for abatement of urban air pollution. 3 Biotech 7(61):1–10. https://doi.org/10.1007/s13205-017-0690-0

Singh H, Verma A, Kumar M, Sharma R, Gupta R, Kaur M, Negi M, Sharma SK (2017b) Phytoremediation: a green technology to clean up the sites with low and moderate level of heavy metals. Austin Biochem 2:1012

Singh H, Sharma R, Savita SMP, Kumar M, Verma A, Ansari MW, Sharma SK (2018) Adaptive physiological response of *Parthenium hysterophorus* to elevated atmospheric CO_2 concentration. Indian Forester 144:1–14

Singh H, Yadav M, Kumar N, Kumar A, Kumar M (2020) Assessing adaptation and mitigation potential of roadside trees under the influence of vehicular emissions: A case study of *Grevillea robusta* and *Mangifera indica* planted in an urban city of India. PLoS ONE 15(1):e0227380. https://doi.org/10.1371/journal.pone.0227380

Singh S, Singh H, Sharma SK, Nautiyal R (2021) Seasonal variation in biochemical responses of bamboo clones in the sub-tropical climate of Indian Himalayan foothills. Heliyon 7(4):e06859. https://doi.org/10.1016/j.heliyon.2021.e06859

Singh H, Bandyopadhyaya S, Kumar A, Singh M, Malik A, Kumari G, Raturi A, Yadav SK, Kumar M, Barthwal S, Thakur A, Rawat PS (2022) Understanding the physiological and biophysical response of urban roadside plantations for assessing adaptation and mitigation mechanisms toward vehicular emissions. Urban Clim 44:101183. https://doi.org/10.1016/j.uclim.2022.101183

Singh M, Singh H, Kumar A, Kumar M, Barthwal S, Thakur A (2024a) Soil nitrogen availability determines the CO_2 fertilization effect on tree species (*Neolamarckia cadamba*): growth and physiological evidence. Environ Sustain 7:53–60. https://doi.org/10.1007/s42398-023-00300-w

Singh H, Kumar N, Kumar A (2024c) Enhancing resource use efficiency in crops through plant functional traits. In: Kumar N, Singh H (eds) Plant functional traits for improving productivity. Springer, Singapore. https://doi.org/10.1007/978-981-97-1510-7_6

Slezina MP, Istomina EA, Korostyleva TV, Kovtun AS, Kasianov AS, Konopkin AA, Shcherbakova LA, Odintsova TI (2021) Molecular insights into the role of cysteine-rich peptides in induced resistance to Fusarium oxysporum infection in tomato based on transcriptome profiling. Int J Mol Sci 22(11):5741. https://doi.org/10.3390/ijms22115741

Sturiale L, Scuderi A (2019) The role of green infrastructures in urban planning for climate change adaptation. Climate 7(10):119. https://doi.org/10.3390/cli7100119

Takahashi M, Feng Z, Mikhailova TA, Kalugina OV, Shergina OV, Afanasieva LV, Heng RK, Abd Majid NM, Sase H (2020) Air pollution monitoring and tree and forest decline in East Asia: a review. Sci Total Environ 10(742):140288. https://doi.org/10.1016/j.scitotenv.2020.140288

Tomson M, Kumar P, Barwise Y, Perez P, Forehead H, French K, Morawska L, Watts JF (2021) Green infrastructure for air quality improvement in street canyons. Environ Int 1(146):106288. https://doi.org/10.1016/j.envint.2020.106288

Trivedi P, Batista BD, Bazany KE, Singh BK (2022) Plant–microbiome interactions under a changing world: responses, consequences and perspectives. New Phytol 234(6):1951–1959. https://doi.org/10.1111/nph.18016

Vahedifard F, Alborzi A, Williams JM, AghaKouchak A (2021) Framework for adaptive design of infrastructure under a changing climate. In: Geo-extreme, pp 267–277

Wang H, Ouyang Z, Chen W, Wang X, Zheng H, Ren Y (2011) Water, heat, and airborne pollutants effects on transpiration of urban trees. Environ Pollut 159(8–9):2127–2137. https://doi.org/10.1016/j.envpol.2011.02.031

Wang XM, Wang XK, Su YB, Zhang HX (2019) Land pavement depresses photosynthesis in urban trees especially under drought stress. Sci Total Environ 25(653):120–130. https://doi.org/10.1016/j.scitotenv.2018.10.281

Wang C, Wang ZH, Kaloush KE, Shacat J (2021a) Cool pavements for urban heat island mitigation: a synthetic review. Renew Sustain Energy Rev 1(146):111171. https://doi.org/10.1016/j.rser.2021.111171

Wang Y, Chang Q, Li X (2021b) Promoting sustainable carbon sequestration of plants in urban greenspace by planting design: a case study in parks of Beijing. Urban for Urban Green 1(64):127291. https://doi.org/10.1016/j.ufug.2021.127291

Wolf KL, Lam ST, McKeen JK, Richardson GR, van Den Bosch M, Bardekjian AC (2020) Urban trees and human health: a scoping review. Int J Environ Res Public Health 17(12):4371. https://doi.org/10.3390/ijerph17124371

Woodward A, Hinwood A, Bennett D, Grear B, Vardoulakis S, Lalchandani N, Lyne K, Williams C (2023) Trees, climate change, and health: an urban planning, greening and implementation perspective. Int J Environ Res Public Health 20(18):6798. https://doi.org/10.3390/ijerph20186798

Yadav SK, Singh H, Nautiyal R, Ginwal HS, Ansari SA, Barthwal S (2019b) Modulation of morpho-physiological responses in *Populus deltoides* by elevated carbon dioxide and temperature. Forest Science 66:105–118. https://doi.org/10.1093/forsci/fxz048

Yadav SK, Singh H, Ginwal HS, Barthwal S (2019a) Elevated CO_2 enhanced growth and physiological process of *Populus deltoides* Bartr. ex Marsh. Indian For 145(1):23–27

Yu Y, Gui Y, Li Z, Jiang C, Guo J, Niu D (2022) Induced systemic resistance for improving plant immunity by beneficial microbes. Plants 11(3):386. https://doi.org/10.3390/plants11030386

Zafar Z, Rasheed F, Atif RM, Maqsood M, Gailing O (2021) Salicylic acid-induced morpho-physiological and biochemical changes triggered water deficit tolerance in *Syzygium cumini* L. saplings. Forests 12(4):491. https://doi.org/10.3390/f12040491

Biological Insights into Forest Adaptation to Climate Change: Ecosystem Resilience and Adaptation

Rishita Pakhira, Subhankar Biswas, C. Manoj, and Hukum Singh

Abstract Climate change presents a significant challenge to global forests, necessitating a deeper understanding of biological mechanisms and adaptive strategies to bolster ecosystem resilience. Understanding biological insights into forest adaptation is vital for long-term sustainability. Genetic, phenotypic, and biochemical adaptation mechanisms underpin forests' resilience to environmental changes. This chapter delves into forest adaptation to climate change, encompassing genetic diversity and adaptive management techniques. Forest's intricate ecosystems display a remarkable ability to adjust to shifting climates. Genetic diversity plays a crucial role, with natural selection favoring individuals possessing advantageous traits, ensuring resilient genotypes persist. Phenotypic plasticity allows trees to adjust their form and function in response to environmental cues, enhancing overall adaptability. Species migration aids forests in tracking suitable climates but often lags behind climate shifts. Assisted migration strategies involving deliberate species transplantation may bridge this gap. Trees utilize physiological mechanisms to combat drought and heat, including protective enzymes and metabolic adjustments. Adaptive forest management enhances resilience through selective breeding, mimicking natural disturbances, and promoting mixed-species forests. Incorporating ecological interactions and functional traits into management approaches fosters holistic adaptation. Habitat restoration efforts aid native species recovery and improve resilience, while soil and water conservation practices are crucial for forest health. Climate-smart forestry bolsters ecosystem robustness while providing valuable services and economic benefits. Adaptive management, habitat restoration, and climate-smart forestry are essential for navigating an uncertain climate, enabling forests to thrive amidst change. This chapter may benefit the scientific community and policymakers by helping them understand forest ecosystem resilience and adaptation to climate change and developing strategies to combat climate change consequences on forest ecosystems.

R. Pakhira (✉) · C. Manoj · H. Singh
ICFRE-Forest Research Institute, PO New Forest, Dehradun, Uttarakhand 248006, India
e-mail: rishitapakhira@gmail.com

S. Biswas
Bidhan Chandra Krishi Viswavidyalaya, P.O. Krishi Viswavidyalaya, Mohanpur, Dist-Nadia, West Bengal 741252, India

H. Singh (ed.), *Urban Forests, Climate Change and Environmental Pollution*,
https://doi.org/10.1007/978-3-031-67837-0_15

Keywords Climate smart forestry · Genetic diversity · Phenotypic plasticity · Resilience · Restoration

1 Introduction

Since the 1850s, the Earth's climate has undergone significant transformations, increasingly influenced by human activities, notably greenhouse gas emissions (IPCC 2021). Elevated greenhouse gases from fossil fuel combustion, deforestation, and land use changes have altered Earth's climate (Tubiello et al. 2013; Kumar et al. 2022, 2024; Singh et al. 2022, 2024a; Singh 2024). Human actions have notably elevated atmospheric CO_2 levels, correlating with heightened occurrences of severe weather events (Anderson-Frey 2017; Dhyani et al. 2021; Kumar et al. 2021e; Prakash et al. 2022). Climate shifts, particularly in temperature ranges, are severe, with projections indicating significant permafrost loss with a 2 °C temperature rise (Wang et al. 2010).

Global climate variations result in diverse outcomes, including rising water levels and escalating flood risks in coastal areas. Drought occurrences increase while permafrost loss threatens stability (Young et al. 2017). The IPCC warns of a potential 1.5 °C temperature rise by 2052, impacting forests and altering species composition (IPCC 2018).

Climate hazards like storms, wildfires, and droughts pose critical challenges, hindering sustainable development (Denton et al. 2014). The IPCC advocates mitigation and adaptation, aiming to reduce greenhouse gas emissions and adapt to climatic changes. Forests are crucial in this framework, acting as carbon stores and providing essential products and services. Climate change alters water cycles and forest dynamics, intensifying floods and droughts globally (Gharbia et al. 2018; Wang et al. 2021).

Indigenous communities in forested areas face heightened vulnerability due to limited adaptive capacity and alternative income sources (Jain and Sajjad 2016). Monitoring climate change's consequences on forests is crucial for mitigation and adaptation strategies (Singh 2021; Sharma and Singh 2021; Singh et al. 2021). This chapter explores forest adaptation to climate change, from genetic diversity to adaptive management techniques.

2 Climate Change and Forests

Forests are vital for stabilizing global temperatures and regulating carbon and nitrogen cycles, with their environmental shifts impacting local and global climates (Singh et al. 2020; Kumar et al. 2021a, b; Barati et al. 2023). Global warming, driven by changing precipitation and temperature patterns, significantly affects the expansion and productivity of transboundary forests, leading to adverse impacts like forest fires, droughts, and pest outbreaks (Kumar et al. 2021c, d). These changes jeopardize the livelihoods of communities heavily reliant on forests for survival (Yadav et al. 2019a, b; Gupta et al. 2019a, b). Droughts, exacerbated by climate change, pose a substantial risk to forest well-being, while intense storms and altered precipitation patterns further stress forest survival (Goldsmith et al. 2022; Tiebel et al. 2023).

Approximately 1.6 billion people worldwide depend on forests for subsistence, including 60 million indigenous people and 1.2 billion in agroforestry-dependent communities (Husain et al. 2018). In Africa, over two-thirds of the population relies on forests for essentials like food and fuelwood, facing disproportionate vulnerability to climate change. Despite this, many forest communities remain unfamiliar with "climate change" due to economic, cultural, and socio-ecological influences. Fluctuations in temperature and precipitation adversely affect agroforestry crops, reducing yields (Jawo et al. 2023).

Rising mosquito populations and new wasp species contribute to increased skin-borne diseases like malaria in the Himalayas. Bangladesh is experiencing rising water-borne illnesses, including diarrhoea and dengue (Rodway 2023). Climate change exacerbates existing challenges forest communities face, extending beyond health concerns.

3 Climate Change Adaptation Strategies

Addressing the challenges of climate change demands a holistic approach that balances mitigation and adaptation strategies. Mitigation involves reducing greenhouse gas (GHG) emissions and enhancing carbon sinks in ecosystems such as forests and soil (Kumar et al. 2019a; b). The IPCC defines adaptation as the 'adjustment in natural or human systems in response to actual or expected climatic stimuli or their effects, which moderates harm or exploits beneficial opportunities.' Adaptation takes various forms, including autonomous, reactive, anticipatory, or planned measures. It plays a crucial role in mitigating the impacts of climate change across all sectors, addressing both immediate and long-term consequences. Even under scenarios with the lowest GHG stabilization, adaptation remains indispensable, considering the associated risks and costs that may surpass the adaptive capacity of natural, managed, and human systems (Sharma et al. 2018; Singh et al. 2018; Gupta et al. 2018a, b).

Historically, mitigation has received more attention than adaptation, primarily due to concerns about the prolonged disruption of global biogeochemical cycles and their impact on the climate system (Cohen et al. 1998). However, it's essential to recognize that a certain degree of climate change is unavoidable due to historical emissions and the inherent inertia of the climate system. Moreover, while mitigation impacts may take decades to materialize, adaptation actions yield almost immediate effects (Singh and Verma 2013; Kumar et al. 2017; Gupta and Singh 2017). Adaptation measures can be implemented at regional or local scales, with efficacy less dependent on the actions of others. Additionally, adaptation often mitigates existing climate variability risks, besides addressing future climate risks (Singh et al. 2017a, b).

Adger et al. (2007) highlight a growing recognition of adaptation as a crucial aspect of climate change response, albeit with significant implementation constraints, especially in vulnerable nations and communities. Adaptation strategies are often integrated into broader risk management, resource planning, and sustainable development initiatives. Proactive management of future climate risks through adaptation approaches holds promise in mitigating economic, social, and environmental costs associated with climate change.

4 Phenotypic Plasticity

Phenotypic plasticity, the ability to undergo phenotypic modifications in response to environmental factors, enhances survival through molecular modifications that alter gene expression. Epigenetic regulation, crucial in phenotypic plasticity, enables gene expression changes without DNA sequence modifications (Iwasaki and Paszkowski 2014). Phenotypic plasticity is pivotal for trees to adapt to rapid environmental changes induced by climate change. Climate change intensifies global drought stress, causing stand-level growth decline and increased mortality. Understanding epigenetic pathways connecting environmental cues, particularly escalating drought stress in adult trees, to subsequent generational phenotypic effects is a pressing scientific challenge (Alberto et al. 2013).

Three significant epigenetic pathways include DNA methylation, histone modifications, and non-coding RNAs (ncRNAs). DNA methylation, extensively studied in model and non-model plants, regulates gene transcription, plant development, and stress responses (Zhang et al. 2018). Histone modifications influence chromatin compaction, affecting transcriptional accessibility. Challenges in studying histone modifications in trees limit research in this area. Non-coding RNAs, mobile within cells and persisting through divisions, play roles in genomic regulation (Yakovlev et al. 2012).

DNA methylation is pivotal in conifers due to extensive transposable element content, influencing gene expression and genome integrity. While CpG methylation's impact on gene expression is complex, CHG and CHH methylation typically impede DNA transcription, preserving genome integrity (Bewick and Schmitz 2017). Stress-induced DNA methylation could underlie phenotypic variation in plants, possibly contributing to "stress memory" and heritability of environmental responses (Kinoshita and Seki 2014).

Studies in various plants suggest inherited epigenetic modifications associated with environmental variations (Boyko et al. 2010; Lang-Mladek et al. 2010), yet mechanisms remain incompletely understood. Understanding epigenetic mechanisms in tree adaptation to environmental changes is essential for future research (Table 1).

Table 1 DNA methylation studies conducted globally to study plant species' response to several abiotic stresses

Plant Species	Technique	Objective	References
Quercus ilex L.	MSAP	Drought stress response	Rico et al. (2014)
Populustrichocarpa Hook	WGBS	Drought stress response	Liang et al. (2014)
Pinus pinea L.	MSAP	Epigenetic variability	Sáez-Laguna et al. (2014)
Medicago sativa L. and *Medicago spp*	MSAP	Salt stress response	Al-Lawati et al. (2016)
Quercus lobata Née	RRBS	Environment response	Gugger et al. (2016)
*Eucalyptus globulus*Labill	MeDIP-seq, MRE-seq	Heteroblastic development	Hasbún et al. (2016)
*Populussimonii*Carr	MSAP	Various abiotic stresses	Song et al. (2016)
*Populuseuphratica*Oliv	WGBS	Salt stress response	Su et al. (2018)
Morus alba L	WGBS	Drought stress response	Li et al. (2020)
Bruguieragymnorhiza (L.) Lamk	WGBS	Salt stress response	Miryeganeh et al. (2022)
Populustremula L. × *Populus alba* L.	RNAi, HPLC, WGBS	Drought stress response	Sow et al. (2018)

5 Genetic Basis of Climate Change Adaptation

Advancements in omics technologies, database availability, and genome editing tools have significantly enhanced the understanding of the genetic basis of complex traits related to climatic adaptation in natural populations (Meena et al. 2011; Kumar et al. 2019c, 2020). This progress aims to mitigate the adverse impacts of climate change and foster the development of climate-adaptive crops (Zaidem et al. 2019). Various methodologies, including quantitative trait locus (QTL) mapping, genome-wide association studies (GWAS), landscape genomics, transcriptomics, and metabolomics, are employed in these endeavors. Utilizing multiple complementary methods allows for a comprehensive understanding of the genes responsible for climate adaptation.

For instance, GWAS examines associations between genome-wide molecular markers (e.g., single nucleotide polymorphisms, SNPs) and ecologically significant phenotypes using hundreds of ecotypes from natural populations. While GWAS provides high resolution due to low levels of linkage disequilibrium resulting from numerous recombination events in natural populations, population structure may confound its results (Pais et al. 2020). Conversely, QTL mapping studies involve mapping populations generated from crosses between individuals with differing traits, providing unbiased results unaffected by population structure but with lower resolution due to fewer recombination events (Zaidem et al. 2019).

Landscape genomics has emerged as a potent approach for examining the connection between genomic variation and environmental diversity among natural populations (Li et al. 2017; Pais et al. 2020). In landscape genomic studies, two prevalent strategies are employed in identifying candidate genes linked to environmental adaptation. One approach involves scanning the genomes of populations inhabiting different habitats to pinpoint outlier loci displaying an excess of population differentiation, while the other entails conducting GWAS between molecular markers and environmental variables treated as phenotypes (Li et al. 2017; Gould et al. 2018; Pais et al. 2020).

Integrating both methods proves valuable in dissecting the genetic basis of climate adaptation, especially for tree species where QTL mapping and reciprocal transplantation face challenges due to extended life cycles. For instance, Pais et al. (2020) utilized both genome scan and GWAS methods to identify genetic variants indicating signals of local adaptation in an endangered flowering dogwood tree (*Cornus florida* L.). Integrating various omics technologies in systems biology has facilitated the identification of candidate genes responsible for complex trait variation. Transcriptomics, proteomics, and metabolomics quantitatively assess the activities of transcripts, proteins, and metabolites, contributing to predicting candidate genes or associating proteins, metabolites, and pathways with traits of interest.

Over the past decade, numerous studies have explored methods for dissecting the genetic basis of climatic adaptation and identifying candidate genes involved in environmental adaptation (Ding et al. 2020). A comprehensive literature review by Franks and Hoffmann (2012) suggested the involvement of candidate genes, genetic regulatory networks, and epigenetic effects in climate change adaptation. Ogura et al. (2019) employed a genomic approach (GWAS) to identify an Arabidopsis gene (EXO70A3) influencing drought tolerance, highlighting the complementary nature of genomic tools and techniques.

6 Assisted Migration

Assisted migration involves intentionally relocating species to simulate or facilitate natural range expansion. This approach can be particularly beneficial for locally adapted and long-lived populations or species, especially when their natural migration capacity is limited or their habitats face threats like fragmentation (Dumroese et al. 2015). A study using genotypes of Fremont cottonwood (*Populusfremontii* Wats.) from 16 populations across its thermal range demonstrated that moving northern populations with beneficial plasticity to warmer regions allowed them to withstand substantial warming before reaching a mortality threshold. Conversely, southern populations relocated to higher elevations and latitudes exhibited non-adaptive, unfavorable plasticity responses (Fang et al. 2017). As a result, active management practices such as assisted migration become essential. Climate change shifts the frost line northward, making it prudent to plant genotypes from lower elevations or latitudes adapted to longer growing seasons, especially in regions with mild winters. Long-lived trees can leverage plasticity to adapt to a rapidly changing environment (Fang et al. 2017).

Ensuring the survival of forest tree species in the face of climate change necessitates matching seed sources to the anticipated climates of the coming decades. Short-distance transfers can help mitigate uncertainty about the scale of climate change by enhancing gene flow among populations and planting more diverse seed sources among and within forest stands. Shifting climates may render current species or populations non-adapted.

Understanding the processes leading to the creation of refugia from contemporary climate change is advantageous for preserving genetic diversity (Keppel et al. 2012; Singh and Kumar 2022; Singh et al. 2024b). Assisted migration represents a departure from traditional natural resource management and conservation biology practices. However, it is approached by considering species and population vulnerability to climate change, selecting options and management targets, and long-term monitoring (Hanewinkel et al. 2013; Yadav et al. 2019a, b; Jangra et al. 2024). These procedures can evolve into effective, cost-efficient, and practical strategies for addressing climate change. Local assisted migration initiatives are underway in Canada and British Columbia (Keppel et al. 2012).

7 Adaptive Forest Management Strategies

Climate change is anticipated to lead to shifts in forest types across most states, causing irreversible damage, potentially driving several species to extinction, and impacting livelihoods and markets (Kanta et al. 2024; Malik et al. 2024). Forest managers and policymakers must address climate change's ecological and socio-economic impacts, emphasizing the need to integrate climate considerations into long-term forest policy and management plans (Ravindranath et al. 2011; Kanta et al. 2024; Malik et al. 2024). Although the Government has implemented progressive forest policies in the past, it is necessary to tailor these policies to incorporate climate change adaptation and ensure effective implementation (Kumar and Singh 2024). The following potential policies, strategies, and practices are discussed.

7.1 *Effective Implementation of Existing Policies/Acts/Guidelines*

Enhance and ensure the effective implementation of existing policies/acts/guidelines such as the Forest Conservation Act of 1980, the Wildlife Protection Act of 1972 and 2002, and wildlife conservation programs like Project Tiger and Project Elephant.

7.2 *Expand Protected Areas*

Increasing the area under Protected Areas connects Protected Areas, Reserve Forests, and Wildlife Reserves, especially in mountainous regions. Involve local communities in addressing their dependence on forests for livelihoods and providing alternatives.

7.3 *Adaptation Plans for Wildlife Species*

Formulate specific plans to help wildlife, including flagship and endangered species, migrate and adapt to the impacts of climate change. Base these plans on existing knowledge of species ecology, distribution, and anticipated climate change.

7.4 *Afforestation and Reforestation Under Climate Change*

Incorporate adaptation practices and strategies into afforestation programs, such as mixed species forestry and planting ecologically compatible species. Consider climate impact studies when undertaking afforestation and reforestation efforts.

7.5 *Improved Silvicultural Practices*

Develop and adopt silvicultural practices that reduce vulnerability and enhance resilience. Promote natural regeneration, mixed species forestry, anticipatory planting, in-situ and ex-situ conservation, short rotation species, and sustainable harvest practices.

7.6 *Promoting Community Forestry*

Encourage community involvement in raising, protecting, and managing plantations and forests. Adopt the Joint Forest Management model to promote community forestry.

7.7 *Forest Fire Management*

Formulate innovative and region-specific fire management strategies, considering the varying fire risks and impacts in different regions. Develop strategies for managing different ecosystems, such as grass production or tree species management, based on the fire risk in each area.

8 Integrative Concept of Adaptive Forest Management

Considering the factors above, we propose a holistic framework for adaptive forest management, recognizing the interconnectedness of actions across various spatial scales concerning greenhouse gas (GHG) mitigation and forest adaptation. Local initiatives significantly impact the effectiveness of global GHG mitigation endeavors, influencing climate change dynamics and local adaptation strategies. To foster effective adaptive forest management, international assessments of species/provenance suitability and modeling activities should be conducted on a large scale. This involves evaluating the distribution ranges of both native and non-native species and their provenances. Establishing a comprehensive trial system spanning temperate and boreal zones across continents like Europe would serve as a valuable platform for experiments, testing novel forest stand types, incorporating diverse tree species and provenances, generating field data crucial for modeling species and provenance suitability, as well as predicting future competition dynamics.

The responsibility of national or regional authorities to optimize suitability and efficacy at the local level is to map out priorities for adaptation strategies. Decision-makers must prioritize, considering the urgency and feasibility of adaptation strategies in different localities. Furthermore, at these levels, potential conflicts between adaptation and mitigation goals can be identified and effectively addressed (Klein et al. 2005).

On the local scale, the implementation of specific on-ground adaptation measures falls under the purview of forest practitioners, a topic not extensively explored here. Stakeholders engage in a dual decision support pathway, involving decision-making from higher to lower levels and knowledge transfer from research to regional and local forest administrations. These administrations, in turn, communicate their findings to the responsible regional and local bodies (Bolte et al. 2009).

9 Forest-Water Nexus into Adaptive Forest Management

Forests are essential for maintaining vital ecosystem services, including water supply regulation, climate stability, erosion mitigation, and sediment retention. These services are the foundation for generating other ecosystem benefits like carbon sequestration, aquatic habitat, and biodiversity. Climate change effects such as heat waves, droughts, wildfires, and disease spread alter forest dynamics, impacting hydrological functions and water-related ecosystem services (Trubin et al. 2022). Consequently, this affects forest composition, structure, distribution, and growth, leading to increased demand for ecosystem services, particularly water production and regulation, with indirect implications for carbon sequestration. Despite the prevalent focus on forest adaptation and resilience, current approaches often prioritize carbon sequestration through management and conservation, overlooking critical water-associated ecosystem services (Sterck et al. 2021; Zhou et al. 2021). Given the impact of climate change on the forest-water nexus, integrating adaptive forest management into broader climate change mitigation strategies is crucial for safeguarding water resources in the future.

10 Approaches for Mitigation

Numerous studies advocate for a more pragmatic approach to mitigation within the forestry sector. Wüstemann et al. (2017) investigated how forest management practices in Germany could contribute to mitigation outcomes from ecosystem-based perspectives. Other research focuses on safeguarding forest resources to enhance carbon sinks and ecosystem services, considering livelihoods, landscape conservation, ecological improvement, agricultural benefits, and biodiversity. Some studies emphasize instruments to facilitate mitigation at the climate project level (Chia et al. 2016; Salvini et al. 2016; Few et al. 2017). Similar to REDD + studies, certain research explores how forests and resources can be optimized for mitigation objectives, yielding benefits in sectors beyond adaptation or mitigation. Failing to incorporate a holistic assessment may lead to trade-offs, exemplified by urban greening. Planting trees in urban areas can sequester carbon and mitigate heat island effects, but choosing inappropriate species may compromise biodiversity and introduce health issues (Salmond et al. 2016).

Similarly, a narrow parameter perspective may obscure or undermine positive synergies. Rahn et al. (2014) demonstrated various expected benefits from an agroforestry system, encompassing productivity improvements and enhanced adaptive capacity to mitigate temperature increases. Shading trees in agroforestry serve multiple purposes, including timber, fruits, firewood, and nitrogen fixation, concurrently strengthening the adaptive capacity of local farmers. The challenge lies in comprehending intricate interplays that might not provide straightforward evidence, where components of the agroforestry system contribute to positive synergies. Acknowledging these effects is crucial for fostering momentum in mitigation approaches.

11 Interaction of Mitigation and Adaptation

Differences between adaptation and mitigation manifest across spatial, temporal, and sectoral scales. Adaptation primarily addresses local vulnerability in the short term, while mitigation is an international concern focused on long-term climate change effects. Efforts to integrate adaptation and mitigation have prompted studies seeking synergy in climate approaches.

The adaptation and mitigation approaches identified in this review exhibit variability in practices contingent on contexts and objectives. Integration development is tied to a vector of integration, indicating that an adaptation-oriented approach requires space for mitigation to play a role and vice versa. For instance, when higher vulnerability prompts a demand for an adaptation approach, it's unlikely that a mitigation approach will be prioritized over adaptation. A mitigation approach must be integrated into an adaptation framework in such cases. While mitigation targets GHG reductions, adaptation approaches vary based on the addressed vulnerabilities. Therefore, mobilizing an adaptation approach based on the local context is crucial to achieving synergy effects (Thornton and Comberti 2017).

To achieve positive synergy, mitigation instruments may be effective when facilitated to encompass adaptation outcomes. Forestry and agriculture, sharing equal benefits in some cases, possess significant capacity to generate synergistic effects, as seen in agroforestry practices, including increased carbon sequestration, agricultural productivity, and improved livelihoods in local communities (Rahn et al. 2014).

Implementation challenges include species selection for shading trees, cost-effectiveness, political arrangements, and funding. Synergy effects also emerge when practices are considered in broader contexts, including social and economic dimensions or cross-sectoral aspects, exploring ways to reduce vulnerability and enhance resilience in forest-dependent communities. This extends to benefits in livelihoods, community progress, poverty reduction, urban development, and transport systems (Rahn et al. 2014; Thornton and Comberti 2017).

12 Forest Resilience and Genetic Diversity

Resilience, perceived as an emerging characteristic in ecosystems, is granted across different scales by genetic diversity, species, functional groups, and system processes. Safeguarding or reinstating the resilience of forests is deemed crucial for societal adaptation to climate change. Forests, by nature, exhibit resistance to certain changes, remaining relatively stable within non-catastrophic disturbances like insect herbivory or minor blowdowns. Due to redundancy among functional species, they may also resist environmental alterations, such as gradual shifts in weather patterns (Drever et al. 2006).

While resilience is evident across various biodiversity levels, genetic composition's role is particularly noteworthy. Molecular genetic diversity within a species, the diversity of species within a forested community, and the broader ecosystem diversity across landscapes all represent facets of biological diversity. At the core of these facets lies the genotypic variation present in populations. Individual organisms within populations contribute to adaptive capacity and resilience through natural selection, ensuring the adaptability of tree species and forest ecosystems. This diversity has historically played a crucial role in forest ecosystems' natural (and artificial) regeneration, allowing for adaptation to significant climate changes during the quaternary period. In the face of anticipated anthropogenic climate warming, preserving this diversity becomes paramount.

The genetic diversity within a species, encompassing additive genetic variance, is pivotal as it forms the foundation for natural selection and adaptation to environmental changes. Long-term programs grounded in artificial selection have illustrated the potential for evolutionary change. Genetic strategies aimed at reforestation amidst rapid climate change must prioritize maintaining both species and genetic diversity within species. It is crucial to acknowledge that forested ecosystems' genetic diversity and adaptive capacity rely significantly on in situ genetic variation within each species' population. Populations exposed to environmental changes surpassing their adaptive or dispersal capabilities may confront the threat of extinction. Genetic diversity dictates a species' range of fundamental eco-physiological tolerances, influencing inter-specific competitive interactions, which, in conjunction with dispersal mechanisms, are fundamental determinants of potential species responses to change. Throughout history, plants have navigated significant climate changes through adaptation and migration (Thompson et al. 2009).

13 Forest Restoration

Forest restoration is a set of forest management activities geared towards increasing forest cover, structure, function, and/or species structure. The objectives and approaches of forest restoration include strategies like "reclamation" to establish forests on land lacking vegetation, "rehabilitation" to achieve desired structure or species composition, and "reconstruction" to re-establish native plant communities on land formerly used for other purposes (McDonald et al. 2016).

Forest restoration encompasses diverse management practices involving the planting or natural regeneration of native or non-native tree species, with or without tree harvest or removal. Examples of forest restoration efforts include managed forest plantations of native or non-native species, reforestation with non-native species, and native forest regeneration.

Historically, some of the earliest instances of forest restoration were "reclamation" efforts in regions like Europe, Asia, South America, Africa, and the United States. For example, in the early twentieth century, plantations of non-native Pinus radiata and Eucalyptus globulus were established in Chile to counteract environmental issues caused by 19th-century forest clearing. Similarly, in South Africa, non-native Eucalyptus and Pinus plantations were created on former grassland and shrubland in the early twentieth century. In the southern United States, native pine plantations were established in the 1950s on abandoned cotton and tobacco farmlands after the U.S. Civil War (Fox et al. 2007).

Recent forest restoration efforts include "reclamation" and "rehabilitation," where plantations are established on previously either non-forested or forested land. For instance, from 2000 to 2012, intensively managed forest plantations resulted in both loss and gain of forests in the subtropical climate domain, including regions like the south-eastern United States, South Africa, central Chile, south-eastern Brazil, Uruguay, southern China, Australia, and New Zealand. Forest plantations, when managed to meet certification objectives and standards, contribute to forest restoration (Upton 2019).

Contemporary forest restoration often involves "rehabilitation" and "reconstruction" efforts, focusing on the early stages of native forest restoration. This aligns with the concept of ecological restoration, aiming to assist the recovery of degraded ecosystems and reconstruct them to their pre-degraded state, adjusting for changed conditions (McDonald et al. 2016). Native forest restoration is increasingly considered a "nature-based" strategy to counter the impacts of climate change.

14 Habitat Diversity Restoration

The habitat diversity paradigm posits that maximizing the number of target species in restoration efforts focuses on increasing habitat diversity. Various approaches, both passive and active, have been employed to target habitat diversity, broadly falling into four strategies: (i) varying the starting conditions of secondary succession, (ii) emulating natural disturbances, (iii) establishing or enhancing keystone structures, and (iv) applying ecosystem engineer species. The first two strategies indirectly address habitat diversity by modifying ecosystem processes or states, while the latter two directly target habitat diversity.

i. **Vary the Starting Conditions of Succession**

The most cost-effective approach involves assisting ecological processes like succession, hydrological processes, primary productivity, or decomposition that lead to increased habitat diversity. Preservation or enhancement of naturally occurring ecological processes, particularly succession, plays a crucial role. Examples include the development of meanders in free-flowing rivers and hydro-morphological changes within riverbeds.

ii. **Emulating Natural Disturbances**

Post-restoration succession can lead to reduced species diversity. Emulating natural disturbances becomes essential to maintain intermediate states that support high diversity. For instance, maintaining low-intensity use in temperate montane hay meadows through mowing or grazing prevents succession into closed forests. The 'intermediate landscape diversity hypothesis' suggests that maintaining high-diversity intermediate states requires active habitat management to control the succession rate.

iii. **Establishing Keystone Structures**

Keystone structures are spatial features that provide essential resources for other species. Examples include dolinas in karst areas and wood placement in river and stream restoration. Wood introduces cost-effective habitat diversity, alters flow conditions, retains organic material, and provides surfaces for algae and invertebrates.

iv. **Applying Ecosystem Engineer Species**

Ecosystem engineers, species that create or modify habitats, are valuable tools for increasing habitat diversity. Beavers, for instance, alter flow conditions and create pools or ponds in freshwater ecosystems. Similar impacts are observed with bivalves in marine systems, wild boars in wetlands, and grassland prairie dogs.

15 Climate Smart Forestry

Climate Smart Forestry (CSF) has increased usage in various domains, including academia, policymaking, planning, conservation NGOs, and certification standards. Multiple disciplines, such as forestry and environmental sciences, have adopted CSF, and its principles are well-defined in recent literature, as exemplified by a recent definition proposed by Verkerk et al. (2020). The fundamental tenets of CSF include;

1. **Increasing Carbon Storage**: CSF emphasizes the enhancement of carbon storage in both forests and wood products. This aligns with providing multiple ecosystem services beyond carbon storage.
2. **Enhancing Human Health and Community Resilience**: Adaptive forest management is integral to CSF, focusing on improving human health and fostering community resilience. This reflects a holistic approach that goes beyond traditional forest management practices.
3. **Sustainable Use of Wood Resources**: CSF advocates using wood resources as a substitute for non-renewable and carbon-intensive materials. This underscores the importance of responsible and eco-friendly practices in utilizing wood resources.

The term 'sustainable' is a common thread in CSF applications, prompting a reflection on sustainable forest management (SFM). SFM, a longstanding focus in forestry research since the 1980s, is closely associated with sustainable development. SFM traditionally emphasizes productive forest landscapes, often called 'working' forests, focusing on sustained production to meet societal needs over time. While SFM has evolved to incorporate additional forest values, such as habitat provisioning, some criticisms have been raised. Scholars argue that SFM, rooted in industrialized and development-oriented frameworks, may not adequately encompass socio-cultural values. Critics, particularly political ecologists, contend that industrial forestry prioritizes economic production over other values (Agnoletti and Santoro 2015).

Risks Associated with the Current Framing of CSF

1. **Oversimplified Carbon Sequestration Relationships**

The current CSF framing risks oversimplifying the intricate processes governing forest carbon storage. Simplistic views may lead to undervaluing late succession, primary, or old-growth forests due to misconceptions about carbon sequestration diminishing after an optimal harvest year. Evidence suggests that older trees and forests can continue high-rate carbon sequestration well beyond this presumed threshold.

2. **Aboveground Tree Volume Emphasis**

The focus on aboveground biomass as the primary metric for measuring forest carbon stocks poses challenges. Most forest carbon inventories concentrate on aboveground biomass, potentially biasing estimates towards carbon stored in tree boles. This approach may neglect carbon estimates in other tree parts and forests, resulting in biased model predictions.

3. **'Carbonization' Over Resilience and Biodiversity**

The overemphasis on carbon sequestration in CSF, termed 'carbonization' of forest governance, may neglect other critical forest values. This approach might overlook adaptive mitigation strategies crucial for forest resilience, particularly in the face of climate change. The potential consequences include neglecting biodiversity, essential for maintaining ecological functions and adapting to future challenges.

4. **Misinterpretation of Forest Ecosystem Management and Climate Effects**

The dominance of carbon sequestration as the primary CSF indicator may lead to misinterpretations in understanding non-carbon forest interactions with climate. Forest climate effects are complex and cannot be solely derived from biogenic carbon estimates. Additionally, fossil fuel use in forestry operations and transportation undermines the calculated climate benefits of harvested wood products.

5. **Long-term and End-of-life Carbon Storage Issues**

CSF may overlook challenges related to the long-term and end-of-life carbon storage of wood products. While forests under sustainable management yield crucial commodities, the fate of wood products after use, including incineration and landfill, poses challenges. Carbon released during decomposition in landfills contributes to emissions, emphasizing the need to consider the entire life cycle of wood products in CSF strategies.

Enhanced Framework for CSF

Upon thorough examination, there is a significant opportunity to enhance the CSF concept by including a more comprehensive range of activities that improve climate benefits and contribute to social and environmental co-benefits. In addition to the existing three pillars, we advocate for an expanded definition by introducing two new pillars to shape an 'Enhanced' CSF framework:

1. **Preservation of Natural Places**: This entails preventing the loss of forests, intact ecosystems, forest complexity, biodiversity, or connectivity. The focus is averting conversion to higher management intensity, safeguarding natural landscapes' inherent value and ecological attributes.
2. **Advocacy for Landscape Restoration**: Emphasizing the promotion of landscape restoration, this pillar targets enhancing ecosystem function and connectivity within degraded landscapes. Fostering the recovery of ecosystems aims to elevate overall environmental health and resilience, contributing to the broader goal of sustainable and climate-smart forestry practices.

Components of an Enhanced CSF Framework

The subsequent sections delve into the fundamental components of the proposed Enhanced CSF framework, outlining the intricacies of each phase.

1. **Assessment of Current Forest Condition and Use**

In expanding the CSF paradigm, it is imperative to establish additional connections encompassing diverse forest conditions and their potential climate benefits. While the conventional CSF framework predominantly centres on carbon in productive forests, the enhanced approach extends its scope to include landscapes such as degraded areas, savannas, trees outside of forests, and pristine areas with minimal human intervention.

2. **Calculation of Carbon Storage and Greenhouse Gas Fluxes**

Enhanced CSF interventions must move beyond a singular focus on sequestration rates and carbon storage. A comprehensive evaluation of actual and potential greenhouse gas fluxes is vital. By adopting a broader perspective, the framework can include additional indicators, offering a more nuanced understanding of climate benefits, resilience, and the preservation of substantial carbon pools.

3. **Determination of Strategy and Tactics**

Unlike the traditional focus on 'productive' or 'managed' forests, the Enhanced CSF strategy involves decisions concerning various forested landscapes. The optimal mix of tactics, guided by greenhouse gas information, could encompass afforestation, reforestation, improved forest management, Reduced Impact Logging (RIL), and the avoidance of forested land conversion. These strategies should be explicitly linked within the Enhanced CSF framework to avoid pitfalls and ensure a holistic approach.

4. **Consideration of Feasibility and Implementation**

Beyond the typical feasibility considerations of carbon projects, enhanced CSF broadens its scope to create more inclusive incentive structures. The framework acknowledges social dimensions, governance, participation, and the inclusion of grievance mechanisms as crucial aspects of feasibility. This inclusive approach aims to overcome social barriers, enhance scalability, and incentivize diverse and robust CSF interventions.

5. **Assessment of Broader Impacts of CSF Strategy**

Enhanced CSF operates beyond parcel-level initiatives, viewing them as components of strategies geared towards optimal landscape-scale outcomes. It necessitates a holistic assessment of the broader impacts, including balancing production with protection and the connection of natural areas for restoration. Effective benefit distribution at various levels is crucial, from forest decision-makers to wood users. The framework emphasizes integrated approaches to sustainable land use, improved linkages between carbon stocks and landscape-scale planning, and the incorporation of social and environmental safeguards to ensure equitable and efficient CSF implementation.

16 Conclusion

It concludes by exploring biological insights into forest adaptation to climate change, which reveals the intricate mechanisms governing ecosystem resilience and adaptation. Forests, comprising diverse elements such as genes, species, and ecological processes, exhibit remarkable adaptive capacities, emphasizing the dynamic interplay between biological diversity and environmental challenges. The genetic composition of species plays a pivotal role in resilience, with genetic variability within populations serving as the foundation for adaptation. Biological diversity, from individual species to broader ecosystem diversity, is crucial for rejuvenating and adapting forest ecosystems, particularly amidst rapid climate change. Historical lessons underscore the importance of preserving genetic diversity and capacity as forests face anthropogenic climate warming. Forest restoration strategies emphasize the central role of biological diversity in enhancing forest function and resilience. However, challenges exist, especially in climate-smart forestry, requiring a nuanced approach considering ecological, social, and economic dimensions. A proposed enhanced Climate Smart Forestry framework addresses these challenges, emphasizing a balanced strategy and comprehensive impact assessments. In essence, biological insights emphasize the role of biodiversity in shaping the future of forests amidst climate change, necessitating sustainable management practices and innovative frameworks for their survival and health.

References

Adger WN, Agrawala S, Mirza MMQ, Conde C, Brien KO, Pulhin J, Pulwarty R, Smit B, Takahashi K (2007) Assessment of adaptation practices, options, constraints and capacity, climate change 2007: impacts, adaptation and vulnerability. In: Contribution of working group II to the fourth assessment report of the intergovernmental panel on climate change. WMO, Geneva

Agnoletti M, Santoro A (2015) Cultural values and sustainable forest management: the case of Europe. J for Res 20:438–444

Alberto FJ, Aitken SN, Alía R, González-Martínez SC, Hänninen H, Kremer A, Lefèvre F, Lenormand T, Yeaman S, Whetten R, Savolainen O (2013) Potential for evolutionary responses to climate change–evidence from tree populations. Glob Change Biol 19(6):1645–1661

Al-Lawati A, Al-Bahry S, Victor R, Al-Lawati AH, Yaish MW (2016) Salt stress alters DNA methylation levels in alfalfa (Medicago spp). Genet Mol Res 15(1):15018299

Anderson-Frey AK (2017) Statistical examination of tornado report and warning near-storm environments. Ph.D. Thesis, The Pennsylvania State University, University Park, PA, USA

Barati AA, Zhoolideh M, Azadi H, Lee JH, Scheffran J (2023) Interactions of land-use cover and climate change at global level: How to mitigate the environmental risks and warming effects. Ecol Ind 146:109829

Bewick AJ, Schmitz RJ (2017) Gene body DNA methylation in plants. Curr Opin Plant Biol 36:103–110

Bolte A, Ammer C, Löf M, Madsen P, Nabuurs GJ, Schall P, Spathelf P, Rock J (2009) Adaptive forest management in central Europe: climate change impacts, strategies and integrative concept. Scand J for Res 24(6):473–482

Boyko A, Blevins T, Yao Y, Golubov A, Bilichak A, Ilnytskyy Y, Hollander J, Meins F Jr, Kovalchuk I (2010) Transgenerational adaptation of Arabidopsis to stress requires DNA methylation and the function of Dicer-like proteins. PLoS ONE 5(3):e9514

Chia EL, Fobissie K, Kanninen M (2016) Exploring opportunities for promoting synergies between climate change adaptation and mitigation in forest carbon initiatives. Forests 7(1):24

Cohen S, Demeritt D, Robinson J, Rothman D (1998) Climate change and sustainable development: towards dialogue. Glob Environ Chang 8(4):341–371

Denton F, Wilbanks TJ, Abeysinghe AC, Burton I, Gao Q, Lemos MC, Masui T, O'Brien K, Warner K (2014) Climate-resilient pathways: adaptation, mitigation, and sustainable development. Clim Change, pp 1101–1131

Dhyani K, Kuniyal HB, Singh H, Sobha (2021) Growth and physiological potential of *Terminalia arjuna* under elevated CO_2 levels in Open top chamber condition. J Appl Natural Sci 13(3):1121–1126. https://doi.org/10.31018/jans.v13i3.2490. ISSN 2231-5209

Ding Y, Shi Y, Yang S (2020) Molecular regulation of plant responses to environmental temperatures. Mol Plant 13(4):544–564

Drever CR, Peterson G, Messier C, Bergeron Y, Flannigan M (2006) Can forest management based on natural disturbances maintain ecological resilience? Can J for Res 36(9):2285–2299

Dumroese RK, Williams MI, Stanturf JA, Clair JBS (2015) Considerations for restoring temperate forests of tomorrow: forest restoration, assisted migration, and bioengineering. New for 46:947–964

Fang XW, Zhang JJ, Xu DH, Pang J, Gao TP, Zhang CH, Li FM, Turner NC (2017) Seed germination of Caragana species from different regions is strongly driven by environmental cues and not phylogenetic signals. Sci Rep 7(1):11248

Few R, Martin A, Gross-Camp N (2017) Trade-offs in linking adaptation and mitigation in the forests of the Congo Basin. Reg Environ Change 17:851–863

Fox TR, Jokela EJ, Allen HL (2007) The development of pine plantation silviculture in the southern United States. J Forest 105(7):337–347

Franks SJ, Hoffmann AA (2012) Genetics of climate change adaptation. Annu Rev Genet 46:185–208

Gharbia SS, Smullen T, Gill L, Johnston P, Pilla F (2018) Spatially distributed potential evapotranspiration modeling and climate projections. Sci Total Environ 633:571–592

Goldsmith GR, Allen ST, Braun S, Siegwolf RT, Kirchner JW (2022) Climatic influences on summer use of winter precipitation by trees. Geophys Rese Lett 49(10):e2022GL098323

Gould BA, Chen Y, Lowry DB (2018) Gene regulatory divergence between locally adapted ecotypes in their native habitats. Mol Ecol 27(21):4174–4188

Gugger PF, Fitz-Gibbon S, PellEgrini M, Sork VL (2016) Species-wide patterns of DNA methylation variation in Quercus lobata and their association with climate gradients. Mol Ecol 25(8):1665–1680

Gupta SK, Singh H (2017) Observations on Sporadic Flowering in *Ficus benjamina* L. and *Peltophorum africanum* L. in New Forest. Indian Forester 143:290–291

Gupta SK, Ram J, Singh H (2018a) Comparative study of transpiration in cooling effect of tree species in the atmosphere. J Geosci Environ Prot 6:151–166. https://doi.org/10.4236/gep.2018.68011

Gupta SK, Ram J, Singh H, Kumar P (2018b) Comparative study of cooling under shades of some forestry tree species with respect to ambient temperatures. J Geosci Environ Prot 6:51–66. https://doi.org/10.4236/gep.2018.611004

Gupta SK, Ram J, Singh H (2019a) Relationship between leaf area index and atmospheric cooling of tree species. Indian J for 42(3):203–208

Gupta SK, Singh H, Ram J (2019b) Need for a novel paradigm in forestry research: perspectives of comparative atmospheric cooling by different forestry effect tree species. Indian Forester 145(6):509–515

Hanewinkel M, Cullmann DA, Schelhaas MJ, Nabuurs GJ, Zimmermann NE (2013) Climate change may cause severe loss in the economic value of European forest land. Nat Clim Chang 3(3):203–207

Hasbún R, Iturra C, Bravo S, Rebolledo-Jaramillo B, Valledor L (2016) Differential methylation of genomic regions associated with heteroblasty detected by M&M algorithm in the nonmodel species Eucalyptus globulus Labill. Int J Genom

Husain M, Vishwakarma DK, Rathore JP, Rasool A, Parrey AA, Mahendar K (2018) Local people strategies in biodiversity conservation and sustainable development. Pharma Innov J 7(1):444–450

IPCC (2018) Global warming of 1.5° C. World Meteorological Organization, Geneva, Switzerland

IPCC (2021) Climate change 2021: the physical science basis. In: Contribution of working group I to the sixth assessment report of the intergovernmental panel on climate change. Cambridge University Press. In Press

Iwasaki M, Paszkowski J (2014) Epigenetic memory in plants. EMBO J 33(18):1987–1998

Jain P, Sajjad H (2016) Household dependency on forest resources in the Sariska Tiger Reserve (STR), India: Implications for management. J Sustain for 35(1):60–74

Jangra A, Kumar K, Maikhuri S, Bhandari MS, Pandey S, Singh H, Barthwal S (2024) Unveiling stress-adapted endophytic bacteria: Characterizing plant growth-promoting traits and assessing cross-inoculation effects on *Populus deltoides* under abiotic stress. Plant Physiol Biochem 210:108610. https://doi.org/10.1016/j.plaphy.2024.108610

Jawo TO, Kyereh D, Lojka B (2023) The impact of climate change on coffee production of small farmers and their adaptation strategies: a review. Climate Dev 15(2):93–109

Kanta C, Kumar A, Chauhan A, Singh H, Sharma IP (2024) The interplay between plant functional traits and climate change. In: Kumar N, Singh H (eds) Plant functional traits for improving productivity. Springer, Singapore. https://doi.org/10.1007/978-981-97-1510-7_3

Keppel G, Van Niel KP, Wardell-Johnson GW, Yates CJ, Byrne M, Mucina L, Schut AGT, Hopper SD, Franklin SE (2012) Refugia: identifying and understanding safe havens for biodiversity under climate change. Glob Ecol Biogeogr 21(4):393–404

Kinoshita T, Seki M (2014) Epigenetic memory for stress response and adaptation in plants. Plant Cell Physiol 55(11):1859–1863

Klein RJ, Schipper ELF, Dessai S (2005) Integrating mitigation and adaptation into climate and development policy: three research questions. Environ Sci Policy 8(6):579–588

Kumar N, Singh H (2024) Plant functional traits for improving productivity. Springer Nature, Singapore. https://doi.org/10.1007/978-981-97-1510-7

Kumar N, Suyal DC, Sharma IP, Verma A, Singh H (2017) Elucidating stress proteins in rice (*Oryza sativa* L.) genotype under elevated temperature: a proteomic approach to understand heat stress response. 3 Biotech 7:205. https://doi.org/10.1007/s13205-017-0856-9

Kumar M, Singh MP, Singh H, Dhakate PM, Ravindranath NH (2019a) Forest working plan for the sustainable management of forest and biodiversity in India. J Sustain for 39:1–22. https://doi.org/10.1080/10549811.2019.1632212

Kumar N, Jeena N, Singh H (2019b) Elevated temperature modulates rice pollen structure: a study from the foothill of the Himalayan agro-ecosystem in India. 3 Biotech 9(5):175. https://doi.org/10.1007/s13205-019-1700-1

Kumar N, Jeena N, Kumar N, Gangola S, Singh H (2019c) Phytoremediation facilitating enzymes: an enzymatic approach for enhancing remediation process. In: Bhatt P (ed) Smart bioremediation technologies. Academic Press, Elsevier, pp 289–306. 50 Hampshire Street, 5th Floor, Cambridge, MA 02139, United States

Kumar N, Singh H, Sharma SK (2020) Antioxidants: responses and importance in plant defense system. In: Roychowdhury R, Choudhury S, Hasanuzzaman M, Srivastava S (eds) Sustainable agriculture in the era of climate change. Springer, Cham, pp 251–264. https://doi.org/10.1007/978-3-030-45669-6_11

Kumar A, Kumar P, Singh H, Kumar N (2021a) Modulation of plant functional traits under essential plant nutrients during seasonal regime in natural forests of Garhwal Himalayas. Plant Soil 465:197–212. https://doi.org/10.1007/s11104-021-05003-x

Kumar A, Kumar P, Singh H, Bisht S, Kumar N (2021b) Relationship of physiological plant functional traits with soil carbon stock in the temperate forest of Garhwal Himalaya. Curr Sci 120(8):1368–1373. https://doi.org/10.18520/cs/v120/i8/1368-1373

Kumar A, Kumar P, Singh H, Kumar N (2021c) Adaptation and mitigation potential of roadside trees with bio-extraction of heavy metals under vehicular emissions and their impact on physiological traits during seasonal regimes. Urban for Urban Green 58:126900. https://doi.org/10.1016/j.ufug.2020.126900

Kumar A, Kumar P, Singh H, Kumar N (2021d) Impact of plant functional traits on infiltration rate under different forests of Kempty watershed in Garhwal Himalaya. Indian J Soil Cons 49(1):38–44

Kumar M, Phukon SN, Paygude AC, Tyagi K, Singh H (2021e) Mapping Phenological Functional Types (PhFT) in the Indian Eastern Himalayas using machine learning algorithm in Google Earth Engine. Comput Geosci 158:104982. https://doi.org/10.1016/j.cageo.2021.104982

Kumar A, Singh H, Kumari G, Bisht S, Malik A, Kumar A, Singh M, Raturi BS, Thakur A, Kaushal R (2022) Adaptive resilience of roadside trees to vehicular emissions via leaf enzymatic, physiological, and anatomical trait modulations. Environ Pollut 313:120191. https://doi.org/10.1016/j.envpol.2022.120191

Kumar N, Singh H, Giri K, Kumar A, Joshi A, Yadav A, Singh R, Bisht S, Kumari R, Jeena N, Khairakpam R, Mishra G (2024) Physiological and molecular insights into the allelopathic effects on agroecosystems under changing environmental conditions. Physiol Mol Biol Plants 30:417–433. https://doi.org/10.1007/s12298-024-01440-x

Lang-Mladek C, Popova O, Kiok K, Berlinger M, Rakic B, Aufsatz W, Jonakb C, Hauser M, Luschnig C (2010) Transgenerational inheritance and resetting of stress-induced loss of epigenetic gene silencing in Arabidopsis. Mol Plant 3(3):594–602

Li Y, Zhang XX, Mao RL, Yang J, Miao CY, Li Z, Qiu YX (2017) Ten years of landscape genomics: challenges and opportunities. Front Plant Sci 8:2136

Li R, Hu F, Li B, Zhang Y, Chen M, Fan T, Wang T (2020) Whole genome bisulfite sequencing methylome analysis of mulberry (Morus alba) reveals epigenome modifications in response to drought stress. Sci Rep 10(1):8013

Liang D, Zhang Z, Wu H, Huang C, Shuai P, Ye CY, Tang S, Wang Y, Yang L, Wang J, Yin W, Xia X (2014) Single-base-resolution methylomes of Populustrichocarpa reveal the association between DNA methylation and drought stress. BMC Genet 15(1):1–11

Malik A, Kumar K, Kumari G, Kumar N, Singh H (2024). Application of functional traits in modelling productivity and resilience under climate change. In: Kumar N, Singh H (eds) Plant functional traits for improving productivity. Springer, Singapore. https://doi.org/10.1007/978-981-97-1510-7_5

McDonald T, Gann GD, Jonson J, Dixon K (2016) International standards for the practice of ecological restoration-including principles and key concepts. Society for Ecological Restoration, Washington, DC, USA

Meena D, Singh H, Chaudhari SK (2011) Elucidating strontium response on growth dynamics and biochemical change in *Phaseolus mungo* L. Int J Agric Environ Biotechnol 4:107–113

Miryeganeh M, Marlétaz F, Gavriouchkina D, Saze H (2022) De novo genome assembly and in natura epigenomics reveal salinity-induced DNA methylation in the mangrove tree Bruguieragymnorhiza. New Phytol 233(5):2094–2110

Ogura T, Goeschl C, Filiault D, Mirea M, Slovak R, Wolhrab B, Satbhai SB, Busch W (2019) Root system depth in Arabidopsis is shaped by EXOCYST70A3 via the dynamic modulation of auxin transport. Cell 178(2):400–412

Pais AL, Whetten RW, Xiang QY (2020) Population structure, landscape genomics, and genetic signatures of adaptation to exotic disease pressure in Cornusflorida L.—Insights from GWAS and GBS data. J Systemat Evol 58(5):546–570

Prakash V, Hunney K, Singh H (2022) Effect of elevated carbon dioxide on growth and development of *Santalum album* L. seedlings inoculated with plant growth promoting microorganisms in Open Top Chambers. Int J Environ Agric Biotechnol 7(5):149–159

Rahn E, Läderach P, Baca M, Cressy C, Schroth G, Malin D, van Rikxoort H, Shriver J (2014) Climate change adaptation, mitigation and livelihood benefits in coffee production: where are the synergies? Mitig Adapt Strat Glob Change 19:1119–1137

Ravindranath NH, Murthy IK, Swarnim S (2011).Options for forest management for coping with climate change in South Asia. In: Climate change and food security in South Asia, pp 343–358

Rico L, Ogaya R, Barbeta A, Penuelas J (2014) Changes in DNA methylation fingerprint of Q uercus ilex trees in response to experimental field drought simulating projected climate change. Plant Biol 16(2):419–427

Rodway GW (2023) Climate change in and around the high ranges of Asia: consequences for human health. Wilderness Environ Med 34(1):1–2

Sáez-Laguna E, Guevara MÁ, Diaz LM, Sánchez-Gómez D, Collada C, Aranda I, Cervera MT (2014) Epigenetic variability in the genetically uniform forest tree species Pinus pinea L. PLoS ONE 9(8):e103145

Salmond JA, Tadaki M, Vardoulakis S, Arbuthnott K, Coutts A, Demuzere M, Dirks KN, Heaviside C, Lim S, Macintyre H, McInnes RN, Wheeler BW (2016) Health and climate related ecosystem services provided by street trees in the urban environment. Environ Health 15(1):95–111

Salvini G, Ligtenberg A, Van Paassen A, Bregt AK, Avitabile V, Herold M (2016) REDD+ and climate smart agriculture in landscapes: a case study in Vietnam using companion modelling. J Environ Manage 172:58–70

Sharma R, Singh H, Kaushik M, Nautiyal R, Singh O (2018). Adaptive physiological response, carbon partitioning, and biomass production of *Withania somnifera* (L.) Dunal grown under elevated CO_2 regimes. 3 Biotech 8:267. https://doi.org/10.1007/s13205-018-1292-1

Sharma R, Singh H (2021) Alteration in biochemical constituents and nutrients partitioning of *Asparagus racemosus* in response to elevated atmospheric CO_2 concentration. Environ Sci Pollut Res 29:6812–6821. https://doi.org/10.1007/s11356-021-16050-3

Singh H (2021) An integrated approach considering physiological and biophysical based indicators for assessing tolerance of roadside plantations of *Alstonia scholaris* towards urban roadside air pollution: an assessment of adaptation of plantations for mitigating urban roadside air pollution. Trees. https://doi.org/10.1007/s00468-021-02179-8

Singh H (2024) Exploring adaptive modulation in plant functional traits and its impact on the productivity of *Acacia auriculiformis* under CO_2-enriched environment. Industr Crops Prod 210:118186. https://doi.org/10.1016/j.indcrop.2024.118186

Singh H, Kumar M (2022) Climate change and its impact on Indian Himalayan Forests: current status and research needs. In: Rani S, Kumar R (eds) Climate change. Springer Climate. Springer, Cham, pp 223–242. https://doi.org/10.1007/978-3-030-92782-0_11

Singh H, Verma A (2013) Characterization and screening of high nitrogen efficient rice genotype to curtail environmental pollution. Int J Agric Environ Biotechnol 6(1):23–30

Singh H, Savita Sharma R, Sinha S, Kumar M, Kumar P, Verma A, Sharma SK (2017a) Physiological functioning of *Lagerstroemia speciosa* L. under heavy roadside traffic: an approach to screen potential species for abatement of urban air pollution. 3 Biotech 7(61):1–10. https://doi.org/10.1007/s13205-017-0690-0

Singh H, Verma A, Kumar M, Sharma R, Gupta R, Kaur M, Negi M, Sharma SK (2017b) Phytoremediation: a green technology to clean up the sites with low and moderate level of heavy metals. Austin Biochem 2:1012

Singh H, Sharma R, Savita SMP, Kumar M, Verma A, Ansari MW, Sharma SK (2018) Adaptive physiological response of *Parthenium hysterophorus* to elevated atmospheric CO_2 concentration. Indian Forester 144:1–14

Singh H, Yadav M, Kumar N, Kumar A, Kumar M (2020) Assessing adaptation and mitigation potential of roadside trees under the influence of vehicular emissions: a case study of *Grevillea robusta* and *Mangifera indica* planted in an urban city of India. PLoS ONE 15(1):e0227380. https://doi.org/10.1371/journal.pone.0227380

Singh S, Singh H, Sharma SK, Nautiyal R (2021) Seasonal variation in biochemical responses of bamboo clones in the sub-tropical climate of Indian Himalayan foothills. Heliyon 7(4):e06859. https://doi.org/10.1016/j.heliyon.2021.e06859

Singh H, Bandyopadhyaya S, Kumar A, Singh M, Malik A, Kumari G, Raturi A, Yadav SK, Kumar M, Barthwal S, Thakur A, Rawat PS (2022) Understanding the physiological and biophysical response of urban roadside plantations for assessing adaptation and mitigation mechanisms toward vehicular emissions. Urban Clim 44:101183. https://doi.org/10.1016/j.uclim.2022.101183

Singh M, Singh H, Kumar A, Kumar M, Barthwal S, Thakur A (2024a) Soil nitrogen availability determines the CO_2 fertilization effect on tree species (*Neolamarckia cadamba*): growth and physiological evidence. Environ Sustain 7:53–60. https://doi.org/10.1007/s42398-023-00300-w

Singh H, Kumar N, Kumar A (2024b) Enhancing resource use efficiency in crops through plant functional traits. In: Kumar N, Singh H (eds) Plant functional traits for improving productivity. Springer, Singapore. https://doi.org/10.1007/978-981-97-1510-7_6

Song Y, Ci D, Tian M, Zhang D (2016) Stable methylation of a non-coding RNA gene regulates gene expression in response to abiotic stress in *Populus simonii*. J Exp Bot 67(5):1477–1492

Sow MD, Allona I, Ambroise C, Conde D, Fichot R, Gribkova S, Jorge V, Le-Provost G, Paques L, Plomion C, Salse J, Sanchez-Rodriguez L, Segura V, Tost J, Maury S (2018) Epigenetics in forest trees: state of the art and potential implications for breeding and management in a context of climate change. Adv Bot Res 88:387–453

Sterck FF, Vos MA, Hannula SES, de Goede SS, de Vries WW, den Ouden JJ, Nabuurs GJ, van der Putten WH, Veen CG (2021) Optimizing stand density for climate-smart forestry: a way forward towards resilient forests with enhanced carbon storage under extreme climate events. Soil Biol Biochem 162:108396

Su Y, Bai X, Yang W, Wang W, Chen Z, Ma J, Ma T (2018) Single-base-resolution methylomes of *Populus euphratica* reveal the association between DNA methylation and salt stress. Tree Genet Genomes 14:1–11

Thompson I, Mackey B, McNulty S, Mosseler A (2009) Forest resilience, biodiversity, and climate change. In: A synthesis of the biodiversity/resilience/stability relationship in forest ecosystems. Secretariat of The Convention on Biological Diversity, Montreal. Technical Series, vol 43, no 1, pp 1–67

Thornton TF, Comberti C (2017) Synergies and trade-offs between adaptation, mitigation and development. Clim Change 140:5–18

Tiebel K, Karge A, Wagner S (2023) Does shading and ground cover of moss and litter improve germination and establishment of *Betula pendula* Roth, *Salix caprea* L. and *Populus tremula* L. seedlings during drought stress in climate change?–A greenhouse study. For Ecol Manage 544:121212

Trubin A, Mezei P, Zabihi K, Surový P, Jakuš R (2022) Northernmost European spruce bark beetle Ipstypographus outbreak: modelling tree mortality using remote sensing and climate data. For Ecol Manage 505:119829

Tubiello FN, Salvatore M, Rossi S, Ferrara A, Fitton N, Smith P (2013) The FAOSTAT database of greenhouse gas emissions from agriculture. Environ Res Lett 8(1):015009

Upton C (2019) The forest certification handbook. CRC Press

Verkerk PJ, Costanza R, Hetemäki L, Kubiszewski I, Leskinen P, Nabuurs GJ, Potočnik J, Palahí M (2020) Climate-smart forestry: the missing link. For Policy Econ 115:102164

Wang T, O'Neill GA, Aitken SN (2010) Integrating environmental and genetic effects to predict responses of tree populations to climate. Ecol Appl 20(1):153–163

Wang S, Zhang L, She D, Wang G, Zhang Q (2021) Future projections of flooding characteristics in the Lancang-Mekong River Basin under climate change. J Hydrol 602:126778

Wüstemann H, Bonn A, Albert C, Bertram C, Biber-Freudenberger L, Dehnhardt A, Döring R, Elsasser P, Hartje V, Mehl D, Kantelhardt J, Rehdanz K, Schaller L, Scholz M, Thrän D, Witing F, Hansjürgens B (2017) Synergies and trade-offs between nature conservation and climate policy: Insights from the "Natural Capital Germany–TEEB DE" study. Ecosyst Serv 24:187–199

Yadav SK, Singh H, Nautiyal R, Ginwal HS, Ansari SA, Barthwal S (2019b) Modulation of morpho-physiological responses in *Populus deltoides* by elevated carbon dioxide and temperature. For Sci 66:105–118. https://doi.org/10.1093/forsci/fxz048

Yadav SK, Singh H, Ginwal HS, Barthwal S (2019) Elevated CO_2 enhanced growth and physiological process of *Populus deltoides* Bartr. ex Marsh. Indian For 145(1):23–27

Yakovlev I, Fossdal CG, Skrøppa T, Olsen JE, Jahren AH, Johnsen Ø (2012) An adaptive epigenetic memory in conifers with important implications for seed production. Seed Sci Res 22(2):63–76

Young DJ, Stevens JT, Earles JM, Moore J, Ellis A, Jirka AL, Latimer AM (2017) Long-term climate and competition explain forest mortality patterns under extreme drought. Ecol Lett 20(1):78–86

Zaidem ML, Groen SC, Purugganan MD (2019) Evolutionary and ecological functional genomics, from lab to the wild. Plant J 97(1):40–55

Zhang H, Lang Z, Zhu JK (2018) Dynamics and function of DNA methylation in plants. Nat Rev Mol Cell Biol 19(8):489–506

Zhou G, Xia J, Zhou P, Shi T, Li L (2021) Not vegetation itself but mis-revegetation reduces water resources. Sci China Earth Sci 64:404–411

Unraveling the Tapestry of Plant Phenology in the Face of Climate Change and Environmental Stress

Karuna Gusain, Vivek Chauhan, Hukum Singh, and Mithilesh Singh

Abstract This chapter examines plant phenology's complex realm, tracing its evolution from ancient observations to contemporary precision aided by satellite-based remote sensing. Beyond being a passive recorder, phenology emerges as a dynamic force intricately linked with species-environment symbiosis, influencing spatial species distributions. As climate change accelerates, overshadowing this delicate phenological dance, the narrative expands to depict phenology as a critical driver, offering insights into the pulsating rhythm of ecosystems amid changing environmental dynamics. Drawing from significant reports and expert insights, an overview of climate change contextualizes the intricate relationship between climate change and life event timing, particularly in plant phenology. This exploration underscores the need for adaptive methodologies, blending historical observations with present knowledge. Beyond scientific inquiry, plant phenology holds profound implications for environmental stewardship, conservation, and policy, urging proactive measures to safeguard biodiversity and foster resilient ecosystems. The ongoing dialogue between phenology and climate change and advances in scientific inquiry and technology promises deeper insights into the phenological symphony, guiding informed policy decisions for a sustainable future.

Keywords Climate change · Environmental pollution · Plant phenology · Sustainability · Urban forests

K. Gusain (✉) · M. Singh
G.B. Pant, National Institute of Himalayan Environment, Kosi-Katarmal, Almora, Uttarakhand 263643, India
e-mail: kkarunag97@gmail.com

V. Chauhan
Forest Ecology and Climate Change Division, ICFRE-Forest Research Institute, Dehradun, Uttarakhand 248006, India

H. Singh
Plant Physiology Discipline, ICFRE-Forest Research Institute, P.O. New Forest, Dehradun, Uttarakhand 248006, India

H. Singh (ed.), *Urban Forests, Climate Change and Environmental Pollution*,
https://doi.org/10.1007/978-3-031-67837-0_16

1 Introduction

Phenology, derived from the Greek words "phaino" (to show or appear) and "logos" (study), offers a captivating exploration of the intricate temporal rhythms governing life on Earth. Beyond its role as a passive observer, phenology provides a profound lens through which the interplay between species and their environment unfolds (Yadav et al. 2019). This journey into phenology intertwines scientific curiosity with technological advancement, revealing nature's timekeeping prowess. At its essence, phenology serves as nature's clock, documenting the cyclical cadence of life's pivotal events. From leaf emergence to flower blossoming, migratory journeys to hibernation, each event harmonizes in a ballet choreographed by the changing seasons—a symphony echoing through ecosystems, played in sync with Earth's timeless melodies. Phenology transcends mere observation; it unveils the complex relationships between species and their dynamic environments. Biological timing becomes a language spoken fluently by flora and fauna, revealing dialogues of adaptation and interconnectedness.

Tracing phenology's historical roots reveals an evolutionary tapestry, from ancient civilizations' observations to the meticulous records of naturalists like Gilbert White. Each era contributes to a narrative woven through time, setting the stage for modern inquiries into life's intricate timings. In the contemporary age, technology conducts this symphony, with satellite-based remote sensing elevating phenological studies to new heights. This panoramic view transcends traditional observation, offering unprecedented insights into regional and global phenological patterns (Kumar and Singh 2024). In phenology's lexicon, terms like flowering, migration, and hibernation resonate with life's profound significance, conducting a symphony orchestrated by the environment. Seminal works, such as Chuine's, depict phenology not as a passive recorder but as a dynamic force driving species distributions and shaping biodiversity's fabric in response to environmental cues. In this intricate dance of life, phenology emerges as a dynamic driver intricately woven into ecological systems, influencing the spatial arrangement of species and profoundly impacting biodiversity.

1.1 *Evolution of Phenological Studies*

The historical trajectory of phenological studies, meticulously chronicled by Primack and Gallinat (2017), unfolds as a rich tapestry of evolving comprehension. Beginning with the rudimentary observational records of ancestors, who meticulously noted the timing of natural phenomena, and progressing to the contemporary precision enabled by technologies such as herbarium specimens, the methodologies employed in phenological studies have journeyed through the annals of time, embracing the ever-accelerating wave of technological advancements. This historical evolution deepens scientific insight into the nuanced timings of life events, emphasizing the necessity for adaptive approaches, particularly in light of the looming specter of climate change (Meena et al. 2011; Kumar et al. 2019c).

Exploring the historical importance of phenological studies necessitates recognizing the groundbreaking contributions of Linnaeus in the eighteenth century. Linnaeus established the groundwork for systematic phenological observations through meticulous documentation of flowering times. Amidst the intricate rhythms of seasons, Linnaeus's endeavors remain a cornerstone of phenological research, offering a timeless gauge of ecological vitality (Linnaeus 1751). His legacy reverberates across epochs, underscoring the fundamental connection between meticulous observation and profound ecological understanding.

1.2 Climate Change and Phenological Shifts

The accelerating pace of climate change casts a profound shadow over the delicate ballet of phenology. Parmesan's comprehensive synthesis in 2006 unveils the global fingerprints of climate change on phenological events. Parmesan's work serves as a cornerstone, providing a nuanced understanding of how climatic shifts reverberate through the timing of life events across taxa and geographical regions (Kumar et al. 2020; Singh and Kumar 2022). As the planet warms, the intricate timings of flowering, migration, and hibernation become not just markers of seasons but poignant indicators of the Earth's response to the unfolding climate crisis.

The intertwining threads of history, technology, and climate change weave a narrative that connects past observations, current advancements, and the imperative for future exploration. The definition and significance of phenology expand beyond the realm of temporal events and become an opening to understanding the pulse of ecosystems in the face of environmental changes.

The IPCC Special Report on Global Warming of 1.5 °C, unveiled in 2018, emerges as a pivotal milestone in unraveling the complex tapestry of contemporary climate change. This authoritative document, crafted with meticulous precision, serves as a compendium of scientific knowledge, offering an exhaustive panorama of the repercussions entwined with global warming. As a clarion call for urgent mitigation efforts, the report meticulously outlines the projected impacts of a 1.5 °C temperature elevation above pre-industrial levels. Its scrutiny extends beyond ecological realms, delving into the profound social and economic dimensions that underscore the urgency of collective action (Singh et al. 2024a, b).

In tandem with this global synthesis, the groundbreaking contributions of Diffenbaugh and Field in 2013 cast a focused lens on the shifts transpiring in ecologically critical terrestrial climate conditions. Their seminal work dissects temperature variations and delves into the intricate dance of precipitation patterns and other climatic variables of paramount ecological significance. The research by Diffenbaugh and Field stands as a linchpin, providing nuanced insights that extend beyond global assessments. This work is an indispensable reference, guiding researchers and policymakers through the labyrinthine landscape of ecological responses to the evolving climate (Diffenbaugh and Field 2013; Singh and Verma 2013; Kumar et al. 2017). The convergence of global assessments and specific insights into terrestrial climate conditions is a robust foundation. This nuanced understanding becomes pivotal to understanding the intricate relationship between climate change and the timing of life events, with a particular focus on plant phenology.

2 Urban Phenological Dynamics

The urban landscape, with its concrete jungles and modified microclimates, offers a fascinating area for investigating phenological changes. The urban heat island effect, characterized by elevated temperatures compared to rural areas, significantly influences the timing of plant life events. Schwartz et al. (2006) delve into the lasting impacts of urbanization on phenological patterns, providing valuable insights into how urban heat islands contribute to alterations in plant life cycles. Bolmgren and Lönnberg (2005) examine specific phenological traits within urban environments, uncovering correlations between fruit characteristics and flowering times. McDonnell and Hahs (2008) underscore the necessity for gradient analysis to grasp the intricacies of urban ecosystems.

The urban heat island effect, stemming from increased impervious surfaces and heat-retaining infrastructure, significantly influences temperature regimes within urban environments. Stewart et al. (2005) delved into the spatial and temporal dynamics of temperatures in urban areas, which are crucial for interpreting phenological shifts. Additionally, Huebner (2022) investigates how the urban heat island contributes to changes in the phenological phases of trees, providing valuable insights into the specific impacts on urban vegetation.

Moreover, artificial lighting in urban settings can disrupt natural photoperiodic cues, thereby influencing the phenological responses of plants. Gaston et al. (2015) discuss the broader impacts of artificial light on natural processes, including phenology, shedding light on the consequences of light pollution. Furthermore, Bennie et al. (2016) explore how artificial light affects the phenological patterns of wild plants, offering valuable insights into the effects of light pollution on plant life cycles.

Understanding phenological changes in urban forestry is paramount for addressing their impact on biodiversity and ecosystem services. Baldock et al. (2015) highlight the relationship between urban bee populations and pollination services, underscoring the importance of comprehending urban phenological dynamics for effective pollinator management. Additionally, Melaas et al. (2016) utilized satellite data to quantify the phenology of urban trees, contributing to the understanding of how urban vegetation responds to changing environmental conditions. The different temperature shifts over the past 2000 years are mentioned in Table 1.

Table 1 Temperature shift in the last 2000 years

Year	Temperature anomaly (°C)	Event/period
1 CE	0	Start of common era
400 CE	−0.2	Cooler period, Late Antique Little Ice Age
800 CE	0.1	Medieval warm period
1300 CE	−0.3	Little Ice Age
1750 CE	−0.2 to 0	Industrial revolution begins
1800 CE	−0.3	Industrial era, initial impact of human activities
1850 CE	−0.2	Continued industrialization
1900 CE	0.0	Early twentieth century
1950 CE	0.1	Post-World War II industrialization
2000 CE	0.7	Modern era, significant global warming
2020 CE	1.02	Recent decade

3 Phenological Shifts: Understanding the Basics

Phenological shifts, the delicate temporal adjustments in recurring biological events, serve as crucial indicators of the dynamic interplay between species and their environment. These shifts offer a nuanced understanding of the intricate timings governing life on Earth. Richardson et al. (2013) significantly contribute to this understanding by advocating a more comprehensive representation of vegetation phenology in terrestrial biosphere models. Their research underscores the necessity of capturing the nuanced dynamics of plant life cycles, recognizing the intricate tapestry woven by the timings of flowering, migration, and hibernation (Richardson et al. 2013; Gupta and Singh 2017; Singh et al. 2017a). The significance of phenological shifts extends beyond mere observational records, evolving into a scientific exploration that unveils the forces steering species distributions. Chuine (2010) argues that phenology is not merely a recorder of events but a dynamic driver shaping ecosystems and influencing biodiversity. This profound relationship underscores the integral role of phenological shifts in ecological systems.

4 Factors Influencing Phenological Changes

Understanding the intricate factors that shape and influence phenological changes is pivotal in unraveling the complexities of life events. Phenology is a dynamic field where various environmental, geographical, and ecological elements converge to coordinate the timing of critical biological occurrences.

4.1 Temperature Variations

One of the primary drivers influencing phenological changes is temperature. The relationship between temperature and phenology resembles a symphony echoing through ecosystems. Wolkovich et al. (2012) contribute valuable insights by demonstrating that warming experiments often underpredict plant phenological responses to climate change. This observation prompts a deeper exploration of how temperature variations modulate and potentially disrupt plant life cycles, emphasizing the nuanced nature of plant responses to changing climate conditions and the need for adaptive strategies (Singh et al. 2017b).

Challenges to the conventional understanding of warming experiments suggest inadequate prediction of plant phenological responses to climate change (Zhou et al. 2015). This prompts a deeper examination of temperature variations as a critical driver influencing the intricate timings of plant and animal life cycles.

Temperature is a master orchestrator of phenological events, acting as a fundamental cue regulating critical life stage timing. In contrast, traditional warming experiments provide valuable insights. Anderson et al. (2012) and Li et al. (2019) extend the exploration, highlighting the importance of considering both mean temperature and the frequency and intensity of temperature extremes. These studies underscore the need for a comprehensive understanding of temperature dynamics to predict how phenological events may evolve in response to ongoing climate change.

Furthermore, Garcia and colleagues (2014) research investigates the genetic basis of temperature responses in different plant species. This genetic perspective sheds light on the potential for evolutionary adaptations in the face of changing temperature regimes, providing a more nuanced view of how species may cope with the warming trends associated with climate change.

4.2 Precipitation Patterns and Soil Conditions

Beyond temperature, precipitation patterns and soil conditions join the environmental factors shaping phenological shifts. The availability of water, nutrients, and soil composition intricately influences the timing of events such as flowering and leafing. These factors contribute to the environmental cues that plants rely on to synchronize their life cycles. A comprehensive understanding of how precipitation and soil conditions interact with temperature provides a more holistic perspective on the drivers of phenological changes (Sharma et al. 2018; Singh et al. 2018).

Altered precipitation patterns emerge as significant drivers of phenological changes beyond temperature. Diez et al. (2012) delve into the complex task of forecasting phenology, emphasizing the need to comprehend species variability and community patterns in the face of changing precipitation regimes.

Havard et al. (2016) provide additional insights into the interactive effects of temperature and precipitation on plant phenology. These works emphasize the importance of considering the holistic climate context, as altered precipitation patterns can modulate the responses of plants to temperature cues. Additionally, regional variations in precipitation trends in different ecosystems highlight the context-specific nature of precipitation impacts on phenological events (Chen et al. 2011).

4.3 Geographic and Seasonal Variations

The Earth's landscapes are diverse, exhibiting unique characteristics across different regions. Clewell and Aronson's (2006) insights into the positive responses of species to climate change challenge conventional assumptions of uniformly negative impacts. The research reveals species' adaptability to changing environmental cues by examining phenological tracking. This adaptability introduces a layer of complexity, suggesting that the effects of climate change on phenology may vary across ecosystems. Understanding this heterogeneity is crucial for predicting and managing the consequences of phenological shifts (Gupta et al. 2018a, b).

Furthermore, seasonal variations, a constant in the natural world, further contribute to the intricate dance of life events. The changing length of days, sunlight exposure, and temperature fluctuations during different seasons create a dynamic backdrop for phenological shifts. Deciphering how these seasonal dynamics interact with environmental cues is essential for understanding the timing of events such as flowering, fruiting, and migration.

4.4 Species Interactions and Responses

Species interactions and responses introduce complexity to phenological changes. Diez et al. (2012) delve into forecasting phenology, emphasizing the need to understand species variability and community patterns. The cascading effects within ecosystems, triggered by phenological changes, underscore the interconnectedness of species in response to shifting environmental cues. This research highlights the importance of considering individual species and the broader ecological context when studying phenological shifts.

The intricate web of species interactions and responses to climate-induced phenological changes adds another layer of complexity to our understanding. Examining changes in predator–prey relationships and the adaptation strategies employed by different species provides valuable insights into the broader ecological consequences of phenological shifts.

McDowell et al. (2020) explored the cascading effects of phenological changes on species interactions, exploring how altered timing of key life events can disrupt established predator–prey relationships, potentially leading to resource availability and utilization mismatches. Additionally, studies by Martin et al. (2023) explore the adaptability of different species to changing phenological cues, revealing the diverse strategies employed to cope with or capitalize on shifting ecological dynamics.

Species' responses to climate-induced phenological shifts are not isolated; they are part of a broader interconnected web. Bogdziewicz et al. (2020) explore how changes in the phenology of one species cascade through entire ecosystems, influencing the timing of critical ecological events. This interconnectedness underscores the need for a holistic understanding of phenological shifts, considering the ripple effects on ecological processes.

Li et al. (2009) delve into the cascading effects of phenological changes on species interactions, including predator–prey relationships and plant-pollinator dynamics. These studies shed light on the intricate ways in which phenological shifts reverberate through trophic levels, emphasizing the importance of considering the broader ecological consequences.

The mechanisms driving phenological changes are intricate and multifaceted. Temperature variations, altered precipitation patterns, and the complex interplay of species interactions and responses collectively shape the evolving tapestry of life events. Hence, it becomes increasingly evident that a holistic and interdisciplinary approach is essential for unraveling the complexities of phenological shifts in a changing climate.

4.5 Human-Induced Plant Phenological Consequences

Human activities have significantly altered natural ecosystems, leading to profound consequences for the timing of biological events, or phenology, in plant species worldwide. The activities such as deforestation, urbanization, and agricultural expansion have profoundly altered natural habitats and landscapes, leading to shifts in plant phenological patterns. Habitat loss and fragmentation disrupt the environmental cues that plants rely on to time their life cycle events, resulting in mismatches between resources and needs.

Clewell and Aronson (2006) and Jones et al. (2012) have shown how land use change influences plant phenology by altering habitat availability and microclimatic conditions. Urbanization, in particular, creates heat islands and modifies local microclimates, accelerating phenological changes in urban environments compared to surrounding natural areas.

Human-induced pollution from sources such as air and water impact plant phenology. Air pollutants like nitrogen oxides and sulfur dioxide directly affect plant physiology and growth, changing flowering and leafing times. Similarly, water pollution from industrial runoff, agricultural runoff, and urban wastewater can alter aquatic ecosystems' phenological patterns. Parmesan and Yohe (2003) and Garcia et al. (2012) have demonstrated the link between pollution and changes in plant phenology, highlighting the need for comprehensive strategies to mitigate pollution's impacts on ecosystems.

Introducing invasive species by human activities poses a significant threat to native ecosystems and can disrupt plant phenological patterns. Invasive species often have different phenological cues and requirements than native species, leading to mismatches in timing and interactions. Diez et al. (2012) and Bogdziewicz et al. (2020) have shown how invasive species alter plant phenology by outcompeting native species for resources or disrupting plant-pollinator interactions. Effective management strategies are needed to control invasive species and minimize their impacts on native ecosystems.

Human-induced plant phenological shifts have far-reaching consequences for ecosystems and society. Disruptions to plant-pollinator interactions may reduce fruit and seed production, affecting wildlife populations and ecosystem functioning. Changes in the timing of agricultural activities impact human communities' food production, livelihoods, and food security. Furthermore, phenological changes affect human health, economics, and culture. Shifts in the timing of allergenic plant species' flowering, for example, can exacerbate seasonal allergies and respiratory illnesses. Changes in the timing of cultural events tied to natural phenomena, such as harvest festivals or religious ceremonies, disrupt traditional practices and societal rhythms.

5 Historical Context of Climate Change and Phenology

Understanding the intricate relationship between climate change and phenological shifts requires a comprehensive exploration of the historical context, empirical evidence, and the interconnectedness of species in responding to environmental cues. A crucial aspect in unraveling the influence of climate change on phenology is to delve into the historical context spanning the past century, as mentioned in Table 2. Thackeray et al. (2016) offer a nuanced perspective by comprehensively analyzing phenological sensitivity across taxa and trophic levels. Their work not only elucidates the individual responses of species but also reveals the interconnectedness of entire ecosystems as they adapt to changing environmental cues. Further, they contribute essential insights into the temporal dynamics of phenological shifts by examining historical trends and showcasing how various species have responded to the evolving climate.

Acknowledging the broader scientific discourse on the historical trajectory of climate change is imperative. Pioneering studies by Jones et al. (2012) and Ward et al. (2018) investigate the long-term temperature records and climatic patterns, providing a foundation for understanding the broader climatic context within which phenological shifts unfold.

Phenological changes are not speculative and form a robust foundation for understanding the impacts of climate change on the timing of life events. Parmesan and Yohe (2003) contribute significantly by presenting a globally coherent fingerprint of climate change impacts across natural systems. Their work synthesizes data from

Table 2 Historical temperature shifts affecting plant phenology (Last 2000 years)

Causes of phenological shifts	Description	Examples
Temperature changes	Shifts in average temperature affecting plant life cycles	Early flowering in cherry blossoms, altered budburst in deciduous trees
Precipitation patterns	Changes in rainfall patterns impacting plant phenology	Delayed flowering in response to reduced rainfall
Climate variability	General changes in climate conditions influencing phenology	Altered fruiting patterns due to overall climate warming
Habitat disturbance	Disruptions to plant habitats influencing phenological events	Accelerated flowering in response to habitat clearing
Urbanization	Urban development impacting local microclimates and phenology	Early leafing in urban areas due to the urban heat island effect

diverse ecosystems, offering a comprehensive overview of how climatic shifts reverberate through the timing of life events across taxa and geographical regions. Amalgamating evidence from the different areas and species solidifies the understanding that phenological changes are a tangible manifestation of a warming climate.

Building upon this empirical foundation, Garcia et al. (2012) employ advanced monitoring techniques, including satellite observations and citizen science initiatives, to capture fine-scale phenological changes. These studies corroborate earlier findings and provide detailed insights into the specific mechanisms driving phenological responses in different ecosystems.

6 Ecological Consequences and Biodiversity Implications

The ecological ramifications of phenological shifts reverberate throughout diverse ecosystems, profoundly impacting biodiversity and instigating trophic cascade effects. Thackeray et al. (2010) laid foundational insights into the consequences of phenological shifts on biodiversity by introducing the concept of trophic level asynchrony. This seminal study illuminated how disparate trophic levels may respond to climate-induced phenological changes at varying rates, resulting in temporal mismatches in resource availability. Vrieling et al. (2017) delved into the ramifications of phenological mismatches on species interactions in marine ecosystems, shedding light on the broader implications for biodiversity conservation.

Investigations into species diversity responses to phenological shifts unveil nuanced patterns across ecosystems (Jochner et al. 2016; Smith and Ramsay 2020). While some species may exhibit adaptive responses to changing phenological cues, others face heightened decline risks. These studies underscore the importance of considering overall biodiversity and the specific responses of individual species to phenological changes (Yadav et al. 2019; Gupta et al. 2019).

Bartomeus et al. (2013) provided crucial insights into trophic cascade effects triggered by phenological changes, particularly in plant-pollinator interactions. Understanding how shifts in flowering timing impact pollinators and subsequently influence plant reproduction is pivotal for predicting broader ecosystem disruptions. Investigations by Peralta et al. (2020) and Gaiarsa et al. (2021) explored the cascading effects of phenological changes on higher trophic levels, elucidating implications for predator–prey dynamics and overall ecosystem stability.

Suding et al. (2015) delved into how phenological shifts influence critical ecological processes like nutrient cycling and carbon sequestration. Their study underscored the intricate interconnectedness of phenology with fundamental ecosystem functions, highlighting the potential for cascading effects that permeate entire landscapes (Singh et al. 2024a, b).

The ecological ramifications of phenological shifts are multifaceted, shaping biodiversity patterns and instigating trophic cascade effects across diverse ecosystems. Examining trophic-level asynchrony, species-specific responses, and ecosystem-level dynamics aids in understanding how phenological changes shape the intricate web of life. These insights are indispensable for informing conservation strategies and mitigating the potential impacts of ongoing climate-induced phenological shifts (Gupta et al. 2019; Kumar et al. 2019a, b).

7 Adaptation and Resilience Strategies

Understanding and fostering adaptation and resilience strategies are essential to mitigating the impacts of phenological shifts on ecosystems (Meena et al. 2011; Kumar et al. 2019c). Valuable insights and exploring strategies are required to provide comprehensive knowledge of how organisms cope with changing phenology (Kumar et al. 2020; Singh and Kumar 2022). Considering these strategies may help conservationists and policymakers to develop targeted approaches to enhance the resilience of ecosystems in the face of ongoing environmental changes.

7.1 Species-Level Adaptations

The researchers have delved into evolutionary adaptations in response to changing climatic conditions (Walther et al. 2002; Hoffmann and Sgrò 2011; Chevin et al. 2015). Their work underscores that certain species may undergo genetic changes over generations, allowing them better to synchronize their life cycles with altered phenological cues. Understanding the genetic basis of these adaptations is crucial for predicting the long-term resilience of populations (Gupta and Singh 2017; Singh et al. 2017a, b).

Behavioural adaptations also play a crucial role in species responding to phenological shifts. Scientists have explored the concept of behavioural plasticity, illustrating how individuals within a population can adjust their behaviours in response to changing environmental conditions (Sih et al. 2011; Capilla-Lasheras et al. 2022). These studies underscore the importance of considering physiological adaptations and the flexible behaviours that enable species to cope with phenological uncertainties.

Additionally, phenotypic plasticity offers another dimension to species-level adaptations (Pigliucci 2001; Nicotra et al. 2010). Understanding how organisms exhibit different phenotypes in response to environmental cues provides insights into their capacity to cope with phenological changes. This plasticity may involve changes in morphology, physiology, or behavior, enabling species to adjust to variable conditions (Singh and Verma 2013; Kumar et al. 2017).

7.2 Ecological Resilience Initiatives

Beyond the individual level, the conservation of keystone species becomes crucial for maintaining ecological resilience. Ecological resilience underscores the significance of critical species in stabilizing ecosystems (Holling 1973). Scientists have illustrated that certain species, due to their ecological roles, influence the timing of essential events in ecosystems, contributing to overall stability (Paine 1969; Power et al. 1996).

Conservation efforts extend beyond individual species to encompass entire habitats. Habitat preservation and restoration are vital for enhancing ecological resilience (Hobbs and Harris 2001; Suding et al. 2015). These initiatives focus on creating and maintaining diverse habitats that can support a variety of species and their interconnected phenological relationships.

Examining community-level adaptations provides insights into how ecological communities can adapt to changing phenology (Peralta et al. 2020; Elias et al. 2021). Their work emphasizes the intricate relationships between species within communities and the potential for community-level adaptations to shape the resilience of entire ecosystems.

8 Mitigation and Management Strategies

Mitigating the impacts of phenological shifts requires a multi-faceted approach that spans conservation practices, fine-scale processes, and global mitigation policies. Further exploration of strategies and their associated studies offers a more nuanced understanding of managing and mitigating the effects of changing phenology (Singh et al. 2020; Kumar et al. 2021a, b).

Expanding on the concept of predictive phenology emphasizes the integration of multiple disciplines, including ecology, evolution, and climate science (Morin et al.

2010; Miller-Rushing et al. 2010). Their efforts advocate for developing predictive models that consider the synergistic effects of various factors on phenological events. Such models enhance the ability to forecast changes and implement proactive conservation practices.

Engaging the public in phenological observations through citizen science initiatives contributes valuable data for conservation efforts. The importance of citizen science in monitoring phenological shifts has been recognized globally (Dickinson et al. 2010; McKinley et al. 2017). These initiatives provide large-scale data and foster public awareness and participation in conservation practices.

Incorporating principles of restoration ecology highlights the potential for actively restoring ecosystems to enhance their resilience to phenological changes. Reintroducing native species and restoring natural habitats assist conservationists in creating conditions that support historical phenological relationships (Kumar et al. 2021c, d).

9 Global and Local Mitigation Policies

International collaborations are paramount in addressing the global dimensions of climate change. Studies have examined the importance of international agreements, such as the Convention on Biological Diversity, and emphasize the need for collaborative strategies to mitigate the impacts of climate change on biodiversity and phenology (Settele et al. 2014; Sala et al. 2000).

Tailoring mitigation strategies to regional contexts is crucial. Scientists have explored the effectiveness of regional adaptation strategies in mitigating the impacts of climate change on ecosystems (Opdam et al. 2013; Seidl et al. 2016). Their studies advocate context-specific approaches considering different regions' unique phenological patterns and ecological dynamics.

Shifting towards ecosystem-based adaptation highlights the importance of considering entire ecosystems in mitigation policies. Addressing the broader ecological context may enhance ecosystems' adaptive capacity to change phenological patterns (CBD 2010; Reyers et al. 2013).

10 Future Research Needs

Future research endeavors on phenological shifts in the face of climate change must focus on identifying gaps in understanding, leveraging technological advancements, and addressing the evolving challenges posed by shifting plant phenology (Yadav et al. 2019; Gupta et al. 2019). The following are suggestive aspects that should be undertaken to address phenological shifting under environmental and climate change scenarios.

10.1 Global Patterns of Plant Phenology

Global assessments are essential for capturing broader trends in plant phenology and offer insights into shifting plant phenology patterns across diverse ecosystems. These studies emphasize the need for a comprehensive understanding of global phenological dynamics and the factors influencing these patterns (Gupta et al. 2019; Kumar et al. 2019a).

10.2 Integrating Social-Ecological Systems

Future research should explore integrating social-ecological systems to address the impacts of phenological shifts holistically. Such studies provide frameworks for understanding the interplay between human communities and changing plant phenology (Kumar et al. 2019b; Gupta et al. 2018b). Investigating the socio-economic implications of altered phenological patterns and the potential for adaptive governance strategies is crucial for developing resilient systems.

10.3 Microbial Interactions and Phenology

The role of microbial communities in shaping plant phenology remains an understudied aspect that offers glimpses into the intricate relationships between plants and soil microbes. Future investigations should explore how microbial communities influence plant phenological events and vice versa.

11 Technological Advancements

11.1 High-Resolution Remote Sensing

Integrating high-resolution remote sensing technologies provides an avenue for detailed and accurate plant phenology monitoring. Such studies explore the potential of satellite imagery and unmanned aerial vehicles (UAVs) in capturing fine-scale phenological variations. Future research should examine the applications of these technologies across different ecosystems and scales.

11.2 Advances in Phenological Modelling

Machine learning algorithms present a promising frontier in advancing phenological modeling. These approaches demonstrate machine learning approaches' efficacy in predicting phenological events with high precision. Future research should focus on refining these models, incorporating additional variables, and assessing their applicability across diverse plant communities (Sharma et al. 2018; Singh et al. 2018; Gupta et al. 2018a).

11.3 Citizen Science Integration

Harnessing the power of citizen science is pivotal for large-scale phenological observations. Researchers have emphasized citizen science initiatives' potential in collecting real-time data on plant phenology. Future research should explore innovative ways to integrate citizen science data into existing phenological databases and enhance the collaborative nature of phenological research.

12 Conclusion

The symphony of plant phenology serves as a captivating approach to the interconnected dance between species and their environment, resonating throughout ecological history. From Linnaeus's pioneering observations to contemporary satellite-based remote sensing, the dynamic interplay of phenological events unveils the evolving tapestry of life. This exploration underscores the imperative for adaptive methodologies in climate change, emphasizing the rich historical context that intertwines past observations with present understanding. Beyond scientific inquiry, plant phenology holds profound implications for environmental stewardship, conservation, and policy considerations. The historical context weaves a narrative thread connecting the past to the present, urging us to recognize our role as custodians of the Earth. Continued research and technological innovation ensure the harmonious symphony of plant phenology endures for future generations. The implications for conservation and policy underscore the need for proactive measures to preserve biodiversity, understand species' responses, and foster resilient ecosystems. Ongoing scientific inquiry, coupled with emerging technologies like machine learning algorithms, promises to unravel more layers of the phenological symphony. Integrating these advancements into future research will likely open new avenues for understanding and predicting phenological shifts more precisely. Informed policy decisions, aligned with scientific findings, become crucial in addressing the cascading effects of climate change on plant life cycles, contributing to the overall health of ecosystems. The policies crafted with a deep understanding of the intricate relationships between climate dynamics and plant life cycles can pave the way for resilient ecosystems and sustainable coexistence.

References

Anderson JT, Inouye DW, McKinney AM, Colautti RI, Mitchell-Olds T (2012) Phenotypic plasticity and adaptive evolution contribute to advancing flowering phenology in response to climate change. Proc Roy Soc B Biol Sci 279(1743):3843–3852

Baldock KCR, Goddard MA, Hicks DM, Kunin WE, Mitschunas N, Osgathorpe LM, Potts SG, Robertson KM, Scott AV, Stone GN (2015) Where is the UK's pollinator biodiversity? The importance of urban areas for flower-visiting insects. Proc Roy Soc B Biol Sci 282(1803):20142849

Bartomeus I, Ascher JS, Gibbs J, Danforth BN, Wagner DL, Hedtke SM, Winfree R (2013) Historical changes in northeastern US bee pollinators related to shared ecological traits. Proc Natl Acad Sci 110(12):4656–4660

Bennie J, Davies TW, Cruse D, Gaston KJ (2016) Ecological effects of artificial light at night on wild plants. J Ecol 104(3):611–620

Bogdziewicz M, Szymkowiak J, Bonal R, Hacket-Pain A, Espelta JM, Pesendorfer M, Grewling L, Kasprzyk I, Belmonte J, Kluska K (2020) What drives phenological synchrony? Warm springs advance and desynchronize flowering in oaks. Agric for Meteorol 294:108140

Bolmgren K, Lönnberg K (2005) Herbarium data reveal an association between fleshy fruit type and earlier flowering time. Int J Plant Sc 166(4):663–670

Capilla-Lasheras P, Thompson MJ, Sánchez-Tójar A, Haddou Y, Branston CJ, Réale D, Charmantier A, Dominoni DM (2022) A global meta-analysis reveals higher variation in breeding phenology in urban birds than in their non-urban neighbours. Ecol Lett 25(11):2552–2570

CBD (Convention on Biological Diversity) (2010) Connecting biodiversity and climate change mitigation and adaptation: report of the Second Ad Hoc technical expert group on biodiversity and climate change. In: CBD Tech Ser 53

Chen I-C, Hill JK, Ohlemüller R, Roy DB, Thomas CD (2011) Rapid range shifts of species associated with high levels of climate warming. Science 333(6045):1024–1026

Chevin L-M, Visser ME, Tufto J (2015) Estimating the variation, autocorrelation, and environmental sensitivity of phenotypic selection. Evolution 69(9):2319–2332

Chuine I (2010) Why does phenology drive species distribution? Philos Trans Roy Soc B Biol Sci 365(1555):3149–3160

Clewell AF, Aronson J (2006) Motivations for the restoration of ecosystems. Conserv Biol 20(2):420–428

da Martin RA, Silva CRB, Moore MP, Diamond SE (2023) When will a changing climate outpace adaptive evolution? Wiley Interdisc Rev Clim Change 14(6):e852

Dickinson JL, Zuckerberg B, Bonter DN (2010) Citizen science as an ecological research tool: challenges and benefits. Annu Rev Ecol Evol Syst 41:149–172

Diez JM, Ibáñez I, Miller-Rushing AJ, Mazer SJ, Crimmins TM, Crimmins MA, Bertelsen CD, Inouye DW (2012) Forecasting phenology: from species variability to community patterns. Ecol Lett 15(6):545–553

Diffenbaugh NS, Field CB (2013) Changes in ecologically critical terrestrial climate conditions. Science 341(6145):486–492

Elias JPC, Vergne DC, Zanatta MP, Campos CCF, Ramos FN (2021) Edge creation changes the timing and intensity of phenological reproductive patterns and species activities in forest tree communities. Rodriguésia 72:e01882019

Gaiarsa MP, Kremen C, Ponisio LC (2021) Pollinator interaction flexibility across scales affects patch colonization and occupancy. Nat Ecol Evol 5(6):787–793
Garcia RA, Burgess ND, Cabeza M, Rahbek C, Araújo MB (2012) Exploring consensus in 21st century projections of climatically suitable areas for African vertebrates. Glob Change Biol 18(4):1253–1269
Garcia RA, Cabeza M, Rahbek C, Araújo MB (2014) Multiple dimensions of climate change and their implications for biodiversity. Science 344(6183):1247579
Gaston KJ, Duffy JP, Bennie J (2015) Quantifying the erosion of natural darkness in the global protected area system. Conserv Biol 29(4):1132–1141
Gupta SK, Singh H (2017) Observations on Sporadic Flowering in *Ficus benjamina* L. and *Peltophorum africanum* L. in New Forest. Indian for 143:290–291
Gupta SK, Ram J, Singh H (2018a) Comparative study of transpiration in cooling effect of tree species in the atmosphere. J Geosci Environ Prot 6:151–166. https://doi.org/10.4236/gep.2018.68011
Gupta SK, Ram J, Singh H, Kumar P (2018b) Comparative study of cooling under shades of some forestry tree species with respect to ambient temperatures. J Geosci Environ Prot 6:51–66. https://doi.org/10.4236/gep.2018.611004
Gupta SK, Ram J, Singh H (2019a) Relationship between leaf area index and atmospheric cooling of tree species. Indian J for 42(3):203–208
Havard V, Smith B, Raupach M, Briggs P, Nieradzik L, Beringer J, Hutley L, Trudinger CM, Cleverly J (2016) Coupling carbon allocation with leaf and root phenology predicts tree–grass partitioning along a savanna rainfall gradient. Biogeosciences 13(3):761–779
Hobbs RJ, Harris JA (2001) Restoration ecology: repairing the earth's ecosystems in the new millennium. Restor Ecol 9(2):239–246
Hoffmann AA, Sgrò CM (2011) Climate change and evolutionary adaptation. Nature 470(7335):479–485
Holling CS (1973) Resilience and stability of ecological systems. Annu Rev Ecol Syst 4(1):1–23
Huebner CD (2022) Effects of global climate change on regeneration of invasive plant species from seeds. In: Plant regeneration from seeds. Elsevier, pp 243–257
Jochner S, Sparks TH, Laube J, Menzel A (2016) Can we detect a nonlinear response to temperature in European plant phenology? Int J Biometeorol 60:1551–1561
Jones PD, Lister DH, Osborn TJ, Harpham C, Salmon M, Morice CP (2012) Hemispheric and large-scale land-surface air temperature variations: an extensive revision and an update to 2010. J Geophys Res Atmos 117(D5)
Kumar M, Singh MP, Singh H, Dhakate PM, Ravindranath NH (2019a) Forest working plan for the sustainable management of forest and biodiversity in India. J Sustain for 39:1–22. https://doi.org/10.1080/10549811.2019.1632212
Kumar M, Phukon SN, Paygude AC, Tyagi K, Singh H (2021a) Mapping Phenological Functional Types (PhFT) in the Indian Eastern Himalayas using machine learning algorithm in Google Earth Engine. Comput Geosci 158:104982. https://doi.org/10.1016/j.cageo.2021.104982
Kumar A, Kumar P, Singh H, Kumar N (2021b) Modulation of plant functional traits under essential plant nutrients during seasonal regime in natural forests of Garhwal Himalayas. Plant Soil 465:197–212. https://doi.org/10.1007/s11104-021-05003-x
Kumar A, Kumar P, Singh H, Bisht S, Kumar N (2021c) Relationship of physiological plant functional traits with soil carbon stock in the temperate forest of Garhwal Himalaya. Curr Sci 120(8):1368–1373. https://doi.org/10.18520/cs/v120/i8/1368-1373
Kumar A, Kumar P, Singh H, Kumar N (2021d) Adaptation and mitigation potential of roadside trees with bio-extraction of heavy metals under vehicular emissions and their impact on physiological traits during seasonal regimes. Urban for Urban Green 58:126900. https://doi.org/10.1016/j.ufug.2020.126900
Kumar N, Suyal DC, Sharma IP, Verma A, Singh H (2017) Elucidating stress proteins in rice (*Oryza sativa* L.) genotype under elevated temperature: a proteomic approach to understand heat stress response. 3 Biotech 7:205. https://doi.org/10.1007/s13205-017-0856-9

Kumar N, Jeena N, Singh H (2019b) Elevated temperature modulates rice pollen structure: a study from the foothill of the Himalayan agro-ecosystem in India. 3 Biotech 9(5):175. https://doi.org/10.1007/s13205-019-1700-1

Kumar N, Jeena N, Kumar N, Gangola S, Singh H (2019c) Phytoremediation facilitating enzymes: an enzymatic approach for enhancing remediation process. In: Bhatt P (ed) Smart bioremediation technologies. Academic Press, Elsevier, pp 289–306. 50 Hampshire Street, 5th Floor, Cambridge, MA 02139, United States

Kumar N, Singh H, Sharma SK (2020) Antioxidants: responses and importance in plant defense system. In: Roychowdhury R, Choudhury S, Hasanuzzaman M, Srivastava S (eds) Sustainable agriculture in the era of climate change. Springer, Cham, pp 251–264. https://doi.org/10.1007/978-3-030-45669-6_11

Li C, Wu C, Duan B, Korpelainen H, Luukkanen O (2009) Age-related nutrient content and carbon isotope composition in the leaves and branches of *Quercus aquifolioides* along an altitudinal gradient. Trees 23:1109–1121

Li X, Zhou Y, Meng L, Asrar GR, Lu C, Wu Q (2019) A dataset of 30 m annual vegetation phenology indicators (1985–2015) in urban areas of the conterminous United States. Earth Syst Sci Data 11(2):881–894

Linnaeus C (1751) Philosophia Botanica. Impensis Direct. LaurentiiSalvii

McDonnell MJ, Hahs AK (2008) The use of gradient analysis studies in advancing our understanding of the ecology of urbanizing landscapes: current status and future directions. Landsc Ecol 23:1143–1155

McDowell NG, Allen CD, Anderson-Teixeira K, Aukema BH, Bond-Lamberty B, Chini L, Clark JS, Dietze M, Grossiord C, Hanbury-Brown A (2020) Pervasive shifts in forest dynamics in a changing world. Science 368(6494):eaaz9463

McKinley DC, Miller-Rushing AJ, Ballard HL, Bonney R, Brown H, Cook-Patton SC, Evans DM, French RA, Parrish JK, Phillips TB (2017) Citizen science can improve conservation science, natural resource management, and environmental protection. Biol Cons 208:15–28

Meena D, Singh H, Chaudhari SK (2011) Elucidating strontium response on growth dynamics and biochemical change in *Phaseolus mungo* L. Int J Agric Environ Biotechnol 4:107–113

Melaas EK, Friedl MA, Richardson AD (2016) Multiscale modeling of spring phenology across Deciduous Forests in the Eastern United States. Glob Change Biol 22(2):792–805

Miller-Rushing AJ, Høye TT, Inouye DW, Post E (2010) The effects of phenological mismatches on demography. Philos Trans Roy Soc B Biol Sci 365(1555):3177–3186

Morin X, Roy J, Sonié L, Chuine I (2010) Changes in leaf phenology of three European oak species in response to experimental climate change. New Phytol 186(4):900–910

Nicotra AB, Atkin OK, Bonser SP, Davidson AM, Finnegan EJ, Mathesius U, Poot P, Purugganan MD, Richards CL, Valladares F (2010) Plant phenotypic plasticity in a changing climate. Trends Plant Sci 15(12):684–692

Opdam P, Nassauer JI, Wang Z, Albert C, Bentrup G, Castella J-C, McAlpine C, Liu J, Sheppard S, Swaffield S (2013) Science for action at the local landscape scale. Landsc Ecol 28:1439–1445

Paine RT (1969) A note on trophic complexity and community stability. Am Nat 103(929):91–93

Parmesan C, Yohe G (2003) A globally coherent fingerprint of climate change impacts across natural systems. Nature 421(6918):37–42

Peralta G, Vazquez DP, Chacoff NP, Lomáscolo SB, Perry GLW, Tylianakis JM (2020) Trait matching and phenological overlap increase the spatio-temporal stability and functionality of plant–pollinator interactions. Ecol Lett 23(7):1107–1116

Pigliucci M (2001) Phenotypic plasticity: beyond nature and nurture. JHU Press

Power ME, Tilman D, Estes JA, Menge BA, Bond WJ, Mills LS, Daily G, Castilla JC, Lubchenco J, Paine RT (1996) Challenges in the quest for keystones: identifying keystone species is difficult—but essential to understanding how loss of species will affect ecosystems. Bioscience 46(8):609–620

Primack RB, Gallinat AS (2017) Insights into grass phenology from herbarium specimens. New Phytol 213(4):1567–1568

Reyers B, Biggs R, Cumming GS, Elmqvist T, Hejnowicz AP, Polasky S (2013) Getting the measure of ecosystem services: a social–ecological approach. Front Ecol Environ 11(5):268–273

Richardson AD, Keenan TF, Migliavacca M, Ryu Y, Sonnentag O, Toomey M (2013) Climate change, phenology, and phenological control of vegetation feedbacks to the climate system. Agric for Meteorol 169:156–173

Sala OE, Stuart Chapin F, Armesto JJ, Berlow E, Bloomfield J, Dirzo R, Huber-Sanwald E, Huenneke LF, Jackson RB, Kinzig A (2000) Global biodiversity scenarios for the year 2100. Science 287(5459):1770–1774

Schwartz MD, Ahas R, Aasa A (2006) Onset of spring starting earlier across the Northern Hemisphere. Glob Change Biol 12(2):343–351

Seidl R, Spies TA, Peterson DL, Stephens SL, Hicke JA (2016) Searching for resilience: addressing the impacts of changing disturbance regimes on forest ecosystem services. J Appl Ecol 53(1):120–129

Settele J, Scholes R, Betts R, Bunn S, Leadley P, Nepstad D, Overpeck JT, Taboada MA, Fischlin A, Moreno JM (2014) Terrestrial and inland water systems. Clim Change 260:271–359

Sharma R, Singh H, Kaushik M, Nautiyal R, Singh O (2018) Adaptive physiological response, carbon partitioning, and biomass production of *Withania somnifera* (L.) Dunal grown under elevated CO_2 regimes. 3 Biotech 8:267. https://doi.org/10.1007/s13205-018-1292-1

Sih A, Ferrari MCO, Harris DJ (2011) Evolution and behavioural responses to human-induced rapid environmental change. Evol Appl 4(2):367–387

Singh H, Kumar M (2022) Climate change and its impact on Indian Himalayan Forests: current status and research needs. In: Rani S, Kumar R (eds) Climate change. Springer Climate. Springer, Cham, pp 223–242. https://doi.org/10.1007/978-3-030-92782-0_11

Singh H, Verma A (2013) Characterization and screening of high nitrogen efficient rice genotype to curtail environmental pollution. Int J Agric Environ Biotechnol 6(1):23–30

Singh H, Savita Sharma R, Sinha S, Kumar M, Kumar P, Verma A, Sharma SK (2017a) Physiological functioning of *Lagerstroemia speciosa* L. under heavy roadside traffic: an approach to screen potential species for abatement of urban air pollution. 3 Biotech 7(61):1–10. https://doi.org/10.1007/s13205-017-0690-0

Singh H, Verma A, Kumar M, Sharma R, Gupta R, Kaur M, Negi M, Sharma SK (2017b) Phytoremediation: a green technology to clean up the sites with low and moderate level of heavy metals. Austin Biochem 2:1012

Singh H, Sharma R, Savita SMP, Kumar M, Verma A, Ansari MW, Sharma SK (2018) Adaptive physiological response of *Parthenium hysterophorus* to elevated atmospheric CO_2 concentration. Indian Forester 144:1–14

Singh H, Yadav M, Kumar N, Kumar A, Kumar M (2020) Assessing adaptation and mitigation potential of roadside trees under the influence of vehicular emissions: A case study of Grevillea robusta and Mangifera indica planted in an urban city of India. PLoS ONE 15(1):e0227380. https://doi.org/10.1371/journal.pone.0227380

Singh M, Singh H, Kumar A, Kumar M, Barthwal S, Thakur A (2024a) Soil nitrogen availability determines the CO_2 fertilization effect on tree species (*Neolamarckia cadamba*): growth and physiological evidence. Environ Sustain 7:53–60. https://doi.org/10.1007/s42398-023-00300-w

Singh H, Kumar N, Kumar A (2024c) Enhancing resource use efficiency in crops through plant functional traits. In: Kumar N, Singh H (eds) Plant functional traits for improving productivity. Springer, Singapore. https://doi.org/10.1007/978-981-97-1510-7_6

Smith AM, Ramsay PM (2020) A comparison of ground-based methods for obtaining large-scale, high-resolution data on the spring leaf phenology of temperate tree species. Int J Biometeorol 64:521–531

Stewart IT, Cayan DR, Dettinger MD (2005) Changes toward earlier streamflow timing across western North America. J Clim 18(8):1136–1155

Suding K, Higgs E, Palmer M, Callicott JB, Anderson CB, Baker M, Gutrich JJ, Hondula KL, LaFevor MC, Larson BMH (2015) Committing to ecological restoration. Science 348(6235):638–640

Thackeray SJ, Sparks TH, Frederiksen M, Burthe S, Bacon PJ, Bell JR, Botham MS, Brereton TM, Bright PW, Carvalho L (2010) Trophic level asynchrony in rates of phenological change for marine, freshwater and terrestrial environments. Glob Change Biol 16(12):3304–3313

Thackeray SJ, Henrys PA, Hemming D, Bell JR, Botham MS, Burthe S, Helaouet P, Johns DG, Jones ID, Leech DI (2016) Phenological sensitivity to climate across taxa and trophic levels. Nature 535(7611):241–245

Vrieling A, Skidmore AK, Wang T, Meroni M, Ens BJ, Oosterbeek K, O'Connor B, Darvishzadeh R, Heurich M, Shepherd A (2017) Spatially detailed retrievals of spring phenology from single-season high-resolution image time series. Int J Appl Earth Obs Geoinf 59:19–30

Walther G-R, Post E, Convey P, Menzel A, Parmesan C, Beebee TJC, Fromentin J-M, Hoegh-Guldberg O, Bairlein F (2002) Ecological responses to recent climate change. Nature 416(6879):389–395

Ward DS, Shevliakova E, Malyshev S, Rabin S (2018) Trends and variability of global fire emissions due to historical anthropogenic activities. Glob Biogeochem Cycles 32(1):122–142

Wolkovich EM, Cook BI, Allen JM, Crimmins TM, Betancourt JL, Travers SE, Pau S, Regetz J, Davies TJ, Kraft NJB (2012) Warming experiments underpredict plant phenological responses to climate change. Nature 485(7399):494–497

Yadav SK, Singh H, Nautiyal R, Ginwal HS, Ansari SA, Barthwal S (2019) Modulation of morpho-physiological responses in Populus deltoides by elevated carbon dioxide and temperature. For Sci 66:105–118. https://doi.org/10.1093/forsci/fxz048

Zhou H, Xu M, Pan H, Yu X (2015) Leaf-age effects on temperature responses of photosynthesis and respiration of an alpine oak, Quercus aquifolioides, in southwestern China. Tree Physiol 35(11):1236–1248

Role of Urban Forests in Modulating Urban Climate

Anita Rawat and Hukum Singh

Abstract Rapid urbanization has profoundly impacted cities worldwide in recent years, presenting significant challenges for their microclimates. The migration of populations from rural to urban areas has led to the loss of natural vegetation and degradation of urban ecosystems, resulting in a phenomenon known as the Urban Heat Island (UHI) effect. Higher temperatures in urban areas characterize this effect compared to surrounding rural regions. Consequently, urban communities face many environmental issues, including poor air quality, acid rain, smog, flooding, landslides, and a lack of recreational spaces, threatening residents' health and well-being. The projected effects of climate change and continued urbanization are expected to exacerbate the UHI effect and its associated problems, highlighting the urgent need for climate-conscious urban planning. Urban greening emerges as a nature-based solution to enhance quality of life, preserve biodiversity, and bolster climate resilience. Urban forests, in particular, play a crucial role in the ecology of human habitats by regulating the urban microclimate and mitigating the impacts of UHI and climate change. Through processes such as evapotranspiration and shading, urban forests provide cooling, carbon storage, runoff reduction, and improved air quality. Additionally, they offer habitats for wildlife and serve as recreational areas for city dwellers, contributing to stress reduction and mental well-being. Addressing the multifaceted challenge of optimizing urban microclimates requires an integrated, multilevel, systematic approach. This entails developing and refining comprehensive research methodologies that combine field measurements with numerical simulations to better understand the role of urban forests in the sustainable development of cities. Effective planning and managing green spaces may maximize the ecosystem services urban forests provide, ultimately enhancing the resilience and livability of urban environments.

Keywords Urban heat island · Landscape planning · Ecosystem services · Green-blue-grey infrastructure · Thermal comfort · Urban policy · Built environment

---

A. Rawat (✉) · H. Singh
ICFRE-Forest Research Institute, PO New Forest, Dehradun, Uttarakhand 248006, India
e-mail: anita81rawat@gmail.com

H. Singh (ed.), *Urban Forests, Climate Change and Environmental Pollution*,
https://doi.org/10.1007/978-3-031-67837-0_17

1 Introduction

The global population is anticipated to rise from 7.9 billion in 2022 to approximately 8.5 billion in 2030, 9.7 billion in 2050, and 10.4 billion in 2100 (UNDESA 2022). This growth exerts significant pressure on natural resources and dramatically strains natural ecosystems. For instance, more than 13 million hectares of forest land are converted annually for agricultural, urban, and industrial purposes (Zipperer et al. 2020). Urbanization has seen a remarkable surge, with the urban population doubling since 1985 to reach 4.3 billion, representing over half of the global population (UNDESA 2019). According to projections from the UN World Urbanization Prospects, over two-thirds of the global population will reside in urban areas by 2050, signifying a substantial increase (Ritchie et al. 2024).

Urbanization leads to habitat destruction and fragmentation and triggers environmental alterations. The widespread replacement of natural ecosystems with built-up areas intensifies pressure on cities and their surroundings, resulting in resource scarcity, compromised air and water quality, and diminished green spaces. Moreover, escalating global temperatures have heightened the frequency and severity of natural disasters such as floods, droughts, and heatwaves, posing heightened risks to densely populated areas and critical infrastructure (IPCC 2023; Kumar et al. 2024; Singh et al. 2024b; Singh 2024).

Furthermore, the combustion of fossil fuels by power plants emits greenhouse gases, primarily CO_2, exacerbating global climate change (EPA 2010). Average annual greenhouse gas emissions during 2010–2019 surpassed those of any previous decade, with a growth rate of 1.3% per year. In 2019, approximately 79% of global greenhouse gas emissions originated from the energy, industry, transport, and buildings sectors, with 22% stemming from agriculture, forestry, and other land use (IPCC 2023).

These urban and demographic challenges contribute to the degradation of the urban microclimate, characterized by rising urban temperatures. The Urban Heat Island (UHI) phenomenon exacerbates this issue, intensifying under the influence of global warming. UHI alters precipitation patterns, enhances ozone production, modifies biogeochemical processes, and poses health risks to humans and native species, particularly during heatwaves (Akbari et al. 1992; EPA 2010).

Mitigating the UHI effect is imperative for achieving sustainability in cities, necessitating improvements in the urban microclimate. Microclimates, which encompass fine-scale climate variations distinct from the broader atmosphere or macroclimate, are influenced by various factors, including topography, land cover, and proximity to water bodies. Addressing these issues demands adopting innovative, cost-effective measures to enhance the urban microclimate and revitalize urban areas (Kumar et al. 2022; Singh 2021).

The imperative to plan for sustainable urban growth is evident, considering the need to balance environmental preservation with economic development and social progress. Effective urban planning should prioritize long-term sustainability over short-term gains, anticipating future challenges and implementing strategies to mitigate environmental impacts while accommodating population growth and economic expansion. This chapter underscores the significance of microclimate modification in urbanization, the role of urban forests in regulating microclimates, and the various factors influencing microclimate regulation. Additionally, it discusses assessment methods and implications for urban planning and policy and outlines future challenges and opportunities in this domain.

2 Urbanization and Microclimate Modification

According to the ‘2018 Revision of World Urbanization Prospects’, it is projected that by 2050, approximately 68% of the global population will reside in urban areas, with India contributing around 416 million individuals to this urban demographic. As urbanization progresses, cities evolve into dense conglomerations of high-rise structures, with dwindling pockets of green spaces serving as vital relief zones amidst the concrete sprawl. However, the presence of Urban Heat Islands (UHIs) significantly alters the microclimate of these open areas, undermining their intended functionality and value (Srivastav et al. 2023).

The relentless expansion of urban areas accelerates climate change, transforming natural landscapes into artificial built environments and exacerbating various environmental issues. Zipperer et al. (2020) underline that while cities serve as hubs of economic and cultural activity, they also profoundly impact the environment and the ecological services derived from natural systems. For instance, de Souza et al. (2015) studied the urban growth of Manaus and its effects on microclimates, highlighting the alteration of Surface Energy Balance (SEB) and local atmospheric processes due to urbanization.

The formation of UHIs, resulting from reduced vegetation, increased non-reflective surfaces, and anthropogenic emissions, exacerbates environmental degradation and poses risks to human health and native species (Zipperer et al. 2020). The consequences include altered precipitation patterns, intensified heat waves, and worsening air quality, with vulnerable populations facing increased health risks (Akbari et al. 1992; EPA 2010). Eurosurveillance (2005) reports a stark toll on human life during heatwaves, emphasizing the urgent need for mitigation strategies.

Moreover, UHIs pose energy efficiency challenges, escalating energy consumption, infrastructure vulnerability, and ecological footprints (Akbari et al. 1992; Santamouris et al. 2007). Swain et al. (2016) investigated the impact of urbanization on Bhubaneshwar’s microclimate, identifying land use and land-cover changes as significant drivers of UHI effects.

Large-scale urbanization exacerbates habitat loss, pollution, and the deterioration of urban dwellers’ quality of life (United Nations). Amirtham et al. (2009) underscore the role of urbanization in altering surface characteristics and atmospheric properties, leading to microclimate changes and UHI formation. Urban development modifies natural relief, increases surface roughness, and influences precipitation patterns

(Bherwani et al. 2020; Dhyani et al. 2021; Kumar et al. 2021a, b, c, d, e; Prakash et al. 2022).

Urban geometry and building density further shape the microclimate, affecting aerodynamics, air quality, and thermal comfort (Giyasova 2021). The configuration of streets, buildings, and open spaces, along with environmental factors like solar radiation and wind speed, play crucial roles in urban climate dynamics (Jamei et al. 2017; Singh 2021).

3 Characteristics of Urban Forests

Urban forest ecosystems have many unique characteristics that, in combination, distinguish them from other forest types. These characteristics include (1) close proximity to large or dense human populations, (2) relatively high diversity of species and forest patch structures, (3) multiple public and private ownership types, and (4) management often geared toward sustaining tree health and ecosystem services. Urban forests can include trees planted in many different types of growing spaces: in the parkways, along streets, and next to sidewalks in cities; in public parks or cemeteries; trees in front or back yards on residential property; riparian trees along creeks and lowlands; on institutional properties such as school or hospital campuses or business or industrial parks; along highways; growing spontaneously in vacant lots and along alleys; trees in remnant natural areas or planted woodlands; green roofs, green walls, landscaping and more (Vogt 2020). It includes land cover such as golf course fairways and greens, grassy medians and shoulders along roads, lands under electrical transmission lines, and lawns. Because they are co-located in places inhabited by people, urban forests include the trees, soil, and other abiotic and biotic features in the surrounding growing environment and the people living in cities who impact the urban forest. There may be some remnant patches of native forests that have been enclosed by urban development and set aside as parks, nature preserves, or backyard woodlands, or there may also be areas of planted forest within a city that share more traditional forest characteristics. Whether planted or remnant, these areas are sometimes called urban forest natural areas, naturalized areas, forest patches, urban woodlands or woodlots, or forested parks (Vogt 2020). These areas include not just trees but also shrubs and herbs as ground cover, and depending on quality, may contain habitat for urban wildlife, including small mammals, insects, and birds. Urban forest natural areas may abut bodies of water near or within urban areas, such as streams, rivers, ponds, lakes, or canals. The growing conditions for trees, soils, hydrology, nutrient flows, microclimate, and ecology in these areas are more likely to mimic those for trees in natural areas outside of cities. However, urban forests growing outside natural areas such as parks, the front or back yards of residential properties, along streets, etc., face many types of growing conditions from the semi-natural conditions. Both above and belowground growing spaces may be more constrained, especially for trees growing in highly constructed spaces (Sharma and

Singh 2021; Singh et al. 2021). For instance, trees growing in a tree pit are exceptionally constrained: tree pits are small growing spaces, often found along streets, in plazas, parking lots, etc.

The base of a tree in a tree pit is likely surrounded on all four sides by concrete paving impervious to water, and the area immediately adjacent to the tree trunk may be covered with a metal grate that, if not correctly sized, may interfere with the root flare and healthy trunk growth. Belowground, tree roots must compete with urban infrastructure such as buried power and cable lines, storm and sanitary sewer pipes, gas lines, building, and road foundations, etc., cope with high levels of soil compaction, soil nutrient imbalance, and inadequate soil drainage (or conversely, poor water retention). Aboveground, these trees may be heavily pruned so that they do not interfere with urban infrastructure such as buildings, fences, signage, street lights, aboveground power, and cable lines, etc., or may have limited exposure to light due to shading by tall urban buildings (Vogt 2020; Singh et al. 2020; Kumar et al. 2021a). Because of this suite of constraints, trees in constrained urban areas may be in poor condition, of small size and sub-optimal form, and experience high mortality rates and short life spans. On the other hand, trees growing in high-quality urban parks may have relatively unlimited growing space, be well-cared for (through regular pruning and watering, pest management, etc.), and even may have access to additional nutrients (either through direct fertilizing of the tree or as a byproduct of fertilizing the nearby turf grass). They may grow in a wide-open area without competition with other trees or other urban infrastructure. These high-quality growing conditions may lead to long-lived trees that grow large and provide significant benefits.

4 Role of Urban Forests in Microclimate Regulation

In the past, urban tree selection for plantings was primarily based on aesthetic demands, while nowadays, the ecosystem services of green infrastructure are getting more and more attention (Gómez-Baggethun and Barton 2013). In many US cities, planting programs are implemented to include environmental and health benefits for urban trees in species selection (McPherson and Simpson 2003). The benefits which arise from appropriate tree selection can be various, starting from the reduction in air pollution, reducing carbon emission and carbon sequestration, reducing surface runoff, natural disaster mitigation, combating UHI effect, regulating air temperature and humidity, air cooling and shading, increased biodiversity, noise reduction to a better health of pedestrians (Kumar et al. 2021b, c, d).

Reducing air pollution: Trees influence air quality in several ways. Trees remove pollution from the air by intercepting airborne particles on their leaves and branches and absorbing gaseous pollutants into their leaves via stomata. Pollution removal by trees within a city can be thousands of tons annually, with air-quality improvement typically less than 1 percent. Trees reduce pollutant emissions from power plants by reducing building energy requirements, thereby improving air quality (Nowak et al.

2006; Yadav et al. 2019a, b; Gupta et al. 2019). Coniferous trees have the highest filtering capacity compared with deciduous trees due to the larger total surface area of leaves. However, an assortment of coniferous and deciduous trees is much better for reducing air pollution (Bolund and Hunhammar 1999).

Reducing carbon dioxide emissions: Trees can reduce the emission of CO_2 from power plants by reducing energy needs for cooling or heating buildings by providing shade in the summer, reducing summer air temperatures (primarily through transpirational cooling), and blocking winter winds. Urban trees reduce carbon dioxide (CO_2), a significant greenhouse gas, by directly removing it from the atmosphere and storing ("sequestering") the carbon in the trees as biomass.

Reducing surface runoff: By intercepting and retaining or slowing the flow of precipitation reaching the ground, urban forests can play an essential role in urban hydrologic processes. They can reduce the rate and volume of stormwater runoff, flooding damage, and stormwater treatment costs and enhance water quality (Gupta et al. 2019; Kumar et al. 2019a). Tree roots reduce erosion and create soil conditions that allow rainwater to infiltrate the ground and recharge the groundwater supply. The root systems of trees and shrubs further act as a filter system for rainwater, catching pollutants and helping purify the water.

Combating UHI effect: Vegetated landscapes in and around an urban area can significantly reduce the urban heat island effect. Trees can influence thermal comfort by reflecting, transmitting, and absorbing solar radiation, modifying air temperature, increasing air humidity through evapotranspiration, and reducing wind speed (Streiling and Matzarakis 2003). In a study, augmentation of the green area from 2487 to 4398 m^2 led to an overall temperature decrease of approximately 0.45 °C, highlighting the favorable influence of heightened green space on microclimate conditions (Adıgüzel 2023). Appropriate tree species and their features, such as treetop density, size and shape of leaves, tree age, and planting arrangement greatly impact their efficiency in regulating air temperature and humidity (de Abreu-Harbich et al. 2015). Coniferous trees were more effective than broadleaf trees in reducing temperature and thermal comfort levels, which were most improved when planted in paired settings. Focusing on planting arrangement, Miloševic et al. (2017) tested tree location's impact on thermal comfort on street parking lots. Stojakovic et al. (2020) developed an approach based on the evolutionary algorithm for insulation mitigation by changing tree locations.

Cooling effect: Plants exert a cooling effect in two ways, first, evapotranspiration, which is attributed to features such as leaf shapes, stomatal density, canopy structures, foliage density, water use strategies, etc., and second, shading, which is influenced by tree characteristics like crown diameter, canopy height, canopy shape etc. which may affect their cooling efficiency along with the climatic conditions and soil moisture content (Richter et al. 2022). Broadleaved species, such as *Eucalyptus* and *Olea, Ficus* and *Tipuana tipu* trees, and *Ulmus parvifolia* are reported to have a better cooling effect compared to trees with larger leaves, which tend to have higher crown temperatures (Leuzinger et al. 2010).

5 Effects of Microclimate on Urban Ecosystems

Urban ecosystems interface biological communities of plants and animals and urban infrastructure. They can include urban forests, native plantings, or green infrastructure such as green roofs, community gardens, and rain gardens. These ecosystems provide various services for cities and residents, including reducing energy consumption, improving air and water quality, and lowering ambient temperatures (Kumar et al. 2019b). While these systems can aid in mitigating the effects of climate change, they are also threatened by the extreme events brought on by climate change. Some of the impacts of urbanization and microclimate modification on urban ecosystems are as follows:

Survival threat to vegetation and wildlife: Urban vegetation is at risk of exposure to increased stormwater runoff and pollutants because of climate change. Severe flooding and high salinity could also kill many plants within an urban ecosystem. Extreme heat waves can cause stress in trees and damage their ability to uptake carbon dioxide or ultimately kill the tree, leading to further increases in atmospheric carbon dioxide. Extended periods of drought also result in less carbon dioxide uptake since trees halt gas exchange to protect water storage functions. Urban trees provide shelter and forage to animals like squirrels, chipmunks, and birds, while prairie plantings and shrubs provide cover for rabbits and food for pollinators. Without a consistent source of shelter and food, these animals are not likely to survive or will be forced to relocate.

Adverse impact on human health: Pollution from vehicles, factories, and power plants can cause respiratory problems, heart disease, and even cancer. Poor air quality, contaminated water supplies, and natural disasters such as floods cause diseases' rapid spread and evolution. Extreme heat events, which occur more frequently in UHIs, can and do result in deaths, cardiopulmonary diseases, reduced capacity for outdoor labor, mental health concerns, and kidney disease (Sharma et al. 2018; Singh et al. 2018). The demographics most vulnerable to the negative impacts of UHIs are senior citizens and those without resources to cool off, such as air conditioners (Tong et al. 2021). Elevated urban heat island (UHI) temperatures, especially during the summer months, can precipitate various health issues, including but not limited to general discomfort, respiratory challenges, heat cramps, exhaustion, non-fatal heat stroke, and heat-related fatalities. Eurosurveillance (2005) reported approximately 22,080 excess deaths in regions including England, Wales, France, Italy, and Portugal during and in the aftermath of the heat waves experienced in the summer of 2003.

Loss of biodiversity: Habitat fragmentation caused by urban development leads to loss of connectivity, which makes it impossible for some species to migrate across the landscape, thereby inhibiting genetic drift, a process essential for maintaining the genetic diversity needed for species survival (Santangelo et al. 2022) and ecosystem stabilization. This results in the extinction of threatened species. Species diversity is also impacted by introducing non-native and invasive species from travel and shipping processes, which can out-compete essential native species, cause biotic homogenization, and introduce new disease vectors (Gaertner et al. 2017). Elevated temperature stimulates the decline of many beneficial insects, such as wild bees and butterflies (Gupta et al. 2018a, b). The insect populations adapted to cold climates have declined (e.g., dragonflies, stoneflies, and bumblebees), showing a general reduction in the population density of pollinators such as wasps, ants, and beetles in Mediterranean regions. In urban settings, industrial sites, transportation, and sewage increase soil contamination by the heavy metals in green infrastructures, which can reduce not only belowground biodiversity but also influence the vegetation structure of lawns and grassland patches (Sánchez-Bayo and Wyckhuys 2019). The change in habitat structure of the natural landscape, caused by industrialization and agricultural intensification, has reduced the food resources and shelter sites for many specialist species, thus leading to the disappearance of many habitat specialist species and their replacement with a few generalist species adapted to the artificial human environment (Singh and Verma 2013; Kumar et al. 2017).

Effect on microbiome: Subsurface microbial communities are associated with major biochemical processes that support plant growth and ensure key ecosystem services involving nitrogen cycling, biodegradation, and decomposition. Edaphic variables such as physical (moisture and texture) and chemical properties (pH, solid minerals, and organic matter) of soil can influence the diversity and geographical distribution of microorganism communities (Joyner et al. 2019). In urban areas, the presence of heavy metals and other pollutants such as pesticides, fertilizers, salt, etc., in soil and exposure to petroleum products impact the soil ecosystem, as they can alter the structure of soil bacteria communities and have a substantial effect on their abundance and diversity (Delgado-Balbuena et al. 2016).

6 Factors Influencing Microclimate Regulation

The four main factors influencing an area's microclimate are air temperature, humidity, wind speed, and solar radiation. The landscape elements in an urban form that regulate these factors can potentially affect the urban microclimate regulation. The major landscape elements in an urban setting are the green, grey, and blue infrastructure. The characteristic features of these elements, such as morphological and physiological traits of trees, their configuration and arrangement, composition and geometry of built environment, and distribution of blue spaces drive microclimate regulation in urban areas as follows (Gupta and Singh 2017; Singh et al. 2017a).

Canopy geometry: The critical factor for trees to exert a cooling effect resides in the canopy (Cheung et al. 2021), which blocks direct sunlight on the surface, while the leaves of the canopy can perform photosynthesis and transpiration (Kantor et al. 2016). Canopies with different three-dimensional characteristics have different cooling capacities. Pileated canopy shapes have the most robust cooling and humidification effects, while the shape of the columnar canopy is the weakest. Increasing the Tree Crown Diameter (TCD) has better cooling and humidification effects than increasing the trunk height (Speak et al. 2020) because a larger TCD reduces the average Sky View Factor (Oshio et al. 2021; Karimi et al. 2020), and is more conducive in improving the thermal comfort.

Tree community configuration: Plant community configurations such as grasslands, woodlands, and ornamental shrubs also play an essential role in cooling and humidification effects. The tree–grassland and tree–shrub–grass types significantly improved comfort (Bao et al. 2022).

Tree physical characteristics: When analyzing the thermal effect from the physical attributes of a tree, crown density, tree height, trunk height, and crown diameter are the main analytical factors (Singh et al. 2017b). Crown density is the primary determinant of a tree's heat reduction potential, contributing about 60% of Ta reduction. Generally, a tree with high foliage density is a high heat mitigator and vice-versa, as confirmed by previous studies (Morakinyo et al. 2018, 2020). In reality, the number, type, size, and arrangement of leaves affect crown density collectively. In ENVI-met, this combined metric is generally represented by LAI (Fabbri et al. 2017). The strong positive correlations between LAI and solar radiation attenuation and reduction in Ta, Tmrt, PET, and TS (Sodoudi et al. 2018; Morakinyo et al. 2018; Wai et al. 2020) under the canopy have been confirmed. However, a tree with a dense crown may block the wind and increase Ta (Aboelata and Sodoudi 2019), an effect that should be assessed critically.

Tree height was an essential tree parameter for human thermal comfort improvement. Weaker than LAI, a strong and positive correlation between tree height and Tmrt (Morakinyo et al. 2018) and PET reduction has been confirmed. When trees have the same LAI values, the tallest tree with a broad and scattered crown recorded the most solar attenuation.

Trunk height is the distance between the lower surface of the tree crown and the ground. A crown at a greater height can bring better ventilation. In contrast, for more shading beneath the tree, the crown should be at a lower height (Morakinyo et al. 2018). Due to the "umbrella effect" (Brown and Gillespie 1995), broader tree crowns can provide more shading, more TS reduction (Karimi et al. 2020), and more UHI depression than narrower tree crowns. However, a broad tree crown may obstruct wind and ventilation.

Planting arrangement: The tree-planting design includes key attributes: number of trees (tree coverage ratio) and tree planting patterns (including tree arrangement, planting orientation, intervals among trees, etc.). Areas with similar tree cover but different planting designs produce different thermal performance due to effects on wind speed, wind direction, and shading pattern. A good tree-planting design embodies the thermal benefits of both ventilation and shading (Abdi et al. 2020). Additionally, some green indices could reflect the planting pattern quantitatively, such as the landscaping deviation index, landscaping isolation index (Rui et al. 2019), and land shape index (Sodoudi et al. 2018).

Tree coverage: The number of trees in a particular place is quantitatively represented by tree coverage ratio (TCR) (Ouyang et al. 2020) or green coverage ratio

(GCR) (Morakinyo et al. 2018). Many reviewed studies have confirmed a higher TCR lowering Ta (Morakinyo et al. 2018) and PET, as well as raising RH in the daytime. However, some reviewed studies found that the correlations were non-linear because of WS and RH variations, indicator selection, tree planting pattern, background urban environment, and ENVI-met version (Ouyang et al. 2020). Overall, the impact of trees cannot be considered as the more, the better, especially in high-density urban areas.

If planted incorrectly, a tree's thermal contributions may be underutilized, leading to wasted overlapped shadows or airflow blocking (Morakinyo et al. 2020). Also, urban morphology regulates the background thermal environment in an urban environment, which may strengthen or weaken the thermal effects of trees. Therefore, planting design should incorporate appropriate tree planting location, arrangement, orientation, and inter-tree interval to optimize the shaded area and improve ventilation.

The potential of trees to regulate the energy demand for cooling a building is mainly associated with radiation shading and a corresponding decrease in the increased solar heat of open facades of buildings. It is directly related to their foliage density and planting patterns (Meena et al. 2011; Kumar et al. 2019c; Giyasova 2021). Higher energy savings of up to 54% have been achieved with the trees forming a continuous shade canopy with a leaf area density of 2.5 m^2/m^3. However, the cooling potential of the street trees was quite negligible when they were not tall enough to obscure most of the building's outer façade. The more dispersed the vegetation, the better the thermal comfort achieved. Scattered trees can provide shade, avoid unnecessary overlapped tree crowns, and interact more with the surrounding environment. Also, every tree's full transpirational cooling potential can be achieved due to the "oasis effect" (Sodoudi et al. 2018).

Grey infrastructure: In the more complex geometry of the urban environment, different building densities, variations in heights, surface materials, issues of thermal response of buildings, cooling energy needs, etc., influence the microclimate regulation. With the increase in building densities, the sky view factor decreases, which limits the radiation cooling of the sky. In addition, solar and infrared interflections increase with the surfaces of neighboring buildings (Lauzet et al. 2019). The escalation of building surfaces and the proliferation of solid floors further drive temperature elevation.

Blue spaces: Blue spaces in the urban ecosystem are essential in regulating urban microclimate and human thermal comfort (Cao et al. 2022). Adding water bodies can improve the thermal comfort of residential areas as water both loses and gains heat slower than the surrounding land. Doing so helps bring moisture into the air, building a more humid microclimate. The water body absorbs and accumulates heat in summer, and through evaporation, it absorbs heat from the surrounding environment and keeps it cool. The cooling effect is the best when the water bodies are added dispersedly.

Thus, blue-green-grey infrastructure needs to be established carefully to mitigate the urban heat island effect and get the maximum thermal comfort for the well-being of the urban ecosystem. Using more scientific approaches, designers and planners can optimize site design to reduce urban heat, including building shape, street geometry, and landscape design standards (Heris et al. 2020).

7 Techniques for Enhancing Microclimate Regulation

The idea of local climate management involves altering the energy balance of local climatic systems and, through that, the local climate. The fundamental principle behind this is energy conservation, which dictates that all incoming energy from the sun must be accounted for within the system. The incoming solar energy is diverted into two energy fluxes: the latent heat flux, which can contribute to evapotranspiration, and the sensible heat flux, which contributes to heating the air and soil (Kumar et al. 2020; Singh and Kumar 2022). How the incoming solar energy is diverted into these two energy fluxes depends on landscape characteristics. The essence of microclimate management is to maximize the latent heat flux and minimize the sensible heat flux to achieve a cooler local climate. The presence of blue-green infrastructure directs solar energy mainly towards evapotranspiration, thereby reducing the amount of heat available for heating the air and soil. It thus leads to a cooling effect on the local climate. We can alter the local energy balance by implementing wind management, shading, mulching practices, water harvesting, and regreening, creating more favorable microclimatic conditions.

Climate-sensitive urban design offers a sustainable solution to urban overheating. It involves a combination of innovative choices, including urban fabric, urban morphology, and re-integration of urban green and blue infrastructures, the so-called 'cool islands' of trees, shrubs, herbs, green roofs, vertical greenings and water bodies, including streams, lakes, ponds, artificial swales and stormwater retention ponds (Singh et al. 2024a). The following measures can enhance microclimate regulation:

- Augmentation of green space by converting vacant urban spaces into verdant zones while concurrently safeguarding existing green expanses can curtail urban heat influence (Adıgüzel 2023).
- Green facades and roofs entail incorporating vegetation onto building exteriors, and roofs manifest as efficacious measures for mitigating heat impact within densely populated cities where open space is a significant constraint (Xing et al. 2018).
- Grass paving is more effective than stone in lowering air temperature and improving thermal comfort at the near-surface level (Choi et al. 2021).
- The urban surface materials should undergo reevaluation and refurbishment, incorporating more reflective alternatives. Technological advancements can focus on the design and materials of cool roofs, such as single-ply or liquid-applied options, including white paints, elastomeric, polyurethane, or acrylic coatings (Santamouris 2014).

- To mitigate the adverse effects of heat on outdoor thermal conditions, it's crucial to control the albedo of building surfaces and road pavement materials. Additionally, reconfiguring urban forms to enhance city ventilation can be effective (Zhang et al. 2019).
- Cool pavements, characterized by higher albedo and infrared emittance values, demonstrate lower surface temperatures, often up to 10 °C than conventional pavements. This contributes to reduced heat convection to the ambient air (Gaitani et al. 2011).
- Implementing ponding and circulation zones can significantly mitigate the urban heat effect by fostering cooling through water evaporation.
- There's a necessity to adopt the concept of "zero-heating buildings" or "neutral microclimate buildings" due to their energy efficiency, carbon emission reduction potential, and ability to improve indoor microclimate quality on-site (He 2019).
- Advancements in HVAC standards are crucial for enhancing energy efficiency. Technological upgrades, such as improved insulation and tightness to limit heat loss, can facilitate waste heat collection (He 2019).

8 Methods for Assessing Microclimate Regulation

Different approaches have been used for understanding the change in UMC with the increase in urbanization. Satellite observations, field observations, Computational fluid dynamics (CFD), and Weather Research and Forecasting (WRF) simulations are usually adopted to determine urban heat islands and other related parameters of UMC.

Field monitoring and analysis: Field observations were one of the best ways to prove the scientific hypothesis during a lack of sophisticated instrumentation and computers. Temperature was one of the earliest parameters explored through field observations and wind. Later on, diurnal variations were recorded along with energy balance between the canyon (Ambrosini et al. 2014). Albedo and urban canopy effects were studied by the end of the nineteenth century. With the establishment of the concept of urban heat island, observations using other methods like remote sensing were relied upon. Field monitoring is now mainly used to validate results generated by models or remotely sensed observations and interpretations. Peripheral parameters like air quality are becoming mainstream and measured. Field measurements using sophisticated methods, including sensors and monitoring networks, are now being used to record and analyze finer observations, like the impact of vehicular movement on UMC (Shahrestani et al. 2015).

Remote sensing and GIS-based assessments: In remote sensing technology, multiple parameter analysis is possible as various mathematical models can get numerous data from reflected light. Earlier, the measurements were made for temperature, albedo, and humidity parameters. However, with advancements, variation in the height of roughness and the planetary boundary was incorporated in measurements along with a diurnal variation in temperature humidity measurements. These measurements were done using sensors and satellites, like advanced very high-resolution radiometer (AVHRR), Moderate Resolution Imaging Spectroradiometer (MODIS), Landsat Thematic Mapper (TM), Landsat Enhanced Thematic Mapper (ETM), Advanced Spaceborne Thermal Emission and Reflection Radiometer (ASTER), etc. Although multiple measurements are made using satellites, predictions and decision-making are still lacking in this field as the research is restricted to measurements and observations (Bherwani et al. 2020).

Modeling and simulation-based assessments: UMC research has shifted towards modeling and simulation-based assessments. Computational fluid dynamics (CFD) is widely used to assess UMC due to the presence of many in-built solver packages. CFD-based assessments also started with temperature and wind as crucial model parameters. Weather Research and Forecasting (WRF) is an adaptable model that studies the effects of mesoscale interactions at the micro level. Another added versatility is that it can simulate coupled models with other models like single/multilayer urban canopy models, building energy models, and regional climate models (Jin et al. 2010). Over the years, CFD has gradually upgraded for parameters including latent heat, air quality, urban canopy, and 3-dimensional flows. Openfoam and ENVI-met are based on CFD models and are used widely nowadays (Jandaghian 2018).

In recent years, ENVI-met (Windows-based) and OpenFOAM (Linux/Ubuntu-based) tools and related solver packages have been widely used for simulating urban canopy models with three-dimensional complex geometries (Ambrosini et al. 2014). Programming languages and indigenous software packages have been developed recently, like the Surface Urban Energy and Water Balance Scheme (SUEWS) for UMC simulations. The use of SUEWS to discover the effects of designing sustainable cities or to measure the advantages of more planning scenarios, such as including green cover and water bodies to alleviate heat stress, has also been studied in recent years. Nowadays, Python programming is widely used by the atmospheric sciences community for numerical modeling, simulation, and analysis of data. Various datasets, structures, and program development tools are used to interconnect high-level languages with low-level ones like Fortran and Python. Real-time simulation of whole building performance with the actual on-field measurement by Energy Plus is also being carried out. For data collection and monitoring, many sensor deployments have been proposed in urban areas that can feed into the latest models (Yang and Bou-Zeid 2019).

9 Implications for Urban Planning and Policy

Urban systems are critical for achieving deep emissions reductions and advancing climate-resilient development (IPCC 2023). Considering UHI mitigation measures early in the design process, policymakers, involving multi-disciplinary teams, including built environment professionals, economists, and climate scientists, who improve the UHI policy response regarding materials, technologies, and practices can contribute to the cooling of cities (Hamilton et al. 2014). Some of the strategies that can be developed and well-integrated into urban planning and policy for the development of resilient cities in the face of climate change are:

- Policy instruments such as tree and wetland protection policies are essential in maintaining quality green and blue spaces.
- Government policy intervention by enforcing land use regulations to compel adherence to proper housing plans and construction.
- By-law enforcement mechanisms within local regulatory regimes to ensure the sustainable management and protection of natural spaces.
- LEED certification (Leadership in Energy and Environmental Design, a green building certification program) to establish green infrastructure practices in individual buildings.
- Asset management systems for urban ecosystems and natural landscapes are used similarly to engineered infrastructure.
- Zoning to protect environmentally sensitive areas from urban development and sprawl.
- Mainstreaming green infrastructure into road planning and design for a "smarter" and more sustainable city.
- Adopting sustainable public transport: The saving potential for CO_2 emissions is highest for electric vehicles with renewable electricity generated from wind turbines, closely followed by hydrogen vehicles fueled by hydrogen generated from water electrolysis using wind electricity.
- Implementation of the walkable city model in urban planning to reduce the distance traveled to access resources such as food and jobs.
- Implement large-scale landscaping projects such as natural heat sinks (C40 2015).
- Explore and identify incentive or subsidy programs to encourage developers to integrate natural features (e.g., wetlands) and green infrastructure (e.g., green roofs) into new development, e.g., Toronto's Eco-Roof Incentive Program.
- Exploring albedo enhancing projects through the generation of multi-criteria 'cool city model' when the variables of cool pavements, cool roofs, and vegetation variables are considered. Such simulation models can support urban planners and policymakers to justify the investment in such projects.

10 Challenges and Opportunities

Current urban microclimate restoration strategies, such as integrating green infrastructure, de-paving surfaces, and implementing environmental design interventions, have demonstrated efficacy in alleviating urban heat island effects and enhancing thermal comfort (Hashemi et al. 2023; Bherwani et al. 2020). These approaches harness vegetation coverage, albedo adjustments, and porous surfaces to diminish heat absorption, boost evapotranspiration, and lessen air quality (Yadav et al. 2019a, b; Jangra et al. 2024). However, obstacles to their implementation persist, including structural limitations, constraints on available space for de-paving measures, and the imperative for policy backing and cohesive urban planning to facilitate widespread adoption (Sylliris et al. 2023). Overcoming these hurdles necessitates the development of frameworks, streamlining decision-making processes, and prioritizing integrated assessments of parameters to bridge the chasm between research insights and practical deployment in urban settings.

Two pivotal global events, the 1st and 2nd World Forums on Urban Forests, have markedly enhanced our comprehension of the role of urban forestry in crafting healthier and more resilient cities. Pioneering concepts, such as the 3-30-300 rule, envisage a scenario where individuals can view at least three trees from their window, inhabit neighborhoods with a minimum of 30 percent tree cover, and reside no more than 300 m away from high-quality urban green spaces. Swift advancements in technology facilitate the collection of spatial data for urban forestry planning and monitoring, while governments worldwide are allocating funds to urban forestry programs, acknowledging trees' indispensable role in fortifying urban resilience and enhancing human well-being (Borelli et al. 2023). The Kunming-Montreal Global Biodiversity Framework (GBF), ratified in 2022 by 196 countries, is a noteworthy example of the global discourse surrounding nature-based solutions, with urban forestry as a pivotal component. This framework aims to halt and reverse biodiversity loss by 2030 and advocates for the application of nature-based solutions to attain this ambitious goal, notably Target 12, which seeks to augment the area of, access to, and benefits from green and blue spaces for human health and well-being in urban and densely populated areas.

Furthermore, prominent global city networks such as C40 Cities, the International Council for Local Environmental Initiatives (ICLEI), and United Cities and Local Governments (UCLG) have recognized the pivotal role of nature and green spaces in enhancing urban quality of life. In 2021, C40 Cities endorsed the C40's Urban Nature Declaration, pledging increased large-scale investments in urban nature to shield cities from the repercussions of climate change and ensure universal access to green spaces. Cities worldwide engage in initiatives like the Tree Cities of the World, FAO, and the Arbor Day Foundation collaboration that acknowledges cities committed to managing their urban forests and trees. In 2023, four years post-launch, the program lauded 170 cities from 20 countries for their endeavors (Borelli et al. 2023).

11 Future Directions

Efforts aimed at enhancing urban forests globally encompass a multitude of strategies. These include raising awareness through educational campaigns, fostering research and collaboration among stakeholders, actively engaging the public, and ensuring equitable access to green spaces. Moreover, providing technical assistance and capacity development is essential, integrating urban forestry into policy-development plans and fostering partnerships and knowledge sharing among professionals and institutions (Borelli et al. 2023).

However, it's crucial to acknowledge that urban forestry is not a one-size-fits-all solution. Each city and region faces unique challenges and opportunities, demanding tailored strategies. The bottom line for cities is that addressing environmental degradation and biodiversity loss due to urbanization must be prioritized alongside economic and social issues (Kanta et al. 2024; Malik et al. 2024). Without native biodiversity, the ecosystem services vital to human well-being will be severely diminished or lost entirely. Therefore, mitigating the environmental effects of urbanization requires proactive planning and decision-making.

Urban ecosystem management is inherently multidisciplinary, necessitating diverse, flexible social, environmental, economic, and decision-making tools and institutions (Kumar N and Singh 2024). These tools and institutions must be adaptable to changes within one or more systems, ensuring effective management and sustainable urban development.

12 Conclusions

Urban ecosystems and green infrastructure offer vital ecosystem services that enhance human health and well-being while mitigating the impacts of natural disturbances and extreme weather events. Upholding a sustainable and robust supply of these services is imperative for bolstering resilience in urban areas. However, it's crucial to recognize that only healthy, well-functioning ecosystems can furnish these invaluable services and contribute positively to a city's resilience. Climate change poses a significant threat to the future viability of urban ecosystems, emphasizing the urgent need to safeguard and enhance their resilience. Thus, it is imperative not only to ensure the continued existence of urban ecosystems and the services they provide to fortify the resilience of cities but also to bolster the resilience of these ecosystems themselves in the face of climate change. This dual approach is essential for fostering sustainable urban development and ensuring the well-being of both urban residents and the environment.

References

Abdi B, Hami A, Zarehaghi D (2020) Impact of small-scale tree planting patterns on outdoor cooling and thermal comfort. Sustain Cities Soc 56:102085

Aboelata A, Sodoudi S (2019) Evaluating urban vegetation scenarios to mitigate urban heat island and reduce buildings' energy in dense built-up areas in Cairo. Build Environ 166:106407

Adıgüzel F (2023) Effects of green spaces on microclimate in sustainable urban planning. Int J Environ Geoinform 10(3):124–131

Akbari H, Davis S, Dorsano S, Huang J, Winert S (1992) Cooling our communities—a guidebook on tree planting and white coloured surfacing. US Environmental Protection Agency, Office of Policy Analysis, Climate Change Division, USA

Ambrosini D, Galli G, Mancini B, Nardi I, Sfarra S (2014) Evaluating mitigation effects of urban heat islands in a historical small center with the ENVI-Met® climate model. Sustain 6(10):7013–7029

Amirtham LR, Devadas MD, Perumal M (2009) Mapping of micro-urban heat islands and land cover changes: a case in Chennai City, India. Int J Clim Change: Impacts Resp 1(2):71

Bao Y, Gao M, Luo D, Zhou X (2022) The influence of plant community characteristics in urban parks on the microclimate. Forests 13(9):1342

Bherwani H, Singh A, Kumar R (2020) Assessment methods of urban microclimate and its parameters: a critical review to take the research from lab to land. Urban Clim 34:100690

Bolund P, Hunhammar S (1999) Ecosystem services in urban areas. Ecol Econ 29(2):293–301

Borelli S, Conigliaro M, Di Cagno F (2023) Urban forests: a global perspective. FAO, Rome

Brown RD, Gillespie TJ (1995) Microclimatic landscape design: creating thermal comfort and energy efficiency, vol 1. Wiley, New York

C40 (2015) Cities100: Paris—green spaces keep the city cool. https://www.c40.org/case_studies/cities100-paris-green-spaces-keep-the-city-cool

Cao S, Wang Y, Ni Z, Xia B (2022) Effects of blue-green infrastructures on the microclimate in an urban residential area under hot weather. Front Sustain Cities 4:824779

Cheung PK, Jim CY, Hung PL (2021) Preliminary study on the temperature relationship at remotely-sensed tree canopy and below-canopy air and ground surface. Build Environ 204:108169

Choi GY, Kim HS, Kim H, Lee JS (2021) How do paving and planting strategies affect microclimate conditions and thermal comfort in apartment complexes? Int J Clim Change Strat Manag 13(2):97–119

De Abreu-Harbich LV, Labaki LC, Matzarakis A (2015) Effect of tree planting design and tree species on human thermal comfort in the tropics. Landsc Urban Plan 138:99–109

de Souza MOD, Maciel Filho R, Mantelatto PE, Cavalett O, Rossell CEV, Bonomi A, Leal MRLV (2015) Sugarcane processing for ethanol and sugar in Brazil. Environ Develop 15:35–51

Delgado-Balbuena L, Bello-López JM, Navarro-Noya YE, Rodríguez-Valentín A, Luna-Guido ML, Dendooven L (2016) Changes in the bacterial community structure of remediated anthracene-contaminated soils. PLoS ONE 11(10):e0160991

Dhyani K, Kuniyal HB, Singh H and Sobha (2021). Growth and physiological potential of *Terminalia arjuna* under elevated CO_2 levels in Open top chamber condition. J Appl Nat Sci 13(3):1121–1126. https://doi.org/10.31018/jans.v13i3.2490. ISSN 2231-5209

EPA (2010) Reducing urban heat islands: compendium of strategies—urban heat island basics. Available at: http://www.epa.gov/hiri/resources/compendium.htm

Eurosurveillance (2005) Workshops sustainable & energy efficient building in Africa, vol 10. Johannesburg, South Africa, pp 7–9

Fabbri K, Canuti G, Ugolini A (2017) A methodology to evaluate outdoor microclimate of the archaeological site and vegetation role: a case study of the Roman Villa in Russi (Italy). Sustain Cities Soc 35:107–133

Gaertner M, Wilson JR, Cadotte MW, MacIvor JS, Zenni RD, Richardson DM (2017) Non-native species in urban environments: patterns, processes, impacts and challenges. Biol Invasions 19:3461–3469

Gaitani N, Spanou A, Saliari M, Synnefa A, Vassilakopoulou K, Papadopoulou K, Pavlou K, Santamouris M, Papaioannou M, Lagoudaki A (2011) Improving the microclimate in urban areas: a case study in the centre of Athens. Build Serv Eng Res Technol 32(1):53–71

Giyasova I (2021) Factors affecting microclimatic conditions in urban environment. E3S Web Conf 244:06010

Gómez-Baggethun E, Barton DN (2013) Classifying and valuing ecosystem services for urban planning. Ecol Econ 86:235–245

Gupta SK, Singh H (2017) Observations on sporadic flowering in *Ficus benjamina* L. and *Peltophorum africanum* L. in new forest. Indian Forest 143:290–291

Gupta SK, Ram J, Singh H (2018a) Comparative study of transpiration in cooling effect of tree species in the atmosphere. J Geosci Environ Protect 6:151–166. https://doi.org/10.4236/gep.2018.68011

Gupta SK, Ram J, Singh H, Kumar P (2018b) Comparative study of cooling under shades of some forestry tree species with respect to ambient temperatures. J Geosci Environ Protect 6:51–66. https://doi.org/10.4236/gep.2018.611004

Gupta SK, Ram J, Singh H (2019) Relationship between leaf area index and atmospheric cooling of tree species. Indian J Forest 42(3):203–208

Hamilton I, Stocker J, Evans S, Davies M, Carruthers D (2014) The impact of the London Olympic Parkland on the urban heat island. J Build Perform Simul 7(2):119–132

Hashemi F, Poerschke U, Iulo LD, Chi G (2023) Urban microclimate, outdoor thermal comfort, and socio-economic mapping: a case study of Philadelphia, PA. Build 13(4):1040

He BJ (2019) Towards the next generation of green building for urban heat island mitigation: zero UHI impact building. Sustain Cities Soc 50:101647

Heris MP, Middel A, Muller B (2020) Impacts of form and design policies on urban microclimate: assessment of zoning and design guideline choices in urban redevelopment projects. Landsc Urban Plan 202:103870

IPCC (2023) Summary for policymakers. In: Climate change 2023: synthesis report. Contribution of working groups I, II and III to the sixth assessment report of the intergovernmental panel on climate change [Core Writing Team, H. Lee and J. Romero (eds.)]. IPCC, Geneva, Switzerland, pp 1–34. https://doi.org/10.59327/IPCC/AR6-9789291691647.001

Jamei E, Ossen DR, Rajagopalan P (2017) Investigating the effect of urban configurations on the variation of air temperature. Int J Sustain Built Environ 6(2):389–399

Jandaghian Z (2018) Flow and pollutant dispersion model in a 2D urban street canyons using computational fluid dynamics. Comp Eng Phys Model 1(1):83–93

Jangra A, Kumar K, Maikhuri S, Bhandari MS, Pandey S, Singh H, Barthwal S (2024) Unveiling stress-adapted endophytic bacteria: characterizing plant growth-promoting traits and assessing cross-inoculation effects on *Populus deltoides* under abiotic stress. Plant Physiol Biochem 210:108610. https://doi.org/10.1016/j.plaphy.2024.108610

Jin M, Shepherd JM, Zheng W (2010). Urban surface temperature reduction via the urban aerosol direct effect: a remote sensing and WRF model sensitivity study. Adv Meteorol

Joyner JL, Kerwin J, Deeb M, Lozefski G, Prithiviraj B, Paltseva A, McLaughlin J, Groffman P, Cheng Z, Muth TR (2019) Green infrastructure design influences communities of urban soil bacteria. Front Microbiol 10:433434

Kanta C, Kumar A, Chauhan A, Singh H, Sharma IP (2024). The interplay between plant functional traits and climate change. In: Kumar N, Singh H (eds) Plant functional traits for improving productivity. Springer, Singapore. https://doi.org/10.1007/978-981-97-1510-7_3

Kántor N, Kovács A, Takács Á (2016) Small-scale human-biometeorological impacts of shading by a large tree. Open Geosci 8(1):231–245

Karimi A, Sanaieian H, Farhadi H, Norouzian-Maleki S (2020) Evaluation of the thermal indices and thermal comfort improvement by different vegetation species and materials in a medium-sized urban park. Energy Rep 6:1670–1684

Kumar N, Suyal DC, Sharma IP, Verma A and Singh H (2017) Elucidating stress proteins in rice (*Oryza sativa* L.) genotype under elevated temperature: a proteomic approach to understand heat stress response. 3 Biotech 7:205. https://doi.org/10.1007/s13205-017-0856-9

Kumar M, Singh MP, Singh H, Dhakate PM, Ravindranath NH (2019a) Forest working plan for the sustainable management of forest and biodiversity in India. J Sustain Forest 39:1–22. https://doi.org/10.1080/10549811.2019.1632212

Kumar N, Jeena N and Singh H (2019b) Elevated temperature modulates rice pollen structure: a study from the foothill of the Himalayan agro-ecosystem in India. 3 Biotech 9(5):175. https://doi.org/10.1007/s13205-019-1700-1

Kumar N, Jeena N, Kumar N, Gangola S and Singh H (2019c) Phytoremediation facilitating enzymes: an enzymatic approach for enhancing remediation process. In: Bhatt P (ed) Smart bioremediation technologies. Academic Press, Elsevier. 50 Hampshire Street, 5th Floor, Cambridge, MA 02139, United States, pp 289–306

Kumar N, Singh H and Sharma SK (2020). Antioxidants: responses and importance in plant defense system. In: Roychowdhury R, Choudhury S, Hasanuzzaman M, Srivastava S (eds) Sustainable agriculture in the era of climate change. Springer, Cham, pp 251–264. https://doi.org/10.1007/978-3-030-45669-6_11

Kumar A, Kumar P, Singh H, Kumar N (2021a) Modulation of plant functional traits under essential plant nutrients during seasonal regime in natural forests of Garhwal Himalayas. Plant Soil 465:197–212. https://doi.org/10.1007/s11104-021-05003-x

Kumar A, Kumar P, Singh H, Kumar N (2021b) Adaptation and mitigation potential of roadside trees with bio-extraction of heavy metals under vehicular emissions and their impact on physiological traits during seasonal regimes. Urban Forest Urban Green 58:126900. https://doi.org/10.1016/j.ufug.2020.126900

Kumar A, Kumar P, Singh H, Kumar N (2021c) Impact of plant functional traits on infiltration rate under different forests of Kempty watershed in Garhwal Himalaya. Indian J Soil Conserv 49(1):38–44

Kumar A, Kumar P, Singh H, Bisht S, Kumar N (2021d) Relationship of physiological plant functional traits with soil carbon stock in the temperate forest of Garhwal Himalaya. Curr Sci 120(8):1368–1373. https://doi.org/10.18520/cs/v120/i8/1368-1373

Kumar M, Phukon SN, Paygude AC, Tyagi K and Singh H (2021e). Mapping phenological functional types (PhFT) in the Indian Eastern Himalayas using machine learning algorithm in Google Earth Engine. Comp Geosci 158. https://doi.org/10.1016/j.cageo.2021.104982

Kumar A, Singh H, Kumari G, Bisht S, Malik A, Kumar A, Singh M, Raturi , Barthwal S, Thakur A and Kaushal R (2022). Adaptive resilience of roadside trees to vehicular emissions via leaf enzymatic, physiological, and anatomical trait modulations. Environ Pollut 313. https://doi.org/10.1016/j.envpol.2022.120191

Kumar N, Singh H, Giri K, Kumar A, Joshi A, Yadav A, Singh R, Bisht S, Kumari R, Jeena N, Khairakpam R, Mishra G (2024) Physiological and molecular insights into the allelopathic effects on agroecosystems under changing environmental conditions. Physiol Mol Biol Plants 30:417–433. https://doi.org/10.1007/s12298-024-01440-x

Lauzet N, Rodler A, Musy M, Azam MH, Guernouti S, Mauree D, Colinart T (2019) How building energy models take the local climate into account in an urban context—a review. Renew Sustain Energy Rev 116:109390

Leuzinger S, Vogt R, Körner C (2010) Tree surface temperature in an urban environment. Agric for Meteorol 150(1):56–62

Malik A, Kumar K, Kumari G, Kumar N, Singh H (2024) Application of functional traits in modelling productivity and resilience under climate change. In: Kumar N, Singh H (eds) Plant functional traits for improving productivity. Springer, Singapore. https://doi.org/10.1007/978-981-97-1510-7_5

McPherson EG, Simpson JR (2003) Potential energy savings in buildings by an urban tree planting programme in California. Urban Forest Urban Green 2:73–86

Meena D, Singh H, Chaudhari SK (2011) Elucidating strontium response on growth dynamics and biochemical change in *Phaseolus mungo* L. Int J Agricult Environ Biotech 4:107–113

Milošević DD, Bajšanski IV, Savić SM (2017) Influence of changing trees locations on thermal comfort on street parking lot and footways. Urban Forest Urban Green 23:113–124

Morakinyo TE, Lau KKL, Ren C, Ng E (2018) Performance of Hong Kong's common trees species for outdoor temperature regulation, thermal comfort and energy saving. Build Environ 137:157–170

Morakinyo TE, Ouyang W, Lau KKL, Ren C, Ng E (2020) Right tree, right place (urban canyon): tree species selection approach for optimum urban heat mitigation-development and evaluation. Sci Total Environ 719:137461

Nowak DJ, Crane DE, Stevens JC (2006) Air pollution removal by urban trees and shrubs in the United States. Urban Forest Urban Green 4(3–4):115–123

Oshio H, Kiyono T, Asawa T (2021) Numerical simulation of the nocturnal cooling effect of urban trees considering the leaf area density distribution. Urban Forest Urban Green 66:127391

Ouyang W, Morakinyo TE, Ren C, Ng E (2020) The cooling efficiency of variable greenery coverage ratios in different urban densities: A study in a subtropical climate. Build Environ 174:106772

Prakash V, Hunney K and Singh H (2022) Effect of elevated carbon dioxide on growth and development of *Santalum album* L. seedlings inoculated with plant growth promoting microorganisms in Open Top Chambers. Int J Environ, Agricult Biotech 7(5):149–159

Richter R, Ballasus H, Engelmann RA, Zielhofer C, Sanaei A, Wirth C (2022) Tree species matter for forest microclimate regulation during the drought year 2018: disentangling environmental drivers and biotic drivers. Sci Rep 12(1):17559

Ritchie H, Rosado P, Roser M (2024) Energy production and consumption. Our world in data

Rui L, Buccolieri R, Gao Z, Gatto E, Ding W (2019) Study of the effect of green quantity and structure on thermal comfort and air quality in an urban-like residential district by ENVI-met modelling. Build Simul 12:183–194

Sánchez-Bayo F, Wyckhuys KA (2019) Worldwide decline of the entomofauna: a review of its drivers. Biol Cons 232:8–27

Santamouris M (2014) Cooling the cities–a review of reflective and green roof mitigation technologies to fight heat island and improve comfort in urban environments. Sol Energy 103:682–703

Santamouris M, Paraponiaris K, Mihalakakou G (2007) Estimating the ecological footprint of the heat island effect over Athens, Greece. Climatic Change 80(3):265–276

Santangelo JS, Roux C, Johnson MT (2022) The effects of environmental heterogeneity within a city on the evolution of clines. J Ecol 110(12):2950–2959

Shahrestani M, Yao R, Luo Z, Turkbeyler E, Davies H (2015) A field study of urban microclimates in London. Renew Energy 73:3–9

Sharma R, Singh H (2021) Alteration in biochemical constituents and nutrients partitioning of *Asparagus racemosus* in response to elevated atmospheric CO_2 concentration. Environ Sci Pollut Res 29:6812–6821. https://doi.org/10.1007/s11356-021-16050-3

Sharma R, Singh H, Kaushik M, Nautiyal R, Singh O (2018) Adaptive physiological response, carbon partitioning, and biomass production of *Withania somnifera* (L.) Dunal grown under elevated CO_2 regimes. 3 Biotech 8:267. https://doi.org/10.1007/s13205-018-1292-1

Singh H (2021) An integrated approach considering physiological and biophysical based indicators for assessing tolerance of roadside plantations of *Alstonia scholaris* towards urban roadside air pollution: an assessment of adaptation of plantations for mitigating urban roadside air pollution. Trees. https://doi.org/10.1007/s00468-021-02179-8

Singh H (2024). Exploring adaptive modulation in plant functional traits and its impact on the productivity of *Acacia auriculiformis* under CO_2-enriched environment. Indust Crops Products 210:118186. https://doi.org/10.1016/j.indcrop.2024.118186

Singh H, Verma A (2013) Characterization and screening of high nitrogen efficient rice genotype to curtail environmental pollution. Int J Agricult, Environ Biotech 6(1):23–30

Singh H, Kumar M (2022) Climate change and its impact on Indian Himalayan forests: current status and research needs. In: Rani S, Kumar R (eds) Climate change. Springer Climate. Springer, Cham, pp 223–242. https://doi.org/10.1007/978-3-030-92782-0_11

Singh H, Sharma RS, Sinha S, Kumar M, Kumar P, Verma A, Sharma SK (2017a) Physiological functioning of *Lagerstroemia speciosa* L. under heavy roadside traffic: an approach to screen potential species for abatement of urban air pollution. 3 Biotech 7(61):1–10. https://doi.org/10.1007/s13205-017-0690-0

Singh H, Verma A, Kumar M, Sharma R, Gupta R, Kaur M, Negi M, Sharma SK (2017b) Phytoremediation: a green technology to clean up the sites with low and moderate level of heavy metals. Austin Biochem 2:1012

Singh H, Sharma R, Savita SMP, Kumar M, Verma A, Ansari MW, Sharma SK (2018) Adaptive physiological response of *Parthenium hysterophorus* to elevated atmospheric CO_2 concentration. Indian Forest 144:1–14

Singh H, Yadav M, Kumar N, Kumar A, Kumar M (2020) Assessing adaptation and mitigation potential of roadside trees under the influence of vehicular emissions: A case study of *Grevillea robusta* and *Mangifera indica* planted in an urban city of India. PLoS ONE 15(1):e0227380. https://doi.org/10.1371/journal.pone.0227380

Singh S, Singh H, Sharma SK, Nautiyal R (2021) Seasonal variation in biochemical responses of bamboo clones in the sub-tropical climate of Indian Himalayan foothills. Heliyon 7(4):e06859. https://doi.org/10.1016/j.heliyon.2021.e06859

Singh H, Kumar N, Kumar A (2024a). Enhancing resource use efficiency in crops through plant functional traits. In: Kumar N, Singh H (eds) Plant functional traits for improving productivity. Springer, Singapore. https://doi.org/10.1007/978-981-97-1510-7_6

Singh M, Singh H, Kumar A, Kumar M, Barthwal S, Thakur A (2024b) Soil nitrogen availability determines the CO_2 fertilization effect on tree species (*Neolamarckia cadamba*): growth and physiological evidence. Environ Sustain 7:53–60. https://doi.org/10.1007/s42398-023-00300-w

Sodoudi S, Zhang H, Chi X, Müller F, Li H (2018) The influence of spatial configuration of green areas on microclimate and thermal comfort. Urban Forest Urban Green 34:85–96

Speak A, Montagnani L, Wellstein C, Zerbe S (2020) The influence of tree traits on urban ground surface shade cooling. Landsc Urban Plan 197:103748

Srivastav VT, Sharma A, Jadon SS (2023) Microclimate analysis of high-density urban residential open enclosures: a case of Thane, India. Environ Conservat J 24(2):434–447

Stojakovic V, Bajsanski I, Savic S, Milosevic D, Tepavcevic B (2020) The influence of changing location of trees in urban green spaces on insolation mitigation. Urban Forest Urban Green 53:126721

Streiling S, Matzarakis A (2003) Influence of single and small clusters of trees on the bioclimate of a city: a case study. Arboricult Urban Forest (AUF) 29(6):309–316

Swain D, Roberts GJ, Dash J, Vinoj V, Lekshmi K, Tripathy S (2016) May. Impact of rapid urbanization on the microclimate of Indian cities: a case study for the city of Bhubaneswar. In: Land surface and cryosphere remote sensing III, vol 9877. SPIE, pp 366–372

Sylliris N, Papagiannakis A, Vartholomaios A (2023) Improving the climate resilience of urban road networks: a simulation of microclimate and air quality interventions in a typology of streets in Thessaloniki historic centre. Land 12(2):414

Tong S, Prior J, McGregor G, Shi X, Kinney P (2021) Building healthy communities: urban heat: an increasing threat to global health. The BMJ 375

United Nations Department of Economic and Social Affairs, Population Division (2022) World population prospects 2022: summary of results. UN DESA/POP/2022/TR/NO. 3

United Nations Department of Economic and Social Affairs, Population Division (2019) World urbanization prospects: the 2018 revision (ST/ESA/SER.A/420). United Nations, New York

Vogt J (2020) Urban forests: biophysical features and benefits. Encycl. World's Biomes 5:48–57

Wai KM, Tan TZ, Morakinyo TE, Chan TC, Lai A (2020) Reduced effectiveness of tree planting on micro-climate cooling due to ozone pollution—a modeling study. Sustain Cities Soc 52:101803

Xing QW, Hao XL, Yang K, Huang MH, Zhang KT et al (2018) Energy saving mechanism of vertical greening and its potential of energy saving and emission reduction in Changsha area. Build Energy Conservat 46:111–114

Yadav SK, Singh H, Ginwal HS, Barthwal S (2019a) Elevated CO_2 enhanced growth and physiological process of *Populus deltoides* Bartr. ex Marsh. Indian Forest 145(1):23–27

Yadav SK, Singh H, Nautiyal R, Ginwal HS, Ansari SA, Barthwal S (2019b) Modulation of morpho-physiological responses in *Populus deltoides* by elevated carbon dioxide and temperature. Forest Sci 66:105–118. https://doi.org/10.1093/forsci/fxz048

Yang J, Bou-Zeid E (2019). Designing sensor networks to resolve spatio-temporal urban temperature variations: fixed, mobile or hybrid? Environ Res Lett 14(7):074022

Zhang G, He BJ, Zhu Z, Dewancker BJ (2019) Impact of morphological characteristics of green roofs on pedestrian cooling in subtropical climates. Int J Environ Res Public Health 16(2):179

Zipperer WC, Northrop R, Andreu M (2020) Urban development and environmental degradation. In: Oxford research encyclopedia of environmental science

Urban Forests and Carbon Sequestration

Lopamudra Nayak, Sandeep B. Adavi, Priyanka Lal, Laxmipriya Behera, Ujala Rashmi Sahu, Muhammad Ahsan Altaf, Awadhesh Kumar, Ravinder Kumar, Rahul Kumar Tiwari, and Milan Kumar Lal

Abstract Both rapid urbanization and carbon dioxide (CO_2) emissions substantially contribute to climate change. Carbon dioxide is the most frequently released greenhouse gas by humans. The built environment, which includes activities like cooking, urban transit, and building heating, is mostly to blame for this. The results show an inverted U-shaped relationship between CO_2 emissions and urbanization. Several factors, including transportation density and atmospheric stability, influence the concentration of near-surface CO_2 in cities. Urban forests are essential for maintaining the basic biological and environmental services that cities offer for the survival of people, animals, and plants as the world's cities become increasingly urbanized. Because plants can sequester CO_2 through photosynthesis, they can store carbon in the soil and plant biomass. Cities' green spaces may significantly influence local atmospheric CO_2 concentrations, as seen by comparisons between urban

L. Nayak · L. Behera · U. R. Sahu · A. Kumar · M. K. Lal (✉)
Division of Crop Physiology and Biochemistry, ICAR-National Rice Research Institute (ICAR-NRRI), Cuttack 753006, Odisha, India
e-mail: milan2925@gmail.com

S. B. Adavi
School of Crop Resistance System Research, ICAR-National Institute of Biotic Stress Management, Baronda, Raipur 493225, Chhattisgarh, India

P. Lal
Lovely Professional University, Jalandhar GT Road (NH1), Phagwara, Punjab, India

M. A. Altaf
Key Laboratory for Quality Regulation of Tropical Horticultural Crops of Hainan Province, School of Breeding and Multiplication (Sanya Institute of Breeding and Multiplication), Hainan University, Sanya 572025, China

R. Kumar
ICAR-Indian Agricultural Research Institute, New Delhi, India

R. Kumar · R. K. Tiwari · M. K. Lal
ICAR-Central Potato Research Institute (ICAR-CPRI), Shimla 171001, Himachal Pradesh, India

R. K. Tiwari
ICAR-Indian Institute of Sugarcane Research, Lucknow, Uttar Pradesh, India

H. Singh (ed.), *Urban Forests, Climate Change and Environmental Pollution*,
https://doi.org/10.1007/978-3-031-67837-0_18

and rural areas, which show lower CO_2 concentrations in the presence of vegetation. The amount of CO_2 sequestered over the "urban forest" varies throughout the growth season, with nighttime emissions in response to respiration and absorption taking place during the day when plants are photosynthetically active. Increased atmospheric CO_2 concentrations stimulate more efficient photosynthesis in plants by acting as a fertilizer. However, the photosynthetic system of urban plants is regularly hampered by environmental stresses, which, in extreme conditions, can lead plants to switch from being carbon sinks to carbon sources. The many functions, benefits, and advantages that urban woods provide are essential to the sustainable development of metropolitan regions. This chapter analyzed the most recent studies to understand better the roles that urban trees play today and in the future in removing CO_2 from the atmosphere, and we highlight the need for more study.

Keywords Urbanization · Carbon dioxide (CO_2) emissions · Urban forests · Photosynthesis · Sustainable development

1 Introduction

The centres of economic development and expansion are cities. Nearly half of India's GDP is currently generated by urban regions. Still, the country's fast urbanization is also a significant cause of pollution both within and outside of cities, changes in land use, habitat loss, and biodiversity reduction (Satterthwaite et al. 2010; Kumar et al. 2022; Singh 2021). According to a survey by the International Council of Local Environmental Initiatives (ICLEI), South Asia, the average per capita carbon emissions in India's metropolitan cities are 1.19 tons, whereas the average per capita carbon emissions in the country's non-metropolitan cities are just 0.90 tons. There are three ways to lower the amount of carbon dioxide in the atmosphere: by using less energy, changing how energy is used, or increasing the rate at which carbon dioxide is removed from the atmosphere by trees through a process called carbon sequestration, which naturally lowers atmospheric carbon dioxide concentrations (Sharma et al. 2020; Kumar et al. 2024; Singh et al. 2024b). According to research by many academics, urban green spaces can be crucial in reducing a city's carbon footprint (Strohbach et al. 2012). In addition to directly lowering atmospheric CO_2 concentration through sequestering carbon, a greenspace's vegetation and soil indirectly influence the carbon balance by affecting the urban energy balance and, consequently, CO_2 emissions associated with energy consumption (Churkina 2016; Singh 2024). Over half of the world's population lives in urban regions due to significant urbanization. Urban outdoor activity and human health are impacted by the urban thermal environment and the complete urban eco-environment, which includes phenology, biodiversity, and water and air quality. The pace of deterioration of the urban thermal environment has grown due to the pressure from rapid urbanization (Huang et al. 2018; Ren et al. 2022). Anthropogenic activities have had a major impact on the global climate, which has resulted in decreased agricultural output and climate change

(Ortiz-Bobea et al. 2021). According to reports, anthropogenic climate change (ACC) has contributed to a 21% decrease in global agricultural production since 1961. The impact of ACC has been worse in warm regions like Africa (− 34%) than in cooler regions like Europe and Central Asia (− 7.1%). The world's supply of crops is predicted to be further impacted by climate change in the coming decades, both in terms of quantity and quality, because of the intricate interactions between temperature changes and increased atmospheric CO_2(Raza et al. 2019; Dhyani et al. 2021; Kumar et al. 2021e).

It is noted that urban forestry contributes significantly to cities' ecosystems. According to Christopoulou et al. (2007), the urban forest is a complex ecosystem tightly linked to the urban ecology (Christopoulou et al. 2007; Prakash et al. 2022). It is known to be a forest and flora-like forest that exists in diverse compositions in the periphery of urban environments. They can be found in open areas, parks, gardens, streams, and forest stands. It is impossible to ignore the positive effects urban woods have on human health and the environment, which include preserving biodiversity and assisting in the adaptation and mitigation of climate change (Singh 2021; Sharma and Singh 2021; Smith et al. 2022). Few attempts have addressed the potential of trees in carbon sequestration in urban environments despite the well-established and confirmed relevance of wooded areas in carbon sequestration (Singh et al. 2020, 2021; Kumar et al. 2021a). Urban green spaces play an essential role in offsetting carbon emissions locally; therefore, it is crucial to establish how much carbon is sequestered by urban centers. Huge university campuses offer ample space for urban tree plantings, which may help reduce climate change's effects. Knowing how much carbon an urban green space can store is helpful because it allows an organization or institution to value and offset its emissions from the space (Kumar et al. 2021b, c). Therefore, the significance, application, and health of urban forests, which are crucial in removing CO_2 from the atmosphere, are covered in this work.

2 Environment-Urbanization Relationships

The United Nations Population Division predicts that, in October 1999, the global population will surpass the landmark six billion. The UN just released long-term estimates that extend to 2150. The "most likely" medium-fertility scenario predicts that by 2200, the world's population will have stabilized at slightly around 11 billion. According to the United Nations, there has been a significant increase in the number and size of cities worldwide, indicating a rapid rise in urbanization over the past century (Kacyira 2016; Kumar et al. 2021d). The global population was estimated to reach 4 billion in 2015, up from 746 million in 1950, a five-fold increase. This trend is expected to continue in the coming decades, especially in low- and middle-income countries, which will present significant challenges for them in managing urbanization (Yadav et al. 2019a, b; Gupta et al. 2019a). The demographics of cities worldwide threaten the sustainability of their environments and the welfare of their citizens in the modern world. Over 80% of Americans live in metropolitan settings, according to Nowak, who highlights the significant impact of urban woods on daily life (Nowak 2016). The rapid loss of trees and forest cover in and around urban areas has been dramatically exacerbated by city growth and intensification, which has occurred without proper consideration for the land use capacity and local demands for food, woody building materials, and wood energy (Fang et al. 2018; Gupta et al. 2019a; Kumar et al. 2019a). Cities suffer enormous losses from destroyed infrastructure and fatalities from landslides, floods, dust invasions, water shortages, and soil erosion. However, their naturally occurring environment, which is significantly impacted by the growing effects of urbanization, is crucial to maintaining and expanding biodiversity. Rapid economic growth can harm the environment by leading to changes in land use, increased resource use, and pollution (Kumar et al. 2019b; Schwela 2000). A substantial body of research emphasizes how urban woods can lower pollution levels and balance out city greenhouse gas emissions (McPherson 1998). Given their capacity to offer several advantages, proper management and upkeep of urban forests and green spaces are vital to maximizing the health benefits (Nowak et al. 2014). Humans have significantly changed the global ecosystem, altering land, increasing biota movement, and affecting global biogeochemical cycles (Singh et al. 2022). Since the Industrial Revolution, it has been steadily rising in the atmosphere, which is thought to be the cause of disturbances in the planet's energy balance and accumulation of energy that leads to ice melting, sea level rise, changes in the circulation of the ocean and atmosphere, extreme weather, flooding, and other events (Skofronick-Jackson et al. 2017). These events significantly impact food production and have other irreversible effects. Urban areas' cities promote themselves as the centers of socioeconomic progress. However, maintaining equity and sustainable urban growth is hampered by the planet's fast urbanization, which lacks the resources to meet the world's expanding needs for food and other necessities.

3 Changes in Climate and Carbon Emissions

Carbon is ubiquitous in rocks, soil, oceans, atmosphere, fossil fuels, and living things. It is constantly recycled throughout the earth's system. Greenhouse gases produced by humans are considered the primary cause of climate change and global warming, with CO_2 being the most significant (Sharma et al. 2018; Singh et al. 2018). Carbon dioxide is crucial to the Earth's atmosphere because it absorbs infrared radiation at 4.26 mm (asymmetric stretching vibrational mode) and 14.99 mm (bending vibrational mode). The greenhouse effect, which helps to keep our world reasonably mild, is mainly caused by CO_2. Still, if the concentration of CO_2 increases dramatically, it can also cause global warming and climate change, which is hostile (*AR5 Climate Change 2014: Mitigation of Climate Change—IPCC*, n.d.). Without CO_2 in the atmosphere, Earth's temperatures would drastically drop, and life on the planet would be very different from what it is today (Kuittinen 2015; Gupta et al. 2018a, b; Gupta et al. 2019b). Even while the global temperature only slightly rises in response to an increase in CO_2 concentration in the atmosphere, this change is sufficient to cause an increase in the amount of water vapor in the atmosphere due to ocean evaporation. Global warming is the outcome of this significant increase in the greenhouse effect caused by the increased amount of water vapor in the atmosphere. The most critical impact on the earth's temperature comes from this particular phenomenon. Strangely enough, the concentration of CO_2 controls the amount of water vapor in the atmosphere and, as a result, determines the Earth's equilibrium temperature. The amount of CO_2 in the atmosphere has increased by over 50% during the 1950s, which has increased the blanketing effect and raised global temperatures by 1 °C. The relative ratios of carbon isotopes provide evidence that human activity is the cause of the CO_2 emissions. The ratio of 13C in CO_2 derived from fossil fuels is significantly lower than in CO_2 created from modern decaying plants, which is why the relative amount of 13C in the atmosphere has been dropping (Craig 1954).

4 Ecosystem Benefits Provided by Urban Forests

In 2023, Castro et al. divided the ecosystem services provided by urban forests into three primary categories: regulating the air, water, soil, and climate, providing the standard of the habitat, and supplying more commodities and services (Singh and Verma 2013; Kumar et al. 2017; Castro et al. 2023). While air, water, soil, and climate ecosystem services also have an impact on the delivery of other goods and services, the above and below-ground space inhabited by urban trees is essential for the provision of all other ecosystem services (Chen et al. 2016; Giannico et al. 2016). Along a rural-urban transect, there are differences in the relative importance of ecosystem services, i.e., the impact on the environment and the overall community perception. Given the density of the population and built infrastructure, improving the microclimate and air quality and preserving buildings are critical in cities (Gupta and Singh 2017; Singh et al. 2017a). In contrast, in rural areas, forests are quantitatively more important for their ability to sequester CO_2, provide energy, nutrients, and resources, and restore soil and water (Singh et al. 2017b). The quality of the habitat is crucial for biodiversity in urban woods. Most services are indirect and intangible, yet they significantly impact urban society's and local ecosystems' well-being. These services have long been acknowledged, and numerous studies have tried to define and measure them.

5 Advantages for the Environment

Urban woods offer essential environmental services such as noise reduction, microclimate adjustment, pollution absorption and filtration, and precipitation retention. However, because of their ethereal nature, such functions are complex for the common layperson to understand. Numerous pollutants, including sulfur dioxide gas, nitrogen oxide gas particles, and noise, have been brought into cities by human activity. In many places, these toxins have led to severe environmental and public health issues (Meena et al. 2011; Perera 2018; Kumar et al. 2019c). The heat island effect makes cities hot and miserable for residents and work. Thousands of dollars must be invested annually to reduce the excessive temperatures and shift the structures' bioclimatic regime into a pleasant range. Numerous studies have shown that urban woods are an excellent way to mitigate these issues.

5.1 *Reduced Noise*

Humans may experience physical and psychological stress as a result of noise. Cities are known for producing a lot of unwanted noise due to the dense population and machinery. Noise levels above 70 decibels (dBA) are usually considered annoying. Sound attenuation occurs over distance and involves absorption, deflection, reflection, refraction, and masking. However, barriers can be used to absorb sound across shorter distances. Artificial structures and living flora reduce noise levels (Peckham 2010; Kumar et al. 2020; Singh and Kumar 2022). Urban woodlands help reduce noise levels. The leaves, branches, and twigs of urban forest trees and shrubs might absorb, divert, or reflect the most irritating frequencies with the right design (Jim and Chen 2008).

Additionally, trees filter noise by creating soothing noises from the wind, rustling tree leaves, or birds singing in the branches. According to research, vegetation may significantly reduce noise when planted with sufficient width and density. Densely crowned trees and plants should be planted near the noise locations to mitigate noise. JE Kuser claims that soft ground surfaces plus a 30-m-wide tree belt may reduce loud noise by 50% or more, or 6–10 dBA. Noise levels can be lowered by three to five decibels even inside a three-meter-wide belt of thick shrubs and trees with dense foliage at their base. According to Fang and Ling (2003), using vegetation in urban areas to screen noise at its source would be a more practical approach than attenuating it at the recipient sites (Fang and Ling 2003).

5.2 *Functions of Biomass*

In urban forests, plants use photosynthesis to produce oxygen and serve as the primary producers by absorbing carbon dioxide. The release rate of oxygen and CO_2 sequestration in urban vegetation is determined annually by the photosynthetic capability of plants, which is affected by the age structure and species composition (Fares et al. 2017; Grote et al. 2016). As a greenhouse gas linked to a higher risk of global warming, CO_2 can be directly and indirectly reduced by vegetation (Ali et al. 2020; Moll and Kollin 1996). Throughout their growth, trees actively sequester and store carbon. Additionally, trees cool cities by providing shade and evapotranspiration.

This lowers the need for air conditioning and prevents CO_2 emissions linked to electricity production. Trees are more successful in recollecting carbon and cooling the air than other plant life forms. In addition, they have more incredible lifespans, more biomass, and a more extensive leaf area index.

5.3 Air Pollution Reduction

In cities, air pollution is a significant and prevalent issue, mainly because it can harm people's health (Wang et al. 2015). Additional issues brought on by air pollution include acidic deposition, diminished visibility, and harm to artificial materials and vegetation (Kojima and Lovei 2001). Ozone (O_3), nitrogen oxides (NOx), sulfur dioxide (SO_2), and particulates—which are typically described as tiny particulate matter (PM) with a diameter of less than 10 μm (PM10)are the main air pollutants found in modern cities. Automobiles and stationary fossil fuel combustion sources are the primary sources of SO_2 and NOx; chemical reactions between NOx and volatile organic compounds form O_3 (Fenger 2002), and road-transport emissions are the main source of PM10 (Chow et al. 1996; Singh et al. 2024a). PM10 can also be produced by other human activities, such as building, industrial processes, burning fossil fuels in power plants, and chemical reactions involving gaseous pollutants. In urban green spaces, trees remove air pollutants primarily from the atmosphere through a process known as dry deposition, in which gaseous and particulate pollutants are carried to and absorbed mainly through the surfaces of the plants (Grote et al. 2016; Ysebaert et al. 2021). The kind and features of air pollutants, the local weather conditions, the canopy area, and the species of plants all affect this ecosystem service. Particles that are later washed off by precipitation may be trapped by leaves, branches, stems, and related surface structures (e.g., pubescence on leaves). Furthermore, tree transpiration can raise air humidity, which helps particles in the air settle (Andresen 1989). Because urban tree canopies have a rougher surface than other vegetation species, they are better at collecting particles. Several researchers have assessed and calculated the ecosystem service that urban woods provide in eliminating air pollutants (Akbari et al. 2001; Nowak et al. 2006). For example, particle removal for the St. Louis Street tree-covered area reached about 2.75 lb/acre/day (3.08 kg/ha/day) (Jim and Chen 2008). The removal rates for particles, NOx, SO_2, and O_3 were found to be 1.3–3.9 lb/acre/day (1.4–4.4 kg/ha/day), 1.2–3.6 lb/acre/day (1.3–4.0 kg/ha/day), 20.2–66.3 lb/acre/day (22.7–74.4 kg/ha/day), and 30.9–99.5 lb/acre/day (34.7–111.6 kg/ha/day) in other tree-covered areas within the same city (Schmid 2015).

5.4 *Improvement of the Microclimate*

It is commonly known that, in temperate latitudes, urban areas are typically 0.5–1.5 °C warmer than the surrounding countryside (Waffle et al. 2017). In tropical locations, this difference can reach up to 3 °C. High air temperatures and low relative humidity characterize these urban heat islands. City centers and other highly populated regions may become unbearably hot for people due to changes in the urban microclimate. It takes a lot of energy to artificially alter the indoor microclimate to a comfortable level (Moonen et al. 2012). People in Guangzhou, China, strongly believe urban woods may improve the microclimate (Jim and Chen 2008).

Three processes allow urban forests to improve thermal comfort and efficiently affect the microclimate throughout the summer (Lai et al. 2019; Lin et al. 2021; Sodoudi et al. 2018). First, strategically placed trees can block part of the sun's rays from reaching structures. This lowers the energy needed for building cooling and initial heating and heat storage (Simpson et al. 1996). Second, urban woods can change the surrounding environment of buildings by serving as windbreaks. It was discovered that a neighborhood's dispersed trees increased surface roughness, which decreased wind speeds (Beckett et al. 1998). Third, evapotranspiration from urban trees may decrease summertime air temperatures (Akbari et al. 2005). To cool the surrounding air on a hot summer day, a mature tree may transpire up to 100 gallons (or 378.5 kg) of water into the earth's atmosphere (French 2004). Because latent heat of vaporization from the surrounding air is absorbed to turn liquid water into water vapour, transpired water from the surfaces of leaves can chill the air.

6 Carbon Sequestration and Urban Forests

All plant species aid in carbon sequestration, with varying amounts of contribution depending on the availability of water, sufficient sunshine, and inorganic nutrients (Ramaiah and Avtar 2019). Reforestation of defunct pasture and agricultural land results in carbon sequestration, a mitigation strategy to reduce atmospheric concentrations of CO_2 and other greenhouse gases (Silver et al. 2000; Yadav et al. 2019a, b; Jangra et al. 2024). Urban woods offer a multitude of co-benefits to cities and play a crucial role in mitigating the consequences of climate change through significant carbon sequestration efficiency (Nero et al. 2017). One criterion used to assess ecological benefits is carbon sequestration, which quantifies a tree's ability to fix carbon. Climate change, energy consumption, carbon cycles, and local climate are all impacted by urban forests (Gill et al. 2007; Kanta et al. 2024; Malik et al. 2024). Urban woods are essential for the globe, as they are human-dominated ecosystems.

As per a study, the existence and management of green spaces in urban areas are determined by the population's educational attainment and environmental consciousness, as their views towards greenery are vital in urban greenery growth (Chaurasia and Munoth 2022). For instance, the designation of a city as a "National Garden

City" in China and as "Tree City USA" in the USA were applied using the same fundamental standards, i.e., green space canopy coverage ratio, green space to land ratio, and per-capita green space to award these recognitions to cities in various countries, which designated more than 3000 towns, cities, and villages as "National Garden Cities" and more than 100 cities as "Tree City USA" Singapore's "Urban Green Model" is another comparable program that is worthwhile to imitate. India is also involved in these initiatives to inform the public and policymakers about the advantages of urban green spaces because there is a shortage of information regarding the relationship between ecosystem services and human well-being. Because of their large size, the high surface-to-volume ratio of foliage, hairy or rough leaf and bark surfaces, and their ability to effectively filter airborne particles, shelter-belt plants are needed in urban areas, parks, residential areas, streets, road avenues, and industrial sites. They also help clean up airborne particulate pollution in the environment, which benefits other essential ecosystem services (Bolund and Hunhammar 1999).

India's metropolitan areas have also been negatively impacted by the global phenomena of urbanization brought on by development (Chadchan and Shankar 2012). The detrimental impacts of rising carbon emissions serve as a stark reminder of the urgent need to plan the entire area covered by urban forests and tree cover, with the primary goal of management being the mitigation of environmental harm. Government-maintained mini forests and enormous, fragmented, tree-dominated, non-forested areas are also considered urban forests, as are some larger, recognized green spaces that are identified as patches, such as parks, gardens, avenues, historical monuments, parks, and areas of beautification that are valued for their social, cultural, and aesthetic qualities as well as their tribal significance. For example, dividing a potent biodiversity hotspot in Arunachal Pradesh, India, also aids carbon sequestration (Ugle et al. 2010).

7 Urban Forest in Indian Cities

Urban cities require natural environments to escape their hectic, demanding, frequently dull, and depressing daily lives. Local government is responsible for bringing a portion of nature closer to city residents, as poor and middle-class people cannot afford to regularly travel to hill stations and far-off national parks to enjoy nature (Kumar and Singh 2024). Creating parks and gardens in and around metropolitan areas can help achieve this. Because of the continuous emphasis on growing more trees and the rigorous oversight of tree-cutting permits in New Delhi, India has developed into one of the greenest capitals in the world. This has been made possible despite the infrastructure projects that arose due to the demands of the 2010 Commonwealth Games. Delhi's total area is 20% covered with greenery, translating to an availability of roughly 21.43 m^2 of green space per person. The Delhi Development Authority (DDA) has numerous gardens, forests in the city, biodiversity parks, and other green belts; the Municipal Corporation of Delhi (MCD) manages around 14,000 parks; and the New Delhi Municipal Council (NDMC) has roughly

1,000 parks and gardens. The Parks and Garden Society was established recently to organize Delhi's greening initiatives. Delhi is home to 2 biodiversity parks and nine forests. There are plans to develop nine more city forests.

Nevertheless, it is necessary to locate empty spaces that can be covered with greenery. The ridge's whole expanse (about 6,000 hectares) must be planted. This is crucial since many people are moving to Delhi for work, study, and housing.

To encourage the planting of trees in urban areas, the Indian state of Gujarat has launched a novel scheme that links tree planting to religious rituals from different religions. Every *Grah* (planet), *Nakshatra* (constellation), and *Rashi* (zodiac) has a favorite tree, according to the Hindu sacred text Puranas. Such tree planting and preservation have a beneficial impact on human life. Punit Van, an urban forest spanning six hectares in Gandhinagar, is a holy site that promotes a love and appreciation for trees. It encompasses all these qualities. The Gandhinagar capital project spans 57 km^2. As of 2005, the city's tree cover accounted for 57.13% of its 32.56 km^2 total geographical area. With a population of about 0.2 million in 2001, more than 160 m2 of green space was available per person (Chaudhry et al. 2011). The state government employed remote sensing satellite imagery to evaluate city tree cover alterations in 1979, 1986, 1999, and 2005 (Rawat and Kumar 2015). A few other urban forests, or Vans, have been established in other Gujarati cities along the same tree-religion theme. These include Kailash Van, which is close to Ahmedabad, Mangalya Van in Ambaji, Tirthankar Van in Mehsana district, Harihar Van in Somnath, Bhakti Van in Chotila of Surendranagar district, and Shamal Van in Shamlaji of Sabarkantha district.

Bangalore, India, is called the "Garden City of India" because of its many parks and private gardens, trees that line the roads and avenues, and stunning Lalbagh and Cubbon Parks. There are 705 parks in the city, which are divided into large parks and small and medium-sized parks. In addition to these standard parks, about 200 undeveloped open spaces and green places are just waiting to be turned into parks. According to Gowda et al. (2008), these are essentially community amenity plots designated for constructing community infrastructure, such as parks and gardens (Gowda et al. 2008). The authors have offered a few specific recommendations for revitalizing the city's urban green spaces, including the creation of additional large regional parks similar to Cubbon Park and Lal Bagh, the use of NGOs and multinational corporations for park development and upkeep, and the effective use of rainwater harvesting techniques. Most of Bangalore's avenue species are exotic, mostly planted for their eye-catching aesthetic and rapid growth. Multiple-use trees, such as *Mangifera indica, Murraya koenigii, Moringa oleifera, Tamarindus indica, Artocarpus integrifolia, Phyllanthus embelica*, and *Syzygium cumini*, should be planted, especially in residential areas where they will yield edible pods, flowers, fruits, leaves (Sudha and Ravindranath 2000). In certain parks and gardens, encroachments urgently need to be removed. Such encroachers are typically wealthy and lack concern for the environment. The city's estimated crown cover is 19.9% of the geographical area (Raj et al. 2021).

8 Potential for Sequestering Carbon

Trees employ photosynthesis to change carbon dioxide and water into sugar molecules and oxygen. While some of the sugar is stored, the tree uses the majority for structural and other functions. The tree's structure is derived from cellulose, mainly composed of sugar bonded together. Carbon makes up a significant portion of the bulk of this sugar. Carbon makes up the majority of the mass of sugar molecules since it has an atomic mass of 12, hydrogen has an atomic mass of 1, and oxygen has an atomic mass of 16. Accordingly, one tonne of carbon stored in the tree is equivalent to the emission of 2.67 tonnes of oxygen and the removal of 44/12, or 3.67, tonnes of carbon from the atmosphere. In the Conterminous United States, urban trees sequester 700 million tonnes of carbon at a gross carbon sequestration rate of 22.8 million tC. Compared to 53.5 tC/ha in forest stands, urban forests' average national carbon storage density is 25.1 tC/ha (Nowak and Crane 2002). In addition to removing significant amounts of air pollutants, these urban trees help enhance the air quality in cities. According to a few studies, one acre filled with 600 tropical trees could store up to 15 tonnes of CO_2 annually. Forty trees can store one tonne of CO_2 annually, and one million trees covering 1667 acres could store 25,000 tonnes of CO_2 annually. It is anticipated that planting 400,000 trees in Canberra between 2008 and 2012 will save energy, mitigate pollution, and sequester carbon at a cost of between $20 and $67 million (Brack 2002).

Similarly, between 2002 and 2008, 115,200 native street trees were planted in the City of Tshwane Metropolitan Municipality in South Africa. According to estimates, planting trees will reduce CO_2 equivalent by 200,492 tonnes and sequester 54,630 tonnes of carbon. According to Beijing's research (2002), central trees helped to clean the air by removing 1261.4 tonnes of pollutants. PM10, or particulate matter with an aerodynamic diameter of 10 μm, was the most reduced air pollutant, with a reduction of 772 tons. Furthermore, according to Yang et al. (2005), the urban forest stored roughly 0.2 million tons of carbon dioxide (CO_2) in biomass form (Yang et al. 2005). From an estimated 7.79 million ha urban area, the total amount of carbon stored by the urban trees is expected to be 23.8 million tonnes or 3.01 tonnes of carbon/ha. Only 2.21% of the carbon stock comes from urban forests, compared to 17.11 tons of carbon per hectare from all other forests and tree cover (Ugle et al. 2010). As a result, there is plenty of room for urban forests to contribute more to global carbon reserves. According to the National Mission (2009) for a Green India whitepaper, the Sub Mission may increase carbon stocks by 0.06 million tons annually. A few studies conducted in India include "Sequestered standing carbon stock in selective tree species of University campus at Aurangabad, Maharashtra, India" (G. Flora), which revealed the above-ground biomass for trees as follows: *Mangifera indica* 3.13 t/tree, *Ficus religiosa* is 4.27 t/tree, *Delonix regia* 2.12 t/tree, *Ficus benghalensis* 3.89 t/tree, *Butea monosperma* 2.10 t/tree, *Azadirachta indica* 1.91 t/tree, *Pongamia pinnata* 1.57 t/tree, and *Peltophorum pterocarpum* 2.01 t/tree, respectively (Ugle et al. 2010). In another study, a 20-year-old Silver oak shade tree is thought to store up to 41.8 Mg/ha of carbon (Bherwani et al. 2022). The study highlights that the standing carbon stock of young urban trees is not significant, but as the trees mature, there is a possibility that biomass will increase. As a result, the growth rate affects carbon sequestration. Another Indian instance supporting several vulnerable and endangered plant, animal, and bird species is the Kerwa urban forest region in Bhopal. In addition, it is an important carbon sink, storing roughly 19.5 thousand tons of carbon above ground (Lahoti et al. 2020).

9 Conversations and Conclusion

This research demonstrates that it is undeniably true that urban forests and the ecosystem services they provide enormous advantages. Urban woods are believed to improve the standard of living in metropolitan areas. More precisely, ecosystem services include mitigation of air pollution, energy conservation, rainfall, sequestration and storage of carbon, interception, reduction of the urban heat island effect, and adaptability to climate change. Urban forest ecosystem evaluations are a vital tool for increasing our knowledge of these priceless resources and helping quantify the advantages of urban forests. It is evident that urban forests, particularly carefully designed and maintained urban forests, have numerous positive effects on the environment, society, and human psychology. To develop good management concerns in urban ecosystems, a thorough study of the current circumstances of urban forests is necessary. Research has shown how important it is to carry out more studies on the functions that trees play in sequestering carbon in urban environments with a high concentration of trees, such as public parks, avenues, streets, and university campuses. Previous research has also emphasized the importance of estimating biomass to determine the capacity for CO_2 sequestration from the chosen tree species. In addition to forests, the function of urban forests in carbon sequestration promotes the assessment of the quantity of carbon lost and captured as well as the estimation of the carbon pool in urban forests because it has been determined that biomass estimation is required. However, further research is needed to investigate specific techniques and approaches to quantify and estimate the biomass of urban forests more precisely and their potential for sequestering carbon.

References

Akbari H, Pomerantz M, Taha H (2001) Cool surfaces and shade trees to reduce energy use and improve air quality in urban areas. Elsevier. Retrieved December 11, 2023, from https://www.sciencedirect.com/science/article/pii/S0038092X0000089X

Akbari A, Org E, Akbari H (2005) Title energy saving potentials and air quality benefits of urban heat island mitigation permalink. https://escholarship.org/uc/item/4qs5f42s. Publication Date Energy Saving Potentials and Air Quality Benefits of Urban Heat Island Mitigation 1. http://heatisland.lbl.gov/

Ali KA, Ahmad MI, Yusup Y (2020) Issues, impacts, and mitigations of carbon dioxide emissions in the building sector. Sustain (Switzerland) 12(18). https://doi.org/10.3390/SU12187427

Andresen JW (1989) Tree city USA: volunteer urban forestry. Arboricult J 13(4):333–343. https://doi.org/10.1080/03071375.1989.9746813

AR5 Climate Change 2014: Mitigation of Climate Change—IPCC (n.d.) Retrieved January 26, 2024, from https://www.ipcc.ch/report/ar5/wg3/

Beckett KP, Freer-Smith PH, Taylor G (1998) Urban woodlands: their role in reducing the effects of particulate pollution. Environ Pollut 99(3):347–360. https://doi.org/10.1016/S0269-7491(98)00016-5

Bherwani H, Banerji T, Menon R (2022) Role and value of urban forests in carbon sequestration: review and assessment in Indian context. Environ, Develop Sustain, 1–24. https://doi.org/10.1007/S10668-022-02725-5/METRICS

Bolund P, Hunhammar S (1999) Ecosystem services in urban areas. Ecol Econ 29(2):293–301. https://doi.org/10.1016/S0921-8009(99)00013-0

Brack CL (2002) Pollution mitigation and carbon sequestration by an urban forest. Environ Pollut 116:S195–S200. https://doi.org/10.1016/S0269-7491(01)00251-2

Castro BB, Cotas J, Gomes L, Pacheco D, Pereira L (2023) Ecosystem services provided by seaweeds. Hydrobiol 2:75–96. https://doi.org/10.3390/HYDROBIOLOGY2010006

Chadchan J, Shankar R (2012) An analysis of urban growth trends in the post-economic reforms period in India. Int J Sustain Built Environ 1(1):36–49. https://doi.org/10.1016/J.IJSBE.2012.05.001

Chaudhry P, Bagra K, Singh B (2011) Urban greenery status of some Indian cities: a short communication. Int J Environ Sci Develop, 98–101. https://doi.org/10.7763/IJESD.2011.V2.104

Chaurasia S, Munoth N (2022) Carbon sequestration potential of urban trees: a case of Kolar area in Bhopal City, India. J Institut Eng (India): Series A 103(2):359–374. https://doi.org/10.1007/S40030-022-00621-9/METRICS

Chen J, Zhu L, Fan P, Tian L, Lafortezza R (2016) Do green spaces affect the spatiotemporal changes of PM2.5 in Nanjing? Ecol Process 5(1). https://doi.org/10.1186/s13717-016-0052-6

Chow J, Watson J, Lowenthal DH, Countess RJ (1996) Sources and chemistry of PM10 aerosol in Santa Barbara County, CA. Elsevier. Retrieved December 11, 2023, from https://www.sciencedirect.com/science/article/pii/1352231095003630

Christopoulou O, Polyzos S, Minetos D (2007) Peri-urban and urban forests in Greece: obstacle or advantage to urban development? Manage Environ Qual: An Int J 18(4):382–395. https://doi.org/10.1108/14777830710753794/FULL/XML

Churkina G (2016) The role of urbanization in the global carbon cycle. Front Ecol Evol 3:168382. https://doi.org/10.3389/FEVO.2015.00144/BIBTEX

Craig H (1954) Carbon 13 in plants and the relationships between carbon 13 and carbon 14 variations in nature. J Geol 62(2):115–149. https://doi.org/10.1086/626141

Dhyani K, Kuniyal HB, Singh H, Sobha (2021). Growth and physiological potential of *Terminalia arjuna* under elevated CO_2 levels in Open top chamber condition. J Appl Nat Sci 13(3):1121–1126. https://doi.org/10.31018/jans.v13i3.2490. ISSN 2231-5209

Fang C, Ling D-L (2003) Investigation of the noise reduction provided by tree belts. Landscap Urban Plan 63:187–195. https://www.sciencedirect.com/science/article/pii/S0169204602001901

Fang J, Yu G, Liu L, Hu S, Stuart Chapin F (2018) Climate change, human impacts, and carbon sequestration in China. Proc Natl Acad Sci USA 115(16):4015–4020. https://doi.org/10.1073/PNAS.1700304115/ASSET/8C10F6EC-061F-4F95-8060-C5627FAC84EA/ASSETS/GRAPHIC/PNAS.1700304115FIG04.JPEG

Fares S, Paoletti E, Calfapietra C, Mikkelsen TN, Samson R, Le Thiec D (2017) Carbon sequestration by urban trees, 31–39. https://doi.org/10.1007/978-3-319-50280-9_4

Fenger, J. (2002). Chapter 1: urban air quality. developments in environmental science 1(C):1–52. https://doi.org/10.1016/S1474-8177(02)80004-3

French CJ (2004) Tree planting for phytoremediation: the fate of soil contaminants on brownfield sites

Giannico V, Lafortezza R, John R, Sanesi G, Pesola L, Chen J (2016) Estimating stand volume and above-ground biomass of urban forests using LiDAR. Mdpi.com. Retrieved December 15, 2023, from https://www.mdpi.com/2072-4292/8/4/339

Gill SE, Handley J, Ennos AR, Pauleit S (n.d.) Adapting cities for climate change: the role of the green infrastructure. Ingentaconnect.Com. Retrieved December 22, 2023, from https://www.ingentaconnect.com/content/alex/benv/2007/00000033/00000001/art00008

Gowda K, Sridhara MV, Rajan S (2008) Planning and management of parks and green areas: The case of Bangalore metropolitan area. Manag Environ Qual: Int J 19(3):270–282. https://doi.org/10.1108/14777830810866419

Grote R, Samson R, Alonso R, Amorim JH, Cariñanos P, Churkina G, Fares S, Thiec DL, Niinemets Ü, Mikkelsen TN, Paoletti E, Tiwary A, Calfapietra C (2016) Functional traits of urban trees: air pollution mitigation potential. Front Ecol Environ 14(10):543–550. https://doi.org/10.1002/FEE.1426

Gupta SK, Singh H (2017) Observations on Sporadic Flowering in *Ficus benjamina* L. and *Peltophorum africanum* L. in New Forest. Indian Forest 143:290–291

Gupta SK, Ram J, Singh H (2018a) Comparative study of transpiration in cooling effect of tree species in the atmosphere. J Geosci Environ Protect 6:151–166. https://doi.org/10.4236/gep.2018.68011

Gupta SK, Ram J, Singh H, Kumar P (2018b) Comparative study of cooling under shades of some forestry tree species with respect to ambient temperatures. J Geosci Environ Protect 6:51–66. https://doi.org/10.4236/gep.2018.611004

Gupta SK, Ram J, Singh H (2019a) Relationship between leaf area index and atmospheric cooling of tree species. Indian J Forest 42(3):203–208

Gupta SK, Singh H, Ram J (2019b) Need for a novel paradigm in forestry research: perspectives of comparative atmospheric cooling by different forestry effect tree species. Indian Forest 145(6):509–515

Huang KT, Yang SR, Matzarakis A, Lin TP (2018) Identifying outdoor thermal risk areas and evaluation of future thermal comfort concerning shading orientation in a traditional settlement. Sci Total Environ 626:567–580. https://doi.org/10.1016/J.SCITOTENV.2018.01.031

Jangra A, Kumar K, Maikhuri S, Bhandari MS, Pandey S, Singh H, Barthwal S (2024) Unveiling stress-adapted endophytic bacteria: characterizing plant growth-promoting traits and assessing cross-inoculation effects on Populus deltoides under abiotic stress. Plant Physiol Biochem 210:108610. https://doi.org/10.1016/j.plaphy.2024.108610

Jim CY, Chen WY (2008) Assessing the ecosystem service of air pollutant removal by urban trees in Guangzhou (China). J Environ Manage 88(4):665–676. https://doi.org/10.1016/J.JENVMAN.2007.03.035

Kacyira AK (2016) Urbanization: emerging challenges and a new global urban agenda. Brown J World Aff 23. https://heinonline.org/HOL/Page?handle=hein.journals/brownjwa23&id=389&div=&collection

Kanta C, Kumar A, Chauhan A, Singh H, Sharma IP (2024) The interplay between plant functional traits and climate change. In: Kumar N, Singh H (eds) Plant functional traits for improving productivity. Springer, Singapore. https://doi.org/10.1007/978-981-97-1510-7_3

Kojima M, Lovei M (2001) Urban air quality management: coordinating transport, environment, and energy policies in developing countries 56. https://books.google.com/books/about/Urban_Air_Quality_Management.html?id=muZAtgWs6AoC

Kuittinen M (2015) Setting the carbon footprint criteria for public construction projects. Proced Econ Fin 21:154–161. https://doi.org/10.1016/S2212-5671(15)00162-8

Kumar N, Singh H (2024) Plant functional traits for improving productivity. Springer Nature, Singapore. https://doi.org/10.1007/978-981-97-1510-7

Kumar N, Suyal DC, Sharma IP, Verma A, Singh H (2017) Elucidating stress proteins in rice (*Oryza sativa* L.) genotype under elevated temperature: a proteomic approach to understand heat stress response. 3 Biotech 7:205. https://doi.org/10.1007/s13205-017-0856-9
Kumar M, Singh MP, Singh H, Dhakate PM, Ravindranath NH (2019a) Forest working plan for the sustainable management of forest and biodiversity in India. J Sustain Forest 39:1–22. https://doi.org/10.1080/10549811.2019.1632212
Kumar N, Jeena N, Singh H (2019b) Elevated temperature modulates rice pollen structure: a study from the foothill of the Himalayan agro-ecosystem in India. 3 Biotech 9(5):175. https://doi.org/10.1007/s13205-019-1700-1
Kumar N, Jeena N, Kumar N, Gangola S, Singh H (2019c) Phytoremediation facilitating enzymes: an enzymatic approach for enhancing remediation process. In: Bhatt P (ed) Smart bioremediation technologies. Academic Press, Elsevier, 50 Hampshire Street, 5th Floor, Cambridge, MA 02139, United States, pp 289–306
Kumar N, Singh H, Sharma SK (2020) Antioxidants: responses and importance in plant defense system. In: Roychowdhury R, Choudhury S, Hasanuzzaman M, Srivastava S (eds) Sustainable agriculture in the era of climate change. Springer, Cham, pp 251–264. https://doi.org/10.1007/978-3-030-45669-6_11
Kumar A, Kumar P, Singh H, Kumar N (2021a) Modulation of plant functional traits under essential plant nutrients during seasonal regime in natural forests of Garhwal Himalayas. Plant Soil 465:197–212. https://doi.org/10.1007/s11104-021-05003-x
Kumar A, Kumar P, Singh H, Bisht S, Kumar N (2021b) Relationship of physiological plant functional traits with soil carbon stock in the temperate forest of Garhwal Himalaya. Curr Sci 120(8):1368–1373. https://doi.org/10.18520/cs/v120/i8/1368-1373
Kumar A, Kumar P, Singh H, Kumar N (2021c) Adaptation and mitigation potential of roadside trees with bio-extraction of heavy metals under vehicular emissions and their impact on physiological traits during seasonal regimes. Urban Forest Urban Green 58:126900. https://doi.org/10.1016/j.ufug.2020.126900
Kumar A, Kumar P, Singh H, Kumar N (2021d) Impact of plant functional traits on infiltration rate under different forests of Kempty watershed in Garhwal Himalaya. Indian J Soil Conserv 49(1):38–44
Kumar M, Phukon SN, Paygude AC, Tyagi K, Singh H (2021e) Mapping phenological functional types (PhFT) in the Indian Eastern Himalayas using machine learning algorithm in Google Earth engine. Comp Geosci 158. https://doi.org/10.1016/j.cageo.2021.104982
Kumar A, Singh H, Kumari G, Bisht S, Malik A, Kumar A, Singh M, Raturi , Barthwal S, Thakur A, Kaushal R (2022) Adaptive resilience of roadside trees to vehicular emissions via leaf enzymatic, physiological, and anatomical trait modulations. Environ Pollut 313. https://doi.org/10.1016/j.envpol.2022.120191
Kumar N, Singh H, Giri K, Kumar A, Joshi A, Yadav A, Singh R, Bisht S, Kumari R, Jeena N, Khairakpam R, Mishra G (2024) Physiological and molecular insights into the allelopathic effects on agroecosystems under changing environmental conditions. Physiol Mol Biol Plants 30:417–433. https://doi.org/10.1007/s12298-024-01440-x
Lahoti S, Lahoti A, Joshi RK, Saito O (2020) Vegetation structure, species composition, and carbon sink potential of urban green spaces in Nagpur City, India. Land 9:107. https://doi.org/10.3390/LAND9040107
Lai D, Liu W, Gan T, Liu K, Chen Q (2019) A review of mitigating strategies to improve the thermal environment and thermal comfort in urban outdoor spaces. Sci Total Environ 661:337–353. https://doi.org/10.1016/J.SCITOTENV.2019.01.062
Lin W, Zeng C, Lam NSN, Liu Z, Tao J, Zhang X, Lyu B, Li N, Li D, Chen Q (2021) Study of the relationship between the spatial structure and thermal comfort of a pure forest with four distinct seasons at the microscale level. Urban Forest Urban Green 62:127168. https://doi.org/10.1016/J.UFUG.2021.127168
Malik A, Kumar K, Kumari G, Kumar N, Singh H (2024) Application of functional traits in modelling productivity and resilience under climate change. In: Kumar N, Singh H (eds) Plant

functional traits for improving productivity. Springer, Singapore. https://doi.org/10.1007/978-981-97-1510-7_5

McPherson EG (1998) Atmospheric carbon dioxide reduction by Sacramento's urban forest. J Arboricult 24(4):215–223. https://doi.org/10.48044/jauf.1998.026

Meena D, Singh H, Chaudhari SK (2011) Elucidating strontium response on growth dynamics and biochemical change in *Phaseolus mungo* L. Int J Agricult Environ Biotechnol 4:107–113

Moll G, Forests CKA (1996) Urban trees and carbon. Go.Gale.Com. Retrieved December 11, 2023, from https://go.gale.com/ps/i.do?id=GALE%7CA18646225&sid=googleScholar&v=2.1&it=r&linkaccess=abs&issn=00028541&p=AONE&sw=w

Moonen P, Defraeye T, Dorer V, Blocken B, Carmeliet J (2012) Urban physics: effect of the micro-climate on comfort, health and energy demand. Front Architect Res 1(3):197–228. https://doi.org/10.1016/j.foar.2012.05.002

Nero BF, Callo-Concha D, Anning A, Denich M (2017) Urban green spaces enhance climate change mitigation in cities of the global South: the case of Kumasi, Ghana. Proced Eng 198:69–83. https://doi.org/10.1016/J.PROENG.2017.07.074

Nowak D (2016) Definitions and characteristics of urban forest ecosystems of agricultural and urban forests in the United States

Nowak DJ, Crane DE (2002) Carbon storage and sequestration by urban trees in the USA. Environ Pollut 116(3):381–389. https://doi.org/10.1016/S0269-7491(01)00214-7

Nowak DJ, Crane DE, Stevens JC (2006) Air pollution removal by urban trees and shrubs in the United States. Urban Forest Urban Green 4(3–4):115–123. https://doi.org/10.1016/J.UFUG.2006.01.007

Nowak DJ, Hirabayashi S, Bodine A, Greenfield E (2014) Tree and forest effects on air quality and human health in the United States. Environ Pollut (Barking, Essex : 1987) 193:119–129. https://doi.org/10.1016/J.ENVPOL.2014.05.028

Ortiz-Bobea A, Ault TR, Carrillo CM, Chambers RG, Lobell DB, Ortiz-Bobea A, Ault TR, Carrillo CM, Chambers RG, Lobell DB (2021) Anthropogenic climate change has slowed global agricultural productivity growth. NatCC 11(4):306–312. https://doi.org/10.1038/S41558-021-01000-1

Peckham SC (2010) Nature in the city: ecological consciousness development associated with naturalized urban spaces and urban forest values in Calgary, AB and Halifax, NS. https://dalspace.library.dal.ca//handle/10222/13180

Perera F (2018) Pollution from fossil-fuel combustion is the leading environmental threat to global pediatric health and equity: Solutions exist. Int J Environ Res Publ Heal 15(1). https://doi.org/10.3390/ijerph15010016

Prakash V, Hunney K, Singh H (2022) Effect of elevated carbon dioxide on growth and development of *Santalum album* L. seedlings inoculated with plant growth promoting microorganisms in open top chambers. Int J Environ Agricult Biotech 7(5):149–159

Raj KG, Trivedi S, Ramesh KS, Sudha R, Subramoniam SR, Ravishankar HM, Vidya A (2021) Assessment of vegetation cover of Bengaluru City, India, using geospatial techniques. J Indian Soc Remote Sens 49(4):747–758. https://doi.org/10.1007/S12524-020-01259-5

Ramaiah M, Avtar R (2019) Urban green spaces and their need in cities of rapidly urbanizing India: a review. Urban Sci 3:94. https://doi.org/10.3390/URBANSCI3030094

Rawat JS, Kumar M (2015) Monitoring land use/cover change using remote sensing and GIS techniques: a case study of Hawalbagh block, district Almora, Uttarakhand, India. Egyptian J Remote Sens Space Sci 18(1):77–84. https://doi.org/10.1016/J.EJRS.2015.02.002

Raza A, Razzaq A, Mehmood SS, Zou X, Zhang X, Lv Y, Xu J (2019) Impact of climate change on crops adaptation and strategies to tackle its outcome: a review. Plants 8(2). https://doi.org/10.3390/PLANTS8020034

Ren Z, Fu Y, Dong Y, Zhang P, He X (2022) Rapid urbanization and climate change significantly contribute to worsening urban human thermal comfort: a national 183-city, 26-year study in China. Urban Clim 43:101154. https://doi.org/10.1016/J.UCLIM.2022.101154

Satterthwaite D, McGranahan G, Tacoli C (2010) Urbanization and its implications for food and farming. Philosop Trans Royal Soc B: Biol Sci 365(1554):2809–2820. https://doi.org/10.1098/RSTB.2010.0136

Schmid JA (2015) Vegetation types, functions, and constraints in metropolitan environments. Plan Uses Manage Land, 499–528. https://doi.org/10.2134/AGRONMONOGR21.C21

Schwela D (2000) Air pollution and health in urban areas. Rev Environ Health 15(1–2):13–42. https://doi.org/10.1515/REVEH.2000.15.1-2.13

Sharma R, Singh H (2021) Alteration in biochemical constituents and nutrients partitioning of *Asparagus racemosus* in response to elevated atmospheric CO_2 concentration. Environ Sci Pollut Res 29:6812–6821. https://doi.org/10.1007/s11356-021-16050-3

Sharma R, Singh H, Kaushik M, Nautiyal R, Singh O (2018) Adaptive physiological response, carbon partitioning, and biomass production of *Withania somnifera* (L.) Dunal grown under elevated CO_2 regimes. 3 Biotech 8:267. https://doi.org/10.1007/s13205-018-1292-1

Sharma R, Pradhan L, Kumari M, Bhattacharya P (2020) Assessment of carbon sequestration potential of tree species in Amity University Campus Noida. Environ Sci Proceed 3:52. https://doi.org/10.3390/IECF2020-08075

Silver WL, Ostertag R, Lugo AE (2000) The potential for carbon sequestration through reforestation of abandoned tropical agricultural and pasture lands. Restor Ecol 8(4):394–407. https://doi.org/10.1046/J.1526-100X.2000.80054.X

Simpson J, McPherson EG (1996) Potential of tree shade for reducing residential energy use in California. Fs.Usda.GovJR Simpson, EG McPherson. J Arboricult. Retrieved December 21, 2023, from https://www.fs.usda.gov/research/treesearch/download/61736.pdf

Singh H (2021) An integrated approach considering physiological and biophysical based indicators for assessing tolerance of roadside plantations of Alstonia scholaris towards urban roadside air pollution: an assessment of adaptation of plantations for mitigating urban roadside air pollution. Trees. https://doi.org/10.1007/s00468-021-02179-8

Singh H (2024). Exploring adaptive modulation in plant functional traits and its impact on the productivity of *Acacia auriculiformis* under CO_2-enriched environment. Indust Crop Prod 210:118186. https://doi.org/10.1016/j.indcrop.2024.118186

Singh H, Verma A (2013) Characterization and screening of high nitrogen efficient rice genotype to curtail environmental pollution. Int J Agricult, Environ Biotech 6(1):23–30

Singh H, Kumar M (2022) Climate change and its impact on Indian Himalayan forests: current status and research needs. In: Rani S, Kumar R (eds) Climate change. Springer Climate, Springer, Cham, pp 223–242. https://doi.org/10.1007/978-3-030-92782-0_11

Singh H, Sharma RS, Sinha S, Kumar M, Kumar P, Verma A, Sharma SK (2017a). Physiological functioning of *Lagerstroemia speciosa* L. under heavy roadside traffic: an approach to screen potential species for abatement of urban air pollution. 3 Biotech 7(61):1–10. https://doi.org/10.1007/s13205-017-0690-0

Singh H, Verma A, Kumar M, Sharma R, Gupta R, Kaur M, Negi M, Sharma SK (2017b) Phytoremediation: a green technology to clean up the sites with low and moderate level of heavy metals. Austin Biochem 2:1012

Singh H, Sharma R, Savita SMP, Kumar M, Verma A, Ansari MW, Sharma SK (2018) Adaptive physiological response of *Parthenium hysterophorus* to elevated atmospheric CO_2 concentration. Indian Forest 144:1–14

Singh H, Yadav M, Kumar N, Kumar A, Kumar M (2020) Assessing adaptation and mitigation potential of roadside trees under the influence of vehicular emissions: a case study of Grevillea robusta and Mangifera indica planted in an urban city of India. PLoS ONE 15(1):e0227380; https://doi.org/10.1371/journal.pone.0227380

Singh S, Singh H, Sharma SK, Nautiyal R (2021) Seasonal variation in biochemical responses of bamboo clones in the sub-tropical climate of Indian Himalayan foothills. Heliyon 7(4):e06859. https://doi.org/10.1016/j.heliyon.2021.e06859

Singh H, Bandyopadhyaya S, Kumar A, Singh M, Malik A, Kumari G, Raturi A, Yadav SK, Kumar M, Barthwal S, Thakur A and Rawat PS (2022) Understanding the physiological and biophysical response of urban roadside plantations for assessing adaptation and mitigation mechanisms toward vehicular emissions. Urban Clim 44. https://doi.org/10.1016/j.uclim.2022.101183

Singh H, Kumar N, Kumar A (2024a) Enhancing resource use efficiency in crops through plant functional traits. In: Kumar N, Singh H (eds) Plant functional traits for improving productivity. Springer, Singapore. https://doi.org/10.1007/978-981-97-1510-7_6

Singh M, Singh H, Kumar A, Kumar M, Barthwal S, Thakur A (2024b) Soil nitrogen availability determines the CO_2 fertilization effect on tree species (*Neolamarckia cadamba*): growth and physiological evidence. Environ Sustain 7:53–60. https://doi.org/10.1007/s42398-023-00300-w

Skofronick-Jackson G, Petersen WA, Berg W, Kidd C, Stocker EF, Kirschbaum DB, Kakar R, Braun SA, Huffman GJ, Iguchi T, Kirstetter PE, Kummerow C, Meneghini R, Oki R, Olson WS, Takayabu YN, Furukawa K, Wilheit T (2017) The global precipitation measurement (GPM) mission for science and society. Bull Am Meteor Soc 98(8):1679–1695. https://doi.org/10.1175/BAMS-D-15-00306.1

Smith P, Arneth A, Barnes DKA, Ichii K, Marquet PA, Popp A, Pörtner HO, Rogers AD, Scholes RJ, Strassburg B, Wu J, Ngo H (2022) How do we best synergize climate mitigation actions to co-benefit biodiversity? Glob Change Biol 28(8):2555–2577. https://doi.org/10.1111/GCB.16056

Sodoudi S, Zhang H, Chi X, Müller F, Li H (2018) The influence of spatial configuration of green areas on microclimate and thermal comfort. Urban Forest Urban Green 34:85–96. https://doi.org/10.1016/J.UFUG.2018.06.002

Strohbach MW, Arnold E, Haase D (2012) The carbon footprint of urban green space—a life cycle approach. Landsc Urban Plan 104(2):220–229. https://doi.org/10.1016/J.LANDURBPLAN.2011.10.013

Sudha P, Ravindranath NH (2000) A study of Bangalore urban forest. Landsc Urban Plan 47(1–2):47–63. https://doi.org/10.1016/S0169-2046(99)00067-5

Ugle P, Rao S, RamaChandra TV (2010) Carbon sequestration potential of urban trees. In: Proceedings of the Lake 22. https://wgbis.ces.iisc.ac.in/energy/lake2010/Theme1/prachi_u.pdf

Waffle AD, Corry RC, Gillespie TJ, Brown RD (2017) Urban heat islands as agricultural opportunities: an innovative approach. Landsc Urban Plan 161:103–114. https://doi.org/10.1016/j.landurbplan.2017.01.010

Wang L, Gong H, Liao W, Wang Z (2015) Accumulation of particles on the surface of leaves during leaf expansion. Sci Total Environ 532:420–434. https://doi.org/10.1016/J.SCITOTENV.2015.06.014

Yadav SK, Singh H, Ginwal HS and Barthwal S (2019a) Elevated CO_2 enhanced growth and physiological process of *Populus deltoides* Bartr. ex Marsh. Indian Forest 145(1):23–27

Yadav SK, Singh H, Nautiyal R, Ginwal HS, Ansari SA, Barthwal S (2019b) Modulation of morpho-physiological responses in *Populus deltoides* by elevated carbon dioxide and temperature. Forest Sci 66:105–118. https://doi.org/10.1093/forsci/fxz048

Yang J, McBride J, Zhou J, Sun Z (2005) The urban forest in Beijing and its role in air pollution reduction. Urban Forest Urban Green 3(2):65–78. https://doi.org/10.1016/J.UFUG.2004.09.001

Ysebaert T, Koch K, Samson R, Denys S (2021) Green walls for mitigating urban particulate matter pollution—a review. Urban Forest Urban Green 59:127014. https://doi.org/10.1016/J.UFUG.2021.127014

Urban Tree Selection and Management Strategies for Climate Adaptation

Vikas Verma Patel, Sagar Vishwakarma, and Reetu Gangwar

Abstract Urban areas have experienced rapid urbanization, which involves unplanned and uncontrolled growth and development, demographic expansion, and other anthropogenic activities. Rapid urbanization has occurred in the cities, resulting in unplanned and uncontrollably growing populations, demographic shifts, and other human activities. For this reason, climate change's potential biotic and abiotic effects may impact urban forests and trees. Tree species that exhibit strong stress tolerance and high local climatic adaptability should be discovered and chosen to increase urban forests' resilience to climate change's effects. Urban forests and trees are directly or indirectly impacted by climate phenomena in urban environments, such as frost, cold, heat waves, drought, thunderstorms, waterlogging, soil salinization, pests, and disease. An evaluation framework can be proposed and used to evaluate the climate change adaptation of about 70 urban tree species using a method that combines quantitative data and qualitative description to choose urban tree species in the context of climate change. The most important prevalent tree species viz. *Azadirachtaindica, Terminaliacatappa, Peltophorumpterocarpum, Delonixregia, Annonasquamosa, Acacia auriculiformis, Prosopisjuliflora, Tamarindusindica, Tecomastans, Leucaenaleucocephala, Borassusflabellifer, Sennasiamea, Delonixregia* and *Eucalyptus tereticornis,* etc., in urban areas used to deal with the impact of climate change in urban areas. In light of the effects of climate change, warm subtropical, and highly tolerant species make the best selections for urban tree species. The need to create new techniques, such as using digital geotechnology for urban tree care, is becoming more pressing due to the growing demand to conserve trees and the limited availability of resources (workforce and water). Developing an urban forest is a long-term project that needs ongoing

V. V. Patel (✉)
Department of Botany, Veerangna Rani Avanti Bai Lodhi Government Girls Degree College, Bareilly 243005, Uttar Pradesh, India
e-mail: vikasbotany@gmail.com

S. Vishwakarma
Department of Botany and Microbiology, Gurukula Kangri (Deemed to Be University), Haridwar 249204, Uttarakhand, India

R. Gangwar
Department of Botany, Bareilly College, Bareilly 243001, Uttar Pradesh, India

H. Singh (ed.), *Urban Forests, Climate Change and Environmental Pollution*,
https://doi.org/10.1007/978-3-031-67837-0_19

support and dedication from urban communities to achieve the full availability and equitable distribution of numerous advantages. It is proposed that following guidelines can help make tree-planting projects more successful. The following principles are adhered to: (1) provide basic knowledge of urban tree species; (2) facilitate community engagement on tree planting and maintenance; (3) focus more emphasis on tree canopy cover goals than planting quantity of trees; (4) focus on post-planting care and encouraging retention of existing trees; (5) increasing species diversity, through careful species selection to enhance resilience of urban trees.

Keywords Urban tree species · Management strategies · Climatic adaptability

1 Introduction

Cities serve as a domicile and habitat for a developing proportion of the global population while also performing an essential role in preserving ecological, economic, and social welfare. India's most advanced financial system has experienced the most rapid growth globally. The city is projected to accommodate an astounding 60 crore individuals at the current expansion rate. On the pretext that urban forests provide purported environmental and social benefits, a number of the largest cities in the world have established tree-planting initiatives (Stephanie et al. 2013; Yadav et al. 2019a, b; Jangra et al. 2024; Kanta et al. 2024; Malik et al. 2024; Kumar and Singh 2024). Recent studies have evaluated the impacts of green space and trees on individuals' physical, mental, and social well-being, employing the World Health Organization's conceptualization of health (Kondo et al. 2018; Wolf et al. 2020). Urban trees are of significant ecological value in human habitats due to their multifaceted functions, which encompass air, water, and sunlight filtration, provision of shelter for avian and animal species that contribute to food webs, and establishment of recreational areas for humans (Nowak et al. 2013; Meena et al. 2011; Kumar et al. 2019c, 2020; Singh and Kumar 2022; Singh et al. 2024). Compared to the amount of attention devoted to the forests, urban forests effectively reduce stormwater discharge, mitigate urban pollution, and provide energy-conserving shade for residential areas (Nagendra and Gopal 2010).

Indigenous species are often displaced, while non-native ones are encouraged to take root, potentially leading to invasive outcomes. This decline in native species coincides with heightened human activity. Research indicates that non-native tree (NNT) species can disrupt ecosystem functioning, diminish native biodiversity, and alter local climate (Bowler et al. 2010). Over the last two decades, scientific evidence has emphasized the vital role of green spaces in urban ecological systems. However, urban planners and administrators frequently underestimate the importance of trees (Nagendra and Gopal 2010).

Consequently, there's a significant risk of losing green spaces as urban areas expand to accommodate growing populations, which could threaten the quality of life for future generations. Urbanization leads to substantial changes in land cover and use, affecting ecosystems' configuration, functionality, and structure (Singh and Verma 2013; Kumar et al. 2017; Gupta and Singh 2017; Singh et al. 2017a, b). These changes raise concerns about their impact on daily life and the sustainability of future generations.

Pavement, buildings, and vegetation canopy significantly influence the development of urban ecosystems. Urban forests, consisting of trees in yards, along roads, and parks within urban areas, are essential to these ecosystems. Urban trees enhance landscape aesthetics, mitigate air pollution, and help regulate urban energy consumption, water usage, and stormwater runoff. Expansion of pavement areas notably affects energy exchange, hydrology, and microclimate in urban areas (Arnold and Gibbons 1996). Several challenges in urban ecosystem management stem from these factors, including increased stormwater runoff, urban heat islands, and reduced air quality associated with impervious surfaces. There's a known correlation between water use, air quality, tree density, and crown density (Nowak et al. 2018). Understanding air quality, energy dynamics, and hydrological processes in urban ecosystems relies on factors such as tree species, leaf and stem surface areas, tree size, and pavement coverage percentages. Identifying forest structure characteristics is crucial for assessing the environmental benefits of urban forests and informing management decisions, such as leaf removal, tree pruning, and pest control (Korkmaz et al. 2018; Sharma et al. 2018; Singh et al. 2018; Gupta et al. 2018a, b). Urban administrators can leverage this understanding to plan and implement effective urban forest management strategies.

2 Basic Principles of Trees in Urban Design

Beautiful trees make great cities because a dynamic urban forest supports a healthy community, economy, and environment. Moreover, urban design incorporates trees as an essential element (Arnold 1980). Nonetheless, numerous urban design experts lack the fundamental understanding that forms the foundation of the science and art of tree integration into the urban fabric. In addition, countless urban designs, codes, and public works standards demonstrate the lack of knowledge regarding the fundamental requirements of trees and the tenets of tree planting. This dearth of comprehension is a significant contributor to the demise of trees. Seven to ten years is the average lifespan of a boulevard tree, which yields a negative return on investment (Moll 1989). In light of the increasing demand for urban living, the significance of green infrastructure, and the completion of thoroughfares, it is becoming more critical than ever for the prosperity of any municipality or city to comprehend how to design appropriately for trees. Deciding on an appropriate spatial configuration that fosters the robust growth of a tree in an urban setting can be a formidable challenge, considering the biological nature of trees (Yadav et al. 2019a, b; Gupta et al. 2019a, b; Kumar et al. 2019a, b). This article presents a collection of tools utilized to create aesthetically pleasing urban environments while also being cost-effective by integrating several fundamental principles. In addition to framing space and humanizing the urban environment, trees serve as living construction materials that are indispensable to the urban form of cities. The urban forest restores order to an urban environment that is frequently disorderly. Trees are inseparable from the urban structure. The tree is the one architectural feature that can unite an entire city. Therefore, it is necessary to pay attention to some unique points for urban plantation design;

2.1 Establish Large Trees

In urban design, planting trees creates and reinforces space, not to adorn it. The flowering trees (smaller trees) will never attain sufficient size to furnish the requisite shading canopy for the comfort and proportion of this entire paved area.

2.2 Limit Tree Species

The urban environment presents a formidable challenge for tree survival, characterized by inadequate drainage, disturbed and compacted soils, limited sunlight, animal refuse accumulation, polluted stormwater runoff, and human abuse. Forced diversity (of trees) results in the planting of unsuitable species of trees that cannot even survive, let alone thrive. Furthermore, integrating various tree species into a design compromises its fundamental objective.

2.3 Plant Trees Close Together Using a Geometric Grid

The value of a tree in a city is its contribution to the more extensive urban forest. Arnold (1980) describes how trees collaborate to "create entrance canopies for buildings, separate areas with different scales or geometric configurations, and form arcades that connect buildings." A haphazard arrangement of tree plantings contributes to the sense of disorder. However, trees arranged nearby in a geometric or grid-like formation establish a visual syncopation that enhances human perception of the area. This is analogous to how metrical form in poetry directs measurement and comprehension. To adequately characterize the structure of an urban forest, it is imperative to possess fundamental data encompassing tree counts, spatial distributions, species composition, dimensions, and growing conditions. Choosing the appropriate tree species is critical for the success of tree planting initiatives in urban areas. Trees that are ideally adapted for urban environments should be selected, including those that can withstand drought, provide tree canopy, and contribute to aesthetic and environmental design.

2.4 *Design the Tree from the Roots Up*

Urban streets are highly competitive for space, and trees planted will not perform as intended if they are not incorporated into the design from the project's outset. Trees require soil for optimal development and survival, particularly for large shade trees. Many methodologies exist for "Creating Space for Roots," encompassing suspended sidewalk systems, engineered soil solutions, root paths, trenches, and vaults (Urban 2008). Although these solutions initially incur higher expenses, they ultimately result in long-term cost savings (Fowler 2011).

2.5 *Respect the Base of the Tree Because Trees Grow.*

A tree will develop if properly designed, and any design must account for the expansion of the trunk flare and roots. Refraining from employing tree grates and trunk protectors is advisable due to their high cost and propensity to encircle the tree. Designating an open area encircling its base is essential to facilitate the tree's development, enclosing low seating walls or a fence. Additionally, using flexible paving materials in combination with engineered soil designs is considered a best practice.

2.6 *Plant the Right Tree in the Right Place*

Trees are living things; each species has different requirements and capacities for acclimating to various urban environments. Understanding and considering individual tree species' diverse requirements and adaptations is paramount when formulating a tree palette for a given project.

3 Benefits of Planting Urban Trees

The advantages of tree sowing in urban environments are manifold and extensive. Trees improve by absorbing pollutants, including nitrogen dioxide, sulfur dioxide, and fine particulate matter. Air quality. Furthermore, their contribution to carbon sequestration is crucial in alleviating climate change's consequences (Nowak 2000). Moreover, trees contribute to the mitigation of the urban heat island phenomenon, which transpires when the temperature of cities surpasses that of adjacent rural regions owing to the absorption and reflection of heat by structures, roads, and hardscaped surfaces. Trees mitigate heat gain by providing shade and evapotranspiration, thereby contributing to the cooling of urban environments and diminishing the demand for air conditioning energy. Urban trees support local ecosystems, increase biodiversity, provide habitat for small mammals, insects, and birds, and purify the air and soil of contaminants and particulates. This is particularly crucial in urban regions, where natural vegetation is frequently scarce, and the community is adversely affected by effluent from constructed structures, automobiles, and hardscaped surfaces.

4 Site Evaluation for Tree Plantation

The initial step in selecting appropriate trees for a planting site involves thorough site evaluation, encompassing both above-ground and below-ground considerations. Neglecting this crucial assessment often leads to the premature demise of trees in urban settings. Understanding the hardiness zones of trees is pivotal in gauging their suitability for urban environments. The hardiness zone map, devised by the United States Department of Agriculture (USDA), delineates regions based on their average lowest winter and highest summer temperatures. Before tree selection, it's imperative to ascertain the hardiness zone number corresponding to the planting site on the map.

4.1 Above-Ground Site Analysis

Several crucial factors should be taken into account during the above-ground assessment. Urban environments pose various challenges, including overhead wires, streetlights, buildings, signage, vandalism, and regulatory requirements. Additionally, environmental factors like light exposure, slope, wind patterns, salt levels, and the presence of existing trees play significant roles. The following are critical considerations in this regard.

4.1.1 Light Exposure

Note the duration of direct sunlight the planting site receives during summer hours. When evaluating sites across different seasons, it's essential to consider the seasonal shifts in solar angle. Crape myrtle and other full-sun trees typically necessitate a minimum of six hours of direct sunlight daily for optimal growth. However, all-day sun exposure fosters the best development and form. Trees adaptable to full sun to partial shade can thrive with three to six hours of direct sunlight.

Conversely, trees with shade preferences are better suited to locations receiving less than three hours of direct sunlight. Most large trees flourish in full-light conditions. Trees may experience heightened heat loads when positioned near structures due to sunlight reflecting off glass or walls. Optimal trees for such locations exhibit drought tolerance and thrive in full light. Additionally, trees with expansive root systems benefit from increased soil volume, aiding their ability to withstand reflected light. Ensuring adequate irrigation further supports the vitality of these trees.

4.1.2 Slope Exposure

Transplanting trees with thin bark, like maples, cherries, and plums, presents challenges, especially on western and southern slopes. These slopes experience higher transpiration and evaporation rates, making it harder to maintain soil moisture. To mitigate this, we intend to provide additional irrigation to trees on southern and western exposures to prevent desiccation, favoring drought-tolerant species for these areas. Conversely, northern slopes retain moisture better and receive less direct sunlight, making them suitable for a broader range of tree species.

4.1.3 Wind

Trees experience increased water loss to the atmosphere in windy conditions. Therefore, it's advisable to opt for drought-tolerant tree species, especially in areas prone to strong winds like coastal regions. Alternatively, additional precautions such as expanding irrigation capabilities or providing shading can be implemented. Additionally, trees selected for such locations should be capable of enduring both drought and wet conditions, particularly if the soil has inadequate drainage.

4.1.4 Salt

When airborne salt settles on the ground and infiltrates the soil, it has the potential to scorch twigs and foliage or permeate the roots of trees. Despite the deformities often inflicted by direct exposure to saline air, salt-tolerant trees exhibit resilience and expansive growth. Conversely, salt-sensitive trees suffer foliage burns and stunted growth when subjected to salty air. Trees with one-sided canopies near coastal areas are particularly vulnerable to hurricane-force winds, which can disproportionately affect the canopy from the windward side.

4.1.5 Other Trees

Young trees needing full or partial sunlight, often developing broad canopies, tend to lean towards the sun, resulting in one-sided growth. This phenomenon commonly occurs beneath the canopy of mature trees. The presence of these mature trees can hinder or even halt the development of new trees due to various factors such as root competition, insufficient water, and shading.

Table 1 Recommendations for planting trees 40 feet away from power lines or street lamps

Distance from power lines or lights	Tree size (at maturity)
0–6 feet	Planting is not recommended unless trees remain 20 feet tall
6–40 feet	Height should be 10 feet or shorter than wire/light, or canopy diameter should be less than twice the distance to wire/light
More than 40 feet	Any tree can be planted

4.1.6 Overhead Wires and Street/Security Lights

Before planting, it's crucial to survey the area. Often, trees end up too close to utility lines and security lights, requiring utility companies to regularly prune branches that encroach on wires, incurring significant expenses annually. To alleviate these costs and reduce hurricane damage, it's advisable to limit planting trees near power lines to those of suitable size (refer to Table 1). Optimal placement involves situating trees as far from power lines as possible.

Frequently, trees are planted in the same areas as overhead security lighting, resulting in the canopy eventually overshadowing the lights. This necessitates consistent pruning, leading to misshapen canopies. Effective planning involves separating trees from security lights, positioning lights approximately 12 feet above the ground to accommodate tree canopy growth. Alternatively, selecting trees that remain small at maturity allows them to coexist beneath the lights. While some trees can adjust their canopy growth away from buildings, narrow-canopied trees within 10 feet of a structure can be a suitable choice. If shade is desired, consider planting several small-statured trees to create a dense canopy.

4.1.7 Signs

Conflicts between signs and trees frequently arise due to poor planning. To mitigate this, planting large trees near short signs and small trees near tall signs is recommended. Over time, large-maturing trees may obstruct lower signs. However, maintaining these trees with a single trunk enables the pruning of lower branches, ensuring sign visibility. Numerous municipalities have implemented landscape regulations requiring signs to be positioned within 8 ± 10 feet from the ground, addressing this potential issue. For optimal compliance with ordinances, signs should ideally be placed at a distance of 3 ± 4 feet above the ground.

4.1.8 Vandalism

Sometimes, trees face deliberate harm or destruction, while others may suffer unintentional damage, mainly when situated in vulnerable spots like sidewalk cutouts, where pedestrian traffic is common. To minimize the risk of vandalism, it's prudent to refrain from planting species with thin bark, like red maple, or to opt for trees with a trunk diameter of at least 4 inches.

4.2 Below-Ground Site Analysis

Soil characteristics, including pH, drainage, depth, salinity, proximity to the water table, and obstacles to root growth, such as barriers, play pivotal roles in determining tree suitability. Neglecting these factors often leads to the failure of many plantations. The value of high-quality soil cannot be overstated; it must be preserved diligently. After completion of the task, soil can be reinstated at the site to promote tree growth. Furthermore, thorough pre-construction planning enables collaboration with contractors to prevent soil over-compaction in areas designated for tree preservation and planting (Hall et al. 2010). Implementing reinforced fencing to separate these zones from heavy machinery and vehicles and imposing penalties on contractors for violations helps ensure compliance.

4.2.1 Rooting Space Restrictions

Align the tree's eventual size with the soil volume accessible for root development. This technique maintains the stability and health of trees during cyclones. Additionally, it safeguards adjacent pavement, curbing, and footpaths from harm. Soil compacted and inadequately aerated beneath pavement is generally unsuitable for sustaining roots unless the soil consists of coarse sediment and is well-drained (Fig. 1). The primary habitat of roots will be the unpaved soil space or the area between the soil and the pavement base. This can impede the establishment a robust root system, potentially leading to the tree's vulnerability during hurricanes. Certain trees that can tolerate damp sites (such as bald cypress) have acclimated to grow roots beneath pavement and can withstand high gusts.

4.2.2 Soil pH

Soil pH plays a critical role in regulating soil microorganisms' activity and the availability of nutrients essential for plant growth. To accurately assess soil pH across the site, it's imperative to conduct pH tests wherever there are noticeable differences in soil color or texture. This ensures a comprehensive understanding of the soil conditions, which may vary significantly within the project area, affecting the suitability of planting certain species.

For sampling, employ a trowel or shovel to create ten small excavations spaced five to ten feet apart in an open area, such as a lawn, where soil uniformity is expected. For analysis, extract a sediment section extending from the surface to a depth of

Fig. 1 Stormwater management. This figure displays how to reduce onsite flooding and municipality load while increasing tree growth by integrating stormwater into your Strata vault tree pits

12 inches from each excavation. The optimal soil pH range for fostering healthy tree growth typically falls between 4.8 and 7.2. If the soil pH drops below 4.8, prioritize trees known for their resilience to acidic conditions. Conversely, when soil pH exceeds 7.2, trees that demonstrate tolerance to alkaline environments should be chosen. Remember that only a few tree species can thrive in soils with pH levels exceeding 9.0. This tailored approach ensures the selection of trees best suited to the specific soil conditions encountered on the site, promoting successful growth and development.

4.2.3 Inadequate Drainage, Compacted Soil, and Low Oxygen

Soil preparation and type play pivotal roles in the success of urban trees, necessitating the presence of nutrient-rich, well-draining, and uncompacted soil. Adequate irrigation is essential to ensure newly planted trees receive ample water for developing robust root systems. City Green offers specialized soil tailored to your trees' specific needs (Singh et al. 2020; Kumar et al. 2021a, b, c, d).

Urban areas often witness soil consolidation, mainly due to poor drainage and compaction, even with alluvial soil. Such conditions deprive tree roots of oxygen for survival and growth. While challenging soils may limit options to drought-tolerant species like bald cypress, sweet bay, or pond apple, trees with vigorous root systems like ficus and oak are worth considering. However, their extensive surface roots may pose risks to nearby structures such as curbs, sidewalks, and pavements and disrupt lawn maintenance activities.

For better hurricane resistance, opt for small- to medium-sized trees not exceeding 40 feet tall at maturity. To assess drainage and compaction effectively, excavate multiple trenches around the site, each at least 18 inches deep. Breaking up the compacted surface layer within a 15-foot radius around the tree before planting can significantly enhance drainage and promote tree growth. In such instances, tree selection should prioritize site suitability over tolerance to damp conditions.

4.2.4 Subsurface Compacted Layers

Loose soil layered atop compacted subsoil presents unique challenges. Often, roots struggle to penetrate compacted subsoil, preferring porous soil for development. In cases where compacted subsurface is exposed, and less than two feet of porous soil is available, opting for small to medium-sized trees is recommended. Due to their limited root systems, large-maturing trees may pose stability and safety risks.

4.2.5 Soil Depth and Distance to the Water Table

Planting trees near shallow bedrock or limited soil depth requires careful consideration. Mature trees in such conditions may experience storm-induced toppling due to their shallow root systems. However, certain tree species can adapt by sending roots into solution cavities within limestone for stability. To assess the site's suitability, dig several holes two to three feet deep and observe for any water accumulation. Trees can be planted once the cavity remains dry after two to four hours. Opt for trees capable of enduring damp conditions if water begins pooling in the cavity. When the distance to the water table is less than two feet, planting medium- to small-sized trees is advisable. However, exceptions like bluff cypress and tupelos may thrive even in such conditions, especially when clustered. Throughout the year, the water table's depth fluctuates due to factors like transpiration and seasonal changes. Therefore, it's crucial to consider these variations, as extreme weather events can also affect the water table depth.

5 Choose Desirable Tree Attributes

It's crucial to establish and uphold a variety of tree species within the community. While the selection of trees has focused mainly on their suitability for the site (Almas and Conway 2018), it's essential to consider additional factors beyond mere growth capability. Attributes such as function, size, form, and longevity play pivotal roles in determining the suitability of a species for planting.

5.1 Function

Healthy trees offer a multitude of benefits to our environment and well-being. They serve as shelters, oxygen generators, erosion regulators, and water resource protectors, contributors to bitumen durability, wildlife sanctuaries, and stabilizers of stream banks. Moreover, a tree's specific purpose can influence a range of its characteristics, such as size, shape, lifespan, canopy density, color, growth rate, and fruit attributes.

5.2 Mature Size

Optimal choices for street planting and covering vast open spaces are mature trees surpassing 50 feet in height, provided ample space above and below ground. Strategically placing medium to large trees can offer substantial shelter for structures, potentially reducing reliance on air conditioning. Nonetheless, it's crucial to acknowledge that larger trees are more vulnerable to damage and destruction. In urban settings with constrained soil space, planting smaller trees (around 30 feet at maturity) is often advised despite their limited sheltering capabilities.

5.3 Form

The form of a tree plays a crucial role in determining its maintenance requirements. Trees situated near pavements contribute significantly to urban landscapes. Conversely, larger or smaller upright trees can be shaped to extend their canopy over walkways or streets. Typically, pyramidal trees need less pruning to establish sturdy, wind-resistant branches than trees with different shapes (Sivarajah et al. 2018). Regular pruning is required within the first 25 years post-planting for trees with rounded, oval, or spreading canopies to ensure proper structure and clearance.

5.4 Longevity

In most landscape situations, enormous, long-lasting trees would seem to be the most logical selection for planting due to the enduring impact they would produce. Many of today's streets, residences, and buildings will likely be surpassed in age, however, if long-lasting trees are planted and maintained with reasonable diligence. Due to this rationale, the significance of tree longevity concerns may be diminished in densely populated areas unless specific measures are implemented to safeguard the tree.

5.5 Desirable Plant Species

The relationship between the location of trees and the relative stresses they experience has recently received attention in India. Among the stresses are a variety of pollutants, mechanical damage, extremes in ambient temperature, de-icing salt, limited root volume for crown development, poor quality growing substrates, and inadequate root access to oxygen and water. Nonetheless, the trees are frequently subjected to toxins and human and animal stress. Lastly, trees in urban woodlands may live long and experience relatively low stress levels. The stress in urban woodlands is primarily determined by long-distance air pollution, biotic damages, recreational activities that increase fire risk, and meteorological growth conditions. Urban areas can likely see some success using plant materials selected for urban forestry. If planners and practitioners are to expand the range of species used in cities successfully, then some key factors should be considered when selecting species and genotypes. For plantations in urban areas, such plants that require less water and are drought-resistant should be chosen. A list (Table 2) of some selected plants in the facility has been made, which can be helpful for plantation in urban areas and on the roadside.

6 Engaging the Community in the Tree-Planting Process

Community engagement is pivotal in urban tree-planting endeavors. Active involvement of residents not only garners support for environmental conservation (Vogt et al. 2015) but also fosters a sense of ownership and responsibility. Collaborative efforts with local organizations, hosting public events, and distributing educational materials elucidating the benefits of trees and proper maintenance can bolster community participation. Moreover, involving community members in selecting tree species and planting locations further enhances engagement and ensures alignment with local needs and preferences.

The societal benefits of tree planting extend far beyond environmental improvements. Trees play a transformative role, turning barren landscapes into vibrant, green spaces that serve as focal points for community interaction and cohesion. They provide gathering spots for various activities, from children's play to neighborhood gatherings, fostering social bonds and unity among residents. This communal aspect contributes significantly to social cohesion and enhances the overall quality of life in neighborhoods.

Furthermore, tree-planting initiatives address disparities in tree distribution, often termed "tree equity" or "tree justice" issues. Research indicates that lower-income communities in urban areas tend to have fewer trees, exacerbating environmental and health inequalities. Prioritizing tree planting in underserved communities may help to improve local environments and distribute vital benefits such as green spaces and cleaner air to those who need them most.

Table 2 Common list of diverse trees suitable for urban planting

S. no.	Family	Species name	Common name	Deciduous/ evergreen
1	Anacardiaceae	Lanneacoromandelica (Houtt.) Merr	Indian ash	Deciduous
2	Anacardiaceae	*Mangiferaindica* L	Mango	Evergreen
3	Annonaceae	*Polyalthialongifolia* (Sonner.) Thw	False ashoka	Evergreen
4	Apocynaceae	*Tabernaemontanadivaricata* (L.) R.Br. ex Roem. & Schult	Chandni/ pinwheel flower	Evergreen
5	Apocynaceae	*Thevetiaperuviana*L	Kaner	Evergreen
6	Apocynaceae	*Alstoniascholaris* (L.) R.Br	Scholar tree	Evergreen
7	Apocynaceae	*Plumeria alba* L	White frangipani	Deciduous
8	Apocynaceae	*Plumeriapudica*Jacq	Golden arrow	Deciduous
9	Arecaceae	*Caryotaurens* L	Fishtail palm	Evergreen
10	Arecaceae	*Borassusflabellifer* L	Palmyra palm	Evergreen
11	Bignoniaceae	*Jacaranda mimosifolia* D.Don	Blue jacaranda	Deciduous
12	Bignoniaceae	*Kigeliaafricana* (Lam.) Benth	Sausage tree	Evergreen
13	Bignoniaceae	*Spathodeacampanulata* P.Beauv	Squirt tree	Deciduous
14	Boraginaceae	*Cordiasebestena* L	Scarlet cordia	Deciduous
15	Combretaceae	*Terminaliaarjuna* (Roxb.) Wight & Arn	Arjun tree	Deciduous
16	Combretaceae	*Terminaliabellirica* (Gaertn.) Roxb	Baheda	Evergreen
17	Combretaceae	*Terminaliacatappa* L	Indian-almond	Deciduous
18	Euphorbiaceae	*Phyllanthusemblica* L	Indian gooseberry	Evergreen
19	Fabaceae	*Acacia nilotica* (L.) Willd. ex Del	Babul	Deciduous
20	Fabaceae	*Dalbergiasissoo* DC	Indian rosewood	Evergreen
21	Fabaceae	*Adenantherapavonina* L	Coral wood	Evergreen
22	Fabaceae	*Cassia fistula* L	Indian laburnum	Deciduous
23	Fabaceae	*Buteamonosperma* (Lam.) Taub	Flame of the forest	Deciduous
24	Fabaceae	*Hardwickiabinata*Roxb	Indian black wood	Deciduous
25	Fabaceae	*Saracaasoca* (Roxb.)Willd	Ashok tree	Evergreen
26	Fabaceae	*Sesbaniagrandiflora* (L.) Pers	Vegetable hummingbird	Deciduous
27	Fabaceae	*Albiziasaman* (Jacq.) Merr	Rain tree	Evergreen
28	Fabaceae	*Prosopisjuliflora* (Sw.) Dc	Mesquite	Evergreen
29	Fabaceae	*Acacia leucophloea* (Roxb.) Willd	White bark acacia	Deciduous

(continued)

Table 2 (continued)

S. no.	Family	Species name	Common name	Deciduous/ evergreen
30	Fabaceae	*Bauhinia acuminata* L	White orchid-tree	Evergreen
31	Fabaceae	*Bauhinia purpurea* L	Orchid tree	Deciduous
32	Fabaceae	*Dalbergialatifolia*Roxb	Indian rosewood	Evergreen
33	Fabaceae	*Pongamiapinnata* (L.)	Pierre Indian beech	Evergreen
34	Fabaceae	*Sennasiamea* (Lam.) H.S. Irwin & Barneby	Siamese cassia	Evergreen
35	Fabaceae	*Delonixregia* (Boj. ex Hook) Rafin	Royal poinciana	Deciduous
36	Fabaceae	*Tamarindusindica* L	Indian date	Evergreen
37	Fabaceae	*Parkinsoniaaculeata* L	Palo verde	Evergreen
38	Lythraceae	*Lagerstroemia indica* L	Bonnet flower	Evergreen
39	Lythraceae	*Lagerstroemia reginae*Roxb	Queen of flowers	Evergreen
40	Lythraceae	*Lagerstroemia speciosa* (L.) Pers	Pride of India	Evergreen
41	Magnoliaceae	*Magnolia champaca* (L.) Baill. ex Pierre	Champak	Evergreen
42	Malpighiaceae	*Malpighiaemarginata* DC	Acerola cherry	Evergreen
43	Malvaceae	*Hibiscus tiliaceus* L	Beach hibiscus	Deciduous
44	Malvaceae	*Bombaxceiba* L	Cotton tree	Deciduous
45	Malvaceae	*Adansoniadigitata* L	Baobab	Deciduous
46	Meliaceae	*Azadirachtaindica* A. Juss	Neem tree	Deciduous
47	Meliaceae	*Meliaazedarach* L	Chinaberry	Deciduous
48	Meliaceae	*Swieteniamahagoni* (L.) Jacq	Mahogany	Evergreen
49	Moraceae	*Ficusracemosa* L	Cluster fig	Evergreen
50	Moraceae	*Artocarpusheterophyllus* L	Jackfruit	Evergreen
51	Moraceae	*Ficusamplissima* J.E. Smith	Indian bat tree	Evergreen
52	Moraceae	*Ficusbenghalensis* L	Indian banyan	Evergreen
53	Moraceae	*Ficusreligiosa* L	Sacred fig	Evergreen
54	Moraceae	*Ficuselastica*Roxb. exHornem	Indian rubber tree	Evergreen
55	Myrtaceae	*Callistemon citrinus* (Curtis) Skeels	Bottlebrush	Evergreen
56	Myrtaceae	*Globuluslabill*	Eucalyptus	Evergreen
57	Myrtaceae	*Syzygiumcumini* (L.) Skeels	Jamun	Evergreen
58	Myrtaceae	*Psidiumguajava* L	Guava	Deciduous
59	Nyctaginaceae	*Pisoniagrandis* R. Br	Bird-catcher tree	Evergreen

(continued)

Table 2 (continued)

S. no.	Family	Species name	Common name	Deciduous/ evergreen
60	Nyctanthaceae	*Nyctanthesarbor-tristis* L	Night-flowering jasmine	Evergreen
61	Poaceae	*Bambusabambos* (L.) Voss	Indian thorny bamboo	Evergreen
62	Rhamnaceae	*Ziziphusjujuba* Mill	Indian jujube	Evergreen
63	Rubiaceae	*Morindapubescens* Sm	Morinda tree	Evergreen
64	Rubiaceae	*Hamelia patens* L	Firebush	Evergreen
65	Rutaceae	*Murrayakoenigii* (L.) Spreng	Curry tree	Evergreen
66	Rutaceae	*Aeglemarmelos* (L.) Correa	Bael	Evergreen
67	Sapotaceae	*Madhucalongifolia* (J. Koenig ex L.) J.F. Macbr	Mahua	Deciduous
68	Simaroubaceae	*Ailanthus excelsa*Roxb	Indian tree of heaven	Deciduous
69	Verbenaceae	*Tectonagrandis* L.f	Teak	Deciduous
70	Verbenaceae	*Durantaerecta* L. Golden Dewdrop Evergreen	Golden Dewdrop	Evergreen
71	Cupressaceae	*Taxodiumdistichum* L	Bald cypress	Deciduous
72	Magnoliaceae	*Magnolia virginiana* L	Sweetbay	Evergreen
73	Annonaceae	*Annonaglabra* L	Pond apple	Evergreen
74	Fagaceae	*Quercusrobur*L	Oak tree	Evergreen

7 Tree Canopy Cover Goals

Preserving the extant tree stock and establishing new trees are two essential components of urban forest management. The primary objectives of management practices are to enhance ecosystem services by increasing tree size, species diversity, and overall canopy cover (Fotis et al. 2018). Consequently, increasing the extent of the urban canopy has emerged as a prevalent objective among municipalities seeking to alleviate the impacts of urban expansion. Given that under 20% of the population resides in urban regions with inadequate canopy cover, establishing and attaining a minimum target of 20% would yield significant societal advantages. When viewed from above, urban canopy cover or tree canopy cover is the area or proportion of the ground covered by the healthy tree crown, including leaves, branches, and trunks. According to Walters and Sinnett (2021), urban canopy cover is widely regarded as a benefit or amenity for the environment because it is linked to numerous ecosystem services. The advantages above encompass regional cooling, mitigation of floods, reduction of air pollutants, and enhancements to local biodiversity and human welfare. Improved aesthetic value, pollution reduction, and social cohesion have all been linked to urban canopy cover. The assessable metric of canopy cover is employed to quantify the overall percentage of tree canopy cover that encompasses a spatially delineated urban region. To expedite the process of augmenting canopy cover, some cities in India have established ambitious targets for canopy cover and integrated them into strategies that advocate for sustainability and livability. Numerous canopy cover objectives are relatively long-term; therefore, it is critical to comprehend the optimal planting schedule and tree stock size to maximize returns on investment and ensure feasibility.

8 Post-planting Care and Encouraging Retention of Existing Trees

Proper maintenance is essential for the health and endurance of trees planted in urban areas (Ordonez 2015). Regular irrigation is critical for supplying trees with adequate moisture, particularly during high temperatures and arid conditions. It is essential to prune to eliminate damaged or diseased branches and encourage healthy growth. Resolving problems immediately is critical to prevent their spread and subsequent tree injury. This entails the surveillance of various indicators, including thermal measurements, property values, air pollution reduction, tree health and biodiversity enhancement, and increased community engagement (Nagendra et al. 2013). You can determine the financial and ecological benefits that these trees provide to the community by monitoring their development.

Site modifications may facilitate the accommodation of a greater diversity of tree species. Site modifications, including but not limited to relocating cables or street lamps, grading, enhancing drainage, and applying soil amendments to expansive regions, can significantly influence soil conditions before planting. These following modifications can subsequently impact tree growth and species selection:

8.1 Moving Lights and Wires

To accommodate trees, overhead power lines and street lighting may be relocated or altered. While this behavior is not customary, it is unexpectedly frequent when it occurs in response to a suggestion. This approach frequently serves as a more enduring resolution to a design dilemma, permitting planting trees along a street in a prohibited area without necessitating the relocation or alteration of fixtures.

8.2 Changing Soil pH

Planting trees that have adapted to the soil's current pH is preferable to changing the pH of the soil. Introducing sulfur or limestone into soil typically results in a transient alteration of pH levels. Consistent application is required to modify the pH levels. It is preferable to replace the soil or plant species that are tolerant of the pH of the site.

8.3 Improving Drainage and Reducing Runoff

Water dripping from a site can transport soil, fertilizers, and pesticides, contributing to environmental degradation (Urban 2008). Often, soil grading is performed to optimize water retention on-site. Redirected water must remain on-site to prevent discharge and sedimentation rather than being directed into streams or streets. In cases where drainage is adequate, engineered solutions such as curbing can be employed to retain water in roadways and linear areas. For optimal root development in compacted areas, it's recommended to dig four or more trenches from the planting pit using a trencher, followed by backfilling with the excavated soil. This technique creates channels that facilitate root growth, marginally improving drainage, enhancing water percolation, and reducing discharge.

8.4 Other Soil Improvements

Certain techniques of soil modification can enhance the conditions of the soil to facilitate root development. These include replacing soil, adding fill soil, and applying inorganic or organic matter widely. Trees do not benefit from amending the tiny backfill soil adjacent to the root ball.

8.5 Evaluate Maintenance Practices

It's crucial to carefully consider the management of the site post-sowing when choosing trees. Maintenance tasks such as irrigation, pruning, fertilization, and insect control play a significant role in a tree's ability to thrive. For example, suppose it's anticipated that trees won't need frequent pruning once they're established. In that case, preference may be given to trees with inherently strong structures, like those with an excurrent growth habit.

Table 3 Irrigation schedules depend on the size of nursery stock and desired objective; establishment takes approximately 3–5 months per inch of trunk caliper

Size of Nursery stock	Irrigation schedule for vigor	Irrigation schedule for survival
< 2 inch caliper	Daily: 2 weeks Alternate day: 2 months Weekly: until established	Twice weekly for 2–3 months
2–4 inch caliper	Daily: 1 month Alternate day: 3 month Weekly: until established	Twice weekly for 3–4 months
> 4-inch caliper	Daily: 6 weeks Alternate day: 5 months Weekly: until established	Twice weekly for 4–5 months

8.6 Irrigation

The capability of delivering irrigation determines nursery stock sizes and species most suitable for a given site. When trees can be consistently irrigated only once established, selecting drought-tolerant varieties and sowing nursery stock of any magnitude is advisable. Trees should receive consistent irrigation from establishment to maturity or be planted in their native range and soil type to ensure successful cultivation. This approach enables the cultivation of any tree, regardless of its tolerance to arid conditions. If irrigation isn't feasible during specific periods outlined in Table 3, consider using smaller nursery stock for planting.

Although rainwater management is a secondary benefit of planting trees, capturing rainwater by large trees is an essential function that should not come at the expense of creating aesthetically pleasing and functional urban spaces.

8.7 Pruning

Regular pruning is necessary to preserve the health and longevity of trees. Trees that are pruned effectively are more resistant to hurricane-force gusts. A proficient urban forestry program allocates a specific portion of the budget for this essential pruning. Consider planting species that require only moderate pruning to develop and sustain excellent structure if this is the case. This is a summary! If pruning is to occur infrequently or not at all, or if the timing and method of pruning are unknown, then avoid planting mature trees near structures (such as streetlights) that could potentially obstruct tree growth.

8.8 *Fertilization*

Most micronutrient deficiencies in alkaline soils are attributable to fertilization issues. If a historically significant or other exceptional circumstance necessitates planting a tree that cannot tolerate alkaline soil (e.g., queen palm) in such soil, be ready to implement a routine monitoring and treatment regimen aimed at averting micronutrient deficiencies.

8.9 *Cleanup*

Under the canopy, trees bearing fruit that is excessively succulent (fig, seagrape, cocoplum, or queen palm), large (Royal poinciana), or hard (Mahogany) may cause an annoyance or hazard on the sidewalks and pavement. The fruit may cause pedestrians to trip and fall and be unappealing. In areas with heavy pedestrian traffic, if funds for cleanup are limited, consider planting trees devoid of this type of unsightly fruit, such as the fruitless 'Rotundiloba' cultivar of sweet gum.

9 Urban Forest Diversity

Urban forest diversity is described as "the complexity of tree species composition, the size distribution of trees, and the patterns of distribution within an urban forest" by Professor Emeritus Joe McBride of the University of California–Berkeley. In light of the ongoing expansion of urban areas and the challenges posed by climate change, tree diversification has become an increasingly vital concept and practice. Diversity in natural forests comprises both species diversity and forest structure. As illustrated here, one example is the vertical stratification of a forest and the life that each stratum sustains (Singh 2021; Sharma and Singh 2021; Singh et al. 2021). Certain animal species exhibit specific preferences for particular trees, as exemplified by the phenomenon of association; in contrast, others may rely on the dimensions or age of trees to fulfill their requirements (Aronson et al. 2014). Temperature, space, light conditions, forest density, and extent determine certain animal niches. For instance, a forest adjacent to a meadow offers a variety of wildlife habitats, and certain species' ability to migrate from one to the other is crucial. Undeniably, climate also exerts an influence. Habitat modifications disrupt equilibrium and may engender novel and occasionally undesirable associations, whether the result of natural occurrences or anthropogenic activities (Dhyani et al. 2021; Kumar et al. 2021e; Prakash et al. 2022).

9.1 *Reasons for Lack of Urban Diversity*

Dr. Joe McBride has identified several factors that contribute to the lack of tree diversity in communities:

- Regional circumstances include the arid climate of the harshest hardiness zones, exceedingly high temperatures, and a shortage of species diversity within the local biosphere.
- Past inclinations towards specific introduced tree species, sometimes dating back to pioneer days.
- The expertise or absence of local arborists or others entrusted with tree planting and maintenance responsibility.
- Psychological and spiritual elements, including formative encounters with trees during childhood or religious symbolism.
- The availability of nursery stock for sowing.

Insufficient urban tree diversity exposes a community to the peril of substantial tree loss in the event of an unforeseen catastrophe and hinders the optimization of ecosystem services. As a result, to increase the resilience of the urban forest, the selection of trees should consistently strive for a high degree of diversity, achieved through the selection of species suitable for each site. Enhanced biodiversity in urban forests contributes to their capacity to mitigate pest and disease threats, bolster resistance to future climate conditions, and consequently augment trees' long-term benefits (Kondo et al. 2018). As an illustration, an urban forest with a varied age structure can serve as a protective barrier against stressors that affect trees at distinct life phases, thereby diminishing the probability of substantial tree canopy depletion at the local level and enabling administrators to distribute maintenance expenses consistently over an extended period.

10 Conclusion

Planting trees undoubtedly yields numerous social and economic advantages. Beyond enhancing environmental health, trees foster community cohesion and drive economic growth. By recognizing these benefits, communities can make informed decisions about investing in green initiatives. Responsibility for planting and maintaining trees lies with individuals, organizations, and governments, ensuring a sustainable future for future generations. Every effort to create a greener, healthier, and more prosperous world is worthwhile, whether it involves reforesting deforested regions, planting trees in urban areas, or prioritizing underserved communities. Together, we can leverage the power of trees to shape a better, more socially dynamic, economically stable, and environmentally sustainable future. Urban areas, characterized by extensive impervious cover, increased pollution, and higher population densities, face heightened susceptibility to climate change. These factors, compounded by a concentration of built structures, amplify the impacts of urban heat, drought, and extreme weather events. As a result, urban residents are particularly vulnerable to the physical and mental health effects of climate stressors. This chapter equips professionals at the intersection of urban forestry, climate, and public health with resources to support climate adaptation planning and activities. It offers insights and strategies for optimizing urban forestry projects to enhance climate resilience and human health outcomes.

References

Almas AD, Conway TM (2018) Resident attitudes and actions toward native tree species: a case study of residents in four Southern Ontario municipalities. Arboricult Urban Forest 44(2):101–115

Arnold HF (1980) Trees in urban design. Van Nostrand Reinhold Company, New York

Arnold CL, Gibbons CJ (1996) Impervious surface coverage—the emergence of a key environmental indicator. J Am Plan Assoc 62:243–258

Aronson M, Sorte F, Nilon C, Katti M, Goddard M (2014) A global analysis of the impacts of urbanization on bird and plant diversity reveals key anthropogenic drivers. Proc Soc B: Biol Sci 281(1780):20133330

Bowler DE, Buyung-Ali L, Knight TM, Pullin AS (2010) Urban greening to cool towns and cities: a systematic review of the empirical evidence. Landsc Urban Plan 97:147–155

Dhyani K, Kuniyal HB, Singh H, Sobha (2021) Growth and physiological potential of *Terminalia arjuna* under elevated CO_2 levels in open top chamber condition. J Appl Nat Sci 13(3):1121–1126. https://doi.org/10.31018/jans.v13i3.2490. ISSN 2231-5209

Duany A, Plater-Zyberk E, Falk B (2014) Center for applied transect studies. Retrieved from http://www.transect.org/index.htmlUrban

Fotis A, Morin T, Fahey R, Hardiman B, Bohrer G, Curtis P (2018) Forest structure in space and time: biotic and abiotic determinants of canopy complexity and their effects on net primary productivity. Agric For Meteorol 250:181–191

Fowler D (2011) Achieving the goal of 25% canopy coverage in phoenix by 2030 (Unpublished Master of Landscape Architecture Thesis). Arizona State University, Tempe Arizona

Gupta SK, Singh H (2017) Observations on sporadic flowering in *Ficus benjamina* L. and *Peltophorum africanum* L. in New Forest. Indian Forest 143:290–291

Gupta SK, Ram J, Singh H (2018a) Comparative study of transpiration in cooling effect of tree species in the atmosphere. J Geosci Environ Protect 6:151–166. https://doi.org/10.4236/gep.2018.68011

Gupta SK, Ram J, Singh H, Kumar P (2018b) Comparative study of cooling under shades of some forestry tree species with respect to ambient temperatures. J Geosci Environ Protect 6:51–66. https://doi.org/10.4236/gep.2018.611004

Gupta SK, Ram J, Singh H (2019a) Relationship between leaf area index and atmospheric cooling of tree species. Indian J For 42(3):203–208

Gupta SK, Singh H, Ram J (2019b) Need for a novel paradigm in forestry research: perspectives of comparative atmospheric cooling by different forestry effect tree species. Indian Forest 145(6):509–515

Hall L, Vonderscher K, Adkins R (2010) City of phoenix tree and shade master plan. Retrieved from http://phoenix.gov/parks/shade.html

Jangra A, Kumar K, Maikhuri S, Bhandari MS, Pandey S, Singh H, Barthwal S (2024) Unveiling stress-adapted endophytic bacteria: Characterizing plant growth-promoting traits and assessing cross-inoculation effects on *Populus deltoides* under abiotic stress. Plant Physiol Biochem 210:108610. https://doi.org/10.1016/j.plaphy.2024.108610

Kanta C, Kumar A, Chauhan A, Singh H, Sharma IP (2024) The interplay between plant functional traits and climate change. In: Kumar N, Singh H (eds) Plant functional traits for improving poductivity. Springer, Singapore. https://doi.org/10.1007/978-981-97-1510-7_3

Kaspar J, Kendal D, Sore R, Livesley S (2017) Random point sampling to detect gain and loss in tree canopy cover in response to urban densification. Urban Forest Urban Green 24:26–34

Kondo MC, Fluehr JM, McKeon T, Branas CC (2018) Urban green space and its impact on human health. Int J Environ Res Public Heal 15:445

Korkmaz M, Fakir H, Alkan H (2018) Effects of nature training projects on environmental perception and attitudes. Appl Ecol Environ Res 16(1):359–369. https://doi.org/10.15666/aeer/1601_359369

Kumar N, Singh H (2024) Plant functional traits for improving productivity. Springer Nature, Singapore. https://doi.org/10.1007/978-981-97-1510-7

Kumar N, Suyal DC, Sharma IP, Verma A, Singh H (2017) Elucidating stress proteins in rice (*Oryza sativa* L.) genotype under elevated temperature: a proteomic approach to understand heat stress response. 3 Biotech 7:205. https://doi.org/10.1007/s13205-017-0856-9

Kumar M, Singh MP, Singh H, Dhakate PM, Ravindranath NH (2019a) Forest working plan for the sustainable management of forest and biodiversity in India. J Sustain Forest 39:1–22. https://doi.org/10.1080/10549811.2019.1632212

Kumar N, Jeena N, Singh H (2019b) Elevated temperature modulates rice pollen structure: a study from the foothill of the Himalayan agro-ecosystem in India. 3 Biotech 9(5):175. https://doi.org/10.1007/s13205-019-1700-1

Kumar N, Jeena N, Kumar N, Gangola S, Singh H (2019c) Phytoremediation facilitating enzymes: an enzymatic approach for enhancing remediation process. In: Bhatt P (ed) Smart bioremediation technologies. Academic Press, Elsevier, 50 Hampshire Street, 5th Floor, Cambridge, MA, 02139, United States, pp 289–30

Kumar N, Singh H, Sharma SK (2020) Antioxidants: responses and importance in plant defense system. In Roychowdhury R, Choudhury S, Hasanuzzaman M, Srivastava S (eds) Sustainable agriculture in the era of climate change. Springer, Cham, pp 251–264. https://doi.org/10.1007/978-3-030-45669-6_11

Kumar A, Kumar P, Singh H, Kumar N (2021a) Modulation of plant functional traits under essential plant nutrients during seasonal regime in natural forests of Garhwal Himalayas. Plant Soil 465:197–212. https://doi.org/10.1007/s11104-021-05003-x

Kumar A, Kumar P, Singh H, Bisht S, Kumar N (2021b) Relationship of physiological plant functional traits with soil carbon stock in the temperate forest of Garhwal Himalaya. Curr Sci 120(8):1368–1373. https://doi.org/10.18520/cs/v120/i8/1368-1373

Kumar A, Kumar P, Singh H, Kumar N (2021c) Adaptation and mitigation potential of roadside trees with bio-extraction of heavy metals under vehicular emissions and their impact on physiological traits during seasonal regimes. Urban Forest Urban Green 58:126900. https://doi.org/10.1016/j.ufug.2020.126900

Kumar A, Kumar P, Singh H, Kumar N (2021d) Impact of plant functional traits on infiltration rate under different forests of Kempty watershed in Garhwal Himalaya. Indian J Soil Conserv 49(1):38–44

Kumar M, Phukon SN, Paygude AC, Tyagi K, Singh H (2021e) Mapping phenological functional types (PhFT) in the Indian Eastern Himalayas using machine learning algorithm in Google Earth engine. Comp Geosci 158. https://doi.org/10.1016/j.cageo.2021.104982

Malik A, Kumar K, Kumari G, Kumar N, Singh H (2024) Application of functional traits in modelling productivity and resilience under climate change. In: Kumar N, Singh H (eds) Plant functional traits for improving productivity. Springer, Singapore. https://doi.org/10.1007/978-981-97-1510-7_5

Meena D, Singh H, Chaudhari SK (2011) Elucidating strontium response on growth dynamics and biochemical change in *Phaseolus mungo* L. Int J Agricult Environ Biotech 4:107–113

Moll CW (1989) The state of our urban forest. Am Forest 95:61–64
Nagendra H, Gopal D (2010) Street trees in Bangalore—density, diversity, composition and distribution. Urban Forest Urban Green 9:129–137
Nagendra H, Lucas R, Honrado JP, Jongman RHG, Tarantino C, Adamo M, Mairota P (2013) Remote sensing for conservation monitoring: assessing protected areas, habitat extent, habitat condition, species diversity and threats. Ecol Ind 33:45–59
Neumann M, Mues V, Moreno A, Hasenauer H, Seidl R (2017) Climate variability drives recent tree mortality in Europe. Glob Change Biol 23(11):4788–4797
Nowak D (2000) The interactions between urban forests and global climate change. Retrieved 22 November 2019 from. https://www.researchgate.net/publication/288254909_The_interactions_between_urban_forests_and_global_climate_change
Nowak DJ, Hoehn RE, Bodine AR, Greenfield EJ, O'Neil-Dunne J (2013) Urban forest structure, ecosystem services and change in Syracuse, NY. Urban Ecosyst, 1–23
Nowak D, Hirabayashi S, Doyle M, McGovern M, Pasher J (2018) Air pollution removal by urban forests in Canada and its effect on air quality and human health. Urban Forest Urban Green 29:40–48
Ordonez C (2015) Adopting public values and climate change adaptation strategies in urban forest management: a review and analysis of the relevant literature. J Environ Manage 164:215–221
Prakash V, Hunney K, Singh H (2022) Effect of elevated carbon dioxide on growth and development of *Santalum album* L. seedlings inoculated with plant growth promoting microorganisms in Open Top Chambers. Int J Environ, Agricult Biotech 7(5):149–159
Sharma R, Singh H (2021) Alteration in biochemical constituents and nutrients partitioning of *Asparagus racemosus* in response to elevated atmospheric CO_2 concentration. Environ Sci Pollut Res 29:6812–6821. https://doi.org/10.1007/s11356-021-16050-3
Sharma R, Singh H, Kaushik M, Nautiyal R, Singh O (2018) Adaptive physiological response, carbon partitioning, and biomass production of *Withania somnifera* (L.) Dunal grown under elevated CO_2 regimes. 3 Biotech 8:267. https://doi.org/10.1007/s13205-018-1292-1
Singh H (2021) An integrated approach considering physiological and biophysical based indicators for assessing tolerance of roadside plantations of *Alstonia scholaris* towards urban roadside air pollution: an assessment of adaptation of plantations for mitigating urban roadside air pollution. Trees. https://doi.org/10.1007/s00468-021-02179-8
Singh H, Verma A (2013) Characterization and screening of high nitrogen efficient rice genotype to curtail environmental pollution. Int J Agricult, Environ Biotech 6(1):23–30
Singh H, Kumar M (2022) Climate change and its impact on Indian Himalayan Forests: current status and research needs. In: Rani S, Kumar R (eds) Climate change. Springer Climate. Springer, Cham, pp 223–242. https://doi.org/10.1007/978-3-030-92782-0_11
Singh H, Sharma RS, Sinha S, Kumar M, Kumar P, Verma A, Sharma SK (2017a) Physiological functioning of *Lagerstroemia speciosa* L. under heavy roadside traffic: an approach to screen potential species for abatement of urban air pollution. 3 Biotech, 7(61):1–10. https://doi.org/10.1007/s13205-017-0690-0
Singh H, Verma A, Kumar M, Sharma R, Gupta R, Kaur M, Negi M, Sharma SK (2017b) Phytoremediation: a green technology to clean up the sites with low and moderate level of heavy metals. Austin Biochem 2:1012
Singh H, Sharma R, Savita SMP, Kumar M, Verma A, Ansari MW, Sharma SK (2018) Adaptive physiological response of *Parthenium hysterophorus* to elevated atmospheric CO_2 concentration. Indian Forest 144:1–14
Singh H, Yadav M, Kumar N, Kumar A, Kumar M (2020) Assessing adaptation and mitigation potential of roadside trees under the influence of vehicular emissions: a case study of *Grevillea robusta* and *Mangifera indica* planted in an urban city of India. PLoS One 15(1):e0227380. https://doi.org/10.1371/journal.pone.0227380
Singh S, Singh H, Sharma SK, Nautiyal R (2021) Seasonal variation in biochemical responses of bamboo clones in the sub-tropical climate of Indian Himalayan foothills. Heliyon 7(4):e06859. https://doi.org/10.1016/j.heliyon.2021.e06859

Singh H, Kumar N, Kumar A (2024) Enhancing resource use efficiency in crops through plant functional traits. In: Kumar N, Singh H (eds) Plant functional traits for improving productivity. Springer, Singapore. https://doi.org/10.1007/978-981-97-1510-7_6

Sivarajah S, Smith SM, Thomas SC (2018) Tree cover and species composition effects on academic performance of primary school students. PLoS One 13(2):e0193254. https://doi.org/10.1371/journal.pone.0193254

Stephanie P, Thomas G, Diane EP, Sassan S, Jean-Daniel S (2013) Urban tree planting programs, function or fashion? Los Angeles and urban tree planting campaigns. Geo J 78:475–493

The International Plant Names Index (2012) Published on the internet. http://www.ipni.org [accessed 18 October 2018]

The Plant List (2018) Version 1.1. Published on the Internet; http://www.theplantlist.org/ (accessed 18 October)

Urban J (2008) Up by the roots, healthy soils and trees in the built environment. Internal Soc Arboricult, Champaign, IL

Vogt J, Watkins S, Mincey S, Patterson M, Fischer B (2015) Explaining planted-tree survival and growth in urban neighborhoods: a social–ecological approach to studying recently-planted trees in Indianapolis. Landsc Urban Plan 136:130–143

Walters M, Sinnett D (2021) Achieving tree canopy cover targets: a case study of Bristol, UK. Urban Forest Urban Green 65(2021):127296. https://doi.org/10.1016/j.ufug.2021.127296

Wolf KL, Lam ST, McKeen JK, Richardson GRA, Van den Bosch M, Bardekjian AC (2020) Urban trees and human health: a scoping review. Int J Environ Res Public Health 17:4371

Yadav SK, Singh H, Ginwal HS, Barthwal S (2019a) Elevated CO_2 enhanced growth and physiological process of *Populus deltoides* Bartr. ex Marsh. Indian Forest 145(1):23–27

Yadav SK, Singh H, Nautiyal R, Ginwal HS, Ansari SA, Barthwal S (2019b) Modulation of morpho-physiological responses in *Populus deltoides* by elevated carbon dioxide and temperature. Forest Sci 66:105–118. https://doi.org/10.1093/forsci/fxz048

Ecosystem Services Provided by Urban and Peri-Urban Forests

Rupali Singh, S. B. Chavan, V. D. Kakade, A. S. Morade, Alok Kumar Singh, G. B. Rawale, A. R. Uthappa, A. Keerthika, A. R. Chichaghare, Shubham Gurav, and K. Sammi Reddy

Abstract Urban and peri-urban forests (UPFs) are critical in addressing climate change challenges and fostering urban sustainability. The world faces escalating climate threats, with human activities causing historic CO_2 emissions and rising atmospheric CO_2 concentration. Tree planting and reforestation are crucial to limit global warming to 1.5 °C, aligning with global efforts. As urbanization surges, UPFs emerge as essential components for climate change mitigation and adaptation, often called "city lungs. Recognized in Sustainable Development Goal-11", UPFs offer diverse ecosystem services, contributing to urban quality of life. However, these forests face threats from unregulated urban development and inadequate management, often undervalued by decision-makers. Urban forests offer numerous benefits, addressing various urban issues such as food security, poverty, pollution, and climate change. They are pivotal in advancing Sustainable Development Goals (SDGs), impacting nine specific SDGs. Understanding urban ecosystem services (UES) becomes imperative as more than half the global population resides in cities. In this chapter, the concept of ecosystem services is divided into four categories: provisioning, regulating, supporting, and cultural. This paper focuses on provisioning and regulating services provided by urban trees, detailing their role in food security, pollution mitigation, and climate regulation. Specifically, it explores the regulating services of heat mitigation, dust retention, carbon sequestration, and noise reduction, emphasizing the impact of tree species, environmental factors, and their role

R. Singh
ICFRE-Forest Research Institute, P. O. New Forest, Dehradun 248006, Uttarakhand, India

R. Singh · S. B. Chavan (✉) · V. D. Kakade · A. S. Morade · G. B. Rawale · S. Gurav · K. S. Reddy
ICAR-National Institute of Abiotic Stress Management, Baramati 413115, Maharashtra, India
e-mail: sangramc8@gmail.com

A. K. Singh
Rajiv Gandhi South Campus, Banaras Hindu University, Mirzapur 231001, India

A. R. Uthappa
ICAR-Central Coastal Agricultural Research Institute, Ela, Old Goa, India

A. Keerthika · A. R. Chichaghare
ICAR-Central Arid Zone Research Institute, Jodhpur, Rajasthan, India

H. Singh (ed.), *Urban Forests, Climate Change and Environmental Pollution*,
https://doi.org/10.1007/978-3-031-67837-0_20

in enhancing the urban quality of life. Recognizing and incorporating their diverse ecosystem services into comprehensive climate change mitigation strategies is crucial for building healthier, more resilient cities.

Keywords Urban forestry · Sustainability · Urban heat islands · Ecosystem services · Carbon sequestration · Climate change

1 Introduction

Urban and peri-urban forests (UPFs) are crucial in addressing the pressing challenges of climate change and enhancing urban sustainability. In a world where climate change threatens food and water security, with billions of people living in climate-vulnerable areas, it is evident that human activities, driven primarily by greenhouse gas emissions, have propelled global warming. Historical net carbon dioxide (CO_2) emissions from 1850 to 2019 have reached alarming levels, resulting in a rise in atmospheric CO_2 concentration to 421 parts per million (ppm), with projections indicating further increases to 700–900 ppm due to industrialization (Kumar et al. 2022; Singh 2021; Kumar et al. 2024; Singh et al. 2024a, b; Stallinga 2023). The consequences of this warming, including a one-degree Celsius increase since pre-industrial times, are being felt worldwide, with countries like India ranking among the top emitters of greenhouse gases (GHGs). India's commitment to mitigating climate change is reflected in its participation in global initiatives like the Nationally Determined Contributions (NDCs) under the United Nations Framework on Climate Change (UNFCC). However, the escalating emissions have far-reaching implications for land use systems, accounting for significant portions of GHG emissions and hindering efforts to achieve the targets set by the Paris Agreement (IPCC).

To combat the adverse effects of climate change, limiting global warming to 1.5 °C is imperative, necessitating a 45% reduction in emissions by 2030 and achieving net-zero emissions by 2050 (Olhoff and Christensen 2020). The urgency is further underscored by the rising heat-related mortality rates in India, with a 55% increase in heat-related deaths from 2017 to 2021 compared to the early 2000s (D'Souza et al. 2022). One contributing factor to this issue is the phenomenon of urban heat islands, where metropolitan areas experience higher temperatures than their suburban counterparts, resulting in increased energy consumption and air pollution (Varol 2022). In response to this crisis, tree planting and reforestation have gained prominence as crucial strategies aligned with global initiatives such as the Bonn Challenge, Trillion Trees vision, and the World Economic Forum's 1t.org project, which aims to remove up to 10 gigatonnes of CO_2 equivalent by 2050 (Artz 2021). However, as urbanization continues to surge, with over half the global population residing in cities and urban areas, urban forests have become vital components of climate change mitigation and adaptation. They are often called "city lungs" and function as carbon reservoirs, absorbing and storing CO_2 while providing multiple benefits to urban dwellers (K'A Kumu and Olima 2007).

The importance of urban and peri-urban forests (UPFs) has gained recognition from international entities, and science-policy initiatives have been integrated into the global sustainability framework. This acknowledgment is particularly evident within the United Nations General Assembly's Sustainable Development Goal-11, which focuses on creating inclusive, safe, resilient, and sustainable cities and human settlements. Two key developments have driven the growing attention to UPFs in science and policy agendas. Firstly, UPFs offer various ecosystem services to urban inhabitants, including climate regulation, air purification, flood control, food provision, energy and water supply, disaster risk reduction, and cultural values. These services are essential for enhancing urban quality of life and addressing climate-related challenges (Jim and Chen 2009; Dou et al. 2017).

In addition to their environmental services, urban forests significantly impact human well-being by providing shade, air purification, and aesthetically pleasing landscapes that enhance the overall quality of life (Alizadeh and Hitchmough 2020). In the face of mounting climate concerns, urban forests emerge as a powerful strategy for enhancing city sustainability and resilience. However, their integration should be part of a comprehensive climate change mitigation approach, encompassing emission reduction, renewable energy adoption, and climate adaptation (Dhyani et al. 2021; Kumar et al. 2021a, b, c, d, e, f; Prakash et al. 2022). Collectively, the diverse benefits provided by urban forests contribute to healthier and more sustainable cities, highlighting the harmonious coexistence of urban environments and the natural world.

Forests within and around cities face threats from unregulated urban development, limited investment, and inadequate management. Despite their potential to create a healthier environment, urban forests are often undervalued, primarily appreciated for aesthetics rather than their vital ecosystem functions (Singh 2021; Sharma and Singh 2021; Singh et al. 2021). Many urban decision-makers, including mayors and planners, remain unaware of urban forests' substantial economic, social, and environmental benefits. Consequently, these green spaces are often deprioritized, with budget allocations favoring health, welfare, and safety areas (Singh et al. 2020; Kumar et al. 2021a, b, c, d). The untapped potential of urban forests to enhance the quality of life for urban and peri-urban residents remains a missed opportunity. Here are some potential benefits from urban forests (Table 1).

Urban forests are pivotal in advancing sustainable development goals (SDGs), with nine specific SDGs substantially impacted by their presence. These green spaces address various urban challenges while enhancing environmental quality, social well-being, and economic prosperity (Fig. 1).

Table 1 Potential benefits of urban forests

Urban issue	Potential benefits of urban forests
Food security	Provide food, clean water, and wood fuel
Urban poverty	Create jobs and increase income
Soil and landscape degradation	Improve soil conditions and prevent erosion
Reduced biodiversity	Preserve and increase biodiversity
Air and noise pollution	Remove air pollutants and buffer noise
Greenhouse gas emissions	Sequester carbon and mitigate climate change, improve local climate, and build resilience
Extreme weather events	Mitigate local climate and build resilience
Energy shortage	Save energy through shading/cooling and grow wood fuel
Heat island effect	Cool the built environment through shade and evapotranspiration
Limited accessible green space	Provide more accessible natural green spaces
Public health	Improve the physical and mental health of residents
Flooding	Mitigate stormwater runoff and reduce flooding
Limited recreational opportunities	Provide opportunities for recreation and environmental education
Exposure	Provide shelter
Limited water resources	Enable infiltration and the reuse of wastewater
Lack of community and social cohesion	Provide distinctive places for formal and informal outdoor interaction

Source Salbitano et al. (2016)

2 Classifying Urban Ecosystem Services

The concept of ecosystem services was introduced to delineate the invaluable advantages natural ecosystems bestow upon society. This term was instrumental in increasing consciousness regarding the importance of biodiversity and the preservation of ecosystems (Birkhofer et al. 2015; Yadav et al. 2019a, b; Gupta et al. 2019a; Kumar et al. 2019a, b). Cities confront multifaceted challenges in our contemporary world, where a significant majority of the global population resides in urban areas, exceeding the 50% mark. These include urban heat island effects, food security concerns, flood risks, and pollution issues. The concept of ecosystem services has become intrinsically linked to the sustainability and liveability of urban environments (Russo and Cirella 2019).

For a long time, cities have been relatively understudied ecosystems in this context (Stott et al. 2015), leading to a notable knowledge gap. Consequently, there is a mounting interest in urban ecology and the notion of urban ecosystem services (UES), which offer innovative solutions to the aforementioned urban challenges (Russo

Fig. 1 Contribution of urban forests to SDGs

and Cirella 2021; Stott et al. 2015). Research and understanding of these "novel" ecosystems are imperative to mitigate the impacts of a burgeoning urban population and the ongoing rural-to-urban transition, ensuring the well-being and resilience of urban communities (Russo and Cirella 2021).

A growing corpus of literature has helped us comprehend the biophysical, economic, and sociocultural aspects of urban ecosystem services in recent years. The Economics of Ecosystems and Biodiversity (TEEB 2011) and the Millennium Ecosystem Assessment (2005) addressed urban ecosystem services and have also drawn more attention in the policy discussion on ecological infrastructure. In contrast to other ecosystems like wetlands or forests, urban ecosystems have received comparatively little attention in the literature on ecosystem services, even though more than half of the world's population now lives in cities. The purpose of this section is to categorize and describe the ecosystem services offered in urban areas and how they may improve the quality of life in these regions.

Ecosystem services were divided into four main categories (Fig. 2) by The Economics of Ecosystem Services and Biodiversity (TEEB 2010).

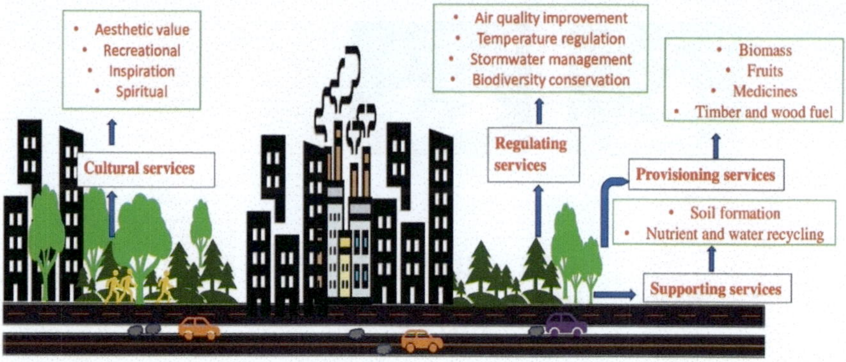

Fig. 2 Diagrammatic view of ecosystem services from Urban Forests. Modified and adapted from Abdel (2021)

3 Provisioning Service

Ecosystem services are vital in meeting fundamental human needs and sustaining environmental health. They directly contribute to food security by supplying crops, livestock, fisheries, and edible fruits and nuts from forests (MEA 2005). Additionally, ecosystems regulate water flow and quality, acting as natural reservoirs and purifying water in wetlands (Brauman et al. 2007). Furthermore, ecosystems provide medicinal resources, such as quinine from the cinchona tree, which is historically significant in traditional medicines (CBD 2004). They also offer fuel, construction materials, genetic resources for breeding programs, and raw materials for pharmaceuticals and biochemicals (TEEB 2010).

Research on provisioning ecosystem services from urban ecosystems, particularly within the framework of green infrastructure, has gained prominence. Notably, there's a growing acknowledgment of the importance of domesticated fruit trees in urban areas. For instance, studies in Berlin underscore the historical legacy and low risk associated with consuming fruit grown in urban settings (Larondelle and Strohbach 2016; von Hoffen and Säumel 2014). Various countries have initiated urban food forestry projects, focusing on publicly owned greenspaces planted with apple trees to bolster food security (Clark and Nicholas 2013). Nonetheless, there remain unanswered questions concerning the complete range of ecosystem services provided by urban forests, their significance to different city residents, and their impact on residents' practices (Haase et al. 2014).

Urban environments are recognized as crucial sources of provisioning ecosystem services across diverse cultural contexts. For instance, Maori elders in New Zealand gather culturally important medicines in urban areas to safeguard conservation areas (Wehi and Wehi 2010). In South Africa, wild tree fruits and herbaceous species in peri-urban townships uphold cultural heritage and supply tangible goods like food, medicines, fodder, and fuel wood (Shackleton et al. 2011). Similarly, foraging supports cultural practices for marginalized communities in rapidly urbanizing cities like Bangalore, India (D'Souza and Nagendra 2011; Unnikrishnan and Nagendra 2015). In the United States, foragers harvest various items from diverse urban ecosystems for food, medicine, crafting, and other purposes (McLain et al. 2012; Poe et al. 2013, 2014; Hurley et al. 2015). However, it's worth noting that foraging from ornamental plantings represents an unforeseen and unauthorized use. Semeraro et al. (2021) emphasize the importance of ensuring the sustainability of these ecosystem services for both human well-being and environmental health. Numerous studies have explored the provisioning of ecosystem services from various ecosystems, often employing comparative analyses across different subtypes. Table 2 summarizes studies conducted worldwide, focusing on multiple provisioning services and their contributions to the well-being of communities and the environment.

Table 2 List of provisioning services reported in the literature from urban green spaces

S. no.	Location	Key findings	References
1	Western Australia	Irrigated urban lawns offer a critical foraging habitat for kangaroos by providing a consistent food source, reducing conflicts with agriculture, and supplying much-needed water during dry periods	Ignatieva et al. (2020)
2	Europe	Harvesting plant biomass from lawns and meadows offers valuable resources with economic benefits when used for livestock feed or biofuel production	Brunzel et al. (2018); Sikorska et al. (2020)
3	Temperate Europe	Proposed that urban meadows, transformed from traditional lawns and cultivated with bioenergy crops in city parks, could yield economic benefits through biomass and biofuel production	Sikorska et al. (2020)
4		The sustainable management and upkeep of urban green spaces are crucial, as they play a significant role in preserving freshwater resources by facilitating groundwater infiltration and removing nutrients and impurities	Lundy and Wade (2011)
5	Tuscany, Italy	It showed reduced maintenance costs and increased water savings when urban UGS were managed as wildflower meadows	Bretzel et al. (2007)
6	United States	creating meadow-like vegetation using drought-tolerant native plants in arid urban areas can conserve much freshwater compared to lawns	Smith and Hilaire (1999)
7	USA	Some residential lawns have also been removed or replaced in southern California, USA, to reduce water use and management costs	Hartin et al. (2022)
8	USA	Highlights the potential of urban green biomass as a valuable resource for bioenergy production, constituting a form of provisioning services	Erickson et al. (2001) and Sobol et al. (2021)

4 Regulating Services

Regulating ecosystem services (RES) refers to the benefits derived from ecosystem processes that serve to control and stabilize natural environments (MEA 2005). These services are essential in mitigating the adverse impacts of natural occurrences and human actions, safeguarding human health and ecosystem integrity. RES plays a crucial role in reducing risks associated with environmental disruptions, making them indispensable for a sustainable and healthy planet. Water purification and waste treatment, air quality maintenance, soil erosion control, flood protection, climate regulation, pest and disease regulation, pollination, and regulation of the frequency and intensity of natural hazards' flow are all mechanisms used by RES to protect the natural environment (MEA 2005; Kandziora et al. 2013; Smith et al. 2013; Villamagna et al. 2013; Sutherland et al. 2018;). Urban people benefit from various regulating functions trees provide, including carbon sequestration, oxygen release, water containment, nitrogen oxide absorption, and heat reduction. Tree species and features precisely determine the type and amount of these services. For example, leaf area and crown size determine a tree's ability to manage microclimate, reduce noise, and intercept rainfall. Because of their wider canopy breadth, deciduous trees typically outperform evergreen ones in microclimate regulation (Hong et al. 2011; Sharma et al. 2018; Singh et al. 2018; Gupta et al. 2018a, b). On a larger scale, the surrounding green space's characteristics, such as vegetation structure and arrangement, influence trees' regulating services. We presented findings for four regulating services in the following text: heat mitigation, dust retention, carbon sequestration, and noise reduction.

(1) **Heat mitigation**

Trees have a higher capacity to mitigate heat compared to shrubs and grasses. On clear summer days, trees dominated areas were 0.5–1 °C cooler than shrub—dominated areas during the day and at night (Giridharan et al. 2008). In addition to reducing residential thermal comfort, urban trees can effectively reduce air conditioning usage and save energy (Akbari 2002). As increase in urbanization causes urban heat islands when cities replace natural land cover with dense concentrations of pavement, buildings, and other surfaces that absorb and retain heat (Zölch et al. 2019).

The urban heat island effect can result in temperatures up to 12 °C higher than in surrounding less-developed places (Fig. 3). Trees cool the environment through shading and transpiration. Shading reduces heat from the sun, cutting energy use by about 12% in residential and 1.3% in commercial areas (Li 2005). Transpiration involves heat absorption and water release, lowering air temperature. Studies have assessed various tree species for their heat mitigation capabilities and linked traits like canopy size, transpiration rate, leaf density, and metabolic cycle to cooling effects (Singh and Verma 2013; Kumar et al. 2017; Gupta and Singh 2017; Singh et al. 2017a, b; Li et al. 2018). Heat mitigation capacity in trees is associated with specific traits such as canopy size, transpiration intensity, leaf density, and the metabolic cycle of

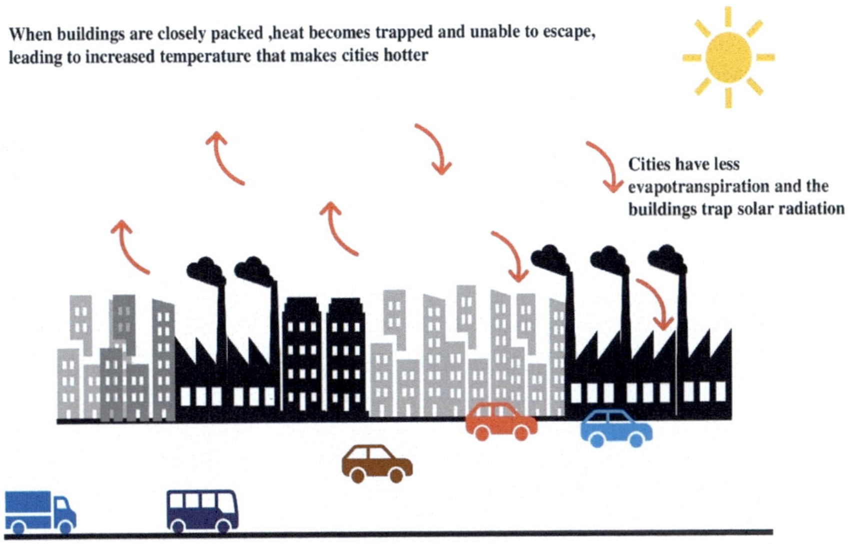

Fig. 3 Thermal heat effect and increase in temperature between the closed-packed buildings. Modified and adapted from Abdel (2021)

the tree (Li et al. 2018). During the peak of summer, a comprehensive study involved 332 individual trees across 85 different species in sub-canopy areas, comparing them to the adjacent unshaded ground. Among these traits, the leaf area index (LAI) and the overall size of the canopy emerged as pivotal factors. Notably, hawthorn trees excelled in providing effective cooling due to their substantial LAI (Speak et al. 2020).

A separate investigation conducted in Chengdu revealed that weeping willow trees exhibited the highest transpiration rate, resulting in approximately double the cooling effect on the surrounding 1000 cubic meters of air when compared to the osmanthus tree. This observation underscores the importance of tree selection and its specific traits in mitigating heat and enhancing local environmental comfort.

Researchers have identified specific urban tree species that effectively combat heat in cities. In Beijing, China, a study revealed that dry willow (*Salix matsudana*) excels in cooling and humidifying capacity per unit of leaf area, with elm (*Prunus triloba*) and plane tree (*Platanus occidentalis*) also demonstrating high cooling potential (Yin et al. 2021). Apart from tree characteristics, the cooling impact of trees is influenced by environmental factors like the ground surface. For instance, studies indicate that grass reduces surface temperature by 3 °C per unit of LAI (Leaf Area Index). At the same time, asphalt contributes to a more substantial 6 °C reduction in surface temperature (Speak et al. 2020).

(2) **Dust retention potential of tree**

Urban air pollution is a significant concern due to its adverse impact on human health and productivity (Ding et al. 2019). Urban trees mitigate this problem by effectively capturing atmospheric particulate matter. Several studies have investigated the dust retention capabilities of common urban tree species and the factors influencing these variations (Meena et al. 2011; Kumar et al. 2019c; Kumar et al. 2020; Singh and Kumar 2022; Singh et al. 2024a).

Research focusing on different green space structures has shown that trees and shrubs exhibit a greater capacity to aggregate and capture particulate matter than grass, making them more effective in dust deposition. Notably, passing through a forested area can result in a remarkable 75% reduction in atmospheric dust concentration compared to an unvegetated residential area (Qi et al. 2013) Even in severe dust sources like open pit coal mines, a 15-m-wide birch forest strip reduced dust levels by more than 50% (Qi et al. 2013).

In the specific context of Beijing, measurements have indicated that mature woodlands can achieve an impressive dust retention rate of up to 61.1% during summer and around 20% during winter. These findings highlight the valuable role of urban trees in mitigating air pollution and enhancing the quality of urban environments. The ability of trees to retain dust is primarily influenced by various leaf characteristics, canopy structure, branch, and leaf density, as well as the angle of the leaves (Chang et al. 2015). Studies concentrating on these tree traits have revealed that species with larger leaf areas and higher leaf densities are more effective in dust retention. *Cedrusdeodara* has the highest dust retention ability of 75.4 $\mu g\ cm^{-2} d^{-1}$, followed by *Neriumindicum* and Acer*palmatum*, while *Platamusorientalis* exhibits the lowest dust capture ability of 9.19 $\mu g\ cm^{-2} d^{-1}$, as reported by Jia et al. (2021). For instance, research conducted in Beijing identified tree species like downy ash (*Fraxinus tomentosa*), spruce (*Piceaasperata*), white pine (*Pinusbungeana*), lateral cypress (*Platycladusorientalis*), and cypress (*Juniperussabina*) as particularly adept at adsorbing and deterring PM 2.5 pollutants (Chang et al. 2015).

Comparative studies have also indicated that coniferous trees have superior dust retention capacity compared to broad-leaved trees (Xia 2015). Additionally, tree species with hairy and intricately folded leaves exhibit enhanced dust retention abilities. For example, species such as water cotoneaster (*Cotoneaster multiflorus*), hairy cherry (*Prunus tomentosa*), and lonicera (*Lonicera japonica*) are more effective in dust capture due to the presence of grooved and downy microstructures on their leaf surfaces (Yan et al. 2021).

It was also observed that torch trees (*Rhustyphina*) often keep their stomata closed, leading to a lower dust retention capacity (Yang et al. 2015). On the canopy level, trees with larger canopies and finer structures tend to have a higher capacity for retaining dust particles. Several environmental factors significantly impact the dust retention ability of urban trees. Temperature, maintenance, irrigation, precipitation, wind speed, and pollution levels are crucial. Elevated temperatures and strong winds can accelerate the movement of pollutants, making it more challenging for trees to adsorb and capture them effectively (Prusty et al. 2005) (Fig. 4).

High levels of environmental pollution can quickly saturate a tree's dust retention capacity, diminishing its overall effectiveness in capturing airborne particles (Wei 2021). These environmental factors collectively influence the ability of urban trees to serve as effective barriers against dust and pollutants, emphasizing the importance of

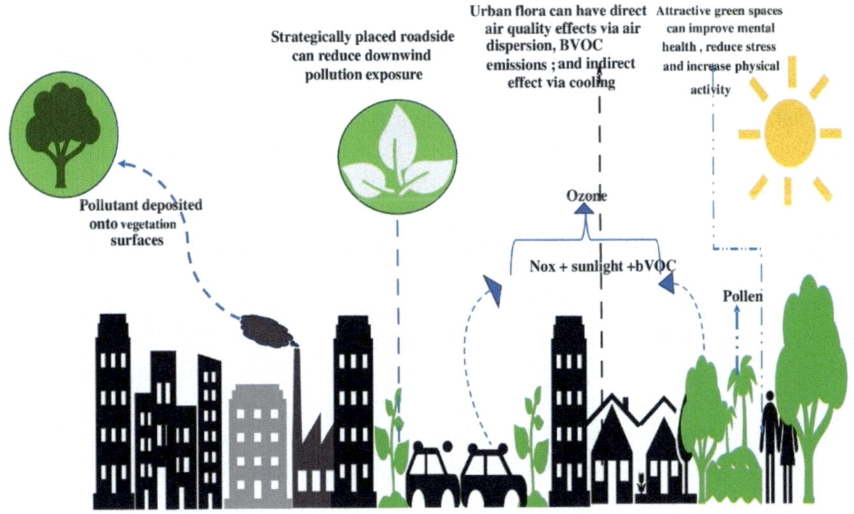

Fig. 4 Schematic diagram of air quality benefits and downsides of green infrastructure in the built environment. *Source*: Modified and adapted from Abdel (2021)

considering these elements in urban planning and tree management strategies (Yadav et al. 2019a, b; Jangra et al. 2024; Kanta et al. 2024; Malik et al. 2024; Kumar and Singh 2024).

An essential indicator of a plant's resilience to air pollution is the Air Pollution Tolerance Index (APTI). Plants with lower APTI values can indicate higher pollution levels, while those with higher APTI values can significantly reduce pollution. A study published in Current Science by Dubey et al. (2023) identified trees like pipal, neem, and mangoes, as well as crops like maize, pigeon pea, and safflower, as the best choices for areas with elevated air pollution.

The presence of greenery around industrial buildings provides a practical and environmentally friendly means to combat air pollution. In a study conducted in the Talkatora Industrial Area in Lucknow, Uttar Pradesh, India, the APTI of 25 plant species was determined. *Ficus bengalensis, Ficusreligiosa, Eucalyptus globulus,* and *Azadirachtaindica* were the most air pollution-tolerant, while *Polythalialongifolia* was the most sensitive. *Heveabrasiliensis* demonstrated tolerance to air pollution (Dash and Dash 2018).

Moreover, the dust-capturing ability of plant leaves was assessed, with *Moringa oleifera* found to be the most effective (5.7 mg/cm^2), and *Acacia nilotica* (0.10 mg/cm2) the least effective (Bharti et al. 2018). Plant species with APTI values below 11 serve as reliable bio-indicators of air quality, signifying higher pollution levels. Conversely, species with APTI values equal to or greater than 17 are suitable for creating green belts, as they exhibit a high degree of air pollution tolerance and can effectively contribute to improving environmental conditions. This classification based on APTI values facilitates informed decisions regarding plant selection for specific environmental and air quality management purposes (Bharti et al. 2018).

Pandey et al. (2015) evaluated nine plant species commonly found in urban areas of Varanasi. They assessed these species for their Air Pollution Tolerance Index (APTI) and Anticipated Performance Index (API) to determine the most suitable plants for developing long-term air pollution reduction in urban forests in Varanasi. The study identified *Ficusbenghalensis* (26.01) and *Ficus religiosa* (23.35) as excellent performers based on these indices. Additionally, *Polyalthialongifolia* (22.88), *Ficus-glomerata* (*Roxb.*) (19.22), *Anthocephalus indicus* (17.36) and *Mangifera indica* (16.89*)* were recognized as very good performers. Similarly, *Cassia fistula* (24.52), *Drypetesroxburghii* (22.11), *Terminalia arjuna* (20.54), *Psidiumguajava* (20.38), *Millingtoniahortensis* (19.22), *and Dalbergia sissoo* (15.95) were estimated to be good performers in terms of their anticipated performance in air pollution reduction. This research provides valuable insights for selecting and promoting specific plant species in urban forest development to enhance air quality in Varanasi.

(3) **Carbon Sequestration Potentials Trees and Influencing Factors**

The rapid increase in atmospheric carbon is a crucial driver of climate change, and there are two primary strategies to mitigate this issue: reducing carbon emissions and enhancing carbon fixation. Urban forests and plantations have the potential to make a substantial contribution to carbon storage and sequestration efforts. (Table 3) By storing carbon in terrestrial carbon sinks such as plants, plant products, and soils over extended periods, they can effectively address the challenges of climate change. These green spaces act as valuable reservoirs for capturing and storing carbon dioxide, a major greenhouse gas responsible for global warming. Through photosynthesis, trees and other vegetation absorb carbon dioxide from the atmosphere and convert it into organic matter, locking carbon away in their biomass. Trees play a vital role in combating the rise of atmospheric carbon dioxide by capturing, fixing, and storing it above and below the ground (Lewis et al. 2019). Therefore, planting trees in urban areas has the potential to reduce the concentration of CO_2 in the atmosphere by accumulating it in the form of biomass. As trees grow, they absorb carbon dioxide from the air and store it within their plant tissues, leading to the growth of various tree parts. This process helps mitigate climate change by removing CO_2 from the atmosphere and storing it in a natural carbon sink. (Matthews et al. 2000).

According to Nowak and Crane (2002), urban trees across the contiguous United States currently store approximately 700 million tonnes of carbon, with a gross carbon sequestration rate of 22.8 million tonnes per year. The CO_2 fix model, as suggested by Kaul et al. (2010), indicates that the carbon storage potential in soil and products is less sensitive than trees' carbon stock. Furthermore, studies have found that tropical urban forests are carbon-rich ecosystems, contributing up to 55% of terrestrial carbon sinks. Within these ecosystems, aboveground biomass serves as the primary carbon pool (Gibbs et al. 2007; Vashum and Jayakumar 2012). Similarly, Keerthika and Chavan (2022) conducted a study to estimate the carbon sequestration capacity of various tree species at different ages. They concluded that among the tree species, *Populusdeltoides* stands out as the species with the maximum carbon

Table 3 Analysis of carbon sequestration by dominant trees in urban areas

S. no.	Name of city and country	Dominant tree species	Carbon sequestration	References
1	Mumbai, Thane	*Azardirachtaindica, Ficusbenghalensis, Tamarindiusindica, Casurinaequisetifolia, Pongamiapinnata*	3289.8 kg C/tree, 2375.5 kg C/ tree,1666.9 kg C/ tree,1530.5 kg C/ tree 865.9 kgC/ tree	Tak and Kakde (2020)
2	Sambalpur, Odisha	*Ficusbenghalensis, Ficus religiosa*	3355.06 kgC/ tree/yr 1645.33 kgC/ tree/yr	Sahu et al. (2020)
3	Amity University, Noida	*Ficusbenjamina Alstoniascholaris*	30.53 tons 16.38 tons	Sharma et al. (2020)
5	Mumbai, India	*Ficusbenghalensis, Caryota mentis, Roystoniaregia, Alstoniascholaris, Syzygium cumini*	25,463.0 kgC/tree 3568.5 kg C/tree, 2418.3 kg C/tree, 2792.92 kg C/tree 1906.7 kg C/tree	Mithbavkar et al. (2022)
6	Jammu & Kashmir	*Ficusbenghalensis, Eucalyptus species, Grevillearobusta, Platanusorientalis, AlstoniaScholaris*	15,349.08 tonnes 1703.157 tonnes 1024.976 tonnes 1021.493 tonnes 580.624 tonnes	Devi (2017)
7	Nigeria	*Azadirachtaindica, Albizialebbeck*	5448.8 kg C 1040.4 kg C	Eneji et al. (2014)
8	Pakistan	*Azadirachtaindica, Conocarpus erectus*	662.3 kg C 192.7 kg C	Ajani and Shams (2016)
9	Oman	*Ficusbenghalensis, Azadirachta indicia, Conocarpus erectus*	3399.3 tonnes 2845.2 tonnes 2286 tonnes	Al-Nadabi and Sulaiman (2023)
11	Tamil Nadu	*Peltophorum, Ferrogenium, Millingtonia, Hortensis Cyatheadealbata*	1601.635 kg C/ tree 1104.243 kg C/ tree 436.475 kg C/tree	Saral et al. (2017)

sequestration potential (Fig. 5). It can sequester 12.61 tons of carbon per year per hectare at the age of 8 years. On the other hand, *Mangifera indica* and *Prosopis cineraria* exhibit the lowest carbon sequestration capacity among the listed species, with *Mangifera indica* sequestering 0.38 tons per year per hectare and *Prosopis cineraria* sequestering 0.46 tons per year per hectare at their respective ages.

Young secondary and plantation forests, characterized by fast-growing trees, play a significant role in carbon absorption, with a substantial portion of carbon stored in various tree components such as stems, roots, leaves, and branches (Ali et al. 2023).

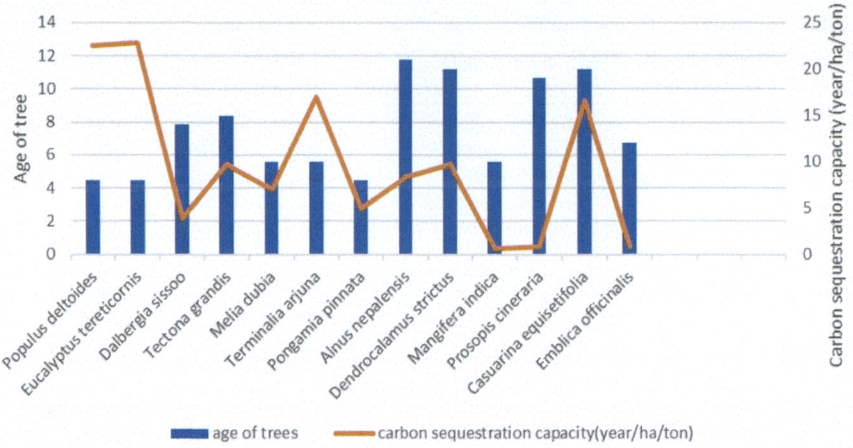

Fig. 5 Carbon sequestration capacity of various tree species at different ages. *Source* Keerthika and Chavan (2022)

The carbon sequestration rate in these forests is influenced by factors including plant growth, specific traits of tree species, wood density, and prevailing environmental conditions.

Research has demonstrated a connection between soil carbon concentration and tree roots, indicating that soil carbon levels increase with higher tree productivity, root diameter, and specific root length (SRL), decreasing as the C: N content rises (Sun et al. 2020). Elevated SRL, associated with greater fine root density, increases root carbon exudation and turnover (Bergmann et al. 2020). Additionally, the mycorrhizal associations of tree roots, including clumping and ectomycorrhizal fungi, are crucial factors in soil carbon sequestration (Averill and Hawkes 2016). Empirical studies have revealed positive relationships between canopy area, diameter at breast height, tree height, and carbon sequestration. Similarly, soil carbon storage positively correlates with photosynthesis rate, chlorophyll a, b, and carotenoid content. Various regression models have indicated that water use efficiency and stomatal conductance are reliable predictors of soil carbon stocks in bristlecone pine forests (Kumar et al. 2021a, b, c, d, e, f).From an environmental perspective, factors like elevation and precipitation indirectly influence carbon sequestration by affecting tree growth. These environmental variables play a significant role in determining the effectiveness of trees in capturing and storing carbon in various ecosystems.

(4) **Noise Reduction Effect of Trees and Influencing Factors**

Sound attenuation, which refers to the reduction in the intensity of sound waves, occurs when these waves pass through thick clusters of plants taller than 1 m. This phenomenon happens due to the blocking effect of the plants. Often, trees are strategically planted alongside roads to help mitigate traffic noise.

This sound attenuation process can relate to the capillary absorption coefficient formula. According to this formula, the coefficient of absorption for capillaries is inversely proportional to the radius of the tube and directly proportional to the square root of the frequency. In simpler terms, the absorption coefficient increases as the tube radius decreases. Moreover, higher-frequency sounds are absorbed more effectively by the capillaries.

In essence, this relationship explains how sound is attenuated when passing through plant clusters and provides insights into how different factors influence this phenomenon.

$$\alpha = \frac{2}{ac0}\sqrt{\gamma\omega\eta - \rho o} \tag{1}$$

In this equation, α is the sound absorption coefficient, a is the radius of the tube, and the $\sqrt{\gamma\omega\eta - \rho o}$ is frequency (Yuan 2009).

Studies have previously indicated that when it comes to tree characteristics, features like dense foliage with intricate branches, larger leaf size, and leaf surfaces covered with fine hairs can be likened to fine capillaries. This analogy means that such tree traits are more effective at reducing noise due to their ability to enhance the absorption coefficient. In simpler terms, trees with bushy branches, large leaves, and hairy leaf surfaces are better at providing sound insulation against higher-frequency noise, reinforcing the idea that finer capillaries and higher noise frequencies result in more excellent sound absorption (Watts et al. 1998). Trees and other plants, when strategically planted, can also act as an effective barrier against unwanted noise pollution. These natural barriers can be placed alongside roads, in residential areas, or in public spaces to help alleviate the impact of bothersome noises by absorbing, deflecting, refracting, or even masking sound waves. As stated by the United States Department of Agriculture, a well-planned tree barrier spanning 100 feet in width has been found to decrease noise levels by about 5 to 8 decibels (USDA). It's important to highlight that some researchers have emphasized the significance of factors beyond just the physical characteristics of the plants in achieving noise reduction benefits. Elements such as the length, width, height, density, and arrangement of green belts play pivotal roles (Cool and Haverbrake 1974).

Studies have consistently demonstrated the effectiveness of vegetation belts in reducing noise levels, especially when they are sufficiently dense and wide. For example, Leeuwen (2016) found that bamboo plants outperformed conventional barriers in mitigating road traffic noise when arranged as a noise barrier. This underscores the superior noise reduction capabilities of vegetation barriers like bamboo. Furthermore, research by Wang et al. (2013) indicates that green belts ranging from 16 to 70 m in width are particularly effective in reducing noise levels. Additionally, the composition and density of vegetation within these green belts play a crucial role. It's been observed that a combination of trees and shrubs within the green belt yields better noise reduction compared to having only one type of vegetation (Wang 2007). Moreover, a higher density of trees and shrubs further enhances the noise reduction potential (Wang 2007).

Moreover, Zhang et al. (2009) suggest that planting green belts closer to the noise source is more effective in reducing noise than placing them near the area requiring protection. Onder and Kocbeker (2012) conducted a study in Konya, Turkey, investigating the effectiveness of bushes in reducing roadside noise. Different plant species were categorized into three groups based on their distance from the noise source (7, 11, and 20 m) and plant width (5, 9, and 20 m). The study measured and calculated the decrease in noise levels in decibels. Results showed that the highest attenuation obtained was 6.3 dB(A) from the first group of plant species, 5.5 dB(A) from the second group and 6.2 dB(A) from the third group. The study concluded that the reduction in noise levels depended on the species of plants used as barriers, the distance of the barrier from the noise source, and the width of the plants. Certainly, Nasiri et al. (2015) investigated the impact of broad-leaved and coniferous tree species on reducing noise pollution along forest roads. The study found significant reductions in sound levels at various distances from the sound source. Broad-leaved tree species showed reductions of 10.6 dB at 20 m, 14.5 dB at 100 m, and 19.4 dB at 300 m.

Similarly, coniferous tree species exhibited reductions of 10.4 dB, 14.3 dB, and 16.8 dB at the same distances. The study emphasized the importance of tree species with a lower Height to Diameter at Breast Height (H:D) ratio and a broad crown in effectively mitigating noise pollution. These findings underscore the significance of vegetation belts and their configurations in noise reduction.

Moreover, the ground covered by branches, leaves, and dead branches effectively absorbs low and medium-frequency noise (Zhou et al. 2005). However, it's important to note that in urban settings, foliage is often regularly cleared to prevent fire hazards and maintain a neat environment, which can disrupt this natural noise reduction process.

(5) **Supporting services**

Supporting ecosystem services constitute the fundamental functions that underpin all other ecosystem services and maintain ecological processes. These services include nutrient cycling, soil formation, and primary production. Nutrient cycling, facilitated by microorganisms and decomposers, ensures the recycling of essential elements in ecosystems. Soil formation involves accumulating organic matter and the development of fertile soils, crucial for agriculture and plant growth (Six et al. 2004). Primary production, mainly driven by photosynthesis in plants, forms the basis of food chains and sustains life on Earth (Elser et al. 2007). These services are fundamental to ecosystems' functionality, highlighting nature's interconnectedness. They support the overall health and productivity of natural systems and, in turn, human well-being.

In the context of India, several critical supporting ecosystem services are vital for maintaining ecological stability and sustainability. These services include primary production, soil formation, nutrient cycling, and oxygen production. (Gogoi et al. 2021). Notably, these services exhibit substantial variability seasonally and inter-annually within the country (Nayak et al. 2016, 2019). From 1981 to 2006, India's total Net Primary Productivity (NPP) was estimated to be 1.42 Peta grams of carbon.

Effectively managed urban green spaces sequester more soil organic carbon than native prairies, particularly in drier temperate North American regions (Pouyat et al. 2009). Additionally, both lawns and meadows play a pivotal role in retaining nutrients, minimizing sediment transport, enhancing water infiltration and quality, and facilitating aquifer recharge when compared to impervious surfaces, thereby mitigating issues related to soil erosion, flooding, and nutrient leaching into adjacent water bodies (Beard and Green 1994; Groffman et al. 2009; Monteiro 2017). Petrovic's (1990) review underscored that well-maintained urban turfgrass lawns can retain up to 90% nitrogen, preventing nitrogen loss into groundwater and downstream water systems. Furthermore, Monteiro's (2017) review emphasized the role of urban turfgrass lawns in reducing water runoff intensity, minimizing flooding, mitigating soil erosion, and curtailing soil and water pollution. Enhancing soil carbon content and urban green spaces increases soil aggregate stability and improves soil quality (Bronick and Lal 2005).

6. **Cultural Services**

Cultural ecosystem services are the non-material benefits people obtain from nature, enriching their quality of life and well-being. These services include spiritual and aesthetic values, recreation, and opportunities for inspiration and education. Natural landscapes, such as serene forests or scenic coastlines, offer aesthetic pleasure and spiritual connection (Haines-Young and Potschin 2013). Recreation in natural settings, like parks and wildlife reserves, enhances physical and mental health, promoting relaxation and stress relief (Keniger et al. 2013). Moreover, contact with nature can inspire creativity, serve as a valuable educational resource, and foster a sense of community and place. Urban ecosystems offer cultural services and evidence supporting urban biodiversity's role in enhancing well-being. Research by Fuller et al. (2007) highlights that green spaces' psychological benefits correlate with higher biodiversity, and a green view from windows reduces job stress and increases job satisfaction (Sop 2009). Multiple studies confirm property value appreciation in proximity to green areas (Brander and Koetse 2011).

Urban biodiversity also informs eco-design and bio-mimicry in architecture and urban planning, providing valuable design insights (Blanco et al. 2021). These cultural services demonstrate the profound influence of nature on human culture and social interactions, contributing to holistic human development. Numerous research studies have delved into wooded landscapes' beauty and leisure aspects. For instance, a wealth of evidence supports the idea that people typically favour forests with a more natural appearance and more considerable, mature trees.

Researchers such as (Blanco et al. 2021; Gao et al. 2007) contributed to this body of knowledge, highlighting the widespread preference for forests that exhibit these characteristics. A significant portion of the scientific literature has predominantly focused on ecosystem services related to providing resources and regulating environmental processes, while cultural services have not received as much attention (Keerthika and Chavan 2022). Dasgupta et al. (2022) conducted a study investigating six non-material values associated with landscapes in the Sundarbans delta. These values included spiritual, recreational, heritage, aesthetic, educational, and negative aspects. Their research revealed that in the Indian context, much of the cultural ecosystem studies focus on recreational value, ecotourism, and traditional forest-based livelihoods. However, only recent studies have begun to explore the impact of green spaces on the well-being of urban residents, as exemplified by Gandherva et al. (2019).

Furthermore, there has been a growing interest in understanding place-based attachment and the psychological benefits people derive from natural environments, as demonstrated by Sen and Guchhait (2021). This became particularly relevant during the COVID-19 pandemic, as people were encouraged to stay home, increasing stress levels. Nature was found to offer valuable benefits in helping individuals cope with this stress, as Basu et al. indicated (2021). Urban forests are undeniably valued for their aesthetic appeal, with people deriving enjoyment from the diverse colour combinations, structures, and textures of urban trees and plants (Tyrväinen et al. 2005). The aesthetic appreciation of urban forests is closely linked to visitors' education, culture, age, gender, and personal interests. Numerous prior studies have affirmed that the aesthetic experiences in green spaces positively impact individuals' mental and emotional well-being (Kaplan and Kaplan 1989).

Moreover, the presence of fresh air and open green spaces further contributes to the enhancement of human mental health and positive moods. Preference studies, especially in Europe, have demonstrated that urban green spaces significantly benefit human mental health (Barton and Pretty 2010). Urban parks, in particular, possess substantial educational potential for individuals of all ages (Carter 1993). Among these, urban parks, including botanical gardens, provide diverse educational opportunities (Krasny and Tidball 2009). Urban forests, therefore, serve as ideal locations for field trips, allowing students to gain valuable scientific knowledge in areas like plant and tree identification and insights into management and policy issues (Krasny and Tidball 2009).

The framework for non-material benefits of Cultural and Ecosystem Services (CEA), as outlined by the Millennium Ecosystem Assessment, encompasses a wide array of services, including spiritual, religious, traditional knowledge, education, inspiration, aesthetics, scenic beauty, social connections, a sense of place, cultural heritage, and recreational and ecotourism services. These services are entirely shaped by human perceptions, consciousness, knowledge, and socio-cultural developments connected to the natural world. Understanding how people perceive and appreciate these benefits is crucial because effectively managing them is essential for sustainability (Gould et al. 2018).

It's also important to note that opinions regarding cultural services can vary among individuals and change over time, highlighting the dynamic nature of these services and the need for adaptable management approaches to ensure their long-term viability. (Gould et al. 2018). Keerthika and Chavan (2022) conducted a study to quantify the cultural services of the MFA model in Tamil Nadu. They reported the relationship between WTP and income, which increases as income increases. Platania and Rizzo (2018) and Nie et al. (2019) have also reported similar studies. Cultural Ecosystem Services (CES) encompass the direct and indirect use of natural resources, such as appreciating scenic walks or recognizing green spaces' cultural heritage and spiritual significance. Recognizing the cultural benefits of ecosystem services is essential for achieving socio-ecological sustainability and making informed management decisions (Table 4).CES encompasses spiritual enrichment, recreation, mental and physical well-being, aesthetic enjoyment, and cognitive development, as acknowledged by the Millennium Ecosystem Assessment (MEA 2005).

There is also a growing interest in studying the role of recreational value in urban green spaces. These spaces offer residents a wide range of benefits, including ecological, health, and social advantages. For instance, in Nagpur city, the average per capita availability of recreational urban green spaces is 3.65 m^2, although this value varies across different city zones (Lahoti et al. 2019). In a study examining the emotional perception of visitors in an urban park in Jammu, it was found that 39% of respondents described their experience as one of "happiness," closely followed by feelings of "freedom" and being "closer to nature" as reported by Duggal and Chib (2012). This reflects the positive emotional and psychological aspects of urban green spaces. Cultural ecosystem services enhance human well-being and enrich our connection with nature, offering spiritual, aesthetic, recreational, and educational benefits. Despite their importance, these services have historically received less attention in the literature, possibly due to the quantification challenges. Recognizing their dynamic and multifaceted nature is crucial for effective management and sustainable integration into human culture and society, as demonstrated by a growing body of research in this field.

Table 4 International Insights on Urban Green Space Perceptions and Benefits

Location	Key findings	References
Australia	Emphasizes the pivotal role of social perceptions in urban green space planning, particularly in the global South, highlighting the need for better communication of community benefits and context-based valuation to promote greener cities	Cilliers (2023)
Australia	Reviewed how social practices, usage patterns, and people's perceptions and preferences interrelate with urban green spaces	Farahani and maller (2018)
China	Reviewed and suggested that people perceive urban green spaces and infrastructure as practical tools in mitigating climate-related risks. This perception reduces the stress associated with these risks and contributes to overall well-being	Duan et al. (2018)
Beijing	Reviewed and suggested enhancing the well-being benefits of urban green spaces, directing their usage, and fostering an appreciation for nature among users, emphasizing the value of spending time in such environments	Priego et al. (2008), Chen et al. (2020), and Xu et al. (2021)
South Bengal	Suggested that perceptions and usage patterns of urban green spaces (UGSs) play a significant role in fostering social cohesion, supporting child development, and reducing stress among urban residents	Macintyre et al. (2019) and Sen and Guchhait (2021)

5 Conclusion

Urban and peri-urban forests (UPFs) play a crucial role in enhancing the livability and resilience of cities by providing various ecosystem services. These services encompass provisioning functions like ensuring food security, mitigating pollution, and regulating climate. The multifaceted benefits offered by UPFs, including heat reduction, dust control, carbon storage, and noise mitigation, significantly contribute to improving the quality of life in urban areas. Cities must recognize and integrate these ecosystem services into their climate change mitigation strategies and urban planning endeavors. Incorporating UPFs into decision-making processes may benefit cities in effectively tackling climate change challenges, promoting sustainability, and building resilience for current and future generations. Acknowledging the pivotal role of UPFs in urban environments is imperative for creating cities that can thrive amidst environmental pressures and meet the evolving needs of society.

References

Abdel AFG (2021) Role of urban trees and forests in solving the problem of air pollution. In: Hassan MM, Elsayed AE (eds) Proceedings of ICFD14; Fourteenth international Conference of Fluid Dynamics 2–3 April. Fairmont Nile City, Egypt, pp 1–2

Ajani A, Shams ZI (2016) Comparative status of sequestered carbon stock of *Azadirachtaindica* and *Conocarpus erectus* at the University of Karachi Campus, Pakistan. Int J Environ 5:89–97

Akbari H (2002) Shade trees reduce building energy use and CO_2 emissions from power plants. Environ Pollut 116:119–126

Ali L, Khan SM, Ahmad Z, Siddiq Z, Ullah A, Yoo S, Han HS, Raposo A (2023) Carbon sequestration potential of different forest types in Pakistan and its role in regulating services for public health. Front Public Health 10:1064586

Alizadeh B, Hitchmough JD (2020) How will climate change affect future urban naturalistic herbaceous planting? the role of plant origin and fitness. Urban Forest Urban Green 56:126786

Al-Nadabi AS, Sulaiman H (2023) Screening tree species for carbon storage potential through urban tree inventory in planted vegetation. Open J Forest 13:61–74

Artz R (2021) Reforestation and climate change mitigation: a review of international initiatives. Environ Manage 42(5):836–846

Averill C, Hawkes CV (2016) Ectomycorrhizal fungi slow soil carbon cycling. Ecol Lett 19:937–947

Barton J, Pretty J (2010) What is the best dose of nature and green exercise for improving mental health? Amulti-Study Anal Environ Sci Tech 44(10):3947–3955

Basu M, Dasgupta R, Kumar P, Dhyani S (2021) Home gardens moderate the relationship between COVID-19-induced stay-at-home orders and mental distress: a case study with urban residents of India. Environ Res Comm 3:105002

Beard JE, Green RL (1994) The role of turfgrasses in environmental protection and their benefits to humans. J Environ Qual 23(3):452–460

Bergmann J, Weigelt A, Plas FVD, Laughlin DC, Kuyper TW, Guerrero-Ramirez NR, Valverde-Barrantes OJ, Bruelheide H, Freschet GT, Iversen CM et al (2020) The fungal collaboration gradient dominates the root economics space in plants. Sci Adv 6:3756

Bharti SK, Trivedi A, Kumar N (2018) Air pollution tolerance index of plants growing near an industrial site. Urban Clim 24:820–829

Birkhofer K, Diehl E, Andersson J, Ekroos J, Früh-Müller A, Machnikowski F, Smith HG (2015) Ecosystem services—current challenges and opportunities for ecological research. Front Ecol Evol 2:87

Blanco E, Pedersen Zari M, Raskin K, Clergeau P (2021) Urban ecosystem-level biomimicry and regenerative design: Linking ecosystem functioning and urban built environments. Sustain 13(1):404

Brander LM, Koetse MJ (2011) The value of urban open space: meta-analyses of contingent valuation and hedonic pricing results. J Environ Manage 92(10):2763–2773

Brauman KA, Daily GC, Duarte TK, Mooney HA (2007) Ecosystem services: water and food. Science 315(5817):1575–1576

Bretzel F, Pezzarossa B, Carrai C, Malorgio F (2007) Wildflower plantings to reduce the management costs of urban gardens and roadsides. VI International Symposium Floriculture Crops 813:263–270

Bronick CJ, Lal R (2005) Soil structure and management: a review. Geoderma 124:3–22

Brunzel S, Kellermann J, Nachev M, Sures B, Hering D (2018) Energy crop production in an urban area: a comparison of habitat types and land use forms targeting economic benefits and impact on species diversity. Urban Ecosyst 21:615–623

Buffers for Noise Control, Conservation buffers. United States Department of Agriculture. Guidelines/6.0/Aesthetics&Visualquality. https://www.fs.usda.gov/nac/buffers/docs/6/6.4ref.pdf

Carter EJ (1993) The potential of urban forestry in developing countries: a concept paper. In: Prepared on Behalf of the Food and Agriculture Organization, Rome, June, Potential Benefits and Problems, Chapter 4, pp 21–25

Chang YM, Yan PB, Yang J (2015) Suggestions on tree species selection for urban greening in Beijing to control PM 2.5 pollution. J Landscape Architect 31:69–73

Chen C, Luo W, Li H, Zhang D, Kang N, Yang X et al (2020) Impact of perception of green space for health promotion on willingness to use parks and actual use among young urban residents. Int J Environ Res Public Health 17:5560

Cilliers EJ (2023) Social perceptions of the value of green spaces: a view from the South. Front Sustain Cities 4:1037123

Clark KH, Nicholas KA (2013) Introducing urban food forestry: a multifunctional approach to increase food security and provide ecosystem services. Landscape Ecol 28(9):1649–1669. https://doi.org/10.1007/s10980-013-9903-z

Convention on Biological Diversity (CBD) (2004) Ecosystem services: benefits supplied to human societies by natural ecosystems. Retrieved from https://www.cbd.int/ecosystem/serv.shtml

Cool DL, Haverbeke DFV (1974) Trees and shrubs for noise abatement. University of Nebraska College of Agricultural Experimental Station Bulletin, Lincoln, NE, USA

D'Souza P (2022) Times of India, https://timesofindia.indiatimes.com/india/a-recent-study-has-shown-that-heat-related-deaths-in-the-country-have-gone-up-by-more-than50percent/articleshow/95089618.cms?utm_source=contentofinterest&utm_medium=text&utm_campaign=cppst

D'Souza R, Nagendra H (2011) Changes in public commons as a consequence of urbanization: the Agara Lake in Bangalore, India. Environ Manage 47(5):840–850. https://doi.org/10.1007/s00267-011-9658-8

Dasgupta R, Hashimoto S, Basu M, Okuro T, Johnson BA, Kumar P, Dhyani S (2022) Spatial characterization of non-material values across multiple coastal production landscapes in the Indian Sundarban delta. Sustaina Sci, 1–14

Dash SK, Dash AK (2018) Air pollution tolerance index to assess the pollution tolerance level of plant species in industrial areas. Asian J Chem 29(12):219–222

Devi R (2017) Carbon storage by trees in urban parks: a case study of Jammu, Jammu and Kashmir, India. Int J Adv Res Develop 2(4):250–253

Dhyani K, Kuniyal HB, Singh H, Sobha (2021) Growth and physiological potential of *Terminalia arjuna* under elevated CO_2 levels in Open top chamber condition. J Appl Nat Sci 13(3):1121–1126. https://doi.org/10.31018/jans.v13i3.2490 (ISSN 2231-5209)

Ding HJ, Jiang J, Zhang JC (2019) Regional variation and factor analysis of soil permeability. J Soil Water Conserv 33:51–56

Dobbs C, Escobedo FJ, Zipperer WC (2011) A framework for developing urban forest ecosystem services and goods indicators. Landsc Urban Plan 99:196–206

Dou Y, Zhen L, De Groot R, Du B, Yu X (2017) Assessing the importance of cultural ecosystem services in urban areas of Beijing municipality. Ecosyst Serv 24:79–90

Duan J, Wang Y, Fan C, Xia B, de Groot R (2018) Perception of urban environmental risks and the effects of urban green infrastructures (UGIs) on human well-being in four public green spaces of Guangzhou, China. Environ Manag 62:500–517

Dubey R, Choudhary AK, Singh S, Ajay A, Kumar S, Kumar R, Singh VK (2023) Assessing the impact of air pollution of trees and crops in Eastern Gangetic plains of India. Curr Sci 124(8):956

Duggal A, Chib A (2012) The role of urban green spaces for the sustainable city, Jammu (J&K). Paripex Indian J Res 3:92–94

Elser JJ, Bracken ME, Cleland EE, Gruner DS, Harpole WS, Hillebrand H, Smith JE (2007) Global analysis of nitrogen and phosphorus limitation of primary producers in freshwater, marine and terrestrial ecosystems. Ecol Lett 10(12):1135–1142

Eneji IS, Obinna O, Azua ET (2014) Sequestration and carbon storage potential of tropical forest reserve and tree species located within Benue State of Nigeria. J Geosci Environ Protect 2:157–166

Erickson JE, Cisar JL, Volin JC, Snyder GH (2001) Comparing nitrogen runoff and leaching between newly established St. Augustine grass turf and an alternative residential landscape. Crop Sci 41:1889–1895

Farahani LM, Maller C (2018) Perception and preferences of urban greenspaces: a literature review and framework for policy and practice. Landscape Online 2:1–22

Fuller RA, Irvine KN, Devine-Wright P, Warren PH, Gaston KJ (2007) Psychological benefits of greenspace increase with biodiversity. Biol Let 3(4):390–394

Gandherva D, Bhattacharya R, Bhattacharya P (2019) Assessment of user's perception towards urban green spaces: a case study of Delhi India. J Ecol Nat Res 3(1):15

Gao RR, Wu MX, Zhou CL (2007) Scenic beauty estimation of street trees landscape in Qingdao. J Shandong Univ Sci Tech 37:16–19

Gibbs HK, Brown S, Niles JO, Jonathan A, Foley JA (2007) Monitoring and estimating tropical forest carbon stocks: making REDD a reality. Environ Res Lett 2:1–13

Giridharan R, Lau S, Ganesan S, Givoni B (2008) Lowering the outdoor temperature in high-rise high-density residential developments of coastal Hong Kong: The vegetation influence. Build Environ 43:1583–1595

Gogoi B, Borah N, Baishya A, Nath D, Dutta S, Das R, Bhattacharyya D, Sharma K, Valente D, Petrosillo I (2021) Enhancing soil ecosystem services through sustainable integrated nutrient management in double rice-cropping system of North-East India. Ecol Ind 132:108262

Gould RK, Coleman K, Gluck SB (2018) Exploring dynamism of cultural ecosystems services through a review of environmental education research. Ambio 47(8):869–883

Groffman PM, Williams CO, Pouyat RV, Band LE, Yesilonis ID (2009) Nitrate leaching and nitrous oxide flux in urban forests and grasslands. J Environ Qual 38:48–1860. https://doi.org/10.2134/jeq2008.0521

Gupta SK, Singh H (2017) Observations on sporadic flowering in *Ficus benjamina* L. and *Peltophorum africanum* L. in New Forest. Indian Forest 143:290–291

Gupta SK, Ram J, Singh H (2018a) Comparative study of transpiration in cooling effect of tree species in the atmosphere. J Geosci Environ Protect 6:151–166. https://doi.org/10.4236/gep.2018.68011

Gupta SK, Ram J, Singh H, Kumar P (2018b) Comparative study of cooling under shades of some forestry tree species with respect to ambient temperatures. J Geosci Environ Protect 6:51–66. https://doi.org/10.4236/gep.2018.611004

Gupta SK, Ram J, Singh H (2019a) Relationship between leaf area index and atmospheric cooling of tree species. Indian J for Forest 42(3):203–208

Gupta SK, Singh H, Ram J (2019b) Need for a novel paradigm in forestry research: perspectives of comparative atmospheric cooling by different forestry effect tree species. Indian Forest 145(6):509–515

Haase D, Frantzeskaki N, Elmqvist T (2014) Ecosystem services in urban landscapes: practical applications and governance implications. Ambio 43:407–412. https://doi.org/10.1007/s13280-014-0503-1

Haines-Young R, Potschin M (2013) Common international classification of ecosystem services (CICES): consultation on Version 4, August–December 2012, p 34

Hartin JS, Surls RA, Bush JP (2022) Lawn removal motivation, satisfaction, and landscape maintenance practices of Southern Californians. Hort Tech 32(1):57–66

Hong B, Lin BR, Hu LH, Li SH (2011) Optimal tree design for sunshine and ventilation in residential district using geometrical models and numerical simulation. Build Simul 4:351–63

Hurley P, Emery M, McLain R, Poe M, Grabbatin B, Goetcheus C (2015) Whose urban forest? the political ecology of gathering urban Non timber forest products. In Isenhour C, Checker M, McDonogh G (eds). Sustainability in the global city: myth and practice. Cambridge University Press. https://doi.org/10.1017/CBO9781139923316.014

Ignatieva M, Haase D, Dushkova D, Haase A (2020) Lawns in cities: from a globalised urban green space phenomenon to sustainable nature-based solutions. Land 9(3):73

Jangra A, Kumar K, Maikhuri S, Bhandari MS, Pandey S, Singh H, Barthwal S (2024) Unveiling stress-adapted endophytic bacteria: characterizing plant growth-promoting traits and assessing cross-inoculation effects on *Populus deltoides* under abiotic stress. Plant Physiol Biochem 210:108610. https://doi.org/10.1016/j.plaphy.2024.108610

Jia M, Zhou D, Lu S, Yu J (2021) Assessment of foliar dust particle retention and toxic metal accumulation ability of fifteen roadside tree species: relationship and mechanism. Atmos Pollut Res 12(1):36–45

Jim CY, Chen WY (2009) Ecosystem services and valuation of urban forests in China. Cities 26(4):187–194

K'A Kumu OA, Olima WH (2007) The dynamics and implication of residential segregation in Nairobi. Habitat Internat 31:87–99

Kandziora M, Burkhard B, Müller F (2013) Interactions of ecosystem properties, ecosystem integrity and ecosystem service indicators-a theoretical matrix exercise. Ecol Ind 28:54–78

Kanta C, Kumar A, Chauhan A, Singh H, Sharma IP (2024) The interplay between plant functional traits and climate change. In: Kumar N, Singh H (eds) Plant functional traits for improving productivity. Springer, Singapore. https://doi.org/10.1007/978-981-97-1510-7_3

Kaplan R, Kaplan S (1989) The experience of nature—a psychological perspective, Vol 7. Cambridge University Press, Cambridge, pp 12–17

Kaul M, Mohren GMJ, Dadwal VK (2010) Carbon storage and sequestration potential of selected tree species in India. Mitigation Adap Strat Global Change 15:489–510

Keerthika A, Chavan SB (2022) Oxygen production potential of trees in India. Curr Sci 122(7):850

Keerthika A, Parthiban KT, Kumar DS (2021) Understanding cultural ecosystem services of multifunctional agroforestry: a study from the foothills of the Nilgiris, Western Ghats, India. Current Sci 121(12):1610

Keniger LE, Gaston KJ, Irvine KN, Fuller RA (2013) What are the benefits of interacting with nature? Int J Environ Res Public Health 10(3):913–935

Krasny ME, Tidball KG (2009) Community gardens as contexts for science, stewardship, and civic action learning. Cities Environ 2:1–18

Kumar N, Singh H (2024) Plant Functional traits for improving productivity. Springer Nature, Singapore. https://doi.org/10.1007/978-981-97-1510-7

Kumar N, Suyal DC, Sharma IP, Verma A, Singh H (2017) Elucidating stress proteins in rice (*Oryza sativa* L.) genotype under elevated temperature: a proteomic approach to understand heat stress response. 3 Biotech 7:205. https://doi.org/10.1007/s13205-017-0856-9

Kumar M, Singh MP, Singh H, Dhakate PM, Ravindranath NH (2019a) Forest working plan for the sustainable management of forest and biodiversity in India. J Sustain Forest 39:1–22. https://doi.org/10.1080/10549811.2019.1632212

Kumar N, Jeena N, Singh H (2019b). Elevated temperature modulates rice pollen structure: a study from the foothill of the Himalayan agro-ecosystem in India. 3 Biotech 9(5):175. https://doi.org/10.1007/s13205-019-1700-1

Kumar N, Jeena N, Kumar N, Gangola S, Singh H (2019c) Phytoremediation facilitating enzymes: an enzymatic approach for enhancing remediation process. In: Bhatt P (ed) Smart bioremediation technologies. Academic Press, Elsevier, 50 Hampshire Street, 5th Floor, Cambridge, MA 02139, United States, pp 289–306

Kumar N, Singh H, Sharma SK (2020) Antioxidants: responses and importance in plant defense system. In: Roychowdhury R, Choudhury S, Hasanuzzaman M, Srivastava S (eds) Sustainable agriculture in the era of climate change. Springer, Cham, pp 251–264. https://doi.org/10.1007/978-3-030-45669-6_11

Kumar A, Kumar P, Singh H, Kumar N (2021a) Modulation of plant functional traits under essential plant nutrients during seasonal regime in natural forests of Garhwal Himalayas. Plant Soil 465:197–212. https://doi.org/10.1007/s11104-021-05003-x

Kumar A, Kumar P, Singh H, Kumar N (2021b) Adaptation and mitigation potential of roadside trees with bio-extraction of heavy metals under vehicular emissions and their impact on physiological traits during seasonal regimes. Urban Forest Urban Green 58:126900. https://doi.org/10.1016/j.ufug.2020.126900

Kumar A, Kumar P, Singh H, Kumar N (2021c) Impact of plant functional traits on infiltration rate under different forests of Kempty watershed in Garhwal Himalaya. Indian J Soil Conserv 49(1):38–44

Kumar A, Kumar P, Singh H, Bisht S, Kumar N (2021d) Relationship of physiological plant functional traits with soil carbon stock in the temperate forest of Garhwal Himalaya. Curr Sci 120(8):1368–1373. https://doi.org/10.18520/cs/v120/i8/1368-1373

Kumar M, Phukon SN, Paygude AC, Tyagi K and Singh H (2021e) Mapping phenological functional types (PhFT) in the Indian Eastern Himalayas using machine learning algorithm in Google Earth Engine. Comp Geosc 158. https://doi.org/10.1016/j.cageo.2021.104982

Kumar A, Kumar P, Singh H, Bisht S, Kumar N (2021f) Relationship of physiological plant functional traits with soil carbon stock in the temperate forest of Garhwal Himalaya. Curr Sci 120:1368–1373

Kumar A, Singh H, Kumari G, Bisht S, Malik A, Kumar A, Singh M, Raturi , Barthwal S, Thakur A and Kaushal R (2022) Adaptive resilience of roadside trees to vehicular emissions via leaf enzymatic, physiological, and anatomical trait modulations. Environ Pollut 313. https://doi.org/10.1016/j.envpol.2022.120191

Kumar N, Singh H, Giri K, Kumar A, Joshi A, Yadav A, Singh R, Bisht S, Kumari R, Jeena N, Khairakpam R, Mishra G (2024) Physiological and molecular insights into the allelopathic effects on agroecosystems under changing environmental conditions. Physiol Mol Biol Plants 30:417–433. https://doi.org/10.1007/s12298-024-01440-x

Lahoti S, Lahoti A, Saito O (2019) Benchmark assessment of recreational public urban green space provisions: a case of typical urbanizing Indian City, Nagpur. Urban Forest Urban Green 44:126424

Larondelle N, Strohbach MW (2016) A murmur in the trees to note: urban legacy effects on fruit trees in Berlin, Germany. Urban Forest Urban Green 17:11–15. https://doi.org/10.1016/j.ufug.2016.03.005

Lewis S, Wheeler C, Mitchard ETA, Koch A (2019) Restoring natural forests is the best way to remove atmospheric carbon. Nature 568:25–28

Li ZG, Chen D, Cai SZ, Che SQ (2018) The ecological services of plant communities in parks for climate control and recreation—a case study in Shanghai, China. Plos One 13:e0196445

Li XG (2005) Observations and assessments of mitigation effects of the urban forests on atmospheric pollutants. Master's Thesis, Central South University of Forestry and Technology, Changsha, China, pp 23–29
Lundy L, Wade R (2011) Integrating sciences to sustain urban ecosystem services. Progress in Phys Geograp 35:653–669
Macintyre VG, Cotterill S, Anderson J, Phillipson C, Benton JS, French DP (2019) I would never come here because I've got my own garden: Older adults' perceptions of small urban green spaces. Int J Environ Res Public Health 16:1994
Malik A, Kumar K, Kumari G, Kumar N, Singh H (2024) Application of functional traits in modelling productivity and resilience under climate change. In: Kumar N, Singh H (eds) Plant functional traits for improving productivity. Springer, Singapore. https://doi.org/10.1007/978-981-97-1510-7_5
Matthews E, Payne R, Rohweder M, Murray S (2000) Forest ecosystem: carbon storage sequestration. Carbon sequestration in soil. Global Clim Change Digest 12:174–180
McLain R, Poe M, Hurley P, Lecompte-Mastenbrook J, Emery M (2012) Producing edible landscapes in Seattle's urban forest. Urban Forest Urban Green 11:187–194. https://doi.org/10.1016/j.ufug.2011.12.002
Millennium Ecosystem Assessment (MEA) (2005) Ecosystems and human well-being, vol 5. Island Press, Washington, DC, p 563
Meena D, Singh H, Chaudhari SK (2011) Elucidating strontium response on growth dynamics and biochemical change in *Phaseolus mungo* L. Int J Agricult Environ Biotech 4:107–113
Mithbavkar V, Jadhav M, Ganekar D, Gavas S (2022) Carbon sequestration potential of trees in urban areas; a case study. Int J Adv Innov Research 9(3):2394–7780
Monteiro JA (2017) Ecosystem services from turfgrass landscapes. Urban Forest Urban Green 26:151–157
Nasiri M, Fallah A, Nasiri B (2015) The effects of tree species on reduction of the rate of noise pollution at the edge of hyrcanian forest roads. Environ Eng Manage J 14(5)
Nayak AK, Shahid M, Nayak AD, Dhal B, Moharana KC (2019) Assessment of ecosystem services of rice farms in Eastern India. Ecol Proc 8:35
Nayak R, Chandra A, Patel N, Dadhwal V (2016) Terrestrial net primary productivity and net ecosystem productivity over India. National Remote Sensing Centre, Balanagar, Hyderabad. https://bhuvanapp3.nrsc.gov.in/data/download/tools/document/CASAtechnicalreport_final.pdf
Nie X, Chen Q, Xiao T, Wang H (2019) Willingness to pay for ecological function regions protection based on a choice experiment method: a case study of the Shiwandashan Nature Reserve. Qual Quant 53(2):813–829
Nowak DJ, Crane DE (2002) Carbon storage and sequestration by urban trees in the USA. Environ Pollut 116(3):381–389
Olhoff A, Christensen JM (2020) Emissions gap report. UNEP DTU Partnership. https://www.unep.org/emissions-gap-report-2020
Önder S, Akay A (2015) Reduction of traffic noise pollution effects by using vegetation, Turkey' sample. J Eng Econ Develop 2(2):23
Onder S, Kocbeker Z (2012) Importance of the green belts to reduce noise pollution and determination of roadside noise reduction effectiveness of bushes in Konya, Turkey. Int J Agricult Biosyst Eng 6(6):373–376
Pandey AK, Pandey M, Mishra A, Tiwary SM, Tripathi BD (2015) Air pollution tolerance index and anticipated performance index of some plant species for development of urban forest. Urban Forest Urban Green 14(4):866–871
Petrovic AM (1990) The fate of nitrogenous fertilizers applied to turfgrass. J Environ Qual 19(1):1–14
Platania M, Rizzo M (2018) Willingness to pay for protected areas: a case of Etna Park. Ecological Indicator 93:201–206

Poe M, LeCompte-Mastenbrook J, McLain R, Hurley P (2014) Urban foraging andrelational ecologies of belonging. Soc Cult Geogr 15(8):901–919. https://doi.org/10.1080/14649365.2014.908232

Poe MR, McLain RJ, Emery M, Hurley PT (2013) Urban forest justice and therights to wild foods, medicines, and materials in the city. Human Ecol 41(3):409–422. https://doi.org/10.1007/s10745-013-9572-1

Pouyat R, Groffman P, Yesilonis I, Hernandez L (2009) Soil carbon pools and fluxes in urban ecosystems. Environ Pollut 116:107–118

Prakash V, Hunney K, Singh H (2022) Effect of elevated carbon dioxide on growth and development of *Santalum album* L. seedlings inoculated with plant growth promoting microorganisms in open top chambers. Int J Environ, Agricult Biotech 7(5):149–159

Priego C, Breuste J, Rojas J (2008) Perception and value of nature in urban landscapes: comparative analysis in cities in Germany, Chile and Spain. Landscape Online 7:22

Prusty BA, Mishra PC, Azeez PA (2005) Dust accumulation and leaf pigment content in vegetation near the national highway at Sambalpur, Orissa, India. Ecotoxicol Environ Safety 60:228–235

Qi JZ, Wei JH, Zhang Q, Liu RH (2013) Problems and countermeasures in research on urban trees' dust retaining ability. World J Forest Res 26:52–57

Russo A, Cirella GT (2019) Edible urbanism 5.0. Palgrave Comm 5(1):1–9

Russo A, Cirella GT (2021) Urban sustainability: integrating ecology in city design and planning. In: Sustainable human–nature relations: environmental scholarship, economic evaluation, urban strategies. Springer, pp 187–204

Sahu C, Nayak HN, Sahu SK (2020) Carbon sequestration potential of trees in an urban area: a case study of sambalpur town in eastern india. Fresenius Environ Bull 29(10):8757–8766

Salbitano F, Borelli S, Conigliaro M, Chen Y (2016) Guidelines on urban and peri-urban forestry. FAO Forestry Paper 178

Saral AM, SteffySelcia S, Devi K (2017) Carbon storage and sequestration by trees in VIT University campus. IOP Conf Series: Mat Sci Eng 263(2):022008. IOP Publishing

Semeraro T, Scarano A, Buccolieri R, Santino A, Aarrevaara E (2021) Planning of urban green spaces: an ecological perspective on human benefits. Land 10:105

Sen S, Guchhait SK (2021) Urban green space in India: perception of cultural ecosystem services and psychology of situatedness and connectedness. Ecol Ind 123:107338

Shackleton C, Shackleton S, Shanley P (2011) Building a holistic picture: an integrative analysis of current and future prospects for non-timber forest products in a changing world. Springer, Berlin, Germany. https://doi.org/10.1007/978-3-642-17983-9

Sharma R, Singh H (2021) Alteration in biochemical constituents and nutrients partitioning of *Asparagus racemosus* in response to elevated atmospheric CO_2 concentration. Environ Sci Pollut Res 29:6812–6821. https://doi.org/10.1007/s11356-021-16050-3

Sharma R, Pradhan L, Kumari M, Bhattacharya P (2020) Assessment of carbon sequestration potential of tree species in Amity University Campus Noida. Environ Sci Proceed 3(1):52

Sharma R, Singh H, Kaushik M, Nautiyal R and Singh O (2018) Adaptive physiological response, carbon partitioning, and biomass production of *Withania somnifera* (L.) Dunal grown under elevated CO_2 regimes. 3 Biotech 8:267. https://doi.org/10.1007/s13205-018-1292-1

Sikorska D, Macegoniuk S, Łaszkiewicz E, Sikorski P (2020) Energy crops in urban parks as a promising alternative to traditional lawns—perceptions and a cost-benefit analysis. Urban Forest Urban Green 49:126579

Singh H (2021) An integrated approach considering physiological and biophysical based indicators for assessing tolerance of roadside plantations of *Alstonia scholaris* towards urban roadside air pollution: an assessment of adaptation of plantations for mitigating urban roadside air pollution. Trees. https://doi.org/10.1007/s00468-021-02179-8

Singh H (2024). Exploring adaptive modulation in plant functional traits and its impact on the productivity of *Acacia auriculiformis* under CO_2-enriched environment. Indust Crops Prod 210:118186. https://doi.org/10.1016/j.indcrop.2024.118186

Singh H, Verma A (2013) Characterization and screening of high nitrogen efficient rice genotype to curtail environmental pollution. Int J Agricult, Environ Biotech 6(1):23–30

Singh H, Kumar M (2022) Climate change and its impact on Indian Himalayan forests: current status and research needs. In: Rani S, Kumar R (eds) Climate change. Springer climate. Springer, Cham, pp 223–242. https://doi.org/10.1007/978-3-030-92782-0_11

Singh H, Savita Sharma R, Sinha S, Kumar M, Kumar P, Verma A, Sharma SK (2017a). Physiological functioning of *Lagerstroemia speciosa* L. under heavy roadside traffic: an approach to screen potential species for abatement of urban air pollution. 3 Biotech 7(61):1–10. https://doi.org/10.1007/s13205-017-0690-0

Singh H, Verma A, Kumar M, Sharma R, Gupta R, Kaur M, Negi M, Sharma SK (2017b) Phytoremediation: a green technology to clean up the sites with low and moderate level of heavy metals. Austin Biochem 2:1012

Singh H, Sharma R, Savita SMP, Kumar M, Verma A, Ansari MW, Sharma SK (2018) Adaptive physiological response of *Parthenium hysterophorus* to elevated atmospheric CO_2 concentration. Indian Forest 144:1–14

Singh H, Yadav M, Kumar N, Kumar A, Kumar M (2020) Assessing adaptation and mitigation potential of roadside trees under the influence of vehicular emissions: a case study of *Grevillea robusta* and *Mangifera indica* planted in an urban city of India. PLoS One 15(1):e0227380. https://doi.org/10.1371/journal.pone.0227380

Singh S, Singh H, Sharma SK, Nautiyal R (2021) Seasonal variation in biochemical responses of bamboo clones in the sub-tropical climate of Indian Himalayan foothills. Heliyon 7(4):e06859. https://doi.org/10.1016/j.heliyon.2021.e06859

Singh H, Bandyopadhyaya S, Kumar A, Singh M, Malik A, Kumari G, Raturi A, Yadav SK, Kumar M, Barthwal S, Thakur A, Rawat PS (2022) Understanding the physiological and biophysical response of urban roadside plantations for assessing adaptation and mitigation mechanisms toward vehicular emissions. Urban Climate 44. https://doi.org/10.1016/j.uclim.2022.101183

Singh H, Kumar N, Kumar A (2024a). Enhancing resource use efficiency in crops through plant functional traits. In: Kumar N, Singh H (eds) Plant functional traits for improving productivity. Springer, Singapore. https://doi.org/10.1007/978-981-97-1510-7_6

Singh M, Singh H, Kumar A, Kumar M, Barthwal S, Thakur A (2024b) Soil nitrogen availability determines the CO_2 fertilization effect on tree species (*Neolamarckia cadamba*): growth and physiological evidence. Environ Sustain 7:53–60. https://doi.org/10.1007/s42398-023-00300-w

Six J, Bossuyt H, Degryze S, Denef K (2004) A history of research on the link between (micro) aggregates, soil biota, and soil organic matter dynamics. Soil Tillage Res 79(1):7–31

Smith P, Ashmore MR, Black HI (2013) The role of ecosystems and their management in regulating climate, and soil, water and air quality. J Appl Ecol 50(4):812–829

Smith C, Hilaire RS (1999) Xeriscaping in the urban environment, New Mexico. J Sci, 241–242

Sobol Ł, Dyjakon A, Suardi A, Preißmann R (2021) Analysis of the possibility of energetic utilization of biomass obtained from grass mowing of a large-area golf course—a case study of Tuscany. Energies 14(17):5520

Sop SW (2009) The influence of forest view through a window on job satisfaction and jobstress. Scand J for Res 22(3):248–253

Speak A, Montagnani L, Wellstein C, Zerbe S (2020) The influence of tree traits on urban ground surface shade cooling. Landsc Urban Plan 197:103748

Springer TL (2012) Biomass yield from an urban landscape biomass. Bioenergy 37:82–87

Stallinga P (2023) Residence time versus adjustment time of carbon dioxide in the atmosphere. Entropy 25(2):384

Stott I, Soga M, Inger R, Gaston KJ (2015) Land sparing is crucial for urban ecosystem services. Front Ecol Environ 13(7):387–393

Sun LJ, Ataka M, Han MG, Han YF, Gan DY, Xu T, Guo YP, Zhu B (2020) Root exudation as a major competitive fine-root functional trait of 18 coexisting species in a subtropical forest. New Phytol 229:259–271

Sutherland IJ, Villamagna AM, Dallaire CO, Bennett EM, Chin AT, Yeung AC, Cormier R (2018) Undervalued and under pressure: a plea for greater attention toward regulating ecosystem services. Ecol Ind 94:23–32

Tak AA, Kakde UB (2020) Analysis of carbon sequestration by dominant trees in urban areas of Thane city. Int J Global Warm 20(1):1–11

TEEB The Economics of Ecosystems and Biodiversity (TEEB) (2011) Rewarding benefits through payments and markets, chapter 5. In: Ten Brink P (ed) TEEB in national and international policy making. Earthscan, London, pp 177–257

The Economics of Ecosystems and Biodiversity (TEEB) (2010) TEEB Manual for cities: ecosystem services in urban management. www.teebweb.org

Tyrväinen L, Pauleit S, Seeland K, de Vries S (2005) Benefits and uses of urban forests and trees. In: Urban forests and trees. Springer, Berlin, Heidelberg, pp 81–114

Uniyal A, Rawat GS (2018) Energy-Food-Water; the fundamental provisioning services from the Himalayan forests: a case study from Dhauladhar mountain range, North-West Himalaya. Indian J Forest 41(1):17–26

Unnikrishnan H, Nagendra H (2015) Privatizing the commons: impact on ecosystem services in Bangalore's lakes. Urban Ecosyst 18(2):613–632. https://doi.org/10.1007/s11252-014-0401-0

Van Leeuwen HJ (2016) Bamboo plants as a noise barrier to reduce road traffic noise. In INTER-NOISE and NOISE-CON Congress and Conference Proceedings 253(4):4481–4487

Varol C (2022) Impacts of urban heat islands on human well-being and animal behavior: a review. Environ Sci Pollut Res 29(6):6082–6097

Vashum KT, Jayakumar S (2012) Methods to estimate above-ground biomass and carbon stock in natural forests—a review. J Ecosyst Ecograp 2:4. https://doi.org/10.4172/2157-7625.1000116

Villamagna AM, Angermeier PL, Bennett EM (2013) Capacity, pressure, demand, and flow: a conceptual framework for analyzing ecosystem service provision and delivery. Ecol Complex 15:114–121

Von Hoffen LP, Säumel I (2014) Orchards for edible cities: cadmium and lead contentin nuts, berries, pome and stone fruits harvested within the inner city neighbourhoods in Berlin, Germany. Ecotoxicol Environ Saf 101:233–239. https://doi.org/10.1016/j.ecoenv.2013.11.023

Wang WL, Guo XP, Wang MY, Zhang P (2013) Research progress of traffic noise attenuation by green belts. J Northwest Univ 28:240–244

Wang CM (2007) The analysis of traffic noise characteristic and the effect of afforested belt noise reduction. Master's Thesis, Northwest A&F University, Yangling, China, p 43

Watts GR, Chinn L, Godfrey N (1998) The effects of vegetation on the perception of traffic noise. Appl Acoust 56:39–56

Wehi PM, Wehi WL (2010) Traditional plant harvesting in contemporary fragmented and urban landscapes. Conserv Biol 24(2):594–604. https://doi.org/10.1111/j.1523-1739.2009.01376.x

Wei WJ (2021) Mechanisms of particulate matters capture capacity and response of leaf functional traits for urban tree species. Ph.D. Thesis, Chinese Academy of Forestry, Beijing, China, pp 24–28

Xia HZ (2015) Study on the dust-retention effect of the 18 kind common urban greening trees in Dalian. Master's Thesis, Liaoning Normal University, Dalian, China

Xu J, Wang F, Chen L, Zhang W (2021) Perceived urban green and residents' health in Beijing. SSM Population Heal J 14:100790

Yadav SK, Singh H, Ginwal HS and Barthwal S (2019a) Elevated CO_2 enhanced growth and physiological process of *Populus deltoides* Bartr. ex Marsh. Indian For 145(1):23–27

Yadav SK, Singh H, Nautiyal R, Ginwal HS, Ansari SA, Barthwal S (2019b) Modulation of morpho-physiological responses in *Populus deltoides* by elevated carbon dioxide and temperature. Forest Sci 66:105–118. https://doi.org/10.1093/forsci/fxz048

Yan Q, Xu LS, Duan YH (2021) Dust retention capacity and dust particle size of 20 commonly used greening tree species. Chinese J Plant Ecol 40:3259–3267

Yang J, Wang HX, Xie BZ, Shi H, Wang YH (2015) Accumulation of particulate matter on leaves of nine urban greening plant species with different micromorphological structures in Beijing. Research J Environ Sci 28:384–392
Yin YJ, Chen QH, Zhao R, Sun M, Chen LQ (2021) Transpiration characteristics and temperature & humidity benefits of common greening tree species in Beijing. J Northwest Univ 36:31–36
Yuan L (2009) Study on noise attenuation and it's application by tree belts along highway. Ph.D. Thesis, Chang'an University, Xi'an, China
Zhang WQ, Liu JW, Hu HY, Xiao LE, Li X (2009) Effect of the construction of green belts on the attenuation of traffic noise along the urban trunk roads in Xiamen City. Subtropical Plant Sci 38:74–78
Zhou JX, Ding YC, Li H, Li YP (2005) Investigation of traffic noise attenuation provides by green belts. Environ Eng 4:48–51
Zölch T, Rahman MA, Wagner G, Pfleiderer EJ, Pauleit S (2019) Designing public squares with green infrastructure to optimize thermal comfort. Build Environ 149:640–654

Ecosystem Services and Maintaining Their Flow in Urban Environments: Nurturing Sustainable Flow in Urban Environments

Sarita Bisht, Hukum Singh, and Amit Kumar

Abstract Urban forests provide invaluable contributions to society and economies through ecosystem services across four key domains: provisioning, regulating, cultural, and supporting services. However, the global surge of urbanization poses profound challenges to these services, disrupting natural ecosystems vital for clean air, water, and sustenance. This chapter examines ecosystem services' critical relevance in urban landscapes, emphasizing the imperative of maintaining energy flow amidst urbanization. It examines urban areas' obstacles in safeguarding these services, exploring effective strategies and case studies. Urbanization often triggers habitat loss, pollution, and disrupted hydrological cycles, jeopardizing essential life-supporting systems. Illuminating these challenges' interconnectedness may underscore the urgent need to preserve ecological services in urban environments. The chapter advocates for prioritizing ecosystem services in urban planning by highlighting tangible approaches, such as integrating green infrastructure like parks and urban trees, community engagement, and sustainable policies. Valuing urban ecosystem services offered by urban forests and conserving resources may pave the way for sustainable urban environments, benefiting present and future generations. Ultimately, the knowledge generated in this chapter serves as a vital resource for researchers, policymakers, and urban planners committed to embedding ecosystem services at the core of urban development, fostering healthier, more equitable, and sustainable cities.

Keywords Ecosystem services · Urban environment · Urban sustainability · Green infrastructure · Environmental education and policy

S. Bisht (✉) · H. Singh
Plant Physiology Discipline, ICFRE-Forest Research Institute, P.O. New Forest, Dehradun 248006, Uttarakhand, India
e-mail: saritabisht015@gmail.com

A. Kumar
Department of Forestry, North Eastern Hill University, Tura Campus, Tura, Meghalaya, India

H. Singh (ed.), *Urban Forests, Climate Change and Environmental Pollution*,
https://doi.org/10.1007/978-3-031-67837-0_21

1 Introduction

An ecosystem constitutes a dynamic community of plants, animals, microorganisms, and the abiotic environment, functioning together as a cohesive unit. Various factors such as soil composition, topography, climate, and water availability shape the distribution of ecosystems (Sweeney and Mitchell 2019; Yadav et al. 2019a, b; Jangra et al. 2024; Kanta et al. 2024; Malik et al. 2024; Kumar and Singh 2024). Within these ecosystems, energy, and materials continuously cycle through living organisms, a process honed over millions of years in response to disturbances, stresses, and ecological interactions (Meena et al. 2011; Kumar et al. 2019a, 2020; Upadhyay 2020). However, the escalating pace of urbanization underscores the imperative to comprehend and prioritize ecosystem services within urban areas. Ecosystem services encompass many benefits crucial to human well-being, including access to clean water, fertile soil, pure air, and recreational spaces. Yet, rapid urban expansion often comes at the cost of depleting these invaluable services (Millennium Ecosystem Assessment 2005).

Urban forests, encompassing the tree cover within cities, play a crucial role in improving environmental quality and urban life by offering many ecosystem services. Apart from their aesthetic and cultural significance, urban trees enhance biodiversity, mitigate heat island effects, lessen stormwater runoff, sequester carbon, and purify the air (McPherson et al. 1997; Nowak and Crane 2002; Tyrväinen et al. 2005). With more than half of the global population residing in urban areas—expected to increase to 68% by 2050—urbanization presents both opportunities and challenges. Habitat destruction, pollution, and altered hydrology are among the adverse consequences linked to urban expansion, threatening the sustainability of ecosystem services (Bebbington and Unerman 2018).

Rapid urbanization reshapes the relationship between human societies and the environment, profoundly affecting ecosystem services and urban resilience (Gomez 2017). This transformation underscores the urgent need to safeguard ecosystem services and strengthen urban ecosystems and their associated infrastructures (Talubo et al. 2022). The complexity of maintaining ecosystem services in urban settings stems from the interconnectedness of various challenges, including habitat loss, pollution, and disruptions in hydrological cycles, which directly endanger ecological systems supporting human well-being (Pickett et al. 2001). Prioritizing and integrating ecosystem services into urban planning and development processes is crucial to ensuring cities remain livable and sustainable (Andersson et al. 2014).

This chapter explores the pivotal role of ecosystem services within urban environments and assesses diverse strategies and policy instruments to sustain urban vegetation's societal contributions. Investing in urban ecosystem services, generating essential data for educational and research purposes, and evaluating their economic value can enhance cities' capacity for informed decision-making in urban planning and management, benefiting both present and future generations. The chapter may advance scientific understanding and application of ecosystem-based approaches to urban sustainability by addressing these pressing concerns.

2 Ecosystem Services in Urban Environments

Ecosystem services, encompassing nature's various benefits to human society and the economy, play a crucial role in urban environments. Urbanization, a global phenomenon characterized by rapid population growth and expansion of cities, profoundly impacts the supply and maintenance of ecosystem services (Singh and Kumar 2022; Singh et al. 2024). As more people migrate to urban areas, cities often encroach upon natural ecosystems, disrupting the flow of these vital services. Different habitats, such as parks, private gardens, and street trees, generate urban ecosystem services. They are created by engineered green infrastructure, which includes vertical greening systems, green roofs, and sustainable urban drainage systems.

2.1 Classification of Ecosystem Services

Ecosystem services encompass four main categories: provisioning, regulating, cultural, and supporting services. Provisioning services involve the tangible products and materials obtained from ecosystems. Regulating services refers to the role of ecosystems in regulating environmental processes. Cultural services provide non-material benefits such as cultural and spiritual values. Finally, supporting services encompass the foundational structures, processes, and functions that define ecosystems (Masiero et al. 2019).

2.1.1 Provisioning Services

Provisioning services encompass ecosystems' direct and tangible benefits to meet essential human needs and enhance well-being. These services encompass diverse vital resources derived from the natural environment. A close relationship exists between provisioning services and producing goods, including food, fuel, water, and genetic resources. Urban environments, in particular, excel in water provision due to their impermeable surfaces, which facilitate rainwater retention and accumulation (Shafique and Kim 2017; Semadeni-Davies et al. 2008). Ecosystem services provide sustenance such as food and water and serve as raw materials for various production supply chains, including timber, fuels, and fibers. While many of these services originate outside urban areas, there is an increasing focus on in situ provisioning (Bauhus et al. 2010). Various provisioning services are discussed in detail below.

Food

Ecosystems are primary sources of food production. They provide various crops, fruits, vegetables, grains, fish, meat, and other agricultural products essential for human consumption. For instance, forests, grasslands, and marine ecosystems contribute significantly to food production by yielding fruits, grains, seafood, and livestock (Hajer et al. 2016).

Water

Natural ecosystems, especially freshwater ecosystems like rivers, lakes, wetlands, and aquifers, provide clean and accessible water. They contribute to water purification, regulation, and storage, ensuring a sustainable supply of freshwater for human consumption, agriculture, and industrial purposes (MEA 2005).

Raw Materials

Urban forests are primary sources of raw materials such as timber, fibers (cotton, wool, and wood pulp), medicinal plants, biofuels, and natural resources like minerals. These resources support diverse production supply chains, including construction, manufacturing, and pharmaceutical industries (MEA 2005).

Genetic Resources

Urban forests harbor vast genetic diversity in plants, animals, and microorganisms. This diversity provides a rich pool of genetic resources crucial for breeding programs in agriculture, medicine, and biotechnology, contributing to developing new crops, livestock breeds, medicines, and drugs (Kremsa 2021).

2.1.2 Regulating Services

Regulating services are vital functions provided by ecosystems that play a crucial role in regulating and maintaining environmental processes. These services directly influence human well-being by ensuring ecosystems' stability, balance, and resilience. Urban ecosystems contribute significantly to mitigating environmental challenges such as air pollution, climate regulation, and flood control, as highlighted in Table 1 (Salmond et al. 2016). In post-industrial cities, unmanaged vacant land harbors flora that may serve as refugia within highly developed pavement landscapes akin to forests and parks. Furthermore, arboriculture initiatives on managed vacant lots and public right-of-ways have effectively reduced air pollution, absorbed greenhouse gases, and decreased ambient air temperatures by providing shade on asphalt, concrete, and surrounding structures. This, in turn, enhances energy efficiency for local buildings (McPherson et al. 1997).

Climate Regulation

Ecosystems are crucial in regulating the Earth's climate. Forests, oceans, and other ecosystems absorb carbon dioxide (CO_2) through photosynthesis, mitigating the effects of climate change by acting as carbon sinks (Singh and Verma 2013; Kumar et al. 2017; Ussiri and Lal 2017). Additionally, they regulate temperature through evapotranspiration, which influences local and regional climates.

Natural Hazard Regulation

Ecosystems protect against natural disasters and hazards. Coastal ecosystems like mangroves and coral reefs act as natural barriers against storm surges and tsunamis, reducing the impact of these events on human communities. Wetlands and forests can help control flooding by absorbing excess water and slowing its flow (Ussiri and Lal 2017; Gupta and Singh 2017; Singh et al. 2017a, b).

Water Purification

Ecosystems, particularly wetlands, play a significant role in purifying water. They act as natural filters by trapping and breaking down pollutants, improving water quality. Wetlands and riparian zones also aid in sedimentation, nutrient removal, and the breakdown of organic matter in water bodies, contributing to cleaner water (Powlson et al. 2011).

Table 1 Classification of ecosystem services

Provisioning services	Regulating services	Cultural services	Supporting services
Products obtained from ecosystems	Benefits obtained from regulation of ecosystem processes	Non-material ecosystems that enrich lives	Underpinning services that enable other services to function
• Food: crops, wild plants, tea, honey, etc. • Fresh wood • Fiber • Biochemicals • Genetic resources	• Climate regulation • Disease regulation • Water regulation • Water purification • Pollution	• Spiritual and religious • Recreation and ecotourism • Aesthetic • Educational sense of place • Cultural heritage	• Soil formation • Nutrient cycling • Primary production

Pollination

Ecosystems support pollination services, which are vital for agricultural productivity. Bees, birds, bats, and other pollinators facilitate the reproduction of many plants, including numerous crops. This process is essential for producing fruits, vegetables, and seeds, directly impacting food security (Sharma et al. 2018; Singh et al. 2018; Gupta et al. 2018a, b).

Disease Regulation

Natural ecosystems can regulate diseases by controlling disease vectors or providing natural habitats that suppress disease-causing organisms. Biodiverse ecosystems often have lower incidences of certain diseases due to the balance between species that can limit disease transmission (Wood et al. 2014).

Hence, the regulating services are critical for maintaining the balance and functionality of ecosystems, directly impacting human societies and their resilience to environmental changes and hazards.

2.1.3 Cultural Services

Cultural services encompass the intangible benefits ecosystems offer humans, enriching their overall well-being, cultural identity, and quality of life. Within urban areas, green spaces and natural areas play a pivotal role in providing cultural and recreational advantages that contribute to residents' social well-being, mental health, and quality of life. These urban green spaces serve as vital outlets for relaxation and recreational activities and foster a sense of connection to nature (Barton and Pretty 2010). However, urban residents often face challenges in appreciating nature due to limited exposure to quality habitats, which can lead to undervaluing the importance of conservation efforts both within and beyond the city (Green et al. 2016). Efforts to make parks more accessible are underway, facilitating urban dwellers' ability to access and enjoy these benefits more effectively. Furthermore, street trees play a significant role in enhancing the psychosocial well-being of communities, contributing to reduced crime rates and improved neighborhood walkability (Lee 2021).

Aesthetic and Spiritual Values

Natural landscapes, such as forests, mountains, and water bodies, offer aesthetic beauty and spiritual significance to human societies. Scenic views, biodiversity, and serene environments contribute to emotional and psychological well-being, inspiring art, literature, and cultural practices (MEA 2005).

Cultural Heritage and Identity

Ecosystems often hold cultural significance and heritage for indigenous communities and societies. Sacred sites, traditional knowledge, rituals, and cultural practices are intricately woven into ecosystems and biodiversity, forming a profound connection between human societies and nature. These sites and practices contribute to the preservation of cultural identity and heritage.

Recreational and Tourism Opportunities

Natural environments offer recreational opportunities such as hiking, birdwatching, camping, and eco-tourism. These activities promote physical health, leisure, and tourism revenue, benefiting local economies and communities.

Educational and Inspirational Values

Ecosystems serve as living laboratories, providing education, research, and inspiration opportunities. They offer insights into ecological processes, biodiversity, and the interconnectedness of life, fostering environmental awareness and inspiring future generations of scientists, conservationists, and nature enthusiasts.

2.1.4 Supporting Services

Supporting services are foundational functions ecosystems provide to sustain and facilitate the production of other ecosystem services. They encompass vital environmental processes like soil formation, nutrient cycling, and plant growth (Table 1), as well as the provision of suitable habitats for beneficial arthropods and animals that contribute to regulating services. Urban agriculture's effectiveness hinges on various supporting services, such as soil structure, nutrient cycling, and water availability (Powlson et al. 2011; Yadav et al. 2019a, b; Gupta et al. 2019a, b). These services, categorized within the Millennium Ecosystem Assessment (MEA) framework, are indispensable for ecosystem functionality and comprise several essential components.

Nutrient Cycling

Ecosystems play a crucial role in regulating the cycling and distribution of essential nutrients such as nitrogen, phosphorus, and carbon. Microorganisms, plants, and decomposers are instrumental in breaking down organic matter, recycling nutrients, and sustaining nutrient cycles vital for plant growth and overall ecosystem productivity (Kumar et al. 2019a, b).

Soil Formation

Ecosystems contribute to the formation and maintenance of soils through processes like weathering, organic matter accumulation, and microbial activity. Healthy soils are crucial for supporting plant growth, regulating water flow, and providing habitat for soil organisms.

Primary Production (Photosynthesis)

Photosynthesis, primarily by plants and phytoplankton, converts sunlight into chemical energy, supporting the foundation of food webs and providing ecosystem energy. It sustains the production of biomass, fuels ecosystem productivity, and is essential for all life forms (Singh et al. 2020; Kumar et al. 2021a, b).

Habitat Provision

Ecosystems provide diverse habitats that support a wide range of species. These habitats, ranging from forests and grasslands to wetlands and coral reefs, offer shelter, breeding grounds, and resources for numerous plants and animals, contributing to biodiversity and ecosystem resilience.

3 Factors Affecting the Flow of Ecosystem Services

Various factors influence the flow of ecosystem services. These factors shape the resilience and sustainability of ecosystems, emphasizing the need for comprehensive understanding and effective management to ensure the continual provision of essential services for human societies. These services, including provisioning (food, water), regulating (pollination, climate regulation), supporting (soil formation, nutrient cycling), and cultural (recreation, aesthetic values), are impacted by various factors.

3.1 Biodiversity Loss

Biodiversity loss diminishes ecosystem resilience and the ability to provide services. Ecosystems with higher species diversity tend to be more stable and productive, thus influencing the flow of services (Cardinale et al. 2012). One of the primary drivers of biodiversity loss is the destruction, fragmentation, and degradation of habitats due to human activities such as deforestation, urbanization, agriculture expansion, and infrastructure development. Loss of habitats directly affects various species and their ability to survive.

3.2 Land Use Changes

Alterations in land use, such as deforestation, urbanization, and agricultural expansion, can disrupt ecosystems and their services. Conversion of forests to agriculture, for instance, can reduce water regulation and carbon storage capacities (Foley et al. 2005).

3.3 Climate Change

Alterations in climate patterns due to human-induced factors like greenhouse gas emissions led to temperature changes, altered precipitation, rising sea levels, and extreme weather events. These changes impact ecosystems and habitats, causing species distribution shifts and threatening survival (Kumar et al. 2021c, d). Changing climatic patterns impact the distribution and functioning of ecosystems, affecting their ability to provide services. Altered precipitation patterns, temperature shifts, and extreme weather events can disrupt services like crop pollination and flood regulation (IPCC 2021).

3.4 Pollution and Contamination

Pollution from various industrial, agricultural, or domestic sources degrades ecosystems and hinders their capacity to provide services. Water pollution, for example, affects aquatic ecosystems' ability to purify water and support fisheries (Kremen and Ostfeld 2005). Chemical pollutants from industrial, agricultural, and domestic sources contaminate air, water, and soil. These pollutants disrupt ecosystems, harm species, and accumulate in the food chain, leading to health issues and population declines.

3.5 Overexploitation and Resource Depletion

Unsustainable resource extraction depletes ecosystems, diminishing their capacity to deliver services. This results in the collapse of soil degradation, affecting provisioning services (Costanza et al. 2017). Overexploitation destabilizes ecosystems and jeopardizes the survival of several species.

3.6 *Invasive Species*

The introduction of invasive species disrupts ecosystems, altering nutrient cycles, predation patterns, and species interactions, thus affecting ecosystem services (Vila et al. 2011).

3.7 *Lack of Awareness and Conservation Efforts*

Insufficient public awareness, inadequate conservation policies, weak enforcement of regulations, and a lack of sustainable practices contribute to the ongoing loss of biodiversity.

3.8 *Socioeconomic Drivers*

Economic activities and societal demands influence land management practices, affecting ecosystem services. Policies, market dynamics, and governance structures shape how ecosystems are utilized and conserved, impacting service flows (Daily et al. 2000). The globalization of trade and markets can influence land use change, affecting the provision of ecosystem services (Lambin et al. 2001). Understanding these factors and their interconnectedness is crucial for sustainable ecosystem management and preserving the flow of ecosystem services for future generations.

4 Maintaining Ecosystem Services in Urban Environments

Maintaining ecosystem services in urban environments is crucial for the well-being of both human populations and the natural world. Ecosystem services encompass a wide range of benefits that ecosystems provide to society, including clean air and water, pollination of crops, climate regulation, and cultural and recreational opportunities. Urbanization often poses significant challenges to these services. However, careful planning and management may support thriving ecosystems that contribute to their residents' overall quality of life. Urban areas support these services through various strategies.

4.1 Green Infrastructure and Urban Planning

Urban planning is pivotal in maintaining ecosystem services. Incorporating green infrastructure, such as parks, green roofs, and urban forests, can enhance biodiversity, improve air quality, and mitigate the urban heat island effect. Zoning regulations that protect natural habitats and promote sustainable land use also contribute to the preservation of ecosystem services. Green infrastructure in urban settings involves strategically integrating natural elements such as parks, green roofs, trees, and permeable surfaces into the built environment to provide various ecosystem services while enhancing the overall liability of cities (Plieninger and Bieling 2013). Creating and maintaining parks, community gardens, and green spaces in cities offers recreational opportunities, enhances biodiversity, supports mental health, and provides spaces for social interactions (Shanahan et al. 2016). Further, increasing tree canopy cover in urban areas helps mitigate air pollution, reduce energy consumption through shading, regulate temperature, sequester carbon, and provide habitats for wildlife (Escobedo et al. 2011). Using porous surfaces and nature-based drainage systems like bioswales or rain gardens is an approach that allows water to infiltrate the ground, reducing stormwater runoff, mitigating flooding, and improving water quality by filtration. (Orta-Ortiz and Geneletti 2021).

Urban biodiversity conservation enhances sustainable ecological services. Various vegetations is maintained by planting native plants, constructing animal corridors, and establishing green areas that resemble natural habitats. Insect management and general ecosystem resilience biodiversity increase a city's capacity to adapt to changing environmental conditions. The green infrastructure supports urban biodiversity and fosters ecological resilience by giving different species habitats and corridors (Goddard et al. 2010). Creating ecological corridors or greenways in urban landscapes facilitates wildlife movement, enhances genetic diversity, and helps maintain functional ecosystems even in fragmented urban settings (Bennett 2003).

Green infrastructure significantly enhances climate resilience by mitigating extreme weather impacts, reducing heat stress, and improving urban adaptation to climate change (Wamsler et al. 2016). Successful implementation requires collaboration among urban planners, policymakers, communities, and stakeholders to design, maintain, and optimize these natural features within cities (Cardinale et al. 2012). Incorporating vegetation on rooftops and walls serves various purposes, such as stormwater management, mitigating the urban heat island effect, improving air quality, conserving energy through insulation, and creating habitats for biodiversity (Joshi and Teller 2021). By integrating nature-based solutions into urban planning, cities can seamlessly weave ecosystem services into the urban fabric. Strategies like mixed land-use planning, compact city design, and incorporating green spaces into urban landscapes significantly promote sustainable urban ecosystems (Niemelä et al. 2011).

4.2 Sustainable Water Management

Sustainable water management is critical in urban areas. Implementing green stormwater infrastructure, such as porous pavements and green swales, helps absorb and filter rainwater, reducing the risk of floods and improving water quality. Protecting and restoring urban wetlands and riparian zones also contribute to water purification and flood control.

4.3 Community Engagement and Education

Involving communities in conservation efforts, promoting awareness about the value of ecosystems, and fostering stewardship among urban residents lead to better management and preservation of ecosystem services. (Andersson et al. 2014). Urban areas harness the benefits of ecosystem services by implementing these strategies, improving residents' quality of life while promoting sustainability and resilience in urban environments. Educating and involving the community is critical to maintaining ecological services. Citizen science projects, community-based conservation projects, and environmental education programs help residents feel responsible. Informed and involved communities are more likely to support policies and practices that protect ecosystem services.

5 Urban Planning and Policy

Urban planning and policy are pivotal in the sustainable development and maintenance of ecosystem services in urban environments. Ecosystem services, encompassing nature's benefits to human societies, are critical for the well-being of urban populations (Singh 2021; Sharma and Singh 2021; Singh et al. 2021). In urban planning, land use, infrastructure, and services are organized deliberately to maximize resource usage, reduce environmental impact, and improve habitat. Integrated planning considers social, economic, and environmental dimensions that lead to the creation of sustainable cities (Jordán and Infante 2012).

In addition, zoning laws and regulations guide the spatial distribution of activities within cities, balancing residential, commercial, industrial, and green spaces and the protection of natural habitats. Well-designed zoning regulations promote mixed land use, walkability, and access to amenities (Levy 2016). Curitiba, Brazil, is an example where land use planning has integrated green spaces and preserved ecological corridors, improving air quality, enhanced aesthetics, and increased resilience to climate change (Balaban 2017).

Planning policies prioritizing public transportation, non-motorized transport, and efficient road networks reduce traffic congestion, air pollution, and carbon emissions.

Transit-oriented development encourages compact, mixed-use areas near transit hubs, enhancing accessibility and reducing reliance on private vehicles (Litman 2015). Policies promoting creating and preserving green spaces, parks, and urban forests contribute to biodiversity, mitigate urban heat islands, improve air quality, and offer recreational opportunities. Urban green spaces foster community cohesion and mental well-being (Rakhshandehroo et al. 2017).

Besides, policies addressing affordable housing, social housing programs, and equitable access to services and opportunities are crucial for inclusive cities. Urban planning should aim for mixed-income neighbourhoods to foster social integration and reduce spatial inequalities (Thurber et al. 2018). Urban planning policies must integrate climate change considerations by promoting resilient infrastructure, floodplain management, green infrastructure, and regulations that minimize vulnerability to extreme weather events (Ahern 2011).

5.1 Community Engagement and Participation

Effective urban planning involves engaging communities in decision-making. Participatory approaches foster a sense of ownership, understanding of local needs, and aligning policies with community aspirations (Bai et al. 2016). Urban planning and policy frameworks must be adaptive, responsive to evolving challenges, and inclusive, considering diverse perspectives to create sustainable, resilient, and equitable cities for present and future generations. Here are some examples of Showcasing cities with successful urban planning for ecosystem services:

5.1.1 Singapore: The Garden City

Singapore has pioneered green infrastructure in its urban planning, earning it the title "Garden City". It has implemented initiatives such as the "Bishan-Ang Mo Kio Park", where a concrete canal was transformed into a natural river, enhancing water quality and providing recreational spaces. Additionally, Singapore's commitment to vertical greenery, seen in iconic structures likes the "Gardens by the Bay" and the "Super Tree Grove", contributes to improved air quality and biodiversity. The government's strategic urban planning and policies promoting sustainable development serve as a model for other cities (Yuen 1996).

5.1.2 Portland, Oregon, USA

Stormwater Management and Green Infrastructure Portland has excelled in incorporating green infrastructure to manage stormwater, addressing ecological and urban challenges. The "Grey to Green" initiative focuses on transforming conventional grey infrastructure into green solutions. Bioswales, permeable pavements, and green roofs are implemented to manage stormwater, enhance water quality, and provide habitat for urban wildlife. The emphasis on sustainable stormwater management underscores the city's commitment to preserving and improving ecosystem services (Shandas 2015).

5.2 Capacity Building and Cultural Practices

There is a need to provide training and resources to community members, local organizations, and leaders to equip them with the skills necessary for sustainable practices, effective advocacy, and project implementation within their neighbourhoods. In addition, it is essential to recognize and integrate diverse cultural perspectives and social dynamics within communities to enhance the effectiveness and inclusivity of urban ecosystem initiatives. Cultural events, celebrations, and community networks contribute to a sense of belonging and collective responsibility (Archer et al. 2014).

5.3 Stakeholders Engagement

Building partnerships between communities, local governments, NGOs, businesses, and academia is advised to foster collective action and knowledge sharing. Collaborative efforts leverage diverse expertise and resources toward shared goals of sustainable urban development. Further, urban planners, authorities, and policymakers should establish transparent communication channels that allow for ongoing dialogue, feedback mechanisms, and evaluation of community initiatives. This practice fosters trust, promotes accountability, and ensures community perspectives are continuously considered in decision-making processes (Cornwall 2008).

Authority must engage youth through educational programs, mentorship opportunities, and involvement in community projects, educate them about sustainability, and harness their innovative ideas and energy toward building resilient urban ecosystems (Russ and Krasny 2017).

6 Challenges and Future Directions

As urbanization continues to accelerate globally, managing and preserving ecosystem services within urban areas pose significant challenges. Ecosystem services, essential for human well-being, are under increasing pressure due to rapid urban expansion, pollution, and changing land-use patterns. Maintaining ecosystem services in urban environments faces several challenges due to the complex interactions between natural and anthropogenic environments (Birkhofer et al. 2015). Following challenges obstruct the development and maintenance of healthy urban ecosystems;

- Urbanization and land use change
- Biodiversity loss and habitat fragmentation
- Climate change impacts
- Pollution and environmental degradation
- Limited green spaces and access to nature
- Lack of integrated planning and policy
- Socioeconomic inequalities
- Capacity and knowledge gaps

Various directions may shape the future of urban ecosystems and sustainability, such as

- Biodiversity conservation and restoration
- Smart and sustainable technologies
- Circular economy initiatives
- Social inclusion in planning and decision-making
- Resilience and adaptation
- Nature-centric design
- Green infrastructure policy integration with developmental projects
- Creating awareness among communities regarding nature conservation

Collectively, the abovementioned initiatives enhance urban ecosystems resilience, ensure the supply of essential services, and promote more sustainable and resilient cities. These future directions seek to direct urban development toward sustainability, resilience, and improved quality of life for every citizen as cities evolve and face new challenges. Combining innovative strategies, technological advancements, and a commitment to equitable and resilient urban growth will shape the future of urban ecosystems and sustainability. As cities expand, these primary directions provide a road map for designing urban environments that integrate human needs with ecological health and longevity. Maintaining ecosystem services in urban areas is a complex challenge, but proactive measures and sustainable urban design can help establish a more resilient and ecologically balanced metropolitan landscape. Addressing the challenges and implementing future-oriented strategies may help cities strive to ensure the continued provision of essential ecosystem services for the benefit of both current and future generations.

7 Conclusion

The preservation and enhancement of ecosystem services within urban areas are paramount for the well-being of humans and the environment. As urbanization continues to reshape landscapes and communities, maintaining the flow of ecosystem services becomes increasingly crucial. Through strategic urban planning, green infrastructure implementation, biodiversity conservation, and community engagement, cities can transcend the conventional paradigm of environmental degradation associated with rapid urban growth. Recognizing the intrinsic value of these services is essential for informed decision-making in urban planning and development. A holistic approach that integrates green infrastructure, sustainable land use practices, and community engagement is imperative to sustain ecosystem services in urban contexts. Striking a balance between urban expansion and ecological preservation ensures that essential services promoting human health, improving quality of life, and mitigating the effects of climate change continue to be delivered. Understanding the reciprocal relationship between urban environments and ecological services is crucial for fostering resilient and thriving cities. As we confront the challenges of urbanization, adopting sustainable behaviors becomes both a responsibility and an investment in a healthier, more harmonious future for urban residents and the ecosystems they rely on. Nature-based solutions, technological innovations, and circular economy principles are emerging as vital pillars in guiding urban development towards sustainability. However, achieving resilient urban ecosystems requires collaboration. Equitable access to nature, participatory governance, and environmental justice are fundamental elements empowering communities to actively contribute to preserving ecosystem services. Experiences from cities worldwide illustrate that informed and engaged communities are potent catalysts for positive change. The future of urban environments hinges on a collective commitment to creating spaces that meet the needs of present generations and safeguard future generations' well-being. Implementing these principles and fostering a holistic approach to urban development pave the way for cities to thrive as innovation, sustainability, and resilience centers, ensuring the enduring flow of essential ecosystem services for generations.

References

Ahern J (2011) From fail-safe to safe-to-fail: sustainability and resilience in the new urban world. Landsc Urban Plan 100(4):341–343

Almusaed A, Almssad A (eds) (2023) Sustainable smart cities: a vision for tomorrow. BoD–Books on Demand

Andersson E, Barthel S, Borgström S, Colding J, Elmqvist T, Folke C, Gren A (2014) Reconnecting cities to the biosphere: stewardship of green infrastructure and urban ecosystem services. Ambio 43:445–453

Archer D, Almansi F, DiGregorio M, Roberts D, Sharma D, Syam D (2014) Moving towards inclusive urban adaptation: approaches to integrating community-based adaptation to climate change at city and national scale. Climate Dev 6(4):345–356

Bai X, Surveyer A, Elmqvist T, Gatzweiler FW, Güneralp B, Parnell S, Webb R (2016) Defining and advancing a systems approach for sustainable cities. Current Opinion Environ Sustain 23:69–78

Balaban O (2017) Land use. In: Urbanization and climate co-benefits. Routledge, pp 67–87

Barton J, Pretty JULES (2010) Urban ecology and human health and wellbeing. Urban Ecol 12(1):202–229

Bauhus J, Pokorny B, van der Meer PJ, Kanowski P, Kanninen M (2010) Ecosystem goods and services–the key for sustainable plantations. In: Ecosystem goods and services from plantation forests, pp 205–227

Beatley T (2011) Basophilic cities: integrating nature into urban design and planning. Island Press

Bebbington J, Unerman J (2018) Achieving the United Nations sustainable development goals: an enabling role for accounting research. Account, Audit Account J 31(1):2–24
Bennett AF (2003) Linkages in the landscape: the role of corridors and connectivity in wildlife conservation. IUCN
Betsill MM, Bulkeley H (2006) Cities and the multilevel governance of global climate change. Glob Gov 12:141
Birkhofer K, Diehl E, Andersson J, Ekroos J, Früh-Müller A, Machnikowski F, Smith HG (2015) Ecosystem services—current challenges and opportunities for ecological research. Front Ecol Evol 2:87
Caragliu A, Del B, C–Nijkamp P (2011) Smart cities. Europe J Urban Tech 18(2):65–82
Cardinale BJ, Duffy JE, Gonzalez A, Hooper DU, Perrings C, Venail P, Naeem S (2012) Biodiversity loss and its impact on humanity. Nature 486(7401):59–67
Cornwall A (2008) Unpacking 'Participation': models, meanings and practices. Community Dev J 43(3):269–283
Costanza R, De Groot R, Braat L, Kubiszewski I, Fioramonti L, Sutton P, Grasso M (2017) Twenty years of ecosystem services: how far have we come and how far do we still need to go? Ecosyst Serv 28:1–16
Daily GC, Söderqvist T, Aniyar S, Arrow K, Dasgupta P, Ehrlich PR, Walker B (2000) The value of nature and the nature of value. Science 289(5478):395–396
Daily GC, Polasky S, Goldstein J, Kareiva PM, Mooney HA, Pejchar L, Shallenberger R (2009) Ecosystem services in decision making: time to deliver. Front Ecol Environ 7(1):21–28
Escobedo FJ, Kroeger T, Wagner JE (2011) Urban forests and pollution mitigation: analyzing ecosystem services and disservices. Environ Pollut 159(8–9):2078–2087
Foley JA, DeFries R, Asner GP, Barford C, Bonan G, Carpenter SR, Snyder PK (2005) Global consequences of land use. Science 309(5734):570–574
Gaur N, Sharma S, Yadav N (2024) Environmental pollution. In: Green chemistry approaches to environmental sustainability. Elsevier, pp 23–41
Goddard MA, Dougill AJ, Benton TG (2010) Scaling up from gardens: biodiversity conservation in urban environments. Trends Ecol Evol 25(2):90–98
Gomez JEA Jr (2017) The size of cities: a synthesis of multi-disciplinary perspectives on the global megalopolis. Prog Plan 116:1–29
Green OO, Garmestani AS, Albro S, Ban NC, Berland A, Burkman CE, Shuster WD (2016) Adaptive governance to promote ecosystem services in urban green spaces. Urban Ecosyst 19:77–93
Gupta SK, Singh H (2017) Observations on Sporadic Flowering in *Ficus benjamina* L. and *Peltophorum africanum* L. in New Forest. Indian Forest 143:290–291
Gupta SK, Ram J, Singh H (2018a) Comparative study of transpiration in cooling effect of tree species in the atmosphere. J Geosci Environ Protect 6:151–166. https://doi.org/10.4236/gep.2018.68011
Gupta SK, Ram J, Singh H, Kumar P (2018b) Comparative study of cooling under shades of some forestry tree species with respect to ambient temperatures. J Geosci Environ Protect 6:51–66. https://doi.org/10.4236/gep.2018.611004
Gupta SK, Ram J, Singh H (2019a) Relationship between leaf area index and atmospheric cooling of tree species. Indian J Forest 42(3):203–208
Gupta SK, Singh H, Ram J (2019b) Need for a novel paradigm in forestry research: perspectives of comparative atmospheric cooling by different forestry effect tree species. Indian Forester 145(6):509–515
Hajer MA, Westhoek H, Ingram J, Van Berkum S, Ozay L (2016) Food systems and natural resources. United Nations Environmental Programme
IPCC (2021) Summary for policymakers. In: Climate change 2021: the physical science basis. Contribution of working group I to the sixth assessment report of the intergovernmental panel on climate change
Jangra A, Kumar K, Maikhuri S, Bhandari MS, Pandey S, Singh H, Barthwal S (2024) Unveiling stress-adapted endophytic bacteria: characterizing plant growth-promoting traits and assessing

cross-inoculation effects on *Populus deltoides* under abiotic stress. Plant Physiol Biochem 210:108610. https://doi.org/10.1016/j.plaphy.2024.108610

Jordán R, Infante B (2012) A strategic planning approach for developing eco-efficient and socially inclusive urban infrastructure. Local Environ 17(5):533–544

Joshi MY, Teller J (2021) Urban integration of green roofs: current challenges and perspectives. Sustainability 13(22):12378

Kabisch N, Frantzeskaki N, Pauleit S, Naumann S, Davis M, Artmann M, Bonn A (2016) Nature-based solutions to climate change mitigation and adaptation in urban areas: perspectives on indicators, knowledge gaps, barriers, and opportunities for action. Ecol Soc 21(2)

Kanta C, Kumar A, Chauhan A, Singh H, Sharma IP (2024) The interplay between plant functional traits and climate change. In: Kumar N, Singh H (eds) Plant functional traits for improving productivity. Springer, Singapore. https://doi.org/10.1007/978-981-97-1510-7_3

Kremen C, Ostfeld RS (2005) A call to ecologists: measuring, analyzing, and managing ecosystem services. Front Ecol Environ 3(10):540–548

Kremsa VS (2021) Sustainable management of agricultural resources (agricultural crops and animals). In: Sustainable resource management. Elsevier, pp 99–145

Kumar N, Singh H (2024) Plant functional traits for improving productivity. Springer Nature, Singapore. https://doi.org/10.1007/978-981-97-1510-7

Kumar N, Suyal DC, Sharma IP, Verma A, Singh H (2017) Elucidating stress proteins in rice (*Oryza sativa* L.) genotype under elevated temperature: a proteomic approach to understand heat stress response. 3 Biotech 7:205. https://doi.org/10.1007/s13205-017-0856-9

Kumar M, Singh MP, Singh H, Dhakate PM, Ravindranath NH (2019a) Forest working plan for the sustainable management of forest and biodiversity in India. J Sustain Forest 39:1–22. https://doi.org/10.1080/10549811.2019.1632212

Kumar N, Jeena N, Singh H (2019b). Elevated temperature modulates rice pollen structure: a study from the foothill of the Himalayan agro-ecosystem in India. 3 Biotech 9(5):175. https://doi.org/10.1007/s13205-019-1700-1

Kumar N, Jeena N, Kumar N, Gangola S, Singh H. (2019c). Phytoremediation facilitating enzymes: an enzymatic approach for enhancing remediation process. In: Bhatt P (ed) Smart bioremediation technologies. Academic Press, Elsevier. 50 Hampshire Street, 5th Floor, Cambridge, MA 02139, United States, pp 289–306

Kumar N, Singh H, Sharma SK (2020) Antioxidants: responses and importance in plant defense system. In: Roychowdhury R, Choudhury S, Hasanuzzaman M, Srivastava S (eds) Sustainable agriculture in the era of climate change. Springer, Cham, pp 251–264. https://doi.org/10.1007/978-3-030-45669-6_11

Kumar A, Kumar P, Singh H, Kumar N (2021a) Modulation of plant functional traits under essential plant nutrients during seasonal regime in natural forests of Garhwal Himalayas. Plant Soil 465:197–212. https://doi.org/10.1007/s11104-021-05003-x

Kumar A, Kumar P, Singh H, Bisht S, Kumar N (2021b) Relationship of physiological plant functional traits with soil carbon stock in the temperate forest of Garhwal Himalaya. Curr Sci 120(8):1368–1373. https://doi.org/10.18520/cs/v120/i8/1368-1373

Kumar A, Kumar P, Singh H, Kumar N (2021c) Adaptation and mitigation potential of roadside trees with bio-extraction of heavy metals under vehicular emissions and their impact on physiological traits during seasonal regimes. Urban Forest Urban Green 58:126900. https://doi.org/10.1016/j.ufug.2020.126900

Kumar A, Kumar P, Singh H, Kumar N (2021d) Impact of plant functional traits on infiltration rate under different forests of Kempty watershed in Garhwal Himalaya. Indian J Soil Conserv 49(1):38–44

Lambin EF, Turner BL, Geist HJ, Agbola SB, Angelsen A, Bruce JW, Xu J (2001) The causes of land-use and land-cover change: moving beyond the myths. Glob Environ Chang 11(4):261–269

Lee S (2021) Does tree canopy moderate the association between neighborhood walkability and street crime? Urban Forest Urban Green 65:127336

Levy JM (2016) Contemporary urban planning. Taylor & Francis

Lin BB, Philpott SM, Jha S, Liere H (2017) Urban agriculture as a productive green infrastructure for environmental and social well-being. In: Greening cities: forms and functions, pp 155–179
Litman T (2015) Evaluating public transit benefits and costs. Victoria transport policy institute, Victoria, BC, Canada
MacArthur E (2013) Towards the circular economy. J Ind Ecol 2(1):23–44
Malik A, Kumar K, Kumari G, Kumar N, Singh H (2024) Application of functional traits in modelling productivity and resilience under climate change. In: Kumar N, Singh H (eds) Plant functional traits for improving productivity. Springer, Singapore. https://doi.org/10.1007/978-981-97-1510-7_5
Masiero M, Pettenella D, Boscolo M, Barua SK, Animon I, Matta JR (2019) Valuing forest ecosystem services: a training manual for planners and project developers. Forestry Working Paper No. 11. Rome, FAO. 216 pp. Licence: CC BY-NC-SA 3.0 IGO
McPherson EG, Nowak D, Heisler G, Grimmond S, Souch C, Grant R, Rowntree R (1997) Quantifying urban forest structure, function, and value: the Chicago urban forest climate project. Urban Ecosyst 1:49–61
MEA (2005) Ecosystems and human well-being: wetlands and water. World resources institute
Meena D, Singh H, Chaudhari SK (2011) Elucidating strontium response on growth dynamics and biochemical change in *Phaseolus mungo* L. Int J Agricult Environ Biotech 4:107–113
Niemelä J, Breuste JH, Guntenspergen G, McIntyre NE, Elmqvist T, James P (eds) (2011) Urban ecology: patterns, processes, and applications. Oxford University Press, Oxford
Nowak DJ, Crane DE (2002) Carbon storage and sequestration by urban trees in the USA. Environ Pollut 116(3):381–389
Orta-Ortiz MS, Geneletti D (2021) Reviewing the performance of nature-based solutions for stormwater management in urban areas. International Conference on Innovation in Urban and Regional Planning. Springer International Publishing, Cham, pp 15–22
Pickett ST, Cadenasso ML, Grove JM, Nilon CH, Pouyat RV, Zipperer WC, Costanza R (2001) Urban ecological systems: linking terrestrial ecological, physical, and socioeconomic components of metropolitan areas. Annu Rev Ecol Syst 32(1):127–157
Plieninger T, Bieling C (2013) Resilience-based perspectives to guiding high-nature-value farmland through socioeconomic change. Ecol Soc 18(4)
Power AG (2010) Ecosystem services and agriculture: tradeoffs and synergies. Philos Trans Royal Soc B: Biol Sci 365(1554):2959–2971
Powlson DS, Gregory PJ, Whalley WR, Quinton JN, Hopkins DW, Whitmore AP, Goulding KW (2011) Soil management in relation to sustainable agriculture and ecosystem services. Food Policy 36:S72–S87
Rakhshandehroo M, Yusof MJM, Arabi R, Parva M, Nochian A (2017) The environmental benefits of urban open green spaces. Alam Cipta 10(1):10–16
Reed MS (2008) Stakeholder participation for environmental management: a literature review. Biol Cons 141(10):2417–2431
Reta Z, Adgo Y, Girum T, Mekonnen N (2020) Assessment of contribution of non-timber forest products in the socio-economic status of peoples in Eastern Ethiopia. Open Access J Biogener Sci Res
Roslan AF, Fernando T, Biscaya S, Sulaiman N (2021) Transformation towards risk-sensitive urban development: a systematic review of the issues and challenges. Sustainability 13(19):10631
Russ A, Krasny ME (eds) (2017) Urban environmental education review. Cornell University Press
Salmond JA, Tadaki M, Vardoulakis S, Arbuthnott K, Coutts A, Demuzere M, Wheeler BW (2016) Health and climate related ecosystem services provided by street trees in the urban environment. Environ Health 15(1):95–111
Semadeni-Davies A, Hernebring C, Svensson G, Gustafsson LG (2008) The impacts of climate change and urbanisation on drainage in Helsingborg, Sweden: suburban stormwater. J Hydrol 350(1–2):114–125
Shafique M, Kim R (2017) Green stormwater infrastructure with low impact development concept: a review of current research. Desalin Water Treat 83(7):16–29

Shanahan DF, Franco L, Lin BB, Gaston KJ, Fuller RA (2016) The benefits of natural environments for physical activity. Sports Med 46(7):989–995

Shandas V (2015) Neighborhood change and the role of environmental stewardship: a case study of green infrastructure for stormwater in the City of Portland, Oregon, USA. Ecol Soc 20(3)

Sharma R, Singh H (2021) Alteration in biochemical constituents and nutrients partitioning of *Asparagus racemosus* in response to elevated atmospheric CO_2 concentration. Environ Sci Pollut Res 29:6812–6821. https://doi.org/10.1007/s11356-021-16050-3

Sharma R, Singh H, Kaushik M, Nautiyal R, Singh O (2018) Adaptive physiological response, carbon partitioning, and biomass production of *Withania somnifera* (L.) Dunal grown under elevated CO_2 regimes. 3 Biotech 8:267. https://doi.org/10.1007/s13205-018-1292-1

Singh H (2021) An integrated approach considering physiological and biophysical based indicators for assessing tolerance of roadside plantations of *Alstonia scholaris* towards urban roadside air pollution: an assessment of adaptation of plantations for mitigating urban roadside air pollution. Trees. https://doi.org/10.1007/s00468-021-02179-8

Singh H, Verma A (2013) Characterization and screening of high nitrogen efficient rice genotype to curtail environmental pollution. Int J Agricult, Environ Biotech 6(1):23–30

Singh H, Kumar M (2022) Climate change and its impact on Indian Himalayan forests: current status and research needs. In: Rani S, Kumar R (eds) Climate change. Springer Climate. Springer, Cham, pp 223–242. https://doi.org/10.1007/978-3-030-92782-0_11

Singh H, Savita Sharma R, Sinha S, Kumar M, Kumar P, Verma A, Sharma SK (2017a) Physiological functioning of *Lagerstroemia speciosa* L. under heavy roadside traffic: an approach to screen potential species for abatement of urban air pollution. 3 Biotech 7(61):1–10. https://doi.org/10.1007/s13205-017-0690-0

Singh H, Verma A, Kumar M, Sharma R, Gupta R, Kaur M, Negi M, Sharma SK (2017b) Phytoremediation: a green technology to clean up the sites with low and moderate level of heavy metals. Austin Biochem 2:1012

Singh H, Sharma R, Savita SMP, Kumar M, Verma A, Ansari MW, Sharma SK (2018) Adaptive physiological response of *Parthenium hysterophorus* to elevated atmospheric CO_2 concentration. Indian Forester 144:1–14

Singh H, Yadav M, Kumar N, Kumar A, Kumar M (2020). Assessing adaptation and mitigation potential of roadside trees under the influence of vehicular emissions: a case study of *Grevillea robusta* and *Mangifera indica* planted in an urban city of India. PLoS One 15(1):e0227380. https://doi.org/10.1371/journal.pone.0227380

Singh S, Singh H, Sharma SK, Nautiyal R (2021) Seasonal variation in biochemical responses of bamboo clones in the sub-tropical climate of Indian Himalayan foothills. Heliyon 7(4):e06859. https://doi.org/10.1016/j.heliyon.2021.e06859

Singh H, Kumar N, Kumar A (2024). Enhancing resource use efficiency in crops through plant functional traits. In: Kumar N, Singh H (eds) Plant functional traits for improving productivity. Springer, Singapore. https://doi.org/10.1007/978-981-97-1510-7_6

Sweeney F, Mitchell R (2019) Ecology, environment and pollution. Scientific e-Resources

Talubo JP, Morse S, Saroj D (2022) Whose resilience matters? a socio-ecological systems approach to defining and assessing disaster resilience for small islands. Environ Challenges 7:100511

Thurber A, Bohmann CR, Heflinger CA (2018) Spatially integrated and socially segregated: the effects of mixed-income neighbourhoods on social well-being. Urban Studies 55(9):1859–1874

Tyrväinen L, Pauleit S, Seeland K, De Vries S (2005) Benefits and uses of urban forests and trees. In: Urban forests and trees: a reference book, pp 81–114

Upadhyay RK (2020) Markers for global climate change and its impact on social, biological and ecological systems: a review. Am J Clim Chang 9(03):159

Ussiri DA, Lal R (2017) Carbon sequestration for climate change mitigation and adaptation. Springer International Publishing, Cham, pp 287–325

Vilà M, Espinar JL, Hejda M, Hulme PE, Jarosík V, Maron JL, Pyšek P (2011) Ecological impacts of invasive alien plants: a meta-analysis of their effects on species, communities and ecosystems. Ecol Lett 14(7):702–708

Wamsler C, Niven L, Beery TH, Bramryd T, Ekelund N, Jönsson KI, Stalhammar S (2016) Operationalizing ecosystem-based adaptation: harnessing ecosystem services to buffer communities against climate change. Ecol Soc 21(1)

Wolch JR, Byrne J, Newell JP (2014) Urban green space, public health, and environmental justice: the challenge of making cities 'just green enough.' Landsc Urban Plan 125:234–244

Wood CL, Lafferty KD, DeLeo G, Young HS, Hudson PJ, Kuris AM (2014) Does biodiversity protect humans against infectious disease? Ecology 95(4):817–832

Yadav SK, Singh H, Ginwal HS and Barthwal S (2019a). Elevated CO_2 enhanced growth and physiological process of *Populus deltoides* Bartr. ex Marsh. Indian Forester 145(1):23–27

Yadav SK, Singh H, Nautiyal R, Ginwal HS, Ansari SA, Barthwal S (2019b) Modulation of morpho-physiological responses in *Populus deltoides* by elevated carbon dioxide and temperature. Forest Sci 66:105–118. https://doi.org/10.1093/forsci/fxz048

Yuen B (1996) Creating the Garden City: the Singapore experience. Urban Studies 33(6):955–970

Advancing Urban Forest Resilience: Strategies, Challenges, and Innovations in the Face of Climate and Environmental Change

Priyanka Lal, Pankaj Thakur, Lopamudra Nayak, Sandeep B. Adavi, Laxmipriya Behera, Muhammad Ahsan Altaf, Awadhesh Kumar, Ravinder Kumar, Rahul Kumar Tiwari, and Milan Kumar Lal

Abstract In urban sustainability, the intersection of climate change, environmental shifts, and the resilience of urban ecosystems emerges as a pivotal focal point. This chapter delves into the forefront of this intricate and dynamic landscape. As cities worldwide confront escalating impacts of climate change, surging population densities, and mounting environmental pressures, the urgency to enhance the resilience of urban forests grows ever more critical. This chapter explores state-of-the-art strategies, inherent challenges, and innovative solutions that improve urban forest resilience. Moreover, it analyzes the complex terrain of urban forestry, including

P. Lal
Lovely Professional University, Jalandhar GT Road (NH1), Phagwara, Punjab, India

P. Thakur
Chaudhary Charan Singh National Institute of Agricultural Marketing (CCS NIAM), Jaipur 302033, India

L. Nayak · L. Behera · A. Kumar · M. K. Lal (✉)
Division of Crop Physiology and Biochemistry, ICAR-National Rice Research Institute (ICAR-NRRI), Cuttack 753006, Odisha, India
e-mail: milan2925@gmail.com

S. B. Adavi
School of Crop Resistance System Research, ICAR-National Institute of Biotic Stress Management, Baronda, Raipur 493225, Chhattisgarh, India

M. A. Altaf
Key Laboratory for Quality Regulation of Tropical Horticultural Crops of Hainan Province, School of Breeding and Multiplication (Sanya Institute of Breeding and Multiplication), Hainan University, Sanya 572025, China

R. Kumar
ICAR-Indian Agricultural Research Institute, New Delhi, India

R. Kumar · R. K. Tiwari · M. K. Lal
ICAR-Central Potato Research Institute (ICAR-CPRI), Shimla 171001, Himachal Pradesh, India

R. K. Tiwari
ICAR-Indian Institute of Sugarcane Research, Lucknow, Uttar Pradesh, India

H. Singh (ed.), *Urban Forests, Climate Change and Environmental Pollution*,
https://doi.org/10.1007/978-3-031-67837-0_22

tactics utilized to bolster resilience, obstacles encountered during implementation, and transformative innovations shaping the trajectory of urban ecosystems. The insights gleaned may aid urban planners and policymakers in advancing urban forest resilience and providing sustainable solutions that harmonize with the evolving dynamics of our urban environments.

Keywords Urban sustainability · Climate change · Environmental transformations · Urban ecosystems · Resilience

1 Introduction

Urban areas face significant vulnerability to climate change impacts due to increased population densities, pollution, and construction activities. These factors exacerbate urban heat, drought, and extreme weather events, amplifying the challenges of climate change in urban environments (Kanta et al. 2024; Malik et al. 2024; Kumar and Singh 2024). Urban forest cover is crucial in helping cities mitigate these challenges by reducing greenhouse gas emissions and mitigating the adverse effects of extreme heat and climate variations on human health and the ecosystem (Yadav et al. 2019a, b; Jangra et al. 2024). Strengthening urban ecosystems has become increasingly vital amidst the escalating repercussions of climate change and environmental alterations.

Adopting a participatory and inclusive approach to urban forest planning, decision-making, and implementation is essential, albeit challenging. Understanding the specific services desired by communities and mitigating adverse effects on health, the environment, and economic opportunities is crucial (Meena et al. 2011; Kumar et al. 2019c, 2020). Whether initiatives stem from top-down directives or grassroots movements, they necessitate a multi-level holistic approach engaging all stakeholders, incorporating diverse interventions with the potential for substantial benefits.

This chapter delves into urban forest resilience, offering an in-depth exploration of the techniques and challenges involved. It highlights innovative approaches contributing to overall well-being, drawing from contemporary research to elucidate tactics enhancing urban forest resilience. It illustrates how cities strategically plan and manage their urban forests to withstand climate change impacts, emphasizing the importance of diverse tree species, adaptive management practices, and green infrastructure integration. Furthermore, it critically examines challenges cities face in sustainable urban forestry, including limited resources, conflicting urban planning priorities, and potential socio-environmental disparities. This examination offers insights into crafting effective and inclusive strategies for resilient urban forest management.

2 Vulnerabilities of Urban Regions to Climate Change

Urban areas, housing half of the global population, are particularly vulnerable to climate change, contributing to over 70% of Greenhouse Gas (GHG) emissions, a percentage that continues to rise with population growth. Consequently, climate change significantly stresses urban areas, resulting in frequent floods and other natural disasters. Coastal urban areas, in particular, face heightened risks due to climate change, impacting habitats, infrastructure, property, and ecosystems (Pelling and Blackburn 2014; Singh and Kumar 2022; Singh et al. 2024b).

The composition of urban construction, including elements like roads and infrastructure, plays a crucial role in GHG emissions and exacerbates the effects of climate change. Urban areas constructed with non-sustainable materials increase the likelihood of heat accumulation, contributing to global warming (Singh and Verma 2013; Kumar et al. 2017). Building design, composition, form, and alignment influence air conditioning and heating demand, further exacerbating climate impacts. Impermeable surfaces in urban regions worsen flooding and contribute to the Urban Heat Island effect, where metropolitan areas experience higher temperatures than surrounding rural areas due to reduced evaporation and increased sensible heat release (Grimmond 2011; Gupta and Singh 2017; Singh et al. 2017a, b).

The limited number of trees in urban environments disrupts natural processes like evapotranspiration, exacerbating the impacts of droughts and floods. However, cities have the advantage of assuming a prominent role in predicting and adapting to climate fluctuations. By integrating resilience into their development processes, cities can create more resilient projects that effectively address the needs of their populations.

A conceptual framework of urban vulnerability is presented in Fig. 1, illustrating how individuals and civilizations can leverage favorable circumstances to mitigate potential hazards through understanding, strategizing, and adapting to a dynamic environment. Immediate action is required to establish frameworks for Urban forests, enhancing the environment's capacity to withstand and adapt to climate change.

3 Role of Urban Forest Cover in Climate Change Mitigation

As discussed in the previous section on the susceptibility of urban areas to climate change, expanding urban forests would improve aesthetic enjoyment and serve as a defense against climate-related hazards. It has a pivotal function in mitigating climate change by offering a variety of environmental, social, and economic advantages (Sharma et al. 2018; Singh et al. 2018). The importance of urban forest cover in climate change mitigation encompasses numerous crucial issues.

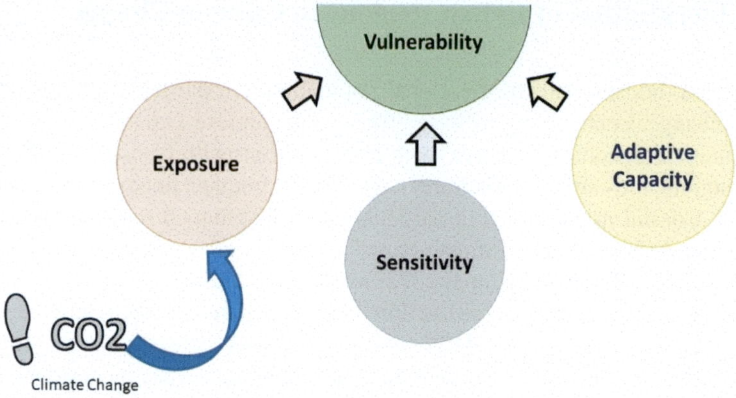

Fig. 1 Concept of urban vulnerability

3.1 Carbon Sequestration and Carbon Sink

The increase in population and industrial activity has resulted in a rise in pollution levels in metropolitan areas. Consequently, trees play a crucial role in mitigating this issue by absorbing carbon dioxide (CO_2) through photosynthesis and storing carbon in their biomass (Gupta et al. 2018a, b). Carbon sequestration is a method that reduces the concentration of greenhouse gases in the atmosphere. Urban areas benefit from a collective green cover that acts as a carbon sink, effectively absorbing and retaining carbon from the atmosphere (Yadav et al. 2019a, b; Gupta et al. 2019a, b). This aids in alleviating the impacts of increasing CO_2 concentrations, a significant catalyst of climate change.

3.2 Temperature Regulation and Air Quality Improvement

Trees offer shade and lower surface temperatures by evapotranspiration. The cooling effect is particularly crucial in urban heat islands, characterized by extensive areas covered with concrete and asphalt, which can result in higher temperatures (Kumar et al. 2019a, b). Urban woods enhance localized climate resilience by reducing heat. Trees and vegetation serve as natural purifiers, capturing and eliminating atmospheric contaminants (Singh et al. 2020; Kumar et al. 2021a). This initiative enhances air quality by mitigating the influence of pollutants that contribute to climate change and adversely harm human well-being.

3.3 Reduced Runoff

Urban forests contribute to stormwater management by absorbing and attenuating rainfall. This mitigates the potential for flooding and erosion, which are intensified by extreme weather events caused by climate change. Urban woods offer a habitat and refuge for diverse species of plants and animals. Biodiversity is crucial for the well-being of ecosystems, and urban forests play a significant role in preserving equilibrium within local ecosystems.

3.4 Energy Conservation

Properly placed trees can reduce the need for air conditioning by providing shade to buildings. Additionally, they can act as windbreaks, reducing heat loss during colder months thereby contributing to energy conservation.

3.5 Urban Planning and Climate Resilience

Urban forests should be integrated into urban planning strategies to maximize their benefits. Adaptive solutions and well-designed green spaces are effective strategies to enhance a city's resilience to climate change (Fig. 2). Urban forests serve as educational tools, raising awareness about the importance of sustainable practices, including tree planting, maintenance, and conservation (Kumar et al. 2021b, c). Urban forests are integral components of climate change mitigation efforts. Recognizing their multifaceted contributions and incorporating them into urban planning and development can significantly enhance the overall resilience of cities in the face of climate challenges (Kumar et al. 2021d).

3.6 Challenges in Urban Forest Resilience

Urban forests face various challenges, and addressing these challenges is vital for effective urban forest management, resilience, and long-term sustainability. A comprehensive list of challenges has been presented in Fig. 3, recognizing and addressing what is crucial for effective urban forest management and conservation.

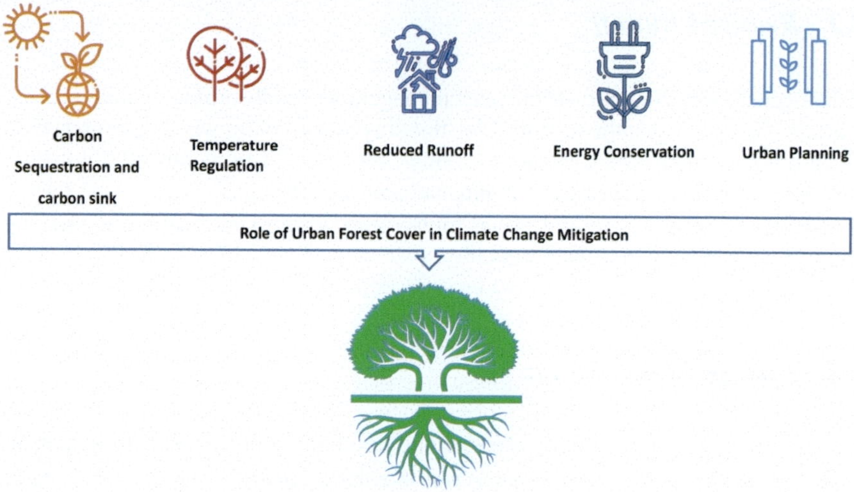

Fig. 2 Role of urban forests in climate change mitigation

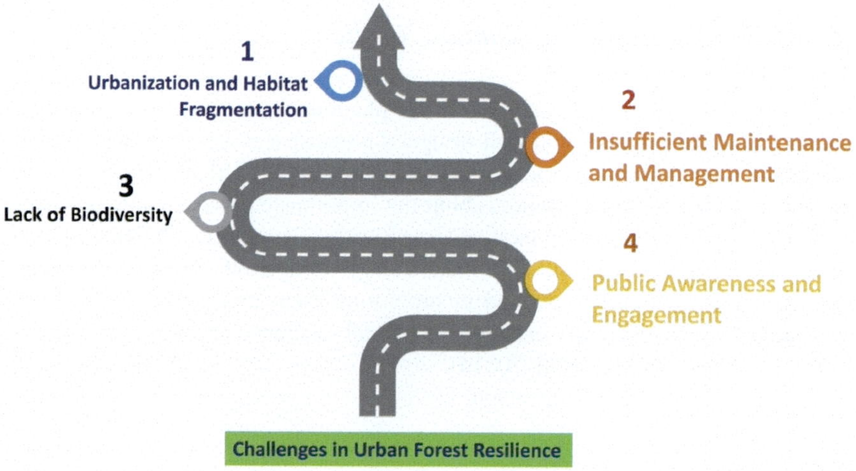

Fig. 3 Challenges in urban forest resilience

3.7 *Urbanization and Habitat Formation*

Urban environments are not always suitable for tree species growth. Exposure to pollutants, high temperatures, extreme drought, inundation, and limited space above and below ground inhibit proper root and crown formation (Singh 2021; Sharma and Singh 2021). Rapid urbanization and shrinking land size can often lead to losing natural habitats and green spaces, reducing urban forests' overall size and connectivity (Singh et al. 2021). This fragmentation can isolate populations of flora and fauna, impacting biodiversity.

3.8 Insufficient Maintenance and Management

Effective maintenance is essential for a flourishing urban forest, but municipal authorities must allocate time, resources, and coordination. Urban forests are vulnerable to inadequate resources for adequate maintenance. Financially constrained private property owners may find it challenging to prioritize. Insufficient resources also impede endeavors to catalog both public and private green areas (Dhyani et al. 2021). Uncertainty over the care of trees and forests might arise when there is inadequate information about land ownership or tenure. Furthermore, urban development leads to encroachment, which in turn causes the depletion of urban forest canopy and other green areas (van den Bosch et al. 2017; Kumar et al. 2021e; Prakash et al. 2022). The public and decision-makers often perceive urban trees as a financial liability or potential danger, although their advantages may be inadequately comprehended and underestimated (Escobedo et al. 2011).

4 Innovative Approaches for Urban Forest Resilience

In the contemporary urban landscape, environmental challenges, population growth, and diminishing green spaces necessitate a comprehensive exploration of strategies to fortify urban forest resilience (Kumar et al. 2022; Singh et al. 2022). This section delves into scientifically grounded approaches to address the challenges confronting trees and green spaces in urban environments.

4.1 Green Infrastructure Planning

Urban areas, characterized by spatial constraints, require meticulous planning of green spaces to fulfill multifaceted roles encompassing stormwater management, air purification, and biodiversity conservation. Integrating green infrastructure into urban planning emerges as a strategic imperative for shaping resilient and adaptive landscapes (Monteiro and Ferreira 2020; Benedict and McMahan 2002; Demuzere et al. 2014; Gómez-Baggethun and Barton 2013). It can be executed in different ways:

4.1.1 Green Roofs and Vertical Forests

Implementing vertical forests and green walls on urban structures presents a promising avenue to optimize green space utilization in densely populated areas. Beyond aesthetic enhancement, these interventions contribute significantly to mitigating heat island effects, providing habitat, and managing stormwater effectively (Kumar et al. 2024; Singh et al. 2024a).

4.1.2 Permeable Pavements and Tree Pits

Combining permeable pavements with strategically placed tree pits fosters water infiltration and root growth. This integrated approach serves dual purposes by supporting stormwater management objectives and promoting the health of urban trees.

5 Smart Urban Forest Management

The recognition of green spaces as integral components of sustainable cities has precipitated the emergence of Smart Forest Management. This innovative approach leverages cutting-edge technologies to enhance urban forests' planning, monitoring, and maintenance, ensuring their pivotal role in ecological balance, air quality improvement, and overall community well-being (Singh 2024). Utilize sensor networks to monitor tree health, soil conditions, and environmental factors. This data can inform adaptive management strategies, allowing for timely interventions to address stressors or threats to the urban forest (Pelling and Blackburn 2014). Furthermore, urban forest diversity improves aesthetics, providing residents with a visually appealing and engaging landscape. Beyond aesthetics, a diverse range of tree species supports a broader array of wildlife and promotes ecological balance within urban ecosystems. Use Geographic Information Systems (GIS) and data analytics to analyze urban forest health and vulnerabilities. This data-driven approach helps identify priority areas for intervention and guides long-term planning strategies. Incorporating green technologies, such as solar-powered sensors, into urban forest monitoring provides real-time data for informed decision-making, enhancing adaptive management strategies (Thakur et al. 2023).

6 Urban Forest Diversity and Tree Selection

Promoting species diversity within urban forests is an aesthetic consideration and a resilience strategy against pests, diseases, and climate variability. Thoughtful selection of tree species that are well-adapted to local conditions contributes substantially to the overall robustness of urban forests. The selection of well-suited species for the local environment may help urban planners enhance the urban forest's overall health and longevity while promoting biodiversity. Proactively selecting climate-resilient tree species tailored to prevailing and anticipated climatic conditions is foundational in enhancing urban forest resilience against changing environmental dynamics.

7 Community Engagement and Citizen Science

The involvement of communities in tree planting and maintenance activities may empower residents to become stewards of their local urban forests, foster a sense of ownership, and strengthen the resilience of green spaces. Community involvement can be beneficial as the urban forests are common property resources and can be managed well. Additionally, implementing adaptive pruning schedules and care practices that respond dynamically to changing environmental conditions is essential for ensuring urban trees' long-term health and resilience. Educational programs targeted at urban planners, arborists, and the public play a pivotal role in disseminating awareness about the importance of urban forest resilience. Such initiatives foster a culture of sustainability and resilience within urban ecosystems.

Along with this, collaborative governance can be beneficial for urban forest management. Promoting multi-stakeholder collaboration among government agencies, community groups, businesses, and researchers is crucial for developing and implementing robust urban forest strategies. Collective efforts are indispensable for addressing the multifaceted challenges facing urban green spaces.

8 Conclusion

Globally, urban areas face escalating vulnerabilities to the impacts of climate change and environmental pollution, necessitating proactive and multidimensional strategies to foster resilience. The insights into various techniques in this chapter may enhance urban forest resilience and achieve the goals of sustainable urban forestry. This comprehensive understanding of these challenges would help craft effective and inclusive strategies, from resource constraints to conflicting planning priorities and potential socio-environmental disparities. Interventions such as incorporating suitable tree species, adaptive management techniques, and green infrastructure integration are valuable for withstanding the impacts of a changing climate. The inclusive approach informs planning, education, outreach, research, and potential changes in policy or infrastructure, fostering resilient urban forest management that aligns with the diverse and dynamic nature of urban environments. Implementation of innovative approaches establish and maintain resilient urban forests capable of withstanding current challenges and adapting to future climate change and urban development uncertainties.

References

Benedict MA, McMahon ET (2002) Green infrastructure: smart conservation for the 21st century. Renew Resour J 20:12–17

Demuzere M, Orru K, Heidrich O, Olazabal E, Geneletti D, Orru H, Bhave AG, Mittal N, Feliu E, Faehnle M (2014) Mitigating and adapting to climate change: Multi-functional and multi-scale assessment of green urban infrastructure. J Environ Manage 146:107–115

Dhyani K, Kuniyal HB, Singh H, Sobha (2021) Growth and physiological potential of *Terminalia arjuna* under elevated CO_2 levels in Open top chamber condition. J Appl Nat Sci 13(3):1121–1126. https://doi.org/10.31018/jans.v13i3.2490. ISSN 2231-5209

Escobedo FJ, Kroeger T, Wagner JE (2011) Urban forests and pollution mitigation: analyzing ecosystem services and disservices. Environ Pollut (Barking, Essex: 1987) 159(8–9):2078–2087

Gómez-Baggethun E, Barton DN (2013) Classifying and valuing ecosystem services for urban planning. Ecol Econ 86:235–245

Grimmond CSB (2011) Urban climate in the coastal zone: resilience to weather extremes. Megacities and the coast: transformation for resilience. A preliminary review of knowledge, practice and future research. LandOcean Interactions in the Coastal Zone (LOICZ), Chapter 7

Gupta SK, Singh H (2017) Observations on Sporadic Flowering in *Ficus benjamina* L. and *Peltophorum africanum* L. in New Forest. Indian Forest 143:290–291

Gupta SK, Ram J, Singh H (2018a) Comparative study of transpiration in cooling effect of tree species in the atmosphere. J Geosci Environ Protect 6:151–166. https://doi.org/10.4236/gep.2018.68011

Gupta SK, Ram J, Singh H, Kumar P (2018b) Comparative study of cooling under shades of some forestry tree species with respect to ambient temperatures. J Geosci Environ Protect 6:51–66. https://doi.org/10.4236/gep.2018.611004

Gupta SK, Ram J, Singh H (2019a) Relationship between leaf area index and atmospheric cooling of tree species. Indian J Forest 42(3):203–208

Gupta SK, Singh H, Ram J (2019b) Need for a novel paradigm in forestry research: perspectives of comparative atmospheric cooling by different forestry effect tree species. Indian Forester 145(6):509–515

Jangra A, Kumar K, Maikhuri S, Bhandari MS, Pandey S, Singh H, Barthwal S (2024) Unveiling stress-adapted endophytic bacteria: Characterizing plant growth-promoting traits and assessing cross-inoculation effects on *Populus deltoides* under abiotic stress. Plant Physiol Biochem 210:108610. https://doi.org/10.1016/j.plaphy.2024.108610

Kanta C, Kumar A, Chauhan A, Singh H, Sharma IP (2024) The interplay between plant functional traits and climate change. In: Kumar N, Singh H (eds) Plant functional traits for improving productivity. Springer, Singapore. https://doi.org/10.1007/978-981-97-1510-7_3

Kumar N, Singh H (2024) Plant functional traits for improving productivity. Springer Nature, Singapore. https://doi.org/10.1007/978-981-97-1510-7

Kumar N, Suyal DC, Sharma IP, Verma A, Singh H (2017) Elucidating stress proteins in rice (*Oryza sativa* L.) genotype under elevated temperature: a proteomic approach to understand heat stress response. 3 Biotech 7:205. https://doi.org/10.1007/s13205-017-0856-9

Kumar M, Singh MP, Singh H, Dhakate PM, Ravindranath NH (2019a) Forest working plan for the sustainable management of forest and biodiversity in India. J Sustain Forest 39:1–22. https://doi.org/10.1080/10549811.2019.1632212

Kumar N, Jeena N, Singh H (2019b). Elevated temperature modulates rice pollen structure: a study from the foothill of the Himalayan agro-ecosystem in India. 3 Biotech 9(5):175. https://doi.org/10.1007/s13205-019-1700-1

Kumar N, Jeena N, Kumar N, Gangola S, Singh H (2019c) Phytoremediation facilitating enzymes: an enzymatic approach for enhancing remediation process. In: Bhatt P (ed) Smart bioremediation technologies. Academic Press, Elsevier. 50 Hampshire Street, 5th Floor, Cambridge, MA 02139, United States, pp 289–306

Kumar N, Singh H, Sharma SK (2020) Antioxidants: responses and importance in plant defense system. In: Roychowdhury R, Choudhury S, Hasanuzzaman M, Srivastava S (eds) Sustainable agriculture in the era of climate change. Springer, Cham, pp 251–264. https://doi.org/10.1007/978-3-030-45669-6_11

Kumar A, Kumar P, Singh H, Kumar N (2021a) Modulation of plant functional traits under essential plant nutrients during seasonal regime in natural forests of Garhwal Himalayas. Plant Soil 465:197–212. https://doi.org/10.1007/s11104-021-05003-x

Kumar A, Kumar P, Singh H, Bisht S, Kumar N (2021b) Relationship of physiological plant functional traits with soil carbon stock in the temperate forest of Garhwal Himalaya. Curr Sci 120(8):1368–1373. https://doi.org/10.18520/cs/v120/i8/1368-1373

Kumar A, Kumar P, Singh H, Kumar N (2021c) Adaptation and mitigation potential of roadside trees with bio-extraction of heavy metals under vehicular emissions and their impact on physiological traits during seasonal regimes. Urban Forest Urban Green 58:126900. https://doi.org/10.1016/j.ufug.2020.126900

Kumar A, Kumar P, Singh H, Kumar N (2021d) Impact of plant functional traits on infiltration rate under different forests of Kempty watershed in Garhwal Himalaya. Indian J Soil Conserv 49(1):38–44

Kumar M, Phukon SN, Paygude AC, Tyagi K, Singh H (2021e) Mapping phenological functional types (PhFT) in the Indian Eastern Himalayas using machine learning algorithm in Google Earth Engine. Comp Geosci 158. https://doi.org/10.1016/j.cageo.2021.104982

Kumar A, Singh H, Kumari G, Bisht S, Malik A, Kumar A, Singh M, Barthwal SR, Thakur A, Kaushal R (2022) Adaptive resilience of roadside trees to vehicular emissions via leaf enzymatic, physiological, and anatomical trait modulations. Environ Pollut 313. https://doi.org/10.1016/j.envpol.2022.120191

Kumar N, Singh H, Giri K, Kumar A, Joshi A, Yadav A, Singh R, Bisht S, Kumari R, Jeena N, Khairakpam R, Mishra G (2024) Physiological and molecular insights into the allelopathic effects on agroecosystems under changing environmental conditions. Physiol Mol Biol Plants 30:417–433. https://doi.org/10.1007/s12298-024-01440-x

Malik A, Kumar K, Kumari G, Kumar N, Singh H (2024) Application of functional traits in modelling productivity and resilience under climate change. In: Kumar N, Singh H (eds) Plant functional traits for improving productivity. Springer, Singapore. https://doi.org/10.1007/978-981-97-1510-7_5

Meena D, Singh H, Chaudhari SK (2011) Elucidating strontium response on growth dynamics and biochemical change in *Phaseolus mungo* L. Int J Agricult Environ Biotech 4:107–113

Monteiro R, Ferreira JC (2020) Green infrastructure planning as a climate change and risk adaptation tool in coastal urban areas. J Coast Res 95:889–893

Pelling M, Blackburn S (2014) Megacities and the coast: risk, resilience and transformation. Earthscan from Routledge. ISBN: 9780415815123

Prakash V, Hunney K, Singh H (2022). Effect of elevated carbon dioxide on growth and development of *Santalum album* L. seedlings inoculated with plant growth promoting microorganisms in open top chambers. Int J Environ, Agricult Biotech 7(5):149–159

Sharma R, Singh H (2021) Alteration in biochemical constituents and nutrients partitioning of *Asparagus racemosus* in response to elevated atmospheric CO_2 concentration. Environ Sci Pollut Res 29:6812–6821. https://doi.org/10.1007/s11356-021-16050-3

Sharma R, Singh H, Kaushik M, Nautiyal R, Singh O (2018) Adaptive physiological response, carbon partitioning, and biomass production of *Withania somnifera* (L.) Dunal grown under elevated CO_2 regimes. 3 Biotech 8:267. https://doi.org/10.1007/s13205-018-1292-1

Singh H (2021) An integrated approach considering physiological and biophysical based indicators for assessing tolerance of roadside plantations of *Alstonia scholaris* towards urban roadside air pollution: an assessment of adaptation of plantations for mitigating urban roadside air pollution. Trees. https://doi.org/10.1007/s00468-021-02179-8

Singh H (2024). Exploring adaptive modulation in plant functional traits and its impact on the productivity of *Acacia auriculiformis* under CO_2-enriched environment. Indust Crops Prod 210:118186. https://doi.org/10.1016/j.indcrop.2024.118186

Singh H, Verma A (2013) Characterization and screening of high nitrogen efficient rice genotype to curtail environmental pollution. Int J Agricult, Environ Biotech 6(1):23–30

Singh H, Kumar M (2022) Climate change and its impact on Indian Himalayan Forests: current status and research needs. In: Rani S, Kumar R (eds) Climate change. Springer Climate. Springer, Cham, pp 223–242. https://doi.org/10.1007/978-3-030-92782-0_11

Singh H, Bandyopadhyaya S, Kumar A, Singh M, Malik A, Kumari G, Raturi A, Yadav SK, Kumar M, Barthwal S, Thakur A and Rawat PS (2022) Understanding the physiological and biophysical response of urban roadside plantations for assessing adaptation and mitigation mechanisms toward vehicular emissions. Urban Clim 44. https://doi.org/10.1016/j.uclim.2022.101183

Singh H, Sharma RS, Sinha S, Kumar M, Kumar P, Verma A, Sharma SK (2017a) Physiological functioning of *Lagerstroemia speciosa* L. under heavy roadside traffic: an approach to screen potential species for abatement of urban air pollution. 3 Biotech 7(61):1–10. https://doi.org/10.1007/s13205-017-0690-0

Singh H, Verma A, Kumar M, Sharma R, Gupta R, Kaur M, Negi M, Sharma SK (2017b) Phytoremediation: a green technology to clean up the sites with low and moderate level of heavy metals. Austin Biochem 2:1012

Singh H, Sharma R, Savita SMP, Kumar M, Verma A, Ansari MW, Sharma SK (2018) Adaptive physiological response of *Parthenium hysterophorus* to elevated atmospheric CO_2 concentration. Indian Forester 144:1–14

Singh H, Yadav M, Kumar N, Kumar A, Kumar M (2020). Assessing adaptation and mitigation potential of roadside trees under the influence of vehicular emissions: a case study of *Grevillea robusta* and *Mangifera indica* planted in an urban city of India. PLoS ONE 15(1):e0227380. https://doi.org/10.1371/journal.pone.0227380

Singh S, Singh H, Sharma SK, Nautiyal R (2021) Seasonal variation in biochemical responses of bamboo clones in the sub-tropical climate of Indian Himalayan foothills. Heliyon 7(4):e06859. https://doi.org/10.1016/j.heliyon.2021.e06859

Singh M, Singh H, Kumar A, Kumar M, Barthwal S, Thakur A (2024a) Soil nitrogen availability determines the CO_2 fertilization effect on tree species (*Neolamarckia cadamba*): growth and physiological evidence. Environ Sustain 7:53–60. https://doi.org/10.1007/s42398-023-00300-w

Singh H, Kumar N, Kumar A (2024b). Enhancing Resource use efficiency in crops through plant functional traits. In: Kumar N, Singh H (eds) Plant functional traits for improving productivity. Springer, Singapore. https://doi.org/10.1007/978-981-97-1510-7_6

Thakur P, Mehta P, Guleria A, Divyanshu, Singh P, Sharma P (2020) Study on marketing performance and constraints of pea (*Pisum sativum*) output in high hills wet temperate zone of Himachal Pradesh, India. Indian J Ecol 50:129–135. https://doi.org/10.55362/ije/2023/3865

van den Bosch CCK, Ferrini F, Fini A, Ferrini F (2017) Introduction. In: Routledge handbook of urban forestry

Yadav SK, Singh H, Ginwal HS, Barthwal S (2019a). Elevated CO_2 enhanced growth and physiological process of *Populus deltoides* Bartr. ex Marsh. Indian Forest 145(1):23–27

Yadav SK, Singh H, Nautiyal R, Ginwal HS, Ansari SA, Barthwal S (2019b) Modulation of morpho-physiological responses in *Populus deltoides* by elevated carbon dioxide and temperature. Forest Sci 66:105–118. https://doi.org/10.1093/forsci/fxz048

Urban Forests: Benefits, Barriers, and Sustainable Solutions

Aman Mahajan, Bill Nelson Paul, Ankush Mehra, Hitesh Gupta, and S. Sarath

Abstract The significance of urban forests and green spaces has increased recently as more people flock to cities and urban areas for opportunities, convenience, education, etc. The benefits of urban forests are multifaceted and conducive to enhancing resilience, livability, and the health of cities. Urban forests reduce urban areas' carbon footprint and alleviate climate change. They improve the interception and infiltration of precipitation and act as natural filters, removing pollutants from the air and water. Urban forests provide essential habitat for wildlife and contribute to biodiversity in cities. They contribute to the health and welfare of the community through a reduction in pollution and improved aesthetics. Moreover, they provide direct economic benefits such as timber, fruits, etc., as well as indirect benefits such as property value appreciation, tourism, and a reduction in the cost of temperature management. Despite the numerous advantages of urban forests, they face challenges, such as continuous exposure to pollutants and susceptibility to pests and diseases, impacting their growth and development. Urban expansion generally prioritizes infrastructure over green spaces, creating space and resource constraints and leading to the decline of urban forest cover. Nonetheless, the existence of urban forests highlights the value of maintaining and growing city green areas. This chapter further dwells on urban forest's ecological, climatic, and social aspects of well-being and also describes the barriers and sustainable solutions in such efforts.

Keywords Urban forests · Green spaces · Air pollution · Climate change · Carbon footprint

A. Mahajan (✉) · B. N. Paul · A. Mehra
ICFRE-Forest Research Institute, PO New Forest, Dehradun 248006, Uttarakhand, India
e-mail: aman.mahajan58@gmail.com

H. Gupta
College of Forestry, Veer Chandra Singh Garhwali Uttarakhand University of Horticulture and Forestry, Ranichauri, Tehri Garhwal 249199, Uttarakhand, India

S. Sarath
ICFRE-Institute of Wood Science and Technology, Bengaluru 560003, Karnataka, India

H. Singh (ed.), *Urban Forests, Climate Change and Environmental Pollution*,
https://doi.org/10.1007/978-3-031-67837-0_23

1 Introduction

Urban forestry is managing and caring for trees and forests within urban and peri-urban areas. It encompasses the planning, planting, maintaining, and preserving of trees in cities, towns, and other developed areas. This includes street trees, parks, green belts, and other green spaces within urban landscapes. Urban forestry aims to enhance the quality of life for urban residents, improve environmental conditions, mitigate the urban heat island effect, and promote sustainable development (Costanza et al. 2017; Malik et al. 2024; Kumar and Singh 2024). Urban forestry plays a crucial role in addressing various environmental, social, and economic challenges faced by cities worldwide. Trees and urban green spaces provide many direct and indirect benefits in improving the ecology and human well-being. The significant benefits of urban forests are pollution reduction, carbon sequestration, reduction in air temperature, reduction of surface runoff, improving health, conservation of biodiversity, increasing property values, and reduction in energy consumption (Allen et al. 2017; Yadav et al. 2019a, b; Jangra et al. 2024; Kanta et al. 2024).

The concept of urban forestry has its roots in ancient civilizations where trees were valued for their cultural, aesthetic, and practical significance. However, the formal recognition of urban forestry as a discipline emerged in the nineteenth and early twentieth centuries with the rise of urbanization and industrialization. During this period, the rapid expansion of urban areas led to the degradation of urban environments, prompting the need for systematic tree planting and management. One of the earliest examples of urban forestry initiatives is the creation of parks and boulevards in cities such as Paris, London, and New York in the nineteenth century. These efforts were driven by landscape architects, urban planners, and conservationists who recognized the importance of green spaces in improving public health and well-being (Berland et al. 2017).

In the United States, establishing the National Urban Forestry Program in the 1970s marked a significant milestone in developing urban forestry as a professional discipline. Since then, cities worldwide have implemented urban forestry programs and policies to address urban challenges and harness the benefits of trees in urban environments. Overall, the historical evolution of urban forestry reflects society's evolving understanding of the value of trees and green spaces in creating liveable, sustainable, and resilient cities. This explores the pivotal importance of urban forests, illuminating the opportunities for sustainable urban development while delving into the challenges that confront their preservation and expansion.

2 Classification of Urban Forests

In a conventional format, the urban forests are divided based on their origin into natural and artificial. However, the Food and Agricultural Organization (FAO) has classified urban forests based on their significance and location into peri-urban forests, city parks, pocket parks, street trees, and other green spaces.

2.1 *Based on the Origin*

Based on the origin of the urban forests, they are divided into natural and artificial urban forests. Natural Urban Forest (NUF) is the fragmented patches or pockets of the natural forests or woodlands within the urban or peri-urban periphery that get protection or are saved for any reason. Aarey Forest (Mumbai) and Aravalli Biodiversity Park (Delhi) of India are examples of Natural Urban Forest (NUF). Artificial Urban Forest (AUF) are the forests that have been raised or developed through planned action on land where there was no sign of forests in the past or on tree-less land abandoned a long back within the urban or peri-urban periphery can be grouped under this category of artificial urban forest. There are numerous examples from almost all the urban and peri-urban areas nationwide. The trees raised in or around the governmental public or private land with the institutional or residential setup are good examples. Other examples of AUF are the public gardens in residential societies or the green belts developed in the industrial zones.

2.2 *Based on Location and Importance*

The FAO (2016) adopted five streamlined categories to classify urban forests based on their significance in addressing particular difficulties. For instance, peri-urban forests and woodlands are crucial to the supply of forest products such as fuel wood. However, urban forests and city parks are more important for recreation. (1) Peri-urban forests and woodlands forests surrounding towns and cities provide goods and services such as wood, fibre, fruit, other non-wood forest products, clean water, recreation, and tourism. (2) City parks and urban forests are large urban or district parks that usually have more than 0.5 hectares. They generally have a varying land cover and are at least partly equipped with facilities for recreation and leisure. (3) Pocket parks and gardens with trees are small district parks that usually have an area of less than 0.5 hectares. These are equipped with facilities for recreation and leisure and include private gardens and green spaces. (4) Trees on streets or in public squares are individual trees in parking lots and squares, small groups of trees on streets, linear tree populations, etc. (5) Other green spaces with trees are green areas that do not fall into the four categories listed above. These include river banks, urban agricultural plots, vacant lands, lawns, sports grounds, botanical gardens, open fields, and cemeteries.

3 Importance and Benefits

Urban forests provide many benefits, such as improving the community environment even with city space limitations. The benefits include enhanced biodiversity, human health, air quality, food production, and ecosystem services (Fig. 1).

3.1 *Social and Aesthetic Benefits*

Urban forestry significantly contributes to the social well-being of urban residents by providing opportunities for recreation, relaxation, and connection with nature. Moreover, urban forests serve as natural sanctuaries, offering respite from the hustle and bustle of urban life and promoting mindfulness and mental rejuvenation (Maas et al. 2009). Access to green spaces and trees has been linked to numerous physical and psychological health benefits, including stress reduction, improved mood, and enhanced overall well-being. Cities with trees and urban forests are healthier places to

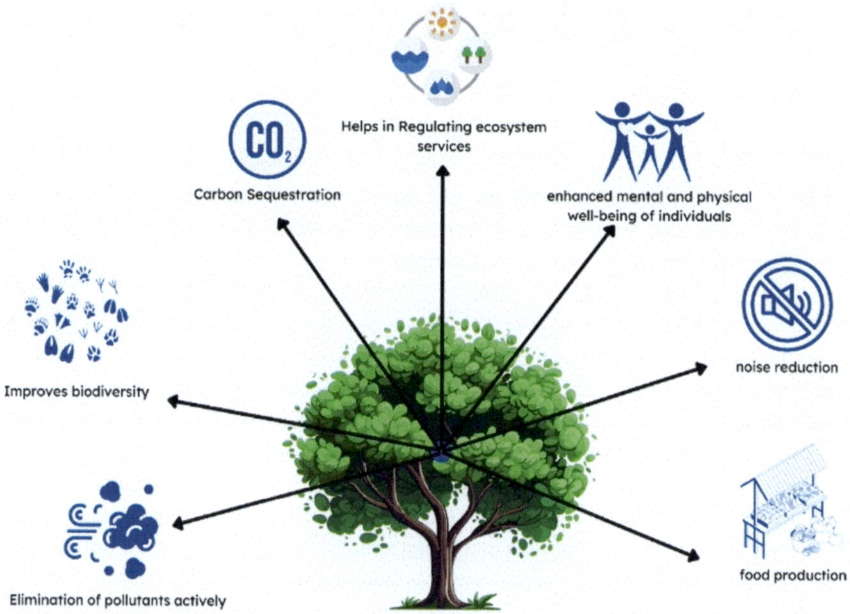

Fig. 1 The benefits derived from urban forests

live, thanks partly to cleaner air, which helps people breathe more easily (Eisenman et al. 2019). The leaves of the trees absorb or capture pollutants such as nitrogen dioxide, sulphur dioxide, ozone, ammonia, and particulate material. Some of these pollutants can cause serious health problems.

Additionally, the shade of the trees protects people from skin damage caused by ultraviolet light and other ways urban forests may improve public health. Urban forests indirectly help the community's well-being by promoting leisure activities and exercise. These forests can calm the mind and reduce stress. According to Morita et al. (2007), forest immersion led to a decrease in depression and an increase in feelings of vitality. This effect was even more vital for participants who started with higher stress levels. Another way urban forests improve health is by enhancing immunity from visiting urban forests. The reduced pro-inflammatory factors and increased Killer cells were retained even after a week (Wolf et al. 2020). Urban forest exposure also reduces the mortality rate due to cardiovascular disease and stroke. Nowak and Dwyer (2007) reported a decrease in mortality of 16% due to stroke and 13% for cardiovascular disease. The link between urban forests and fewer deaths from cardiovascular disease suggests that planting trees could be a supplementary tool for policymakers looking to tackle such a problem.

Urban woodlands and parks provide people with places for recreation. Greenery in urban areas creates comfortable environments for outdoor activities, exercise, and social gatherings, fostering community cohesion and social interaction. Hence, sharing environmental information and education programs may raise public awareness and foster an appreciation for the city's wildlife and plant life. Urban forests are precious as they provide residents with convenient, high-quality recreation options, which reduce disturbances in ecologically sensitive areas further away. Even when designing urban forests for people's enjoyment, it's essential to consider some ecological factors to ensure staying healthy and enjoyable places to visit in the long run (Kumar et al. 2022; Singh 2021; Kumar et al. 2024; Singh et al. 2024b; Singh 2024).

Urban forestry contributes to urban landscapes' aesthetic quality and visual appeal, enhancing cities' and neighborhoods' overall ambiance and character. Urban green spaces soften the built environment, contrast hard surfaces, and create visually pleasing settings, attracting residents, visitors, and tourists (Kumar et al. 2021e; Prakash et al. 2022). Additionally, integrating landscaping features like flower beds, shrubs, and ground cover complements trees and elevates overall aesthetic quality. Aesthetics are often subjective, and the preferences may vary from person to person. Age, occupation, educational qualification, and individual nature affect a person's aesthetic preference. Well-maintained street trees, parks, and urban forests contribute to a sense of place and identity, instilling pride and attachment among residents and fostering a positive urban image. Moreover, aesthetically pleasing urban environments have been shown to enhance psychological well-being, promote social interaction, and attract investment and economic development opportunities (Locke et al. 2015).

3.2 Climatic and Environmental Benefits

The urban forest significantly enhances urban environmental conditions and preserves biodiversity. These ecological benefits extend beyond large wooded areas; even smaller clusters, tree-lined streets, and individual trees improve the overall environmental quality in urban settings. Urban areas are characterized by high level of air pollution due to vehicular exhausts, industrial emissions, and the day-to-day activities of the high population concentrated in small areas. In this context, the role of urban green spaces, mainly the urban forest, in mitigating air pollutants has garnered attention. Trees most effectively capture air pollutants when planted near emission sources. Increased leaf surface roughness and leaf hairs improve the capture of particulate pollutants. The higher leaf roughness is why woodland canopies outperform other vegetation in capturing pollutants (Singh 2021; Sharma and Singh 2021; Singh et al. 2021).

Cities usually have a lot of impervious surfaces, preventing infiltration of water and increasing runoff. Incorporating trees and vegetation may improve the hydrology of urban areas. Cites with impervious surfaces negatively affect the local hydrology, soil moisture, and nutrient uptake (Revelli and Porporato 2018). Urban forests are crucial in mitigating surface runoff, easing the burden on urban sewage systems, and moderating peak stream flows. Open areas adorned with trees and other porous surfaces facilitate rainwater infiltration, recharging groundwater in an otherwise impervious urban environment. The overall hydrology of the area is improved due to urban forestry practices (Singh et al. 2020; Kumar et al. 2021a, b).

Urban woodlands and trees consistently exhibit cooler surfaces during hot summer days due to shade evapotranspiration cooling (Yang et al. 2005; Paschalis et al. 2021). Introducing vegetation along with built areas may reduce the temperature inside the buildings. Konijnendijk et al. (2005) reported temperature reduction in a transition zone outside the parks and the built-up regions up to a distance of approximately 200–400 m on the windward side. Therefore, it is crucial to preserve vegetation within urban land uses and create a dense network of publicly accessible green spaces. Tree plantings significantly reduce airflow, which may lower the energy demand of nearby buildings for heating and air conditioning. To strike a balance, planting schemes are necessary to mitigate overheating while ensuring adequate ventilation to prevent the accumulation of pollutants under tree canopies.

3.3 Habitat for Wildlife

Trees in urban parks serve as crucial habitats for birds, bats, and invertebrates. The tree canopies offer food, shelter, and nesting for birds and invertebrates, supporting urban biodiversity and ecosystem health (McPherson et al. 2018). Some urban forests are fragmented remains of the original forest before the establishment of the cities. These forest fragments also double as conservation sites for threatened species in their native range. Native trees are often better adapted to local environmental conditions and support a broader range of species than non-native species (Conway et al. 2019). Factors such as tree density, overall tree stand extent, and age support urban biodiversity. Interestingly, urban areas exhibit a relatively high biodiversity due to human influence and the presence of many exotic species. This rich diversity of species within urban ecosystems is utilized for educational purposes, allowing residents to appreciate the unique features and diverse values that trees bring to city environments (Kumar et al. 2021c, d).

3.4 Economic Benefits

Urban forestry generates significant economic benefits for communities and individuals by enhancing property values, reducing energy costs, and lowering infrastructure maintenance expenses. The values attached to urban forests are broadly classified as use and non-use values. Use values are categorized into two main types: consumptive and non-consumptive use values. On the other hand, non-use values are typically divided into several subcategories, including option values, quasi-option values, bequest values, and existence values. Consumptive use values include the value of forest products that may be sold, such as wood, fruit, etc. However, there is no direct monetary value for the primary benefits i.e., non-consumptive use values such as decreased wind speed, a balanced microclimate, shade and erosion management, calm environments, clean air, and beautiful landscapes of urban and peri-urban forests (Tyrväinen 1999). The economic worth can be assessed by considering benefits such as lower energy bills for heating and cooling or alternative environmental control expenses. Further, urban forests help manage stormwater runoff, mitigate flood risks, and reduce the need for costly engineered drainage systems and flood control measures (Ernstson and Sörlin 2013). The non-use values, such as cultural and historical value and willingness to pay for preservation for the future or for benefiting from existing facilities, are usually less prevalent in urban forests.

There are various methods for estimating the economic worth of urban forests. Opportunity costs, maintenance cost estimation, and production value estimation of forests are examples of traditional methodologies. However, newer methods include hedonic pricing, tree pricing continuous evaluation, travel cost, and environment benefit evaluation (Tyrväinen 1999). Because these techniques are predicated on valuing commodities at market prices, their applicability in urban forests is constrained. Another complexity in evaluating urban forest amenity values is that they are not purchased and sold through markets. The hedonic pricing estimates the value of urban forest in conjunction with the increase in housing premiums due to it. The continuous evaluation method estimates the willingness to pay to improve and maintain urban forests for their perceived benefits. The tree pricing method is used only in small groups of trees or individual trees, where the individual value of a tree is estimated based on factors such as age, health, form, aesthetics, etc. The travel cost method estimates the value of the urban forest based on the money spent to travel to the area. The goal of estimating the environmental benefits of urban forests is to determine how much shade, evapotranspiration, and altered air movement impact urban climates. The benefits of urban forests as amenities are often not sufficiently considered when deciding land use and green space management. Urban forest conversion to other uses is under rising pressure in many cities, and financial difficulties beset the management of these lands. Existing green spaces play a crucial role in the policy-making process if the socio-economic benefit of ecological aspects is proven, for instance, through a premium on home prices. Economic analysis is applied in decision-making at different levels, and at the very least, it improves the stakeholders' comprehension of potential values.

4 Challenges in Urban Forestry

Urbanization is increasing globally, with more than half of the world's population presently living in cities, with expectations for cities to reach two-thirds of the world population by 2050 (Dye 2008). Thus, rapid urbanization has brought about a wide range of challenges. As they grow, cities are losing green spaces and are likely to continue with mounting anthropogenic pressures.

4.1 Land Use Change and Space Constraints

The primary challenge in urban forests is the encroachment and change in land use. Land use change is one of the primary challenges for urban forestry, resulting in the loss, fragmentation, or degradation of green spaces and natural habitats. Rapid urbanization demands land for various projects, such as housing, infrastructure, industrial, etc., for the growing population. Many cities globally have experienced the destruction of recreational parks, tree gardens, and peri–urban plantations in the bid to create space for establishing housing and infrastructure development. Intensive pressures by governments and land speculators cause gentrification and the destruction of facilities used by low-income residents. People who cannot afford conventional cooking facilities have been reported to illegally cut down avenue trees along the street and botanical gardens for fuel. Additionally, urban sprawl and unplanned development exacerbate urban heat island effects, alter microclimates, and disrupt ecosystem processes, posing significant challenges to urban forest management and sustainability (Lyytimäki and Sipilä 2009). Balancing competing land use priorities and ensuring the preservation and protection of green spaces are critical challenges for urban forest management and planning (Konijnendijk and Gauthier 2006).

4.2 Air Pollution

Air pollution poses a significant threat to the health and vitality of urban forests and the overall ecosystem health. Urban areas are often characterized by high levels of air pollution from vehicle emissions, industrial activities, and other sources, which can negatively affect tree health and growth. Air pollutants such as sulphur dioxide, nitrogen oxides, ozone, and particulate matter can damage tree foliage, impair photosynthesis, and weaken tree defences against pests and diseases (Singh and Kumar 2022; Singh et al. 2024a). Even though it improves the overall air quality, introducing vegetation can also have some adverse effects. The pollen of some species of trees and plants increases the particulate pollutants in the air (Escobedo et al. 2011). Moreover, vegetation increases the secondary ozone formation at ground level and increases the release of some volatile organic compounds (Gao et al. 2022).

4.3 Soil Contamination

Soil contamination is another issue when implementing urban forestry initiatives. Common contamination by construction rubble containing cementitious substances raises alkalinity, inducing deficiency in some plant nutrients (Ware 1990; Jim 1998). Moreover, a pH increase over 9.0 causes alkaline toxicity in plants. Additionally, soil pollution from contaminants such as heavy metals, chemicals, and pesticides can impair root growth, nutrient uptake, and soil microbial activity, compromising tree health and ecosystem functioning. Addressing pollution sources, implementing pollution control measures, and promoting sustainable land use practices are essential for mitigating the impacts of pollution on urban forests (Livesley et al. 2016). Urban forests become a dumping ground for household and industrial waste in areas with inadequate waste management. Dumping waste in these areas has two main consequences, i.e., landfill effect and health hazards. Wastes occupy space that could be used for planting more trees, hindering the growth of the urban forest (Meena et al. 2011; Kumar et al. 2019b).

Moreover, industrial waste poses a health risk to residents, potentially causing various illnesses due to contamination. Generally, urban soils of cut-and-fill sites often have coarse texture due to contamination by construction rubble (Jim 1998). The excessive coarse fragments from construction wastes obstruct root growth and reduce water and nutrient-holding capacity.

4.4 Soil Compaction

Soil is a critical factor in urban forests, but its importance is mainly neglected. The soils in urban environments are disturbed in ways that are not conducive to tree growth (Scharenbroch et al. 2005). The inherited site soil usually has limited organic matter, nutrient stock, and colloidal materials, suppressing nutrient-cation exchange capacity and nutrient-holding potential. Deficiency in organic and inorganic aggregating agents creates a weak and unstable structure, thus predisposing the soil to structural failure under pressure to cause compaction. The bare soil surface is susceptible to compaction by trampling, vehicular runs, or rain splash, which may form a surface-sealing soil crust. Roadside soils under the pavement are compacted to meet engineering load-bearing specifications (Jim 1998). Existing roots can wither in compacted, and new root growth is often hampered. Moreover, compacted soils are unfavorable to soil organisms involved in nutrient cycling, thus dampening organic decomposition and nutrient release.

4.5 Financial Constraints

Effective urban forest management and governance require supportive policies, regulations, and adequate funding mechanisms. However, many cities face challenges in implementing and enforcing urban forestry policies due to limited resources, competing priorities, and institutional barriers. Urban forestry initiatives conducted on a large scale require much money to implement and sustain. Costs incurred for establishing and maintaining activities such as irrigation, tending operations, waste removal, and protection can often be very high. This is mainly in the case of instant results being wanted, such as in amenity plantings, where large saplings are planted, which require intensive aftercare. Another issue is that urban forests require continuous management for many years, while the support from government bodies might decline over time. The extremely high demand for rapid urban development has also negatively affected governments' proportional allocation of funds for urban forest establishment and management.

4.6 Threats to Human Safety

Poorly planted or inappropriate tree species are a hazard to urban inhabitants and property through tree and branch falls. This may be particularly likely in countries where cyclones or hurricanes are regularly experienced. There might be an increasing occurrence of over-mature and damaged trees that need replacement. The regular management and upkeep of urban forests are not adequately conducted in many areas, hindering the identification of dangerous trees, removal, and replacement. Urban forests represent deleterious effects such as allergenic pollen produced by certain trees and provide a refuge for vectors of human pathogens (Escobedo et al. 2011).

4.7 Climate Change

Urban forests are vulnerable to extreme weather and climate change, with some species being more vulnerable. Thus, incorporating the role of climate change as a driver of urban tree dieback and mortality into adaptive management practices may help policymakers and urban forest managers reduce risks and economic losses (Yadav et al. 2019a, b; Gupta et al. 2019a, b). Urban forests are affected by gradual changes in weather parameters (such as changing precipitation patterns and rising temperatures). Extreme weather (e.g., heat waves, flooding, and storms) or drought causes mortality and catalyzes other factors contributing to tree decline (Brando et al. 2014; Brandt et al. 2016; Jahani and Saffariha 2021). Levels of soil, water, and atmospheric pollution are often high in urban areas, and climatic conditions are harsh due to the artificial environment created by human structures. Additionally, the stress from the urban environment might make the trees susceptible to pests and diseases, causing damage and mortality. Climate change often increases vegetation stress, pest and disease incidence, and their impact on trees (Tubby and Webber 2010).

4.8 Design and Policy

The integration and planning of urban forestry with regional policies, climate, and ecology are other problems. With limited resources and more societal demand, properly designing urban forests with a strong emphasis on sustainability, functionality, and management is crucial. In landscape and garden architecture, aesthetics plays a significant role. However, the functional aspects should be considered in their design. Trees are living organisms; therefore, a thorough knowledge of their nature and life cycle is indispensable for designing urban green spaces. Introducing species unsuitable for a locality can spread out of its intended range, becoming invasive and displacing local species (McMillen et al. 2019; Singh and Verma 2013). Some species, even though providing many benefits, might not be suitable for the region's climate, thereby increasing the resources required for managing them. Another challenge for urban forestry is maintaining and developing sustainable urban forest resources that meet social needs. This elucidates why multi-purpose tree species are widely favored for urban forestry. Urban forest structure significantly impacts the mix of outputs, and determining the best mix requires technical information on trade-offs between the different outputs, costs, and values. Moreover, gaps in coordination and collaboration among government agencies, stakeholders, and community groups may lead to fragmented or ineffective urban forest management. Strengthening policy frameworks, securing sustainable funding sources, and fostering multi-sectoral partnerships are critical for overcoming policy and funding challenges and promoting the sustainable management and conservation of urban forests (Vogt et al. 2015; Gupta and Singh 2017; Singh et al. 2017a, b).

5 Opportunities and Policy Changes

Urban forestry presents multiple opportunities that contribute to the well-being of both communities and the urban environment. Looking forward, urban forestry provides even more opportunities in the future as cities grapple with pressing challenges such as urbanization, climate change, and public health. As cities become more densely populated, innovative solutions like vertical forests on buildings will become more prominent. These create green spaces while maximizing land use in cities where land is scarce. Urban forestry can also be integrated with green infrastructure projects, using trees and vegetation to reduce pollution and create natural cooling systems (Kumar et al. 2019a). Urban forests regulate climate change by absorbing carbon dioxide from the atmosphere and storing it in their biomass. Strategies for urban forestry must be developed and adapted to meet the challenges posed by climate change. As climate change can exacerbate the stress on vegetation, the future will likely focus on planting climate-resilient species specifically chosen for their resistance to heat, drought, and pests. Urban forestry initiatives often do not reap their full benefits without the active involvement of local communities. Therefore, there is a growing recognition of the importance of public participation in planning, establishing, and maintaining urban forests. Enhanced public awareness will help increase the engagement of citizens in tree-planting events, community gardens, and urban forest education programs. This collaborative approach fosters a sense of ownership and responsibility for the community of the urban green spaces, making their management much more accessible (Sharma et al. 2018; Singh et al. 2018; Gupta et al. 2018a, b). Urban forestry has to be incorporated into all tiers of development planning, including local, regional, and national development plans. Moreover, the empowerment of regional government bodies with sufficient autonomy and resources to allocate space, design, and execute urban forestry initiatives is essential. Integrating technologies like remote sensing and big data analysis can offer valuable insights for managing urban forests. Satellite imagery helps determine the change in vegetation cover, health, and composition and optimizes maintenance practices in urban forests and green spaces. Non-destructive methods such as tree tomography can detect damaged trees using ultrasonic waves and decide on further management practices. Moreover, modeling software may help determine the value of services provided by urban forests and express them in monetary terms. Strategies such as selecting climate-resilient species and nature-based solutions, adopting green infrastructure into all tiers of urban planning, and ensuring community participation and technology integration are the way forward for sustainable and resilient cities.

5.1 Innovative Urban Forestry Partnerships

Engaged citizens and community-based organizations are vital in preserving and managing urban parks worldwide. This collaborative effort has been evident in various cities such as Delhi, India; Manila, the Philippines; Mexico City, Mexico; and New York, United States (Kuchelmeister 1999a). For instance, in Yokohama, Japan, the city's Board of Parks collaborates with several citizen associations to operate an ecological park. These associations include specialists interested in specific aspects of nature, such as bird-watchers, as well as generalists interested in social activities during leisure time. Similarly, in Zurich, Switzerland, the municipal administration has actively encouraged citizens to assume responsibility for specific urban green areas.

Moreover, initiatives involving vulnerable groups as partners and clients have demonstrated significant impact. For example, in Bombay, India, a park was developed for and with street children on a former garbage dump site amidst sprawling slums and a polluted creek (Pye-Smith 1996). Likewise, in São Paulo, Brazil, the One Million Trees project engaged street children in various activities, including tree watering, protection against vandalism, and conducting door-to-door educational campaigns (Kuchelmeister 1999b).

Public–private partnerships have become prevalent in various countries, where businesses contribute to the upkeep of street trees or sponsor tree planting in front of their establishments in exchange for promotional opportunities, such as advertising on tree protectors or attracting customers. In Chile, certain companies go a step further by financing and overseeing entire public parks (IDB 1997). In Sacramento, California, the concern over declining funds for municipal tree care has led to innovative partnerships involving the Sacramento Tree Foundation, the Municipal Utility District, and trained volunteers. They collaborate on initiatives like Dutch elm disease control and encouraging residents to plant energy-efficient trees in their yards (McPherson and Luttinger 1998). Furthermore, domestic and international city partnerships are emerging as significant contributors to development efforts. These partnerships often involve collaborations between municipal associations, NGOs, and urban forestry initiatives. For example, Singapore has supported urban forestry projects in Manila, the Philippines. Additionally, there are instances of North–South urban forestry cooperation, such as the partnership between Guelph, Canada, and Jinja, Uganda, as well as the collaboration between León, Nicaragua, and Utrecht, the Netherlands (Kuchelmeister 1999b).

5.2 *Urban Forestry Policies in a Global Perspective*

Urban forestry policies vary across countries depending on environmental concerns, urbanization rates, cultural values, and governance structures. Here's an overview of urban forestry policies in a global perspective.

5.2.1 The FAO Guidelines on Urban and Peri-Urban Forestry

The guidelines, developed by the Food and Agriculture Organization of the United Nations (FAO), provide a comprehensive framework for planning, managing, and monitoring urban forests. These guidelines cover many topics, including tree selection, planting and maintenance, pest and disease control, and urban and peri-urban forestry's social and economic benefits (UPF). Guidelines promote awareness of the contributions of UPF to environmental sustainability, improved quality of life, and resilience in urban areas. Overall, the FAO Guidelines on Urban and Peri-urban Forestry serve as a valuable resource for promoting urban forests' establishment, management, and appreciation, contributing to developing sustainable and resilient cities worldwide.

5.2.2 The UN Sustainable Development Goals (SDGs)

The United Nations set up Sustainable Development Goals or Global Goals in 2015, a collection of 17 interconnected goals acting as a guideline to achieve a more sustainable and better future for all. SDGs address challenges such as inequality, climate change, environmental degradation, peace, and justice to create a peaceful and prosperous world for everyone by 2030. Several SDGs directly or indirectly promote urban forestry, including SDG 11 on Sustainable Cities and Communities, SDG 13 on Climate Action, and SDG 15 on Life on Land. These goals provide a global framework for urban development that recognizes the importance of green spaces, including urban forests.

5.2.3 The Tree Cities of the World Program

The Tree Cities of the World program is a global initiative recognizing cities and towns committed to caring for and celebrating their urban forests. It's jointly managed by the Food and Agriculture Organization (FAO) of the United Nations and the Arbor Day Foundation. It recognizes cities that are committed to sustainably managing their urban forests. The program provides participating cities with technical assistance, training, and networking opportunities. Acknowledging and supporting cities committed to urban forestry in the program helps create greener, healthier, and more liveable communities around the globe.

5.3 Initiatives from Individual Countries

5.3.1 The Indian Nagar Van Yojana (NVY)

Nagar Van Yojana (NVY) is a program formulated to create and promote urban forests and the greening of cities in India. The program, funded by the compensatory afforestation program (CAMPA), plans to develop 1000 'Nagar Van' (urban forests) with an area of 10 to 50 hectares and 'Nagar Vatika' of 1–10 hectares. The initiative's primary focus is to improve green cover, biodiversity, air quality, and quality of life in urban areas. Moreover, it encourages community participation involving NGOs, educational institutions, and local bodies. The scheme is envisioned to create cleaner, greener, and healthier cities nationwide.

5.3.2 Green Vancouver

In 2010, Vancouver, Canada, introduced the 'Green Vancouver' initiative, an ambitious program designed to tackle the city's environmental issues comprehensively. This strategy comprises ten objectives covering aspects such as fostering a green economy, demonstrating climate leadership, promoting green buildings and transportation, achieving zero waste, enhancing access to nature, reducing ecological footprint, ensuring clean water and air, and promoting local food production. Notably, Target 1, under the "access to nature" goal, aims to ensure that in 2020, every resident will be within a five-minute walk of a park, greenway, or similar green space. Additionally, Target 2 sets a goal of planting 150,000 trees within the city between 2010 and 2020, with 37,000 trees already planted by 2014 (City of Vancouver 2012).

5.3.3 Green Infrastructure and Biodiversity Plan

In May 2013, Barcelona, Spain, introduced its Green Infrastructure and Biodiversity Plan, aligning with the strategy outlined by the European Commission to advance the adoption of green infrastructure across Europe. The plan defines green infrastructure as a network of areas featuring public or private agricultural or naturally landscaped vegetation, serving multiple functions by providing ecological, environmental, social, and economic benefits. Moreover, it emphasizes enhancing these benefits by establishing connectivity within the green infrastructure network. Additionally, the plan seeks to integrate green infrastructure seamlessly with other urban infrastructure elements (Ajuntament de Barcelona 2013).

5.3.4 Ghana's National Urban Policy

The Ministry of Local Government and Rural Development initiated Ghana's national urban policy in 2012. Known as the National Urban Policy Framework, it aims to foster sustainable, well-organized urban development. This entails ensuring adequate housing and services, establishing efficient institutions, and creating a conducive living and working environment for all citizens. The framework addresses various urban challenges, including land-use disorder, uncontrolled urban sprawl, environmental degradation, insufficient urban infrastructure and services, urban poverty, slums, squatter settlements, and weak rural–urban connections (Ministry of Local Government and Rural Development of Ghana 2012).

5.3.5 Green Infrastructure in Philadelphia

Philadelphia is working to make green infrastructure a regular procedure to enhance local stormwater management and reduce strain on the city's sewer system. Forecasts indicate that adopting grey infrastructure to solve the sewerage problem will cost the city $8 billion more over ten years than utilizing green infrastructure, which includes wetlands rehabilitation and protection, bioswales, rain barrels, and permeable pavements. Additionally, the city would benefit from green infrastructure through better water quality, increased carbon sequestration, enhanced wildlife habitat, and more accessible recreational open spaces. To bolster the green infrastructure strategy, the city updated its stormwater charging structure and extended discounts to clients that used green infrastructure solutions to lower the impervious cover on their properties (Valderrama et al. 2013).

6 Case Studies

Most of the world's population lives in urban areas and cities, serving as centers for innovation, trade, culture, scientific advancement, and productivity. As discussed previously, urban forests have numerous benefits that help improve the liveability of modern cities. Several case studies on the environmental, social, aesthetic, and economic benefits of urban forests and green spaces are described below.

6.1 Climatic and Environmental Perspective

Reducing air pollution, summer temperatures, surface runoff, and carbon sequestration of carbon and conservation are some of the environmental benefits of urban forests. Air pollution in cities is caused by particulate matter and harmful gases that affect human health and well-being. Yang et al. (2005) conducted a study to evaluate the efficacy of Urban forests in alleviating pollution in Beijing, China. Urban forests reduce particulate matter, ozone, sulphur dioxide and nitrous oxide reduction. The annual air pollutant absorption by trees from the atmosphere was estimated to be 1261.4 tons. The amount of pollution removed was seasonal, with the highest removal occurring in summer and the lowest in winter. The seasonal change in pollution reduction is due to varying emissions in different seasons and the deciduous nature of most trees in the city. Of the pollutants removed by urban forests, particulate matter accounted for most of the total pollutants (61%). The rate of pollution removal was estimated to be about 27.5 g/year per square meter of the canopy. The trees reduce the air temperature by shade and evapotranspiration and assimilate it into their biomass. An estimated 41,701 tonnes of carbon are stored annually as biomass throughout the region, and the overall drop in air temperature of 1.6 °C was observed.

Moreover, as most of the energy requirement for the city comes from coal power, the reduction in energy requirement due to urban forests reduces the release of pollutants and carbon dioxide. The carbon dioxide prevented from entering the atmosphere amounts to 44% of the biomass sequestered. However, the study found that the tree size and species composition already present in the region were not optimal for removing pollutants and carbon, and improvement was suggested.

Similarly, Yang et al. (2005) stated that the effects of urban forests in reducing air pollutants such as particulate matter, carbon monoxide, nitrous oxide, sulphur dioxide, and carbon monoxide were estimated in 88 cities in Canada (Nowak et al. 2018). Out of all the pollutants, the highest reduction was observed for particulate material. The average air quality improvement due to removing all the contaminants was 1 percent. Among the different pollutants, sulphur dioxide had the highest percentage reduction while carbon dioxide had the lowest. The removal of gaseous pollutants was higher during the day, whereas particulate pollutants were captured at a similar rate during the day and night.

Urban forests have a defined role in controlling throughfall, surface runoff, and maintaining the hydrological balance. Inkiläinen et al. (2013) evaluated the role of urban vegetation in controlling throughfall and surface runoff during storms in Raleigh, North Carolina, USA. The cumulative throughfall in the urban forest with an average canopy cover of 67% was 88.9% of the gross precipitation. The throughfall was observed to be less in urban forest patches that had a higher canopy cover. However, increasing the canopy cover could not fully mitigate the incidence of stormwater damage. Reducing impervious surfaces, improving vertical stratification of vegetation, and maintaining good soil cover were suggested to reduce the runoff damage.

6.2 Social and Aesthetic Perspective

Urban forests provide many social benefits, such as areas for recreation, stress reduction, health benefits, and improving the overall aesthetics of an area. A survey was conducted by Shin et al. (2005) among visitors to 12 urban forests in Korea to examine the perceived psychosocial benefits. Among the different psychosocial outcomes, the sense of peace and tranquillity, a break from challenges, and an opportunity to exercise were ranked highest on average. The perceived psychosocial outcomes differed based on age and educational qualifications but did not differ by gender. Higher importance was given to psychosocial benefits by older people and people with a higher level of education. Managing trees and green spaces in urban environments has increasingly been recognized for positively impacting mental health. An investigation into the mental health benefits provided by urban forests to students preparing for exams was conducted in Belgrade, Serbia (Vujcic and Tomicevic-Dubljevic 2018). Mental health parameters such as depression, stress, and anxiety were estimated using the 'Depression Anxiety Stress Scale'. The students were divided into two groups, one of which was made to spend 2 h every day for five weeks in an urban forest setting while the control group went on with their everyday activities. The result shows a significant improvement in these parameters compared to the control group. However, it should be noted that the scale used for estimating the mental health parameters is self-reported. The aesthetics of urban forests are difficult to quantify as they are subjective and are affected by a person's social, cultural, and personality traits. Therefore, the effect of aesthetics on other benefits of urban forests, such as increased tourism or property value appreciation, can provide a better estimate of its intrinsic value. A study was conducted in Savannah, Georgia, to establish the contribution of urban forests and green spaces to the city's beauty and tourism (Deng et al. 2010). The importance of urban forests for tourism was estimated with a questionnaire concerning parameters such as beauty, image, attractiveness, and tourism experience, which were measured with a percentage scale and attribute ranking from relative importance. The responses from the study imply that urban forests contributed more than 65% to the study parameters. The urban forests were found to be a popular attraction for tourists (74.1%), second only to historical attractions.

6.3 Economical Perspective

The consumptive benefits of urban forests are more accessible to estimate. However, they don't provide the complete picture, as the primary value of urban forests is from non-consumptive use and non-use values. The potential benefits from energy savings due to temperature control, property value appreciation, etc., are much higher than the direct economic benefits. Payton et al. (2008) tried to estimate the benefits of urban forests on their potential improvement in the price of housing units in Marion

County, Indiana, USA. The Normalized Difference Vegetation Index (NDVI) was used to measure the greenness of the property and surrounding areas.

In contrast, the housing prices were estimated using the existing sales database. The value of the properties increased with a corresponding increase in vegetation in all the models used in the study. The study concluded that for every percent increase in NDVI, the households in the region are willing to pay $26 to $56 per year.

A study in Sacramento County by Simpson (1998) evaluated the effects of changes in air temperature, solar radiation, and wind speed due to trees planted on the energy requirement for heating, cooling, and peak electricity use. The effects of urban forests on climate and their subsequent impact on changes in energy requirements were simulated. The total savings in heating and cooling were estimated to be about 20 million US dollars. The urban forest in the county contributes to an annual air-conditioning savings of approximately 157 Giga Watt hour, which amounts to about $18.5 million (energy costs based on market rates at the time of the study). This savings in energy represents about 12% of the air conditioning usage and 1.5% of overall electrical consumption. Even though trees increased the heating costs due to shade in winter, the overall energy consumption is still lesser due to the positive impact of reduced wind speed. The annual energy savings from wind reduction was estimated to be 145 terajoules (TJ), equivalent to $1.3 million. Furthermore, peak energy reductions lead to an additional savings of $6 million.

The cooling effects produced by trees are not only contributed by shade but also by evapotranspiration from the foliage, which can also reduce the temperature. The impact of cooling demand and savings associated with evapotranspiration and subsequent cooling effects in 3 cities in the UK (Moss et al. 2019). Even though all trees provide a cooling effect through shade and evapotranspiration, they can drastically differ depending on the species. Therefore, the study also considered the individual evapotranspiration of each species. The energy saving was calculated by comparing the cooling achieved through tree transpiration to standard cooling methods and by incorporating evapotranspiration into dynamic building airflow and thermal modeling programs. The study uncovered that evaporative cooling has the potential to achieve up to a 13.4% reduction in energy usage, specifically for sensible heat cooling. Total savings were estimated from 83.9 to 321£ across the cities when compared with the power requirement to produce cooling similar to that of an air conditioner. The cooling value per individual tree was estimated to range from 0.03 to 0.05 ($\pm$ 0.01) £/hr/tree for the three cities. The total economic value of evapotranspiration cooling for the study area was estimated to be more than £10 million.

As pollution affects human health and well-being, the potential health benefits associated with reducing pollution due to urban forests can be expressed in monetary terms. In a study, the effect of pollution and the monetary value related to its removal by urban forests across cities in Canada were estimated (Nowak et al. 2018). The health hazards associated with the levels of different pollutants and the values associated with their change were calculated using the 'Environmental Benefits Mapping and Analysis' program developed by the United States Environmental Protection Agency. The estimated total human health value was $227.2 million for a corresponding reduction in pollution of 16,500 tons across all cities. The value per hectare of tree cover for the pollution reduction was estimated at $511.

7 Conclusion

As the world rapidly urbanizes, the importance of urban forests becomes increasingly clear, serving as vital hubs for environmental preservation and human well-being amidst the concrete landscapes of modern cities. More than just clusters of trees, urban forests act as the lungs of urban areas, cleansing the air and offering a tranquil escape from urban chaos. They foster biodiversity, providing sanctuary and sustenance for numerous plant and animal species. Furthermore, these green havens are crucial in combatting climate change by absorbing carbon dioxide and alleviating the urban heat island effect, contributing to a more sustainable and pleasant urban environment. Additionally, they enhance city aesthetics and serve as spaces for recreation, relaxation, and community interaction. Despite their recognized significance, urban forests encounter various challenges in planning, establishment, and maintenance due to pressures from development, pollution, funding shortages, and climate change. Socio-economic factors such as unequal access to green spaces and environmental injustice further compound these challenges, disproportionately affecting marginalized communities. Exploring the potential of urban forests through innovative planning, collaborative partnerships, and community engagement may guide the development of cities that coexist harmoniously with nature, enriching the lives of current and future generations. Embracing these opportunities is both a choice and a responsibility toward building resilient and equitable urban environments for all. Numerous urban forestry initiatives have emerged worldwide, reflecting the global recognition of the importance of urban forests at various scales.

References

Ajuntament de Barcelona (2013) Barcelona green infrastructure and biodiversity plan 2020. Barcelona, Spain. https://ajuntament.barcelona.cat/ecologiaurbana/en/what-we-do-and-why/green-city-and-biodiversity/green-and-biodiversityplan

Allen KS, Harper RW, Bayer A, Brazee NJ (2017) A review of nursery production systems and their influence on urban tree survival. Urban Forest Urban Green 21:183–111. https://doi.org/10.1016/j.ufug.2016.12.002

Berland A, Shiflett SA, Shuster WD, Garmestani AS, Goddard HC, Herrmann D L and Hopton ME (2017) The role of trees in urban stormwater management. Lands Urban Plann 162:167–177. https://doi.org/10.1016/j.landurbplan.2017.02.017

Brando PM, Balch JK, Nepstad DC, Morton DC, Putz FE, Coe MT, Silvério D, Macedo MN, Davidson EA, Nobrega CC (2014) Abrupt increases in Amazonian tree mortality due to drought–fire interactions. Proceed Natl Acad Sci 111(17):6347–6352. https://doi.org/10.1073/pnas.1305499111

Brandt L, Lewis AD, Fahey R, Scott L, Swanston CDL (2016) A framework for adapting urban forests to climate change. Environ Sci Pol 66:393–402. https://doi.org/10.1016/j.envsci.2016.06.005

City of Vancouver (2012) Greenest City 2020 action plan. http://vancouver.ca/files/cov/Greenest-city-action-plan.pdf

Conway TM, Almas AD, Coore D (2019) Ecosystem services, ecological integrity, and native species planting: how to balance these ideas in urban forest management? Urban Forest Urban Green 41:1–5. https://doi.org/10.1016/j.ufug.2019.03.006

Costanza R, DeGroot R, Braat L, Kubiszewski I, Fioramonti L, Sutton P, Farber S, Grasso M (2017) Twenty years of ecosystem services: how far have we come and how far do we still need to go? Ecosyst Serv 28:1–16. https://doi.org/10.1016/j.ecoser.2017.09.008

Deng J, Arano KG, Pierskalla C, McNeel J (2010) Linking urban forests and urban tourism: a case of Savannah, Georgia. Tour Anal 5(2):167–181. https://doi.org/10.3727/108354210X12724863327641

Dye C (2008) Health and urban living. Sci e319:766–769. https://doi.org/10.1126/science.1150198

Eisenman TS, Churkina G, Jariwala SP, Kumar P, Lovasi GS, Pataki DE, Weinberger KR, Whitlow TH (2019) Urban trees, air quality, and asthma: an interdisciplinary review. Lands Urban Plan 187:47–59. https://doi.org/10.1016/j.landurbplan.2019.02.010

Ernstson H, Sörlin S (2013) Ecosystem services as the technology of globalization: on articulating values in urban nature. Ecol Econ 86:274–284

Escobedo FJ, Kroeger T, Wagner JE (2011) Urban forests and pollution mitigation: analyzing ecosystem services and disservices. Environ Pollut 159(8–9):2078–2087. https://doi.org/10.1016/j.envpol.2011.01.010

FAO (2016) Guidelines on urban and peri-urban forestry, by F. Salbitano, S. Borelli, M. Conigliaro and Y. Chen. FAO Forestry Paper No. 178. Rome, Food and Agriculture Organization of the United Nations

Gao Y, Ma M, Yan F, Su H, Wang S, Liao H, Zhao B, Wang X, Sun Y, Hopkins JR, Chen Q (2022) Impacts of biogenic emissions from urban landscapes on summer ozone and secondary organic aerosol formation in megacities. Sci Total Environ 814:152654. https://doi.org/10.1016/j.scitotenv.2021.152654

Gupta SK, Singh H (2017) Observations on Sporadic Flowering in *Ficus benjamina* L. and *Peltophorum africanum* L. in New Forest. Indian Forester 143:290–291

Gupta SK, Ram J, Singh H (2018a) Comparative study of transpiration in cooling effect of tree species in the atmosphere. J Geosci Environ Protect 6:151–166. https://doi.org/10.4236/gep.2018.68011

Gupta SK, Ram J, Singh H, Kumar P (2018b) Comparative study of cooling under shades of some forestry tree species with respect to ambient temperatures. J Geosci Environ Protect 6:51–66. https://doi.org/10.4236/gep.2018.611004

Gupta SK, Ram J, Singh H (2019a) Relationship between leaf area index and atmospheric cooling of tree species. Indian J Forest 42(3):203–208

Gupta SK, Singh H, Ram J (2019b) Need for a novel paradigm in forestry research: perspectives of comparative atmospheric cooling by different forestry effect tree species. Indian Forester 145(6):509–515

IDB (1997) Good practices for urban greening. Env-109. Washington, DC, Inter-American Development Bank. https://publications.iadb.org/en/publication/11417/good-practices-urban-greening. Accessed 10 Mar 2024

Inkiläinen EN, McHale MR, Blank GB, James AL, Nikinmaa E (2013) The role of the residential urban forest in regulating throughfall: A case study in Raleigh, North Carolina, USA. Lands Urban Plann 119:91–103. https://doi.org/10.1016/j.landurbplan.2013.07.002

Jahani A, Saffariha M (2021) Modeling of trees failure under windstorm in harvested Hyrcanian forests using machine learning techniques. Scient Rep 11(1):1–13. https://doi.org/10.1038/s41598-020-80426-7

Jangra A, Kumar K, Maikhuri S, Bhandari MS, Pandey S, Singh H, Barthwal S (2024) Unveiling stress-adapted endophytic bacteria: Characterizing plant growth-promoting traits and assessing cross-inoculation effects on *Populus deltoides* under abiotic stress. Plant Physiol Biochem 210:108610. https://doi.org/10.1016/j.plaphy.2024.108610

Jim CY (1998) Physical and chemical properties of a Hong Kong roadside soil about urban tree growth. Urban Ecosyst 2:171–181. https://doi.org/10.1023/A:1009585700191

Kanta C, Kumar A, Chauhan A, Singh H, Sharma IP (2024). The interplay between plant functional traits and climate change. In: Kumar N, Singh H (eds) Plant functional traits for improving productivity. Springer, Singapore. https://doi.org/10.1007/978-981-97-1510-7_3

Konijnendijk CC and Gauthier M (2006) Urban forestry for multifunctional urban land use. Cities farming for the future: agriculture for green and productive cities (R. van Veenhuizen, ed). RUAF Foundation, Rome, pp 414–416

Konijnendijk C, Nilsson K, Randrup T, Schipperijn J (eds) (2005) Urban forests and trees: a reference book Springer, Heidelberg, Berlin

Kuchelmeister G (1999a) Urbanization in developing countries—time for action for national forest programs and international development cooperation for the urban millennium. Forest Policy Research Forum: The Role of National Forest Programs to Ensure Sustainable Forest Management, 14–17 June, Joensuu, Finland

Kuchelmeister G (1999b) Innovative partnerships for sustainable urban forestry in developing countries. Prepared for 1999 National Urban Forest Conference, Building Cities of Green, Seattle, Washington, USA

Kumar N, Singh H (2024) Plant functional traits for improving productivity. Springer Nature, Singapore. https://doi.org/10.1007/978-981-97-1510-7

Kumar M, Singh MP, Singh H, Dhakate PM, Ravindranath NH (2019a) Forest working plan for the sustainable management of forest and biodiversity in India. J Sustain Forest 39:1–22. https://doi.org/10.1080/10549811.2019.1632212

Kumar N, Jeena N, Kumar N, Gangola S, Singh H (2019b) Phytoremediation facilitating enzymes: an enzymatic approach for enhancing remediation process. In: Bhatt P (ed) Smart bioremediation technologies. Academic Press, Elsevier, Cambridge, MA, pp 289–306

Kumar A, Kumar P, Singh H, Kumar N (2021a) Modulation of plant functional traits under essential plant nutrients during seasonal regime in natural forests of Garhwal Himalayas. Plant Soil 465:197–212. https://doi.org/10.1007/s11104-021-05003-x

Kumar A, Kumar P, Singh H, Bisht S, Kumar N (2021b) Relationship of physiological plant functional traits with soil carbon stock in the temperate forest of Garhwal Himalaya. Curr Sci 120(8):1368–1373. https://doi.org/10.18520/cs/v120/i8/1368-1373

Kumar A, Kumar P, Singh H, Kumar N (2021c) Adaptation and mitigation potential of roadside trees with bio-extraction of heavy metals under vehicular emissions and their impact on physiological traits during seasonal regimes. Urban Forest Urban Green 58:126900. https://doi.org/10.1016/j.ufug.2020.126900

Kumar A, Kumar P, Singh H, Kumar N (2021d) Impact of plant functional traits on infiltration rate under different forests of Kempty watershed in Garhwal Himalaya. Indian J Soil Conserv 49(1):38–44

Kumar M, Phukon SN, Paygude AC, Tyagi K, Singh H (2021e) Mapping phenological functional types (PhFT) in the Indian Eastern Himalayas using machine learning algorithm in Google Earth Engine. Comp Geosci 158. https://doi.org/10.1016/j.cageo.2021.104982

Kumar A, Singh H, Kumari G, Bisht S, Malik A, Kumar A, Singh M, Raturi , Barthwal S, Thakur A, Kaushal R (2022) Adaptive resilience of roadside trees to vehicular emissions via leaf enzymatic, physiological, and anatomical trait modulations. Environ Pollut 313. https://doi.org/10.1016/j.envpol.2022.120191

Kumar N, Singh H, Giri K, Kumar A, Joshi A, Yadav A, Singh R, Bisht S, Kumari R, Jeena N, Khairakpam R, Mishra G (2024) Physiological and molecular insights into the allelopathic effects on agroecosystems under changing environmental conditions. Physiol Mol Biol Plants 30:417–433. https://doi.org/10.1007/s12298-024-01440-x

Livesley SJ, McPherson EG and Calfapietra C (2016) The urban forest and ecosystem services: Impacts on urban water, heat, and pollution cycles at the tree, street, and city scale. J Environ Qual 45(1):119–124. https://doi.org/10.2134/jeq2015.11.0567

Locke DH, Roman LA, Murphy-Dunning C (2015) Why opt-in to a planting program? long-term residents value street tree aesthetics. Arboricult Urban Forest 41(6):324–333. https://www.isa-arbor.com/Publications/Arboriculture-Urban-Forestry. Accessed Mar 10 2024

Lyytimäki J, Sipilä M (2009) Hopping on one leg—the challenge of ecosystem disservices for urban green management. Urban Forest Urban Green 8(4):309–315. https://doi.org/10.1016/j.ufug.2009.09.003

Maas J, Van Dillen SM, Verheij RA, Groenewegen PP (2009) Social contacts as a possible mechanism behind the relation between green space and health. Heal Place 15(2):586–595. https://doi.org/10.1016/j.healthplace.2008.09.006

Malik A, Kumar K, Kumari G, Kumar N, Singh H (2024) Application of functional traits in modelling productivity and resilience under climate change. In: Kumar N, Singh H (eds) Plant functional traits for improving productivity. Springer, Singapore. https://doi.org/10.1007/978-981-97-1510-7_5

McMillen H, Campbell LK, Svendsen ES (2019) Weighing values and risks of beloved invasive species: the case of the survivor tree and conflict management in urban green infrastructure. Urban Forest Urban Green 40:44–52. https://doi.org/10.1016/j.ufug.2018.06.023

McPherson EG, Berry AM, van Doorn NS (2018) Performance testing to identify climate-ready trees. Urban Forest Urban Green 29:28–39. https://doi.org/10.1016/j.ufug.2018.07.023

Meena D, Singh H, Chaudhari SK (2011) Elucidating strontium response on growth dynamics and biochemical change in *Phaseolus mungo* L. Int J Agricult Environ Biotech 4:107–113

Ministry of Local Government and Rural Development (2012) Ghana National Urban Policy Framework. https://www.nalagghana.org/wp-content/uploads/2019/03/national_urban_policy_framework_ghana_2012.pdf

Morita E, Fukuda S, Nagano J, Hamajima N, Yamamoto H, Iwai Y, Nakashima T, Ohira H, Shirakawa TJ (2007) Psychological effects of forest environments on healthy adults: Shinrin-yoku (forest-air bathing, walking) as a possible method of stress reduction. Public Heal 121(1):54–63. https://doi.org/10.1016/j.puhe.2006.05.024

Moss JL, Doick KJ, Smith S, Shahrestani M (2019) Influence of evaporative cooling by urban forests on cooling demand in cities. Urban Forest Urban Green 37:65–73

Nowak DJ, Dwyer JF (2007) Understanding the benefits and costs of urban forest ecosystems. In: Urban and community forestry in the northeast. Springer Netherlands, Dordrecht, pp 25–46. https://doi.org/10.1007/978-1-4020-4289-8_2

Nowak DJ, Hirabayashi S, Doyle M, McGovern M, Pasher J (2018) Air pollution removal by urban forests in Canada and its effect on air quality and human health. Urban Forest Urban Green 29:40–48. https://doi.org/10.1016/j.ufug.2017.10.019

Paschalis A, Chakraborty TC, Fatichi S, Meili N, Manoli G (2021) Urban forests as main regulator of the evaporative cooling effect in cities. AGU Adv 2(2):e2020AV000303. https://doi.org/10.1029/2020AV000303

Payton S, Lindsey G, Wilson J, Ottensmann JR, Man J (2008) Valuing the benefits of the urban forest: a spatial hedonic approach. J Environ Planning Manage 51(6):717–736. https://doi.org/10.1080/09640560802423509

Prakash V, Hunney K, Singh H (2022) Effect of elevated carbon dioxide on growth and development of *Santalum album* L. seedlings inoculated with plant growth promoting microorganisms in Open Top Chambers. Int J Environ, Agricult Biotech 7(5):149–159

Pye-Smith C (1996) Building green islands in Bombay. People & The Planet 8(4). https://peopleandplanet.net/doc.php?id=1040. Accessed 10 Mar 2024

Revelli R, Porporato A (2018) Ecohydrological model for the quantification of ecosystem services provided by urban street trees. Urban Ecosyst 21:489–504. https://doi.org/10.1007/s11252-018-0741-2

Scharenbroch BC, Lloyd JE, Johnson-Maynard JL (2005) Distinguishing urban soils with physical, chemical, and biological properties. Pedobiologica 49(4):283–296. https://doi.org/10.1016/j.pedobi.2004.12.002

Sharma R, Singh H (2021) Alteration in biochemical constituents and nutrients partitioning of *Asparagus racemosus* in response to elevated atmospheric CO_2 concentration. Environ Sci Pollut Res 29:6812–6821. https://doi.org/10.1007/s11356-021-16050-3

Sharma R, Singh H, Kaushik M, Nautiyal R, Singh O (2018) Adaptive physiological response, carbon partitioning, and biomass production of *Withania somnifera* (L.) Dunal grown under elevated CO_2 regimes. 3 Biotech 8:267. https://doi.org/10.1007/s13205-018-1292-1.

Shin WS, Kwon HG, Hammitt WE, Kim BS (2005) Urban Forest Park use and psychosocial outcomes: a case study in six cities across South Korea. Scand J For Res. 20(5):441–447. https://doi.org/10.1080/02827580500339930

Simpson JR (1998) UrbanForest impacts on regional cooling and heating energy use: Sacramento County case study. J Arboricult 24:201–214. https://doi.org/10.48044/jauf.1998.025

Singh H (2021) An integrated approach considering physiological and biophysical based indicators for assessing tolerance of roadside plantations of *Alstonia scholaris* towards urban roadside air pollution: an assessment of adaptation of plantations for mitigating urban roadside air pollution. Trees. https://doi.org/10.1007/s00468-021-02179-8

Singh H (2024) Exploring adaptive modulation in plant functional traits and its impact on the productivity of *Acacia auriculiformis* under CO_2-enriched environment. Indust Crops Prod 210:118186. https://doi.org/10.1016/j.indcrop.2024.118186

Singh H, Verma A (2013) Characterization and screening of high nitrogen efficient rice genotype to curtail environmental pollution. Int J Agricult, Environ Biotech 6(1):23–30

Singh H, Kumar M (2022) Climate change and its impact on Indian Himalayan Forests: current status and research needs. In: Rani S, Kumar R (eds) Climate change. Springer Climate. Springer, Cham, pp 223–242. https://doi.org/10.1007/978-3-030-92782-0_11

Singh H, Sharma RS, Sinha S, Kumar M, Kumar P, Verma A, Sharma SK (2017a) Physiological functioning of *Lagerstroemia speciosa* L. under heavy roadside traffic: an approach to screen potential species for abatement of urban air pollution. 3 Biotech 7(61):1–10. https://doi.org/10.1007/s13205-017-0690-0

Singh H, Verma A, Kumar M, Sharma R, Gupta R, Kaur M, Negi M, Sharma SK (2017b) Phytoremediation: a green technology to clean up the sites with low and moderate level of heavy metals. Austin Biochem 2:1012

Singh H, Sharma R, Savita SMP, Kumar M, Verma A, Ansari MW, Sharma SK (2018) Adaptive physiological response of *Parthenium hysterophorus* to elevated atmospheric CO_2 concentration. Indian Forester 144:1–14

Singh H, Yadav M, Kumar N, Kumar A, Kumar M (2020) Assessing adaptation and mitigation potential of roadside trees under the influence of vehicular emissions: a case study of *Grevillea robusta* and *Mangifera indica* planted in an urban city of India. PLoS One 15(1): e0227380. https://doi.org/10.1371/journal.pone.0227380

Singh S, Singh H, Sharma SK, Nautiyal R (2021) Seasonal variation in biochemical responses of bamboo clones in the sub-tropical climate of Indian Himalayan foothills. Heliyon 7(4):e06859. https://doi.org/10.1016/j.heliyon.2021.e06859

Singh H, Kumar N, Kumar A (2024a) Enhancing resource use efficiency in crops through plant functional traits. In: Kumar N, Singh H (eds) Plant functional traits for improving productivity. Springer, Singapore. https://doi.org/10.1007/978-981-97-1510-7_6

Singh M, Singh H, Kumar A, Kumar M, Barthwal S, Thakur A (2024b) Soil nitrogen availability determines the CO_2 fertilization effect on tree species (*Neolamarckia cadamba*): growth and physiological evidence. Environ Sustain 7:53–60. https://doi.org/10.1007/s42398-023-00300-w

Tyrväinen L (1999) Monetary valuation of urban forest amenities in Finland. Academic dissertation. Research Papers 739. Finnish Forest Research Institute, Vantaa.

Tubby K, Webber J (2010) Pests and diseases threatening urban trees under a changing climate. Forestry: An Int J Forest Res 83(4):451–459. https://doi.org/10.1093/forestry/cpq027

Valderrama A, Levine L, Bloomgarden E, Bayon R, Wachowicz K, Kaiser C, Holland C, Ranney N, Scott J, Kerr O, DePhilip M (2013) Creating clean water cash flows developing private markets for green stormwater infrastructure in Philadelphia. Natural Resources Defense Council. https://encouragecapital.com/wp-content/uploads/2015/09/StormWater_Report_11.pdf

Vogt J, Hauer RJ, Fischer BC (2015) The costs of maintaining and not maintaining the urban forest: a review of the urban forestry and arboriculture literature. Arboricult Urban Forest 41(6):293–323. https://doi.org/10.48044/jauf.2015.027

Vujcic M, Tomicevic-Dubljevic J (2018) Urban forest benefits to the younger population: the case study of the city of Belgrade, Serbia. For Policy Econ 96:54–62. https://doi.org/10.1016/j.forpol.2018.08.006

Ware G (1990) Constraints to tree growth imposed by urban soil alkalinity. J Arboricult 16(1):35–38. https://doi.org/10.48044/jauf.1990.011

Wolf KL, Lam ST, McKeen JK, Richardson GR, Van Den Bosch M, Bardekjian AC (2020) Urban trees and human health: a scoping review. Int J Environ Res Public Heal (12):4371. https://doi.org/10.3390/ijerph17124371

Yadav SK, Singh H, Ginwal HS, Barthwal S (2019a) Elevated CO_2 enhanced growth and physiological process of *Populus deltoides* Bartr. ex Marsh. Indian Forest 145(1): 23–27

Yadav SK, Singh H, Nautiyal R, Ginwal HS, Ansari SA, Barthwal S (2019b) Modulation of morpho-physiological responses in *Populus deltoides* by elevated carbon dioxide and temperature. Forest Sci 66:105–118. https://doi.org/10.1093/forsci/fxz048

Yang J, McBride J, Zhou J, Sun Z (2005) The urban forest in Beijing and its role in air pollution reduction. Urban Forest Urban Green 3(2):65–78. https://doi.org/10.1016/j.ufug.2004.09.001

Stubble Burning and Urban Environmental Pollution: Impacts and Control Measures

Anusreeta Dutta, Hukum Singh, Md. Quasid Qamar, Tanmay Mathews Kottuppalli, Falguni Upreti, Smita Ghosh, K. H. Anagha, Riyanka Saha, and Ankit Kumar

Abstract Urban air pollution poses a significant challenge in India's major cities, including Delhi, Noida, and Gurgaon, exacerbated by rapid urbanization and industrialization. Stubble burning, a prominent contributor to air pollution, adds to the woes of the National Capital Territory. This process accounts for a substantial portion of global biomass-burning incidents, with India alone burning 14 million metric tonnes of stubble annually. Other factors, such as coal plants, vehicular emissions, and deforestation, exacerbate the issue. Indirect causes, including historical agricultural practices and inadequate government policies, compound the problem. The resultant primary pollutants degrade air quality and disrupt the climate system by releasing harmful greenhouse gases and adversely affecting soil productivity and carbon–nitrogen equilibrium. Health impacts are severe, with Delhi residents experiencing a significant reduction in life expectancy due to air pollution-related ailments such as asthma and acute respiratory infections. The economic toll is also substantial, with air pollution's cumulative impact rising to 7.7% of GDP in 2018. Nature-based solutions, such as urban forestry and green spaces, offer a promising avenue for

A. Dutta (✉)
Department of Humanities and Social Sciences, Birla Institute of Technology and Science, Hyderabad, Telangana 500078, India
e-mail: anusreetadutta.23@gmail.com

School of Property, Construction and Project Management, Royal Melbourne Institute of Technology, Melbourne, VIC 3004, Australia

H. Singh · Md. Q. Qamar · F. Upreti · S. Ghosh
ICFRE-Forest Research Institute, PO New Forest, Dehradun, Uttarakhand 248006, India

T. M. Kottuppalli
Department International Relations, Pondicherry Central University, Pondicherry 605 014, India

K. H. Anagha
Indian Institute of Technology, Roorkee, Uttarakhand 247667, India

R. Saha
West Bengal State University, Berunanpukhuria, West Bengal 700126, India

A. Kumar
Department of Vectorology, ICMR-Vector Control Research Centre, Puducherry 605006, India

H. Singh (ed.), *Urban Forests, Climate Change and Environmental Pollution*,
https://doi.org/10.1007/978-3-031-67837-0_24

combating air pollution in densely populated cities like Delhi. These interventions may support urban agriculture, mitigate urban heat islands, and reduce soil erosion, noise pollution, and energy consumption. The Delhi government has initiated several measures to increase urban greenery, including biodiversity parks and green ambulance services. However, more concerted efforts and policy interventions are needed to effectively address the multifaceted challenge of urban air pollution.

Keywords Air pollution · Stubble burning · Industrialization · Agriculture

1 Introduction

Rapid urbanization and careless industrialization are directly linked to city environmental stress. Delhi, the national capital of India, along with its neighboring industrial hubs like Noida and Gurgaon, are hotspots for pollution, especially during the winter months. With a daunting population of twenty-five million (United Nations 2018) and experiencing extreme climatic conditions, Delhi is one of the most polluted cities globally (Rizwan et al. 2013). The primary pollutant, PM2.5, significantly degrades the air quality.

While industrialization and urbanization are significant causes, stubble burning from nearby agrarian states exacerbates the issue, particularly during winter. Farmers burning crop residue, or 'parali', after harvest contributes to nearly 80% of global biomass burning, with a significant portion occurring in tropical regions like Southeast Asia (Sahu et al. 2015). Although stubble burning might seem minor compared to other pollution sources in Delhi, it underscores the government's neglect of farmers' needs in a country heavily reliant on agriculture.

Delhi releases nearly three thousand metric tonnes of air pollutants annually (Rizwan et al. 2013). Vehicular pollution is the major contributor (67%), followed by coal-thermal power plants (12%). Secondary pollutants like NOx and PM2.5 also degrade air quality. Daily PM2.5 concentrations in Delhi, Noida, and Gurgaon often exceed national standards (Kumar et al. 2015). Winter haze, exacerbated by Diwali festivities, significantly reduces visibility and worsens air quality due to high particulate matter concentrations.

Biomass burning contributes significantly to PM2.5 levels in Delhi, coupled with desert aerosols from the Thar Desert (Kirpa et al. 2012). Anthropogenic VOC emissions in India are estimated at twenty-one million metric tonnes annually, with biomass burning accounting for 60% (Varshney and Padhy 1998). India ranks second in stubble burning, contributing 33% of global incidents (Kant et al. 2022). The practice releases vast amounts of CO_2, CO, SO_2, and black carbon, harming the environment and soil productivity.

Stubble burning impacts soil health by depleting essential nutrients and raising soil temperatures, affecting microbial balance (Jangra et al. 2024; Kanta et al. 2024; Kumar and Singh 2024; Malik et al. 2024; Singh et al. 2018; Yadav et al. 2019a, b). Human health is also severely affected, with increased rates of respiratory and cardiovascular diseases (Kulshreshtha et al. 2021). The issue is so severe that it frequently reaches the Supreme Court for hearings.

The adverse effects of air pollution on public health are well-documented, with twenty-two Indian cities among the top fifty most polluted globally in 2019 (Bera et al. 2021). Urban green spaces offer a sustainable solution to mitigate pollution. They help regulate urban temperatures, reduce air pollutants, and provide ecosystem services like carbon sequestration (Kumar et al. 2022, 2024; Nowak et al. 2006; Singh 2024; Singh et al. 2022, 2024). Urban green infrastructure is increasingly promoted to enhance resilience and adapt to climate change.

Delhi hosts fourteen city forests managed by various agencies, providing crucial ecosystem services. The Okhla Wildlife Sanctuary (OBS) is a significant urban green space offering groundwater recharge, soil erosion control, and aesthetic value. Despite its importance, research on carbon sequestration by urban trees is limited.

This chapter emphasizes the need for urban forestry to mitigate long-term air quality degradation in Delhi-NCR. We highlight the need for urban green spaces as a nature-based solution to protect against incoming pollutants from neighboring states due to stubble burning. The ongoing conflict over stubble burning underscores the importance of sustainable urban planning and inter-state cooperation to address air pollution effectively.

2 Methodology

In this study, we endeavor to underscore the significance of urban spaces as potential barriers to mitigating incoming air pollution in the Delhi-NCR region during the winter months, offering valuable insights for policymakers. The chapter is meticulously crafted based on a thematic literature review encompassing journals, articles, and pertinent documents. A substantial portion of the reviewed research articles was sourced from platforms like Google Scholar and Scopus and references from existing literature. Our search was guided by primary keywords such as 'air pollution', 'urban forestry', 'urban green spaces', 'stubble burning', 'winter pollution', 'North India', 'Delhi', 'pollution', and 'gas chamber'. We identified 567 relevant literatures in Google Scholar and 82 in Scopus, which were systematically categorized according to our thematic focus for a comprehensive review.

Additionally, we referred to websites, press releases, and real-time incidents to enrich our understanding of the subject matter. Our analysis delves into the approaches adopted to establish the interrelationship between urban green spaces acting as a barrier to mitigate the impact of stubble-burning pollution, drawing upon insights gleaned from previous studies. Furthermore, we meticulously highlight the connections between the selected papers, offering a coherent narrative synthesizing the existing body of knowledge on the topic.

3 Urban Green Spaces

To comprehend the significance of urban green spaces and their potential to mitigate environmental challenges, it is imperative to grasp the current landscape and the associated challenges and how these relate to human well-being. This understanding can inform why preserving urban greenery is crucial. The interdependence of humans and the environment underscores the importance of maintaining ecosystem services, as human life is intricately tied to the health of its surroundings. Sustainable living necessitates increased investment in environmental conservation (Dhyani et al 2021; Kumar et al. 2021a, b, c, d, e; Prakash et al. 2022). From an anthropological standpoint, ecosystem services are defined as the unidirectional benefits that connect ecosystems to human society.

Urban landscapes possess a unique blend of natural and built environments. The concept of urban spaces with vegetation originates from the natural areas within cities, primarily covered with vegetation (Singh 2021; Singh et al. 2021; Sharma and Singh 2021). Urban green infrastructure encompasses roadside plantations, remnant forests, and vegetation patches. Trees in urban areas include those in parks, gardens, and along streets, with land use and the natural environment being pivotal factors influencing the extent of urban tree cover (Nowak et al. 2006).

The benefits of urban tree cover are manifold, ranging from temperature reduction to energy conservation. Mature trees near buildings have been shown to reduce energy consumption for cooling purposes. Moreover, urban tree cover enhances commercial and residential areas' aesthetic appeal and economic value. However, rapid industrialization and urbanization have significantly impacted designated vegetation, leading to increased air pollution and noise emissions.

The rise in population and vehicular traffic has exacerbated air pollution in metropolitan areas like Delhi-NCR, especially during winter, posing a serious global concern. Natural disasters increasingly stem from anthropogenic actions, emphasizing the vital role of vegetation in environmental sustainability. Stricter policies are needed to preserve vegetation cover for development projects to mitigate this issue.

Research underscores the potential of urban green spaces as a solution to declining greenery, particularly in urban areas. Urban forests and trees provide various ecosystem services, including air purification, microclimate regulation, and noise reduction (Bolund and Hunhammar 1999). They also improve public health and enhance environmental quality.

Urban green spaces are crucial in combating air pollution by absorbing pollutants and improving air quality (Kumar et al. 2021a, b, c, d; Singh et al. 2020). The study in China and Portugal has demonstrated the negative correlation between air pollution and green spaces, emphasizing the urgent need to create and maintain green spaces to address air pollution issues. Forest floors in urban green spaces can trap air pollutants, while tree canopies help reduce dust and prevent secondary pollution (Suhane et al. 2023).

In India, research should prioritize urban green spaces and their potential benefits. Successful initiatives in cities like Pune and Bangalore have highlighted the importance of urban trees in carbon sequestration and the need for well-planned urban green spaces. Consolidated data on urban tree distribution and vegetation can inform better management and policymaking, fostering public involvement in sustainability efforts (Nagendra and Gopal 2010).

Overall, understanding the role of urban green spaces and implementing effective measures to preserve and enhance them is critical for addressing environmental challenges and promoting sustainable urban development.

4 The Need for Urban Green Spaces

Urbanization has swept across developing countries like India in recent decades, accompanied by significant demographic shifts (Avtar et al. 2019). Currently, India, China, and Nigeria stand at the forefront, experiencing rapid urban migration, projected to contribute to approximately 35% of urban population growth between 2018 and 2050 (Avtar et al. 2019). The bulk of the world's rural populace, approximately 90%, resides in regions such as Asia and Africa, with a notable migration trend towards urban centers. Projections suggest that by 2050, the urban population in Asia and Africa will surge by 64% and 56%, respectively.

Economic prosperity is a primary driver of urbanization, with labor migration occurring nationwide. A poignant quote from Hindi poet Sudhama Pandey's 'Dhoomil', as cited by 'The Wire' media house, encapsulates the harsh reality faced by many: "admi ko peeth nahi, pet girata hain" (Pande 2023), signifying the struggle for sustenance. However, the process of unplanned urbanization raises serious sustainability concerns. The plight of migrant laborers, exacerbated by events such as the COVID-19 pandemic, underscores the pressing need for governmental and societal attention to this issue. Tragic incidents like the recent train collision in Orissa, claiming the lives of around 270 migrant laborers en route to employment opportunities, highlight the disconnect between urban and rural policy frameworks.

By 2030, it is estimated that nearly 60% of the global population will inhabit urban areas, with India's urbanization rate projected to double by 2050. Amidst escalating concerns over deteriorating air quality in urban centers, urgent measures are imperative to foster sustainable urbanization practices. Overcrowding in metropolitan hubs like Delhi, Kolkata, Chennai, and Mumbai strains energy, transportation, and water resources and exacerbates climatic and health challenges (Hasan Rashed 2023). Addressing these issues is crucial to enhancing air purification and overall air quality.

Amidst these challenges, urban green spaces are a potential solution to mitigate environmental degradation and enhance public well-being. These spaces offer invaluable ecosystem services and foster public engagement with nature-based solutions (Gupta et al. 2019; Khera et al. 2009; Kumar et al. 2019b, c; Yadav et al. 2019a, b). Particularly in countries like India and China, where pollution levels are high, urban green spaces play a vital role in combating environmental degradation. Moreover, they support urban agriculture, mitigate microclimatic effects, and reduce the prevalence of urban heat islands (Jennings et al. 2016).

Despite their potential benefits, challenges persist in effectively managing and utilizing urban green spaces. Issues such as inadequate species selection and lack of maintenance hinder their efficacy (Alvey 2006; Chacalo et al. 1994). Addressing these challenges requires comprehensive planning, management, and policy interventions (Bhalla and Bhattacharya 2015). Strengthening research efforts in water resource management is also crucial to support the establishment of urban green spaces (Ramaiah and Avtar 2019).

The government has made significant strides in enhancing urban green spaces in Delhi. Initiatives such as establishing biodiversity parks and tree conservation measures underscore the commitment to environmental sustainability. The 'City Plants Million Trees' campaign exemplifies proactive measures to preserve ecological diversity (Imam and Banerjee 2016). Therefore, fostering sustainable urbanization practices and promoting the development of urban green spaces is paramount to mitigating environmental degradation and improving public health and well-being in rapidly urbanizing regions like Delhi.

5 Stubble Burning and Its Environmental Impacts in the Northwestern States

India, as a rural nation, generates a significant amount of agricultural waste annually. The Indo-Gangetic Plain (IGP), covering 20% of the country's geographical area, contributes substantially to food grain production, catering to 40% of the nation's population. Approximately twelve million hectares of land in the IGP region are dedicated to wheat-rice crop rotation. After harvest, wheat and rice are predominantly used, and substantial crop residues and biomass remain in the fields during the paddy season. It is customary for farmers in the IGP to burn agricultural crop residues in October and November each year, considering it a convenient method to clear the fields for the upcoming Rabi season.

In 2012–2013, India produced 361 million tonnes of sugarcane, 94 million tonnes of wheat, and 105 million tonnes of rice, as the Directorate of Economics and Statistics reported. A report from the Indian Agricultural Research Institute (IARI) revealed that fourteen million metric tonnes of stubble were burnt out of the twenty-two million metric tonnes produced.

Most crop residues from rice, wheat, and sugarcane are subjected to burning due to the lucrative returns associated with these crops, discouraging farmers from exploring alternative practices. Traditionally, Haryana and Punjab account for approximately 48% of stubble burning (Gadde et al. 2009). This high prevalence of burning can be attributed to socioeconomic factors, as farmers perceive it as a convenient and cost-effective method to prepare fields for the next sowing season. This sentiment has been echoed in research by (Ravindra et al. 2019), highlighting stubble burning as a significant loophole in India's agricultural history. Additionally, the association of stubble burning with the Diwali festival exacerbates air pollution levels significantly. Studies conducted in Delhi, Noida, Gurgaon, and Ghaziabad have demonstrated hazardous air quality levels during the overlap of stubble burning and Diwali festivities (Barman et al. 2008). Ghei and Sane (2018) reported a stark increase of 40 $\mu g/m^3$ in PM2.5 concentration post-Diwali in 2018, further underscoring the detrimental impact of these events on air quality.

This chapter emphasizes the persistent pollution from stubble burning, which has garnered attention from academic researchers and policymakers. Despite efforts to introduce biochemical and physiological interventions, the efficacy remains questionable. We contend that the focus of research should shift towards sustainable solutions to mitigate airborne pollution during winters in the Delhi-NCR region, mainly through the development of green spaces. This section comprehensively analyzes the winter pollution caused by stubble burning, urging for concerted efforts toward sustainable interventions.

5.1 On-Field Impact

5.1.1 Effect on Air Quality

In India, agriculture sustains 80% of the population, with the rice–wheat system being a prevalent cropping pattern and a significant contributor to stubble burning (Sain 2020). A grim episode in Delhi's winter of 2017 saw 2.5 million farmers burning 23 million tonnes of paddy residue, exacerbating stubble pollution (Moussa et al. 2017). Stubble burning is a major activity that deteriorates air quality, releasing aerosols and gaseous pollutants. (Mittal et al. 2009)Highlights that 90% of biomass burning is attributed to human activities, with natural fibers contributing negligibly.

Although burning crop residues is not the sole source of pollution, it significantly adds to environmental risks alongside various other factors. Stubble burning emerges as a prominent contributor to harmful air pollutants, elevating levels of particulate matter in the atmosphere and impacting climate change dynamics. Particulate matter, classified into PM2.5 and PM10 based on aerodynamic diameter and chemical composition, is crucial in air quality degradation. The research underscores stubble burning as a primary contributor to Delhi's air pollution crisis. Light particulate matter, carried by winds, prolongs atmospheric residence and travel, intensifying the smog effect (Zhao and Luo 2017).

5.1.2 Climatic Effect

Stubble burning emits several harmful greenhouse gases, including methane and carbon dioxide, contributing significantly to air pollution during winters as these

Table 1 A comparative analysis of residue generated and burned agricultural residues of Indian States

Name of the states	Generated residue	Burned residue
Uttar Pradesh	59.8	21.3
Punjab	50.8	19.8
Maharashtra	46.5	7.6
Andhra Pradesh	43.9	2.8
West Bengal	35.8	4.5
Karnataka	33.8	5.7
Madhya Pradesh	33.2	1.8
Rajasthan	29.4	1.8
Gujrat	28.6	3.9
Haryana	27.8	9.06
Orissa	20.1	1.4
Tamil Nadu	19.8	4.1
Assam	11.4	0.74
Chattisgarh	11.3	0.9

pollutants form smog when mixed in the atmosphere. Agriculture alone accounts for 17–32% of total global emissions (Ravindra et al. 2019), with India's estimated emissions at approximately 658.823 Tg, representing about 12.2% of global greenhouse gas emissions. According to a UNEP report from October 2022, India's per capita carbon emission is around 280 million tonnes, lagging behind the world average but still posing significant concerns. A study by the Ministry of Agriculture and Farmers' Welfare outlined the recent trend of emissions released by stubble burning across different states (Table 1).

5.1.3 Effect on Edaphic Conditions

Crop residue burning not only depletes the organic content of the soil but also diminishes the population of crucial microorganisms residing in the topsoil layer. The ramifications of stubble burning on-field are manifold, including a reduced capacity of the soil to absorb organic matter, a decline in nutrient richness, and impaired nitrogen and phosphorus cycling. Studies reported that stubble burning can elevate the temperature of the landscape by 33.8 °C to 44.2 °C, leading to a significant depletion of soil nitrogen by up to 73% within 2.5 cm of the soil surface. This drastic reduction in nitrogen content adversely affects the populations of essential bacterial and fungal communities crucial for soil fertility, with stubble burning estimated to decrease soil bacteria content by 50%.

Furthermore, the loss of beneficial pests due to stubble burning renders crops more susceptible to diseases, exacerbating the proliferation of harmful pests. The solubility capacity of the upper soil layers diminishes post-burning, resulting in decreased crop yields and soil fertility. Additionally, the open burning of residue threatens surrounding vegetation and standing crops while decimating soil microflora and fauna. The leftover residue from burning is an excellent absorbent, potentially compromising the efficacy of applied weedicides if left untreated (Jain et al. 2014).

The Department of Soils, Punjab Agricultural University (PAU) 2006 revealed that Punjab produces approximately seventeen million tonnes of wheat straw and about twenty-three million tonnes of paddy straw annually (Kumar et al. 2015). When straw is burnt, the soil's top three inches are subjected to extreme temperatures, disrupting the delicate carbon–nitrogen equilibrium. This leads to the release of carbon into the atmosphere in the form of CO_2, exacerbating the already precarious situation of soil organic carbon levels in Punjab (Reddy et al. 1984). Despite previous warnings and studies, Punjab continues to overlook the detrimental consequences of crop residue burning.

The rice–wheat cropping sequence prevalent in Punjab results in the removal of over three hundred kilograms of nitrogen per hectare, significantly disrupting nutrient cycles and leading to an approximate loss of 0.824 million tonnes of NPK from the soil (Aulakh et al. 2001). These findings underscore the urgent need for sustainable agricultural practices and effective measures to curb crop residue burning to safeguard soil health and agricultural sustainability.

5.2 *Off-Field Impacts*

5.2.1 Mortality Rates

The impact of stubble burning on air pollution has led to countless deaths worldwide since the onset of industrialization. In India, the toll of air pollution-related fatalities has surged from 1990 to 2015 due to escalating pollution levels (Sharma et al. 2019). In particular, residents of the Indo-Gangetic Plain face dire consequences, with life expectancy shortened by an average of seven years compared to residents of other regions in India. Globally, studies estimate that approximately six hundred thousand people suffer premature deaths annually due to air pollution in India alone (Ghude et al. 2016). Research indicates that Delhi residents have experienced a reduction in life expectancy by 6.4 years attributable to air pollution (Ghude et al. 2016). Furthermore, projections suggest that adhering to World Health Organization (WHO) and National Ambient Air Quality Standards could extend Delhiites' lifespan by fifteen years.

In India, particulate pollution alone has claimed the lives of 0.67 million individuals, while overall air pollution-related mortality stands at 1.24 million. Alarmingly, over half of these deaths, approximately 51%, involve individuals below the age of seventy (Liu et al. 2018). These figures underscore the urgent need for effective measures to curb air pollution and safeguard public health.

5.2.2 Effect on Human Health and Well-Being

Agricultural residue burning is associated with many health ailments, from asthma and bronchial asthma to acute respiratory infections and eye irritation. It has been noted that India leads in the number of cataract cases annually, possibly due to the pervasive air pollution (Arbex et al. 2004). Pollution incidents can have fatal consequences, particularly for individuals with pre-existing respiratory conditions. Even brief exposure to carbon monoxide release can disrupt blood flow and pose immediate health risks. Moreover, chronic exposure to elevated levels of air pollution can lead to long-term health complications, including bronchitis, emphysema, asthma, and chronic obstructive pulmonary disease.

The impact of pollution extends across various landscapes, affecting farmers and urban residents in the Delhi-NCR region during burning incidents in the field. Farmers exposed to stubble smoke commonly report eye and lung irritation (Kumar et al. 2015), highlighting the pervasive health risks associated with agricultural residue burning.

5.3 *Socio-economic Development*

5.3.1 The Effect of Air Pollution on Economic Development

The country's economy faces a severe threat from air pollution, with developing countries significantly contributing to the global economic burden. According to a 2013 World Bank report, air pollution cost the global economy $225 billion in 2013. This impact is significant, particularly for India, where the cumulative cost of air pollution on the economy escalated from 4.5 to 7.7% of its GDP by 2018, with projections indicating a potential increase to 15% by 2050 (Kumar et al. 2015).

Furthermore, the economic implications of managing air pollution extend to local economies. Studies suggest that maintaining good air quality in households could yield substantial economic benefits. For instance, it is estimated that the economic benefit of ensuring clean air in households in Kolkata and Delhi could amount to approximately 950 rupees (equivalent to 12.7 USD, assuming 1 USD = 75 INR) and Rs 2086 (27.8 USD) per annum, respectively (Kumar et al. 2015). These findings underscore the significant economic stakes of addressing air pollution at both national and local levels.

5.4 *Geographical Barriers*

The National Capital Region (NCR), encompassing Delhi and its surrounding suburbs such as Noida, Gurgaon, Ghaziabad, and Faridabad, is shrouded in a dense blanket of toxic smog for approximately 10–12 weeks during the fall and winter months. This prolonged exposure to hazardous air pollutants seriously threatens the local population's health. The escalation in respirable airborne dust levels profoundly impacts air quality, thereby jeopardizing the well-being of residents.

6 The Major Factors Intensifying Air Pollution: Sources and Determinants

6.1 Continentality

Due to stubble burning, emitted pollutants disperse across the plains, while the towering Himalayas act as a formidable barrier to the north. The mountain range impedes the natural dispersion of pollutants, causing them to be trapped within the plains. In particular, the air becomes stagnant during the colder winter months due to the mountain barrier. The frigid air, combined with high mountains, causes smog to concentrate heavily in the plains, inhibiting its vertical ascent (as warm air rises and cool air settles). This atmospheric phenomenon, known as temperature inversion, is well-documented (Mukai 2018).

During the warmer months, the planetary boundary layer, the lowest layer of the atmosphere, contains warmer and lighter air, facilitating vertical mixing of pollutants into the upper atmosphere. This process allows pollutants to disperse and combine with cleaner air (Kahn et al. 2007). However, in winter, the planetary boundary layer becomes thinner due to the prevalence of cooler air near the Earth's surface. The warm air above acts as a lid, trapping colder air below and restricting vertical mixing. This winter inversion phenomenon limits the space for pollutants to disperse vertically, exacerbating air pollution levels (Chen et al. 2017; Kahn et al. 2007).

Furthermore, the coupling of primary and secondary pollutants with vehicular exhaust, particularly in conditions of high humidity (40–70%), light wind (1–2 m/sec), and temperatures ranging between 20–30 °C, exacerbates winter haze, especially in landlocked national capitals like Delhi. In the National Capital Region (NCR), which encompasses Delhi and its surrounding areas, industrial activities further exacerbate pollution levels, contributing to increasingly severe winters.

6.2 Insufficient Winds Stopping Pollutant Dispersion

The wind is pivotal when considering winter pollution resulting from stubble burning. Winds carrying pollutants from stubble-burning sites often encounter the imposing barrier of the Himalayas, where they become trapped due to atmospheric dynamics. As air pressure forces pollutants in one direction and hinders their rapid dispersion in the opposite direction, particulate matter accumulates over the northern plains. Crop burning typically occurs from September to November, coinciding with the onset of winter. During this period, gentle northwest winds gently push the polluted air towards the Delhi-NCR plains, exacerbating the region's smoggy conditions. A report by Down To Earth highlights that scientists from the USA and India have observed that the winter menace of stubble pollution extends beyond Maharashtra, Telangana, Chhattisgarh, and parts of Orissa.

6.3 Additional Factors Contributing to Air Pollution in Delhi-NCR

A report by The Times of India highlighted research findings from the Centre for Science and Environment, shedding light on the significant contribution of eleven coal-based thermal power plants in the Delhi-NCR region to pollution, accounting for 8% of the total pollution burden. The study, based on an evaluation of environmental status reports available on the power plants' websites, was conducted by the Central Electricity Authority, the technical arm of the Union Power Ministry, using data from April 2022 to August 2023. The report delves into a detailed analysis of the roadmap for coal power plants within the 2060 timeline. In 2015, the Union Ministry directed the eleven studied power plants to comply with emission norms for PM 2.5 and SO_2 by 2040, a deadline recently extended to 2060 due to the failure of seven out of eleven plants to meet the prescribed standards, attributed to poor maintenance of electrostatic precipitators and other operational parameters.

Automobile pollution is recognized as a significant contributor to air pollution in Delhi and surrounding heavily industrialized areas (Tripathi et al. 2019). The government implemented the odd–even vehicle policy to mitigate pollution levels in 2016. However, environmental experts suggest that while the policy has reduced road congestion, its impact on winter pollution remains limited. The Delhi government has also implemented measures like the Graded Action Plan to address pollution effectively.

Deforestation and construction activities further exacerbate air pollution levels. Extensive infrastructure development aimed at creating a visually appealing capital often leads to removing trees, significantly impacting air quality. It is imperative for nations to assign value to trees and involve public participation to ensure optimal outcomes. Construction activities also emit harmful pollutants, prompting regulatory measures like Phase III of the Graded Action Plan, implemented in 2023 following a severe spike in the air quality index of the national capital. While the regulation initially faced criticism for halting major public development projects, the government later relaxed restrictions in response to public concerns.

7 Depicting Other Associated Causes of Stubble Burning

7.1 Green Revolution

Farmers in Punjab and Haryana have been encouraged to adopt the environmentally unsustainable wheat-rice crop rotation system for over three decades. The Green Revolution, initiated in the 1970s, significantly impacted India's food security by using fertilizers, pesticides, and high-yielding crop varieties to increase agricultural productivity and ensure food security (Rathna Priya et al. 2019). Despite their inherently dry climate for paddy-based agriculture, Punjab and Haryana, the leading states for rice cultivation, have experienced adverse environmental effects due to this agricultural exploitation, leading to the menace of winter pollution in Delhi-NCR. The prevailing mindset of burning crop residues in the fields represents the easiest, fastest, and least expensive approach, emblematic of agro-exploitation.

7.2 The Insufficient MSP Regime

The Minimum Support Price (MSP) is an agricultural incentive provided by the Indian government to farmers to guarantee a minimum profit for their crops. Designed to secure livelihoods, especially for 86% of small farmers, the MSP regime aims to shield farmers from market price fluctuations arising from climate change, market integration, and misinformation (Sharma et al. 2019). The MSP policy is also crucial for incentivizing investment and modern agricultural practices among farmers. However, historical policies such as free electricity, fertilizer subsidies, and rising MSPs have led to the proliferation of paddy cultivation (Panigrahi 2021). Empirical data from various news outlets indicate a significant increase in paddy cultivation at the expense of other crops, exacerbating stubble-burning incidents (Parikh 2020). Despite the intention to benefit farmers and promote agriculture, the MSP regime's combination of bias and inadequacy has contributed significantly to the plight of farmers. While there have been discussions about revising the MSP policy based on the recommendations of the National Commission on Farmers, led by MS Swaminathan, the current MSP calculation formula lacks comprehensive cost coverage, resulting in underpayment to farmers (Singh and Bhogal 2021).

7.3 Lack of Access to Mechanization

India's enormous population of small farmers faces numerous challenges, including limited access to mechanization due to financial constraints. Punjab, Haryana, Uttar Pradesh, and Rajasthan governments have initiated schemes to provide machinery at subsidized rates, yet equitable distribution remains challenging. Policymakers should prioritize further mechanization development and ensure equitable access for all farmers.

7.4 *Lack of Monitoring*

The Central Pollution Control Board has established the National Ambient Air Quality Standards, which are mandatory for all Indian states (Ganguly et al. 2020; Gulia et al. 2015). Analysis utilizing satellite data has revealed significant stubble-burning incidents in Haryana and Punjab, particularly during the post-harvest months of October and November (Das et al. 2023; Thumaty et al. 2015). Shady monitoring practices have normalized excessive risk over time, and necessitating improved oversight and enforcement of environmental regulations.

Given the issues above, urban green spaces emerge as a potential mitigation tool against stubble-burning pollution in the Delhi-NCR region. The pollutants from stubble burning exacerbate air quality in the national capital, posing significant health risks. Trees act as resistant barriers against incoming pollutants, underscoring the need to intensify research and policy efforts to increase urban green spaces, including urban forests and parks, to mitigate the impacts of stubble-burning pollution.

8 Discussion

Delhi is the foremost city grappling with winter pollution from vehicular emissions, municipal waste, industrial activities, and biomass/crop residue burning (Kumar et al. 2015). On a global scale, India ranks eighth among the most polluted countries. It is fourth in the list of fifty of the most polluted cities worldwide, according to IQAIR's 2018 air quality report. The gravity of air pollution is underscored by its lethal impact. Delhi, Noida, and Gurgaon are identified as the leading contributors to premature deaths attributed to air pollution, as highlighted in the Global Burden of Disease Report. The dire consequences of toxic air inhalation are projected to curtail life expectancy by a decade, as predicted by the University of Chicago Energy Policy Institute's Air Quality Life Index 2023. PM2.5, a hazardous particulate matter, is estimated to slash life expectancy by five years, surpassing the detrimental effects of smoking, alcohol consumption, unsafe water, and transportation issues. Alarmingly, studies suggest that air pollution could potentially diminish the life expectancy of 40% of Indians within the next decade (Pandey et al. 2021).

Amidst burgeoning urbanization, strategic planning and incorporating urban green spaces emerge as imperative components for city development. The surge in population migration to Delhi-NCR from rural areas in 2021, surpassing the natural population growth rate, underscores the pressing need for urban vegetation cover (Fig. 1). Urban green spaces serve as vital buffers against airborne pollutants, effectively mitigating the impacts of urbanization-induced environmental degradation. As air

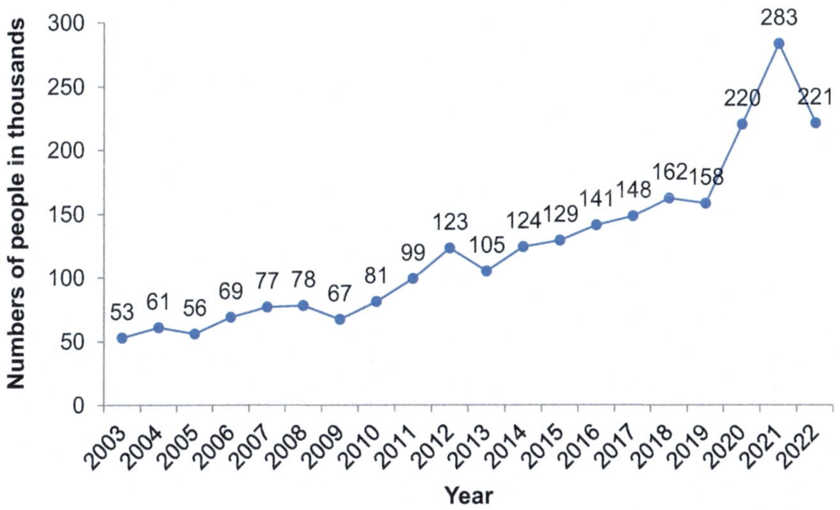

Fig. 1 Urban migration population in the national capital of Delhi in the period (2003–2021). *Source* Statista 2023

currents disperse pollutants, the significance of green spaces becomes pronounced in combating localized pollution sources, such as stubble-burning events. Additionally, urban greenery aids in ameliorating microclimate variations induced by land use changes, vehicular emissions, and industrial activities, thereby alleviating the urban heat island effect prevalent during Delhi-NCR's winter months.

Efforts towards integrating green spaces into urban landscapes align with the 11th Sustainable Development Goal of creating green, safe, and inclusive cities and communities. Moreover, the imperative to limit global warming to 1.5 °C, as advocated by the Intergovernmental Panel on Climate Change (IPCC), necessitates transformative actions in energy, land use, and urban infrastructure, wherein urban green spaces play a pivotal role (Kabisch and Haase 2014). Green spaces offer tangible benefits in tackling air pollution and climate change by actively reducing PM2.5 concentrations and mitigating the urban heat island effect (Gago et al. 2013).

Beyond physical health benefits, urban green spaces foster mental and emotional well-being, as the World Health Organization endorsed. Regional policymakers are urged to leverage urban green spaces to attract investments from national budgets, NGOs, and private enterprises. Private sector initiatives, such as Mamma Earth's 'Plant Goodness and Plastic' policy, underscore the potential for private players to drive sustainability agendas and engage consumers in green initiatives. However, despite strides in green management strategies, challenges persist due to inadequate urban planning and governance structures. Delhi's journey towards implementing urban green spaces has been marred by top-down policy approaches and governance hurdles, as highlighted by (Rana et al. 2019). Overcoming these obstacles necessitates a concerted effort towards inclusive and sustainable urban development, with urban green spaces serving as indispensable components in fostering healthier, more resilient cities.

According to the United Nations Development Programme (UNDP), cities, though occupying only 3% of the Earth's land, consume 70% of energy and are responsible for 70% of carbon emissions, contributing to 90% of the world's GDP. Sustainable growth and environmental conservation are paramount for the survival of humankind. Maintaining natural vegetation cover is crucial for achieving sustainability, as economic production cannot come to a standstill.

Kumar et al. (2019a, b, c) emphasize the necessity of smart planning and the integration of vegetation cover in development initiatives (Fig. 2). It underscores the significance of green infrastructure in mitigating air pollution. It stresses the need for further research in developing urban green spaces to address this issue effectively. Prioritizing sustainable urban development and the preservation of natural ecosystems may guide us toward a healthier and more resilient environment for present and future generations.

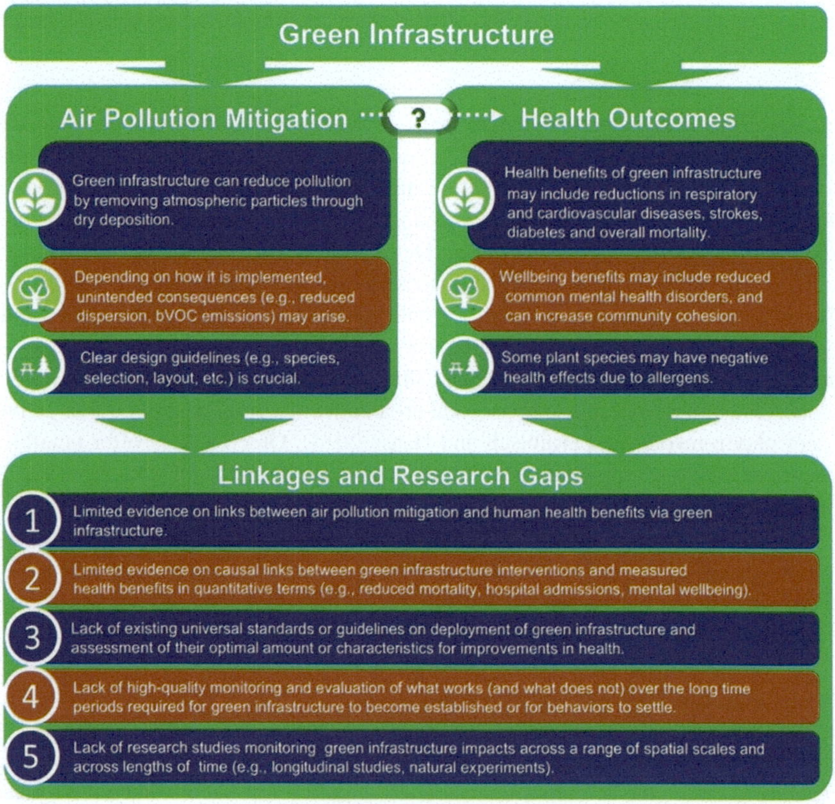

Fig. 2 The correlation between air pollution, green infrastructure, and health outcomes, highlighting research gaps in related design (Kumar et al. 2019a, b, c)

The Supreme Court has issued mandates to Punjab, Haryana, and Rajasthan governments, urging them to address the primary causes of Delhi's inflowing stubble pollution, which has earned the city the ominous moniker of a 'gas chamber'. Criticizing government policymakers for suggesting alternatives to farmers without adequate measures, is not a comprehensive solutions. Punjab has successfully reduced stubble pollution emissions by 40%, showcasing the effectiveness of ongoing policies. However, despite significant subsidies of nearly 60–70% for alternative tools in various states, the issue persists, underscoring the need for a bottom-up approach to address these challenges (Khundrakpam and Sarmah 2023).

Our chapter advocates for a bottom-up approach to tackle these issues. We contend that green infrastructures, such as urban green spaces and forests, can be practical barriers to absorbing pollutants and preventing incoming stubble pollution (Gupta et al. 2018a, b; Sharma et al. 2018; Singh et al. 2018). Implementing such approaches is particularly crucial in Delhi-NCR, where proper urban green spaces are lacking. Furthermore, we emphasize the importance of species selection in addressing heavily polluted environments, encompassing air, water, and soil quality.

The Yamuna River, a major water source for Delhi, is severely polluted, with several parts exhibiting high levels of heavy metals in groundwater. The Delhi Development Authority's 2017 report underscores the depleted groundwater levels due to extensive migration, compounding environmental stresses from agricultural, industrial, and domestic sectors (Gupta and Singh 2017; Kumar et al. 2017; Singh and Verma 2013; Singh et al. 2017a, b). If urban greening measures are not swiftly adopted, Delhi risks facing a situation akin to the 2006 London smog, posing significant public health and safety threats.

9 Conclusion

The chapter provides a unique perspective on tackling the pervasive air pollution issue stemming from stubble burning by advocating for urban forestry as a viable solution. The pollution emanating from stubble burning, particularly during the winter months, exacerbates the air quality index in Delhi, posing significant challenges to the city's residents. We commenced our discussion by introducing various facets of urban green spaces, emphasizing their pivotal role in urban planning. As the narrative unfolded, we underscored the imperative for urban green spaces as essential human and environmental protection tools. Through meticulous analysis, we elucidated the causal mechanisms of stubble pollution and how the geography of Delhi-NCR amplifies the concentration of incoming pollution in the city. Our comprehensive review aimed to construct a compelling argument advocating using urban spaces, including parks and forests, as natural solutions to combat incoming stubble pollution in India's capital. However, we also identified a significant gap in research databases on urban forestry in the country, which poses a substantial challenge in the field. To bridge this gap, policymakers need to prioritize the contributions of urban forestry to public welfare. Researchers and practitioners must quantify the benefits urban vegetation provides in alleviating environmental stress and fostering community engagement. Additionally, policymakers need to translate learnings from international platforms into actionable policies that address local environmental challenges. Looking ahead, the focus on urban green space research should encompass both the environmental needs, such as mitigating air pollution caused by stubble burning and addressing the needs of the people. This dual focus will be instrumental in shaping policies and interventions that promote sustainable urban development while enhancing the quality of life for urban residents.

References

Alvey AA (2006) Promoting and preserving biodiversity in the urban forest. Urban Forest Urban Green 5(4):195–201. https://doi.org/10.1016/j.ufug.2006.09.003

Arbex MA, Cançado JED, Pereira LAA, Braga ALF, Saldiva PHDN (2004) Queima de biomassa e efeitos sobre a saúde. J Bras Pneumol 30(2):158–175. https://doi.org/10.1590/S1806-37132004000200015

Aulakh M, Khera T, Doran J, Bronson K (2001) Denitrification, N_2O and CO_2 fluxes in rice-wheat cropping system as affected by crop residues, fertilizer N and legume green manure. Biol Fertil Soils 34(6):375–389. https://doi.org/10.1007/s003740100420

Avtar, Sahu, Aggarwal, Chakraborty, Kharrazi, Yunus, Dou, Kurniawan (2019) Exploring renewable energy resources using remote sensing and GIS—a review. Resources 8(3):149. https://doi.org/10.3390/resources8030149

Barman SC, Singh R, Negi MPS, Bhargava SK (2008) Ambient air quality of Lucknow City (India) during use of fireworks on Diwali Festival. Environ Monit Assess 137(1–3):495–504. https://doi.org/10.1007/s10661-007-9784-1

Bera B, Bhattacharjee S, Shit PK, Sengupta N, Saha S (2021) Significant impacts of COVID-19 lockdown on urban air pollution in Kolkata (India) and amelioration of environmental health. Environ Dev Sustain 23(5):6913–6940. https://doi.org/10.1007/s10668-020-00898-5

Bhalla P, Bhattacharya P (2015) Urban biodiversity and green spaces in Delhi: a case study of new settlement and Lutyens' Delhi. J Hum Ecol 52(1–2):83–96. https://doi.org/10.1080/09709274.2015.11906933

Bolund P, Hunhammar S (1999) Ecosystem services in urban areas. Ecol Econ 29(2):293–301. https://doi.org/10.1016/S0921-8009(99)00013-0

Chacalo A, Aldama A, Grabinsky J (1994) Street tree inventory in Mexico City. Arboricult Urban Forest 20(4):222–226. https://doi.org/10.48044/jauf.1994.040

Chen J, Li C, Ristovski Z, Milic A, Gu Y, Islam MS, Wang S, Hao J, Zhang H, He C, Guo H, Fu H, Miljevic B, Morawska L, Thai P, Lam YF, Pereira G, Ding A, Huang X, Dumka UC (2017) A review of biomass burning: emissions and impacts on air quality, health and climate in China. Sci Total Environ 579:1000–1034. https://doi.org/10.1016/j.scitotenv.2016.11.025

Das P, Behera M, Abhilash P (2023) A rapid assessment of stubble burning and air pollutants from satellite observations. Trop Ecol. https://doi.org/10.1007/s42965-022-00291-5

Dhyani K, Kuniyal HB, Singh H, Sobha (2021) Growth and physiological potential of *Terminalia arjuna* under elevated CO_2 levels in open top chamber condition. J Appl Nat Sci 13(3):1121–1126. https://doi.org/10.31018/jans.v13i3.2490 (ISSN 2231-5209)

Gadde B, Bonnet S, Menke C, Garivait S (2009) Air pollutant emissions from rice straw open field burning in India, Thailand and the Philippines. Environ Pollut 157(5):1554–1558. https://doi.org/10.1016/j.envpol.2009.01.004

Gago EJ, Roldan J, Pacheco-Torres R, Ordóñez J (2013) The city and urban heat islands: a review of strategies to mitigate adverse effects. Renew Sustain Energy Rev 25:749–758. https://doi.org/10.1016/j.rser.2013.05.057

Ganguly T, Selvaraj KL, Guttikunda SK (2020) National clean air programme (NCAP) for Indian cities: review and outlook of clean air action plans. Atmosp Environ: X 8:100096. https://doi.org/10.1016/j.aeaoa.2020.100096

Ghei D, Sane R (2018) Correction: estimates of air pollution in Delhi from the burning of firecrackers during the festival of Diwali. PLoS ONE 13(12):e0205131. https://doi.org/10.1371/journal.pone.0205131

Ghude SD, Chate DM, Jena C, Beig G, Kumar R, Barth MC, Pfister GG, Fadnavis S, Pithani P (2016) Premature mortality in India due to PM $_{2.5}$ and ozone exposure. Geophys Res Lett 43(9):4650–4658. https://doi.org/10.1002/2016GL068949

Gulia S, Shiva Nagendra SM, Khare M, Khanna I (2015) Urban air quality management-a review. Atmos Pollut Res 6(2):286–304. https://doi.org/10.5094/APR.2015.033

Gupta SK, Singh H (2017) Observations on sporadic flowering in *Ficus benjamina* L. and *Peltophorum africanum* L. in new forest. Indian Forest 143:290–291

Gupta SK, Ram J, Singh H (2018a) Comparative study of transpiration in cooling effect of tree species in the atmosphere. J Geosci Environ Protect 6:151–166. https://doi.org/10.4236/gep.2018.68011

Gupta SK, Ram J, Singh H, Kumar P (2018b) Comparative study of cooling under shades of some forestry tree species with respect to ambient temperatures. J Geosci Environ Protect 6:51–66. https://doi.org/10.4236/gep.2018.611004

Gupta SK, Ram J, Singh H (2019) Relationship between leaf area index and atmospheric cooling of tree species. Indian J Forest 42(3):203–208

Hasan Rashed A (2023) The impacts of unsustainable urbanization on the environment. In: Almusaed A, Almssad A (eds) Sustainable regional planning. IntechOpen. https://doi.org/10.5772/intechopen.110089

Imam AUK, Banerjee UK (2016) Urbanisation and greening of Indian cities: problems, practices, and policies. Ambio 45(4):442–457. https://doi.org/10.1007/s13280-015-0763-4

Jain N, Bhatia A, Pathak H (2014) Emission of air pollutants from crop residue burning in India. Aerosol Air Q Res 14(1):422–430. https://doi.org/10.4209/aaqr.2013.01.0031

Jangra A, Kumar K, Maikhuri S, Bhandari MS, Pandey S, Singh H, Barthwal S (2024) Unveiling stress-adapted endophytic bacteria: characterizing plant growth-promoting traits and assessing cross-inoculation effects on *Populus deltoides* under abiotic stress. Plant Physiol Biochem 210:108610. https://doi.org/10.1016/j.plaphy.2024.108610

Jennings V, Larson L, Yun J (2016) Advancing sustainability through urban green space: cultural ecosystem services, equity, and social determinants of health. Int J Environ Res Public Health 13(2):196. https://doi.org/10.3390/ijerph13020196

Kabisch N, Haase D (2014) Green justice or just green? Provision of urban green spaces in Berlin, Germany. Landsc Urban Plan 122:129–139. https://doi.org/10.1016/j.landurbplan.2013.11.016

Kahn RA, Li W-H, Moroney C, Diner DJ, Martonchik JV, Fishbein E (2007) Aerosol source plume physical characteristics from space-based multiangle imaging. J Geophys Res Atmosp 112(D11):2006JD007647. https://doi.org/10.1029/2006JD007647

Kanta C, Kumar A, Chauhan A, Singh H, Sharma IP (2024) The interplay between plant functional traits and climate change. In: Kumar N, Singh H (eds) Plant functional traits for improving productivity. Springer, Singapore. https://doi.org/10.1007/978-981-97-1510-7_3

Khera N, Mehta V, Sabata BC (2009) Interrelationship of birds and habitat features in urban greenspaces in Delhi, India. Urban Forest Urban Green 8(3):187–196. https://doi.org/10.1016/j.ufug.2009.05.001

Khundrakpam P, Sarmah JK (2023) Stubble burning in India: politics and policy responses. Indian J Public Adm 69(2):303–316. https://doi.org/10.1177/00195561221149934

Kumar N and Singh H (2024) Plant functional traits for improving productivity. Springer Nature, Singapore. https://doi.org/10.1007/978-981-97-1510-7

Kumar R, Barth MC, Pfister GG, Nair VS, Ghude SD, Ojha N (2015) What controls the seasonal cycle of black carbon aerosols in India? J Geophys Res Atm 120(15):7788–7812. https://doi.org/10.1002/2015JD023298

Kumar N, Suyal DC, Sharma IP, Verma A, Singh H (2017) Elucidating stress proteins in rice (*Oryza sativa* L.) genotype under elevated temperature: a proteomic approach to understand heat stress response. 3 Biotech 7:205. https://doi.org/10.1007/s13205-017-0856-9

Kumar P, Druckman A, Gallagher J, Gatersleben B, Allison S, Eisenman TS, Hoang U, Hama S, Tiwari A, Sharma A, Abhijith KV, Adlakha D, McNabola A, Astell-Burt T, Feng X, Skeldon AC, De Lusignan S, Morawska L (2019a) The nexus between air pollution, green infrastructure and human health. Environ Int 133:105181. https://doi.org/10.1016/j.envint.2019.105181

Kumar M, Singh MP, Singh H, Dhakate PM, Ravindranath NH (2019b) Forest working plan for the sustainable management of forest and biodiversity in India. J Sustain forest 39:1–22. https://doi.org/10.1080/10549811.2019.1632212

Kumar N, Jeena N and Singh H (2019c) Elevated temperature modulates rice pollen structure: a study from the foothill of the Himalayan agro-ecosystem in India. 3 Biotech 9(5):175. https://doi.org/10.1007/s13205-019-1700-1.

Kumar M, Phukon SN, Paygude AC, Tyagi K, Singh H (2021a) Mapping phenological functional types (PhFT) in the Indian Eastern Himalayas using machine learning algorithm in Google Earth engine. Comput Geosci 158:104982. https://doi.org/10.1016/j.cageo.2021.104982

Kumar A, Kumar P, Singh H, Kumar N (2021b) Modulation of plant functional traits under essential plant nutrients during seasonal regime in natural forests of Garhwal Himalayas. Plant Soil 465:197–212. https://doi.org/10.1007/s11104-021-05003-x

Kumar A, Kumar P, Singh H, Bisht S, Kumar N (2021c) Relationship of physiological plant functional traits with soil carbon stock in the temperate forest of Garhwal Himalaya. Curr Sci 120(8):1368–1373. https://doi.org/10.18520/cs/v120/i8/1368-1373

Kumar A, Kumar P, Singh H, Kumar N (2021d) Adaptation and mitigation potential of roadside trees with bio-extraction of heavy metals under vehicular emissions and their impact on physiological traits during seasonal regimes. Urban Forest Urban Green 58:126900. https://doi.org/10.1016/j.ufug.2020.126900

Kumar A, Kumar P, Singh H, Kumar N (2021e) Impact of plant functional traits on infiltration rate under different forests of Kempty watershed in Garhwal Himalaya. Indian J Soil Conser 49(1):38–44

Kumar A, Singh H, Kumari G, Bisht S, Malik A, Kumar A, Singh M, Raturi BS, Thakur A, Kaushal R (2022) Adaptive resilience of roadside trees to vehicular emissions via leaf enzymatic, physiological, and anatomical trait modulations. Environ Pollut 313:120191. https://doi.org/10.1016/j.envpol.2022.120191

Kumar N, Singh H, Giri K, Kumar A, Joshi A, Yadav A, Singh R, Bisht S, Kumari R, Jeena N, Khairakpam R, Mishra G (2024) Physiological and molecular insights into the allelopathic effects on agroecosystems under changing environmental conditions. Physiol Mol Biol Plants 30:417–433. https://doi.org/10.1007/s12298-024-01440-x

Liu T, Marlier ME, DeFries RS, Westervelt DM, Xia KR, Fiore AM, Mickley LJ, Cusworth DH, Milly G (2018) Seasonal impact of regional outdoor biomass burning on air pollution in three Indian cities: Delhi, Bengaluru, and Pune. Atmos Environ 172:83–92. https://doi.org/10.1016/j.atmosenv.2017.10.024

Malik A, Kumar K, Kumari G, Kumar N, Singh H (2024) Application of functional traits in modelling productivity and resilience under climate change. In: Kumar N, Singh H (eds) Plant functional traits for improving productivity. Springer, Singapore. https://doi.org/10.1007/978-981-97-1510-7_5

Mittal SK, Singh N, Agarwal R, Awasthi A, Gupta PK (2009) Ambient air quality during wheat and rice crop stubble burning episodes in Patiala. Atmos Environ 43(2):238–244. https://doi.org/10.1016/j.atmosenv.2008.09.068

Moussa DT, El-Naas MH, Nasser M, Al-Marri MJ (2017) A comprehensive review of electrocoagulation for water treatment: potentials and challenges. J Environ Manage 186:24–41. https://doi.org/10.1016/j.jenvman.2016.10.032

Mukai S (2018) Remote sensing of heavy aerosol pollution episodes: smoke and dust. In: Remote sensing of aerosols, clouds, and precipitation. Elsevier, pp 85–108. https://doi.org/10.1016/B978-0-12-810437-8.00004-9

Nagendra H, Gopal D (2010) Street trees in Bangalore: density, diversity, composition and distribution. Urban Forest Urban Green 9(2):129–137. https://doi.org/10.1016/j.ufug.2009.12.005

Nowak DJ, Crane DE, Stevens JC (2006) Air pollution removal by urban trees and shrubs in the United States. Urban Forest Urban Green 4(3–4):115–123. https://doi.org/10.1016/j.ufug.2006.01.007

Pandey A, Brauer M, Cropper ML, Balakrishnan K, Mathur P, Dey S, Turkgulu B, Kumar GA, Khare M, Beig G, Gupta T, Krishnankutty RP, Causey K, Cohen AJ, Bhargava S, Aggarwal AN, Agrawal A, Awasthi S, Bennitt F, Dandona L (2021) Health and economic impact of air pollution in the states of India: the global burden of disease study 2019. Lancet Planet Health 5(1):e25–e38. https://doi.org/10.1016/S2542-5196(20)30298-9

Panigrahi R (2021) Economics of subsidies in a welfare state: dynamics of populist policies and farm profitability. Glob Bus Rev 22(3):705–717. https://doi.org/10.1177/0972150918816902

Prakash V, Hunney K and Singh H (2022). Effect of elevated carbon dioxide on growth and development of *Santalum album* L. seedlings inoculated with plant growth promoting microorganisms in open top chambers. Int J Environ Agric Biotechnol 7(5):149–159

Ramaiah M, Avtar R (2019) Urban green spaces and their need in cities of rapidly urbanizing India: a review. Urban Sci 3(3):94. https://doi.org/10.3390/urbansci3030094

Rana NP, Luthra S, Mangla SK, Islam R, Roderick S, Dwivedi YK (2019) Barriers to the development of smart cities in Indian context. Inf Syst Front 21(3):503–525. https://doi.org/10.1007/s10796-018-9873-4

Rathna Priya TS, Eliazer Nelson ARL, Ravichandran K, Antony U (2019) Nutritional and functional properties of coloured rice varieties of South India: a review. J Ethnic Foods 6(1):11. https://doi.org/10.1186/s42779-019-0017-3

Ravindra K, Singh T, Mor S, Singh V, Mandal TK, Bhatti MS, Gahlawat SK, Dhankhar R, Mor S, Beig G (2019) Real-time monitoring of air pollutants in seven cities of North India during crop residue burning and their relationship with meteorology and transboundary movement of air. Sci Total Environ 690:717–729. https://doi.org/10.1016/j.scitotenv.2019.06.216

Reddy KR, Patrick WH, Broadbent FE (1984) Nitrogen transformations and loss in flooded soils and sediments. C R C Critic Rev Environ Control 13(4):273–309. https://doi.org/10.1080/10643388409381709

Rizwan S, Nongkynrih B, Gupta S (2013) Air pollution in Delhi: its magnitude and effects on health. Indian J Commun Med 38(1):4. https://doi.org/10.4103/0970-0218.106617

Sahu LK, Sheel V, Pandey K, Yadav R, Saxena P, Gunthe S (2015) Regional biomass burning trends in India: analysis of satellite fire data. J Earth Syst Sci 124(7):1377–1387. https://doi.org/10.1007/s12040-015-0616-3

Sain M (2020) Production of bioplastics and sustainable packaging materials from rice straw to eradicate stubble burning: a mini-review. Environ Conser J 21(3):1–5. https://doi.org/10.36953/ECJ.2020.21301

Sharma R, Singh H (2021) Alteration in biochemical constituents and nutrients partitioning of *Asparagus racemosus* in response to elevated atmospheric CO_2 concentration. Environ Sci Pollut Res 29:6812–6821. https://doi.org/10.1007/s11356-021-16050-3

Sharma R, Singh H, Kaushik M, Nautiyal R, Singh O (2018) Adaptive physiological response, carbon partitioning, and biomass production of *Withania somnifera* (L.) Dunal grown under elevated CO_2 regimes. 3 Biotech 8:267. https://doi.org/10.1007/s13205-018-1292-1

Sharma R, Choudhary D, Kumar P, Venkateswaran J, Solanki CS (2019) Do solar study lamps help children study at night? Evidence from rural India. Energy Sustain Dev 50:109–116. https://doi.org/10.1016/j.esd.2019.03.005

Singh H (2021) An integrated approach considering physiological and biophysical based indicators for assessing tolerance of roadside plantations of *Alstonia scholaris* towards urban roadside air pollution: an assessment of adaptation of plantations for mitigating urban roadside air pollution. Trees. https://doi.org/10.1007/s00468-021-02179-8

Singh H (2024) Exploring adaptive modulation in plant functional traits and its impact on the productivity of *Acacia auriculiformis* under CO_2-enriched environment. Ind Crops Prod 210:118186. https://doi.org/10.1016/j.indcrop.2024.118186

Singh H, Verma A (2013) Characterization and screening of high nitrogen efficient rice genotype to curtail environmental pollution. Int J Agric Environ Biotechnol 6(1):23–30

Singh P, Bhogal S (2021) Interrogating the MSP regime, farm laws and agrarian future in India. Millennial Asia 12(3):332–349. https://doi.org/10.1177/09763996211056996

Singh H, Savita Sharma R, Sinha S, Kumar M, Kumar P, Verma A and Sharma SK (2017a) Physiological functioning of *Lagerstroemia speciosa* L. under heavy roadside traffic: an approach to screen potential species for abatement of urban air pollution. 3 Biotech 7(61):1–10. https://doi.org/10.1007/s13205-017-0690-0

Singh H, Verma A, Kumar M, Sharma R, Gupta R, Kaur M, Negi M, Sharma SK (2017b) Phytoremediation: a green technology to clean up the sites with low and moderate level of heavy metals. Austin Biochem 2:1012

Singh H, Sharma R, Savita SMP, Kumar M, Verma A, Ansari MW, Sharma SK (2018) Adaptive physiological response of *Parthenium hysterophorus* to elevated atmospheric CO_2 concentration. Indian Forest 144:1–14

Singh H, Yadav M, Kumar N, Kumar A and Kumar M (2020) Assessing adaptation and mitigation potential of roadside trees under the influence of vehicular emissions: a case study of *Grevillea robusta* and *Mangifera indica* planted in an urban city of India. PLoS ONE 15(1):e0227380. https://doi.org/10.1371/journal.pone.0227380

Singh S, Singh H, Sharma SK, Nautiyal R (2021) Seasonal variation in biochemical responses of bamboo clones in the sub-tropical climate of Indian Himalayan foothills. Heliyon 7(4):e06859. https://doi.org/10.1016/j.heliyon.2021.e06859

Singh H, Bandyopadhyaya S, Kumar A, Singh M, Malik A, Kumari G, Raturi A, Yadav SK, Kumar M, Barthwal S, Thakur A, Rawat PS (2022) Understanding the physiological and biophysical response of urban roadside plantations for assessing adaptation and mitigation mechanisms toward vehicular emissions. Urban Climate 44:101183. https://doi.org/10.1016/j.uclim.2022.101183

Singh M, Singh H, Kumar A, Kumar M, Barthwal S, Thakur A (2024) Soil nitrogen availability determines the CO_2 fertilization effect on tree species (*Neolamarckia cadamba*): growth and physiological evidence. Environ Sustain 7:53–60. https://doi.org/10.1007/s42398-023-00300-w

Suhane S, Polara R, Gajjar A, Agrawal R, Sharma U (2023) The effect of urban green spaces on air pollution reduction: a case of Ahmedabad. E3S Web Conf 436:10009. https://doi.org/10.1051/e3sconf/202343610009

Thumaty KC, Rodda SR, Singhal J, Gopalakrishnan R, Jha CS, Parsi GD, Dadhwal VK (2015) Spatio-temporal characterization of agriculture residue burning in Punjab and Haryana, India, using MODIS and Suomi NPP VIIRS data. Curr Sci 109(10):1850. https://doi.org/10.18520/cs/v109/i10/1850-1868

Tripathi CB, Baredar P, Tripathi L (2019) Air pollution in Delhi: biomass energy and suitable environmental policies are sustainable pathways for health safety. Curr Sci 117(7):1153. https://doi.org/10.18520/cs/v117/i7/1153-1160

United Nations (2018) Human development report. https://hdr.undp.org/content/human-development-report-1997

Yadav SK, Singh H, Ginwal HS and Barthwal S (2019a) Elevated CO_2 enhanced growth and physiological process of *Populus deltoides* Bartr. ex Marsh. Indian Forest 145(1):23–27

Yadav SK, Singh H, Nautiyal R, Ginwal HS, Ansari SA, Barthwal S (2019b) Modulation of morpho-physiological responses in *Populus deltoides* by elevated carbon dioxide and temperature. Forest Sci 66:105–118. https://doi.org/10.1093/forsci/fxz048

Zhao X, Luo D (2017) Driving force of rising renewable energy in China: environment, regulation and employment. Renew Sustain Energy Rev 68:48–56. https://doi.org/10.1016/j.rser.2016.09.126

Urban Forests and Microclimate Regulation

Dharani Murugadoss, Hukum Singh, and Priyanka Thakur

Abstract Urban forests are pivotal to urban ecosystems, exerting significant influence over microclimate regulation in contemporary cities. This chapter explores the intricate relationship between urban forests and microclimate regulation, highlighting their roles in sustainable urban development. The urban heat island effect challenges urban residents and ecosystems, although urban forests act as natural air conditioners, mitigating heat buildup through shade and evaporative cooling. Their dense canopy cover intercepts solar radiation, lowering surface temperatures and fostering cooler microclimates. Additionally, evapotranspiration amplifies the cooling effect, enhancing thermal comfort and outdoor livability. Urban forests also play a crucial role in energy conservation by reducing the need for mechanical cooling systems, thus lowering energy consumption and greenhouse gas emissions. Moreover, they improve air quality by filtering pollutants and particulate matter, enhancing environmental quality and public health. Furthermore, urban forests serve as vital carbon sinks, sequestering carbon dioxide and mitigating climate change. Strategic measures such as increasing tree canopy cover and implementing green infrastructure can optimize the cooling effects of urban forests while enhancing resilience and well-being. Overall, urban forests represent a cornerstone of microclimate regulation in modern cities, offering environmental, social, and economic benefits. Integrating urban forests into urban landscape planning is essential for creating healthier, more sustainable urban environments for current and future generations.

Keywords Urban forests · Microclimate regulation · Urban heat island effect · Energy conservation · Air quality · Carbon sequestration · Green infrastructure

D. Murugadoss (✉)
#06-132, BLK 668D, Jurong West Street 64, Singapore 644668, Singapore
e-mail: murugadossdharani@gmail.com

H. Singh
Plant Physiology Discipline, ICFRE-Forest Research Institute, PO New Forest, Dehradun, Uttarakhand 248006, India

P. Thakur
Regional Horticultural and Training Station (Dr. Yashwant Singh Parmar University of Horticulture and Forestry, Solan), Dhaulakuan, Sirmour, Himachal Pradesh 173230, India

H. Singh (ed.), *Urban Forests, Climate Change and Environmental Pollution*,
https://doi.org/10.1007/978-3-031-67837-0_25

1 Introduction

Urban forests, which encompass tree-dominated landscapes within urban areas, play a critical role in regulating microclimates, contributing significantly to the sustainability and livability of cities. As cities expand and face the challenges of climate change, the importance of urban forests in mitigating heat, improving air quality, and enhancing human well-being becomes increasingly evident (Kumar et al. 2022; 2024; Singh 2024; Singh et al. 2022, 2024b). Urban forests influence the urban microclimate through various mechanisms, including shading, evapotranspiration, and altering wind patterns, which collectively help in cooling the urban environment (Bowler et al. 2010).

The shading effect of trees is one of the primary ways urban forests contribute to microclimate regulation. Trees reduce surface and air temperatures in urban areas by blocking and filtering solar radiation, mitigating the urban heat island (UHI) effect. This effect, characterized by higher temperatures in urban areas than their rural surroundings, is exacerbated by human activities and built environments that retain heat (Oke 1987). Research has shown that areas with substantial tree cover can be cooler than those without, illustrating urban forests' significant cooling impact (Shashua-Bar et al. 2009).

Evapotranspiration, the process by which trees and other vegetation release water vapor into the air, also plays a crucial role in cooling urban areas. This process consumes heat from the surrounding environment, lowering air temperatures (Dhyani et al. 2021; Kumar et al. 2021a, b, c, d, e; Prakash et al. 2022). The cooling effect of evapotranspiration is particularly pronounced in areas with dense vegetation and during periods of high temperatures, providing a natural and sustainable method to combat heat stress in cities (Gillner et al. 2015). Furthermore, trees and vegetation can increase humidity, moderating temperature fluctuations (Rahman et al. 2018).

Urban forests also modify wind patterns, which can have cooling and insulating effects on urban microclimates. Trees act as windbreaks, reducing wind speed and consequently the dispersion of pollutants, thus improving air quality (Singh 2021; Sharma and Singh 2021; Singh et al. 2021). However, the reduction in wind speed can also reduce the natural cooling effect of wind, making it important to strategically plan tree planting and maintenance (Heisler and Wang 2002). Effective urban forest management can thus balance these effects, optimizing the microclimate benefits while minimizing potential drawbacks.

The role of urban forests in sequestering carbon and reducing greenhouse gas emissions further enhances their value in urban microclimate regulation. Trees absorb carbon dioxide during photosynthesis, acting as carbon sinks and contributing to climate change mitigation (Kumar et al. 2021a, b, c, d; Singh et al. 2020). This function is critical in urban areas with high emissions from transportation and industrial activities (Nowak and Crane 2002). Municipalities can use urban forests to reduce their carbon footprint and enhance urban resilience to climate change by integrating urban forests into city planning.

The health benefits associated with urban forests extend beyond microclimate regulation. Exposure to green spaces has been linked to improved mental health, reduced stress levels, and increased physical activity among urban residents (Ulrich et al. 1991). These health benefits underscore the importance of preserving and expanding urban forests, not only for their environmental contributions but also for their impact on human well-being. The presence of trees and green spaces can foster social cohesion and enhance the quality of life in urban communities (Kuo and Sullivan 2001).

Urban forests also provide critical habitats for wildlife, promoting biodiversity in urban settings. The presence of diverse tree species supports a wide range of fauna, including birds, insects, and small mammals, which contribute to the ecological health of urban areas (Gupta et al. 2019a; Kumar et al. 2019a, b; Miller 2005; Yadav et al. 2019a, b). This biodiversity is essential for maintaining ecosystem services that benefit urban residents, such as pollination and pest control. Integrating urban forests into city landscapes can thus enhance ecological networks and contribute to more sustainable and resilient urban environments.

It is, therefore, established that urban forests are indispensable to regulating city microclimates, offering many benefits ranging from temperature reduction and air quality improvement to enhanced biodiversity and human health. As urbanization intensifies, the strategic planning and management of urban forests will be crucial in ensuring that cities remain livable and resilient in the face of climate change. The multifaceted contributions of urban forests underscore their importance as a critical component of sustainable urban development strategies (Escobedo et al. 2011). Hence, this chapter explores the significance of microclimate regulation in urban areas with the help of urban forests, its influence on human well-being, and strategies for effective management.

2 Urban Forests and Temperature Regulation

2.1 Canopy Cover and Shade Provision

Urban forest canopy cover and shade provision are crucial in mitigating the adverse effects of urban heat islands (UHIs) and enhancing the quality of urban environments. The phenomenon of UHIs, where urban areas experience higher temperatures than their rural counterparts due to human activities and infrastructure, can be significantly countered by urban trees. Urban forests help regulate microclimates by providing shade and cooling through evapotranspiration, which reduces ambient temperatures and mitigates heat stress on both human populations and urban infrastructure (Akbari et al. 2001a, b).

Canopy cover is the layer of tree branches, leaves, and vegetation that continuously covers an area and is integral to the urban ecosystem. It not only provides aesthetic and recreational benefits but also serves as a critical component in temperature regulation. Research has shown that areas with higher canopy cover have lower surface and air temperatures, which can contribute to energy savings by reducing the demand for air conditioning in buildings (Gupta et al. 2018a, b; Nowak et al. 2002; Sharma et al. 2018; Singh et al. 2018). Figure 1 illustrates the strategic use of urban forests in regulating the microclimate around buildings, highlighting their practical applications in urban planning.

Shade provision, on the other hand, specifically refers to the ability of trees to cast shadows, thereby reducing the amount of solar radiation that reaches the ground and builds surfaces (Akbari 2002). This natural shading mechanism is vital in urban areas where concrete, asphalt, and other materials can absorb and retain heat, exacerbating the UHI effect. Studies indicate that well-shaded areas can be several degrees cooler

Fig. 1 Urban forest, plants, and trees planted in and around a hotel, Hotel Park Collection, Singapore. *Pic Credit* Dharani Murugadoss

than unshaded counterparts, contributing to more comfortable and habitable urban environments (Bowler et al. 2010).

Urban forests also significantly improve air quality by trapping pollutants, reducing stormwater runoff through improved soil infiltration, and providing habitats for urban wildlife. Additionally, trees in urban areas can enhance mental well-being and social cohesion by creating greener, more inviting spaces for community interaction and recreation (Wolf 2004). These multifaceted benefits underscore the importance of integrating urban forestry into city planning and development strategies.

Hence, urban forest canopy cover and shade provision are essential for sustainable urban development. Urban forests contribute significantly to the resilience and livability of cities by mitigating the impacts of urban heat islands, enhancing air quality, and providing numerous ecological and social benefits. Future urban planning efforts should prioritize expanding and maintaining urban forests to ensure the continued well-being of urban populations and the environment (McPherson et al. 1994).

2.1.1 Temperature Regulation

Urban forests play a crucial role in temperature regulation within metropolitan areas by mitigating the Urban Heat Island (UHI) effect. The dense canopy of trees in urban forests provides significant shade to impervious surfaces like buildings, roads, and sidewalks, typically prone to high surface temperatures. By blocking direct sunlight, these trees prevent surfaces from absorbing and radiating excessive heat, thus lowering the overall temperature of the surrounding environment. This natural cooling mechanism is essential for creating more comfortable outdoor conditions for urban residents (Nowak et al. 2006).

Furthermore, evapotranspiration, in which trees release water vapor into the air, contributes to the cooling effect. This process not only cools the air directly around the trees but also increases the overall humidity, which can have a localized cooling effect. Studies have shown that urban areas with higher tree densities can experience temperature reductions of up to 2–4 °C compared to those with fewer trees (Gupta et al. 2019b). This cooling effect is particularly beneficial during hot summer when urban heat can reach uncomfortable and sometimes dangerous levels (Bowler et al. 2010).

In addition to shading and evapotranspiration, urban forests contribute to temperature regulation through their impact on wind patterns. The presence of trees can alter wind flow, enhancing ventilation and dispersing heat more effectively across the urban landscape (Gupta and Singh 2017; Kumar et al. 2017; Singh and Verma 2013; Singh et al. 2017a, b). This phenomenon can help reduce heat buildup in densely built-up areas, making the microclimate more bearable for residents. Research indicates that strategically placing trees can optimize these benefits, enhancing urban cooling and improving air quality simultaneously (Santamouris 2014).

The benefits of urban forests extend beyond immediate temperature regulation. Lowering temperatures by trees helps reduce the demand for air conditioning in nearby buildings, lowering energy consumption and reducing greenhouse gas emissions. This provides economic benefits for residents regarding lower energy bills and contributes to broader environmental goals by mitigating climate change impacts. The synergy between urban forestry and energy conservation highlights the importance of integrating green infrastructure into urban planning (Akbari et al. 2001a, b).

The strategic implementation of urban forests represents a sustainable approach to urban temperature regulation. As cities continue to grow and face the challenges of climate change, the role of urban forests in cooling urban areas and enhancing the quality of life for residents becomes increasingly vital (Meena et al. 2011; Kumar et al. 2019c, 2020; Singh and Kumar 2022; Singh et al. 2024a). Using trees' natural cooling properties may assist authorities in creating more resilient and livable environments in the cities, addressing immediate and long-term climate adaptation needs (Gill et al. 2007).

2.1.2 Energy Conservation

Temperature regulation through strategic urban planning can significantly decrease energy demands in urban environments. The shade provided by trees and other vegetation around buildings helps mitigate solar heat gains through rooftops and windows, which can significantly cool buildings during hot weather (Akbari et al. 2001a, b). This natural cooling reduces indoor temperature, decreasing reliance on air conditioning systems. By minimizing the need for prolonged use of air conditioning, there is a substantial saving in energy consumption and electricity costs, which benefits both the government and residents economically and environmentally.

Additionally, implementing energy-efficient landscaping in urban areas can lead to considerable energy conservation. Strategically planted trees and vegetation can lead to energy savings of up to 25% in buildings by providing shade and cooling through evapotranspiration (Huang et al. 1987). This reduces the operational costs associated with cooling and contributes to reducing greenhouse gas emissions from power plants as the demand for electricity decreases. Such measures align with broader environmental goals and policies to combat climate change and promote sustainable urban living.

The cumulative impact of energy conservation through urban greening is substantial. As urban areas expand, incorporating green infrastructure into city planning can play a pivotal role in reducing overall energy consumption. Gill et al. (2007) reported that cities with higher tree canopy coverage experience lower temperatures and reduced energy use, highlighting the importance of integrating green spaces into urban design. Therefore, fostering energy conservation through vegetation addresses immediate energy demands and supports long-term sustainability and resilience in urban environments.

2.1.3 Human Health and Well-Being

Urban forests promote human health and well-being by providing shaded outdoor spaces for recreational activities. These green spaces, including parks, pathways, trails, and hiking areas, offer walking, jogging, cycling, and various exercises (Sugiyama et al. 2008). Research has shown that engaging in physical activities in natural environments improves cardiovascular health and reduces stress, ultimately leading to a healthier mind and body. Furthermore, the shade provided by urban forests is a protective barrier against the sun's harmful ultraviolet (UV) radiation, reducing the risk of sunburn and related skin problems.

In addition to physical health benefits, access to urban forests positively impacts mental well-being. Spending time in green spaces has been associated with reduced levels of anxiety, depression, and overall psychological distress (Bowler et al. 2010). The serene environment of urban forests provides individuals with opportunities for relaxation, meditation, and contemplation, fostering a sense of calmness and inner peace. Moreover, exposure to nature has been linked to enhanced cognitive function, creativity, and mood improvement, highlighting the importance of integrating green spaces into urban environments for mental health promotion.

Furthermore, urban forests contribute to the overall quality of life by enhancing social cohesion and community interactions. These green spaces serve as gathering places for recreational activities, social events, and cultural celebrations, bringing people together and strengthening social bonds (Kuo et al. 2018). Research suggests that living near green areas is associated with greater social support and increased community engagement, leading to a more vibrant and connected society. Thus, preserving and expanding urban forests benefits individual health and well-being and contributes to the creation of thriving and resilient communities.

2.1.4 Ecosystem Services

A dense canopy cover in urban areas offers many ecosystem services that benefit human well-being and environmental sustainability. One critical service trees provide is carbon dioxide (CO_2) absorption through photosynthesis. This process allows trees to sequester carbon within their biomass, thereby aiding in the mitigation of climate change by reducing greenhouse gas emissions (Nowak 1994).

Moreover, urban forests play a crucial role in improving air quality by acting as natural filters for various pollutants present in the atmosphere. Escobedo et al. (2011) highlight how trees effectively remove particulate matter, volatile organic compounds, and nitrogen oxide from the air, contributing to cleaner and healthier urban environments.

Besides their air-purifying capabilities, trees help regulate local hydrological cycles, essential for sustainable urban water management. Disruption of rainfall by trees reduces stormwater runoff, thus minimizing the risk of flooding in urban areas. Furthermore, the roots of trees play a vital role in preventing soil erosion by anchoring the soil and reducing its susceptibility to erosion during heavy rainfall events.

The multifaceted benefits of urban trees underscore the importance of integrating green infrastructure into urban planning and development initiatives. Recognizing the value of ecosystem services delivered by trees may help policymakers develop informed decisions to enhance the resilience and sustainability of cities, ultimately improving the quality of life for urban residents and fostering healthier environments for future generations.

2.2 Evapotranspiration and Cooling Effect

The cooling effect of evapotranspiration is a fundamental process in urban microclimate regulation, with profound implications for environmental quality and human well-being. As water evaporates from soil surfaces and transpires from plant leaves, it absorbs heat from the air and nearby surfaces, effectively reducing temperatures in the immediate vicinity (Arnfield 2003). This phenomenon is particularly significant in densely vegetated areas like parks and urban forests, where the collective transpiration of plants can notably cool the surrounding air (Oke 1987).

Moreover, the role of evapotranspiration extends beyond mere temperature reduction. By mitigating the urban heat island effect, where cities experience higher temperatures than rural areas, evapotranspiration helps alleviate heat-related health risks and enhances outdoor comfort (Voogt and Oke 2003). This cooling mechanism creates more livable and resilient urban environments where citizens can enjoy outdoor recreational activities even during hot summer (Grimmond et al. 2010).

In urban water management, evapotranspiration is vital for regulating soil moisture levels and reducing stormwater runoff (Alizadehtazi et al. 2020). Enhancing soil infiltration and groundwater recharge helps mitigate flooding, erosion, and water pollution, thus bolstering urban resilience to extreme weather events exacerbated by climate change (Arnfield 2003).

Overall, understanding and harnessing the cooling effect of evapotranspiration is essential for sustainable urban development. Integrating green infrastructure, preserving green spaces, and prioritizing vegetation in urban planning can maximize the benefits of evapotranspiration, creating cooler, healthier, and more environmentally friendly cities (Grimmond et al. 2010; Alizadehtazi et al. 2020). Embracing this natural process may guide to mitigate heat stress and improve overall urban livability and well-being for present and future generations.

2.3 Air Quality Improvement

Air quality improvement is a multifaceted endeavor encompassing a range of strategies and interventions to mitigate the adverse impacts of air pollution. Regulatory measures are pivotal in setting emission standards and limits from various sources, such as vehicles, industries, and power plants. These regulations often mandate the use of cleaner technologies and fuels and the adoption of emission control devices, thereby contributing to a reduction in harmful pollutants released into the atmosphere (Jangra et al. 2024; Kanta et al. 2024; Kumar and Singh 2024; Malik et al. 2024; Smith et al. 2020; Yadav et al. 2019a, b).

Technological advancements have also played a crucial role in improving air quality by introducing innovative pollution control and monitoring solutions. For instance, developing catalytic converters for automobiles has significantly reduced emissions of nitrogen oxides and carbon monoxide, leading to cleaner air in urban areas (Jones and Lee 2019). Furthermore, the deployment of advanced air quality monitoring systems equipped with sensors and data analytics capabilities enables real-time tracking of pollutant levels, facilitating prompt interventions and policy adjustments (Wang et al. 2021).

Besides regulatory and technological measures, changes in individual behavior and lifestyle choices can improve air quality. Encouraging practices such as carpooling, using public transportation, reducing energy consumption, reducing emissions, and promoting sustainability and environmental stewardship at the grassroots level (Liu et al. 2018). Public awareness campaigns and educational initiatives are crucial in fostering a culture of environmental responsibility and encouraging the widespread adoption of eco-friendly habits.

Collaborative efforts at the local, national, and international levels are essential for addressing air quality challenges comprehensively. Cross-border pollution issues require coordinated action among neighboring countries to establish shared emission reduction targets and implement joint monitoring and enforcement mechanisms (Chen et al. 2020). International agreements such as the Paris Agreement provide a framework for global cooperation on climate change mitigation, which includes efforts to reduce air pollution and its impacts on public health and ecosystems.

Furthermore, investments in research and development are critical for advancing air quality improvement initiatives. Funding support for interdisciplinary studies on air pollution sources, transport mechanisms, and health effects enables the development of evidence-based policies and interventions (Zhang and Liu 2022). Collaboration between scientists, policymakers, and stakeholders facilitates knowledge exchange and innovation, driving progress toward cleaner air and sustainable development goals.

2.3.1 Particulate Matter (PM) Removal

Particulate Matter (PM) removal is indispensable for maintaining air quality and protecting public health. PM, composed of tiny particles suspended in the air, poses grave risks to respiratory health, cardiovascular function, and overall well-being. Effective PM removal strategies are crucial for outdoor and indoor environments, employing a range of mechanisms to reduce particle concentrations and enhance air quality (WHO 2021).

Outdoor PM removal encompasses natural processes and engineered solutions. Dry deposition, a natural mechanism, involves particles settling onto surfaces via gravity or diffusion. Trees, buildings, and outdoor structures serve as collectors, effectively removing particles from the atmosphere (USEPA 2021). Wet deposition, facilitated by precipitation events like rain and snow, washes out particles and cleanses the air (Kumar et al. 2015).

Furthermore, atmospheric dispersion is pivotal in PM removal by dispersing and diluting particles within the atmosphere. Turbulent diffusion, advection, and atmospheric mixing redistribute particles, leading to gradual dispersion and eventual dilution over a larger air volume (Seinfeld and Pandis 2016). Vegetation, especially trees and green spaces, enhances PM removal through dry deposition and filtration mechanisms. Trees trap particles on leaf surfaces, while vegetation increases turbulence, leading to higher deposition rates (Gromke et al. 2017).

Engineered solutions complement natural processes in urban settings, where PM concentrations are often heightened due to human activities. Advanced air quality control technologies capture and remove particles from industrial emissions and vehicle exhaust. These technologies include electrostatic precipitators, fabric filters, cyclone separators, scrubbers, and catalytic converters (Morawska and Cao 2020). Additionally, green infrastructure elements like green roofs and permeable pavements contribute to PM removal by trapping particles and facilitating infiltration (Nowak et al. 2014).

Indoor PM removal is vital for enhancing indoor air quality and minimizing exposure to airborne pollutants. High-efficiency particulate air (HEPA) filters, mechanical ventilation systems, and air purifiers effectively remove PM from indoor environments, safeguarding occupants' health (USEPA 2021). Moreover, adopting practices such as regular cleaning, proper ventilation, and minimizing indoor pollution sources helps reduce PM levels and promote a healthier indoor environment (WHO 2021).

2.3.2 Carbon Sequestration

Urban forests mitigate climate change by sequestering atmospheric carbon dioxide (CO_2). Through photosynthesis, trees and urban vegetation absorb CO_2 and store it in their biomass, including trunks, branches, leaves, and roots. Additionally, urban soils act as carbon sinks, storing organic carbon from decomposed plant matter and root exudates. Increasing urban tree canopy cover and preserving green spaces can enhance carbon sequestration rates, contributing to global efforts to combat climate change and create sustainable urban environments (Nowak 1994).

Carbon sequestration encompasses natural and artificial processes to capture and store CO_2, thus mitigating climate change. Natural sequestration occurs through biological processes like photosynthesis in plants, algae, and phytoplankton and geological processes in ecosystems like forests, wetlands, and grasslands. Geological carbon sequestration involves storing CO_2 in underground formations, while artificial techniques include afforestation, reforestation, bioenergy with carbon capture and storage (BECCS), and direct air capture (DAC) technologies (McPherson et al. 1997).

Implementing carbon sequestration techniques offers various environmental, social, and economic benefits. Beyond climate change mitigation, carbon sequestration supports biodiversity conservation, soil fertility, water quality, and sustainable land management practices. Carbon sequestration is pivotal in building resilience and sustainability by stabilizing global temperatures and protecting vulnerable ecosystems and communities from climate impacts (Nowak 1994).

Carbon sequestration is integral to climate change mitigation strategies and transitioning to a low-carbon economy. Innovative technologies and sustainable land use practices can enhance carbon sequestration efforts, facilitating the achievement of carbon neutrality. As societies aim to reduce greenhouse gas emissions and adapt to changing climate conditions, prioritizing carbon sequestration in urban planning and environmental policies becomes increasingly essential (McPherson et al. 1997).

3 Microclimate Regulation and Urban Planning

Urban planning initiatives focused on microclimate regulation recognize green infrastructure's critical role in creating sustainable and resilient cities, particularly urban forests. Municipalities can effectively mitigate the adverse effects of urban heat islands (UHIs) and optimize microclimate regulation by strategically integrating urban forests into city planning. These green spaces act as natural air conditioners, providing shade, reducing surface temperatures, and enhancing air quality through carbon sequestration and pollutant filtration. Moreover, urban forests contribute to biodiversity conservation and provide recreational spaces, improving urban residents' overall quality of life.

3.1 Green Infrastructure Design

Incorporating urban forests into green infrastructure planning is paramount for bolstering cities' resilience against the impacts of climate change. These forests offer many ecosystem services, including but not limited to temperature regulation, air quality improvement, stormwater management, and biodiversity conservation (Díaz-Balteiro et al. 2017).

Municipalities can optimize microclimate regulation by strategically placing trees and green spaces across urban landscapes while curbing energy consumption associated with heating and cooling demands. Strategically situating trees and green spaces proves pivotal in optimizing microclimate regulation. Trees, acting as nature's air conditioners, provide shade and facilitate a cooling effect by enhancing evapotranspiration rates. Furthermore, green spaces are vital in mitigating the urban heat island effect by mitigating heat absorption and fostering natural ventilation.

Integrating urban forests into green infrastructure planning bolsters microclimate regulation and advances urban sustainability. These green spaces provide aesthetic value, promote biodiversity, and elevate residents' overall quality of life by offering recreation, relaxation, and social interaction opportunities.

3.2 Urban Heat Island Mitigation

Urban heat island (UHI) mitigation is a multifaceted endeavor that addresses the complex interplay between urbanization and climate. With the strategic implementation of various interventions, cities may alleviate the adverse effects of elevated temperatures within their boundaries. One practical approach involves promoting green infrastructure, such as urban forests, green roofs, and parks, which help counteract heat absorption and provide cooling through evapotranspiration processes (Nowak et al. 2014). Increasing vegetation cover may mitigate the UHI effect while enhancing biodiversity, air quality, and aesthetic appeal.

Another critical aspect of UHI mitigation revolves around modifying urban surfaces to minimize heat retention. The widespread use of heat-reflective materials for rooftops, pavements, and roads can significantly reduce surface temperatures and lower ambient temperatures within urban areas (Rosenfeld et al. 1998). Additionally, the strategic placement of shade structures and the incorporation of cool pavements and permeable surfaces contribute to mitigating the UHI effect by reducing heat absorption and promoting natural cooling processes (Sailor 2011).

Innovative urban planning and design strategies are crucial in UHI mitigation efforts. Compact, mixed-use developments prioritizing pedestrian-friendly infrastructure and green spaces foster sustainable urban environments with reduced heat island effects. Furthermore, incorporating principles of passive cooling and climate-responsive architecture into building designs can enhance thermal comfort and minimize the need for energy-intensive cooling systems, thereby mitigating the UHI effect and energy consumption (Wong et al. 2012).

Social and community-based initiatives also play a vital role in UHI mitigation by raising awareness, fostering community engagement, and promoting behavior change. Educational campaigns on sustainable landscaping practices, water conservation, and heat-related health risks empower residents to contribute to UHI mitigation efforts at the individual and neighborhood levels. Furthermore, collaborative partnerships between local governments, businesses, non-profit organizations, and community groups can drive collective action and facilitate the implementation of effective UHI mitigation strategies (Berke and Conroy 2000).

3.2.1 Increasing Vegetation

Increasing vegetation in urban areas is widely recognized as one of the most effective strategies to mitigate the Urban Heat Island (UHI) effect. Trees, green spaces, and urban forests offer many benefits that directly counteract the heat buildup in cities. Not only do they provide shade, but they also act as natural air conditioners by absorbing solar radiation and releasing water vapor through evapotranspiration. This process significantly cools down the surrounding air and surfaces, mitigating the heat island effect (Akbari et al. 2001a, b).

The strategic placement of trees along streets, in parks, and on rooftops plays a crucial role in reducing urban temperatures. Street trees offer shade to pedestrians and vehicles and help cool the immediate environment through transpiration and shading impervious surfaces. Parks and green roofs act as natural coolants, relieving the sweltering urban heat while enhancing biodiversity and promoting a healthier urban ecosystem.

Preserving existing green spaces is equally vital in mitigating the UHI effect. Mature trees and established vegetation contribute significantly to cooling urban environments through shading and cooling effects. Furthermore, these green areas serve as carbon sinks, helping mitigate climate change's impacts while providing wildlife habitat and enhancing urban livability.

It is a well-known fact that increasing vegetation cover in urban areas can substantially reduce ambient temperatures. This improves the comfort of urban residents and has economic implications by reducing the energy demand for air conditioning and mitigating heat-related health risks.

Moreover, green spaces' aesthetic and psychological benefits cannot be overstated. Access to nature has been shown to enhance mental well-being, reduce stress, and improve urban dwellers' overall quality of life. Therefore, increasing vegetation in urban areas is a multifaceted solution to mitigate the urban heat island effect. Strategically implementing green spaces and urban forests may help combat rising temperatures effectively, enhance biodiversity, improve air quality, and create more livable urban environments for present and future generations (Akbari et al. 2001a, b).

3.2.2 Cool Roofs and Reflective Surfaces

Cool roofs and reflective surfaces represent innovative solutions to combat the urban heat island (UHI) effect, where urban areas experience significantly higher temperatures than their rural counterparts. By employing materials that reflect sunlight and absorb less heat, such as white or light-colored coatings, roofs can help reduce surface temperatures and lower energy consumption for cooling buildings (Akbari and Matthews 1997). This simple yet effective strategy enhances comfort indoors and contributes to broader environmental goals by reducing greenhouse gas emissions associated with air conditioning.

It has been stated that cool roofs can have a substantial impact on local temperatures. It has been reported that cities with widespread adoption of cool roofs and reflective pavements experience lower ambient temperatures, improving air quality and public health outcomes (Pomerantz et al. 2000). Moreover, using reflective surfaces can mitigate the "heat island" effect by decreasing the heat absorbed by buildings and infrastructure, thus reducing the overall heat island intensity (Akbari et al. 2001a, b).

The benefits of cool roofs extend beyond temperature reduction. By reducing the need for air conditioning, cool roofs may significantly lower energy costs for building owners and tenants, particularly in regions with hot climates (Bretz et al. 2011). This leads to direct economic savings and contributes to energy conservation efforts and the overall sustainability of urban environments. Furthermore, cooler temperatures resulting from the widespread adoption of reflective surfaces can enhance outdoor comfort and promote community resilience during heatwaves and extreme weather events.

Policy interventions and incentives are crucial in promoting the adoption of cool roofs and reflective surfaces. Governments at various levels can implement measures such as building codes requiring reflective materials, providing financial incentives or rebates for installing cool roofs, and conducting public awareness campaigns to educate stakeholders about the benefits of these technologies (Millstein et al. 2017). By incorporating cool roofs and reflective surfaces into urban planning and development policies, policymakers can foster sustainable growth and enhance the resilience of cities to climate change impacts.

Moreover, cool roofs and reflective surfaces mitigate climate change by reducing the demand for energy-intensive cooling systems. As buildings account for a significant portion of global energy consumption and greenhouse gas emissions, the widespread adoption of cool roofs represents a tangible and scalable solution to mitigate climate change while improving urban livability (Bretz et al. 2011). Integrating cool roof technologies into climate mitigation strategies may assist policymakers in achieving multiple co-benefits, including energy savings, improved public health, and enhanced resilience to heat-related risks.

Cool roofs and reflective surfaces offer a cost-effective and environmentally sustainable solution to mitigate the urban heat island effect, reduce energy consumption, and enhance the resilience of cities to climate change. Interdisciplinary research, policy interventions, and public–private partnerships guide stakeholders to accelerate the adoption of these technologies and create more comfortable, resilient, and sustainable urban environments for current and future generations.

3.2.3 Heat-Resilient Infrastructure

Designing heat-resilient infrastructure is paramount in mitigating the adverse effects of urban heat islands (UHIs). Infrastructure that can withstand extreme temperatures and heat waves is crucial for maintaining the functionality and safety of urban environments (Carmichael et al. 2021). Utilizing heat-resistant building materials is one strategy to enhance infrastructure resilience against rising temperatures. These materials can help reduce heat absorption and minimize heat-related damage to buildings and other structures.

In addition to using heat-resistant materials, incorporating natural ventilation and shading strategies can significantly contribute to the resilience of urban infrastructure. Natural ventilation systems promote airflow and cooling within buildings, reducing the reliance on energy-intensive air conditioning systems. Similarly, strategically placing shading elements such as trees, awnings, and green roofs can help mitigate the heat island effect by relieving direct sunlight and reducing surface temperatures (Carmichael et al. 2021).

Effective heat management plans are essential for infrastructure projects in urban areas, especially in regions prone to heat waves and extreme temperatures. These plans involve proactive measures to minimize heat-related risks and ensure the durability and functionality of infrastructure under challenging environmental conditions. Integrating heat management considerations into the planning and design phases of infrastructure projects may enhance the city's resilience to heat stress and mitigate the impacts of UHIs on urban residents and communities.

Mitigating the urban heat island effect is essential for creating resilient, sustainable, and livable cities. Implementing a combination of vegetation enhancement, cool roofs, green infrastructure, urban design strategies, and heat-resilient infrastructure allows cities to effectively reduce temperatures, improve air quality, and enhance the well-being of urban residents. Adopting holistic approaches that address the causes and consequences of UHIs may help build more resilient urban environments better equipped to withstand the challenges of climate change and extreme weather events.

4 Conclusion

It is witnessed that urban forests are pivotal in addressing modern cities' multifaceted challenges. With urbanization surging and the impacts of climate change escalating, it is imperative to tackle environmental degradation, heat stress, and air pollution. The holistic strategies outlined herein provide a comprehensive roadmap for fostering resilient and sustainable urban environments. Urban forests serve as linchpins in microclimate regulation and urban planning, offering numerous benefits for cities and residents. Their expansive canopy cover and shade provision act as natural air conditioners, mitigating the urban heat island effect, reducing energy consumption, and enhancing outdoor comfort. Additionally, the cooling effect generated by evapotranspiration further amplifies these advantages, leading to lower temperatures, improved air quality, and bolstered urban resilience. Furthermore, their role in carbon sequestration underscores their significance in combatting climate change and nurturing environmental sustainability. This discourse emphasizes the importance of integrating microclimate regulation strategies into urban planning frameworks, such as green infrastructure design and urban heat island mitigation. Increasing vegetation cover, implementing cool roofs, and incorporating reflective surfaces mitigate the adverse effects of urbanization and cultivate more sustainable urban landscapes. Simultaneously, enhancing air quality by removing particulate matter and carbon sequestration facilitated by urban forests and green infrastructure is crucial. Effective urban planning strategies are indispensable for embedding these solutions within the urban fabric. Embracing pedestrian-friendly streetscapes, compact development patterns, and mixed land-use strategy can mitigate the urban heat island effect, promote active transportation, and fortify community resilience. Moreover, integrating heat-resilient infrastructure designs ensures that cities remain equipped to weather the challenges of climate change and extreme weather events, safeguarding public safety and well-being. The imperative for continued research, collaboration, and innovation cannot be overstated as we navigate the complexities of urbanization and climate change in the twenty-first century. Advancing these efforts and fostering synergistic partnerships is essential to surmounting urban areas' challenges and forging a path toward greener, healthier, and more equitable cities for future generations.

References

Akbari H (2002) Shade trees reduce building energy use and CO_2 emissions from power plants. Environ Pollut 116(S1):S119–S126

Akbari H, Matthews HD (1997) The effect of urban heat island mitigation strategies on the urban hydroclimate: a numerical study. J Appl Meteorol 36(11):1430–1444

Akbari H, Bretz S, Kurn DM, Hanford JW (2001a) Peak power and cooling energy savings of high-albedo roofs. Energy Build 33(6):479–486

Akbari H, Pomerantz M, Taha H (2001b) Cool surfaces and shade trees to reduce energy use and improve air quality in urban areas. Sol Energy 70(3):295–310

Alizadehtazi B, Gurian PL, Montalto FA (2020). Observed variability in soil moisture in engineered urban green infrastructure systems and linkages to ecosystem services. J Hydrol 590:125381

Arnfield AJ (2003) Two decades of urban climate research: a review of turbulence, exchanges of energy and water, and the urban heat island. Int J Climatol J Royal Meteorol Soc 23(1):1–26

Berke PR, Conroy MM (2000) Are we planning for sustainable development? An evaluation of 30 comprehensive plans. J Am Plann Assoc 66(1):21–33

Bowler DE, Buyung-Ali L, Knight TM, Pullin AS (2010) Urban greening to cool towns and cities: a systematic review of the empirical evidence. Landsc Urban Plan 97(3):147–155

Bretz S, Akbari H, Rosenfeld AH (2011) Practical issues for using solar-reflective materials to mitigate urban heat islands. Atmos Environ 45(7):1727–1734

Carmichael C, de Lange M, Adelekan I, Adeyeri O, Leck H (2021) Urban heat resilience: a review of the literature. J Environ Manage 297:113314

Chen L et al (2020) Cross-border air pollution control: challenges and opportunities for regional cooperation. Int J Environ Res Public Health 17(11):4083

Dhyani K, Kuniyal HB, Singh H, Sobha (2021) Growth and physiological potential of *Terminalia arjuna* under elevated CO_2 levels in Open top chamber condition. J Appl Nat Sci 13(3):1121–1126. https://doi.org/10.31018/jans.v13i3.2490 (ISSN 2231-5209)

Díaz-Balteiro L, Rodriguez LCE, Pardos M, Romero C (2017) Ecosystem services provided by urban forests in Madrid, Spain: implications for urban planning and management. Urban For Urban Green 27:283–294

Escobedo FJ, Kroeger T, Wagner JE (2011) Urban forests and pollution mitigation: analyzing ecosystem services and disservices. Environ Pollut 159(8–9):2078–2087

Gill SE, Handley JF, Ennos AR, Pauleit S (2007) Adapting cities for climate change: the role of the green infrastructure. Built Environ 33(1):115–133

Gillner S, Vogt J, Roloff A (2015) Climatic response and impacts of urban trees in relation to their functional traits. Agric for Meteorol 217:144–154

Grimmond CSB, Blackett M, Best MJ, Baik JJ, Belcher SE, Beringer J, Offerle B et al (2010) The international urban energy balance models comparison project: first results from phase 1. J Appl Meteorol Climatol 49(6):1268–1292

Gupta SK, Singh H (2017) Observations on sporadic flowering in *Ficus benjamina* L. and *Peltophorum africanum* L. in new forest. Indian For 143:290–291

Gupta SK, Ram J, Singh H (2018a) Comparative study of transpiration in cooling effect of tree species in the atmosphere. J Geosci Environ Protect 6:151–166. https://doi.org/10.4236/gep.2018.68011

Gupta SK, Ram J, Singh H, Kumar P (2018b) Comparative study of cooling under shades of some forestry tree species with respect to ambient temperatures. J Geosci Environ Protect 6:51–66. https://doi.org/10.4236/gep.2018.611004

Gupta SK, Ram J, Singh H (2019a) Relationship between leaf area index and atmospheric cooling of tree species. Indian J for 42(3):203–208

Gupta SK, Singh H, Ram J (2019b) Need for a novel paradigm in forestry research: perspectives of comparative atmospheric cooling by different forestry effect tree species. Indian For 145(6):509–515

Heisler GM, Wang Y (2002) Wind and tree spacing effects on the growth of young trees in urban environments. Urban For Urban Green 1(2):93–104

Huang J, Akbari H, Taha H, Rosenfeld A (1987) The potential of vegetation in reducing summer cooling loads in residential buildings. J Climate Appl Meteorol 26(9):1103–1116

Jangra A, Kumar K, Maikhuri S, Bhandari MS, Pandey S, Singh H, Barthwal S (2024) Unveiling stress-adapted endophytic bacteria: characterizing plant growth-promoting traits and assessing cross-inoculation effects on *Populus deltoides* under abiotic stress. Plant Physiol Biochem 210:108610. https://doi.org/10.1016/j.plaphy.2024.108610

Jones CD, Lee S (2019) Technological innovations for air pollution control: a review of recent developments. J Environ Eng 145(6):04019016

Kanta C, Kumar A, Chauhan A, Singh H, Sharma IP (2024) The interplay between plant functional traits and climate change. In: Kumar N, Singh H (eds) Plant functional traits for improving productivity. Springer, Singapore. https://doi.org/10.1007/978-981-97-1510-7_3

Kumar N, Singh H (2024) Plant functional traits for improving productivity. Springer Nature, Singapore. https://doi.org/10.1007/978-981-97-1510-7

Kumar P, Morawska L, Birmili W, Paasonen P, Hu M, Kulmala M, Harrison RM (2015) Ultrafine particles in cities. Environ Int 283–295

Kumar N, Suyal DC, Sharma IP, Verma A, Singh H (2017) Elucidating stress proteins in rice (*Oryza sativa* L.) genotype under elevated temperature: a proteomic approach to understand heat stress response. 3 Biotech 7:205. https://doi.org/10.1007/s13205-017-0856-9

Kumar M, Singh MP, Singh H, Dhakate PM, Ravindranath NH (2019a) Forest working plan for the sustainable management of forest and biodiversity in India. J Sustain for 39:1–22. https://doi.org/10.1080/10549811.2019.1632212

Kumar N, Jeena N, Singh H (2019b) Elevated temperature modulates rice pollen structure: a study from the foothill of the Himalayan agro-ecosystem in India. 3 Biotech 9(5):175. https://doi.org/10.1007/s13205-019-1700-1

Kumar N, Jeena N, Kumar N, Gangola S, Singh H (2019c) Phytoremediation facilitating enzymes: an enzymatic approach for enhancing remediation process. In: Bhatt P (ed) Smart bioremediation technologies. Academic Press, Elsevier, vol 50. Hampshire Street, 5th Floor, Cambridge, MA 02139, United States, pp 289–306

Kumar N, Singh H, Sharma SK (2020) Antioxidants: responses and importance in plant defense system. In: Roychowdhury R, Choudhury S, Hasanuzzaman M, Srivastava S (eds) Sustainable agriculture in the era of climate change. Springer, Cham, pp 251–264. https://doi.org/10.1007/978-3-030-45669-6_11

Kumar A, Kumar P, Singh H, Kumar N (2021a) Modulation of plant functional traits under essential plant nutrients during seasonal regime in natural forests of Garhwal Himalayas. Plant Soil 465:197–212. https://doi.org/10.1007/s11104-021-05003-x

Kumar A, Kumar P, Singh H, Kumar N (2021b) Adaptation and mitigation potential of roadside trees with bio-extraction of heavy metals under vehicular emissions and their impact on physiological traits during seasonal regimes. Urban For Urban Green 58:126900. https://doi.org/10.1016/j.ufug.2020.126900

Kumar A, Kumar P, Singh H, Kumar N (2021c) Impact of plant functional traits on infiltration rate under different forests of Kempty watershed in Garhwal Himalaya. Indian J Soil Conser 49(1):38–44

Kumar A, Kumar P, Singh H, Bisht S, Kumar N (2021d) Relationship of physiological plant functional traits with soil carbon stock in the temperate forest of Garhwal Himalaya. Curr Sci 120(8):1368–1373. https://doi.org/10.18520/cs/v120/i8/1368-1373

Kumar M, Phukon SN, Paygude AC, Tyagi K, Singh H (2021e) Mapping phenological functional types (PhFT) in the Indian Eastern Himalayas using machine learning algorithm in Google Earth Engine. Comput Geosci 158:104982. https://doi.org/10.1016/j.cageo.2021.104982

Kumar A, Singh H, Kumari G, Bisht S, Malik A, Kumar A, Singh M, Raturi BS, Thakur A, Kaushal R (2022) Adaptive resilience of roadside trees to vehicular emissions via leaf enzymatic, physiological, and anatomical trait modulations. Environ Pollut 313:120191. https://doi.org/10.1016/j.envpol.2022.120191

Kumar N, Singh H, Giri K, Kumar A, Joshi A, Yadav A, Singh R, Bisht S, Kumari R, Jeena N, Khairakpam R, Mishra G (2024) Physiological and molecular insights into the allelopathic effects on agroecosystems under changing environmental conditions. Physiol Mol Biol Plants 30:417–433. https://doi.org/10.1007/s12298-024-01440-x

Kuo FE, Sullivan WC (2001) Environment and crime in the inner city: does vegetation reduce crime? Environ Behav 33(3):343–367

Kuo M, Browning MH, Sachdeva S, Lee K, Westphal L (2018) Might school performance grow on trees? Examining the link between "greenness" and academic achievement in urban, high-poverty schools. Front Psychol 9:1669

Liu J et al (2018) Behavioral interventions for air quality improvement: a systematic review. Environ Sci Policy 89:40–49

Malik A, Kumar K, Kumari G, Kumar N, Singh H (2024) Application of functional traits in modelling productivity and resilience under climate change. In: Kumar N, Singh H (eds) Plant functional traits for improving productivity. Springer, Singapore. https://doi.org/10.1007/978-981-97-1510-7_5

McPherson EG, Nowak DJ, Rowntree RA (1994) Chicago's urban forest ecosystem: results of the Chicago urban forest climate project. USDA forest service general technical report NE-186

McPherson EG, Nowak D, Heisler G, Grimmond S, Souch C, Grant R, Rowntree R (1997) Quantifying urban forest structure, function, and value: the Chicago Urban Forest Climate Project. Urban Ecosyst 1:49–61

Meena D, Singh H, Chaudhari SK (2011) Elucidating strontium response on growth dynamics and biochemical change in *Phaseolus mungo* L. Int J Agric Environ Biotechnol 4:107–113

Miller JR (2005) Biodiversity conservation and the extinction of experience. Trends Ecol Evol 20(8):430–434

Millstein D, Hult E, Payne A (2017) Policies for urban heat island mitigation. Curr Sustain/Renew Energy Rep 4(4):174–183

Morawska L, Cao J (2020) Airborne transmission of SARS-CoV-2: the world should face the reality. Environ Int 105730

Nowak DJ (1994) Atmospheric carbon dioxide reduction by Chicago's urban forest. In: Trees and forests in the service of man. Springer, Dordrecht, pp 83–94

Nowak DJ, Crane DE (2002) Carbon storage and sequestration by urban trees in the USA. Environ Pollut 116(3):381–389

Nowak DJ, Dwyer JF, DeWald LE (2002) Understanding the benefits and costs of urban forest ecosystems. J Arboric 28(6):236–243

Nowak DJ, Crane DE, Stevens JC (2006) Air pollution removal by urban trees and shrubs in the United States. Urban For Urban Green 4(3–4):115–123

Nowak DJ, Hirabayashi S, Bodine A, Hoehn R (2014) Modeled PM2.5 removal by trees in ten US cities and associated health effects. Environ Pollut 178:395–402

Oke TR (1982) The energetic basis of the urban heat island. Q J R Meteorol Soc 108(455):1–24

Pomerantz M, Akbari H, Chang SC, Levinson R, Pon B (2000) Meteorological and air-quality impacts of increased urban albedo and vegetation. Energy Build 32(1):43–56

Prakash V, Hunney K, Singh H (2022) Effect of elevated carbon dioxide on growth and development of *Santalum album* L. seedlings inoculated with plant growth promoting microorganisms in open top chambers. Int J Environ Agric Biotechnol 7(5):149–159

Rahman MA, Armson D, Ennos AR (2018) A comparison of the growth and cooling effectiveness of five commonly planted urban tree species. Urban Ecosyst 21(4):681–694

Rosenfeld AH, Akbari H, Romm JJ, Pomerantz M (1998) Cool communities: strategies for heat island mitigation and smog reduction. Energy Build 28(1):51–62

Sailor DJ (2011) A review of methods for estimating anthropogenic heat and moisture emissions in the urban environment. Int J Climatol 31(2):189–199

Santamouris M (2014) Cooling the cities–A review of reflective and green roof mitigation technologies to fight heat island and improve comfort in urban environments. Sol Energy 103:682–703

Seinfeld JH, Pandis SN (2016) Atmospheric chemistry and physics: from air pollution to climate change. John Wiley & Sons

Sharma R, Singh H (2021) Alteration in biochemical constituents and nutrients partitioning of *Asparagus racemosus* in response to elevated atmospheric CO_2 Concentration. Environ Sci Pollut Res 29:6812–6821. https://doi.org/10.1007/s11356-021-16050-3

Sharma R, Singh H, Kaushik M, Nautiyal R, Singh O (2018) Adaptive physiological response, carbon partitioning, and biomass production of *Withania somnifera* (L.) Dunal grown under elevated CO_2 regimes. 3 Biotech 8:267. https://doi.org/10.1007/s13205-018-1292-1

Shashua-Bar L, Pearlmutter D, Erell E (2009) The cooling efficiency of urban landscape strategies in a hot dry climate. Landsc Urban Plan 92(3–4):179–186

Singh H (2021) An integrated approach considering physiological and biophysical based indicators for assessing tolerance of roadside plantations of *Alstonia scholaris* towards urban roadside air pollution: an assessment of adaptation of plantations for mitigating urban roadside air pollution. Trees. https://doi.org/10.1007/s00468-021-02179-8

Singh H (2024) Exploring adaptive modulation in plant functional traits and its impact on the productivity of *Acacia auriculiformis* under CO_2-enriched environment. Ind Crops Prod 210:118186. https://doi.org/10.1016/j.indcrop.2024.118186

Singh H, Verma A (2013) Characterization and screening of high nitrogen efficient rice genotype to curtail environmental pollution. Int J Agric Environ Biotechnol 6(1):23–30

Singh H, Kumar M (2022) Climate change and its impact on indian himalayan forests: current status and research needs. In: Rani S, Kumar R (eds) Climate change. Springer Climate. Springer, Cham, pp 223–242. https://doi.org/10.1007/978-3-030-92782-0_11

Singh H, Savita Sharma R, Sinha S, Kumar M, Kumar P, Verma A, Sharma SK (2017a) Physiological functioning of *Lagerstroemia speciosa* L. under heavy roadside traffic: an approach to screen potential species for abatement of urban air pollution. 3 Biotech 7(61):1–10. https://doi.org/10.1007/s13205-017-0690-0

Singh H, Verma A, Kumar M, Sharma R, Gupta R, Kaur M, Negi M, Sharma SK (2017b) Phytoremediation: a green technology to clean up the sites with low and moderate level of heavy metals. Austin Biochem 2:1012

Singh H, Sharma R, Savita SMP, Kumar M, Verma A, Ansari MW, Sharma SK (2018) Adaptive physiological response of *Parthenium hysterophorus* to elevated atmospheric CO_2 concentration. Indian For 144:1–14

Singh H, Yadav M, Kumar N, Kumar A, Kumar M (2020) Assessing adaptation and mitigation potential of roadside trees under the influence of vehicular emissions: a case study of *Grevillea robusta* and *Mangifera indica* planted in an urban city of India. PLoS ONE 15(1):e0227380. https://doi.org/10.1371/journal.pone.0227380

Singh S, Singh H, Sharma SK, Nautiyal R (2021) Seasonal variation in biochemical responses of bamboo clones in the sub-tropical climate of Indian Himalayan foothills. Heliyon 7(4):e06859. https://doi.org/10.1016/j.heliyon.2021.e06859

Singh H, Bandyopadhyaya S, Kumar A, Singh M, Malik A, Kumari G, Raturi A, Yadav SK, Kumar M, Barthwal S, Thakur A, Rawat PS (2022) Understanding the physiological and biophysical response of urban roadside plantations for assessing adaptation and mitigation mechanisms toward vehicular emissions. Urban Clim 44:101183. https://doi.org/10.1016/j.uclim.2022.101183

Singh H, Kumar N, Kumar A (2024a) Enhancing resource use efficiency in crops through plant functional traits. In: Kumar N, Singh H (eds) Plant functional traits for improving productivity. Springer, Singapore. https://doi.org/10.1007/978-981-97-1510-7_6

Singh M, Singh H, Kumar A, Kumar M, Barthwal S, Thakur A (2024b) Soil nitrogen availability determines the CO_2 fertilization effect on tree species (*Neolamarckia cadamba*): growth and physiological evidence. Environ Sustain 7:53–60. https://doi.org/10.1007/s42398-023-00300-w

Smith AB et al (2020) Regulatory approaches to air quality improvement: a comparative analysis. Environ Policy Gov 30(3):185–201

Sugiyama T, Leslie E, Giles-Corti B, Owen N (2008) Associations of neighborhood greenness with physical and mental health: do walking, social coherence and local social interaction explain the relationships? J Epidemiol Community Health 62(5):e9

Ulrich RS, Simons RF, Losito BD, Fiorito E, Miles MA, Zelson M (1991) Stress recovery during exposure to natural and urban environments. J Environ Psychol 11(3):201–230

USEPA (2021) Particulate matter (PM) basics. U.S. Environmental Protection Agency. Retrieved from https://www.epa.gov/pm-pollution/particulate-matter-pm-basics

Voogt JA, Oke TR (2003) Thermal remote sensing of urban climates. Remote Sensing of Environment, 86(3):370–384. https://doi.org/10.1016/S0034-4257(03)00079-8

Wang Y et al (2021) Real-time air quality monitoring systems: advances and challenges. Atmos Environ 244:117957

WHO (2021) Ambient (outdoor) air pollution. World Health Organization. Retrieved from https://www.who.int/news-room/fact-sheets/detail/ambient-(outdoor)-air-quality-and-health

Wolf KL (2004) Trees and business district preferences: a case study of Athens, Georgia, US. J Arboricult 30(6):336–346

Wong NH, Jusuf SK, Tan AY, Sia A, Ong CL (2012) Mitigation of urban heat island by landscape design. Landsc Urban Plan 104(3–4):198–205

Yadav SK, Singh H, Ginwal HS, Barthwal S (2019a) Elevated CO_2 enhanced growth and physiological process of *Populus deltoides* Bartr. ex Marsh. Indian For 145(1):23–27

Yadav SK, Singh H, Nautiyal R, Ginwal HS, Ansari SA, Barthwal S (2019b) Modulation of morpho-physiological responses in *Populus deltoides* by elevated carbon dioxide and temperature. For Sci 66:105–118. https://doi.org/10.1093/forsci/fxz048

Zhang Q, Liu H (2022) Research and development priorities for air quality improvement: a perspective from emerging economies. Environ Sci Technol 56(1):15–27

The Potential of People's Engagement by Green Taxation to Maintain Urban Green Space

Anusreeta Dutta, Anugrah Tripathi, Md. Quasid Qamar, Smita Ghosh, Falguni Upreti, and Himanshi Payal

Abstract Green taxes hold significant potential as effective tools for financing crucial initiatives, promoting sustainable urban design, and addressing the challenges of urbanization and climate change. This chapter explores citizen engagement in green taxation decision-making, from levying taxes to mitigate pollution impacts to allocating funds for citizen-centric urban forestry management. It emphasizes the role of taxation policies in shaping public behavior and awareness regarding environmental conservation and the adverse effects of pollution on forests and biodiversity. The chapter examines successful global green taxation initiatives, such as Amsterdam's green restoration project, Sweden's NOx emission tax, and France's forest tax, illustrating how environmental incentives can positively impact the environment. It underscores the need for moderation and a balanced approach between environmental and economic factors to implement circular economy and green taxation policies effectively. The economic evaluation of urban green infrastructure is discussed, highlighting methods like Discounted Cash Flow analysis and LiDAR remote sensing to measure ecosystem services and understand the financial benefits of urban greenery. The chapter highlights global successes in urban forestry through citizen participation and green taxation. Additionally, it focuses on the status of urban forestry in India, policies to enhance citizen engagement in green taxation, and the need to address social and economic concerns to improve participation in climate mitigation measures. It also examines reasons for past failures in mitigating climate change, such as decreased funding and continued investment in fossil fuels,

---

A. Dutta (✉)
Department of Humanities and Social Sciences, Birla Institute of Technology and Science, Hyderabad, Telangana 500078, India
e-mail: anusreetadutta.23@gmail.com

School of Property, Construction, and Project Management, Royal Melbourne Institute of Technology, Melbourne, VIC 3004, Australia

A. Tripathi
Directorate of Research, Indian Council of Forestry Research and Education, PO New Forest, Dehradun, Uttarakhand 248006, India

A. Tripathi · Md. Q. Qamar · S. Ghosh · F. Upreti · H. Payal
ICFRE-Forest Research Institute, PO New Forest, Dehradun, Uttarakhand 248006, India

H. Singh (ed.), *Urban Forests, Climate Change and Environmental Pollution*,
https://doi.org/10.1007/978-3-031-67837-0_26

which hinder effective policy implementation. The chapter argues for the importance of public participation in green taxation measures, underscoring its critical role in fostering sustainable urban environments and successful climate mitigation efforts.

Keywords Green taxation · Green bonds · Sustainability · Governance · Polluter pay

1 Introduction

In recent years, sustainability has gained widespread recognition, emphasizing integrating economic, environmental, and social concerns to enhance opportunities for people to live in a cleaner, healthier environment (Kumar et al. 2019c, 2020; Meena et al. 2011; Singh and Kumar 2022; Singh et al. 2024b; United Nations and United Nations 2008;. This chapter explores the role of individual contributions, mainly through monetary means, to enrich urban greenery. As pollution levels rise, individuals are motivated to reside in a purified environment in their communities, surrounded by gardens, parks, and fields, desiring pollution-free spaces (Kumar et al. 2022; 2024; Singh 2024; Singh et al. 2022; 2024a). Addressing global environmental concerns, such as pollution caused by rapid development, is imperative. The continuous and unchecked development has become detrimental to human health additionally. The industrial sector is a major pollution emitter, followed by the energy, food, agriculture, and transport sectors (IPCC 2018).

Environmental issues have affected the natural world and created economic disparities in urban landscape valuation. The economic valuation of urban landscapes can enhance property values around urban areas (Banday and Aneja 2019; Dhyani et al. 2021; Kumar et al. 2021a, b, c, d, e; Prakash et al. 2022). The aesthetic appeal of a property significantly influences buyer interest, reflecting individuals' willingness to pay for environmental quality.

The concept is particularly relevant now as it has a wide trade-off with the principle of "polluter pays" for environmental damage. Effective governance is crucial for implementing green taxes (Atalla et al. 2022), acting as an intermediary to facilitate the development of urban green spaces funded by public payments. In developing countries like India, a consolidated green tax policy is absent. However, initiatives like the Graded Response Action Plan (GRAP) in Delhi, which addresses air pollution through graded measures based on the Air Quality Index (AQI), exemplify direct environmental taxation (NCAP 2019).

India's long-term neglect of pollution has led to severe environmental impacts, necessitating stringent measures that might face public resistance (Ghorani-Azam et al. 2016). An inclusive policy encouraging public participation in environmental accountability is urgently needed (US-EPA 2023). Environmental taxes serve as incentive-based tools, compelling people to pay for the environmental damage they cause ("Governance by Green Taxes" 1994). As a form of environmental taxation

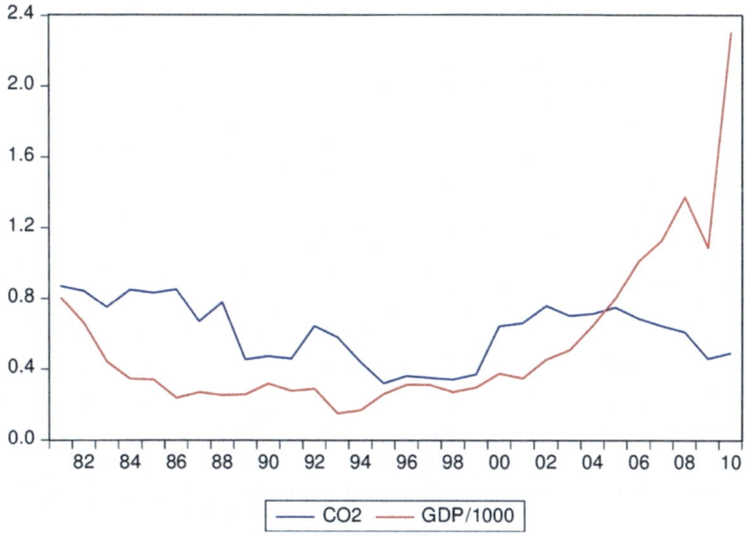

Fig. 1 The relationship between CO_2 emissions and economic growth (GDP growth)—Environmental Kuznet curve (EKC)—A case of Nigeria (Osu et al. 2016)

on pollution, green taxation raises awareness of the environmental harm caused by individual choices (Oates and Portney 2003).

Policymakers view this willingness to pay as indicative of potential economic growth contributing to GDP (OECD 2021). Studies on the correlation between GDP growth and CO_2 emissions have confirmed a positive relationship, represented by the Environmental Kuznets Curve (EKC) (Osu et al. 2016) (Fig. 1).

In countries like India, the immediate need is not solely a policy-maker's perspective but active public engagement through taxation. Taxation, a fundamental civic duty, strengthens national revenue and is a potent measure for environmental preservation. Its flexibility allows equitable access to resources in return. A green tax regime operates on the principle that polluters should bear the full social cost of their actions (Pigou and Aslanbeigui 2017), promoting an eco-friendly infrastructure (Gramkow 2020). This regime is more effective when individuals understand and appreciate the benefits of urban greenery, such as trees and parks, which enhance their quality of life.

A study in Palermo, Italy, indicated residents' willingness to pay for urban vegetation due to the benefits they derived from nearby green spaces (Galati et al. 2023). Similarly, a survey of 3009 households in Mandeville, Louisiana, USA, showed a positive correlation between willingness to pay for tree preservation and the benefits provided by trees, such as urban cooling, pollution filtration, and carbon sequestration. This evidence underscores the success of public willingness to fund environmental greenery. To encourage contributions to urban greening, it is crucial to consider the financial capacities of different demographic groups, including variations in class, gender, age, literacy, income, and awareness (Sharma and Singh

2021; Singh 2021; Singh et al. 2021). This chapter extends a critical perspective that connects the public's willingness to pay for environmental damage with green taxation, justifying environmental taxation as a motivational scheme where people pay for the pollution they cause.

A green approach should prioritize urban green finance in urban budgets, implementing property taxes, transportation fees to reduce car traffic, and charges to promote responsible water consumption (Kumar et al. 2021a, b, c, d; Singh et al. 2020; Sub-National Finances and Fiscal Consolidation 2013). Rapid urbanization has exceeded cities' carrying capacities, leading to air, water, and noise pollution, which stresses urban infrastructure and residents' quality of life. Studies show that urban parks and forests provide significant benefits, helping alleviate everyday stress.

Effective global governance must ensure the implementation of cost-effective, climate-smart policies and institutionalize climate action based on public involvement. The success of these policies depends on proper implementation at all governance levels—local, regional, and transboundary (Carattini et al. 2017). A balance between environmental improvement and economic concerns is necessary to foster a clean environment (Tietenberg 1990). Initiatives such as congestion charges, flexible parking fees, toll lanes, and split-rate property taxes are important steps toward this goal. Citizens with secure employment are more likely to accept green taxation as part of their environmental responsibility.

Internationally, the Nationally Determined Contributions (NDCs) post-Paris Agreement drive governments to implement domestic policies to tackle climate change (Carattini et al. 2017). Although green taxes have reduced pollution and improved economic performance, they can disrupt production and consumption processes in the short term (OECD 2021). This disruption highlights the complex relationship between economic development and environmental awareness (Gramkow 2020). It underscores the need for an effective circular economy and green taxation policies. Concerns about the economic impact on low-income consumers who might resist higher taxes complicate the implementation of green tax policies.

In India, environmental taxes are incorporated into energy and motor transport taxes for administrative convenience and political feasibility (Tietenberg 1990). For instance, Delhi requires owners of vehicles older than eight years to pay an annual green tax of Rs 3800 (Yadav 2021). The Ministry of Road Transport and Highways (MoRTH) has successfully implemented this taxation based on the 'polluter pays' principle in the National Capital.

However, revenue streams in Indian urban areas can either support or hinder green policy priorities. Fiscal policies at all governance levels should be refined to achieve green growth and reassessed to identify unintended consequences that might impede sustainable development. This includes addressing anti-green biases in local tax provisions and mitigating counterproductive incentives from environmentally harmful subsidies (Yadav 2021). We highlight various green taxation mechanisms used worldwide for promoting and conserving urban forestry, including roadside greening, urban parks, and peri-urban forestry. Further, we discuss the need for a comprehensive green taxation policy in developing countries like India, providing empirical evidence of its economic and socio-political benefits in enhancing climate resilience. We also emphasize the importance of an indigenous policy that shares the accountability of impact and benefits with public participation, aiding in meeting regional, national and international climate targets.

2 Methodological Framework

The methodological framework employed in this chapter adopts a comprehensive approach to investigate the intricate interplay between green taxation, citizen engagement, and urban forestry initiatives. Our research began with an exhaustive thematic literature review, incorporating research articles, journal publications, policy documents, and other pertinent sources. We refined our focus by employing primary keywords in Google Scholar, targeting themes such as "urban green finance," "urban forestry," "urban parks," "green taxation," "economic valuation," "green incentive," "emission," "contingent valuation method," and "green revenue." This extensive search yielded over 700 relevant pieces of literature, which were meticulously categorized according to the chapter's theme and objectives.

To explore global initiatives, we highlighted success stories of urban forestry projects and assessed the impact of taxation policies on public behavior and environmental awareness. Specifically, in the context of India, we thoroughly examined the status of urban forestry, existing policies, and the challenges and successes in aligning social and economic concerns to enhance citizen engagement in climate mitigation measures.

The analysis extended to a global scale, investigating climate finance trends to understand the funding dynamics and challenges countries face in their efforts to mitigate climate change. This included assessing how climate finance is allocated, the effectiveness of these allocations, and the barriers to accessing such funds.

By combining a thematic literature review with a targeted investigation of both global and Indian contexts, this chapter provides a nuanced understanding of how green taxation and urban forestry initiatives can be leveraged to foster sustainable urban environments. This comprehensive methodological approach ensures that the findings are grounded in robust evidence, offering valuable insights for policymakers and stakeholders engaged in environmental conservation and urban planning.

3 Green Taxation: Encouraging Urban Forestry Initiatives

Urbanization, fuelled by population influx and compounded by factors like population growth and climate change, has led to the degradation of forests and a reduction in green spaces in major cities worldwide. This has compromised crucial ecosystem services such as carbon sequestration, biodiversity conservation, and regulation of nutrient cycles, posing risks to both macroeconomic stability and ecological balance (Estevão and Kemper 2021; Gupta et al. 2019a; Kumar et al. 2019a, b; Yadav et al.

2019). The relationship between population growth and the availability of green spaces should be directional. Achieving long-term climate goals necessitates substantial investment in climate action, estimated at approximately $1 trillion annually over the next 15 years, to meet the Nationally Determined Contributions (NDCs) outlined in the Paris Agreement.

To finance urban green initiatives, the FAO has proposed a sustainable financing mechanism that combines various revenue sources, including general taxes, green-specific taxes, municipal bonds, fines, and fee revenues, to incentivize urban green planning. Policymakers can explore funding from industries contributing to emissions or pollution through pollution taxes, green bonds, and beautification taxes to establish urban parks, green spaces, and tree plantation drives near industrial areas. Furthermore, forestry policy can involve reallocating taxes to support forestry-related activities and providing subsidies to landowners for preserving forested lands.

In Yokohama, Japan, policymakers established the Yokohama Green-Up Plan and implemented the Yokohama green tax to address climate risks, encompassing forested land within its scope. This initiative legalized private ownership of such lands. It imposed a green tax on forestry products and green spaces while creating additional incentives for private landowners interested in sustainable forest management and green projects. Similar success was achieved through the Payment for Ecosystem Services scheme under the Forest and Ecosystem Taxation in Japan (FETT), facilitating the transfer of budgets allocated to privately owned urban green spaces (Kohsaka and Uchiyama 2022). Public–private partnerships, exemplified by California's Landscape and Lighting Assessment District (LLAD), have also played a crucial role in funding public projects related to urban greenery.

Post-Paris Agreement, private sector involvement in emission reduction targets has become increasingly important, with initiatives like the Landscape and Lighting Assessment District (LLAD) in San Francisco and the TREE CITY initiative in Illertissen, Germany, demonstrating successful models of public–private collaboration (Pérez et al. 2022)These initiatives utilize a combination of funding mechanisms, including parcel taxes, general obligation bonds, and public–private partnerships, to finance urban forestry projects and ensure sustainable urban development.

Green bonds have emerged as a critical financing tool for environmentally sustainable projects such as urban forestry, afforestation activities, and waste management (World Bank 2019). Issued by institutions like the World Bank and the European Union, green bonds offer advantages such as broadening investor outreach, reducing capital costs, and enhancing brand reputation through rigorous impact measurement and reporting (Gupta et al. 2018a, b; Sharma et al. 2018; Singh et al. 2018).

Hence, green taxation and innovative financing mechanisms like green bonds are essential for incentivizing urban forestry initiatives. Public–private partnerships and supranational funding frameworks are crucial in mobilizing resources for sustainable urban development. Emphasizing community participation and environmental responsibility, these initiatives pave the way for a greener and more sustainable urban future.

4 The Need for Green Taxation to Protect Urban Forests

The overarching objective of the sustainable development framework is to safeguard public health by prioritizing cleaner and sustainable technologies, supported by structured environmental policies akin to carbon taxes, which play a pivotal role globally (Oates and Portney 2003). Addressing environmental degradation is a crucial challenge within the development scenario. Economic growth across various sectors is fundamentally linked to energy utilization, particularly by urban utilities such as industry, transportation, and housing. A study conducted by Yousaf Raza et al. (2023) for the period 1980–2018 in Bangladesh established a positive correlation between fossil fuel energy use and GDP growth, indicating a causal connection that results in increased CO_2 emissions attributable to fossil fuel use within each sector.

A crucial driver for advancing green growth is the heightened costs of using environmentally harmful commodities (Di Maria et al. 2017). The Urban Environmental Transformation Hypothesis indicates that industrial activities intrinsic to urban development significantly contribute to various forms of pollution, including water and sewage pollution, air pollution, and land degradation (Zhang et al. 2017). This hypothesis underscores the interconnectedness of urbanization and industrialization, emphasizing their collective role in amplifying energy consumption and emissions, thus causing substantial environmental harm. Notably, nations that have relied on fossil fuels as their primary energy source since the industrial era—particularly the USA, China, and European countries—are major contributors to carbon emissions.

Extensive studies reveal that coniferous forests in the Slovak Republic, Czech Republic, Poland, and Germany face severe damage due to significant land use changes, including converting forestry land for agricultural expansion aimed at regional and global food security. This alteration in land use patterns contributes to heightened climatic concerns (Di Maria et al. 2017). There is a growing global emphasis on environmental preservation in response to escalating environmental degradation. Countries worldwide are formulating and implementing environmental conservation policies as proactive measures to mitigate and combat the adverse effects of environmental damage.

The International Maritime Organization (IMO) has outlined an ambitious plan to achieve a 50% reduction in emissions compared to 2008 levels by 2050, aligning with the overarching objectives of the Paris Agreement. Implementing a well-structured carbon pricing mechanism effectively is crucial to achieving this goal. The United Nations Framework Convention for Climate Change underscores carbon taxes as the most precise means to curtail future carbon emissions and create employment opportunities, fostering a robust carbon market. A comprehensive carbon pricing framework empowers polluting entities to disregard environmental considerations, incur financial costs, or proactively engage in emission reduction measures (World Bank 2022).

In nations at the forefront of global economic activity, such as the United Kingdom and the United States, integrating public opinion through referendums strengthens democratic principles. It enhances the likelihood of successful green taxation implementation. Bridging the gap between the populace and their elected government is pivotal, necessitating identifying and incorporating public perspectives in policy frameworks to ensure successful and impactful outcomes. However, this theory does not hold for India, where the democratic structure keeps the government as the highest decision-making institution, often widening the gap between people and their representatives and causing structural injustice. Poor policy implementation in India can be attributed to political factors, policy overlap, and improper budget allocation (Khan et al. 2019).

Creating urban green spaces in developing countries is considered an ambitious goal due to a severe lack of capital for investment in developing urban infrastructure, including green parks and urban forests. Green taxation, a pivotal tool in combating environmental degradation, operates on the principle of "polluter pays," with its tax base rooted in substances causing environmental harm. While effective in revenue generation, green taxation also instigates public concern and induces behavioural change, focusing on transforming the sustainability landscape to combat air pollution. However, the success of green taxation in the global north cannot dictate its fate in the global south, owing to distinct geographical, social, and political factors that shape policy frameworks.

The pressing need for green taxation in developing nations like India can help reduce pollution in metropolitan cities with high pollution levels and environmental degradation incidents. However, challenges arise due to a predominantly centralized administrative structure and limited public participation in decision-making processes. Participatory governance is essential to make the people of India feel less neglected about decisions made behind closed doors. This necessitates re-evaluation and restructuring of the current system to adopt a more inclusive and efficient approach to addressing environmental concerns. Local green taxes, such as vehicular taxes, reveal notable public behaviour transitions towards sustainable transport options. However, enforcement challenges persist, particularly when private car owners or corporate entities seek cost-minimization strategies to evade taxation.

In India, the evolution of environmental tax has undergone various phases, yet greater efforts are needed to promote widespread awareness and acceptance among the general population. Educating people about their capacity to contribute to environmental protection at the grassroots level, including minimizing the open burning of non-biodegradable waste, is crucial for the success of such initiatives (Pal 2020). Given the global outcry regarding climate change and environmental degradation, translating international treaties and conventions into tangible actions is urgently needed. The enforcement of green taxation is a pivotal measure to swiftly and effectively address this pressing need.

5 The Rise of Green Taxation

Environmental concerns gained significant attention with Rachel Carson's *Silent Spring*, sparking global dialogue on environmental issues. This led to the first major international environmental conference, the Stockholm Conference, in 1972. Following this, the Kyoto Protocol, established in 1997 and effective in 2005, marked a pivotal shift in global climate action efforts. The Protocol categorized countries into developed (Annex I) and developing (Annex II) groups and introduced three carbon trading mechanisms: Emission Trading, Joint Implementation, and the Clean Development Mechanism (CDM). These mechanisms allowed companies to exceed or reduce government-set carbon emission standards to balance emissions by purchasing credits from other industries (Kuyper et al. 2018).

Despite these mechanisms, developed countries benefited more than developing countries under the CDM due to internal barriers within the latter. For instance, Africa's CDM implementation fell short of expectations, with only a few active projects out of 200 registered ones a decade after the mechanism's introduction. The carbon capture and sequestration method was recommended as a potential method for CDM projects.

In India, a 2013 TERI report indicated that 60% of CO_2 emissions were concentrated in 64 specific districts, posing a significant challenge to sustainable socio-economic development. Although India has significant geological sequestration potential and engages in international collaborations, no operational Carbon Capture and Storage (CCS) projects exist in the country, highlighting the critical importance of advancing technology and implementing large-scale CCS initiatives in the next 10–15 years (Shaw and Mukherjee 2022).

The European Union's Emission Trading System is part of the European Green Deal, which includes the 'Fit for 55' package, the 'Think 2030' Program, the European Climate Law, and the 2030 Biodiversity Strategy. These initiatives collectively aim to achieve at least a 55% reduction in greenhouse gas emissions by 2030, addressing climate change, biodiversity loss, and the overexploitation of natural resources. Environmental-related taxation constitutes 5.9% of total taxes within the EU, reflecting necessary tax reforms to achieve green economy goals, eliminate fossil fuel subsidies, and adhere to the Polluter Pays Principle, which shifts the tax burden to those causing pollution.

Environmental monetary reforms in the EU tax polluting industries while providing subsidies through social benefits and investment grants for environmental protection (CEPA) and resource management (CReMA). These financial transactions are recorded using the System of Environmental-Economic Accounting (SEEA), which also identifies Potential Environmentally Damaging Subsidies (PEDS), including fossil fuel subsidies and tax relaxations for car purchases and housing constructions.

Several global initiatives highlight the role of green taxation in environmental protection. Finland's Biodiversity Programme subsidizes landowners for growing forest crops on private land, emphasizing compensation based on positive environmental practices (METSO 2008). Central Africa's "forest fees" implementation encourages private land conservation. Environmental Tax on Wood (ETW) and Non-Timber Forest Products (NTFPs) under the harvester pays principle to fund forestry development is important. Pennsylvania's Clean and Green Program provides tax relief to private forest landowners. Japan's Kochi City introduced the Forest Environmental Transfer Tax under the Payment for Ecosystem Services (PES) framework to compensate for ecosystem services provided by forestlands.

According to the OECD's Policy Instrument for the Environment (PINE) 2023 database, 1871 tax policy instruments are operational in 129 countries, with Finland, Norway, and Sweden initiating carbon taxes in the early 1990s (OECD 2023).

In India, the disconnect between the political structure and social concerns poses a significant obstacle to effective policy implementation. For instance, despite having the highest renewable capacity in Southeast Asia, Rajasthan faces implementation conflicts in renewable energy due to land disputes with indigenous people. Socioeconomic challenges, including job scarcity, low education levels, and caste discrimination, contribute to long-standing economic crises in households. The government's ambitious international commitments and insensitivity to local concerns exacerbate conflicts. This chapter contends that the socio-political discourse profoundly impacts the eco-environmental framework, perpetuating poverty and burdening environmental concerns.

Overall, green taxation is a critical tool for addressing environmental degradation, operating on the principle that the polluter pays. While effective in revenue generation, it also induces behavioral change, aiming to transform the sustainability landscape to combat air pollution. However, the success of green taxation in the global north cannot dictate its fate in the global south due to distinct geographical, social, and political factors shaping policy frameworks. The pressing need for green taxation in developing nations like India can help reduce pollution in metropolitan areas. Still, challenges arise due to a centralized administrative structure and limited public participation in decision-making. Therefore, a re-evaluation and restructuring of the current system are necessary to adopt a more inclusive and efficient approach to addressing environmental concerns.

6 Success Stories in Green Taxation

Initiating environment-related cesses as early as 1950, Finland emerged as a pioneer in environmental taxation. One of its most successful programs to reduce packaging pollution was the deposit and refund scheme for beverage packaging. This initiative peaked in 2015 with a 95% return rate, targeted single-use, refillable PET (Polyethylene Terephthalate) and glass bottles, each with a fixed return rate ranging from €0.10 to €0.40. Incentives were provided to manufacturers and importers to encourage participation in the deposit return system, illustrating a collaborative effort between the government, civic society, retailers, and the beverage sector.

Costa Rica's response to extensive deforestation between the 1900s and 1990s involved addressing issues such as taxes on barren land, low livestock rates, and transport system expansion (Karousakis 2007). In 1996, the country enacted the Forest Law, introducing Payment for Ecosystem Services (PES) to support landowners engaging in responsible forest management financially. Administered by the National Forest Financial Fund (FONAFIFO) and funded by tax revenues, national and international loans, and private market instruments, the program successfully paid $120 million to 4400 landowners from 1997 to 2005, enrolling half a million hectares into the PES system (Pagiola 2008).

France implemented its first forest tax in 1946 and established the National Forestry Fund (FFN) in 2000 to promote the regeneration of forest farms (Tissot and Kohler 2013). This initiative significantly influenced French forests, increasing the population of coniferous trees and benefiting biodiversity and the environment in various forest locations (Tissot and Kohler 2013). Subsequently, in 1969, the country imposed a tax on the extraction of natural resources from forest lands, demonstrating active development in environmental protection (Tissot and Kohler 2013).

In 1992, the Swedish government introduced a tax on nitrogen (NOx) emissions, analysing the potential of 97 plants to assist in directing monetary compensation for improving environmental quality. This tax led to a 40% reduction in emissions between 1997 and 2001 (Höglund-Isaksson and Sterner 2009). Additionally, Sweden imposed a tax on energy used for space heating, electricity generation, and industrial activities to counter soil acidification. The initial rate was fixed at SEK 40/kg NOx emitted and later raised to SEK 50/kg in 2008. The revenue generated from this initiative supported industrial plants that emit low levels of NOx, promote energy efficiency, and mitigate potential adverse effects on competitiveness (Höglund-Isaksson and Sterner 2009).

In the UK, the introduction of a landfill tax in 1996 aimed to reflect the environmental cost of landfilling, reduce waste output, and increase recycling (Seely 2009). The tax rates were £2.65 (€2.96) per tonne for inert waste and £84.40 (€94.21) per tonne for biodegradable waste. This resulted in a significant decline in waste sent to landfill, from 50 million tonnes in 2001 to approximately 12 million tonnes in 2015 (Elliott 2016). Ireland, post-Kyoto Protocol, implemented a Plastic Bag Levy in 2001, charging an extra €0.22 per plastic bag at the point of sale. This initiative substantially decreased the contribution of discarded plastic bags to overall litter pollution, from 5% in 2001 to 0.13% in 2015, raising €200 million over twelve years. The funds were allocated to nationwide environmental projects (Anastasio and Nix 2022).

Under the visionary Nagar Van Scheme, India planned to develop 200 urban forests at the municipality level across the country in five years, financed by the Compensatory Afforestation Fund Management and Planning Authority (CAMPA). The Warje Urban Forest in Pune, an exemplary urban forestry project initiated through a Public–Private Partnership (PPP) by the Technology, Education, Research, and Rehabilitation for the Environment (TERRE) policy center and Tata Motors, serves as an inspiring model. Transforming a 16-hectare barren land into a biodiverse oasis set the stage for the nationwide 'Nagar Van' project, hosting over 10,000 indigenous plant species and diverse wildlife. The project, initially funded by the Pune Municipal Corporation and the CSR of Tata Motors, later collaborated with the CAMPA fund. Similar projects include the Asola Bhatti Wildlife Sanctuary, the Aravalli project in Delhi, and Yamuna Biodiversity Park.

The rise of environmental taxes is evident in India and globally, indicating a gradual acceptance by the public. This acceptance is notably more robust when taxes are applied to producers rather than consumers. Despite the imposition of small taxes, numerous exemptions exist, even when they fall short of reaching their optimal level, resulting in minimal environmental improvement. The success of green taxation schemes largely depends on public acceptance and involvement in the process, highlighting the importance of transparent and participatory governance in achieving environmental goals.

7 Economic Valuation of Urban Forestry and Green Taxation

The economic valuation of urban green infrastructure plays a pivotal role in achieving future targets and goals. Mainstreaming economic valuation requires a comprehensive data inventory, which presents a significant challenge. Research utilizing LiDAR remote sensing data to establish a consistent forest-related database provides insights into successful measures, such as Japan's Forest Environmental Transfer Tax (FETT) (Kohsaka and Uchiyama 2022). The authors elaborate that the collaboration between state and municipal levels of governance in allocating tax budgets for forest data development has been instrumental (Kohsaka and Uchiyama 2022) (Fig. 2).

The structure of forests is characterized by factors such as leaf area index, age, diameter, species composition, diversity, and the natality and mortality rates of forest flora. Forests are crucial in moderating air quality, temperature, hydrology, microclimate, and energy use efficiency. McPherson et al. (1997) highlighted that establishing structural and functional relationships in the ecological processes of forests provides a foundation for measuring and predicting economic benefits. Traditional approaches for estimating forest goods, such as maintenance, opportunity costs, and

Fig. 2 Steps of evaluating urban forests

production value, face challenges in urban forests. Emerging technological estimation approaches, like i-Tree developed by the USDA Forest Service and the Biome-BGC ecosystem process model, offer efficient ways to assess and quantify ecosystem services, including water flux, carbon, and nitrogen analysis. These tools categorize ecosystem services into regulatory, provisioning, supporting, and cultural services, aiding forestry and arboriculture researchers in their synthesis efforts (Nowak and Crane 2002).

Discounted Cash Flow (DCF) analysis of social perceptions regarding tree cover and urban forests involves calculating economic components based on trees' present and future property value. This analysis suggests that investing in fuller and larger trees with high trade value can be a strategic approach, mainstreaming urban forestry practices with community participation. The Council of Tree and Landscape Appraisers (CTLA) developed a single tree valuation method, assigning a compensatory or replacement value to each tree and considering them structural assets (Nowak and Crane 2002). Other studies, such as the one in Ribnjak Park in Croatia, estimate the value of individual trees based on the Danish model for tree value determination, considering dimensions, social values, expected age, location, and health of the trees (Beljan et al. 2015).

Studies using the contingent valuation method (CVM) reveal that consumers are willing to pay more for goods in stores near urban green spaces, emphasizing the economic value of such environments. In Hawassa City, Ethiopia, a study estimated the economic cost of recreation, conservation, and development of the Urban Forest Park, revealing community-based willingness to pay for urban forest development (Zegeye et al. 2023). The contingent valuation method also demonstrates that consumers are willing to pay more for goods in stores near urban green spaces than those without trees. Travel cost methodology studies found that housing near parks and green spaces correlates with growing vegetables locally using urban agroforestry systems. Various economic methodologies, including the hedonic pricing method, have also successfully assigned value to urban green in the USA. In North American cities, the hedonic pricing method measures the economic value of urban parks, reflecting an increase in property values near parks of commercial properties (Zambrano-Monserrate et al. 2021). Evaluating the economic value of urban forests requires a blend of advanced technologies like i-Tree and economic methodologies such as Discounted Cash Flow and contingent valuation methods. These approaches highlight the importance of community participation and societal perceptions, emphasizing the dual benefits of environmental enhancement and economic well-being in urban forestry practices.

8 Global Urban Green Policy

Understanding the complex environment, mainly the pivotal role of green taxation in addressing urbanization, industrialization, and climate change challenges, is imperative (Gupta and Singh 2017; Kumar et al. 2017; Singh and Verma 2013; Singh et al.

2017a, b). Drawing insights from diverse global perspectives, including experiences from Nigeria, Italy, and India, the study underscores the integration of environmental, economic, and social dimensions through green taxation strategies, specifically emphasizing urban forestry conservation. The "polluter pays" and "user pays" principles emerge as fundamental incentives driving a green agenda in urban finance, deemed essential by policymakers for fostering future economic growth while tackling environmental issues, notably the preservation of urban green infrastructure (Zegeye et al. 2023).

Success stories of environmental taxation initiatives worldwide further highlight the global acknowledgment and impact of such measures. Examples include Finland's pioneering deposit refund scheme, Costa Rica's innovative forest law and payment for ecosystem services program, and France's enduring commitment to forest conservation. Similarly, Sweden's effective reduction in nitrogen emissions, the UK's successful landfill tax, and Ireland's plastic bag levy underscore the potential of targeted taxation in achieving positive environmental outcomes. The green renovation projects of Amsterdam, particularly in Amsterdam Noord, aim to create an age-friendly green environment featuring elements like urban forests, waste management green networks, and sustainable buildings. The initiative envisions transforming Amsterdam Noord into a future-oriented garden city, emphasizing sustainability and connectivity with its dike city heritage.

Furthermore, the American Public Works Association's (APWA) recommendations on financing systems and revenue sources for urban forestry, encompassing environmental taxation mechanisms such as payments for ecosystem services (REDD +), tradable permits, and biodiversity offsets, contribute significantly to the discourse.

9 Gaps and Challenges in Green Taxation and Environmental Protection

The Adaptation Gap Report 2023 highlighted a significant issue: despite intergovernmental commitments to double climate finance for developing and underdeveloped nations, funding has decreased by 15%. Moreover, countries receiving climate finance from developed nations continue to invest in fossil fuels, exacerbating the climate crisis. This underscores the negative implications of the current climate finance flow and highlights several key challenges associated with implementing environmental or green taxes.

9.1 Lack of Standardized Policies

There is no universal standard for green taxation policies. For example, European Union member countries levy taxes on carbon, NOX, and SOX emissions, while the United States has a specific carbon tax. In India, green taxes primarily focus on vehicles, particularly those older than ten years without a Pollution under Control (PUC) certificate, which are subject to a tax.

9.2 Corporate Social Responsibility (CSR)

The European Union defines Corporate Social Responsibility (CSR) as enterprises' voluntary inclusion of social and environmental concerns (EC 2001). However, CSR initiatives often overlook environmental degradation due to market-driven foreign policy regulations, data inefficiencies, and lack of positive communication regarding environmental achievements (EC 2001). Companies globally increasingly incorporate sustainability into their CSR efforts, aligning with evolving Environmental Social Governance (ESG) schemes. For instance, ITC Ltd.'s e-Choupal program has empowered millions of farmers and promoted social and farm forestry, greening extensive land areas and generating substantial employment.

9.3 Need for Real-Time Climate Data and Modelling

Effective green tax reforms require robust climate data and modelling to establish relationships between emissions, environmental degradation, and economic impacts. Accurate economic valuation of environmental services is essential to raise public awareness. The Contingent Valuation Method (CVM) is one such method, as demonstrated by a study in Joensuu, Finland, which assessed the benefits of urban forests and residents' willingness to pay for them (Tyrväinen and Väänänen 1998). Similar research is needed in India to engage the public in environmental management effectively.

9.4 Broader Societal Implications

The convergence of technical and environmental challenges necessitates coordinated efforts across economics, management, sustainability politics, and research and development. Sustainable technology development involves a dynamic interplay of radical and incremental changes. A comprehensive approach is needed to address the relationship between technological advancements and environmental considerations (Hauschild et al. 2020).

9.5 Policy Interaction and Conflicts

Environmental conservation policies often conflict with fiscal policies within forestry and other land-use systems, necessitating a re-evaluation of economic perspectives. The intensity of fiscal reform varies across states and local policies, leading to inconsistencies and lack of coordination.

9.6 Institutional and Regulatory Frameworks

India's rich biodiversity and varied carbon storage potential across cities require comprehensive inventorying of forest data and institutional methodologies. The lack of systematic approaches poses challenges in assessing and managing urban ecosystems (Sahoo et al. 2021). The absence of a regulatory framework for different governance zones further complicates urban forest management, potentially leading to data miscalculations and inequitable Payment for Ecosystem Services distribution. Reforming sectoral policies with a whole-economy perspective can reduce the cost of achieving sustainability objectives.

10 Recommendations and Future Directions

The study has provided valuable insights into the intricate balance between economic growth and environmental conservation, particularly in urban and peri-urban areas. We emphasize raising awareness, considering regional priorities, and incorporating public participation in decision-making processes. Future research directions could include conducting in-depth analyses of the socio-economic impacts of green taxation policies on the conservation of urban forestry spaces. Additionally, case studies on successful implementations of such projects and strategies for overcoming challenges associated with global variations in governance structures and public attitudes are warranted.

Furthermore, real-time climate data and the economic valuation of environmental degradation need to be addressed. Understanding the potential economic burden on industries, especially in fossil fuel-dependent economies, is crucial for effective policy formulation and implementation. In the context of India, it is essential to highlight the challenges posed by the gap between social concerns and political structures in cities and suburban areas. Therefore, we advocate for a deeper understanding of socio-political and socio-environmental dynamics to ensure the successful implementation of green taxation in urban green planning policies, including urban and peri-urban forestry, park development, and the promotion of green buildings.

11 Conclusion

The intricate interplay between economic development, urbanization, environmental concerns, and green taxation underscores the imperative of community participation in promoting sustainable practices. Green taxation is an incentive-based tool to deter harmful activities and foster responsible behaviour, aligning individual actions with broader sustainability objectives. However, the challenges and gaps identified in implementing green taxation highlight the need for standardized policies, enhanced corporate social responsibility, and access to real-time climate data in urban forestry. The socio-political analysis, particularly within the context of India, emphasizes the significance of understanding local dynamics to ensure effective policy implementation, especially in densely populated urban areas with limited green infrastructure. A balanced approach is consistently advocated, acknowledging the delicate equilibrium between economic concerns and environmental benefits. Economic valuation of urban green infrastructure emerges as a central theme, showcasing technology integration, economic analysis, and environmental policy. The diverse methodologies discussed, ranging from LiDAR remote sensing to Discounted Cash Flow analysis, offer valuable insights into quantifying ecosystem services and understanding the economic benefits associated with urban greenery. The success stories highlighted, such as Finland's innovative deposit refund scheme and Amsterdam's green renovation project funded by a green bond scheme, are tangible examples of how green taxation and environmental incentives can drive positive environmental outcomes. The gained knowledge contributes to a nuanced understanding of the complexities and potential solutions for equitable and effective environmental policies in developing urban green spaces. Green taxation emerges as a powerful tool to incentivize sustainable urban planning, finance essential projects, and address the challenges of urbanization and climate change. Incorporating global initiatives, successful models, and financing mechanisms provides a comprehensive perspective on the intersection of economics and environmental conservation in urban environments.

References

Anastasio M, Nix J (2022) Plastic bag Levy in Ireland. https://ieep.eu/wp-content/uploads/2022/12/IE-Plastic-Bag-Levy-final-1-1.pdf

Atalla G, Mills M, McQueen J (2022) Six ways that governments can drive the green transition. https://www.ey.com/en_us/insights/government-public-sector/six-ways-that-governments-can-drive-the-green-transition

Banday UJ, Aneja R (2019) Energy consumption, economic growth and CO_2 emissions: evidence from G7 countries. World J Sci Technol Sustain Dev 16(1):22–39. https://doi.org/10.1108/WJSTSD-01-2018-0007

Beljan K, Posavec S, Jerčić K (2015) University of Zagreb, Faculty of Forestry, Department of Forest Inventory and Management, Svetošimunska 25, HR-10000 Zagreb, Croatia, & University of Zagreb, Faculty of Forestry, Department of Forest Inventory and Management, Svetošimunska 25, HR-10000 Zagreb, Croatia (2015). Economic valuation of urban trees: Ribnjak Park Case Study, Zagreb. South-East Euro For 6(1). https://doi.org/10.15177/seefor.15-08

Carattini S, Baranzini A, Thalmann P, Varone F, Vöhringer F (2017) Green taxes in a post-Paris world: are millions of nays inevitable? Environ Resource Econ 68(1):97–128. https://doi.org/10.1007/s10640-017-0133-8

Dhyani K, Kuniyal HB, Singh H, Sobha (2021) Growth and physiological potential of *Terminalia arjuna* under elevated CO_2 levels in Open top chamber condition. J Appl Nat Sci 13(3):1121–1126. https://doi.org/10.31018/jans.v13i3.2490, ISSN 2231-5209

Di Maria C, Smulders S, Van Der Werf E (2017) Climate policy with tied hands: optimal resource taxation under implementation lags. Environ Res Econ 66(3):537–551. https://doi.org/10.1007/s10640-016-0091-6

EC (2001) Promoting a European framework for corporate social responsibility. https://www.europarl.europa.eu/meetdocs/committees/deve/20020122/com%282001%29366_en.pdf

Elliott T (2016) Landfill tax in the United Kingdom. https://ieep.eu/wp-content/uploads/2022/12/UK-Landfill-Tax-final.pdf

Estevão M, Kemper K (2021) How fiscal policy can help save forests. https://blogs.worldbank.org/en/climatechange/how-fiscal-policy-can-help-save-forests

Galati A, Coticchio A, Peiró-Signes Á (2023) Identifying the factors affecting citizens' willingness to participate in urban forest governance: evidence from the municipality of Palermo, Italy. For Policy Econ 155:103054. https://doi.org/10.1016/j.forpol.2023.103054

Ghorani-Azam A, Riahi-Zanjani B, Balali-Mood M (2016) Effects of air pollution on human health and practical measures for prevention in Iran. J Res Med Sci 21(1):65. https://doi.org/10.4103/1735-1995.189646

Governance by Green Taxes (1994) Making pollution prevention pay author: Mikael Skou Andersen £40.00, hardback, pp247 ISBN 0 7190 4231 3 £14.99, paperback ISBN 0 7190 4232 1 Manchester University Press, 1994. Euro Environ 4(6):27–27. https://doi.org/10.1002/eet.3320040607

Gramkow C (2020) Green fiscal policies: an armoury of instruments to recover growth sustainably. https://repositorio.cepal.org/server/api/core/bitstreams/6dd8170b-a96b-4682-b40e-9b7ed7a93749/content

Gupta SK, Singh H (2017) Observations on sporadic flowering in *Ficus benjamina* L. and *Peltophorum africanum* L. in new forest. Indian For 143:290–291

Gupta SK, Ram J, Singh H (2018a) Comparative study of transpiration in cooling effect of tree species in the atmosphere. J Geosci Environ Protect 6:151–166. https://doi.org/10.4236/gep.2018.68011

Gupta SK, Ram J, Singh H, Kumar P (2018b) Comparative study of cooling under shades of some forestry tree species with respect to ambient temperatures. J Geosci Environ Protect 6:51–66. https://doi.org/10.4236/gep.2018.611004

Gupta SK, Ram J, Singh H (2019a) Relationship between leaf area index and atmospheric cooling of tree species. Indian J for 42(3):203–208

Gupta SK, Singh H, Ram J (2019b) Need for a novel paradigm in forestry research: perspectives of comparative atmospheric cooling by different forestry effect tree species. Indian For 145(6):509–515

Hauschild MZ, Kara S, Røpke I (2020) Absolute sustainability: challenges to life cycle engineering. CIRP Ann 69(2):533–553. https://doi.org/10.1016/j.cirp.2020.05.004

Höglund-Isaksson L, Sterner T (2009) Innovation effects of the Swedish Nox charge. https://www.oecd.org/env/consumption-innovation/43211635.pdf

IPCC (2018) Summary for Policymakers, in: Global Warming of 1.5 C. An IPCC Special Report on the impacts of global warming of 1.5 C above pre-industrial levels and related global greenhouse gas emission pathways, in the context of strengthening the global response to the threat of climate change, sustainable development, and efforts to eradicate poverty, 2018. https://www.ipcc.ch/sr15/chapter/spm/

Karousakis K (2007) Incentives to reduce GHG emissions from deforestation: lessons learned from Costa Rica and Mexico. https://redd.unfccc.int/uploads/2_126_redd_20081208_oecd_incentives.pdf

Khan MD, Shakya S, Vu HHT, Ahn JW, Nam G (2019) Water environment policy and climate change: a comparative study of India and South Korea. Sustainability 11(12):3284. https://doi.org/10.3390/su11123284

Kohsaka R, Uchiyama Y (2022) Use of the forest environment transfer tax for forest data development and exchange: evidence from all 47 prefectures in Japan. For Sci Technol 18(4):201–212. https://doi.org/10.1080/21580103.2022.2133017

Kumar N, Suyal DC, Sharma IP, Verma A, Singh H (2017) Elucidating stress proteins in rice (*Oryza sativa* L.) genotype under elevated temperature: a proteomic approach to understand heat stress response. 3 Biotech 7:205. https://doi.org/10.1007/s13205-017-0856-9

Kumar M, Singh MP, Singh H, Dhakate PM, Ravindranath NH (2019a) Forest working plan for the sustainable management of forest and biodiversity in India. J Sustain for 39:1–22. https://doi.org/10.1080/10549811.2019.1632212

Kumar N, Jeena N, Singh H (2019b) Elevated temperature modulates rice pollen structure: a study from the foothill of the Himalayan agro-ecosystem in India. 3 Biotech 9(5):175. https://doi.org/10.1007/s13205-019-1700-1

Kumar N, Jeena N, Kumar N, Gangola S, Singh H (2019c) Phytoremediation facilitating enzymes: an enzymatic approach for enhancing remediation process. In: Bhatt P (ed) Smart bioremediation technologies. Academic Press, Elsevier. 50 Hampshire Street, 5th Floor, Cambridge, MA 02139, United States, pp 289–306

Kumar N, Singh H, Sharma SK (2020) Antioxidants: responses and importance in plant defense system. In: Roychowdhury R, Choudhury S, Hasanuzzaman M, Srivastava S (eds) Sustainable agriculture in the era of climate change. Springer, Cham, pp 251–264. https://doi.org/10.1007/978-3-030-45669-6_11

Kumar M, Phukon SN, Paygude AC, Tyagi K, Singh H (2021a) Mapping phenological functional types (PhFT) in the Indian Eastern Himalayas using machine learning algorithm in Google Earth Engine. Comput Geosci 158:104982. https://doi.org/10.1016/j.cageo.2021.104982

Kumar A, Kumar P, Singh H, Kumar N (2021b) Modulation of plant functional traits under essential plant nutrients during seasonal regime in natural forests of Garhwal Himalayas. Plant Soil 465:197–212. https://doi.org/10.1007/s11104-021-05003-x

Kumar A, Kumar P, Singh H, Bisht S, Kumar N (2021c) Relationship of physiological plant functional traits with soil carbon stock in the temperate forest of Garhwal Himalaya. Curr Sci 120(8):1368–1373. https://doi.org/10.18520/cs/v120/i8/1368-1373

Kumar A, Kumar P, Singh H, Kumar N (2021d) Adaptation and mitigation potential of roadside trees with bio-extraction of heavy metals under vehicular emissions and their impact on physiological traits during seasonal regimes. Urban For Urban Green 58:126900. https://doi.org/10.1016/j.ufug.2020.126900

Kumar A, Kumar P, Singh H, Kumar N (2021e) Impact of plant functional traits on infiltration rate under different forests of Kempty watershed in Garhwal Himalaya. Indian J Soil Conser 49(1):38–44

Kumar A, Singh H, Kumari G, Bisht S, Malik A, Kumar A, Singh M, Raturi BS, Thakur A, Kaushal R (2022) Adaptive resilience of roadside trees to vehicular emissions via leaf enzymatic, physiological, and anatomical trait modulations. Environ Pollut 313:120191. https://doi.org/10.1016/j.envpol.2022.120191

Kumar N, Singh H, Giri K, Kumar A, Joshi A, Yadav A, Singh R, Bisht S, Kumari R, Jeena N, Khairakpam R, Mishra G (2024) Physiological and molecular insights into the allelopathic effects on agroecosystems under changing environmental conditions. Physiol Mol Biol Plants 30:417–433. https://doi.org/10.1007/s12298-024-01440-x

Kuyper J, Schroeder H, Linnér B-O (2018) The evolution of the UNFCCC. Annu Rev Environ Res 43(1):343–368. https://doi.org/10.1146/annurev-environ-102017-030119

McPherson EG, Nowak D, Heisler G, Grimmond S, Souch C, Grant R, Rowntree R (1997) No title found. Urban Ecosyst 1(1):49–61. https://doi.org/10.1023/A:1014350822458

Meena D, Singh H, Chaudhari SK (2011) Elucidating strontium response on growth dynamics and biochemical change in *Phaseolus mungo* L. Int J Agric Environ Biotechnol 4:107–113

METSO (2008) Forest biodiversity programme for Southern Finland (METSO). https://mmm.fi/en/forests/biodiversity-and-protection/metso-programme

NCAP (2019) National clean air programme. Ministry of environment, forest and climate change. https://moef.gov.in/wp-content/uploads/2019/05/NCAP_Report.pdf

Nowak DJ, Crane DE (2002) Carbon storage and sequestration by urban trees in the USA. Environ Pollut 116(3):381–389. https://doi.org/10.1016/S0269-7491(01)00214-7

Oates WE, Portney PR (2003) The political economy of environmental policy. In: Handbook of environmental economics, vol 1. Elsevier, pp 325–354. https://doi.org/10.1016/S1574-0099(03)01013-1

OECD (2021) Financial markets and climate transition opportunities, challenges and policy implications. https://www.oecd.org/finance/Financial-Markets-and-Climate-Transition-Opportunities-challenges-and-policy-implications.htm

OECD (2023) OECD (Organ. Econ. Co-op. Dev.) OECD sustainable manufacturing indicators. Organisation for economic co-operation and development. https://www.oecd.org/innovation/green/toolkit/oecdsustainablemanufacturingindicators.htm

Osu BO, Okonkwo CU, Ujumadu RN (2016) The relationship between GDP and CO_2 emission in Nigeria using the least square polynomials. https://ideas.repec.org/a/arp/ajoams/2016p51-55.html

Pagiola S (2008) Payments for environmental services in Costa Rica. Ecol Econ 65(4):712–724. https://doi.org/10.1016/j.ecolecon.2007.07.033

Pal A (2020) Green taxation—its impact and necessity in India. https://doi.org/10.13140/RG.2.2.36376.62724

Pérez L, Hunt V, Samandari H, Nuttall R, Biniek K (2022) Does ESG really matter—and why? https://www.mckinsey.com/~/media/mckinsey/business%20functions/sustainability/our%20insights/does%20esg%20really%20matter%20and%20why/does-esg-really-matter-and-why-vf.pdf

Pigou AC, Aslanbeigui N (2017) The economics of welfare, 1st ed. Routledge. https://doi.org/10.4324/9781351304368

Prakash V, Hunney K, Singh H (2022) Effect of elevated carbon dioxide on growth and development of *Santalum album* L. seedlings inoculated with plant growth promoting microorganisms in open top chambers. Int J Environ Agric Biotechnol 7(5):149–159

Sahoo UK, Tripathi OP, Nath AJ, Deb S, Das DJ, Gupta A, Devi NB, Charturvedi SS, Singh SL, Kumar A, Tiwari BK (2021) Quantifying tree diversity, carbon stocks, and sequestration potential for diverse land uses in Northeast India. Front Environ Sci 9:724950. https://doi.org/10.3389/fenvs.2021.724950

Seely A (2009) Landfill tax: introduction and early history. https://researchbriefings.files.parliament.uk/documents/SN00237/SN00237.pdf

Sharma R, Singh H, Kaushik M, Nautiyal R, Singh O (2018). Adaptive physiological response, carbon partitioning, and biomass production of *Withania somnifera* (L.) Dunal grown under elevated CO_2 regimes. 3 Biotech 8:267. https://doi.org/10.1007/s13205-018-1292-1

Sharma R, Singh H (2021) Alteration in biochemical constituents and nutrients partitioning of *Asparagus racemosus* in response to elevated atmospheric CO_2 concentration. Environ Sci Pollut Res 29:6812–6821. https://doi.org/10.1007/s11356-021-16050-3

Shaw R, Mukherjee S (2022) The development of carbon capture and storage (CCS) in India: a critical review. Carbon Capture Sci Technol 2:100036. https://doi.org/10.1016/j.ccst.2022.100036

Singh H (2021) An integrated approach considering physiological and biophysical based indicators for assessing tolerance of roadside plantations of *Alstonia scholaris* towards urban roadside air pollution: an assessment of adaptation of plantations for mitigating urban roadside air pollution. Trees. https://doi.org/10.1007/s00468-021-02179-8

Singh H, Verma A (2013) Characterization and screening of high nitrogen efficient rice genotype to curtail environmental pollution. Int J Agric Environ Biotechnol 6(1):23–30

Singh H, Kumar M (2022) Climate change and its impact on Indian Himalayan Forests: current status and research needs. In: Rani S, Kumar R (eds) Climate change. Springer climate. Springer, Cham, pp 223–242. https://doi.org/10.1007/978-3-030-92782-0_11

Singh H, Savita Sharma R, Sinha S, Kumar M, Kumar P, Verma A, Sharma SK (2017a) Physiological functioning of *Lagerstroemia speciosa* L. under heavy roadside traffic: an approach to screen potential species for abatement of urban air pollution. 3 Biotech 7(61):1–10. https://doi.org/10.1007/s13205-017-0690-0

Singh H, Verma A, Kumar M, Sharma R, Gupta R, Kaur M, Negi M, Sharma SK (2017b) Phytoremediation: a green technology to clean up the sites with low and moderate level of heavy metals. Austin Biochem 2:1012

Singh H, Sharma R, Savita SMP, Kumar M, Verma A, Ansari MW, Sharma SK (2018) Adaptive physiological response of *Parthenium hysterophorus* to elevated atmospheric CO_2 concentration. Indian For 144:1–14

Singh H, Yadav M, Kumar N, Kumar A, Kumar M (2020) Assessing adaptation and mitigation potential of roadside trees under the influence of vehicular emissions: a case study of *Grevillea robusta* and *Mangifera indica* planted in an urban city of India. PLoS ONE 15(1):e0227380. https://doi.org/10.1371/journal.pone.0227380

Singh S, Singh H, Sharma SK, Nautiyal R (2021) Seasonal variation in biochemical responses of bamboo clones in the sub-tropical climate of Indian Himalayan foothills. Heliyon 7(4):e06859. https://doi.org/10.1016/j.heliyon.2021.e06859

Singh H, Bandyopadhyaya S, Kumar A, Singh M, Malik A, Kumari G, Raturi A, Yadav SK, Kumar M, Barthwal S, Thakur A, Rawat PS (2022) Understanding the physiological and biophysical response of urban roadside plantations for assessing adaptation and mitigation mechanisms toward vehicular emissions. Urban Clim 44:101183. https://doi.org/10.1016/j.uclim.2022.101183

Singh H (2024) Exploring adaptive modulation in plant functional traits and its impact on the productivity of *Acacia auriculiformis* under CO_2-enriched environment. Ind Crops Prod 210:118186. https://doi.org/10.1016/j.indcrop.2024.118186

Singh M, Singh H, Kumar A, Kumar M, Barthwal S, Thakur A (2024a) Soil nitrogen availability determines the CO_2 fertilization effect on tree species (*Neolamarckia cadamba*): growth and physiological evidence. Environ Sustain 7:53–60. https://doi.org/10.1007/s42398-023-00300-w

Singh H, Kumar N, Kumar A (2024b) Enhancing resource use efficiency in crops through plant functional traits. In: Kumar N, Singh H (eds) Plant functional traits for improving productivity. Springer, Singapore. https://doi.org/10.1007/978-981-97-1510-7_6

Sub-national Finances and Fiscal Consolidation (2013) Walking on thin ice (OECD regional development working papers 2013/02; OECD regional development working papers, vol 2013/02). https://doi.org/10.1787/5k49m8cqkcf3-en

Tietenberg TH (1990) Economic instruments for environmental regulation. Oxf Rev Econ Policy 6(1):17–33. https://doi.org/10.1093/oxrep/6.1.17

Tissot W, Kohler Y (2013) Integration of nature protection in forest policy in France. https://efi.int/sites/default/files/files/publication-bank/projects/france.pdf

Tyrväinen L, Väänänen H (1998) The economic value of urban forest amenities: an application of the contingent valuation method. Landsc Urban Plan 43(1–3):105–118. https://doi.org/10.1016/S0169-2046(98)00103-0

United Nations and United Nations (eds) (2008) Achieving sustainable development and promoting development cooperation: dialogues at the economic and social council. United Nations

US-EPA (2023) Economic incentives. https://www.epa.gov/environmental-economics/economic-incentives

World Bank (2019) World development report. http://refhub.elsevier.com/S0301-4215(22)00457-8/sref56

World Bank (2022) Poverty and equity briefs. Poverty and equity brief India. https://databankfiles.worldbank.org/public/ddpext_download/poverty/987B9C90-CB9F-4D93-AE8C-750588BF00QA/current/Global_POVEQ_IND.pdf

Yadav N (2021) Old vehicle owners may have to pay up to ₹3800 per year as 'Green tax'. https://www.businessinsider.in/business/auto/news/old-vehicle-owners-may-have-to-pay-up-to-3800-per-year-as-green-tax/articleshow/80473971.cms

Yadav SK, Singh H, Nautiyal R, Ginwal HS, Ansari SA, Barthwal S (2019) Modulation of morpho-physiological responses in *Populus deltoides* by elevated carbon dioxide and temperature. For Sci 66:105–118. https://doi.org/10.1093/forsci/fxz048

Yousaf Raza M, Hasan MM, Chen Y (2023) Role of economic growth, urbanization and energy consumption on climate change in Bangladesh. Energ Strat Rev 47:101088. https://doi.org/10.1016/j.esr.2023.101088

Zambrano-Monserrate MA, Ruano MA, Yoong-Parraga C, Silva CA (2021) Urban green spaces and housing prices in developing countries: a two-stage quantile spatial regression analysis. For Policy Econ 125:102420. https://doi.org/10.1016/j.forpol.2021.102420

Zegeye G, Erifo S, Addis G, Gebre GG (2023) Economic valuation of urban forest using contingent valuation method: the case of Hawassa city, Ethiopia. Trees, For People 12:100398. https://doi.org/10.1016/j.tfp.2023.100398

Zhang J, Zeng W, Wang J, Yang F, Jiang H (2017) Regional low-carbon economy efficiency in China: analysis based on the super-SBM model with CO_2 emissions. J Clean Prod 163:202–211. https://doi.org/10.1016/j.jclepro.2015.06.111

Biotechnological Innovations in Urban Forestry: Adapting and Mitigating Climate Change and Environmental Pollution

Komal Rani and Abhishek Kumar

Abstract Urbanization has placed significant anthropogenic pressure on environmental health, impacting forests and green spaces. As cities confront the challenges of climate change and environmental degradation, urban forestry becomes increasingly vital. This chapter examines the integration of biotechnology in urban forestry to enhance the sustainability and resilience of urban environments. Biotechnological approaches provide innovative solutions for mitigating environmental stresses, improving air and water quality, and promoting biodiversity conservation. The key strategies discussed include the genetic modification of trees, DNA storage technology, RNA interference, biomimicry, phytoremediation, biodiversity assessment, precision tree farming, rhizosphere engineering, and bio-based materials. Policy recommendations for incorporating biotech-driven solutions into urban planning frameworks are also highlighted. Leveraging biotechnology interventions may guide urban planners and policymakers to develop more sustainable, greener, and resilient urban landscapes, fostering healthier environments and improving residents' quality of life.

Keywords Urban green spaces · Urban cities · Sustainable cities · Biotechnology · Biomimicry · Urban greening · Urban forestry · Genetic engineering

K. Rani (✉)
Genetics and Tree Improvement Division, ICFRE-Tropical Forest Research Institute, Jabalpur 482021, Madhya Pradesh, India
e-mail: komalbiotech96@gmail.com

A. Kumar
Forest Ecology and Climate Change Division, ICFRE-Forest Research Institute, Dehradun, Uttarakhand 248006, India

H. Singh (ed.), *Urban Forests, Climate Change and Environmental Pollution*,
https://doi.org/10.1007/978-3-031-67837-0_27

1 Introduction

According to the World Bank, Urban population of India is about 36% of the total population. Due to rapid urbanization, anthropogenic pressure has been developed on the country's health and the quality of forests. The land has been diverted for agriculture to meet the food demands, and altered temperature and precipitation events have led to deteriorated environmental conditions at a very unprecedented rate (Kumar et al. 2022, 2024; Singh et al. 2022). Urban forestry is crucial in augmenting the quality of urban spaces and the people's health. During these altering climatic conditions, urban forestry is pivotal in maintaining sustainable urban spaces worldwide. Urban spaces (USs) include avenue plantations, recreational parks, community forests, and residential and commercial landscaping (Singh et al. 2024a; b). Apart from having environmental benefits and aesthetic values, the USs also plays a significant role in health and community (Fig. 1). Cities worldwide increasingly adopt greenification and urban forestry initiatives in response to socio-ecological challenges. This trend underscores the growing global focus on Biophilia, which fosters human-nature interaction to improve urban life quality (Lefosse et al. 2023). The gradual integration of natural elements within urban environments, facilitated by Nature-Based Solutions (NBSs), involves transforming urban areas from predominantly concrete landscapes into urban forests, reestablishing connections between human-made habitats and the biosphere (Kanta et al. 2024). In NBS, Green Infrastructures have garnered significant attention for their strategic importance, extending landscape planning towards a more comprehensive approach to urban design (Dhyani et al. 2021; Kumar et al. 2021a, b, c, d, e; Quattrone 2023). These initiatives act as proactive ecosystems composed of green spaces, effectively mitigating environmental degradation and fulfilling ecosystem service objectives.

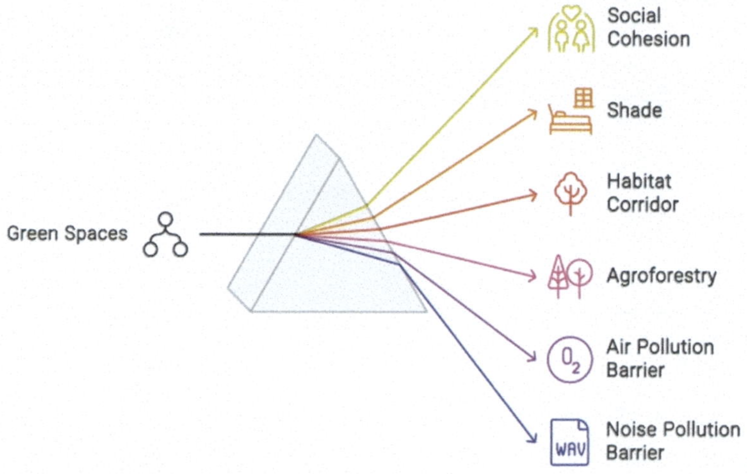

Fig. 1 Advantages of green spaces

Strategic urban planning, utilizing various tools and policies, is essential for developing future climate-resilient and green-smart cities. In response to these challenges, biotechnology is anticipated to spearhead the next innovation wave within urban environments (Gladkov and Gladkova 2021; Prakash et al. 2022). Current biotechnology research is exploring avenues such as energy conservation using plant-based methods and data storage through artificially synthesized DNA strands (Wang et al. 2021). Integrating these biotechnological advancements with smart city technologies allows urban cities to evolve into self-sustaining entities. These bio-smart cities would be capable of generating, storing, and analysing data through organic processes seamlessly integrated with Artificial Intelligence (AI). This symbiosis of biology and technology can transform urban areas into dynamic, living organisms that intelligently adapt and respond to their inhabitants' needs. The convergence of IoT and biotechnology promises to create smarter, greener, and more resilient urban landscapes (Vermesan and Friess 2013) This chapter effectively details the role of biotechnology in creating smart urban green spaces (UGSs) through interventions in urban forestry and making them climate resilient.

2 Need for Biotechnology in Urban Forestry

Biotechnology has paved the way for advanced farming, bioremediation, production of plants with desired traits, editing the genome of plants, and production of bio-based material for green infrastructure (Sharma and Singh 2021; Singh 2021). UGS is influenced by noise and light pollution to a severe extent, not only by residents but also by plants. Light pollution hinders the clear view of the skies and affects the circadian rhythm of plants. The remote sensing technique offers a solution to alleviate the artificial light at night (ALAN) by delineating city spaces and promoting sustainable development (Ye et al. 2024). Elsahragty and Kim (2015) suggest increasing the spacing between the light poles and decreasing their heights to reduce light pollution. Besides light pollution, the US also suffers from noise pollution, which harms residents' health. Plants absorb light and convert it into heat energy, and there are numerous examples of trees that can reduce noise pollution. Evergreen trees like *Dalbergia sissoo, Eucalyptus, Bamboo, Pinus* spp., and *Terminalia arjuna*have thick foliage and absorb noise efficiently (Hannan et al. 1999; Pandey et al. 2015).

Allam (2020) first depicted urban resilience using biophilic and smart city indicators. Tarek and Ouf (2021) suggested the possible indicators for biophilic and smart cities that may be implemented jointly to attain city resilience. When integrated into urban resilience principles, the synthesized indicators of smart and biophilic approaches offer valuable insights for urban planning (Tarek and Ouf 2021). Firstly, smart indicators provide tools and technologies to monitor, analyze, and respond to various environmental, social, and economic pressures (Agbali et al. 2017; Singh et al. 2021). These indicators enable data-driven decision-making processes, facilitating proactive responses to growing issues such as climate change impacts, resource management, and infrastructure resilience (Bi and Little 2022). By leveraging smart solutions, urban planners can optimize resource allotment, efficient service delivery, and enhance overall urban efficiency (Agbali et al. 2017; Singh et al. 2021). Secondly, biophilic indicators emphasize integrating nature and natural elements into the urban environment (Hussein and Al-Khafaji 2023). Incorporating green spaces, biodiversity corridors, and natural ecosystems within cities enhances aesthetic appeal and delivers tangible benefits for urban resilience (Daniels et al. 2020). Biophilic design can mitigate environmental risks, improve air and water quality, reduce urban heat island effects, and promote community well-being. Prioritizing biophilic solutions may guide urban planners to create more sustainable, resilient, and livable cities (Kumar et al. 2021a; Quattrone 2023; Singh et al. 2020).

In urban planning, these indicators serve as guiding principles for designing resilient and adaptive cities (Figueiredo et al. 2018). By considering the multifaceted interactions between urban systems, natural environments, and human communities, planners can develop holistic strategies that address the interconnected challenges of urbanization (Bi and Little 2022), as depicted in Fig. 2. Furthermore, by adopting smart and biophilic approaches, cities can foster innovation and inclusivity and enhance the quality of life for residents (Daniels et al. 2020). Integrating these indicators into urban planning processes is essential for building resilient cities capable of thriving in the face of future uncertainties (Figueiredo et al. 2018). An illustrative instance demonstrating the utilization of biotechnological products inspired by biomimicry is found in the City Tree projects (Allam 2020). These projects draw inspiration from nature and are developed by the start-up Green City Solutions (Osborne et al. 2013). Biotechnology offers exciting possibilities for enhancing urban trees, making them more resilient, productive, and adaptable to urban environments. However, ecological concerns limit the use of hybrid or genetically modified trees for urban plantations (Brennan et al. 2023).

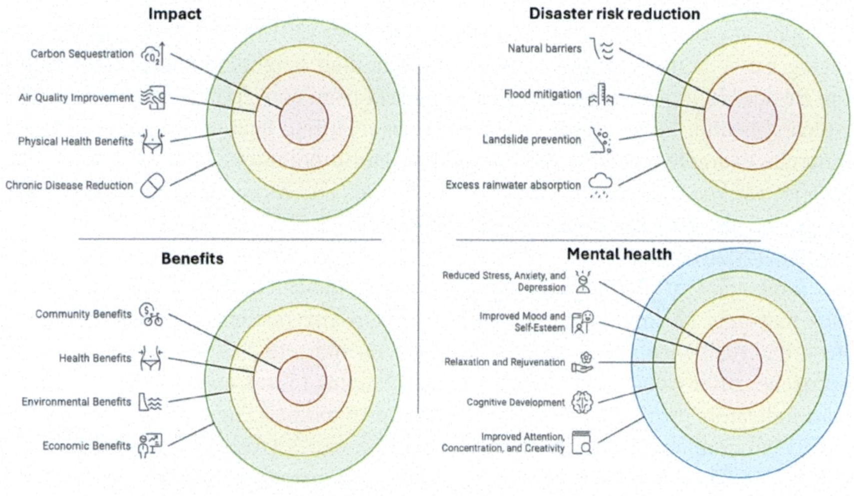

Fig. 2 Urban spaces and their multifaceted role

Adaptation and mitigation strategies should go hand in hand because mitigation is essential for sustaining life under these altering conditions, and without mitigation, adaptation will not be feasible. This chapter discusses some mitigation and adaptation strategies to combat climate change and environmental pollution utilizing advanced biotechnological interventions in UF.

3 Utilizing Biotechnological Approaches for Climate Change

3.1 Genetic Modification of Trees for Resilience

Genetic engineering (GE) is another powerful tool that manipulatesan organism's genome by adding or deleting the required traits. This approach is widely used to modify genes for better adaptation to salinity, heat, drought, and biotic stress management. However, there are ethical, environmental, and social concerns over using GE in forestry. Species that proliferate, sequester more carbon, and resist pests and disease can be developed using GE. Confined field trials of many species like Poplar, Eucalyptus, and Pinus have been established with insect-resistant, fast-growing, and drought-resistant traits in countries like the US, Canada, China, and Brazil (Haggman et al. 2013). Eucalyptus is an important raw material for the paper and pulp industries. It is a fast-growing tree with more pulping efficiency (20%). Brazil-based company FuturaGene developed the GM eucalyptus plant in May 2015 and is currently under confined trials in Brazil (Germantown and Paulo 2015).

Similarly, Canada approved a GM apple tree with a non-browning fruit trait that does not turn brown after cutting for several hours (Baker and Horin 2018). Rubber plant, also used as avenue plantation in urban spaces, is GM and planted in June 2021 in the Assam state of India. This GM tree will have more productivity and be suitable for the cold climatic conditions of the geographical region (Parashar 2021). In February 2023, US-based company Living Carbon planted the first genetically modified Poplar plant in the forest at the commercial level. This Poplar tree is a hybrid of *Populus tremula* and *P. alba* and sequesters more carbon than the ordinary Poplar. All the trees commercially planted are female, so there is no concern about pollen escape (Popkin 2023). These are some of the few examples of GM trees produced using a biotechnological approach for combating climate change to secure food security and create resilient urban trees. However, there are concerns about the lack of genetic diversity in the GM plants and trees that grow rapidly and generally possess low wood density and strength. This area of research is to shorten the life span of trees, which sequester more carbon and grow quickly, as in the battle against climate change, these technologies can be a game changer.

3.2 DNA as a Storage Tool

DNA is emerging as a groundbreaking medium for data storage, offering a solution to the limitations of conventional storage devices like hard drives and SSDs (Song 2021). Its immense storage capacity, high density, resilience under harsh conditions, and low power consumption position DNA as a superior alternative for long-term data archiving (Cevallos et al. 2022). Pioneering projects like Microvenus and Genesis have laid the groundwork for DNA data storage, demonstrating the feasibility of encoding information into the fabric of life itself (Demidov 2020). Although challenges such as the cost of DNA synthesis, slower retrieval speeds, and current limitations in rewriting and access remain, ongoing research addresses these issues (Doricchi et al. 2022). Moreover, DNA's potential extends beyond mere storage; it holds promise for secret communication, cryptography, and even computing applications, making it a versatile and future-proof technology (Doricchi et al. 2022).

3.3 RNA Interference (RNAi)

Due to increasing food demand and changes in climatic patterns like altered precipitation, rising temperatures impacted the distribution and behavior of pathogens and pests (Fladung et al. 2021). Therefore, trees must be more climate-resilient to resist biotic and abiotic stress and ensure ecosystem sustainability and long-term health. Genetic engineering approaches like double-stranded RNA mediated interference (RNAi) have proved to be a significant tool in managing forest tree disease (Cagliari et al. 2019). RNAi is a cellular process that degrades the target gene by suppressing its transcription (Svoboda 2020). RNAi has been used in many vulnerable tree species and devastated by insect pests and pathogens. Virus-resistant varieties of Plum were developed against the Plum pox virus disease of plum (Scorza et al. 1994) fungal pathogen-resistant apple variety (Bhagta et al. 2023). Resistance against the stem and bark borer of important tree species like Pine spp was developed against *Dendroctonus frontalis*, commonly known as the Southern pine beetle *D. ponderosae*, mountain pine beetle (Kyre et al. 2019). *Agrilus planipennis*, is an invasive pest of the Ash tree and has damaged a vast population of Ash trees in Europe. RNAi has been used to silence the genes crucial for the survival of this insect, and resistant Ash trees have been produced against EAB (Emerald ash borer) (Rodrigues et al. 2018). RNAi is a potential tool to combat pest pathogen infestation and increase tree yield by silencing the target genes.

3.4 Phytoremediation and Bioremediation

Phytoremediation is a method to eradicate and degrade pollutants using green plants from the environment (McGrath and Zhao 2003). Post-COVID-19 pandemic, the focus shifted to Indoor air quality (IAQ), which was often neglected. Indoor plants like *Ficus benjamina, Dracaenatrifasciata,* Areca palm, Indian rubber plant, *Aglaonema,* are some plants that effectively reduce the pollutants from the air (Ravindra and Mor 2022). Urban spaces are the major sources of greenhouse gas emissions, and some of the hazardous contaminants in the air include PM10, CO_2, NO_2, O_3. Pinus sppis efficient in removing PM10, and Aesculus hippocastanum is highly efficient in CO_2 removal in urban spaces (Araminiene et al. 2023; Kumar et al. 2021b; c). *Pteris vittate,* a hyperaccumulator fern, is known for its absorbing capacity of arsenic (Todeschini et al. 2007), which is also aided by arbuscular mycorrhizal fungi (AMF) by increasing Arsenic translocator factor in the fern (Trotta et al. 2006). *Moringa oleifera* is known as a superfood, and its bark and leaf extracts biosorption of heavy metals in the polluted lake of metro city Bangaluru, India (George et al. 2016) effectively. Microbial bioremediation for pollution control is imperative to curb the pollution level in urban spaces. The microflora associated with the phyllosphere and endosphere of the trees help not only in the removal of toxic pollutants but also promote the growth and development of trees. *Suillusbovinus,* is Zinc tolerant fungi and bacteria like *Pseudomonas sp. Klebsiella* sp. and *Enterobacter* sp. are highly efficient in alleviating zinc toxicity from the soil (Hussain et al. 2022; Jayan and Bhatlu 2021).

3.5 Biodiversity Assessment and Conservation

Urban green spaces (UGSs) harbormore biodiversity. Recent biotechnology approaches like DNA barcoding and genome sequencing are utilized to assess the native micro biodiversity in and around urban ecosystems. Environmental DNA (eDNA) metabarcoding is an emerging, non-invasive, and efficient technology where biodiversity is evaluated by collecting the sample and analyzing the genetic material of different species in the soil and water samples (Bae et al. 2023; Kumar et al. 2021d). eDNA metabarcoding and camera traps revealed the presence of 43 species in the water bodies of the UGS and urban ecological corridors (Kim et al. 2022). Taking many low-volume samples from the water bodies of UGS helped determine the water body's community structure by performing eDNA barcoding (Yang et al. 2021). A study conducted in South Africa in an urban ecosystem by Webster et al. (2020) documented some rare and endangered species in the water management area of the urban ecosystems, implying UGS are also playing a crucial part in biodiversity conservation. This advanced technology, aided by visual surveys and camera traps, gave an authentic picture of the diversity of urban ecosystems.

3.6 *Precisiontree Farming for Urban Spaces*

Precision agriculture, tree farming, and vertical gardens are some frontier technologies that are product-oriented (Sishodia et al. 2020). As land is a limited source for any country, productivity must be increased for available land resources. It involves Global Positioning System (GPS), Geographic information systems (GIS), remote sensing (RS), sensors, and data analytics to get a clear picture of the targeted site to deliver water, fertilizers, and other nutrients in an efficient, productive, and sustainable way (Vuran et al. 2018). Precision farming is changing the paradigm of farming in Indian rural and urban spaces, and it utilizes less resources and gives a targeted result efficiently (Shafi et al. 2019). Biotechnology is transforming precision farming by enhancing productivity regarding food supply and carbon sequestration by trees. Gene editing technologies like Clustered Regularly Interspaced Short Palindromic Repeat (CRISPR)- Cas-9 technology help edit the genes within an organism with no foreign DNA involvement (Deltcheva et al. 2011). This is known for its precision editing of genes. The newest buzz in science is about luminescent plants, these plants shine at night and can be distinguished from weeds. This makes crop management easier in urban areas with limited territory for tree farming (Sanchez et al. 2018). Luminescence is observed naturally in bacteria, fungi, protozoans, and animals but never in plants (Burford and Robison 2020). It was made possible to observe luminescence in plants because of genetic engineering. The first report of luminescence in plants was generated by Hu and Cheng (1995) when green fluorescent protein (GFP) from *Aequorea Victoria* was successfully expressed in *Arabidopsis.* Since then, numerous ornamental plants that are integral plants of urban planning have been engineered to be bioluminescent, like *Chrysanthemum morifolium* (Kishi-Kaboshi et al. 2017), *Dahlia pinnata* (Ke and Tsai 2022) and *Torenia fournieri* (Sasaki et al. 2014). These glow-in-dark ornamental flowers are of aesthetic value to the urban landscapes and depict a fascinating amalgamation of technology and nature.

3.7 *Rhizosphere Engineering*

The rhizosphere involves the roots, soil, and microbial flora, and rhizosphere engineering manipulates all these components to enhance productivity, carbon sequestration, and sustainability. This region is influenced by root exudates, plant growth-promoting bacteria (PGPR), and soil health (Paredes-Mendoza and Espinosa-Victoria 2010). Some nutrients are present in nature in bound form, and it is difficult for plants to assess them in this form. When stressed, PGPRs maintain nutrient availability and phytohormone production, inducing systemic resistance in plants.

Iron is an essential micronutrient crucial in various physiological and biochemical processes of the plant system. Its availability to the plants is bound by factors like high pH, low iron solubility, and its bound form. The bound form of iron is released by forming siderophores with the help of PGPRs. These siderophores are high-affinity, chelating molecules that solubilize iron, zinc, and phosphate into the soil and fulfill the nutritional requirements of the plants (Khan et al. 2023).

Urbanization puts pressure on the microbial communities associated with the trees. Rosier et al. (2021) found that trees in rural ecosystems have distinct microbiota. In contrast, urban trees share common microbial flora, as there has been a more significant overlap of taxa in urban environments. Petroleum products have contaminated the soil of UGS, rhizosphere engineering using bacteria like *Pseudomonadata, Bacteroidetes, Actinobacteria, and Nocardioides*reclaimed land by remediating the hydrocarbons accumulated into the soil with the burning of petroleum products (Singha and Pandey 2021).

3.8 Development of Bio-based Materials for Urban Infrastructure

Smart cities can make their inhabitants living environments healthier and more enjoyable by imitating natural patterns, materials, and processes using biomimicry. It involves emulating nature's strategies and designs to solve human challenges. Urban greening can lead to innovative designs that are efficient, sustainable, and resilient, much like natural ecosystems. For example, natural ventilation systems, living walls, and green roofs are designed according to ecological systems (Gupta et al. 2019a; Tarek and Ouf 2021; Yadav et al. 2019). Biomimicry helps design energy-efficient buildings by studying how organisms regulate temperature, capture sunlight, and optimize energy use (Chayaamor-Heil and Hannachi-Belkadi 2017). One iconic example of this biomimicry is the building of the Ministry of Environment, Forest, and Climate Change, Government of India, New Delhi. It is India's highest-rated green building with net zero emissions (Khandelwal et al. 2020). Biochar promotes plant growth and development and is prepared from many feedstocks like wood, food waste, and weeds. Apart from its nutritional usage to plants, it can be utilized in forming green buildings and increasing the quality of soil (Kumar et al. 2019a, b; Novotný et al. 2023). Cooking oil, wood-based oil, and manure-based bio-bitumen are used nowadays in place of bitumen in green buildings as modifiersand binders, thus promoting sustainable infrastructure (Zhang et al. 2024).

4 Policy Recommendations for Integrating Biotech-Driven Solutions into Urban Planning Frameworks

Biotechnology and urban forestry policies encompass a range of measures to promote sustainable urban development while addressing environmental, social, and economic challenges. These policies often include regulations governing the use of biotechnology in urban forestry projects, ensuring safety, efficacy, and ethical considerations (Holzinger et al. 2023; Sharma et al. 2018; Singh et al. 2018). Additionally, policies may focus on biodiversity conservation, climate change mitigation and adaptation, and integrating green infrastructure into urban planning efforts (Gupta et al. 2018a, b; Ramyar et al. 2021). Biotechnology can play a crucial role in these policies by developing tree varieties that are more resilient to climate stressors and enhancing ecosystem services delivery. These policies strive to balance the potential benefits of biotechnological advancements with considerations for environmental protection, public health, and social equity, promoting sustainable and resilient urban environments. Different tools and methods for sustainable UGS are depicted in Fig. 3.

These policies recognize the importance of addressing disparities in the distribution of urban tree canopy (UTC) to ensure that all communities benefit from the ecosystem services provided by urban forests and to promote climate resilience (Englund et al. 2021; Kumar et al. 2017; Singh and Verma 2013). Socioeconomic and racial factors often contribute to inequities in UTC distribution, with historically marginalized communities bearing a disproportionate burden of environmental challenges such as heat islands and air pollution (Gupta and Singh 2017; Hansmann et al. 2016; Singh et al. 2017a). Equity-focused policies aim to rectify these disparities by prioritizing tree planting and maintenance efforts in underserved neighborhoods, providing resources and support for community-led greening initiatives, and incorporating principles of environmental justice into urban planning processes (Dutta et al. 2020).

Integrating social equity into urban resilience planning ensures that resilience strategies prioritize the needs of communities (Rigolon et al. 2020; Singh et al.

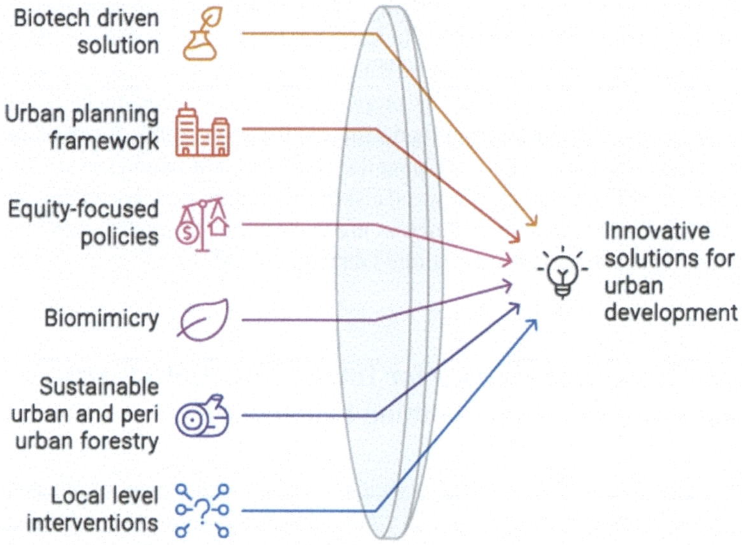

Fig. 3 Different tools and methods for urban development

2017b) Adequate funding allocation supports equitable tree planting initiatives while creating pathways to green in urban forestry (Hansmann et al. 2016). Preventing gentrification ensures greening efforts benefit existing communities (Rigolon et al. 2020). Establishing dedicated committees or task forces clarifies roles and responsibilities while shifting tree maintenance responsibility from property owners to local government ensures equitable access to green spaces (Carter et al. 2015). Collaboration across agencies and sectors aligns efforts towards common goals, and integrating tree planting with broader climate adaptation strategies enhances cross-sectoral climate resiliency planning.

4.1 Adverse Impact of Policies

Some policies can unintentionally harm urban forests, compromising their health and sustainability (Jim 2013). One example is road and sidewalk design policies that do not allocate adequate space for tree roots or canopy growth. Inadequate space restricts the ability of trees to access essential nutrients and water, leading to stunted growth, increased susceptibility to pests and diseases, and premature death. Furthermore, national or regional urban expansion and development policies can significantly impact tree cover (Daniel et al. 2016). Rapid urbanization often results in clearing green spaces and forests to make way for infrastructure and housing projects, leading to the loss of valuable tree canopy and biodiversity (Singh 2024). These policies can exacerbate issues such as heat islands, air pollution, and wildlife habitat loss, undermining the ecological and socio-economic benefits that urban forests provide (Kumar et al. 2019c; Meena et al. 2011; Sanesi et al. 2011).

4.2 Sustainable Urban and Peri-Urban Forestry

The United Nations Economic Commission for Europe (UNECE) underscores the significance of sustainable urban and peri-urban forestry as a pivotal nature-based solution for fostering healthier and climate-resilient cities (Amaratunga et al. 2023). Integrating sustainable forestry practices into urban and peri-urban landscapes may enhance urban environmental quality, mitigate the adverse effects of climate change, and promote the well-being of urban residents (Kumar et al. 2020; Singh and Kumar 2022). Sustainable forestry initiatives in urban areas involve the strategic planning, management, and preservation of trees and green spaces to optimize ecosystem services such as air purification, carbon sequestration, and temperature regulation. Moreover, peri-urban forestry initiatives extend these benefits to areas adjacent to urban centers, addressing land-use challenges and promoting biodiversity conservation. A "3–30-300 rule" highlights the significance of equity in urban forestry to ensure that all inhabitants benefit from our urban trees and forests (Policy Brief 2024). The "3–30-300 rule" stipulates that a person should be able to see three trees or more from their home, that each neighborhood should have a minimum of 30% tree canopy cover, and that the closest excellent public green space should be 300 m away.

4.3 Local-Level Interventions

Cities worldwide implementvarious strategies to enhance urban forests through city-specific plans, regulations, and institutional arrangements (Pretzhch et al. 2017; Singh et al. 2024b). These efforts aim to increase tree canopy coverage, promote biodiversity, and improve environmental quality. Urban forestry master plans provide a roadmap for managing and expanding tree cover, while zoning and land-use regulations integrate tree protection and planting requirements into development projects (Carreiro and Zipperer 2008). Tree preservation ordinances safeguard significant trees, and cities often set tree canopy goals as part of their sustainability agendas. Community engagement and education initiatives empower residents to participate in tree planting and maintenance, enabling a sense of ownership and stewardship. Interagency collaboration ensures coordinated efforts across city departments, while public–private partnerships leverage resources and expertise from diverse stakeholders. Incentive programs incentivize private investment in urban forestry, encouraging property owners to contribute to greener, more resilient urban environments.

5 Conclusion

Urban greening plays a crucial role in creating healthier, more sustainable cities. Integrating green spaces, trees, and natural elements into urban environments offers numerous benefits, including enhanced air quality by filtering pollutants, regulating temperature, and mitigating the urban heat island effect. Trees sequester carbon, contributing to climate change mitigation, and access to parks and green areas positively impacts human well-being. Biotechnology allows us to create UGS that are aesthetically pleasing, sequester carbon, improve air quality, and enhance overall sustainability. Over the past decade, significant developments in biotechnology have advanced the cultivation of genetically modified tree species, leading to improved resilience to pests and diseases, increased tolerance to environmental stresses such as salinity or drought, and faster growth rates. Many of these biotechnologically engineered trees are now grown worldwide, both in confined areas and commercially. Hence, harnessing biotechnological advancements may enhance urban forests' health, resilience, and sustainability in metropolitan regions. Techniques such as genetic engineering and tissue culture allow for the propagation of trees with desired traits, facilitating the rapid establishment of diverse and resilient urban forests. Furthermore, biotechnology interventions may restore degraded urban ecosystems by enhancing soil fertility, promoting biodiversity, and improving ecosystem services. Moreover, biotech-led micro-enterprises may create employment opportunities in urban forestry and encourage community engagement in tree planting and maintenance efforts. Integrating biotechnology into urban forestry promises to create greener, healthier, and more livable cities, benefiting current and future generations.

References

Agbali M, Trillo C, Fernando TP, Arayici Y (2017) Creating smart and healthy cities by exploring the potentials of emerging technologies and social innovation for urban efficiency: lessons from the innovative city of Boston. Int J Energy Power Eng 11(5)

Allam Z (2020) Biotechnology to render future cities as living and intelligent organisms. In: Allam Z (Ed) Biotechnology and future cities: towards sustainability, resilience and living urban organisms. Springer International Publishing, pp 1–15. https://doi.org/10.1007/978-3-030-43815-9_1

Amaratunga D, Anzellini V, Guadagno L, Hagen JS, Komac B, Krausmann E, Linsen M, Pescaroli G, Rossi JL, Samuel K (2023) United nations office for disaster risk reduction regional assessment report on disaster risk reduction 2023: Europe and Central Asia

Araminienė V, Sicard P, Černiauskas V, Coulibaly F, Varnagirytė-Kabašinskienė I (2023) Estimation of air pollution removal capacity by urban vegetation from very high-resolution satellite images in Lithuania. Urban Clim 51:101594

Bae S, Kim P, Yi CH (2023) Biodiversity and spatial distribution of ascidian using environmental DNA metabarcoding. Mar Environ Res 185:105893

Baker A, Horin L (2018) Arctic apples: a fresh new take on genetic engineering. Harvard University blog. Available on https://sitn.hms.harvard.edu/flash/2018/arctic-apples-fresh-new-take-genetic-engineering/

Bhagta S, Bhardwaj V, Kant A (2023) Exogenous dsRNA trigger RNAi in venturia inaequalis resulting in down regulation of target genes and growth reduction. Mol Biol Rep 50(10):8421–8429

Bi C, Little JC (2022) Integrated assessment across building and urban scales: a review and proposal for a more holistic, multi-scale, system-of-systems approach. Sustain Cities Soc 82:103915

Brennan AN, Ma Z, Jacobs DF (2023) Perceptions of land managers towards using hybrid and genetically modified trees. New For 54(4):605–636. https://doi.org/10.1007/s11056-021-09895-6

Burford BP, Robison BH (2020) Bioluminescent backlighting illuminates the complex visual signals of a social squid in the deep sea. Proc Natl Acad Sci 117(15):8524–8531

Cagliari D, Dias NP, Galdeano DM, Dos Santos EÁ, Smagghe G, Zotti MJ (2019) Management of pest insects and plant diseases by non-transformative RNAi. Front Plant Sci 10:483210

Carreiro MM, Zipperer WC (2008) Urban forestry and the eco-city: today and tomorrow. In: Carreiro MM, Song Y-C, Wu J (eds) Ecology, planning, and management of urban forests: international perspectives. Springer, pp 435–456. https://doi.org/10.1007/978-0-387-71425-7_27

Carter JG, Cavan G, Connelly A, Guy S, Handley J, Kazmierczak A (2015) Climate change and the city: building capacity for urban adaptation. Prog Plan 95:1–66

Cevallos Y, Nakano T, Tello-Oquendo L, Rushdi A, Inca D, Santillán I, Shirazi AZ, Samaniego N (2022) A brief review on DNA storage, compression, and digitalization. Nano Commun Netw 31:100391

Chayaamor-Heil N, Hannachi-Belkadi N (2017) Towards a platform of investigative tools for biomimicry as a new approach for energy-efficient building design. Buildings 7(1):19

Daniel C, Morrison TH, Phinn S (2016) The governance of private residential land in cities and spatial effects on tree cover. Environ Sci Policy 62:79–89

Daniels P, El Baghdadi O, Desha C, Matthews T (2020) Evaluating net community benefits of integrating nature within cities. Sustain Earth 3:1–15

Deltcheva E, Chylinski K, Sharma CM, Gonzales K, Chao Y, Pirzada ZA, Eckert MR, Vogel J, Charpentier E (2011) CRISPR RNA maturation by trans-encoded small RNA and host factor RNase III. Nature 471(7340):602–607

Demidov VV (2020) DNA beyond genes. Springer

Dhyani K, Kuniyal HB, Singh H, Sobha (2021) Growth and physiological potential of *Terminalia arjuna* under elevated CO_2 levels in open top chamber condition. J Appl Nat Sci 13(3):1121–1126. https://doi.org/10.31018/jans.v13i3.2490 (ISSN 2231-5209)

Doricchi A, Platnich CM, Gimpel A, Horn F, Earle M, Lanzavecchia G, Cortajarena AL, Liz-Marzán LM, Liu N, Heckel R (2022) Emerging approaches to DNA data storage: challenges and prospects. ACS Nano 16(11):17552–17571

Dutta A, Bouri E, Das D, Roubaud D (2020) Assessment and optimization of clean energy equity risks and commodity price volatility indexes: implications for sustainability. J Clean Prod 243:118669

Elsahragty M, Kim JL (2015) Assessment and strategies to reduce light pollution using geographic information systems. Procedia Eng 118:479–488

Englund M, Gonzalez L, Seto K, Skinner K (2021) Assessment of the urban forestry landscape for promising equity-focused policies and practices

Figueiredo L, Honiden T, Schumann A (2018) Indicators for resilient cities

Fladung M, Häggman H, Sutela S (2021) Application of RNAi technology in forest trees. RNAi for plant improvement and protection. CABI, Wallingford UK, pp 54–71

George KS, Revathi KB, Deepa N, Sheregar CP, Ashwini TS, Das S (2016) A study on the potential of moringa leaf and bark extract in bioremediation of heavy metals from water collected from various lakes in Bangalore. Procedia Environ Sci 35:869–880

Germantown MD, Paulo S (2015) Futuragene release: driving sustainable agriculture through improved qualities in eucalyptus and poplar trees. biospace. https://www.biospace.com/article/releases/-b-futuragene-b-release-driving-sustainable-agriculture-through-improved-qualities-in-eucalyptus-and-poplar-trees-/?keywords=crops

Gladkov EA, Gladkova OV (2021) New directions of biology and biotechnology in urban environmental sciences. HemijskaIndustrija 75(6):365–368

Gupta SK, Singh H (2017) Observations on sporadic flowering in *Ficus benjamina* L. and *Peltophorum africanum* L. in new forest. Indian For 143:290–291

Gupta SK, Ram J, Singh H (2018a) Comparative study of transpiration in cooling effect of tree species in the atmosphere. J Geosci Environ Protect 6:151–166. https://doi.org/10.4236/gep.2018.68011

Gupta SK, Ram J, Singh H, Kumar P (2018b) Comparative study of cooling under shades of some forestry tree species with respect to ambient temperatures. J Geosci Environ Protect 6:51–66. https://doi.org/10.4236/gep.2018.611004

Gupta SK, Ram J, Singh H (2019a) Relationship between leaf area index and atmospheric cooling of tree species. Indian J for 42(3):203–208

Häggman H, Raybould A, Borem A, Fox T, Handley L, Hertzberg M, Lu MZ, Macdonald P, Oguchi T, Pasquali G, Pearson L, McLean M (2013) Genetically engineered trees for plantation forests: key considerations for environmental risk assessment. Plant Biotechnol J 11(7):785–798

Hannan MO, Roy AK, Hasnin SM (1999) Comparative study on the constitutents of sound and disease affected sissoo (*Dalbergia sissoo*roxb) wood. Khulna University Studies, pp 139–142

Hansmann R, Whitehead I, Krajter Ostoić S, Živojinović I, Stojanovska M, Jones N, Bernasconi A, Benamar S, Lelieveld C, Barstad J (2016) Partnerships for urban forestry and green infrastructure delivering services to people and the environment: a review on what they are and aim to achieve. South-East Euro For: SEEFOR 7(1):9–19

Holzinger A, Keiblinger K, Holub P, Zatloukal K, Müller H (2023) AI for life: trends in artificial intelligence for biotechnology. New Biotechnol 74:16–24

Hu W, Cheng CL (1995) Expression of Aequorea green fluorescent protein in plant cells. FEBS Lett 369(2–3):331–334
Hussein WA, Al-Khafaji AS (2023) Integrating the biophilia concept into urban planning: a case study of Kufa City, Iraq. J Urban Dev Manag 2(3):125–134
Hussain S, Khan M, Sheikh TMM, Mumtaz MZ, Chohan TA, Shamim S, Liu Y (2022) Zinc essentiality, toxicity, and its bacterial bioremediation: a comprehensive insight. Front Microbiol 13:900740
Introduction to Policy for Urban Forests (n.d.) Cities4Forests. Retrieved May 19, 2024, from https://cities4forests.com/resource/urban-forests-for-healthier-cities-policy-planning-regulations-and-institutional-arrangements/introduction-to-policy-for-urban-forests/
Jayan N, Bhatlu MLD (2021) Isolation and studies on zinc removal using microorganism from contaminated soil. Mater Today: Proc 44:1892–1897
Jim CY (2013) Sustainable urban greening strategies for compact cities in developing and developed economies. Urban Ecosyst 16:741–761
Kanta C, Kumar A, Chauhan A, Singh H, Sharma IP (2024) The interplay between plant functional traits and climate change. In: Plant functional traits for improving productivity. Springer Nature Singapore, Singapore, pp 41–58
Ke HM, Tsai IJ (2022) Understanding and using fungal bioluminescence–recent progress and future perspectives. Curr Opin Green Sustain Chem 33:100570
Khan S, Ambika, Rani K, Sharma S, Kumar A, Singh S, Thapliyal M, Rawat P, Thakur A, Pandey S, Thapliyal A, Singh Y (2023) Rhizobacterial mediated interactions in curcuma longa for plant growth and enhanced crop productivity: a systematic review. Fro Plant Sci 14:1231676
Khandelwal R, Jain R, Gupta M (2020) Case study: India's first net-zero energy building—Indira Paryavaran Bhavan. Int J Sci Technol Res 9(11):353–357
Kim WM, Lee HJ, Song W (2022) Environmental DNA metabarcoding effectively monitors terrestrial species by using urban green spaces. Urban For Urban Green 78:127782
Kishi-Kaboshi M, Aida R, Sasaki K (2017) Generation of gene-edited chrysanthemum morifolium using multicopy transgenes as targets and markers. Plant Cell Physiol 58(2):216–226
Kumar N, Suyal DC, Sharma IP, Verma A, Singh H (2017) Elucidating stress proteins in rice (*Oryza sativa* L.) genotype under elevated temperature: a proteomic approach to understand heat stress response. 3 Biotech 7:205. https://doi.org/10.1007/s13205-017-0856-9
Kumar M, Singh MP, Singh H, Dhakate PM, Ravindranath NH (2019a) Forest working plan for the sustainable management of forest and biodiversity in India. J Sustain for 39:1–22. https://doi.org/10.1080/10549811.2019.1632212
Kumar N, Jeena N, Singh H (2019b) Elevated temperature modulates rice pollen structure: a study from the foothill of the Himalayan agro-ecosystem in India. 3 Biotech 9(5):175. https://doi.org/10.1007/s13205-019-1700-1
Kumar N, Jeena N, Kumar N, Gangola S, Singh H (2019c) Phytoremediation facilitating enzymes: an enzymatic approach for enhancing remediation process. In: Bhatt P (ed) Smart bioremediation technologies. Academic Press, Elsevier. 50 Hampshire Street, 5th Floor, Cambridge, MA 02139, United States, pp 289–306
Kumar N, Singh H, Sharma SK (2020) Antioxidants: responses and importance in plant defense system. In: Roychowdhury R, Choudhury S, Hasanuzzaman M, Srivastava S (eds) Sustainable agriculture in the era of climate change. Springer, Cham, pp 251–264. https://doi.org/10.1007/978-3-030-45669-6_11
Kumar A, Kumar P, Singh H, Kumar N (2021a) Modulation of plant functional traits under essential plant nutrients during seasonal regime in natural forests of Garhwal Himalayas. Plant Soil 465:197–212. https://doi.org/10.1007/s11104-021-05003-x
Kumar A, Kumar P, Singh H, Kumar N (2021b) Adaptation and mitigation potential of roadside trees with bio-extraction of heavy metals under vehicular emissions and their impact on physiological traits during seasonal regimes. Urban For Urban Green 58:126900. https://doi.org/10.1016/j.ufug.2020.126900

Kumar A, Kumar P, Singh H, Kumar N (2021c) Impact of plant functional traits on infiltration rate under different forests of Kempty watershed in Garhwal Himalaya. Indian J Soil Conser 49(1):38–44

Kumar A, Kumar P, Singh H, Bisht S, Kumar N (2021d) Relationship of physiological plant functional traits with soil carbon stock in the temperate forest of Garhwal Himalaya. Curr Sci 120(8):1368–1373. https://doi.org/10.18520/cs/v120/i8/1368-1373

Kumar M, Phukon SN, Paygude AC, Tyagi K, Singh H (2021e) Mapping phenological functional types (PhFT) in the Indian Eastern Himalayas using machine learning algorithm in Google Earth Engine. Comput Geosci 158:104982. https://doi.org/10.1016/j.cageo.2021.104982

Kumar A, Singh H, Kumari G, Bisht S, Malik A, Kumar A, Singh M, Raturi BS, Thakur A, Kaushal R (2022) Adaptive resilience of roadside trees to vehicular emissions via leaf enzymatic, physiological, and anatomical trait modulations. Environ Pollut 313:120191. https://doi.org/10.1016/j.envpol.2022.120191

Kumar N, Singh H, Giri K, Kumar A, Joshi A, Yadav A, Singh R, Bisht S, Kumari R, Jeena N, Khairakpam R, Mishra G (2024) Physiological and molecular insights into the allelopathic effects on agroecosystems under changing environmental conditions. Physiol Mol Biol Plants 30:417–433. https://doi.org/10.1007/s12298-024-01440-x

Kyre BR, Rodrigues TB, Rieske LK (2019) RNA interference and validation of reference genes for gene expression analyses using qPCR in southern pine beetle, dendroctonus frontalis. Sci Rep 9(1):5640

Lefosse D, van Timmeren A, Ratti C (2023) Biophilia upscaling: a systematic literature review based on a three-metric approach. Sustainability 15(22), Article 22. https://doi.org/10.3390/su152215702

McGrath SP, Zhao FJ (2003) Phytoextraction of metals and metalloids from contaminated soils. Curr Opin Biotechnol 14(3):277–282

Meena D, Singh H, Chaudhari SK (2011) Elucidating strontium response on growth dynamics and biochemical change in *Phaseolus mungo* L. Int J Agric Environ Biotechnol 4:107–113

Novotný M, Marković M, Raček J, Šipka M, Chorazy T, Tošić I, Hlavínek P (2023) The use of biochar made from biomass and biosolids as a substrate for green infrastructure: a review. Sustain Chem Pharm 32:100999

Osborne M, Kearns P, Yang J (2013) Learning cities: developing inclusive, prosperous and sustainable urban communities. Int Rev Educ 59:409–423

Pandey AK, Pandey M, Mishra A, Tiwary SM, Tripathi BD (2015) Air pollution tolerance index and anticipated performance index of some plant species for development of urban forest. Urban For Urban Green 14(4):866–871

Parashar U (2021) World's first genetically modified rubber planted in Guwahati. Hindustan Times. https://www.hindustantimes.com/india-news/worlds-first-genetically-modified-rubber-planted-in-guwahati-101624429542403.html

Paredes-Mendoza M, Espinosa-Victoria D (2010) Organic acids produced by phosphate solubilizing rhizobacteria: a critical review. Terra Latinoamericana 28(1):61–70

Popkin G (2023) For the first time, genetically modified trees have been planted in a U.S. Forest. The New York Times. Available on https://www.nytimes.com/2023/02/16/science/genetically-modified-trees-livingcarbon.html#:~:text=The%20poplars%20may%20be%20the,Monday%20hope%20to%20transform%20forestry

Prakash V, Hunney K, Singh H (2022) Effect of elevated carbon dioxide on growth and development of *Santalum album* L. seedlings inoculated with plant growth promoting microorganisms in open top chambers. Int J Environ Agric Biotechnol 7(5):149–159

Pretzsch H, Biber P, Uhl E, Dahlhausen J, Schütze G, Perkins D, Rötzer T, Caldentey J, Koike T, Con TV, Chavanne A, Lefer B (2017) Climate change accelerates growth of urban trees in metropolises worldwide. Sci Rep 7(1):15403

Quattrone G (2023) Green and blue infrastructures and nature-based solutions to reduce pollutant emissions and make transitioning urban ecosystems more climate change-adaptive. Green Build Constr Econ 80–89

Ramyar R, Ackerman A, Johnston DM (2021) Adapting cities for climate change through urban green infrastructure planning. Cities 117:103316

Ravindra K, Mor S (2022) Phytoremediation potential of indoor plants in reducing air pollutants. Fro Sustain Cities 4:1039710

Rigolon A, Keith SJ, Harris B, Mullenbach LE, Larson LR, Rushing J (2020) More than "just green enough": helping park professionals achieve equitable greening and limit environmental gentrification. J Park Recreat Adm 38(3)

Rodrigues TB, Duan JJ, Palli SR, Rieske LK (2018) Identification of highly effective target genes for RNAi-mediated control of emerald ash borer, Agrilus Planipennis. Sci Rep 8(1):5020

Rosier CL, Polson SW, D'Amico V III, Kan J, Trammell TL (2021) Urbanization pressures alter tree rhizosphere microbiomes. Sci Rep 11(1):9447

Sanchez E, Artuso E, Lombardi C, Visentin I, Lace B, Saeed W, Lolli ML, Kobauri P, Ali Z, Spyrakis F, Cubas P, Prandi C (2018) Structure–activity relationships of strigolactones via a novel, quantitative in planta bioassay. J Exp Bot 69(9):2333–2343

Sanesi G, Gallis C, Kasperidus HD (2011) Urban forests and their ecosystem services in relation to human health. In: Forests, trees and human health, pp 23–40

Sasaki K, Kato K, Mishima H, Furuichi M, Waga I, Takane KI, Yamaguchi H, Ohtsubo N (2014) Generation of fluorescent flowers exhibiting strong fluorescence by combination of fluorescent protein from marine plankton and recent genetic tools in *Torenia fournieri* Lind. Plant Biotechnol 31(4):309–318

Scorza R, Ravelonandro M, Callahan AM, Cordts JM, Fuchs M, Dunez J, Gonsalves D (1994) Transgenic plums (*Prunus domestica* L.) express the plum pox virus coat protein gene. Plant Cell Rep 14:18–22

Shafi U, Mumtaz R, García-Nieto J, Hassan SA, Zaidi SAR, Iqbal N (2019) Precision agriculture techniques and practices: from considerations to applications. Sensors 19(17):3796

Sharma R, Singh H (2021) Alteration in biochemical constituents and nutrients partitioning of asparagus racemosus in response to elevated atmospheric CO_2 concentration. Environ Sci Pollut Res 29:6812–6821. https://doi.org/10.1007/s11356-021-16050-3

Sharma R, Singh H, Kaushik M, Nautiyal R, Singh O (2018) Adaptive physiological response, carbon partitioning, and biomass production of *Withania somnifera* (L.) Dunal grown under elevated CO_2 regimes. 3 Biotech 8:267. https://doi.org/10.1007/s13205-018-1292-1

Singh H (2021) An integrated approach considering physiological and biophysical based indicators for assessing tolerance of roadside plantations of *Alstonia scholaris* towards urban roadside air pollution: an assessment of adaptation of plantations for mitigating urban roadside air pollution. Trees. https://doi.org/10.1007/s00468-021-02179-8

Singh H (2024) Exploring adaptive modulation in plant functional traits and its impact on the productivity of Acacia auriculiformis under CO_2-enriched environment. Ind Crops Prod 210:118186. https://doi.org/10.1016/j.indcrop.2024.118186

Singh H, Verma A (2013) Characterization and screening of high nitrogen efficient rice genotype to curtail environmental pollution. Int J Agric Environ Biotechnol 6(1):23–30

Singh H, Kumar M (2022) Climate change and its impact on indian himalayan forests: current status and research needs. In: Rani S, Kumar R (eds) Climate change. Springer Climate. Springer, Cham, pp 223–242. https://doi.org/10.1007/978-3-030-92782-0_11

Singh H, Savita Sharma R, Sinha S, Kumar M, Kumar P, Verma A and Sharma SK (2017a) Physiological functioning of *Lagerstroemia speciosa* L. under heavy roadside traffic: an approach to screen potential species for abatement of urban air pollution. 3 Biotech 7(61):1–10. https://doi.org/10.1007/s13205-017-0690-0

Singh H, Verma A, Kumar M, Sharma R, Gupta R, Kaur M, Negi M, Sharma SK (2017b) Phytoremediation: a green technology to clean up the sites with low and moderate level of heavy metals. Austin Biochem 2:1012

Singh H, Sharma R, Savita SMP, Kumar M, Verma A, Ansari MW, Sharma SK (2018) Adaptive physiological response of *Parthenium hysterophorus* to elevated atmospheric CO_2 concentration. Indian For 144:1–14

Singh H, Yadav M, Kumar N, Kumar A, Kumar M (2020) Assessing adaptation and mitigation potential of roadside trees under the influence of vehicular emissions: a case study of Grevillea robusta and Mangifera indica planted in an urban city of India. PLoS ONE 15(1):e0227380. https://doi.org/10.1371/journal.pone.0227380

Singh S, Singh H, Sharma SK, Nautiyal R (2021) Seasonal variation in biochemical responses of bamboo clones in the sub-tropical climate of Indian Himalayan foothills. Heliyon 7(4):e06859. https://doi.org/10.1016/j.heliyon.2021.e06859

Singh H, Bandyopadhyaya S, Kumar A, Singh M, Malik A, Kumari G, Raturi A, Yadav SK, Kumar M, Barthwal S, Thakur A, Rawat PS (2022) Understanding the physiological and biophysical response of urban roadside plantations for assessing adaptation and mitigation mechanisms toward vehicular emissions. Urban Clim 44:101183. https://doi.org/10.1016/j.uclim.2022.101183

Singh M, Singh H, Kumar A, Kumar M, Barthwal S, Thakur A (2024a) Soil nitrogen availability determines the CO_2 fertilization effect on tree species (*Neolamarckia cadamba*): growth and physiological evidence. Environ Sustain 7:53–60. https://doi.org/10.1007/s42398-023-00300-w

Singh H, Kumar N, Kumar A (2024b) Enhancing resource use efficiency in crops through plant functional traits. In: Kumar N, Singh H (eds) Plant functional traits for improving productivity. Springer, Singapore. https://doi.org/10.1007/978-981-97-1510-7_6

Singha LP, Pandey P (2021) Rhizosphere assisted bioengineering approaches for the mitigation of petroleum hydrocarbons contamination in soil. Crit Rev Biotechnol 41(5):749–766

Sishodia RP, Ray RL, Singh SK (2020) Applications of remote sensing in precision agriculture: a review. Remote Sens 12(19):3136

Song X (2021) Novel techniques and applications in molecular computing, data storage, diagnostics, and fabrication. Duke University

Svoboda P (2020) Key mechanistic principles and considerations concerning RNA interference. Front Plant Sci 11:1237

Tarek S, Ouf ASE-D (2021) Biophilic smart cities: the role of nature and technology in enhancing urban resilience. J Eng Appl Sci 68(1):40. https://doi.org/10.1186/s44147-021-00042-8

Todeschini V, Franchin C, Castiglione S, Burlando B, Biondi S, Torrigiani P, Berta P, Lingua G (2007) Responses to copper of two registered poplar clones inoculated or not with arbuscular mycorrhizal fungi. Caryologia 60(1–2):146–155

Trotta A, Falaschi P, Cornara L, Minganti V, Fusconi A, Drava G, Berta G (2006) Arbuscular mycorrhizae increase the arsenic translocation factor in the as hyperaccumulating fern Pteris vittata L. Chemosphere 65(1):74–81

Vermesan O, Friess P (2013) Internet of things: converging technologies for smart environments and integrated ecosystems. River publishers

Vuran MC, Salam A, Wong R, Irmak S (2018) Internet of underground things in precision agriculture: architecture and technology aspects. Ad Hoc Netw 81:160–173

Wang J, Qian F, Huang S, Lv Z, Wang Y, Xing X, Chen M, Han S-T, Zhou Y (2021) Recent progress of protein-based data storage and neuromorphic devices. Adv Intell Syst 3(1):2000180

Webster HJ, Emami-Khoyi A, van Dyk JC, Teske PR, Jansen van Vuuren B (2020) Environmental DNA metabarcoding as a means of estimating species diversity in an urban aquatic ecosystem. Animals 10(11):2064

Yadav SK, Singh H, Nautiyal R, Ginwal HS, Ansari SA, Barthwal S (2019) Modulation of morpho-physiological responses in *Populus deltoides* by elevated carbon dioxide and temperature. For Sci 66:105–118. https://doi.org/10.1093/forsci/fxz048

Yang J, Zhang X, Jin X, Seymour M, Richter C, Logares R, Khim JS, Klymus K (2021) Recent advances in environmental DNA-based biodiversity assessment and conservation. Divers Distrib 27(10):1876–1879

Ye Y, Tong C, Dong B, Huang C, Bao H, Deng J (2024) Alleviate light pollution by recognizing urban night-time light control area based on computer vision techniques and remote sensing imagery. Ecol Ind 158:111591

Zhang Y, Ding P, Zhang L, Luo X, Cheng X, Zhang H (2024) Green roads ahead: a critical examination of bio-bitumen for sustainable infrastructure. Fro Mater 11:1382014

Molecular and Gene Expression in Plants Under Climate Change and Urban Environment

Sandeep B. Adavi, B. Jagadhesan, Sonali Kadam, Sagar Karande, Lopamudra Nayak, Priyanka Lal, Laxmipriya Behera, Ravinder Kumar, Rahul Kumar Tiwari, Awadhesh Kumar, and Milan Kumar Lal

Abstract The sessile plants must reprogram themselves physiologically, metabolically, and at the molecular level to overcome the impacts of changing climatic conditions. Climatic adversities such as drought, salinity, elevated $CO_{2,}$ and elevated temperatures may occur independently or more often in combination. Numerous studies indicate that plants, including crucial crops, mobilize their starch reserves to release sugars, energy, and related metabolites, aiding in stress mitigation. Maintaining ion homeostasis and mineral uptake is vital for plants under stress, with various mechanisms employed to regulate homeostasis. Urban environments compound these difficulties, further impeding plant survival and growth. The diminishing number of plant species resilient to climate and urban conditions directly results from escalating stresses. Recent studies have effectively harnessed 'omic'

S. B. Adavi
School of Crop Resistance System Research, ICAR-National Institute of Biotic Stress Management, Baronda, Raipur, Chhattisgarh 493 225, India

B. Jagadhesan · R. Kumar
ICAR-Indian Agricultural Research Institute, New Delhi, India

S. Kadam · S. Karande
School of Soil Stress Management, ICAR-National Institute of Abiotic Stress Management, Baramati, Maharashtra, India

S. Karande
Department of Plant Pathology and Agricultural Microbiology, Mahatma Phule Krishi Vidyapeeth, Rahuri, India

L. Nayak · L. Behera · A. Kumar · M. K. Lal (✉)
Crop Physiology and Biochemistry Division, ICAR-National Rice Research Institute, Cuttack 753006, India
e-mail: milan2925@gmail.com

P. Lal
Lovely Professional University, Jalandhar GT Road (NH1) Phagwara, Punjab, India

R. K. Tiwari
ICAR-Central Potato Research Institute, Shimla, Himachal Pradesh 171001, India

ICAR-Indian Institute of Sugarcane Research, Lucknow, Uttar Pradesh, India

H. Singh (ed.), *Urban Forests, Climate Change and Environmental Pollution*,
https://doi.org/10.1007/978-3-031-67837-0_28

methodologies to uncover interconnected transcriptional, proteomic, and metabolic networks associated with stress perception and response. These investigations have extended beyond the model plant Arabidopsis (*Arabidopsis thaliana*) to encompass various crop, garden, and woody species. Plants sense and respond to these climate adversities at phenological, physiological, metabolical, and molecular levels. At the transcriptomic level, changes can be observed in photosynthesis, photorespiration, light-harvesting complex, electron transport, genes linked to cellulose synthesis, carbon metabolism, glycolysis, and starch-sucrose metabolism. However, future research directions appear to be more challenging due to the unpredictable changes in our global climate. It is still a significant task to establish a complete predictive model that would enable us to easily examine the complex biological networks involved in plant responses to coupled climate change problems. A model like this would provide insightful information and help crop plants become more resilient to changing weather patterns. Ultimately, this might help us get closer to sustainable agriculture, guaranteeing the ability of food, fiber, fuel, oil, and other necessary agricultural plants to withstand changing environmental conditions.

Keywords Sessile plants · Climatic adversities · Urban environments · Omic methodologies · Sustainable agriculture

1 Introduction

Climate change is characterized by alterations in climate patterns primarily driven by the release of greenhouse gases. The emissions of these gases result in heat accumulation in the Earth's atmosphere, acting as the primary contributor to global warming. These emissions have dual origins, stemming from natural occurrences comprising earthquakes, volcanoes, wetlands, forest fires, and human activities encompassing land alterations, industrialization, deforestation, generation of power, etc. (Yue and Gao 2018). The statistical analysis carried out by Yue and Gao of global greenhouse gas emissions asserted that the Earth's natural system tends to self-balance. At the same time, extra stress is added by anthropogenic activities on earth (Yue and Gao 2018).

The primary greenhouse gases include carbon dioxide (CO_2), nitrous oxide (N_2O), methane (CH_4), and fluorinated gases viz., HFCs (hydrofluorocarbons), PFCs (perfluorocarbons), and SF6(sulfur hexafluoride) (UNFCCC 2008). It was reported that a total of 55.3 $GtCO_2$ greenhouse gases had been amounted in 2018, with CO_2 emitted from fossils for industry and energy production contributing to 37.5 $GtCO_2$ (UNEP 2019), which is a 2% increase compared to 1.5% increase in earlier decades. It was also reported that 3.5 $GtCO_2$ emission is attributed to land-use change (UNEP 2019). Collectively, 74% of greenhouse gas emissions came from land use and fossil-based emissions 2018. In 2018 the emission of methane (CH_4), nitrous oxide (N_2O), and fluorinated gases showed 1.7%, 0.8% and 6.1% increase which was 1.3, 1 and 4.6% in past decades respectively (UNEP 2019). Intergovernmental Panel on Climate Change

(IPCC) report delineated that 1.0 °C rise in global temperature is mainly caused by anthropogenic activities and is expected to reach 1.5 °C by 2052 compared to the pre-industrial era (IPCC 2018).

Although the literature has not always clearly stated how they relate, global warming and urban warming happen simultaneously. The measurement of temperature differences between urban and non-urban areas can be used to calculate the extent of urban warming. For example, over the past ten years, average surface air temperatures in several urbanized cities have shown notable increases; US cities have seen rises of 0.27 °C, Chinese cities have seen increases of 0.5 °C, and Japanese cities have seen increases of 2 °C (Huang et al. 2009; Kalnay and Cai 2003) These increases are ascribed to the synergistic impacts of urban and global warming. The warming of surface air temperatures is influenced differently by different characteristics of urbanization (Kumar et al. 2022, 2024 Singh et al. 2022, 2024a, b). Changes in the ground-level energy budget and altered air circulation patterns brought on by buildings like skyscrapers, power generators, industrial operations, and automobiles are major factors influencing heat generation in metropolitan areas (Hinkel et al. 2003; Huang et al. 2009). The concentrations of various aerosol pollutants, the thermal inertia of solid surfaces, and the lack of vegetative cover all contribute to urban warming (Velazquez-Lozada et al. 2006). Urbanization is known to impact the local climate, although its exact magnitude is yet unknown (Dhyani et al. 2021; Kumar et al. 2021a, b, c, d, e; Prakash et al. 2022). 'Urban Heat Island Effect (UHIE)' or 'Heat Island Intensity' describes the build-up of heat caused by land use patterns, like densely populated areas and larger impermeable areas, combined with ongoing energy use in commercial, industrial, and residential (Huang et al. 2009; Kataoka et al. 2009; Narumi et al. 2009). It is measured as the temperature differential between urban and rural areas (Kataoka et al. 2009).

As vegetation affects climate, hydrology, the carbon and nitrogen cycles, and other aspects of terrestrial ecosystem function, it becomes an essential component of the biosphere (Nesbitt et al. 2017). Maintaining the health and functioning of terrestrial ecosystems requires understanding the mechanisms underlying vegetation growth. Climate-related factors, including temperature, precipitation, and radiation, which provide the conditions needed for vegetation growth (de Jong et al. 2013; Pearson et al. 2013), and disturbances brought about by human activity and natural phenomena are the two main driving factors influencing vegetation growth (Brando et al. 2014; Choat et al. 2012; Zhang et al. 2014). In recent decades, there have been significant changes in vegetation growth at regional and global scales due to human activity and climate change (Jiang et al. 2017; Liu et al. 2018). Nevertheless, our knowledge of vegetation dynamics and the mechanisms behind the climate change and urbanization effect is limited as only a few studies have been conducted to look into these two factors' as combined effect on urban vegetation.

2 Reprogramming at Multiple Levels Under Abiotic Stress

2.1 *Plant Response to Drought*

Drought is a meteorological phenomenon characterized by a prolonged period of low precipitation, resulting in decreased soil moisture and increased evaporation, leading to ongoing water loss through transpiration. It is widely regarded as the most common limitation caused by climate in numerous regions across the globe (Rojas 2020; Sharma and Singh 2021; Singh 2021; Singh et al. 2021). This stress will lead to dehydration or water deficit and negatively impact various aspects of plant physiology, development, reproduction, and crop productivity (Claeys and Inzé 2013). Relative water content (RWC) is a widely used and appropriate indicator for assessing the degree of dehydration. RWC of leaves in well-irrigated plants is approximately 90%. However, the RWC ranges from 60 to 70% with mild drought stress. With moderate stress, the RWC is between 40 and 60%. In the case of severe stress, the RWC drops below 40% (Laxa et al. 2019). Isohydric species, also known as "water savers," are able to maintain a nearly constant RWC by exerting fine control over their stomatal behaviour. These plants exhibit a quick decline in stomatal conductance (gs) and restrict excessive loss of water without reducing leaf area in response to dryness. However, they also experience a reduced photosynthetic rate.

Conversely, anisohydric plants, also known as "water wasters," experience a decline in the amount of water in their leaves and a significant loss in leaf size. However, they retain their stomata open and sustain a higher photosynthetic rate (Kumar et al. 2021a, b, c, d; Singh et al. 2020; Skelton et al. 2015). ABA synthesis mainly occurs in roots and is transported to shoots by xylem vessels. ABA is a signal for soil water scarcity synthesized in the root and transported to the shoot. In *Arabidopsis thaliana*, it was shown that the CLE-25 peptide, which comes from the roots, is transported through the vasculature to the leaves. This peptide acts as a signal for water deficiency, leading to the manufacture of ABA by activating biosynthetic enzymes (Gupta et al. 2020). ABA also contributes to various subsequent processes that preserve tissue moisture (a strategy to prevent dehydration). These processes involve osmotic adjustment, which consists of accumulating organic osmotic substances (such as carbohydrates, glycine-betaine, soluble proteins and proline) in shoots and roots (Takahashi et al. 2018).

Dehydration adversely affects membrane integrity and impedes numerous physiological and biochemical processes. The membrane stability index or cell membrane damage index is often used to measure dehydration tolerance. This index indicates capacity to sustain membrane integrity under a specific dehydration level (Gupta et al. 2019a; Kumar et al. 2019a, d; Yadav et al. 2019a, b). Plants utilise dehydration tolerance systems to preserve the integrity of their cell membranes and maintain cell homeostasis. These mechanisms also allow plants to restore their physiological functions once the stress has ended (de Carvalho 2008). The synthesis of compatible solutes (mannitol, proline, raffinose, pinitol, glycine-betaine, trehalose) and protective proteins (chaperons, dehydrins and LEA proteins) for membrane and enzyme protection is regulated by both ABA independent and ABA-dependent pathways.

To sustain photosynthesis, plants use a range of defence mechanisms to safeguard their photosynthetic organs from harm while adjusting to water deprivation. Moderate and severe drought stress increases many nonstomatal variables, such as chlorophyll concentration, photosynthetic enzyme activity, and active oxygen metabolism (Gupta et al. 2018a, b; Sharma et al. 2018; Singh et al. 2018). Drought not only directly hampers chlorophyll production but also hinders the absorption of mineral elements from the soil, resulting in leaf nutritional deficit (such as leaf etiolation). The control of photosynthetic enzymes is a highly intricate process. Minor drought stress can have a negligible impact on the efficiency of photosynthetic carboxylation. However, it can significantly hinder the activity of RuBPCase, leading to a notable decrease in photosynthetic carboxylation efficiency during severe drought stress (Parry et al. 2001).

In drought-stressed plants, ROS buildup is increased due to photosynthetic disruption and increased photorespiration due to limited CO_2 uptake (You and Chan 2015). Excessive ROS generation, including superoxide radicals (O_2), hydrogen peroxide (H_2O_2), hydroxyl radical (OH^-), and singlet oxygen ($^1O_2$), damages organelles by oxidizing lipids and damaging nucleic acids and proteins (Gupta and Singh 2017; Kumar et al. 2017; Singh and Verma 2013; Singh et al. 2017a, 2017b). Plants have both enzymatic and non-enzymatic ROS-scavenging mechanisms to combat oxidative damage. Some antioxidants are enzymatic, such as peroxidases, catalases and superoxide dismutase. Non-enzymatic components such as proline, glutathione, ascorbic acid, flavonoids, carotenoids, phenolic substances and alpha-tocopherol reduce oxidative damage by reducing ROS activity and cooperating with antioxidant enzymes. Furthermore, the alternative oxidase (AOX) plays a role in preventing the excessive production of reactive oxygen species (ROS) in mitochondrial electron transport chains (Mahmood et al. 2022).

3 Coping with Elevated Temperatures

The incidence of global warming, variable weather patterns, and high temperatures has increased, posing increasingly challenging environmental stress for plants. The stress of this nature hampers the regular growth and development of plants, posing a significant risk to food safety and security (Jagadish et al. 2021). Thermal fluctuation directly impacts photosynthesis, which is the first metabolic process affected. As the temperature rises above the optimal level for a particular species, the rates of photosynthesis decline (Gupta and Singh 2017; Kumar et al. 2017; Mathur et al. 2014;Singh and Verma 2013; Singh et al. 2017a, b). The Photosystem II (PSII) exhibits a pronounced susceptibility to elevated temperatures, is composed of a complex of integral proteins, and plays a vital role in facilitating electron transport during the photochemical phase of photosynthesis. Photosystem II (PSII) damage becomes apparent within a timeframe ranging from several minutes to a few hours of heat exposure (Allakhverdiev et al. 2008). Elevated temperatures modify the characteristics of lipids, resulting in increased fluidity of membranes and subsequently interfering with membrane functions. Proteins exhibit ideal temperature ranges for their action. Elevated temperatures disrupt enzyme activity, causing metabolic pathways to become imbalanced. Ultimately, proteins undergo denaturation at high temperatures. Active oxygen species (AOS) occurrence results from membrane and protein damage (Hall and André 2001).

Cells acquire thermo tolerance spontaneously after exposure to a conditioning pretreatment, such as a brief, sub-lethal high temperature or moderate stress therapy. As in nature, a slow temperature increase before the deadly temperature might produce thermo tolerance (Kumar et al. 2019c; Kumar et al. 2020; Meena et al. 2011; Singh and Kumar 2022; Singh et al. 2024a; Vierling 1991). Synthesis of Heat Shock Proteins (HSPs) is the well-defined response of plants and other organisms to elevated temperatures. HSP transcriptions are triggered within a few seconds when the temperature rises to a crucial level of 5–10 °C higher than the ideal temperature for growth. Five distinct classes of Heat Shock Proteins (HSPs) have been identified in both plants and other organisms: The proteins Hsp60/GroE, Hsp70/DnaK, Hsp90, Hsp100/ClpB and small HSPs (sHSPs). The Hsp70 and Hsp60 proteins exhibit remarkable conservation across different species, indicating their essential function in cellular stress response. Chaperones are important for many biological activities, including folding proteins, moving them across membranes, regulating protein breakdown, preventing irreversible protein aggregation, and modifying protein activity. The latter task is essential for surviving high-temperature stress and understanding how high temperatures induce these proteins (Bourgine and Guihur 2021).

Heat induces the upregulation of several non-HSP genes in plants. Lipid peroxidation (LPO) is a significant consequence of ROS, involving the oxidation of polyunsaturated fatty acids, subsequently leading to membrane lipids' degradation (Repetto et al. 2012). The non-enzymatic components of plant antioxidants comprising glutathione, ascorbate, carotenoids and tocopherols (vitamin E), and enzymatic complex system comprising guaiacol peroxidase, glutathione reductase (GR), catalase (CAT), ascorbate peroxidase (APX), and superoxide dismutase (SOD) will help is scavenging ROS (Hajiboland 2014). Superoxide dismutase (SOD) is the initial enzyme responsible for the detoxification of ROS. It accomplishes this by turning the superoxide radical into H_2O_2. Subsequently, H_2O_2 is further transformed into water by the enzymes CAT, ascorbate peroxidase (APX), and guaiacol peroxidase (Gupta et al. 2016). In addition to having a functional heat shock element (HSE) in its 5′ promoter region, the Arabidopsis cytosolic ascorbate peroxidase gene (*APX1*) is activated by heat (Storozhenko et al. 1998).

4 Tolerance Mechanisms for Elevated CO_2

The main reactions of plants to elevated carbon dioxide levels e[CO_2] are an augmentation in the rate of photosynthesis (A) and a decrease in stomatal conductance (gs) (Ainsworth and Rogers 2007; Padhan et al. 2020a). The rise in A occurs because Rubisco is not saturated at ambient [CO_2] in C3 plants (Drake et al. 1997). The effects of e[CO_2] on plant growth, development, and morphology have been extensively studied in both C3 and C4 species (Adavi and Sathee 2019; Sathee et al. 2022a, b; Sinto et al. 2022a, b). In general, e(CO_2) enhances the efficiency of leaf photosynthesis, leading to increased plant height, thicker stems, and a more significant number of branches and leaves (Antoo et al. 2022; Padhan et al. 2020a, b). The study conducted by Ainsworth and Long (2005) found that these plants tend to be consistently larger in size. The growth rate often exhibits non-uniform patterns, with alterations in the ratio of roots to shoots. This ratio rose in 60% of the species examined, whereas it either decreased or remained stable in the other species in the remaining studies (Rogers et al. 1996).

Ainsworth and Long (2005) determined that the number of leaves increased in C3 grasses cultivated at elevated carbon dioxide levels while the leaf area index remained the same. The rate at which leaves expand may be greater during the initial leaf growth stages (Pritchard et al. 1999; Seneweera and Conroy 2005). According to Tricker et al. (2005), the density of stomata decreases as the concentration of carbon dioxide [CO_2] increases. The primary direct impacts of e[CO_2] include an augmentation in carbohydrate accessibility and a decrease in water consumption. These combined effects would promote cell growth and phenological advancement (Padhan et al. 2020b). Additionally, there would be an enhancement in light and nutrient availability to moderate the reaction. Under e[CO_2], the dry mass per leaf nitrogen (N) unit increases. For instance, in the case of rice, the crucial nitrogen concentrations were 4.0% under current atmospheric carbon dioxide levels. However, when the carbon dioxide levels were doubled, the critical concentration was reduced to 2.8% (Aben et al. 1999). Despite the decrease in concentrations (by 30% in the given example), the e[CO_2] can nevertheless enhance the overall N need of the crops, acting as a "fertilizer". Species exhibiting the greatest increase in biomass growth under elevated CO_2 typically possess the largest carbon sinks. The grain protein concentration (GPC) often decreases with elevated carbon dioxide levels e[CO_2] (Högy and Fangmeier 2008; Kimball et al. 2002; Sathee et al. 2022b; Sinto et al. 2022b).

5 Plant Tolerance to Salinity

There are three different types of processes underlying salinity tolerance. (1) Osmotic stress tolerance; (2) Na^+ exclusion from leaf blades; and (3) Tissue tolerance, or the ability of tissue to withstand accumulations of Na^+ or, in some instances, Cl^-.

The diminished photosynthetic rate amplifies the generation of ROS and enhances the enzymatic activity that cleanse these organisms (Logan 2005). The hormone ABA facilitates a transfer of salt stress signals from the roots to the shoot over long distances, for which cells in the roots must first detect both the osmotic and ionic aspects of introducing Na^+ and then promptly react to it. The reactions of root cells are essential for maintaining their proper function in response to the increased external Na^+ concentration. Additionally, these responses serve as signals to the shoot, indicating that the shoot function must also be adjusted. The increase in Na^+ concentration will lead to a rise in the concentration of free Ca^{2+} in the cytosol ($[Ca^{2+}]_{cyt}$). The introduction of Na^+ into the extracellular environment seems capable of stimulating the movement of Ca^{2+} into the cytosol through the plasma membrane and the tonoplast, which is rather intriguing. Leaf cells can handle high levels of Na^+ and Cl^- through morphological modifications and intracellular partitioning (Zhao et al. 2020).

By partitioning within cells, the cytosolic Na^+ concentrations should be as low as 10–30 mM. Na^+ inhibits most enzyme concentrations of about 100 mM *in vitro*, although some are sensitive to lower values (Flowers et al. 1991). The concentration at which Cl^- becomes harmful is unclear but likely comparable to that of Na^+. Ideally, Na^+ and Cl^- should be primarily stored in cell vacuoles. The high Na^+ contents in healthy leaves confirm this sequestering. Many tissues have values over 200 mM, however these identical amounts ultimately inhibit enzyme function in vitro. Differences in *AtNHX1* or *AtAVP1* expression levels may impact leaf vacuole Na^+ sequestration. Overexpression of *NHX* genes or *AtAVP1* in various plant species increases salt tolerance. Increasing vacuole Na^+ concentrations necessitates synchronized osmotic pressure increases in other subcellular compartments, including the cytoplasm, to maintain volume. This can be achieved by increasing K^+ and compatible solute concentrations to sub-toxic levels (Zhao et al. 2020). When Na^+ and Cl^- are stored in the vacuole, 'compatible solutes' must accumulate in the cytosol and organelles to balance the osmotic pressure of the ions in the vacuole. Sucrose, proline, and glycine betaine are the most typically accumulated chemicals; however, their molecules can also accumulate in large amounts in some species (Munns and Tester 2008).

6 Molecular Reprogramming

HSFs, a conserved transcription factors, bind to conserved cis-elements (HSEs) in gene promoters, including HSP (Nover et al. 2001). 5′-AGAAnnTTCT-3 ′ is the palindromic HSE sequence. When the plant senses a temperature rise, HSFs transition from a cytosolic monomeric state to a trimeric state in the nucleus, allowing them to bind HSEs. Recruitment of transcription components by HSF leads to gene expression within minutes of temperature increase. Possible signalling molecules for heat stress tolerance include calcium ions, abscisic acid (ABA), SA, ethylene (or ACC), and AOS. Pre-treatment with any of these compounds makes plants thermotolerant. Sphingolipids, phosphatidic acid, membrane fluidity changes, and several kinases and phosphatases are thought to have a role in heat sensing. Furthermore, it was observed that NADPH oxidase, which is found in the apoplastic fluid, generates ROS to facilitate the integration of signalling networks associated with stress response mechanisms. Reduced turgor pressure causes tension changes in plasma membranes, which are detected by osmotic stress sensors such as receptor-like kinases (RLKs), integrin-like proteins (ILPs) and histidine kinases (HKs). The drought-responsive TFs include Zip (AREB/ABF), MYB/MYC, AP2/ERF (DREB/CBF), NAC and WRKY. Water deficiency causes membrane destabilization and cytoplasmic Ca^{2+} influx (ABA independent). Calcium signals are detected and transduced by CaM, CDPK, and CBLs, interacting with the MAPK cascade to activate TFs (NAC, DREB) and express functional protein genes (LEA proteins, chaperones, dehydrins, osmolyte biosynthesis enzymes). On the other hand, JA activates the MYC2 TF, leading to stress-responsive gene expression (Mahmood et al. 2022).

Various plant species have exhibited alterations in DNA methylation as a result of drought-induced stress. These modifications were associated with the activation of genes that encode transcription factors and were implicated in mechanisms that enhance resistance to drought or were connected to susceptibility to drought (Kakoulidou et al. 2021). The study found that drought stress induced histone H3 lysine 4 tri methylation (H3K4me3) in the gene body region of NCED3, a critical enzyme in the synthesis of ABA. In addition, some studies have documented an elevation in H3K4me3 and H3 lysine 9 acetylation (H3K9Ac) in the promoter region of specific genes, including RD29A, RD29B, RD22, and RELATED TO AP2.4 (RAP2.4), which are responsible for the production of LEA proteins. The level of histone modification and the extent of gene expression are contingent upon both the length and intensity of stress. Non-coding short RNAs (miRNAs, siRNAs) are essential elements of the stress response at the molecular level that are also engaged in memory pathways. They can initiate DNA methylation and histone changes (Thiebaut et al. 2019).

7 Transcriptomic and Genomic Insights into Plant Stress Responses Under Abiotic Stress

Plants exhibit diverse transcriptomic responses to adapt to challenging environmental situations like salinity, drought, and heavy metal exposure. These abiotic stresses induce oxidative stress, nutritional imbalances, and osmotic disturbances, disrupting the photosynthesis mechanism and consequently impeding plant growth and development (Jangra et al. 2024; Kanta et al. 2024; Kumar and Singh 2024; Malik et al. 2024; Yadav et al. 2019a, b). To counteract the detrimental effects of these stresses, plants employ various strategies, activating secondary messengers that stimulate cell signalling and initiate additional transcriptomic responses, thereby fortifying their protective mechanisms. Integrating modern biological sciences branches, including transcriptomics, metabolomics, and proteomics, enhances our comprehension of the intricate stress-tolerant mechanisms plants employ (Javaid et al. 2022).

7.1 Photosynthesis, Photorespiration, Light-Harvesting Complex and Electron Transport

Understanding the interplay between abiotic stressors, photosynthetic processes, and gene regulation mechanisms provides valuable insights into improving plant resilience and productivity under challenging environmental conditions. The process of photosynthesis, crucial for both plants and certain bacteria, extends beyond the leaves, occurring in various organisms and environments (Taiz and Zeiger 2010). Chloroplasts, the sites of photosynthesis, are highly sensitive to multiple abiotic stresses such as anoxia, heavy metals, UV radiation, and fluctuations in light intensity (Demmig-Adams et al. 2018; Kohli et al. 2018; Paunov et al. 2018; Sharma et al. 2019).

Heat stress exacerbates the sensitivity of the photosynthetic system, particularly affecting Rubisco, the key enzyme in photosynthesis, and its activase proteins (Anjana Jajoo and Allakhverdiev 2017; Crafts-Brandner and Salvucci 2004; Mathur et al. 2014). Differential thermostability of Rubisco activase isoforms underlines their roles in acclimation to heat stress (Wang et al. 2010). Light and high temperatures often coincide, leading to photoinhibition of PSII, especially under elevated temperatures, affecting chlorophyll content and the oxygen-evolving complex (Janka et al. 2013; Jordan and Ogren 1984; Yamori et al. 2014). Additionally, salt stress and drought significantly impair photosynthesis, affecting yield and quality through various mechanisms (Hui et al. 2018; Tanveer et al. 2019; Tattini et al. 2014; Teixeira et al. 2016).

Heavy metals induce oxidative stress, impacting chloroplast structure and function, while extreme light stress damages the photosynthetic machinery (Elsheery and Cao 2008; Shahzad et al. 2018). Transcription factors (TFs) regulate gene expression in response to stress, affecting photosynthesis directly or indirectly (Gururani et al. 2015). WRKY and BZR1 TFs influence cell wall and chloroplast genes, while bZIP-type TFs and HY5 regulate CAB gene expression for stress tolerance (Saibo et al. 2009). TFs also modulate Rubisco expression and structure and genes related to stress response and adaptation (Ashraf and Harris 2013; Khafagy et al. 2009).

Under stress conditions, TFs regulate genes associated with CAM, glycolysis, and ABA signaling pathways, enhancing stress tolerance and maintaining photosynthetic efficiency (Saibo et al. 2009; Schaeffer et al. 1995; Zhang et al. 2008). CBF/DREB TFs orchestrate stress-responsive gene networks, which are crucial for maintaining normal photosynthesis and growth under adverse conditions (Agarwal et al. 2017). Understanding the molecular mechanisms underlying gene expression and TF regulation provides insights into enhancing photosynthetic activity and stress tolerance in plants (Nawaz et al. 2018).

Furthermore, specific genes like HRD and ABR17 contribute to stress resistance and improved photosynthetic performance when overexpressed, while mitochondrial PPR proteins play roles in post-transcriptional regulation and embryonic development (Karaba et al. 2007; Manna 2015; Srivastava et al. 2004). Enhanced salt tolerance in tobacco is linked to regulating specific genes involved in inositol metabolism, resulting in improved growth and photosynthesis under stress (Patra et al. 2010).

Mitogen-activated protein kinases (MAPKs) facilitate signal transduction from chloroplasts to the nucleus, regulating metabolic processes and gene expression in response to stress (Dietz et al. 2016; Raja et al. 2017). Genes involved in sugar regulation also influence photosynthetic capacity by modulating leaf sugar levels and activating photosynthetic components (Pego et al. 1999).

7.2 *Starch-Sucrose Conversion*

Starch, a glucose homopolymer, is stored within plastids as discrete granules and serves as the primary storage carbohydrate in numerous plant species, comprising up to 90% of the total dry weight in perennating organs (Martin and Smith 1995; Streb and Zeeman 2012). Recent studies have underscored the significance of starch pools produced in non-storage sinks and photosynthetic source tissues as pivotal components of overall plant production (Sulpice et al. 2009). When carbohydrate absorption is impeded during environmental stress, starch acts as a crucial buffer against the adverse effects of stress-induced carbon depletion, serving as a carbon and energy reservoir (Krasavina et al. 2014).

The dynamic functionality of starch optimizes carbon utilization, acting as a sugar sink during sugar surpluses or as a sugar source during carbon demand (MacNeill et al. 2017). Recent reviews emphasize the critical role of starch in regulating source-sink interactions (MacNeill et al. 2017) and enhancing abiotic stress tolerance (Lal et al. 2022; Thalmann et al. 2016). The modulation of starch metabolism can either facilitate starch accumulation or elevate cellular sugar levels in response to stress, leading to significant biological implications due to alterations in the proportions of these molecules. Unfavorable conditions often hinder photosynthesis, reducing sugar concentrations and triggering a sugar-deficiency response, potentially resulting in sugar depletion over prolonged exposure.

Conversely, restricted growth under stress conditions can lead to decreased sugar utilization, causing sugar overaccumulation, which provides osmoprotection against excessive sugar levels. Under various environmental stressors, starch accumulation has been observed, with water deprivation inhibiting starch biosynthesis in multiple plant species (Geigenberger et al. 1997). Although less documented, salinity-induced constraints on starch production have been noted, with high sodium chloride concentrations causing a reduction in rice seedlings' starch content (Chen et al. 2008). Extreme heat stress inhibits enzymes in starch metabolism, reducing starch biosynthesis and increasing sugar accumulation in cereal grains. Conversely, cold stress activates amylase isoforms, leading to starch breakdown, while drought induces enzymes that degrade starch, raising sugar levels.

The precise mechanisms governing starch metabolism regulation remain incompletely understood, although it is widely recognized that the circadian rhythm and light intensity are crucial regulatory factors (Graf and Smith 2011). Many starch-degrading enzymes exhibit transcript levels in Arabidopsis leaves that typically peak at the end of the day and decrease at night. However, inconsistent correlations have been reported between this gene expression pattern and associated protein quantities or enzyme activity levels (Smith et al. 2003).

Transcriptional modulation appears to play a role in regulating starch metabolism in response to stress, with numerous genes involved in primary glucose metabolism induced by ABA in several transcriptome studies (Yoshida et al. 2015). Stress-induced changes in starch metabolism have been linked to ABA production, a crucial event in abiotic stress tolerance (Kempa et al. 2008). For example, *P. patens* exhibited enhanced freezing tolerance due to exogenous ABA-induced starch breakdown (Nagao et al. 2005). Recent research by Thalmann et al. (2016) demonstrated that exogenous ABA in Arabidopsis stimulated the expression of *AMY3* and *BAM1*, increasing starch breakdown and BAM1 enzyme activity. Molecular genetic analyses suggest that the canonical ABA signaling pathway, involving SNF1-related kinase 2 s (SnRK2s) and ABA-responsive element (ABRE) binding factors AREB/ABFs, may partially mediate the ABA effect on starch metabolism (Thalmann et al. 2016). This mechanism appears to be conserved among eudicots, as ABREs are found in the promoters of *BAM1* ortholog genes. However, ABA is unlikely to be the sole regulatory mechanism reprogramming carbohydrate metabolism under stress, as several starch-degrading genes can be induced to transcribe by dehydration-responsive element Binding protein 1A and 2A (DREB1A and DREB2A) (Maruyama et al. 2009).

Further investigation is needed to fully elucidate the regulatory network controlling transcriptional changes in genes associated with carbohydrates during abiotic stress. Enzyme variations can be attributed to both the encoded enzymes' characteristics and gene expression patterns. For example, two closely related glucan hydrolytic enzymes, BAM1 and BAM3, exhibit molecular differences. BAM1 prefers higher temperatures and more alkaline pH values, while BAM3 is more active at lower temperatures and slightly acidic pH values (Monroe et al. 2014).

Given that daytime temperatures are higher and that photosynthesis simultaneously raises the pH in the stroma, it is conceivable that BAM1 and BAM3 underwent separate molecular evolution to fulfill their distinct roles. Furthermore, post-translational modifications such as disulfide-bridge formation, phosphorylation, or complex formation can occur in numerous enzymes involved in starch metabolism (Kötting et al. 2010). Disulfide-bridge formation reversibly inactivates BAM1 and AMY3 (Seung et al. 2013); however, the importance of this post-translational modification for plant stress tolerance remains unclear.

Although much remains to be learned about how starch metabolism integrates into stress signal transduction pathways to alter source-sink relationships, emerging evidence is gradually bridging these gaps. The T6P/SnRK1 signaling cascade appears particularly significant in linking differences in starch metabolism to appropriate stress response (Jamsheer and Laxmi 2015). This cascade, with its antagonistic effects between the metabolite T6P and the kinase SnRK1 (Nunes et al. 2013), influences changes in source-sink allocation (Griffiths et al. 2016; Paul et al. 2017; Yu et al. 2015). High T6P levels relative to SnRK1 in sink tissues activate storage pathways, including starch production, possibly in a tissue-specific manner (Wurzinger et al. 2018). Conversely, reduced T6P levels indicate "starvation," activating SnRK1 to mobilize starch and other reserves and produce sucrose for needy cells (Bledsoe et al. 2017; Griffiths and Paul 2017; Yu et al. 2015).

Ávila-Castañeda et al. (2014) reported that SnRK1 subunits bind to starch granules and interact with maltose, suggesting their potential role as cytosolic sensors for released glucose. Several studies demonstrate that T6P/SnRK1 responds to environmental stress (Lin et al. 2017; Paul et al. 2017), affects source-sink relations (Bledsoe et al. 2017; Henry et al. 2015), and induces changes in starch metabolism (Ávila-Castañeda et al. 2014; Lin et al. 2017; Loreti et al. 2018). Further research is needed to strengthen the connection between these physiological and biochemical events (Griffiths and Paul 2017; Henry et al. 2015; Paul et al. 2017; Yu et al. 2015). Nevertheless, T6P/SnRK1 activity appears to be a critical regulator of assimilate distribution between sink and source, enhancing plant survival under abiotic stress; starch metabolism, including accelerated synthesis or remobilization, is pivotal in this process (Yu et al. 2015).

7.3 *Plant Nutrition*

Plant development and survival depend heavily on sixteen chemical components. The inorganic elements contained in food, known as mineral nutrients, are crucial for the growth of plants and cannot be produced by the plant itself (Duman 2012). Salinity, acidity, water scarcity, and xenobiotics are some environmental conditions that affect the uptake, translocation, and exclusion of these mineral nutrients. For example, plants' biological characteristics are altered by salinity, which impacts water acquisition, ion homeostasis, and retention. Salt-treated soils frequently have high Na^+/Ca^{2+}, Na^+/K^+, Ca^{2+}/Mg^{2+}, and Cl^-/NO_3^- ratios, which can cause ionic imbalances and toxicities (Parida and Das 2005). On the other hand, plants cannot withstand elevated levels of salt within their cytoplasm. As a coping technique, they consequently compartmentalize ions in various tissues or store surplus salts in vacuoles (Zhu 2003). Plants under salt stress show higher concentrations of Na^+ and Cl^- and lower concentrations of Ca^{2+}, K^+, and Mg^{2+}(MA 2001). For many biological processes, such as enzyme activity, membrane potential maintenance, and cell volume management, the cytosolic K^+/Na^+ balance must be maintained. By competing with sodium ions for membrane-binding sites, calcium ions (Ca^{2+}) help mitigate salt stress's effect (Tuna et al. 2007). Calcium sulfate ($CaSO_4$) was found to raise the Ca^{2+}, N, and K^+ while decreasing the concentration of Na^+ in the leaves of tomato plants under salt stress (Tuna et al. 2007). Increasing external Ca concentrations often cause an increase in Ca uptake and a decrease in Na uptake and concentrations.

On the other hand, ion interactions, precipitation, and elevated ionic strength provide difficulties since they can lower the activity of Ca^{2+} in the soil solution and restrict its availability (Grattan and Grieve 1998). According to Siddiqui et al. (2010), salinity frequently causes reduced nitrogen (N) buildup, which results in nitrogen deficit in plants. Further studies have demonstrated that ionic strength and soil salinity determine the availability of phosphorus (P), with salinity influencing the solubility of Ca–P minerals (Violante et al. 2002).

Another environmental element that affects plant nutrition is drought, which modifies the concentration of ions in plant tissues. According to Hu et al. (2007), decreased soil moisture during drought stress restricts the diffusion of nutrients to the absorbing root surface and lowers the amount of nutrients transported from roots to shoots. By promoting membrane hydration, cell wall cohesiveness, and cell resilience to dehydration, calcium (Ca^{2+}) helps plants tolerate drought better (Ma et al. 2009). In plant cells under drought stress, potassium (K) is essential for maintaining osmoregulation, and K transport rises in the early phases of stress adaptation (Zhang et al. 2006). Drought-related soil nitrogen deficiencies are common, and there's a strong correlation between water availability and nitrogen absorption. Increased availability, absorption, and utilization of nitrogen by plants are often correlated with higher soil moisture content (Albrizio et al. 2010). However, research by Payne et al. (1995) revealed that pearl millet (*Pennisetum glaucum*) drought stress had higher levels of total nitrogen and NO_3^-. During drought stress, magnesium (Mg) levels in leaves can rise, helping to preserve water content. Chloroplasts depend on Mg^{2+}, and a lack of it can lower photosynthetic activity. A concentration impact brought on by decreased soil moisture or ion transfer from old to new leaves could be the cause of the elevated ion concentration in the root zone during dry stress (Mahouachi 2009).

The phytotoxic element cadmium (Cd) can change the content and make-up of plant nutrients and interfere with mineral feeding. Cd^{2+} modifies protein shape, interferes with membrane permeability, and competes with vital proteins. Cd's interaction with minerals depends on its concentration; at low concentrations, Cd may hyperpolarize root plasma membranes, elevating transmembrane potential (Gonçalves et al. 2009). Changes in nutrient intake can result from substituting Cd for certain nutritional elements, such as Mg, Zn, and Mn (Patra et al. 2004). Studies on *Mirabilis palaja* exhibited that Phosphorus (P) and Cd have an antagonistic interaction (Yu and Zhou 2009). Arsenic (As) may cause direct nutritional competition or change how plants metabolize their resources. Because As and phosphate share comparable chemical properties, As can function as a transportable phosphate analog. Research on the As hyperaccumulator *Pteris vittata* revealed a negative relationship between As accumulation and Ca levels, suggesting Ca can avert As poisoning (Tu and Ma 2005). Another dangerous element, mercury (Hg), has a strong affinity for sulfhydryl groups, which can cause disruptions to vital proteins and change the permeability of membranes. When some dietary elements like Mg, Zn, and Mn are replaced with mercury, the absorption of those nutrients is impacted (Patra et al. 2004; Rodriguez et al. 2009). Widely employed in many different contexts, herbicides can also affect how well plants absorb mineral nutrients. As an illustration, certain herbicides, such as 2,4-dichlorophenol (2,4-DCP), may raise the concentration of Cu and Zn that are soluble in water, while other herbicides, like glyphosate, may form complexes with divalent ions that may lower the availability of nutrients (Chen et al. 2004; Sheals et al. 2003).

8 Conclusion and Future Prospect

Anthropogenic activities and natural events have induced climate change, resulting in a notable surge in greenhouse gas emissions that fuel global warming and disrupt ecosystems. This climate shift manifests in urban areas as the Urban Heat Island Effect (UHIE), presenting distinctive challenges to local climates. Urbanization, a major driver of land-use change, exacerbates UHIE and impacts vegetation dynamics. However, our comprehension of the concurrent impacts of climate change and urbanization on vegetation is insufficient, emphasizing the necessity for further research.

This chapter delves into the molecular mechanisms governing plant stress responses, highlighting the significance of transcriptomics, genomics, and proteomics in unraveling how plants adapt to environmental challenges. The photosynthetic apparatus, crucial in plant biology, is highly susceptible to various abiotic stresses such as drought, heat, and heavy metal exposure. Understanding regulatory networks involving transcription factors and candidate genes offers insights into enhancing stress tolerance. Additionally, the discussion on starch-sucrose metabolism underscores the pivotal role of starch as a dynamic carbohydrate reserve influencing plant responses to stress conditions. The T6P/SnRK1 signaling cascade emerges as a central regulator connecting starch metabolism, assimilate distribution, and plant survival under abiotic stress. Exploring nutrient uptake and metabolism sheds light on how essential mineral nutrients are impacted by environmental stressors such as salinity, drought, and toxic elements like cadmium and arsenic. The intricate relationships between these elements underscore the importance of understanding plant nutrition amid environmental challenges.

It is imperative to bridge gaps in understanding the simultaneous effects of urbanization and climate change on vegetation dynamics. Advanced technologies like remote sensing and high-throughput omics analyses can comprehensively view these interactions across different spatial and temporal scales. Molecular insights into plant stress responses create opportunities for developing stress-tolerant crops through genetic engineering. Given the mounting challenges climate change poses, future studies should explore the potential applications of these findings in agriculture. Furthermore, unraveling the intricacies of starch metabolism and the T6P/SnRK1 signaling cascade presents opportunities for targeted interventions in crop improvement programs. Identifying and manipulating key regulators can contribute to developing crops with enhanced resilience to environmental stressors. Understanding the crosstalk between stress pathways and nutrient uptake systems will be crucial in designing sustainable agricultural practices resilient to changing environmental conditions.

References

Aben SK, Seneweera SP, Ghannoum O, Conroy JP (1999) Nitrogen requirements for maximum growth and photosynthesis of rice, *Oryza sativa* L. cv. Jarrah grown at 36 and 70 Pa CO_2. Funct Plant Biol 26:759–766

Adavi SB, Sathee L (2019) Elevated CO 2-induced production of nitric oxide differentially modulates nitrate assimilation and root growth of wheat seedlings in a nitrate dose-dependent manner. Protoplasma 256:147–159

Agarwal PK, Gupta K, Lopato S, Agarwal P (2017) Dehydration responsive element binding transcription factors and their applications for the engineering of stress tolerance. J Exp Bot 68:2135–2148

Ainsworth EA, Long SP (2005) What have we learned from 15 years of free-air CO_2 enrichment (FACE)? A meta-analytic review of the responses of photosynthesis, canopy properties and plant production to rising CO_2. New Phytol 165:351–372

Ainsworth EA, Rogers A (2007) The response of photosynthesis and stomatal conductance to rising [CO2]: mechanisms and environmental interactions. Plant Cell Environ 30:258–270

Albrizio R, Todorovic M, Matic T, Stellacci AM (2010) Comparing the interactive effects of water and nitrogen on durum wheat and barley grown in a mediterranean environment. Field Crops Res 115:179–190

Allakhverdiev SI, Kreslavski VD, Klimov VV, Los DA, Carpentier R, Mohanty P (2008) Heat stress: an overview of molecular responses in photosynthesis. Photosynth Res 98:541–550

Anjana Jajoo AJ, Allakhverdiev SI (2017) High-temperature stress in plants: consequences and strategies for protecting photosynthetic machinery. In: Plant stress physiology. CABI Wallingford UK, pp 138–154

Antoo S, Singh D, Jha SK, Adavi S, Kumar RR, Chinnusamy V, Singh MP, Sathee L (2022) Effect of optimum/high nitrogen supply and elevated CO_2 on yield and nitrogen accumulation in bread wheat

Ashraf M, Harris PJC (2013) Photosynthesis under stressful environments: an overview. Photosynthetica 51:163–190

Ávila-Castañeda A, Gutiérrez-Granados N, Ruiz-Gayosso A, Sosa-Peinado A, Martínez-Barajas E, Coello P (2014) Structural and functional basis for starch binding in the SnRK1 subunits AKINβ2 and AKINβγ. Front Plant Sci 5:199

Bledsoe SW, Henry C, Griffiths CA, Paul MJ, Feil R, Lunn JE, Stitt M, Lagrimini LM (2017) The role of Tre6P and SnRK1 in maize early kernel development and events leading to stress-induced kernel abortion. BMC Plant Biol 17:1–17

Bourgine B, Guihur A (2021) Heat shock signaling in land plants: from plasma membrane sensing to the transcription of small heat shock proteins. Front Plant Sci 12

Brando PM, Balch JK, Nepstad DC, Morton DC, Putz FE, Coe MT, Silvério D, Macedo MN, Davidson EA, Nóbrega CC et al (2014) Abrupt increases in Amazonian tree mortality due to drought–fire interactions. Proc Natl Acad Sci 111:6347–6352

Chen YX, Lin Q, He YF, Tian GM (2004) Behavior of Cu and Zn under combined pollution of 2, 4-dichlorophenol in the planted soil. Plant Soil 261:127–134

Chen H-J, Chen J-Y, Wang S-J (2008) Molecular regulation of starch accumulation in rice seedling leaves in response to salt stress. Acta Physiol Plant 30:135–142

Choat B, Jansen S, Brodribb TJ, Cochard H, Delzon S, Bhaskar R, Bucci SJ, Feild TS, Gleason SM, Hacke UG et al (2012) Global convergence in the vulnerability of forests to drought. Nature 491:752–755

Claeys H, Inzé D (2013) The agony of choice: how plants balance growth and survival under water-limiting conditions. Plant Physiol 162:1768–1779

Crafts-Brandner SJ, Salvucci ME (2004) Analyzing the impact of high temperature and CO_2 on net photosynthesis: biochemical mechanisms, models and genomics. Field Crops Res 90:75–85

de Carvalho MH (2008) Drought stress and reactive oxygen species: production, scavenging and signaling. Plant Signal Behav 3:156–165

de Jong R, Schaepman ME, Furrer R, De Bruin S, Verburg PH (2013) Spatial relationship between climatologies and changes in global vegetation activity. Glob Chang Biol 19:1953–1964

Demmig-Adams B, Stewart JJ, Baker CR, Adams WW III (2018) Optimization of photosynthetic productivity in contrasting environments by regulons controlling plant form and function. Int J Mol Sci 19:872

Dhyani K, Kuniyal HB, Singh H and Sobha (2021) Growth and physiological potential of *Terminalia arjuna* under elevated CO_2 levels in open top chamber condition. J Appl Nat Sci 13(3):1121–1126. https://doi.org/10.31018/jans.v13i3.2490 (ISSN 2231-5209)

Dietz K-J, Turkan I, Krieger-Liszkay A (2016) Redox-and reactive oxygen species-dependent signaling into and out of the photosynthesizing chloroplast. Plant Phys 171:1541–1550

Drake BG, Gonzàlez-Meler MA, Long SP (1997) More efficient plants: a consequence of rising atmospheric CO_2? Annu Rev Plant Biol 48:609–639

Duman F (2012) Uptake of mineral elements during abiotic stress. In: Abiotic stress responses in plants: metabolism, productivity and sustainability, pp 267–281

Elsheery NI, Cao K-F (2008) Gas exchange, chlorophyll fluorescence, and osmotic adjustment in two mango cultivars under drought stress. Acta Physiol Plant 30:769–777

Flowers TJ, Hajibagherp MA, Yeo AR (1991) Ion accumulation in the cell walls of rice plants growing under saline conditions: evidence for the Oertli hypothesis. Plant Cell Environ 14:319–325

Geigenberger P, Reimholz R, Geiger M, Merlo L, Canale V, Stitt M (1997) Regulation of sucrose and starch metabolism in potato tubers in response to short-term water deficit. Planta 201:502–518

Gonçalves JF, Antes FG, Maldaner J, Pereira LB, Tabaldi LA, Rauber R, Rossato LV, Bisognin DA, Dressler VL, de Moraes Flores EM et al (2009) Cadmium and mineral nutrient accumulation in potato plantlets grown under cadmium stress in two different experimental culture conditions. Plant Physiol Biochem 47:814–821

Graf A, Smith AM (2011) Starch and the clock: the dark side of plant productivity. Trends Plant Sci 16:169–175

Grattan SR, Grieve CM (1998) Salinity–mineral nutrient relations in horticultural crops. Sci Hortic 78:127–157

Griffiths CA, Paul MJ, Foyer CH (2016) Metabolite transport and associated sugar signalling systems underpinning source/sink interactions. Biochimica et Biophysica Acta (BBA)-Bioenergetics 1857:1715–1725

Griffiths CA, Paul MJ (2017) Targeting carbon for crop yield and drought resilience. J Sci Food Agric 97:4663–4671

Gupta SK, Singh H (2017) Observations on sporadic flowering in *Ficus benjamina* L. and *Peltophorum africanum* L. in new forest. Ind Forest 143:290–291

Gupta DK, Palma JM, Corpas FJ (2016) Redox state as a central regulator of plant-cell stress responses. Springer

Gupta SK, Ram J, Singh H (2018a) Comparative study of transpiration in cooling effect of tree species in the atmosphere. J Geosci Environ Protect 6:151–166. https://doi.org/10.4236/gep.2018.68011

Gupta SK, Ram J, Singh H, Kumar P (2018b) Comparative study of cooling under shades of some forestry tree species with respect to ambient temperatures. J Geosci Environ Protect 6:51–66. https://doi.org/10.4236/gep.2018.611004

Gupta SK, Ram J, Singh H (2019a) Relationship between leaf area index and atmospheric cooling of tree species. Ind J for 42(3):203–208

Gupta A, Rico-Medina A, Caño-Delgado AI (2020) The physiology of plant responses to drought. Science 368(1979):266–269
Gururani MA, Venkatesh J, Tran LSP (2015) Regulation of photosynthesis during abiotic stress-induced photoinhibition. Mol Plant 8:1304–1320
Hajiboland R (2014) Reactive oxygen species and photosynthesis. In: Oxidative damage to plants. Elsevier, pp 1–63
Hall K, André M-F (2001) New insights into rock weathering from high-frequency rock temperature data: an Antarctic study of weathering by thermal stress. Geomorphology 41:23–35
Henry C, Bledsoe SW, Griffiths CA, Kollman A, Paul MJ, Sakr S, Lagrimini LM (2015) Differential role for trehalose metabolism in salt-stressed maize. Plant Physiol 169:1072–1089
Hinkel KM, Nelson FE, Klene AE, Bell JH (2003) The urban heat island in winter at Barrow, Alaska. Int J Climatol 23:1889–1905
Högy P, Fangmeier A (2008) Effects of elevated atmospheric CO_2 on grain quality of wheat. J Cereal Sci 48:580–591
Hu Y, Burucs Z, von Tucher S, Schmidhalter U (2007) Short-term effects of drought and salinity on mineral nutrient distribution along growing leaves of maize seedlings. Environ Exp Bot 60:268–275
Huang S, Taniguchi M, Yamano M, Wang C (2009) Detecting urbanization effects on surface and subsurface thermal environment—a case study of Osaka. Sci Total Environ 407:3142–3152
Hui D, Yu C-L, Deng Q, Dzantor EK, Zhou S, Dennis S, Sauve R, Johnson TL, Fay PA, Shen W et al (2018) Effects of precipitation changes on switchgrass photosynthesis, growth, and biomass: a mesocosm experiment. PLoS ONE 13:e0192555
IPCC (2018) Summary for policymakers. Global warming of 1.5°C. IPCC, Geneva
Jagadish SV, Way DA, Sharkey TD (2021) Scaling plant responses to high temperature from cell to ecosystem. Plant Cell Environ 44
Jamsheer KM, Laxmi A (2015) Expression of Arabidopsis FCS-Like Zinc finger genes is differentially regulated by sugars, cellular energy level, and abiotic stress. Front Plant Sci 6:746
Jangra A, Kumar K, Maikhuri S, Bhandari MS, Pandey S, Singh H, Barthwal S (2024) Unveiling stress-adapted endophytic bacteria: characterizing plant growth-promoting traits and assessing cross-inoculation effects on Populus deltoides under abiotic stress. Plant Physiol Biochem 210:108610. https://doi.org/10.1016/j.plaphy.2024.108610
Janka E, Körner O, Rosenqvist E, Ottosen C-O (2013) High temperature stress monitoring and detection using chlorophyll a fluorescence and infrared thermography in chrysanthemum (Dendranthema grandiflora). Plant Physiol Biochem 67:87–94
Javaid MH, Khan AR, Salam A, Neelam A, Azhar W, Ulhassan Z, Gan Y (2022) Exploring the adaptive responses of plants to abiotic stresses using transcriptome data. Agriculture 12:211
Jiang L, Bao A, Guo H, Ndayisaba F et al (2017) Vegetation dynamics and responses to climate change and human activities in Central Asia. Sci Total Environ 599:967–980
Jordan DB, Ogren WL (1984) The CO 2/O 2 specificity of ribulose 1, 5-bisphosphate carboxylase/oxygenase. Planta 161:308–313
Kakoulidou I, Avramidou EV, Baránek M, Brunel-Muguet S, Farrona S, Johannes F, Kaiserli E, Lieberman-Lazarovich M, Martinelli F, Mladenov V et al (2021) Epigenetics for crop improvement in times of global change. Biology (basel) 10:766
Kalnay E, Cai M (2003) Impact of urbanization and land-use change on climate. Nature 423:528–531
Kanta C, Kumar A, Chauhan A, Singh H, Sharma IP (2024) The interplay between plant functional traits and climate change. In: Kumar N, Singh H (eds) Plant functional traits for improving productivity. Springer, Singapore. https://doi.org/10.1007/978-981-97-1510-7_3
Karaba A, Dixit S, Greco R, Aharoni A, Trijatmiko KR, Marsch-Martinez N, Krishnan A, Nataraja KN, Udayakumar M, Pereira A (2007) Improvement of water use efficiency in rice by expression of HARDY, an Arabidopsis drought and salt tolerance gene. Proc Natl Acad Sci 104:15270–15275
Kataoka K, Matsumoto F, Ichinose T, Taniguchi M (2009) Urban warming trends in several large Asian cities over the last 100 years. Sci Total Environ 407:3112–3119

Kempa S, Krasensky J, Dal Santo S, Kopka J, Jonak C (2008) A central role of abscisic acid in stress-regulated carbohydrate metabolism. PLoS ONE 3:e3935

Khafagy MA, Arafa AA, El-Banna MF (2009) Glycinebetaine and ascorbic acid can alleviate the harmful effects of NaCl salinity in sweet pepper. Aust J Crop Sci 3:257–267

Kimball BA, Kobayashi K, Bindi M (2002) Responses of agricultural crops to free-air CO_2 enrichment. Adv Agron 77:293–368

Kohli SK, Handa N, Sharma A, Gautam V, Arora S, Bhardwaj R, Wijaya L, Alyemeni MN, Ahmad P (2018) Interaction of 24-epibrassinolide and salicylic acid regulates pigment contents, antioxidative defense responses, and gene expression in Brassica juncea L. seedlings under Pb stress. Environ Sci Pollut Res 25:15159–15173

Kötting O, Kossmann J, Zeeman SC, Lloyd JR (2010) Regulation of starch metabolism: the age of enlightenment? Curr Opin Plant Biol 13:320–328

Krasavina MS, Burmistrova NA, Raldugina GN (2014) The role of carbohydrates in plant resistance to abiotic stresses. In: Emerging technologies and management of crop stress tolerance. Elsevier, pp 229–270

Kumar N and Singh H (2024). Plant functional traits for improving productivity. Springer Nature, Singapore. https://doi.org/10.1007/978-981-97-1510-7

Kumar N, Suyal DC, Sharma IP, Verma A, Singh H (2017) Elucidating stress proteins in rice (*Oryza sativa* L.) genotype under elevated temperature: a proteomic approach to understand heat stress response. 3 Biotech 7:205. https://doi.org/10.1007/s13205-017-0856-9

Kumar M, Singh MP, Singh H, Dhakate PM, Ravindranath NH (2019a) Forest working plan for the sustainable management of forest and biodiversity in India. J Sustain for 39:1–22. https://doi.org/10.1080/10549811.2019.1632212

Kumar N, Jeena N and Singh H (2019b) Elevated temperature modulates rice pollen structure: a study from the foothill of the Himalayan agro-ecosystem in India. 3 Biotech 9(5):175. https://doi.org/10.1007/s13205-019-1700-1

Kumar N, Jeena N, Kumar N, Gangola S, Singh H (2019c) Phytoremediation facilitating enzymes: an enzymatic approach for enhancing remediation process. In: Bhatt P (ed) Smart bioremediation technologies. Academic Press, Elsevier, p 50; Hampshire Street, 5th Floor, Cambridge, MA 02139, United States, pp 289–306

Kumar N, Singh H and Sharma SK (2020) Antioxidants: responses and importance in plant defense system. In: Roychowdhury R, Choudhury S, Hasanuzzaman M, Srivastava S (eds) Sustainable agriculture in the era of climate change. Springer, Cham, pp 251–264. https://doi.org/10.1007/978-3-030-45669-6_11

Kumar A, Kumar P, Singh H, Kumar N (2021a) Modulation of plant functional traits under essential plant nutrients during seasonal regime in natural forests of Garhwal Himalayas. Plant Soil 465:197–212. https://doi.org/10.1007/s11104-021-05003-x

Kumar A, Kumar P, Singh H, Kumar N (2021b) Adaptation and mitigation potential of roadside trees with bio-extraction of heavy metals under vehicular emissions and their impact on physiological traits during seasonal regimes. Urban Forest Urban Green 58:126900. https://doi.org/10.1016/j.ufug.2020.126900

Kumar A, Kumar P, Singh H, Kumar N (2021c) Impact of plant functional traits on infiltration rate under different forests of Kempty watershed in Garhwal Himalaya. Ind J Soil Conser 49(1):38–44

Kumar A, Kumar P, Singh H, Bisht S, Kumar N (2021d) Relationship of physiological plant functional traits with soil carbon stock in the temperate forest of Garhwal Himalaya. Curr Sci 120(8):1368–1373. https://doi.org/10.18520/cs/v120/i8/1368-1373

Kumar M, Phukon SN, Paygude AC, Tyagi K, Singh H (2021e) Mapping phenological functional types (PhFT) in the Indian Eastern Himalayas using machine learning algorithm in Google Earth Engine. Comput Geosci 158:104982. https://doi.org/10.1016/j.cageo.2021.104982

Kumar A, Singh H, Kumari G, Bisht S, Malik A, Kumar A, Singh M, Raturi BS, Thakur A, Kaushal R (2022) Adaptive resilience of roadside trees to vehicular emissions via leaf enzymatic, physiological, and anatomical trait modulations. Environ Pollut 313:120191. https://doi.org/10.1016/j.envpol.2022.120191

Kumar N, Singh H, Giri K, Kumar A, Joshi A, Yadav A, Singh R, Bisht S, Kumari R, Jeena N, Khairakpam R, Mishra G (2024) Physiological and molecular insights into the allelopathic effects on agroecosystems under changing environmental conditions. Physiol Mol Biol Plants 30:417–433. https://doi.org/10.1007/s12298-024-01440-x

Lal MK, Sharma N, Adavi SB, Sharma E, Altaf MA, Tiwari RK, Kumar R, Kumar A, Dey A, Paul V, Singh B (2022) From source to sink: mechanistic insight of photoassimilates synthesis and partitioning under high temperature and elevated ¬ [CO 2]. Plant Mol Biol. https://doi.org/10.1007/s11103-022-01274-9

Laxa M, Liebthal M, Telman W, Chibani K, Dietz K-J (2019) The role of the plant antioxidant system in drought tolerance. Antioxidants 8:94

Lin Y, Zhang J, Gao W, Chen Y, Li H, Lawlor DW, Paul MJ, Pan W (2017) Exogenous trehalose improves growth under limiting nitrogen through upregulation of nitrogen metabolism. BMC Plant Biol 17:1–13

Liu R, Xiao L, Liu Z, Dai J (2018) Quantifying the relative impacts of climate and human activities on vegetation changes at the regional scale. Ecol Indic 93:91–99

Logan BA (2005) Reactive oxygen species and photosynthesis. In: Antioxidants and reactive oxygen species in plants, pp 250–267

Loreti E, Valeri MC, Novi G, Perata P (2018) Gene regulation and survival under hypoxia requires starch availability and metabolism. Plant Physiol 176:1286–1298

MA K (2001) Experimental assessment of salinity tolerance of Ceriops tagal seedlings and saplings from the Indus delta, Pakistan. Aquat Bot 70:259–268

Ma Y, Song W, Liu Z, Zhang H, Guo X, Shao H, Ni F (2009) The dynamic changing of Ca2+ cellular localization in maize leaflets under drought stress. C R Biol 332:351–362

MacNeill GJ, Mehrpouyan S, Minow MAA, Patterson JA, Tetlow IJ, Emes MJ (2017) Starch as a source, starch as a sink: the bifunctional role of starch in carbon allocation. J Exp Bot 68:4433–4453

Mahmood T, Iqbal MS, Li H, Nazir MF, Khalid S, Sarfraz Z, Hu D, Baojun C, Geng X, Tajo SM et al (2022) Differential seedling growth and tolerance indices reflect drought tolerance in cotton. BMC Plant Biol 22:331

Mahouachi J (2009) Changes in nutrient concentrations and leaf gas exchange parameters in banana plantlets under gradual soil moisture depletion. Sci Hortic 120:460–466

Malik A, Kumar K, Kumari G, Kumar N, Singh H (2024) Application of functional traits in modelling productivity and resilience under climate change. In: Kumar N, Singh H (eds) Plant functional traits for improving productivity. Springer, Singapore. https://doi.org/10.1007/978-981-97-1510-7_5

Manna S (2015) An overview of pentatricopeptide repeat proteins and their applications. Biochimie 113:93–99

Martin C, Smith AM (1995) Starch biosynthesis. Plant Cell 7:971

Maruyama K, Takeda M, Kidokoro S, Yamada K, Sakuma Y, Urano K, Fujita M, Yoshiwara K, Matsukura S, Morishita Y et al (2009) Metabolic pathways involved in cold acclimation identified by integrated analysis of metabolites and transcripts regulated by DREB1A and DREB2A. Plant Physiol 150:1972–1980

Mathur S, Agrawal D, Jajoo A (2014) Photosynthesis: response to high temperature stress. J Photochem Photobiol B 137:116–126

Meena D, Singh H, Chaudhari SK (2011) Elucidating strontium response on growth dynamics and biochemical change in *Phaseolus mungo* L. Int J Agric Environ Biotechnol 4:107–113

Monroe JD, Storm AR, Badley EM, Lehman MD, Platt SM, Saunders LK, Schmitz JM, Torres CE (2014) b-Amylase1 and b-amylase3 are plastidic starch hydrolases in arabidopsis that seem to be adapted for different thermal, pH, and stress conditions1 [W][OPEN]

Munns R, Tester M (2008) Mechanisms of salinity tolerance. Annu Rev Plant Biol 59:651–681

Nagao M, Minami A, Arakawa K, Fujikawa S, Takezawa D (2005) Rapid degradation of starch in chloroplasts and concomitant accumulation of soluble sugars associated with ABA-induced freezing tolerance in the moss Physcomitrella patens. J Plant Physiol 162:169–180

Narumi D, Kondo A, Shimoda Y (2009) Effects of anthropogenic heat release upon the urban climate in a Japanese megacity. Environ Res 109:421–431

Nawaz MA, Jiao Y, Chen C, Shireen F, Zheng Z, Imtiaz M, Bie Z, Huang Y (2018) Melatonin pretreatment improves vanadium stress tolerance of watermelon seedlings by reducing vanadium concentration in the leaves and regulating melatonin biosynthesis and antioxidant-related gene expression. J Plant Physiol 220:115–127

Nesbitt L, Hotte N, Barron S, Cowan J, Sheppard SRJ (2017) The social and economic value of cultural ecosystem services provided by urban forests in North America: a review and suggestions for future research. Urban Urban Green 25:103–111

Nover L, Bharti K, Döring P, Mishra SK, Ganguli A, Scharf K-D (2001) Arabidopsis and the heat stress transcription factor world: how many heat stress transcription factors do we need? Cell Stress Chaperones 6:177

Nunes C, Primavesi LF, Patel MK, Martinez-Barajas E, Powers SJ, Sagar R, Fevereiro PS, Davis BG, Paul MJ (2013) Inhibition of SnRK1 by metabolites: tissue-dependent effects and cooperative inhibition by glucose 1-phosphate in combination with trehalose 6-phosphate. Plant Physiol Biochem 63:89–98

Padhan BK, Sathee L, Meena HS, Adavi SB, Jha SK, Chinnusamy V (2020b) CO_2 elevation accelerates phenology and alters carbon/nitrogen metabolism vis-à-vis ROS abundance in bread wheat. Front Plant Sci 11:1061

Padhan BK, Sathee L, Meena HS, Adavi SB (2020a) CO 2 elevation accelerates phenology and alters carbon/nitrogen metabolism vis-à-vis ROS abundance in bread wheat 11:1–18

Parida AK, Das AB (2005) Salt tolerance and salinity effects on plants: a review. Ecotoxicol Environ Saf 60:324–349

Parry M, Griffiths H, Jones HG (2001) Plant responses water stress. J Exp Bot 52:22–32

Patra M, Bhowmik N, Bandopadhyay B, Sharma A (2004) Comparison of mercury, lead and arsenic with respect to genotoxic effects on plant systems and the development of genetic tolerance. Environ Exp Bot 52:199–223

Patra B, Ray S, Richter A, Majumder AL (2010) Enhanced salt tolerance of transgenic tobacco plants by co-expression of PcINO1 and McIMT1 is accompanied by increased level of myo-inositol and methylated inositol. Protoplasma 245:143–152

Paul MJ, Oszvald M, Jesus C, Rajulu C, Griffiths CA (2017) Increasing crop yield and resilience with trehalose 6-phosphate: targeting a feast–famine mechanism in cereals for better source–sink optimization. J Exp Bot 68:4455–4462

Paunov M, Koleva L, Vassilev A, Vangronsveld J, Goltsev V (2018) Effects of different metals on photosynthesis: cadmium and zinc affect chlorophyll fluorescence in durum wheat. Int J Mol Sci 19:787

Payne WA, Hossner LR, Onken AB, Wendt CW (1995) Nitrogen and phosphorus uptake in pearl millet and its relation to nutrient and transpiration efficiency. Agron J 87:425–431

Pearson RG, Phillips SJ, Loranty MM, Beck PSA, Damoulas T, Knight SJ, Goetz SJ (2013) Shifts in Arctic vegetation and associated feedbacks under climate change. Nat Clim Chang 3:673–677

Pego JV, Weisbeek PJ, Smeekens SCM (1999) Mannose inhibits Arabidopsis germination via a hexokinase-mediated step. Plant Physiol 119:1017–1024

Prakash V, Hunney K and Singh H (2022) Effect of elevated carbon dioxide on growth and development of *Santalum album* L. seedlings inoculated with plant growth promoting microorganisms in open top chambers. Int J Environ Agric Biotechnol 7(5):149–159

Pritchard SG, Rogers HH, Prior SA, Peterson CM (1999) Elevated CO_2 and plant structure: a review. Glob Chang Biol 5:807–837

Raja V, Majeed U, Kang H, Andrabi KI, John R (2017) Abiotic stress: interplay between ROS, hormones and MAPKs. Environ Exp Bot 137:142–157

Repetto M, Semprine J, Boveris A (2012) Lipid peroxidation: chemical mechanism, biological implications and analytical determination. Lipid Peroxidation 1:3–30

Rodríguez E, Peralta-Videa JR, Israr M, Sahi S V, Pelayo H, Sánchez-Salcido B, Gardea-Torresdey JL (2009) Effect of mercury and gold on growth, nutrient uptake, and anatomical changes in Chilopsis linearis. Environ Exp Bot 65:253–262

Rogers GS, Milham PJ, Gillings M, Conroy JP (1996) Sink strength may be the key to growth and nitrogen responses in N-deficient wheat at elevated CO_2. Funct Plant Biol 23:253–264

Rojas O (2020) Agricultural extreme drought assessment at global level using the FAO-agricultural stress index system (ASIS). Weather Clim Extrem 27:100184

Saibo NJM, Lourenço T, Oliveira MM (2009) Transcription factors and regulation of photosynthetic and related metabolism under environmental stresses. Ann Bot 103:609–623

Sathee L, Adavi SB, Padhan BK, Sinto A, Anand A, Sailo N, Jain V, Jha SK (2022a) Chapter 3. In: Elevated atmospheric CO_2 induced changes in nitrogen metabolism and crop quality. Academic Press

Sathee L, Adavi SB, Padhan BK, Sinto A, Anand A, Sailo N, Jain V, Jha SK (2022b) Chapter 3. Elevated atmospheric CO_2 induced changes in nitrogen metabolism and crop quality. Academic Press

Schaeffer HJ, Forsthoefel NR, Cushman JC (1995) Identification of enhancer and silencer regions involved in salt-responsive expression of Crassulacean acid metabolism (CAM) genes in the facultative halophyte Mesembryanthemum crystallinum. Plant Mol Biol 28:205–218

Seneweera SP, Conroy JP (2005) Enhanced leaf elongation rates of wheat at elevated CO_2: is it related to carbon and nitrogen dynamics within the growing leaf blade? Environ Exp Bot 54:174–181

Seung D, Thalmann M, Sparla F, Abou Hachem M, Lee SK, Issakidis-Bourguet E, Svensson B, Zeeman SC, Santelia D (2013) Arabidopsis thaliana AMY3 is a unique redox-regulated chloroplastic α-amylase. J Biol Chem 288:33620–33633

Shahzad B, Tanveer M, Che Z, Rehman A, Cheema SA, Sharma A, Song H, ur Rehman S, Zhaorong D (2018) Role of 24-epibrassinolide (EBL) in mediating heavy metal and pesticide induced oxidative stress in plants: a review. Ecotoxicol Environ Saf 147:935–944

Sharma R, Singh H (2021) Alteration in biochemical constituents and nutrients partitioning of *Asparagus racemosus* in response to elevated atmospheric CO_2 concentration. Environ Sci Pollut Res 29:6812–6821. https://doi.org/10.1007/s11356-021-16050-3

Sharma R, Singh H, Kaushik M, Nautiyal R, Singh O (2018) Adaptive physiological response, carbon partitioning, and biomass production of *Withania somnifera* (L.) Dunal grown under elevated CO_2 regimes. 3 Biotech 8:267. https://doi.org/10.1007/s13205-018-1292-1

Sharma A, Shahzad B, Kumar V, Kohli SK, Sidhu GPS, Bali AS, Handa N, Kapoor D, Bhardwaj R, Zheng B (2019) Phytohormones regulate accumulation of osmolytes under abiotic stress. Biomolecules 9:285

Sheals J, Granström M, Sjöberg S, Persson P (2003) Coadsorption of Cu (II) and glyphosate at the water–goethite (α-FeOOH) interface: molecular structures from FTIR and EXAFS measurements. J Colloid Interface Sci 262:38–47

Singh H and Kumar M (2022) Climate change and its impact on indian himalayan forests: current status and research needs. In: Rani S, Kumar R (eds) Climate change. Springer Climate. Springer, Cham, pp 223–242. https://doi.org/10.1007/978-3-030-92782-0_11

Siddiqui MH, Mohammad F, Khan MN, Al-Whaibi MH, Bahkali AHA (2010) Nitrogen in relation to photosynthetic capacity and accumulation of osmoprotectant and nutrients in Brassica genotypes grown under salt stress. Agric Sci China 9:671–680

Singh H (2021) An integrated approach considering physiological and biophysical based indicators for assessing tolerance of roadside plantations of *Alstonia scholaris* towards urban roadside air pollution: an assessment of adaptation of plantations for mitigating urban roadside air pollution. Trees. https://doi.org/10.1007/s00468-021-02179-8

Singh H, Verma A (2013) Characterization and screening of high nitrogen efficient rice genotype to curtail environmental pollution. Int J Agric Environ Biotech 6(1):23–30

Singh H, Savita Sharma R, Sinha S, Kumar M, Kumar P, Verma A, Sharma SK (2017a) Physiological functioning of *Lagerstroemia speciosa* L. under heavy roadside traffic: an approach to screen potential species for abatement of urban air pollution. 3 Biotech 7(61):1–10. https://doi.org/10.1007/s13205-017-0690-0

Singh H, Verma A, Kumar M, Sharma R, Gupta R, Kaur M, Negi M, Sharma SK (2017b) Phytoremediation: a green technology to clean up the sites with low and moderate level of heavy metals. Austin Biochem 2:1012

Singh H, Sharma R, Savita SMP, Kumar M, Verma A, Ansari MW, Sharma SK (2018) Adaptive physiological response of *Parthenium hysterophorus* to elevated atmospheric CO_2 concentration. Ind Forest 144:1–14

Singh H, Yadav M, Kumar N, Kumar A, Kumar M (2020) Assessing adaptation and mitigation potential of roadside trees under the influence of vehicular emissions: a case study of Grevillea robusta and Mangifera indica planted in an urban city of India. PLoS ONE 15(1):e0227380. https://doi.org/10.1371/journal.pone.0227380

Singh S, Singh H, Sharma SK, Nautiyal R (2021) Seasonal variation in biochemical responses of bamboo clones in the sub-tropical climate of Indian Himalayan foothills. Heliyon 7(4):e06859. https://doi.org/10.1016/j.heliyon.2021.e06859

Singh H, Bandyopadhyaya S, Kumar A, Singh M, Malik A, Kumari G, Raturi A, Yadav SK, Kumar M, Barthwal S, Thakur A, Rawat PS (2022) Understanding the physiological and biophysical response of urban roadside plantations for assessing adaptation and mitigation mechanisms toward vehicular emissions. Urban Climate 44:101183. https://doi.org/10.1016/j.uclim.2022.101183

Singh H, Kumar N, Kumar A (2024a) Enhancing resource use efficiency in crops through plant functional traits. In: Kumar N, Singh H (eds) Plant functional traits for improving productivity. Springer, Singapore. https://doi.org/10.1007/978-981-97-1510-7_6

Singh M, Singh H, Kumar A, Kumar M, Barthwal S, Thakur A (2024b) Soil nitrogen availability determines the CO_2 fertilization effect on tree species (*Neolamarckia cadamba*): growth and physiological evidence. Environ Sustain 7:53–60. https://doi.org/10.1007/s42398-023-00300-w

Sinto A, Sathee L, Singh D, Jha SK, Adavi SB, Kumar RR, Chinnusamy V (2022a) Elevated CO_2 and Nitrogen dose affect grain ionome, grain morphology and associated gene expression in wheat (Triticum aestivum L.) 82:143–152

Sinto A, Sathee L, Singh D, Jha SK, Adavi SB, Kumar RR, Chinnusamy V (2022b) Elevated CO_2 and nitrogen dose affect grain ionome, grain morphology and associated gene expression in wheat (Triticum aestivum L.) 82:143–152

Skelton RP, West AG, Dawson TE (2015) Predicting plant vulnerability to drought in biodiverse regions using functional traits. Proc Natl Acad Sci 112:5744–5749

Smith AM, Zeeman SC, Thorneycroft D, Smith SM (2003) Starch mobilization in leaves. J Exp Bot 54:577–583

Srivastava S, Fristensky B, Kav NNV (2004) Constitutive expression of a PR10 protein enhances the germination of *Brassica napus* under saline conditions. Plant Cell Physiol 45:1320–1324

Storozhenko S, De Pauw P, Van Montagu M, Inzé D, Kushnir S (1998) The heat-shock element is a functional component of the Arabidopsis APX1 gene promoter. Plant Physiol 118:1005–1014

Streb S, Zeeman SC (2012) Starch metabolism in Arabidopsis. In: The arabidopsis book/American society of plant biologists, p 10

Sulpice R, Pyl E-T, Ishihara H, Trenkamp S, Steinfath M, Witucka-Wall H, Gibon Y, Usadel B, Poree F, Piques MC et al (2009) Starch as a major integrator in the regulation of plant growth. Proc Natl Acad Sci 106:10348–10353

Taiz L, Zeiger E (2010) Plant physiology, 5th ed. Sinauer Associates, Sunderland, MA, p 464

Takahashi F, Kuromori T, Sato H, Shinozaki K (2018) Regulatory gene networks in drought stress responses and resistance in plants. In: Survival strategies in extreme cold and desiccation: adaptation mechanisms and their applications, pp 189–214

Tanveer M, Shahzad B, Sharma A, Khan EA (2019) 24-Epibrassinolide application in plants: an implication for improving drought stress tolerance in plants. Plant Physiol Biochem 135:295–303

Tattini M, Velikova V, Vickers C, Brunetti C, Di Ferdinando M, Trivellini A, Fineschi S, Agati G, Ferrini F, Loreto F (2014) Isoprene production in transgenic tobacco alters isoprenoid, non-structural carbohydrate and phenylpropanoid metabolism, and protects photosynthesis from drought stress. Plant Cell Environ 37:1950–1964

Teixeira RN, Ligterink W, França-Neto J de B, Hilhorst HWM, da Silva EAA (2016) Gene expression profiling of the green seed problem in soybean. BMC Plant Biol 16:1–15

Thalmann M, Pazmino D, Seung D, Horrer D, Nigro A, Meier T, Kölling K, Pfeifhofer HW, Zeeman SC, Santelia D (2016) Regulation of leaf starch degradation by abscisic acid is important for osmotic stress tolerance in plants. Plant Cell 28:1860–1878

Thiebaut F, Hemerly AS, Ferreira PCG (2019) A role for epigenetic regulation in the adaptation and stress responses of non-model plants. Front Plant Sci 10:246

Tricker PJ, Trewin H, Kull O, Clarkson GJJ, Eensalu E, Tallis MJ, Colella A, Doncaster CP, Sabatti M, Taylor G (2005) Stomatal conductance and not stomatal density determines the long-term reduction in leaf transpiration of poplar in elevated CO_2. Oecologia 143:652–660

Tu C, Ma LQ (2005) Effects of arsenic on concentration and distribution of nutrients in the fronds of the arsenic hyperaccumulator *Pteris vittata* L. Environ Pollut 135:333–340

Tuna AL, Kaya C, Ashraf M, Altunlu H, Yokas I, Yagmur B (2007) The effects of calcium sulphate on growth, membrane stability and nutrient uptake of tomato plants grown under salt stress. Environ Exp Bot 59:173–178

UNEP (2019) Emissions gap report. UN environment program, Nairobi. https://wedocs.unep.org/bitstream/handle/20.500.11822/30797/EGR2019.pdf?sequence=1&isAllowed=y. Accessed 22 Dec 2019

UNFCCC (2008) Kyoto protocol reference manual on accounting of emissions and assigned amount. https://unfccc.int/resource/docs/publications/08_unfccc_kp_ref_manual.pdf. Accessed 22 Dec 2019

Velazquez-Lozada A, Gonzalez JE, Winter A (2006) Urban heat island effect analysis for San Juan, Puerto Rico. Atmos Environ 40:1731–1741

Vierling E (1991) The roles of heat shock proteins in plants. Annu Rev Plant Biol 42:579–620

Violante A, Pigna M, Ricciardella M, Gianfreda L (2002) Adsorption of phosphate on variable charge minerals and soils as affected by organic and inorganic ligands. In: Developments in soil science. Elsevier, pp 279–295

VM-D, Zhai P, Pörtner H-O, Roberts D, Skea J, Shukla PRP, A, Moufouma-Okia W, Péan C, Pidcock R, Connors S, Matthews JBR, Chen Y, Zhou X, Gomis MI, Lonnoy EM, T, Tignor MWT (2018) Global warming of 1.5 °C. In: An IPCC special report on the impacts of global warming of 1.5 °C above pre-industrial levels and related global greenhouse gas emission pathways, in the context of strengthening the global response to the threat of climate

Wang D, Li X-F, Zhou Z-J, Feng X-P, Yang W-J, Jiang D-A (2010) Two Rubisco activase isoforms may play different roles in photosynthetic heat acclimation in the rice plant. Physiol Plant 139:55–67

Wurzinger B, Nukarinen E, Nägele T, Weckwerth W, Teige M (2018) The SnRK1 kinase as central mediator of energy signaling between different organelles. Plant Physiol 176:1085–1094

Yadav SK, Singh H, Ginwal HS, Barthwal S (2019) Elevated CO_2 enhanced growth and physiological process of *Populus deltoides* Bartr. ex Marsh. Ind Forest 145(1):23–27

Yadav SK, Singh H, Nautiyal R, Ginwal HS, Ansari SA, Barthwal S (2019b) Modulation of morpho-physiological responses in *Populus deltoides* by elevated carbon dioxide and temperature. Forest Sci 66:105–118. https://doi.org/10.1093/forsci/fxz048

Yamori W, Hikosaka K, Way DA (2014) Temperature response of photosynthesis in C 3, C 4, and CAM plants: temperature acclimation and temperature adaptation. Photosynth Res 119:101–117

Yoshida T, Fujita Y, Maruyama K, Mogami J, Todaka D, Shinozaki K, Yamaguchi-Shinozaki K (2015) Four A rabidopsis AREB/ABF transcription factors function predominantly in gene expression downstream of SnRK2 kinases in abscisic acid signalling in response to osmotic stress. Plant Cell Environ 38:35–49

You J, Chan Z (2015) ROS regulation during abiotic stress responses in crop plants. Front Plant Sci 6:1092

Yu Z, Zhou Q (2009) Growth responses and cadmium accumulation of *Mirabilis jalapa* L. under interaction between cadmium and phosphorus. J Hazard Mater 167:38–43

Yu S-M, Lo S-F, Ho T-HD (2015) Source–sink communication: regulated by hormone, nutrient, and stress cross-signaling. Trends Plant Sci 20:844–857

Yue XL, Gao QX (2018) Contributions of natural systems and human activity to greenhouse gas emissions. Adv Clim Change Res 9:243–252

Zhang Y, Wang Z, Zhang L, Cao Y, Huang D, Tang K (2006) Molecular cloning and stress-dependent regulation of potassium channel gene in Chinese cabbage (Brassica rapa ssp. Pekinensis). J Plant Physiol 163:968–978

Zhang X, Wollenweber B, Jiang D, Liu F, Zhao J (2008) Water deficits and heat shock effects on photosynthesis of a transgenic *Arabidopsis thaliana* constitutively expressing ABP9, a bZIP transcription factor. J Exp Bot 59:839–848

Zhang J, Wang X, Wang J, Wang W (2014) Carbon and nitrogen contents in typical plants and soil profiles in Yanqi Basin of Northwest China. J Integr Agric 13:648–656

Zhao C, Zhang H, Song C, Zhu J-K, Shabala S (2020) Mechanisms of plant responses and adaptation to soil salinity. The Innovation 1

Zhu J-K (2003) Regulation of ion homeostasis under salt stress. Curr Opin Plant Biol 6:441–445

Molecular Insights into the Modulation of Gene Expression in Trees Under Abiotic Stresses

Namo Dubey, Subhash Kumar, and Swati Verma

Abstract Trees are constantly exposed to environmental stresses due to climatic variations over time across geographical places. Trees have evolved various molecular processes to govern growth and development under challenging conditions, reducing the impact on their growth and development. Over time, stress measurement tools and software advanced dramatically, increasing our understanding of how woody plants respond physiologically to various stresses. However, it is only recently that advances in biochemical and next-generation sequencing approaches enabled us to investigate the complicated molecular networks that underpin the stress physiology in woody trees. This chapter looks at recent breakthroughs in understanding the molecular mechanisms underlying how trees adapt to different abiotic stresses. We further focus on functional, hormonal cascades, epigenetic, and small RNA-mediated responses in trees against abiotic stresses.

Keywords Forest trees · Abiotic stresses · Stress tolerance · Gene expression · Gene regulation

1 Introduction

Forest trees account for 82% of the total biomass on land and are essential for maintaining more than half of the biodiversity on Earth (Neale and Kremer 2011; Roy et al. 2001). In addition to their ecological importance, trees contribute significantly

N. Dubey
Department of Biotechnology, College of Life and Applied Sciences, Yeungnam University, Gyeongsan, Gyeongsangbuk-do, Republic of Korea

S. Kumar
Department of Food Science and Technology, Yeungnam University, Gyeongsan, Gyeongsangbuk-do, Republic of Korea

S. Verma (✉)
College of Horticulture and Forestry, Thunag, Dr. Y. S. Parmar University of Horticulture and Forestry, Nauni, Solan, Himachal Pradesh 173 230, India
e-mail: swativerma@yspuniversity.ac.in

H. Singh (ed.), *Urban Forests, Climate Change and Environmental Pollution*,
https://doi.org/10.1007/978-3-031-67837-0_29

to mitigating the adverse effects of climate change. Trees supply a range of critical resources, including timber, paper, biomass, fuel, and biomaterials (Harfouche et al. 2012; Kumar et al. 2022, 2024; Singh 2024; Singh et al., 2022, 2024). Nonetheless, the ecological strain, specifically the influence of climate change on worldwide forest productivity, presents a substantial peril to the growth and survival of trees.

The IPCC (Intergovernmental Panel on Climate Change) assessments predict that field-grown trees will face more stress due to current and future global climate change. The efficiency of traditional breeding methods in improving specific tree growth characteristics is limited by factors such as large genomes, long generation times, and a lack of well-defined mutations (Dhyani et al. 2021; Kumar et al. 2021a, b, c, d, e; Neale and Kremer 2011; Prakash et al. 2022). However, progress in genomic-based methods and genetic engineering, specifically in forest-tree species, has broadened breeding opportunities in the past few decades (White et al. 2014). Understanding how forest trees acclimate to unfavorable environmental circumstances is essential for maintaining productivity and fulfilling forthcoming commercial requirements (Kumar et al. 2021a, b, c, d; Singh et al. 2020). The continuous endeavors to enhance the ability of plants to withstand abiotic stress employing molecular analysis and genetic engineering rely on a thorough comprehension of plant signaling networks and the genes that control crucial proteins (Singh 2021; Sharma and Singh 2021; Singh et al. 2021). The successful poplar genome sequencing and the imminent availability of genome sequences for other tree species are driving advancements in forest-tree genetics.

Omics investigations, such as transcriptomic, proteomic, and metabolomic analyses, provide valuable information about the regulatory networks involved in plant responses to abiotic stress. All research areas, including gene networks, epigenetic regulation of gene expression, signaling networks and systems, and biology, contribute to our comprehension of complex, polygenic features in plant genetic enhancement. Although the discussion has covered potential and challenges in creating trees that can tolerate abiotic stress, recent technologies such as next-generation Ecotilling and genomic selection (GS) offer other ways to comprehend the processes in forest trees. Although there has been advancement in the genetic engineering of poplar trees to enhance their ability to withstand stress, there is minimal focus on applying techniques similar to those of other forest trees. Additionally, there is a lack of field testing of genetically modified trees in real-world scenarios when there is a water shortage.

Further implementation of tree transgenesis is required to establish a connection between systems biology, physiology, and field performance. This chapter explores recent breakthroughs in comprehending molecular reactions to drought, salinity, heat, and cold stress in trees. It encompasses sensing, subsequent signaling cascades, and signaling outputs that respond to stress. In addition, we discuss strategies for developing trees that have improved tolerance to environmental circumstances. These strategies include studying the epigenetic responses to abiotic stress, using genomic selection techniques, and researching quantitative trait loci (QTL).

2 General Concept and Critical Characteristics of the Physiology of Plant Stress

Plant stress encompasses adverse conditions that impede plants' growth and metabolic processes (Lichtenthaler 1996). However, from a functional perspective, physiologists often define stress as a deviation from the equilibrium state induced by various factors (Kranner et al. 2010). This chapter seeks to distinguish between internal and external stressors whenever possible.

Abiotic stress denotes environmental conditions that provoke physiological changes in organisms. The stress response describes the physiological states arising from stressors. Thermal stress arises from temperature fluctuations, while drought stress results from water scarcity. The plant's physiological response to stress can vary based on the intensity and frequency of the stressors and the plant's developmental stage. It's noteworthy that different tree species exhibit varying capacities to withstand stress.

Moreover, individual plants may display significant variations in stress resistance throughout the year, particularly in response to cold stress (Harrison et al. 1978; Weiser 1970). This suggests temporal fluctuations in tree stress responses, which can be managed daily (Hüve et al. 2006) or yearly (Harrison et al. 1978). Hardening or acclimation refers to the plant's physiological adjustments as it gradually encounters a stressful environment. These adjustments protect the plant from growth inhibition and damage caused by sudden environmental changes. Notably, such responses do not involve alterations in the DNA sequence (Crisp et al. 2016).

Adaptive transgenerational plasticity refers to the persistence of hardening effects across generations (Herman and Sultan 2011). Under persistent and challenging conditions, genetic changes may become permanent across generations due to selection pressures, leading to population adaptability (Mitchell-Olds et al. 2007). The plant's overall tolerance to adverse abiotic conditions can be influenced by adaptive transgenerational plasticity, acclimation, and genetic adaptation, all of which may coincide (Fig. 1).

2.1 Drought Stress

Drought stress contributes to tree extinction (McDowell et al. 2018; Williams et al. 2012). Trees employ diverse strategies to cope with water scarcity, involving physiological, biochemical, and morphological adjustments categorized as drought avoidance and drought tolerance. Drought avoidance primarily entails regulating water status to minimize water loss and maintain continuous water absorption. This includes both constitutive mechanisms, such as constructing barriers to reduce water evaporation (Aharoni et al. 2004), and inducible responses, like reducing leaf area through

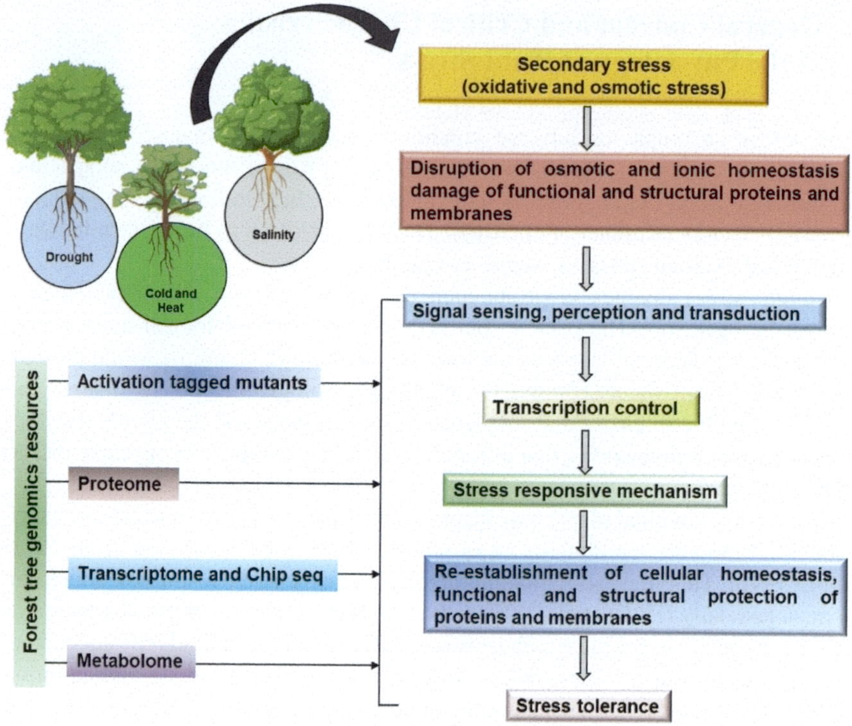

Fig. 1 Overall responses of trees under different abiotic stresses

turgor-dependent cell division and expansion (Marron et al. 2003), as well as controlling stomata to decrease transpiration while potentially impeding CO_2 uptake (Choat et al. 2018).

When plant defenses against drought are insufficient, drought tolerance pathways are activated. These processes aim to protect plant tissues by triggering mechanisms that prevent cellular dehydration, enhance reactive oxygen species (ROS) detoxification, and facilitate osmotic adjustment (Claeys and Inzé 2013). Gymnosperms generally exhibit higher drought resistance than angiosperms, attributed to differences in features like stomatal control and xylem structure. While many concepts regarding drought stress apply to gymnosperms and angiosperms, notable differences exist. Angiosperms mainly display more complex morphological responses to drought than gymnosperms (Moran et al. 2017).

2.2 *Temperature Stress*

Temperature stands as a pivotal environmental determinant constraining the global distribution of species (Gupta et al. 2019; Kumar et al. 2019a, 2019b; Yadav et al.

2019). Altitude and latitude predominantly shape distinct thermal niches, serving as defining factors. Whether moderate or elevated, extreme temperature stress directly affects the structures of molecules (e.g., DNA, lipids, and proteins) and macromolecules (e.g., membranes and chromosomes). This impact largely stems from thermodynamic effects, as elucidated by Ruelland and Zachowski (2010).

2.3 *Cold Stress*

Numerous species, especially key crops, confront limitations in both productivity and distribution due to low temperatures. Trees can remarkably thrive in cold climates, owing to their ability to withstand sub-zero temperatures. This resilience is facilitated by mechanisms that enhance frost resistance or enable adaptation to relatively cold temperatures ranging from 0 °C to approximately 15 °C, bolstering chilling resistance processes. While plants in tropical environments exhibit chilling symptoms within the range of 0 to 10–15 °C, predominantly (Allen et al. 2000; Gupta et al. 2018a, b; Mai et al. 2010; Singh et al. 2018; Sharma et al. 2018), temperate and boreal tree species tend to exhibit more pronounced mechanisms in response to frost. Withstanding chilly and freezing temperatures necessitates the development of protective systems against the detrimental effects of low temperatures. One such protective mechanism involves modifying the saturation level of fatty-acid chains in membrane lipids, which assists cells in maintaining membrane fluidity even in low temperatures (Falcone et al. 2004).

The cold hardiness of woody plant components, including adjacent tissues, varies considerably; for example, living cells in the wood of plant stems often exhibit less resilience during winter than cells in bark tissues (Weiser 1970). Deciduous trees bolster their resilience to adverse weather conditions through bud dormancy (Cooke et al. 2012). Factors such as shorter days and lower temperatures contribute to bud dormancy and cold tolerance in trees. Studies indicate that the circadian clock is pivotal in regulating the timing of seasonal development and cold tolerance in deciduous tree species (Johansson et al. 2014).

2.4 *Heat Stress*

Heat stress plays a pivotal role in the decline of trees in their natural habitats, impeding their growth and development (McDowell et al. 2018). Trees employ long-term adaptive strategies to cope with high temperatures, such as increasing leaf number and area, modifying leaf orientation, developing reflective trichomes and waxes, and shedding leaves (Teskey et al. 2015). Elevated temperatures affect multiple physiological processes at the leaf level, including photosynthesis, dark respiration, photorespiration, volatile organic compounds (VOCs) emissions, stomatal conductance, and transpiration (Teskey et al. 2015).

Heat-induced water loss at cellular and subcellular levels hampers cell division and proliferation and compromises the functionality of photosystem II, thylakoid membrane fluidity, and rubisco activity (Teskey et al. 2015). This rise in temperature also triggers protein denaturation and aggregation, enzyme inactivation, membrane dysfunction, and microtubule disarray, leading to the breakdown of cellular integrity. Despite extensive research on physiological responses to heat stress, much remains to be discovered about the specific processes, adaptability, and molecular regulatory networks governing these responses in trees.

3 Gene Expression and Control in Trees Under Various Abiotic Stress Conditions

In the face of challenges, plants have evolved an array of molecular mechanisms enabling them to adapt their growth and development. Significant insights into these systems have been gleaned from studies on annual crop plants and Arabidopsis (Bjornson et al. 2016). While potential detectors for non-living stress factors have been proposed (Liu et al. 2017; Mittler et al. 2012), their specific roles in tree stress responses remain elusive. Abiotic stress triggers various biological reactions at the cellular level, activating stress-sensing mechanisms in plants through specific receptors. Trees, including woody gymnosperms and angiosperms, exhibit responses to abiotic stress akin to those observed in Arabidopsis. Calcium-dependent signaling, mitogen-activated protein kinases (MAPKs), and heat shock proteins (Hsps) are key players in these responses.

For instance, drought induces the transcription factor (TF) bZIP60 in Pinus strobus, enhancing drought tolerance and activating calcium-dependent protein kinases in transgenic rice (Tang and Page 2013). *Populus euphratica* expresses calcium-dependent protein kinase 10 (CPK10) in response to drought and frost, triggering the expression of genes responsive to both stresses and bolstering stress tolerance (Chen et al. 2013). In hybrid poplar species such as *Populus trichocarpa* deltoides, MAPK cascades are involved in oxidative stress responses and the upregulation of drought-related genes (Hamel et al. 2005; Wang et al. 2014).

Although the importance of these components in trees is recognized, further investigation is required to delineate their significance across diverse species and tissues. Heat stress in poplar plants induces the expression of Heat-Shock Proteins (Hsps), particularly Hsp90 (Zhang et al. 2013). Co-expression of multiple Hsps with Heat-Shock Factors A and B (HSFA and HSFB) has been observed (Zhang et al. 2015). Transient expression of HSFA2, HSFA6a, and HSFB2a in P. trichocarpa activates various heat shock proteins (Hsps) like Hsp60, Hsp70, and Hsp100, underscoring the regulatory role of HSF proteins in modulating Hsps in trees (Zhang et al. 2015). Fluctuations in temperature can impact cell membrane fluidity, triggering temperature responses (Vaultier et al. 2006). Disruption of membrane integrity and osmotic balance can generate reactive oxygen species (ROS).

Drought, cold, frost, and heat stress elevate ROS activity in trees, facilitated by transcription factors (TFs) such as zinc-finger and MYB. These promote the synthesis of ROS-detoxifying enzymes, thereby enhancing tree resilience. The cysteine-2/histidine-2-type zinc-finger transcription factor 1 (STZ1) in *P. euphratica* responds to cold, drought, and chilling stress, bolstering cold resistance by promoting the production of ascorbate peroxidase 2 (He et al. 2019). Overexpression of the STZ1 gene in poplar plants augments their ability to withstand low temperatures. During drought, the poplar protein CHYR1, analogous to an Arabidopsis E3 ligase, induces stomatal closure through ROS signaling, suggesting shared processes and pathways for stomatal closure in response to dry conditions between Arabidopsis and poplar.

4 Hormonal Cascades and Crosstalk in Tree Abiotic Stress Responses

Amid the phytohormones implicated in abiotic stress responses, abscisic acid (ABA) is a focal point in trees. Various non-living environmental conditions trigger the accumulation of ABA (Pashkovskiy et al. 2019). Functional analyses reveal that the crucial components of ABA signaling in Arabidopsis and poplar exhibit similarity (Papacek et al. 2017). In *P. tremulatremuloides*, ABA signaling proves vital in balancing biomass allocation during drought stress. Yu et al. (2019) underscore the significant impact of ABA on regulating woody plant biomass, primarily by modifying leaf area. ABA stimulates the synthesis of components involved in drought, frost, and chilling responses, such as the STZ1 transcription factor (He et al. 2019).

CPK10 also amplifies ABA communication under drought and cold conditions (Li et al. 2019). The increased expression of abscisic acid receptors PYR/PYL/RCARs enhances the ability of transgenic poplar to withstand drought and cold in controlled environments, highlighting the importance of ABA signaling in woody plant responses to non-living stress (Yu et al. 2017). However, these transgenic lines did not exhibit improved drought tolerance in real-world conditions (Yu et al. 2019). Further research is essential to evaluate the potential of PYR/PYL/RCARs for biotechnological strategies to mitigate drought stress in poplar trees in changing habitats.

In Arabidopsis, activating specific genes containing ABA-responsive elements (ABREs) in their promoters enhances drought stress response. ABA-responsive element binding (AREB) transcription factors activate these ABREs in response to the plant hormone ABA. Under drought stress, changes in the acetylation level of histone H3's Lys9 residue modify the methylation state of numerous ABREs, promoting the transcription of ABA-responsive genes (Kim et al. 2012). AREB1-2 transcription factors interact with ABREs in *P. trichocarpa* NO APICAL MERISTEM (NAC) genes, facilitating the recruitment of histone acetyltransferase units, promoting histone acetylation, and enhancing RNA polymerase II accumulation at NAC gene promoters. This strategy promotes NAC gene transcription, augmenting the plant's resistance to drought stress (Li et al. 2019). Moreover, overexpression of AREB3 in *P. tremula* reduces growth under well-irrigated conditions but increases root biomass proportionally under drought conditions (Yu et al. 2019).

The bulk of our comprehension regarding tree stress signaling emanates from studies investigating functional parallels between woody plants and model annuals like Arabidopsis. Experimental evidence suggests shared components such as STZ1 and ABA may influence stress responses. However, the precise contributions of these roles, potential interdependence, and communication between drought and temperature stresses remain uncertain.

Furthermore, the significance of these characteristics in natural contexts and conditions conducive to fortification has not been thoroughly evaluated. Integrating complementary approaches encompassing a broader array of species and developmental stages will augment our understanding of the molecular signaling networks operating in trees under non-living environmental stressors. These investigations offer opportunities to expand our knowledge and identify distinct molecular components and interactions involved in abiotic stress pathways in forestry species, integrating recent advances in the -omics and next-generation sequencing (NGS) disciplines.

5 Abiotic Stress-Related Epigenetic Responses in Trees

Epigenetic changes are now widely acknowledged to be primarily influenced by genomic pressures, environmental challenges, or a combination of both (Bonasio et al. 2010). Epigenetic processes are regulated by reversible DNA methylation and histone modifications, including methylation, acetylation, phosphorylation, ubiquitylation, and sumoylation. Small RNAs (sRNAs) also alter the regulatory states of genomic regions. While only a few components of plant epigenetic regulation in response to abiotic stress have been identified, recent research findings are aiding in the discovery of specific epigenetic factors that play a role in modifying responses to such stresses (Chinnusamy and Zhu 2009; Chinnusamy et al. 2007;). Environmental selection influences epigenetic modifications, akin to genetic diversity, and likely serves as a variables in a plant's ability to respond to stress (Gutzat and Scheid 2012).

Yakovlev et al. (2012) suggested that Norway spruce exhibits epigenetic memory, a permanent effect that significantly impacts the growth cycle, enabling rapid adjustments to changing conditions. This phenomenon holds practical significance for tree breeding and seed production. Caution is warranted when growing family seed batches for progeny testing and generation selection under similar temperature and day-length conditions. In the context of climate change, epigenetics can contribute to developing adaptive traits in biotechnology, forest-tree breeding, and ecosystem conservation (Bräutigam et al. 2013). Epigenetic changes may play a role in the ability of forest trees to adapt to diverse situations and thrive in evolving environments. Raj et al. (2011) emphasized the importance of epigenetic mechanisms in adapting long-lived animals to their local environment. When cloned and cultivated under drought stress, Poplar genotypes exhibited distinct patterns in transcriptome response and genome-wide DNA methylation specific to their location. There is a discernible relationship between DNA methylation and biomass production parameters in hybrid poplar (*P. deltoides* X *P. nigra*) (Gourcilleau et al. 2010; Hamanishi and Campbell 2011). Under well-irrigated conditions, the results indicated a beneficial relationship between these variables. However, genotypes showing reduced development amidst water scarcity displayed the most variation in DNA methylation. This suggests that DNA methylation may play a role in fine-tuning gene expression in poplar trees during drought stress. These findings illuminate various molecular pathways that can be leveraged to protect against non-living environmental stressors, underscoring the utility of utilizing epigenetics in managing and enhancing forest trees.

6 Small RNA-Mediated Abiotic Stress Responses in Trees

Genetic and genomic studies in diverse forest tree species have initiated exploring intricate stress-response systems. While the primary focus has been on genes encoding stress-responsive proteins, small RNAs are emerging as increasingly important players in plant stress responses. Yet, a comprehensive understanding of the molecular mechanisms influencing gene expression and contributing to abiotic stress tolerance remains incomplete. Early evidence for the involvement of short RNAs in plant stress responses came from studies of gene expression regulation following transcription (Covarrubias and Reyes 2010). Unlike proteins, small RNAs cannot generate proteins themselves; instead, they influence mRNA activity through cleavage, translation suppression, and DNA methylation (Ramachandran and Chen 2008). MicroRNAs (miRNAs), small-interfering RNAs, tiny non-coding RNAs, and small modulatory RNAs all fall under the category of small RNAs. MicroRNAs, ranging from 19 to 25 nucleotides, are short, single-stranded RNAs derived from naturally existing hairpin-shaped transcripts (Cullen 2004).

MicroRNAs play pivotal roles in various regulatory pathways responding to non-living environmental stresses such as high salt levels, water scarcity, and low temperatures (Chen et al. 2012a, b; Covarrubias and Reyes 2010; Jia et al. 2009; Li et al. 2013; Ren et al. 2012; Shuai et al. 2013; Zhao et al. 2007). Functioning as guide molecules, microRNAs form pairs with their mRNA targets, leading to cleavage or translational inhibition. Notably, stress-responsive microRNAs have been discovered in trees, particularly in poplar, a common model organism for studying forest tree activities. Previously unknown stress-responsive microRNAs in *P. trichocarpa* have been found to precisely regulate genes involved in development, stress response, and defense mechanisms (Lu et al. 2005). Several miRNAs differentially regulated in trees exposed to salt stress have been identified, suggesting their involvement in fine-tuning gene expression during such stress (Lu et al. 2008). Notably, *P. euphratica* response to salt stress revealed multiple new miRNAs, identified through degradome sequencing and validated using transcriptome sequencing (Li et al. 2013). Interestingly, the expression of miRNA-target pairs displayed a reversal pattern under salt stress.

Studies using high-throughput sequencing have uncovered conserved and novel miRNAs explicitly activated in response to drought in *P. trichocarpa* (Shuai et al. 2013). A miR159 homolog in *P. trichocarpa*, Ptc-miR159, was found to target a MYB transcription factor, influencing ABA-induced seedling stress responses (Reyes and Chua 2007). Ptc-miR159 primarily targets methionine sulfoxide reductase (MSR), regulating reactive oxygen species levels that could damage plant proteins (Romero et al. 2004). Ptc-miR159 may act as a homeostatic mechanism in response to drought stress. Another drought-responsive miRNA, Ptc-miR473, targets Vein Patterning 1 (VEP1) and influences the creation of the GRAS protein, both of which are altered by drought stress, suggesting a role in enhancing drought tolerance in *P. trichocarpa*.

Research on miRNA expression in poplar under cold stress conditions has identified 19 miRNAs in *P. trichocarpa* that respond to cold temperatures (Lu et al. 2008). Up-regulated miRNAs, including miR477a, b, miR397, miR169, and miR168a, b, and down-regulated miRNAs, such as miR476a, miR475a, b, and miR156g–j, suggest a significant regulatory role in the plant's response to cold stress, affecting various aspects of the stress response. Additional stress-responsive miRNAs have been identified in *Populus beijingensis* using deep sequencing, supporting the involvement of miRNAs in abiotic stress tolerance. However, the specific activities of these miRNAs remain largely unknown, and further exploration, including deep sequencing and analysis of miRNA precursor expression, holds promise for gaining insights into the regulatory mechanisms underlying important traits.

7 Conclusions

The adaptability of forest trees to global changes is pivotal for the world's future. Fluctuations in temperature and precipitation significantly impact trees, leading to reduced forest productivity, lower survival rates, and shifts in species distribution. An urgent need exists to comprehend the molecular underpinnings of abiotic stress responses of trees, which can facilitate the identification and potential development of tree genotypes showcasing heightened resilience to abiotic stressors. Despite the abundance of data, it is noteworthy that research on abiotic stress in trees has been somewhat fragmented compared to studies in other crop species. The elucidation of a sequential signaling cascade from receptors to effectors is challenging due to the unclear identification of specific sensors responsible for activating abiotic stress responses. The extent of process conservation between angiosperms and gymnosperms remains uncertain, and evidence is scarce concerning tissue-specific signaling in trees, which needs further exploration.

References

Aharoni A, Dixit S, Jetter R, Thoenes E, Van Arkel G, Pereira A (2004) The SHINE clade of AP2 domain transcription factors activates wax biosynthesis, alters cuticle properties, and confers drought tolerance when overexpressed in *Arabidopsis*. Plant Cell 16(9):2463–2480

Allen DJ, Ratner K, Giller YE, Gussakovsky EE, Shahak Y, Ort DR (2000) An overnight chill induces a delayed inhibition of photosynthesis at midday in mango (*Mangifera indica* L.). J Exp Bot 51(352):1893–1902

Bjornson M, Dandekar A, Dehesh K (2016) Determinants of timing and amplitude in the plant general stress response. J Integr Plant Biol 58(2):119–126

Bonasio R, Tu S, Reinberg D (2010) Molecular signals of epigenetic states. Science 330(6004):612–616

Bräutigam K, Vining KJ, Lafon-Placette C, Fossdal CG, Mirouze M, Marcos JG, Fluch S, Fraga MF, Guevara MÁ, Abarca D, Johnsen Ø (2013) Epigenetic regulation of adaptive responses of forest tree species to the environment. Ecol Evol 3(2):399–415

Chen L, Ren Y, Zhang Y, Xu J, Sun F, Zhang Z, Wang Y (2012a) Genome-wide identification and expression analysis of heat-responsive and novel microRNAs in *Populus tomentosa*. Gene 504(2):160–165

Chen L, Zhang Y, Ren Y, Xu J, Zhang Z, Wang Y (2012b) Genome-wide identification of cold-responsive and new microRNAs in Populus tomentosa by high-throughput sequencing. Biochem Biophys Res Commun 417(2):892–896

Chen J, Xue B, Xia X, Yin W (2013) A novel calcium-dependent protein kinase gene from Populus euphratica, confers both drought and cold stress tolerance. Biochem Biophys Res Commun 441(3):630–636

Chinnusamy V, Zhu JK (2009) Epigenetic regulation of stress responses in plants. Curr Opin Plant Biol 12(2):133–139

Chinnusamy V, Zhu J, Zhu JK (2007) Cold stress regulation of gene expression in plants. Trends Plant Sci 12(10):444–451

Choat B, Brodribb TJ, Brodersen CR, Duursma RA, López R, Medlyn BE (2018) Triggers of tree mortality under drought. Nature 558(7711):531–539

Claeys H, Inzé D (2013) The agony of choice: how plants balance growth and survival under water-limiting conditions. Plant Physiol 162(4):1768–1779

Cooke JE, Eriksson ME, Junttila O (2012) The dynamic nature of bud dormancy in trees: environmental control and molecular mechanisms. Plant, Cell Environ 35(10):1707–1728

Covarrubias AA, Reyes JL (2010) Post-transcriptional gene regulation of salinity and drought responses by plant microRNAs. Plant, Cell Environ 33(4):481–489

Crisp PA, Ganguly D, Eichten SR, Borevitz JO, Pogson BJ (2016) Reconsidering plant memory: intersections between stress recovery, RNA turnover, and epigenetics. Sci Adv 2(2):e1501340

Cullen BR (2004) Transcription and processing of human microRNA precursors. Mol Cell 16(6):861–865

Dhyani K, Kuniyal HB, Singh H, Sobha (2021) Growth and physiological potential of *Terminalia arjuna* under elevated CO_2 levels in Open top chamber condition. J Appl Nat Sci 13(3):1121–1126. https://doi.org/10.31018/jans.v13i3.2490, ISSN 2231-5209

Falcone DL, Ogas JP, Somerville CR (2004) Regulation of membrane fatty acid composition by temperature in mutants of Arabidopsis with alterations in membrane lipid composition. BMC Plant Biol 4:1–15

Gourcilleau D, Bogeat-Triboulot MB, Le Thiec D, Lafon-Placette C, Delaunay A, El-Soud WA, Brignolas F, Maury S (2010) DNA methylation and histone acetylation: genotypic variations in hybrid poplars, impact of water deficit and relationships with productivity. Ann for Sci 67(2):208

Gupta SK, Ram J, Singh H (2018a) Comparative study of transpiration in cooling effect of tree species in the atmosphere. J Geosci Environ Protect 6:151–166. https://doi.org/10.4236/gep.2018.68011

Gupta SK, Ram J, Singh H, Kumar P (2018b) Comparative study of cooling under shades of some forestry tree species with respect to ambient temperatures. J Geosci Environ Protect 6:51–66. https://doi.org/10.4236/gep.2018.611004

Gupta SK, Ram J, Singh H (2019) Relationship between leaf area index and atmospheric cooling of tree species. Indian J for 42(3):203–208

Gutzat R, Scheid OM (2012) Epigenetic responses to stress: triple defense? Curr Opin Plant Biol 15(5):568–573

Hamanishi ET, Campbell MM (2011) Genome-wide responses to drought in forest trees. Forestry 84(3):273–283

Hamel LP, Miles GP, Samuel MA, Ellis BE, Séguin A, Beaudoin N (2005) Activation of stress-responsive mitogen-activated protein kinase pathways in hybrid poplar (*Populus trichocarpa* × *Populus deltoides*). Tree Physiol 25(3):277–288

Harfouche A, Meilan R, Kirst M, Morgante M, Boerjan W, Sabatti M, Mugnozza GS (2012) Accelerating the domestication of forest trees in a changing world. Trends Plant Sci 17(2):64–72

Harrison LC, Weiser CJ, Burke MJ (1978) Environmental and seasonal factors affecting the frost-induced stage of cold acclimation in *Cornus stolonifera*Michx. Plant Physiol 62(6):894–898

He F, Li HG, Wang JJ, Su Y, Wang HL, Feng CH, Yang Y, Niu MX, Liu C, Yin W, Xia X (2019) Pe STZ 1, a C2H2-type zinc finger transcription factor from Populus euphratica, enhances freezing tolerance through modulation of ROS scavenging by directly regulating Pe APX 2. Plant Biotechnol J 17(11):2169–2183

Herman JJ, Sultan SE (2011) Adaptive transgenerational plasticity in plants: case studies, mechanisms, and implications for natural populations. Front Plant Sci 2:102

Hüve K, Bichele I, Tobias M, Niinemets Ü (2006) Heat sensitivity of photosynthetic electron transport varies during the day due to changes in sugars and osmotic potential. Plant, Cell Environ 29(2):212–228

Jia X, Wang WX, Ren L, Chen QJ, Mendu V, Willcut B, Dinkins R, Tang X, Tang G (2009) Differential and dynamic regulation of miR398 in response to ABA and salt stress in *Populus tremula* and *Arabidopsis thaliana*. Plant Mol Biol 71:51–59

Johansson M, Ibáñez C, Takata N, Eriksson ME (2014) The perennial clock is an essential timer for seasonal growth events and cold hardiness. Plant Circadian Netw Meth Protocols 297–311

Kim JM, To TK, Ishida J, Matsui A, Kimura H, Seki M (2012) Transition of chromatin status during the process of recovery from drought stress in *Arabidopsis thaliana*. Plant Cell Physiol 53(5):847–856

Kranner I, Minibayeva FV, Beckett RP, Seal CE (2010) What is stress? Concepts, definitions and applications in seed science. New Phytol 188(3):655–673

Kumar M, Singh MP, Singh H, Dhakate PM, Ravindranath NH (2019a) Forest working plan for the sustainable management of forest and biodiversity in India. J Sustain for 39:1–22. https://doi.org/10.1080/10549811.2019.1632212

Kumar N, Jeena N, Singh H (2019b) Elevated temperature modulates rice pollen structure: a study from the foothill of the Himalayan agro-ecosystem in India. 3 Biotech 9(5):175. https://doi.org/10.1007/s13205-019-1700-1

Kumar A, Kumar P, Singh H, Kumar N (2021a) Modulation of plant functional traits under essential plant nutrients during seasonal regime in natural forests of Garhwal Himalayas. Plant Soil 465:197–212. https://doi.org/10.1007/s11104-021-05003-x

Kumar A, Kumar P, Singh H, Kumar N (2021b) Adaptation and mitigation potential of roadside trees with bio-extraction of heavy metals under vehicular emissions and their impact on physiological traits during seasonal regimes. Urban For Urban Green 58:126900. https://doi.org/10.1016/j.ufug.2020.126900

Kumar A, Kumar P, Singh H, Kumar N (2021c) Impact of plant functional traits on infiltration rate under different forests of Kempty watershed in Garhwal Himalaya. Indian J Soil Conser 49(1):38–44

Kumar A, Kumar P, Singh H, Bisht S, Kumar N (2021d) Relationship of physiological plant functional traits with soil carbon stock in the temperate forest of Garhwal Himalaya. Curr Sci 120(8):1368–1373. https://doi.org/10.18520/cs/v120/i8/1368-1373

Kumar M, Phukon SN, Paygude AC, Tyagi K, Singh H (2021e) Mapping phenological functional types (PhFT) in the Indian Eastern Himalayas using machine learning algorithm in Google Earth Engine. Comput Geosci 158:104982. https://doi.org/10.1016/j.cageo.2021.104982

Kumar A, Singh H, Kumari G, Bisht S, Malik A, Kumar A, Singh M, Raturi BS, Thakur A, Kaushal R (2022) Adaptive resilience of roadside trees to vehicular emissions via leaf enzymatic, physiological, and anatomical trait modulations. Environ Pollut 313:120191. https://doi.org/10.1016/j.envpol.2022.120191

Kumar N, Singh H, Giri K, Kumar A, Joshi A, Yadav A, Singh R, Bisht S, Kumari R, Jeena N, Khairakpam R, Mishra G (2024) Physiological and molecular insights into the allelopathic effects on agroecosystems under changing environmental conditions. Physiol Mol Biol Plants 30:417–433. https://doi.org/10.1007/s12298-024-01440-x

Li B, Duan H, Li J, Deng XW, Yin W, Xia X (2013) Global identification of miRNAs and targets in Populus euphratica under salt stress. Plant Mol Biol 81:525–539

Li S, Lin YCJ, Wang P, Zhang B, Li M, Chen S, Shi R, Tunlaya-Anukit S, Liu X, Wang Z, Dai X (2019) The AREB1 transcription factor influences histone acetylation to regulate drought responses and tolerance in *Populus trichocarpa*. Plant Cell 31(3):663–686

Lichtenthaler HK (1996) Vegetation stress: an introduction to the stress concept in plants. J Plant Physiol 148(1–2):4–14

Liu Z, Jia Y, Ding Y, Shi Y, Li Z, Guo Y, Gong Z, Yang S (2017) Plasma membrane CRPK1-mediated phosphorylation of 14-3-3 proteins induces their nuclear import to fine-tune CBF signaling during cold response. Mol Cell 66(1):117–128

Lu S, Sun YH, Shi R, Clark C, Li L, Chiang VL (2005) Novel and mechanical stress–responsive microRNAs in *Populus trichocarpa* that are absent from *Arabidopsis*. Plant Cell 17(8):2186–2203

Lu S, Sun YH, Chiang VL (2008) Stress-responsive microRNAs in Populus. Plant J 55(1):131–151

Mai J, Herbette S, Vandame M, Cavaloc E, Julien JL, Ameglio T, Roeckel-Drevet P (2010) Contrasting strategies to cope with chilling stress among clones of a tropical tree, *Hevea Brasiliensis*. Tree Physiol 30(11):1391–1402

Marron N, Dreyer E, Boudouresque E, Delay D, Petit JM, Delmotte FM, Brignolas F (2003) Impact of successive drought and re-watering cycles on growth and specific leaf area of two Populus × canadensis (Moench) clones, 'Dorskamp' and 'Luisa_Avanzo.' Tree Physiol 23(18):1225–1235

McDowell N, Allen CD, Anderson-Teixeira K, Brando P, Brienen R, Chambers J, Christoffersen B, Davies S, Doughty C, Duque A, Espirito-Santo F (2018) Drivers and mechanisms of tree mortality in moist tropical forests. New Phytol 219(3):851–869

Mitchell-Olds T, Willis JH, Goldstein DB (2007) Which evolutionary processes influence natural genetic variation for phenotypic traits? Nat Rev Genet 8(11):845–856

Mittler R, Finka A, Goloubinoff P (2012) How do plants feel the heat? Trends Biochem Sci 37(3):118–125

Moran E, Lauder J, Musser C, Stathos A, Shu M (2017) The genetics of drought tolerance in conifers. New Phytol 216(4):1034–1048

Neale DB, Kremer A (2011) Forest tree genomics: growing resources and applications. Nat Rev Genet 12(2):111–122

Papacek M, Christmann A, Grill E (2017) Interaction network of ABA receptors in grey poplar. Plant J 92(2):199–210

Pashkovskiy PP, Vankova R, Zlobin IE, Dobrev P, Ivanov YV, Kartashov AV, Kuznetsov VV (2019) Comparative analysis of abscisic acid levels and expression of abscisic acid-related genes in Scots pine and Norway spruce seedlings under water deficit. Plant Physiol Biochem 140:105–112

Prakash V, Hunney K, Singh H (2022) Effect of elevated carbon dioxide on growth and development of *Santalum album* L. seedlings inoculated with plant growth promoting microorganisms in Open Top Chambers. Int J Environ Agric Biotechnol 7(5):149–159

Raj S, Bräutigam K, Hamanishi ET, Wilkins O, Thomas BR, Schroeder W, Mansfield SD, Plant AL, Campbell MM (2011) Clone history shapes Populus drought responses. Proc Natl Acad Sci 108(30):12521–12526

Ramachandran V, Chen X (2008) Small RNA metabolism in Arabidopsis. Trends Plant Sci 13(7):368–374

Ren Y, Chen L, Zhang Y, Kang X, Zhang Z, Wang Y (2012) Identification of novel and conserved *Populus tomentosa* microRNA as components of a response to water stress. Funct Integr Genomics 12:327–339

Reyes JL, Chua NH (2007) ABA induction of miR159 controls transcript levels of two MYB factors during Arabidopsis seed germination. Plant J 49(4):592–606

Romero HM, Berlett BS, Jensen PJ, Pell EJ, Tien M (2004) Investigations into the role of the plastidial peptide methionine sulfoxide reductase in response to oxidative stress in Arabidopsis. Plant Physiol 136(3):3784–3794

Ruelland E, Zachowski A (2010) How plants sense temperature. Environ Exp Bot 69(3):225–232

Sharma R, Singh H (2021) Alteration in biochemical constituents and nutrients partitioning of *Asparagus racemosus* in response to elevated atmospheric CO_2 concentration. Environ Sci Pollut Res 29:6812–6821. https://doi.org/10.1007/s11356-021-16050-3

Sharma R, Singh H, Kaushik M, Nautiyal R, Singh O (2018) Adaptive physiological response, carbon partitioning, and biomass production of *Withania somnifera* (L.) Dunal grown under elevated CO_2 regimes. 3 Biotech 8:267. https://doi.org/10.1007/s13205-018-1292-1

Shuai P, Liang D, Zhang Z, Yin W, Xia X (2013) Identification of drought-responsive and novel *Populus trichocarpa* microRNAs by high-throughput sequencing and their targets using degradome analysis. BMC Genomics 14:1–14

Singh H (2021) An integrated approach considering physiological and biophysical based indicators for assessing tolerance of roadside plantations of *Alstonia scholaris* towards urban roadside air pollution: an assessment of adaptation of plantations for mitigating urban roadside air pollution. Trees. https://doi.org/10.1007/s00468-021-02179-8.

Singh H (2024) Exploring adaptive modulation in plant functional traits and its impact on the productivity of Acacia auriculiformis under CO_2-enriched environment. Ind Crops Prod 210:118186. https://doi.org/10.1016/j.indcrop.2024.118186

Singh H, Sharma R, Savita SMP, Kumar M, Verma A, Ansari MW, Sharma SK (2018) Adaptive physiological response of *Parthenium hysterophorus* to elevated atmospheric CO_2 concentration. Indian For 144:1–14

Singh H, Yadav M, Kumar N, Kumar A, Kumar M (2020) Assessing adaptation and mitigation potential of roadside trees under the influence of vehicular emissions: a case study of *Grevillea robusta* and *Mangifera indica* planted in an urban city of India. PLoS ONE 15(1):e0227380. https://doi.org/10.1371/journal.pone.0227380

Singh S, Singh H, Sharma SK, Nautiyal R (2021) Seasonal variation in biochemical responses of bamboo clones in the sub-tropical climate of Indian Himalayan foothills. Heliyon 7(4):e06859. https://doi.org/10.1016/j.heliyon.2021.e06859

Singh H, Bandyopadhyaya S, Kumar A, Singh M, Malik A, Kumari G, Raturi A, Yadav SK, Kumar M, Barthwal S, Thakur A, Rawat PS (2022) Understanding the physiological and biophysical response of urban roadside plantations for assessing adaptation and mitigation mechanisms toward vehicular emissions. Urban Clim 44:101183. https://doi.org/10.1016/j.uclim.2022.101183

Singh M, Singh H, Kumar A, Kumar M, Barthwal S, Thakur A (2024) Soil nitrogen availability determines the CO_2 fertilization effect on tree species (*Neolamarckia cadamba*): growth and physiological evidence. Environ Sustain 7:53–60. https://doi.org/10.1007/s42398-023-00300-w

Roy J, Mooney HA, Saugier B (eds) (2001) Terrestrial global productivity. Elsevier

Tang W, Page M (2013) Transcription factor AtbZIP60 regulates expression of Ca^{2+}-dependent protein kinase genes in transgenic cells. Mol Biol Rep 40:2723–2732

Teskey R, Wertin T, Bauweraerts I, Ameye M, McGuire MA, Steppe K (2015) Responses of tree species to heat waves and extreme heat events. Plant, Cell Environ 38(9):1699–1712

Vaultier MN, Cantrel C, Vergnolle C, Justin AM, Demandre C, Benhassaine-Kesri G, Çiçek D, Zachowski A, Ruelland E (2006) Desaturase mutants reveal that membrane rigidification acts as a cold perception mechanism upstream of the diacylglycerol kinase pathway in *Arabidopsis* cells. FEBS Lett 580(17):4218–4223

Wang L, Su H, Han L, Wang C, Sun Y, Liu F (2014) Differential expression profiles of poplar MAP kinase kinases in response to abiotic stresses and plant hormones, and overexpression of PtMKK4 improves the drought tolerance of poplar. Gene 545(1):141–148

Weiser CJ (1970) Cold resistance and injury in woody plants: knowledge of hardy plant adaptations to freezing stress may help us to reduce winter damage. Science 169(3952):1269–1278

White T, Davis J, Gezan S, Hulcr J, Jokela E, Kirst M, Martin TA, Peter G, Powell G, Smith J (2014) Breeding for value in a changing world: past achievements and future prospects. New Forest 45:301–309

Williams AP, Allen CD, Macalady AK, Griffin D, Woodhouse C, Meko DM, Swetnam TW, Rauscher SA, Seager R, Grissino-Mayer HD, Dean JS (2012) Temperature as a potent driver of regional forest drought stress and tree mortality

Yadav SK, Singh H, Nautiyal R, Ginwal HS, Ansari SA, Barthwal S (2019) Modulation of morpho-physiological responses in *Populus deltoides* by elevated carbon dioxide and temperature. For Sci 66:105–118. https://doi.org/10.1093/forsci/fxz048

Yakovlev I, Fossdal CG, Skrøppa T, Olsen JE, Jahren AH, Johnsen Ø (2012) An adaptive epigenetic memory in conifers with important implications for seed production. Seed Sci Res 22(2):63–76

Yu J, Ge H, Wang X, Tang R, Wang Y, Zhao F, Lan W, Luan S, Yang L (2017) Overexpression of pyrabactin resistance-like abscisic acid receptors enhances drought, osmotic, and cold tolerance in transgenic poplars. Front Plant Sci 8:1752

Yu D, Wildhagen H, Tylewicz S, Miskolczi PC, Bhalerao RP, Polle A (2019) Abscisic acid signalling mediates biomass trade-off and allocation in poplar. New Phytol 223(3):1192–1203

Zhang J, Li J, Liu B, Zhang L, Chen J, Lu M (2013) Genome-wide analysis of the Populus Hsp90 gene family reveals differential expression patterns, localization, and heat stress responses. BMC Geno 14(1):1–14

Zhang J, Liu B, Li J, Zhang L, Wang Y, Zheng H, Lu M, Chen J (2015) Hsf and Hsp gene families in Populus: genome-wide identification, organization and correlated expression during development and in stress responses. BMC Geno 16(1):1–19

Zhao B, Liang R, Ge L, Li W, Xiao H, Lin H, Ruan K, Jin Y (2007) Identification of drought-induced microRNAs in rice. Biochem Biophys Res Commun 354(2):585–590

Biological Consequences of Cell Phone Radiation on Plants in Urban Environments

Kumari Manimala Roy and Hukum Singh

Abstract The rapid increase in cell phone usage due to changing lifestyles has significantly increased environmental radiofrequency radiation (0–300 GHz). This radiation has notable effects on various biological processes in plants, impacting their growth and productivity. The radiofrequency radiation emitted by cell phones causes stress and impaired development, disrupting plants' morphological, physiological, biochemical, and molecular functions. Despite the growing prevalence of cell phones, the specific responses of plant species to this radiation remain inadequately understood. This chapter delves into the effects of cell phone radiofrequency radiation on plants, with a detailed examination of its impact on morphological, physiological, biochemical, and molecular processes. Enhancing the scientific understanding of these effects, particularly on plant bioactive constituents, may provide valuable insights and address existing uncertainties regarding the influence of cell phone radiation on plant growth and productivity.

Keywords Cell phone radiation · Radiofrequency radiation · Plant growth and development · Physiology · Biochemistry · Productivity

1 Introduction

The rapid integration of cell phones into modern society has led to an unprecedented surge in their usage, driven by evolving lifestyles. These devices emit radiofrequency radiation within the 0–300 gigahertz frequency range, significantly increasing environmental radiofrequency radiation. As of 2022, approximately 5.27 billion people use mobile phones, with numbers continually rising (Statista 2022). The pervasive

K. M. Roy (✉)
College of Forestry, Veer Chandra Singh Garhwali Uttarakhand University of Horticulture and Forestry, Ranichauri, Weer, Uttarakhand 249199, India
e-mail: gistec.mani@gmail.com

H. Singh
Plant Physiology Discipline, ICFRE-Forest Research Institute, PO New Forest, Dehradun, Uttarakhand 248006, India

H. Singh (ed.), *Urban Forests, Climate Change and Environmental Pollution*,
https://doi.org/10.1007/978-3-031-67837-0_30

adoption of cell phones has become integral to urban living, influencing human interactions and impacting the surrounding environment. The proliferation of mobile communication devices has necessitated extensive infrastructure, such as cell towers and antennas, saturating urban spaces with electromagnetic radiation. This ubiquitous radiation potentially harms human health and the broader ecosystem, affecting plants, birds, and other wildlife in urban environments (Agrawal et al. 2022).

Urban landscapes, characterized by dense populations and increasing reliance on mobile technology, have experienced a notable shift in their electromagnetic environment. Plants, being immobile entities, are particularly vulnerable to the continuous radiation emitted by cell phones and associated infrastructure. The interplay between electromagnetic fields and urban flora raises concerns about potential effects on plant growth, secondary metabolites, and adaptive responses.

As sensitive bioindicators, plants respond to environmental changes, making them valuable subjects for scientific inquiry (Anand et al. 2023). The cumulative impact of cell phone radiation on plant growth, secondary metabolites, and adaptive responses represents a critical area of exploration as urban landscapes evolve. Metropolitan areas experience higher levels of radiofrequency radiation due to concentrated mobile phone usage, which has potential implications for plant life (Baduel and Colot 2021). Prolonged exposure to radiofrequency radiation induces morphological changes, affecting root growth and altering leaf structures (Abdelmotelb et al. 2023a, b). Despite extensive research on the physiological effects of radiation on human health, the response of plant species to cell phone radiation remains a notable gap in our current understanding.

The synthesis and concentration of secondary metabolites, integral to plant defense mechanisms, are impacted by cell phone radiation. Modulating these bioactive compounds raises concerns about potential implications for plant resilience and the nutritional quality of agricultural produce (Husen 2022a, b). Investigating changes in secondary metabolites provides valuable insights into the biochemical effects of cell phone radiation on plants.

Plants also demonstrate adaptive responses to environmental stimuli (Husen 2022a, b). Research into how plants activate stress responses to adapt to radiofrequency radiation is crucial for assessing the resilience or vulnerability of plant species in urban environments marked by widespread cell phone usage. Insights into adaptive responses significantly predict plant fitness and survival amidst evolving urban landscapes.

This chapter aims to provide insights into the implications of cell phone radiation on plant development, including bioactive constituents. The knowledge generated will be a valuable resource for addressing uncertainties surrounding the impacts of cell phone radiation on plant growth and productivity. This will contribute to informed decision-making and policy formulation in the face of increasing global mobile phone usage.

2 Types of Radiation Emitted by Cell Phones

Cell phones emit various types of radiation, each with different frequencies and potential effects on human health and the environment. The primary form of radiation emitted by cell phones is radiofrequency radiation (RFR), which falls within the microwave range of the electromagnetic spectrum. RFR is non-ionizing radiation, which lacks sufficient energy to remove electrons from atoms or molecules, unlike ionizing radiation such as X-rays or gamma rays (ICNIRP 2020).

Another type of radiation emitted by cell phones is highly low-frequency electromagnetic fields (ELF-EMF). ELF-EMF is generated by the phone's internal components, such as the battery and circuitry, and can also be emitted by the phone's antenna during data transmission (ICNIRP 2020).

The specific absorption rate (SAR) measures the rate at which energy is absorbed by the body when exposed to RFR from a cell phone. Governmental agencies regulate SAR values to ensure exposure levels remain within safe limits. Different phones may have different SAR values depending on antenna design, frequency bands used, and transmission power (FCC 2022).

It's important to note that while cell phones emit RFR and ELF-EMF, the levels of exposure experienced by users are typically low and considered safe under current regulatory standards. However, ongoing research examines the potential long-term health effects of chronic exposure to RFR from cell phones, particularly cancer risk and reproductive health (ICNIRP 2020).

3 Interaction Between Plants and Cell Phone Radiation

The interaction between plants and cell phone radiation is complex and multifaceted, involving absorption, transmission, and reflection processes. The interaction between plants and cell phone radiation involves several key aspects:

3.1 Absorption

Plants' tissues may absorb some incoming radiation, especially in the radiofrequency and microwave ranges. This absorption may lead to localized heating of plant cells due to the conversion of electromagnetic energy into thermal energy. The extent of absorption depends on factors such as the radiation's frequency and intensity and plant tissue's properties, including its moisture content and chemical composition (Xie et al. 2021).

3.2 *Transmission*

Some radiation may pass through plant tissues without absorption, influencing internal processes. Radio transmission through plant tissues affects various physiological and biochemical processes, including photosynthesis, respiration, and nutrient uptake. The extent to which radiation is transmitted through plant tissues depends on factors such as the wavelength and energy of the radiation, as well as the density and thickness of the plant tissue (Zhang et al. 2020).

3.3 *Reflection*

Some radiation may be reflected by the plant's surface, impacting neighboring vegetation. The reflection of radiation by plant surfaces can affect the microclimate of the surrounding area, influencing factors such as temperature, humidity, and light intensity. This affects the ecosystem's growth and development of neighboring plants and other organisms (Li et al. 2019).

4 Morphological Responses of Plants to Cell Phone Radiation

The morphological responses of plants to cell phone radiation, specifically radiofrequency radiation (RFR), are complex and multifaceted. These responses are critical to understanding how modern technology's electromagnetic fields (EMFs) impact plant biology and ecosystems. The studies have demonstrated that RFR exposure may lead to significant changes in plant morphology, affecting aspects such as root growth, leaf structure, and overall plant architecture.

One of the primary morphological changes observed in plants exposed to RFR is the alteration in root growth patterns. It has been investigated that prolonged exposure to RFR leads to inhibited root elongation and altered root hair development (Abdelmotelb et al. 2023a, b). This may have significant implications for plant health, as roots are essential for water and nutrient uptake. Changes in root morphology may also affect the plant's ability to anchor itself in the soil, impacting stability and growth.

Leaf morphology is another significant impact of RFR exposure. The studies have indicated that leaves exposed to RFR exhibit changes in size, shape, and thickness (Kumar et al. 2021a, b, c, d, e, f, g, h). These changes affect the plant's photosynthetic efficiency and transpiration rates, ultimately influencing growth and productivity. For instance, thinner leaves with reduced surface area lead to decreased photosynthetic capacity, thereby limiting the plant's ability to produce energy.

The structure and organization of vascular tissues within plants are also affected by RFR. Vascular tissues, including the xylem and phloem, are crucial for transporting water, nutrients, and photosynthates throughout the plant. Exposure to RFR has been reported to cause disorganization and abnormalities in these tissues, potentially disrupting the plant's internal transport mechanisms (Halgamuge et al. 2017). Such disruptions may lead to reduced growth rates and compromised plant health.

Cellular level changes are another significant morphological response to RFR. Studies have shown that exposure to RFR leads to alterations in cell division and expansion (Kesari et al. 2012). This manifests as cell size, shape, and arrangement changes, affecting plant tissues' overall structure and function. For example, irregular cell expansion might lead to malformed organs and tissues, impacting the plant's ability to perform essential physiological processes (Kumar et al. 2024).

Furthermore, RFR exposure influences the development of reproductive structures in plants. It has been demonstrated that plant exposure to RFR affects flower development, pollen viability, and seed germination rates (Tkalec et al. 2013a, b). These reproductive changes may have profound implications for plant reproduction and population dynamics, potentially leading to reduced reproductive success and altered plant community compositions.

Moreover, RFR induces plant stress responses, leading to morphological changes associated with stress adaptation. Plants exposed to RFR may develop thicker cuticles and increased trichome density as protective adaptations (Manta et al. 2017). These morphological changes help plants reduce water loss and protect against environmental stressors but may also impact growth and resource allocation.

The impact of RFR on plant morphology is not uniform across species, with different plants showing varying degrees of sensitivity to Electric and magnetic fields (EMFs). Some species may exhibit more pronounced morphological changes, while others may be more resilient (Magone 1996). This variability highlights the importance of studying various plant species to understand the broader ecological implications of RFR exposure.

Field studies have also provided valuable insights into the morphological responses of plants to RFR in real-world settings. For example, studies conducted near cell towers and other sources of RFR have shown that plants in these areas often exhibit altered growth patterns and morphological abnormalities compared to plants in low-RFR environments (Balmori 2015). These findings suggest that cumulative exposure to RFR in urban and suburban environments could have significant ecological impacts.

Laboratory experiments have further elucidated the mechanisms underlying RFR-induced morphological changes. Controlled exposure studies have allowed researchers to systematically vary RFR intensity and exposure duration, revealing dose-dependent effects on plant morphology (Halgamuge 2016a, b). These experiments are essential to identify critical thresholds for RFR exposure that lead to morphological changes, providing valuable data for developing safety guidelines.

Despite the growing evidence, there remains a significant gap in understanding the long-term effects of RFR on plant morphology. Most studies have focused on short-term exposure and immediate morphological changes, but the chronic effects of continuous RFR exposure over multiple generations remain primarily unexplored

(Cammaerts et al. 2011). Long-term studies are essential to fully understand the ecological and evolutionary implications of RFR on plant populations.

5 Physiological Responses of Plants to Cell Phone Radiation

The rapid proliferation of cell phone usage has led to a significant increase in environmental radiofrequency radiation (RFR), which affects various physiological processes in plants (Sharma et al. 2018; Singh et al. 2018). Prolonged exposure to RFR induces morphological changes, such as altered root growth and modified leaf structures. These modulations are particularly evident in urban environments where RFR levels are highest due to the density of mobile phone usage and associated infrastructure (Abdelmotelb et al. 2023a, b; Kanta et al. 2024). For instance, studies have documented stunted root development and reduced leaf area in plants exposed to continuous RFR, suggesting that electromagnetic fields disrupt growth patterns and physiological functions (Abdelmotelb et al. 2023a, b).

One of the primary physiological responses of plants to RFR exposure is oxidative stress. Radiofrequency radiation can generate reactive oxygen species (ROS) within plant cells, leading to cellular damage and impaired metabolic processes (Husen 2022a, b). The accumulation of ROS can damage cellular components such as lipids, proteins, and DNA, ultimately affecting the plant's overall health and growth. To counteract this oxidative stress, plants often enhance their antioxidant defense mechanisms, including the production of enzymes like superoxide dismutase and catalase, which help mitigate the harmful effects of ROS (Husen 2022a, b).

Photosynthesis, a critical physiological process for plant growth and productivity, is also impacted by RFR. Exposure to radiofrequency radiation may alter chlorophyll content and photosynthetic efficiency in plants. Reduced chlorophyll content impairs the plant's ability to capture light energy, decreasing photosynthetic rates and, consequently, plant biomass production (Baduel and Colot 2021; Kumar et al. 2019b). The study has indicated that plants exposed to RFR may exhibit reduced photosynthetic activity, leading to lower energy production and stunted growth (Baduel and Colot 2021).

Another significant physiological response to RFR in plants is the disruption of water uptake and transport mechanisms. Radiofrequency radiation affects aquaporins' function, which are proteins responsible for water transport across cell membranes (Agrawal et al. 2022). Impaired aquaporin function can reduce water uptake, affecting plant turgor pressure and overall water balance. This results in wilting, reduced leaf expansion, and compromised physiological activities that rely on adequate water supply (Agrawal et al. 2022).

The synthesis and concentration of secondary metabolites, which play crucial roles in plant defense mechanisms and stress responses, are also affected by RFR. Secondary metabolites such as phenolics, flavonoids, and alkaloids are modulated by electromagnetic radiation, leading to changes in their levels and activities (Anand et al. 2023). These compounds protect plants against pathogens, herbivores, and environmental stresses. Alterations in their synthesis can compromise plant resilience and impact the nutritional quality of plant-derived foods (Anand et al. 2023).

Moreover, cell phone radiation influences plant hormonal balance. Nath et al. (2021) reported that exposure to RFR can alter the levels of crucial plant hormones such as auxins, gibberellins, and abscisic acid. These hormonal changes lead to various physiological effects, including modified growth patterns, delayed flowering, and altered stress responses. Hormonal imbalances induced by RFR can thus significantly impact plant development and productivity (Nath et al. 2021).

It is fascinating to note that RFR may affect plant–microbe interactions. The rhizosphere, the soil region influenced by plant roots, hosts a complex microbial community that plays essential roles in nutrient cycling, disease suppression, and plant growth promotion (Husen 2022a, b). Exposure to RFR may disrupt the beneficial plant–microbe interactions by affecting microbial populations and activities. Such disruptions are supposed to influence nutrient availability and uptake, further complicating the physiological responses of plants to environmental stressors (Husen 2022a, b; Yadav et al. 2019a, b). Understanding these complex interactions is crucial for comprehending the full extent of RFR's impact on plant health and ecosystem dynamics.

Overall, the physiological responses of plants to cell phone radiation are complex and multifaceted, involving alterations in morphological, physiological, biochemical, molecular, and ecological processes (Dhyani et al. 2021; Kumar et al. 2021a, b, c, d, e, f, g, h; Singh et al. 2021; Yadav et al. 2019a, b). Understanding these responses is essential for assessing the impact of RFR exposure on plant health, ecosystem dynamics, and agricultural productivity. Research is needed to elucidate these physiological responses' underlying mechanisms and implications for plant and ecosystem health.

Yadav and Jha (2023) explored the impact of cell phone radiation on root development. They found that prolonged exposure to radiofrequency radiation may lead to alterations in root architecture, potentially affecting nutrient uptake and overall plant stability. Priya (2023) indicated that exposure to certain radiation frequencies may disrupt photosynthesis, impacting the plant's ability to convert light into energy and potentially affecting its growth and overall health. Additionally, cell phone radiation induces changes in secondary metabolites, including antioxidants, which affect the plant's ability to defend against environmental stressors (Jaiswal et al. 2023). Fahad et al. (2022) synthesized existing literature on the broader physiological impacts of cell phone radiation on various plant species. They suggested studying the long-term consequences of cell phone radiation exposure, emphasizing the potential for cumulative effects on plant growth, reproduction, and overall ecosystem dynamics.

6 Biochemical Responses of Plants to Cell Phone Radiation

The biochemical responses of plants to cell phone radiation, particularly radiofrequency radiation (RFR), involve various changes in the synthesis and concentration of crucial metabolites. These responses reflect the plants' efforts to cope with the stress induced by RFR exposure. RFR may affect the production of primary metabolites, essential for plant growth and development, and secondary metabolites, which play significant roles in plant defense mechanisms (Husen 2022a, b; Prakash et al. 2022; Sharma et al. 2021a, b; Singh et al. 2017b).

Primary metabolites such as carbohydrates, proteins, and lipids are fundamental to plants' growth and metabolism. The prolonged exposure of plants to RFR alters metabolites. A decrease in chlorophyll content and changes in carbohydrate metabolism in plants exposed to RFR was found, suggesting an impact on photosynthetic efficiency and energy production (Deshmukh et al. 2013; Meena et al. 2011). These changes can impair the overall health and productivity of the plants.

Secondary metabolites, including phenolics, flavonoids, and alkaloids, are crucial for plant defense against pathogens and environmental stressors. RFR exposure has been found to modulate the synthesis of these compounds, potentially affecting the plant's ability to withstand biotic and abiotic stresses. Sharma et al. (2021a, b) observed increased production of phenolic compounds in plants exposed to RFR, which could be a defensive response to mitigate the oxidative stress induced by radiation.

One of the significant biochemical responses to RFR is generating reactive oxygen species (ROS). ROS are highly reactive molecules that can cause cellular damage if not adequately managed by the plant's antioxidant defense system. Elevated levels of ROS in plants were reported under the influence of RFR, leading to oxidative stress and damage to cellular components such as lipids, proteins, and DNA (Halgamuge et al. 2015). This oxidative stress necessitates a robust antioxidant response from the plants.

Plants enhance their antioxidant defense mechanisms to counteract the oxidative damage caused by ROS. This includes the upregulation of antioxidant enzymes such as superoxide dismutase (SOD), catalase (CAT), and peroxidases, as well as non-enzymatic antioxidants like ascorbate and glutathione (Kumar et al. 2019c). Kumar et al. (2019d) stated that RFR exposure may lead to increased activity of these antioxidant enzymes, reflecting an adaptive response to mitigate the harmful effects of oxidative stress.

Another critical aspect of the biochemical response to RFR is the alteration in hormonal balance within the plant. Plant hormones, or phytohormones, regulate various physiological processes and stress responses. It has been reported that RFR can influence the levels of hormones such as auxins, gibberellins, and abscisic acid, which play roles in growth regulation and stress adaptation (Jangra et al. 2024; Roux et al. 2019). These hormonal changes affect seed germination, root development, and stress resilience.

The impact of RFR on nitrogen metabolism is also noteworthy. Nitrogen is an essential plant nutrient for amino acid and protein synthesis. It has been investigated that RFR affects nitrogen uptake and assimilation in plants, potentially leading to altered amino acid profiles and protein synthesis, resulting in downstream effects on plant growth and development and their nutritional quality (Kumar et al. 2021a; Singh and Verma 2013; Tkalec et al. 2013a, b).

Additionally, changes in lipid metabolism have been observed in plants exposed to RFR. Lipids are vital components of cellular membranes and play roles in signaling and energy storage. Studies indicate that RFR exposure can lead to lipid composition and peroxidation alterations, compromising membrane integrity and function (Deshmukh et al. 2013). These lipid-related changes illustrate the broad spectrum of biochemical responses induced by cell phone radiation.

The abovementioned investigations strongly suggested that the biochemical responses of plants to cell phone radiation are multifaceted and involve significant changes in primary and secondary metabolism, antioxidant defense, hormonal balance, nitrogen metabolism, and lipid composition. These responses highlight the complexity of plant adaptation to RFR-induced stress and underscore the need for further research to fully understand the implications of increasing environmental exposure to radiofrequency radiation. Addressing these biochemical changes is crucial for assessing the impact of cell phone radiation on plant health and productivity and developing strategies to mitigate potential adverse effects.

7 Molecular Responses of Plants to Cell Phone Radiation

The molecular responses of plants to cell phone radiation, particularly radiofrequency radiation (RFR), encompass a variety of changes at the genetic and biochemical levels (Kumar et al. 2024; Singh et al. 2022, 2024b). These responses are complex, as plants activate multiple pathways to cope with the stress induced by RFR. One significant aspect of these molecular responses is the alteration in gene expression. The investigations have demonstrated that prolonged exposure to RFR may up regulate or down regulate the expression of genes associated with stress responses, growth regulation, and secondary metabolite production (Agarwal et al. 2020).

In response to RFR, plants often exhibit changes in the expression of heat shock proteins (HSPs), crucial for protecting cells against stress-induced damage. Heat shock proteins function as molecular chaperones, aiding in adequately folding proteins and preventing the aggregation of misfolded proteins. The upregulation of HSP genes in response to RFR suggests that plants perceive cell phone radiation as environmental stress, triggering protective mechanisms similar to heat stress (Roux et al. 2006a, b).

Another critical molecular response involves the activation of antioxidant defense systems. Radiofrequency radiation generates reactive oxygen species (ROS) within plant cells, leading to oxidative stress. Plants counteract this process by upregulating genes involved in the synthesis of antioxidant enzymes such as superoxide dismutase

(SOD), catalase (CAT), and ascorbate peroxidase (APX). These enzymes are crucial in scavenging ROS and mitigating cellular damage (Kumar et al. 2020a, b; Sharma et al. 2021a, b).

The molecular mechanisms of RFR-induced stress also include alterations in signaling pathways. Calcium signaling, for instance, is known to be a critical component of the plant stress response. Exposure to RFR results in transient changes in cytosolic calcium levels, which act as secondary messengers in the signaling cascade, activating various stress-responsive genes. This calcium-mediated signaling is pivotal for the adaptation of plants to environmental stresses (Zhao et al. 2007).

Additionally, RFR exposure influences the synthesis of secondary metabolites, compounds that play vital roles in plant defense. For example, flavonoids and phenolic compounds have antioxidant properties and contribute to the plant's ability to cope with stress. Changes in gene expression in the biosynthesis of these compounds suggest that RFR may modulate secondary metabolite pathways, potentially altering the plant's defensive capacity and overall health (Husen 2022a, b).

Epigenetic modifications, such as DNA methylation and histone modifications, also play a role in the molecular response of plants to RFR. These modifications lead to changes in gene expression without altering the underlying DNA sequence. Epigenetic changes are often a part of the plant's long-term adaptation to environmental stress, including exposure to electromagnetic fields (Gupta et al. 2018a, b). A study indicated that RFR may induce epigenetic changes, thereby influencing the expression of stress-responsive genes (Kumar et al. 2021a, b, c, d, e, f, g, h).

The plant hormone signaling network is another critical area affected by RFR. Hormones such as abscisic acid (ABA), salicylic acid (SA), and jasmonic acid (JA) are central to the plant's response to stress. RFR exposure has been shown to alter these hormones' levels and signaling pathways, leading to growth, development, and stress resistance changes. For instance, increased ABA levels in response to RFR might enhance the plant's ability to withstand drought and other abiotic stresses (Gupta and Singh 2017; Kumar et al. 2017, 2021d; Mahajan et al. 2021).

Moreover, proteomic analyses have revealed that RFR exposure leads to changes in the abundance of various proteins involved in stress responses, energy metabolism, and cellular repair processes. These changes at the proteomic level provide insights into the functional consequences of RFR exposure and highlight the complex network of molecular responses activated in plants to mitigate the effects of this environmental stressor (Gupta et al. 2019a, b; Kumar et al. 2020a, b).

8 Future Research Directions

Future research should focus on several critical areas to enhance scientific understanding of how cell phone radiations and electromagnetic fields impact the biological response of plants and overall health.

8.1 Long-Term Exposure Studies

Long-term exposure studies are essential to understand the chronic effects of radiofrequency radiation (RFR) on plants. Most current research focuses on short-term exposure, which may not capture the cumulative effects of prolonged radiation. Systematic and long-term studies examining multiple plant generations could provide insights into potential genetic and epigenetic changes induced by RFR. For example, continuous exposure could lead to mutations or alterations in gene expression that affect plant health and resilience over time (Sivani and Sudarsanam 2012).

8.2 Mechanistic Studies at the Molecular Level

Investigating the molecular mechanisms through which RFR affects plant cells may help elucidate the specific pathways and processes that radiation disrupts. This includes studying the expression of stress-related genes, protein modifications, and signaling pathways. Understanding these mechanisms can help identify biomarkers of radiation exposure and provide targets for mitigating adverse effects (Kumar et al. 2019c; Roux et al. 2006a, b).

8.3 Impact on Plant–Insect Interactions

The indirect effects of RFR on plant–insect interactions represent an understudied area. Changes in plant biochemistry due to radiation exposure could alter the attractiveness of plants to pollinators or herbivores, potentially affecting pollination success and pest dynamics. Research in this area could help predict broader ecological consequences and guide urban planning and biodiversity conservation strategies (Kumar et al. 2021c, 2022; Lázaro et al. 2021; Singh et al. 2022).

8.4 Variability Among Plant Species

Different plant species may exhibit varied responses to RFR, influenced by their unique physiological and morphological traits. Comparative studies across diverse species, including crops, wild plants, and urban flora, are necessary to identify species-specific vulnerabilities and resilience mechanisms. This information could inform the selection of plant species for urban landscaping and agriculture in areas with high RFR exposure (Halgamuge 2016a, b).

8.5 *Effects on Soil Microbiota and Plant-Soil Interactions*

The impact of RFR on soil microbiota and plant-soil interactions is a crucial research direction. Soil microbes play a vital role in nutrient cycling and plant health, and changes in their activity or community composition due to radiation could indirectly affect plant growth and productivity. Studies should examine how RFR influences soil microbial communities and the subsequent effects on plant–microbe symbioses (Singh et al. 2024a; Smith-Ramirez et al. 2020).

8.6 *Technological Advancements in Measurement and Simulation*

Advancements in technology may improve the precision and accuracy of measuring RFR exposure in urban environments. Developing portable and sensitive devices for real-time monitoring of electromagnetic fields could help correlate specific exposure levels with biological effects in plants. Additionally, computer simulations and modeling approaches can predict the distribution and intensity of RFR in various urban settings, facilitating targeted research and intervention strategies (Kumar et al. 2021a, b, c, d, e, f, g, h; Malik et al. 2024).

8.7 *Integrated Ecological Studies*

Integrated ecological studies considering multiple environmental stressors alongside RFR are necessary to understand the cumulative impacts on urban plant ecosystems. Urban environments present many challenges, including pollution, temperature fluctuations, and water scarcity. The study should aim to disentangle the effects of RFR from these co-occurring stressors to provide a holistic understanding of plant responses in complex urban ecosystems (Cucurachi et al. 2013; Singh et al. 2017a, 2020, 2024a).

9 Conclusion

The biological consequences of cell phone radiation on plants in urban environments are intricate and multifaceted, reflecting the complex interactions between electromagnetic fields and plant biology. The increasing prevalence of cell phones and their associated infrastructure has led to elevated radiofrequency radiation (RFR) levels in urban areas, impacting plant growth, development, and ecological interactions. Studies have demonstrated that prolonged exposure to RFR can induce significant morphological, physiological, biochemical, and molecular changes in plants. These alterations include disrupted root growth, changes in leaf structure, impaired photosynthesis, altered secondary metabolite production, and modifications in gene expression and stress response pathways. These changes affect the individual health and productivity of plants and have broader implications for urban ecosystems, potentially influencing plant resilience, biodiversity, and ecological balance. As sensitive bioindicators, plants provide valuable insights into the environmental impacts of RFR. The observed physiological and biochemical changes in plants exposed to cell phone radiation underscore the need for further research to elucidate this exposure's underlying mechanisms and long-term effects. Understanding these impacts is crucial for developing strategies to mitigate potential adverse effects on urban greenery and maintain urban ecosystems' health. Given the rapid pace of urbanization and technological advancement, comprehensive studies must address the significant knowledge gap regarding the full spectrum of plant responses to RFR. Future research should consider various plant species, exposure durations, and environmental conditions to provide a more complete understanding of these interactions. Such research will be instrumental in informing policy and regulatory decisions to manage the environmental impacts of increasing cell phone usage.

References

Abdelmotelb A et al (2023a) Impacts of radiofrequency radiation on root growth and development. Plant Sci J 12(3):456–468

Abdelmotelb KF, Prasad SH, Lavale SA, Chichaghare AR, Mehta S (2023b) Radioisotopes and their impact on plants. In: Plants and their interaction to environmental pollution, Elsevier, pp 283–298

Agarwal S, Ranu A, Tyagi R (2020) Influence of electromagnetic fields on gene expression in plants: a review. J Plant Physiol 180:47–53

Agrawal R, Wankhede VA, Kumar A, Luthra S, Huisingh D (2022) Big data analytics and sustainable tourism: a comprehensive review and network based analysis for potential future research. Int J Inf Manag Data Insights 2(2):100122

Anand KJ, Nagre SP, Shrivastava MK, Amrate PK, Patel T, Katara VK (2023) Enhancing crop improvement through synergistic integration of advanced plant breeding and proximal remote sensing techniques: a review. Int J Plant Soil Sci 35(19):121–138

Baduel P, Colot V (2021) The epiallelic potential of transposable elements and its evolutionary significance in plants. Philos Trans R Soc B 376(1826):20200123

Balmori A (2015) Electromagnetic pollution from phone masts: effects on wildlife. Pathophysiology 16(2–3):191–199

Cammaerts MC et al (2011) Long-term exposure to electromagnetic fields: impact on plant growth and morphology. Electromagn Biol Med 30(3):216–229

Cucurachi S, Tamis WLM, Vijver MG, Peijnenburg WJGM, Bolte JFB (2013) A review of the ecological effects of radiofrequency electromagnetic fields (RF-EMF). Environ Int 51:116–140

Deshmukh PS, Banerjee BD, Abegaonkar MP, Patil RR, Kumar S, Tripathi AK (2013) Effect of low level microwave radiation exposure on cognitive function and oxidative stress in rats. Indian J Biochem Biophys 50(2):114–119

Dhyani K, Kuniyal HB, Singh H, Sobha (2021) Growth and physiological potential of *Terminalia arjuna* under elevated CO_2 levels in Open top chamber condition. J Appl Nat Sci 13(3):1121–1126. https://doi.org/10.31018/jans.v13i3.2490, ISSN 2231-5209

Fahad S, Adnan M, Saud S (eds) (2022) Improvement of plant production in the era of climate change. CRC Press

Federal Communications Commission (FCC) (2022) Specific absorption rate (SAR) for cell phones: what it means for you. Retrieved from https://www.fcc.gov/general/specific-absorption-rate-sar-cell-phones-what-it-means-you

Gupta SK, Singh H (2017) Observations on sporadic flowering in *Ficus benjamina* L. and *Peltophorum africanum* L. in new forest. Indian For 143:290–291

Gupta SK, Ram J, Singh H (2018a) Comparative study of transpiration in cooling effect of tree species in the atmosphere. J Geosci Environ Protect 6:151–166. https://doi.org/10.4236/gep.2018.68011

Gupta SK, Ram J, Singh H, Kumar P (2018b) Comparative study of cooling under shades of some forestry tree species with respect to ambient temperatures. J Geosci Environ Protect 6:51–66. https://doi.org/10.4236/gep.2018.611004

Gupta SK, Ram J, Singh H (2019a) Relationship between leaf area index and atmospheric cooling of tree species. Indian J for 42(3):203–208

Gupta SK, Singh H, Ram J (2019b) Need for a novel paradigm in forestry research: perspectives of comparative atmospheric cooling by different forestry effect tree species. Indian For 145(6):509–515

Halgamuge MN (2016a) Critical thresholds for RFR-induced morphological changes in plants. Bioelectromagnetics 37(5):327–337

Halgamuge MN (2016b) Weak radiofrequency radiation exposure from mobile phone radiation on plants. Electromagn Biol Med 36(2):213–235

Halgamuge MN, Yak SK, Eberhardt JL (2015) Reduced growth of soybean seedlings after exposure to weak microwave radiation from GSM 900 mobile phone and base station. Bioelectromagnetics 36(2):87–95

Halgamuge MN et al (2017) Disruption of vascular tissue organization by electromagnetic fields. Bot Stud 58(1):7–19

Husen A (2022a) Plant responses to radiofrequency electromagnetic fields: a review. Environ Exp Bot 190:104611

Husen A (ed) (2022b) Plants and their interaction to environmental pollution: damage detection, adaptation, tolerance, physiological and molecular responses. Elsevier

International Commission on Non-Ionizing Radiation Protection (ICNIRP) (2020) ICNIRP guidelines for limiting exposure to electromagnetic fields. Retrieved from https://www.icnirp.org/en/frequencies/index.html

Jaiswal D, Pandey A, Agrawal M, Agrawal SB (2023) Photosynthetic, biochemical and secondary metabolite changes in a medicinal plant Chlorophytum borivillianum (Safed musli) against low and high doses of UV-B radiation. Photochem Photobiol 99(1):45–56

Jangra A, Kumar K, Maikhuri S, Bhandari MS, Pandey S, Singh H, Barthwal S (2024) Unveiling stress-adapted endophytic bacteria: characterizing plant growth-promoting traits and assessing cross-inoculation effects on *Populus deltoides* under abiotic stress. Plant Physiol Biochem 210:108610. https://doi.org/10.1016/j.plaphy.2024.108610

Kanta C, Kumar A, Chauhan A, Singh H, Sharma IP (2024) The interplay between plant functional traits and climate change. In: Kumar N, Singh H (eds) Plant functional traits for improving productivity. Springer, Singapore. https://doi.org/10.1007/978-981-97-1510-7_3

Kesari KK et al (2012) Cellular effects of radiofrequency electromagnetic fields. J Environ Biol 33(2):231–237

Kumar N, Suyal DC, Sharma IP, Verma A, Singh H (2017) Elucidating stress proteins in rice (*Oryza sativa* L.) genotype under elevated temperature: a proteomic approach to understand heat stress response. 3 Biotech 7:205. https://doi.org/10.1007/s13205-017-0856-9

Kumar M, Singh MP, Singh H, Dhakate PM, Ravindranath NH (2019a) Forest working plan for the sustainable management of forest and biodiversity in India. J Sustain for 39:1–22. https://doi.org/10.1080/10549811.2019.1632212

Kumar N, Jeena N, Singh H (2019b) Elevated temperature modulates rice pollen structure: a study from the foothill of the Himalayan agro-ecosystem in India. 3 Biotech 9(5):175. https://doi.org/10.1007/s13205-019-1700-1

Kumar N, Jeena N, Kumar N, Gangola S, Singh H (2019c) Phytoremediation facilitating enzymes: an enzymatic approach for enhancing remediation process. In: Bhatt P (ed) Smart bioremediation technologies. Academic Press, Elsevier. 50 Hampshire Street, 5th Floor, Cambridge, MA 02139, United States, pp 289–306

Kumar S, Deshmukh PS, Banerjee BD, Abegaonkar MP, Tripathi AK (2019d) Effect of mobile phone radiation on antioxidant enzymes and its amelioration by vitamin C. Indian J Exp Biol 57(6):337–342

Kumar N, Singh H, Sharma SK (2020) Antioxidants: responses and importance in plant defense system. In: Roy chowdhury R, Choudhury S, Hasanuzzaman M, Srivastava S (eds) Sustainable agriculture in the era of climate change. Springer, Cham, pp 251–264. https://doi.org/10.1007/978-3-030-45669-6_11

Kumar D, Kumar R, Singh R (2020b) Proteomic responses of plants to electromagnetic fields. Plant Proteomics 7:212–226

Kumar A, Kumar P, Singh H, Kumar N (2021a) Modulation of plant functional traits under essential plant nutrients during seasonal regime in natural forests of Garhwal Himalayas. Plant Soil 465:197–212. https://doi.org/10.1007/s11104-021-05003-x

Kumar A, Kumar P, Singh H, Kumar N (2021b) Adaptation and mitigation potential of roadside trees with bio-extraction of heavy metals under vehicular emissions and their impact on physiological traits during seasonal regimes. Urban For Urban Green 58:126900. https://doi.org/10.1016/j.ufug.2020.126900

Kumar A, Kumar P, Singh H, Kumar N (2021c) Impact of plant functional traits on infiltration rate under different forests of Kempty watershed in Garhwal Himalaya. Indian J Soil Conser 49(1):38–44

Kumar A, Kumar P, Singh H, Bisht S, Kumar N (2021d) Relationship of physiological plant functional traits with soil carbon stock in the temperate forest of Garhwal Himalaya. Curr Sci 120(8):1368–1373. https://doi.org/10.18520/cs/v120/i8/1368-1373

Kumar M, Phukon SN, Paygude AC, Tyagi K, Singh H (2021e) Mapping phenological functional types (PhFT) in the Indian Eastern Himalayas using machine learning algorithm in Google Earth Engine. Comput Geosci 158:104980. https://doi.org/10.1016/j.cageo.2021.104982

Kumar S, Gupta S, Rattan S (2021f) Measurement of electromagnetic fields in urban environment: challenges and solutions. J Electromag Waves Appl 35(11):1492–1506

Kumar V et al (2021g) Effects of electromagnetic radiation on leaf morphology and photosynthesis. J Plant Physiol 189:34–45

Kumar V, Sharma R, Agrawal S (2021h) Epigenetic modifications and plant adaptation to radiofrequency radiation. Plant Sci 309:110948

Kumar A, Singh H, Kumari G, Bisht S, Malik A, Kumar A, Singh M, Raturi A, Barthwal S, Thakur A, Kaushal R (2022) Adaptive resilience of roadside trees to vehicular emissions via leaf enzymatic, physiological, and anatomical trait modulations. Environ Pollut 313:120191. https://doi.org/10.1016/j.envpol.2022.120191

Kumar N, Singh H, Giri K, Kumar A, Joshi A, Yadav A, Singh R, Bisht S, Kumari R, Jeena N, Khairakpam R, Mishra G (2024) Physiological and molecular insights into the allelopathic effects on agroecosystems under changing environmental conditions. Physiol Mol Biol Plants 30:417–433. https://doi.org/10.1007/s12298-024-01440-x

Lázaro A, Lundgren R, Totland Ø (2021) The effect of radiofrequency electromagnetic fields on wild pollinators: recent knowledge and future directions. Environ Pollut 273:116451

Li J, Lu J, Zhang Y, Wang Z, Zhang W, He Z (2019) Electromagnetic radiation of mobile phone affects growth, yield and fruit quality of strawberry (Fragaria × ananassa). Plant Physiol Biochem 139:646–652

Magone I (1996) Plant sensitivity to electromagnetic fields. Electromagn Biol Med 15(2):121–134

Mahajan S, Yadav SK, Kumar R (2021) Plant hormone signaling under electromagnetic field stress. Plant Signal Behav 16(4):1865896

Malik A, Kumar K, Kumari G, Kumar N, Singh H (2024) Application of functional traits in modelling productivity and resilience under climate change. In: Kumar N, Singh H (eds) Plant functional traits for improving productivity. Springer, Singapore. https://doi.org/10.1007/978-981-97-1510-7_5

Manta D et al (2017) Stress adaptation in plants exposed to radiofrequency radiation. Environ Stress Biol 15(4):299–312

Meena D, Singh H, Chaudhari SK (2011) Elucidating strontium response on growth dynamics and biochemical change in *Phaseolus mungo* L. Int J Agric Environ Biotechnol 4:107–113

Nath A, Singha H, Lahkar BP (2021) Urbanization, electromagnetic radiation and sparrows: a case study from Guwahati, India

Prakash V, Hunney K, Singh H (2022) Effect of elevated carbon dioxide on growth and development of *Santalum album* L. seedlings inoculated with plant growth promoting microorganisms in open top chambers. Int J Environ Agric Biotechnol 7(5):149–159

Priya M (2023) Effect of cell phones on growth and metabolism of *Vigna unguiculata* L. Walp. Dogo Rangsang Res J 13:34–37

Roux D, Priem F, Perrot-Rechenmann C (2006a) Heat shock proteins in plants: mechanisms of induction and function. J Plant Physiol 163(3):324–330

Roux D, Vian A, Girard S, Bonnet P, Paladian F, Davies E, Gaba V (2006b) Electromagnetic fields (900 MHz) evoke consistent molecular responses in tomato plants. Physiol Plant 128(2):283–288

Roux D, Vian A, Girard S, Bonnet P, Paladian F, Davies E, Gendraud M (2019) Electromagnetic fields (900 MHz) evoke consistent molecular responses in tomato plants. Physiol Plant 166(2):289–302

Sharma R, Singh H, Kaushik M, Nautiyal R, Singh O (2018) Adaptive physiological response, carbon partitioning, and biomass production of *Withania somnifera* (L.) Dunal grown under elevated CO_2 regimes. 3 Biotech 8:267. https://doi.org/10.1007/s13205-018-1292-1

Sharma P, Jha AB, Dubey RS (2021a) Reactive oxygen species, oxidative stress, and antioxidative defense mechanism in plants under stressful conditions. J Bot 2021:2170371

Sharma VP, Singh HP, Kohli RK, Batish DR (2021b) Mobile phone radiation inhibits Vigna radiata (mung bean) root growth by inducing oxidative stress. Sci Rep 11(1):13917

Singh H, Verma A (2013) Characterization and screening of high nitrogen efficient rice genotype to curtail environmental pollution. Int J Agric Environ Biotechnol 6(1):23–30

Singh H, Savita Sharma R, Sinha S, Kumar M, Kumar P, Verma A, Sharma SK (2017a) Physiological functioning of *Lagerstroemia speciosa* L. under heavy roadside traffic: an approach to screen potential species for abatement of urban air pollution. 3 Biotech 7(61):1–10. https://doi.org/10.1007/s13205-017-0690-0

Singh H, Verma A, Kumar M, Sharma R, Gupta R, Kaur M, Negi M, Sharma SK (2017b) Phytoremediation: a green technology to clean up the sites with low and moderate level of heavy metals. Austin Biochem 2:1012

Singh H, Sharma R, Savita SMP, Kumar M, Verma A, Ansari MW, Sharma SK (2018) Adaptive physiological response of *Parthenium hysterophorus* to elevated atmospheric CO_2 concentration. Indian For 144:1–14

Singh H, Yadav M, Kumar N, Kumar A, Kumar M (2020) Assessing adaptation and mitigation potential of roadside trees under the influence of vehicular emissions: a case study of *Grevillea robusta* and *Mangifera indica* planted in an urban city of India. PLoS ONE 15(1):e0227380. https://doi.org/10.1371/journal.pone.0227380

Singh S, Singh H, Sharma SK, Nautiyal R (2021) Seasonal variation in biochemical responses of bamboo clones in the sub-tropical climate of Indian Himalayan foothills. Heliyon 7(4):e06859. https://doi.org/10.1016/j.heliyon.2021.e06859

Singh H, Bandyopadhyaya S, Kumar A, Singh M, Malik A, Kumari G, Raturi A, Yadav SK, Kumar M, Barthwal S, Thakur A, Rawat PS (2022) Understanding the physiological and biophysical response of urban roadside plantations for assessing adaptation and mitigation mechanisms toward vehicular emissions. Urban Climate 44:101183. https://doi.org/10.1016/j.uclim.2022.101183

Singh M, Singh H, Kumar A, Kumar M, Barthwal S, Thakur A (2024a) Soil nitrogen availability determines the CO_2 fertilization effect on tree species (*Neolamarckia cadamba*): growth and physiological evidence. Environ Sustain 7:53–60. https://doi.org/10.1007/s42398-023-00300-w

Singh H, Kumar N, Kumar A (2024c) Enhancing resource use efficiency in crops through plant functional traits. In: Kumar N, Singh H (eds) Plant functional traits for improving productivity. Springer, Singapore. https://doi.org/10.1007/978-981-97-1510-7_6

Sivani S, Sudarsanam D (2012) Impacts of radio-frequency electromagnetic field (RF-EMF) from cell phone towers and wireless devices on biosystem and ecosystem—a review. Biol Med 4(4):202–216

Smith-Ramirez C, Montalba R, Soto R (2020) Influence of electromagnetic fields from high voltage power lines on the microbiota of rhizosphere and plants. J Environ Manage 259:110011

Statista (2022) Number of mobile phone users worldwide from 2013 to 2022

Tkalec M et al (2013a) Effects of radiofrequency radiation on plant reproduction. Biol Plant 57(3):601–608

Tkalec M, Malarić K, Pavlica M, Pevalek-Kozlina B, Vidaković-Cifrek Ž (2013b) Effects of radiofrequency electromagnetic fields on seed germination and root meristematic cells of Allium cepa L. Mutation Res/Gen Toxicol Environ Mutagenesis 743(1–2):77–81

Xie Y, Wei M, Wang Z, Wang L, Sun H (2021) Effects of electromagnetic radiation from mobile phone on growth and metabolism of rice. Environ Pollut 268(Pt A):115769

Yadav A, Jha S (2023) Effects of cell phone radiation on plants growth, active constituents and production. In: Plants and their interaction to environmental pollution. Elsevier, pp 299–307

Yadav SK, Singh H, Ginwal HS, Barthwal S (2019a) Elevated CO_2 enhanced growth and physiological process of *Populus deltoides* Bartr. ex Marsh. Indian For 145(1):23–27

Yadav SK, Singh H, Nautiyal R, Ginwal HS, Ansari SA, Barthwal S (2019b) Modulation of morpho-physiological responses in *Populus deltoides* by elevated carbon dioxide and temperature. For Sci 66:105–118. https://doi.org/10.1093/forsci/fxz048

Zhang Y, He Z, Zhang W, He M, Lu J (2020) Effects of electromagnetic radiation on growth and physiological characteristics of tea plant (*Camellia sinensis* L.) seedlings. Environ Sci Pollut Res 27(8):7933–7941

Zhao F, Wang P, Zhao Y (2007) Calcium signaling in plant stress response. Plant Cell Rep 26:771–782

Consequences of Artificial Light at Night (ALAN) on Urban Tree Functioning: A Case Study

Amit Kumar and Hukum Singh

Abstract Plants use light as a source of both energy and signalling. Street lighting is integral to urban infrastructure, providing safety and visibility during night hours. However, the widespread use of artificial lighting in urban areas has raised concerns about its potential impact on the physiological functioning of urban plants. This study aimed to understand the physiological responses of tree species at the seedling stage under exposure to artificial light at night (ALAN) provided by 50-W LEDs. The findings revealed that ALAN significantly affected chlorophyll content, stomatal conductance, transpiration, air and leaf temperature, leaf area, and soil moisture. All measured physiological functional traits declined under ALAN exposure compared to control plants. Further, long-term studies are recommended to draw comprehensive conclusions that will assist urban planners and policymakers in designing landscapes for developing green spaces in urban areas.

Keywords Artificial light · Plant growth and development · Morphology and biochemical processes

1 Introduction

Light is a crucial abiotic factor that provides the energy necessary for the growth and development of plants. Sunlight regulates numerous vital functions in living organisms and serves as the planet's primary energy source. Primary producers like plants, blue-green algae, and photoautotrophs absorb light, particularly in the 400–700 nm range, known as photosynthetically active radiation, and convert it into energy through photosynthesis, producing starch, sucrose, and other complex organic compounds as food reserves (Sharkey 2020).

A. Kumar (✉) · H. Singh
Plant Physiology Discipline, ICFRE-Forest Research Institute, PO New Forest, Dehradun, Uttarakhand 248006, India
e-mail: amiitkr.7@gmail.com

H. Singh (ed.), *Urban Forests, Climate Change and Environmental Pollution*,
https://doi.org/10.1007/978-3-031-67837-0_31

In addition to regulating plant growth and development, light plays a significant role in signaling processes such as photomorphogenesis (light-regulated morphogenesis), phototropism (light-mediated movements), circadian rhythms (light-regulated biological rhythms), and many other processes (Kami et al. 2010; Sharma et al. 2018; Singh et al. 2018; Gupta et al. 2018a, b). Light properties, including intensity, duration, and wavelength, can positively and negatively affect plants and other organisms (Nozue and Maloof 2006).

However, recent human developments, such as the construction of highways, buildings, LED lighting technology, product advertisements, and industrial growth, have led to increased artificial light in the environment, contributing to anthropogenic light pollution (Svechkina et al. 2020). The widespread use of artificial light at night (ALAN) poses severe and widespread effects on living organisms by altering life cycles and impacting biological processes, including photosynthetic activity, stomatal movements, enzyme activity, flower opening, and other critical functions (Gaston et al. 2014).

The interaction between plants and their environment is vital for determining current and future plant growth. ALAN has been shown to impact various micro environmental characteristics, including light, soil biology, humidity, and the biotic community, affecting interactions with living organisms, including plants (Haynes et al. 2020). Urbanization, electrification, population growth, and socioeconomic development have significantly expanded ALAN's density and spatiotemporal distribution. From 2012 to 2016, ALAN increased globally by approximately 1.8% per year, a trend likely to continue (Kyba et al. 2017). By disrupting circadian rhythms and altering plant perceptions of day length, ALAN can affect phenology, potentially leading to changes in species abundance, distribution, ecological community organization, and ecosystem functioning (Davies et al. 2012; Lewansik and Voigt 2014).

The long evolutionary history of life on Earth, which has relied heavily on light as an environmental factor, underscores the importance of understanding these impacts (Singh and Verma 2013; Kumar et al. 2017; Gupta and Singh 2017; Singh et al. 2017a, b). Plants have evolved complex pigment systems and photoreceptors to absorb and respond to specific wavelengths of light for growth regulation and hormonal functions (Argüello-Astorga and Herrera-Estrella 1998; Möglich et al. 2010). Unfortunately, the abundance of artificial light from urbanization and modernization in the twentieth century can significantly disrupt these natural cycles, a phenomenon referred to as "photopollution" (Verheijen 1984).

Night light pollution, defined as "the modulation of night natural glows by anthropogenic light sources," causes artificial skyglow and disrupts natural darkness (Falchi et al. 2016). This disruption affects approximately 80% of flowering plants sensitive to photoperiodism, influencing processes like leaf fall, bud-burst, and blossoming (Klimešová and Středa 2016). Some tree species have adapted by altering stomatal behavior to conserve water, but artificial lighting can delay these adaptations, impacting protein content and seed size (Chaney 2002).

Light intensity and spectral composition, influenced by the moon phase, season, and weather, provide crucial cues for biological processes like seedling development, photosynthesis, growth, and budding (Briggs and Christie 2002). Phototropism, chloroplast movements, and stomatal opening are all affected by light intensity and direction, impacting photosynthetic efficiency and water use (Celaya and Liscum 2005; Wada et al. 2003; Dietrich et al. 2001).

Light pollution also affects pollination, with studies showing reduced nocturnal pollinator visits and crop yields in areas with ALAN (Knop et al. 2017). Abiotic and biotic stresses, including those induced by ALAN, can alter plant physiological, biochemical, and molecular processes, causing oxidative stress and reducing overall plant health (Kwak et al. 2018). Additionally, introducing excessive artificial light has decreased diurnal relative humidity and temperature differences, potentially affecting fruit development, seed set, and yield potential (Gipson and Joham 1968).

Given the significance of these impacts, understanding the intricate mechanisms of plant growth, photosynthesis, and physiological responses to artificial light is crucial for optimizing landscape planning and ensuring successful plant cultivation in urban environments (Kumar et al. 2022; Singh et al. 2022; Kumar et al. 2024; Singh et al. 2024a; Singh 2024). Further research is necessary to establish precise guidelines for utilizing artificial light in different landscapes and mitigate ALAN's adverse effects on plant ecosystems. Therefore, this study aims to understand the physiological modulations and biophysical changes in the leaves of tree species exposed to street lighting.

2 Materials and Methodology

2.1 Study Area

The Forest Research Institute (FRI) is located in Dehradun, nestled in the lush green mountains of the lower Shiwalik range (Fig. 1). Covering 450 ha, it is set against the backdrop of the outer Himalayas. The FRI and College Area Campus, a census-designated town, lies between Kaulagarh in the north and the Indian Military Academy in the south, with the Tons River forming its western boundary. The area boasts immense biodiversity in Dehradun's subtropical climate, which experiences significant seasonal variations, including some endangered flora and fauna.

Established in the 1920s by the British for forest management practices, FRI is primarily an artificial ecosystem. The land, previously used for agriculture, was acquired in the 1920s and transformed into forested areas. Initially, the forest management practices focused on timber production. However, the focus moved towards conservation following India's Forest Policy shift in the 1970s. The entire New Forest area is classified as a reserve forest under section 26 of the Indian Forest Act, 1927 (IFA) and under the administrative control of the Director of FRI, Dehradun. The FRI campus was officially declared a reserve forest in 1942.

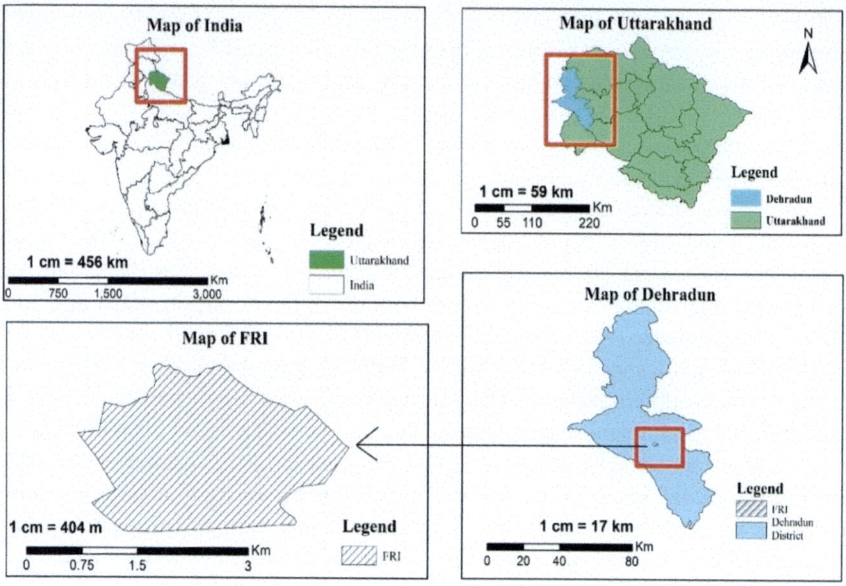

Fig. 1 Location map of FRI, Dehradun study area

2.2 Climate

The climate at the Forest Research Institute (FRI) campus is conducive to plant development, with most trees planted for timber and non-wood forest products. The mean yearly air temperature recorded at the FRI observatory from 1980 to 2010 is 22.73 °C. The mean monthly temperature varies between 32.23 and 33.37 °C. May and June are the hottest, while December and January are the coldest. The monsoon typically starts in mid-June and lasts until mid-September.

2.3 Vegetation

The vegetation at FRI is a mix of deciduous and evergreen species, rich in floral diversity. The area hosts trees, shrubs, bamboo, climbers, grasses, and herbs. The habitats include Chir-pine forest, Sal forest, Teak plantation, mixed forest, Botanical Garden, Arboretum, and Bamboosetum. The vegetation primarily consists of mixed plantations typical of a subtropical forest.

2.4 Experimental Plant Species

The selected plant species for the study are commonly found in urban areas and are frequently exposed to street light. The chosen species are Eucalyptus spp., Paulownia (*Paulownia tomentosa*), and Indian butter tree (*Diploknema butyracea*). Eighteen seedlings of each species were used in the study, ensuring robust results through replication.

2.5 Fitting of Street Lights

Three LED lights, each of 50W, were installed on a pole in three directions to ensure consistent light exposure for the plants during the study (Fig. 2).

2.6 Exposure of Plants to Different Conditions

The plants were divided into two groups for the study:

1. Controlled (Natural) Condition: Plants received natural daylight and no light at night.
2. ALAN Condition: Plants were exposed to light for 24 h, receiving natural sunlight during the day and artificial light at night, simulating continuous light exposure.

This setup allowed the study to compare the effects of artificial light at night (ALAN) on the physiological responses of the selected plant species.

2.7 Management of Seedlings

The management of seedlings for the selected species involved several methods to ensure optimal growth:

Watering: The plants were watered to maintain adequate soil moisture.
Sunlight Exposure: Efforts were made to ensure the seedlings received proper sunlight, avoiding any shadow zones that could impede growth.
Weeding: Weeding was carried out weekly to prevent competition for nutrients and water.
Grouting: Grouting was performed weekly to support the seedlings and promote healthy growth.

Fig. 2 Exposure of plant species to ALAN condition

2.8 Morphological and Biophysical Monitoring of Plants

2.8.1 Plant Height Measurement

The height of the treated plant species was measured using a meter tape. Measurements were taken from the base of the stem (at the soil surface) to the top of the canopy (tip of the apical bud) and recorded in centimeters (cm).

2.8.2 Seedling Diameter Measurement

The diameter of the seedlings was measured using a digital Vernier caliper, with the results expressed in millimeters (mm).

2.8.3 Leaf Area Estimation

The leaf area of the selected species was estimated using the graph paper method:

The leaf was placed on graph paper.
The outer boundary of the leaf was traced onto the paper with a pencil.
The leaf area was calculated by counting the squares on the graph paper covered by the leaf outline.
The total number of squares counted was considered the leaf's area.

2.8.4 Leaf Water Content (LWC)

Fresh leaves were weighed immediately after collection. They were then oven-dried at 80 °C until a constant weight was achieved. The leaf water content was calculated using the following equation and expressed as a percentage:

$$LWC = \frac{(Fresh\,weight - Dry\,weight)}{Fresh\,weight} * 100$$

where, LWC is leaf water content, and weight is in gram (gm).

2.8.5 Monitoring Physiological Response

The physiological response of seedlings was assessed using a portable photosynthesis system (CIRAS-3 PP System). We monitored several physiological traits, including leaf temperature, air temperature, stomatal conductance, leaf transpiration, and leaf chlorophyll content. Measurements were taken under clear sky conditions from well-expanded and healthy leaves. Leaf chlorophyll content was estimated from fresh, healthy leaves using a chlorophyll concentration meter (Apogee Instruments), and the chlorophyll content was expressed as the Chlorophyll Content Index (CCI).

2.8.6 Soil Moisture Content Analysis

Soil samples were collected from the growing seedlings' bags to monitor soil moisture content. A known weight of soil samples was collected and then oven-dried at 80 °C until the weight remained constant. The soil moisture content (SMC) was calculated using the equation below and expressed as a percentage:

$$SWC = \frac{(Fresh\, weight - Dry\, weight)}{Dry\, weight} * 100$$

3 Results

3.1 Chlorophyll Content

The effect of artificial light at night (ALAN) on the chlorophyll content of plant species differs from that of plants under controlled conditions (Fig. 3). ALAN significantly impacted chlorophyll content in the selected species. Under controlled conditions, the maximum chlorophyll content was observed in the Eucalyptus species, followed by the Paulownia species, with the Indian butter tree having the least. Specifically, Eucalyptus exhibited a chlorophyll content index (CCI) of 23.16, Paulownia had a CCI of 13.2, and the Indian butter tree had a CCI of 10.3.

In contrast, the chlorophyll content decreased across all species under ALAN conditions (Fig. 3). Eucalyptus had a CCI of 15.3, Paulownia had a CCI of 11.4, and the Indian butter tree had a CCI of 10.14. This indicates that ALAN has a significant adverse effect on the chlorophyll content of these plant species, with Eucalyptus showing the most substantial reduction.

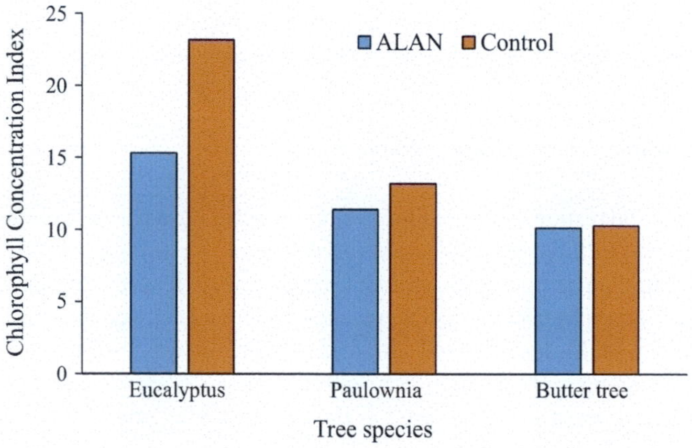

Fig. 3 The impact of ALAN on chlorophyll content of different plant species

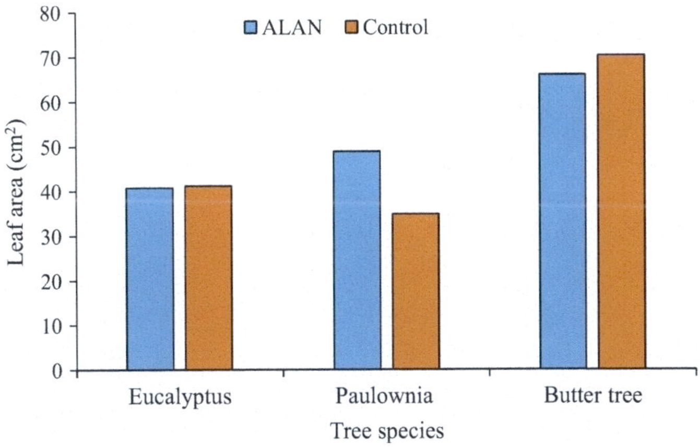

Fig. 4 Leaf area of different plant species under ALAN and controlled condition

3.2 *Leaf Area*

The average leaf area was higher in controlled conditions than in ALAN plantations for most species (Fig. 4). In controlled conditions, the average leaf area of the Indian butter tree was 70.3 cm^2, while at an ALAN site, it was 66 cm^2. For Paulownia, the leaf area was 48.9 cm^2 under ALAN conditions and 34.8 cm^2 in controlled conditions. Eucalyptus showed a minimal difference between the two conditions, with a leaf area of 41.1 cm^2 in controlled conditions and 40.7 cm^2 under ALAN conditions, indicating that the impact of ALAN on the Eucalyptus leaf area was negligible.

3.3 *Plant Height*

The average plant height was more in controlled conditions than ALAN plants (Fig. 5). Specifically, in the controlled condition, the average plant height of Eucalyptus species was 99.4 cm, whereas in the ALAN site, it measured 96.8 cm. Similarly, in the Indian butter tree, the height of plants slightly increased in controlled conditions, measuring 45.6 cm compared to 44 cm in ALAN plantations. The height of Paulownia species remained consistent at 46 cm in ALAN and controlled conditions.

3.4 *Plant Girth*

The average plant dbh was greater in plantations under ALAN conditions than controlled conditions (Fig. 6). Specifically, in ALAN, the average dbh of Indian butter

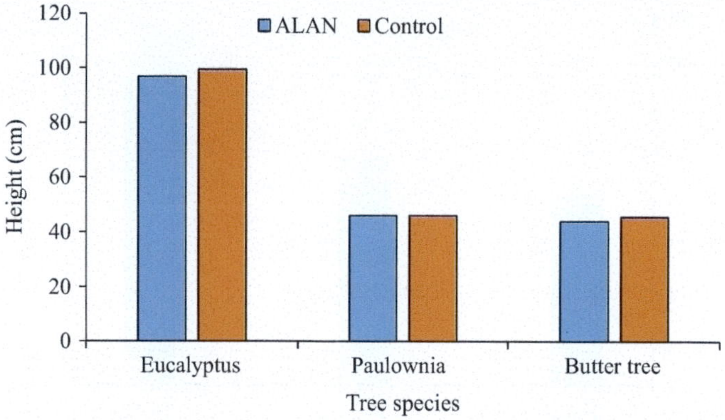

Fig. 5 Effect of ALAN on plant height of different plant species

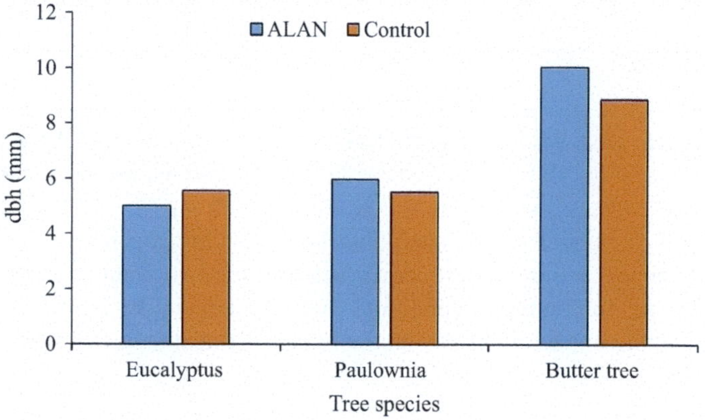

Fig. 6 Effect of ALAN on dbh of different plant species

trees was measured at 10.046 mm, whereas in controlled sites, it was 8.86 mm. For Paulownia species, the dbh of plants in ALAN conditions was slightly higher than controlled conditions, with measurements of 5.97 mm and 5.54 mm, respectively. Conversely, in Eucalyptus species, it was noted that the dbh in controlled plantations was slightly more than in ALAN plantations, with average girths of 5.55 mm and 5.01 mm, respectively.

3.5 *Soil Moisture*

The average soil moisture content was higher for all species under ALAN conditions than in controlled conditions (Fig. 7). In the ALAN condition, the average soil

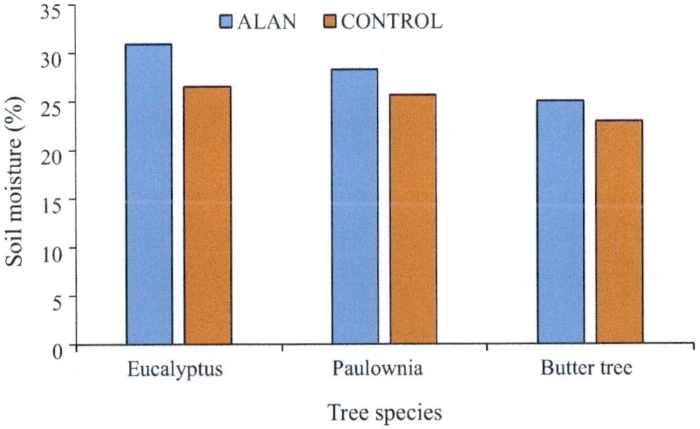

Fig. 7 Effect of ALAN on soil moisture content of different plant species

moisture content for Eucalyptus species was 30.88%, whereas it was 26.5% under controlled conditions. Similarly, for Paulownia species, the soil moisture content under ALAN conditions was higher at 28.26%, compared to 25.6% in controlled conditions. In Indian butter trees, the soil moisture content was also higher under ALAN conditions, measuring 25%, while it was 22.9% under controlled conditions.

3.6 *Leaf Water Content*

The leaf water content varied among species under different conditions (Fig. 8). The average water content in Eucalyptus species was 59.16% under ALAN conditions, slightly higher than the 58.3% than controlled conditions. Conversely, in Paulownia species, the difference in leaf water content between ALAN and controlled conditions was minimal, with values of 70.98% and 70.19%, respectively. However, in the Indian butter tree, the leaf water content was higher under controlled conditions, measuring 51.5%, compared to 48.7% under ALAN conditions.

3.7 *Physiological Response of Eucalyptus Species*

The average physiological characteristics were higher in the ALAN condition plants compared to the control plants (Fig. 10). In the ALAN condition, the ambient temperature was measured at 28.97 °C, whereas in the control condition, it was 27.19 °C (Fig. 9). Similarly, the leaf temperature was notably higher in the ALAN condition, reaching 31.12 °C, compared to 27.47 °C in the control condition. Stomatal conductance was also higher in the ALAN condition (5.16 mol H_2O m^{-2} s^{-1}) compared to

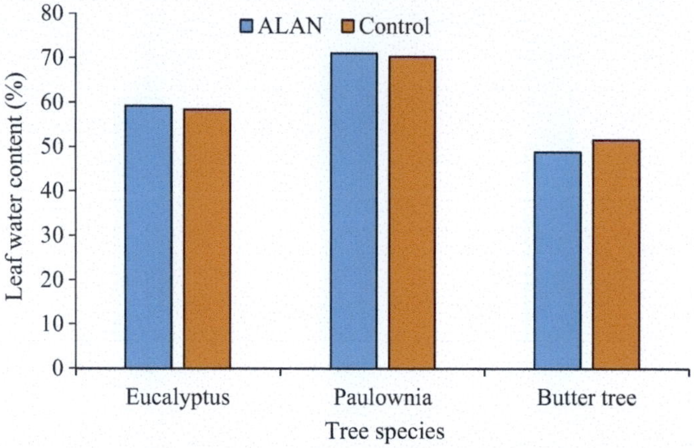

Fig. 8 Effect of ALAN on leaf water content of different plant species

Fig. 9 Effect of ALAN on air and leaf temperature of Eucalyptus species

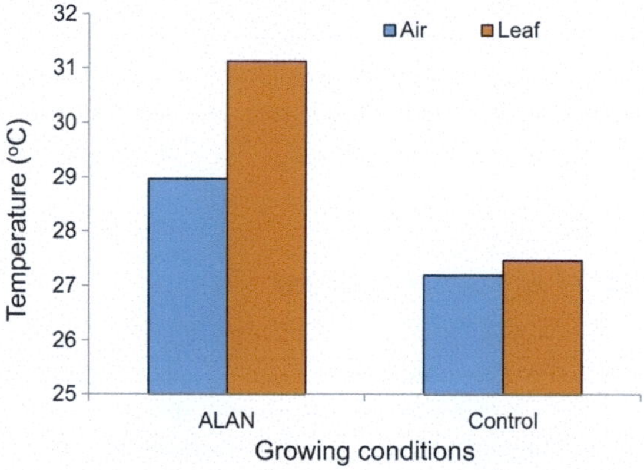

control (4.12 mol H_2O m^{-2} s^{-1}). This trend persisted in transpiration rates, where the ALAN condition exhibited greater values (0.26 mmol H_2O m^{-2} s^{-1}) than the control condition (0.18 mmol H_2O m^{-2} s^{-1}).

3.8 *Physiological Response of Paulownia Species*

The average physiological characteristics were found to be higher in the plants under ALAN conditions compared to the control plants (Fig. 12). In the ALAN condition, the ambient temperature was measured at 29.01 °C, while in the control condition, it was 27.55 °C (Fig. 11). Leaf temperature was also higher in the ALAN condition,

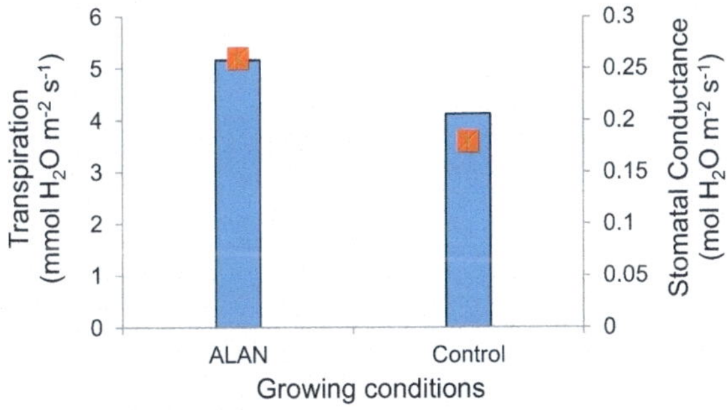

Fig. 10 Effect of ALAN on leaf transpiration and stomatal conductance of Eucalyptus species

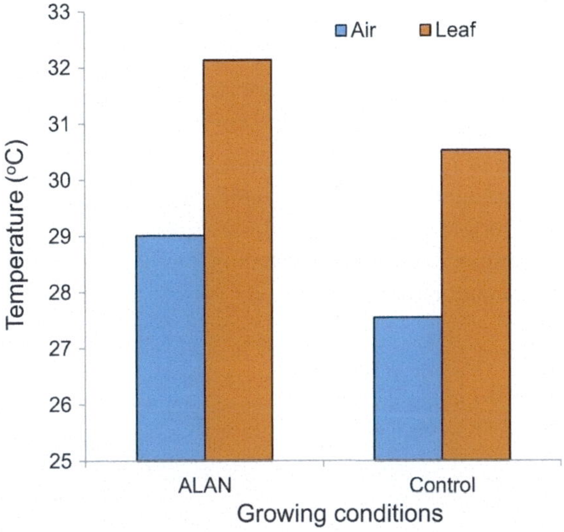

Fig. 11 Effect of ALAN on air and leaf temperature of Paulownia species

reaching 32.13 °C, compared to 30.52 °C in the control condition. Stomatal conductance was higher in the ALAN condition (8.42 mol H_2O m^{-2} s^{-1}), whereas it was 5.42 mol H_2O m^{-2} s^{-1} in the control condition (Fig. 12). Similarly, transpiration rates were more excellent in the ALAN condition at 0.55 mmol H_2O m^{-2} s^{-1} compared to 0.35 mmol H_2O m^{-2} s^{-1} in the control condition.

3.9 *Physiological Response of Indian Butter Tree*

The average physiological characteristics were observed to be higher in the ALAN condition compared to the control plants (Fig. 14). In the ALAN condition, the ambient temperature was measured at 28.83 °C, whereas in the control condition, it was 27.52 °C. Leaf temperature was also higher in the ALAN condition (30.98 °C),

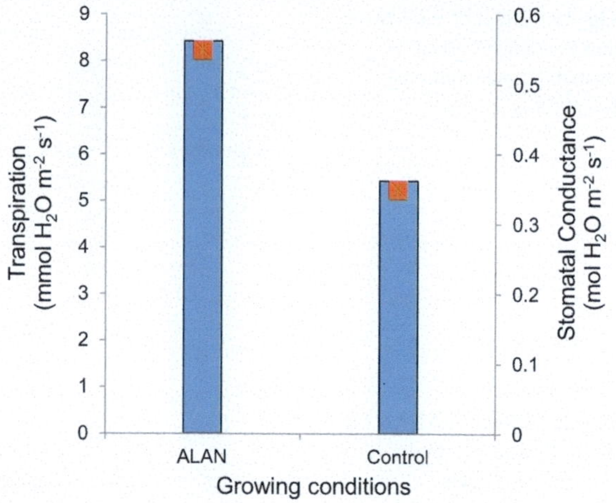

Fig. 12 Effect of ALAN on leaf transpiration and stomatal conductance of Paulownia species

compared to 25.17 °C the control condition (Fig. 13). Stomatal conductance was notably higher in the ALAN condition (measured at 3.03 mol H_2O m^{-2} s^{-1}) compared to 1.61 mol H_2O m^{-2} s^{-1} in the control condition. Similarly, transpiration rates were greater in the ALAN condition (6.05 mmol H_2O m^{-2} s^{-1}) than the control condition (2.17 mmol H_2O m^{-2} s^{-1}) (Fig. 14).

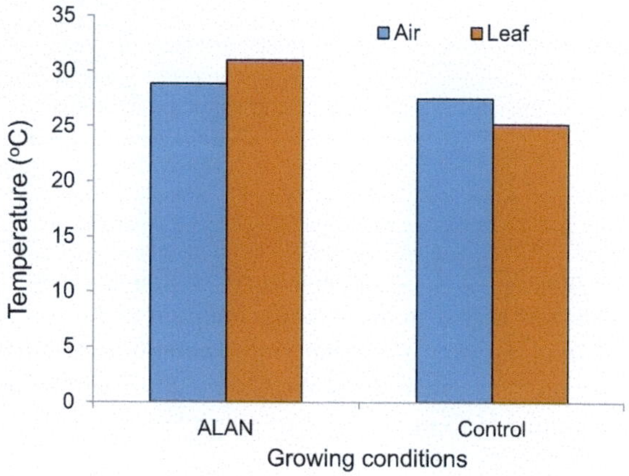

Fig. 13 Effect of ALAN on air and leaf temperature of Indian butter tree

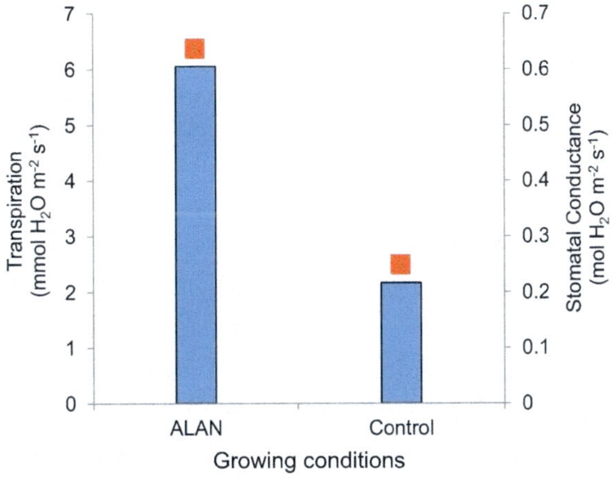

Fig. 14 Effect of ALAN on leaf transpiration and stomatal conductance of Indian butter tree

4 Discussion

4.1 *Chlorophyll Content*

Chlorophyll is crucial for plant health and photosynthesis, with higher concentrations indicating active photosynthetic activity and overall plant health (Palta 1990; Meena et al. 2011; Kumar et al. 2019c; Kumar et al. 2020;Singh and Kumar 2022; Singh et al. 2024a). The chlorophyll content was significantly increased in all three species under controlled conditions. Eucalyptus species exhibited the highest chlorophyll content in the control condition, followed by the Paulownia species and the Indian butter tree. The most remarkable increment in total chlorophyll was observed in Eucalyptus, with approximately 51.3% increase in the control condition compared to 15.3% in the ALAN condition. Paulownia species showed a variation in chlorophyll content from control to ALAN condition, with an increase of approximately 15.7%. Conversely, the chlorophyll content of the Indian butter tree varied from 10.14 in the ALAN condition to 10.3 in the control condition. With its higher chlorophyll content, Eucalyptus appears well-suited for roadside plantations, even under artificial light conditions, as it demonstrates robust growth and development.

4.2 *Leaf Area*

The leaf area directly impacts light interception and photoperiodism in plants (Bréda 2003). Eucalyptus species exhibited a minimal change in leaf area (approx. 0.98%) from control to ALAN condition. In contrast, Paulownia species showed a substantial increase in leaf area (40.5%) under the ALAN condition. Indian butter tree increased leaf area (6.5%) from ALAN to control condition. These variations indicate potential adaptations of the selected species to street light conditions, influencing their growth patterns.

Given its smaller leaf area and open canopy, Eucalyptus species seems suitable for roadside plantation or under street light conditions, facilitating optimal growth and development even under artificial light.

4.3 Plant Height

Plant height significantly increased under controlled conditions, with the Indian butter tree exhibiting the highest height in the control condition, followed by Eucalyptus species and Paulownia species. Indian butter tree showed the most incredible height increment (approx. 3.6%) in the control condition compared to the ALAN condition. Eucalyptus species demonstrated a height increment of approximately 2.68%, while Paulownia species exhibited no change in height between conditions.

Ideal plant height for roadside plantations should surpass the height of street lights, minimizing light interference (Bose et al. 2018; Dhyani et al 2021; Kumar et al. 2021a, b, c, d, e; Prakash et al. 2022). Indian butter tree, with its faster growth and height exceeding street lights, is suitable for roadside plantation, followed by Eucalyptus species.

4.4 Plant Girth

Plant girth, influenced by artificial light, varies among species (Dorais and Gosselin 2000; Haynes et al. 2020). Indian butter tree exhibited the most tremendous girth increase (approx. 13.6%) under ALAN condition, followed by Paulownia species. Conversely, Eucalyptus species demonstrated a decrease in circumference (approx. 9%) under ALAN condition. Thinner girth, as observed in Eucalyptus species, suggests suitability for growth under artificial light conditions, particularly for roadside plantations.

4.5 Soil Moisture

Soil moisture, crucial for plant growth (Haynes et al. 2020), was significantly increased under ALAN conditions across all species. Eucalyptus species exhibited the highest soil moisture content (approx. 16.2%) under ALAN conditions, followed by Paulownia species and the Indian butter tree. Eucalyptus's higher soil moisture content suggests efficient nutrient uptake even under artificial light conditions.

4.6 Leaf Water Content

Leaf water content, indicative of plant health, varied among species (Attia et al. 2015). Paulownia species demonstrated the highest leaf water content (approx. 70.9%), while Indian butter trees exhibited the lowest (approx. 48.71%). Paulownia's minimal water loss under artificial light conditions suggests its suitability for growth without hindrance.

4.7 Physiological Characteristics of Different Plant Species

Physiological characteristics, including ambient temperature, leaf temperature, stomatal conductance, and transpiration, were significantly higher under ALAN conditions across all species. Eucalyptus species exhibited the most pronounced changes, followed by the Paulownia species and the Indian butter tree. The interrelationship of physiological characteristics influences plant growth and development under continuous light conditions (Singh 2021; Sharma and Singh 2021; Singh et al. 2021). Eucalyptus species, with minimal changes in physiological responses, appear well-adapted to artificial light conditions, making them more suitable for plantation in such environments. Conversely, Indian butter trees, with substantial changes in physiological characteristics, may not be ideal for plantation under artificial light conditions (Singh et al. 2020; Kumar et al. 2021a, 2021b, 2021c, 2021d).

Understanding the physiological responses of different plantation species to artificial light conditions is crucial for informed landscape planning and ensuring optimal growth and development in urban environments. Further research is necessary to explore the underlying mechanisms and establish precise guidelines for utilizing artificial light in various planting scenarios (Yadav et al. 2019; Gupta et al. 2019a; Gupta et al. 2019a; Kumar et al. 2019a, b).

5 Conclusion

The impact of street lights on plant physiology encompasses a broad and significant field of study. Prolonged exposure to artificial lighting, such as that emitted by street lights, can disrupt the intricate balance of plant responses to light cues, altering plant phenology and morphology. The altered spectra of street lights may interfere with essential plant processes mediated by light receptors like phytochromes and cryptochromes, potentially leading to abnormal growth patterns and physiological imbalances. Various measures can be implemented, such as using shielded lighting fixtures to direct light downward, minimizing light spills, and adjusting street light intensity during off-peak hours to mitigate these adverse effects. However, our research indicates that Artificial Light At Night (ALAN) significantly affects various physiological parameters, including soil moisture, transpiration, stomatal conductance, and chlorophyll concentration. Compared to control plants, those exposed to ALAN exhibited decreased physiological responses. Further investigation is essential to deepen our understanding of street light impacts on plant physiology. This includes studying species-specific responses to artificial lighting, examining the effects of different light spectra and intensities, and conducting long-term assessments of light pollution's ecological consequences on plant communities and their interactions with other organisms. Advancing knowledge in this area may help to develop more effective strategies to mitigate the adverse effects of street lights on plant health and ecosystem dynamics.

References

Argüello-Astorga G, Herrera-Estrella L (1998) Evolution of light-regulated plant promoters. Annu Rev Plant Biol 1:525–555

Attia Z, Domec JC, Oren R, Way DA, Moshelion M (2015) Growth and physiological responses of isohydric and anisohydric poplars to drought. J Exp Bot 66(14):4373–4381

Bose B, Pant B, Singhal RK, Kumar M, Mondal S (2018) Phytochrome: physiology molecular aspects and sustainable crop production. In: Emerging trends of plant physiology for sustainable crop production

Bréda GNJJ (2003) Ground-based measurements of leaf area index: a review of methods, instruments and current controversies. J Exp Bot 54:2403–2417

Briggs WR, Christie JM (2002) Phototropins 1 and 2: versatile plant blue-light receptors. Trends Plant Sci 7:204–210. https://doi.org/10.1016/s1360-1385(02)02245-8

Briggs WR (2006) Physiology of plant responses to artificial lighting. In: Rich C, Longcore T (eds) Ecological consequences of artificial night lighting. Island Press, Washington, pp 389–411

Celaya RB, Liscum E (2005) Phototropins and associated signaling: providing the power of movement in higher plants. Photochem Photobiol 81:73–80. https://doi.org/10.1111/j.1751-1097.2005.tb01524.x

Chaney W (2002) Does night lighting harm trees? Circular FNR-FAQ-17, Department of Forestry and Natural Resources, Purdue University. Accessed 13 Nov 2016

Davies TW, Bennie J, Gaston KJ (2012) Street lighting changes the composition of invertebrate communities. Biol Lett 8:764–767. https://doi.org/10.1098/rsbl.2012.0216

Dhyani K, Kuniyal HB, Singh H, Sobha (2021). Growth and physiological potential of Terminalia arjuna under elevated CO_2 levels in Open top chamber condition. J Appl Nat Sci 13(3):1121-1126. https://doi.org/10.31018/jans.v13i3.2490. ISSN 2231-5209

Dietrich P, Sanders D, Hedrich R (2001) The role of ion channels in light dependent stomatal opening. J Exp Bot 52:1959–1967. https://doi.org/10.1093/jexbot/52.363.1959

Dorais M, Gosselin A (2000) Physiological response of greenhouse vegetable crops to supplemental lighting. Acta Hort 580:59–67

Falchi F, Cinzano P, Duriscoe D, Kyba CC, Elvidge CD, Baugh K, Portnov BA, Rybnikova NA, Furgoni R (2016) The new world atlas of artificial night sky brightness. Sci Adv 2:e1600377

Gaston KJ, Bennie J (2014) Demographic effects of artificial nighttime lighting on animal populations. Environ Rev 22:1–8. https://doi.org/10.1139/er-2014-0041

Gaston KJ, Bennie J, Davies TW, Hopkins J (2013) The ecological impacts of night time light pollution: a mechanistic appraisal. Biol Rev 88:912–927

Gaston KJ, Duffy JP, Gaston S, Bennie J, Davies TW (2014) Human alteration of natural light cycles: causes and ecological consequences. Oecologia 176(4):917–931

Gipson JR, Joham HE (1968) Influence of night temperature on growth and development of cotton (*Gossypium hirsutum* L.). I. Fruiting and boll development. Agron J 60:292–295

Gupta SK, Singh H (2017) Observations on sporadic flowering in *Ficus benjamina* L. and *Peltophorum africanum* L. in new forest. Indian for 143:290–291

Gupta SK, Ram J, Singh H (2018a) Comparative study of transpiration in cooling effect of tree species in the atmosphere. J Geosci Environ Protect 6:151–166. https://doi.org/10.4236/gep.2018.68011

Gupta SK, Ram J, Singh H, Kumar P (2018b) Comparative study of cooling under shades of some forestry tree species with respect to ambient temperatures. J Geosci Environ Protect 6:51–66. https://doi.org/10.4236/gep.2018.611004

Gupta SK, Ram J, Singh H (2019a) Relationship between leaf area index and atmospheric cooling of tree species. Indian J for 42(3):203–208

Gupta SK, Singh H, Ram J (2019b) Need for a novel paradigm in forestry research: perspectives of comparative atmospheric cooling by different forestry effect tree species. Indian for 145(6):509–515

Haynes KJ, DiBiase E, Roach DA, Carr DE (2020) Interactions between artificial light at night soil moisture and plant density affect the growth of a perennial wildflower. Oecologia 1–8

Johnston TJ, Pendleton JW, Peters DB, Hicks DR (1969) Influence of supplemental light on apparent photosynthesis, yield, and yield components of soybeans (*Glycine max* L.) 1. Crop Sci 9:577–581

Kami C, Lorrain S, Hornitschek P, Fankhauser C (2010) Light-regulated plant growth and development. In Current topics in developmental biology. CRC Academic Press, pp 29–66.

Klimešová J, Středa T (2016) Agrometeorological and biological aspects of maize transpiration. In: Brzezina et al (eds) Mendel and bioclimatology. Mendel University in Brno, Brno, pp 150–156

Knop E, Zoller L, Ryser R, Gerpe C, Hörler M, Fontaine C (2017) Artificial light at night as a new threat to pollination. Nature 548(7666):206–209

Kumar M, Singh MP, Singh H, Dhakate PM, Ravindranath NH (2019a) Forest working plan for the sustainable management of forest and biodiversity in India. J Sustain for 39:1–22. https://doi.org/10.1080/10549811.2019.1632212

Kumar A, Kumar P, Singh H, Kumar N (2021a) Modulation of plant functional traits under essential plant nutrients during seasonal regime in natural forests of Garhwal Himalayas. Plant Soil 465:197–212. https://doi.org/10.1007/s11104-021-05003-x

Kumar A, Kumar P, Singh H, Kumar N (2021b) Adaptation and mitigation potential of roadside trees with bio-extraction of heavy metals under vehicular emissions and their impact on physiological traits during seasonal regimes. Urban Fo Urban Greening 58:126900. https://doi.org/10.1016/j.ufug.2020.126900

Kumar A, Kumar P, Singh H, Kumar N (2021c) Impact of plant functional traits on infiltration rate under different forests of Kempty watershed in Garhwal Himalaya. Indian J Soil Conserv 49(1):38–44

Kumar A, Kumar P, Singh H, Bisht S, Kumar N (2021d) Relationship of physiological plant functional traits with soil carbon stock in the temperate forest of Garhwal Himalaya. Curr Sci 120(8):1368–1373. https://doi.org/10.18520/cs/v120/i8/1368-1373

Kumar M, Phukon SN, Paygude AC, Tyagi K, Singh H (2021e) Mapping Phenological Functional Types (PhFT) in the Indian Eastern Himalayas using machine learning algorithm in Google Earth Engine. Comput Geosci 158:104982. https://doi.org/10.1016/j.cageo.2021.104982

Kumar A, Singh H, Kumari G, Bisht S, Malik A, Kumar A, Singh M, Raturi BS, Thakur A, Kaushal R (2022) Adaptive resilience of roadside trees to vehicular emissions via leaf enzymatic, physiological, and anatomical trait modulations. Environ Pollut 313:120191. https://doi.org/10.1016/j.envpol.2022.120191

Kumar N, Singh H, Giri K, Kumar A, Joshi A, Yadav A, Singh R, Bisht S, Kumari R, Jeena N, Khairakpam R, Mishra G (2024) Physiological and molecular insights into the allelopathic effects on agroecosystems under changing environmental conditions. Physiol Mol Biol Plants 30:417–433. https://doi.org/10.1007/s12298-024-01440-x

Kumar N, Suyal DC, Sharma IP, Verma A, Singh H (2017) Elucidating stress proteins in rice (*Oryza sativa* L.) genotype under elevated temperature: a proteomic approach to understand heat stress response. 3 Biotech 7:205. https://doi.org/10.1007/s13205-017-0856-9

Kumar N, Jeena, Singh H (2019b) Elevated temperature modulates rice pollen structure: a study from the foothill of the Himalayan agro-ecosystem in India. 3 Biotech 9(5):175. https://doi.org/10.1007/s13205-019-1700-1

Kumar N, Jeena N, Kumar N, Gangola S and Singh H. (2019c). Phytoremediation facilitating enzymes: an enzymatic approach for enhancing remediation process. In: Bhatt P (ed) Smart bioremediation technologies. Academic Press, Elsevier, 50 Hampshire Street, 5th Floor, Cambridge, MA 02139, United States, pp 289–306

Kumar N, Singh H, Sharma SK (2020) Antioxidants: responses and importance in plant defense system. In: Roychowdhury R, Choudhury S, Hasanuzzaman M, Srivastava S (eds) Sustainable agriculture in the era of climate change. Springer, Cham, pp 251–264. https://doi.org/10.1007/978-3-030-45669-6_11.

Kwak MJ, Je SM, Cheng HC, Seo SM, Park JH, Baek SG et al (2018) Night light adaptation strategies for photosynthetic apparatus in yellow-poplar (*Liriodendron tulipifera* L.) exposed to artificial night lighting. Forests 9(2):74

Kyba CCM, Tong KP, Bennie J, Birriel I, Birriel JJ, Cool A et al (2015) Worldwide variations in artificial skyglow. Sci Rep 5:8409

Kyba CC et al (2017) Artificially lit surface of Earth at night increasing in radiance and extent. Sci Adv 3:e1701528

Lewansik D, Voigt CC (2014) Artificial light puts ecosystem services of frugivorous bats at risk. J Appl Ecol 52:388–394. https://doi.org/10.1111/1365-2664.12206

Meena D, Singh H, Chaudhari SK (2011) Elucidating strontium response on growth dynamics and biochemical change in *Phaseolus mungo* L. Int J Agric Environ Biotechnol 4:107–113

Möglich A, Yang X, Ayers RA, Moffat K (2010) Structure and function of plant photoreceptors. Annu Rev Plant Biol 61:21–47

Nozue K, Maloof JN (2006) Diurnal regulation of plant growth. Plant, Cell Environ 29(3):396–408

Palta JP (1990) Leaf chlorophyll content. Remote Sens Rev 5(1):207–213. https://doi.org/10.1080/02757259009532129

Prakash V, Hunney K, Singh H (2022) Effect of elevated carbon dioxide on growth and development of *Santalum album* L. seedlings inoculated with plant growth promoting microorganisms in Open Top Chambers. Int J Environ Agric Biotechnol 7 (5):149–159

Sharkey TD (2020) Emerging research in plant photosynthesis. In: Emerging topics in life sciences

Sharma R, Singh H, Kaushik M, Nautiyal R, Singh O (2018) Adaptive physiological response, carbon partitioning, and biomass production of *Withania somnifera* (L.) Dunal grown under elevated CO_2 regimes. 3 Biotech 8:267. https://doi.org/10.1007/s13205-018-1292-1

Sharma R, Singh H (2021) Alteration in biochemical constituents and nutrients partitioning of *Asparagus racemosus* in response to elevated atmospheric CO_2 concentration. Environ Sci Pollut Res 29:6812–6821. https://doi.org/10.1007/s11356-021-16050-3

Singh H (2021) An integrated approach considering physiological and biophysical based indicators for assessing tolerance of roadside plantations of *Alstonia scholaris* towards urban roadside air pollution: an assessment of adaptation of plantations for mitigating urban roadside air pollution. Trees. https://doi.org/10.1007/s00468-021-02179-8

Singh H, Verma A (2013) Characterization and screening of high nitrogen efficient rice genotype to curtail environmental pollution. Int J Agric Environ Biotechnol 6(1):23–30

Singh H, Verma A, Kumar M, Sharma R, Gupta R, Kaur M, Negi M, Sharma SK (2017b) Phytoremediation: a green technology to clean up the sites with low and moderate level of heavy metals. Austin Biochem 2:1012

Singh H, Sharma R, Savita SMP, Kumar M, Verma A, Ansari MW, Sharma SK (2018) Adaptive physiological response of *Parthenium hysterophorus* to elevated atmospheric CO_2 concentration. Indian Forester 144:1–14

Singh S, Singh H, Sharma SK, Nautiyal R (2021) Seasonal variation in biochemical responses of bamboo clones in the sub-tropical climate of Indian Himalayan foothills. Heliyon 7(4):e06859. https://doi.org/10.1016/j.heliyon.2021.e06859

Singh H, Bandyopadhyaya S, Kumar A, Singh M, Malik A, Kumari G, Raturi A, Yadav SK, Kumar M, Barthwal S, Thakur A, Rawat PS (2022) Understanding the physiological and biophysical response of urban roadside plantations for assessing adaptation and mitigation mechanisms toward vehicular emissions. Urban Climate 44:101183. https://doi.org/10.1016/j.uclim.2022.101183

Singh M, Singh H, Kumar A, Kumar M, Barthwal S, Thakur A (2024b) Soil nitrogen availability determines the CO_2 fertilization effect on tree species (*Neolamarckia cadamba*): growth and physiological evidence. Environ Sustain 7:53–60. https://doi.org/10.1007/s42398-023-00300-w

Singh H, Kumar M (2022) Climate change and its impact on Indian Himalayan forests: current status and research needs. In: Rani S, Kumar R (eds) Climate change. Springer climate. Springer, Cham, pp 223–242. https://doi.org/10.1007/978-3-030-92782-0_11

Singh H, Savita Sharma R, Sinha S, Kumar M, Kumar P, Verma A, Sharma SK (2017a) Physiological functioning of *Lagerstroemia speciosa* L. under heavy roadside traffic: an approach to screen potential species for abatement of urban air pollution. 3 Biotech 7(61):1–10. https://doi.org/10.1007/s13205-017-0690-0

Singh H, Yadav M, Kumar N, Kumar A, Kumar M (2020) Assessing adaptation and mitigation potential of roadside trees under the influence of vehicular emissions: a case study of *Grevillea robusta* and *Mangifera indica* planted in an urban city of India. PLoS ONE 15(1):e0227380. https://doi.org/10.1371/journal.pone.0227380

Singh H, Kumar N, Kumar A (2024a). Enhancing Resource Use Efficiency in Crops through Plant Functional Traits. In: Kumar, N., Singh, H. (eds) Plant Functional Traits for Improving Productivity. Springer, Singapore. https://doi.org/10.1007/978-981-97-1510-7_6

Singh H (2024). Exploring adaptive modulation in plant functional traits and its impact on the productivity of Acacia auriculiformis under CO_2-enriched environment. Ind Crops Prod 210:118186. https://doi.org/10.1016/j.indcrop.2024.118186

Singhal RK, Kumar V, Kumar M, Bose B. Responses of different light intensities and continue light during dark period on rice (*Oryza sativa* L.) seed germination and seedling development. J Pharmacognosy Phytochem 8(4):2602–2609

Svechkina A, Portnov BA, Trop T (2020) The impact of artificial light at night on human and ecosystem health: a systematic literature review. Landscape Ecol 1–18

Velez-Ramirez AI, Heuvelink E, vanIeperen W, Vreugdenhil D, Millenaar FF (2012) Continuous light as a way to increase greenhouse tomato production: expected challenges. Acta Horticulturae VII international symposium on light in horticultural systems

Verheijen FJ, Photopollution: artificial light optic spatial control systems fail to cope with. Incidents, causation, remedies, Exp Biol 44:1–18

Wada M, Kagawa T, Sato Y (2003) Chloroplast movement. Annu Rev Plant Biol 54:455–468

Yadav SK, Singh H, Nautiyal R, Ginwal HS, Ansari SA, Barthwal S (2019) Modulation of morpho-physiological responses in *Populus deltoides* by elevated carbon dioxide and temperature. For Sci 66:105–118. https://doi.org/10.1093/forsci/fxz048

Understanding the Carbon Sequestration Potential of Urban Gardens and Parks: Implications for Sustainable Urban Planning and Environmental Conservation

Bijaya Mishra, Tej Kumar Shrestha, Rajeev Joshi, Bijay Maharjan, Pushpa Rana Magar, and Kamal Raj Gosai

Abstract The primary objective of this study was to assess the carbon sequestration in the above-ground vegetation of urban gardens and parks within the Kathmandu Valley. To identify potential urban gardens and parks in the Kathmandu Valley, we cross-referenced GIS research with input from local government officials. Subsequently, we randomly selected 30 urban gardens and parks, each covering at least 0.5 hectares, across three districts: Kathmandu (18), Bhaktapur (7), and Lalitpur (5). These selected urban gardens encompassed various types, including government, public, and institutional. We conducted a complete tree enumeration and sampling based on the size and density of the urban parks. Complete tree enumeration was carried out in urban gardens smaller than one hectare or those with scattered trees. In contrast, sampling was employed in larger gardens with substantial forest cover. A non-destructive method was adopted for estimating the biomass of individual trees in the selected sample sites, measuring tree height and Diameter at Breast Height (DBH) at 1.3 m. Trees with a DBH greater than or equal to 5 cm were calculated to estimate tree biomass. Tee biomass was computed using the equation developed by

B. Mishra
Pristine Services Pvt. Ltd., Bhaktapur, Nepal

T. K. Shrestha
School of Natural Sciences, Massey University, Albany Campus, New Zealand

R. Joshi (✉) · P. R. Magar
Faculty of Forestry, College of Natural Resource Management, Agriculture and Forestry University, Katari-56310, Udayapur, Nepal
e-mail: joshi.rajeev20@gmail.com

K. R. Gosai
Tri-Chandra Multiple Campus, Ghantaghar, Kathmandu, Nepal

B. Maharjan
Lumbini Environmental Services Pvt. Ltd, NewBaneshwor, Kathmandu, Nepal

B. Mishra
Nature First Nepal, Kathmandu, Nepal

H. Singh (ed.), *Urban Forests, Climate Change and Environmental Pollution*,
https://doi.org/10.1007/978-3-031-67837-0_32

Chave et al. (2005), which considers climate and forest stand types. In a Kathmandu Valley urban garden, the number of stems with a DBH greater than 5 cm was found to be 418.08 per hectare. The government-run urban garden exhibited the highest number of stems per acre (673 stems per acre), followed by the institutional urban garden (433.53 stems per acre). The above-ground tree biomass carbon levels in institutional, government, and public urban gardens were 50.10, 72.64, and 13.70 tons per hectare, respectively. The below-ground tree biomass carbon content in institutional, government, and public urban gardens was 10, 14.53, and 2.70 tons per hectare, respectively. The total carbon content originating from trees in institutional, government, and public urban gardens were 60.10, 87.17, and 16.40 tons per hectare, respectively. The calculated carbon level averaged 54.56 tons per hectare, which is lower than the national carbon level, likely attributed to the lower occurrence of stems per hectare in public parks.

Keywords Carbon sequestration · Environmental conservation · Urban green space · Sustainable urban planning

1 Introduction

Climate change is a critical global environmental, economic, and security challenge (IPCC 2007; Karagiannis & Soldatos 2010). One promising strategy to mitigate and adapt to its impacts is the Reducing Emissions from Deforestation and Forest Degradation (REDD+) initiative, which focuses on the conservation and sustainable management of forests in developing countries (Angelsen 2008; FAO 2011; Yadav et al. 2019; Gupta et al. 2019a). The Intergovernmental Panel on Climate Change (IPCC) reported that the forest sector contributes significantly to greenhouse gas emissions, primarily due to deforestation and forest degradation (IPCC 2007; Gupta et al. 2019a; Kumar et al. 2019a). Stern (2006) noted that curbing these practices is a cost-effective means of reducing emissions.

With urbanization on the rise globally, it's projected that by 2030, 60% of the world's population will reside in cities (Rydin et al. 2012; Singh et al. 2021). Despite being one of the least urbanized countries, Nepal is experiencing rapid urbanization, leading to various environmental and societal challenges (Bakrania 2015; Nepali 2017; Kumar et al. 2022; Singh et al. 2022). Recognizing the importance of urban green spaces, the Environment-Friendly Local Governance Framework (2013) advocates for household-level greenery, suggesting planting at least two trees per household (MFALD 2019).

Urban green spaces encompass trees in gardens, parks, and along streets, providing a range of ecosystem services such as air quality improvement, noise pollution reduction, biodiversity conservation, and microclimate regulation (Ugle et al. 2010; Singh et al. 2018; Shah & Gavali 2017). These spaces play a vital role in carbon sequestration, mitigating the urban carbon footprint (Strohbach et al. 2012; Nowak & Crane 2002). Vegetation and soil in urban greenspaces sequester carbon, directly reducing atmospheric CO_2 concentrations and indirectly affecting carbon balance through their impact on urban energy balance (Churkina 2016; Kumar et al. 2024; Singh et al. 2024; Singh 2024).

Urban trees, by absorbing CO_2 through photosynthesis and providing shade that reduces energy consumption, contribute significantly to carbon sequestration in cities (Martin et al. 2012; Akbari et al. 2001; Kumar et al. 2019b). While urban greenspace carbon sequestration is modest compared to city emissions, proper tree species selection and management can optimize their benefits (Singh et al. 2017; Dhyani et al. 2021).

Assessing carbon storage and sequestration in urban greenspaces is crucial, as it impacts urban carbon balances and national carbon accounting (Singh et al. 2020; Kumar et al. 2021a). However, this aspect has been underestimated, highlighting the need for further investigation and management of urban trees (Brown et al. 2012; Kumar et al. 2021a, b, c, d, e; Prakash et al. 2022).

Climate change poses socioeconomic challenges globally, with cities major contributors to anthropogenic CO_2 emissions (Seto et al. 2014; Kumar et al. 2021b, c, d). Fast urbanization exacerbates environmental and health issues, emphasizing the urgency of urban green initiatives (Gong et al. 2012; Zhou et al. 2014; Young et al. 2012). Trees are invaluable in cities, offering numerous benefits such as carbon sequestration, air pollutant filtration, and aesthetic enhancement. Recognizing their importance, efforts should focus on native biodiversity and ecosystem services, including carbon sequestration, within urban landscapes (Singh 2021; Sharma and Singh 2021).

This study aims to evaluate the carbon sequestration potential of urban gardens and parks in Kathmandu Valley. Objectives include assessing these spaces' distribution, types, and conditions, estimating carbon stocks, comparing stocks among different green spaces, and identifying factors influencing carbon stocks. Findings will inform sustainable urban planning and conservation efforts.

2 Material and Method

2.1 Study Site

The research was conducted within the confines of Kathmandu Valley (as illustrated in Fig. 1). Located in the northeastern region of Nepal, Kathmandu Valley sits at an average elevation of 1350 m above sea level. The valley encompasses three major cities: Kathmandu, Lalitpur, and Bhaktapur. Kathmandu district consists of one Metropolitan City and ten Municipalities, comprising a total of 135 wards covering an area of 433.61 km^2, with a population of 1,699,288. Bhaktapur district includes four Municipalities, encompassing 38 wards and covering an area of 123.79 km^2, with a population of 298,684. Lalitpur district comprises one Metropolitan City, two

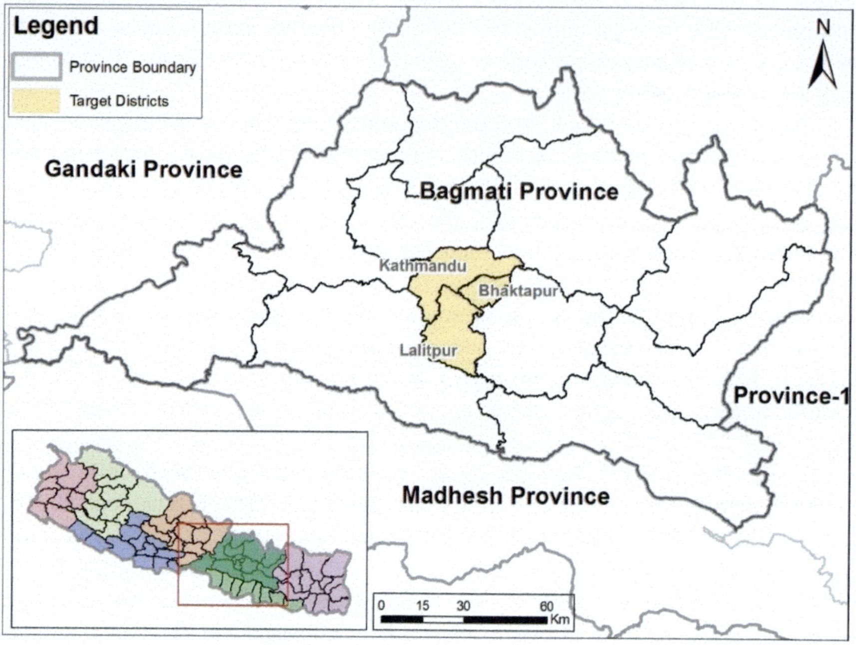

Fig. 1 Location map of study site

Municipalities, and three Rural Municipalities, with 71 wards, covering an area of 396.83 km^2, and a population of 457,606 (as depicted in Fig. 2).

Literature review

We relied on secondary data sources to acquire comprehensive information about the status, distribution, and various forms of urban forestry in the Kathmandu Valley. We thoroughly reviewed published and unpublished materials, including documents, journals, reports, and books, to gather relevant and essential information on the subject. We utilized search engines such as Google Scholar, ISI Web of Science, Science Direct, Scopus by Elsevier, ResearchGate, and JSTOR. We employed specific keywords such as "carbon sequestration," "carbon stock," and "biomass estimation" to search for articles that align with the objectives of this research task.

Site selection

We identified prospective urban gardens using geographic information system (GIS) analysis and consulting with the relevant local authorities.

Preparatory works on GIS

We searched for moderate-resolution satellite images and secured images free from cloud cover. Subsequently, we performed essential image pre-processing tasks, such

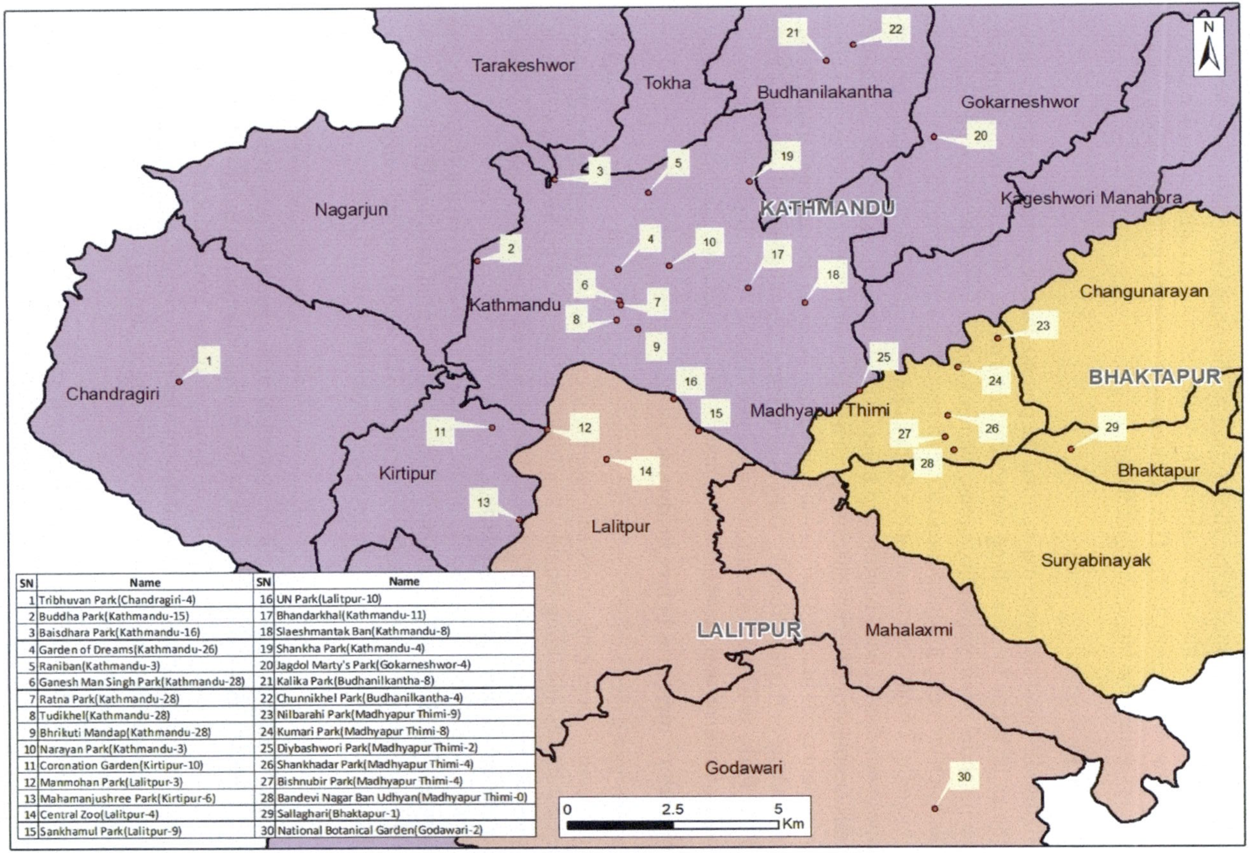

Fig. 2 Selected urban gardens distribution in three districts

as stacking layers and carrying out geometric and radiometric correction/rectification as needed. All GIS data and satellite images were transformed into a uniform geographic projection system using the re-projection tool to ensure compatibility. Furthermore, we selected a specific portion of the image encompassing the urban parks and gardens within the Kathmandu Valley area by creating a mask based on the location of interest (AOI). This allowed us to identify potential urban gardens and parks within the larger satellite scene.

Field survey

Utilizing GIS analysis, we cross-verified the potential urban gardens and parks in the Kathmandu Valley by consulting with local government officials. After identifying these potential urban garden locations, we randomly selected 30 urban gardens and parks, each covering a minimum area of 0.5 hectares, across three districts: 18 from Kathmandu, 7 from Bhaktapur, and 5 from Lalitpur. This selection included municipalities with district headquarters and others with diverse characteristics as required for our sample size. The chosen urban gardens represent a variety of factors, including government types like the Coronation Garden and National Botanical Garden, public types such as Sallaghari Temple, Dibyaswori Park, Shankhadar Park, Buddha Park, and UN Park, as well as institutional types like Ganesh Man Singh Park, Garden of Dreams, Raniban, and Tudikhel (see Annexes 1–3 for details).

Preparatory measures before fieldwork

The GPS-based method for locating the sample urban parks and gardens was supplemented with topographic maps and satellite images created by the GIS specialist. The coordinates, including latitude, longitude, and the WGS UTM44 N system coordinates, were documented and printed. The team used these coordinates to pinpoint the exact locations of the urban parks.

Before embarking on fieldwork, the team made necessary preparations, including obtaining checklists, field data entry forms, and essential field equipment such as DBH tape, measuring tape, and GPS devices. The team received training on data collection tools, methods, and the proper completion of data forms.

Inventory design

After identifying and confirming the potential urban gardens, they were classified based on their types and size. To assess the tree population, we employed a method that depended on the size and density of the urban parks. We conducted a comprehensive tree enumeration for urban gardens smaller than one hectare or with dispersed trees. In contrast, we resorted to sampling for larger gardens with dense forest cover to estimate the tree population.

Sample plots and sampling design for larger parks

In the case of larger gardens, we employed a random sampling approach to choose the sample plots. We used Concentric Circular Sample Plots (CCSP) with radii of 20 m, 15 m, 8 m, and 4 m for sampling trees with DBH measurements falling into the ranges of $\geq$ 30 cm, 20–29.9 cm, 10–19.9 cm, and 5–9.9 cm, respectively (Table 1).

Table 1 Inventory methods for different Parks

S. No.	Park name	District	Area	Inventory method	Management	Remarks
1	Sallaghari	Bhaktapur	3.27	Sample plot	Public	Dense tree
2	BandeviNagar Ban Udhyan	Bhaktapur	0.797	Sample plot	Institutional	Dense tree
3	Bishnubir Park	Bhaktapur	2.71	Sample plot	Institutional	Dense tree
4	Dibyaswori Park	Bhaktapur	0.65	Total enumeration	Public	Small area
5	Kumari Park	Bhaktapur	1.5	Total enumeration	Public	Scattered tree
6	Nilbarahi Park	Bhaktapur	20.5	Sample plot	Institutional	Dense tree
7	Shankhadar Park	Bhaktapur	0.69	Total enumeration	Public	Small area
8	Chunnikhel Park	Kathmandu	0.52	Total enumeration	Public	Small area
9	Kalika Park	Kathmandu	0.56	Total enumeration	Public	Small area
10	Tribhuvan Park	Kathmandu	7.83	Sample plot	Public	Dense tree
11	Jagdol Marty's Park	Kathmandu	1.31	Sample plot	Institutional	Dense tree
12	Baisdhara Park	Kathmandu	1.2	Sample plot	Public	Dense tree
13	Bhandarkhal	Kathmandu	8.28	Sample plot	Public	Dense tree
14	Bhrikuti Mandap	Kathmandu	1.38	Total enumeration	Public	Scattered tree
15	Buddha Park	Kathmandu	0.51	Total enumeration	Public	Small area
16	Ganesh Man Singh Park	Kathmandu	0.5	Total enumeration	Institutional	Small area
17	Garden of Dreams	Kathmandu	0.59	Total enumeration	Institutional	Small area
18	Narayan Park	Kathmandu	0.95	Total enumeration	Public	Small area
19	Raniban	Kathmandu	7.6	Sample plot	Institutional	Dense tree
20	Ratna Park	Kathmandu	2.34	Total enumeration	Public	Scattered tree
21	Shankha Park	Kathmandu	1.44	Total enumeration	Public	Scattered tree

(continued)

Table 1 (continued)

S. No.	Park name	District	Area	Inventory method	Management	Remarks
22	Slaeshmantak Ban	Kathmandu	17.4	Sample plot	Institutional	Dense tree
23	Tudikhel	Kathmandu	14.4	Total enumeration	Institutional	Scattered tree
24	Coronation Garden	Kathmandu	16.1	Sample plot	Government	Dense tree
25	Mahamanjushree Park	Kathmandu	1	Total enumeration	Institutional	Small area
26	National Botanical Garden	Lalitpur	23	Sample plot	Government	Dense tree
27	UN Park	Lalitpur	5.11	Sample plot	Public	Dense tree
28	Central Zoo	Lalitpur	5.18	Total enumeration	Institutional	Scattered tree
29	Manmohan Park	Lalitpur	1.86	Total enumeration	Public	Scattered tree
30	Sankhamul Park	Lalitpur	0.58	Total enumeration	Public	Small area

2.2 *Data Collection*

2.2.1 Field Observation

Field observations were carried out to evaluate the distribution, types, and conditions of various urban gardens and parks in the Kathmandu Valley. This study employed a linear transect method to identify vegetation types through visual assessments. The critical parameters considered included the predominant tree species, vegetation characteristics, tree height, tree density, understorey density, environmental topography, and the origin and extent of disturbances. We also documented the presence of different plant species within the parks.

2.2.2 Measurement of Tree Height and Diameter at Breast Height (DBH)

We employed a non-invasive approach for estimating biomass to measure the tree height and DBH (Diameter at Breast Height) of individual trees within the selected sample sites of the urban garden. To estimate the tree biomass, we enumerated individual trees with a DBH greater than or equal to 5 cm, measured at 1.3 m above the ground. The tree height and diameter were determined using a clinometer and measuring tape. Each tree was individually documented, along with its species name.

Field data were recorded and organized in spreadsheets. Tree species were identified primarily through visual observation, and any uncertain samples were collected and later verified by a botanist.

2.3 Data Analysis

2.3.1 Estimation of Above-Ground and Below-Ground Biomass (AGB and BGB)

Above-ground and below-ground biomasses were estimated using allometric equations based on field measurements of the tree's diameter at breast height (DBH).

Above-ground tree biomass (AGTB)

The equation (Chave et al. 2005) used to estimate the tree biomass based on climate and forest stand types was used. The allometric equations (models) used were

$$\mathrm{AGTB} = 0.0509 * \rho \mathrm{D}^2\mathrm{H}$$

where,

$AGTB$ = aboveground tree biomass (kg)

ρ = wood specific gravity (gm cm^{-3}) (Referred global wood density database of South Esat Asia (Tropical))

D = tree diameter at breast height (DBH) [cm]; and

H = tree height (m).

Below Ground Tree Biomass (BGTB)

Estimating below-ground tree biomass is much more complicated than AGTB's. So, MacDicken's (1997) root-to-shoot ratio value of 1:5 was followed to estimate below-ground biomass as 20% of above-ground tree biomass.

Estimation of Total Biomass (TB)

The total biomass of individual trees is the sum of their above and below-ground biomasses, respectively, given by the following equation:

$$\textbf{Total Biomass} = \textbf{AGTB} + \textbf{BGTB}$$

Estimation of Carbon Content

The biomass stock was converted into carbon stock after multiplication with the (IPCC 2006) default carbon fraction of 0.47.

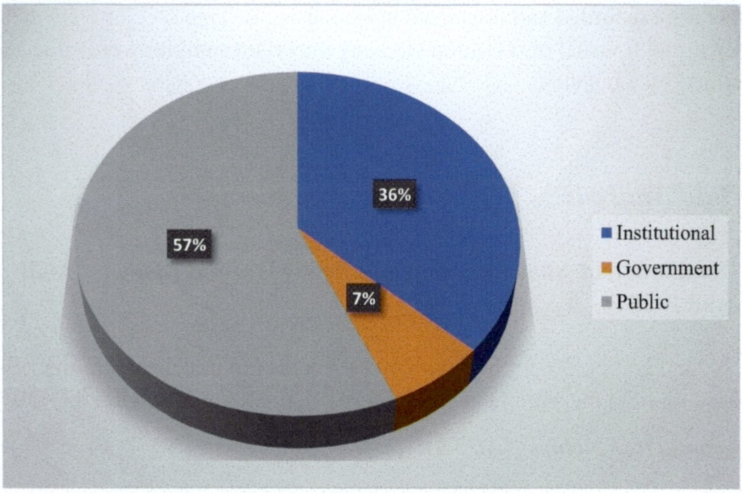

Fig. 3 Types of sampled urban garden

$$\textbf{Total carbon stock} = \textbf{Total Biomass} \times \textbf{0.47}$$

3 Results and Discussion

3.1 Status of Urban Garden

3.1.1 Distribution and Types of Urban Garden

Urban gardens have been classified into three categories: government, public, and institutional. The study reveals that the majority of the urban gardens (57%) fall under the public category, followed by institutional gardens (36%), with government gardens representing the smallest proportion (7%) (as depicted in Fig. 3).

3.1.2 Plant Species

Various plant species were documented across the different categories of urban gardens. In total, 62, 46, and 54 distinct tree species were observed in the Institutional urban garden (IUG), Government urban garden (GUG), and Public urban garden (PUG), respectively (Table 2).

Table 2 Plant species occurrence in different urban gardens

S. No.	Species	Family	Local name	Occurrence
1	*Acer oblongum*	Aceraceae	Firfire	IUG, PUG
2	*Alangium chinense*	Alangiaceae	–	GUG
3	*Albizia chinensis*	Leguminosae	Siran	IUG, PUG
4	*Albizia julibrissin*	Leguminosae	Rato siris	IUG
5	*Albizzia* spp.	Leguminosae	Siris	IUG, PUG
6	*Alnus nepalensis*	Betulaceae	–	IUG, GUG, PUG
7	*Araucaria bidwilli*	Araucariaceae	–	IUG
8	*Bauhinia purpurea*	Leguminosae	Tanki	GUG
9	*Bauhinia variegata*	Leguminosae	Koiralo	PUG
10	*Bombax ceiba*	Bombacaceae	Simal	PUG
11	*Brassiopsis hainla*	Araliaceae	Seto chuletro	IUG
12	*Buddleja asiatica*	Buuddlejaceae	–	IUG, PUG
13	*Calistemon citrinus*	Myrtaceae	Kalkiful	IUG, PUG
14	*Castanopsis indica*	Fagaceae	Dhalekatus	IUG, GUG
15	*Castanopsis tribuloides*	Fagaceae	Musurekatus	IUG
16	*Casuarina equisetifolia*	Casuarinaceae	–	GUG
17	*Cedrus deodara*	Pinaceae	Devdar	GUG
18	*Celtis australis*	Ulmaceae	Khari	IUG, GUG, PUG
19	*Choerospondias axillaris*	Anacardiaceae	Lapsi	IUG, GUG, PUG
20	*Cinnamomum camphora*	Lauraceae	Kapur	IUG, GUG, PUG
21	*Cinnamomum tamala*	Lauraceae	Tejpat	IUG
22	*Citrus maxima*	Rutaceae	–	IUG, GUG, PUG
23	*Cordia dichotoma*	Cordiaceae	Bohari	GUG
24	*Crateva unilocularis*	Capparaceae	–	IUG
25	*Cupressus torulosa*	Cupressaceae	Dhupisalla	IUG, GUG, PUG
26	*Dalbergia sissoo*	Leguminosae	Sissoo	IUG, GUG
27	*Diospyros* spp.	Ebenaceae	–	GUG
28	*Elaeocarpus serratus*	Elaeocarpaceae	–	IUG, PUG
29	*Engelhardtia spicata*	Juglandaceae	–	IUG
30	*Eucalyptus citriodora*	Myrtaceae	–	GUG
31	*Eucalyptus grandis*	Myrtaceae	–	PUG
32	*Eucalyptus* spp.	Myrtaceae	Masala	IUG, GUG, PUG
33	*Eurya acuminata*	Pentaphylacaceae	–	IUG
34	*Ficus auriculata*	Moraceae	Nemaro	GUG
35	*Ficus benghalensis*	Moraceae	Bar	IUG, PUG
36	*Ficus benjamina*	Moraceae	Sami	IUG, PUG
37	*Ficus elastica*	Moraceae		GUG, PUG

(continued)

Table 2 (continued)

S. No.	Species	Family	Local name	Occurrence
38	*Ficus lacor*	Moraceae	Kavro	IUG, GUG, PUG
39	*Ficus religiosa*	Moraceae	Dudhilo	IUG, PUG
40	*Fraxinus floribunda*	Oleaceae	Lakuri	IUG, PUG
41	*Grevillea robusta*	Proteaceae	Kangiyoful	IUG, GUG, PUG
42	*Grewia optiva*	Tiliaceae	Bhimal	IUG
43	*Hymenodictyon excelsum*	Rubiaceae	–	IUG
44	*Ilex excelsa*	Aquifoliaceae	Puwale	IUG, PUG
45	*Jacaranda mimosifolia*	Bignoniaceae	Birendraful	GUG, PUG
46	*Juglans regia*	Juglandaceae	Okhar	GUG, PUG
47	*Juniperus* spp.	Juglandaceae	–	GUG
48	*Largerstroemia indica*	Lythraceae	–	IUG, GUG
49	*Lagerstroemia parviflora*	Lythraceae	Botdhayaro	PUG
50	*Leucaena leucocephala*	Leguminosae	Ipil ipil	GUG, PUG
51	*Ligustrum confusum*	Lauraceae		IUG, PUG
52	*Litsea oblonga*	Lauraceae		IUG, GUG, PUG
53	*Magnifera indica*	Anacardiaceae		IUG, PUG
54	*Magnolia grandiflora*	Magnoliaceae		IUG, PUG
55	*Melia azedarach*	Meliaceae	Bakaino	IUG, GUG, PUG
56	*Michelia champaca*	Magnoliaceae	Champ	IUG, GUG, PUG
57	*Mimosa* spp.	Fabaceae		GUG
58	*Morus alba*	Moraceae	Kimbu	IUG, GUG, PUG
59	*Myrica esculenta*	Myricaceae	Kafal	IUG, GUG
60	*Myrsine capitellata*	Primulaceae	Seti Kath	IUG, PUG
61	*Myrsine semiserrata*	Myrsinaceae		IUG, PUG
62	*Neolitsea cuipala*	Lauraceae		IUG, PUG
63	*Osmanthus fragrance*	Oleaceae		PUG
64	*Paulownia tomentosa*	Paulowniaceae		GUG
65	*Perseaduthiei*	Lauraceae		IUG, PUG
66	*Persea odoratissima*	Lauraceae		IUG
67	*Phyllanthus emblica*	Phyllanthaceae		IUG
68	*Pinus roxburghii*	Pinaceae	Khotesalla	IUG, GUG, PUG
69	*Pinus* spp.	Pinaceae		IUG, PUG
70	*Populus deltoides*	Salicaceae	Laharepipal	PUG
71	*Populus euramericana*	Salicaceae		GUG
72	*Prunus cerasoides*	Rosaceae	Paiyu	IUG, GUG, PUG
73	*Pyrus communis*	Rosaceae		IUG
74	*Pyrus pashia*	Rosaceae	Mayal	IUG, GUG, PUG

(continued)

Table 2 (continued)

S. No.	Species	Family	Local name	Occurrence
75	*Quercus glauca*	Fagaceae	Falat	IUG
76	*Quercus lanata*	Fagaceae	Thulo bajha	GUG
77	*Rhododendron arboretum*	Ericaceae	Laliguras	IUG, GUG, PUG
78	*Salix babylinica*	Salicaceae	Bains	IUG, GUG
79	*Sapindus mukorossi*	Sapindaceae	Rittha	IUG, GUG
80	*Sequoia sempervirens*	Cupressaceae		GUG
81	*Schima wallichii*	Theaceae	Chilaune	IUG, GUG, PUG
82	*Syzygium cumini*	Myrtaceae	Jamun	IUG, PUG
83	*Syzygium jambos*	Myrtaceae		PUG
84	*Tamarindus indica*	Leguminosae	Imli	GUG
85	*Terminalia chebula*	Combretaceae	Harro	PUG
86	*Thuja orientalis*	Cupressaceae		GUG, PUG
87	*Zanthoxylum armatum*	Rutaceae	Timur	PUG
88	*ZizyphusincurvaRoxb*	Rhamnaceae		IUG, GUG, PUG

3.1.3 Number of Stems in the Urban Garden of Kathmandu Valley (DBH ≥ 5 cm)

In the urban gardens of the Kathmandu Valley, the count of tree stems with a diameter greater than 5 cm at breast height (DBH) averaged 418.08 ha^{-1}. Notably, government urban gardens had the highest stem density per hectare, with institutional urban gardens following closely behind (as illustrated in Table 3 and Fig. 4).

Threats to park

Parks are often marred by litter and have become gathering spots for vagrants. Many parks in the capital city lack fundamental amenities such as restrooms and water sources. Even those with restroom facilities are often in poor condition. The parks' beauty is tarnished by the careless trash disposal by visitors. In some parks, maize cobs are roasted and sold, but the husks and cobs are discarded. Fruit peels are scattered throughout, further diminishing the aesthetic appeal of the parks.

Table 3 Number of stems (≥ 5 cm DBH) in the urban garden of Kathmandu Valley

S. No.	Types	Stem/ha
1	Institutional	433.53
2	Government	673
3	Public	147.71
Average		418.08

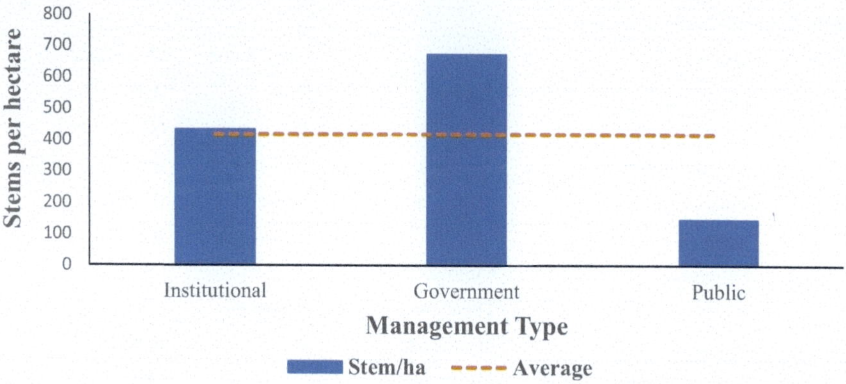

Fig. 4 Number of stems per hectare in different types of urban gardens of Kathmandu Valley

3.2 *Carbon Stock*

3.2.1 AboveGround Tree Biomass Carbon (AGTB)

The Above Ground Tree Biomass and Carbon in institutional, government, and public urban gardens were determined to be 106.6 tons ha^{-1} and 50.1 tons ha^{-1}, 154.55 tons ha^{-1} and 72.64 tons ha^{-1}, 29.1 tons ha^{-1} and 13.7 tons ha^{-1}, respectively (as presented in Table 4).

Above-ground tree biomass (AGTB) in the urban garden of Kathmandu Valley

The AGTB and carbon of Kathmandu Valley were found to be 96.75 tons ha^{-1} and 45.48 tons ha^{-1}, respectively. The total AGTB carbon stock in the urban garden of Kathmandu Valley was 6976.70 tons. Government urban garden has the highest AGTB and AGTB carbon stock (Table 5 and Fig. 5).

Below Ground Tree Biomass (BGTB) Carbon

The BGTB and BGTB Carbon in institutional, government, and public urban gardens were found to be 21.3 tons ha^{-1} and 10.0 tons ha^{-1}, 30.91 tons/ha and 14.53 tons/ha, 5.8 tons/ha and 2.7 tons ha^{-1} respectively (Table 6). The Below Ground Tree Biomass (BGTB) and Below Ground Tree Biomass Carbon in institutional, government, and public urban gardens were determined to be 21.3 tons ha^{-1} and 10.0 tons ha^{-1}, 30.91 tons ha^{-1} and 14.53 tons ha^{-1}, 5.8 tons ha^{-1} and 2.7 tons ha^{-1}, respectively (as outlined in Table 6).

Below Ground Tree Biomass (BGTB) in the urban garden of Kathmandu Valley

The BGTB and Carbon in the Kathmandu Valley were measured at 19.34 tons ha^{-1} and 9.07 tons ha^{-1}, respectively. The combined BGTB Carbon stock in the urban gardens of the Kathmandu Valley amounted to 1392.43 tons. Notably, government

Table 4 AGTB and AGTB carbon in different urban gardens

S. No.	Species	AGTB (tons ha^{-1})			AGTB carbon (tons ha^{-1})		
		IUG	GUG	PUG	IUG	GUG	PUG
1	*Acer oblongum*	0.016		0.188	0.008		0.089
2	*Alangiumchinense*		0.048			0.022	
3	*Albizia chinensis*	0.288		0.059	0.135		0.028
4	*Albizia julibrissin*	0.002			0.001		
5	*Albizzia* spp.	0.917		0.027	0.431		0.013
6	*Alnus nepalensis*	2.526	5.329	0.175	1.187	2.505	0.082
7	*Araucaria bidwilli*	0.292			0.137		
8	*Bauhinia purpurea*		0.067			0.031	
9	*Bauhinia variegata*			0.005			0.002
10	*Bombax ceiba*			2.479			1.165
11	*Brassiopsishainla*	0.003			0.001		
12	*Buddleja asiatica*	0.004		0.008	0.002		0.004
13	*Calistemoncitrinus*	0.006		0.054	0.003		0.025
14	*Castanopsis indica*	1.622	3.731		0.763	1.754	
15	*Castanopsis tribuloides*	0.19			0.089		
16	*Casuarina equisetifolia*		0.637			0.299	
17	*Cedrus deodara*		0.021			0.01	
18	*Celtis australis*	0.155	3.115	1.807	0.073	1.464	0.85
19	*Choerospondias axillaris*	0.121	7.008	0.224	0.057	3.294	0.105
20	*Cinnamomum camphora*	1.554	25.175	0.858	0.73	11.832	0.403
21	*Cinnamomum tamala*	0.005		0.012	0.002		0.006
22	*Citrus maxima*	0.007	0.002	0.007	0.003	0.001	0.003
23	*Cordia dichotoma*		0.002			0.001	
24	*Crateva unilocularis*	0.001			0.001		
25	*Cupressus torulosa*	0.013	2.905	0.115	0.006	1.366	0.054
26	*Dalbergia sissoo*	0.067	0.001		0.031	0.001	
27	*Diospyros* spp.		1.769			0.831	
28	*Elaeocarpus serratus*	0.042		0.015	0.02		0.007
29	*Engelhardtia spicata*	0.023			0.011		
30	*Eucalyptus citriodora*		47.151			22.161	
31	*Eucalyptus grandis*			0.077			0.036

(continued)

Table 4 (continued)

S. No.	Species	AGTB (tons ha^{-1})			AGTB carbon (tons ha^{-1})		
		IUG	GUG	PUG	IUG	GUG	PUG
32	*Eucalyptus sps*	0.101	0.71	0.429	0.048	0.333	0.202
33	*Eurya acuminata*	0.076			0.036		
34	*Ficus auriculata*		0.001			0	
35	*Ficus benghalensis*	0.003		0.02	0.001		0.009
36	*Ficus benjamina*	0.017		0.011	0.008		0.005
37	*Ficus elastica*		0.008	0.006		0.004	0.003
38	*Ficus lacor*	0.015	0.001	0.027	0.007	0	0.013
39	*Ficus religiosa*	0.061		0.502	0.029		0.236
40	*Fraxinus floribunda*	0.002		0.341	0.001		0.16
41	*Grevillea robusta*	0.443	7.977	3.547	0.208	3.749	1.667
42	*Grewia optiva*	0.122			0.057		
43	*Hymenodictyon excelsum*	0.769			0.361		
44	*Ilex excels*	0.006		1.822	0.003		0.856
45	*Jacaranda mimosifolia*		3.073	0.067		1.444	0.031
46	*Juglans regia*		0.009	0.001		0.004	0
47	*Juniperus* spp.		0.004			0.002	
48	*Lagerstroemia parviflora*			0.225			0.106
49	*Largerstroemia indica*	0.003	1.827	0.015	0.002	0.859	0.007
50	*Leucaena leucocephala*		0.025	0.006		0.012	0.003
51	*Ligustrum confusum*	0.006		0.152	0.003		0.071
52	*Litsea oblonga*	0.01	0.01		0.005	0.005	
53	*Magnifera indica*	0.705		0.017	0.331		0.008
54	*Magnolia grandiflora*	0		0.157	0		0.074
55	*Melia azedarach*	0.558	0.05	0.068	0.262	0.023	0.032
56	*Micheliachampaca*	5.29	1.072	0.122	2.486	0.504	0.057
57	*Mimosa* spp.		0.001			0	
58	*Morus alba*	0.013	0.059	0.692	0.006	0.028	0.325
59	*Myrica esculenta*	2.54	0.001		1.194	0.001	
60	*Myrica esculenta*	0.033			0.016		
61	*Myrsine capitellata*	0.006		0.288	0.003		0.136
62	*Myrsine semiserrata*	0.675		0.036	0.317		0.017
63	*Neolitsea cuipala*	0.002		0.056	0.001		0.026

(continued)

Table 4 (continued)

S. No.	Species	AGTB (tons ha^{-1})			AGTB carbon (tons ha^{-1})		
		IUG	GUG	PUG	IUG	GUG	PUG
64	*Osmanthus fragrance*			1.854			0.871
65	*Paulownia tomentosa*		1.135			0.533	
66	*Perseaduthiei*	0.011		2.95	0.005		1.387
67	*Persea odoratissima*	0.433			0.204		
68	*Phyllanthus emblica*	1.577			0.741		
69	*Pinus roxburghii*	40.814	14.658	3.189	19.183	6.889	1.499
70	*Pinus* spp.	0.241		1.007	0.113		0.473
71	*Populus deltoides*			0.141			0.066
72	*Populus euramericana*		1.683			0.791	
73	*Prunus cerasoides*	0.069	1.744	0.76	0.032	0.82	0.357
74	*Pyrus communis*	0.345			0.162		
75	*Pyrus pashia*	0.022	0.005	0.033	0.01	0.002	0.015
76	*Pyrus pashia*	1.324			0.622		
77	*Quercus glauca*	2.632			1.237		
78	*Quercus lanata*		0.2			0.094	
79	*Rhododendron arboretum*	0.504	3.479	0.026	0.237	1.635	0.012
80	*Salix babylinica*	0.003	12.166		0.001	5.718	
81	*Sapindus mukorossi*	0.551	0.009		0.259	0.004	
82	*Schima wallichii*	0.36	5.285	0.051	0.169	2.484	0.024
83	*Schima wallichii*	37.207			17.488		
84	*Sequoia sempervirens*		0.032			0.015	
85	*Syzygium cumini*	0.309		0.079	0.145		0.037
86	*Syzygium jambos*			0.003			0.002
87	*Tamarindus indica*		0.09			0.042	
88	*Terminalia chebula*			0.007			0.003
89	*Thuja orientalis*		0.255	0.023		0.12	0.011
90	*Zanthoxylum armatum*			0.013			0.006
91	*Zizyphus incurva Roxb*	0.661	0.015	2.485	0.311	0.012	1.168
92	Others	0.267	1.997	1.746	0.125	0.938	0.821
	Total	106.6	154.5	29.1	50.1	72.6	13.7

Table 5 AGTB and AGTB carbon in the urban garden of Kathmandu Valley

S. No.	Types	AGTB biomass (tons ha^{-1})	AGTB carbon (tons ha^{-1})	Total AGTB carbon (tons)
1	Institutional	106.60	50.10	3606.70
2	Government	154.55	72.64	2840.22
3	Public	29.10	13.70	529.78
Average		96.75	45.48	6976.70

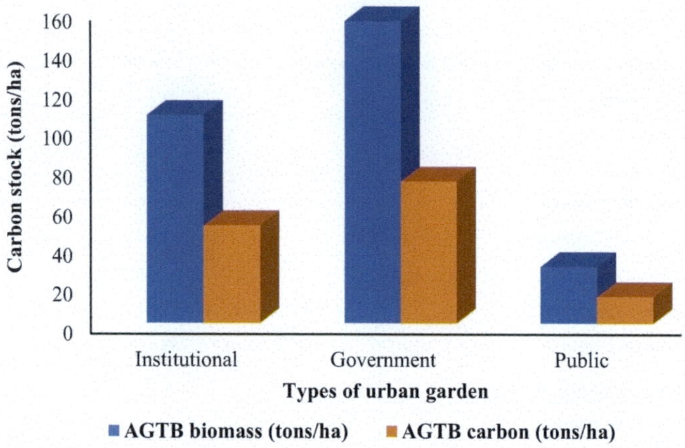

Fig. 5 AGTB and AGTB carbon in different types of urban gardens in Kathmandu Valley

urban gardens had the highest Below Ground Tree Biomass and Below Ground Tree Biomass Carbon stock (as depicted in Table 7 and Fig. 6).

Total carbon stock

The overall carbon stock in the urban gardens of the Kathmandu Valley amounted to 8369.13 tons. Government urban gardens exhibited the highest carbon stock, ranking institutional urban gardens second (Table 8 and Fig. 7).

Table 6 BGTB and BGTB carbon in different urban gardens

S. No.	Species	BGTB (tons ha^{-1})			BGTB carbon (tons ha^{-1})		
		IUG	GUG	PUG	IUG	GUG	PUG
1	*Acer oblongum*	0.003		0.038	0.002		0.018
2	*Alangium chinense*		0.01			0.004	
3	*Albizia chinensis*	0.058		0.012	0.027		0.006
4	*Albizia julibrissin*	0			0		
5	*Albizzia* spp.	0.183		0.005	0.086		0.003
6	*Alnus nepalensis*	0.505	1.066	0.035	0.237	0.501	0.016
7	*Araucaria bidwilli*	0.058			0.027		
8	*Bauhinia purpurea*		0.013			0.006	
9	*Bauhinia variegata*			0.001			0
10	*Bombax ceiba*			0.496			0.233
11	*Brassiopsis hainla*	0.001			0		
12	*Buddleja asiatica*	0.001		0.002	0		0.001
13	*Calistemon citrinus*	0.001		0.011	0.001		0.005
14	*Castanopsis indica*	0.324	0.746		0.153	0.351	
15	*Castanopsis tribuloides*	0.038			0.018		
16	*Casuarina equisetifolia*		0.127			0.06	
17	*Cedrus deodara*		0.004			0.002	
18	*Celtis australis*	0.031	0.623	0.361	0.015	0.293	0.17
19	*Choerospondiasaxillaris*	0.024	1.402	0.045	0.011	0.659	0.021
20	*Cinnamomum camphora*	0.311	5.035	0.172	0.146	2.366	0.081
21	*Cinnamomum tamala*	0.001		0.002	0		0.001
22	*Citrus maxima*	0.001	0	0.001	0.001	0	0.001
23	*Cordia dichotoma*		0			0	
24	*Crateva unilocularis*	0			0		
25	*Cupressus torulosa*	0.003	0.581	0.023	0.001	0.273	0.011
26	*Dalbergia sissoo*	0.013	0		0.006	0	
27	*Diospyros* spp.		0.354			0.166	
28	*Elaeocarpus serratus*	0.008		0.003	0.004		0.001
29	*Engelhardtia spicata*	0.005			0.002		
30	*Eucalyptus citriodora*		9.43			4.432	
31	*Eucalyptus grandis*			0.015			0.007
32	*Eucalyptus* spp.	0.02	0.142	0.086	0.01	0.067	0.04
33	*Eurya acuminata*	0.015			0.007		
34	*Ficus auriculata*		0			0	
35	*Ficus benghalensis*	0.001		0.004	0		0.002
36	*Ficus benjamina*	0.003		0.002	0.002		0.001

(continued)

Table 6 (continued)

S. No.	Species	BGTB (tons ha^{-1})			BGTB carbon (tons ha^{-1})		
		IUG	GUG	PUG	IUG	GUG	PUG
37	*Ficus elastica*		0.002	0.001		0.001	0.001
38	*Ficus lacor*	0.003	0	0.005	0.001	0	0.003
39	*Ficus religiosa*	0.012		0.1	0.006		0.047
40	*Fraxinus floribunda*	0		0.068	0		0.032
41	*Grevillea robusta*	0.089	1.595	0.709	0.042	0.75	0.333
42	*Grewia optiva*	0.024			0.011		
43	*Hymenodictyon excelsum*	0.154			0.072		
44	*Ilex excels*	0.001		0.364	0.001		0.171
45	*Jacaranda mimosifolia*		0.615	0.013		0.289	0.006
46	*Juglans regia*		0.002	0		0.001	0
47	*Juniperus* spp.		0.001			0	
48	*Lagerstroemia parviflora*			0.045			0.021
49	*Largerstroemia indica*	0.001	0.365	0.003	0	0.172	0.001
50	*Leucaena leucocephala*		0.005	0.001		0.002	0.001
51	*Ligustrum confusum*	0.001		0.03	0.001		0.014
52	*Litsea oblonga*	0.002	0.002		0.001	0.001	
53	*Magnifera indica*	0.141		0.003	0.066		0.002
54	*Magnolia grandiflora*	0		0.031	0		0.015
55	*Melia azedarach*	0.112	0.01	0.014	0.052	0.005	0.006
56	*Micheliachampaca*	1.058	0.214	0.024	0.497	0.101	0.011
57	*Mimosa* spp.		0			0	
58	*Morus alba*	0.003	0.012	0.138	0.001	0.006	0.065
59	*Myrica esculenta*	0.508	0		0.239	0	
60	*Myrica esculenta*	0.007			0.003		
61	*Myrsine capitellata*	0.001		0.058	0.001		0.027
62	*Myrsine semiserrata*	0.135		0.007	0.063		0.003
63	*Neolitsea cuipala*	0		0.011	0		0.005
64	*Osmanthus fragrance*			0.371			0.174
65	*Paulownia tomentosa*		0.227			0.107	
66	*Persea duthiei*	0.002		0.59	0.001		0.277
67	*Persea odoratissima*	0.087			0.041		
68	*Phyllanthus emblica*	0.315			0.148		
69	*Pinus roxburghii*	8.163	2.932	0.638	3.837	1.378	0.3
70	*Pinus* spp.	0.048		0.201	0.023		0.095
71	*Populus deltoides*			0.028			0.013

(continued)

Table 6 (continued)

S. No.	Species	BGTB (tons ha^{-1})			BGTB carbon (tons ha^{-1})		
		IUG	GUG	PUG	IUG	GUG	PUG
72	*Populus euramericana*		0.337			0.158	
73	*Prunus cerasoides*	0.014	0.349	0.152	0.006	0.164	0.071
74	*Pyrus communis*	0.069			0.032		
75	*Pyrus pashia*	0.004	0.001	0.007	0.002	0	0.003
76	*Pyrus pashia*	0.265			0.124		
77	*Quercus glauca*	0.526			0.247		
78	*Quercus lanata*		0.04			0.019	
79	*Rhododendron arboretum*	0.101	0.696	0.005	0.047	0.327	0.002
80	*Salix babylinica*	0.001	2.433		0	1.144	
81	*Sapindus mukorossi*	0.11	0.002		0.052	0.001	
82	*Schima wallichii*	0.072	1.057	0.01	0.034	0.497	0.005
83	*Schima wallichii*	7.441			3.498		
84	*Sequoia sempervirens*		0.006			0.003	
85	*Syzygium cumini*	0.062		0.016	0.029		0.007
86	*Syzygium jambos*			0.001			0
87	*Tamarindus indica*		0.018			0.008	
88	*Terminalia chebula*			0.001			0.001
89	*Thuja orientalis*		0.051	0.005		0.024	0.002
90	*Zanthoxylum armatum*			0.003			0.001
91	*Zizyphus incurvaRoxb*	0.132	0.005	0.497	0.062	0.002	0.234
92	Others	0.053	0.399	0.349	0.125	0.188	0.164
	Total	21.3	30.9	5.8	10	14.5	2.7

Table 7 BGTB and BGTB carbon in urban garden of Kathmandu Valley

S. No	Types	BGTB biomass (tons ha^{-1})	BGTB carbon (tons ha^{-1})	Total BGTB carbon (tons)
1	Institutional	21.30	10.00	719.90
2	Government	30.91	14.53	568.12
3	Public	5.80	2.70	104.41
Average		19.34	9.07	1392.43

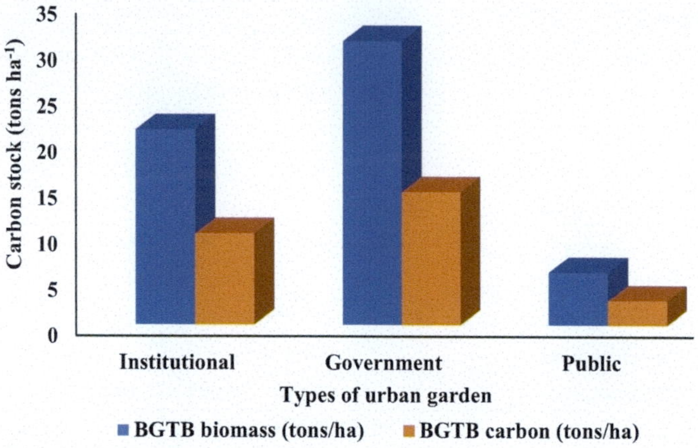

Fig. 6 BGTB and BGTB carbon in different types of urban gardens in Kathmandu Valley

Table 8 Total carbon stock in the urban garden of Kathmandu Valley

Types	AGTB carbon (tons ha^{-1})	BGTB carbon (tons ha^{-1})	Total carbon (tons ha^{-1})	Total carbon (tons)
Institutional	50.10	10	60.10	4326.599
Government	72.64	14.53	87.17	3408.347
Public	13.70	2.70	16.40	634.188
Average	45.48	9.07	54.56	8369.13

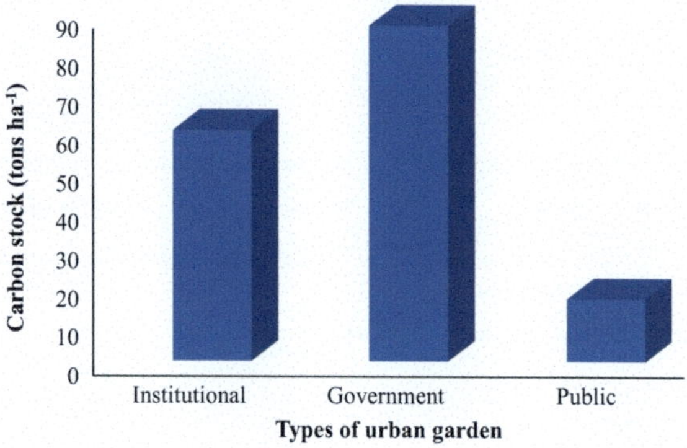

Fig. 7 Total Carbon in different types of urban garden in Kathmandu Valley

4 Conclusion

Assessing the ecosystem services of urban forests and green spaces hinges on vital measures such as carbon storage and sequestration. In the context of the Kathmandu Valley, urban gardens fall into three distinct categories: government, public, and institutional. The density of tree stems with a diameter exceeding 5 cm at breast height within the urban gardens of the Kathmandu Valley was documented at 418.08 ha^{-1}. Notably, government urban gardens display the highest stem density, with 673 stems per hectare, closely followed by institutional urban gardens, with 433.53 stems per hectare. As for above-ground total biomass (AGTB) carbon, institutional, government, and public urban gardens store 50.10, 72.64, and 13.70 tons per hectare, respectively. Additionally, these gardens' below-ground total biomass (BGTB) carbon content amounts to 10, 14.53, and 2.70 tons per hectare, respectively. When considering both AGTB and BGTB, the total carbon content in institutional, government, and public urban gardens reaches 60.10, 87.17, and 16.40 tons per hectare, respectively. Significantly, the carbon stored in the trees, totaling 54.56 tons per hectare, is relatively lower than the national carbon standard, possibly due to the lower stem density in public parks. During field visits and through input from park visitors, we gathered recommendations for effective urban park management. These recommendations include establishing collection points for biodegradable waste, public education initiatives regarding park property and environmental preservation, creating informative materials for park visitors, taking care of trees, tree plantation, and identifying new garden and park development areas. Additionally, we advocated for recognizing gardens at the national level and formulating associated policies.

Annexes

Annex 1: Details of Sample Sites in Kathmandu District

S. No.	Park name	Municipality/ Metropolitan	Ward	Latitude	Longitude	Area (ha)
1	Chunnikhel Park	Budhanilkantha	4	27°45′46.4″	85°21′54.3″	0.52
2	Kalika Park	Budhanilkantha	8	27°45′34.22″	85°21′33.89″	0.56
3	Tribhuvan Park	Chandragiri	4	27°41′23.75″	85°13′14.15″	7.83
4	Jagdol Marty's Park	Gokarneshwor	4	27°44′33.28″	85°22′55.89″	1.31
5	Baisdhara Park	Kathmandu	16	27°44′1.79″	85°18′4.09″	1.2
6	Bhandarkhal	Kathmandu	11	27°42′34.65″	85°20′32.15″	8.28
7	Bhrikuti Mandap	Kathmandu	28	27°42′2.7″	85°19′6.8″	1.38
8	Buddha Park	Kathmandu	15	27°42′57″	85°17′3.8″	0.51
9	Ganesh Man Singh Park	Kathmandu	28	27°42′25.43″	85°18′53.29″	0.5
10	Garden of Dreams	Kathmandu	26	27°42′50″	85°18′52.5″	0.59
11	Narayan Park	Kathmandu	3	27°42′52.87″	85°19′31.82″	0.95

(continued)

(continued)

S. No.	Park name	Municipality/ Metropolitan	Ward	Latitude	Longitude	Area (ha)
12	Raniban	Kathmandu	3	27°43′50.94″	85°19′15.77″	7.6
13	Ratna Park	Kathmandu	28	27°42′21.96″	85°18′54.21″	2.34
14	Shankha Park	Kathmandu	4	27°43′59.61″	85°20′33.78″	1.44
15	Slaeshmantak Ban	Kathmandu	8	27°42′23.02″	85°21′15.72″	17.4
16	Tudikhel	Kathmandu	28	27°42′9.97″	85°18′51.07″	14.4
17	Coronation Garden	Kirtipur	10	27°40′46.47″	85°17′14.63″	16.1
18	Mahamanjushree Park	Kirtipur	6	27°39′29.65″	85°17′38.35″	1

Annex 2: Details of Sample Sites in Bhaktapur District

S. No.	Park name	Municipality/ Metropolitan	Ward	Latitude	Longitude	Area (ha)
1	Sallaghari	Bhaktapur	1	27°40′26.89″	85°24′39.83″	3.27
2	Bandevi Nagar Ban Udhyan	Madhyapur Thimi	12	27°40′26.91″	85°23′9.46″	0.797
3	Bishnubir Park	Madhyapur Thimi	4	27°40′26.91″	85°23′2.8″	2.71
4	Dibyaswori Park	Madhyapur Thimi	3	27°41′13.61″	85°21′57.26″	0.65
5	Kumari Park	Madhyapur Thimi	8	27°41′32″	85°23′12.9″	1.5
6	Nilbarahi Park	Madhyapur Thimi	9	27°41′54.50″	85°23′43.62″	20.5
7	Shankhadar Park	Madhyapur Thimi	4	27°40′54″	85°23′5.20″	0.69

Annex 3: Details of Sample Sites in Lalitpur District

S. No.	Park name	Municipality/ Metropolitan	Ward	Latitude	Longitude	Area (ha)
1	National Botanical Garden	Godawari	2	27°35′45.45″	85°22′53.07″	23
2	UN Park	Lalitpur	10	27°41′8.10″	85°19′34.56″	5.11
3	Central Zoo	Lalitpur	4	27°40′21.27″	85°18′42.66″	5.18
4	Manmohan Park	Lalitpur	3	27°40′44.45″	85°17′57.10″	1.86
5	Sankhamul Park	Lalitpur	9	27°40′43.08″	85°19′53.55″	0.58

Annex 4: Institutional Type Urban Garden

S. No.	Park name	District	Municipality/Metropolitan	Area (ha)
1	BandeviNagar Ban Udhyan	Bhaktapur	Madhyapur Thimi	0.797
2	Bishnubir Park	Bhaktapur	Madhyapur Thimi	2.71
3	Nilbarahi Park	Bhaktapur	Madhyapur Thimi	20.5
4	Jagdol Marty's Park	Kathmandu	Gokarneshwor	1.31
5	Ganesh Man Singh Park	Kathmandu	Kathmandu	0.5
6	Garden of Dreams	Kathmandu	Kathmandu	0.59
7	Raniban	Kathmandu	Kathmandu	7.6
8	Slaeshmantak Ban	Kathmandu	Kathmandu	17.4
9	Tudikhel	Kathmandu	Kathmandu	14.4
10	Mahamanjushree Park	Kathmandu	Kirtipur	1
11	Central Zoo	Lalitpur	Lalitpur	5.18
	Total			71.99

Annex 5: Government Type Urban Garden

S. No.	Park name	District	Municipality/Metropolitan	Area (ha)
1	Coronation Garden	Kathmandu	Kritipur	16.1
2	National Botanical Garden	Lalitpur	Godawari	23
	Total			39.1

Annex 6: Public Type Urban Garden

S. No	Park name	District	Municipality/Metropolitan	Area (ha)
1	Sallaghari	Bhaktapur	Bhaktapur	3.27
2	Dibyaswori Park	Bhaktapur	Madhyapur Thimi	0.65
3	Kumari Park	Bhaktapur	Madhyapur Thimi	1.5
4	Shankhadar Park	Bhaktapur	Madhyapur Thimi	0.69
5	Chunnikhel Park	Kathmandu	Budhanilkantha	0.52
6	Kalika Park	Kathmandu	Budhanilkantha	0.56
7	Tribhuvan Park	Kathmandu	Chandragiri	7.83
8	Baisdhara Park	Kathmandu	Kathmandu	1.2
9	Bhandarkhal	Kathmandu	Kathmandu	8.28
10	Bhrikuti Mandap	Kathmandu	Kathmandu	1.38
11	Buddha Park	Kathmandu	Kathmandu	0.51
12	Narayan Park	Kathmandu	Kathmandu	0.95
13	Ratna Park	Kathmandu	Kathmandu	2.34
14	Shankha Park	Kathmandu	Kathmandu	1.44
15	UN Park	Lalitpur	Lalitpur	5.11
16	Manmohan Park	Lalitpur	Lalitpur	1.86
17	Sankhamul Park	Lalitpur	Lalitpur	0.58
	Total			38.67

References

Akbari H, Pomerantz M, Taha H (2001) Cool surfaces and shade trees to reduce energy use and improve air quality in urban areas. Solar Egergy 70(3):295–310

Anglesen A (2008) Moving ahead with REDD: issues, options and implications. CIFOR, Bogor

Bakrania S (2015) Urbanization and urban growth in Nepal. GSDRC helpdesk research report 1294. University of Birmingham, Birmingham, UK

Brown S, Miltner E, Cogger C (2012) Carbon sequestration potential in urban soils. Springer

Chang J et al (2017) Assessing the ecosystem services provided by urban green spaces along urban center-edge gradients. Nat Sci Rep 7:11226

Chave J, Andalo C, Brown S, Cairns MA, Chambers JQ, Eamus D, Fölster H, Fromard F, Higuchi N, Kira T, Lescure J-P, Nelson BW, Ogawa H, Puig H, Riéra B, Yamakura T (2005) Tree allometry and improved estimation of carbon stocks and balance in tropical forests. Oecologia 145:87–99

Churkina G (2016) The role of urbanization in the global carbon cycle. Front Ecol Evol 3:144

Davies Z et al (2011) Mapping an urban ecosystem service:quantifying above-ground carbon storage at a city-wide scale. J Appl Ecol 48(5):1125–1134

Dhakal S (2010) GHG emissions from urbanization and opportunities for urban carbon mitigation, s.l.: Curr Opin Environ Sustain

Dhyani K, Kuniyal HB, Singh H, Sobha (2021) Growth and physiological potential of Terminalia arjuna under elevated CO_2 levels in Open top chamber condition. J Appl Nat Sci 13(3):1121–1126. https://doi.org/10.31018/jans.v13i3.2490. ISSN 2231-5209
FAO (2011) State of the world's forests. FAO, Rome
Gong P et al (2012). Urbanisation and health in China. Lancet
Grimm NB et al (2008) Global change and the ecology of cities. Science 756–760
Gupta SK, Ram J, Singh H (2019a) Relationship between leaf area index and atmospheric cooling of tree species. Indian J for 42(3):203–208
Gupta SK, Singh H, Ram J (2019b) Need for a novel paradigm in forestry research: perspectives of comparative atmospheric cooling by different forestry effect tree species. Indian Forester 145(6):509–515
IPCC (2006) Guidelines for national greenhouse gas inventories. Prepared by the National Greenhouse Gas Inventories Programme. In: Eggleston HS, Buendia L, Miwa K, Ngara T, Tanabe K (eds). IGES, Japan
IPCC (2007) Climate change 2007—The physical science basis: working group I contribution to the fourth assessment report of the IPCC (climate change 2007), s.l.: s.n.
Karagiannis I, Soldatos P (2010) Estimation of critical CO_2 values when planning the power source in water desalination: the case of the small Aegean islands. Energy Policy 3891–3897
Keeley M et al (2013) Perspectives on the use of green infrastructure for stormwater management in Cleveland and Milwaukee. Environ Manage 51(6):1093–1108
Kumar M, Singh MP, Singh H, Dhakate PM, Ravindranath NH (2019a) Forest working plan for the sustainable management of forest and biodiversity in India. J Sustain for 39:1–22. https://doi.org/10.1080/10549811.2019.1632212
Kumar A, Kumar P, Singh H, Kumar N (2021a) Modulation of plant functional traits under essential plant nutrients during seasonal regime in natural forests of Garhwal Himalayas. Plant Soil 465:197–212. https://doi.org/10.1007/s11104-021-05003-x
Kumar A, Kumar P, Singh H, Kumar N (2021b) Adaptation and mitigation potential of roadside trees with bio-extraction of heavy metals under vehicular emissions and their impact on physiological traits during seasonal regimes. Urban Forestry Urban Greening 58:126900. https://doi.org/10.1016/j.ufug.2020.126900
Kumar A, Kumar P, Singh H, Kumar N (2021c) Impact of plant functional traits on infiltration rate under different forests of Kempty watershed in Garhwal Himalaya. Indian J Soil Conserv 49(1):38–44
Kumar A, Kumar P, Singh H, Bisht S, Kumar N (2021d) Relationship of physiological plant functional traits with soil carbon stock in the temperate forest of Garhwal Himalaya. Curr Sci 120(8):1368–1373. https://doi.org/10.18520/cs/v120/i8/1368-1373
Kumar M, Phukon SN, Paygude AC, Tyagi K, Singh H (2021e) Mapping phenological functional types (PhFT) in the Indian Eastern Himalayas using machine learning algorithm in Google Earth Engine. Comput Geosci 158:104982. https://doi.org/10.1016/j.cageo.2021.104982
Kumar A, Singh H, Kumari G, Bisht S, Malik A, Kumar A, Singh M, Raturi BS, Thakur A, Kaushal R (2022) Adaptive resilience of roadside trees to vehicular emissions via leaf enzymatic, physiological, and anatomical trait modulations. Environ Pollut 313:120191. https://doi.org/10.1016/j.envpol.2022.120191
Kumar N, Singh H, Giri K, Kumar A, Joshi A, Yadav A, Singh R, Bisht S, Kumari R, Jeena N, Khairakpam R, Mishra G (2024) Physiological and molecular insights into the allelopathic effects on agroecosystems under changing environmental conditions. Physiol Mol Biol Plants 30:417–433. https://doi.org/10.1007/s12298-024-01440-x
Kumar N, Jeena N, Singh H (2019d). Elevated temperature modulates rice pollen structure: a study from the foothill of the Himalayan agro-ecosystem in India. 3 Biotech, 9(5):175. https://doi.org/10.1007/s13205-019-1700-1.
MacDicken KG (1997) A guide to monitoring carbon storage in forestry and Agro forestry projects. Winrock International Institute for Agricultural Development

Martin N, Chappelka A, Leowenstein E, Keever G (2012) Comparison of carbon storage, carbon sequestration, and air pollution removal by protected and maintained urban forests in Alabama, USA. Int J Biodivers Sci Ecosyst Services Manage 8(3):265–272

MFALD (2019) Environment-friendly Local Governance Framework, 2013. Ministry of Federal Affairs and Local Development, Government of Nepal, Kathmandu

Nowak D, Crane D (2002) Carbon sequestration and storage by urban trees in the USA. Environ Pollut 116(3):381–389

Paloheimo E, Salmi O (2013) Evaluating the carbon emissions of the low carbon city: a novel approach for consumer based allocation. Cities 30:233–239

Pimpalkhute P, Rao S, Ramachandra T (2010) Carbon sequestration potential of urban trees. s.l., s.n.

Prakash V, Hunney K, Singh H (2022) Effect of elevated carbon dioxide on growth and development of *Santalum album* L. seedlings inoculated with plant growth promoting microorganisms in Open Top Chambers. Int J Environ Agric Biotechnol 7(5):149–159.

Roy S, Byrne J, Pickering C (2012) A systematic quantitative review of urban tree benefits, costs, and assessment methods across cities in different climatic zones. Urban Forestry Urban Greening 11(4):351–363

Rydin Y et al (2012) Shaping cities for health: complexity and the planning of urban environments in the 21st century. The lancet, pp. 2079–2108.

Seto K, Guneralp B, Hutyra L (2012) Global forecasts of urban expansion to 2030 and direct impacts on biodiversity and carbon pools. Proceedings of National Academy of Sciences of the United States of America, pp 16083–16088

Seto K et al (2014) Human settlements, infrastructure and spatial planning. In: Intergovernmental panel on climate change. Working group III—Mitigation of climate change. Cambridge University Press, New York

Shah D, Gavali D (2017) Floral diversity in Vadodara gardens, Gujarat, India. Int J Conserv Sci 8:113–120

Sharma R, Singh H (2021) Alteration in biochemical constituents and nutrients partitioning of asparagus racemosus in response to elevated atmospheric CO_2 concentration. Environ Sci Pollut Res 29:6812–6821. https://doi.org/10.1007/s11356-021-16050-3

Singh H (2021) An integrated approach considering physiological and biophysical based indicators for assessing tolerance of roadside plantations of Alstonia scholaris towards urban roadside air pollution: an assessment of adaptation of plantations for mitigating urban roadside air pollution. Trees. https://doi.org/10.1007/s00468-021-02179-8

Singh H (2024) Exploring adaptive modulation in plant functional traits and its impact on the productivity of Acacia auriculiformis under CO_2-enriched environment. Ind Crops Prod 210:118186. https://doi.org/10.1016/j.indcrop.2024.118186

Singh S, Bhattacharya P, Gupta N (2018) Dust particles characterization and innate resistance for Thevetia peruviana in different land-use pattern of urban area. Int J Environ Sci Technol 15:1061–1072

Singh S, Singh H, Sharma SK, Nautiyal R (2021) Seasonal variation in biochemical responses of bamboo clones in the sub-tropical climate of Indian Himalayan foothills. Heliyon 7(4):e06859. https://doi.org/10.1016/j.heliyon.2021.e06859

Singh H, Bandyopadhyaya S, Kumar A, Singh M, Malik A, Kumari G, Raturi A, Yadav SK, Kumar M, Barthwal S, Thakur A, Rawat PS (2022) Understanding the physiological and biophysical response of urban roadside plantations for assessing adaptation and mitigation mechanisms toward vehicular emissions. Urban Climate 44:101183. https://doi.org/10.1016/j.uclim.2022.101183

Singh M, Singh H, Kumar A, Kumar M, Barthwal S, Thakur A (2024) Soil nitrogen availability determines the CO2 fertilization effect on tree species (*Neolamarckia cadamba*): growth and physiological evidence. Environ Sustain 7:53–60. https://doi.org/10.1007/s42398-023-00300-w

Singh S, Gupta N, Bhattacharya P (2017) Assessment of air pollution responses through biochemical and physical traits observed for four plant species in an urban area, Delhi, India. In: Climate change, resource conservation and sustainability strategies

Singh H, Yadav M, Kumar N, Kumar A, Kumar M (2020) Assessing adaptation and mitigation potential of roadside trees under the influence of vehicular emissions: a case study of Grevillea robusta and Mangifera indica planted in an urban city of India. PLoS ONE 15(1): e0227380. https://doi.org/10.1371/journal.pone.0227380

Stern N (2006) Stern review on the economics of climate change. London, October

Strohbach M, Haase D (2012) Above-ground carbon storage by urban trees in Leipzig, Germany: Analysis of patterns in a European city. Landsc Urban Plan 104(1):95–104

Strohbach M, Arnold E, Haase D (2012) The carbon footprint of urban green space—A life cycle approach. Landsc Urban Plan 104:220–229

Ugle, P, Roa S, Ramachandra T (2010) Carbon sequestration potential of urban trees. In: Proceedings of the Lake 2010: Wetlands, biodiversity and climate change, 22–24 Dec, pp 1–12

Velasco E, Roth M, Norford L, Molina L (2016) Does urban vegetation enhance carbon sequestration? Landsc Urban Plan 148:99–107

Yadav SK, Singh H, Nautiyal R, Ginwal HS, Ansari SA, Barthwal S (2019) Modulation of morpho-physiological responses in Populus deltoides by elevated carbon dioxide and temperature. Forest Sci 66:105–118. https://doi.org/10.1093/forsci/fxz048

Young G et al (2012) Differential exposure to hazardous air pollution in the United States: a multilevel analysis of urbanization and neighborhood socioeconomic deprivation. Int J Environ Res Public Health 9(6):2204–2225

Zhou D et al (2014) Surface urban heat island in China's 32 major cities: spatial patterns and drivers. Remote Sens Environ 152:51–61

Role of Urban Trees in Minimizing the Atmospheric Carbon Dioxide Level of the Megacity of Kolkata, India

Pallavi Dutta, Sana Ahmed, Prosenjit Pramanick, Sufia Zaman, and Abhijit Mitra

Abstract The megacity of Kolkata (22°33′36.00″N and 88°25′12.00″ E) in the maritime state of West Bengal (India), with a population of some 15 million and population density of about 24,000 individuals per square km, is noted for its high population density, large number of vehicles and several industrial units at the outskirt of the city. However, due to unplanned urban expansion at the cost of trees and wetlands areas in the city, the use of low-grade fuel to run the vehicles, and emissions from industrial units (mostly without proper treatment), the near-surface atmospheric carbon dioxide level has hiked up and crossed the threshold value. In this study, we assessed the level of atmospheric CO_2 in five major locations of the city for more than three decades (1990–2022) through three seasons (premonsoon, monsoon, and post-monsoon). The near-surface atmospheric CO_2 level showed the highest value of 427 ppm (during March 2019 at Moulali; Stn. 1) and the lowest value of 326 ppm (during September 2020 at Park Circus; Stn. 5). The stored carbon in the Above Ground Stem Biomass (AGSB) referred to as Above Ground Stem Carbon (AGSC) of the dominant tree species and common to all these sites was estimated during March 2022 to identify the species-wise carbon storage potential of the planted trees. The mean values of AGSC exhibit the order *Azadirachta indica* > *Tamarindus indica* > *Aegle marmelos* > *Delonix regia* > *Mangifera indica* > *Ficus religiosa* > *Acacia auriculiformis* > *Cocos nucifera* > *Bombax ceiba* > *Tectona grandis* > *Terminalia arjuna* > *Albizia saman* > *Ziziphus mauritiana* > *Artocarpus heterophyllus* > *Alstonia scholaris* in all the selected sites. The present study highlights the potential of urban trees to store carbon within the stem biomass, which may serve as a practical roadmap to combat the rising trend of carbon dioxide in the near-surface atmosphere of the city.

Keywords Kolkata · Urban trees · Near-surface atmospheric CO_2 level · Above Ground Stem Biomass (AGSB) · Above Ground Stem Carbon (AGSC)

P. Dutta · S. Ahmed · P. Pramanick · S. Zaman
Department of Oceanography, Techno India University, West Bengal, EM 4 Salt Lake, Sector V, Kolkata 700091, India

A. Mitra (✉)
Department of Marine Science, University of Calcutta, 35 B.C. Road, Kolkata 700019, India
e-mail: abhijit_mitra@hotmail.com

H. Singh (ed.), *Urban Forests, Climate Change and Environmental Pollution*,
https://doi.org/10.1007/978-3-031-67837-0_33

1 Introduction

One of our most pressing environmental challenges is the escalating carbon dioxide levels in urban near-surface atmospheres worldwide. This surge can be attributed to the rapidly expanding population's substantial electricity consumption from coal and other fossil fuels and the widespread use of electrical devices across various sectors, including offices, academic institutions, hospitals, markets, shopping malls, recreational centers, and industries. Additionally, the transportation sector of cities significantly contributes to carbon dioxide emissions. The proliferation of carbon dioxide in the atmosphere poses numerous adverse impacts at both local and regional levels (Mitra 2013; Meena et al. 2011; Kumar et al. 2019c; Kumar et al. 2020;Singh and Kumar 2022; Singh et al. 2024).

These sectors collectively demand massive amounts of energy, predominantly supplied by coal in many countries. The combustion of coal for energy production releases substantial quantities of carbon dioxide into the atmosphere (Singh et al. 2020; Kumar et al. 2021a, b, c, d). Studies have shown that burning 1 kg of bituminous coal yields approximately 2.42 kg of carbon dioxide. According to the Global Electricity Review by EMBER (2021), India stands out as one of the leading countries, ranking second, in generating electricity from coal (Table 1).

Kolkata, formerly known as Calcutta, is the capital of West Bengal, a maritime state in India. Renowned as one of India's major metropolitan cities alongside Delhi, Mumbai, Bengaluru, and Chennai, Kolkata boasts a population of approximately 15.5 million (https://www.newgeography.com). Notably, the city is characterized by its high population density, reaching 24,000 persons per square kilometer (source). Moreover, Kolkata hosts a multitude of industries both within its bounds and in the surrounding areas. These industries contribute significantly to the city's electricity consumption and serve as crucial sources of employment across various sectors (Table 2).

These industries, coupled with unplanned urban development and the plying of many diesel and petrol-based vehicles, are the sources of the rise of carbon dioxide levels in the city's atmosphere. In this research, we have attempted to scan the town's near-surface atmospheric carbon dioxide level at five sites, namely Moulali, Shyambazar, Manicktala, Rajabazar, and Park Circus. Data on atmospheric carbon dioxide were collected from the sampling stations located in these sites over more than three decades (1990–2022) through three seasons, namely premonsoon (March-June), monsoon (July–October), and postmonsoon (November-February). Simultaneously, the dominant trees common to these five sites were also monitored during March 2022 to evaluate the carbon storage potential of the tree species. This approach has been adopted primarily because the producer communities on the planet Earth scrub the carbon dioxide from the atmosphere to carry out the process of photosynthesis irrespective of the habitat types (Mitra et al. 2012, 2015; Zaman et al. 2014; Bhattacharyya et al. 2015; Mitra 2015; Chakraborty et al. 2016; Banerjee et al. 2016, 2017a, 2017b, 2018, 2020; Agarwal et al. 2017, 2021; Pal et al. 2020). Thus, the present study was undertaken to construct a baseline dataset of the dominant tree

Table 1 Ranking of top 25 countries in the World generating coal-fire electricity

Rank	Country	Coal generation (TWh)
1	China	4631
2	India	947
3	United States	774
4	Japan	274
5	South Korea	192
6	South Africa	191
7	Indonesia[a]	168
8	Russia	155
9	Vietnam	141
10	Australia	135
11	Germany	134
12	Taiwan	117
13	Poland	110
14	Turkey	99
15	Kazakhstan	72
16	Malaysia[a]	67
17	Philippines	49
18	Canada	45
19	Ukraine	38
20	Thailand	35
21	Czechia	32
22	Pakistan	29
23	Serbia	25
24	Brazil	22
25	Israel*	22

Source: Ember Global Electricity Review (2021); [a] For Indonesia, Malaysia, and Israel, 2019 has been reflected as 2020 data sets could not be obtained

species in the city of Kolkata that have the potential to retard the rise of carbon dioxide levels in the city's atmosphere. However, we took a conservative approach and estimated the Above Ground Stem Biomass (AGSB), which is a relatively permanent storehouse of carbon, unlike the branches and leaves that often fall during strong wind action or seasonal effects (Singh and Verma 2013; Agarwal et al. 2017; Kumar et al. 2017; Gupta and Singh 2017; Singh et al. 2017a, b).

Table 2 "Major large-scale industries/public sector undertakings in and around the megacity of Kolkata

S. No	Name of the unit
1	M/s. Garden Reach Ship Builders', Garden Reach, Kolkata
2	M/s. India Tobacco Co. Ltd., Garden Reach, Kolkata
3	M/s. Paharpur Cooling Tower Ltd., Garden Reach, Kolkata
4	M/s. General Electricals Co. Ltd., Paharpur, Kolkata
5	M/s. India Foils Ltd., Taratala, Kolkata
6	M/s. Hindustan Development Corpn. Ltd., Tiljala, Kolkata
7	M/s. Kilburn Engg. Co. Ltd., Majherhat, Kolkata
8	M/s. Siel Hardmetal Ltd., Behala, Kolkata
9	M/s. Indian Oxygen Ltd., Taratala, Kolkata
10	M/s. Willamson Magor Ltd., Majherhat, Kolkata
11	M/s. Modern Bread Industry Ltd., Taratala, Kolkata
12	M/s. Eveready Battery, Taratala, Kolkata
13	M/s. Britania Biscuit Co. Ltd., Hide Road, Kolkata
14	M/s. Balmer Lawrie & Co Ltd., Hide Road, Kolkata
15	M/s. Calcutta Chemical Co. Ltd., Bandel Road, Kolkata
16	M/s. Boroline Industries, Chiriamore, Kolkata
17	M/s. Usha Fan Industry Ltd., Bansdroni, Kolkata
18	M/s. Bharat Brake & Valves, S.M.Avenue, Kolkata
19	M/s. Rescon India (P) Ltd., Behala, Kolkata
20	M/s. Hindustan Levers Ltd., Garden Reach, Kolkata
21	M/s. Polar Fan Industry Ltd., Behala, Kolkata
22	M/s. Central Inland Water Transport Corpn., Dock Yard Road, Kolkata
23	M/s. British Engg. Pumps Ltd., Taratala, Kolkata
24	M/s. Bharat Process & Mechanical Engg., Ultadanga, Kolkata
25	M/s. Greaves Foseco Ltd.,Taratala, Kolkata
26	Downstream units of HPL (10 Units)
27	Easy Fit Jewellery Pvt. Ltd
28	Jute Mills (2 Units)
29	Mizan & Co
30	Sree Ganesh Jewellery House Ltd
31	HM Biscuits Industries
32	Gitanjali Gems Ltd.

Source Ahmed et al. (2022a)

2 Materials and Methods

2.1 Site Selection

Five sites were selected in Kolkata based on the degree of anthropogenic activities (Table 3 and Fig. 1).

Table 3 Geographical location of the selected sites

Stations	Coordinates
Moulali (Station 1)	22°33′36.0″ N; 88°22′00.3″ E
Shyambazar (Station 2)	22°36′06.5″ N; 88°22′26.4″ E
Manicktala (Station 3)	22°35′08.0″ N; 88°22′29.5″ E
Rajabazar (Station 4)	22°34′31.7″ N; 88°22′24.5″ E
Park Circus (Station 5)	22°32′40.1″ N; 88°22′09.4″ E

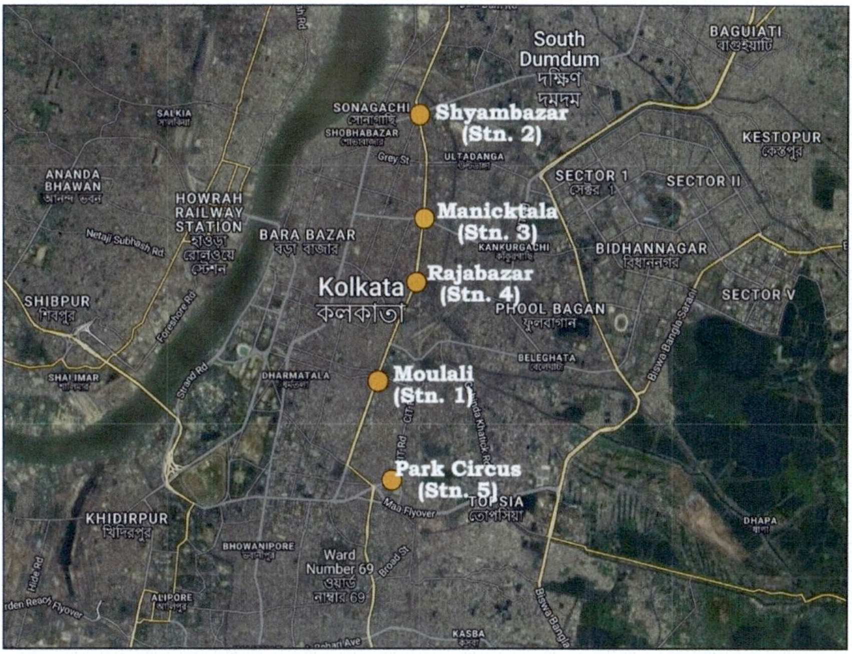

Fig. 1 Map showing the selected five sites for the present study

2.2 Analysis of Near-Surface Atmospheric CO_2

A portable CO_2 analyzer (Lutron CO_2 meter, GCH-2018) was used to measure the near-surface atmospheric carbon dioxide concentrations in and around the selected stations (10 in numbers per site) during the study period (1990–2022) through three seasons. As a part of quality assurance, ten readings were taken during the afternoon hours at each site at 15 m apart. Finally, the average value for every selected site was considered for further statistical analysis.

2.3 Stem Biomass Estimation

The non-destructive method was used to determine the stem biomass for each tree species in which the measurement of the Diameter at the Breast Height (DBH) was done manually after evaluating the circumference with a measuring tape and height using a laser beam (BOSCH DLE 70 Professional model)." The Form Factor was determined according to the method outlined by Koul and Panwar (2008). The following expression determined the stem volume (V),

$$V = FH\Pi r^2$$

where H is the height of the target tree, F is the form factor, and r is the tree's radius derived from its DBH.

The stem core of each species was used to estimate the species-wise specific gravity (G), and using the expression AGSB = GV, the Above Ground Stem Biomass (AGSB) for each species was calculated.

The evaluation of AGSB per hectare was estimated by counting the population of each tree species in the 10 m × 10 m quadrant and finally extrapolated the same to know the counts in 10,000 sq. m (1 hectare) area. To ensure the reliability of the output data on AGSB per hectare, the mean value of 10 quadrants was considered for species-wise population count in each site.

2.4 Stem Carbon Estimation

Direct estimation of percent carbon was done by CHN analyzer (Vario MACRO elementar). The analysis was carried out with a portion of fresh stem samples of the target trees of each species after oven drying at 70 °C, mixed randomly, and grounded to pass through a mesh size of 0.5 mm. The value obtained was converted to stored carbon per species (in tonnes) considering the AGSB (Above Ground Stem Biomass) for each species and their respective number (in the 10 m × 10 m quadrant).

The flowchart for species-wise stem carbon estimation is presented in Fig. 2.

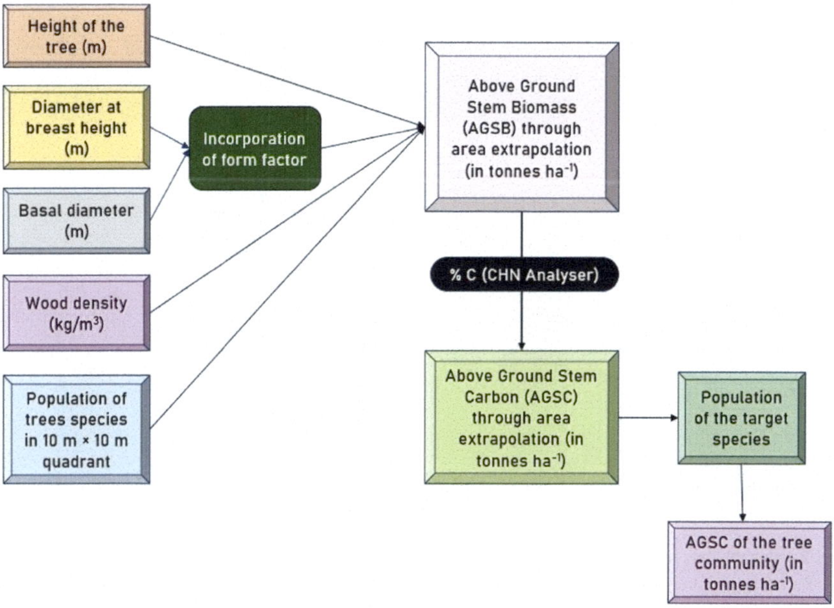

Fig. 2 Flowchart of the estimation of Above Ground Stem Carbon (AGSC)

3 Results

The near-surface atmospheric carbon dioxide in the study sites varied as per the order Moulali (400 ppm) > Rajabazar (397 ppm) > Manicktala (395 ppm) > Shyambazar (392 ppm) > Park Circus (386 ppm). These values are the mean of three seasons (Fig. 3).

The seasonal variation of the carbon dioxide level exhibited the order premonsoon > postmonsoon > monsoon irrespective of sites (Figs. 4, 5, 6, 7 and 8).

It is evident from the mean CO_2 level that the ratios of CO_2 level in the selected sites to the safe level of CO_2 in the atmosphere (350 ppm) are 1.143:1, 1.120:1, 1.129:1, 1.134:1, 1.103:1 at Stns. 1, 2, 3, 4 and 5 respectively. A lower ratio value is a proxy for better carbon dioxide gas air quality.

Our aim to identify and rank the species-wise stored carbon potential of the common tree species was carried out through the evaluation of the biomass of the tree where the rank varied as per the sequence *Azadirachta indica* > *Tamarindus indica* > *Aegle marmelos* > *Delonix regia* > *Mangifera indica* > *Ficus religiosa* > *Acacia auriculiformis* > *Cocos nucifera* > *Bombax ceiba* > *Tectona grandis* > *Terminalia arjuna* > *Albizia saman* > *Ziziphus mauritiana* > *Artocarpus heterophyllus* > *Alstonia scholaris* in all the selected sites (Figs. 9, 10, 11, 12 and 13).

The readings obtained from species-wise carbon percentage analysis using the CHN analyzer and converted to AGSC (Figs. 14, 15, 16, 17 and 18) showed a similar

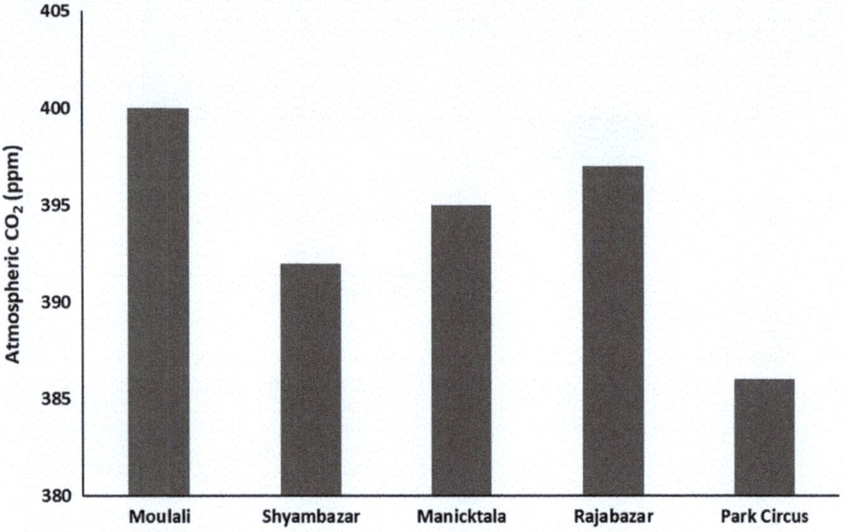

Fig. 3 "Near-surface atmospheric carbon dioxide (in ppm) level in the selected sites (these values are the mean of three seasons)"

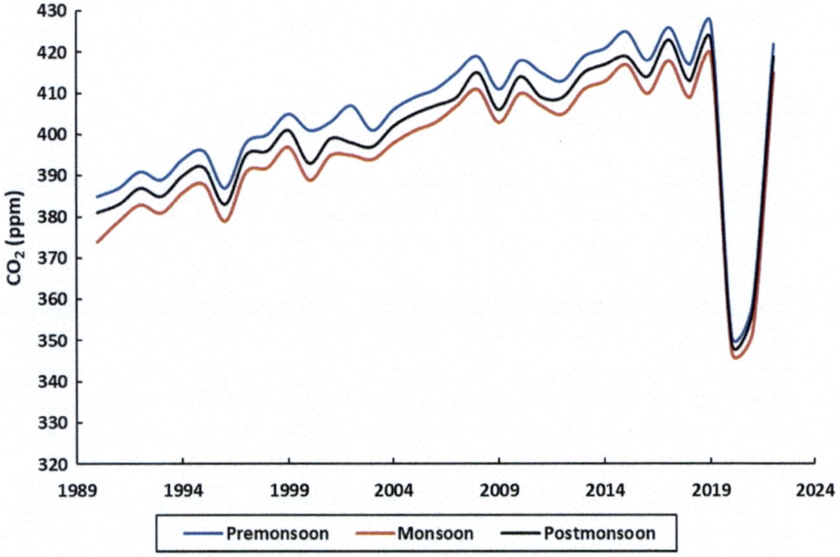

Fig. 4 "Near-surface atmospheric carbon dioxide (in ppm) value through three seasons at Moulali (Stn. 1)"

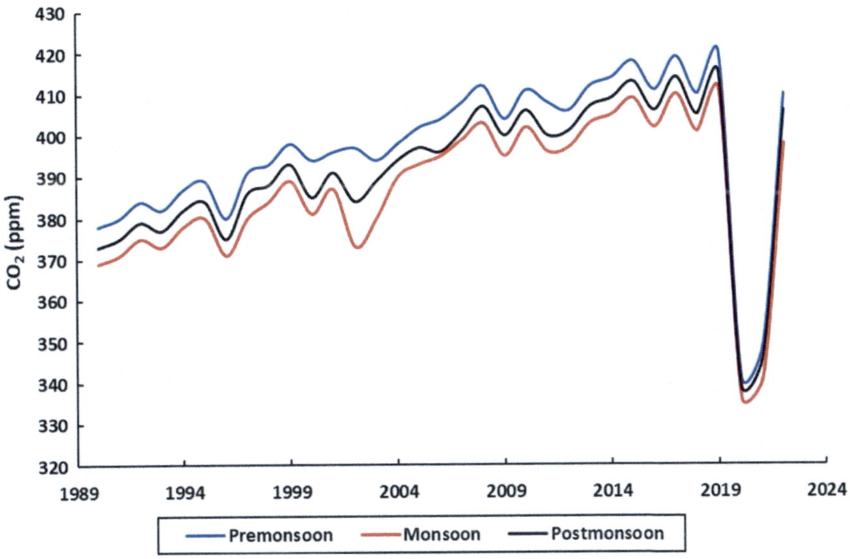

Fig. 5 "Near-surface atmospheric carbon dioxide (in ppm) value through three seasons at Shyambazar (Stn. 2)"

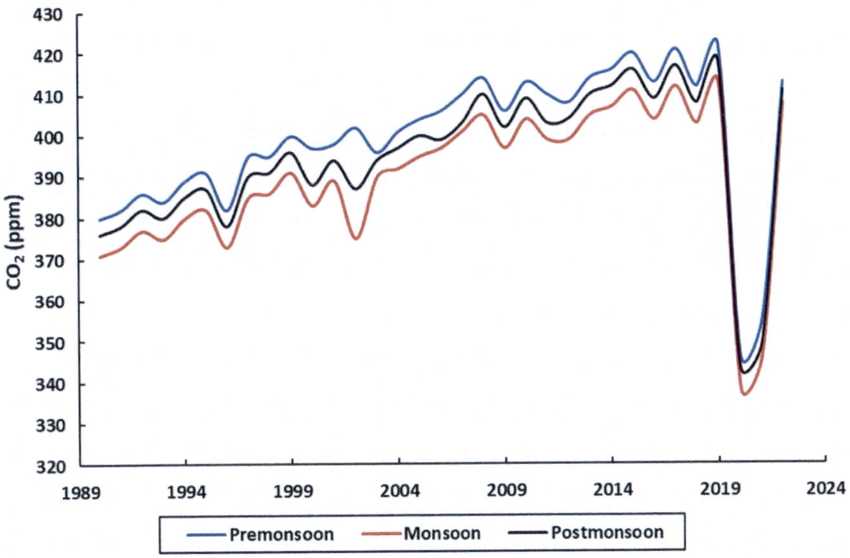

Fig. 6 "Near-surface atmospheric carbon dioxide (in ppm) value through three seasons at Manicktala (Stn. 3)"

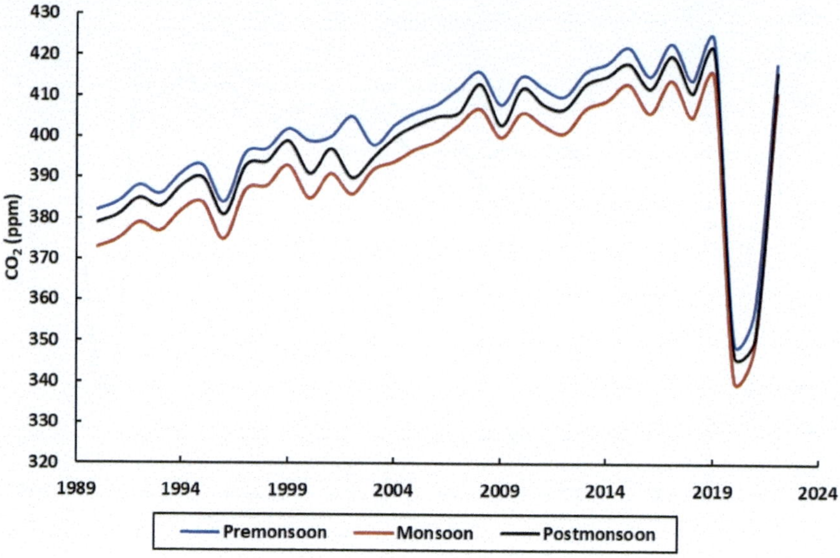

Fig. 7 "Near-surface atmospheric carbon dioxide (in ppm) value through three seasons at Rajabazar (Stn. 4)"

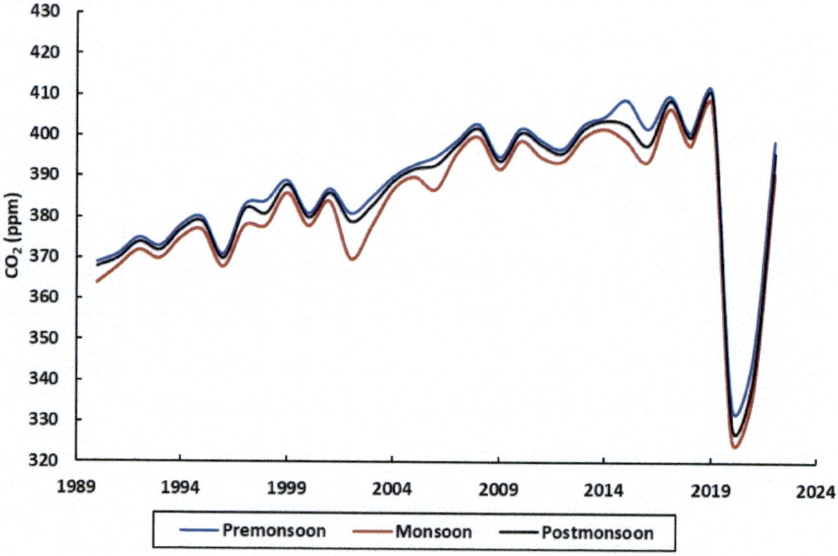

Fig. 8 "Near-surface atmospheric carbon dioxide (in ppm) value through three seasons at Park Circus (Stn. 5)"

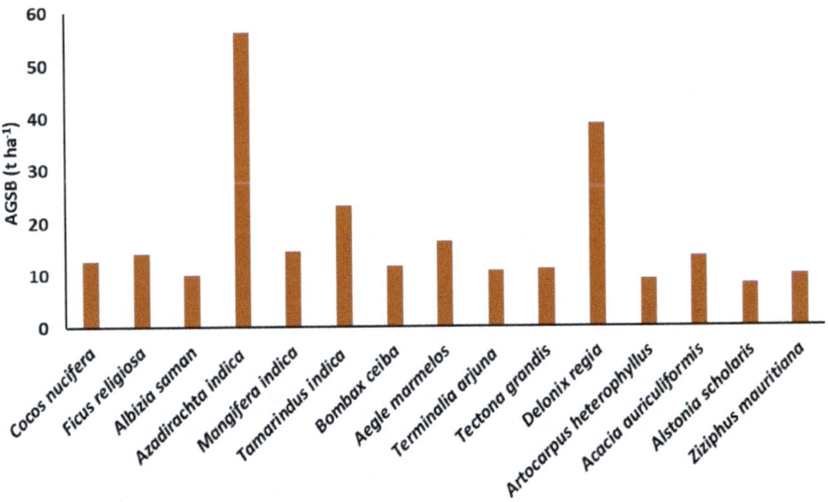

Fig. 9 "AGSB (in t ha^{-1}) of the common tree species at Moulali (Stn. 1) during March 2022"

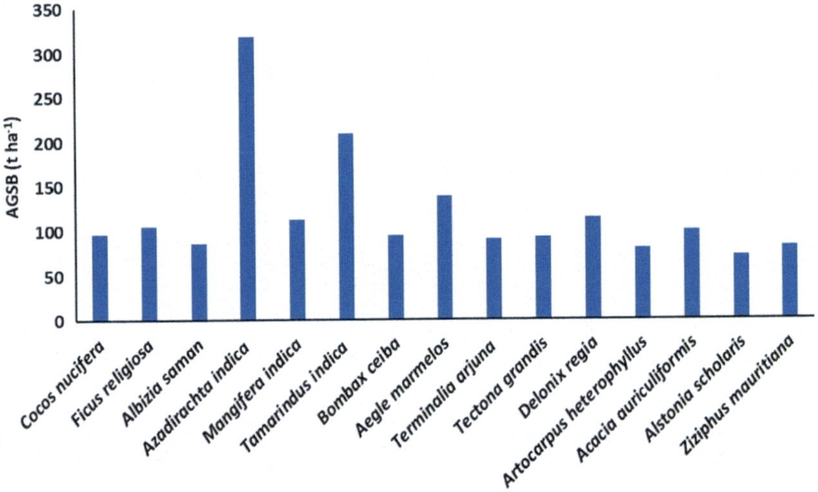

Fig. 10 "AGSB (in t ha^{-1}) of the common tree species at Shyambazar (Stn. 2) during March 2022"

trend to that of the AGSB from which carbon dioxide equivalent (CO_2–e) was estimated through the incorporation of the conversion factor 3.67 (Petersson et al. 2012). Table 4 presents the site-wise CO_2-e considering the most common dominant tree species.

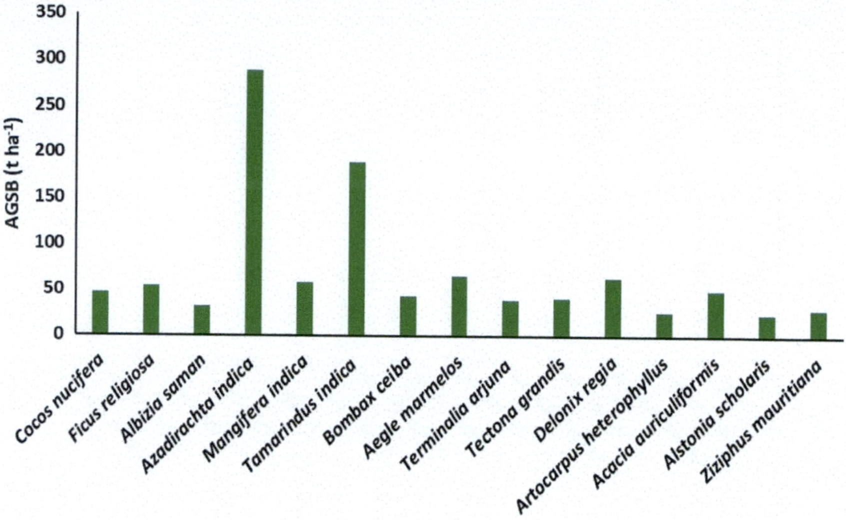

Fig. 11 "AGSB (in t ha^{-1}) of the common tree species at Manicktala (Stn. 3) during March 2022"

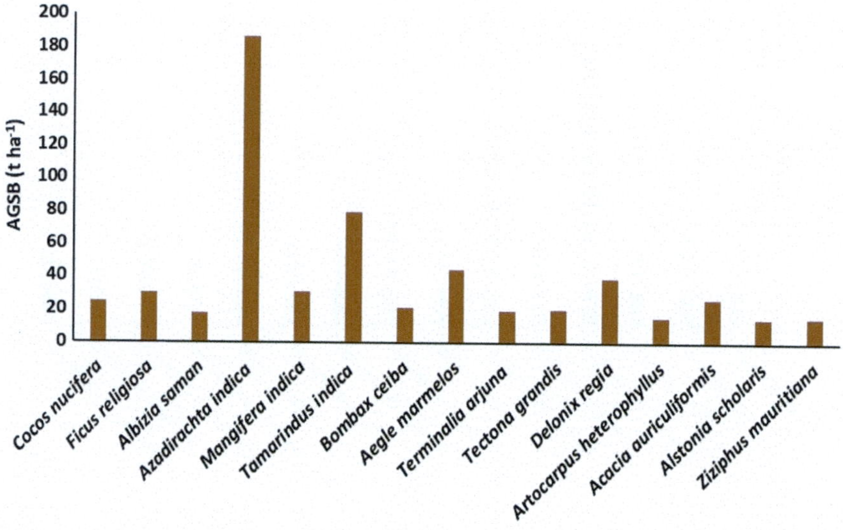

Fig. 12 "AGSB (in t ha^{-1}) of the common tree species at Rajabazar (Stn. 4) during March 2022"

4 Discussion

The city of Kolkata, with a population of about 15 million, supports several small and medium-scale industries that are the point sources of carbon dioxide emission. Also, many vehicles ply over the city, producing carbon dioxide emissions. According to

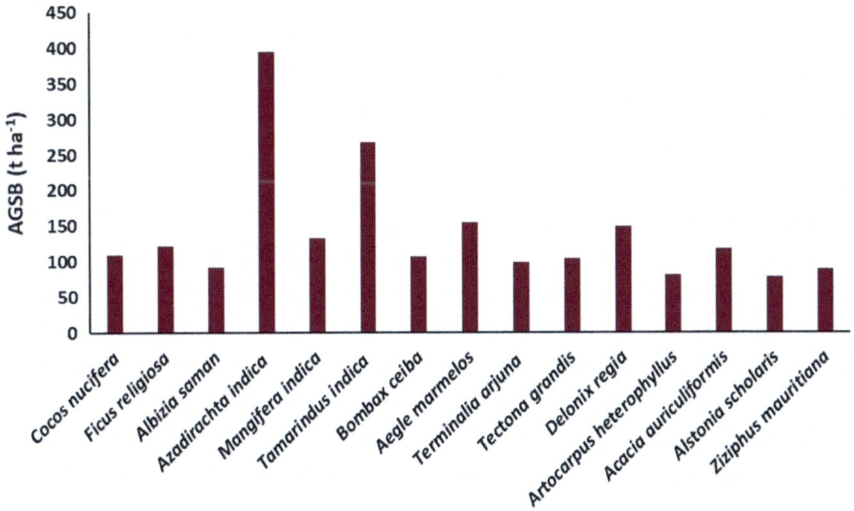

Fig. 13 "AGSB (in t ha^{-1}) of the common tree species at Park Circus (Stn. 5) during March 2022"

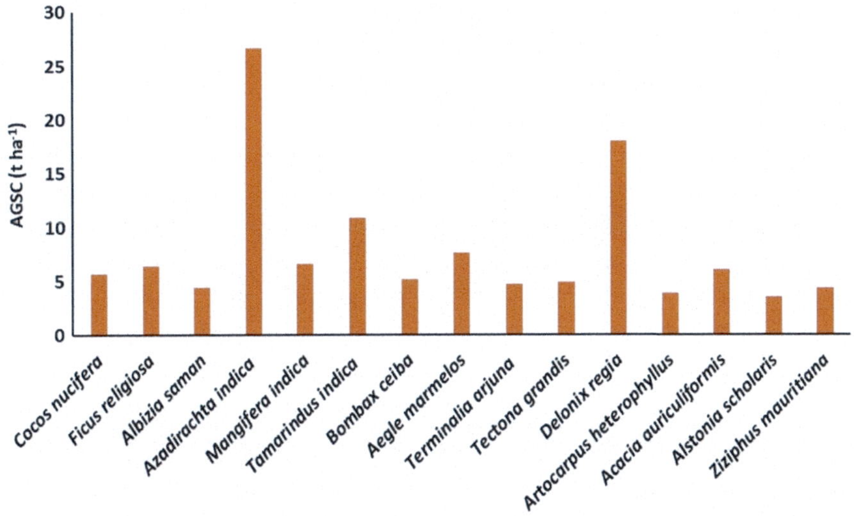

Fig. 14 "AGSC (in t ha^{-1}) of the common tree species at Moulali (Stn. 1) during March 2022"

Chowdhury (2015), the "vehicular population of Kolkata is near about 12.04 lakhs and total road space within the KMC area 7.0%. From these data sets, it can be inferred that a negative correlation exists between vehicle growth and limited road space because traffic congestion and automobile emissions are expected in Kolkata.

The main reasons behind the maximum vehicular emission are the high average age of the vehicles, shallow road surface area, road surface conditions such as

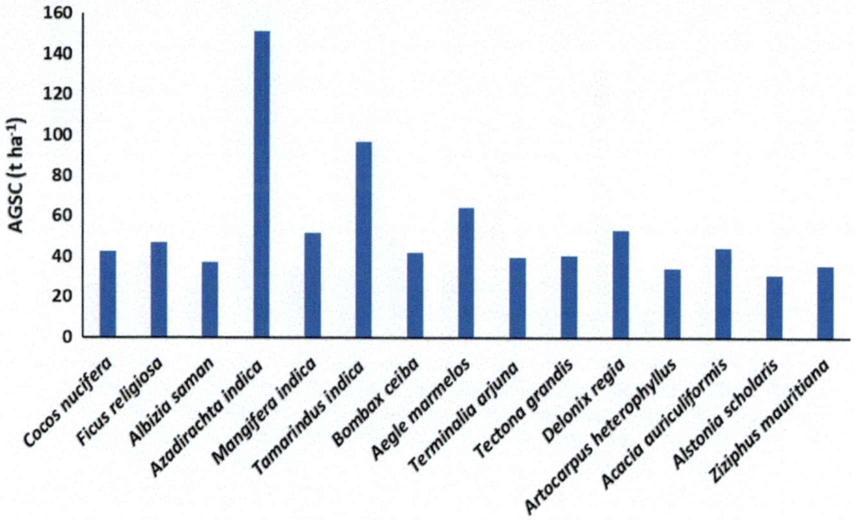

Fig. 15 "AGSC (in t ha^{-1}) of the common tree species at Shyambazar (Stn. 2) during March 2022"

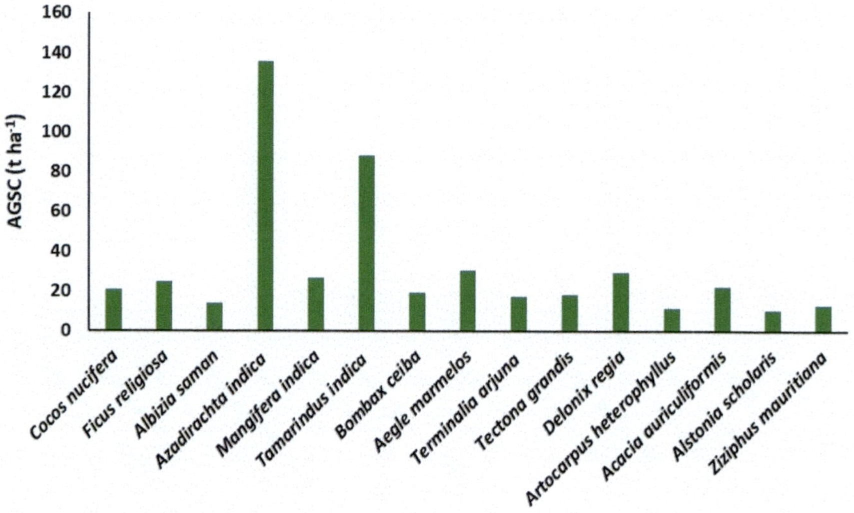

Fig. 16 "AGSC (in t ha^{-1}) of the common tree species at Manicktala (Stn. 3) during March 2022"

damaged roads, speed breakers, etc., high density of population using the same road space, mixed vehicular mode along with the slow-moving vehicles that are primarily responsible for high-level emission from the transport sector of the city (Sharma et al. 2018; Singh et al. 2018; Gupta et al. 2018a, 2018b). The major causes of air pollution associated with the transport sector of Kolkata are shown in Fig. 19.

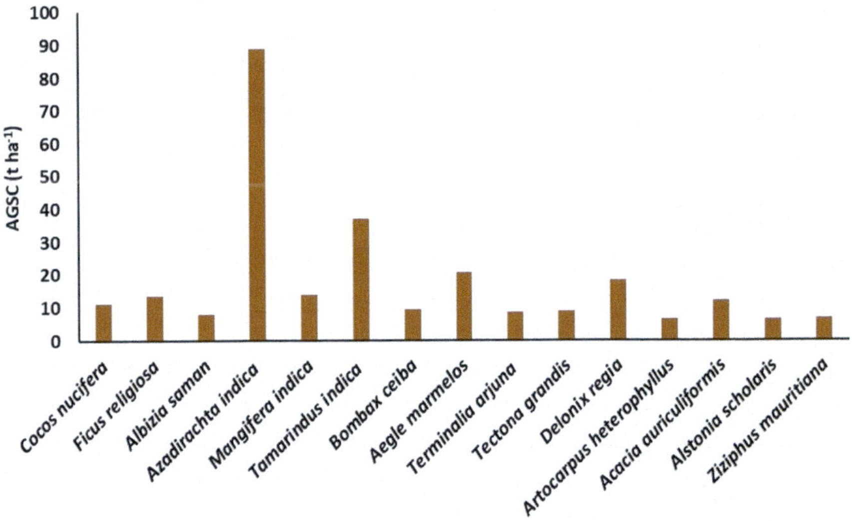

Fig. 17 "AGSC (in t ha^{-1}) of the common tree species at Rajabazar (Stn. 4) during March 2022"

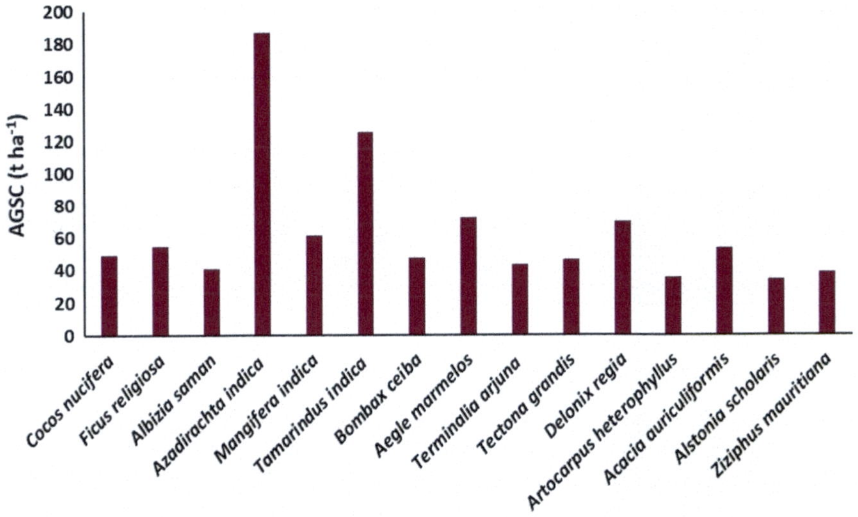

Fig. 18 "AGSC (in t ha^{-1}) of the common tree species at Park Circus (Stn. 5) during March 2022

The city of Kolkata is the capital of the state of West Bengal, which is a maritime state on the northeast coast of the Indian subcontinent. The city has recently become highly congested as it has attracted populations from all over the country for employment and better livelihood. The numbers of vehicles, offices, academic institutes, and industrial activities have also increased simultaneously to meet the growing demand of this rising population. All these sectors require enormous energy to run their

Table 4 Station-wise carbon dioxide - equivalent (in ha^{-1}) of the common tree species

S. No	Species	Moulali	Shyambazar	Manicktala	Rajabazar	Park Circus
1	 *Cocos nucifera*	20.97	157.13	76.76	40.68	179.69
2	 *Ficus religiosa*	23.54	173.54	89.99	49.51	201.37
3	 *Albizia saman*	16.18	138.17	51.12	28.58	148.49
4	 *Azadirachta indica*	97.90	556.27	497.72	325.61	685.65
5	 *Mangifera indica*	24.27	189.73	98.61	50.84	224.11

(continued)

Table 4 (continued)

S. No	Species	Moulali	Shyambazar	Manicktala	Rajabazar	Park Circus
6	 *Tamarindus indica*	40.02	356.82	323.52	135.88	459.82
7	 *Bombax ceiba*	19.04	155.88	71.15	34.73	174.37
8	 *Aegle marmelos*	28.02	236.95	112.36	76.30	265.29
9	 *Terminalia arjuna*	17.42	147.04	63.30	31.55	157.71
10	 *Tectona grandis*	17.92	150.50	67.40	32.60	169.11
11	 *Delonix regia*	65.90	196.33	109.18	66.90	255.44

(continued)

Table 4 (continued)

S. No	Species	Moulali	Shyambazar	Manicktala	Rajabazar	Park Circus
12	 *Artocarpus heterophyllus*	14.41	127.41	42.01	23.80	128.66
13	 *Acacia auriculiformis*	22.26	163.38	81.50	44.03	195.06
14	 *Alstonia scholaris*	12.87	113.84	39.01	23.61	124.68
15	 *Ziziphus mauritiana*	15.97	131.78	47.29	24.61	141.54

respective activities, generating considerable carbon dioxide in the ambient media (Sengupta et al. 2022; Ahmed et al. 2022b). The trees of this city have also been deforested in many places to provide space for the city's horizontal or road expansion (Singh 2021; Sharma and Singh 2021; Singh et al. 2021). Hence, it has become necessary to probe the carbon dioxide level in different zones of the city and the carrying capacity of the tree species to combat the rise of carbon dioxide in the city atmosphere. The present study identified a few core findings like (1) significant spatiotemporal variations of the near-surface carbon dioxide level in the city

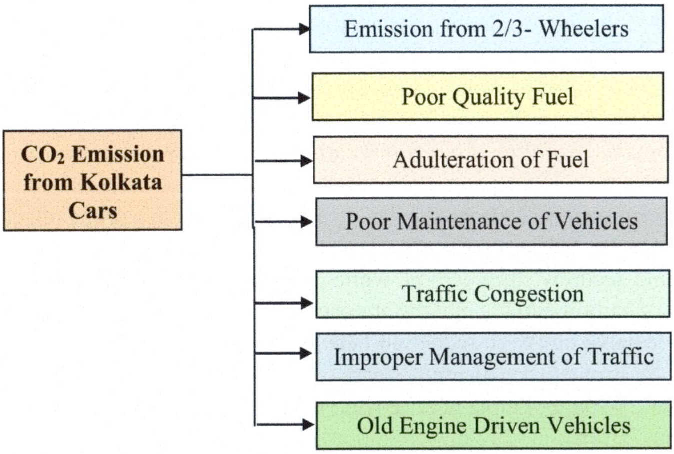

Fig. 19 Major Sources of CO_2 Emission from the Cars in the City of Kolkata

atmosphere (Table 5). (2) *Azadirachta indica*, *Tamarindus indica*, *Aegle marmelos*, *Delonix regia,* and *Mangifera indica* are the most critical trees in terms of carbon storage, and (3) carbon credit can be achieved for the city trees in terms of monetary benefits (provided the criteria for carbon credit are satisfied) that can be further utilized for better management, maintenance and expansion of the trees.

In the context of global environmental sustainability, using carbon credits for urban trees remains underdeveloped across many nations despite its potential to yield significant benefits.

Forestry initiatives to reduce greenhouse gas emissions are gaining traction on local, regional, and global scales (Sedjo et al. 2001; Bigsby 2009). Wofsy (2001) notes that global forests absorb approximately a quarter of the carbon emitted from fossil fuel combustion annually. Carbon markets commonly recognize five categories of forestry projects: afforestation, reforestation, avoided deforestation, intensive forest management, and urban forestry.

Urban forestry, though still in its early stages, has been the subject of some research focusing on the measurement and quantification of carbon sequestration in urban trees

Table 5 Spatio-temporal variation of near-surface carbon dioxide

Source of variation	SS	df	MS	F	*P*-value	F crit
Between stations	362.9333	4	90.7333	80.0588	1.73E-06	3.8379
Between seasons	176.9333	2	88.4667	78.0588	5.65E-06	4.4590
Error	9.0667	8	1.1333			
Total	548.9333	14				

(Birdsey 1992; Jo and McPherson 1995; Hoover et al. 2000; Nowak and Crane 2002; Myeong et al. 2006; Pouyat et al. 2006; Smith et al. 2006), as well as the evaluation of ecological and other associated benefits (Rowntree and Nowak 1991; McPherson 1994, 1998; Jo and McPherson 2001; Brack 2002). However, a significant knowledge gap remains regarding urban forest carbon credit projects' market potential and quality criteria.

Various criteria have been established to assess the merit of carbon offset projects. For instance, the effectiveness of a forestry project depends on its capacity to sequester and securely store carbon, while projects like tidal energy aim to prevent the release of carbon dioxide into the atmosphere. Some projects are highly regarded for their verifiable emission reduction benefits, while others are esteemed for their adherence to stringent verification standards, ensuring the permanency of credits (Yadav et al. 2019a, b; Gupta et al. 2019a; Gupta et al. 2019a; Kumar et al. 2019c; Kumar et al. 2019b).

Tansey (2009) highlighted the challenge nascent carbon credit markets faced where marketing often precedes quality. Carbon markets enforce rigorous policies, standards, and independent verification mechanisms to validate the authenticity of carbon credits, reflecting the growing demand for high-quality credits in voluntary markets (Dhyani et al 2021; Kumar et al. 2021a, b, c, d, e; Prakash et al. 2022). Corporations engaging in GHG reduction schemes as part of their Corporate Social Responsibility (CSR) actively seek reliable credit suppliers with robust scientific backing and adherence to carbon credit-related policies.

Several organizations have published consumer guidelines and technical reports to assist purchasers in thoroughly evaluating carbon credits. For example, Carbon Concierge (2008) developed a criteria-based matrix for assessing the quality of carbon credit providers. Beane et al. (2008) devised a scorecard evaluating forest carbon credit quality based on eight general components. These initiatives signify a burgeoning market for high-quality credits, suggesting limited marketing scope for projects failing to meet stringent quality criteria.

The United States has undertaken extensive research in the field of urban forestry. Nowak and Walton (2005) estimated that urban land use constituted slightly over 3% of the total land area in 2000, projected to reach 8% by 2050. Additionally, Nowak et al. (2001) reported that urban areas in the U.S. have an average tree coverage of approximately 27%, with an estimated 4 billion urban trees and an additional 70 billion trees growing in metropolitan areas nationwide (Bratkovich et al. 2008). Studies such as Nowak and Crane (2002) provide insights into current carbon storage and potential annual sequestration rates in major U.S. metropolitan areas. With urban land expected to expand rapidly, there is ample opportunity to establish new urban forestry projects, including plantations on marginal land and acquiring land for urban parks and public open spaces.

In this study, we investigate the extent to which common urban trees in Kolkata meet quality criteria. These trees are commonly found in urban settings such as streets, parks, riparian buffers, and public and private spaces. Our analysis focuses on three key criteria:

1. Assessing site-specific variations in near-surface atmospheric levels to determine vulnerability scales.
2. Examining site-specific variations in Aboveground Carbon Storage Capacity (AGSC) among dominant tree species in selected sites to identify species with high carbon storage potential.
3. Evaluating the correlation between tree species and carbon storage by analyzing the correlation coefficient between stored carbon in trees and atmospheric carbon dioxide levels, supported by statistical significance.

Our findings reveal a significant negative correlation between these variables ($r_{AGSC \times CO2} = -0.96$; $p < 0.1$), indicating that urban trees play a crucial role as a sink for atmospheric carbon dioxide.

The AGSC value of each species was transformed to respective CO_2-e for all the five sites and the observed order (in tonnes ha^{-1}) is Park Circus; Stn. 5 (3511.01) > Shyambazar; Stn.2 (2994.77) > Manicktala; Stn.3 (1770.93) > Rajabazar; Stn. 4 (989.23) > Moulali; Stn. 1 (436.71) (Fig. 20).

The reason for estimating the CO_2–e values is to evaluate the possibility of acquiring carbon credit provided all the criteria for a high-quality carbon credit project are fulfilled.

The price of CO_2-e varies by region and is determined by the carbon trading market, and it also exhibits temporal variation. As of March 31, 2023, the European Union Emission Trading Scheme (EU-ETS) carbon price was above US $96 per metric tonne of CO_2-e (https://www.statista.com). Some examples of the oscillation of the CO_2-e prices are US $7.60 in the Regional Greenhouse Gas Initiative (RGGI) auction in March 2021 (https://www.eia.gov), US$185 as the suggested price by researchers in Nature (https://phys.org/news/2022-09-carbon-dioxide-price.html)

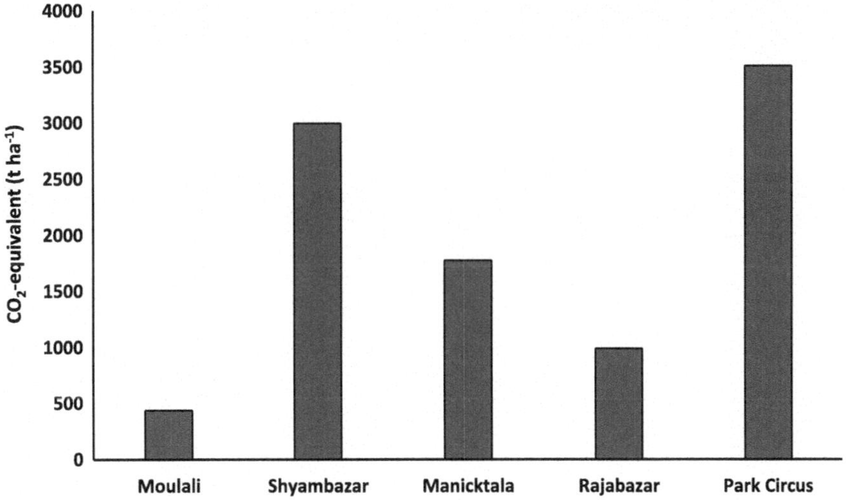

Fig. 20 Total CO_2-e (in t ha^{-1}) in the selected stations

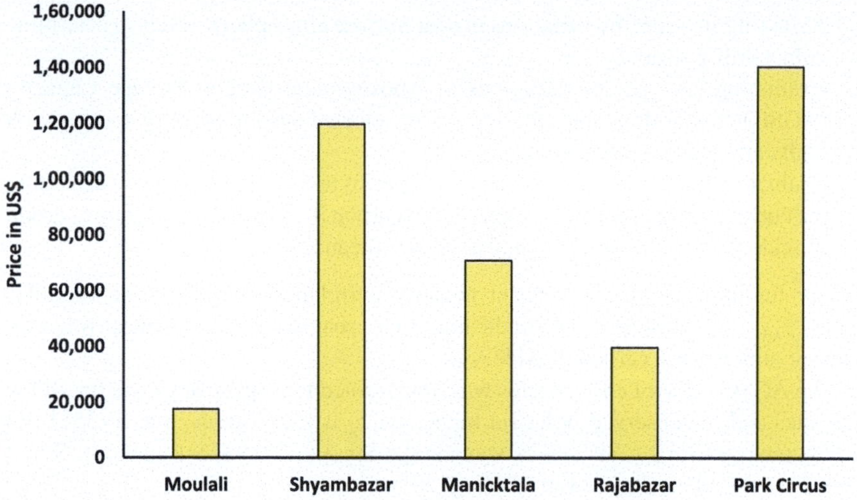

Fig. 21 Total price in US$ of the total CO_2-e in the selected stations

etc. In this study, we have considered the social cost of CO_2 emission in India, which is about $86 per tonne of CO_2 (Ricke et al. 2018). Based on this, we have stepped down the value to US $ 40, which is equivalent to INR 3334.40 (as per the date of 7th December 2023, 1 US $ = INR 83.36; https://www.xe.com/currencyconverter; calculated on 7th December 2023). The main reasons for this step down are our study is (i) not emission related rather storage related by dominant tree species, (ii) devoid of plantation and maintenance, instead it involves the survey, sampling, and analysis cost only, and (iii) more of a voluntary nature, an initiative of Techno India University, West Bengal. On this foundation, the price of CO_2–e per hectare in the five selected sites of Kolkata is presented in Figs. 21 and 22.

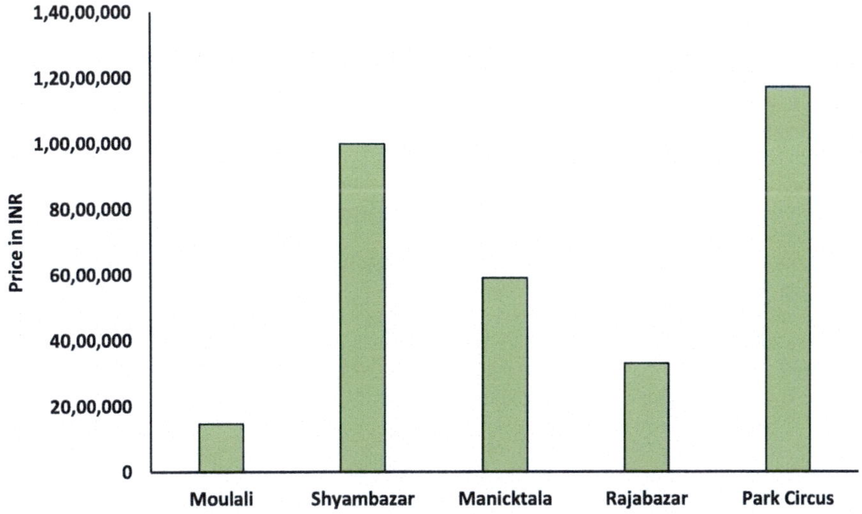

Fig. 22 Total price in INR of the total CO_2-e in the selected stations

5 Conclusion

The present study delves into the significance of urban forests in mitigating the escalating carbon dioxide levels in Kolkata's atmosphere. It endeavours to establish a connection between urban forestry and the carbon trading sector, although it lacks third-party verification and is thus unregistered according to prevailing norms. Despite its limited scope as a research endeavour, it serves as a valuable blueprint for Indian cities, offering insights to embrace, customize, and expand sustainable urban forestry strategies. This approach holds substantial merit and has garnered recognition on the international stage. The Sixth Assessment Report (2022) of the Intergovernmental Panel on Climate Change (IPCC) underscored the importance of ecologically sustainable urban planning and infrastructure design. Noteworthy concepts such as rooftop gardens, park networks, urban forests, urban agriculture, and rainwater harvesting emerged as pivotal tools for climate change mitigation and adaptation within urban landscapes.

References

Agarwal S, Banerjee K, Pal N, Mallik K, Bal G, Pramanick P, Mitra A (2017) Carbon sequestration by mangrove vegetations: A case study from Mahanadi mangrove wetland. J Environ Sci Comp Sci Eng Tech 7(1):16–29

Agarwal S, Chakrabarty SP, Zaman S, Pramanick P, Biswas P, Mukhopadhyay N, Dutta J, Mitra A (2021) Assessment of carbon storage potential of dominant tree species in an educational campus, Kolkata, West Bengal. Ela J for Wildl 10(2):869–878

Ahmed S, Trivedi S, Pal N, Pramanick P, Zaman S, Mitra A (2022a) Study on floral carbon storage potential in the megacity of kolkata: a roadmap towards net-zero carbon emission. J Environ Sociobiol 19(2):113–123

Ahmed S, Zaman S, Gobato R, Mitra A (2022b) Pulse of acidification in the Gangetic water in and around the megacity of Kolkata, India. Par J Sci Edu 8(2):12–15. https://tiny.cc/PJSE24476153v8i2p012-015

Banerjee K, Amin G, Fazli P, Pramanick P, Zaman S, Mitra A (2016) *Excoecaria agallocha*: a potential Mangrove species in context to carbon storage in high Saline zone. J Environ Sci Comp Sci Eng Tech 5(2):115–120

Banerjee K, Agarwal S, Pal N, Mitra A (2017a) Bhitarkanika Mangrove forest: a potential sink of carbon. Int J Res Cult Soc 1(7):280–286

Banerjee K, Bal G, Pal N, Amin G, Mitra A (2017b) Carbon sequestration by Mangrove vegetation: a case study from Odisha Bhitarkanika Wild Life Sanctuary (BWLS). Par J Sci Edu 3(4):1–9

Banerjee K, Sahoo CK, Bal G, Mallik K, Pal R, Mitra A (2020) High blue carbon stock in mangrove forests of Eastern India. Trop Ecol (springer) 61(1):150–167. https://doi.org/10.1007/s42965-020-00072-y

Banerjee K, Khemendu GR, Paul R, Mitra A (2018) Carbon storage potential in dominant trees of Koraput District of Odisha. In: Singh VP et al (eds) Impact of climate change on marine environments. Energy and environment, water science and technology library, vol 80. Springer, pp 235–243. https://doi.org/10.1007/978-981-10-5798-4_22

Beane JL, Hagan JM, Whitman AA, Gunn JS (2008) Forest carbon offsets: a scorecard for evaluating project quality. Manomet Center for Conservation Sciences Report MCCS NCI 2008-1, Brunswick, Maine

Bhattacharyya S, Mitra A, Raha AK (2015) Stored carbon in above ground biomass of dominant mangrove floral species in Sagar Island of Indian Sundarbans. J Chem Biol Phys Sci 5(4):4664–4672

Bigsby H (2009) Carbon banking: creating flexibility for forest owners. For Ecol Manag 257(1):378–383

Birdsey RA (1992) Methods to estimate forest carbon storage. Forests and global change, opportunities for increasing forest cover, vol 1. American Forests, Washington, DC, pp 255–261

Brack CL (2002) Pollution mitigation and carbon sequestration by an urban forest. Environ Pollut 116:195–200

Bratkovich S, Bowyer J, Fernholz K, Lindburg A (2008) Urban tree utilization and why it matters. Dovetail Partners, Inc. Retrieved from http://www.dovetailinc.org/files/DovetailUrban0108ig.pdf. Accessed 12 Dec 2010

Chakraborty A, Datta U, Zaman S, Pal N, Pramanick P, Mitra A, Mitra A (2016) Spatio-temporal variation of stored carbon in *Sonneratia apetala*: a case study from Indian Sundarbans. Int J Innov Res Sci Eng Tech 5(3):2975–2980

Chowdhury IR (2015) Scenario of vehicular emissions and its effect on human health in Kolkata City. Int J Humanit Soc Sci Invent 4(5):1–9

Carbon Concierge (2008) Carbon offset provider evaluation matrix. Carbon Concierge. Retrieved from http://www.carbonconcierge.com/. Accessed 15 Mar 2011

Dhyani K, Kuniyal HB, Singh H, Sobha (2021) Growth and physiological potential of *Terminalia arjuna* under elevated CO_2 levels in Open top chamber condition. J Appl Nat Sci 13(3):1121-1126. https://doi.org/10.31018/jans.v13i3.2490. ISSN 2231-5209

EMBER Global Electricity Review Report (2021)

Gupta SK, Singh H (2017) Observations on sporadic flowering in *Ficus benjamina* L. and *Peltophorum africanum* L. in new forest. Indian Forester 143:290–291

Gupta SK, Ram J, Singh H (2018a) Comparative study of transpiration in cooling effect of tree species in the atmosphere. J Geosci Environ Prot 6:151–166. https://doi.org/10.4236/gep.2018.68011

Gupta SK, Ram J, Singh H, Kumar P (2018b) Comparative study of cooling under shades of some forestry tree species with respect to ambient temperatures. J Geosci Environ Protect 6:51–66. https://doi.org/10.4236/gep.2018.611004

Gupta SK, Ram J, Singh H (2019a) Relationship between leaf area index and atmospheric cooling of tree species. Indian J for 42(3):203–208

Gupta SK, Singh H, Ram J (2019b) Need for a novel paradigm in forestry research: perspectives of comparative atmospheric cooling by different forestry effect tree species. Indian Forester 145(6):509–515

Hoover CM, Birdsey RA, Health LS, Stout SL (2000) How to estimate carbon sequestration on small forest tracts. J for 98(9):13–19

https://phys.org/news/2022-09-carbon-dioxide-price.html

https://www.eia.gov/todayinenergy/detail.php?id=47256

https://www.newgeography.com/content/002620-the-evolving-urban-form-kolkata-50-mile-city

https://www.populationu.com/cities/kolkata-population

https://www.statista.com/statistics/1241719/carbon-trading-prices-worldwide-by-select-country/#:~:text=As%20of%20March%2031%2C%202023,the%20world's%20first%20international%20ETS

https://www.xe.com/currencyconverter/convert/?Amount=1&From=USD&To=INR

IPCC Climate Change (2022) Impacts, adaptation and vulnerability. In: Pörtner H-O, Roberts DC, Tignor M, Poloczanska ES, Mintenbeck K, Alegría A, Craig M et al. (eds) Contribution of working group II to the sixth assessment report of the intergovernmental panel on climate change. Chapter 9: Africa. Cambridge University Press, pp 1285–1456, 3056 p. https://report.ipcc.ch/ar6/wg2/IPCC

Jangra A, Kumar K, Maikhuri S, Bhandari MS, Pandey S, Singh H, Barthwal S (2024) Unveiling stress-adapted endophytic bacteria: Characterizing plant growth-promoting traits and assessing cross-inoculation effects on *Populus deltoides* under abiotic stress. Plant Physiol Biochem 210:108610. https://doi.org/10.1016/j.plaphy.2024.108610

Jo H, McPherson E (1995) Carbon storage and flux in urban residential green space. J Environ Manag 45:109–133. https://doi.org/10.1006/jema.1995.0062

Jo H, McPherson E (2001) Indirect carbon reduction by residential vegetation and planting strategies in Chicago, USA. J Environ Manage 61(2):165–177. https://doi.org/10.1006/jema.2000.0393

Kanta C, Kumar A, Chauhan A, Singh H, Sharma IP (2024). The Interplay between plant functional traits and climate change. In: Kumar N, Singh H (eds) Plant functional traits for improving productivity. Springer, Singapore. https://doi.org/10.1007/978-981-97-1510-7_3.

Koul DN, Panwar P (2008) Prioritizing land-management options for carbon sequestration potential. Curr Sci 95(5):658–663

Kumar M, Singh MP, Singh H, Dhakate PM, Ravindranath NH (2019a) Forest working plan for the sustainable management of forest and biodiversity in India. J Sustain for 39:1–22. https://doi.org/10.1080/10549811.2019.1632212

Kumar A, Kumar P, Singh H, Kumar N (2021a) Modulation of plant functional traits under essential plant nutrients during seasonal regime in natural forests of Garhwal Himalayas. Plant Soil 465:197–212. https://doi.org/10.1007/s11104-021-05003-x

Kumar A, Kumar P, Singh H, Kumar N (2021b) Adaptation and mitigation potential of roadside trees with bio-extraction of heavy metals under vehicular emissions and their impact on physiological traits during seasonal regimes. Urban for Urban Greening 58:126900. https://doi.org/10.1016/j.ufug.2020.126900

Kumar A, Kumar P, Singh H, Kumar N (2021c) Impact of plant functional traits on infiltration rate under different forests of Kempty watershed in Garhwal Himalaya. Indian J Soil Conserv 49(1):38–44

Kumar A, Kumar P, Singh H, Bisht S, Kumar N (2021d) Relationship of physiological plant functional traits with soil carbon stock in the temperate forest of Garhwal Himalaya. Curr Sci 120(8):1368–1373. https://doi.org/10.18520/cs/v120/i8/1368-1373

Kumar M, Phukon SN, Paygude AC, Tyagi K, Singh H (2021e) Mapping phenological functional types (PhFT) in the Indian Eastern Himalayas using machine learning algorithm in Google Earth Engine. Comput Geosci 158:104982. https://doi.org/10.1016/j.cageo.2021.104982

Kumar N, Singh H (2024) Plant functional traits for improving productivity. Springer, Singapore. https://doi.org/10.1007/978-981-97-1510-7

Kumar N, Suyal DC, Sharma IP, Verma A, Singh H (2017) Elucidating stress proteins in rice (*Oryza sativa* L.) genotype under elevated temperature: a proteomic approach to understand heat stress response. 3 Biotech 7:205. https://doi.org/10.1007/s13205-017-0856-9

Kumar N, Jeena N, Singh H (2019b) Elevated temperature modulates rice pollen structure: a study from the foothill of the Himalayan agro-ecosystem in India. 3 Biotech 9(5):175. https://doi.org/10.1007/s13205-019-1700-1

Kumar N, Jeena N, Kumar N, Gangola S, Singh H (2019c) Phytoremediation facilitating enzymes: an enzymatic approach for enhancing remediation process. In: Bhatt P (ed) Smart bioremediation technologies. Academic Press, Elsevier, 50 Hampshire Street, 5th Floor, Cambridge, MA 02139, United States, pp 289–306

Kumar N, Singh H, Sharma SK (2020) Antioxidants: responses and importance in plant defense system. In: Roychowdhury R, Choudhury S, Hasanuzzaman M, Srivastava S (eds) Sustainable agriculture in the era of climate change. Springer, Cham, pp 251–264. https://doi.org/10.1007/978-3-030-45669-6_11

Malik A, Kumar K, Kumari G, Kumar N, Singh H (2024). Application of functional traits in modelling productivity and resilience under climate change. In: Kumar N, Singh H (eds) Plant functional traits for improving productivity. Springer, Singapore. https://doi.org/10.1007/978-981-97-1510-7_5

McPherson GE (1994) Using urban forests for energy efficiency and carbon storage. J for 92(10):36–41

McPherson GE (1998) Atmospheric carbon dioxide reduction by Sacramento's urban forest. J Arbor 24(4):215–223

Meena D, Singh H, Chaudhari SK (2011) Elucidating strontium response on growth dynamics and biochemical change in *Phaseolus mungo* L. Int J Agric Environ Biotechnol 4:107–113

Mitra A (2015) Study on stored carbon in dominant tress species in two major cities of India; an approach for a carbon free urban zone. EC Chem 1(2):63–66

Mitra A, Sengupta K, Banerjee K (2012) Spatial and temporal trends in biomass and carbon sequestration potential of *Sonneratia apetala* Buch-Ham in Indian Sundarbans. Proc Natl Aca Sci Ind Sect b: Biol Sci 82(2):317–323. https://doi.org/10.1007/s40011-012-0021-5

Mitra A, Bagchi J, Thakur S, Parkhi US, Debnath S, Pramanick P, Zaman S (2015) Carbon sequestration in Bhubaneswar city of Odisha, India. Int J Innov Res Sci Eng Tech 4(8):6942–6947

Mitra A (2013) Sensitivity of mangrove ecosystem to changing climate. Springer, New Delhi, Heidelberg, New York, Dordrecht, London. ISBN-10: 8132215087; ISBN-13: 978-8132215080. ISBN 978-81-322-1509-7 (eBook), XIX, p 323. https://doi.org/10.1007/978-81-322-1509-7

Myeong S, Nowak DJ, Duggin MJ (2006) A temporal analysis of urban forest carbon storage using remote sensing. Remote Sens Environ 101(2):277–282

Nowak DJ, Crane DE (2002) Carbon storage and sequestration by urban trees in the USA. Environ Pollut 116(3):381–389

Nowak DJ, Walton JT (2005) Projected urban growth (2000–2050) and its estimated impact on the US forest resource. J for 103(8):383–389

Nowak DJ, Noble MH, Sisinni SM, Dwyer JF (2001) Assessing the US urban forest resource. J for 99(3):37–42

Pal N, Zaman S, Mitra A (2020) Carbon sequestration by Mangrove seedlings: A case study from Indian Sundarbans. In: Patra SC, Dr. Mitra A (eds) Climate change & blue carbon; webinar articles on "blue carbon domain: a potential regulator of climate change" under RUSA 2.0, abstract # 15. Centre of Excellence Environment, Climate Change and Public Health, Utkal University, pp 81–85

Petersson H, Holm S, Ståhl G, Alger D, Fridman J, Lehtonen A, Lundström A, Mäkipää R (2012) Individual tree biomass equations or biomass expansion factors for assessment of carbon stock changes in living biomass—A comparative study. For Ecol Mange 270:78–84

Pouyat RV, Yesilonis ID, Nowak DJ (2006) Carbon storage by urban soils in the United States. J Environ Qual 35:1566–1575

Prakash V, Hunney K, Singh H (2022) Effect of elevated carbon dioxide on growth and development of *Santalum album* L. seedlings inoculated with plant growth promoting microorganisms in Open Top Chambers. Int J Environ Agric Biotechnol 7(5):149–159

Ricke K, Drouet L, Caldeira K, Tavoni M (2018) Country level social cost of carbon. Nat Clim Change 8:895–900. https://doi.org/10.1038/s41558-018-0282-y

Rowntree RA, Nowak DJ (1991) Quantifying the role of urban forests in removing atmospheric carbon dioxide. J Arbori 17(10):269–275

Sedjo RA, Marland G, Fruit K (2001) Renting carbon offsets: the question of permanence. Resources for the Future, Washington, DC

Sengupta S, Ray Chaudhuri T, Ahmed S, Agarwal S, Zaman S, Mitra A (2022) Status of the near surface atmospheric CO_2 around the healthcare units in the megacity of Kolkata during two different COVID-19 lockdown phases. Par J Sci Edu 8(2):16–23. http://tiny.cc/PJSE24476153 v8i2p016-023

Sharma R, Singh H, Kaushik M, Nautiyal R, Singh O (2018) Adaptive physiological response, carbon partitioning, and biomass production of *Withania somnifera* (L.) dunal grown under elevated CO_2 regimes. 3 Biotech 8:267. https://doi.org/10.1007/s13205-018-1292-1

Sharma R, Singh H (2021) Alteration in biochemical constituents and nutrients partitioning of *Asparagus racemosus* in response to elevated atmospheric CO_2 concentration. Environ Sci Pollut Res 29:6812–6821. https://doi.org/10.1007/s11356-021-16050-3

Singh H (2021) An integrated approach considering physiological and biophysical based indicators for assessing tolerance of roadside plantations of *Alstonia scholaris* towards urban roadside air pollution: An assessment of adaptation of plantations for mitigating urban roadside air pollution. Trees. https://doi.org/10.1007/s00468-021-02179-8

Singh H, Verma A (2013) Characterization and screening of high nitrogen efficient rice genotype to curtail environmental pollution. Int J Agric Environ Biotechnol 6(1):23–30

Singh H, Verma A, Kumar M, Sharma R, Gupta R, Kaur M, Negi M, Sharma SK (2017b) Phytoremediation: a green technology to clean up the sites with low and moderate level of heavy metals. Austin Biochem 2:1012

Singh H, Sharma R, Savita SMP, Kumar M, Verma A, Ansari MW, Sharma SK (2018) Adaptive physiological response of *Parthenium hysterophorus* to elevated atmospheric CO_2 concentration. Indian Forester 144:1–14

Singh S, Singh H, Sharma SK, Nautiyal R (2021) Seasonal variation in biochemical responses of bamboo clones in the sub-tropical climate of Indian Himalayan foothills. Heliyon 7(4):e06859. https://doi.org/10.1016/j.heliyon.2021.e06859

Singh H, Kumar M (2022) Climate change and its impact on Indian Himalayan forests: current status and research needs. In: Rani S, Kumar R (eds) Climate change. Springer climate. Springer, Cham, pp 223–242. https://doi.org/10.1007/978-3-030-92782-0_11

Singh H, Savita Sharma R, Sinha S, Kumar M, Kumar P, Verma A Sharma SK (2017a) Physiological functioning of *Lagerstroemia speciosa* L. under heavy roadside traffic: an approach to screen potential species for abatement of urban air pollution. 3 Biotech 7(61):1–10. https://doi.org/10.1007/s13205-017-0690-0

Singh H, Yadav M, Kumar N, Kumar A, Kumar M (2020) Assessing adaptation and mitigation potential of roadside trees under the influence of vehicular emissions: a case study of *Grevillea robusta* and *Mangifera indica* planted in an urban city of India. PLoS ONE 15(1):e0227380. https://doi.org/10.1371/journal.pone.0227380

Singh H, Kumar N, Kumar A (2024). Enhancing resource use efficiency in crops through plant functional traits. In: Kumar N, Singh H (eds) Plant functional traits for improving productivity. Springer, Singapore. https://doi.org/10.1007/978-981-97-1510-7_6

Smith JE, Heath LS, Skog KE, Birdsey RA (2006) Methods for calculating forest ecosystem and harvested carbon with standard estimates for forest types of the United States. General technical report, NE-343. USDA Forest Service, Northeastern Research Station, Newtown Square, PA
Tansey J (2009) Carbon offsets then and now. A white paper in the sustainable enterprise report, vol 1. Deloitte and Kyoto Publishing, Vancouver
Wofsy SC (2001) Climate change: where has all the carbon gone? Sci 292(5525):2261–2263
Yadav SK, Singh H, Nautiyal R, Ginwal HS, Ansari SA, Barthwal S (2019b) Modulation of morpho-physiological responses in *Populus deltoides* by elevated carbon dioxide and temperature. Forest Sci 66:105–118. https://doi.org/10.1093/forsci/fxz048
Yadav SK, Singh H, Ginwal HS Barthwal S (2019) Elevated CO_2 enhanced growth and physiological process of *Populus deltoides* Bartr. ex Marsh. Indian Forester 145(1):23–27
Zaman S, Pramanick P, Pal N, Biswas P, Biswas S, Bera N, Mitra A (2014) Stored carbon in dominant trees of urban area of Kolkata, India. Ideas Ideol Int e J 2(7):1–7

Adaptive Strategies of Quercus leucotrichophora in Different Environmental Gradients: Modulations in Leaf Biophysical and Biochemical Functional Traits

Priyanka Mishra and Hukum Singh

Abstract Understanding the adaptive strategies of plants through leaf functional trait modulations is critical for assessing the impacts of environmental changes. Biophysical and biochemical functional traits are crucial characteristics of plant leaves that provide valuable insights into their ecological and physiological roles. The variations in these traits to environmental gradients influence critical ecosystem services, including carbon sequestration, nutrient recycling, and productivity. This study evaluated the biophysical and biochemical characteristics of *Quercus leucotrichophora* in two different environments to understand their adaptive responses to environmental conditions. Twelve leaf traits were examined: six biochemical, four biophysical, and two stoichiometric traits. The findings indicated significant variations in leaf dry matter, carbon and nitrogen, specific leaf area, proline, carotenoids, and chlorophyll between Mussoorie and FRI Dehradun. However, leaf thickness was higher at FRI Dehradun than at Mussoorie. These variations highlight the sensitivity of leaf traits to environmental gradients, underscoring their importance in ecosystem services such as carbon sequestration and nutrient cycling. The study provides insights into the adaptive mechanisms of *Q. leucotrichophora*, emphasizing the role of leaf trait in plant adaptation to environmental conditions.

Keywords Adaptive resilience · Environmental gradients · Leaf functional traits · *Quercus leucotrichophora*

P. Mishra (✉) · H. Singh
Plant Physiology Discipline, ICFRE-Forest Research Institute, PO New Forest, Dehradun, Uttarakhand 248006, India
e-mail: mishra44pri@gmail.com

H. Singh (ed.), *Urban Forests, Climate Change and Environmental Pollution*,
https://doi.org/10.1007/978-3-031-67837-0_34

1 Introduction

Functional traits encompass the anatomical, structural, and functional characteristics vital for a plant's survival, growth, and development, ultimately shaping the ecosystem dynamics (Garnier et al. 2015). Leaf attributes are crucial for plant adaptation and species distribution (Diaz et al. 2007), providing insights into forest functioning, including nutrient dynamics and energy flux (Cai et al. 2020). Functional traits are pivotal for understanding community and ecosystem dynamics, guiding predictions of ecosystem functioning and species coexistence along environmental gradients (Ackerly and Cornwell 2007).

Plant functional traits are integral to global vegetation, landscape, and trait-based models, facilitating vegetation dynamics and biochemical cycle predictions (Albert et al. 2008; Norberg et al. 2001). Leaf traits, particularly in the temperate Himalayas, are underexplored concerning their interactions with the environment (Singh et al. 2017b). Yet, they offer predictive insights into plant performance, resource utilization, and ecosystem processes. Biophysical leaf traits such as leaf area and specific leaf area influence canopy structure, light interception, and photosynthetic productivity, further modulated by elevation and microclimate (Qin and Shangguan 2019; Li et al. 2021).

Leaves orchestrate various ecological functions, including photosynthesis, gas exchange, carbon storage, and nutrient cycling (Gupta et al. 2018a; Kumar et al. 2021a), significantly influencing forest structure, biomass, carbon storage, and productivity (Tian et al. 2016).

Altitudinal gradients induce changes in vegetation, temperature, and soils, with higher elevations exhibiting distinct climatic conditions (Kumar et al. 2017). Climate variations, including shifts in direct photosynthetic radiation, humidity, and intraspecific variation in leaf traits, shape vegetation patterns along altitudinal gradients (Wright et al. 2017; Yadav et al. 2019a, b). Standardized protocols for measuring leaf traits help mitigate intraspecific variation, facilitating trait-environment research (Cornelissen et al. 2003).

Urbanization and vehicular emissions pose significant challenges to air quality, with roadside plants acting as natural filters for pollutants (Gupta et al. 2018b; Kumar et al. 2022; Joshi and Swami 2009). Exposure to vehicular emissions alters plant physiology and growth (Singh et al. 2020, 2022), highlighting the importance of urban vegetation in mitigating air pollution effects. Certain tree species exhibit resilience to air pollutants, serving as bio-indicators and playing crucial roles in urban air quality management. Hence, understanding plant functional traits and their responses to environmental changes is vital for predicting ecosystem dynamics, managing forest health, and mitigating urban air pollution (Kumar et al. 2019c; Singh et al. 2017a; Gupta and Singh 2017).

1.1 The Ecological and Environmental Significance of *Q. leucotrichophora*

It is also known as banj oak extends far beyond its physical presence in the landscape. A member of the Fagaceae family, Quercus spp., a significant group of evergreen or deciduous trees is found in temperate and tropical climates worldwide, with over 450 species exhibiting diverse flowering and fruiting patterns (Tejerina et al. 2011).

Banj oak, scientifically known as *Quercus leucotrichophora* A. Camus (formerly *Quercus oblongata* D. Don), typically grows to a height of 15 m, occasionally reaching 24 m. It displays a notable preference for partial shade during its initial two years, transitioning to a requirement for ample sunlight afterward. Thriving best in clays derived from shale and clay loams, this species thrives in fertile soils but struggles in arid environments. The tree's seed typically ripens between November and March, with mature seeds persisting on the tree for several months. Banj oak is indigenous to the Himalayan region, thriving at altitudes ranging from 1500 to 2400 m above sea level.

The canopy of *Q. leucotrichophora*is characterized by its full and rounded shape, while its bark evolves from a smooth tan-brown to a somewhat corky and furrowed texture as the tree matures. Its leaves characterized by jagged teeth and elongated oval shapes, alternate along the branches.

Ecologically, banj oak plays a crucial role in ecosystem services, serving as a fundamental component of the biogeochemical cycle and environmental regulation. Its acorn fruit is a vital food source for squirrels and insects, sustaining diverse food webs. Additionally, banj oak trees form symbiotic relationships with saprophytic and ectomycorrhizal fungi, facilitating carbon cycling and nitrogen fixation. Notably, oak tree is renowned for its carbon storage capacity and air pollutant removal efficacy. Moreover, banjoak forests exhibit resilience to environmental stressors such as flooding, hurricane-force winds, fires, and seawater erosion (Kumar et al. 2020; Kumar et al. 2019c).

Q. leucotrichophora holds significant cultural, economic, and ecological value in the Himalayan region. Serving as a keystone species across various habitats, banj oak forests provide vital ecosystem services and livelihoods for local communities (Singh and Rawat 2012). The leaves of banj oak are utilized for composting and animal fodder due to their nutrient-rich composition, contributing to soil fertility and agricultural productivity. Furthermore, banj oak forests support a rich biodiversity of lichens, insects, avifauna, pteridophytes, and bryophytes, underscoring their ecological importance (Oldfield and Eastwood 2007).

However, banj oak populations face threats from overexploitation, habitat degradation, and climate change. The overutilization of banj oak for fuel and fodder has led to habitat loss and ecosystem disturbance, affecting various associated flora and fauna (Singh et al. 2016). Climate variability and seasonality also significantly influence interspecific variation in leaf characteristics, impacting plant adaptation strategies and ecosystem functions (Ordonez et al. 2009; Jangra et al. 2024). Given the vulnerability of the Himalayan region to the short- and long-term effects of climate change, urgent conservation efforts are needed to safeguard the ecological integrity and cultural heritage associated with banj oak forests (Walker et al. 2006).

1.2 Impact of Changing Environmental Conditions on Leaf Biophysical Traits

Leaf biophysical traits encompass morphological or physical characteristics of leaves, such as leaf thickness (LT), leaf dry matter content (LDMC), leaf area (LA), specific leaf area (SLA), and more. While many leaf biophysical properties are influenced by species characteristics, changes in environmental components also play a significant role. Variables such as height, distance from the sea, and average annual precipitation have been identified as substantial influencers of leaf shape. In species exhibiting high environmental plasticity, distinct ranges or trends in biophysical variation can often be observed (Kanta et al. 2024). Environmental factors profoundly affect leaf biophysical characteristics, including rising atmospheric carbon dioxide levels and changing climatic conditions. Predicting plant responses to environmental changes necessitates a thorough understanding of these effects. For instance, alterations in temperature and precipitation can lead to changes in leaf area index (LAI), a critical biophysical characteristic indicating the amount of leaf cover in plants (Myneni et al. 1997). Rising atmospheric CO_2 levels impact plant water consumption efficiency by affecting stomatal conductance, influencing ecosystem water cycles and plant physiology (Ainsworth and Rogers 2007; Prakash et al. 2022).

According to Cornelissen et al. (2003), a Specific Leaf Area (SLA) is defined as the area of a fresh leaf on one side divided by the leaf's oven-dry mass. Specific Leaf Area is often positively correlated with the relative growth rate in herbaceous species. Leaf dry matter content (LDMC) is associated with the average density of leaf tissues and typically scales with 1/SLA (Cornelissen et al. 2003). It is calculated as the oven-dried mass of a leaf divided by its water-saturated fresh mass of leaf samples. The partial rehydration method, shown to yield results similar to the complete rehydration method, is commonly used to measure LDMC (Vaieretti et al. 2007). Leaf size is frequently quantified using leaf area, the projected or one-sided area of a single leaf, measured in square millimeters. Various factors such as climate fluctuations, geology, height, and latitude have been linked to interspecific variation in leaf area. Variation in leaf area among climatic zones may also be related to allometric parameters, ecological responses to nutritional stress and environmental perturbations, and evolutionary variables, including phylogeny.

1.3 Impact of Changing Environmental Conditions on Leaf Biochemical Traits

Leaf biochemical attributes, such as proline, carotenoids, chlorophyll content, leaf nitrogen concentration (LNC), leaf carbon concentration (LCC), and leaf phosphorus concentration (LPC), among others, are influenced by variations in the surrounding environment. LNC, for instance, refers to the total quantity of nitrogen per unit of dry leaf mass and is strongly linked to the mass-based maximum photosynthetic rate (Cornelissen et al. 2003 and Kumar et al. 2021b). Similarly, LCC represents the total quantity of carbon per unit of leaf dry mass, indicating the amount invested in structural tissues. Moreover, LNC and LPC, which denote the total amounts of nitrogen and phosphorus per unit of dry leaf mass, often vary significantly depending on the availability of nitrogen or phosphorus in the environment. In some cases, the ratio of LNC to LPC is used as a tool for analysis, mainly when nitrogen or phosphorus availability is more limiting for plant growth. Like many legumes species that actively fix nitrogen typically exhibit higher LNC ratios than other plants growing in similar locations (Singh et al. 2021; Singh and Verma 2013; Sharma and Singh 2021).

Functional traits, including leaf biochemical attributes, have been utilized to comprehend and predict community and ecosystem functioning, such as net primary productivity and biochemical cycles, and to understand patterns of diversity and species coexistence along environmental gradients (Ackerly and Cornwell 2007; De Bello et al. 2010; Singh et al. 2024b). Moreover, through the parameterization of Plant Functional Traits (PFTs) in dynamic global vegetation models (Kumar et al. 2024; Sitch et al. 2003; Malik et al. 2024), landscape models (Albert et al. 2008), and trait-based models (Norberg et al. 2001), plant functional traits form the foundation for modeling tools to predict vegetation dynamics and biochemical cycles.

1.4 Adaptive Responses of Functional Traits to Changing Environments

Studying the functional characteristics of leaves has significantly advanced scientific understanding of how plants adapt to their ever-changing environment. Climate change, stemming from various natural and urban sources, has led to a notable variation in functional traits across different environments (Kumar et al. 2024). Specific leaf area tends to increase with decreasing light exposure conditions (Samuelson and Stokes 2012; Catoni et al. 2015; Singh et al. 2024b), while leaf thickness typically increases in response to intense irradiance (Yang et al. 2014). Under weak lighting conditions, the thickness of spongy tissue increases while that of palisade tissue decreases (Gangrong and Qingsheng 2006). Maximum stomatal conductance may benefit species under low CO_2 conditions, nutrient supply limitations, and higher irradiance (Taylor et al. 2012; Sharma et al. 2018; Kumar et al. 2021a, b, c, d, e).

Trait correlations can exhibit changes in response to environmental shifts, suggesting that several trait combinations may not be necessary in stable environments (Yin et al. 2018). Vehicular emissions heavily influence urban vegetation, increasing carbon presence in the atmosphere and subsequent alterations in plants' biophysical and biochemical traits (Singh et al. 2017a, 2018). Even in natural environments, plants face stressors that prompt adaptations. For instance, higher specific leaf area (SLA) may increase water loss, while higher leaf dry matter content (DMC) and leaf thickness (LT) may enhance moisture diffusion resistance and distance. SLA negatively correlates with LDMC and LT, indicating varying water retention abilities.

The functional biodiversity of plants may be influenced and reshaped by temperature, implying that trait correlations could vary across different climatic conditions (Kumar et al. 2019d).This variability is crucial to consider in discussions concerning global climate change. The waxy texture of the cuticle and sticky leaves of roadside trees contribute to thicker leaves, which tend to attract dust particles. The transpiration rate is directly impacted by the thickness of the cuticle layers on the leaf surface, thereby influencing leaf stomatal behavior. Moreover, the thickening leaves in polluted areas signify a species' high tolerance to intense vehicle pollution, serving as an adaptive mechanism to abiotic stress (Singh et al. 2017a, b, c; Kumar et al. 2021c).

2 Case Study

2.1 Study Sites

The present study was conducted at two different altitudes in the Himalayan foothills i.e. Dehradun (FRI; Forest Research Institute) and Mussoorie. FRI, Dehradun, is situated between latitude 30°20′44.6″ N to 30°19′59.7″ N and longitude 77°59′55.7″ E to 77°59′43.2″ E, with an average elevation of 650 m. Similarly, Mussoorie lies between latitude N30°28.013′ to N30°28.538′ and longitude E78°1.501′ to E78°2.307′, with an average elevation of 1880 m.

Dehradun and Mussoorie are located in the mountain ranges of the Himalayan foothills. The area falls under the subtropical region and experiences characteristics of both temperate and tropical climates. The minimum air temperature ranges from 3.6 to 29.4 °C, while the maximum ranges between 19.3 and 35.3 °C. Over the past 20 years, the average air temperature has varied between 10.9 and 27.10 °C, with the lowest temperatures recorded in January and the highest in June. Relative humidity levels range between 52.11 and 85.16% from April to August. Based on 25 years of rainfall data, the maximum rainfall occurs in July (630.70 mm), while the minimum is recorded in December (2.80 mm). On average, the highest number of sunshine hours per day, calculated over the past 20 years, is observed in May (9.34 h/day). The natural vegetation in Dehradun and Mussoorie predominantly consists of tropical moist deciduous and Himalayan moist temperate types.

2.2 Analysis of Leaf Functional Traits

2.2.1 Leaf Biophysical Functional Traits

Fresh leaves were collected from the Mussoorie and the controlled site at FRI Dehradun. Each leaf's thickness (LT) was measured in millimetres using a Digital Vernier Caliper. To estimate Leaf Dry Matter Content (LDMC), 30 leaf samples were randomly collected from the selected trees. The fresh mass of each leaf sample was measured in the field, and then they were oven-dried at 70 °C for at least 72 h or until a constant weight was achieved. The formula developed by Tecco et al. (2013) was used to estimate LDMC.

Leaf area (LA), expressed in square centimetres per plant, and represents the photosynthetic surface area produced by an individual plant over a specific time interval. It was quantified using ImageJ software. Specific Leaf Area (SLA), expressed in square centimetres per gram, was calculated as the ratio of a plant's leaf area to its dry weight.

2.2.2 Leaf Biochemical Functional Traits

The leaves were detached from the twigs and taken to the laboratory to estimate biochemical parameters such as total chlorophyll content, proline, carotenoid, leaf carbon content, leaf nitrogen content, and leaf phosphorus content of the selected species (Singh et al. 2021; Meena et al. 2011).

The leaf chlorophyll content was estimated by adopting the protocol of Hiscox and Israeistam (1979). Fifty milligrams of fresh leaves were finely chopped and transferred into vials containing 7 ml of Dimethyl Sulphoxide (DMSO). The vials were incubated at 67 °C for 30 min and filtered into graduated test tubes. The volume was adjusted to 10 ml using DMSO, and Optical Density (O.D.) values were recorded for chlorophyll and carotenoid at specific wavelengths using a Spectrophotometer (Hiscox and Israelstam 1979). Carotenoid content was computed per the equation of Singh et al. (2010).

Proline content was determined by homogenizing 0.5 g of the sample in 10 ml of 3% aqueous Sulphosalicylic acid. The filtrate was mixed with glacial acetic acid and ninhydrin and heated in a boiling water bath. The reaction was terminated, and the intensity of the red color was measured at 520 nm. The amount of proline in the sample was estimated using a standard curve (Bates et al. 1973).

Leaf samples were digested with salicylic acid using a digestion mixture, and leaf nitrogen content (LNC) was determined by the Micro-Kjeldahl method (Singh et al. 2013; Singh & Verma, 2013a, b). Leaf carbon concentration (LCC) was calculated by burning dry leaf samples in an electric furnace at 400 °C. The ash content left after burning was weighed, and the carbon concentration was calculated as per the equation of Negi et al. (2003). Phosphorus content was estimated using dry tissue as per the standard method.

2.3 Statistical Analysis

Statistical analysis of the result obtained for all the parameters was done using the SPSS software to determine the *p*-value of the readings obtained in the FRI campus and Mussoorie. The *p*-value was calculated to determine whether the means obtained for the two data sets were significantly different.

3 Result and Discussion

3.1 Adaptive Response of Biophysical Functional Traits

The adaptive response of the biophysical functional traits of the leaf is presented in Fig. 1. The data shows that the leaf dry matter content (LDMC) of *Quercus leucotrichophora* was significantly higher in Mussoorie (0.61 ± 0.01 mg g^{-1}) compared to FRI (0.48 ± 0.01 $^{-1}$). This difference may be attributed to variations in temperature, annual rainfall, and windfall. Higher altitudes typically experience lower temperatures, higher windfall, and more significant yearly rain, contributing to increased litter fall. This, in turn, leads to more decomposition and subsequent soil enrichment. Similar results were reported by Kumar et al. (2019a, b, c, d) and Kumar et al. (2021d).

Specific leaf area (SLA), a fundamental parameter for studying plant growth that denotes leaf surface area per unit mass, was significantly higher in Mussoorie (12.25 ± 0.07 cm^2 g^{-1}) than in FRI (9.97 ± 0.25 cm^2 g^{-1}). This difference might be due to plant adaptation. In FRI, *Q. leucotrichophora* receives less sunlight than Mussoorie, resulting in a lower rate of photosynthesis. To compensate, the plant adapts by expanding its leaves, increasing the specific leaf area and enhancing photosynthesis (Auslander et al. 2003).

Leaf area (LA), which represents the photosynthetic surface area and is measured in cm^2, was significantly larger in Mussoorie (35.05 ± 0.01 cm^2) compared to FRI (24.84 ± 1.11 cm^2). This may be due to lower temperatures at higher altitudes, which can cause organ sizes to decrease, thereby increasing leaf area (Korner and Diemer 1987; Gupta et al. 2019a, b).

Leaf thickness (LT) is associated with the plant's adaptive response to pollutants. The data showed no significant difference in leaf thickness between the samples, with FRI at 0.52 ± 0.019 mm and Mussoorie at 0.42 ± 0.0011 mm. This slight difference might be due to the accumulation of dust particles on leaf surfaces, leading to thicker waxy cuticles and sticky leaves. As the cuticle layer thickens, transpiration rates decrease, and stomatal behavior is affected (Sarma et al. 2017; Singh et al. 2021).

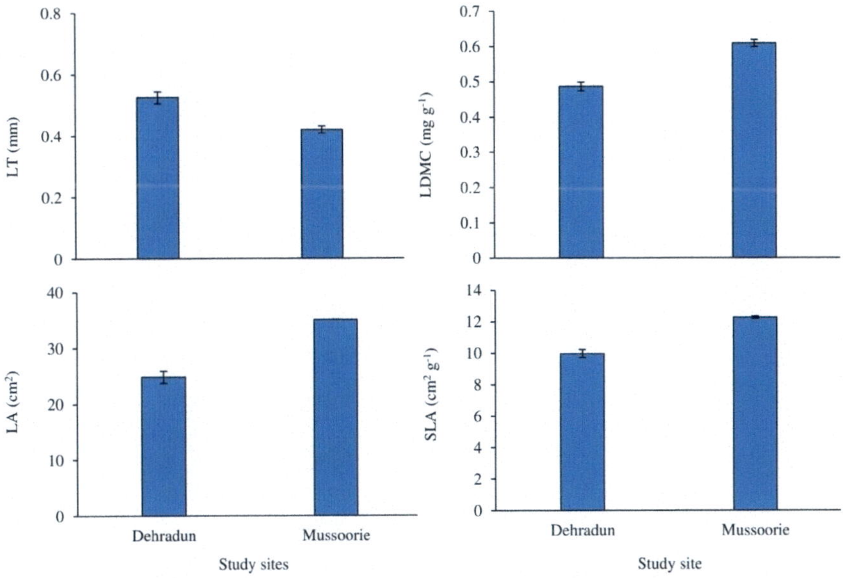

Fig. 1 Adaptive response of leaf biophysical functional traits of *Quercusleucotricophora*growing in Dehradun and Mussoorie

3.2 *Adaptive Response of Leaf Biochemical Functional Traits*

The variation in the adaptive response of leaf biochemical functional traits of *Quercus leucotricophora* growing in Dehradun and Mussoorie is depicted in Fig. 2. Chlorophyll and carotenoids are crucial pigments for photosynthesis. The chlorophyll content is significantly higher in the Mussoorie (1.63 ± 0.019 mg g^{-1}) compared to the FRI (1.51 ± 0.019 mg g^{-1}). Carotenoid content is statistically similar between the two samples, with only a slight difference observed. The slightly lower levels in the FRI sample may be due to reduced radiation, as both pigments are radiation-dependent, and higher radiation levels enhance photosynthesis (Singh et al. 2017a).

Proline is vital in the stress response. The proline content in the leaf is significantly higher in the Mussoorie (0.25 ± 0.015 mg g^{-1}) compared to the FRI (0.17 ± 0.011 mg g^{-1}). This difference might be due to lower stress levels in FRI, where the plant benefits from higher temperatures due to vehicular emissions and other factors. In contrast, Mussoorie experiences low temperatures due to snowfall, leading to increased stress and, consequently, higher proline content (Singh et al. 2017a, 2020).

Leaf carbon content (LCC) is a crucial plant trait and was statistically similar between the two sites i.e. FRI ($46.06 \pm 0.55\%$) and Mussoorie ($46.86 \pm 0.78\%$). The slight difference might be due to the structural components of the leaves; an increase in structural components correlates with higher carbon content (Yadav et al. 2019a, b; Dhyani et al. 2021; Han et al. 2005). The slightly higher LCC in Mussoorie could

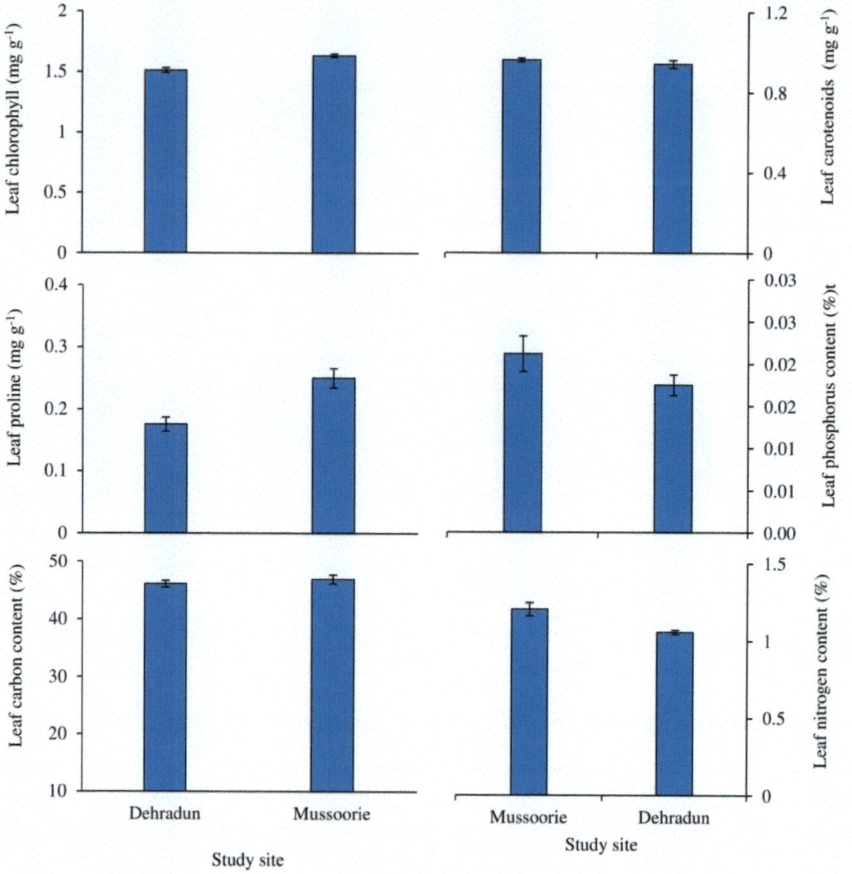

Fig. 2 Adaptive response of leaf biochemical functional traits of *Quercus leucotricophora* growing in Dehradun and Mussoorie

be due to more favorable natural conditions, whereas FRI, located on the roadside, experiences more disturbances.

The leaf phosphorus content (LPC) of both sites was non-significant. The LPC of the FRI plants was $0.017 \pm 0.001\%$, while Mussoorie had $0.021 \pm 0.002\%$. This difference may be attributed to the higher elevation in Mussoorie, where the weathering of rocks releases phosphorus, making it biologically available in the soil and subsequently absorbed by plants (Han et al. 2005; Singh et al. 2017a, b, c).

The data also show that leaf nitrogen content is significantly higher in the Mussoorie (1.20 ± 0.042 mg g^{-1}) compared to the FRI (1.05 ± 0.012 mg g^{-1}). This difference might be due to higher annual rainfall in Mussoorie, which increases humidity, accelerates decomposition, enriches soil nutrients, and improves nitrogen accumulation (Liu et al. 2018). The impact of environmental gradients on carbon and

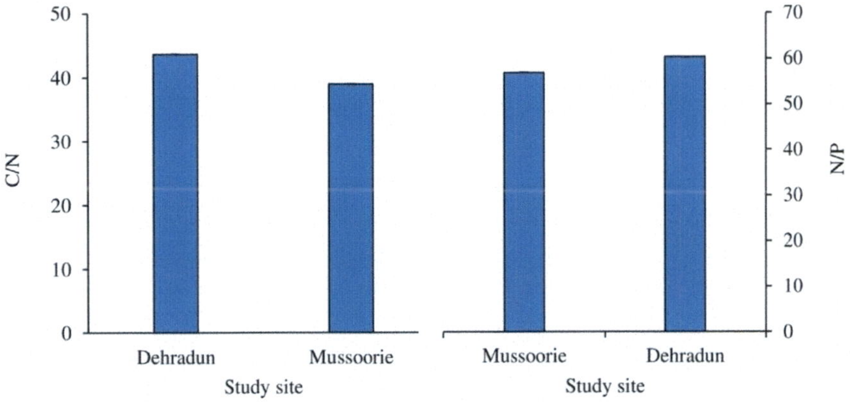

Fig. 3 The figure displays carbon and nitrogen ratio (C/N) and nitrogen and phosphorus ratio (N/P) in the leaf of *Quercus leucotricophora* growing in Dehradun and Mussoorie

nitrogen ratio (C/N) and nitrogen and phosphorus ratio (N/P) is presented in Figs. 2 and 3.

4 Conclusion

We analyzed the variation in leaf functional traits of *Quercus leucotricophora* from two different environmental gradients (Dehradun and Mussoorie). The results indicate that environmental gradients significantly influences leaf nitrogen concentration, dry matter content, and thickness. Some traits, such as chlorophyll and carotenoid content, decreased under unfavorable conditions than natural. Proline content was notably higher in Mussoorie, indicating higher stress levels in this region due to climatic factors. Leaf carbon content and phosphorus were also slightly higher at the Mussoorie. Leaf thickness was significantly impacted by environmental conditions. Overall, the study reveals that environmental variations at altitudinal gradients influence the biophysical and biochemical responses of *Q. leucotricophora.* The variations in biophysical and biochemical attributes highlight the species' adaptive potential. These findings support that *Q. leucotricophora* may alter its morphology and physiology in response to changing environmental conditions, demonstrating its capacity for adaptation.

References

Ackerly DD, Cornwell WK (2007) A trait-based approach to community assembly: partitioning of species trait values into within- and among-community components. Ecol Lett 10:135–145

Ainsworth EA, Rogers A (2007) The response of photosynthesis and stomatal conductance to rising [CO_2]: mechanisms and environmental interactions. Plant Cell Environ 30(3):258–270

Albert CH, Thuiller W, Lavorel S, Davies ID, Garbolino E (2008) Land-use change and subalpine tree dynamics: colonization of Larix decidua in French subalpine grasslands. J Appl Ecol 45:659–669

Auslander M, Nevo E, Inbar M (2003) The effects of slope orientation on plant growth, developmental instability and susceptibility to herbivores. J Arid Environ 55(3):405–416

Bates LS, Waldren RPA, Teare ID (1973) Rapid determination of free proline for water-stress studies. Plant Soil 39:205–207

Cai H, Li F, Jin G (2020) Forest strata-dependent effects of vegetation attributes and soil nutrients on decadal changes in aboveground net carbon stock in two temperate forests. CATENA 194:104776. https://doi.org/10.1016/j.catena.2020.104776

Catoni R, Gratani L, Sartori F, Varone L, Granata MU (2015) Carbon gain optimization in fve broadleaf deciduous trees in response to light variation within the crown: correlations among morphological, anatomical and physiological leaf traits. Acta Bot Croat 74:71–94. https://doi.org/10.1515/botcro-2015-0010

Cornelissen JHC, Lavorel S, Garnier E, Díaz S, Buchmann N, Gurvich DE, Reich PB, TerSteege H, Morgan HD, Van Der Heijden MGA, Pausas JG, Poorter H (2003) Handbook of protocols for standardised and easy measurement of plant functional traits worldwide. Aust J Bot 51:335–380

De Bello F, Lavorel S, Díaz S, Harrington R, Cornelissen JH, Bardgett RD, Harrison PA (2010) Towards an assessment of multiple ecosystem processes and services via functional traits. Biodivers Conserv 19:2873–2893

Dhyani K, Kuniyal HB, Singh H, Sobha (2021) Growth and physiological potential of *Terminalia arjuna* under elevated CO_2 levels in Open Top Chamber condition. J Appl Nat Sci 13(3):1121–1126. https://doi.org/10.31018/jans.v13i3.2490. ISSN 2231-5209

Díaz S, Lavorel S, de Bello F, Quétier F, Grigulis K, Robson TM (2007) Incorporating plant functional diversity effects in ecosystem service assessments. Proc Natl Acad Sci 104(52):20684–20689. 10.1073/ pnas.0704716104

Flückiger W, Oertli JJ, Flückiger H (1979) Relationship between stomatal diffusive resistance and various applied particle sizes on leaf surfaces. Z Pflanzenphysiol 91(2):173–175

Gangrong S, Qingsheng C (2006) Leaf anatomic plasticity of white clover and its response to diferent light intensities. Acta Agrestia Sinica 14:301–305, 1007-0435(2006)04-030105

Garnier E, Cortez J, Billès G, Navas M-L, Roumet C, Debussche M, Laurent G, Blanchard A, Aubry D, Bellmann A, Neill C, Toussaint J-P (2004) Plant functional markers capture ecosystem properties during secondary succession. Ecology 85:2630–2637

Garnier E, Navas M-L, Grigulis K (2015) Plant functional diversity: organism traits, community structure, and ecosystem properties. Oxford University Press. https://doi.org/10.1093/acprof:oso/9780198757368.001.0001

Gupta SK, Singh H (2017) Observations on Sporadic Flowering in *Ficus benjamina* L. and *Peltophorum africanum* L. in New Forest. Ind for 143:290–291

Gupta SK, Ram J, Singh H (2018a) Comparative study of transpiration in cooling effect of tree species in the atmosphere. J Geosci Environ Protect 6:151–166. https://doi.org/10.4236/gep.2018.68011

Gupta SK, Ram J, Singh H, Kumar P (2018b) Comparative study of cooling under shades of some forestry tree species with respect to ambient temperatures. J Geosci Environ Protect 6:51–66. https://doi.org/10.4236/gep.2018.611004

Gupta SK, Ram J, Singh H (2019a) Relationship between leaf area index and atmospheric cooling of tree species. Indian J for 42(3):203–208

Gupta SK, Singh H, Ram J (2019b) Need for a novel paradigm in forestry research: perspectives of comparative atmospheric cooling by different forestry effect tree species. Indian Forester 145(6):509–515

Han W, Fang J, Guo D, Zhang Y (2005) Leaf nitrogen and phosphorus stoichiometry across 753 terrestrial plant species in China. New Phytol 168(2):377–385. https://doi.org/10.1111/j.1469-8137.2005.01530.x

Hiscox JD, Israelstam GF (1979) A method for extraction of chlorophyll from leaf tissue without Maceration. Can J Bot 57:1332–1334. https://doi.org/10.1139/b79-163

Jangra A, Kumar K, Maikhuri S, Bhandari MS, Pandey S, Singh H, Barthwal S (2024) Unveiling stress-adapted endophytic bacteria: Characterizing plant growth-promoting traits and assessing cross-inoculation effects on *Populus deltoides* under abiotic stress. Plant Physiol Biochem 210:108610. https://doi.org/10.1016/j.plaphy.2024.108610

Joshi PC, Swami A (2009) Air pollution induced changes in the photosynthetic pigments of selected plant species. J Environ Biol 30:295–298

Kanta C, Kumar A, Chauhan A, Singh H, Sharma, IP (2024) The interplay between plant functional traits and climate change. In: Kumar N, Singh H (eds) Plant functional traits for improving productivity. Springer, Singapore. https://doi.org/10.1007/978-981-97-1510-7_3

Korner C, Diemer M (1987) In situ photosynthetic responses to light, temperature and carbon dioxide in herbaceous plants from low and high altitude. Funct Ecol 179–194

Kumar N, Singh H (2024) Plant functional traits for improving productivity. Springer, Singapore. https://doi.org/10.1007/978-981-97-1510-7

Kumar N, Suyal DC, Sharma IP, Verma A, Singh H (2017) Elucidating stress proteins in rice (*Oryza sativa* L.) genotype under elevated temperature: a proteomic approach to understand heat stress response. 3 Biotech 7:205. https://doi.org/10.1007/s13205-017-0856-9

Kumar M, Singh MP, Singh H, Dhakate PM, Ravindranath NH (2019a) Forest working plan for the sustainable management of forest and biodiversity in India. J Sustain for 39:1–22. https://doi.org/10.1080/10549811.2019.1632212

Kumar N, Jeena N, Singh H (2019b) Elevated temperature modulates rice pollen structure: a study from the foothill of the Himalayan agro-ecosystem in India. 3 Biotech 9(5):175. https://doi.org/10.1007/s13205-019-1700-1

Kumar N, Jeena N, Kumar N, Gangola S, Singh H (2019c) Phytoremediation facilitating enzymes: an enzymatic approach for enhancing remediation process. In: Bhatt P (ed) Smart bioremediation technologies. Academic Press, Elsevier. 50 Hampshire Street, 5th Floor, Cambridge, MA 02139, United States, pp 289–306

Kumar N, Singh H, Sharma SK (2020) Antioxidants: responses and importance in plant defense system. In: Roy chowdhury R, Choudhury S, Hasanuzzaman M, Srivastava S (eds) Sustainable agriculture in the era of climate change. Springer, Cham, 251–264. https://doi.org/10.1007/978-3-030-45669-6_11

Kumar A, Kumar P, Singh H, Kumar N (2021a) Modulation of plant functional traits under essential plant nutrients during seasonal regime in natural forests of Garhwal Himalayas. Plant Soil 465:197–212. https://doi.org/10.1007/s11104-021-05003-x

Kumar A, Kumar P, Singh H, Kumar N (2021b) Adaptation and mitigation potential of roadside trees with bio-extraction of heavy metals under vehicular emissions and their impact on physiological traits during seasonal regimes. Urban for Urban Greening 58:126900. https://doi.org/10.1016/j.ufug.2020.126900

Kumar A, Kumar P, Singh H, Kumar N (2021c) Impact of plant functional traits on infiltration rate under different forests of Kempty watershed in Garhwal Himalaya. Indian J Soil Conserv 49(1):38–44

Kumar A, Kumar P, Singh H, Bisht S, Kumar N (2021d) Relationship of physiological plant functional traits with soil carbon stock in the temperate forest of Garhwal Himalaya. Curr Sci 120(8):1368–1373. https://doi.org/10.18520/cs/v120/i8/1368-1373

Kumar M, Phukon SN, Paygude AC, Tyagi K, Singh H (2021e) Mapping Phenological Functional Types (PhFT) in the Indian Eastern Himalayas using machine learning algorithm in Google Earth Engine. Comput Geosci 158:104982. https://doi.org/10.1016/j.cageo.2021.104982

Kumar A, Singh H, Kumari G, Bisht S, Malik A, Kumar A, Singh M, Raturi A, Barthwal S, Thakur A, Kaushal R (2022) Adaptive resilience of roadside trees to vehicular emissions via leaf enzymatic, physiological, and anatomical trait modulations. Environ Pollut 313:120191. https://doi.org/10.1016/j.envpol.2022.120191

Kumar N, Singh H, Giri K, Kumar A, Joshi A, Yadav A, Singh R, Bisht S, Kumari R, Jeena N, Khairakpam R, Mishra G (2024) Physiological and molecular insights into the allelopathic effects on agroecosystems under changing environmental conditions. Physiol Mol Biol Plants 30:417–433. https://doi.org/10.1007/s12298-024-01440-x

Li Y et al (2020) Spatiotemporal variation in leaf size and shape in response to climate. J Plant Ecol 13:87–96

Li Zhu S, Lian J, Zhang H, Liu H, Ye W, Ye Q (2021) Functional traits are good predictors of tree species abundance across 101 subtropical forest species in China. Front Plant Sci 12. https://www.frontiersin.org/articles/10.3389/fpls.2021.541577

Liu Y, Ding HUI (2008) Variation in air pollution tolerance index of plants near a steel factory: implications for landscape-plant species selection for industrial areas. WSEAS Trans Environ Devel 4:24–32

Liu S, Zamanian K, Schleuss PM, Zarebanadkouki M, Kuzyakov Y (2018) Degradation of Tibetan grasslands: consequences for carbon and nutrient cycles. Agric Ecosyst Environ 252:93–104

Lourens P, Frans B (2006) Leaf traits are good predictors of plant performance across 53 rain forest species. Ecology 87:1733–1743

Malik A, Kumar K, Kumari G, Kumar N, Singh H (2024). Application of Functional Traits in Modelling Productivity and Resilience under Climate Change. In: Kumar, N., Singh, H. (eds) Plant Functional Traits for Improving Productivity. Springer, Singapore. https://doi.org/10.1007/978-981-97-1510-7_5

Meena D, Singh H, Chaudhari SK (2011) Elucidating strontium response on growth dynamics and biochemical change in Phaseolus mungo L. Int J Agric Environ Biotechnol 4:107–113

Myneni RB, Keeling CD, Tucker CJ, Asrar G, Nemani RR (1997) Increased plant growth in the northern high latitudes from 1981 to 1991. Nature 386(6626):698–702

Negi JDS, Manhas RK, Chauhan PS (2003) Carbon allocation in different components of some tree species of India: a new approach for carbon estimation. Curr Sci 85(11):1528–1531

Norberg J, Swaney DP, Dushoff J, Lin J, Casagrandi R, Levin SA (2001) Phenotypic diversity and ecosystem functioning in changing environments: a theoretical framework. Proc Natl Acad Sci USA 98:11376–11381

Oldfield S, Eastwood A (2007) The red list of oaks. Fauna and Flora of International Cambridge, UK. ISBN 9781 903703250

Ordoñez JC, Van Bodegom PM, Witte JPM, Wright IJ, Reich PB, Aerts R (2009) A global study of relationships between leaf traits, climate and soil measures of nutrient fertility. Glob Ecol Biogeogr 18(2):137–149

Prakash V, Hunney K, Singh H (2022) Effect of elevated carbon dioxide on growth and development of Santalum album L. seedlings inoculated with plant growth promoting microorganisms in Open Top Chambers. Int J Environ Agric Biotechnol 7(5):149–159

Qin J, Shangguan Z (2019) Effects of forest types on leaf functional traits and their interrelationships of *Pinus massoniana* coniferous and broad-leaved mixed forests in the subtropical mountain. Southeastern China Ecol Evol 9(12):6922–6932. https://doi.org/10.1002/ece3.5259

Samuelson LJ, Stokes TA (2012) Leaf physiological and morphological responses to shade in grass-stage seedlings and young trees of long leaf pine. Forests 3:684–699. https://doi.org/10.3390/F3030684

Sarma B, Chanda SK, Bhuyan M (2017) Impact of dust accumulation on three roadside plants and their adapative responses at National Highway 37, Assam, India. Trop Plant Res 4 161–167. https://doi.org/10.22271/tpr.2017.v4.i1.023

Sharma R, Singh H, Kaushik M, Nautiyal R and Singh O (2018). Adaptive physiological response, carbon partitioning, and biomass production of *Withania somnifera* (L.) Dunal grown under elevated CO_2 regimes. 3 Biotech 8 267. https://doi.org/10.1007/s13205-018-1292-1

Sharma R, Singh H (2021) Alteration in Biochemical constituents and nutrients partitioning of *Asparagus racemosus* in response to elevated atmospheric CO_2 concentration. Environ Sci Pollut Res 29:6812–6821. https://doi.org/10.1007/s11356-021-16050-3

Singh H (2021) An integrated approach considering physiological and biophysical based indicators for assessing tolerance of roadside plantations of *Alstonia scholaris* towards urban roadside air pollution: an assessment of adaptation of plantations for mitigating urban roadside air pollution. Trees. https://doi.org/10.1007/s00468-021-02179-8

Singh H (2024) Exploring adaptive modulation in plant functional traits and its impact on the productivity of *Acacia auriculiformis* under CO_2-enriched environment. Ind Crops Prod 210. 118186. https://doi.org/10.1016/j.indcrop.2024.118186

Singh G, Rawat GS (2012) Quantitative analysis of tree species diversity in different Oak (Quercus spp.) dominated forests in Garhwal Himalaya, India. Notulae Scientia Biologicae. J Mt Sci 15(2):307–318, 3184:132–140. https://doi.org/10.15835/nsb448200

Singh H, Verma A (2013a) Characterization and screening of high nitrogen efficient rice genotype to curtail environmental pollution. Int J Agric Environ Biotechnol 6(1):23–30

Singh H, Verma A (2013b) Physiological responses of rice cultivars to various nitrogen levels. Int J Agric Environ Biotechnol 6(3):383–388

Singh H, Kumar M (2022) Climate change and its impact on Indian Himalayan forests: current status and research needs. In Rani S, Kumar R (eds) Climate change. Springer climate. Springer, Cham, pp 223–242. https://doi.org/10.1007/978-3-030-92782-0_11

Singh H, Verma A, Krishnamoorthy M, Shukla A (2010) Consequence of diverse nitrogen levels on leaf pigments in five rice genotypes under field emergent circumstance. Int J Bio-resource Stress Manag 1:189–193

Singh G, Rai ID, Rawat GS (2012) The mortality of banj oak (*Quercus leucotrichophora A. Camus*) trees in Mussoorie, Uttarakhand: is it an alarming call for rapid degradation. Current Sci 102 (12):1622–1623

Singh H, Verma A, Rai SK (2013) Biochemical evaluation of different rice genotypes grown at various nitrogen levels using SDS-PAGE. Curr Adv Agric Sci 5(1):144–146

Singh G, Padalia H, Rai ID et al (2016) Spatial extent and conservation status of Banj oak (*Quercus leucotrichophora* A. Camus) forests in Uttarakhand, Western Himalaya. Trop Ecol 57(2):255–262

Singh H, Savita Sharma R, Sinha S, Kumar M, Kumar P, Verma A, Sharma SK (2017a) Physiological functioning of *Lagerstroemia speciosa* L. under heavy roadside traffic: an approach to screen potential species for abatement of urban air pollution. 3 Biotech 7(61):1–10. https://doi.org/10.1007/s13205-017-0690-0

Singh H, Verma A, Kumar M, Sharma R, Gupta R, Kaur M et al (2017b) Phytoremediation: a green technology to clean up the sites with low and moderate level of heavy metals. Austin Biochem 2:1012

Singh H, Verma A, Kumar M, Sharma R, Gupta R, Kaur M, Negi M, Sharma SK (2017c) Phytoremediation: a green technology to clean up the sites with low and moderate level of heavy metals. Austin Biochem 2:1012

Singh H, Sharma R, Savita SMP, Kumar M, Verma A, Ansari MW, Sharma SK (2018) Adaptive physiological response of *Parthenium hysterophorus* to elevated atmospheric CO_2 concentration. Indian Forester 144:1–14

Singh H, Yadav M, Kumar N, Kumar A, Kumar M (2020) Assessing adaptation and mitigation potential of roadside trees under the influence of vehicular emissions: a case study of *Grevillea robusta* and *Mangifera indica* planted in an urban city of India. PLoS ONE 15(1):e0227380. https://doi.org/10.1371/journal.pone.0227380

Singh S, Singh H, Sharma SK, Nautiyal R (2021) Seasonal variation in biochemical responses of bamboo clones in the sub-tropical climate of Indian Himalayan foothills. Heliyon 7(4):e06859. https://doi.org/10.1016/j.heliyon.2021.e06859

Singh H, Bandyopadhyaya S, Kumar A, Singh M, Malik A, Kumari G, Raturi A, Yadav SK, Kumar M, Barthwal S, Thakur A, Rawat PS (2022) Understanding the physiological and biophysical response of urban roadside plantations for assessing adaptation and mitigation mechanisms toward vehicular emissions. Urban Climate 44:101183. https://doi.org/10.1016/j.uclim.2022.101183

Singh M, Singh H, Kumar A, Kumar M, Barthwal S, Thakur A (2024a) Soil nitrogen availability determines the CO_2 fertilization effect on tree species (*Neolamarckia cadamba*): growth and physiological evidence. Environ Sustain 7:53–60. https://doi.org/10.1007/s42398-023-00300-w

Singh H, Kumar N, Kumar A (2024b) Enhancing resource use efficiency in crops through plant functional traits. In: Kumar N, Singh H (eds) Plant functional traits for improving productivity. Springer, Singapore. https://doi.org/10.1007/978-981-97-1510-7_6

Smart SM et al (2017) Leaf dry matter content is better at predicting above-ground net primary production than specific leaf area. Funct Ecol 31:1336–1344

Taylor SH et al (2012) Photosynthetic pathway and ecological adaptation explain stomatal trait diversity amongst grasses. New Phytol 93:387–396

Tecco PA, Urcelay C, Diaz S, Cabido M, Pérez-Harguindeguy N (2013) Contrasting functional trait syndromes underlay woody alien success in the same ecosystem. Austral Ecol 38(4):443–451

Tejerina D, Garcia TS, Vaca MC, Vasquez FM, Cava R (2011) Acorns *Quercus rotundifolia* Lam and grass as natural sources of antioxidants and fatty acids in the "montanera" feeding of Iberian pig: intra- and inter-annual variations. Food Chem 124:997–1004

Tian M, Yu G, He N, Hou J (2016) Leaf morphological and anatomical traits from tropical to temperate coniferous forests Mechanisms and influencing factors. Sci Rep 6:19703

Vaieretti MV, Diaz S, Vile D, Garnier E (2007) Two measurement methods of leaf dry matter content produce similar results in a broad range of species. Ann Bot 99(5):955–958

Walker M, Wahren CH, Hollister RD (2006) Plant community responses to experimental warming across the tundra biome. Proc Natl Acad Sci USA 103:1342–1346

Wright IJ, Dong N, Maire V, Prentice IC, Westoby M, Diaz S et al (2017) Global climatic drivers of leaf size. Science 357:917–921. https://doi.org/10.1126/science.aal4760

Yadav SK, Singh H, Ginwal HS, Barthwal S (2019a) Elevated CO_2 enhanced growth and physiological process of *Populus deltoides* Bartr. ex Marsh. Indian Forester 145(1):23–27

Yadav SK, Singh H, Nautiyal R, Ginwal HS, Ansari SA, Barthwal S (2019b) Modulation of morpho-physiological responses in *Populus deltoides* by elevated carbon dioxide and temperature. Forest Sci 66:105–118. https://doi.org/10.1093/forsci/fxz048

Yang SJ, Sun M, Zhang YJ, Cochard H, Cao KF (2014) Strong leaf morphological, anatomical, and physiological responses of a subtropical woody bamboo (*Sinarundinarianitida*) to contrasting light environments. Plant Ecol 215:97–109. https://doi.org/10.1007/s11258-013-0281-z

Yang X, Chi XL, Ji CJ, Liu HY, Ma WH, Mohhammat A et al (2016) Variations of leaf N and P concentrations in shrubland biomes across northern China: phylogeny, climate, and soil. Biogeosciences 13:4429–4438. https://doi.org/10.5194/bg-13-4429-2016

Yin Q et al (2018) Te relationships between leaf economics and hydraulic traits of woody plants depend on water availability. Sci Total Environ 621:245–252. https://doi.org/10.1016/j.scitotenv.2017.11.171

Examining Turgor Maintenance Mechanisms in Rhododendron arboreum and Quercus lanata Within Shivapuri Nagarjun National Park, Nepal

Ganga Bahadur Khatri, Kanta Poudyal, Sarita Dhakal, Rajeev Joshi, and Bhuwan Bohara

Abstract This study aims to analyze the maintenance of turgor and phenological activities in two dominant evergreen trees, *Q. lanata,* and *R. arboreum*, as well as a shrub, all of which are found within Shivapuri Nagarjun National Park. Several parameters, including pressure–volume analysis, xylem conductance, twig starch content, specific leaf mass (SLM), and phenomorphological cycles, were measured every month from February 2019 to January 2020. Noteworthy fluctuations were observed in the water relations of the tissues, indicating variations in turgor maintenance for both species. Osmotic potential at zero turgor (Ψ_{sz}) and osmotic potential at full turgor (Ψ_{sf}) decreased during dry months, suggesting that *Q. lanata* and *R. arboreum* exhibited elastic adjustment to preserve turgor and displayed a considerable level of drought tolerance. Xylem conductance strongly correlated with phenological activities, as indicated by increased Hv value during leaf senescence for both species. Starch content (starch0 and starch1) displayed significant seasonal variations and showed a notable association with phenological activities. SLM decreased during senescence and increased during leaf expansion and maturity. Moisture content (%) at a depth of 15 cm showed the lowest and highest values in August and from March to September, respectively. However, *Q. lanata* maintained similar moisture content to that at 15 cm at a depth of 30 cm, while *R. arboreum* displayed the lowest value in February. Soil moisture content exhibited significant variations across different months but not between the sampled species.

Keywords *Rhododendron arboreum* · *Quercus lantata* · Turgor maintenance · Osmotic potential · Elastic adjustment

G. B. Khatri · K. Poudyal · S. Dhakal
Department of Botany, Amrit Campus, Tribhuvan University, Kathmandu, Nepal

R. Joshi (✉)
College of Natural Resource Management, Faculty of Forestry, Agriculture and Forestry University, Katari-56310, Udayapur, Nepal
e-mail: joshi.rajeev20@gmail.com

B. Bohara
Doon(PG) College of Agriculture Science and Technology, Dehradun, India

H. Singh (ed.), *Urban Forests, Climate Change and Environmental Pollution*,
https://doi.org/10.1007/978-3-031-67837-0_35

1 Introduction

Plant growth is primarily associated with maintaining favorable water conditions (DaMatta and Ramalho 2006; Kumar et al. 2022; Singh et al. 2022; Kumar et al. 2024). The plant water status is determined by examining cell and tissue water content (Kozlowski 1971). Turgor maintenance is crucial for leaf expansion in higher plants, even when water availability is low (Kramer and Boyer 1995; Singh et al. 2024a, b). The adjustment of internal water status is achieved through turgor maintenance, which involves elastic and osmotic adjustments to prevent damage to the plasma membrane caused by low tissue water potential (Galindo et al. 2004; Dhyani et al 2021; Kumar et al. 2021). Trees in the central Himalayas experience a distinct seasonal climate, with only three rainy months followed by limited precipitation throughout the year (Singh et al. 2009). This extended dry period reduces tree water potential (ψ), alters tree phenology, and potentially limits tree species distribution (Zobel and Singh 1995; Singh et al. 2000; Poudyal et al. 2004). Dehydration-tolerant species adapt to low plant water potentials by maintaining low osmotic potentials (Morgan 1984; Prakash et al. 2022), which enables them to uptake water from dry soils and recover after drought. Studies have shown that the response of tropical trees to seasonal drought is more complex than their response to low temperatures, as drought effects are often mitigated by soil water reserves and various drought avoidance mechanisms in trees (Borchert 1994a; Nepstad et al. 1994).

Turgor maintenance is crucial for these two species' phenological processes of leaf fall, flowering, and emergence during the dry summer (Poudyal et al. 2012). Tree species with different drought coping strategies often coexist in tropical dry forests and regions with seasonal water limitations (Yadav et al. 2019; Jangra et al. 2024). Cell water relations depend on water potential, which is determined by the osmotic potential of the vacuolar sap and the turgor or pressure potential. Most studies on osmotic adjustment, turgor loss, and the role of the elastic modulus in sensing water deficits focus on leaves and utilize the Pressure–Volume technique (Pallardy et al. 1991; Kanta et al. 2024; Malik et al. 2024). Osmotic adjustment enables water absorption and maintains cell turgor under drought conditions, allowing the plants to sustain higher photosynthetic rates and growth (Cattivelli et al. 2008; Kumar and Singh 2024). As water availability decreases, cells adjust their water status by accumulating osmotically active compounds, lowering the osmotic potential and thus maintaining turgor. Elastic adjustment occurs through modifications in cell walls, making them more elastic and facilitating tissue shrinkage during dehydration (Blake and Tschaplinski 1992; Meena et al. 2011; Kumar et al. 2019). Elastic adjustment plays a vital role in maintaining essential processes. Water conductivity through the soil–plant continuum is crucial for controlling water loss (Aranda et al. 2001), and the structure and size of water-conducting vessels in the xylem play a significant role in the adaptive physiology of plants under water stress (Tyree and Ewers 1991).

Therefore, the water conductive system's structure limits leaf water potential, stomatal behavior, and gas exchange (Tyree and Ewers 1996; Kumar et al. 2020; Singh and Kumar 2022). Water storage in stems and leaves significantly affects trees' seasonal and daily water balance. Seasonal variation in wood water storage is a crucial factor influencing phenology and tree species distribution in tropical dry forests (Borchert 1994b, 1998; Butz 2019). Organic osmolytes, especially carbohydrates, play an important role in osmotic adjustment by protecting subcellular structures from dehydration stress (Hare et al. 1998; Singh et al. 2024a). During the dormant season, reserves are allocated to essential processes such as cold season acclimation, cellular respiration, and defense, enabling trees to survive in changing environmental conditions. Specific leaf mass (SLM), correlated with ecological conditions, including PAR, is widely used to indicate plant functioning, including photosynthetic and respiratory rates, chemical composition, and resistance to herbivory (Lugg and Sinclair 1979; Singh and Verma 2013; Kumar et al. 2017). Species with higher specific leaf mass (SLM) have higher energy costs (Griffin 1994; Gupta and Singh 2017; Singh et al. 2017a, b) and require more time for leaf expansion. Evergreen species generally have higher SLM values than deciduous species in natural settings and controlled conditions (Sobrado 1991; Villar and Merino 2001). Specific leaf mass (SLM) and leaf thickness significantly increase drought tolerance in early and middle successional species compared to late successional species (Abrams and Kubiske 1990; Abrams et al. 1994). In climates with distinct seasonality, vegetation adapts its phenology accordingly (Lieth 1974).

In the central Himalayas, deciduous leaves complete expansion late in the dry season, while evergreen leaves reach full size during the wet season (Negi and Singh 1992). Leaffall is seasonal in the central Himalayas, with the peak month generally being April or May, when 37–53% of the annual leaf fall occurs (Singh and Singh 1987). This pattern of leaf shedding is similar to dry tropical forests but differs from evergreen tropical forests, where leaf fall is relatively consistent throughout the year (Reich 1995). The study was conducted at Shivapuri Nagarjun National Park (SNNP), located in the mid-hills on the northern fringe of the Kathmandu valley, where there is significant variation in annual temperature and precipitation. During the summer months of May and June, the maximum temperature at Kakani weather station ranges from 22.4 to 25.4 °C, while in the coldest months of December and January, the minimum temperature ranges from 2.3 to 5.6 °C. The vegetation in the central Himalayas adapts to selective environmental conditions compared to other parts of the world. The monsoon season from mid-June to mid-September brings heavy rainfall. The concentrated warm season followed by 8–9 months of dry periods is likely to affect plant adaptation and ecosystem processes profoundly (Singh and Singh 1992; Zobel and Singh 1995). The physiological dynamics and adaptive strategies of *R. arboreum* and *Q. lanata* in response to their specific environment have not been extensively studied. Recent research suggests that forests in certain areas may face increased vulnerability to drought shortly due to water depletion and the amplified effects of climate variability and extremes associated with global warming. The role of the environment, particularly drought, in shaping the distribution and performance of Himalayan tree species is not well understood (Zobel and Singh 1995, Poudyal 2004). A comprehensive study on these ecologically significant plants, namely *R. arboreum* and *Q. lanata,* could contribute to their effective management within the ecosystem, considering the current context of escalating deforestation and global warming. This study aims to thoroughly investigate the responses of *R. arboreum* and *Q. lanata* to water deficits to phenological events, examining seasonal changes in various parameters such as pressure–volume analysis, xylem conductance, starch content in twigs, specific leaf mass, and soil moisture content percentage. The research seeks to evaluate the pressure–volume curve to understand the effective mechanism of turgor maintenance in *R. arboreum* and *Q. lanata,* explore the monthly influence of twig conductance (Huber value, hydraulic conductivity, specific conductivity, leaf-specific conductivity) and leaf dynamics (specific leaf mass) on turgor maintenance, investigate the relationship between starch reserves in twigs and leaf development, and assess the potential monthly phenological shifts during dry periods and the underlying mechanisms involved in plant rehydration.

2 Materials and Methods

2.1 *Study Area*

The study was conducted in ShivapuriNagarjun National Park (SNNP), located in the country's mid-hills on the northern outskirts of the Kathmandu valley. Initially designated as Shivapuri National Park in 2002, it covered an area of 144 km^2 (Bhuju et al. 2007). Later 2009, it was expanded by incorporating the Nagarjun Forest Reserve, adding 15 km^2. According to the Tree Improvement and Silvicultural Component (TIS) report 2002, SNNP comprises four distinct forest types: Pine, Oak, Rhododendron, and *Schima-Castanopsis.* Subtropical Schima-Castanopsis forests, with dominant tree species such as *Schima wallichii, Castanopsis indica, Castanopsis tribuloides, Alnus nepalensis,* and *Eurya acuminata* characterize the lower region (up to 1900 m). *Alnus nepalensis* thrives in moist and shaded areas. Moving to the middle belt (2000–2300 m), mixed Oak forests are prevalent, with key tree species including *Quercus lanata, Rhododendron arboreum, and Cyclobalanopsis glauca.* Oak forests dominate the upper belt (2300–2732 m), with *Quercus semecarpifolia, Quercus lamellosa,* and *Rhododendron arboreum* as the primary tree species. Shivapuri experiences a subtropical to warm temperate climate characterized by considerable variation in annual temperature and precipitation levels.

The study area has a typically warm and temperate monsoon climate with three distinct seasons around the year: cold and dry winter (October to February), hot and dry summer (March to May), and monsoon (June to September). The study area does not have a meteorological station, so climatic data were collected from the Department of Hydrology and Meteorology, Naxal, Kathmandu, Nepal. The highest rainfall was recorded in August, i.e., 558.2 mm, and the maximum and minimum temperatures were recorded as the highest at 30.5 and 3.6 °C respectively (Fig. 1).

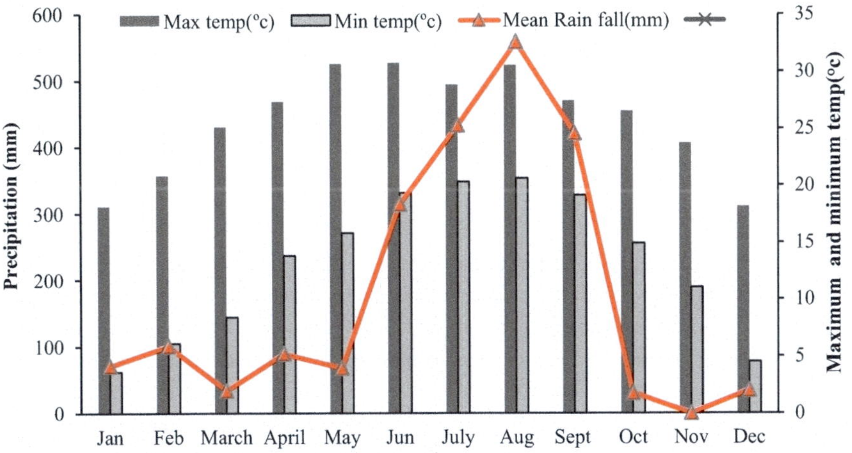

Fig. 1 Ombrothermic graph for the year 2019–2020

2.2 *Sampling Design*

Quercus lanata and Rhododendron arboretum were taken as experimental plants. P-V analysis, xylem conductance, starch percentage, specific leaf mass, soil moisture content at 15 and 30 cm depth, and phenological activities were measured in three representative trees of each species. All these parameters were studied at a monthly interval while phenological data were measured monthly, whose average data was considered. The measurements of the parameters such as pressure–volume analysis, xylem conductance, twig starch content, SLM, and phenomorphological cycles were taken from February 2019 to January 2020 (except June, October, and December). For various reasons, measurements could not be taken in June, October, and December. Each sampled tree was marked, and all measurements were taken on the same marked trees on the same date throughout the observation period.

2.3 *Soil Moisture Content*

The moisture content of the soil was determined by the gravimetric method. Two hundred grams of the inner layer of soil were extracted at a depth of 15 and 30 cm from the study site of both species. Two soil samples of each species were analyzed at monthly intervals. The samples were packed into black polythene bags and taken to the laboratory. The collected soil sample was weighed before and after oven drying it for about 36 h at 60 °C, and then its original moisture content was calculated by using the given formula:

$$\text{Soil moisture content } (\%) = \frac{\text{Fresh weight of soil} - \text{Dry weight of soil}}{\text{Dry weight of soil}} \times 100$$

2.4 Pressure–Volume Analysis

Pressure volume analysis was typically performed using a Pressure chamber (PMS instrument company, Corvallis, Oregon, USA) by the bench drying method (Pallardy et al. 1991). The various components of water potential (RWC_z, Ψ_{sz}, Ψ_{sf} and Ψ_{pf}), osmotic adjustment, and the elastic modulus of tissues were determined employing P-V curve analysis. The P-V curves were obtained using the free-drying technique known as bench drying. Leafy twigs were collected from each sample tree after 16 h for the P-V analysis. These twigs were promptly immersed in water at their cut ends upon collection. The samples were then covered with black plastic bags and transported to the laboratory. The twigs were rehydrated overnight and subsequently used for the P-V curve analysis.

The following steps were undertaken: First, the leafy twigs from the most recent growing season were cut underwater to achieve a suitable size for fitting into the pressure chamber, with the cut end inserted into a perforated stopper. The combined weight of the twig and stopper was measured, representing the saturated weight of the tissue. After recording the weight, the water potential (BP) was calculated, along with the corresponding time and fresh weight, which were all documented on a data sheet. The twigs were then hung on a string and left to transpire freely. At 3, 6, 15, 30, and 60 min, tissue mass and water potential measurements were repeated for each sampled twig. The time intervals between measurements were progressively increased to allow the sample's water content to reach equilibrium. These measurements were continued until the samples displayed a consistent decrease in weight and achieved balance pressure (BP). After the final reading, the samples were removed from the stopper, and the weight of the stopper alone was recorded. The samples of each species were wrapped in paper with detailed information written on it and then dried in an oven at 60 °C for 36 h to determine the dry mass. This entire process was carried out for all the studied species' samples. Finally, the relative water content (RWC) was calculated using the provided formula for each water potential and weight measurement.

$$\text{RWC } (\%) = \frac{\text{Fresh weight of the sample (F)} - \text{Dry weight (D)}}{\text{Weight of sample at full saturation (S)} - \text{Dry weight (D)}} \times 100$$

For each measurement, a plot was created by correlating the reciprocal of balance pressure (1/BP) with relative water content (RWC). The linear segment of the curve was identified, and a straight line was drawn through all the data points within this linear region. This line was then extended to RWC = 100%, visualizing the changes in osmotic potential corresponding to varying RWC values. The RWC and osmotic potential at zero turgor were determined by analyzing the P-V curve and the osmotic potential at full turgor (RWC = 100%).

Based on the RWC for each data point, the osmotic potential was derived from the extrapolated straight line. The pressure potential was subsequently calculated as the difference between the water potential, assumed to be -BP, and the osmotic potential.

Furthermore, a Hoffler diagram was constructed utilizing the P-V curve, presenting the relationship between water potential, osmotic potential, pressure potential, and RWC.

Occasionally, newly emerged leaves on the twigs would break due to repeated handling. In such cases, the broken portion was weighed separately while fresh and again when dry, and adjustments were made in the calculations to account for this loss (Zobel 2000).

2.5 *Theoretical Aspect of Pressure–Volume Curve*

Several components determine water potential, and each component affects water potential in specific way. According to Gibbs (1931), water potential (Ψ_w) in a plant cell or tissue is equal to the algebraic sum of the solute potential (Ψ_s) due to dissolved solutes, the pressure potential (turgor pressure) (Ψ_p) due to the pressure developed within cells or tissue, the matric potential (Ψ_m) due to the binding of water molecules to protoplasmic and cell wall constituents and gravitational potential (Ψ_g). Thus, the several components are expressed as:

$$\Psi_w = \Psi_s + \Psi_p + \Psi_m + \Psi_g \tag{1}$$

where s, p, m and g represent the effects of solutes, pressure, matrices and gravity, respectively.

2.5.1 Index of Elasticity (ε)

The elasticity index is the change in Ψ_p for a given fractional change in the weight of symplastic water (Tyree and Jarvis 1982). To calculate the index of elasticity, all values of Ψ_p above zero turgor were plotted against RWC; the slope of a linear regression line through these points was used as ε.

2.5.2 Index of Desiccation Tolerance (IDT)

An index of Desiccation Tolerance (IDT) can estimate the net effect of RWC at turgor loss (RWCz), osmotic potential at zero turgor (Ψ_{sz}), and high tissue elasticity, which in a tissue can often vary in different ways. IDT was calculated using these three parameters to confer desiccation tolerance: low RWCz, Ψ_{sz} and high tissue elasticity

(ε) (Lambers et al. 1998). The lowest end of the range of RWCz, Ψ_{sz}, and low ε was assigned a relative value of 100, indicating high tolerance, and the high end of the range was 0, indicating low tolerance. Each parameter was assigned a score (0–100) based on its relationship to maximal and minimal values, hence taken as relative values.

2.5.3 Xylem (Twig) Conductance

Different attributes of xylem conductance, i.e., specific conductivity (K_s), leaf specific conductivity (K_l), and Huber value (H_V), were measured using gravity flow of a solution of oxalic acid, following the methods described by Ewers and Cruiziat (1991). As a procedure, twigs of at least one-year-old were collected from each sampled tree. These twigs were immediately submerged in water at their cut ends and stored in water overnight before being transported to the laboratory. A segment of the twig was chosen to measure the flow rate, ensuring it was straight and had minimal side branches. All leaves beyond the segment's center were gathered and placed in a labeled polythene bag. The dimensions of these leaves were measured, and their leaf area was calculated by multiplying length, breadth, and a coefficient factor of 0.654 for both species. The basal end of the twig was connected to a tube attached to a funnel, with the connection point wrapped with parafilm to prevent leakage. The tubes were filled with solution and cleared of bubbles during the connection.

The necessary materials were prepared as follows: an infusion solution of oxalic acid at a concentration of 0.1 M (1.26 g/l) was prepared and filtered using fine filter paper, while a saffranin staining solution with a concentration of 0.1% (weight/volume) was prepared and filtered twice through fine filter paper. An appropriate tube with a clamp at the bottom was fixed to a funnel, and the infusion solution was poured into it. The setup was positioned 50–60 cm above the working table, and any bubbles in the tube were eliminated. A precision balance measuring up to 0.001 g was placed nearby and calibrated to 0.000. A clean and dry vial was taken and filled with loosely packed tissue paper, weighed, and the weight was recorded on a data sheet. The pressure gradient (PG) was defined as the height of the infusion solution divided by 10.2 and the segment length.

The sample was held horizontally at a specified distance from the fluid level for the flow rate measurement. The free end of the sampled segment was pressed against the tissue in the vial to allow the tissue to absorb the fluid passing through the stem segment. As the flow from the twig began, the flow rate per minute was collected in the vial and weighed. The average of the stable weight changes (g/m) was calculated to determine the flow rate of the fluid through the xylem segment. The segment length, diameters, and pith at each end were measured, and the xylem area (Awt) was calculated. These measurements were carried out consistently for all sampled species.

For the measurement of effective xylem area (Awe), the measured sample segment was connected to another tube containing saffranin dye solution (0.1%), and the segment was suspended. The sample remained attached until the dye reached the cut end, after which the tube was clamped, and the sample was removed. The segment was longitudinally split using a sharp blade, and the percentage of stained xylem cross-sectional area was estimated. The effective xylem cross-sectional area (Awe) was then calculated by multiplying the stained xylem percentage by the xylem area (Awt).

All parameters related to xylem conductance, including the Huber value (HV), hydraulic conductivity (K_h), specific conductivity (K_s), and leaf-specific conductivity (K_l), were measured accordingly.

2.5.4 Twig Starch Content

The percentage of starch estimation was done using 1% iodine and potassium iodide solution following Zobel (1996a, b). As the starch level may differ following the growth of the twig, two groups of twigs were measured in three twigs selected randomly from each sampled tree: 0–year twig (0- refers to current year growth) and1–year twig (1—refers to last year growth). The cut surface was flooded with 1% iodine and potassium iodide solution and washed with water after some time. Using a hand lens, the percentage of staining of the cut surface for each twig was visually estimated, and the estimated value was recorded.

2.5.5 Specific Leaf Mass (SLM)

Measurements were done according to Poudyal (2004). A healthy twig comprising enough leaves was taken. The area of each leaf was calculated by multiplying the length, breadth, and coefficient factor. The coefficient factor was used as 0.654 for both species. The leaves were oven-dried at 60–80 °C for at least 36 h, and dry mass was measured for each leaf.

2.6 Phenology

The phenological events (timing of bud break, leaf production, flowering, fruiting, and leaf fall) were recorded in the field by visual observation. In each sampled tree, a single twig was marked, and the length and breadth of each leaf attached to the marked twig were recorded for each month of observation. The obtained values of length and breadth were then multiplied by the coefficient factor (0.654). Overall, nearby non-sampled trees were considered for general understanding. Senescence was recorded when > 25% of the leaves were shed from the trees.

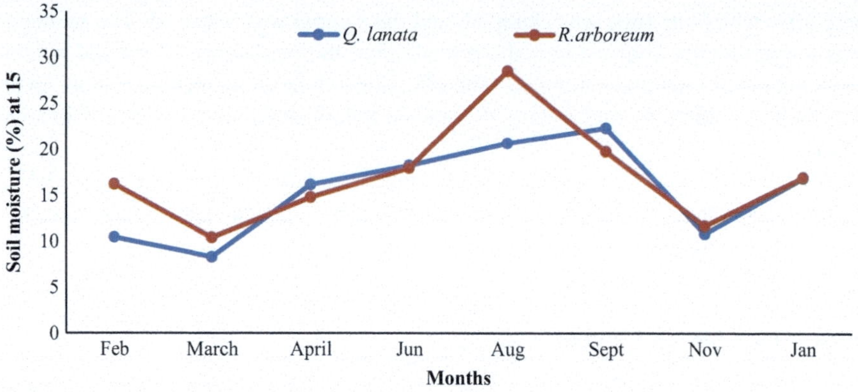

Fig. 2 Variation in soil moisture content (%) at 15 cm depth from February 2019 to January 2020 in *R. arborerum*and *Q. lanata*

2.7 7Statistical Analysis

Mean and standard error values were calculated using descriptive statistics in Microsoft Excel (2010). Variance (ANOVA) analysis with a single factor was performed for each parameter individually. Pearson correlation (r) analysis was conducted to examine the relationships between the studied parameters. Statistical analysis was carried out using SPSS 16.0 for Windows (2010).

3 Results

3.1 Soil Moisture at 15 cm Depth

ANOVA showed that soil moisture at 15 cm exhibited significant variation with sampled months ($P < 0.05$) but not among sampled species ($P = 0.75$).In *R. arboreum*, the moisture percentage ranged from 10.5%(March) to 28.5%(August), whereas in *Q. lanata* the moisture percentage ranged from 8.7%(March) to 22.78% (September) (Fig. 2).

3.2 Soil Moisture at 30 cm Depth

ANOVA showed that soil moisture at 30 cm exhibited significant variation with sampled months ($P < 0.05$) but not among sampled species ($P = 0.94$).In *R. arboreum*, the moisture percentage ranged from 11.09% (February) to 23.60%(August), whereas

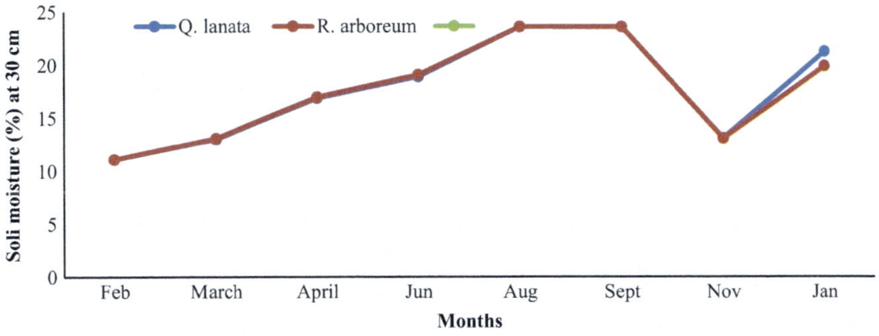

Fig. 3 Variation in soil moisture content (%) at 30 cm depth from February 2019 to January 2020 in *R. arboretum* and *Q. lanata*

in *Q. lanata* moisture, the percentage ranged from 11.094% (March) to 23.61% (September) (Fig. 3).

3.3 Pressure–Volume Analysis

ANOVA showed that RWCz showed significant variation among sampled months ($P < 0.05$) and insignificant variations among species ($P = 0.45$). RWCz in *Q. lanata* ranged between 83.54% (April) to 92.1% (January). RWCz declined during the dry months (April to June) in *R. arboreum*. In *R. arboreum*, RWCz ranged between 83.16% (June) to 92.54% (February) (Fig. 4).

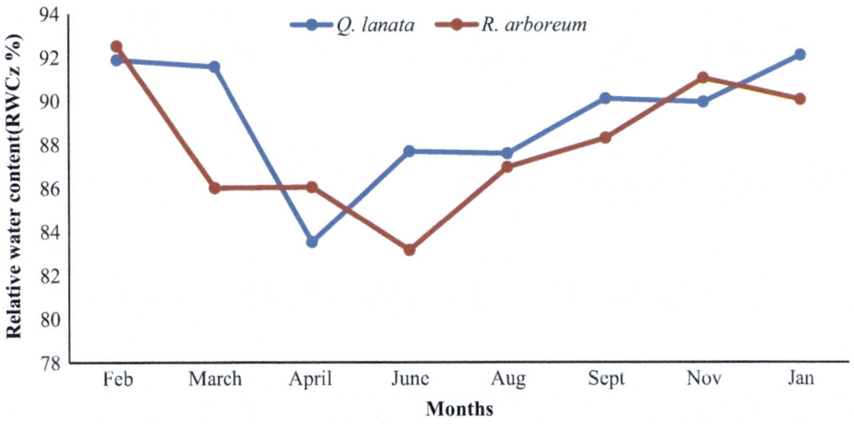

Fig. 4 Variation in relative water content at zero turgor (RWCz,) from February 2019 to January 2020 in *Q. lanata and R. arboreum*

ANOVA showed that the attributes of water relation parameters (ψ_{sz} and ψ_{sf}) varied significantly among months ($P < 0.05$) but insignificantly varied among the species ($P = 0.09$). In *Q. Lanata* mean ψ_{sz} was highest (less negative) in April (− 3.28 MPa) and it was lowest in February (− 5.6 MPa). Mean ψ_{sf} was highest in April (− 2.09 MPa) and lowest in February (− 3.6 MPa). A slightly inconsistent monthly trend was recorded in both ψ_{sz} and ψ_{sf} in *Q. Lanata*. In the case of *R. arboreum,* both ψ_{sz} and ψ_{sf} were highest in September (− 3.06 MPa and −2.16 MPa respectively) and lowest in Feb and March, i.e., 3.78 and − 3.57Mpa (Figs. 5 and 6).

ANOVA showed that the index of tissue elasticity (ε) varied insignificantly among the studied months ($P > 0.05$) but varied significantly between the species ($P = 0.04$) In *Q. Lanata,* the highest and lowest values of ε were recorded in February (5.48Mpa)

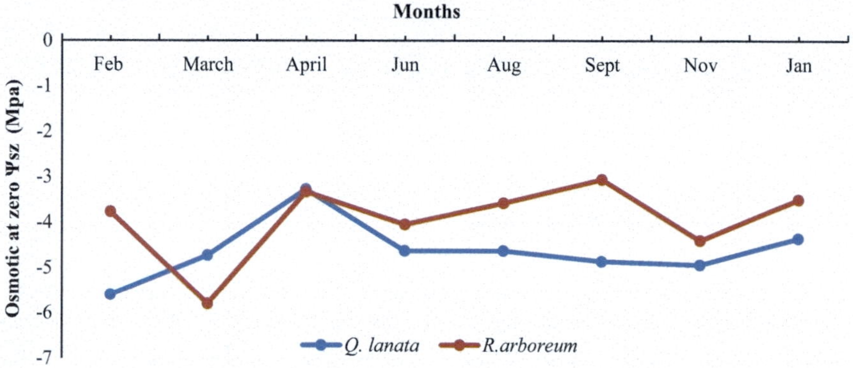

Fig. 5 Variation in osmotic potential at zero turgor from February 2019 to January 2020 in *R. arboreum* and *Q. lanata*

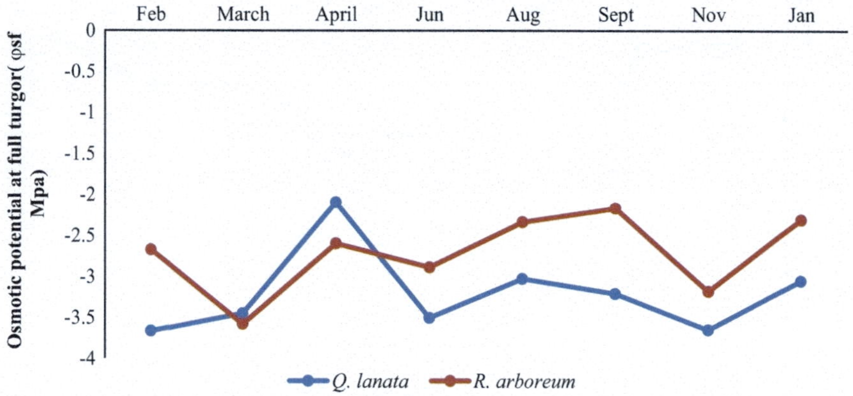

Fig. 6 Osmotic potential variation at full turgor from February 2019 to January 2020 in *R. arboreum* and *Q. lanata*

and April (1.21 MPa), respectively. On the other hand, mean ε ranged from 1.85 MPa (August) and 1.99 Mpa (January) to 4.33 MPa (February) in *R. arborerum* (Fig. 7).

ANOVA showed that the Index of Desiccation Tolerance (IDT) varied significantly among the sampled months ($P < 0.05$) but not with the studied species ($P = 0.17$). The highest and lowest values of IDT were observed in April (88.82) to February (18.14) in *Q. lanata,* whereas in *R. arboretum* the IDT ranged from 22.88 (February) to 71.21 (June) (Fig. 8).

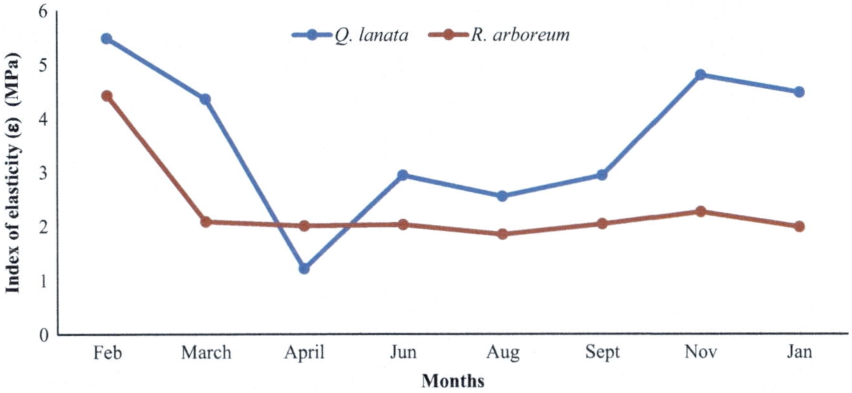

Fig. 7 Variation in index of elasticity (ε) from February 2019 to January 2020 in *R. arboreum* and *Q. lanata*

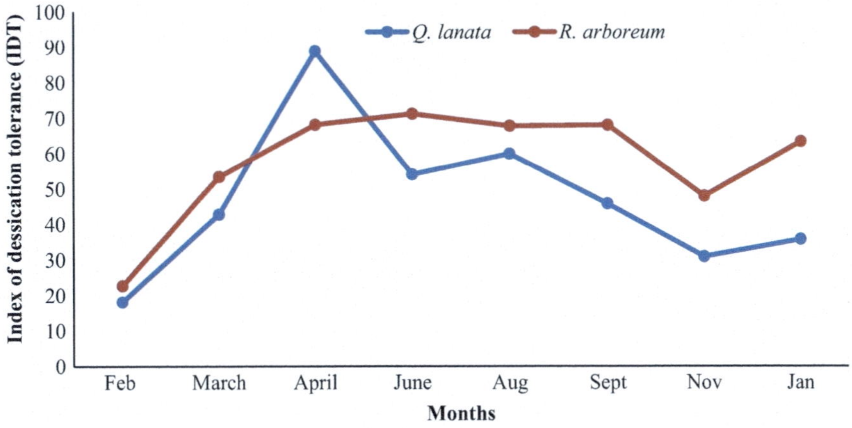

Fig. 8 Variation in index of desiccation tolerance (IDT) from February 2019 to January 2020 in *R. arboreum* and *Q. lanata*

3.4 Xylem Conductance

ANOVA showed that Huber value (H_v), which is a proportionality variation between xylem and leaf surface area, varied significantly among sampled months ($P < 0.05$) and between two species (Annex 1). H_v fluctuated during the sampled months, ranging from 2.1×10^{-4} (March) to 1.11×10^{-4} (April) in *Q. lanata*. In *R. arboretum*, the highest Huber value was in September (3.61×10^{-4}) and the lowest in March (1.80×10^{-4}) (Fig. 9).

ANOVA showed that the specific conductivity (K_s) showed significant variation among the sampled months ($P < 0.05$) but not with the sampled species ($P = 0.39$). In *Q. lanata*, K_s ranged from 0.18 (June) to 1.27 kg m^{-1} s^{-1} MPa^{-1} (August). In *R. arboretum* K_s ranged from 0.22 (February) to 1.8 kg m^{-1} s^{-1} MPa^{-1} (June). An inconsistent trend of monthly variation was recorded in K_s in both species (Fig. 10).

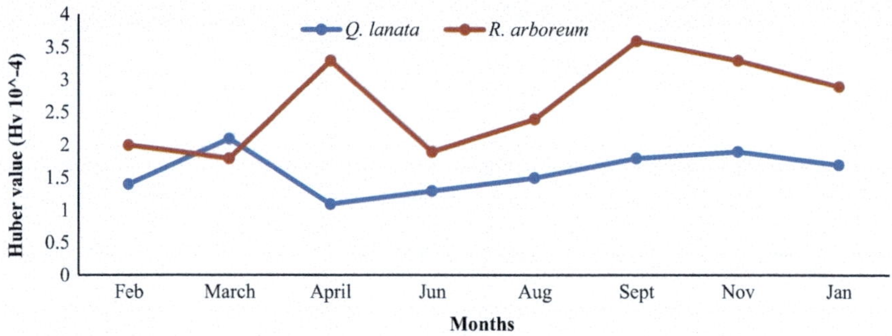

Fig. 9 Variation in Huber value (H_V) during February 2019 to January 2020 in *Q. lanata* and *R. arboreum*

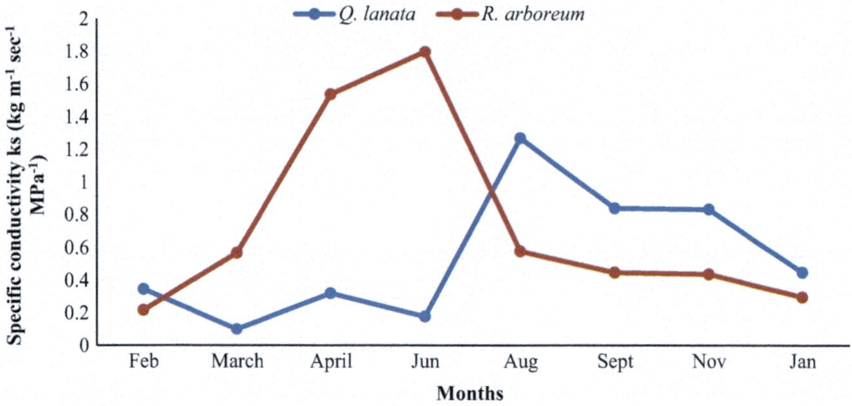

Fig. 10 Variation in specific conductivity (Kh) from February 2019 to January 2020 in *Q. lanata* and *R. arboreum*

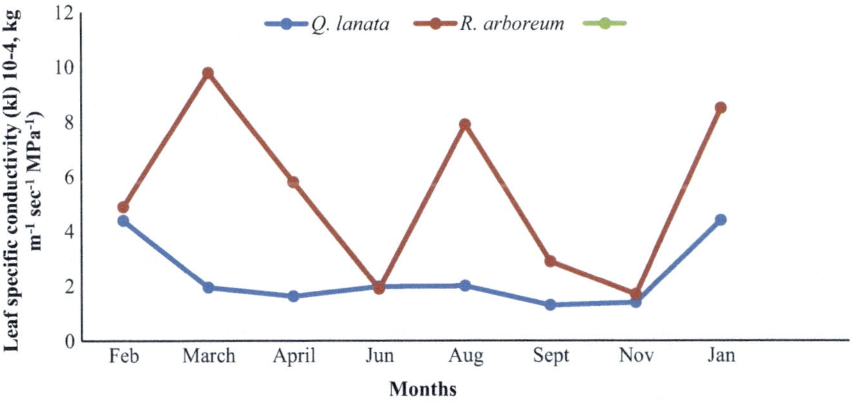

Fig. 11 Variation in leaf specific conductivity (K_l) from February 2019 to January 2020 in *Q. lanata* and *R. arboreum*

ANOVA showed that the leaf-specific conductivity (K_l) varied significantly among the sampled months ($P < 0.05$) but not with the sampled species ($P = 0.27$). In *Q. lanata* K_l was recorded highest during February (4.41×10^{-4} kg m^{-1} s^{-1} MPa^{-1}) and lowest during September (1.30×10^{-4} kg m^{-1} s^{-1} MPa^{-1}). In *R. arboreum* K_l ranged highest in March (9.80×10^{-4} kg m^{-1} s^{-1} MPa^{-1}) and lowest in November (1.20×10^{-4} kg m^{-1} s^{-1} MPa^{-1}). A fluctuating trend of monthly variations was observed in K_l in sampled species (Fig. 11).

3.5 Twig Starch Content

ANOVA analysis showed a significant variation in starch content in zero-year twigs ($starch_0$) ($P < 0.05$) but insignificant variation in one-year twigs ($starch_1$) ($P = 0.31$) among sampled months in *Q. lanata* (Annex 2). However, in the case of *R. arboreum,* significant variation was observed in zero-year twigs, but one-year twigs showed insignificant variation among sampled months ($P = 0.42$). Also, there was negligible variation in starch content in zero-year and one-year twigs among the sampled species ($P = 0.76$) and $P = 0.86$), respectively. $Starch_0$in *Q. lanata* ranged from 0% (September) to 100% (February), and $starch_1$ ranged from 0% (September, November, and January) to 70 to 80% (February and March).In *R. arboreum*starch, the content of the current year ($starch_0$) was highest in April (100%) and lowest in June, August, and September (10%). The previous year, twig ($starch_1$) was highest in April (80%) and lowest in June, August, September, January, and February (0%) (Figs. 12 and 13).

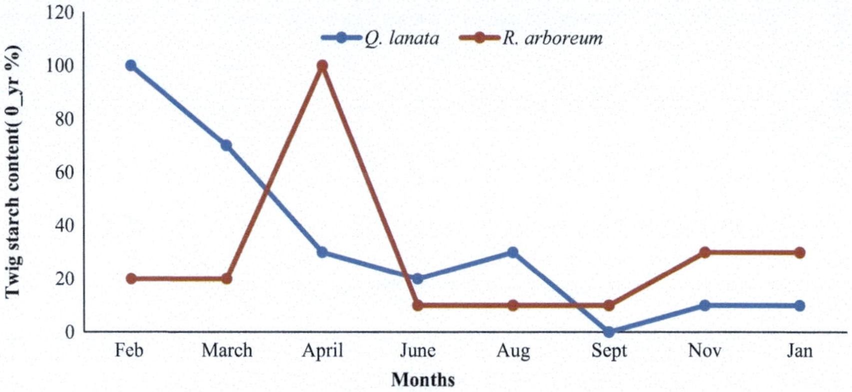

Fig. 12 Variation in starch percentage in current year twig ($starch_0$) from February 2019 to January 2020 in *Q. lanata* and *R. arboreum*

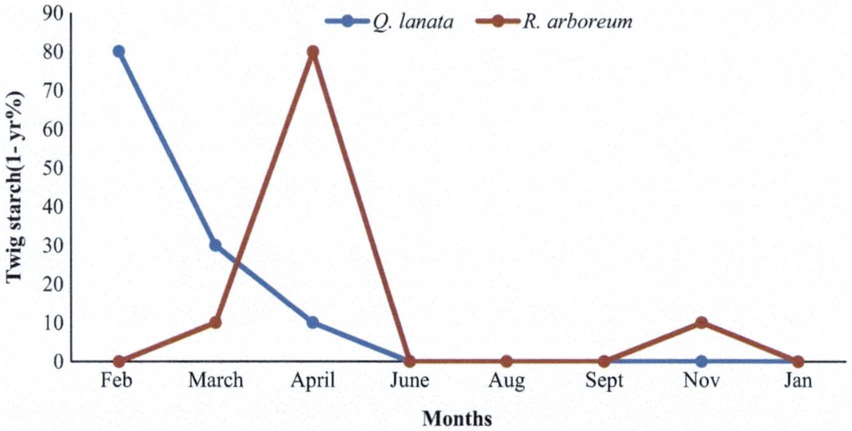

Fig. 13 Variation in starch percentage in one year twig ($starch_1$) from February 2019 to January 2020 in *Q. lanata* and *R. arboreum*

3.6 *Specific Leaf Mass*

ANOVA analysis showed a significant variation in SLM with sampled months ($P < 0.05$) and species ($P = 0.001$). The mean value showed considerable differences among the sampled species. In *Q. lanata,* high and low SLM were observed in January (9.70×10^{-3} g cm^{-2}) and (1.04×10^{-3} g cm^{-2}), respectively. *R. arboreum* had the highest SLM in January (2.27×10^{-3} g cm^{-2}) and lowest in June (1.35×10^{-3} g cm^{-2}). An inconsistent pattern of SLM was observed throughout the studied months in *Q. lanta.*

3.7 Relationship Among the Parameters

3.7.1 Correlation of Relative Water Content at Zero Turgor with Other Variables

RWCz correlated positively with IE, Hv, SLM, K_h and, whereas it showed a negative correlation with K_l, K_s, Hv, ψ_{sz}, ψ_{sf}, IDT, SM at 15 cm, SM at 30 cm in *Q. lanata*. RWCz showed a positively significant correlation with IE and ψ_{sz}, and ψ_{sf} showed a strong negative correlation with RWC_Z. In *R. arboreum,* it exhibited positive correlation with ψ_{sz}, ψ_{sf}, IE, and H_v and correlated negatively with K_h, K_s, K_l, IDT, SM at 15 cm, SM at 30 cm. No strong negative correlation between RWCz and other variables was found in *R. arboretum* except with K_s (Annex 3).

3.7.2 Correlation of ψ_{sz}, ψ_{sf}, and IE with Other Variables

In *R. arboreum,* ψ_{sz} exhibited positive correlations with ψ_{sf}, IE, H_v, K_s, $K_{l,}$ SM at 15 cm, and SM at 30 cm. ψ_{sz} correlated negatively with K_h, IDT. $\boldsymbol{\psi_{sz}}$ showed strong positive correlations with ψ_{sf}. ψ_{sf} correlated positively with K_l, K_h IDT, SM at 15 cm, and SM at 30 cm, as it showed a negative correlation with IE, K_s SLM, K_h. *R. arboretum* IE exhibited negative correlations with K_h, H_v, K_l, K_s IDT, SM at 15 cm, and SM at 30 cm. IE showed strong negative correlations with IDT.ψ_{sz} correlated positively ψ_{sf}, IDT, SM at 15 cm, SM at 30 cm. In contrast, it showed negative correlations with IE, K_s Kh, $H_{v,}$ and K_l in *Q. lanata.* ψ_{sz} showed a strong positive correlation with ψ_{sf} and IDT and a strong negative correlation with IE. In *Q. lanata* Ψ_{sf} exhibited a positive correlation with $K_{s,}$ IDT, SM at 15 cm and SM at 30 cm, whereas it showed a negative correlation with $K_{h,}$ $H_{v,}$ IE. Among them, Ψ_{sf} had a strong positive correlation with IDT and a strong negative correlation with IE. IE correlated positively with $K_{h,}$ $H_{v,}$ and SLM, whereas it negatively correlated with $K_{s,}$ K_l, IDT, and SM at 15vcm and SM at 30 cm. IE showed a strong negative correlation with IDT.

3.7.3 Correlation of Xylem Conductance with Other Variables

In *Q. lanata,* H_v showed a positive correlation with K_l, Ks, and K_h and correlated negatively with IDT, SM at 15 cm, and SM at 30 cm. H_v showed a moderate negative correlation with IDT. K_h showed a strong positive correlation with K_l, $K_{s,}$ and a moderate positive correlation with SLM, SM at 15 cm, and SM at 30 cm, with a negative correlation with IDT. K_s positively correlated with K_l, IDT, and SM at 15 cm and SM at 30 cm. K_s showed a strong positive correlation with K_l. In *Q. lanata,* K_l showed a significant correlation with SM at 15 cm and SM at 30 cm, whereas there was a non-significant correlation with IDT. In *R arboreum,* K_h showed a positive correlation with K_l, Ks, IDT, and SM at 30 cm. In contrast, it negatively correlated with H_v and SM at 15 cm. H_v showed a positive correlation with SLM, IDT, and SM at 15 cm and SM at 30 cm, whereas it showed a negative correlation with K_S. K_S showed a positive correlation with $K_{l,}$ IDT, SM at 30 cm, and SM at 15 cm, whereas a negative correlation with K_l showed a positive correlation with IDT.

3.7.4 Correlation of Specific Leaf Mass (SLM) with Other Variables

SLM showed no significant positive correlation with IDT, SM at 15 cm, and SM at 30 cm in *R. arboretum,* whereas in *Q. lanata* it exhibited a strong negative correlation with IDT.

3.7.5 Correlation of IDT with Other Variables

In *Q. lanata,* IDT showed a positive correlation between SM at 15 cm and SM at 30 cm, whereas, in *R. arboreum,* it exhibited a strong positive correlation between SM at 15 cm and SM at 30 cm.

3.7.6 Correlation of Soil Moisture at 15 cm with Other Variables

In *Q. lanata,* SM at 15 cm showed a strong, significant positive correlation with SM at 30 cm, whereas in n *R arboreum* it showed a significant positive correlation with SM at 30 cm.

3.7.7 Correlation of Starch $_0$ and Starch $_1$ with Other Variables.

In*Q. lanata*starch$_0$ showed significant positive correlation with starch$_l$ like as, positively with RWCz, IE, SLM whereas in *R. arboreum* positively correlated with starch$_{l,}$ K_S, H_v whereas negatively correlated with SLM, SM at 30 cm, SM at 15 cm, RWCz, and K_l (Tables 1 and 2).

3.8 *Phenology*

The phenological activities (bud break, leaf production, flowering, and leaf fall) commenced mainly from March to November. The senescence was confirmed when approximately 25% of the leaves were shed from the sampled and nearby trees. In *Q. lanata,* winter buds matured and broke in late March. Leaf production continued until early January. Leaf fall began in March and was completed in early March. The plant in this habitat is not entirely deciduous. Leaf buds appeared in August. The fruit matured in September. In *Q. lanata,* more than 60% of the leaves were damaged long

Table 1 Pearson correlation coefficient between various parameters shown by *Q. lanata* (Significance values: * = $P < 0.05$)

	(RWCz)	op0-ve	opf -ve	IE	Hv	Ks	Kl	TS₀	TS ₁	IDT	SM (15 cm)	SM (15 cm)
(RWCz)	1											
op0-ve	− 0.76	1										
opf –ve	− 0.75	0.91	1									
IE	0.90	− 0.79	− 0.80	1								
Hv	0.70	− 0.42	− 0.52	0.55	1							
Ks	− 0.09	− 0.13	0.03	− 0.16	0.14	1						
Kl	0.00	− 0.26	− 0.13	− 0.06	0.27	0.98	1					
TS(0)	0.28	− 0.41	− 0.27	0.44	− 0.06	− 0.42	− 0.37	1				
TS(1)	0.38	− 0.52	− 0.33	0.54	− 0.09	− 0.39	− 0.36	0.94	1			
IDT	− 0.92	0.89	0.86	− 0.96	− 0.54	0.02	− 0.08	− 0.33	− 0.48	− 0.65		
SM (15 cm)	− 0.39	0.29	0.39	− 0.65	− 0.35	0.51	0.41	− 0.66	− 0.61	0.20	1	
SM (30 cm)	− 0.23	0.30	0.37	− 0.56	− 0.13	0.54	0.45	− 0.69	− 0.69	0.20	0.95	1

Table 2 Pearson correlation coefficient between various parameters shown by *R. arboretum*. (Significance values: * = $P < 0.05$)

	Rwc 0T	op0-ve	opf -ve	IE	Hv	Ks	Kl	T S (0)	TS (1)	IDT	SM (15cm)	SM (30 cm)
(Rwc 0 T)	1											
op0-ve	0.17	1										
opf -ve	0.14	0.96	1									
IE	0.60	0.03	− 0.07	1								
Hv	0.28	0.59	0.53	− 0.32	1							
Ks	− 0.82	0.07	− 0.03	− 0.34	− 0.10	1						
Kl	− 0.29	0.21	0.17	− 0.35	0.29	0.12	1					
TS (0)	− 0.07	0.21	0.15	− 0.07	0.43	0.41	− 0.44	1				
TS (1)	− 0.25	0.16	0.10	− 0.14	0.38	0.52	− 0.34	0.97	1			
IDT	− 0.72	0.27	0.35	− 0.91	0.31	0.55	0.44	0.13	0.22	1		
SM (15 cm)	− 0.15	0.59	0.68	− 0.18	0.03	− 0.01	0.13	− 0.33	− 0.28	0.38	1	
SM (30 cm)	− 0.35	0.60	0.71	− 0.63	0.35	0.13	0.51	− 0.20	− 0.15	0.21	0.78	1

before senescence. The damages occurred as necrotic patches and holes throughout the leaves.

In *R. arboreum,* leaf fall started in February and maximum, in January. Simultaneously, leaf bud emergence took place in March. The flower bud appeared in February and continued to January. The new leaf emergence took place in April and matured in June. Fruiting occurred in the rainfall period (June) and fell in September. In *R. arboreum,* the senescence rate was slow throughout the sample months, and leaves were damaged as yellowing and holed (Table 3).

Leaf area, which is an important factor controlling water and carbon fluxes by plants, varied insignificantly among sampled months ($P > 0.05$) and between species ($P = 0.27$). In *Q. lanata, the* change of average leaf area in sampled leaves ranged between 17.24 mm^2 (February) to 0.42 mm^2 (August). In *R. arboreum,* it showed an

Table 3 Phenological activities: emergence of leaves (EL)), leaf damage (LD), falling of leaves (FL), coloring of leaves (CL), initiation of flowering (IF), end of flowering (EF), beginning of fruiting (BF), maturing of fruit (MF), leaf bud (LB), Flower bud (FB) and leaf mature (LM) during all sampling months

Species	Monthly observation							
	Feb.	Mar.	Apr.	June	Aug.	Sept.	Nov.	Jan.
Q. lanata	LD, LF	LM, IF, LD EL	LD, LF	BF	LD, MF	LD, LE FF	MF, LD, LB	LD, LF, IF, CL
R. arboreum	IF LD	IF LD LF	EL	LD BF LD	LD LB BF	LD LB	CL LD FB	LM FB LD LF

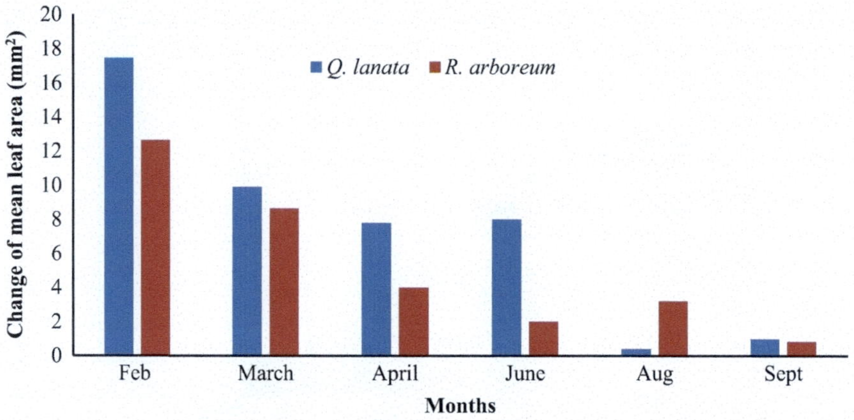

Fig. 14 Variation in change of leaf area from February 2019 to January 2020 in *Q. lanata* and *R. arboreum*

irregular pattern of monthly variation ranging between 12.64 mm^2 (February) and 0.84 mm^2 (September) (Fig. 14).

4 Discussion

In the studied species, soil moisture percentage at 15 cm maintained the lowest and highest values in August and March to September, respectively, in both species. At 30 cm, *Q. lanata* maintained a similar level, but *R. arboreum* had the lowest value in February. The upper soil layers dry out quickly after rainfall stops, and the availability of subsoil moisture varies significantly at different sites, which determines the degree of desiccation observed in trees with low stem water storage. The rate and degree of desiccation are higher in dense hill forests with high root competition compared to widely spaced savanna trees (Borchert 1994a, b, c). These similar results in both species may be attributed to the similar rainfall patterns and soil properties, as the study sites are located within the same altitudinal range. Slightly different results in *R. arboreum* may be due to a relatively moister area compared to the soil around *Q. lanata*. The physical characteristics of the soil largely govern its water-holding capacity and availability. In *Q. lanata*, moisture content at 15 cm and 30 cm significantly correlated with parameters such as Kh, IDT, and ψ_{sz}. However, *R. arboreum* did not significantly correlate with the studied parameters. RWCz, a measure of the plant's ability to maintain turgor under increasing water deficits, reached its maximum value during the beginning of winter (92.1%) in *Q. lanata* (January). The highest RWCz value may be associated with high soil moisture content availability, as it showed a strong positive correlation with moisture content. *Q. lanata* had the lowest RWCz in April (83.54%), positively correlated with IE. The change in tissue elasticity could contribute to turgor maintenance, as leaf damage occurred extensively before senescence, with necrotic patches and holes. The lowest RWCz values were observed in June (83.16%) in *R. arboreum*. This result indicates that both species can maintain a similar turgor level, around 83%. This may be attributed to rainfall throughout the studied months and the evergreen nature of these species. *R. arboreum* exhibited an irregular pattern of RWCz during the studied months, reflecting a strong positive correlation with ψ_{sf}, ψ_{sz}, IE, and Hv, and a strong negative correlation with Ks. This indicates high elastic adjustment and no osmotic adjustment.

Similar findings were reported in M. esculenta, which exhibited an irregular pattern of RWCz during the studied months. It showed a strong positive correlation with ψ_{sf} and ψ_{sz} and a negative correlation with ε, indicating high osmotic adjustment and no elastic adjustment. In *Q. lanata*, ψ_{sf} decreased in April just before leaf flushing, which helped maintain turgidity for new growth. When ψ_{sf} is low, plants can maintain turgidity at low RWCz (Shrestha et al. 2007). In *Q. lanata*, ψ_{sf} reduced after June (− 3.49 MPa) and after leaf emergence (March), which also maintained turgidity. In contrast, in *R. arboreum*, ψ_{sf} was maximum in September (-2.16 MPa) just before leaf bud formation, maintaining turgidity for leaf flushing. Both studied species exhibited no osmotic adjustment, possibly due to precipitation throughout the studied months. The index of elasticity (ε), which represents the flexibility of plant tissues, exhibited an irregular pattern in both species during the sampled months. The role of ε in maintaining turgor is controversial (Cosgrove 1988). Under increasing drought stress, ε has been observed to remain constant (Hsiao et al. 1976), decrease, and increase (Farooq et al. 2012; Mulkey et al. 1991). In February, the highest ε values were recorded in *Q. lanata* and *R. arboreum*, with measurements of 5.48 MPa and 4.33 MPa, respectively. This increase in elasticity was likely due to the onset of the winter season, when plants were less water-stressed. Desiccation tolerance is associated with low ψ_{sz}, low RWCz, and high ε (elasticity) (Lambers et al. 1998). Consequently, both species exhibited high incipient death temperatures (IDT) during the dry months of April to June, coinciding with leaf emergence in *R. arboreum* and the beginning of fruiting and leaf emergence in *Q. lanata*. The mean values of xylem conductance parameters, Hv, Kh, Ks, and Kl, slightly decreased and reached their minimum values from March to June. This decrease might be attributed to low water availability or xylem embolism under dry conditions. Except for Ks in *R. arboreum* and Kh in both species, all the parameters of xylem conductance were highest during late winter due to an adequate amount of stored water in the sub-soil, resulting in less plant stress. The Huber value (Hv), which indicates the relative conducting area, was higher in plants with higher transpiration rates (Larcher 1995). Both species maintained increased Hv values during leaf senescence (January to March) when leaves were damaged by necrosis and holes. *In September, R. arboreum exhibited higher Hv values,* indicating a greater cross-sectional stem area per unit leaf. Hv displayed significant monthly variation but only showed moderate correlation with other water-related parameters such as IDT, soil moisture at 30 cm, and K_h. In *Q. lanata*, H_v reached its highest measurement in February due to a decrease in the desiccation tolerance index, showing a strong negative correlation with IDT. Changes in hydraulic conductivity (K_h) may result from water column interruption in the vessels (embolism) or modifications in xylem vessel size. In *Q. lanata*, K_h exhibited strong seasonal variation, significantly correlating with K_s, K_l, and moisture content at 30 cm, but not other water-related parameters. K_h in *R. arboreum* showed a moderate positive correlation with IDT. Specific conductivity (K_s), which measures the porosity of wood, declined with the onset of dry months. In *Q. lanata*, Ks displayed high seasonal variation, positively correlating with Kl and moderately correlating with soil moisture content at 15 and 30 cm. In *R. arboreum*, K_s exhibited significant monthly variations but did not correlate significantly with other parameters except IDT. Leaf-specific conductivity (K_l), a practical measure of the xylem's hydraulic supply capacity, decreased with reduced water availability. In September and November, K_l reached its lowest levels for *Q. lanata* and *R. arboreum.* Dehydration-induced reduction in K_l may be attributed to decreased permeability of the extra-xylary tissues (Sack and Holbrook 2006, Poudyal 2013) or collapse of xylem cells (Brodribb and Holbrook 2005). K_l in *Q. lanata* exhibited monthly variation but did not significantly correlate with other parameters except SLM. In *R. arboreum*, K_l exhibited higher monthly variation and significantly correlated with SM at 30 cm and IDT. Starch content in twigs displayed a noticeable relationship with leaf phenology. In both species, high starch accumulation was observed in zero-year twigs during leaf emergence (Fig. 14). This high value resulted from the accumulation of reserved carbohydrates, as Poudyal et al. (2012) suggested for Himalayan tree species. As growth commenced, the plants utilized the formed starch. The current year's twigs generally exhibited higher starch staining than the previous year's twigs in these species (Figs. 13 and 14), indicating a quick and efficient transfer of such accumulations to new shoots. Similar results were reported by Poudyal (2014) for *Castanopsis indica* in Phulchowki Hill, suggesting the rapid shift of starch accumulates to new shoots. Zobel (1996a, b) observed a correlation in starch content between 1-year-old and 2-year-old twigs of Pine and *Q. leucotrichophora.*

The decrease in starch staining after leaf emergence also indicated the hydrolysis of photosynthates into soluble carbohydrates. However, a substantial amount of starch remained present in zero-year twigs of *Q. lanata*, except in September. Larcher (1995) stated that about one-third of the reserve material is used during leaf and flower unfolding and fruit development. As flowering and fruiting occurred immediately after leaf flush in the studied species, the existing starch might have been utilized for these reproductive processes. *R. arboreum*, with its multiple leaf flushing behavior, exhibited inconsistent starch staining, where the reserve starch was completely hydrolyzed throughout the monsoon and winter period, except in June, August, and September, in zero-year twigs. This suggests that the reserve starch compensated for all phenological events, indicating continuous depletion that might contribute to osmotic regulation during these months (Poudyal 2004). During senescence, leaves became dry and damaged (necrosis and holes), decreasing leaf density and reducing SLM. SLM in *Q. lanata* did does not show a significant positive correlation with SM at 15 and 30 cm. The results of phenological observations indicate a strong seasonality in the phenological pattern of tree species in sub-tropical forest ecosystems. Abiotic factors such as rainfall periodicity, soil water availability, and variations in photoperiod directly influence plant phenology. Increased water stress during the early dry season generally results in a decline in water potential and induces leaf shedding (Borchert 1995). In *A. nepalensis*, concentrated leaf fall during March–April reduces transpirational water loss and helps rehydrate the tissue (Shrestha et al. 2006). New leaf production and expansion typically start from March onwards, as all phenological processes (leaf emergence, expansion, flowering, and fruiting) occur at the end of the dry season (Ralhan et al. 1985).*Q. lanata* leaves suffered significant damage from insect herbivory long before reaching the senescence stage. The leading cause of defoliation was insect larvae from the Chrysomelidae, Cryptocephalidae, and Scarabaeidae families (Lamichhaney 1995). As a result, the extended period of leaf production in *Q. lanata* may be linked to a reduction in leaf area caused by insect herbivory. Stressed and declining plants are assumed to be more susceptible to insects and diseases than vigorous plants. Therefore, both *Q. lanata* and *R. arboreum*, being evergreen trees and shrubs, are likely to face similar types of stress. Certain diseases and insects specifically affect damaged or stressed plants. In *Q. lanata*, leaf initiation, leaf senescence, and fruiting primarily occurred during the drier months of the year, from January to June. On the other hand, in *R. arboreum*, leaf initiation, leaf senescence, and flowering took place from January to April, with fruiting occurring from June to August (rainy season). Evergreen species shed their leaves throughout the year, but the shedding is relatively more pronounced during the summer (Ralhan et al. 1985). Leaf area, a crucial factor influencing photosynthesis and carbohydrate production, varied among species. In *Q. lanata*, the maximum leaf area was observed in February (average change of 17.33 mm^2) when leaf development peaked, but these leaves were heavily damaged by insect herbivory. The lowest leaf area value might be associated with newly developing leaves, as all the leaves were young in the studied trees during September (0.47 mm^2). Relatively high leaf area can enhance plant photosynthetic capacity and primary productivity. Poudyal et al. (2012) reported a rapid increase in leaf area with lower SLM after the rainy season in *Castanopsis indica* and *Schimawallichii.* Given the adequate soil moisture available then, they attributed this increase to development of more parenchymatous tissues rather than sclerophyllous tissues.

Similarly, the highest and lowest changes in mean leaf area were recorded in February (12.64 cm^2) and September (0.84 mm^2), respectively. This variation could be associated with the timing of leaf maturation, as all leaves were at their peak in January. Fully expanded leaves had a higher leaf area but showed a decline in aging leaves until senescence, indicating a lower rate of photosynthesis. In the study site, *Q. lanata* grew in sun-exposed areas, while *R. arboreum* inhabited slope environments with higher moisture levels and shade. The intensity of light interception by these species was likely due to their high leaf area, leading to a higher photosynthetic rate. Shade leaves have a larger leaf area but lower SLM, indicating smaller sclerenchyma and fewer vascular bundles. Previous studies by Bjorkman (1981), Lichtenthaler (1996), Cao and Booth (2001), and Avramov et al. (2007) reported a lower light saturation point for overall photosynthesis in shade leaves compared to leaves exposed to full sunlight. Furthermore, the variation in leaf shapes and sizes observed in the studied species may better reflect changes in environmental factors such as temperature, light intensity, and water status.

5 Conclusion

The study revealed that sub-tropical forest ecosystems experienced moderate plant stress during the phenologically active periods (dry periods of the year). Both studied species demonstrated various adjustments in turgor maintenance, ranging from temporary responses to low soil moisture to significant survival mechanisms in their natural habitat. Starch accumulation played a crucial role in turgor maintenance, with starch percentage accumulation being a notable factor. During growth and development, accumulated starch could be utilized by converting insoluble starch into soluble carbohydrates (glucose and fructose), potentially contributing to osmotic adjustment during drought conditions. Additionally, mature leaves exhibited adaptive characteristics such as higher specific leaf mass, enabling them to cope with water deficits more effectively than young leaves. *Q. lanata*, which thrived in exposed environments, responded to water stress primarily through drought avoidance mechanisms like leaf shedding, rolling, and partial defoliation. Despite experiencing extensive leaf loss, they demonstrated an adaptive ability to endure prolonged water stress while producing and sustaining leaves for an extended period through efficient tissue elasticity conservation, known as elastic adjustment. The change in tissue elasticity appeared to play a significant role in turgor maintenance for *Q. lanata*, as they suffered significant leaf damage before senescence. Similarly, the evergreen *R. arboretum* exhibited extreme responses to water stress, employing a combination of drought avoidance strategies such as late winter foliage shedding and drought resistance strategies, notably elastic adjustment, in the face of moderate or severe water stress. Elastic adjustment in *R. arboreum* allowed the plant to retain mature leaves and continue growth even during the driest conditions of the year.

6 Annexes

7 Annex 1: Mean Values of Different Parameters of Xylem Conductance; Huber Value (H_v), Specific Conductivity (K_s) and Leaf Specific Conductivity (K_l) During the Sampled months

	Months	Hv (10^{-4})	K_s (kgm^{-1} s^{-1} MPa^{-1}	K_l (10^{-4}) (kg m^{-1} s^{-1} MPa^{-1}
Quercus lanata	Feb	1.4	0.34	4.4
	March	2.1	0.10	1.97
	April	1.1	0.32	1.63
	June	1.3	0.18	1.98
	Aug	1.5	1.27	2
	Sept	1.8	0.84	1.3
	Nov	1.9	0.83	1.4
	Jan	1.7	0.45	4.4
Rhododendron arboreum	Feb	2.0	0.22	4.9
	March	1.8	0.57	9.79
	April	3.3	1.54	5.8
	June	1.9	1.80	1.9
	Aug	2.4	0.58	7.9
	Sept	3.6	0.45	2.9
	Nov	3.3	0.44	1.7
	Jan	2.9	0.30	8.5

8 Annex 2: Mean Values of Starch Percentage in Zero-Year Twig ($Starch_0$) and One year Twig ($Starch_1$) During the Sampled months

	Months	Starch content %	
Q. lanata		$Starch_0$	$Starch_1$
	Feb.	100	80
	March	70	30

(continued)

(continued)

	Months	Starch content %	
	April	30	10
	June	20	0
	Aug.	30	0
	Sept.	0	0
	Nov.	10	0
	Jan.	10	0
R. arboreum	Feb.	20	0
	March	20	10
	April	100	80
	June	10	0
	Aug.	10	0
	Sept.	10	0
	Nov.	30	10
	Jan.	30	0

9 Annex 3: Values of Parameters Obtained from P-V Curve Analysis; Relative Water Content at Zero Turgor (RWCz), Osmotic Potential at Full Turgor (Ψ_{sf}), Osmotic Potential at Zero Turgor (Ψ_{sz}), Index of Tissue Elasticity (ε) and Index of Desiccation Tolerance (IDT) During the Sampled Months

	Parameters	Feb.	March	April	June	Aug.	Sept	Nov.	Jan.
Quercus lanata	RWCz (%)	91.2	91.6	83.54	87.69	87.60	90.1	89.94	92.1
	Ψ_{sf} (MPa)	– 3.66	– 3.44	– 2.09	– 3.49	– 3.02	– 3.02	– 3.64	– 3.04
	Ψ_{sz} (MPa)	– 5.6	– 4.7	– 3.28	– 4.64	– 4.6	– 4.86	– 4.93	– 4.35
	ε (MPa)	5.4	4.35	1.21	2.94	2.55	2.93	4.79	4.47
	IDT	18.14	42.84	88.82	54.09	59.91	45.78	30.9	35.79
Rhododendron arboreum	RWCz (%)	92.54	86.02	86.06	83.16	86.964	88.31	91.05	90.05
	Ψ_{sf} (MPa)	– 2.67	3.57	– 2.59	– 2.88	– 2.33	– 2.16	– 3.17	– 2.30
	Ψ_{sz} (MPa)	– 3.77	– 5.8	– 3.33	– 4.05	– 3.58	– 3.06	– 4.4	– 3.5
	ε (MPa)	4.33	2.09	2.01	2.03	1.85	2.04	2.27	1.99
	IDT	22.78	53.50	68.17	71.21	67.68	68.01	48.01	63.40

References

Abrams MD (1990) Adaptations and responses to drought in *Quercus* species of North, America. Tree Physiol 7:227–238

Abrams MD, Kubiske ME (1990) Leaf structure characteristics of 31 hardwood and conifer tree species in Central Wisconsin: Influence of height regime and shade tolerance rank. Forest Ecol Manage 31:245–253

Bjorkman O (1981) Responses to different quantum flux densities. In: Lange OL, Nobel PS, Osmond CB, Ziegler H (eds) Physiological plant ecology. Encyclopedia of plant physiology, Springer-Verlag, Berlin, pp 57–101

Blake TJ, Tschaplinski TJ (1992) Water relations. In: Mitchell CP, Robertson JB, Hinckley TM, Sennerby-Forsse L (eds) Ecophysiology of short rotation forest crops. Elsevier Applied Science, Amsterdam, pp 66–94

Borchert R (1994a) Water status and development of tropical trees during seasonal drought. Trees 8:115–125

Borchert R (1994b) Induction of rehydration and bud break by irrigation or rain in deciduous trees in tropical forests. Biotropica 24:385–395

Borchert R (1994c) Soil and stem water storage determine phenology and distribution of tropical dry forest trees. Ecology 75:1437–1449

Borchert R (1995) Phenology and flowering periodicity of neotropical dry forest species: evidences from herbarium collections. J Trop Ecol 11:1–8

Borchert R (1998) Responses of tropical trees to rainfall seasonality and its long-term changes. Clim Change 39(2–3):381–393. https://doi.org/10.1023/A:1005383020063

Butz P (2019) Tree water use strategies in a neotropical dry forest. Doctoral dissertation. Universität Göttingen

Cao K, Booth EW (2001) Leaf anatomical structure and photosynthetic induction for seedlings of five dipterocarp species under contrasting light conditions in a Bornean Heath forest. J Trop Ecol 17:163–175

Cattivelli L, Rizza F, Badeck FW, Mazzucotelli E, Mastrangelo AM, Francia E, et al (2008) Drought tolerance improvement in crop plants: an integrated view from breeding to genomics. Field Crop Res 105:1–14

Cosgrove DJ (1988) In defense of the cell volumetric elastic modulus. Plant Cell Environ 11:67–69

DaMatta FM, Ramalho JDC (2006) Impacts of drought and temperature stress on coffee physiology and production: a review. Braz J Plant Physiol 18:55–81. https://doi.org/10.1590/S1677-042020 06000100006

Dhyani K, Kuniyal HB, Singh H, Sobha (2021) Growth and physiological potential of Terminalia arjuna under elevated CO_2 levels in open top chamber condition. J Appl Nat Sci 13(3):1121–1126. https://doi.org/10.31018/jans.v13i3.2490. ISSN 2231-5209

Ewers FW, Cruiziat P (1991) Measuring water transport and storage. In: Lassoie JP, Hinckley TM (eds) Technique and approaches in forest tree ecophysiology. CRC Press, Boca Raton, p 92

Farooq M, Hussain M, Wahid A, Siddique KHM, Aroca R (2012) Plant responses to drought stress: from morphological to molecular features

Galindo FG, Herppich W, Gekas V, Sjöholm I (2004) Factors affecting quality and postharvest properties of vegetables: Integration of water relations and metabolism. Crit Rev Food Sci Nutr 44(3):139–154. https://doi.org/10.1080/10408690490424649

Gibbs JW (1931) The collected works of J. Willard Gibbs, Vol I. Longmans, Green, New York, NY

Griffin KL (1994) Calorimetric estimates of construction cost and their use in ecological studies. Funct Ecol 8:551–562

Gupta SK, Singh H (2017) Observations on Sporadic Flowering in *Ficus benjamina* L. and *Peltophorum africanum* L. in New Forest. Indian Forester 143:290–291

Hare PD, Cress WA, Van Staden J (1998) Dissecting the roles of osmolyte accumulation during stress. Plant Cell Environ 21(6):535–553

Holbrook NM, Putz FE (1996) From epiphyte to tree: Differences in leaf structure and leaf water relation associated with the transition in growth form in eight species of hemiepiphytes. Plant, Cell Environ 19(6):631–642

Hsiao TC, Acevedo E, Fereres E, Henderson WD (1976) Water stress, growth and osmotic adjustment. Philos Trans R Soc London B 273:479–500

Jangra A, Kumar K, Maikhuri S, Bhandari MS, Pandey S, Singh H, Barthwal S (2024) Unveiling stress-adapted endophytic bacteria: Characterizing plant growth-promoting traits and assessing cross-inoculation effects on Populus deltoides under abiotic stress. Plant Physiol Biochem 210:108610. https://doi.org/10.1016/j.plaphy.2024.108610

Kanta C, Kumar A, Chauhan A, Singh H, Sharma IP (2024). The Interplay between plant functional traits and climate change. In: Kumar N, Singh H (eds) Plant Functional traits for improving productivity. Springer, Singapore. https://doi.org/10.1007/978-981-97-1510-7_3.

Kozlowski TT (1971) Growth and development of trees, vol II. Academic Press, New York, NY, p 514

Kramer K, Leinonen I, Loustau D (2000) The importance of phenology for the evaluation of impact of climate change on growth of boreal, temperate and Mediterranean forests ecosystems: an overview. Int J Biometeorol 44:67–75

Kramer PJ, Boyer JS (1995) Water relations of plants and soils. Academic Press, San Diego, p 404

Kumar A, Singh H, Kumari G, Bisht S, Malik A, Kumar A, Singh M, Raturi BS, Thakur A, Kaushal R (2022) Adaptive resilience of roadside trees to vehicular emissions via leaf enzymatic, physiological, and anatomical trait modulations. Environ Pollut 313:120191. https://doi.org/10.1016/j.envpol.2022.120191

Kumar M, Phukon SN, Paygude AC, Tyagi K, Singh H (2021) Mapping phenological functional types (PhFT) in the Indian Eastern Himalayas using machine learning algorithm in Google Earth Engine. Comput Geosci 158:104982. https://doi.org/10.1016/j.cageo.2021.104982

Kumar N, Singh H (2024). Plant functional traits for improving productivity. Springer, Singapore. https://doi.org/10.1007/978-981-97-1510-7

Kumar N, Jeena N, Kumar N, Gangola S, Singh H (2019e) Phytoremediation facilitating enzymes: an enzymatic approach for enhancing remediation process. In: Bhatt P (ed) Smart bioremediation technologies. Academic Press, Elsevier. 50 Hampshire Street, 5th Floor, Cambridge, MA 02139, United States, pp 289–306

Kumar N, Singh H, Sharma SK (2020) Antioxidants: responses and Importance in plant defense system. In: Roychowdhury R, Choudhury S, Hasanuzzaman M, Srivastava S (eds) Sustainable agriculture in the era of climate change. Springer, Cham, pp 251–264. https://doi.org/10.1007/978-3-030-45669-6_11

Kumar N, Singh H, Giri K, Kumar A, Joshi A, Yadav A, Singh R, Bisht S, Kumari R, Jeena N, Khairakpam R, Mishra G (2024) Physiological and molecular insights into the allelopathic effects on agroecosystems under changing environmental conditions. Physiol Mol Biol Plants 30:417–433. https://doi.org/10.1007/s12298-024-01440-x

Kumar N, Suyal DC, Sharma IP, Verma A, Singh H (2017) Elucidating stress proteins in rice (*Oryza sativa* L.) genotype under elevated temperature: a proteomic approach to understand heat stress response. 3 Biotech 7:205. https://doi.org/10.1007/s13205-017-0856-9

Lamichhaney BP (1995) Alnus nepalensis D. Don: a detailed study [Monograph no. 1]. Kathmandu, Nepal

Larcher W (1995) Physiological plant ecology, 3rd edn. Springer, New York, USA, p 448

Lichtenthaler HK (1996) Vegetation stress: An introduction to the stress concept in plants. J Plant Physiol 148:4–14

Lugg DG, Sinclair TG (1979) A survey of soybean cultivars for variability in specific leaf weight. Crop Sci 19:887–892

Malik A, Kumar K, Kumari G, Kumar N, Singh H (2024). Application of functional traits in modelling productivity and resilience under climate change. In: Kumar N, Singh H (eds) Plant functional traits for improving productivity. Springer, Singapore. https://doi.org/10.1007/978-981-97-1510-7_5

McEwen B, Sundqvist C, Younis S (1994) Protochlorophyll (ide) forms in hypocotyls of dark-grown bean (Phaseolus vulgaris). Physiol Plant 90(2):396–407. https://doi.org/10.1111/j.1399-3054.1994.tb00405.x

Meena D, Singh H, Chaudhari SK (2011) Elucidating strontium response on growth dynamics and biochemical change in *Phaseolus mungo* L. Int J Agric Environ Biotechnol 4:107–113

Morgan JM (1984) Osmoregulation and water stress in higher plants. Annu Rev Plant Physiol 35:299–319

Mulkey SS, Wright SJ, Smith AP (1991) Drought acclimation of an understory shrub (*Psychotria limonesis*; Rubiaceae) in a seasonally dry tropical forest in Panama. Am J Bot 78:579–587

Negi GCS, Singh SP (1992) Leaf growth pattern in evergreen and deciduous species of the central Himalaya, India. Int J Biometeorol 36:233–242

Nepstad DC, Carvalho CRde, Davidson EA, Jipp PH, Lefebvre PA, Negreiros GH, Trumbore SE, Vieira S (1994) The role of deep roots in the hydrological and carbon cycles of Amazonian forests and pastures. Natura 372:666–669

Pallardy SG, Pereira JS, Parker WC (1991) Measuring the state of water in tree systems. In: Lassoie JP, Hinckley TM (eds) Techniques and approaches in forest tree ecophysiology. CRC Press, Boca Raton, pp 27–76

Parker WC, Pallardy SG (1987) The influence of resaturation method and tissue type on pressure-volume analysis of *Quercus alba* L. seedlings. J Exp Botany 38(188):535–549

Poudyal K (2004) Plant water relations of some selected Himalayan Tree Species at Phulchowki Hill, Kathmandu. Ph.D. thesis, Tribhuvan University, 167 p

Poudyal K (2013) Plant water relations, phenology and drought adaptation strategy of Himalayan Oak; *Quercus lanata* in Phulchowki Hill, Nepal. Sci Secure J Biotechnol 2(4):99–110

Poudyal K (2014) Plant water relations and phenological shifts in response to drought in *Castanopsis indica* at Phulchowki hill, Kathmandu, Nepal. J Biol Sci Opin 2(1):70–76

Poudyal K, Jha PK, Zobel DB (2003) Seasonal changes in hydraulic conductance of Himalayan trees in response to drought. Ecoprint 10:59–73

Poudyal K, Jha PK, Zobel DB (2012) Role of wood water properties and leaf dynamics in phenology and response to drought in evergreen Himalayan tree species. Ecoprint 19:71–84

Poudyal K, Jha PK, Zobel DB, Thapa CB (2004) Patterns of leaf conductance and water potential of five Himalayan tree species. Trec Physiol 24:689–699

Prakash V, Hunney K, Singh H (2022) Effect of elevated carbon dioxide on growth and development of *Santalum album* L. seedlings inoculated with plant growth promoting microorganisms in Open Top Chambers. Int J Environ Agric Biotechnol 7(5):149–159

Ralhan PK, Khanna RK, Singh SP, Singh JS (1985) Phenological characteristics of the tree layer of Kumaun Himalaya Forests. Vegetation 60:91–101

Shrestha BB, Uprety Y, Jha PK (2006) Wood properties in relation to foliar phenology of some planted tree species at Kirtipur, central Nepal. Trop Ecol 47(2):201–209

Shrestha BB, Uprety Y, Nepal K, Tripathi S, Jha PK (2007) Phenology and water relations of eight woody species in theCoronation Garden of Kirtipur, central Nepal. Himal J Sci 4(6):49–56

Singh H (2024b). Exploring adaptive modulation in plant functional traits and its impact on the productivity of Acacia auriculiformis under CO_2-enriched environment. Ind Crops Prod 210:118186. https://doi.org/10.1016/j.indcrop.2024.118186

Singh H, Kumar M (2022) Climate change and its impact on Indian Himalayan forests: current status and research needs. In: Rani S, Kumar R (eds) Climate change. Springer climate. Springer, Cham, pp 223–242. https://doi.org/10.1007/978-3-030-92782-0_11

Singh H, Bandyopadhyaya S, Kumar A, Singh M, Malik A, Kumari G, Raturi A, Yadav SK, Kumar M, Barthwal S, Thakur A, Rawat PS (2022) Understanding the physiological and biophysical response of urban roadside plantations for assessing adaptation and mitigation mechanisms toward vehicular emissions. Urban Climate 44:101183. https://doi.org/10.1016/j.uclim.2022.101183

Singh H, Kumar N, Kumar A (2024c). Enhancing resource use efficiency in crops through plant functional traits. In: Kumar N, Singh H (eds) Plant functional traits for improving productivity. Springer, Singapore. https://doi.org/10.1007/978-981-97-1510-7_6

Singh H, Savita Sharma R, Sinha S, Kumar M, Kumar P, Verma A, Sharma SK (2017a). Physiological functioning of *Lagerstroemia speciosa* L. under heavy roadside traffic: an approach to screen potential species for abatement of urban air pollution. 3 Biotech 7(61):1–10. https://doi.org/10.1007/s13205-017-0690-0

Singh H, Verma A, Kumar M, Sharma R, Gupta R, Kaur M, Negi M, Sharma SK (2017b) Phytoremediation: a green technology to clean up the sites with low and moderate level of heavy metals. Austin Biochem 2:1012

Singh M, Singh H, Kumar A, Kumar M, Barthwal S, Thakur A (2024b) Soil nitrogen availability determines the CO_2 fertilization effect on tree species (*Neolamarckia cadamba*): growth and physiological evidence. Environ Sustain 7:53–60. https://doi.org/10.1007/s42398-023-00300-w

Singh H, Verma A (2013) Characterization and screening of high nitrogen efficient rice genotype to curtail environmental pollution. Int J Agric Environ Biotechnol 6(1):23–30

Singh JS, Singh SP (1987) Forest vegetation of the Himalaya. Bot Rev 53:80–192

Singh J, Yadav RR, Wilmking M (2009) A 694-year tree-ring based rainfall reconstruction from Himachal Pradesh, India. Clim Dyn 33:1149–1158. https://doi.org/10.1007/s00382-009-0528-5

Singh SP, Tewari A, Singh SK, Pathak GC (2000) Significance of phonologically asynchronous populations of the central Himalayan oaks in drought adaptation. Curr Sci 79:353–357

Tyree MT, Ewers FW (1991) The hydraulic architecture of trees and other woody plants. New Phytologist 119:345–360

Tyree MT, Ewers FW (1996) Hydraulic architecture of woody tropical plants. In: Mulkhey SS, Chazdon RL, Smith APS (eds) Tropical Forest Plants. Chapman and Hall, New York, pp 217–243

Tyree MT, Jarvis PG (1982) Water in tissues and cells. In: Lange OL, Nobel PS, Osmond CB, Ziegler H (eds) Encyclopaedia of plant physiology. New series, vol 12b, Springer, Berlin, pp 35–77

Yadav SK, Singh H, Ginwal HS and Barthwal S (2019). Elevated CO_2 enhanced growth and physiological process of *Populus deltoides* Bartr. ex Marsh. Indian Forester 145(1):23–27

Zobel DB (1974) Local variation in integrating *Abies grandis*—A concolor populations in the central Oregon cascades-II: stomatal reaction to moisture stress. Bot Gaz 135:200–210

Zobel DB (2000) Laboratory manual for plant water relations. Oregon State University, Corvallis

Zobel DB (1996a) Variation of water relations parameters with extended rehydration time, leaf form, season and proportion of leaf. Can J Forest Res 26:175–185

Zobel DB (1996b) Using twig starch levels as an index of tree stress in Himalayan forest. In: Jha PK (ed) Environment and biodiversity: in the context of South Asia

Zobel DB, Singh SP (1995) Tree water relations along the vegetational gradient in the Himalayas. Curr Sci 68:742–745

Zobel DB, Garkoti SC, Singh SP (2001) Leaf conductance of primary and mature leaves of *Pinus roxburghii*: a comparison. J for Res 6:1–5

Zobel DB, Ram J, Bargali SS (1995) Structural and physiological changes in *Quercus leucotrichophora* and *Pinus roxburghii* associated with stand disturbance in a Kumaun Himalaya, India. Int J Ecol Environ Sci 21:45–66

Regulation of Ambient Air Temperature Through Trees: A Comparison Between Industrial and Rural Belt in the State of West Bengal, India

Sana Ahmed, Prosenjit Pramanick, Sufia Zaman, and Abhijit Mitra

Abstract Industrial regions frequently exhibit higher mean air temperature compared to surrounding rural areas. This is often called Urban Heat Island (UHI) of cities and industrial zones. In this research, we present the evidence that in the state of West Bengal (India), the industrial zone like Haldia (22° 06′31.9″ N; 88° 10′19.2″ E) is experiencing higher air temperature compared to the rural area like Jhargram (22° 23′09.7″ N; 86° 52′04.6″ E) after analyzing a long-term data of more than three decades (1990–2022) through three seasons. This research also draws scientific evidence to outline the significant role of trees in regulating the ambient air temperature by using a few relevant tree-based proxies like Population Density (PD), Shannon Weiner Species Diversity Index ($\overline{\mathrm{H}}$), Above Ground Biomass (AGB), and Above Ground Carbon (AGC) estimated for both the selected sites. The ambient air temperature increment at the Haldia industrial belt was 0.19 °C year^{-1}, while for Jhargram, the increment was 0.12 °C year^{-1}. The proxies used to score the potential of the dominant trees (common to both the sites) to regulate the near-surface air temperature demonstrated much better performance and values at Jhargram, the rural belt, compared to the port-cum- industrial complex at Haldia in the state of West Bengal.

Keywords Air temperature · Urban heat island (UHI) · Shannon Weiner Species Diversity Index ($\overline{\mathrm{H}}$) · Above ground biomass (AGB) · Above ground carbon (AGC)

S. Ahmed · P. Pramanick · S. Zaman
Department of Oceanography, Techno India University, West Bengal, EM 4 Salt Lake, Sector V, Kolkata 700091, India

A. Mitra (✉)
Department of Marine Science, University of Calcutta, 35 B.C. Road, Kolkata 700019, India
e-mail: abhijit_mitra@hotmail.com

H. Singh (ed.), *Urban Forests, Climate Change and Environmental Pollution*,
https://doi.org/10.1007/978-3-031-67837-0_36

1 Introduction

The mean global air temperature has increased by 1 °C than witnessed during the pre-industrial times. By 2050, there is a probability that this value will increase by 1.5 °C (IPCC 2021; Yadav et al. 2019a, b; Jangra et al. 2024; Kanta et al. 2024). The rise of air temperature is associated chiefly with unplanned urban expansion, deforestation, and industrial development and is keenly related to consequences like global warming. This may lead to erratic rainfall, more frequent, severe, and intense extreme heat events (Meehl and Tebaldi 2004), and many other adverse impacts on the local scale of weather and climate. The environment in and around the industrial zones and cities is also predicted to become warmer in the coming years (Esperon-Rodriguez et al. 2022). Urban dwellers may face several health issues related to heat waves (Tuholske et al. 2021; Malik et al. 2024; Kumar and Singh 2024). Therefore, an in-depth study is needed to reduce heat energy generated in and around the industrial zones and to provide health security to the population exposed to such an environment of rising temperature. Industrial areas are mosaics of factories (all small, medium, and large scale), buildings, streets, and different types of impervious surfaces, along with infrastructures of various categories (Elmqvist et al. 2013; Meena et al. 2011; Kumar et al. 2019c, 2020; Singh and Kumar 2022; Singh et al. 2024). Impervious surfaces (i.e., man-made surfaces such as concrete, asphalt, and buildings) cover a considerable area in the industrial region, although this coverage varies considerably across regions (Fuller and Gaston 2009). Industrial and urban areas also differ in their radiative, thermal, aerodynamic, and moisture properties relative to surrounding peri-urban and natural areas, absorbing heat and re-radiating longwave radiation within the urban matrix, resulting in higher air and surface temperatures (Oke 1982; Manoli et al. 2019). Therefore, temperatures in industrial and urban areas are relatively higher compared to the rural sectors, leading to the formation of Urban Heat Island (UHI) (Zhao et al. 2014). Burning fossil fuels is a common scenario in the industrial regions, not only from the factories or production units but also from vehicles, offices, and several miscellaneous sources (Fig. 1).

Several strategies to mitigate carbon dioxide levels in the atmosphere and the subsequent rise in temperature have been investigated in different corners of the world. Nature-based solutions like roof gardening, planting trees with high carbon storage potential, expansion of water bodies, etc., are sustainable and cost-effective (Mutani and Todeschi 2020; Ossola and Lin 2021). Trees have the unique potential to reduce local air temperatures (Akbari et al. 2001; Bowler et al. 2010) primarily by way of storing carbon in their biomass (Mitra et al. 2012, 2015, 2016, 2018, 2021a, 2022; Banerjee et al. 2017; Agarwal et al. 2017a, b, b; Ahmed et al. 2021; Dutta et al. 2022; Khokher et al. 2023). Trees cool the surrounding area directly by blocking solar irradiation and transpiring water into the atmosphere (Winbourne et al. 2020).

The temperature in and around the industrial zone can be effectively regulated through alteration of the quantum of heat energy absorbed, stored, and transferred and by adopting eco-friendly and cost-effective nature-based solutions (Singh and Verma 2013; Kumar et al. 2017; Gupta and Singh 2017; Singh et al. 2017a, b).

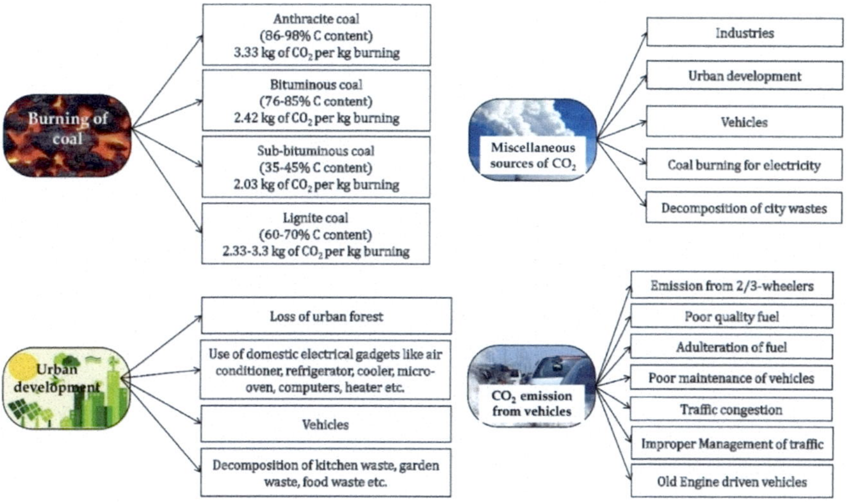

Fig. 1 Sources of carbon dioxide in the atmosphere leading to the rise of temperature in and around the industrial zone

The producer communities of the region offer considerable cooling effects through several mechanisms (Table 1).

Global concern and increasing awareness about climate change and subsequent rise in atmospheric temperature have augmented interest in trees to assist in minimizing the levels of carbon dioxide (Dwyer et al. 1992). One of the most cost-effective and eco-friendly modes of stabilizing the air temperature in industrial and urban areas is through urban forestry, which is keenly associated with the ambient air's carbon dioxide level (Sharma et al. 2018; Singh et al. 2018; Gupta et al. 2018a, 2018b).

Urban forestry includes the trees that grow in and around the houses and other buildings wherein the land use is urban. The trees in such urban areas may either develop from saplings planted after construction or originate as trees spared from the forested area. Trees belonging to older residential neighborhoods usually have extensive tree canopies and comparatively higher density than trees in younger neighborhoods (Singh and Verma 2013; Kumar et al. 2017; Gupta and Singh 2017; Singh et al. 2017a, b).

Urban forestry attains carbon offsets through two critical means. Firstly, they sequester atmospheric carbon into their biomass, and secondly, they offset some of the seasonal energy used for cooling and heating the interior spaces of buildings. Moreover, massive and well-placed trees contribute to a palpable cooling influence through their shades (Yadav et al. 2019a, b; Gupta et al. 2019a, 2019b; Kumar et al. 2019a, b). Transpiration also helps cool ambient air. Trees help reduce heating needs in the winter season by decreasing the wind speeds near buildings.

Trees are the most fundamental determinants that remove carbon from the atmosphere by storing in their biomass for a considerable time. The carbon storage within biomass of trees and soil can be used to mitigate climate change, especially in urban

Table 1 Mechanism of nature-based solution to reduce ambient temperature

S. No.	Mechanism	Description
1	Evaporative cooling and Evapotranspiration	1. In evapotranspiration, the incoming solar radiation converts water into water vapour; thus, solar energy is utilized to drive the evaporation process rather than transfer it into the heat that we feel. This creates a cooling effect in and around the trees 2. In evapotranspiration, water is vaporized from the plant and soil surface. In the industrial zone and thickly populated urban areas, vegetation cover is less, and the soil is mainly covered with concrete, reducing the cooling effect. Hence, the temperature rises
2	Reflectance	In industrial and urban areas, the reflectance is less compared to urban areas. Hence, the reflectance or albedo is less, with more absorption of solar energy that dissipates heat from the surrounding atmosphere
3	Shading	Trees with crowns hinder the incoming solar radiation and create a cooling effect. However, the temperature fall around the tree depends on the shape and dimension of the crown. Dense vegetation is a more effective barrier to prevent incoming solar radiation
4	Carbon storage/ sequestration	Trees absorb carbon dioxide from the atmosphere for photosynthesis and synthesize glucose, leading to carbon storage in the form of starch. Thus, trees act as a natural sink of carbon, which otherwise could increase the surrounding air temperature by creating a carbon dioxide blanket in the atmosphere Considering the potential of trees, we also analyzed the effectiveness of dominant trees in the two regions (industrial and rural) using a few variables like population density, diversity index, and above-ground biomass of the trees common to both areas

settings, and these urban carbon reservoirs can confront several environmental issues (Schimel 1995; Nowak and Crane 2002; Pataki et al. 2006; Grimm et al. 2008).

With this background, the present study also looked into the eco-friendly approach of stabilizing the air temperature and checking the formation of UHI. For this, common trees in both the selected zones were identified. Their efficacy in mitigating the rise of temperature was evaluated through a few relevant proxies like Population Density (PD), Shannon Weiner Species diversity Index ($\overline{H}$), Above Ground Biomass (AGB), and Above Ground Carbon (AGC), which were monitored during 2022, the final year of the present study.

2 Materials and Methods

The study undertaken from 1990 to 2022 can be divided into four phases.

(i) **Study Site**

West Bengal is a maritime state in the northeast sector of the Indian subcontinent. The state has mosaic landscapes like hilly terrain, seas and estuaries, fertile plain land, industrial zones, forests in the rural sector, etc. In the present study, we selected the industrial zone of Haldia (22° 06′ 31.9″ N; 88° 10′ 19.2″ E) and the rural area of Jhargram (22° 23′ 09.7″ N; 86° 52′ 04.6″ E), which is covered mainly by forest vegetation to scan the impact of trees to regulate the ambient air temperature.

(ii) **Estimation of Surface Air Temperature**

Ambient air temperature data were sourced from Climatological Tables of Observatories in India, 9th edition (2022); Data compilation and processing were carried out by the National Data Centre, Office of the Head Climate Research and Services, Pune (India) (https://imdpune.gov.in/library/public/Climate%20of%20WestBengal.pdf). Primary data collection from the sampling sites could not be carried out due to a lack of funding to purchase and install instruments in non-canopy, exposed conditions across the study area, and concern for the risk of vandalism for sensors that are usually visible to the people. A gap in this approach lies in interpreting the canopy cooling services, but considering our constraints, we relied on secondary data sources.

(iii) **Estimation of Tree—Related Parameters**

The sixteen most abundant tree species common to both the sites of Haldia and Jhargram were monitored during the final phase of our survey period in 2022. Population density (PD) of the sixteen tree species was recorded from 10 plots of 10 m × 10 m dimension, and finally Shannon Wiener Species Diversity Index was calculated as per the following expression:

$$\overline{H} = -\sum \frac{n}{N} \log_e \frac{n}{N}$$

where, $\overline{\mathrm{H}}$ = Shannon Wiener Species Diversity Index, n = No. of individuals per species, and N = Total number of individuals of all species.

Above Ground Biomass (AGB) and Above Ground Carbon (AGC) of each of the tree species were analyzed during 2022 as per the procedure of Mitra and Sundaresan (2016) stated in the manual published by CSIR-National Institute of Science Communication And Information Resources (NISCAIR).

(iv) **Statistical Analysis**

To assess whether the abiotic (Air temperature) and biotic (PD, $\overline{\mathrm{H}}$, AGB, and AGC) parameters varied significantly amongst sites and seasons, Analysis of Variance (ANOVA) was performed using SPSS 9.0 for Windows.

3 Results

The air temperature increased in both the industrial zone of Haldia and the rural region of Jhargram irrespective of seasons (Figs. 2 and 3). The average value (of three seasons) is, however, more in Haldia (32.69 °C) compared to Jhargram (31.49 °C).The seasonal trend followed the order premonsoon > monsoon > postmonsoon in both the industrial and rural belts.

We could identify 16 dominant tree species common to both the Haldia industrial and Jhargram rural belts (Fig. 4). The PD, $\overline{\text{H}}$, AGB, and AGC for all the species in

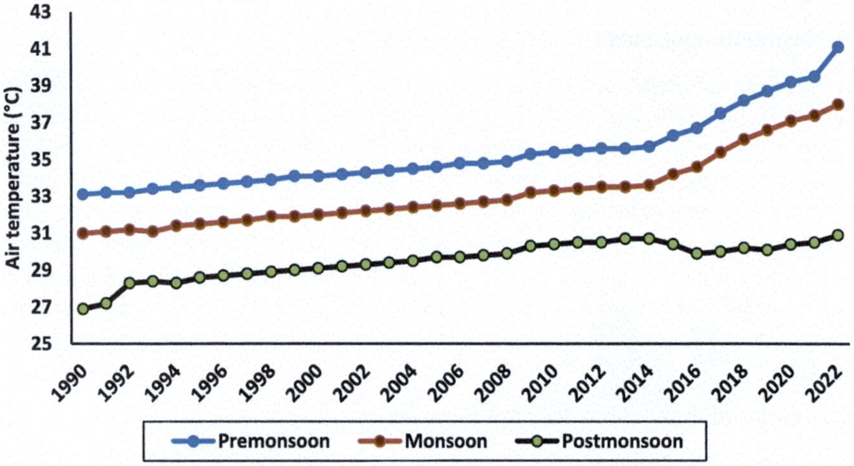

Fig. 2 Temporal variation of air temperature in the industrial zone of Haldia, West Bengal

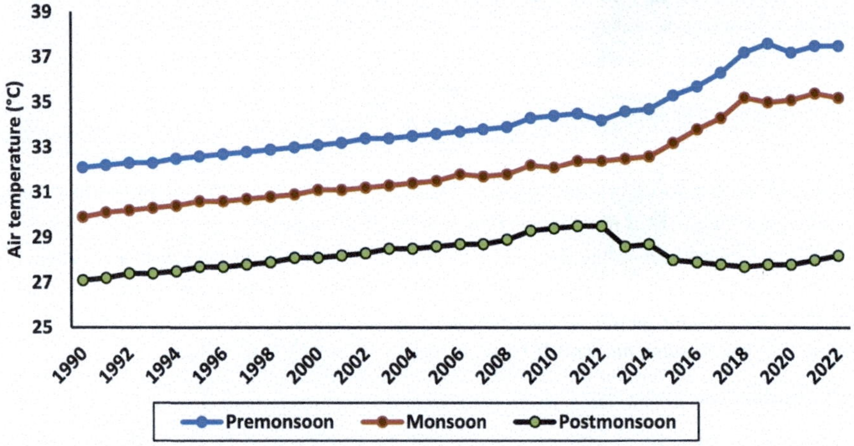

Fig. 3 Temporal variation of air temperature in the rural area of Jhargram, West Bengal

both sites are presented in Table 2. All the biotic parameters are relatively higher in the rural belt of Jhargram, which may be the driver of lowering the temperature in this site compared to the Haldia industrial belt. The ambient air temperature increment at the Haldia industrial belt is 0.19 °C $year^{-1}$, while for Jhargram, the increment is 0.12 °C $year^{-1}$. The chain of industries in the Haldia port-cum-industrial complex, coupled with port activities, may be the possible reason for the higher ambient air temperature. In Jhargram, the vegetation cover and the proxies associated with the tree species in this study scored much better, which may be the factor behind the lower air temperature. The rural area of Jhargram is mainly dominated by the tribal population with minimum industrial activities.

Fig. 4 Common dominant tree species in both study sites: 1. *Cocos nucifera* 2. *Ficus religiosa* 3. *Albizia saman* 4. *Azadirachta indica* 5. *Mangifera indica* 6. *Tamarindus indica* 7. *Bombax ceiba* 8. *Aegle marmelos* 9. *Terminalia arjuna* 10. *Tectona grandis* 11. *Delonix regia* 12. *Artocarpus heterophyllus* 13. *Swietenia mahagoni* 14. *Terminalia catappa* 15. *Psidium guajava* 16. *Acacia auriculiformis*

Table 2 Site-wise parameters monitored during 2022 for the common tree species

Tree species	PD (No./100 m^2)		AGB (tonnes/ha)		AGC (tonnes/ha)	
	Haldia	Jhargram	Haldia	Jhargram	Haldia	Jhargram
Cocos nucifera	17	11	56.24	71.38	27.05	34.48
Ficus religiosa	3	9	103.03	210.56	49.45	101.49
Albizia saman	5	8	49.55	58.72	23.44	27.89
Azadirachta indica	1	6	98.75	228.56	45.23	105.37
Mangifera indica	5	9	56.85	70.20	27.17	33.63
Tamarindus indica	3	8	33.08	69.44	15.12	31.87
Bombax ceiba	1	7	28.39	86.91	13.09	40.33
Aegle marmelos	2	7	98.23	154.65	46.07	72.69
Terminalia arjuna	4	7	46.05	98.37	21.69	46.43
Tectona grandis	3	8	63.28	221.52	30.18	105.89
Delonix regia	5	8	70.29	118.55	32.90	55.96
Artocarpus heterophyllus	1	6	54.56	79.38	26.46	38.82
Swietenia mahagoni	1	5	44.03	88.73	20.96	42.59
Terminalia catappa	1	8	18.52	57.19	8.67	26.94
Psidium guajava	3	8	71.11	113.46	33.71	53.89
Acacia auriculiformis	7	11	66.85	150.98	30.55	69.15
$\overline{H}$	2.42	2.75	Total = 958.81	Total = 1878.60	Total = 28.23	Total = 55.46

4 Discussion

Trees have been observed to buffer the adverse consequences of climate change effectively, preferably the warming and alteration of carbon dioxide levels below the canopy (Bonan 2008; Frey et al. 2016; de Frenne et al. 2019; Zellweger et al. 2020) through creation of a distinct microclimate, which is decoupled from their surrounding environment with substantial deviations between ambient air temperature and temperatures within and below the canopy (Rahman et al. 2017a; b; Davis et al. 2019; de Frenne et al. 2019; Ehbrecht et al. 2019; Richter et al. 2021).

The relatively cooler microclimate beneath the canopy is formed through two fundamental processes, namely by evapotranspiration, where water is converted to water vapor by extracting energy from the local environment (Smithers et al. 2018) and by absorbing and reflecting the incoming shortwave radiation at the top canopy thus preventing the shaded canopy parts from warming (Rahman et al. 2017a). The cooling potential of the forest is a function of the tree density (more preferably population density), diversity, and the AGB, as these biotic variables regulate the process of evapotranspiration and play a significant role in the albedo effect. The present study reflects 13.64% more value of $\overline{\mathrm{H}}$ in the Jhargram rural belt compared to the Haldia Industrial sector. In addition, the AGB and AGC values were also higher by 95.93% and 96.44%, respectively, in Jhargram compared to Haldia, which might be a reason for the lower temperature in the former site, as revealed from our time series analysis of more than three decades. The PD of the tree species in the Haldia Industrial zone is also strikingly lower than the rural zone of Jhargram (Table 2). All these biotic parameters, coupled with high population density and congested buildings, have triggered the formation of UHI in Haldia.

ANOVA computed with the ambient air temperature data sets exhibit significant variations amongst the two sites ($p < 0.01$), irrespective of seasons (Tables 3, 4, and 5). The producer communities seem to be the primary actors for such variation.

Thus, plantation, maintenance of tree species with relatively larger biomass, and management of edaphic factors to promote more diversity and density of the tree species may serve as an effective route to bypass the formation of UHI in the lightly urbanized, thickly populated industrial cities and towns.

Table 3 ANOVA of air temperature during premonsoon

Source of variation	SS	df	MS	F	*P*-value	F crit
Between years	221.3827	32	6.91821	54.4181	2.94E−20	1.804482
Between sites	22.34182	1	22.34182	175.739	1.52E−14	4.149097
Error	4.068182	32	0.127131			
Total	247.7927	65				

Table 4 ANOVA of air temperature during monsoon

Source of variation	SS	df	MS	F	*P*-value	F crit
Between years	209.5894	32	6.549669	78.78403	9.33E−23	1.804482
Between sites	20.8547	1	20.8547	250.855	1.07E−16	4.149097
Error	2.660303	32	0.083134			
Total	233.1044	65				

Table 5 ANOVA of air temperature during postmonsoon

Source of variation	SS	df	MS	F	*P*-value	F crit
Between years	36.24455	32	1.132642	4.185136	5.57E−05	1.804482
Between sites	28.9347	1	28.9347	106.9143	9.94E−12	4.149097
Error	8.660303	32	0.270634			
Total	73.83955	65				

5 Conclusion

The rapid pace of industrial and associated urban expansion causes climate warming relative to the surrounding rural belts, creating a localized high-temperature zone, commonly called Urban Heat Island (UHI). The present study draws scientific evidence to outline the formation of UHI in the Haldia industrial zone of West Bengal by retrieving air temperature data of more than three decades (1990–2022) for two sites, namely the Haldia industrial zone and Jhargram rural belt. The role of trees in combating the rise of temperature has been confirmed through detailed analysis of the population density (PD), diversity index ($\overline{H}$), above-ground biomass (AGB), and above-ground carbon (AGC) content of the most abundant and common trees between these two sites. To sum up, it can be advocated that the selection of tree species based on their AGB and AGC values should be part of the plantation program to combat the rise of temperature due to the increase of anthropogenic activities in the present *era*.

References

Agarwal S, Banerjee K, Pal N, Mallik K, Bal G, Pramanick P, Mitra A (2017a) Carbon sequestration by mangrove vegetations: a case study from Mahanadi mangrove wetland. J Environ Sci Comp Sci Eng Technol 7(1):16–29

Agarwal S, Pal N, Zaman S, Mitra A (2017b) Role of mangroves in carbon sequestration: a case study from Prentice island of Indian Sundarbans. Int J Basic Appl Res 7(12):35–42

Agarwal S, Chakrabarty SP, Zaman S, Pramanick P, Biswas P, Mukhopadhyay N, Dutta J, Mitra A (2021a) Assessment of carbon storage potential of dominant tree species in an Educational Campus, Kolkata, West Bengal. Ela J for Wild 10(2):869–878

Agarwal S, Mitra A, Pramanick P, Mitra A (2021b) Stored carbon in urban trees: ground zero observation from the Konnagar area of West Bengal, India. In: Filho WL, Luetz JM, Ayal DY (eds) Handbook of climate change management (research, leadership, transformation). Published by Springer Nature, pp 1–23, eBook ISBN 978-3-030-22759-3. https://doi.org/10.1007/978-3-030-22759-3_284-1

Ahmed S, Trivedi S, Pal N, Pramanick P, Zaman S, Mitra A (2022) Study on floral carbon storage potential in the megacity of Kolkata: a roadmap towards net-zero carbon emission. J Environ Sociobiol 19(2):113–123

Ahmed S, Zaman S, Mitra A (2021) Carbon sequestration in a common true mangrove flora, *Excoecaria agallocha* of Indian Sundarbans. Chapter 7, In: Mitra A, Calma MM, Chakrabarty SP (eds) Natural resources and their ecosystem services, vol II. Published by HSRA Publication, pp 138–143, ISBN 978-93-5506-064-8

Akbari H, Pomerantz M, Taha H (2001) Cool surfaces and shade trees to reduce energy use and improve air quality in urban areas. Sol Energy 70:295–310

Banerjee K, Bal G, Pal N, Amin G, Mitra A (2017) Carbon sequestration by Mangrove vegetation: a case study from Odisha Bhitarkanika Wild Life Sanctuary (BWLS). Par J Sci Edu 3(4):1–9

Banerjee K, Khemendu GR, Paul R, Mitra A (2018) Carbon storage potential in dominant trees of Koraput District of Odisha. In: Impact of climate change on marine environments, Published by Springer Nature Singapore Pte Ltd., Singh VP et al (eds) Energy and environment, vol 80, Water Science and Technology Library, pp 235–243. https://doi.org/10.1007/978-981-10-5798-4_22

Bonan GB (2008) Forests and climate change: forcings, feedbacks, and the climate benefits of forests. Sci 320:1444–1449

Bowler DE, Buyung-Ali L, Knight TM, Pullin AS (2010) Urban greening to cool towns and cities: a systematic review of theempirical evidence. Landsc Urban Plan 97:147–155

Davis KT, Dobrowski SZ, Holden ZA, Higuera PE, Abatzoglou JT (2019) Microclimatic buffering in forests of the future: the role of local water balance. Ecography 42:1–11

de Frenne P, Zellweger F, Rodríguez-Sánchez F, Scheffers BR, Hylander K, Luoto M, Vellend M, Verheyen K, Lenoir J (2019) Global buffering of temperatures under forest canopies. Nat Ecol Evol 3:744–749

Dutta J, Mitra A, Zaman S, Mitra A (2022) Stored carbon in the mangrove vegetation of Lothian Wild Life Sanctuary of Indian Sundarbans, the designated World Heritage Site. Indian for 148(7):733–741. https://doi.org/10.36808/if/2022/v148i7/151602

Dwyer JF, Mcpherson E, Gregory S, Herbert W, Rowan AR (1992) Assessing the benefits and costs of the urban forest. J Arboric 18:227–234

Ehbrecht M, Schall P, Ammer C, Fischer M, Seidel D (2019) Effects of structural heterogeneity on the diurnal temperature rangein temperate forest ecosystems. For Ecol Manag 432:860–867

Elmqvist T, Fragkias M, Goodness J, Güneralp B, Marcotullio PJ, McDonald RI, Parnell S, Schewenius M, Sendstad M, Seto KC, Wilkinson C (eds) (2013) Urbanization, biodiversity and ecosystem services: challenges and opportunities. Springer, London, UK

Esperon-Rodriguez M, Tjoelker MG, Lenoir J, Baumgartner JB, Beaumont LJ, Nipperess DA, Power SA, Richard B, Rymer PD, Gallagher RV (2022) Climate change increases global risk to urban forests. Nat Clim Chang 12:950–955

Frey SJK, Hadley AS, Johnson SL, Schulze M, Jones JA, Betts MG (2016) Spatial models reveal the microclimatic buffering capacity of old-growth forests. Sci Adv 2:e1501392

Fuller RA, Gaston KJ (2009) The scaling of greenspace coverage in European cities. Biol Lett 5:352–355

Gupta SK, Singh H (2017) Observations on sporadic flowering in *Ficus benjamina* L. and *Peltophorum africanum* L. in new forest. Indian Forester 143:290–291

Gupta SK, Ram J, Singh H (2018a) Comparative study of transpiration in cooling effect of tree species in the atmosphere. J Geosci Environ Prot 6:151–166. https://doi.org/10.4236/gep.2018.68011

Gupta SK, Ram J, Singh H, Kumar P (2018b) Comparative study of cooling under shades of some forestry tree species with respect to ambient temperatures. J Geosci Environ Prot 6:51–66. https://doi.org/10.4236/gep.2018.611004

Gupta SK, Ram J, Singh H (2019a) Relationship between leaf area index and atmospheric cooling of tree species. Indian J for 42(3):203–208

Gupta SK, Singh H, Ram J (2019b) Need for a novel paradigm in forestry research: perspectives of comparative atmospheric cooling by different forestry effect tree species. Indian Forester 145(6):509–515

Grimm NB, Faeth SH, Golubiewski NE, Redman CL, Wu J, Bai X (2008) Global change and the ecology of cities. Sci 319:756

https://imdpune.gov.in/library/public/Climate%20of%20WestBengal.pdf

IPCC Climate Change (2021) The Physical Science Basis. Contribution of Working Group I to the Sixth Assessment Report of the Intergovernmental Panel on Climate Change; Cambridge University Press, Cambridge, UK

Jangra A, Kumar K, Maikhuri S, Bhandari MS, Pandey S, Singh H, Barthwal S (2024) Unveiling stress-adapted endophytic bacteria: characterizing plant growth-promoting traits and assessing cross-inoculation effects on *Populus deltoides* under abiotic stress. Plant Physiol Biochem 210:108610. https://doi.org/10.1016/j.plaphy.2024.108610

Kanta C, Kumar A, Chauhan A, Singh H, Sharma IP (2024) The interplay between plant functional traits and climate change. In: Kumar N, Singh H (eds) Plant functional traits for improving productivity. Springer, Singapore. https://doi.org/10.1007/978-981-97-1510-7_3

Khokher FK, Khan AU,Bargali HS, Ahmed S,Gobato R, Zaman S, Mitra A (2023) Stored carbon in dominant mangrove species in Indian Sundarbans: A situation analysis with two different methods. Par J Sci Edu 9(1):1–9. tiny.cc/PJSE24476153v9i1p001-009

Kumar N, Suyal DC, Sharma IP, Verma A, Singh H (2017) Elucidating stress proteins in rice (*Oryza sativa* L.) genotype under elevated temperature: a proteomic approach to understand heat stress response. 3 Biotech 7:205. https://doi.org/10.1007/s13205-017-0856-9

Kumar M, Singh MP, Singh H, Dhakate PM, Ravindranath NH (2019a) Forest working plan for the sustainable management of forest and biodiversity in India. J Sustain for 39:1–22. https://doi.org/10.1080/10549811.2019.1632212

Kumar N, Jeena N, Singh H (2019b) Elevated temperature modulates rice pollen structure: a study from the foothill of the Himalayan agro-ecosystem in India. 3 Biotech 9(5):175. https://doi.org/10.1007/s13205-019-1700-1

Kumar N, Jeena N, Kumar N, Gangola S, Singh H (2019c) Phytoremediation facilitating enzymes: an enzymatic approach for enhancing remediation process. In: Bhatt P (ed) Smart bioremediation technologies. Academic Press, Elsevier, Cambridge, MA, United States, pp 289–306

Kumar N, Singh H, Sharma SK (2020) Antioxidants: responses and importance in plant defense system. In: Roychowdhury R, Choudhury S, Hasanuzzaman M, Srivastava S (eds) Sustainable agriculture in the era of climate change. Springer, Cham, pp 251–264. https://doi.org/10.1007/978-3-030-45669-6_11

Kumar N, Singh H (2024) Plant functional traits for improving productivity. Springer Nature, Singapore. https://doi.org/10.1007/978-981-97-1510-7

Malik A, Kumar K, Kumari G, Kumar N, Singh H (2024) Application of functional traits in modelling productivity and resilience under climate change. In: Kumar N, Singh H (eds) Plant functional traits for improving productivity. Springer, Singapore. https://doi.org/10.1007/978-981-97-1510-7_5

Meena D, Singh H, Chaudhari SK (2011) Elucidating strontium response on growth dynamics and biochemical change in *Phaseolus mungo* L. Int J Agric Environ Biotechnol 4:107–113

Manoli G, Fatichi S, Schlapfer M, Yu K, Crowther TW, Meili N, Burlando P, Katul GG, Bou-Zeid E (2019) Magnitude of urbanheat islands largely explained by climate and population. Nature 573:55–60

Meehl GA, Tebaldi C (2004) More intense, more frequent, and longer lasting heat waves in the 21st century. Sci 305:994–997

Mitra A, Bagchi J, Thakur S, Parkhi US, Debnath S, Pramanick P, Zaman S (2015) Carbon sequestration in Bhubaneswar city of Odisha, India. Int J Innov Res Sci Eng Tech 4(8):6942–6947

Mitra A, Rudra T, Guha A, Ray A, Pramanick P, Pal N, Zaman S (2016) Ecosystem service of *Avicennia alba* in terms of Carbon sequestration. J Environ Sci Comp Sci Eng Technol 5(1):155–160

Mitra A, Sengupta K, Banerjee K (2012) Spatial and temporal trends in biomass and carbon sequestration potential of *Sonneratia apetala* Buch-Ham in Indian Sundarbans. Proc Natl Aca Sci Ind Section B: Biol Sci 82(2):317–323. https://doi.org/10.1007/s40011-012-0021-5

Mitra A, Sundaresan J (2016) How to study stored carbon in mangroves (2016). Published by CSIR-National Institute of Science Communication and Information Resources (NISCAIR), New Delhi, ISBN: 978-81-7236-349-9

Mutani G, Todeschi V (2020) The effects of green roofs on outdoor thermal comfort, urban heat island mitigation and energy savings. Atmosphere 11:123

Nowak DJ, Crane DE (2002) Carbon storage and sequestration by urban trees in the USA. Environ Pollut 116:381–389. https://doi.org/10.1016/S0269-7491(01)00214-7

Oke TR (1982) The energetic basis of the urban heat island. Q J R Meteorol Soc 108:1–24

Ossola A, Lin BB (2021) Making nature-based solutions climate-ready for the 50 °C world. Environ Sci Policy 123:151–159

Pataki DE, Alig RJ, Fung AS, Golubiewski NE, Kennedy CA, Mcpherson EG, Nowak DJ, Pouyat RV, Romero Lankao P (2006) Urban ecosystems and the North American carbon cycle. Glob Change Biol 12:2092–2102. https://doi.org/10.1111/j.1365-2486.2006.01242.x

Rahman MA, Moser A, Rötzer T, Pauleit S (2017a) Microclimatic differences and their influence on transpirational cooling of *Tilia cordata* in two contrasting street canyons in Munich, Germany. Agric for Meteorol 232:443–456

Rahman MA, Moser A, Rötzer T, Pauleit S (2017b) Within canopy temperature differences and cooling ability of *Tilia cordatatrees* grown in urban conditions. Build Environ 114:118–128

Richter R, Hutengs C, Wirth C, Bannehr L, Vohland M (2021) Detecting tree species effects on forest canopy temperatures withthermal remote sensing: the role of spatial resolution. Remote Sens 13:135. https://doi.org/10.3390/rs13010135

Schimel DS (1995) Terrestrial ecosystems and the carbon cycle. Glob Change Biol 1:77–91

Sharma R, Singh H, Kaushik M, Nautiyal R, Singh O (2018) Adaptive physiological response, carbon partitioning, and biomass production of *Withania somnifera* (L.) Dunal grown under elevated CO_2 regimes. 3 Biotech 8:267. https://doi.org/10.1007/s13205-018-1292-1

Singh H, Verma A (2013) Characterization and screening of high nitrogen efficient rice genotype to curtail environmental pollution. Int J Agric, Environ Biotechnol 6(1):23–30

Singh H, Kumar M (2022) Climate change and its impact on Indian Himalayan forests: current status and research needs. In Rani S, Kumar R (eds) Climate change. Springer Climate. Springer, Cham, pp 223–242. https://doi.org/10.1007/978-3-030-92782-0_11

Singh H, Savita Sharma R, Sinha S, Kumar M, Kumar P, Verma A, Sharma SK (2017a) Physiological functioning of *Lagerstroemia speciosa* L. under heavy roadside traffic: an approach to screen potential species for abatement of urban air pollution. 3 Biotech 7(61):1–10. https://doi.org/10.1007/s13205-017-0690-0

Singh H, Verma A, Kumar M, Sharma R, Gupta R, Kaur M, Negi M, Sharma SK (2017b) Phytoremediation: a green technology to clean up the sites with low and moderate level of heavy metals. Austin Biochem 2:1012

Singh H, Sharma R, Savita SMP, Kumar M, Verma A, Ansari MW, Sharma SK (2018) Adaptive physiological response of *Parthenium hysterophorus* to elevated atmospheric CO_2 concentration. Indian Forester 144:1–14

Smithers RJ, Doick KJ, Burton A, Sibille R, Steinbach D, Harris R, Groves L, Blicharska M (2018) Comparing the relative abilities of tree species to cool the urban environment. Urban Ecosyst 21:851–862

Singh H, Kumar N, Kumar A (2024) Enhancing resource use efficiency in crops through plant functional traits. In: Kumar N, Singh H (eds) Plant functional traits for improving productivity. Springer, Singapore. https://doi.org/10.1007/978-981-97-1510-7_6

Tuholske C, Caylor K, Funk C, Verdin A, Sweeney S, Grace K, Peterson P, Evans T (2021) Global urban population exposure toextreme heat. Proc Natl Acad Sci USA 118:e2024792118

Winbourne JB, Jones TS, Garvey SM, Harrison JL, Wang L, Li D, Templer PH, Hutyra LR (2020) Tree transpiration and urban temperatures: current understanding, implications, and future research directions. Bioscience 70:576–588

Yadav SK, Singh H, Nautiyal R, Ginwal HS, Ansari SA, Barthwal S (2019a) Modulation of morpho-physiological responses in *Populus deltoides* by elevated carbon dioxide and temperature. Forest Sci 66:105–118. https://doi.org/10.1093/forsci/fxz048

Yadav SK, Singh H, Ginwal HS, Barthwal S (2019b) Elevated CO_2 enhanced growth and physiological process of *Populus deltoides* Bartr. ex Marsh. Indian Forester 145(1):23–27

Zellweger F, Frenne PD, Lenoir J, Vangansbeke P, Verheyen K, Bernhardt-Römermann M, Baeten L, Hédl R, Berki I, Brunet J, Van Calster H, Chudomelová M, Decocq G, Dirnböck T, Durak T, Heinken T, Jaroszewicz B, Kopecký M, Máliš F, Macek M, Malicki M, Naaf T, Nagel TA, Ortmann-Ajkai A, Petřík P, Pielech R, Reczyńska K, Schmidt W, Standovár T, Świerkosz K, Teleki B, Vild O, Wulf M, Coomes D (2020) Forest microclimate dynamics drive plant responses to warming. Sci 368:772–775

Zhao L, Lee X, Smith RB, Oleson K (2014) Strong contributions of local background climate to urban heat islands. Nature 511:216–219

Hydrological Variability in Indian Forest Ecosystem: Analysis of Drought Resilience, Recovery and Water Use Efficiency in Moist and Dry Deciduous Forests

Triparna Sett, Bhaskar R. Nikam, Hukum Singh, and Saurabh Purohit

Abstract This study investigates precipitation trends and drought dynamics in Moist Deciduous Forests (MDF) and Dry Deciduous Forests (DDF) over twenty years (2001–2020), with a focus on water use efficiency (WUE). Utilizing the Mann–Kendall test (MK-test), seasonal precipitation trends were analyzed, revealing nuanced patterns across forest types and seasons. The standardized precipitation index (SPI) approach was adopted to identify drought and non-drought/normal conditions. Additionally, MODIS-derived WUE dynamics were scrutinized during drought and non-drought periods to understand ecosystem responses to hydrological challenges. CHIRPS precipitation data was employed for seasonal Mann–Kendall (MK) tests, uncovering distinct trends and identification of drought and non-drought periods in MDF and DDF. The results indicate a decreasing trend in precipitation from January to May in MDF, while DDF shows no significant trend over the twenty years. The drought period (July–December of 2002) and (June-December of 2008) and the non-drought period (July–December of 2012) and (June–December of 2012) were identified for MDF and DDF, respectively. Analysis of WUE during these periods revealed resilient and resistant forest types. This study highlights the importance of understanding ecosystem responses to hydrological variability for informed forest management and conservation. The findings underline that MDF and DDF exhibit different responses to drought periods, reflecting their unique

T. Sett (✉)
Indian Institute of Remote Sensing, ISRO, 4 Kalidas Road, Dehradun, Uttarakhand 248001, India
e-mail: triparna35@gmail.com

B. R. Nikam
EO Applications and DMS Programme Office (EDPO), Indian Space Research Organisation (ISRO), Bengaluru, Karnataka 560094, India

H. Singh
ICFRE-Forest Research Institute, PO New Forest Dehradun, Dehradun, Uttarakhand 248006, India

S. Purohit
Watershed Organisation Trust (WOTR), WOTR-Centre for Resilience Studies, (W-CReS), Pune, Maharashtra 411009, India

H. Singh (ed.), *Urban Forests, Climate Change and Environmental Pollution*,
https://doi.org/10.1007/978-3-031-67837-0_37

ecological attributes and environmental contexts. MDF demonstrates resilience to drought, with decreasing precipitation trends and sustained WUE during dry periods, while DDF displays resistance, characterized by reduced evapotranspiration rates and increased WUE. These empirical insights emphasize the need to comprehend diverse responses and adaptations within forest ecosystems to develop effective management and conservation strategies in the face of evolving climatic dynamics.

Keywords Standardized precipitation index (SPI) · Water use efficiency (WUE) · Mann–Kendall test (MK-test) · Moist Deciduous Forest (MDF) · Dry Deciduous Forest (DDF) · CHIRPS · MODIS

1 Introduction

Forests offer diverse ecological, economic, and social benefits. However, their susceptibility to drought poses significant challenges, impacting water use efficiency (WUE) and ecological resilience. The WUE plays a pivotal role in forest ecosystems and the influencing factors. Exploring vegetation responses to water stress has been a continuous journey, evolving with technological advancements and diverse research endeavors (Kumar et al. 2022, 2024; Singh et al. 2022, 2024; Singh 2024). Beginning with Cohen (1991) laid the groundwork with two laboratory experiments, unraveling the intricate correlations between vegetation indices and water potential in lodge pole pine branches. The significance of indices like NDVI, near-IR to red ratio, Infrared Index, and Moisture Stress Index (MSI) display their responsiveness to varying degrees of water stress. Moving forward, Kogan (1995) highlighted the pressing need for effective drought monitoring. The development of NOAA's Advanced Very High-Resolution Radiometer-based Vegetation Condition Index (VCI) emerges as a tool for accurate and timely drought detection. The VCI, with its adjustments for climate, ecology, and weather conditions, proved instrumental in capturing the onset, intensity, duration, and impact of droughts across large vegetated areas. Peters et al. (2002) reinforced the importance of the VCI in drought monitoring, emphasizing its superiority over conventional ground data. The methodology and technical principles behind the VCI provided a robust framework for its successful application in the United States from 1985 to 1990. The alignment of VCI with precipitation and yield anomaly patterns underscored its efficacy in reflecting real-world drought scenarios. Ji and Peters (2003) further enriched the narrative by exploring the intricate relationship between the Normalized Difference Vegetation Index (NDVI) and moisture-related vegetation conditions in the U.S. Great Plains. Their analysis during the growing season from 1989 to 2000 revealed the 3-month Standardized Precipitation Index (SPI) as a key correlate, emphasizing lag and cumulative precipitation effects on vegetation. This study highlighted the need to consider seasonal effects, particularly in regions with low soil water-holding capacity, when monitoring drought using NDVI. Allen et al. (2010) undertook the first global assessment of tree mortality due to drought and heat stress. Their findings raised alarms about the potential responses of forested ecosystems to climate change, emphasizing increased vulnerability, potential die-off, and associated risks to ecosystem services, including carbon loss. Worrall et al.(2013)showed a decline of *Populus tremuloides* (trembling aspen) across North America, attributing it to severe drought documented in historic climate records. A bio-climate model connected the decline to marginal habitat suitability, particularly in drier regions, emphasizing the sensitivity of ecosystems to drought. Climate projections further indicated a substantial loss of suitable habitat, raising concerns for the future. Assal et al. (2016) extended the exploration to western North American forests, utilizing Landsat time series data of 22 years to examine the impact of severe droughts. The Normalized Difference Moisture Index (NDMI) revealed spatial patterns, with coniferous forests experiencing significant negative trends. This study provided valuable insights for managing regions resistant, persistent, or vulnerable to severe drought. Pei et al. (2018) stressed the importance of accounting for dominant controls on vegetation growth, particularly in regions with complex topography, diverse climate conditions, and different vegetation types. This consideration was crucial for improving the accuracy of vegetation activity analyses using indices like the VCI and the Vegetation Health Index (VHI) in China over the period 1982–2013. Study of drought severity in Mongolia from 2000 to 2016 using NDVI anomalies, identifying periods of mild to severe drought, highlighted specific years; such as 2001, 2007, 2002, 2005, 2004, 2009, and 2006; where significant portions of the country experienced drought conditions (Nanzad et al. 2019). Rousta et al. (2020) analyzed drought effects on Afghanistan's vegetation from 2001 to 2018 using Moderate Resolution Imaging Spectroradiometer (MODIS) and Tropical Rainfall Measuring Mission (TRMM) data/products. Observable declines in vegetation during drought years (2001, 2008) were noted, particularly in the Helmand Watershed. The study utilized the VCI to indicate widespread drought conditions in 2001 and 2008, while 2009 and 2010 exhibited increased vegetation coverage. Javed et al. (2020) employed MODIS and Land Long Term Data Record (LTDR) data to investigate drought evolution in China from 1982 to 2017. Their analysis across four land cover types revealed precipitation trends, with forestland having the highest mean monthly and yearly precipitation. Positive correlations were found between drought indices (NDVI anomaly, VCI) and SPI for different land cover types. VCI outperformed NDVI anomaly correlating with SPI, especially in cropland (3-month SPI) and forestland (6-month SPI). The most recent addition to this extensive exploration comes from (Yu and Guo 2023). Their study focused on monitoring drought in the Canadian prairie region's northern mixed grassland over 36 years using remote sensing inputs. Utilizing a revised vegetation health index (VHIr), the research effectively distinguished between dry and wet years, providing a spatial map of drought severity. VHIr correlated strongly with the dead-to-green biomass ratio and exhibited varying drought sensitivity across grassland landscapes. This study contributed valuable insights into drought severity's spatial and temporal distribution, facilitating improved natural resource management in arid and semi-arid regions.

WUE, the ratio of plant productivity per unit of water use, is commonly employed to measure the connections between forest carbon and water cycles (Hatfield and Dold 2019; Dhyani et al. 2021; Kumar et al. 2021a, b, c, d, e; Prakash et al.

2022).The understanding of WUE has evolved through decades of comprehensive research, employing various methodologies and focusing on diverse ecosystems. Building on the foundations laid by Baldocchi et al.(1987), who utilized the eddy covariance method to investigate WUE dynamics in deciduous forests, subsequent studies have delved into different aspects of this concept. Donovan and Ehleringer (1992)expanded the scope by exploring the eco-physiological characteristics of *Chrysothamnus nauseosus* plants. Their findings indicated a developmental progression in WUE parameters with plant size, shedding light on the nuanced interactions between size, life history, and WUE. Law (2002) broadened the perspective within the FLUXNET network, offering insights into the coupling of carbon and water cycles across diverse ecosystems. Their research highlighted the impact of climate variability on carbon uptake, emphasizing the intricate relationship between environmental factors and ecosystem processes.

Nemani et al. (2002) explored the hydro-climatic changes' influence on the U.S. carbon sink, underscoring the need to consider changes in the hydrologic cycle in global carbon budget analyses. Their study reinforced the intricate connection between climatic variables and ecosystem responses. Yu and Guo (2023) furthered this understanding by examining WUE variations among forest sites. The asynchronous response of Gross Primary Productivity (GPP) and Evapotranspiration (ET) to climatic variables influenced the coupling and decoupling of carbon and water cycles, illustrating the complex dynamics within regional forest ecosystems. Brümmer et al. (2012) provided a continental-scale analysis of WUE values across Canadian ecosystems, revealing a range of WUE levels. The study highlighted the variability among sites and ecosystems, with certain vegetation types exhibiting notably higher WUE. Kauwe et al. (2013) conducted an inter-comparison of ecosystem models, offering valuable insights into the key assumptions influencing model predictions of transpiration and canopy WUE. Their findings underscored the importance of accurately representing stomatal response and other ecosystem processes in enhancing model predictions. Zhang et al. (2015) addressed the accuracy of WUE estimates from MODIS products, emphasizing the need to consider various factors, such as the accuracy of meteorological data and uncertainties in eddy flux data, in quantifying WUE bias. Liu et al. (2017) provided insights into WUE stability over time, with minimal inter-annual variability observed in an old-growth forest in lower subtropical China. Their analysis highlighted the significance of moisture conditions as key drivers of ecosystem WUE. Knauer et al. (2017) investigated the unexpected increase in WUE observed in temperate/boreal forests, challenging previous interpretations. Their study emphasized the need to consider the decreasing CO_2 sensitivity of WUE with increasing scale, providing crucial insights for the physiological understanding of ecosystem WUE changes.

Next, in scientific exploration, a fascinating narrative unfolds as researchers delve into the intricate dynamics of ecosystems, unveiling the secrets that govern the delicate balance between water and carbon cycles. This research spans various landscapes, from arid terrains to lush forests. It traverses continents as scientists strive to decipher the responses of diverse vegetation to the ever-changing climate rhythm, particularly in the face of drought stress.

Following the establishment of fundamental knowledge surrounding forest drought and WUE, elucidating the intricate relationship between these phenomena becomes imperative. These studies offer profound insights, contributing to the scientific discourse on the interplay between forest drought dynamics and WUE.

Jaleel et al. (2008) investigated the complex inter-relationship between forest drought and WUE of *Catharanthus roseus*. This research examines the WUE of two varieties, rosea and alba, subjected to divergent watering regimes. The narrative unfolds as these flowers confront the challenge of drought stress, unraveling a remarkable metamorphosis marked by an augmentation in WUE. Yang et al. (2016) explored the effects of drought on landscapes and ecosystem levels. Their chronicle elucidates the contrasting responses of arid and semi-arid ecosystems to this climatic adversary. In arid regions, a heightened WUE emerges in stark contrast to the diminished efficiency witnessed in the semi-arid areas, underscoring the more significant influence of hydro-climatic conditions on the fate of these ecosystems. Huang et al. (2017) studied the global ecosystem's response to frequent drought events. Legacy effects of drought emerge, revealing ecosystems' resilience to abrupt hydrological changes. The narrative unfolds, emphasizing the elaborate strategies employed by ecosystems during transitions from wet to dry years, shedding light on the wisdom encoded in nature's coping mechanisms. In the northern reaches of Eurasia, Gao et al. (2017) unveil a tale of a boreal Scots pine forest confronting the trials of severe soil moisture drought in 2006. The forest's response, a delicate interplay between carbon assimilation and transpiration, paints a vivid picture of the challenges of changing climatic conditions. Their quest not only exposes vulnerabilities but also underscores the need for refining models to capture the subtleties of this phenomenon. Stan beckon us to the southeastern USA, where the composition of forests, topography, and the relentless force of time shape the relationship between ET and water availability. In the heart of Central China, Niu and Liu (2021) delve into a natural oak forest facing the repetitive cycle of seasonal drought. The forest's anisohydric characteristics defy expectations, showcasing a resilience that challenges the conventional narrative of reduced gross ecosystem production under spring and summer drought. In the dynamic landscape of ecological research, the investigation by Huang et al. (2021) explores the Lancang–Mekong River Basin's spatiotemporal variations in WUE and its responses to drought from 1982 to 2018. Their comprehensive analysis, utilizing GPP and ET data, uncovers a mean yearly WUE of 1.63 g C kg^{-1} H_2O, with distinct variations across land use types, climate regions, and elevations. Continuing this narrative, McQuillan et al. (2022) contribute a novel perspective by quantifying the impact of forest species composition and topography on ET across the Blue Ridge ecoregion. Their findings reveal a trend of decreasing constraints on ET by water availability, with implications for landscape-scale water dynamics, particularly during drought events. In the Yellow River Basin, Qin et al. (2022) undertook an insightful exploration, employing various models to assess the contribution of environmental factors to greenness. Their research underscores the significance of a systematic approach in understanding vegetation changes, emphasizing the interactions between the Temperature Vegetation Dryness Index (TVDI) and WUE. A study of changing dynamics unfolds, raising questions about the impact on down slope water availability and the delicate balance of forested landscapes, unraveling the mysteries of revegetation. The story intricately weaves together environmental factors, unveiling the increasing greenness in the basin. TVDI and WUE emerge as protagonists, steering the narrative towards a profound understanding of vegetation change in vulnerable ecosystems. Luo et al. (2023) extend the temporal dimension, investigating spatiotemporal variations in WUE and its responses to drought in the Lancang–Mekong River Basin from 1982 to 2018. Their work highlights the convoluted relationships between annual WUE and drought across different climatic regions, shedding light on the delayed effects of drought on ecosystem WUE. Zooming into Inner Mongolia, Cheng et al. (2023) delve into the carbon use efficiency (CUE) and WUE of the region's vegetation from 2001 to 2020. Their research identifies shifting drought trends and their impact on carbon and water cycles, providing valuable insights into the complex dynamics of resource use efficiency in response to changing climatic conditions. In a broader ecological context, Hu et al. (2023) explored the role of nitrogen use efficiency (NUE) and WUE in responding to changes in soil nitrogen and water availability in grassland ecosystems. Their findings contribute to a comprehensive understanding of resource use efficiency in different functional groups of plants, shedding light on the interplay between nitrogen deposition, drought, and plant adaptability.

"**Drought-resilient forest type**" and "**drought-resistant forest type**" describe forest types that can withstand and adapt to drought conditions, particularly in climate change. A drought-resistant forest ecosystem has evolved mechanisms to resist and survive periods of prolonged drought or low water availability (Lloret et al. 2011; Singh 2021; Sharma and Singh 2021; Singh et al. 2021). Resilience refers to the ability of both individuals and ecosystems to uphold their functionality in the presence of disturbances (Singh et al. 2020; Kumar et al. 2021a, 2021b, 2021c, 2021d). Resistance refers to the capability to diminish the impact of a disturbance, while recovery denotes the ability to bounce back and restore functionality following the disturbance. This conceptualization, outlined by Lloret et al. (2011), highlights the dynamic nature of resilience and its importance in maintaining the stability and integrity of ecosystems amidst various challenges and disruptions.

Frolking et al. (2009) reviewed forest disturbances and recovery processes, emphasizing the impact on aboveground biomass and canopy structure. They highlighted the need to accurately quantify disturbance impacts using space-borne remote sensing data, which can improve carbon budget estimates and aid in forest carbon cycle modelling. Sohn et al. (2016) conducted a meta-analysis to assess the potential of forest thinning in mitigating drought stress. Their findings suggested that thinning could improve tree performance during and after drought, with varying effects observed between broadleaves and conifers. They emphasized the importance of considering thinning intensity and site characteristics in forest management strategies for drought resilience. Schwarz et al. (2020) critiqued commonly used resilience indices for analyzing tree growth responses to drought. They identified shortcomings in existing indices and proposed a new analytical framework, the "line of full resilience," to improve interpretations of drought responses. Their recommendations aimed to facilitate systematic reviews and the development of evidence-based management strategies for forest resilience. Thong et al. (2020) investigated forest recovery patterns following shifting cultivation in Manipur, North-East India. They assessed tree species composition, diversity, and carbon stock recovery in abandoned fallows and identified environmental factors influencing regeneration. Their study highlighted the potential of certain tree species for restoring degraded forest areas and sequestering carbon in shifting cultivation landscapes. Bohner and Diez (2021) investigated tree resistance and recovery from drought across a large geographic gradient in California. They quantified reductions in tree growth during drought and post-drought recovery for three montane conifers. Their findings highlighted the importance of considering factors at various spatial and temporal scales in understanding the different components of drought responses. Ho et al. (2021) assessed the resistance, recovery, and resilience of two co-occurring palaeotropical pine species in south-central Vietnam. They used dendrochronological methods to examine growth patterns and carbon discrimination against atmospheric CO_2 during the growth decline phase. Their study revealed differences in growth dynamics and resilience between the two species, suggesting varying adaptive strategies to environmental stressors. Shi et al. (2022) studied vegetation resistance to drought in the middle reaches of the Yellow River Basin in China. They analyzed remotely-sensed vegetation indices and Standardized Precipitation Evapotranspiration Index (SPEI) data to assess changes in resistance and resilience to drought. Their findings highlighted the impact of increasing drought intensity on vegetation and provided insights into potential management strategies for enhancing ecosystem stability. Ma et al. (2023) explored changes in vegetation resistance and resilience under different drought disturbances in Jilin Province, China. Using remote sensing data, they developed a framework to analyse vegetation responses to drought and identified spatial and temporal patterns in resistance and resilience across different vegetation types. Their findings provided valuable insights for understanding and managing ecosystem responses to drought. Bargali et al. (2023) investigated the impact of forest fires on tree diversity and regeneration in Uttarakhand, India. Using satellite imagery and field surveys, they assessed tree species diversity and regeneration patterns across different fire frequency classes. Their study underscored the importance of fire management strategies in preserving tree diversity and ecosystem resilience in fire-prone regions.

Though extensively investigated, the patterns of droughts and their inter-links with WUE in the forested ecosystems are highly regional. There are very few studies available in the Indian context. So, our efforts here entail the investigation of hydrological cycle perturbations in Indian forest ecosystems, focusing on drought patterns and WUE. The approach integrates diverse data sources and modeling techniques, including remote sensing data. CHIRPS-gridded rainfall data was employed to compute the SPI and understand the characteristics of meteorological droughts. A big part of the study looks at how WUE and drought patterns change over time and space by using remote sensing data from MODIS-derived WUE data for two selected types of forests in India. Discrepancies in WUE between drought and non-drought periods are scrutinized. Furthermore, the research dives into assessing the ecological resilience of Indian forest ecosystems in response to drought stress. This study highlights the need to understand better the complex interactions between drought intensity, WUE, and ecological resilience and resistance in forest ecosystems and the importance of developing effective strategies for managing these ecosystems in the face of climate change.

2 Study Aims

(a) **Conduct thorough examinations of historical meteorological data to precisely quantify the meteorological drought, with a specific focus** on two different types of forests: Moist Deciduous Forest (MDF) in Uttarakhand and Dry Deciduous Forest (DDF) in between Madhya-Pradesh and Chhattisgarh. The goal is to identify the drought and non-drought period from 2001 to 2020,
(b) **Assess the WUE in the specific forest type**: Examine the water consumption patterns in the identified forest ecosystems during normal and drought situations. MODIS-WUE, offering valuable insights into the adaptive tactics employed by vegetation, and
(c) **Evaluate the ability of ecosystems to withstand and recover from changes in the water cycle:** assess the impact of meteorological droughts on forest ecosystems, focusing on their resistance and resilience mechanisms. Investigate how these ecosystems respond to hydrological disruptions and their implications for long-term ecological stability.

2.1 Scope and Significance

(a) **Ecosystem Conservation:** Directing focused conservation initiatives to save biodiversity, soil productivity, and general ecological balance in forest ecosystems in the Indian climate.
(b) **Empirical Insights for Formulating Adaptive Strategies**: Offering empirically-supported insights to formulate adaptive strategies for Indian forest ecosystems in light of shifting water availability patterns amid climatic fluctuations.
(c) **Scientific Advancement**: Enhancing our comprehension of ecological processes, namely the nexus between hydro-meteorological fluctuations and WUE in forest environments, with potential ramifications for research on global climate change.
(d) **Policy Formulation**: Providing a systematic basis for policymakers to create well-informed policies that tackle the difficulties presented by disturbances in the hydro-meteorological fluctuations.
(e) **Community Resilience**: Enabling local communities to strengthen their ability to withstand and recover from challenges by creating flexible plans that consider the fluctuating availability of water and the ability of ecosystems to bounce back.

3 Study Area and Data

3.1 *Moist Deciduous Forest (MDF)*

The Rajaji National Park spans around 820 km^{2}in the Shivalik foothills, located between 29° 15′ to 30° 31′ North Latitude and 77° 52′ to 78° 22′ East Longitude. This park is in Uttarakhand, India, covering the Haridwar, Dehradun, and Pauri districts. It is situated between the Lesser Himalayas and the upper Gangetic plains. The region has been officially designated as a reserve for "Project Elephant" by the Ministry of Environment, Forests, and Climate Change, Government of India. The primary goal is to ensure the long-term survival of Asian elephants in their natural habitat. Rajaji National Park is classified as a Tropical Moist Deciduous Forest, as Champion and Seth (1968) determined. It is located within the Himalaya Hill biogeography zone. Several indigenous species, such as Mallotus philippensis, *Adina cordifolia, Shorea robusta, Terminalia bellirica, Acacia catechu, Ficus bengalensis*, and *Dalbergia sissoo* inhabit the region. During the hot season, which lasts from May to June, the temperature rapidly rises to 38–40 °C. During the rainy season, which occurs from June to September, there is a high degree of humidity and significant rainfall, with more than 750 mm of precipitation in July and August.

Furthermore, the temperature exhibits low variation. The yearly precipitation in the area fluctuates between 1200 and 1500 mm. The study area for this research is a selected portion of Rajaji National Park (Fig. 1a).

3.2 *Dry Deciduous Forest (DDF)*

The Achanakmar Biosphere Reserve was officially proclaimed India's 14th National Biosphere Reserve on March 30, 2005. It covers a total area of around 3835 km^2. This vast area stretches from the Maikal hill ranges to the point where the Vindhyan and Satpura hill ranges meet, creating a triangle shape over the Indian states of Chhattisgarh and Madhya Pradesh. The biosphere reserve is located between latitudes 22 °15′ to 20° 58′ North and longitudes 81° 25′ to 82° 5′ East. It is within the Malayan realm and represents the Tropical Dry or Deciduous Forest, as classified by Champion and Seth in 1968.

Additionally, it is part of the country's Deccan Peninsular bio-geographic zone. The core zone, situated in the Bilaspur district of Chhattisgarh state (CG), consists of a thick forest that spans mountainous and valley landscapes, encompassing a total area of around 551 km^2. The area comprises ***Dioscore*** Sp., ***Flacourtia indica, Thalictrum*** sp., ***Hypericum japonicum, Talinium portulacifolium, Cocculus hirustus, Dillenia pentagyna***, and ***Curcuma aromatica***. The region has a tropical climate, which is hot and humid. It has three distinct seasons: summer (March-May), rainy season (June–October), and winter (November–February). The average summer temperatures usually vary between 30 and 45 °C, whereas winter temperatures often range

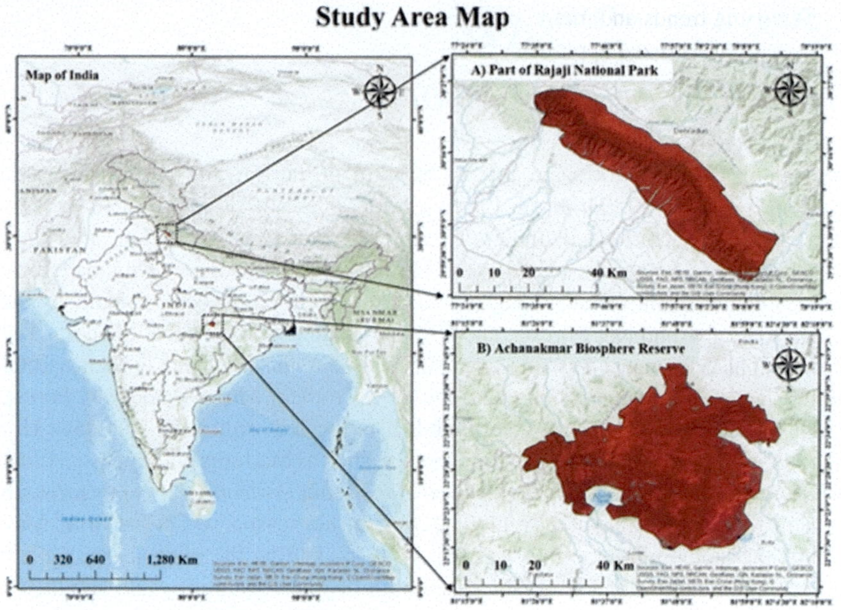

Fig. 1 Study area **a** Part of Rajaji National Park (Moist Deciduous Forest, MDF); **b** Achanakmar Biosphere Reserve Forest (Dry Deciduous Forest, DDF)

from 0 to 25 °C. The region experiences an average annual precipitation of 1292 mm, with the peak rainfall in July and August (Fig. 1b).

3.3 Data Used

3.3.1 CHIRPS Data

Climate Hazards Group InfraRed Precipitation with Station (CHIRPS), developed for the United States Agency for International Development's (USAID's) Famine Early Warning Systems Network (FEWS NET), is a comprehensive global rainfall dataset spanning over 40 years and covering latitudes from 50° S to 50° N. It seamlessly integrates with CHPclim, a gridded data product constructed using a geospatial modeling approach incorporating moving window regressions and inverse distance weighting interpolation. This innovative methodology combines satellite-derived data, gridded physiographic indicators, and climate normals (CNs) from in situ weather stations. As a result, the global 0.05° 12-month precipitation climatology produced by CHIRPS and CHPclim demonstrates superior performance compared to similar global climatology products, particularly in regions characterized by complex terrain and low station densities. This enables the generation of a detailed gridded rainfall time series for analyzing trends and monitoring droughts. CHIRPS's global dataset availability ensures a standardized approach to global precipitation patterns.

3.3.2 MODIS Data

MODIS-derived GPP and ET products are used to compute the WUE from 2001–2020. MOD17 (GPP/NPP)-(Running et al. 2004; Zhao et al. 2005), and MOD16-(ET) (Mu et al. 2011), are two operational MODIS data products that utilize satellite remote sensing data to quantify water and carbon fluxes of vegetated land. MOD17 provides continuous GPP and Net Primary Productivity (NPP) estimates across Earth's entire vegetated land surface. GPP can be input for models that calculate various terrestrial processes, including energy, carbon, water cycle dynamics, and vegetation (Cleugh et al. 2007).On the other hand, MOD16's Collection 6 products include terrestrial ecosystem ET), latent heat flux (LE), and potential ET (PET) data layers. These products are produced at a spatial resolution of 500 m and are available from 2001 onwards. They provide data on an 8-day basis, with resolutions ranging from 500 m to 1 km.

4 Methodology

The study assesses how the forest WUE responded during drought and non-drought periods. Based on this result, resilient and resistant forest types between MDF and DDF were also identified. The SPI was computed using monthly rainfall data to assess the meteorological drought. The MODIS-derived WUEis used in the present analysis.

The daily precipitation data of the CHIRPS was pre-processed from 1981 to 2020 on the Google Earth Engine (GEE) platform, ensuring a consistent spatial resolution of 5.5 km (Didi Sacré Regis et al. 2020; Usman and Nichol 2020; Sa'adi et al. 2024; Funk et al. 2015). The monthly total spatial average rainfall was estimated using daily rainfall products and is used in trend analysis and meteorological drought assessment over two distinct forest types. Similarly, the monthly WUE from 2001–2020 was generated by the equation of GPP/ET (Zhang et al. 2015) with 5.5 km resolution to make consistency with the SPI to analyze the long-term behavior of WUE during drought and non-drought periods.

4.1 Mann–Kendall Trend Analysis on Rainfall Data

We employed trend analysis using the Mann–Kendall test to examine the seasonal and annual rainfall patterns in both study areas. It was developed by Mann (1945) and Kendall (1948) and later refined by Warren and Gilbert (1988). The Mann–Kendall

(MK) test is a non-parametric statistical method. It evaluates whether the data has monotonic increasing or decreasing trends, which may or may not follow a linear pattern. This test serves as an alternative to parametric linear regression analysis, which typically determines if the slope of the linear regression line significantly differs from zero. Unlike regression analysis, the MK test does not assume that the residuals from the fitted regression line adhere to a normal distribution. Therefore, the MK test was instrumental in cases where the normality assumption may not hold (Gadedjisso-tossou and Adjegan 2021).

After ascertaining the positive or negative trend in seasonal and annual rainfall over both study regions, the statistically significant value of trends was derived using Sen's slope estimator (Frimpong et al. 2022).

The MK statistics S can be obtained by

$$S = \sum_{k=1}^{n-1} \sum_{j=k+1}^{n} sgn\left(x_j - x_k\right) \tag{1}$$

$$sgn(x_j - x_k) = \begin{cases} +1, \, if \left(x_j - x_k\right) > 0 \\ 0, \, if \left(x_j - x_k\right) = 0 \\ -1, \, if \left(x_j - x_k\right) < 0 \end{cases} \tag{2}$$

where, n is the length of the sample, $x_k$and x_j *are from* $k = 1, 2 \ldots n{-}1$ *and* $j = k + 1 \ldots., n$. If n is bigger than 8, statistics S approximates the normal distribution. The mean of S is 0, and the variance of S can be acquired as follows:

$$var(S) = \frac{n(n-1)(2n+5)}{18} \tag{3}$$

Then, the test statistics Z is denoted by Eq. (4).

$$Z = \begin{cases} \frac{S-1}{\sqrt{var(S)}}, \, ifS > 0 \\ 0, \, ifS = 0 \\ \frac{S-1}{\sqrt{var(S)}}, \, ifS < 0 \end{cases} \tag{4}$$

If $Z > 0$, it indicates an increasing trend and vice versa. Given a confidence level α, the sequential data would be supposed to experience a statistically significant trend if $|Z| > Z(1-\alpha/2)$, where $Z(1-\alpha/2)$ is the corresponding value of $P = \alpha/2$ following the standard normal distribution.

Further, the magnitude of the statistically significant trend was evaluated using the non-parametric procedure developed by Sen (Frimpong et al. 2022). The trend is calculated by:

$$\beta = Median\left(\frac{x_j - x_i}{j - i}\right), j > i \tag{5}$$

where β is Sen's slope estimate. $\beta > 0$ indicates an upward trend in a time series. Otherwise, the data series presents a downward trend during the period.

4.2 Meteorological Drought Analysis

The SPI, developed by Mckee et al.(1993), serves as a meteorological drought index that assesses the total precipitation at a specific location over a given period compared to the long-term precipitation distribution of the same period. SPI operates as a statistical indicator utilized to estimate the presence of wet or dry conditions based on precipitation levels. The primary purpose of SPIwas to assess drought events and quantify the severity and duration of droughts. It helps understand precipitation variability over different time scales, from short-term to long-term, which is crucial for water resource management, eco-hydrological studies, agriculture, and risk assessment.

4.2.1 Standardized Precipitation Index (SPI)

SPI is a statistical measure (Z-Score) used to assess the deviation of precipitation from the long-term average conditions. It is calculated by taking an individual rainfall value, subtracting the long-term mean, and dividing the difference by the standard deviation. To a gamma probability density function, SPI was computed by fitting precipitation data for specific locations and time periods (such as 1, 3, 6, or 12 months). The cumulative probability of the precipitation distribution was then transformed to a standard normal distribution with a mean of 0 and a variance of 1, following the methodology outlined by Edwards and McKee (1997). The resulting Z value of the normal distribution represents the SPI index, where negative values indicate a rainfall deficit corresponding to dry spells. In contrast, positive values indicate surplus rainfall, indicating wet spells. SPI relies on a predefined non-exceedance rainfall probability for a given time scale.

The formula is as follows:

$$Z = SPI = -\left(t - \frac{c_0 + c_1 t + c_2 t^2}{1 + d_1 t + d_2 t^2 + d_3 t^3}\right) \text{ for } 0 < H(x) \leq 0.5 \quad (6)$$

$$Z = SPI = +\left(t - \frac{c_0 + c_1 t + c_2 t^2}{1 + d_1 t + d_2 t^2 + d_3 t^3}\right) \text{ for } 0.5 < H(x) < 1.0 \quad (7)$$

where,

$$t = \sqrt{ln\left(\frac{1}{H(x))^2}\right)} for 0 < H(x) \leq 0.5$$

Table 1 SPI categorization

SPI values	Categories
≥ 2	Extremely wet
1.50 to 1.99	Severely wet
1.00 to 1.49	Moderately wet
0.00 to 0.99	Mild wet
− 0.99 to 0.00	Mild dry
− 1.49 to − 1.00	Moderately dry
− 1.99 to − 1.50	Severely dry
≤ -2	Extremely dry

$$t = \sqrt{ln\left(\frac{1}{(1.0 - H(x))^2}\right)} for 0.5 < H(x) \leq 1.0$$

And,
$c_0 = 2.515517; c_1 = 0.802853$
$c_2 = 0.010328; d_1 = 1.432788$
$d_2 = 0.189269; d_3 = 0.001308.$

Mckee et al. (1993) specified the criteria for "dry spells" or "wet spells" over any time interval. They classified SPI values to define various dry and wet intensities, as presented in Table 1. This approach provides a standardized and quantitative means of evaluating precipitation patterns and drought severity. SPI can be computed and analyzed for any time scale. For example, the three-month SPI (SPI-3) for June will consider the precipitation total of three months: April, May, and June. It is worth noting that the time sequence of precipitation used for SPI calculation should be at least 30 years, and there should be no missing data in the time series.

SPI typically falls within the range of + 2.0 to − 2.0, with occurrences of extremes beyond this range happening less than 5% of the time. Positive SPI values signify above-average precipitation levels, while negative SPI values indicate below-average precipitation. SPI values near zero suggest precipitation levels close to normal. The magnitude of SPI values reflects the severity of the dry or wet anomaly, with more extreme values indicating more severe precipitation deviations from the norm. SPI values can be interpreted in standardized categories (e.g., mild, moderate, severe, extreme) to facilitate communication and decision-making. Being dimensionless, negative SPI values indicate drought conditions and are associated with descriptors such as 'dry,' 'moderately dry', or 'extremely dry' (Table 1).

4.2.2 Estimation of SPI

The SPI calculation process involved several vital steps. Firstly, the SPIgenerator software received monthly precipitation data in various formats, such as daily, weekly, or monthly, ensuring the availability of a complete dataset for the analysis period. Upon receiving the input data, the software initiated a comprehensive data preprocessing phase, meticulously verifying input parameters to guarantee the accuracy and adequacy of the data for subsequent analysis. For this study, monthly precipitation data was used at the study sites. Subsequently, the software parsed the precipitation data from the input file or directory, systematically extracting precipitation values and their corresponding dates or periods. This step facilitated the organization of the data, laying the groundwork for the SPI calculation process. Following data parsing, the software proceeded with data aggregation, combining precipitation data into larger time intervals based on the specified time scale. This study considered accumulation periods of 1, 3, 6, 9, and 12 months, allowing for a detailed examination of precipitation trends over varying temporal scales. The aggregation process enabled the generation of aggregated precipitation sums for each specified time interval, providing insights into the cumulative precipitation patterns over different periods. This step was crucial for capturing the temporal variability in precipitation and identifying trends or anomalies in the data. Additionally, the software calculated drought periods and frequencies based on the aggregated precipitation data, offering valuable information on the occurrence and severity of drought events.

Following data aggregation, the SPI calculation was performed for each period using the historical precipitation record at the location of interest. This calculation involved comparing the precipitation accumulation over the specified period to the historical distribution of precipitation, typically modeled by a gamma distribution. The SPI provided a standardized measure of drought severity or wetness for each time interval by quantifying the deviation of current precipitation from historical norms. Once the SPI values were computed, the software generated output files containing the SPI values and any additional requested information, such as frequency data or drought periods. Users could specify the output file format, with options including comma-separated values (.csv) files or Excel spreadsheets. This comprehensive output facilitated further analysis and interpretation of the SPI results, enabling researchers to gain valuable insights into the temporal dynamics of precipitation and drought conditions in the study area.

4.2.3 Analysis of Monthly SPI Data

The SPI over the study regions was estimated with 1, 3, 6, 9, and 12-month accumulation periods. Notably, the SPI with a 9-month accumulation period for the present study was used to identify and characterize meteorological droughts over the study area. The 9-month accumulation SPI was trusted to provide better insights into the response of forest ecosystems to drought stress (Jha et al. 2019). The continuous time series of SPI were analyzed, and the rainfall deficit periods (periods with negative SPI) were classified into four categories, indicative of the severity of the rainfall deficit. SPI values ranging from 0.0 to − 0.99 were classified as 'mild drought', values from − 1.00 to − 1.49 were put into 'moderate drought' category, SPI from

– 1.50 to – 1.99 were marked as 'severe drought', and 'extreme drought' was identified as SPI < – 2. The onset of a drought event is marked when the SPI value goes below – 0.99, and the termination criteria of any ongoing drought event was set to SPI > 0.0. The period from onset to termination of an event was the persistence of drought, and the magnitude and duration of precipitation deficits were considered as the intensity of the drought. All the drought events over both the study area were identified, and their characteristics were estimated. Further, these drought events and their characteristics are used to understand and quantify the behaviour of forest WUE during water stress periods.

The intensity and severity of drought are assessed based on the magnitude and duration of precipitation deficits, with negative SPI values indicating more severe drought conditions. Additional indicators were also noted, such as below-average precipitation duration and dry spell frequency. Despite its standardization, flexibility, and sensitivity advantages, SPI relies on long-term, high-quality precipitation data, which may be scarce in some regions. Furthermore, assumptions of precipitation data stationarity pose challenges under changing climate conditions. Careful consideration of local climatic factors and user-specific needs was required for interpreting SPI values, complicating its application in a certain context. SPI is a valuable tool for assessing precipitation anomalies, monitoring drought conditions, and guiding decision-making in climate-sensitive sectors. Its standardized approach and adaptability make it widely applicable across different spatial and temporal scales.

4.3 Application of MODIS-Derived WUE Data

Leveraging raster images from MODIS-derived WUE were used to show how forest WUE behaved during drought and non-drought periods. The spatial average of WUE data was extracted for the specific duration for the exact location from 2001 to 2020 for drought and non-drought/normal periods.

5 Results

5.1 Precipitation Trend Analysis Over MDF and DDF

The monthly precipitation data was classified into four categories based on seasons for the trend analysis using the non-parametric Mann–Kendall test (Fig. 2). The monsoon season (June, July, August, and September—JJAS) shows no trend over the forty years (1980–2020) for MDF and DDF. MDF exhibits a decreasing trend for the winter months (January and February—JF), whereas no significant trend was observed in the DDF. Both the forest types show no trend in the post-monsoon season (October, November, and December—OND), whereas a decreasing trend in MDF was observed in the pre-monsoon season (March, April, and May—MAM) in MDF. DDF for the same period shows no significant trend. The details of the Man-Kendall test are as follows (Table 2).

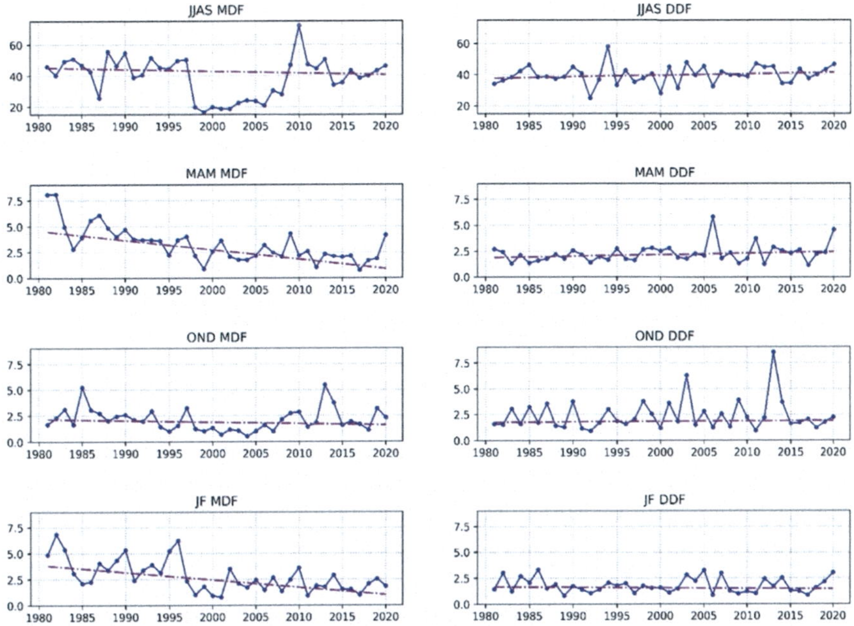

Fig. 2 Seasonal Mann–Kendall trend analysis of precipitation over MDF and DDF

Table 2 Details of Mann–Kendall trend analysis over MDF and DDF

Seasons	Tau	Z value	S value	*P* value	Significance level	Sen's slope	Trend
Moist Deciduous Forest (MDF)							
JJAS	– 0.089	0.803	70.0	0.421	$P > 0.05$	0.116	No trend
MAM	– 0.543	4.928	– 424.0	$8.72e^{-7}$	$P > 0.05$	0.092	Decreasing
OND	– 0.1	0.897	78.0	0.369	$P > 0.05$	0.014	No trend
JF	– 0.382	3.460	– 298.0	0.0005	$P > 0.05$	0.071	Decreasing
Dry Deciduous Forest (DDF)							
JJAS	0.156	1.384	120.0	0.165	$P > 0.05$	0.096	No trend
MAM	0.164	1.479	128.0	0.138	$P > 0.05$	0.013	No trend
OND	0.058	0.524	46.0	0.600	$P > 0.05$	0.005	No trend
JF	– 0.053	0.477	– 42.0	0.632	$P > 0.05$	0.005	No trend

5.2 *Identification of Drought and Non-drought Period*

5.2.1 Drought Period Identification Over DDF

During the 12-month accumulation period, there was a severe drought from November 2015 to December 2015. The drought lasted for one month and had an average value of − 1.54. In a 9-month accumulation period, an extreme drought occurred from September 2000 to November 2000, lasting for two months, with an average SPI value of − 2.23. Additionally, a moderate drought occurred from June 2002 to December 2002 and lastedsix months, with an average SPI value of − 1.25. Another event of moderate drought took place from November 2015 to December 2015, lasting for one month, with an average SPI value of − 1.35. During the 6-month accumulation period, there were several moderate drought events. During a 6-month accumulation period, a severe drought occurred from July 2002 to November 2002, lasting for four months and with an average SPI value of − 1.31. The following severe event was observed from June 2009 to August 2009, lasting two months with an average SPI of − 1.6. Another mild drought event was observed from June 2010 to December 2010, lasting six months with an average SPI value of − 0.48. The following mild drought event occurred from March 2011 to June 2011, lasting for three months, with an average SPI value of − 0.93. The next event lasted from August 2014 to November 2014, lasting for three months, with an average SPI value of − 0.89. Another event occurred from November 2015 to December 2015, lasting one month, with an SPI value of − 1.42. Lastly, an event lasted two months, from June 2016 to August 2016, with an average SPI value of − 0.61.

5.2.2 Non-Drought Period Identification Over DDF

At the 12-month accumulation period, the moderate drought period was observed for May 2003 with an average SPI value of 1.02. At the 9-month accumulation period, a moderate drought was observed from July 2012 to December 2012, which lasted for 5 months with an average SPI value of 1.03. At a six-month accumulation period, mild wetness was observed from January 2007 to July 2007, lasting six months, with an average SPI value of 0.50. Moderate wet condition was seen from March 2013 to June 2013, lasting 3 months with an average SPI value of 1.01. Severe wet condition was observed from January 2020 to June 2020, lasting for 5 months with an average SPI value of 1.10.

5.2.3 Drought Period Identification Over MDF

At 12-month accumulation period, an extreme drought occurred from April 2000 to October 2004, lasting for 54 months. The average SPI value during this period was – 1.62. A moderate drought occurred from October 2005 to September 2009, lasting 47 months, with an average SPI value of – 1.23. A mild drought occurred from September 2006 to September 2009, lasting for a total of 36 months. The average SPI value during this period was – 0.92. At 9-month accumulation period, the severe drought period lasted for 48 months, starting from October 2000 and ending in October 2004, with an average SPI value of – 1.52. The subsequent period of mild drought occurred from July 2005 to July 2009, lasting for 48 months, with an average SPI value of – 1.01. Subsequently, a prolonged severe drought episode occurred from January 2000 to October 2004, lasting 57 months. The average SPI value during this period was – 1.55. A severe drought lasting 48 months, from October 2005 to October 2009, was documented with an average SPI value of – 0.98. At 6 month accumulation period, a mild drought event occurred from May 2017 to November 2017, lasting for 6 months with an average SPI value of – 0.69. A prolonged severe drought occurred from June 2005 to March 2009, lasting 45 months, with an average SPI value of – 0.97. The subsequent extended period of mild drought occurred from May 2016 to November 2017, lasting 6 months, with an average SPI value of – 0.99.

5.2.4 Non-Drought Period Identification Over MDF

At 12 months' accumulation period, the extreme wet condition was seen from July 2010 to January 2011 with 6-month duration with an average SPI value of 2.15. At 9 months' accumulation period, the severe wet period started from January 2011 to March 2010 with 2 months' duration and an average SPI value of 1.63. Moderate wet was observed in July, 2012 December 2012, which lasted five months with an average SPI value of 1.07. Another moderate event was observed from May 2013 to June 2013, which lasted1 month with an average SPI value of 1.04. At 6 months' accumulation period, the moderate wet event was observed from January 2011 to May 2011, which lasted for 4 months with an average SPI value of 1.14; another moderate wet event was observed from January 2018 to July 2018, which lasted for 6 months with SPI value of 1.30.

5.3 Behavior of WUE During Drought and Non-drought Periods

The subsequent goal is to determine the average WUE values for these periods to clarify the fluctuating response of WUE under different hydrological conditions. Stringent selection criteria distinguish their respective occurrences during periods of drought and non-drought. More precisely, the SPI defined drought occurrences with similar duration and intensity. This analysis focuses on the period from July to December 2002, characterized by drought. On the other hand, the period from July to December 2012 represents normal water availability levels or non-drought conditions. The investigation examines explicitly the MDF environment over these timeframes. The temporal classifications are based on SPI levels, which serve as indicators for drought and non-drought situations. The period from June to December 2008 is classified as the drought period for DDF, in contrast to the period from June to December 2012, representing non-drought or normal water availability circumstances. Understanding the relationship between SPI and WUE is extremely important. The SPI was used to measure drought conditions caused by protracted periods of below-average rainfall, whereas WUE represents the ratio of GPP to ET. WUE essentially describes how forest ecosystems adapt to a lack of precipitation.

Within the context of MDF, the behaviour of WUE during drought events can be described as follows: SPI values exhibit initial decline (high drought intensity), and WUE increases until a critical SPI threshold is achieved. An SPI obtained close to or more than − 1 indicates moderate severe drought conditions, decreasing WUE. The reduction in WUE was caused by the severe water shortage that is available for the physiological activities occurring within plant tissues. When the soil moisture levels improve to slightly dry circumstances (i.e., SPI < 0.05), the plant shows an increase in its ability to use the available water resources efficiently. However, because of the likely continuation of less-than-ideal water levels, there was a decrease in WUE under drought circumstances in the MDF ecosystem.

On the other hand, an improvement in WUE was observed in situations that promote carbon sequestration. The inverse relation between WUE and SPI was observed due to decreased soil respiration and canopy transpiration caused by reduced stomatal conductance. This highlights the ability of MDF ecosystems to withstand and adapt to climate disturbances, including periods of long-term drought. However, if the SPI results above a certain threshold in a positive direction, a deviation from the predetermined trajectory of WUE) wasnoticed (Fig. 3a). Unique behavioural traits are evident in DDF. DDFs have increased resistance to extreme climatic variability, such as drought and changes in water availability, during specific seasonal periods. Similar to MDFs, the plant life in DDFs initially increases GPP while SPI values decrease. However, when the SPI reaches or exceeds a threshold of − 1, the WUE shows a negative response, which suggests unfavourable environmental circumstances. Empirical research indicates that WUE increases as SPI values decrease (Hatfield and Dold 2019) (Fig. 3b).

An analysis conducted retrospectively covering the period of drought from 2001 to 2020 reveals frequent incidences of drought, mainly observed between July and December, within the MDF environment. This post-monsoon phase was characterized by decreased soil moisture content and ET due to reduced stomatal conductance. These conditions are typical of a drought. The increase in GPP levels over this period suggests an enhanced WUE within the specified geographical area.

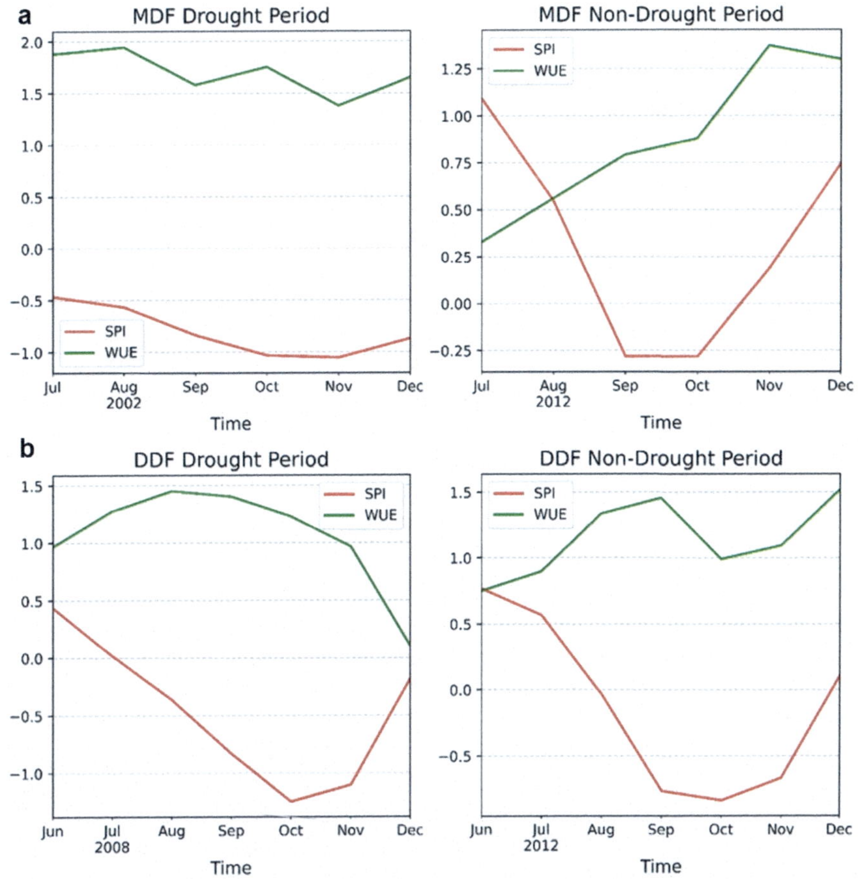

Fig. 3 Behavior of WUE of drought and non-drought period over MDF and DDF

5.4 *Resilience and Recovery of MDF and DDF*

Drought is a very intricate hydrological and climatic phenomenon that leads to detrimental effects on the structure and functioning of ecosystems. Tropical forests are highly susceptible to climate change, which can result in various adverse effects, including forest mortality and increased risk of wildfires (Keenan 2015). Undoubtedly, the impact of climate change could lead to the extinction of numerous plants that cannot acclimate to the changing conditions. These factors can potentially alter the ecosystem's structure and functioning, leading to detrimental effects on floral biodiversity (Memmott et al. 2007). Climate change will cause drought stress, directly affecting the productivity and health of forests. This was due to the increased species mortality caused by disturbances such as insect pests, diseases, and forest fires. The plant's reaction to dryness was intricate, ranging from individual plant death to large-scale forest die-offs. Nevertheless, accurately predicting the species-specific response of plants to dry conditions was a challenge.

This study aimed to determine the forest types resilient and resistant to drought. This was achieved by analyzing the WUE of different forest types throughout drought and non-drought periods. For this estimation, we can rely on the following results. The first test was the MK test, followed by the determination of drought and non-drought periods, and finally, the examination of how WUE behaves during certain periods. According to this study, the MK test on precipitation justifies that the MDF remains stable despite a declining tendency from January to May.

Additionally, it shows a weak link between SPI and WUE during dry periods. In addition, a prolonged spell of drought was also recorded in MDF, lasting around 57 months from January 2000 to October 2004. The severity of the drought ranges from mild to moderate conditions. Therefore, it can be inferred that MDF exhibits drought resilience. However, there was no notable trend in the precipitation pattern in DDF. In addition, the duration of the drought period was less than the MDF. The severity of the drought ranges from moderate to extreme conditions. The most prolonged drought period documented in DDF lasted almost 6 months, from June 2002 to December 2002. The DDF forest exhibits drought resistance as it has a low ET rate, allowing it to sustain itself in a stressed environment.

Consequently, the WUE of the forest was similarly high. The relation between SPI and WUE was higher in DDF compared to MDF. WUE is often highest during periods of extreme water stress, and it falls as water availability increases. The forest system's ability to withstand droughts and water stress demonstrates its resilience. The endurance of this forest type maintains its capacity for survival. The SPI and WUE have a linear relationship. As the SPI, which measures drought, increases, the WUE also increases. Specifically, when the SPI value falls between -1 and -2, the WUE increases. Conversely, when the SPI value falls between 1 and 2, the WUE decreases. This type of condition is seen more prominently in resistant forest type (DDF) than MDF (resilient forest type).

6 Discussion

While forest ecosystems inherently possess resilience, enabling many species and ecosystems to adapt to historical climatic changes, the anticipated scale and pace of future alterations could surpass the natural adaptive capacity of forest species and ecosystems (Keenan 2015). According to this study, MDF and DDF exhibit diverse responses to drought stress, influenced by their ecological characteristics and environmental conditions. MDF, typically found in regions with higher rainfall and moderate temperatures, maintains greater species diversity than DDF, which thrives in drier climates with lower rainfall and higher temperatures. This higher diversity in MDF contributes to more significant ecosystem stability and resilience against drought impacts. Young trees in both forest types are particularly vulnerable to drought. However, well-established DDF trees with deep root systems and efficient nutrient accumulation often exhibit greater tolerance to water deficit conditions. DDF species have developed specific adaptations to cope with drought, such as increased WUE, enhancing their resistance to prolonged dry periods (Hatfield and Dold 2019; Zhao et al. 2020).

Additionally, DDF trees shed their leaves during the dry season as a water conservation strategy, whereas MDF trees retain their leaves year-round. Despite maintaining leaves during drought, MDF may experience reduced photosynthetic capacity, while DDF adjusts leaf phenology to minimize water loss. The resistance of DDF to drought can be attributed to various factors, including increased temperature, radiation, hydraulic conductivity, and adjustments in leaf phenology (Lv et al. 2022). Moreover, the higher species richness in MDF ecosystems influences their resilience rates, allowing them to bounce back more quickly from drought events. Understanding these contrasting responses and adaptations in MDF and DDF ecosystems was crucial for effective forest management and conservation in changing climatic conditions.

7 Challenges

The diverse responses of MDF and DDF during drought period present several challenges for forest management and conservation. One challenge lies in predicting the long-term impacts of climate change on these ecosystems, considering their varying levels of resilience and resistance. Additionally, maintaining the ecological integrity of these forests amidst increasing anthropogenic pressures requires a nuanced understanding of their adaptive capacities and vulnerabilities. Addressing these challenges necessitates interdisciplinary research efforts integrating climatology, ecology, and socio-economic factors to develop robust management strategies that promote forest resilience while ensuring sustainable resource use.

8 Future Directions/Research Needs

Future research endeavors should focus on elucidating the mechanisms underlying the resilience and resistance of MDF and DDF ecosystems during drought periods. This entails investigating physiological, ecological, and genetic factors influencing the adaptive responses of key forest species to changing climatic conditions. Furthermore, there is a critical need for long-term monitoring initiatives to track ecosystem dynamics and assess the effectiveness of management interventions in enhancing forest resilience. Integrating advanced remote sensing techniques, ecological modeling, and field-based experiments can provide valuable insights into the drivers of ecosystem resilience and inform evidence-based conservation practices.

9 Conclusion

The analysis of response of WUE for MDF and DDF in response to drought period and non-drought period, underscores the intricate interplay between ecological dynamics and climatic factors. Our findings reveal that while MDFs maintain greater species diversity and ecosystem stability, DDFs exhibit a high capacity for tolerating water deficit conditions through specific adaptations. These adaptations include physiological adjustments, such as increased WUE and alterations in leaf phenology, which enable DDFs to sustain productivity even amidst prolonged dry periods. MDF, characterized by their year-round leaf retention strategy, struggle to cope with drought conditions. Nonetheless, their response to drought stress is mitigated by their rich species diversity and inherent ecosystem resilience. The variations in response underscore the need for customized management strategies that recognize the unique ecological traits of MDF and DDF. Effective conservation strategies should consider the varying degrees of resilience and resistance exhibited by these forest types, integrating scientific knowledge with community engagement and policy frameworks. Furthermore, interdisciplinary research efforts are crucial for advancing our understanding of the underlying mechanisms driving forest resilience and for developing adaptive management strategies that anticipate and mitigate the impacts of climate change on these vital ecosystems.

References

Allen CD, Macalady AK, Chenchouni H, Bachelet D, McDowell N, Vennetier M, Kitzberger T et al (2010) A global overview of drought and heat-induced tree mortality reveals emerging climate change risks for forests. For Ecol Manage 259(4):660–684. https://doi.org/10.1016/j.foreco.2009.09.001

Assal TJ, Anderson PJ, Sibold J (2016) Spatial and temporal trends of drought effects in a heterogeneous semi-arid forest ecosystem. For Ecol Manage 365(April):137–151. https://doi.org/10.1016/j.foreco.2016.01.017

Baldocchi DD, Verma SB, Anderson DE (1987) Canopy photosynthesis and water-use efficiency in a deciduous forest. J Appl Ecol 24(1):251. https://doi.org/10.2307/2403802

Bargali H, Bhatt D, Sundriyal RC, Uniyal VP, Pandey A, Ranjan R (2023) Effect of forest fire on tree diversity and regeneration in the forests of Uttarakhand, Western Himalaya, India. Front Forests Glob Change 6(July):1–10. https://doi.org/10.3389/ffgc.2023.1198143

Bohner T, Diez J (2021) Tree resistance and recovery from drought mediated by multiple abiotic and biotic processes across a large geographic gradient. Sci Total Environ 789:147744. https://doi.org/10.1016/j.scitotenv.2021.147744

Brümmer C, Andrew Black T, Jassal RS, Grant NJ, Spittlehouse DL, Chen B, Nesic Z et al (2012) How climate and vegetation type influence evapotranspiration and water use efficiency in Canadian Forest, Peatland and Grassland Ecosystems. Agric for Meteorol 153:14–30. https://doi.org/10.1016/j.agrformet.2011.04.008

Champion HG, Seth SK (1968) A revised survey of the forest types of India. The Manager of Publications Delhi 6

Cheng G, Liu T, Wang S, Wu Y, Zhang C (2023) Responses to the impact of drought on carbon and water use efficiency in Inner Mongolia. Land 12(3). https://doi.org/10.3390/land12030583

Cleugh HA, Leuning R, Mu Q, Running SW (2007) Regional evaporation estimates from flux tower and MODIS satellite data. Remote Sens Environ 106(3):285–304. https://doi.org/10.1016/j.rse.2006.07.007

Cohen WB (1991) Response of vegetation indices to changes in three measures of leaf water stress. Photogramm Eng Remote Sens 57(2):195–202

Dhyani K, Kuniyal HB, Singh H, Sobha (2021) Growth and physiological potential of *Terminalia arjuna* under elevated CO_2 levels in Open top chamber condition. J Appl Nat Sci 13(3):1121–1126. https://doi.org/10.31018/jans.v13i3.2490, ISSN 2231-5209

Donovan LA, Ehleringer JR (1992) Contrasting patterns among size and life- shrub history classes of a semi-arid. Funct Ecol 6(4):482–488

Edwards DC, McKee TB (1997) Department of Atmospheric Science. Weather—soundings. Climatology Report No. 97-2, No. 634. http://weather.uwyo.edu/upperair/sounding.html

Frimpong BF, Koranteng A, Molkenthin F (2022) Analysis of temperature variability utilising Mann-Kendall and Sen's slope estimator tests in the Accra and Kumasi Metropolises in Ghana. Environ Syst Res 11(1):1–13. https://doi.org/10.1186/s40068-022-00269-1

Frolking S, Palace MW, Clark DB, Chambers JQ, Shugart HH, Hurtt GC (2009) Forest disturbance and recovery: a general review in the context of spaceborne remote sensing of impacts on aboveground biomass and canopy structure. J Geophys Res: Biogeosci 114(3). https://doi.org/10.1029/2008JG000911

Funk C, Peterson P, Landsfeld M, Pedreros D, Verdin J, Shukla S, Husak G et al (2015) The climate hazards infrared precipitation with stations—a new environmental record for monitoring extremes. Scientific Data 2:1–21. https://doi.org/10.1038/sdata.2015.66

Gadedjisso-tossou A, Adjegan KII (2021) Rainfall and temperature trend analysis by Mann–Kendall test and significance for rainfed cereal yields in. Sci 3(17):1–20

Gao Y, Markkanen T, Aurela M, Mammarella I, Thum T, Tsuruta A, Yang H, Aalto T (2017) Response of water use efficiency to summer drought in a Boreal Scots Pine Forest in Finland. Biogeosciences 14(18):4409–4422. https://doi.org/10.5194/bg-14-4409-2017

Hatfield JL, Dold C (2019) Water-use efficiency: advances and challenges in a changing climate. Front Plant Sci 10(February):1–14. https://doi.org/10.3389/fpls.2019.00103

Hu J, Ma W, Wang Z (2023) Effects of nitrogen addition and drought on the relationship between nitrogen- and water-use efficiency in a temperate grassland. Ecol Processes 12(1). https://doi.org/10.1186/s13717-023-00448-7

Huang, L, He B, Han L, Liu J, Wang H, Chen Z (2017) A global examination of the response of ecosystem water-use efficiency to drought based on MODIS data. Sci Total Environ 601–602:1097–1107. https://doi.org/10.1016/j.scitotenv.2017.05.084

Huang M, Zhai P, Piao S (2021) Divergent responses of ecosystem water use efficiency to drought timing over Northern Eurasia. Environ Res Lett 16(4). https://doi.org/10.1088/1748-9326/abf0d1

Jaleel CA, Gopi R, Sankar B, Gomathinayagam M, Panneerselvam R (2008) Differential responses in water use efficiency in two varieties of *Catharanthus Roseus* under drought stress. C R—Biol 331(1):42–47. https://doi.org/10.1016/j.crvi.2007.11.003

Javed T, Yao N, Chen X, Suon S, Li Y (2020) Drought evolution indicated by meteorological and remote-sensing drought indices under different land cover types in China. Environ Sci Pollut Res 27(4):4258–4274. https://doi.org/10.1007/s11356-019-06629-2

Jha S, Das J, Sharma A, Hazra B, Goyal MK (2019) Probabilistic evaluation of vegetation drought likelihood and its implications to resilience across India. Glob Planet Change 176(January):23–35. https://doi.org/10.1016/j.gloplacha.2019.01.014

Ji L, Peters AJ (2003) Assessing vegetation response to drought in the Northern Great Plains using vegetation and drought indices. Remote Sens Environ 87(1):85–98. https://doi.org/10.1016/S0034-4257(03)00174-3

Kauwe MG, De BE, Medlyn SZ, Walker AP, Dietze MC, Hickler T, Jain AK et al (2013) Forest water use and water use efficiency at elevated CO_2: a model-data intercomparison at two contrasting temperate forest FACE sites. Glob Change Biol 19(6):1759–1779. https://doi.org/10.1111/gcb.12164

Keenan RJ (2015) Climate change impacts and adaptation in forest management: a review. Ann for Sci 72(2):145–167. https://doi.org/10.1007/s13595-014-0446-5

Kendall MG (1948) Rank correlation methods. Griffin, Oxford, England

Knauer J, Zaehle S, Reichstein M, Medlyn BE, Forkel M, Hagemann S, Werner C (2017) The response of ecosystem water-use efficiency to rising atmospheric CO_2 concentrations: sensitivity and large-scale biogeochemical implications. New Phytol 213(4):1654–1666. https://doi.org/10.1111/nph.14288

Kogan FN (1995) Droughts of the late 1980s in the United States as derived from NOAA Polar-Orbiting Satellite Data. Bull—Am Meteorol Soc 76(5):655–668. https://doi.org/10.1175/1520-0477(1995)076%3c0655:DOTLIT%3e2.0.CO;2

Kumar A, Kumar P, Singh H, Kumar N (2021a) Modulation of plant functional traits under essential plant nutrients during seasonal regime in natural forests of Garhwal Himalayas. Plant Soil 465:197–212. https://doi.org/10.1007/s11104-021-05003-x

Kumar A, Kumar P, Singh H, Kumar N (2021b) Adaptation and mitigation potential of roadside trees with bio-extraction of heavy metals under vehicular emissions and their impact on physiological traits during seasonal regimes. Urban for Urban Greening 58:126900. https://doi.org/10.1016/j.ufug.2020.126900

Kumar A, Kumar P, Singh H, Kumar N (2021c) Impact of plant functional traits on infiltration rate under different forests of Kempty watershed in Garhwal Himalaya. Indian J Soil Conserv 49(1):38–44

Kumar A, Kumar P, Singh H, Bisht S, Kumar N (2021d) Relationship of physiological plant functional traits with soil carbon stock in the temperate forest of Garhwal Himalaya. Curr Sci 120(8):1368–1373. https://doi.org/10.18520/cs/v120/i8/1368-1373

Kumar N, Singh H, Giri K, Kumar A, Joshi A, Yadav A, Singh R, Bisht S, Kumari R, Jeena N, Khairakpam R, Mishra G (2024) Physiological and molecular insights into the allelopathic effects on agroecosystems under changing environmental conditions. Physiol Mol Biol Plants 30:417–433. https://doi.org/10.1007/s12298-024-01440-x

Kumar M, Phukon SN, Paygude AC, Tyagi K, Singh H (2021) Mapping Phenological Functional Types (PhFT) in the Indian Eastern Himalayas using machine learning algorithm in Google Earth Engine. Comput Geosci 158. https://doi.org/10.1016/j.cageo.2021.104982

Kumar A, Singh H, Kumari G, Bisht S, Malik A, Kumar A, Singh M, Raturi, Barthwal S, Thakur A, Kaushal R (2022) Adaptive resilience of roadside trees to vehicular emissions via leaf enzymatic, physiological, and anatomical trait modulations. Environ Pollut 313. https://doi.org/10.1016/j.envpol.2022.120191

Law BE (2002) Environmental controls over carbon dioxide and water vapor exchange of terrestrial vegetation. Agric for Meteorol 113:97–120

Ho LT, Hoppe J, Thomas FM (2021) Resistance, recovery and resilience of two co-occurring Palaeotropical Pinus species differing in the sizes of their distribution areas. Forests 12(4):1–16. https://doi.org/10.3390/f12040511

Liu X, Chen X, Li R, Long F, Zhang L, Zhang Q, Li J (2017) Water-use efficiency of an old-growth forest in lower subtropical China. Sci Rep 7(August 2016):1–10. https://doi.org/10.1038/srep42761

Lloret F, Keeling EG, Sala A (2011) Components of tree resilience: effects of successive low-growth episodes in old Ponderosa Pine Forests. Oikos. https://doi.org/10.1111/j.1600-0706.2011.19372.x

Luo X, Wang Y, Li Y (2023) Responses of ecosystem water use efficiency to drought in the Lancang-Mekong River Basin. Front Ecol Evol 11(May):1–13. https://doi.org/10.3389/fevo.2023.1203725

Lv Y, He H, Ren X, Zhang L, Qin K, Wu X, Niu Z, Feng L, Xu Q, Zhang M (2022) High resistance of deciduous forests and high recovery rate of evergreen forests under moderate droughts in China. Ecol Ind 144:109469. https://doi.org/10.1016/j.ecolind.2022.109469

Ma J, Zhang C, Li S, Yang C, Chen C, Yun W (2023) Changes in vegetation resistance and resilience under different drought disturbances based on NDVI and SPEI time series data in Jilin Province, China. Remote Sens 15(13). https://doi.org/10.3390/rs15133280

Mann HB (1945) Non-parametric test against trend. Econometrica 13(3):245–259. http://www.economist.com/node/18330371?story%7B_%7Did=18330371%0A, http://www.jstor.org/stable/1907187

Mckee TB, Doesken NJ, Kleist J (1993) The relationship of drought frequency and duration to time scales. In: 8th Conference on applied climatology, Am Meteorol Soc, 179–184. https://doi.org/10.1002/jso.23002

McQuillan KA, Tulbure MG, Martin KL (2022) Forest water use is increasingly decoupled from water availability even during severe drought. Landscape Ecol 37(7):1801–1817. https://doi.org/10.1007/s10980-022-01425-9

Memmott J, Craze PG, Waser NM, Price MV (2007) Global warming and the disruption of plant-pollinator interactions. Ecol Lett 10(8):710–717. https://doi.org/10.1111/j.1461-0248.2007.01061.x

Mu Q, Zhao M, Running SW (2011) Improvements to a MODIS global terrestrial evapotranspiration algorithm. Remote Sens Environ 115(8):1781–1800. https://doi.org/10.1016/j.rse.2011.02.019

Nanzad L, Zhang J, Tuvdendorj B, Nabil M, Zhang S, Bai Y (2019) NDVI anomaly for drought monitoring and its correlation with climate factors over Mongolia from 2000 to 2016. J Arid Environ 164(February):69–77. https://doi.org/10.1016/j.jaridenv.2019.01.019

Nemani R, White M, Thornton P, Nishida K, Reddy S, Jenkins J, Running S (2002) Recent trends in hydrologic balance have enhanced the terrestrial carbon sink in the United States. Geophys Res Lett 29(10):106-1–106-4. https://doi.org/10.1029/2002gl014867

Niu X, Liu S (2021) Drought affected ecosystem water use efficiency of a natural oak forest in Central China. Forests 12(7):1–19. https://doi.org/10.3390/f12070839

Pei F, Wu C, Liu X, Li X, Yang K, Zhou Y, Wang K, Xu L, Xia G (2018) Monitoring the vegetation activity in China using vegetation health indices. Agric Forest Meteorol 248(101):215–227. https://doi.org/10.1016/j.agrformet.2017.10.001

Peters AJ, Walter-Shea EA, Ji L, Viña A, Hayes M, Svoboda MD (2002) Drought monitoring with NDVI-based standardized vegetation index. Photogramm Eng Remote Sens 68(1):71–75

Prakash V, Hunney K, Singh H (2022) Effect of elevated carbon dioxide on growth and development of Santalum album L. seedlings inoculated with plant growth promoting microorganisms in Open Top Chambers. Int J Environ, Agric Biotechnol 7(5):149–159

Qin G, Meng Z, Fu Y (2022) Drought and water-use efficiency are dominant environmental factors affecting greenness in the Yellow River Basin, China. Sci Total Environ 834:155479. https://doi.org/10.1016/j.scitotenv.2022.155479

Didi Sacré Regis M, Mouhamed L, Kouakou K, Adeline B, Arona D, Saint CHJ, Claude KKA, Jean CTH, Salomon O, Issiaka S (2020) Using the CHIRPS dataset to investigate historical changes in precipitation extremes in West Africa. Climate 8(7). https://doi.org/10.3390/CLI8070084

Rousta I, Olafsson H, Moniruzzaman Md, Zhang H, Liou YA, Mushore TD, Gupta A (2020) Impacts of drought on vegetation assessed by vegetation indices and meteorological factors in Afghanistan. Remote Sens 12(15). https://doi.org/10.3390/RS12152433

Running SW, Nemani RR, Heinsch FA, Zhao M, Reeves M, Hashimoto H (2004) A continuous satellite-derived measure of global terrestrial primary production. Bioscience 54(6):547–560. https://doi.org/10.1641/0006-3568(2004)054[0547:ACSMOG]2.0.CO;2

Sa'adi Z, Alias NE, Yusop Z, Hamed MM, Shukla P, Houmsi MR, Mohamad NA et al (2024) Characterization of the future Northeast Monsoon Rainfall based on the clustered climate zone under CMIP6 in Peninsular Malaysia. Atmos Res 304:107407. https://doi.org/10.1016/j.atmosres.2024.107407

Schwarz J, Skiadaresis G, Kohler M, Kunz J, Schnabel F, Vitali V, Bauhus J (2020) Quantifying growth responses of trees to drought—a critique of commonly used resilience indices and recommendations for future studies. Curr for Rep 6(3):185–200. https://doi.org/10.1007/s40725-020-00119-2

Sharma R, Singh H (2021) Alteration in biochemical constituents and nutrients partitioning of *Asparagus racemosus* in response to elevated atmospheric CO_2 concentration. Environ Sci Pollut Res 29:6812–6821. https://doi.org/10.1007/s11356-021-16050-3

Shi X, Chen F, Ding H, Li Y, Shi M (2022) Quantifying vegetation stability under drought in the middle reaches of Yellow River Basin, China. Forests 13(7). https://doi.org/10.3390/f13071138

Singh H (2021) An integrated approach considering physiological and biophysical based indicators for assessing tolerance of roadside plantations of *Alstonia scholaris* towards urban roadside air pollution: An assessment of adaptation of plantations for mitigating urban roadside air pollution. Trees. https://doi.org/10.1007/s00468-021-02179-8

Singh S, Singh H, Sharma SK, Nautiyal R (2021) Seasonal variation in biochemical responses of bamboo clones in the sub-tropical climate of Indian Himalayan foothills. Heliyon 7(4):e06859. https://doi.org/10.1016/j.heliyon.2021.e06859

Singh M, Singh H, Kumar A, Kumar M, Barthwal S, Thakur A (2024) Soil nitrogen availability determines the CO_2 fertilization effect on tree species (*Neolamarckia cadamba*): growth and physiological evidence. Environ Sustain 7:53–60. https://doi.org/10.1007/s42398-023-00300-w

Singh H, Yadav M, Kumar N, Kumar A, Kumar M (2020) Assessing adaptation and mitigation potential of roadside trees under the influence of vehicular emissions: a case study of Grevillea robusta and Mangifera indica planted in an urban city of India. PLoS ONE 15(1):e0227380. https://doi.org/10.1371/journal.pone.0227380

Singh H, Bandyopadhyaya S, Kumar A, Singh M, Malik A, Kumari G, Raturi A, Yadav SK, Kumar M, Barthwal S, Thakur A, Rawat PS (2022) Understanding the physiological and biophysical response of urban roadside plantations for assessing adaptation and mitigation mechanisms toward vehicular emissions. Urban Climate 44. https://doi.org/10.1016/j.uclim.2022.101183

Singh H (2024) Exploring adaptive modulation in plant functional traits and its impact on the productivity of *Acacia auriculiformis* under CO_2-enriched environment. Ind Crops Prod 210:118186. https://doi.org/10.1016/j.indcrop.2024.118186

Sohn JA, Saha S, Bauhus J (2016) Potential of forest thinning to mitigate drought stress: a meta-analysis. For Ecol Manage 380:261–273. https://doi.org/10.1016/j.foreco.2016.07.046

Thong P, Sahoo UK, Thangjam U, Pebam R (2020) Pattern of forest recovery and carbon stock following shifting cultivation in Manipur, North-East India. PLoS ONE 15(10 October):1–20. https://doi.org/10.1371/journal.pone.0239906

Usman M, Nichol JE (2020) A spatio-temporal analysis of rainfall and drought monitoring in the Tharparkar region of Pakistan. Remote Sens 12(3). https://doi.org/10.3390/rs12030580

Warren J, Gilbert RO (1988) Statistical methods for environmental pollution monitoring. Technometrics 30. https://doi.org/10.2307/1270090

Worrall JJ, Rehfeldt GE, Hamann A, Hogg EH, Marchetti SB, Michaelian M, Gray LK (2013) Recent declines of Populus Tremuloides in North America linked to climate. For Ecol Manage 299:35–51. https://doi.org/10.1016/j.foreco.2012.12.033

Yang Y, Guan H, Batelaan O, McVicar TR, Long D, Piao S, Liang W, Liu B, Jin Z, Simmons CT (2016) Contrasting responses of water use efficiency to drought across global terrestrial ecosystems. Sci Rep 6(November 2015):1–8. https://doi.org/10.1038/srep23284

Yu X, Guo X (2023) Inter-annual drought monitoring in Northern mixed grasslands by a revised vegetation health index from historical Landsat imagery. J Arid Environ 213(May 2022):104964. https://doi.org/10.1016/j.jaridenv.2023.104964

Zhang L, Tian J, He H, Ren X, Sun X, Yu G, Lu Q, Lv L (2015) Evaluation of water use efficiency derived from MODIS products against Eddy variance measurements in China. Remote Sens 7(9):11183–11201. https://doi.org/10.3390/rs70911183

Zhao M, Heinsch FA, Nemani RR, Running SW (2005) Improvements of the MODIS terrestrial gross and net primary production global data set. Remote Sens Environ 95(2):164–176. https://doi.org/10.1016/j.rse.2004.12.011

Zhao J, Xu T, Xiao J, Liu S, Mao K, Song L (2020) Responses of water use efficiency to drought in. Remote Sens 12(199)

Climate Change-Induced Forest Ecosystem Functioning and Alteration in Water-Atmosphere-Soil–Plant (WASP) Relationship

Pooja Kaushik and Hukum Singh

Abstract This chapter focuses on forest ecosystem responses to increasing temperature and shifting precipitation patterns in different magnitudes, such as distressed plants growth, reduced primary productivity, shifting phenological behaviour, unstable soil, and plant carbon sequestration capacity, increasing defoliations in many plants, altered vegetation composition, increased plant mortality rate, distraught water and energy flux, unbalanced nutrient cycling, higher soil acidification, increased loss of biodiversity and many more ecological functions of the forest. These processes are interconnected, emphasizing the value of understanding observations of plants and soil processes, plant-water relationship, water-atmosphere-soil–plant (WASP) interaction, and exclusively the connections among them as part of an enhanced understanding of how forest ecosystems will respond to climate change. In discussing the furthermost expected effects of climate change on the biogeochemical processes in forest soils, it is significant to consider the association among the four compartments of forest ecosystems: the water, the atmosphere, the soil, and the plants (WASP system). However, modification in any of these compartments will influence the others, conferring to the entire transfer of mass and energy and mean residence times established among compartments. We observe that increasing temperatures can stimulate water stress, shifting phenology, and forest species composition, all driving the mean residence time of canopy foliage. The grade of retention of foliage, in turn, influences the nutrient cycles in vegetation and the allocation of nutrients from the canopy to the soil. Notwithstanding constant uncertainty about climate change and forest ecosystem trajectories under global change, our review focuses on the comprehensible form of forest ecological change across WASP systems.

Keywords Nutrient cycle · Phenology · Carbon sequestration · Productivity · Loss of biodiversity · WASP system

---

P. Kaushik (✉)
Lala Kishanchand Government P.G College, Gangoh, Saharanpur, Uttar Pradesh 247341, India
e-mail: pooja.k0409@gmail.com

H. Singh
Plant Physiology Discipline, ICFRE-Forest Research Institute, P.O. New Forest, Dehradun, Uttarakhand 248006, India

H. Singh (ed.), *Urban Forests, Climate Change and Environmental Pollution*,
https://doi.org/10.1007/978-3-031-67837-0_38

1 Introduction

The forest ecosystem is heavily influenced by climate, affecting various physical, chemical, and biological processes that shape the hydrologic cycle, atmospheric conditions, soil dynamics, and plant interactions. Climate change will significantly impact the development and functioning of forest ecosystems worldwide, including tropical deciduous, tropical evergreen, mangrove, boreal, and temperate forests (Kumar et al. 2022; Singh et al. 2022; Kumar et al. 2024). Understanding these impacts requires deeply comprehending how climate influences forest structural complexity, affecting biodiversity, productivity, phenology, carbon sequestration, and other ecosystem functions (Ehbrecht et al. 2021; Fotis et al. 2018; Gough et al. 2019; Gupta et al. 2019b). Forest structural complexity, measured by canopy distribution and tree diversity, is crucial in determining ecosystem responses to climate change (Singh et al. 2024a; Singh 2024). New technologies like LiDAR have enabled better quantification of forest structure, aiding in developing metrics to assess complexity (Dassot et al. 2011; Davies and Asner 2014; Dhyani et al. 2021; Kumar et al. 2021a, b, c, d, e; Prakash et al. 2022). Measures of structural complexity have emerged as strong predictors of key ecosystem processes, such as net primary productivity, driven by factors like canopy occupation and light absorption (Sapijanskas et al. 2014; Seidel et al. 2019).

Additionally, tree species diversity and functional traits shape forest complexity and resilience (Singh et al. 2024b). Despite the critical importance of understanding these dynamics, research on the effects of climate change on forest ecosystems remains limited, partly due to the phenomenon's complexity and the slow response of forests to environmental changes (Harrison et al. 2020; Juchheim et al. 2017, 2020; Singh 2021; Sharma and Singh 2021; Singh et al. 2021). This chapter aims to enhance scientific understanding of how forests adapt to and are impacted by climate change, emphasizing the urgency of addressing this issue through global initiatives aimed at mitigation and adaptation strategies.

2 Impact of Climate Change on Different Magnitudes of Forest Ecosystem

Climate change poses many challenges for forest ecosystems, affecting everything from basic physiological processes in plants to complex interactions within the ecosystem (Fig. 1). It's not just about temperature and precipitation changes but also how they ripple through the entire ecosystem, affecting everything from plant growth and reproduction to the balance of species and nutrient cycling (Singh et al. 2020; Kumar et al. 2021a, b). It's a complex web of interactions that can have far-reaching consequences for the forests and the services they provide to humans and other species.

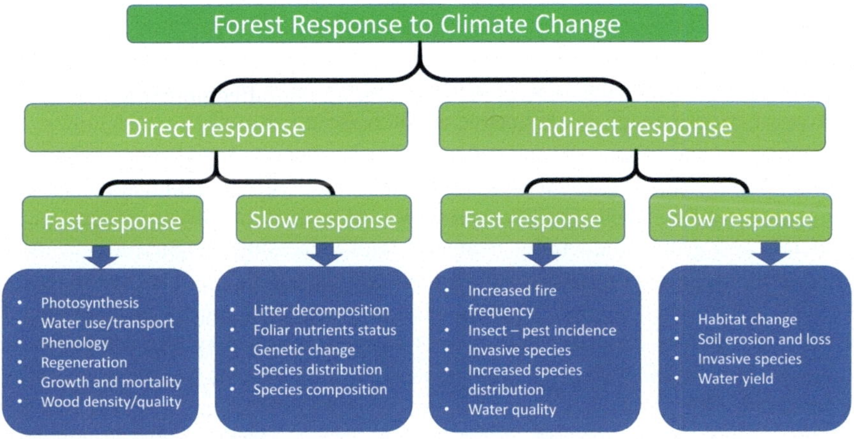

Fig. 1 Direct and indirect forest response to climate change

Climate change has evolved into an all-encompassing catastrophe driven by various anthropogenic activities, notably the excessive consumption of fossil fuels, rampant deforestation, rapid industrialization, population growth, and extensive land-use alterations. As a result, human-induced greenhouse gas (GHG) emissions since the pre-industrial era have led to a significant accumulation of carbon dioxide (CO_2—400 ppm), methane (CH4—1800 ppb), nitrous oxide (N_2O—330 ppb), and other GHGs in the atmosphere (IPCC 2014). Current projections indicate that the Earth's surface temperature is poised to increase by 0.4–0.6 °C on average during the twenty-first century due to the ongoing rate of fossil fuel consumption (Norby et al. 2003). Consequently, numerous facets of the forest ecosystem demand systematic and continuous long-term observations to comprehend the impacts of climate change on forest ecosystems (Ravindranath et al. 1997). These include changes in productivity response, shifts in phenological behavior, disruptions in soil carbon sequestration capacity, imbalances in biogeochemical cycles, alterations in forest biodiversity, disturbances in carbon, water, and energy fluxes, and modifications in water-atmosphere-soil–plant (WASP) interactions within the forest ecosystem (Menzel et al. 2006; Morin 2015; Moser 2010; Penone et al. 2019; Pretzsch 2017).

2.1 Forest Ecosystem Productivity Response Offensively to Climate Change

The impact of climate change on various factors such as topographical area, species diversity, stand composition, tree stage of development, soil type (particularly water holding capacity), CO_2 and nitrogen fertilization, and interactions between any of these factors has been affecting forest productivity (LeBauer and Treseder 2008).

Almost all variations may be temporal, chronological, or relapsing once saturation points have been reached. This is predictable as the circumstance intended for water accessibility, where the decrease of water usually reduces plant development. Still, in areas of excess water, it may primarily increase growth when water logging is concentrated. A similar response to plant growth has been noted for CO_2 enrichment, nitrogen fertilization, and temperature increases (Ollinger et al. 2008; LeBauer and Treseder 2008; Reich and Oleksyn 2008). In wide-range circumstances, plant efficiency and gross productivity increased with increasing temperatures in most forest areas, including the Amazon Forest, possibly due to CO_2 impregnation (Stark et al. 2012). However, in contrast to temperate regions, the tropical forests' temporal productivity increased and decreased once CO_2 saturation levels were reached. Few studies have recorded decreasing growth rates in tropical forests (Feeley et al. 2007). Water and nutrient deficiency over a lengthy period have also been revealed to cause a decline in plant productivity (Malhi et al. 2008). It might be the source designed for the deteriorated productivity and plant efficiency documented in previous studies. Few scientists claim that, founded on paleontological evidence, this may not result in forest dieback frequently declared in connection to anticipated variations in the Amazon region (Mayle and Power 2008). Natural instabilities often reduce forest area, but the damage they cause to standing trees may also decrease productivity (Chakraborty et al. 2018). After long-term observation, many environmentalists projected that climate change would play a very significant role in the alteration of water-atmosphere-soil–plant (WASP) interaction and exclusively the connections among them in the forest ecosystem, which is responsible for tremendous variation in the primary function of the ecosystem, specifically productivity (Albrecht and Kandji 2003).

2.2 Shifting Phenological Behavior of the Forest Ecosystem in Response to Climate Change

In the current scenario, climate change is compelling species diversity to adjust their phenological patterns, such as the timing of seasonal events like leaf sprouting, flowering, and fruiting, or migrating to more suitable ecosystems to avoid extinction (Singh and Kumar 2022). Environmental records indicate significant transformations in plant phenological events, with early occurrences of leaves, flowering, and fruiting, and an increase in the length of the growing season in recent decades. Phenological observations are crucial for understanding how plant species respond to regional climate conditions and changes. Consequently, phenology has emerged as a significant focus of ecological research (Schwartz 1999; Kumar et al. 2021c, d). Over the years, several critical phenological events have advanced, such as leaf emergence and flowering in spring, while events like leaf coloration and fall have been delayed in autumn (Cleland et al. 2007). The consequences of these phenological changes are still largely unknown (Yadav et al. 2019a, b; Gupta et al. 2019a). Still, it is believed to profoundly affect ecosystems, including alterations in species interactions, carbon and water cycles, WASP interactions, and Earth's climate (Piao et al. 2019). In response to this intriguing research area, many ecologists are striving to synthesize and inspire innovative research in plant phenology to address questions about the effects of phenological shifts on ecosystem functioning and local hydrology (Gupta et al. 2019a; Kumar et al. 2019c). This chapter aims to enhance our understanding of the physiological mechanisms driving current phenological changes in spring and fall, shed light on the consequences of these changes on hydrological cycles and ecosystem functioning, and highlight the significant role of climate change in shifting phenological behaviour (Kumar et al. 2019d). In mid-latitudes, the timing of spring events like budding, leafing, flowering, and fruiting depends heavily on atmospheric temperature and precipitation patterns. Warmer temperatures in late winter and early spring generally promote earlier flowering and leafing of plants (Wolkovich et al. 2012). Studies have shown a correlation between temperature increases and earlier flowering times, with higher temperatures leading to progressively earlier flowering in various plant species (Sparks et al. 2000). In India, research has also documented the early flowering, fruiting, and leaf emergence of Himalayan plants due to changing climatic conditions (Negi et al. 2017). Overall, changing climatic variability impacts plant species' phenological behavior, including crops (Kumar et al. 2019a, b, c, d; Kumari and Singh 2018).

2.3 *Unstable Soil Carbon Sequestration Capacity*

Continuous and systematic climate change monitoring has revealed that changing temperatures and precipitation patterns significantly affect the interaction between carbon storage and sequestration in forest ecosystems. As carbon is sequestered in plants and soil within forests, a reduction in atmospheric carbon occurs. By increasing carbon sequestration potential, we can contribute to slowing the rate of global temperature increase (Sharma et al. 2018; Singh et al. 2018; Gupta et al. 2018a, b). This relationship is crucial in climate change discussions, with many tropical nations aiming to reduce carbon emissions and enhance forest carbon sequestration capabilities to access funding for greenhouse gas (GHG) emissions reductions. Increasing CO_2 concentrations, prolonged dry seasons, and rising temperatures are projected to decrease forests' capacity to store and sequester carbon over time, potentially shifting them from carbon sinks to carbon sources in many regions (Ollinger et al. 2008). Carbon sequestration is influenced by productivity and the long-term storage of organic carbon in soil, meaning factors affecting productivity also affect the carbon sequestration potential of forest ecosystems (Atkins et al. 2018; Bushell et al. 2015). Additionally, short-term temperature increases may decrease carbon storage capacity, with outcomes varying based on seasonal variations, precipitation patterns, soil respiration, and human activities like deforestation and forest fires. Early spring warming may intensify carbon sequestration, while early autumn warming may increase respiration more than sequestration (Fig. 2).

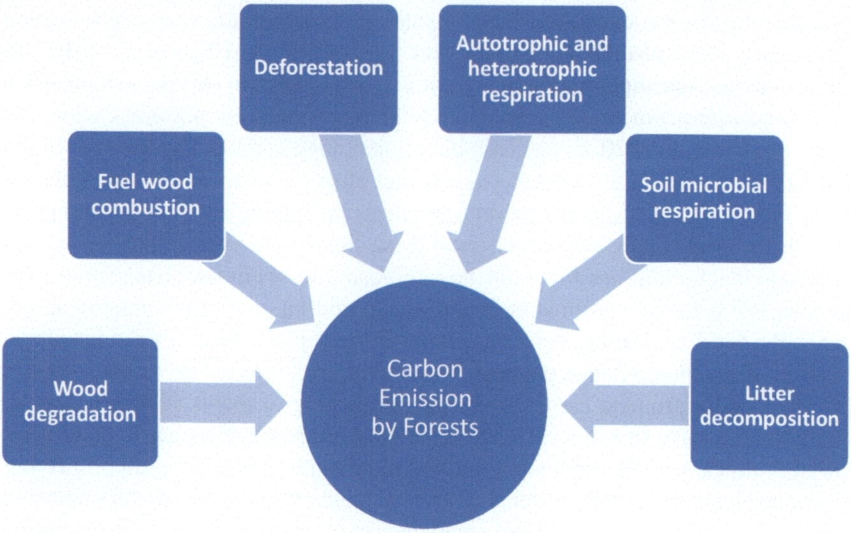

Fig. 2 Factors responsible for carbon emission and climate change in forest ecosystem

2.4 *Climate Change Causes an Imbalance in the Biogeochemical Cycle*

Climate change directly affects the structure and function of forest ecosystems, leading to unstable biogeochemical cycles and nutrient partitioning. Increasing temperatures, changing precipitation patterns, and rising atmospheric carbon dioxide concentrations alter photosynthesis rates, plant growth, and water use efficiency. Plants have developed mechanisms to tolerate these changes to a certain extent. Geochemical processes, such as soil decomposition rates, increase with rising temperatures, enhancing nutrient mineralization and plant availability. However, the interaction between increased greenhouse gas concentrations and temperature alters soil qualities, water availability, and nutrient cycling, disrupting biogeochemical cycles (Singh and Verma 2013; Kumar et al. 2017; Gupta and Singh 2017). This imbalance affects forest productivity by varying the balance between the growth and mortality of producers. Previous studies have shown that climate change directly affects carbon balance and nutrient cycling, impacting forest composition, growing periods, and the hydrological cycle.

The contemporary study aims to understand how climate change affects nutrient accessibility for plant uptake and controls the productivity and survival of forest ecosystems. Organic matter decomposition by microbes releases nutrients into the soil, which is crucial for plant development (Singh et al. 2017a, b). However, increasing atmospheric carbon dioxide levels reduces microbial nitrogen fixation, leading to inadequate nitrogen availability for trees. The productivity of forest ecosystems is influenced by nutrient quantity and availability, with warming temperatures directly affecting photosynthesis rates and nutrient concentrations in soils and leaves (Meena et al. 2011; Kumar et al. 2019d; Kumar et al. 2021a). Ecological studies have shown that rising temperatures can reduce nutrient availability, posing challenges to forest ecosystem functioning (Juneja et al. 2017; Keena 2015; Kumar and Singh 2024; Kumar et al. 2020; Langley 2014).

2.5 *Climate Change Significantly Alters the Forest Biodiversity*

Climate change in forest ecosystems alters the habitats of various species, including plants, animals, microbes, and other organisms. These species must either adapt or migrate to zones with more favorable conditions. Rising atmospheric temperatures, driven by increased greenhouse gas emissions, significantly impact forest ecosystems. Even small fluctuations in average temperatures can alter habitat structure and biodiversity. Climate change contributes to biodiversity degradation, reducing the ecosystem's ability to perform essential functions such as soil binding, nutrient cycling, and water retention (Singh and Kumar 2022).

Approximately 40% of the world's economy and 80% of less industrialized nations' economies depend directly on biological resources and ecosystem services. Climate change is a significant driver of mass extinction, altering species locations, richness, and phenological behavior, leading to declining ecosystem services and disease outbreaks. Other important drivers of biodiversity loss include habitat modification, overexploitation, invasive alien species, and extinction chains. Changes in land and sea use intensity, direct exploitation of species, climate change, and pollution contribute to biodiversity loss. The impact of climate change is expected to surpass other threats in the twenty-first century, both directly and through interactions with different drivers (Singh and Kumar 2022; Singh et al. 2024a; IPCC 2021).

Nature-based solutions such as ecosystem protection, sustainable land management, and habitat restoration are crucial for climate change mitigation and biodiversity protection. However, achieving these goals will be challenging without reducing fossil fuel emissions (Wudu et al. 2023). Additionally, climate change creates favorable conditions for invasive species, leading to allelopathy and suppressing the growth of native plants. Forest pests and pathogens are expected to increase due to climate change, either directly through their richness or dispersal or indirectly through increased water stress or wind damage, making trees more susceptible to attacks. It is likely that global forest disturbances, such as invasive species, insects, and pathogens, will increase in a warming world (Seidl et al. 2017; Turner and Annamalai 2012; Yadav et al. 2019a, b; Jangra et al. 2024; Kanta et al. 2024; Singh and Kumar 2022).

2.6 Distressed Carbon, Water, and Energy Flux in Forest Ecology

The climatic and environmental variability, including rising atmospheric CO_2 concentration, air, and land surface temperature, humidity fluctuations, changes in precipitation patterns, radiation, and other meteorological and land use alterations, have prompted numerous studies on the coupling and response of carbon, water, and energy fluxes in forest/vegetation systems to change climatic conditions (Singh et al. 2022). The studies have led to the developing of various process-based models/methodologies to understand the interlinking of carbon, water, and energy fluxes across different scales, from micrometeorological fluctuations at the leaf level to ecosystem-wide dynamics (Malik et al. 2024; Kumar and Singh 2024). These fluxes, integral components of carbon, water, and energy budgets, are crucial for understanding the interaction between ecosystem functioning and atmospheric systems (Santos and Costa 2004).

Forests influence regional climate through carbon and energy and water flux changes. The Himalayan ecosystem, considered one of the most vulnerable to future climate change, lacks sufficient data despite experiencing warming rates higher than the global average, particularly at higher altitudes (Shukla 2003; Singh et al. 2018). Understanding the functioning of Himalayan forest ecosystems and their response to changing environmental conditions is crucial due to their extensive coverage and presumed sensitivity to climate variability and anthropogenic disturbances (IPCC 2007).

Characterizing carbon, water, and energy flux patterns in forest ecosystems and understanding how environmental changes influence these flux dynamics are essential for computing ecosystem budgets. Cropper et al. (1993) assessed forest standing crops and growth rates. However, few have delved into the process-level understanding necessary to address physiological responses and interactions with climate change and other disturbances (Cropper et al. 1993). Long-term impacts of changes in air temperature, precipitation, atmospheric CO_2, and radiation significantly affect soil–plant-atmosphere exchange processes.

Plant growth depends on capturing and utilizing solar radiation, CO_2, water, and nutrients, with distinct drivers categorized as carbon-driven, water-driven, and solar-driven (Steduto and Albrizio 2005). Forests exchange carbon, water, and energy with the atmosphere, both carbon sinks and sources. Despite the global understanding of carbon, water, and energy flux responses, the Himalayan, highly susceptible to climate change, remains relatively understudied (Fig. 3).

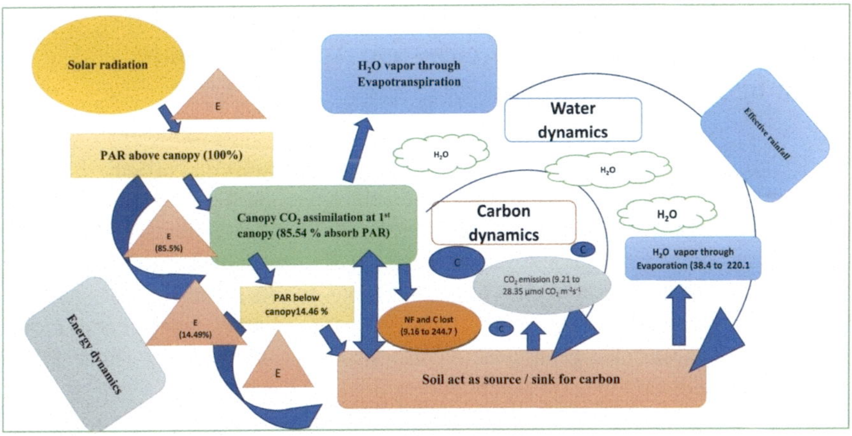

Fig. 3 Conceptual framework for Carbon (C), water (H_2O), and Energy (E) flux dynamics in the soil–plant-atmosphere continuum (SPAC) process of Forest Ecosystem

2.7 *Modification in Water-Atmosphere-Soil–plant (WASP) Interaction in Forest Ecosystem*

Forests have long been acknowledged for their crucial role in water and soil conservation, prompting some countries to institute systems that financially recognize these ecosystem services (Postel and Thompson 2005). Nonetheless, debates persist among foresters and hydrologists regarding the precise impact of forests on water regulation (Innes et al. 2009). This debate gains significance in the context of climate change, as the role of forests in water regulation and soil protection may become increasingly vital. However, changing environmental conditions could compromise their effectiveness in fulfilling this role.

In regions experiencing shifts in precipitation patterns due to climate change, such as reductions in rainy season flows and increases in dry season flows, the efficacy of forests in water regulation may be questioned. This is especially true in areas where total annual rainfall is already low, and a significant portion of it is absorbed or evaporated by forests.

Furthermore, in regions characterized by frequent fog, trees play a crucial role in capturing moisture from the clouds, contributing to what is sometimes referred to as horizontal rain (Stadtmüller 1987). However, there is concern that rising temperatures, as observed in palaeoecological studies of Amazon vegetation changes (Mayle and Power 2008), could lead to a rise in cloud levels, diminishing the occurrence of horizontal precipitation. This potential reduction in horizontal rainfall could significantly affect water availability and ecosystem health in fog forest areas.

3 Coping Strategies and Adaptation Approach to Strengthening Forest Ecosystem Against Climate Change Impact

The depletion of forests in developing and emerging nations, particularly those located in the interior regions of tropical and subtropical latitudes, is widely recognized as a significant contributor to global greenhouse gas (GHG) emissions, thereby exacerbating climate change and global warming. Climate change poses considerable forest risks and presents emerging challenges for forest researchers and the scientific community. Effective adaptation strategies to climate change entail monitoring and anticipating changes and taking responsible actions to mitigate undesirable consequences and capitalize on potential benefits.

The Fourth Assessment Report (AR4) of the Intergovernmental Panel on Climate Change (IPCC 2007) dispelled several doubts about climate change, confirming the certainty of climate warming and its primary association with greenhouse gas emissions, mainly CO_2. Atmospheric carbon dioxide concentrations have risen significantly over the past century, from a pre-industrial level of 278–400 ppm in 2014, and further increased to 423.6 ppm by 2024 (Betts et al. 2024).

Adapting to climate change necessitates regulatory and adaptive measures at all levels, from local communities to national and international spheres. Various communities must develop resilience strategies, encompassing the adoption of appropriate technologies alongside leveraging traditional knowledge and diversifying livelihoods to withstand current and future climatic stresses. Advances in understanding the intricate relationships between forest ecosystems and climate have enhanced our capacity to predict and evaluate ecosystem responses. However, there remains a need for improved guidance for managers in anticipating and responding to potential impacts of climate change and assessing the effectiveness and efficiency of different management responses, given the prevailing inadequate response to potential climate risks. The coping strategies mentioned in Fig. 4 are suggested to strengthen the forest ecosystem.

Coping with climate and weather extremes is indeed intrinsic to forest ecosystems. Numerous events have been attributed to widespread forest degradation, nevertheless the forest system has endeavored to adapt and mitigate climate change vulnerability through various conventional and sophisticated strategies (Albrecht and Kandji 2003). These include promoting intensive afforestation and reforestation programs, fostering more significant involvement of local communities in forest management, curbing settlement expansion and urbanization on forested lands, diminishing reliance on forest fuel, managing invasive alien species, and employing allelopathy techniques, as well as strengthening partnerships, cooperation, and collaboration between local communities, governmental bodies, and private forest managers.

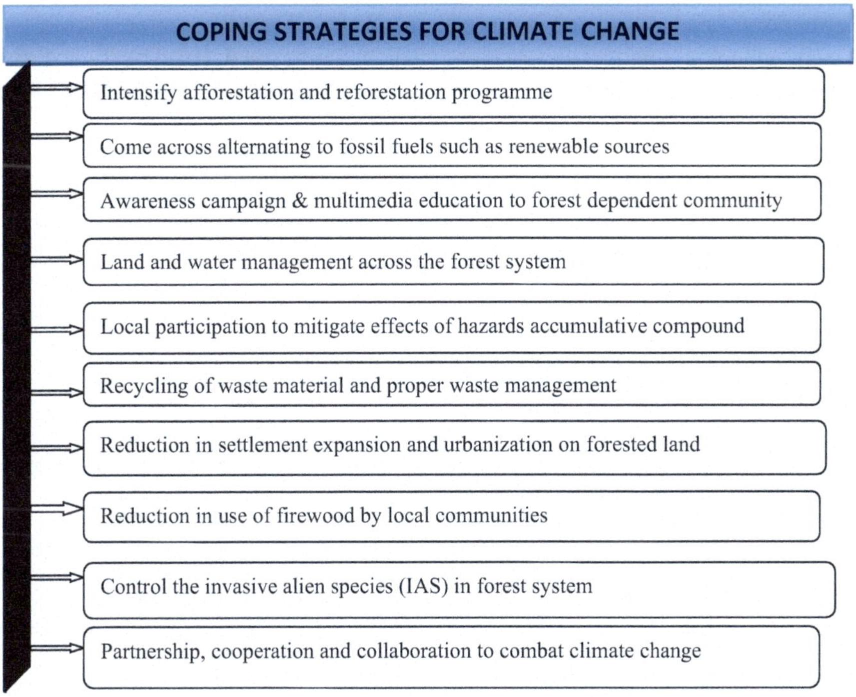

Fig. 4 Coping strategies for climate change in forest ecosystem

4 Conclusion

The contemporary discourse suggests that the forest ecosystem exhibits a nuanced response to the challenges and opportunities of climate change. This study synthesizes the significant impact of climate change across various spheres of the forest ecosystem, including the atmosphere, hydrosphere, lithosphere, and biosphere. Analyzing the intricate relationship between climate dynamics and forest environments identifies the principal challenges that climate variability may impose on the structure and functioning of diverse forest ecosystems. Geographical location, local topography, and land use significantly shape the potential impacts of climate change on forest ecosystems. However, a predominant trend indicates an anticipated rise in temperature, altered precipitation patterns, and increased greenhouse gas emissions. In certain regions, temperature fluctuations could surpass the critical threshold of 2 °C, imperilling the existing ecosystem equilibrium. These temperature shifts are poised to disrupt various ecosystem functions, such as primary productivity, phenological patterns, soil carbon sequestration capacity, carbon–water-energy fluxes, nutrient cycling, biodiversity, and the intricate water-atmosphere-soil–plant (WASP) interactions within forest ecosystems.

Moreover, changes in precipitation patterns and heightened concentrations of CO_2 present additional complexities that are challenging to predict accurately and are primarily influenced by local factors. Recognizing the interconnectedness of these processes underscores the importance of comprehensively understanding WASP interactions. This understanding is pivotal for assessing productivity, carbon sequestration, nutrient cycling, phenological patterns, biodiversity dynamics, and the intricate interplay between these elements. Such insights are crucial for delineating how forest ecosystems will navigate and respond to the multifaceted impacts of climate change.

References

Albrecht A, Kandji ST (2003) Carbon sequestration in tropical agroforestry systems. Agr Ecosyst Environ 99:15–27

Atkins JW, Fahey RT, Hardiman BS, Gough CM (2018) Forest canopy structural complexity and light absorption relationships at the subcontinental scale. J Geophys Res Biogeosci 123:1387–1405

Betts RA, Jones CD, Knight JR, Pope JO, Sandford C (2024) Mauna Loa carbon dioxide forecast for 2024: atmospheric CO_2 rise protected to exceed IPCC 1.5 °C scenarios. Met Office, Hadley Centre

Bushell S, Colley T, Workman M (2015) A unified narrative for climate change. Nat Clim Change 5:971–973

Chakraborty A, Saha S, Sachdeva K, Joshi PK (2018) Vulnerability of forests in the Himalayan region to climate change impacts and anthropogenic disturbances: a systematic review. Reg Environ Change 18:1783–1799. https://doi.org/10.1007/s10113-018-1309-7

Cleland EE, Chuine I, Menzel A, Mooney HA, Schwartz MD (2007) Shifting plant phenology in response to global change. Trends Ecol Evol 22(7):357–365. https://doi.org/10.1016/j.tree.2007.04.003

Cropper Jr, WP, Gholz HL (1993) Constructing a seasonal carbon balance for a forest ecosystem. Clim Res 3:7–12

Dassot M, Constant T, Fournier M (2011) The use of terrestrial LiDAR technology in forest science: application fields, benefits and challenges. Ann Sci 68:959–974

Davies AB, Asner GP (2014) Advances in animal ecology from 3D-LiDAR ecosystem mapping. Trends Ecol Evol 29:681–691

Dhyani K, Kuniyal HB, Singh H, Sobha (2021) Growth and physiological potential of *Terminalia arjuna* under elevated CO_2 levels in Open top chamber condition. J Appl Nat Sci 13(3):1121–1126, ISSN 2231-5209. https://doi.org/10.31018/jans.v13i3.2490

Ehbrecht M, Seidel D, Annighöfer P, Kreft H, Köhler M, Zemp DC, Puettmann K, Nilus R, Babweteera F, Willim K, Stiers M, Soto D, Boehmer HJ, Fisichelli N, Burnett M, Juday G, Stephens SL, Ammer C (2021) Global patterns and climatic controls of forest structural complexity. Nat Commun 12:519. https://doi.org/10.1038/s41467-020-20767-z, www.nature.com/naturecommunications

Feeley KJ, Wright SJ, Nur Supardi MN, Kassim AR, Davies SJ (2007) Decelerating growth in tropical forest trees. Ecol Lett 10:461–469

Fotis AT, Morin TH, Fahey RT, Hardiman BS, Bohrer G, Curtis PS (2018) Forest structure in space and time: biotic and abiotic determinants of canopy complexity and their effects on net primary productivity. Agric Meteorol 250–251:181–191

Gough CM, Atkins JW, Fahey RT, Hardiman BS (2019) High rates of primary production in structurally complex forests. Ecology 100:e02864

Gupta SK, Singh H (2017) Observations on sporadic flowering in *Ficus benjamina* L. and *Peltophorum africanum* L. in new forest. Indian Forester 143:290–291

Gupta SK, Ram J, Singh H (2018a) Comparative study of transpiration in cooling effect of tree species in the atmosphere. J Geosci Environ Prot 6:151–166. https://doi.org/10.4236/gep.2018.68011

Gupta SK, Ram J, Singh H, Kumar P (2018b) Comparative study of cooling under shades of some forestry tree species with respect to ambient temperatures. J Geosci Environ Prot 6:51–66. https://doi.org/10.4236/gep.2018.611004

Gupta SK, Ram J, Singh H (2019a) Relationship between leaf area index and atmospheric cooling of tree species. Indian J for 42(3):203–208

Gupta SK, Singh H, Ram J (2019b) Need for a novel paradigm in forestry research: perspectives of comparative atmospheric cooling by different forestry effect tree species. Indian Forester 145(6):509–515

Harrison S, Spasojevic MJ, Li D (2020) Climate and plant community diversity in space and time. Proc Natl Acad Sci USA 117:4464–4470

Innes J, Joyce LA, Kellomäki S, Louman B, Ogden A, Parrotta J, Thompson I, Ayres M, Ong C, Santoso H, Sohngen B, Wreford A (2009) Management for adapatation. In: Seppala R, Buck A, Katila P (2009) Adaptation of forests and people to climate change. IUFRO World Series 22

IPCC (2007) Climate change 2007: the physical science basis. In: Solomon S, Qin D, Manning M, Chen Z, Marquis M, Averyt KB, Tignor M, Miller HL (eds) Contribution of working group I to the fourth assessment report of the intergovernmental panel on climate change. Cambridge University Press, Cambridge, United Kingdom and New York, NY, USA

IPCC (2014) Climate change 2014: synthesis report. In: Contribution of working groups I, II and III to the fifth assessment report of the intergovernmental panel on climate change

IPCC (2021) Climate change 2021 the physical science basis. In: Working Group I contribution to the sixth assessment report of the intergovernmental panel on climate change

Jangra A, Kumar K, Maikhuri S, Bhandari MS, Pandey S, Singh H, Barthwal S (2024) Unveiling stress-adapted endophytic bacteria: characterizing plant growth-promoting traits and assessing cross-inoculation effects on *Populus deltoides* under abiotic stress. Plant Physiol Biochem 210:108610. https://doi.org/10.1016/j.plaphy.2024.108610

Juchheim J, Ammer C, Schall P, Seidel D (2017) Canopy space filling rather than conventional measures of structural diversity explains productivity of beech stands. Ecol Manag 395:19–26

Juchheim J, Ehbrecht M, Schall P, Ammer C, Seidel D (2020) Effect of tree species mixing on stand structural complexity. Int J Res 93:75–83

Jucker T, Bouriaud O, Coomes DA (2015) Crown plasticity enables trees to optimize canopy packing in mixed species forests. Funct Ecol 29(8):1078–1086

Juneja A, Ceballo RM, Murthy GS (2017) Effects of environmental factors and nutrient availability on the biochemical composition of algae for biofuels production: a review. Energies 6(9):4607–4638

Kanta C, Kumar A, Chauhan A, Singh H, Sharma IP (2024) The interplay between plant functional traits and climate change. In: Kumar N, Singh H (eds) Plant functional traits for improving productivity. Springer, Singapore. https://doi.org/10.1007/978-981-97-1510-7_3

Keena RJ (2015) Climate change impacts and adaptation in forest management: a review. Ann for Sci 72:145–167. https://doi.org/10.1007/s13595-014-0446-

Kumar N, Singh H (2024) Plant functional traits for improving productivity. Springer Nature, Singapore. https://doi.org/10.1007/978-981-97-1510-7

Kumar N, Suyal DC, Sharma IP, Verma A, Singh H (2017) Elucidating stress proteins in rice (*Oryza sativa* L.) genotype under elevated temperature: a proteomic approach to understand heat stress response. 3 Biotech 7:205. https://doi.org/10.1007/s13205-017-0856-9

Kumar M, Kalra N, Khaiter P, Ravindranath NH, Singh V, Singh H, Sharma S, Rahnamayan (2019a) PhenoPine: a simulation model to trace the phenological changes in Pinusroxhburghii in response to ambient temperature rise. Ecol Model 404:12–20

Kumar M, Singh MP, Singh H, Dhakate PM, Ravindranath NH (2019b) Forest working plan for the sustainable management of forest and biodiversity in India. J Sustain for 39:1–22. https://doi.org/10.1080/10549811.2019.1632212

Kumar N, Jeena N, Singh H (2019d) Elevated temperature modulates rice pollen structure: a study from the foothill of the Himalayan agro-ecosystem in India. 3 Biotech 9(5):175; https://doi.org/10.1007/s13205-019-1700-1

Kumar N, Jeena N, Kumar N, Gangola S, Singh H (2019d) Phytoremediation facilitating enzymes: an enzymatic approach for enhancing remediation process. In: Bhatt P (ed) Smart bioremediation technologies. Academic Press, Elsevier, Cambridge, MA, United States, pp 289–306

Kumar N, Singh H, Sharma SK (2020) Antioxidants: responses and importance in plant defense system. In: Roychowdhury R, Choudhury S, Hasanuzzaman M, Srivastava S (eds) Sustainable agriculture in the era of climate change. Springer, Cham, pp 251–264. https://doi.org/10.1007/978-3-030-45669-6_11

Kumar A, Kumar P, Singh H, Kumar N (2021a) Modulation of plant functional traits under essential plant nutrients during seasonal regime in natural forests of Garhwal Himalayas. Plant Soil 465:197–212. https://doi.org/10.1007/s11104-021-05003-x

Kumar A, Kumar P, Singh H, Kumar N (2021b) Adaptation and mitigation potential of roadside trees with bio-extraction of heavy metals under vehicular emissions and their impact on physiological traits during seasonal regimes. Urban Forest Urban Greening 58:126900. https://doi.org/10.1016/j.ufug.2020.126900

Kumar A, Kumar P, Singh H, Kumar N (2021c) Impact of plant functional traits on infiltration rate under different forests of Kempty watershed in Garhwal Himalaya. Indian J Soil Conserv 49(1):38–44

Kumar A, Kumar P, Singh H, Bisht S, Kumar N (2021d) Relationship of physiological plant functional traits with soil carbon stock in the temperate forest of Garhwal Himalaya. Curr Sci 120(8):1368–1373. https://doi.org/10.18520/cs/v120/i8/1368-1373

Kumar M, Phukon SN, Paygude AC, Tyagi K, Singh H (2021e) Mapping phenological functional types (PhFT) in the Indian Eastern Himalayas using machine learning algorithm in Google Earth Engine. Comput Geosci 158. https://doi.org/10.1016/j.cageo.2021.104982

Kumar A, Singh H, Kumari G, Bisht S, Malik A, Kumar A, Singh M, Raturi, Barthwal S, Thakur A, Kaushal R (2022) Adaptive resilience of roadside trees to vehicular emissions via leaf enzymatic, physiological, and anatomical trait modulations. Environ Pollut 313. https://doi.org/10.1016/j.envpol.2022.120191

Kumar N, Singh H, Giri K, Kumar A, Joshi A, Yadav A, Singh R, Bisht S, Kumari R, Jeena N, Khairakpam R, Mishra G (2024) Physiological and molecular insights into the allelopathic effects on agroecosystems under changing environmental conditions. Physiol Mol Biol Plants 30:417–433. https://doi.org/10.1007/s12298-024-01440-x

Kumari A, Singh H (2018) Impact of heat and drought stress on physiological response of *Terminalia arjuna* grown under elevated temperature condition: an adaptive mechanism to climate change. J Pharmacognosy Phytochem 7(1):1374–1378

Langley A (2014) Meetings using results from global change experiments to inform land model development and calibration participants expect a weaker CO_2 response than models or experimental data would suggest, pp 744–746

LeBauer DS, Treseder KK (2008) Nitrogen limitation of net primary productivity in terrestrial ecosystems is globally distributed. Ecology 89(2):371–379

Malhi Y, Timmons Roberts J, Betts RA, Killeen TJ, Li W, Nobre CA (2008) Climate change, deforestation, and the fate of the Amazon. Science 319(5860):169–172

Malik A, Kumar K, Kumari G, Kumar N, Singh H (2024) Application of functional traits in modelling productivity and resilience under climate change. In: Kumar N, Singh H (eds) Plant functional traits for improving productivity. Springer, Singapore. https://doi.org/10.1007/978-981-97-1510-7_5

Mayle FE, Power MJ (2008) Impact of a drier early–mid Holocene climate upon Amazonian forests. Philos Trans R Soc b, Biol Sci 363(1498):1829–1838

Meena D, Singh H, Chaudhari SK (2011) Elucidating strontium response on growth dynamics and biochemical change in *Phaseolus mungo* L. Int J Agric Environ Biotechnol 4:107–113

Menzel A, Sparks TH, Estrella N, Koch E, Aaasa A, Ahas R, Alm-Kübler K, Bissolli P, Braslavská O, Briede A, Chmielewski FM, Crepinsek Z, Curnel Y, Dahl Å, Defila C, Donnelly A, Filella Y, Jatczak K, Måge F, Mestre A, Nordli Ø, Peñuelas J, Pirinen P, Remišová V, Scheifinger H, Striz M, Susnik A, Van Vliet AJH, Wielgolaski FE, Zach S, Zust A (2006) European phenological response to climate change matches the warming pattern. Glob Chang Biol 12:1969–1976

Morin X (2015) (2015) Species richness promotes canopy packing: a promising step towards a better understanding of the mechanisms driving the diversity effects on forest functioning. Funct Ecol 29:993–994

Moser SC (2010) Communicating climate change: history, challenges, process and future directions. Wiley Interdiscip Rev Clim Chang 1:31–53

Negi VS, Maikhuri RK, Pharswan D, Thakur S, Dhyani PP (2017) Climate change impact in the Western Himalayan: 'purple's perception and adaptive strategies. J Mt Sci 14(2):403–416

Norby RJ, DeLucia EH, Gielen B, Calfapietra C, Giardina CP, King JS, Ledford J, McCarthy HR, Moore DJP, Ceulemans R, Angelis PD, Finzi AC, Karnosky DF, Kubiske ME, Lukac M, PregitzerKS S-M, Schlesinger WH, Oren R (2003) Forest response to elevated CO_2 is conserved across a broad range of productivity. Proc Natl Acad Sci 102:18052–18056

Ollinger S, Goodale C, Hayhoe K, Jenkins JP (2008) Potential effects of climate change and rising CO_2 on ecosystem processes in northeastern U.S. forests. Mitig Adapt Strat Glob Change 13(5):467–485

Penone C et al (2019) Specialisation and diversity of multiple trophic groups are promoted by different forest features. Ecol Lett 22:170–180

Piao S, Liu Q, Chen A, Janssens IA, Fu Y, Dai J (2019) Plant phenology and global climate change: current progresses and challenges. Global Change Biol 25(6):1922–1940. https://doi.org/10.1111/gcb.14619

Postel SL, Thompson BH (2005) Watershed protection: capturing the benefits of nature's water supply services. Nat Res Forum 29(2):98–108

Prakash V, Hunney K, Singh H (2022) Effect of elevated carbon dioxide on growth and development of *Santalum album* L. seedlings inoculated with plant growth promoting microorganisms in Open Top Chambers. Int J Environ Agric Biotechnol 7(5):149–159

Pretzsch H (2017) Size-structure dynamics in mixed versus monospecific stands. In: Pretzsch H, Forrester DI, Bauhus J (eds) Mixed-species forests: ecology and management. Springer, pp 211–269

Ravindranath NH, Sukumar R, Deshingkar P (1997) Climate change and forests: impacts and adaptation. Regional Assessment for the Western Ghats, India. Atmospheric Environmental Issues in Developing Countries, Stockholm Environment Institute, Stockholm

Reich PB, Oleksyn J (2008) Climate warming will reduce growth and survival of Scots pine except in the far north. Ecol Lett 11(6):588–597

Santos SNM, Costa MH (2004) A simple tropical ecosystem model of carbon, water and energy fluxes. Ecol Model 176:291–312

Sapijanskas J, Paquette A, Potvin C, Kunert N, Loreau M (2014) Tropical tree diversity enhances light capture through crown plasticity and spatial and temporal niche differences. Ecology 95:2479–2492

Schwartz MD, Reed BC (1999) Surface phenology and satellite sensor-derived onset of greenness: an initial comparison. Int J Remote Sens 20(17):3451–3457

Seidel D, Ehbrecht M, Annighöfer P, Ammer C (2019) From tree to stand-level structural complexity—which properties make a forest stand complex? Agric Meteorol 278:107699

Seidl R, Thom D, Kautz M, Martin-Benito D, Peltoniemi M et al (2017) Forest disturbances under climate change. Nat Clim Chang 7(6):395–402

Sharma R, Singh H, Kaushik M, Nautiyal R, Singh O (2018) Adaptive physiological response, carbon partitioning, and biomass production of *Withania somnifera* (L.) Dunal grown under elevated CO_2 regimes. 3 Biotech 8:267. https://doi.org/10.1007/s13205-018-1292-1

Sharma R, Singh H (2021) Alteration in biochemical constituents and nutrients partitioning of *Asparagus racemosus* in response to elevated atmospheric CO_2 concentration. Environ Sci Pollut Res 29:6812–6821. https://doi.org/10.1007/s11356-021-16050-3

Shukla PR (2003) Climate change and India: vulnerability assessment and adaptation. Orient Blackswan. ISBN 8173714711, 9788173714719

Singh H (2021) An integrated approach considering physiological and biophysical based indicators for assessing tolerance of roadside plantations of *Alstonia scholaris* towards urban roadside air pollution: an assessment of adaptation of plantations for mitigating urban roadside air pollution. Trees. https://doi.org/10.1007/s00468-021-02179-8

Singh H (2024) Exploring adaptive modulation in plant functional traits and its impact on the productivity of *Acacia auriculiformis* under CO_2-enriched environment. Ind Crops Prod 210:118186. https://doi.org/10.1016/j.indcrop.2024.118186

Singh H, Verma A (2013) Characterization and screening of high nitrogen efficient rice genotype to curtail environmental pollution. Int J Agric, Environ Biotechnol 6(1):23–30

Singh H, Verma A, Kumar M, Sharma R, Gupta R, Kaur M, Negi M, Sharma SK (2017b) Phytoremediation: a green technology to clean up the sites with low and moderate level of heavy metals. Austin Biochem 2:1012

Singh H, Sharma R, Savita SMP, Kumar M, Verma A, Ansari MW, Sharma SK (2018) Adaptive physiological response of Parthenium hysterophorus to elevated atmospheric CO_2 concentration. Indian Forester 144:1–14

Singh S, Singh H, Sharma SK, Nautiyal R (2021) Seasonal variation in biochemical responses of bamboo clones in the sub-tropical climate of Indian Himalayan foothills. Heliyon 7(4):e06859. https://doi.org/10.1016/j.heliyon.2021.e06859

Singh M, Singh H, Kumar A, Kumar M, Barthwal S, Thakur A (2024b) Soil nitrogen availability determines the CO_2 fertilization effect on tree species (*Neolamarckia cadamba*): growth and physiological evidence. Environ Sustain 7:53–60. https://doi.org/10.1007/s42398-023-00300-w

Singh H, Kumar M (2022) Climate change and its impact on Indian Himalayan forests: current status and research needs. In: Rani S, Kumar R (eds) Climate change. Springer Climate. Springer, Cham, pp 223–242. https://doi.org/10.1007/978-3-030-92782-0_11

Singh H, Savita Sharma R, Sinha S, Kumar M, Kumar P, Verma A, Sharma SK (2017a) Physiological functioning of *Lagerstroemia speciosa* L. under heavy roadside traffic: an approach to screen potential species for abatement of urban air pollution. 3 Biotech 7(61):1–10. https://doi.org/10.1007/s13205-017-0690-0

Singh H, Yadav M, Kumar N, Kumar A, Kumar M (2020) Assessing adaptation and mitigation potential of roadside trees under the influence of vehicular emissions: a case study of *Grevillea robusta* and *Mangifera indica* planted in an urban city of India. PLoS ONE 15(1):e0227380. https://doi.org/10.1371/journal.pone.0227380

Singh H, Bandyopadhyaya S, Kumar A, Singh M, Malik A, Kumari G, Raturi A, Yadav SK, Kumar M, Barthwal S, Thakur A, Rawat PS (2022) Understanding the physiological and biophysical response of urban roadside plantations for assessing adaptation and mitigation mechanisms toward vehicular emissions. Urban Clim 44. https://doi.org/10.1016/j.uclim.2022.101183

Singh H, Kumar N, Kumar A (2024a) Enhancing resource use efficiency in crops through plant functional traits. In: Kumar N, Singh H (eds) Plant functional traits for improving productivity. Springer, Singapore. https://doi.org/10.1007/978-981-97-1510-7_6

Sparks T (2000) The long-term phenology of woodland species in Britain. In: Kirby KJ, Morecroft MD (eds) Long-term studies in British Woodland. English Nature Science Series, vol 34. English Nature, Peterborough, pp 98–105

Stark SC, Leitold V, Wu JL, Hunter MO, Castilho CVD, Costa FRC, McMahon SM, Parker GG, Shimabukuro MT, Lefsky MA, Keller M, Alves LF, Schietti J, Edemir Y, Shimabukuro, Brandão DO, Woodcock TK, Higuchi N, Camargo PBD, Oliveira R, Saleska SR (2012) Amazon forest carbon dynamics predicted by profiles of canopy leaf area and light environment. Ecol Lett 15:1406–1414

Steduto P, Albrizio R (2005) Resource use efficiency of field-grown sunflower, sorghum, wheat and chickpea: II. Water use efficiency and comparison with radiation use efficiency. Agric For Meteorol 130(3–4):269–281

Stadtmüller T (1987) Cloud forests in the humid tropics: a bibliographic review

Turner AG, Annamalai H (2012) Climate change and the South Asian summer monsoon. Nat Clim Chang 2:587–59

Wolkovich EM, Cook BI, Allen JM, Crimmins TM, Betancourt JL, Travers SE, Pau S, Regetz J, Davies TJ, Kraft NJB, Ault TR, Bolmgren K, Mazer SJ, McCabe GJ, McGill BJ, Parmesan C, Salamin N, Schwartz MD, Cleland EE (2012) Warming experiments underpredict plant phenological responses to climate change. Nature 485:494–497

Wudu K, Abegaz A, Ayele L (2023) The impacts of climate change on biodiversity loss and its remedial measures using nature-based conservation approach: a global perspective. Biodivers Conserv 32:3681–3701. https://doi.org/10.1007/s10531-023-02656-1

Yadav SK, Singh H, Nautiyal R, Ginwal HS, Ansari SA, Barthwal S (2019b) Modulation of morpho-physiological responses in *Populus deltoides* by elevated carbon dioxide and temperature. Forest Sci 66:105–118. https://doi.org/10.1093/forsci/fxz048

Yadav SK, Singh H, Ginwal HS, Barthwal S (2019) Elevated CO_2 enhanced growth and physiological process of *Populus deltoides* Bartr. ex Marsh. Indian Forester 145(1):23–27

Climate-Induced Changes in Forest Ecosystems of Indian Himalayan Region: A Meta-analysis

Shruti Barwa and Hukum Singh

Abstract The evidence of global climate change is now widely acknowledged. Over the past century (1906–2005), the average global temperature has risen by 0.74 °C, with an increase of approximately 0.44 °C in the last 25 years. Consequently, extreme precipitation events, such as heavy rain and severe storms, have become more frequent. Mountain ecosystems worldwide are expected to continue warming at an alarming rate in the coming decades, with the Himalayas experiencing an increase in atmospheric temperature two to three times higher than the global average in recent decades. The fragile landscape of the Indian Himalayan region is particularly susceptible to natural disasters, heightening concerns about the current and potential impacts of climate change. Both anthropogenic activities and global climate change pose significant threats to the Himalayan region, making understanding the magnitude and specifics of climate change effects indispensable. Therefore, a systematic review and meta-analysis of peer-reviewed literature were conducted to assess the impacts of climate change on the ecosystem structure, function, and composition of the Indian Himalayan Region. The study's findings indicate that climate change continues to threaten this region, yet scientific research on vulnerability in social-ecological landscapes remains limited despite the pressing global climate crisis. This underscores the urgent need for comprehensive study and proactive measures to address the challenges of climate change in the Indian Himalayan region.

Keywords Climate change · Indian Himalayan region · Forest ecosystem · Social-ecological landscapes · Vulnerability

S. Barwa (✉)
Indian Institute of Forest Management, Bhadhbhada Road, Nehru Nagar, Bhopal 462003, India
e-mail: shrutibarwa1997@gmail.com

Tata Institute of Social Sciences, Tetelia Road, Jalukbari, Guwahati, Assam 781012, India

H. Singh
ICFRE-Forest Research Institute, P. O. New Forest Dehradun, Dehradun, Uttarakhand 248006, India

H. Singh (ed.), *Urban Forests, Climate Change and Environmental Pollution*,
https://doi.org/10.1007/978-3-031-67837-0_39

1 Introduction

Climate change is a significant concern that the world is currently confronting. It is considered an all-encompassing threat that will pose substantial environmental, economic, social, and political challenges in the coming years and decades. As it affects the global atmosphere, climate change has emerged as a worldwide issue that has captivated global attention (Singh et al 2023). It is the least scientifically predictable, and its repercussions are most likely to harm the most vulnerable and disadvantaged people, who have contributed the least to its major drivers (Singh and Kumar 2022).

Mountains are early indicators of climate change (Singh et al. 2010). River flows are projected to fluctuate as glaciers recede and snowlines rise, leading to numerous societal concerns. These changes affect hydropower generation, biodiversity, forestry, agriculture-based livelihoods, and overall well-being. Mountains cover 24% of the world's land surface and are home to 12% of the global population. They are ecologically, aesthetically, and economically indispensable, not only for the people who rely on mountain resources for day-to-day survival but also for the estimated 40% of the global population who rely on these resources indirectly for water, hydroelectricity, timber, biodiversity, and niche products (Schild 2008; Schild and Sharma 2011).

Mountains are home to over half of the world's 34 biodiversity hotspots and one-third of all terrestrial biodiversity (ICIMOD 2009). They contain water in the form of snow, glaciers, permafrost, wetlands, and rivers, and they also provide groundwater recharge to watersheds (Hua 2009). Major mountain ranges, such as the North American Cordillera, Appalachians, Caledonian Belt, Andes, Urals, Himalayas, Alps, and Tasman Belt, are complex and fragile ecosystems that encompass over a quarter of the earth's land surface and are home to 12% of the world's population (GTOS 2008). These ranges feature diverse vegetation and varied microclimatic and ecological conditions due to significant habitat variability, quick elevational shifts, and fluctuating directional orientation (Sharma et al. 2010). Consequently, mountains have high biodiversity with acute transitions (ecotones) in vegetation sequences and rapid transitions from vegetation and soil to snow and ice (Debarbieux and Price 2012; Jangra et al. 2024; Kanta et al. 2024; Malik et al. 2024; Kumar et al. 2024). Because many species remain isolated at high elevations, mountain ecosystems are generally rich in endemic species, unlike lowland vegetation groups, which occupy climate niches dispersed across wider latitudinal bands. As a result, mountains can be considered the remaining "islands" of wild nature in a sea of altered lowlands, providing essential ecological services.

The Indian Himalayan Region (IHR) is uniquely positioned among mountain ecosystems (Banerji and Basu 2010). These geodynamically young mountains are significant not only for their climate and as a water source for a vast portion of the Indian subcontinent but also their flora, wildlife, human communities, and cultural diversity (Singh 2006; Vaidya and Sharma 2014). The Himalayan mountain range stretches about 2400 kms from east to west in northern South Asia (Roy and Rathore

2019). The range influences regional weather, including the South Asian monsoon, stores water as snow and ice, and serves as a source for major rivers such as the Ganges, Indus, and Brahmaputra, making it the "water tower" for millions in the Indo-Gangetic plains.

However, the Himalayan Mountains are highly vulnerable to the impacts and variability of climate change due to their young and fragile nature and steep gradients (Tewari et al. 2017). The region is under significant threat from increased population pressure and rapid globalization, affecting natural and socioeconomic systems. The climate of the Indian Himalayan region is changing as a result of global warming (Kumar et al. 2020; Kumar et al. 2019c). The most evident impact is the rapid retreat of Himalayan glaciers. Several studies predict that warming will reduce biodiversity in the Himalaya's diverse alpine habitats, such as tundra and rangelands. Endemic species are susceptible to ongoing climate change because of their narrow geographic distribution, the introduction of invasive species, and acute competition. There may be a decline in the niches of endemic species along the elevation gradients. Under global warming, the climatic and topographical barriers of the Himalayas are predicted to stifle species movement along the latitudinal gradient. Bhutiyani et al. (2007) observed that the temperature in the Indian Western Himalayan region has increased by 1.6 °C over the past century, compared to 0.5 °C in the rest of India (Kothawale and Kumar 2005) and 0.76 °C globally. The Himalayan region has seen an increase of 0.98 °C in yearly maximum temperatures.

2 Methodology

2.1 *Study Area*

This meta-analysis focuses on the Indian Himalayan Region (IHR) of Asia. The arc-shaped mountain range in Indian territory spans 12 states: Jammu and Kashmir, Himachal Pradesh, Uttarakhand, Sikkim, Arunachal Pradesh, Nagaland, Manipur, Mizoram, Tripura, Meghalaya, Assam, and the hill districts of West Bengal. This range extends over 2500 km between the Indus and Brahmaputra river systems, covering approximately 16.2% of India's total geographical area.

The Himalayan range is the world's tallest mountain range and borders eight countries: India, Nepal, China, Afghanistan, Pakistan, Bangladesh, Bhutan, and Myanmar. It encompasses an area of about 4.3 million km^2, providing water, food, and energy to nearly 1.5 billion people. In India, the region is broadly divided into the Eastern Himalaya, Western Himalaya, Karakoram Range, and Central Himalaya (Fig. 1). The entire Indian Himalayan Region is crucial for India's ecological security and is a major supplier of timber, medicines, fibers, oils, spices, condiments, firewood, organic manure, fodder, and hydropower.

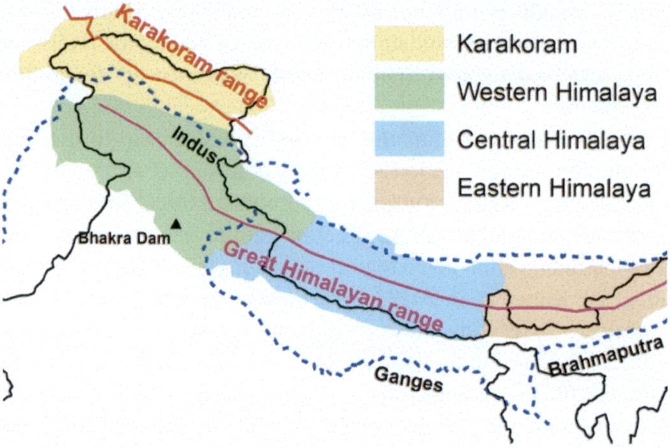

Fig. 1 Broad divisions of Himalayas in India. *Source* Edurev.in

2.2 Literature Gathering Mechanisms

A meta-analysis approach is a powerful and valuable tool for fundamental and applied research, providing a statistical framework for combining and comparing the findings of multiple studies that test the same premise. Meta-analysis is particularly valuable for ecologists and evolutionary biologists, as individual studies often rely on small sample sizes due to time and manpower constraints, resulting in low statistical power (Harrison 2010).

This analysis used a systematic approach to synthesize information, involving a specific step-by-step process for identifying available peer-reviewed literature sources. Systematic reviews in climate science are relatively new but are often regarded as superior to traditional review approaches because they provide quantitative and qualitative literary trend analyses (Berrang-Ford et al. 2015).

Digital libraries such as JSTOR, ResearchGate, Scopus, and Wiley Online Library were utilized to find peer-reviewed publications on the impacts of climate change on the Indian Himalayan Region (IHR), with Google Scholar used to find additional scholarly literature. These digital library search engines were selected for their comprehensive databases covering various interdisciplinary topics, including social and environmental sciences. The study focused on peer-reviewed papers published between 1990 and 2020, documenting climate change implications on the IHR over 30 years. Searches were conducted using factorial combinations of keywords such as (Indian Himalayas and keyword 1—Indian forests), (Indian Himalayas and keyword 2—forest growth dynamics), and so forth for all relevant keywords.

After excluding grey literature (reports, conference proceedings, and notes), articles were selected based on the scope of the study region, resulting in 59 articles from ResearchGate, 54 from JSTOR, 38 from Wiley Online Library, and 52 from Google Scholar, totalling 203 studies for review. While the literature search was not restricted to specific keywords, the primary goal was to document the effects of climate change on the Indian Himalayan Region. Thus, the review procedure focused solely on peer-reviewed scientific literature.

A total of 203 papers were analyzed. The review indicates a considerable increase in the number of studies on the effects of climate change on the Indian Himalayan Region over the last 30 years (Fig. 2). The number of publications increased notably after 2007–2008, with an exponential surge in the previous three to four years. This significant growth in the number of studies highlights an increased interest in understanding the IHR's sensitivity to climate change in recent years.

3 Indian Himalayan Region

3.1 Physiography at a Glance

The Himalayas, formed from the dynamic collision of the Indian and Eurasian plates during the early Eocene and Miocene epochs, have evolved over 45–60 million years, fostering diverse ecosystems by welcoming flora and fauna from all directions. Despite being the youngest mountain range globally, the Himalayas exhibit remarkable complexity and diversity in their biophysical attributes. Geographically and ecologically distinct, the region harbors a unique climate and profoundly influences weather patterns across Asia.

The Indian Himalayan Region, with an average elevation of around 4000 m above sea level, experiences significant climatic influences from the monsoon and westerly winds. The gradient of natural vegetation, ranging from montane forests to alpine deserts, is dictated by topography and atmospheric circulation, with moisture levels decreasing from southeast to northwest. The timberline, marking the upper limit of forests, varies according to local topographic and climatic conditions.

3.2 Biodiversity at a Glance

The Himalayan Mountains are home to some of the world's most diverse ecosystems. With its rich vegetation and various climates and topographies, the Himalayas constitute a complex and dynamic ecosystem that provides numerous ecosystem services (Roy and Rathore 2019). The forest types in the Himalayas range from moist deciduous forests at the base to temperate broadleaf and coniferous forests. At higher elevations and beyond the treeline, alpine meadows host unique species and distinctive biodiversity (Schickhoff et al. 2015; Pandey et al. 2020). The biodiversity in the Himalayan Mountains is significant due to its biogeographic location, habitat heterogeneity, and pronounced endemism (Myers 1988). Notably, the alpine plant diversity recorded in the Himalayan region exceeds the global average (Korner 1999;

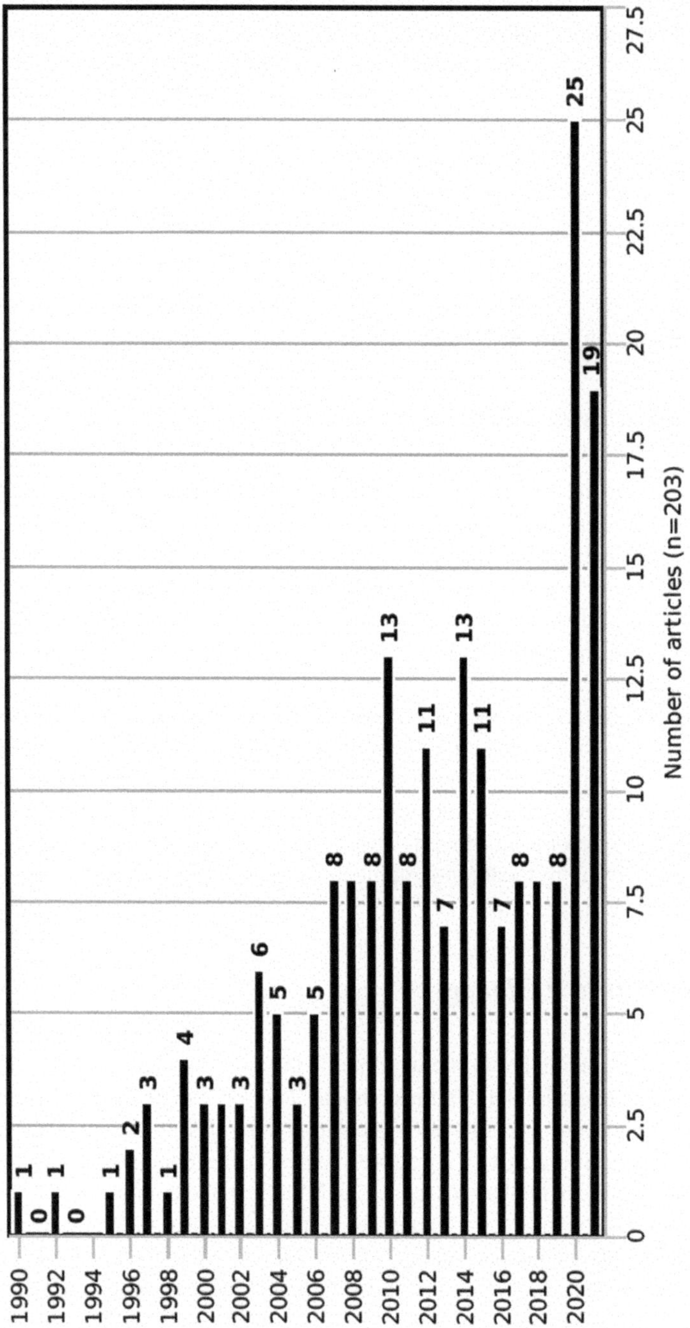

Fig. 2 Year-wise distribution of peer reviewed literature

Table 1 Endemic species distribution of Himalayan flora and fauna

Taxonomic group	Number of species	Endemic species	Endemism percentage
Plants	10,000	3160	31.6
Mammals	300	12	4.0
Birds	977	15	1.5
Reptiles	176	48	27.3
Amphibians	105	42	40.0
Freshwater fish	269	33	12.3

Source http://www.biodiversityhotspots.org/xp/hotspots/himalaya/

Prakash et al. 2022; Dhyani et al. 2021). Overall, the Himalayan region is known to have fundamentally higher plant variety and abundance at elevations between 4200 and 4500 m compared to any other mountain ranges at similar heights.

The Indian Himalayan Region (IHR) harbors about 8000 flowering plant species, accounting for roughly half of India's entire flowering plant population, with nearly a third endemic to the area. The IHR contains around 816 tree species, 675 edible species, and almost 1740 medicinal plants (Singh 2006). The IHR acts as a carbon dioxide "sink" due to its immense green cover, and the annual carbon sequestration by Himalayan trees in the western and northeastern Himalayas is just one of the vital ecosystem services they provide (Upgupta et al. 2015; Watham et al. 2020). In light of the growing threats to the biological diversity of the IHR, its conservation and judicious use can offer substantial economic advantages to local populations while contributing to long-term development (Singh et al. 2024b; Singh et al. 2022).

The number of species in the Indian Himalayan Region exceeds 25,000, and its high endemism (30–40%) adds to its value. The table below (Table 1) details the distribution of endemic species in Himalayan flora and fauna. The IHR includes most of the world's major biome types, including dry steppe and cryptophyte cushions, tropical rainforests, coniferous forests, and deciduous broadleaved forests. More than 60 different ecoregions are supported by a vast range of altitudes, including forests at high altitudes (3000–4000 m). This diversity is even greater than in Amazonian forests (Singh et al. 2011).

3.3 *Timberline at a Glance*

The alpine timberline represents the upper limit of forests on a mountain in a broad sense (Wardle 1974). Above the timberline, dense forests transition to shrubs and/or meadows, either abruptly or gradually. Although the timberline is not a distinct physical line but a transition zone, it often appears sharp from a distance and is commonly perceived as a line (Dutta et al. 2014). The timberline ecotone marks the upper boundary of natural forests, characterized by a steep gradient, increasing stand fragmentation, and stunted growth (Körner 1998).

Research based on satellite imagery and topographic maps indicates that the Himalayan timberline was located between 3900 and 4000 m altitudes in 1960. The current treeline, which usually lies above the timberline, is now at 3542 m, suggesting a downward shift of 350 m (Panigrahy et al. 2010). The average timberline in the Himalayas is approximately 3500 m above sea level. Even minor environmental changes that disrupt the balance between vegetation and atmosphere can lead to significant demographic shifts in these species. Climate changes that disturb the vegetation-climate equilibrium can result in major alterations in species demographics. It is predicted that global warming will cause an upward shift in the timberline, along with changes in habitat orography. This could lead to changes in ecosystem characteristics, the extinction of critical associated flora and fauna, including keystone species, and the invasion of alien species (Navalgund et al. 2019; Singh et al. 2024b; Kumar et al. 2024).

4 Climate Driven Effects on Forest Ecosystems

4.1 Ecosystem Composition and Distribution

The alpine ecosystem is predicted to be significantly impacted by ongoing climate change. Recent studies on alpine vegetation reveal that it will respond quickly and adapt to climate change (Cannone et al. 2007; Singh et al. 2024a), challenging the earlier notion that alpine plants have inherent inertia to temperature increases of 1–2 °C. Unexpected changes in vegetation may occur due to improved hydrological cycles, rapid snowmelt, and higher precipitation resulting from climate change, leading to increased surface instability and disturbance. It should be noted that hydrological disturbances are especially dangerous in alpine areas; for a given rainfall, drainage in these regions is expected to be 4–5 times greater than in lower elevations due to low air temperatures.

According to Beniston (2003), alpine species are susceptible to global warming since they are found near mountain peaks. As the temperature rises, these species will have less room to move upward. Species that rely on snow cover for protection will be vulnerable to frost, and those that require winter chilling for bud break might not survive if temperatures do not remain low enough for sufficient durations (Dimri and Dash 2011; Gaire et al. 2020). Conversely, species that thrive in windy environments may expand their range (Cannone et al. 2007; Kumar et al. 2022; Singh et al. 2022).

Ecological transformations of the landscape and concomitant changes in ecosystem structure and function following the invasion of shrublands into alpine meadows in the Indian Himalayan Region are likely to impact the livelihoods of local human communities significantly. The elimination of alpine meadow ecosystems, in particular, will endanger yak populations that rely on herbaceous plant taxa such as *Heracleum* sp., *Kobresia* sp., *Juncus* sp., and others in the highland Himalaya (Manish et al. 2016). These plant communities provide vital forage meadows for yak populations, which are the Himalayan highlanders' most cherished possessions, integral to their economic and socio-cultural activities. Since shrub species have lesser nutritional value and are unpalatable to most grazers and ungulates, replacing grassland communities with shrublands would negatively impact livestock herding in these locations (Tambe and Rawat 2009).

The Third Assessment Report of the IPCC (2001) stated that climate change might significantly influence forest ecosystems. Most ecosystems and landscapes will be impacted by changes in species composition, productivity, and biodiversity (Leemans and Eickhout 2004), even if global warming is only 1–2 °C, which is far less than the most recent forecasts for warming this century (IPCC 2001).

Climate change will likely alter the composition and distribution of vegetation types such as alpine meadows, steppes, wetlands, peatlands, and forests throughout the Himalayas (Raturi et al. 2024). Alpine meadows, for example, are currently associated with annual precipitation levels exceeding 400 mm, whereas alpine steppe and desert vegetation types are found in areas with annual precipitation levels below 400 mm. Glaciers, snowmelt, and water discharge into wetlands sustain these vegetation types. Although water availability is crucial, overgrazing and climate change are significant factors contributing to rangeland degradation in the region (Zhou et al. 2005; Kumar et al. 2021a, b, c, d, e, f; Singh et al. 2021).

The range of invasive alien plant species (IAPS) in the Kailash Sacred Landscape (KSL), which spans India, Nepal, and China, is expected to expand due to future climate change, posing a threat to native vegetation (Shrestha et al. 2018). The distribution range of these species is predicted to grow vertically and horizontally towards the south and northeast of KSL. In Uttarakhand, rhododendrons and other woody species have already begun to dominate alpine meadows in the Valley of Flowers. As taller woody ones replace herbaceous plants, several changes can occur; for example, there would be fewer lichens and mosses under the shadow of woody plants, and biomass allocation to below-ground components and mycorrhizal fungi would likely shift (Olofsson 2006). Additionally, grazing sheep, goats, horses, and cattle can move to higher altitudes, potentially limiting the establishment of woody plants over time. Furthermore, introducing many species from lower ranges and invasive alien species may alter the composition of plants and animals in meadows, posing a competitive threat to existing species (Singh et al. 2010).

4.2 Ecosystem Phenology

Phenology is one of the most apparent indicators of a plant's response to climate change (Wolkovich et al. 2014). Plants have internal mechanisms that allow them to detect temperature, humidity, and other environmental factors and respond to these changes through life cycle events (Gupta and Singh 2017; Gupta et al. 2018b). Consequently, plant phenology, which is heavily influenced by climate, is one of the most accurate bio-indicators of ongoing climate change (Fitter and Fitter 2002). Changes in phenological events are considered one of the most vulnerable biological responses to climate change (Parmesan and Yohe 2003; Sharma and Singh 2021; Kumar et al. 2021a, b, c, d). Numerous studies, supported by experimental evidence, have shown that plants have shifted their flowering time earlier by several days to weeks in response to increased temperatures attributed to global warming (Parmesan and Yohe 2003; Miller-Rushing and Primack 2008). This shift could lead to a phenological mismatch between flowering plants and pollinators, affecting ecosystem functioning, productivity, and ecological interactions at all levels (Wolkovich et al. 2014; Kumar and Singh 2024; Kumar et al. 2019a).

Notably, phenological patterns in the Himalayas are poorly understood due to the region's diversity of species and a lack of historical data. The Himalayas, for instance, are home to 517 rhododendron species, accounting for about two-thirds of all known species worldwide (Cox and Cox 1997). Few rhododendrons pollinated by wild bees, honey bees, bumblebees, and butterflies are now blooming a month earlier than usual. In the Kumaon Himalayas, it was observed that Rhododendron species began flowering two weeks earlier than usual (Negi and Mukherjee 2020). A Generalized Additive Model (GAM) utilizing real-time field observations (2009–2011) and herbarium records (1893–2003, Hooker 1949) predicted that *Rhododendron arboreum* in the Western Himalayas has been flowering 88–97 days earlier over the past 100 years due to a rise in annual mean maximum temperature (Gaira et al. 2014; Gaira and Dhar 2020; Singh et al. 2020a, b; Singh et al. 2018; Yadav et al. 2019a, b).

4.3 Species Range Shift

A recent study has revealed that the escalating emissions of global greenhouse gases coupled with ongoing climate change are likely to induce significant shifts in the potential habitats of Himalayan Pheasants, pushing them towards higher altitudes and latitudes while concurrently contracting their existing range (Chettri 2014). In extreme future climatic scenarios, there is a possibility that the species may even encroach upon Sky Islands within the Himalayan region, highlighting the urgency of proactive conservation measures. The current network of protected areas (PAs) in the Eastern Himalayas seems inadequate for safeguarding Himalayan Pheasants, as their habitat may rapidly extend beyond these designated zones (Chhetri et al. 2021).

Additionally, a separate investigation indicates that the suitable habitat area for Himalayan Langurs is anticipated to encompass 24,240 km^2 under current conditions. However, projections suggest that Himalayan Langurs will undergo longitudinal and latitudinal shifts in their distribution due to climate change's impacts, further underscoring species' vulnerability to environmental shifts (Bobrowski et al. 2017; Chakraborty et al. 2018; Bhandari et al. 2019; Bagaria et al. 2020; Gupta et al. 2019a, b; Kumar et al. 2019b, c).

Further supporting evidence from Telwala et al. (2013) demonstrates that warming trends have already instigated shifts in the geographic range of a significant majority (87%) of the 124 endemic plant species investigated in the Himalayan region. This conclusion is based on meticulous field experiments utilizing historical data (1849–1950) and contemporary observations (2007–2010) of the elevational ranges of endemic flora.

Moreover, modeling techniques employed by Zomer et al. (2014) have revealed a probable consequence of global warming: a discernible upward shift in the mean elevation of ecoregions (371 m) and bioclimatic zones (357 m). This elevation shift, driven by climate change, may engender intensified competition for resources and land, ultimately leading to a decline in biodiversity (Walker et al. 2006).

4.4 Ecosystem Services

Climate change can profoundly affect ecosystem services and the well-being of people in the Himalayan region, particularly impacting food and feed security, water availability, regulation of natural hazards, spirituality, cultural identity, aesthetics, and recreation.

Few studies highlight the significant impact of climate change on food systems in the challenging environment of the high Himalayan areas. Reduced snow cover and altered precipitation patterns adversely affect subsistence agriculture in high-elevation regions of India and Nepal (Ogra and Badola 2015). Decreased plant density in upper meadows disrupts the livelihoods of yak populations and sheep herders (He and Richards 2014). Climate change also poses significant challenges for cattle management in arid high mountain regions, exacerbating multifaceted poverty in rural mountain communities (Parish and Funnell 1999).

Mountains are often called the "water towers of the world" (Immerzeel et al. 2010), highlighting their crucial role in water provision. Climate change profoundly impacts essential ecosystem services in high-mountain regions, with downstream communities and high-mountain populations experiencing serious consequences (Beniston 2003). Some local communities are witnessing decreased water supplies during dry seasons, while others are experiencing significant streamflow reductions, ultimately affecting water availability.

Furthermore, climate change has additional implications for the regulation of natural disasters. Glacial retreat destabilizes land surfaces, leading to changes in the frequency and location of natural hazards. Among the main hazards are permafrost melting, resulting in landslides and rockfalls, glacial lake outburst floods (GLOFs), and extreme weather events, including violent storms. The studies indicated that permafrost degradation increases the risk of rockfalls in high-alpine areas (Huggel et al. 2012). Intense weather events driven by climate change further contribute to the hazardous conditions in high mountain environments. For example, the severe storm that stranded local mountain guides and visitors on the Himalayan Annapurna trek in 2014 underscores the growing risks associated with climate change (Herring et al. 2018).

4.5 Pollinators

Climate change significantly impacts the biodiversity of pollinators, exerting varied effects on insects based on their biology, current environment, and geographic distribution. Certain species are more vulnerable than others to these effects.

Of particular interest is the Himachal Pradesh economy, which largely relies on apple farming. Despite adequate agronomic inputs and intensive efforts, the decline in natural insect pollinators leads to diminished crop yields and the production of deformed, discolored, and inferior-quality fruits. Insufficient pollination can also trigger pistil senescence and restrict seed production. A study highlights the dwindling diversity of native bees in the region due to habitat loss resulting from land use changes, intensified monoculture, and the adverse impacts of pesticides and herbicides (Pratap 1999). Research by the International Centre for Integrated Mountain Development (ICIMOD) has identified poor pollination as a primary cause of declining apple productivity (Pratap and Partap 2003).

Farmers now resort to renting honey bee colonies for pollinating apple orchards, incurring a cost of Rs. 500 per colony (Ahmad et al. 2002) and adopting temporary measures until the flowering of "polliniser" trees in their orchards. Farmers have resorted to hanging "bouquets" of small flowering branches from pollinizer trees in plastic bags filled with water, strategically placed among the branches of commercially premium apple varieties to attract pollinators (Negi et al. 2012). Additionally, some farmers have removed apple trees altogether due to dissatisfaction with low yields and poor fruit quality resulting from inadequate pollination.

4.6 Insect Pest Incidence

Climate change has direct and indirect impacts on crops and pest populations. Directly, it affects pest reproduction, development, survival, and spread while indirectly influencing the complex relationships among pests, their environment, and other insect species, including natural enemies, competitors, vectors, and mutualistic partners. Insects are poikilothermic organisms, meaning their body temperature aligns with their surroundings. Consequently, temperature is the paramount environmental factor governing insect dispersion, growth, and reproduction (Kocmánková et al. 2009).

Insect physiology is sensitive to temperature fluctuations, with their metabolic rate doubling approximately every 10 °C (Dukes et al. 2009). The studies attest that rising temperatures accelerate insect feeding, development, and mobility, significantly impacting population dynamics through fecundity, survival, generation time, population size, and geographic distribution. Species unable to adapt and evolve in response to escalating temperatures face challenges in maintaining their numbers, while others thrive and reproduce at an accelerated pace. Temperature's influence extends across metabolism, metamorphosis, movement, and host availability, collectively shaping the likelihood of alterations in pest population dynamics (Shrestha 2019).

4.7 *Habitat Suitability*

Climate significantly influences species ranges and shapes overall biodiversity patterns across space and time. Species typically inhabit specific habitats to maintain a climate equilibrium within their range. However, with the onset of climate change, sustaining this equilibrium may necessitate constant shifts in distribution. As the climate rapidly changes in the Himalayas, we anticipate species adapting their distributions accordingly. Unusual precipitation and temperature patterns have been observed in the Himalayas, and these trends are projected to exacerbate (Solomon et al. 2007).

Phenological shifts, alterations in tree lines, invasions by alien species, declines in population and diversity, habitat modifications, and extinctions are all anticipated consequences of climate change, both presently and in the future (Shrestha et al. 2015). Such effects could push populations of endangered species to the brink of extinction. Therefore, understanding the impact of climate change on the distribution and habitats of endangered species is crucial for preserving species diversity, particularly in regions like the Himalayas, where climate change is occurring at an accelerated pace compared to the global average .

For instance, the Kashmir Musk Deer (KMD) is expected to inhabit a high Himalayan band extending from central Nepal to northwestern India, traversing through the Kashmir regions of India and Pakistan en route to Afghanistan. However, suitable habitats do not form a continuous pattern throughout the upper Himalayas. Examination of the expected distribution under future climate change scenarios reveals a northward shift and habitat loss. In the coming decades, habitat is projected to shift northward in regions such as west Nepal and Uttarakhand, resulting in a contiguous distribution of highly suitable habitats in the northern halves of both territories and their adjacent areas in Tibet by the 2050s and 2070s (Jiang et al. 2020).

4.8 *Forest Productivity*

Forests are a crucial component of the terrestrial ecosystem, recognized for their significant role in mitigating global warming by serving as carbon sinks through photosynthesis (Miles and Kapos 2008; Singh 2021; Singh 2024). Therefore, examining future climate conditions' impacts on forest ecosystems is imperative. Forest Net Primary Production (NPP) is highly sensitive to CO_2 concentration and climate change, garnering considerable attention in climate change research . Consequently, forest NPP is a pivotal indicator for studying forest ecosystem responses to future climate change (Sharma and Singh 2021).

Acknowledging that forest productivity is changing due to global warming is crucial. With warmer temperatures persisting for longer periods, plants experience extended growing seasons and increased opportunities for photosynthesis. Consequently, forest growth and carbon sequestration may experience enhancement. However, as temperatures rise, plant respiration also increases (Sendall et al. 2015), which might potentially offset the gains in growth and productivity. Furthermore, temperature elevation, influencing leaf phenology including bud break, leaf development, and leaf senescence, renders trees more susceptible to late spring frosts and soil freezing, thereby inducing stress and reducing their productivity (Adhikari and Kumar 2020; Aryal et al. 2016). Warmer temperatures and drier conditions during the growing season can further induce tree stress, ultimately diminishing overall productivity (Verma et al. 2020).

4.9 Biodiversity

Mountains are home to a diverse array of climatically distinct living zones across short distances and elevations, fostering a multitude of micro-habitats and ecological niches. This biodiversity richness has led to mountains being recognized as biodiversity hotspots and conservation priorities (Körner 2004). However, mountain biodiversity is particularly vulnerable to global warming, with signs of fragmentation and degradation emerging due to exogenous factors such as temperature increases and human activities (Hughes 2000; Houghton et al. 2001; Xu and Wilkes 2004).

The anticipated migration of species to higher elevations in response to climate change, especially with a projected temperature rise of 2 °C, poses significant challenges for alpine plant species, particularly those constrained by limited habitat availability above the tree line (Dirnböck et al. 2003; Singh et al. 2017a; Kumar et al. 2017; Singh et al. 2017b). This could lead to severe fragmentation, habitat loss, or even extinction if these species cannot adapt. Similarly, alpine animals, such as the pika, face lifestyle alterations due to rising temperatures, potentially reducing foraging hours and leading to population declines (Chettri et al. 2010). Other animal species in alpine environments will also contend with shifts in competitors, predators, and food sources in a warmer world.

Climate change projections suggest substantial impacts on India's forests, with consequential alterations to forest types and shifts in plant and animal phenology (Messerli 2009). These changes are expected to negatively affect biodiversity, posing risks to the health of entire ecosystems.

The meta-analysis provided a comprehensive overview of the impacts of climate change on the biodiversity of the Indian Himalayan Region. The findings, systematically summarized in Table 2, offer valuable insights into the current and potential future challenges facing this ecologically sensitive region.

Table 2 Impacts on biodiversity in the Indian Himalayan Region

Biodiversity	Geographical range	Impact	Climate change driver	Source
Snow Leopard (*Panthera uncia*)	Western and Eastern Himalayas	Range shift, Habitat shrinkage	Temperature change	Forrest et al. (2012)
Alpine Plants (Endemic)	IHR as a whole	Upslope shift of elevation range documented for ~ 116 species and median shift of ~ 240 m per 160 years	Higher temperature and more precipitation	Telwala et al. (2013)
Blue Sheep (*Pseudois nayaur*)	Eastern Himalayas	With a decline in grasses and many shrub species at higher elevations, blue sheep move to forage at lower elevations leading to a range shift and decrease in habitat suitability	Temperature Change	Namgail et al. (2010)
Invasion by alien/introduced species and dominance by xeric species (e.g. Mikania, Lantana)	Eastern and Western Himalayas	Species introduction, Land use change	Higher temperature and more precipitation	Pathak et al. (2019)
Alpine Musk Deer (*Moschus chrysogaster*)	Western and Eastern Himalayas	Longitudinal shift of habitat between 2600 and 4600 m	Temperature Change	Lamsal et al. (2018)
Alpine/and subalpine ecosystem	IHR as a whole	Transformation of earlier Quercus-betula forest into the 'Krummholz-type' of vegetation comprising species of Rhododendron, Salix, and Syringa	Temperature change, Declining snowfall, Glacial events	ICIMOD (2017)
Freshwater wetland ecosystem	IHR as a whole	Decrease in population of *Sus salvanius* (Pygmy Hog); beels and associated biodiversity are changing and an impact on large mammals such as crocodiles, river dolphins, wild buffaloes	Temperature change, Variation in precipitation	Chettri et al. (2010)

(continued)

Table 2 (continued)

Biodiversity	Geographical range	Impact	Climate change driver	Source
Increase in forest fires	Western Himalayas	Land degradation	Higher temperature and extreme weather events	Dobriyal and Bijalwan (2017)
Bees and Butterflies	IHR as a whole	Reduced population due to reduced nectar availability resulting from dry spells and drought with phenological changes	Temperature change, Extreme weather events	Rautela and Karki (2015)
Successional shifts from wetlands to terrestrial ecosystem	Eastern Himalayas (loktak lake, deeporbeel)	Habitat alteration, drought and eutrophication	Higher temperature, less precipitation	Sharma et al. (2009)

5 Drivers of Biodiversity Changes in IHR

Climate variability, land use change, and rural migration are pivotal drivers of biodiversity loss in the Indian Himalayan Region. Among these, climate change is a significant force, substantially impacting the region's forests and ecosystems . Highland forest ecosystems have been notably affected by climate variability (Dale et al. 2001). Monitoring the early manifestations of climate change in these ecosystems is imperative, given that alpine environments serve as effective barometers of broader global environmental shifts .

Changes in land cover (the physical attributes of the Earth's surface) and land use (the human activities applied to these attributes) emerge as key drivers of climate change, notably influencing carbon sequestration and nitrogen deposition (Foley et al. 2005). It's essential to recognize that land use and cover alterations contribute to local and regional climate variations and exacerbate global warming, directly impacting biodiversity and resulting in species diversity loss (Sala et al. 2000; Sharma et al. 2018; Singh et al. 2018; Gupta et al. 2018a). Despite numerous land use and cover assessments in the Eastern Himalayas (Champion and Seth 1968), there remains limited documentation of temporal changes. Notably, the conversion of forests to other land uses has accelerated in recent decades, depleting natural resources. Intensified exploitation of forests, driven by rapid economic development and expanding agricultural activities, has led to habitat modification and fragmentation .

A study by Verma et al. (2021) utilized system dynamics techniques in conjunction with causal network analysis to elucidate the interrelationships among causes, actors, and mechanisms of forest change across the Himalayan mountain range, using data from 1984 to 2020. The study identified five proximate drivers, twelve underlying drivers, and five additional factors interconnected by 221 linkages. Noteworthy proximate drivers include smallholder agriculture, extensive non-timber forest product extraction, widespread commercial and non-commercial timber exploitation, and high rates of agricultural abandonment. Underlying factors encompass population growth, low agricultural productivity, international financing for development initiatives, and successful community forest management schemes.

6 India's Efforts to Address Climate Change and Biodiversity in IHR

India has established a National Mission for Sustaining the Himalayan Ecosystem (NSMHE) under the National Action Plan on Climate Change (NAPCC). The mission is designed to delve deeper into understanding the intricate relationship between the Himalayan environment and climate factors, providing invaluable insights for Himalayan sustainable development while safeguarding its fragile ecosystem (https://dst.gov.in). This endeavor will require close collaboration among experts in climatology, glaciology, and related fields. Furthermore, knowledge exchange with South Asian countries and those sharing the Himalayan ecosystem will be crucial.

The mission aims to address several key issues, including:

- Himalayan glaciers and their hydrological implications.
- Natural disaster prediction and control.
- Biodiversity protection and conservation.
- Conservation and protection of wildlife.
- Traditional knowledge societies and their means of subsistence.
- Capacity in science regulation and critical peer review to aid governance challenges relating to the Himalayan Ecosystem's survival.
- Assist in the restoration and rehabilitation process of Uttarakhand.

The mission's primary objective is to establish a long-term national capacity for the continuous assessment of the health of the Himalayan Ecosystem. This capacity will aid policymakers in formulating effective policies and assist the states in the Indian Himalayan Region in implementing sustainable development actions. Achieving this goal will require the development of robust learning mechanisms to balance environmental changes with responsible human activities. The National Mission for the Sustainment of the Himalayan Ecosystem (NSMHE) aims to employ various strategies to cultivate diverse capacities.

Furthermore, the mission will comprehensively assess the Himalayan region's vulnerability to climate change across physical, ecological, and socio-cultural dimensions. This assessment will inform the development of appropriate policy measures and a time-bound action plan to enhance the ecosystem's resilience and ensure the continued provision of essential ecosystem services. Additionally, the mission seeks to bolster national and state-level capacities to analyze climate change and devise suitable responses to Himalayan challenges.

As part of the NSMHE, the Task Force on Micro-flora and Fauna (established in 2014 under DST Grant Number: DST/SPLICE/CCP/NMSHE/TF-2/WII/2014[G] dated 26.08.2014) is dedicated to exploring solutions to mitigate the impacts of climate change on the micro flora and fauna of the Indian Himalayan Region (IHR). The assessments conducted by this Task Force include

- Animal species/communities diversity, distribution, and abundance
- Wildlife habitats, ecosystems, and ecosystem services
- Anthropogenic and climate change impacts on wildlife and ecosystems through scenario building and visualization
- Climate change vulnerability of species/habitats
- Prioritization of species/taxa and monitoring sites.

7 Way Forward

Long-term monitoring of environmental and socio-ecological changes is vital for understanding the impacts of climate change on biodiversity. Establishing permanent plots or units along altitudinal transects spanning tropical and alpine regions are essential for monitoring various ecosystems. However, it is imperative to standardize monitoring parameters to ensure effective monitoring. This requires a consistent, standardized approach and collaborative efforts to gather and analyze data regularly.

Focused research on the impacts of climate change, coping mechanisms, and adaptation strategies is crucial. Prioritizing documentation of changes in crop performance, cropping patterns, and management systems, including shifts in crop nutritional value, is necessary. Institutional support and a reorientation of government research programs toward adaptive research are essential for systematic recording and monitoring.

Assessing critical habitat linkages, protected area effectiveness, and vulnerabilities is imperative. Identifying and strengthening potential habitats and corridors between protected areas are necessary steps. Continuous monitoring of vegetation changes and indicator species within protected areas is essential. Research should focus on evaluating the effectiveness of protected area management and contribute to responsive management approaches and technologies. These findings should inform decision-support systems, prioritizing adaptive reactions and natural system resilience.

The recommendations outlined above are systematically compiled and presented in the accompanying flow chart. This visualization in Fig. 3 offers a clear, structured overview of the proposed strategies for addressing identified issues.

Furthermore, there should be a heightened focus on local initiatives. This includes the establishment of alternative livelihoods, which involves developing competitive advantages through activities that utilize natural resources more productively. This can be achieved by promoting value-added primary produce, which increases

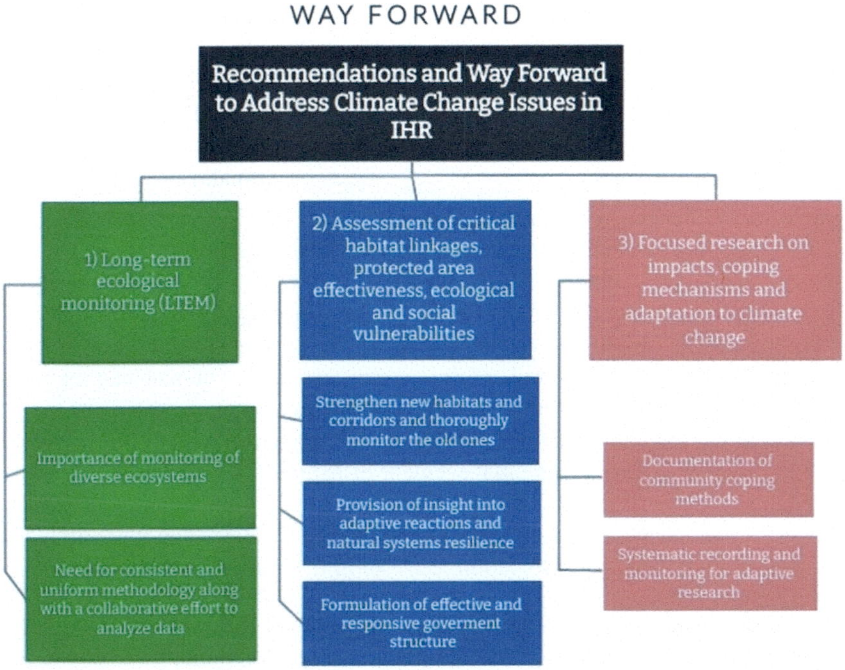

Fig. 3 Recommendations and way forward to address climate issues in IHR

revenues per unit of natural resources. Additionally, adopting alternative materials and technologies that improve resource efficiency or replace resource-intensive practices is crucial. Moreover, creating supplementary sources for flexibility in livelihoods is essential.

In terms of protection and conservation, eco-development interventions are paramount. These initiatives aim to enhance local ecology while improving the welfare of local communities. They achieve this by providing essential services and rights that enable community members to engage in activities that are less dependent on natural resources. Furthermore, these interventions play a vital role in protecting against rapid and slow-onset disasters and facilitating community-based conservation and governance.

8 Why Pursue Long-Term Ecological Research (LTER)?

Long-term ecological research (LTER) is vital for understanding and predicting the dynamics of ecosystems over extended periods, typically spanning decades or even centuries. This extended timeframe enables scientists to capture the complexities of ecological processes, including natural variations, trends, and responses to environmental changes.

LTER provides invaluable insights into the mechanisms underlying ecosystem stability and resilience, biodiversity dynamics, and ecosystem services provisioning. It allows scientists to observe gradual shifts, thresholds, and tipping points that may not be apparent in short-term studies. Moreover, long-term datasets facilitate the detection of trends, anomalies, and unforeseen consequences of human interventions, such as land-use changes, pollution, and habitat fragmentation.

Furthermore, long-term ecological research enhances our capacity for informed decision-making and sustainable management of natural resources. Understanding ecosystems' long-term trajectories and responses to perturbations may help policymakers, land managers, and conservation practitioners to develop more effective strategies for biodiversity conservation, ecosystem restoration, and climate change mitigation and adaptation.

The Intergovernmental Panel on Climate Change (IPCC), in its Fourth Assessment Report (AR4), underscored the inadequacy of climate monitoring data in the Himalayan region. This deficiency is highlighted by the absence of comprehensive scientific reporting on how species respond to shifting global and regional climates, as Pandit (2009) noted, leading the IPCC to characterize it as a "white spot" due to the scarcity of data on natural ecosystems.

Despite the burgeoning interest in climate change research worldwide and the looming threat of global extinction (Xu et al. 2008), it is striking how much of the Himalayan region still lacks sufficient climatological data. High-quality data is essential for assessing the current impacts of climate change and devising reliable predictions to inform planning efforts, particularly in the Himalayan region, where the data collected historically may vary in methodology and instrumentation, thereby introducing uncertainty into its magnitude and quality (Manish et al. 2016). This data deficit poses a significant challenge to global research endeavors, necessitating long-term ecological monitoring (LTEM) initiatives across the Himalayas.

Improving climate data collection in the Indian Himalayan Region is imperative (Negi et al. 2012). The Himalayan highlands, characterized by their cold temperatures, are notably responsive to environmental changes, making them robust indicators of climate change. The biophysical data sets essential for research in the Indian Himalayan Region encompass the cryosphere, hydrosphere, atmospheric pollutants, biodiversity, disasters, and natural hazards.

Notably, the integration of remotely sensed data holds promise for enhancing data collection, particularly in inaccessible terrain; however, ground truthing remains indispensable for validating findings. The long-term research needs in IHR is presented in Fig. 4.

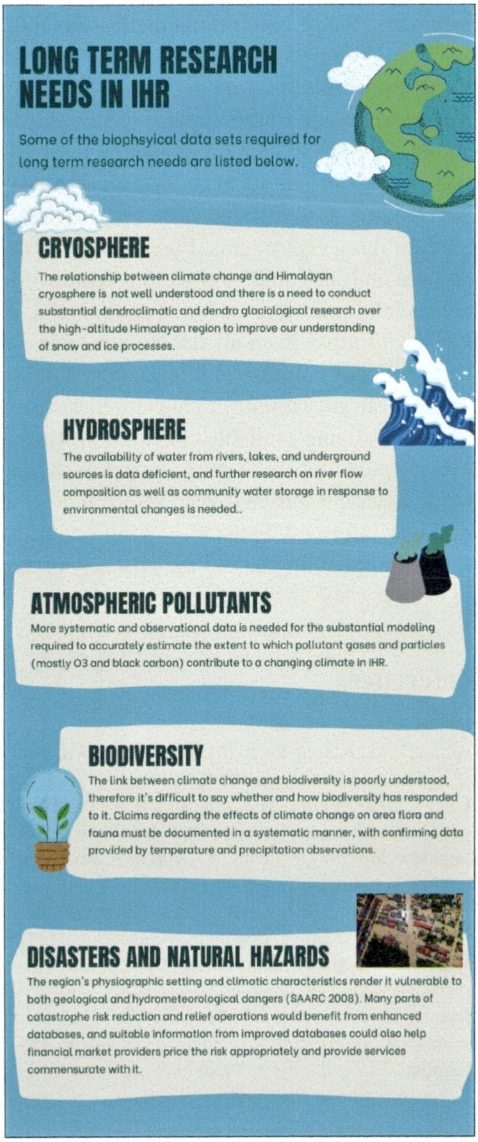

Fig. 4 Long term research needs in IHR

9 Conclusion

This meta-analysis delves into the impacts of climate change on the structure, function, and composition of ecosystems in Indian Himalayan forests, revealing pervasive negative influences across all studied parameters. However, the dearth of data for the Indian Himalayan Region emphasizes the urgent need for sustained, long-term research efforts. While recent years have shown some progress in assessment endeavors, the research remains insufficient, particularly during the critical period between 1990 and 2006. This gap in understanding forest climate change vulnerability, especially in social-ecological landscapes, is striking. Despite advancements in climate change studies regarding Himalayan forests, a considerable lack of knowledge persists concerning regional vulnerabilities. Current forest management initiatives often overlook local ecosystem services crucial for effective adaptation strategies, leading to policy decisions that frequently fall short of addressing the nuanced needs of these ecosystems. Forest policy actions must adopt inclusive approaches at national, sub-national, and local levels to bridge these gaps, integrating ecological and social dimensions of forest ecosystems. Adaptive policies developed collaboratively with input from all stakeholders are crucial for the sustainable harnessing of the benefits of Himalayan forests. Given the scarcity of data in the region, concerted efforts from all Himalayan stakeholders are imperative to conduct comprehensive studies. Utilizing available information, knowledge, and datasets effectively requires an integrated, multi-disciplinary approach that amalgamates contemporary scientific methods with traditional wisdom from diverse communities. Such an approach will facilitate informed decision-making and enhance the efficacy of conservation and adaptation planning across the Himalayan mountain belt in the long term.

References

Adhikari BS, Kumar R (2020) Effect of snowmelt regime on phenology of herbaceous species at and around treeline in Western Himalaya, India. Notulaescientiabiologicae 12(4):901–919

Aryal S, Cockfield G, Maraseni T (2016) Perceived changes in climatic variables and impacts on the transhumance system in the Himalayas. Climate Dev 8:435–446

Bagaria P, Sharma LK, Joshi BD, Kumar H, Mukherjee T, Thakur M, Chandra K (2020) West to east shift in range predicted for Himalayan Langur in climate change scenario. Glob Ecol Conserv

Banerji G, Basu S (2010) Adapting to climate change in Himalayan cold deserts. Int J Clim Change Strat Manag 2(4):426–448. https://doi.org/10.1108/17568691011089945

Beniston M (2003) Climatic change in mountain regions: a review of possible impacts. Clim Change 59(1):5–31

Berrang-Ford L, Pearce T, Ford JD (2015) Systematic review approaches for climate change adaptation research. Reg Environ Change 15:755–769

Bhandari M et al (2019) Prediction mapping through MaxentModeling Paves the way for the conservation of *Rhododendron arboreum* in Uttarakhand Himalayas. J Indian Soc Remote Sens 48:411–422. https://doi.org/10.1007/s12524-019-01089-0

Bhutiyani MR, Kale VS, Pawar NJ (2007) Long-term trends in maximum, minimum and mean annual air temperatures across the Northwestern Himalaya during the twentieth century. Clim Change 85:159–177. https://doi.org/10.1007/s10584-006-9196-1

Bobrowski M, Gerlitz L, Schickhoff U (2017) Modelling the potential distribution of *Betula utilis* in the Himalaya. Glob Ecol Conserv 11:69–83

Cannone N, Sgorbati S, Guglielmin M (2007) Unexpected impacts of climate change on alpine vegetation. Front Ecol Environ 5(7):360–364

Chakraborty A et al (2018) Vulnerability of forests in the Himalayan region to climate change impacts and anthropogenic disturbances: a systematic review. Reg Environ. https://doi.org/10.1007/s10113-018-1309-7

Chettri N (2014) Reconciling mountain biodiversity conservation in a changing climate: a Hindu Kush-Himalayan perspective. Conserv Sci 2:17–27

Chettri N, Sharma E, Shakya B, Thapa R, Bajracharya B, Uddin et al (2010) Biodiversity in the Eastern Himalayas: status, trends and vulnerability to climate change. International Centre for Integrated Mountain Development (ICIMOD)

Chhetri B, Badola HK, Barat S (2021) Modelling climate change impacts on distribution of Himalayan pheasants. Ecol Ind 123. https://doi.org/10.1016/j.ecolind.2021.107368

Cox PA, Cox KN (1997) Encyclopaedia of rhododendron species. Glendoick Publishing, Perth, Australia

Dale VH, Joyce LA, McNulty S, Neilson RP, Ayres MP, Flannigan MD et al (2001) Climate change and forest disturbances: climate change can affect forests by altering the frequency, intensity, duration, and timing of fire, drought, introduced species, insect and pathogen outbreaks, hurricanes, windstorms, ice storms, or landslides. Bio Sci 51(9):723–734

Dhyani K, Kuniyal HB, Singh H, Sobha (2021) Growth and physiological potential of *Terminalia arjuna* under elevated CO_2 levels in open top chamber condition. J Appl Nat Sci 13(3):1121–1126, ISSN 2231-5209. https://doi.org/10.31018/jans.v13i3.2490

Dimri AP, Dash SK (2011) Wintertime climatic trends in the western Himalayas. Clim Change 111:775–800

Dirnböck Ta, Dullinger S, Grabherr G (2003) A regional impact assessment of climate and land-use change on alpine vegetation. J Biogeogr 30(3):401–417

Dukes JS, Pontius J, Orwig D, Garnas JR, Rodgers VL, Brazee N et al (2009) Responses of insect pests, pathogens, and invasive plant species to climate change in the forests of northeastern North America: what can we predict? Can J For Res 39(2):231–248

Dutta P, Dutta B, Das A, Sundriyal RC (2014) Alpine timberline research gap in Himalaya: a literature review. Indian Forester 140:419–427

Fitter AH, Fitter R (2002) Rapid changes in flowering time in British Plants. Science (New York, N.Y.) 296:1689–1691. https://doi.org/10.1126/science.1071617

Foley JA et al (2005) Global consequences of land use. Science 309:570–574

Forrest JL, Wikramanayake E, Shrestha R, Areendran G, Gyeltshen K, Maheshwari A et al (2012) Conservation and climate change: assessing the vulnerability of snow leopard habitat to treeline shift in the Himalaya. Biol Conserv 150(1):129–135

Gaira KS, Dhar U (2020) Phenological change modelling for selected Himalayan medicinal herbs using herbarium records: a case study. Ecol Inf (60). https://doi.org/10.1016/j.ecoinf.2020.101177

Gaira KS, Rawal RS, Rawat B, Bhatt ID (2014) Impact of climate change on the flowering of *Rhododendron arboreum* in central Himalaya, India. Curr Sci 106(12):1735–1738

Gaire N et al (2020) *Abies spectabilis* shows stable growth relations to temperature, but changing response to moisture conditions along an elevation gradient in the central Himalaya. *Dendrochronologia* (60), https://doi.org/10.1016/j.dendro.2020.125675

GTOS (2008) Why focus on mountain regions? www.fao.org/gtos/tems/mod_mou.jsp accessed on 03.09.2009 (accessed 7 Sept 2011)

Gupta SK, Singh H (2017) Observations on sporadic flowering in *Ficus benjamina* L. and *Peltophorum africanum* L. in new forest. Indian Forester 143:290–291

Gupta SK, Ram J, Singh H (2018a) Comparative study of transpiration in cooling effect of tree species in the atmosphere. J Geosci Environ Prot 6:151–166. https://doi.org/10.4236/gep.2018.68011

Gupta SK, Ram J, Singh H, Kumar P (2018b) Comparative study of cooling under shades of some forestry tree species with respect to ambient temperatures. J Geosci Environ Prot 6:51–66. https://doi.org/10.4236/gep.2018.611004

Gupta SK, Ram J, Singh H (2019a) Relationship between leaf area index and atmospheric cooling of tree species. Indian J for 42(3):203–208

Gupta SK, Singh H, Ram J (2019b) Need for a novel paradigm in forestry research: perspectives of comparative atmospheric cooling by different forestry effect tree species. Indian Forester 145(6):509–515

Harrison F (2010) Getting started with meta-analysis. Methods Ecol Evol 2:1–10. https://doi.org/10.1111/j.2041-210X.2010.00056

Herring SC, Christidis N, Hoell A, Kossin JP, Schreck III CJ, Stott PA (2018) Explaining extreme events of 2016 from a climate perspective. Bull Am Meteorol Soc 99(1):S1–S157

Hza O (2009) The Himalayas: water storage under threat. ICIMOD Newslett 56(Winter):3–5

Huggel C, Allen S, Deline P, Fischer L, Noetzli J, Ravanel L (2012) Ice thawing, mountains falling—are alpine rock slope failures increasing? Geol Today 28(3):98–104

Hughes I (2000) Biological consequences of global warming: is the signal already apparent? Trends Ecol Evol 15(2):56–61. https://doi.org/10.1016/s0169-5347(99)01764-4

ICIMOD (2009) Mountain biodiversity and climate change. ICIMOD, Kathmandu, Nepal. http://books.icimod.org/index.php/search/publication/613. Last accessed 13 Oct 2011

ICIMOD W, GBPNIHESD R (2017) Kangchenjunga landscape conservation and development strategy and regional cooperation framework, vol 2. ICIMOD Working Paper 2017

Immerzeel WW, Van Beek LP, Bierkens MF (2010) Climate change will affect the Asian water towers. Science 328(5984):1382–1385

IPCC (2001) Climate change 2001: the scientific basis. In: Houghton JT, Ding Y, Griggs DJ, Noguer M, van der Linden PJ, Dai X, Maskell K, Johnson CA (eds) Contribution of working group I to the third assessment report of the intergovernmental panel on climate change. Cambridge University Press, Cambridge, p 881

IPCC (2001) Climate change 2001: synthesis report. In: A Contribution of Working Groups I, II, and III to the third assessment report of the intergovernmental panel on climate change. Cambridge University Press, Cambridge, United Kingdom, and New York, NY, USA, 398 pp

Jangra A, Kumar K, Maikhuri S, Bhandari MS, Pandey S, Singh H, Barthwal S (2024) Unveiling stress-adapted endophytic bacteria: Characterizing plant growth-promoting traits and assessing cross-inoculation effects on *Populus deltoides* under abiotic stress. Plant Physiol Biochem 210:108610. https://doi.org/10.1016/j.plaphy.2024.108610

Jiang F, Zhang J, Gao H, Cai Z, Zhou X, Li S, Zhang T (2020) Musk deer (Moschus spp.) face redistribution to higher elevations and latitudes under climate change in China. Sci Total Environ 704:135335

Kanta C, Kumar A, Chauhan A, Singh H, Sharma IP (2024) The interplay between plant functional traits and climate change. In: Kumar N, Singh H (eds) Plant functional traits for improving productivity. Springer, Singapore. https://doi.org/10.1007/978-981-97-1510-7_3

Kocmánková E, Trnka M, Juroch J, Dubrovský M, Semerádová D, Možný M, Žalud Z (2009) Impact of climate change on the occurrence and activity of harmful organisms. Plant Prot Sci 45(Special Issue):S48–S52

Körner C (1998) A re-assessment of high elevation treeline positions and their explanation. Oecologia, 115(4):445–459

Körner C (2004) Mountain biodiversity, its causes and function. AMBIO: A J Hum Environ 33:11–17

Kothawale DR, Kumar KR (2005) On the recent changes in surface temperature trends over India. Geophys Res Lett, 32

Kumar N, Singh H (2024) Plant functional traits for improving productivity. Springer Nature, Singapore. https://doi.org/10.1007/978-981-97-1510-7

Kumar N, Suyal DC, Sharma IP, Verma A, Singh H (2017) Elucidating stress proteins in rice (*Oryza sativa* L.) genotype under elevated temperature: a proteomic approach to understand heat stress response. 3 Biotech 7:205. https://doi.org/10.1007/s13205-017-0856-9

Kumar M, Singh MP, Singh H, Dhakate PM, Ravindranath NH (2019a) Forest working plan for the sustainable management of forest and biodiversity in India. J Sustain for 39:1–22. https://doi.org/10.1080/10549811.2019.1632212

Kumar N, Jeena N, Singh H (2019b) Elevated temperature modulates rice pollen structure: a study from the foothill of the Himalayan agro-ecosystem in India. 3 Biotech 9(5):175. https://doi.org/10.1007/s13205-019-1700-1

Kumar N, Jeena N, Kumar N, Gangola S, Singh H (2019c) Phytoremediation facilitating enzymes: an enzymatic approach for enhancing remediation process. In: Bhatt P (ed) Smart bioremediation technologies. Academic Press, Elsevier, Cambridge, MA, United States, pp 289–306

Kumar N, Singh H, Sharma SK (2020) Antioxidants: responses and importance in plant defense system. In: Roy Chowdhury R, Choudhury S, Hasanuzzaman M, Srivastava S (eds) Sustainable agriculture in the era of climate change. Springer, Cham, pp 251–264. https://doi.org/10.1007/978-3-030-45669-6_11

Kumar A, Kumar P, Singh H, Kumar N (2021a) Modulation of plant functional traits under essential plant nutrients during seasonal regime in natural forests of Garhwal Himalayas. Plant Soil 465:197–212. https://doi.org/10.1007/s11104-021-05003-x

Kumar A, Kumar P, Singh H, Kumar N (2021b) Adaptation and mitigation potential of roadside trees with bio-extraction of heavy metals under vehicular emissions and their impact on physiological traits during seasonal regimes. Urban Forest Urban Greening 58:126900. https://doi.org/10.1016/j.ufug.2020.126900

Kumar A, Kumar P, Singh H, Kumar N (2021c) Impact of plant functional traits on infiltration rate under different forests of Kempty watershed in Garhwal Himalaya. Indian J Soil Conserv 49(1):38–44

Kumar A, Kumar P, Singh H, Bisht S, Kumar N (2021d) Relationship of physiological plant functional traits with soil carbon stock in the temperate forest of Garhwal Himalaya. Curr Sci 120(8):1368–1373. https://doi.org/10.18520/cs/v120/i8/1368-1373

Kumar M, Phukon SN, Paygude AC, Tyagi K, Singh H (2021e) Mapping phenological functional types (PhFT) in the Indian Eastern Himalayas using machine learning algorithm in Google Earth Engine. Comput Geosci 158. https://doi.org/10.1016/j.cageo.2021.104982

Kumar A, Kumar P, Singh H, Kumar N (2021f) Modulation of plant functional traits under essential plant nutrients during seasonal regime in natural forests of Garhwal Himalayas. Plant Soil. https://doi.org/10.1007/s11104-021-05003-x

Kumar A, Singh H, Kumari G, Bisht S, Malik A, Kumar A, Singh M, Raturi A, Barthwal S, Thakur A, Kaushal R (2022) Adaptive resilience of roadside trees to vehicular emissions via leaf enzymatic, physiological, and anatomical trait modulations. Environ Pollut 313. https://doi.org/10.1016/j.envpol.2022.120191

Kumar N, Singh H, Giri K, Kumar A, Joshi A, Yadav A, Singh R, Bisht S, Kumari R, Jeena N, Khairakpam R, Mishra G (2024) Physiological and molecular insights into the allelopathic effects on agroecosystems under changing environmental conditions. Physiol Mol Biol Plants 30:417–433. https://doi.org/10.1007/s12298-024-01440-x

Lamsal P, Kumar L, Aryal A, Atreya K (2018). Future climate and habitat distribution of Himalayan Musk Deer (*Moschus chrysogaster*). Ecol Inform 44:101–108

Leemans R, Eickout B (2004) Another reason for concern and global impacts on ecosystems for different levels of climate change. Glob Environ Chang 14:219–228

Malik A, Kumar K, Kumari G, Kumar N, Singh H (2024) Application of functional traits in modelling productivity and resilience under climate change. In: Kumar N, Singh H (eds) Plant functional traits for improving productivity. Springer, Singapore. https://doi.org/10.1007/978-981-97-1510-7_5

Manish K, Telwala Y, Nautiyal DC, Pandit MK (2016) Modelling the impacts of future climate change on plant communities in the Himalaya: a case study from Eastern Himalaya, India. Model Earth Syst Environ 2:1–12

Messerli B (2009) Biodiversity and climate change in the Himalayas. Sustainable Mountain Development No. 55, ICIMOD, Spring 2009

Miles L, Kapos V (2008) Reducing greenhouse gas emissions from deforestation and forest degradation: global land-use implications. Science (New York, N.Y.), 320(5882):1454–1455. https://doi.org/10.1126/science.1155358

Myers N (1988) Threatened biotas: 'hotspots' in tropical forestry. Environmentalist 8:1–20
Namgail T, Van Wieren SE, Mishra C, Prins HHT (2010) Multi-spatial co-distribution of the endangered Ladakh urial and blue sheep in the arid Trans-Himalayan mountains. J Arid Environ 74(10):1162–1169
Navalgund RR, Kumar AS, Nandy S (eds) (2019) Remote sensing of northwest Himalayan ecosystems, vol 1. Springer, Singapore
Negi VS, Maikhuri RK (2012) Socio-ecological and religious perspective of agrobiodiversity conservation: issues, concern and priority for sustainable agriculture, Uttarakhand Himalaya. J Agric Environ Ethics 26:491–512. https://doi.org/10.1007/s10806-012-9386-y
Negi GCS, Mukherjee S (2020) Climate change impacts in the Himalayan Mountain ecosystems. https://doi.org/10.1016/B978-0-12-409548-9.12056-1
Ogra MV, Badola R (2015) Gender and climate change in the Indian Himalayas: global threats, local vulnerabilities, and livelihood diversification at the Nanda Devi Biosphere Reserve, *Earth Syst*. Dynam 6:505–523. https://doi.org/10.5194/esd-6-505-2015
Olofsson J (2006) Short and long-term effects of changes in reindeer grazing pressure on tundra health vegetation. J Ecol 94:431–440
Pandey S et al (2020) Effects of climate change on treeline trees in Sagarmatha (Mt. Everest, Central Himalaya). J Veg Sci. https://doi.org/10.1111/jvs.12921
Panigrahy S, Anitha D, Kimothi MM, Singh SP (2010) Timberline change detection using topographic map and satellite imagery. Trop Ecol 51(1):87–91
Pathak R, Negi VS, Rawal RS, Bhatt ID (2019) Alien plant invasion in the Indian Himalayan Region: state of knowledge and research priorities. Biodivers Conserv 28(12):3073–3102
Prakash V, Hunney K, Singh H (2022) Effect of elevated carbon dioxide on growth and development of *Santalum album* L. seedlings inoculated with plant growth promoting microorganisms in open top chambers. Int J Environ, Agric Biotechnol 7(5):149–159
Pratap U, Partap T (2003) Warning signals from the Apple Valleys of the Hindu Kush—Himalayas: productivity concerns and pollination problems. ICIMOD, Kathmandu
Pratap U (1999) Pollination management of mountain crops through beekeeping: trainers resource book. ICIMOD. Kathmandu, Nepal
Raturi A, Singh H, Kumar P, Chanda A, Raturi A (2024) Spatiotemporal patterns of greenhouse gas fluxes in the subtropical wetland ecosystem of Indian Himalayan foothill. Environ Monit Assess 196(10):882
Rautela P, Karki B (2015) Traditional practices for survival in resource depleted Himalayan region: challenges put forth by climate change and response of local communities. Int J Sci Technol 4(8):395–404
Sala OE, Stuart Chapin FIII, Armesto JJ, Berlow E, Bloomfield J, Dirzo R et al (2000) Global biodiversity scenarios for the year 2100. Science 287(5459):1770–1774
Schickhoff U et al (2015) Do Himalayan treelines respond to recent climate change? An evaluation of sensitivity indicators. Earth Syst Dyn 6:245–265. https://doi.org/10.5194/esd-6-245-2015
Schild A (2008) ICIMOD's position on climate change and mountain systems
Sendall KM, Reich PB, Zhao C, Jihua H, Wei X, Stefanski A et al (2015) Acclimation of photosynthetic temperature optima of temperate and boreal tree species in response to experimental forest warming. Glob Change Biol 21(3):1342–1357
Sharma R, Singh H (2021) Alteration in biochemical constituents and nutrients partitioning of *Asparagus racemosus* in response to elevated atmospheric CO_2 concentration. Environ Sci Pollut Res 29:6812–6821. https://doi.org/10.1007/s11356-021-16050-3
Sharma E, Chettri N, Oli KP (2010) Mountain biodiversity conservation and management: a paradigm shift in policies and practices in the Hindu Kush-Himalayas. Ecol Res 25:909–923. https://doi.org/10.1007/s11284-010-0747-6
Sharma CM, Sharma S, Gurung S, Bajracharya RM, Jüttner I, Pradhan NS (2009) Global climatic change and high altitude lakes: impacts on aquatic biodiversity and pollution status. Nat Resour Manag Rev Res Himal Watersheds 44:103–122

Sharma R, Singh H, Kaushik M, Nautiyal R, Singh O (2018) Adaptive physiological response, carbon partitioning, and biomass production of *Withania somnifera* (L.) Dunal grown under elevated CO_2 regimes. 3 Biotech 8:267. https://doi.org/10.1007/s13205-018-1292-1

Shrestha S (2019) Effects of climate change in agricultural insect pest. Acta Sci Agric 3(12):74–80

Shrestha BB, Budha PB, Wong LJ, Pagad S (2018) Global register of introduced and invasive species—Nepal. Version 2.2. Invasive species specialist group (ISSG). Checklist Dataset. https://doi.org/10.15468/4r0kkr. Accessed via GBIF.org on 2018.03.09

Shrestha KB, Hofgaard A, Vandvik V (2015) Recent treeline dynamics are similar between dry and mesic areas of Nepal, central Himalaya. J Plant Ecol 8(4):347–358

Singh JS (2006) Sustainable development of Indian Himalayan region: linking ecology and economic concerns. Curr Sci 90(6):784–788

Singh H (2021) An integrated approach considering physiological and biophysical based indicators for assessing tolerance of roadside plantations of *Alstonia scholaris* towards urban roadside air pollution: an assessment of adaptation of plantations for mitigating urban roadside air pollution. Trees. https://doi.org/10.1007/s00468-021-02179-8

Singh H (2024) Exploring adaptive modulation in plant functional traits and its impact on the productivity of *Acacia auriculiformis* under CO_2-enriched environment. Ind Crops Prod 210:118186. https://doi.org/10.1016/j.indcrop.2024.118186

Singh SP, Singh V, Skutsch M (2010) Rapid warming in the Himalayas: ecosystem responses and development options. Clim Dev 2(3):221–232

Singh SP, Bassignana-Khadka I, Singh Karky B, Sharma E (2011) Climate change in the Hindu Kush-Himalayas: the state of current knowledge. International Centre for Integrated Mountain Development (ICIMOD)

Singh AK, Bordoloi LJ, Kumar M, Hazarika S, Parmar B (2013) Land use impact on soil quality in eastern Himalayan region of India. Environ Monit Assess. https://doi.org/10.1007/s10661-013-3514-7

Singh H, Kumar N, Singh R, Kumar M (2023) Assessing the climate change impact on the habitat suitability of the range-restricted bird species (Catreus wallichii) in the Indian Himalayan ecosystem. Environ Sci Pollut Res 30(57):121224–121235

Singh H, Verma A, Kumar M, Sharma R, Gupta R, Kaur M, Negi M, Sharma SK (2017b) Phytoremediation: a green technology to clean up the sites with low and moderate level of heavy metals. Austin Biochem 2:1012

Singh H, Sharma R, Savita SMP, Kumar M, Verma A, Ansari MW, Sharma SK (2018) Adaptive physiological response of *Parthenium hysterophorus* to elevated atmospheric CO_2 concentration. Indian Forester 144:1–14

Singh N, Tewari A, Shah S (2020b) Treeline species: regeneration status and seedling dynamics in western Himalayan region. Environ Ecol 38(3B):725–732

Singh S, Singh H, Sharma SK, Nautiyal R (2021) Seasonal variation in biochemical responses of bamboo clones in the sub-tropical climate of Indian Himalayan foothills. Heliyon 7(4):e06859. https://doi.org/10.1016/j.heliyon.2021.e06859

Singh H, Kumar M (2022) Climate change and its impact on Indian Himalayan forests: current status and research needs. In: Rani S, Kumar R (eds) Climate change. Springer Climate. Springer, Cham, pp 223–242. https://doi.org/10.1007/978-3-030-92782-0_11

Singh H, Savita Sharma R, Sinha S, Kumar M, Kumar P, Verma A, Sharma SK (2017a) Physiological functioning of *Lagerstroemia speciosa* L. under heavy roadside traffic: an approach to screen potential species for abatement of urban air pollution. 3 Biotech 7(61):1–10. https://doi.org/10.1007/s13205-017-0690-0

Singh H, Yadav M, Kumar N, Kumar A, Kumar M (2020) Assessing adaptation and mitigation potential of roadside trees under the influence of vehicular emissions: a case study of *Grevillea robusta* and *Mangifera indica* planted in an urban city of India. PLoS ONE 15(1):e0227380. https://doi.org/10.1371/journal.pone.0227380

Singh H, Bandyopadhyaya S, Kumar A, Singh M, Malik A, Kumari G, Raturi A, Yadav SK, Kumar M, Barthwal S, Thakur A, Rawat PS (2022) Understanding the physiological and biophysical

response of urban roadside plantations for assessing adaptation and mitigation mechanisms toward vehicular emissions. Urban Clim 44. https://doi.org/10.1016/j.uclim.2022.101183
Singh M, Singh H, Kumar A, Kumar M, Barthwal S, Thakur A (2024a) Soil nitrogen availability determines the CO_2 fertilization effect on tree species (*Neolamarckia cadamba*): growth and physiological evidence. Environ Sustain 7:53–60. https://doi.org/10.1007/s42398-023-00300-w
Singh H, Kumar N, Kumar A (2024b) Enhancing resource use efficiency in crops through plant functional traits. In: Kumar N, Singh H (eds) Plant functional traits for improving productivity. Springer, Singapore. https://doi.org/10.1007/978-981-97-1510-7_6
Solomon S et al (2007) Climate change 2007: the physical science basis. In: Working Group I contribution to the fourth assessment report of the IPCC
Telwala Y, Brook BW, Manish K, Pandit MK (2013) Climate-induced elevational range shifts and increase in plant species richness in a Himalayan biodiversity epicentre. PloS One 8(2):e57103
Tewari VP, Verma RK, von Gadow K (2017) Climate change effects in the Western Himalayan ecosystems of India: evidence and strategies. Forest Ecosyst 4(13). https://doi.org/10.1186/s40663-017-0100-4
Upgupta S et al (2015) Climate change impact and vulnerability assessment of forests in the Indian western Himalayan region: a case study of Himachal Pradesh, India. Clim Risk Manag 10:63–76
Vaidya RA, Sharma E (2014) Research insights on climate and water in the Hindu Kush Himalayas. ICIMOD, Kathmandu
Verma S et al (2020) Novel approaches to mitigate heat stress impacts on crop growth and development. Plant Physiol Rep. https://doi.org/10.1007/s40502-020-00550-4
Walker MD, Wahren CH, Hollister RD, Henry GH, Ahlquist LE, Alatalo JM et al (2006) Plant community responses to experimental warming across the tundra biome. Proc Nat Acad Sci 103(5):1342–1346
Watham T et al (2020) Environmental control on carbon exchange of natural and planted forests in Western Himalayan foothills of India. Biogeochemistry 151:291–311. https://doi.org/10.1007/s10533-020-00727-x
Wolkovich EM, Cleland EE (2014) Phenological niches and the future of invaded ecosystems with climate change. AoB Plants 6:plu013
Xu J, Wilkes A (2004) Biodiversity impact analysis in northwest Yunnan, southwest China. Biodivers Conserv 13:959–983
Xu X, Chen H, Levy JK (2008) Spatiotemporal vegetation cover variations in the Qinghai-Tibet Plateau under global climate change. Chin Sci Bull 53(6):915–922
Yadav SK, Singh H, Nautiyal R, Ginwal HS, Ansari SA, Barthwal S (2019b) Modulation of morpho-physiological responses in *Populus deltoides* by elevated carbon dioxide and temperature. Forest Sci 66:105–118. https://doi.org/10.1093/forsci/fxz048
Yadav SK, Singh H, Ginwal HS, Barthwal S (2019) Elevated CO_2 enhanced growth and physiological process of *Populus deltoides* Bartr. ex Marsh. Indian Forester 145(1):23–27
Zhou W, Liu G, Pan J, Feng X (2005) Distribution of available soil water capacity in China. J Geogr Sci 15:3–12
Zomer R et al (2014) Projected climate change impacts on spatial distribution of bioclimatic zones and ecoregions within the Kailash Sacred Landscape of China, India, Nepal. Clim Change 125. https://doi.org/10.1007/s10584-014-1176-2

Policies, Strategies, and Regulatory Framework for Sustainable Urban Forest Management

Kundan Burnwal and Nemit Verma

Abstract Urban forests enhance the quality of life through their social, ecological, and economic benefits. This chapter explores the significance of urban forests in mitigating the urban heat island effect, improving air quality, reducing noise pollution, and promoting community well-being. It also highlights the challenges urban forests face, including pollution, developmental pressures, and the expansion of metropolitan areas. Effective governance, supported by policies and strategies, is essential for the sustainable development and management of urban forests. The chapter examines the concept of urban forestry, its multifaceted benefits, and the strategic approaches needed for its development. Additionally, it reviews existing policies and initiatives from various countries, providing insights into different approaches to urban forest management. Emphasis is placed on policy interventions, regulatory frameworks, and community engagement to promote sustainable urban forestry. Addressing urban forests' environmental, social, and economic dimensions provides a comprehensive framework for enhancing urban green spaces, contributing to a healthier and more resilient urban environment.

Keywords Urban forests · Development pressures · Policies · Regulatory frameworks · Community engagement

K. Burnwal
Senior Advisor and Deputy Lead, 'Restore, Conserve, and Protect Forest and Tree Cover for NDC Implementation in India (RECAP4NDC)' Project, Deutsche Gesellschaft Für Internationale Zusammenarbeit (GIZ) GmbH, New Delhi, India
e-mail: kundan.burnwal@giz.de

N. Verma (✉)
Advisor, 'Restore, Conserve, and Protect Forest and Tree Cover for NDC Implementation in India (RECAP4NDC)' Project, Deutsche Gesellschaft Für Internationale Zusammenarbeit (GIZ) GmbH, New Delhi, India
e-mail: nemit.verma@giz.de

H. Singh (ed.), *Urban Forests, Climate Change and Environmental Pollution*,
https://doi.org/10.1007/978-3-031-67837-0_40

1 Introduction

Urban forest plays a vital role in enhancing the quality of life by providing the social, ecological, and economic benefits from nature (Daily 1997; Kumar et al. 2022). A well-managed urban forest contributes to mitigating the heat island effect, improving air quality, reducing runoff and noise pollution, sequestering carbon, and enhancing food security, community cohesion, and community members' physical and mental well-being (Braubach et al. 2017; Kuehler et al. 2017; Lai et al. 2019; Nowak and Dwyer 2007; Sanesi et al. 2011; Sicard et al. 2018; Kumar et al. 2019a, b, c; Weinstein et al. 2015; Wolf 2017, Salbitano et al. 2015). At the same time, urban forests face various challenges due to pollution, ongoing developmental activities, and continuous expansion of the metropolitan areas. Therefore, developing effective strategies for planning, developing, and maintaining urban forests becomes critically important.

The sustainability of any initiative depends on its governance mechanism supplemented by the policies/laws and strategies for achieving fair and equitable outcomes. Specifically in urban forestry, the effectiveness of policies and strategies leads to the sustainable development and management of urban green spaces, strengthening whole ecosystem services in urban areas necessary for city dwellers' mental and physical well-being (Kumar et al. 2024; Singh et al. 2024a; Singh 2024). The legal and institutional framework guides the policy measures to establish a long-term vision towards urban forestry that aligns with international commitments and obligations. To create a healthier and more resilient urban environment and to maximize the benefits of urban green spaces, the policy intervention should focus on effective planning, sustainable management practices, and active community participation (Dhyani et al 2021; Kumar et al. 2021a, b, c, d, e; Prakash et al. 2022).

During urban planning, it is crucial to address green inequality during the integration of urban green spaces as public infrastructure (Sister et al. 2010). There is a need to clearly define the ownership and distribution of benefits, including revenue generation, before establishing urban green spaces. For example, in developing urban green spaces or forest patches, it should be evident who holds the rights and benefits from urban forests/tree resources (Singh 2021; Sharma and Singh 2021; Singh et al. 2021). Moreover, in terms of governance, there should be an institutional mechanism at the national and sub-national levels that provides adaptive measures towards policy interventions based on evolving conditions, priorities, international commitments, obligations, etc. Keeping these facts in mind, this chapter highlights the policy measures, strategies, regulatory framework, challenges, and opportunities associated with urban forest management.

2 Concept of Urban Forest

Urban forest refers to the trees and shrubs in a metropolitan area, including trees in yards and utility corridors, trees along streets, green spaces with trees, and the associated vegetation (Konijnendijk et al. 2006; Escobedo et al. 2011; Miller et al. 2015). In some areas, these urban forests represent the green infrastructure that is most extensive and fully functional, providing ecosystem services such as the abatement of air pollution and the management of stormwater (Lafortezza et al. 2017; Singh et al. 2020; Kumar et al. 2021a, b).

3 Need and Importance of Urban Forests

Rapid urbanization leading to poor air quality has become a major environmental issue for growing cities and city dwellers. Keeping the well-being of the local communities and providing a healthy environment for the citizens, city governments are focusing on clean and green solutions that can reduce air pollution, heat island effects, land degradation, and poor water availability, amongst others (Kumar et al. 2021c, d). Developing urban forests can contribute to environmental sustainability, social cohesion, and economic development. The different aspects involved in designing and managing urban forests are given below:

(a) Environmental Benefits
(b) Social and Health Benefits
(c) Biodiversity Conservation
(d) Green Economic Development
(e) Urban Planning and Sustainable Management of Resources
(f) Community Engagement.

3.1 Environmental Benefits

Trees help improve air quality by absorbing pollutants and releasing oxygen (Nowak et al. 2014). Urban forests provide shade and reduce the urban heat island effect. Cool surfaces and shade trees help to reduce energy use and provide a pleasant environment. Vegetation in urban forests acts as a natural sound barrier, absorbing and deflecting noise (Wang et al. 2014; Yadav et al. 2019; Gupta et al. 2019a). This helps reduce the negative impacts of noise pollution from traffic and other urban sources.

3.2 Social and Health Benefits

Urban forests provide suitable places for recreation, relaxation, and aesthetic enjoyment (Ward Thompson 2012). These activities contribute to improved mental well-being and social interaction. Exposure to nature in urban areas has been linked to improved mental well-being (Bratman et al. 2012; Wolf and Robbins 2015a, b; Braubach et al. 2017). Spending time in natural environments can reduce stress, anxiety, and symptoms of depression.

3.3 Biodiversity Conservation

Urban forests provide diverse habitats for various plant and animal species. Trees, shrubs, and other vegetation offer food, shelter, and breeding sites for birds, insects, mammals, and other wildlife, contributing to urban biodiversity (Nielsen et al. 2014; Kumar et al. 2019c; Kumar et al. 2019b). Urban forests can also act as biodiversity hotspots, supporting a rich variety of plant and animal life. The diverse range of tree species and understory vegetation create microhabitats that attract different species. Moreover, urban forests can be important migration corridors for birds and other wildlife. Green spaces within cities can connect larger natural areas, facilitating the movement of species and promoting genetic diversity.

3.4 Green Economic Development

Properly placed trees can reduce energy consumption by providing shade in the summer and acting as windbreaks in the winter (Akbari et al. 2005). Proximity to green spaces, including urban forests, can positively impact property values (Wolf and Robbins 2015a, b). The planning, development, and maintenance of urban forests create job opportunities. This includes positions related to landscaping, arboriculture, and urban forests, which form part of green infrastructure (Sharma et al. 2018; Singh et al. 2018). Investing in the development and maintenance of this infrastructure provides economic benefits by enhancing the quality of life for residents, attracting businesses, and promoting overall urban resilience, environmental education, park management, and contributing to local employment.

Urban forests contribute to carbon sequestration (Nowak et al. 2002), which can be leveraged in carbon trading and offsetting programs. Businesses and industries seeking to offset their carbon footprint may invest in urban forest conservation or restoration projects.

3.5 Urban Planning and Sustainable Management of Resources

Urban forestry planning by integrating trees into urban planning helps maximize their benefits and mitigate potential conflicts, such as rising temperatures, energy consumption, and poor air quality in a particular area (McPherson et al. 1997). Leveraging funds and partnerships to support green infrastructure projects, grants, public–private partnerships, and community initiatives can help secure resources for the sustainable development and management of urban forests.

3.6 Community Engagement

Involving the community in tree planting and maintenance can create a sense of ownership and environmental stewardship, leading to the sustainable management of resources and improving the whole urban ecosystem (Gobster and Westphal 2004). Community engagement in urban forests brings people together, increasing social cohesion. Shared experiences and collaborative efforts in activities like tree planting or clean-up events strengthen community bonds and create a sense of belongingness.

4 Challenges Faced by Urban Forests

Urban forestry is entirely different from conventional forestry as the urban forests face various challenges such as difficult growing conditions due to pollutants, poor soil quality, human interference, heterogenous light, and water availability, among others that lead to poor health of vegetation, increased susceptibility to diseases and pests, and premature mortality. Consequently, urban forests fail to fulfill the requirements of urban areas in terms of abatement of air pollution, cooling, enhanced scenic quality, and recreational space (Roy and Byrne 2014).

The expansion of cities also threatens the urban green spaces, resulting in the loss of urban forest canopy. Another challenge is the availability of funds, as maintaining urban forests requires more attention, resources, and budget due to harsh growing conditions. Urban forests are undervalued as common people, and decision-makers consider that the financial burden outweighs their benefits (Escobedo et al. 2011; Gupta et al. 2018a, b).

5 Review of Existing Policies in the Context of Green Urban Development in Different Countries

Various countries and cities have implemented policies and initiatives to develop and manage urban forests with different approaches based on the local and regional variations in climate, ecology, and community needs, which often influence the design and implementation of urban forestry policies. Some examples of existing policies in different selected regions to develop and manage urban forests are briefly discussed below:

a. United States—Urban and Community Forestry Program[1]

The U.S. Forest Service administers the Urban and Community Forestry Program, which provides financial assistance, technical support, and educational outreach to enhance urban forestry initiatives nationwide. State forestry agencies often collaborate to implement and tailor these programs to local needs. The program intends to improve more than 140 million acres of urban and community forest across the United States.

b. United Kingdom—England Tree Strategy[2]

The England Tree Strategy aims to expand tree cover in urban and rural areas. It outlines plans to increase tree planting, protect existing trees, and engage local communities. Local authorities play a role in delivering the strategy, and funding is allocated to support tree-planting projects. The England Trees Action Plan 2021–2024 intends to spend over £500 million of the £640 million Nature for Climate Fund on trees and woodlands in England between 2020 and 2025.

c. Canada—Urban Forest Strategy[3]

Toronto has an Urban Forest Strategy focused on maintaining and expanding the city's tree canopy. The strategy includes goals for planting trees, protecting existing trees, and engaging the community in tree stewardship. It addresses issues such as climate change adaptation and biodiversity. The vision of this strategy is to have sustainable, biodiversity, and healthy urban forests that protect and enhance the well-being and prosperity of Canadian communities under five strategic objectives, i.e., national urban forestry capacity building; communications and public education; research; techniques and technologies for urban forest planning and management; and professional development.

d. Australia—Urban Forest Strategy, City of Melbourne and Sydney[4]

---

[1] Urban and Community Forestry Program | US Forest Service (usda.gov) accessed on 22.01.2024.

[2] Consultation launched on the England Tree Strategy—GOV.UK (www.gov.uk) accessed on 22.01.2024.

[3] TC-CUFS-2019-2024-Eng.pdf (treecanada.ca) accessed on 20.01.2024.

[4] Urban Forest Strategy—City of Melbourne accessed on 20.01.2024.
Urban forest strategy—City of Sydney (nsw.gov.au) accessed on 20.01.2024.

The City of Melbourne and Sydney have an Urban Forest Strategy to increase tree canopy cover, improve biodiversity, and create a more sustainable and resilient urban environment. The strategy involves collaborating with local government, community groups, and businesses. It also outlines the targets to increase the canopy of trees and maximize the economic, social, and environmental benefits they can provide. The strategies and targets proposed to achieve this vision include increasing canopy cover, increasing urban forest diversity, improving vegetation health, improving soil moisture and water quality, improving urban ecology, and informing and consulting the community.

e. Singapore—City in a Garden Vision[5]

Singapore's "City in a Garden" vision emphasizes the integration of greenery into the urban landscape. Policies focus on maximizing green spaces, planting trees, and creating a sustainable and livable city. The National Parks Board oversees initiatives to enhance Singapore's green infrastructure.

f. European Union—Biodiversity Strategy for 2030[6]

The EU Biodiversity Strategy for 2030 includes initiatives to enhance green infrastructure, including urban forests. It emphasizes the importance of increasing the quantity and quality of green spaces in urban areas to support biodiversity and ecosystem services.

g. The Million Trees Initiative, Kenya[7]

The Million Trees Initiative in Kenya aims to plant one million trees in urban areas to improve air quality, mitigate urban heat island effects, and enhance urban biodiversity. The initiative collaborates with government agencies, NGOs, community groups, and private sector partners to achieve its goals. A special holiday to each Kenyan has been given to encourage them to plant at least two seedlings to support the achievement of the 100 million targets.

Initiatives Towards Developing Urban Forests—The Indian Context

India has recognized the importance of urban forests and green spaces. Several policies and initiatives have been implemented at the national and local levels to promote and develop urban forests and green spaces. Keeping in mind that policies can evolve, it is advisable to check the latest updates from relevant government agencies. Here are some examples of existing policies related to urban forests in India:

National Forest Policy (NFP) 1988 and Revised National Forest Policy (Draft) 2018

[5] "Garden city" vision is introduced (nlb.gov.sg) accessed on 14.01.2024.

[6] EU biodiversity strategy for 2030 | EUR-Lex (europa.eu) accessed on 14.01.2024.

[7] https://www.aa.com.tr/en/africa/kenyans-granted-special-holiday-to-plant-100m-trees-to-help-fight-climate-crisis/3052991 accessed on 27.02.2024.

While not specific to urban areas, the National Forest Policy provides a framework for the country's conservation, regeneration, and sustainable management of forests. The 2018 draft policy emphasizes the need for urban and peri-urban afforestation to enhance ecosystem services.

Urban Greening Guidelines, Ministry of Housing and Urban Affairs

The Ministry of Housing and Urban Affairs in India has provided guidelines and recommendations for urban greening, including the development of urban forests. These guidelines offer insights into planning, designing, and managing green spaces in urban areas.

National Mission for Green India (GIM)

The National Mission for Green India is one of the major initiatives under the National Action Plan on Climate Change, which primarily focuses on the afforestation being taken up in the forests and non-forest areas. It supports initiatives to increase green cover, improve biodiversity, and enhance environmental sustainability.

Nagar Van Yojna-Urban Forests

National Afforestation and Eco-development Board has been implementing the Nagar Van Yojana (NVY) since 2020, and it envisages developing 1000 Nagar Vans/Vatikas in the country. The support is provided under the scheme to cover mainly the cost of fencing, soil-moisture conservation measures & related activities, administrative activities, plantation, and maintenance in the urban areas.

Smart Cities Mission

The Smart Cities Mission, launched by the Government of India, includes the development of green spaces and parks as one of its key components. Many cities under this mission are working on creating or enhancing urban forests to improve residents' overall quality of life.

State-Level Initiatives

Several states in India have initiated their programs and policies to promote urban forestry. For example, Karnataka's "NammaOoruNamma Mara" (Our City, Our Tree) campaign encourages citizens to participate actively in tree planting and urban forest development.

City-Specific Initiatives

Individual cities in India may have policies and initiatives related to urban forestry. For instance, Bengaluru (Bangalore) has the "Namma Vanam-Namma Dhwani" initiative, focusing on increasing the city's green cover and involving citizens in tree-planting activities.

6 Approach of Policy Interventions in Sustainable Urban Forest Development and Management

Developing an effective policy requires a collaborative planning and management strategy with urban planners, environmental professionals, engineers, architects, and residents. The policies focusing on the development and management of the urban forest vary from region to region and focus majorly on a given area's local context, governance, and socio-ecological conditions. Some of the major considerations for developing the policies in the context of urban forests are given below:

(a) **Urban forestry planning and zoning**

Introducing urban forestry plans and zoning regulations can be important for clean, green, and sustainable urban development. The focus can be quantifying urban forest structure, function, and value.

(b) **Tree canopy goal**

The area-wise tree cover targets can be assigned based on the physical and geographical features, leading to the systematic development of green cover in urban areas.

(c) **Species diversity and resilience**

Promoting planting diverse tree species based on the local floral diversity will be necessary. Priority can be given to fast-growing tree species with shade and higher economic value. Plantation of exotic and invasive species with low bioaccumulation potential must be avoided (Singh and Verma 2013; Kumar et al. 2017).

(d) **Tree preservation during urban development**

Existing policies for urban development should be strengthened with the provision for the preservation and translocation of the existing trees in the area proposed for the construction/developmental activities (Gupta and Singh 2017; Singh et al. 2017a).

(e) **Community engagement and education**

Community-level programs and capacity building for local communities must be introduced. Priorities and designing of initiatives for developing urban green spaces should be given through active community participation by following the suitable model for incentivizing the community towards green development of urban areas (Singh et al. 2017b).

(f) **Planning/Readiness toward green finance**

A separate budget for developing green infrastructure and expanding urban forests can be introduced to help the implementation/development organization shift its approach from traditional urban development to green development.

(g) **Strategy and policy measures to address climate adaptation**

Keeping the vulnerable areas in terms of climate change and existing drivers in the urban areas for environmental degradation, introducing urban forestry as one of the climate change adaptation measures can create positive impacts towards the changing weather and climatic conditions of urban areas. Integrating urban forestry into climate change adaptation plans, considering the role of trees in mitigating heat island effects and reducing vulnerability to extreme weather events, can be one of the major policy measures for developing urban forests.

(h) **Monitoring and evaluation**

Regular/periodic, scientific, and systematic data collection on tree canopy cover, species health, and the socio-economic benefits of urban forests should be done. This will help establish a systematic monitoring and evaluation framework to assess urban forests' health, diversity, and impact.

7 Strategies for Developing Urban Forests

An efficient strategic framework is a critical tool that emphasizes the solutions for existing problems and obtaining maximum benefits from the urban forests' environmental services. The strategic framework should include the following areas when developing an urban forest.

(a) Urban forest inventory: Conduct a comprehensive inventory to assess existing tree cover, identify species diversity, and determine the overall health of the urban forest.
(b) Land suitability analysis: Evaluate land parcels to identify areas with high potential for tree planting, considering soil quality, water availability, and existing infrastructure.
(c) Community engagement and education: Public Awareness Campaigns raise awareness about the importance of urban forests and their benefits to communities through targeted campaigns and educational programs. Community/Citizen Involvement by assigning the targets (can be voluntary) in tree planting initiatives, community clean-up events, and educational workshops to foster a sense of ownership and responsibility.
(d) Green infrastructure planning: Integrate trees into urban planning by collaborating with city planners to incorporate trees into urban designs, ensuring that green spaces are integral to the city's infrastructure.
(e) Multifunctional green spaces: Design urban forests for multiple purposes, such as stormwater management, air purification, and recreational areas.
(f) Species selection and biodiversity: Native species promote and prioritize their planting to enhance biodiversity and ensure the resilience of urban forests against pests and diseases. Creation of diverse ecosystems within urban forests by incorporating a variety of tree species, supporting a healthy and balanced environment.

(g) Maintenance and management: Proactive maintenance followed by regular pruning, disease monitoring, and pest control measures to maintain the health and longevity of urban trees. Adoption/development of comprehensive management plans for urban forest management that outline long-term strategies for urban forest care, growth, and sustainability.
(h) Policy and regulatory framework: Introduction of zoning regulations to protect existing trees and the integration of green spaces in new developments within the city limits.
(i) Incentives for tree planting: Introduce policies that provide incentives for property owners, developers, and businesses to participate in urban forest development actively.
(j) Technology integration: Utilize remote sensing and Geographic Information Systems (GIS) to monitor and assess the health and growth of urban forests. Smart Irrigation Systems to optimize water usage for urban trees, promoting sustainability.
(k) Objective-oriented policy approach: Periodic collection of data on tree canopy cover, species health, and the socio-economic benefits of urban forests to inform future policies and initiatives. The fund should be allocated for developing green infrastructure, including urban forests.

8 Regulatory Framework for Developing Urban Forests

The development of urban forests is subject to various regulatory frameworks that vary across jurisdictions. The regulatory framework typically includes a combination of local, regional, and national laws and policies to promote environmental sustainability, biodiversity conservation, and the overall well-being of urban communities (Meena et al. 2011; Kumar et al. 2019c, 2020). Here are some critical components of a regulatory framework for developing urban green forests;

a. **Land Use Planning and Zoning Regulations**

Municipalities often have land use plans and zoning regulations that designate areas for specific purposes, including green spaces and forests. Zoning codes may include provisions for green belts, parks, and other recreational areas within urban areas (Singh and Kumar 2022; Singh et al. 2024a).

b. **Environmental Impact Assessments (EIAs)**

Before developing urban forests, developers may need to conduct environmental impact assessments to evaluate the project's potential environmental effects. EIAs help identify and mitigate potential negative impacts on biodiversity, soil, water, and air quality.

c. **Biodiversity Conservation Laws**

Laws protecting biodiversity and endangered species may apply, requiring developers to take measures to preserve and enhance local flora and fauna. Protected areas and habitats may have restrictions on development to safeguard biodiversity.

d. **Tree Protection Ordinances**

Many cities have tree protection ordinances that regulate the removal and planting of trees within urban areas. Developers may be required to obtain permits for tree removal and compensate for the loss of trees by planting new ones.

e. **Stormwater Management Regulations**

Regulations may be in place to manage stormwater runoff, which can affect the health of urban green spaces. Green infrastructure solutions, such as porous surfaces and retention ponds, may be required to mitigate the impact of development on water quality.

f. **Public Participation and Consultation**

Some jurisdictions may require public input in the decision-making process for urban green forest development. Public consultation helps ensure that community interests and concerns are considered in the planning and implementation green spaces.

g. **Sustainable Design and Development Standards**

Guidelines and standards promoting sustainable and eco-friendly design practices may be incorporated into the regulatory framework. These may include requirements for using native plants, promoting biodiversity, and adopting energy-efficient and sustainable construction practices.

h. **Community Development Policies**

Urban green forests may be integrated into broader community development policies to enhance residents' quality of life. Policies may encourage the creation of multi-functional green spaces that serve recreational, educational, and ecological purposes.

i. **Monitoring and Enforcement Mechanisms**

Regulatory agencies may have mechanisms in place to monitor compliance with environmental regulations. Enforcement actions, such as fines or project halts, may be imposed for non-compliance with the established regulations.

Urban planners and policymakers must be familiar with the regulations and guidelines applicable in their local area when developing urban forests. Additionally, ongoing collaboration between governmental bodies, environmental organizations, and the local community is crucial for successfully implementing and maintaining these green spaces.

9 Conclusion

Urban forests are indispensable to contemporary urban landscapes, offering myriad benefits that enhance environmental quality, social well-being, and economic vitality. They play a critical role in mitigating the adverse effects of urbanization, such as air pollution, heat islands, and biodiversity loss. This chapter has underscored the necessity for effective governance, robust policies, and strategic planning to ensure urban green spaces' sustainable development and management. Integrating urban forestry into broader urban planning and policy frameworks is vital to achieving these goals. Strategies must encompass comprehensive planning, species diversity, community engagement, and innovative technologies for monitoring and management. Moreover, addressing green inequality and ensuring equitable access to the benefits of urban forests are paramount for fostering inclusive urban environments. This chapter's international case studies and policy reviews demonstrate diverse approaches and successful initiatives worldwide, offering valuable lessons and models for other regions. The emphasis on community participation and integrating urban forestry into climate adaptation strategies highlights the multifaceted benefits and resilience urban forests can provide. As cities grow and evolve, preserving and expanding urban forests will become increasingly critical. Prioritizing urban green spaces in policy agendas, fostering community involvement, and adopting adaptive management practices will guide urban planners and policymakers in creating sustainable, livable environments that support human and ecological health. Through these concerted efforts, urban forests can thrive, ensuring current and future generations enjoy their benefits.

References

Bratman GN, Hamilton JP, Daily GC (2012) The impacts of nature experience on human cognitive function and mental health. Ann N Y Acad Sci 1249:118–136

Braubach M, Egorov A, Mudu P, Wolf T, Ward Thompson C, Martuzzi M (2017) Effects of urban green space on environmental health, equity and resilience. In: Kabisch N, Korn H, Stadler J, Bonn A (eds) Nature-based solutions to climate change adaptation in urban areas: linkages between science, policy and practice. Springer International Publishing, pp 187–205

Daily GC (ed) (1997) Nature's services: societal dependence on natural ecosystems. Island Press, Washington, DC

Dhyani K, Kuniyal HB, Singh H, Sobha (2021) Growth and physiological potential of *Terminalia arjuna* under elevated CO_2 levels in Open top chamber condition. J Appl Nat Sci 13(3):1121–1126, ISSN 2231-5209. https://doi.org/10.31018/jans.v13i3.2490

Escobedo FJ, Kroeger T, Wagner JE (2011) Urban forests and pollution mitigation: analyzing ecosystem services and disservices. Environ Pollut (Barking, Essex: 1987), 159(8–9):2078–2087

Gobster PH, Westphal LM (2004) The human dimensions of urban greenways: planning for recreation and related experiences. Landsc Urban Plan 68(2–3):147–165

Gupta SK, Singh H (2017) Observations on sporadic flowering in *Ficus benjamina* L. and *Peltophorum africanum* L. in new forest. Indian Forester 143:290–291

Gupta SK, Ram J, Singh H (2018a) Comparative study of transpiration in cooling effect of tree species in the atmosphere. J Geosci Environ Prot 6:151–166. https://doi.org/10.4236/gep.2018.68011

Gupta SK, Ram J, Singh H (2019a) Relationship between leaf area index and atmospheric cooling of tree species. Indian J for 42(3):203–208

Gupta SK, Ram J, Singh H, Kumar P (2018b) Comparative study of cooling under shades of some forestry tree species with respect to ambient temperatures. J Geosci Environ Prot 6:51–66. https://doi.org/10.4236/gep.2018.611004

Konijnendijk CC, Ricard RM, Kenney A, Randrup TB (2006) Defining urban forestry—a comparative perspective of North America and Europe. Urban for Urban Greening 4(3–4):93–103

Kuehler E, Hathaway J, Tirpak A (2017) Quantifying the benefits of urban forest systems as a component of the green infrastructure stormwater treatment network: quantifying the benefits of urban forest systems as green infrastructure. Ecohydrology 10(3):e1813

Kumar A, Kumar P, Singh H, Kumar N (2021a) Modulation of plant functional traits under essential plant nutrients during seasonal regime in natural forests of Garhwal Himalayas. Plant Soil 465:197–212. https://doi.org/10.1007/s11104-021-05003-x

Kumar A, Kumar P, Singh H, Kumar N (2021b) Adaptation and mitigation potential of roadside trees with bio-extraction of heavy metals under vehicular emissions and their impact on physiological traits during seasonal regimes. Urban for Urban Greening 58:126900. https://doi.org/10.1016/j.ufug.2020.126900

Kumar A, Kumar P, Singh H, Kumar N (2021c) Impact of plant functional traits on infiltration rate under different forests of Kempty watershed in Garhwal Himalaya. Indian J Soil Conserv 49(1):38–44

Kumar A, Kumar P, Singh H, Bisht S, Kumar N (2021d) Relationship of physiological plant functional traits with soil carbon stock in the temperate forest of Garhwal Himalaya. Curr Sci 120(8):1368–1373. https://doi.org/10.18520/cs/v120/i8/1368-1373

Kumar A, Singh H, Kumari G, Bisht S, Malik A, Kumar A, Singh M, Raturi, Barthwal S, Thakur A, Kaushal R (2022) Adaptive resilience of roadside trees to vehicular emissions via leaf enzymatic, physiological, and anatomical trait modulations. Environ Pollut 313. https://doi.org/10.1016/j.envpol.2022.120191

Kumar M, Phukon SN, Paygude AC, Tyagi K, Singh H (2021) Mapping phenological functional types (PhFT) in the Indian Eastern Himalayas using machine learning algorithm in Google Earth Engine. Comput Geosci 158. https://doi.org/10.1016/j.cageo.2021.104982

Kumar M, Singh MP, Singh H, Dhakate PM, Ravindranath NH (2019a) Forest working plan for the sustainable management of forest and biodiversity in India. J Sustain for 39:1–22. https://doi.org/10.1080/10549811.2019.1632212

Kumar N, Jeena N, Singh H (2019b) Elevated temperature modulates rice pollen structure: a study from the foothill of the Himalayan agro-ecosystem in India. 3 Biotech 9(5):175. https://doi.org/10.1007/s13205-019-1700-1

Kumar N, Jeena N, Kumar N, Gangola S, Singh H. (2019c) Phytoremediation facilitating enzymes: an enzymatic approach for enhancing remediation process. In: Bhatt P (ed) Smart bioremediation technologies. Academic Press, Elsevier, Cambridge, United States, pp 289–306

Kumar N, Singh H, Sharma SK (2020) Antioxidants: responses and importance in plant defense system. In: Roychowdhury R, Choudhury S, Hasanuzzaman M, Srivastava S (eds) Sustainable agriculture in the era of climate change. Springer, Cham, pp 251–264. https://doi.org/10.1007/978-3-030-45669-6_11

Kumar N, Singh H, Giri K, Kumar A, Joshi A, Yadav A, Singh R, Bisht S, Kumari R, Jeena N, Khairakpam R, Mishra G (2024) Physiological and molecular insights into the allelopathic effects on agroecosystems under changing environmental conditions. Physiol Mol Biol Plants 30:417–433. https://doi.org/10.1007/s12298-024-01440-x

Kumar N, Suyal DC, Sharma IP, Verma A, Singh H (2017) Elucidating stress proteins in rice (*Oryza sativa* L.) genotype under elevated temperature: a proteomic approach to understand heat stress response. 3 Biotech 7:205. https://doi.org/10.1007/s13205-017-0856-9

Lafortezza R, Pauleit S, Hansen R, Sanesi G, Davies C (2017) Strategic green infrastructure planning and urban forestry. In: Ferrini F, van den Bosch CCK, Fini A (eds) Routledge handbook of urban forestry. London, UK and New York, NY, pp 179–193

Lai D, Liu W, Gan T, Liu K, Chen Q (2019) A review of mitigating strategies to improve the thermal environment and thermal comfort in urban outdoor spaces. Science Total Environ 661:337–353

McPherson EG, Nowak DJ, Heisler G, Grimmond S, Souch C, Grant R, Rowntree R (1997) Quantifying urban forest structure, function, and value: The Chicago Urban Forest Climate Project. Urban Ecosyst 1:49–61

Meena D, Singh H, Chaudhari SK (2011) Elucidating strontium response on growth dynamics and biochemical change in *Phaseolus mungo* L. Int J Agric Environ Biotechnol 4:107–113

Miller RW, Hauer RJ, Werner LP (2015) Urban forestry: planning and managing urban greenspaces. Waveland Press, Illinois

Nielsen AB, van den Bosch M, Maruthaveeran S, van den Bosch CK, Shrubsole D (2014) Species richness in urban parks and its drivers: a review of empirical evidence. Urban Ecosyst 17(1):305–327

Nowak DJ, Dwyer JF (2007) Understanding the benefits and costs of urban forest ecosystems. In: Kuser JE (ed) Urban and community forestry in the Northeast. Springer, Netherlands, pp 25–46

Nowak DJ, Hirabayashi S, Bodine A, Greenfield E (2014) Tree and forest effects on air quality and human health in the United States. Environ Pollut 159(8–9):2075–2083

Prakash V, Hunney K, Singh H (2022) Effect of elevated carbon dioxide on growth and development of *Santalum album* L. seedlings inoculated with plant growth promoting microorganisms in open top chambers. Int J Environ, Agric Biotechnol 7(5):149–159

Salbitano F, Borelli S, Sanesi G (2015) Urban forestry and agroforestry. In de Zeeuw H, Drechsel P (eds) Cities and agriculture: developing resilient urban food systems. Routledge, Taylor & Francis Group, London; New York

Sanesi G, Gallis C, Kasperidus HD (2011) Urban forests and their ecosystem services in relation to human health. In: Nilsson K, Sangster M, Gallis C, Hartig T, de Vries S, Seeland K, Schipperijn J (eds) Forests, trees and human health. Springer, Netherlands, pp 23–40

Sharma R, Singh H (2021) Alteration in biochemical constituents and nutrients partitioning of *Asparagus racemosus* in response to elevated atmospheric CO_2 concentration. Environ Sci Pollut Res 29:6812–6821. https://doi.org/10.1007/s11356-021-16050-3

Sharma R, Singh H, Kaushik M, Nautiyal R and Singh O (2018) Adaptive physiological response, carbon partitioning, and biomass production of *Withania somnifera* (L.) Dunal grown under elevated CO_2 regimes. 3 Biotech 8:267. https://doi.org/10.1007/s13205-018-1292-1

Sicard P, Agathokleous E, Araminiene V, Carrari E, Hoshika Y, De Marco A, Paoletti E (2018) Should we see urban trees as effective solutions to reduce increasing ozone levels in cities? Environ Pollut 243:163–176

Singh H (2021) An integrated approach considering physiological and biophysical based indicators for assessing tolerance of roadside plantations of *Alstonia scholaris* towards urban roadside air pollution: An assessment of adaptation of plantations for mitigating urban roadside air pollution. Trees. https://doi.org/10.1007/s00468-021-02179-8

Singh H (2024) Exploring adaptive modulation in plant functional traits and its impact on the productivity of *Acacia auriculiformis* under CO_2-enriched environment. Ind Crops Prod 210:118186. https://doi.org/10.1016/j.indcrop.2024.118186

Singh H, Verma A (2013) Characterization and screening of high nitrogen efficient rice genotype to curtail environmental pollution. Int J Agric, Environ Biotechnol 6(1):23–30

Singh H, Savita Sharma R, Sinha S, Kumar M, Kumar P, Verma A and Sharma SK (2017a) Physiological functioning of *Lagerstroemia speciosa* L. under heavy roadside traffic: an approach to screen potential species for abatement of urban air pollution. 3 Biotech 7 (61):1–10. https://doi.org/10.1007/s13205-017-0690-0

Singh H, Verma A, Kumar M, Sharma R, Gupta R, Kaur M, Negi M, Sharma SK (2017b) Phytoremediation: a green technology to clean up the sites with low and moderate level of heavy metals. Austin Biochem 2:1012

Singh H, Sharma R, Savita SMP, Kumar M, Verma A, Ansari MW, Sharma SK (2018) Adaptive physiological response of *Parthenium hysterophorus* to elevated atmospheric CO_2 concentration. Indian Forester 144:1–14

Singh H, Yadav M, Kumar N, Kumar A and Kumar M (2020) Assessing adaptation and mitigation potential of roadside trees under the influence of vehicular emissions: a case study of *Grevillea robusta* and *Mangifera indica* planted in an urban city of India. PLoS ONE 15(1):e0227380. https://doi.org/10.1371/journal.pone.0227380

Singh S, Singh H, Sharma SK, Nautiyal R (2021) Seasonal variation in biochemical responses of bamboo clones in the sub-tropical climate of Indian Himalayan foothills. Heliyon 7(4):e06859. https://doi.org/10.1016/j.heliyon.2021.e06859

Singh H, Bandyopadhyaya S, Kumar A, Singh M, Malik A, Kumari G, Raturi A, Yadav SK, Kumar M, Barthwal S, Thakur A, Rawat PS (2022) Understanding the physiological and biophysical response of urban roadside plantations for assessing adaptation and mitigation mechanisms toward vehicular emissions. Urban Clim 44. https://doi.org/10.1016/j.uclim.2022.101183

Singh H, Kumar N, Kumar A (2024a) Enhancing resource use efficiency in crops through plant functional traits. In: Kumar N, Singh H (eds) Plant functional traits for improving productivity. Springer, Singapore. https://doi.org/10.1007/978-981-97-1510-7_6

Sister C, Wolch J, Wilson J (2010) Got green? Addressing environmental justice in park provision. GeoJournal 75:229–248

Wang Y, Bakker F, De Groot R, Wortche H (2014) Effect of ecosystem services provided by urban green infrastructure on indoor environment: a literature review. Build Environ 77:88–100

Ward Thompson C, Roe J, Aspinall P, Mitchell R, Clow A, Miller D (2012) More green space is linked to less stress in deprived communities: evidence from salivary cortisol patterns. Landsc Urban Plan 105(3):221–229

Weinstein N, Balmford A, DeHaan CR, Gladwell V, Bradbury R, Amano T (2015) Seeing community for the trees: links between contact with natural environments, community cohesion, and crime. Bioscience 65:1141–1153

Wolf KL, Robbins AS (2015) Metro nature, environmental health, and economic value. Environ Health Perspect 123(5):390–398

Wolf KL, Robbins AST (2015b) Metro nature, environmental health, and economic value. Environ Health Perspect 123(5):390–398

Wolf KL (2017) Social aspects of urban forestry and metro nature. In: Ferrini F, van den Bosch CCK, Fini A (eds) Routledge handbook of urban forestry. Routledge, New York, pp 65–81

Yadav SK, Singh H, Nautiyal R, Ginwal HS, Ansari SA, Barthwal S (2019) Modulation of morphophysiological responses in Populus deltoides by elevated carbon dioxide and temperature. Forest Sci 66:105–118. https://doi.org/10.1093/forsci/fxz048

The manufacturer's authorised representative in the EU is Springer Nature Customer Service Centre GmbH, Europaplatz 3, 69115 Heidelberg, Germany. If you have any concerns regarding our products, please contact ProductSafety@springernature.com

Printed and bound by CPI Group (UK) Ltd, Croydon, CR0 4YY
01/10/2026
02225969-0020